HADRON
SPECTROSCOPY

Previous Proceedings in the Series of Conferences on Hadron Spectroscopy

Conference		Held in	Publisher	ISSN/ISBN
HADRON01	9th	Protvino, Russia	AIP Conf. Proceedings Vol. 619	0-7354-0067-9
HADRON99	8th	Beijing, China	North-Holland Nuclear Physics A, Vol. 675	0375-9474
HADRON97	7th	Upton, New York, USA	AIP Conf. Proceedings Vol. 432	1-56396-765-0

Other Related Titles from AIP Conference Proceedings

716   Portable Synchrotron Light Sources and Advanced Applications: International Symposium on Portable Synchrotron Light Sources and Advanced Applications
Edited by Hironari Yamada, Noriko Mochizuki-Oda, and Makoto Sasaki, August 2004, 0-7354-0195-0;
CD-ROM: 0-7354-0196-9

698   Intersections of Particle and Nuclear Physics, 8th Conference; CIPANP 2003
Edited by Zhoreh Parsa, February 2004, CD-ROM included, 0-7354-0169-1

689   Neutrinos, Flavor Physics, and Precision Cosmology: Fourth Tropical Workshop on Particle Physics and Cosmology
Edited by José F. Nieves and Raymond R. Volkas, October 2003, 0-7354-0160-8

688   Scalar Mesons: An Interesting Puzzle for QCD
Edited by Amir H. Fariborz, November 2003, 0-7354-0159-4

687   High Energy Physics: The 25th Annual Montreal-Rochester-Syracuse-Toronto Conference on High Energy Physics, MRST 2003: A Tribute to Joe Schechter
Edited by Amir H. Fariborz, November 2003, 0-7354-0161-6

672   Short Distance Behavior of Fundamental Interactions: 31st Coral Gables Conference on High Energy Physics and Cosmology
Edited by Behram N. Kursunoglu, Metin Camcigil, Stephan L. Mintz, and Arnold Perlmutter, June 2003, 0-7354-0139-X

671   Hydrogen in Materials and Vacuum Systems: First International Workshop on Hydrogen in Materials and Vacuum Systems
Edited by Ganapati Rao Myneni and Swapan Chattopadhyay, July 2003, 0-7354-0137-3

670   Particles and Fields: Tenth Mexican School
Edited by U. Cotti, M. Mondragón, G. Tavares-Velasco, June 2003, 0-7354-0135-7

To learn more about these titles, or the AIP Conference Proceedings Series, please visit the webpage **http://proceedings.aip.org**

HADRON SPECTROSCOPY

Tenth International Conference on
Hadron Spectroscopy

Aschaffenburg, Germany   31 August – 6 September 2003

EDITORS

Eberhard Klempt
Universität Bonn

Helmut Koch
Ruhr-Universität Bochum

Herbert Orth
GSI Darmstadt

SPONSORING ORGANIZATIONS
Bayrisches Staatsministerium für Wissenschaft, Forschung und Kunst
Bundesministerium für Bildung und Forschung
Deutsche Forschungsgemeinschaft
Dynamitron-Tandem-Laboratorium der Ruhr-Universität Bochum
Gesellschaft für Scherionenforschung mbH, Darmstadt

Melville, New York, 2004
AIP CONFERENCE PROCEEDINGS ■ VOLUME 717

Editors:

Eberhard Klempt
Universität Bonn
D-53012 Bonn
GERMANY
E-mail: klempt@iskp.uni-bonn.de

Helmut Koch
Ruhr-Universität Bochum
Universitätsstrasse 150
D-44780 Bochum
GERMANY
E-mail: hkoch@ep1.ruhr-uni-bochum.de

Herbert Orth
Gesellschaft für Schwerionenforschung
Planckstrasse 1
D-64291 Darmstadt
GERMANY
E-mail: h.orth@gsi.de

L.C. Catalog Card No. 2004110100
ISBN 0-7354-0197-7
ISSN 0094-243X
Printed in the United States of America

IN MEMORIAM

Lucien Montanet 1930 - 2003

MESONS

BARYONS

SCALARS

EXOTICS

HEAVY QUARKS

xiv

Preface

The Tenth International Conference on Hadron Spectroscopy was jointly organized by the University of Bochum and the Gesellschaft für Schwerionenforschung (GSI) at Darmstadt. Conferences in this series take place every two years, starting with 1985. The previous conference in 2001 was held in Protvino, Russia. The scope of these conferences is hadron spectroscopy and related aspects of hadron dynamics as well as the understanding the hadronic state of matter from the basics of Quantum Chromodynamics.

The conference took place in the Stadthalle am Schloss of Aschaffenburg, Germany, from 31 August 2003 to 6 September 2003, with about 200 scientists from all over the world attending and contributing. The reason to house the conference in Germany was the decision of GSI to include hadronic physics with antiprotons in its future program.

The timing of the conference was very fortunate as many new and surprising results appeared in the months before its start. The highlights were the discussions about the nature of the recently discovered narrow states. The implications of these findings were discussed in several talks and in an open panel resulting in many ideas for future measurements which will clarify the true nature of these states. Additionally, there were very interesting contributions concerning the properties of hadrons inside nuclear matter, the discovery of baryons with double charm and its implications and the role of the σ/κ structures in low energy $\pi\pi$-and πK-scattering.

The proceedings of the conference will be devoted to Lucien Montanet from CERN, who passed away in 2003 on June 19. He belonged to the pioneers in hadron physics. His eminent role in this field was highlighted during a special plenary session in honor of him.

We would like to thank the International Advisory Committee and the Program Committee for their efforts to guarantee the high scientific level of this conference. We thank the speakers for their excellent presentations. The financial and other support of the sponsors of the conference is gratefully acknowledged. We are indebted to the local organizing committee for pulling the strings during the scientific program and for implementing a enjoyable social program which included a boat trip to the medieval town Seligenstadt, tours of the city and the castle including a wine tasting party and of course the conference diner. The beautiful weather during the whole conference week ensured a very pleasant time in Aschaffenburg and that the HADRON'03 will be remembered.

The next conference of this series is scheduled in August 2005 in Rio de Janeiro. We wish it full success expecting there further exciting results.

Eberhard Klempt
Helmut Koch
Herbert Orth

International Advisory Committee

C. Amsler (Zurich)
P. Barnes (Los Alamos)
E. Berger (Argonne)
T. Bressani (Torino)
S. Chung (BNL)
F. Close (Rutherford)
A. Donnachie (Manchester)
W. Dunwoodie (SLAC)
A. Dzierba (Indiana)
S. Gershtein (Protvino)
F. Gross (JLab)
C. Guaraldo (Frascati)
H. Lipkin (Weizmann)
B. Meadows (FNAL)
V. Metag (Giessen)
L. Montanet (CERN)
S. Nagamiya (KEK)
S. Paul (Munich)
D. Peaslee (Maryland)
K. Seth (Evanston)
A. Skrinsky (Novosibirsk)
K. Takamatsu (KEK)
U. Wiedner (Uppsala)
A. Zaitsev (Protvino)
B. Zou (Beijing)

Scientific Organization

E. Klempt (Bonn)
H. Koch (Bochum)
U. Lynen (GSI)

Local Organization

A. Busch (GSI)
O. Hartmann (GSI)
B. Lewandowski (Bochum)
H. Orth (GSI)
K. Peters (Bochum)
C. Schwarz (GSI)

Conference Secretary

D. Hiltscher (Bochum)

Sponsors

Bayrisches Staatsministerium für Wissenschaft, Forschung und Kunst
Bundesministerium für Bildung und Forschung
Deutsche Forschungsgemeinschaft
Dynamitron-Tandem-Laboratorium der Ruhr - Universität Bochum
Gesellschaft für Scherionenforschung mbG, Darmstadt

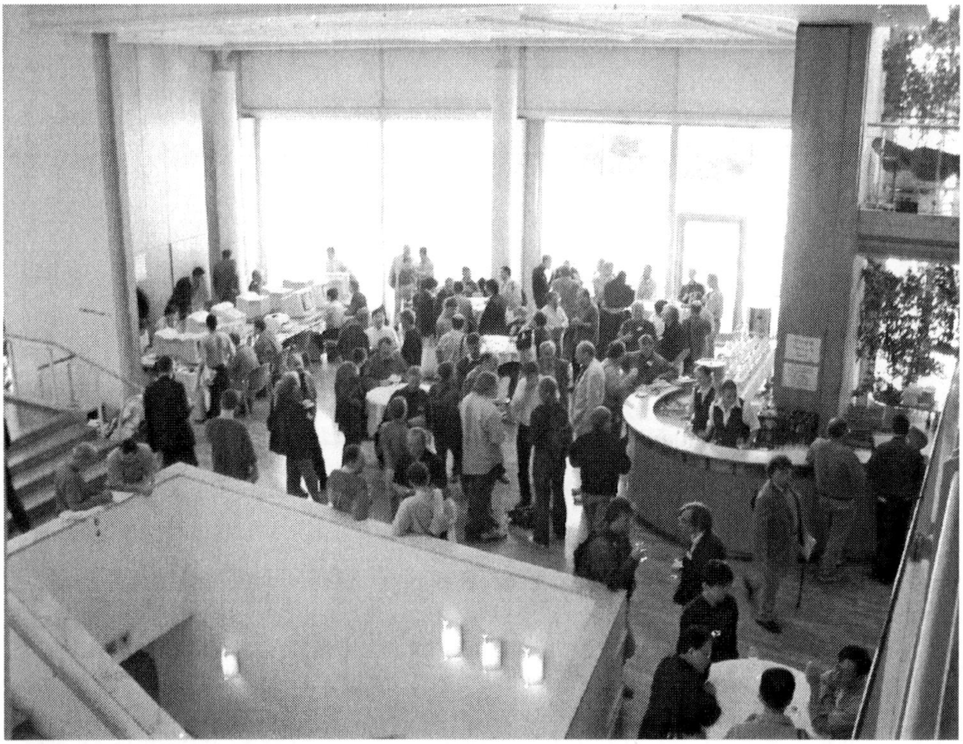

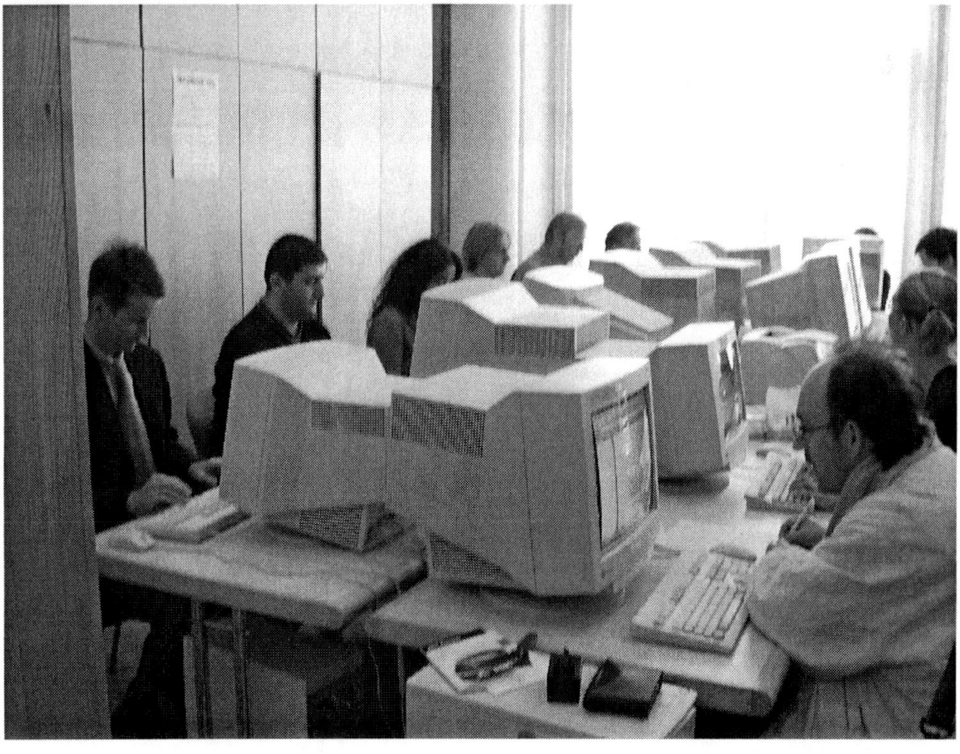

ASCHAFFENBURG

WELCOME ADDRESS

Klaus Herzog, Lord Mayor of the City of Aschaffenburg

I am very pleased to welcome you today in Aschaffenburg. Of course Aschaffenburg is not really the centre of hadron spectroscopy research in the world. And therefore I have to apologize for my little knowledge of your field of activity. This is also the reason that I would not like to tell you something about hadron spectroscopy, but something about our city of Aschaffenburg. As the lord major of Aschaffenburg I am of course very proud that Aschaffenburg is the first German location of your conference history.

It is not just the vicinity to your conference location that the official welcome takes place here in the hall of the "Galerie Jesuitenkirche". It can also be seen as a symbol for the characteristic of our town, which is a town of culture, bringing together a wide variety of cultural life. In this "Galerie Jesuitenkirche" numerous exhibitions of famous and interesting artists have been taken place. At present we are in the lucky position to show some highly interesting pieces of the artist Ernst Barlach. Ms. Ladleif, the manager of this museum will afterwards give you a short impression of this artist and his work. And after the official part you will have the opportunity to get an impression of this exhibition. May be it will also inspire you to give our museum a second visit.

Dear Ladies and Gentlemen,

on the behalf of Aschaffenburg I welcome you very much in our town. Maybe you have already enjoyed the advantages of the location of Aschaffenburg. Coming from the motorway you perhaps had the chance to gain a glimpse to the deep forests of "Odenwald" or "Spessart", which was already mentioned in the Nibelung saga and offers a fantastic chance for hiking, biking and relaxing. Not without good reason, Aschaffenburg is also called „the gate to the Spessart".

Let me take you, Ladies and gentlemen, to a short excursion to the beginning and the rich history of our city. As in many other cases, the beginning of Aschaffenburg`s history remains hidden in the past. One reason may be, that the first settlements were located on German territory outside the boundaries of the Roman Empire reaching the river Main at its peak of expansion.

Has there really existed a Franconian settlement by the name of „Ascis" already 500 years B.C., as mentioned by the unknown geographer from Ravenna? We might never find out. In the year 974 Aschaffenburg is documented for the first time. But parts of the city and the collegiate church "St. Peter and Alexander" were already established then. The arch-chancellor Willigis of Mainz was presented this collegiate and thus began a long period under the electorate of Mainz, reaching all the way into the 19th century.

If you found a little spare time, you should not miss to visit this church with his mixture of architectural styles. Besides, to the most valuable objects among its collection belongs the famous painting "Mourning of Christ" by Matthias Grünewald.

Aschaffenburgs location along the river Main provided excellent conditions for further development. The economical progress began with the construction of the first wooden bridge across the river in the year 989. The toll for the use of the bridge and road may have contributed to receiving the privilege of holding a market and the right of coinage in the 12th century, all infallible signs for the city`s increasing prosperity and independence in the High Middle Ages.

Under the electorate of Mainz, Aschaffenburg grew to become the second residential town besides the city of Mainz itself. The construction of the mighty fortifications built by the elector and arch-bishop Adalbert I. in the early 12th century emphasizes the increasing importance of the city.

Towards the end of the Middle Ages Aschaffenburg developed gradually and its significance is reflected by numerous synods held in the city. Pius II participated in one of this synods shortly before he got elected as pope and was staying in Aschaffenburg.

In the 16th century a very fruitful era began with Albrecht of Brandenburg, elector and arch-bishop of Mainz. As a great supporter of the arts he invited famous painters such as Lukas Cranach (the Elder) and Matthias Grünewald to Aschaffenburg. Nowadays paintings in the church and castle museums still remind of these famous visitors.

But Aschaffenburg was also not spared from the religious and political confusion caused by the schism leading into the so-called "Schmalkaldian War". When emperor Karl V intended to violently solve the religious conflict, Aschaffenburg suffered severely as troops destroyed most of the city. Only five years later the same fate came to pass, when the soldiers of Albrecht of Brandenburg-Kulmbach ravaged and plundered the city. The old gothic castle was destroyed and only the keep remained to this day.

Not until fifty years later, after a long period of stagnation, a new era of prosperity began with the elector Johann Schweickard of Kronberg. He laid the foundation-stone for the new castle, which was completed in 1614.

The Thirty Years War (1618-1648) which was to destroy most of Central Europe, reached Aschaffenburg in 1631, when Gustav Adolf, king of Sweden and leader of the protestant league besieged the city and conquered without much resistance. In the following century Aschaffenburg again became the scenario for martial argumentations.

But nevertheless at the beginning of the 19th century not only in Aschaffenburg the economy, politics and culture were in full bloom. In 1803 the first newspaper, the "Aschaffenburger Zeitung" was published and in 1811 the first theatre, which is now the Municipal Theatre, was opened to the public. The newly founded university invited famous scholars and poets such as Görres, Achim von Arnim and Ludwig Tiek into the city. The great poet of Romanticism, Clemens von Brentano, died in Aschaffenburg and was buried here.

Still, or to be more precise, again Aschaffenburg has its own university: the University of Applied Sciences, a young and very modern campus that hosts more than 1.100 students in a technical and a combined business/law department. It is also due to the university and the quality of the education that high-tech-businesses are a fast-growing sector in our area. We are in the lucky position, that more and more IT-enterprises and network specialists settle down here.

Aschaffenburg and the whole region belongs to the most popular areas to live within Germany. In a recent online survey of 170.000 participants, 82 per cent of the local population rated the region's quality of life as "excellent".

One reason is definitely the comparable healthy economy. The Bavarian Lower Main Area reflects all important segments of the economy: industry, service, high-tech, trade and crafts. The beginning of German fashion industry was near Aschaffenburg, also paper was and is still produced in all its varieties, international companies manufacture important parts for the automotive industry in this region – 90 per cent of all steering wheels installed in German cars are made here. And by far that is not all: in all branches of this region it is not mainly production that counts: service is the name of the game.

You may excuse, that I am quite proud of all these positve aspects. We are happy living in comparably healthy and wealthy conditions which are not longer be taken for granted.

Therefore I am glad that the economy also appreciates our area. At the last conference of the German Chambers of Commerce a survey among 20.000 German companies ranked Aschaffenburg third on the list of Germany's best economic sites!

Among other qualities, "being close to customers", the existing "educational establishments", the "traffic infrastructure" as well as "university co-operations", were mentioned as strong points of this site.

As mentioned, one of the pillars of a vital economy is definitly a functional infrastructure. Profiting by our excellent geographical position we are close to Frankfurt international airport, we have the largest inland port on the River Main in Bavaria and – we have an excellent road network. I can only hope, you will not give the lie, when being in a traffic jam on your way to Aschaffenburg.

But back to the history: Aschaffenburg survived world war I almost fully intact, but the air raids during Worldwar II destroyed nearly 80 % of the city. Especially the old part of the town was almost completely buried under ashes. The greatest loss was the destruction of the castle, landmark and pride of an entire region. In 1950 the reconstruction of the castle began. The city-administration, various entrepreneurs and the residents worked hard and tenacious on rebuilding their city. Thus almost the entire old part of the town could rise again. A new mark within the townscape was set by the city hall, inaugurated in 1959 and building a bridge between the New Functionalism and the traditional architecture of the old part of the city in the immediate vicinity.

Ever since Aschaffenburg developed into a prospering and reputable business city. Meanwhile, round about 70.000 people live in Aschaffenburg. This town, already loved by King Ludwig I and acclaimed by him as the "Bavarian Nice".

The reason for this pet name becomes obvious when strolling through the castle gardens, enjoying the scenery along the river Main and the Pompejanum coming into sight. The Bavarian King, Ludwig I, hat it built during the years 1840 – 1848, as a replica of a villa in Pompeii.

But I don`t want to take your patience any longer. I hope during your stay you will get as many positive impressions as possible from Aschaffenburg by yourselves. If you find a little spare time beside the conference programm, take the chance to walk leisurely through the parks of the city, do a walk through the "Schönbusch Park", one of the most beautiful and the biggest English landscape garden in Bavaria or come to see the art treasures in our numerous museums. Or just enjoy the hospitality of Franconia in one of the typical wine-places in the old town.

And may be we will see the one or other a second time in our "Bavarian Nice". Last but not least I wish the Hadron Conference successful progression and very good conversations.

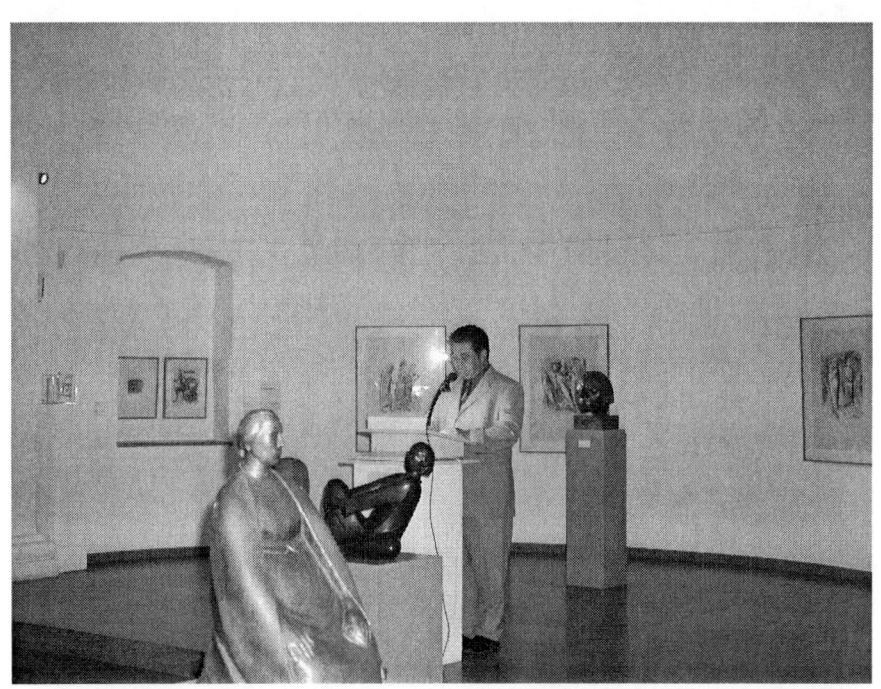

Mesons

Lucien Montanet 1930 – 2003

A bubble chamber physicist

It was with great sadness that we heard that Lucien Montanet had passed away on 19 June. Until quite recently, he had still been here among us at CERN, discussing physics with his usual enthusiasm. A few weeks before he died, he even managed to overcome his exhaustion and pay a warm and eloquent tribute to his friend Charles Peyrou[1].

As a very young graduate from France's Arts et Métiers engineering school, Lucien first extended his knowledge of physics at Paris University before setting out on his physicist's career during his national service at France's ZOE nuclear reactor. In 1957, he was one of the first physicists to take up a position at CERN, which was just starting to be built at the time. To gain experience in particle physics, Lucien went to the Jungfraujoch in Switzerland, where Patrick Blackett from Manchester University had set up a Wilson cloud chamber. There, with other physicists such as Antonino Zichichi and Roberto Salmeron, he analysed cosmic-ray interactions (≈ 100 GeV)[2], the subject of his thesis in 1960[3].

Lucien then worked on images taken from a propane bubble chamber at CERN's synchrocyclotron[4]. Shortly afterwards, he began his long career with liquid hydrogen bubble chambers by working on a 20 cm chamber at the Saclay's Saturne accelerator, which supplied a π^- beam between 500 and 1000 MeV[5]. He then teamed up with Peyrou's group to work on the CERN's 32 cm liquid hydrogen chamber and took part in developing suitable techniques for the analysis of bubble chamber images [6]. In 1961, together with a number of physicists from CERN and Paris (among them Rafael Armenteros and myself), Lucien analysed proton-

CP717, *Hadron Spectroscopy: Tenth International Conference,*
edited by E. Klempt, H. Koch, and H. Orth
© 2004 American Institute of Physics 0-7354-0197-7/04/$22.00

antiproton annihilations on images taken in the 81cm bubble chamber belonging to Saclay and the Ecole Polytechnique, which had just arrived in the CERN-PS South Hall. He co-signed the discovery of the first meson resonance found at CERN and in Europe, the E meson [7] (which later became the E/ι(1440) and is now known as the $\eta^0(1440)$) .

Figure 1: The 2 meter hydrogen bubble chamber at CERN

This marked the start of a long career devoted to meson and baryon spectroscopy. Hard on the heels of the E/ι (η^0(1440) discovery came that of the C (K_1(1270))[8] and D (f_1(1285))[9], all with the 81cm HBC. Thereafter, the images from CERN's 81cm and 2m chambers, fig.1, were used for the analysis of the A_2 into two kaons[10], the saga of the E/ι=(η^0(1440))[7],[16],[17], which is now a glueball candidate, the Q bump, the three-body phase-shift analyses, analysis of the characteristics of numerous resonances produced by π^- at 3.92 GeV and K^- at 4.2 GeV, the masses of the K^0 and of the Λ^0 [11], the width of the ω^0(782)[12], the decay of the K^0, the lifetime of the K^0 short[13] and evidence for the δ (a_0(980))[14], the "Bouddha" (b_1(1235))[15] and the f_1(1420)[17]

Lucien became an eminent specialist in this field. He played a major role in the construction of the present scheme of elementary particles and was a key member of the Particle Data Group, organizing and leading numerous workshops and conferences on hadron spectroscopy. For many years, he was very much in demand as a rapporteur for review talks on these topics.

In his experimental activities, Lucien initiated and coordinated several experiments of central importance at CERN. One of the most illustrious was the EHS (European Hybrid Spectrometer)[18], an elaborate set of particle detectors fed by a high-energy beam at the CERN SPS. Lucien was the spokesman of the large (by the standards of the time) collaboration of institutes working on EHS. The goal of the experiment was to determine, in complex hadronic final states, the dynamic features of strong interactions and to study the associated weak decays. This spectrometer was the first to measure, with excellent precision, the lifetime of the charmed D meson[19],[20]. Among the most remarkable of the sub-detectors that made up the Hybrid Spectrometer were a rapid cycling bubble chamber and a holographic high-resolution bubble chamber[21].

Lucien was adept at attracting young physicists and infusing them with his enthusiasm and experience. In this way, he played a major role in the development of high-energy physics in Spain, which, although not yet a CERN Member State, was an important contributor to EHS.

Lucien played a similar role when, in 1985, he was appointed coordinator of the CERN-USSR Committee, which subsequently became the CERN-Russia Committee. The excellent relations between CERN and Russia are largely due to his skill and devotion to the task. In 1990 he became a member of the Scientific Council of the Joint Institute for Nuclear Research (JINR) at Dubna and played a key role in defining its scientific policy.

Lucien was also a very influential member of the "Crystal Barrel" experiment, which analysed the γ rays and the charged particles produced when the antiprotons from LEAR collided with a liquid hydrogen target[22]. The experiment was known for its precise analyses of the π^0, η and η' final states, its discovery of the new resonances η_2(1685), η_2(1875) [23] and the analysis of many resonances in the $3\pi^0$, $\eta\eta^0\pi$,$\eta\pi^0\pi^0$, $\eta\eta'\pi^0$ states and by finding, in the $3\pi^0$ channel, the f_0(1500)[24] which associated with the E/ι could make a pair of glueballs. It was one of the glories of an already glittering career. The splendid plots representing the three-body reactions are now a reference and could even be described as works of art: fig.2.

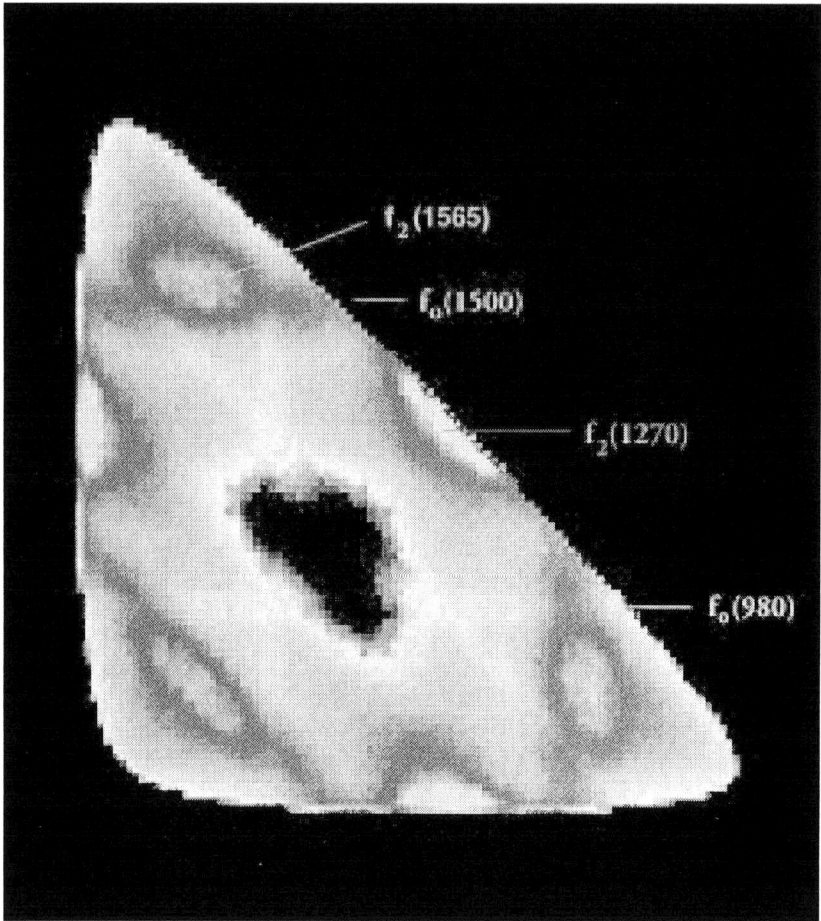

Figure 2: The Dalitz plot from Crystal Barrel for $\bar{p}p \rightarrow 3\pi^0$

In 1973 Lucien became the editor of Physics Letters and continued playing this role competently and efficiently even after his retirement in 1995. This enabled him to remain in close contact with the high-energy physics community and made him a well-known figure for younger generations. Most of the publications of CERN and other European laboratories went under the expert and critical scrutiny of Lucien and his partner Klaus Winter. Up to the last moments of his life, he performed this task with the same diligence.

In Lucien Montanet, we loose one of the pioneers of modern high-energy physics, an inspired, generous and friendly member of our community and a true lover of science.

P. Baillon
CERN CH-1211 Geneva 23

References

1 CERN Courier June 2002 p25
2 L. Montanet et al. A cloud chamber study of nuclear interactions with energies of about 100 GeV. N.C.,17, 166 (1960)
3 L.Montanet "Etude expérimental des intéractions nucléaires de haute énergie dans le rayonnement cosmique" Thesis Paris 31/3/1960
4 A. Loriat et al. The scattering of positive 120 MeV pions on protons N.C., 22, 820 (1961)
5 F.Grard et al. π^- proton scaterring at 516, 616, 710, 887 and 1085 MeV. N.C.,22, 193(1961)
6 M. Bénot et al. La mesure des photographies de chambres à bulles au CERN. Industries Atomiques, 7-8, 54 (1963)
7 R. Amenteros et al. proceeding of the Siena Int. Conf.on elementary part. vol 1 287, (1963) P. Baillon et al. Futher study of the E-mesom in antiproton-proton annihilation at rest. N.C.,50 ,393 (1967)
] R. Amenteros et al. Evidence for a $(K\pi\pi)$ resonance with a mass of 1.230 GeV/c2 PL,9,207 (1964)
9 Ch. D'Andlau et al. Evidence for a non-strange meson of mass 1290 MeV PL,17, 34 (1965)
10 M. Aguilar-Benitez Structure in the $K\overline{K}$ decay mode of the A₂ Meson. PL, 29B,62 (1969)
11 H. Blumenfeld et al. A measurement of the K^0 and Λ^0 masses. Anales de Fisica 68 221 (1972)
12 R.Bizzari et al. $\overline{p}p$ annihilation at rest, the $K\overline{K}$ ω channel. Mass, width and branching ratio of the ω meson NP B27 140 (1971)
13 O. Skjeggestad et al. Measurement of the KSO mean life NP B48 B48 (1972)
14 A. Astier et al. Further study of the I=1 $K\overline{K}$ structure near theshold. PL, 25B, 294 (1967)
15 R.Bizzari et al. Experimental results on the ω πand ππ systems as observed in the $\overline{p}p$ annihilations at rest: $\overline{p}p{\rightarrow}\omega^0\pi^+\pi^-$. NP, B14, 169 (1969)
16 C.Defoix et al. Evidence for decays of the D- and E-mesons into δπ in $\overline{p}p$ annihilation at 700 MeV/c NP B44,125 (1972)
17 C. Dionisi et al. Observation and quantum numbers determination of the E(1420) meson in π^-p interractions at 3.95 GeV/c NP, B169, 1 (1980)
18 M. Aguilar-Benitez et al. The European hybrid spectrometer. A facility to study multihadron events produced in high energy interactions. NIM, 205, 79 (1983)
19 M. Aguilar-Benitez et al. Neutral D-meson properties in 360 GeV/c π^-p interactions. PL B146 266 (1984)
20 M. Aguilar-Benitez et al. D-meson lifetimes. PL, B193 140 (1987)
21 L. Montanet and S. Reucroft High resolution bubble chambers and the observation of short-lived particles. Physics Reports 83 61 (1982)
22 E. Aker et al. The Crystal barrel spectrometer at LEAR. NIM A321 69 (1992)
23 J.Adomeit et al Evidence for two isospin zero J^{PC} = 2⁻⁺mesons at 1645 and 1875 MeV Z. Phys C71 227 (1996)
24 V.V. Anisovich et al. Observation of two J^{PC} = 0⁺⁺ isoscalar resonaces at 1365 and 1520 MeV. PL B323, 233, (1994)

Experimental Approaches in Meson Spectroscopy

Klaus Peters

Ruhr-Universität Bochum, D-44780 Bochum, Universitätsstraße 150, Germany

Abstract. The investigation of the spectrum of mesons from light to heavy is still a very active field. Lattice gauge theory and many effective models provide a lot of insights into the believed structure of the spectrum of mesons. A lot of them have been found but many others remain to be seen. Although many states fit into the proposed scheme, but most mass predictions lack precision and estimates of the widths are usually missing. Some of them are key stone particles since they are exotic in the sense, that their quantum numbers show that they do not contain conventional $q\bar{q}$-pairs or qqq-baryons. This report compares experimental approaches to find and investigate conventional and exotic mesons. It will be discussed how different experimental techniques can contribute to the puzzle and what else is needed to improve on the current situation.

INTRODUCTION - WHY SPECTROSCOPY?

In the last year new important results emerged in the field of hadron spectroscopy which may change our picture of bound quarks. Even in sectors which have been regarded to be well understook like the open charm sector reveal more and more secrets. A complete understanding needs a complete picture of the static properties and the decays of as many hadrons as accessible. Since unflavoured, hidden flavoured and open flavoured mesons and baryons have their particular properties and discovery potential it is vital to survey them in all possible circumstances like in radial but also in gluonic excitations and by exposing these hadrons to the nuclear medium to investigate the chiral effects. By that, the binding in between the quarks is put to the limits which will reveal more about the secrets of the least understood interaction, the strong force. It is this force which generates 98% of the mass of the proton which we not understand at low energies and large distances.

The framework for this investigation are lattice gauge theory and chiral pertubation theory with important predictions within the validity of the respective approaches. But this report is not on theory but the new projects which are running now or have been proposed or already approved. They will allow high precision tests of the models in various different fields. What is common to this generation of experiments is that they provide high-precision and a broad mass range to cover all the sectors mentioned before.

CP717, Hadron Spectroscopy: Tenth International Conference,
edited by E. Klempt, H. Koch, and H. Orth
© 2004 American Institute of Physics 0-7354-0197-7/04/$22.00

Mission statement

The objects of interest are anything which consists of hadrons with or without gluonic degrees of freedom. In the limit of infinite masses a reasonable ordering scheme is the following:

- Mesons and Baryons with the minimal color neutral quark content of $q\bar{q}$ and qqq respectively.
- Multiquarks, sometimes quoted as molecules, which consist of a more complex structure of several mesons and/or baryons like the tetra quark $qq\bar{q}\bar{q}$ or the penta quark $qqqq\bar{q}$. If the quantum numbers allow for it they are just Fock-states of ordinary mesons and baryons.
- Hybrids have instead of a radial excitation of the gluon-exchange potential an excited gluon flux which can be in a pictorial way understood as a vibration of the gluon flux which adds to the quantum number of the final state.
- Glueballs which are a manifest excitation of the strong vaccum. It's a closed gluon flux without any valence quarks.

In summary the objects to deal with, have the principle global structure $q^{3i+n}\bar{q}^{3j+n}g^k$. For example baryons have $i = 1$ and $j, n, k = 0$ ($B = i - j$) while glueballs have $i, j, n = 0$ and $k > 1$.

Hadron spectroscopy provides vital information about the static properties of the hadrons, which are essential to order them into spectra. Since they are the spectra in it's full glory which provide the evidence for understanding the binding of quarks an almost complete coverage of quantum numbers and excitations is necessary to complete the task.

Observables

The important starting point is the knowledge about the mass and the width of any given resonance. In many cases this is easy to measure and can be done also with limited statistics in dirty environments. It is more complicated if the width is extremely small or extremely broad. In the first case it's an experimental in the second case an analysis problem due to resolution and unitarity bounds of the waves respectively. Furthermore the variety of different decay channels offer crucial information about the structure of an object. The same applies to the knowledge about the production modes. The different production and decay modes also manifest the quantum numbers about the minimal quark content, like I^G, c, s etc..

Much more statistics is required to determine properly the spin and parity of the desired state. This needs a well known initial state and/or a well defined reaction plane to use a full spin-parity analysis which depending on the reaction is a dalitzplot-like multi-variant analysis or a partial wave decomposition.

There is a broad spectrum of experiments of the current and next generation in light quark spectroscopy: Hall-D (Jlab), CB-Elsa (Bonn), Compass (CERN), p-Beams (Cosy), Panda (GSI) and others. Unfortunately the light quark sector has always to deal with the many ambiguities for the interpretation of established and new states since the density of states is very large and the width of the states tends to be very broad, thus resulting in a large mixing of states which makes it extremely complicated to clarify the situation. Already in the pure $q\bar{q}$ sector we know the isospin mixing (like ρ and ω), the isoscalar mixing (like f_2 and f_2') and the mixing of kaons from nonets with the same parity but opposite C-parity. In addition there are the many glueball and hybrid states expected which would mix as well as some molecules where the attraction between mesons is strong enough to produce bound states.

A much better situation can be found in the charmed (open and hidden charm) states, where the level density is much smaller and many states are extremely narrow and new states are much easier to find. Any deviation from a mass prediction can easily be measured. The main problem in that sector is statistics, since charm quarks have to be produced either in hadronic processes from gluons (Panda, Compass), or photons (E791, Focus,...) by pair production, or weakly in b-decays (Babar, Belle, Cleo,...) or pair production/fragmentation of electron positron machines (Babar, Belle, Bes, Cleo,...).

Ingredients - Statistics and Resolution

Since it is not sufficient to find a bump and associate with it a mass and some estimate for width it is important to measure the full spectral function and determine the spin-parity. To achieve this, high statistics is a mandatory tool to get this right. For example the complicated S wave structure of $\pi\pi$ interactions – with the $f_0(1500)$ being a major contributor to it – would have never been established without huge statistics in the $p\bar{p} \rightarrow 3\pi^0$ dalitzplot. With more observables the requirement for statistics can be somehow relaxed since the many observables constrain the waves a lot if there are not too many ambiguous broad resonances are involved. The high statistics is needed to disentangle the many overlapping resonances and the crossing interferences which heavily effect the angular distributions. To enhance statistics there are several possibilities apart from just higher luminosity. The other aspects are efficiency of the detector, dead-time due to readout, trigger-efficiency due to electronics and the choice of appropriate channels.

Another aspect is resolution which is important for narrow resonances, since the resolution is folded with the intrinsic width in the experimental spectrum. But resolution is as well necessary for broad structures in multi hadron final states to reduce combinatorics and therefore combinatorial background which is very complicated to handle in spin-paritiy analyses. To reduce this background, apart from detector geometry and layout, can be achieved by using intrinsic properties of the production and/or decay mode and thus something which has to be considered rather early in the design phase of experiments. Such constraints apply for example in the decay of B-mesons in a B-Factory event coming from the $\Upsilon(4s)$ where one can use the knowledge of one B-Meson to constrain the other one, which has been for example successfully used in the search for the $\eta_c(2s)$. Another possibility is the use of formation from the beam and target particle, like in the τc-factories or in $p\bar{p}$ collisions. The resolution is then given just by the properties of the beam and the detector resolution matters only in that point of unambiguous reconstruction of the desired final state. In that respect the $p\bar{p}$ formation is more advantageous, since the hadron doesn't have much bremsstrahlung and an appropriate beam quality is easier to achieve. The electron positron machines always have to suffer from bremsstrahlung which limits any width measurement and usually adds to the systematic error of the mass determination. On the other side $p\bar{p}$ produces a large hadronic background which has to be suppressed using sophisticated detectors and triggers. In any case peaking backgrounds, large combinatorial background and particle mis-identification can be avoided with reasonable resolution which depends strongly on the observable of interest.

PROCESSES

Some process are more suitable for the production of certain hadrons than other. This is obvious in terms of the OZI rule which prevers intermediate states with a similar quark content than the initial and the final state. But as a fact, due to final state interaction of the produced mesons and other suppressed mechanisms it is always possible to create most resonances in any reaction, it's only a matter of cross section/branching fraction. In this section the most desirable cases for exotic and non-exotic resonances are discussed.

Meson and Baryon production

The most promising processes for meson and baryon production are processes where the valence quarks are rearranged and only part of the hadron is produced via hard gluons or from the sea. So pp scattering, $N\bar{N}$ annihilations (like $p\bar{p}$, $n\bar{p}$ + isospin conjugated) or any kind of meson baryon scattering, like πp and Kp are well suited to produce ground state and radially excited conventional hadrons. Nevertheless due to the cleanliness of the reactions also weak decays of heavy mesons like $b \rightarrow c$ ($B \rightarrow DX$) or $c \rightarrow light$ ($D_s \rightarrow K\bar{K}\pi$, $D \rightarrow 3\pi$) can considarably increase the knowledge of certain resonances. Ideally the study of conventional hadrons is performed as close to threshold as possible to reduce the number of partial waves involved but keeping in mind a minimum statistics to achieve an unambiguous solution.

Hybrid and Glueball production

For the production of gluonically excited states like hybrids and glueballs it is important to get the gluon flux excited which will then manifest in the gluonic excitation of the hadron. A natural place to investigate hybrids is in annihilation processes where a lot of gluons are arround to create a coherent gluon flux which adds to the quantum numbers of the final state (Crystal Barrel and Obelix at LEAR). The production of spin-exotic states in such a reaction has been shown to have the same yield as ordinary mesons. A less favorite mechanism is to measure them in hadron hadron scattering as it was done at BNL (E818, E852) and Serpukhov (VES). The problem here is the coupling of the hybrid to the exit channel which is in the $\rho\pi$ case very small and the more suppressed

$b_1\pi$ channel has to act in that way. Due to that, the relative yields between exotic and non-exotic states is at least one order of magnitude smaller than in hadron annihilations. A way out of that is the investigation of photo production at high energies (Hall-D at JLab) where also a reasonable yield of exotics is predicted. The model behind that is the direct vibrational excitation of the gluon flux by the photoinduced hadronemission.

Glueball formation needs the coherent production of several gluons. Very promising processes are the radiative J/ψ decays, since two of the three mandatory vector bosons for it's decay are gluons, while the third is a photon as a trigger. Also ordinary mesons are produced but the rate for glueballs should be high. Other processes with impressive results in the last decade are proton antiproton annihilation (where a lot of gluons are present from the quark antiquark annihilation) and the central production from two hadrons in peripheral collision. This should proceed via pomeron exchange and therefore a strong coupling to coherent glue leads to an enhancement of the formation of glueballs. A nice crosscheck in that respect is the investigation of two photon fusion in electron positron machines where each lepton emits photons from bremsstrahlung which may form hadrons electromagnetically. This production process should favor ordinary mesons and disfavor anything else since it would have to be produced by final state interaction and/or mixing. Nevertheless, since the mixing matrixes are usually badly known this tool can only be used as additional input and not as stand-alone evidence.

Multiquark production

There is no clear picture how molecules are produced. A popular picture is to form them in final state interaction, so most process are suitable to produce molecules. The tetra quarks (mesonic multiquarks, B=0) in the light quark sector should appear naturally together with ordinary mesons by final state interaction. Another possibility is to create directly a system of two quarks and two anti quarks in nucleon antinucleon annihilation and tag this by a Drell-Yan like produced lepton pair. For baryonic multiquarks (B=1) a penta quark system is the simplest case which should be observable in photo production and π,K N scattering. In the heavy quark sector (hidden and open flavour) there are no obvious channels since neither projectile nor target contain the neccessary flavor. Thus it can be investigated in all reactions where (open,hidden)-flavoured mesons are produced.

Determination of quantum numbers

The determination of quatnum numbers is usually the most complicated and tedious procedure, since many aspects are involved. If the resonance is very narrow, then the lineshape doesn't matter, but as soon as the width is large asnd several broad resonances interfere, the problem of maintaining unitarity appears. In that case simple relativistic Breit-Wigner functions are no longer applicable and other approaches which preserve unitiarity must be applied. The resonance itself is only defined by a pole in the complex energy plane of the T-Matrix. So it has a phasemotion which has a complicated pattern and fixed poles, but the couplings may depend on the reaction and therefore the peaks representing the resonance in the real world end up at different energies. So in addition to high statistics and precision also appropriate models have to be used to parametrize the date in order to reveal the resonant structure of the desired wave. But this is not possible in all cases. It is desirable to have the least possible bias due to the analysis model which cannot be achieved in all processes. The next chapter will discuss this in more detail.

An energy independent partial wave analysis is the only unbiased technique to decompose the different waves and impose a dynamical structure afterwards. This is possible in all scattering experiments, like π, K and p scattering on nucleons and photoproduction on nucleons if enough observables are measured to define all planes and axis. But this is only true for infinitely narrow resonances in intermediate states. As soon as broad resonances are involved, decay models (like the isobar model) with or without unitary T-matrix descriptions have to be applied to them which yields systematic bias. Another systematic problem is the maximum angular momentum used for the decomposition and leaves some uncertainties in the analysis like assumption to be made for the t-channel processes.

In other processes with a fixed energy and thus with the need of a recoil system, the initial state can be analysed in terms of an isobar model and therefore has the same limitations in terms of the decay model. An advantage is usually well defined initial state and the possibility of interferences with the recoil system which constrain the dynamical and angular part of the wave description. If the nucleon antinucleon system annihilates in flight this process is again a scattering process which is quite complicated to analyse but has been succesfully performed at Crystal Barrel at LEAR for certain cases. It is important that narrow "final state" particles like ω, ϕ, ... with spin are involved, so that enough constraints for a proper decomposition exist.

The isobar analysis assumes that two particles form always a resonance which completely factorizes with the recoil system. If rescattering is important than this technique is not applicable any longer and one has to live with systematically limited information from that channel.

In some cases which are important for hadron production it is rather complicated or even impossible to do a spin-parity analysis at all. This is for example the case in the electron positron fragmentation into hadrons. There is no well defined initial state and due to the unknown polarization of the mother system a spin-parity determination is unlikeli. The same applies to central production.

Something which is common to all channels is the necessity to have interfering resonances in adition to the desired signal since only interference makes the phase motion observable. To maximise its significance it is desirable to have the signal and the interfering reference resonance with the same yield, thus one should avoid dominance of one or the other which would weaken the evidence.

The case for less favoured modes

In the previous chapter the favourable production channels in terms of rate have been discussed. But since it is important not just to produce, but also to analyse the resonances it is in many cases more appropriate to produce the resonance in a less favoured production mode. This can have many reasons:

- The J^{PC} of the initial state might be well known,
- the reaction involves only low spins which improves the significance of the result conventional
- or exotic states might be supressed respectively or the resolution is extgrmely good due to applicable toplogical constraints.

It is alsways important to compensate the loss in yield by high luminosity and long date taking leading to the many meson factories all around the world.

EXPERIMENTAL APPROACHES

Taking all what has been discussed into account the following picture arises for the different topics for the decade to come.

Hybrid Search

The search for mesonic hybrids has to be divided into two groups, light quark and heavier mesons. There were already some hadrons being discovered with exotic quantum numbers. These and more can be studied at Hall-D (Jlab, if the upgrade gets approved), CB-Elsa, Compass and Panda at GSI. Important topics are the investigation of multiplet partners of the known spin-exotics and additional decay channels as well as a detailed lineshape investigation to prove the resonance nature of all of them.

Charmed hybrids ($c\bar{c}g$) are ideally studied at Panda (GSI) but investiagtions are also going on in B-meson decays at Babar and Belle. Vector hybrids can be studied at Bes. It is also mandatory to look for open charm hybrids which at the moment will only be possible at Panda due to the limited energy range of all other experiments. For the interpretation it is important to complete the spectrum of conventional charmonium and open charm states as a reference.

The natural place to look for baryonic hybrids are photo and meson induced reactions on nucleons like those investigated with Clas and Hall-D (Jlab) and CB-Elsa (Bonn). But since there are no spin-exotics because all spin-parities can be formed in an ordinary way without excited glue it will be very complicated if not impossible to unambiguously identify a state without a better knowledge of the conventional spectrum.

Glueball Search

In the scalar sector the situation between 1.1 and 1.8 GeV/c^2 is experimentally rather clean and many signals like the $f_0(1500)$ and the $f_0(1710)$ show the evidence for an overpopulation and therefore the evidence for the scalar glueball. Nevertheless the heavy mixing complicates the situation and more information is certainly needed. It will gain important input from the radiative ϕ-decays and some hadronic $D_{(s)}$-decays which will help to understand in particular the 1 GeV/c^2 region at the $K\bar{K}$-threshold.

The tensor sector still misses a significant candidate for the tensor glueball and the τc-factories Cleo-c and Bes-III hopefully can clarify the situation in radiative J/ψ decays. Later the Panda-Experiment will contribute with $p\bar{p}$ fine scans between 2.0 and 2.5 GeV/c^2 in many final states.

In the Oddball sector, that means the glueballs with exotic quantum numbers the very heavy ones with $J^{PC} = 0^{+-}$ and 2^{+-} will be looked for in the Panda experiment in parallel to the charmed hybrid search in open and hidden charm decays.

Multiquark Search

As mentioned before multi quarks can be searched in almost all areas. The Θ^+ can be looked for in all reactions with baryonic targets. The knowledge of the width is important for its understanding. But if the width is really very narrow, it can only the measured in scan experiments. If it is a few MeV wide it might be possible to scan it at Cleo-c but due to the dense program it is very unlikly to be performed. If it is even narrower then one must wait until Panda got data on it in $p\bar{p}$ formation at the pair production threshold. Another opportunity is to search for charmed penta quarks which can be done at the B-factories (Babar and Belle) or Panda in fixed target mode.

If the newly discovered D_s states are really tetra quarks, then they may have other charge modes which can be studied in many places which then will clarify the picture if they are ordinary mesons, chiral partners or really multiquark states. Also here the width is an important issue and can only be studied by threshold pair production in $p\bar{p}$ scans since the bremstrahlung effect limits the resolution of any electron positron scan.

SUMMARY

It is important to improve the knowledge of the spectrum of hadrons in several dimensions by putting the hadrons to the limit in terms of excitations and the medium they are inside. Important excitations are the radial and the gluonic excitation of the gluon mediating the strong force between two quarks. Since so many hadronic projects got approved this year (Cleo-c,Bes-III and Panda) most problems can be addressed with the running or approved experiments like:

- Confinement (-potential) by charmonium spectroscopy:
  Panda (all J^{PC}), Cleo-c, Bes-III (only 1^{--})
- Gluonic Components: Cleo-c, Bes-III, Panda (light gg)
  E852, HallD, Panda (uds-Hybrids), Panda (charmed Hybrids)
- Open charm: Panda, Babar, Belle, Cleo-c, Bes-III, Compass
- AntiDecuplet: Clas,CB-Elsa (Θ^+-State), Cleo-c, Panda (Θ^0_c-State)

There is nearly a full coverage of the envisaged sprectrum with mall redundancy which is well suited for maximum output in the next decade. The only aspect which is really missing is a $K^{\pm}$ program in hadron spectroscopy to get a better picture on strange hadrons and strangeness production.

ACKNOWLEDGMENTS

This work is supported by Bundesministerium für Forschung und Bildung (bmb+f) and the Gesellschaft für Schwerionenforschung mbH (GSI).

Threshold Perspectives on Meson Production

M. Wolke

The Svedberg Laboratory, Uppsala University, Box 533, 75121 Uppsala, Sweden *

Abstract. Studies of meson production in nucleon–nucleon collisions at threshold are characterised by few degrees of freedom in a configuration of well defined initial and final states with a transition governed by short range dynamics. Effects from low–energy scattering in the exit channel are inherent to the data and probe the interaction in baryon–meson and meson–meson systems otherwise difficult to access.

From dedicated experiments at the present generation of cooler rings precise data are becoming available on differential and eventually spin observables allowing detailed comparisons between complementary final states. To discuss physics implications of generic and specific properties, recent experimental results on meson production in proton–proton scattering obtained at CELSIUS and COSY serve as a guideline.

INTRODUCTION

High precision data from the present generation of cooler rings, IUCF, CELSIUS, and COSY, have contributed significantly over the last decade to our present knowledge and understanding of threshold meson production (for a recent review see [1]).

Due to the high momentum transfers required to create a meson or mesonic system in production experiments close to threshold the short range part of the interaction is probed. In nucleon–nucleon scattering, for mesons in the mass range up to $1\,\text{GeV}/c^2$ distances from $0.53\,\text{fm}$ (π^0) down to less than $0.2\,\text{fm}$ (ϕ) are involved. At such short distances it is a priori not clear, whether the relevant degrees of freedom are still baryons and mesons, or rather quarks and gluons. As there is no well defined boundary, one goal of the threshold production approach is to explore the limits in momentum transfer for a consistent description using hadronic meson exchange models. Within this framework, questions concerning both the underlying meson exchange contributions and especially the role of intermediate baryon resonances have to be answered.

Another aspect which enriches the field of study arises from the low relative centre–of–mass velocities of the ejectiles: Effects of low energy scattering are inherent to the observables due to strong final state interactions (FSI) within the baryon–baryon, baryon–meson, and meson–meson subsystems. In case of short–lived particles, low energy scattering potentials are otherwise difficult or impossible to study directly.

---

* on leave from Institut für Kernphysik, Forschungszentrum Jülich, 52425 Jülich, Germany

CP717, *Hadron Spectroscopy: Tenth International Conference,*
edited by E. Klempt, H. Koch, and H. Orth
© 2004 American Institute of Physics 0-7354-0197-7/04/$22.00

DYNAMICS OF THE TWO PION SYSTEM

In γ and π induced double pion production on the nucleon the excitation of the $N^*(1440)$ P_{11} resonance followed by its decay to the $N\sigma$ channel, i.e. $N^*(1440) \to p(\pi\pi)_{I=l=0}$, is found to contribute non–negligibly close to threshold [2–4]. Nucleon–nucleon scattering should provide complementary information, eventually on the $\pi\pi$ decay mode of the $N^*(1440)$, which plays an important part in understanding the basic structure of the second excited state of the nucleon [5–7].

Exclusive CELSIUS data from the PROMICE/WASA setup on the reactions $pp \to pp\pi^+\pi^-$, $pp \to pp\pi^0\pi^0$ and $pp \to pn\pi^+\pi^0$ [8–10] are well described by model calculations [11]: For the $\pi^+\pi^-$ and $\pi^0\pi^0$ channels, the reaction preferentially proceeds close to threshold via heavy meson exchange and excitation of the $N^*(1440)$ Roper resonance, with a subsequent pure s–wave decay to the $N\sigma$ channel[1]. While nonresonant contributions are expected to be small, resonant processes with Roper excitation and decay via an intermediate Δ ($pp \to pN^* \to p\Delta\pi \to pp\pi\pi$) and $\Delta\Delta$ excitation ($pp \to \Delta\Delta \to p\pi p\pi$) are strongly momentum dependent and vanish directly at threshold. Double Δ excitation, which is expected to dominate at higher excess energies beyond $Q = 250\,\mathrm{MeV}$ [11] involves higher angular momenta and consequently strongly anisotropic proton and pion angular distributions. On the other hand, the Roper decay amplitude via an intermediate Δ depends predominantly on a term symmetric in the pion momenta (eq.(1)), leading to the $p(\pi^+\pi^-)_{I=l=0}$ channel and an interference with the direct $N\sigma$ decay.

Experimentally, for the reaction $pp \to pp\pi^+\pi^-$ at excess energies of $Q = 64.4\,\mathrm{MeV}$ and $Q = 75\,\mathrm{MeV}$ angular distributions give evidence for only s–waves in the final state, in line with a dominating $pp \to pN^* \to pp(\pi^+\pi^-)_{I=l=0}$ process, with the initial inelastic pp collision governed by heavy meson (σ, ρ) exchange. Roper excitation

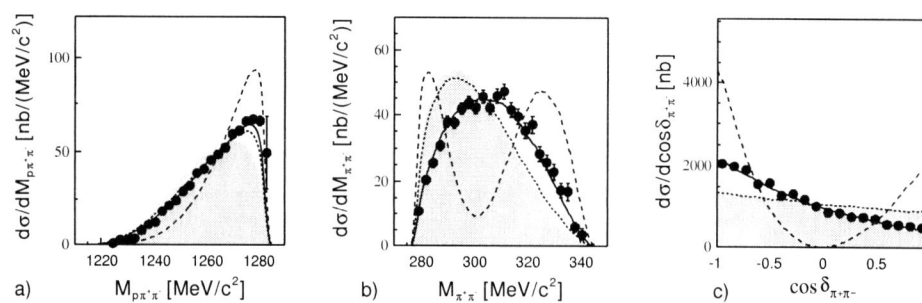

FIGURE 1. Differential cross sections for the reaction $pp \to pp\pi^+\pi^-$ at an excess energy of $Q = 75\,\mathrm{MeV}$. Experimental data (solid circles) for invariant mass distributions of the $(p\pi^+\pi^-)$– (a) and $(\pi^+\pi^-)$–subsystems (b), and the $\pi^+\pi^-$ opening angle (c) are compared to pure phase space (shaded areas) and Monte Carlo simulations for direct decays $N^* \to N\sigma$ (dotted lines), decays via an intermediate Δ resonance $N^* \to \Delta\pi \to N\sigma$ (dashed lines) and an interference of the two decay routes (solid lines) according to eq.(1). Figures are taken from [10].

---

[1] For the $pn\pi^+\pi^0$ final state, this reaction mechanism is trivially forbidden by isospin conservation. An underestimation of the total cross section data [9] by the model predictions [11] might be explained by the neglect of effects from the pn final state interaction in the calculation [12].

is disclosed in the $p\pi^+\pi^-$ invariant mass distribution (Fig.1a), where the data are shifted towards higher invariant masses compared to phase space in agreement with resonance excitation in the low energy tail of the $N^*(1440)$. Compared with Monte Carlo simulations including both heavy meson exchange for N^* excitation, and pp S–wave final state interaction, but only the direct decay $N^* \rightarrow p(\pi^+\pi^-)_{I=l=0}$ (dotted lines), the production process involves additional dynamics, which is apparent from discrepancies especially in observables depending on the π momentum correlation $\vec{k}_1 \cdot \vec{k}_2$, i.e. $\pi^+\pi^-$ invariant mass $M_{\pi\pi}$ (Fig.1b) and opening angle $\delta_{\pi\pi} = \angle(\vec{k}_1 \cdot \vec{k}_2)$ (Fig.1c). A good description of the experimental data is achieved including the $N^*(1440)$ decay via an intermediate Δ and its destructive interference with the direct decay branch to the $N\sigma$ channel (solid lines) in the ansatz for the Roper decay amplitude [11]:

$$\mathcal{A} \propto 1 + c\vec{k}_1 \cdot \vec{k}_2 \left(3D_{\Delta^{++}} + D_{\Delta^0}\right), \tag{1}$$

where the first term describes the direct decay, the parameter c adjusts the relative strengths of the two decay routes, and D_Δ denote the Δ propagators. A fit to the data allows to determine the ratio of partial decay widths $\mathcal{R}(M_{N^*}) = \Gamma_{N^*\rightarrow\Delta\pi\rightarrow N\pi\pi}/\Gamma_{N^*\rightarrow N\sigma}$ at average masses $< M_{N^*} >$ corresponding to excess energies $Q = 64.4$ MeV and $Q = 75$ MeV relative to the $\pi^+\pi^-$ threshold. The numerical results, $\mathcal{R}(1264) = 0.034 \pm 0.004$ and $\mathcal{R}(1272) = 0.054 \pm 0.006$, exhibit the clear dominance of the direct decay to the $N\sigma$ channel in the low energy region of the Roper resonance. On the other hand they indicate the strong energy dependence of the ratio from the momentum dependence in the decay branch via an intermediate Δ, which will surpass the direct decay at higher energies [10]. A model dependent extrapolation based on the validity of ansatz (1) leads to $\mathcal{R}(1440) = 3.9 \pm 0.3$ at the nominal resonance pole in good agreement with the PDG value of 4 ± 2 [13].

Within the experimental programme to determine the energy dependence of the $N^* \rightarrow N\pi\pi$ decay exclusive data (for details see [14]) have been taken simultaneously at the CELSIUS/WASA facility on both the $pp\pi^+\pi^-$ and $pp\pi^0\pi^0$ final states. In case of the $\pi^+\pi^-$ system the preliminary results at an excess energy of $Q = 75$ MeV are in good agreement with the relative strength of the decay routes adjusted to an extrapolated ratio $\mathcal{R}(1440) = 3$. However, at slightly higher excess energy ($Q = 127$ MeV) the data might be equally well described by a value $\mathcal{R}(1440) = 1$, which is noticeably favoured at both excess energies by the data on $\pi^0\pi^0$ production, indicating distinct underlying dynamics in $\pi^0\pi^0$ and $\pi^+\pi^-$ production. One difference becomes obvious from the isospin decomposition of the total cross section [9]: An isospin $I = 1$ amplitude in the $\pi\pi$ system, and accordingly a p–wave admixture, is forbidden by symmetry to contribute to the neutral pion system in contrast to the charged complement. A p–wave component was neglected so far in the analysis, since the unpolarized angular distributions show no deviation from isotropy. However, there is evidence for small, but non–negligible analysing powers from a first exclusive measurement of $\pi^+\pi^-$ production with a polarized beam at the COSY–TOF facility [14, 15], suggesting higher partial waves especially in the $\pi\pi$ system.

At higher energies, i.e. $Q = 208$ MeV and $Q = 286$ MeV with respect to the $\pi^+\pi^-$ threshold, preliminary data for both $\pi^+\pi^-$ and $\pi^0\pi^0$ from CELSIUS/WASA rather follow phase space than expectations based on a dominating $pp \rightarrow pN^* \rightarrow pp\sigma$ reaction mechanism[14]. At these energies, the $\Delta\Delta$ excitation process should influence observables significantly, and, thus, a phase space behaviour becomes even more surprising, unless the $\Delta\Delta$ system is excited in a correlated way.

THE PROTON–PROTON–ETA FINAL STATE

As a general trait in meson production in nucleon–nucleon scattering, the primary production amplitude, i. e. the underlying dynamics can be regarded as energy independent in the vicinity of threshold [16–18]. Consequently, for s–wave production processes, the energy dependence of the total cross section is essentially given by a phase space behaviour modified by the influence of final state interactions. In Fig.2 total cross section data obtained in proton–proton scattering are shown for the pseudoscalar isosinglet mesons η and η' [19]. In both cases, the energy dependence of the total cross sec-

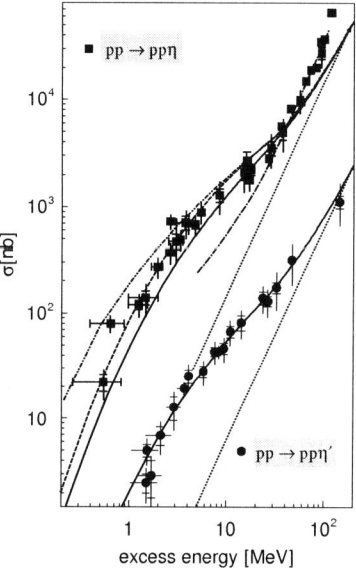

FIGURE 2. Total cross section data for η (squares [20–25]) and η' (circles [24, 26–29]) production in proton–proton scattering versus excess energy Q [19]. In comparison, the energy dependences from a pure phase space behaviour (dotted lines, normalized arbitrarily), from phase space modified by the $^1S_0$ proton–proton FSI including Coulomb interaction (solid lines), and from additionally including the proton–η interaction phenomenologically (dashed line), are shown. Meson exchange calculations for η production including a P–wave component in the proton–proton system [30] are depicted by the dashed–dotted line, while the dashed–double–dotted line corresponds to the arbitrarily normalized energy dependence from a full three–body treatment of the $pp\eta$ final state [31] (see also [32]).

tion deviates significantly from phase space expectations. Including the on–shell $^1S_0$ proton–proton FSI enhances the cross section close to threshold by more than an order of magnitude, in good agreement with data in case of η'. As expected from kinematical considerations [1] the cross section for η production deviates from phase space including the pp FSI at excess energies $Q \geq 40\,\text{MeV}$, where the $^1S_0$ final state is no

20

longer dominant compared to higher partial waves. Deviations at low excess energies seem to be well accounted for by an attractive proton–η FSI (dashed line), treated phenomenologically as an incoherent pairwise interaction [1, 17, 33]. In comparison to the proton–η' (Fig.2) and proton–π^0 systems only the $p\eta$ interaction is strong enough to become apparent in the energy dependence of the total cross section [34]. In differential observables, effects should be more pronounced in the phase space region of low proton–η invariant masses. However, to discern effects of proton–η scattering from the influence of proton–proton FSI, which is stronger by two orders of magnitude, requires high statistics measurements, which have only become available recently [19, 35, 36]: Close to threshold, the distribution of the invariant mass of the proton–proton subsys-

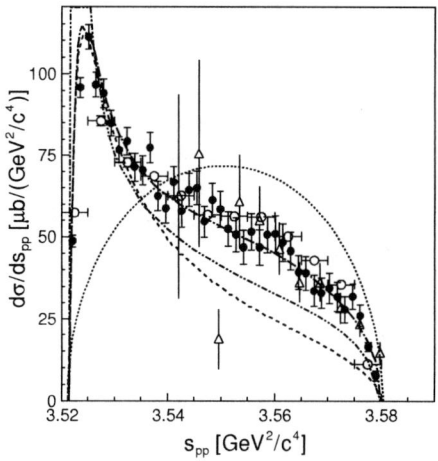

FIGURE 3. Invariant mass squared of the (pp)–subsystem in the reaction $pp \rightarrow pp\eta$ at excess energies of Q = 15.5 MeV (COSY–11, solid circles [19]), Q = 15 MeV (COSY–TOF, open circles [36] and Q = 16 MeV (PROMICE/WASA, open triangles [35]). The dotted and dashed lines follow a pure phase space behaviour and its modification by the phenomenological treatment of the three–body FSI as an incoherent pairwise interaction, respectively. The latter was normalized at small invariant mass values. Effects from including a P–wave admixture in the pp system are depicted by the dashed–dotted line [30], while the dashed–double–dotted line corresponds to a pure s–wave final state with a full three–body treatment [32].

tem is characteristically shifted towards low invariant masses compared to phase space (dotted line in Fig.3). This low–energy enhancement is well reproduced by modifying phase space with the $^1S_0$ pp on–shell interaction. A second enhancement at higher pp invariant masses, i.e. low energy in the $p\eta$ system, is not accounted for even when including additionally the proton–η interaction incoherently (dashed line). However, including a P–wave admixture in the pp system by considering a $^1S_0 \rightarrow {}^3P_0$s transition in addition to the $^3P_0 \rightarrow {}^1S_0$s threshold amplitude, excellent agreement with the experimental invariant mass distribution is obtained (dashed–dotted line [30]). In return, with the P–wave strength adjusted to fit the invariant mass data, the approach fails to reproduce the energy dependence of the total cross section (Fig.2) below excess energies of Q = 40 MeV. Preliminary calculations considering only s–waves in the final state but using a rigorous three–body treatment of the $pp\eta$ final state actually decrease the cross section at large values of the pp invariant mass (dashed–double–dotted lines [32]) compared to an incoherent two–body calculation within the same framework. However, close to threshold the energy dependence of the total cross section is enhanced compared to the phenomenological incoherent treatment and the data (Fig.2). Although part of this enhancement has to be attributed to the neglect of Coulomb repulsion in the pp system, consequently overestimating the pp invariant mass at low values, qualitatively the full three–body treatment has opposite effects compared to a P–wave admixture in the proton–proton system in view of both the total cross section as well as the pp invariant mass distribution. In the approximate description of the total cross section by the phenomenological s–wave approach with an incoherent FSI treatment these two effects seem to cancel casually.

Close to threshold, resonance excitation of the $S_{11}(1535)$ and subsequent decay to the $p\eta$ final state is generally[2] believed to be the dominant η production mechanism [17, 30, 38–43]. In this context, the issue of the actual excitation mechanism of the $S_{11}(1535)$ remains to be addressed. The η angular distribution is sensitive to the underlying dynamics: A dominant ρ exchange favoured in [41] results in an inverted curvature of the η angular distribution compared to π and η exchanges which are inferred to give the largest contribution to resonance excitation in [42]. In the latter approach the interference of the pseudoscalar exchanges in the resonance current with non–resonant nucleonic and mesonic exchange currents turns the curvature to the same angular dependence as expected for ρ exchange. Presently, due to the statistical errors of the available unpolarised data at an excess energy of $Q \approx 40\,\text{MeV}$ [35, 36] it is not possible to differentiate between a dominant ρ or π, η exchange, as discussed in [36]. Data recently taken at the CELSIUS/WASA facility with statistics increased by an order of magnitude compared to the available data might provide an answer in the near future [44].

Spin observables, like the η analyzing power, should even disentangle a dominant ρ meson exchange and the interference of π and η exchanges in resonance excitation with small nucleonic and mesonic currents [42], which result in identical predictions for the unpolarised η angular distribution. First data [45] seem to favour the vector dominance model, but final conclusions both on the underlying reaction dynamics and the admixture of higher partial waves [30] have to await the analysis of data taken with higher statistics for the energy dependence of the η analysing power [46].

ASSOCIATED STRANGENESS PRODUCTION

In elementary hadronic interactions with no strange valence quark in the initial state the associated strangeness production provides a powerful tool to study reaction dynamics by introducing a "tracer" to hadronic matter. Thus, quark model concepts might eventually be related to mesonic or baryonic degrees of freedom, with the onset of quark degrees of freedom expected for kinematical situations with large enough transverse

---

[2] The $pn \to d\eta$ excitation function has been interpreted to provide direct experimental evidence for $S_{11}(1535)$ excitation [23]. It should be noted, however, that in [37] for η production in proton–proton scattering short range nucleonic currents, i.e. σ and ω exchange are found to be much stronger compared to the contribution from resonance currents.

momentum transfer.

First exclusive close–to–threshold data on Λ and Σ^0 production [47, 48] obtained at the COSY–11 facility showed at equal excess energies below $Q = 13\,\mathrm{MeV}$ a cross section ratio of

$$\mathscr{R}_{\Lambda/\Sigma^0} (Q \leq 13\,\mathrm{MeV}) = \frac{\sigma\,(pp \to pK^+\Lambda)}{\sigma\,(pp \to pK^+\Sigma^0)} = 28^{+6}_{-9} \qquad (2)$$

exceeding the high energy value ($Q \geq 300\,\mathrm{MeV}$) of 2.5 [49] by an order of magnitude.

In the meson exchange framework, estimates for π and K exchange contributions based on the elementary scattering processes do not reproduce the experimental value (2) [48, 50]. However, inclusive K^+ production data in pp scattering at an excess energy of $Q = 252\,\mathrm{MeV}$ with respect to the $pK^+\Lambda$ threshold show enhancements at the Λp and ΣN thresholds of similar magnitude [51]. Qualitatively, a strong $\Sigma^0 N \to \Lambda p$ final state conversion might account for both the inclusive SATURNE results as well as the Σ^0 depletion in the COSY–11 data. Evidence for such conversion effects is known e. g. from fully constrained kaon absorption on deuterium via $K^- d \to \pi^- \Lambda p$ [52].

In exploratory calculations performed within the framework of the Jülich meson exchange model [50], taking into account both π and K exchange diagrams in a coupled channel approach, a final state conversion is rather excluded as origin of the experimentally observed ratio: While Λ production is found to be dominated by kaon exchange, both π and K exchange turn out to contribute to the Σ^0 channel with similar strength. Qualitatively, this result is experimentally confirmed at higher excess energies between $Q = 200\,\mathrm{MeV}$ and $Q = 430\,\mathrm{MeV}$ from polarization transfer measurements from the DISTO experiment [53–55]. It is concluded in [50], that only a destructive interference of π and K exchange might explain the experimental value (2). Σ production in different isospin configurations should provide a crucial test for this interpretation, since for the reaction $pp \to nK^+\Sigma^+$ the interference pattern is found to be opposite compared to the $pK^+\Sigma^0$ channel. Data close to threshold have recently been taken at the COSY–11 facility [56].

However, within an effective Lagrangian approach [57] both Λ and Σ^0 production channels are concluded to be dominated by π exchange and excitation of the $S_{11}(1650)$ close to threshold, while at excess energies above $Q = 300\,\mathrm{MeV}$ the $N^*(1710)$ governs strangeness production[3]. In this energy range the influence of resonances becomes evident from recent data on invariant mass distribution determined at COSY–TOF [59].

To study the transition region between the low–energy enhancement (2) and the high energy value measurements have been extended up to excess energies of $Q = 60\,\mathrm{MeV}$ [58, 60]: In order to describe the energy dependence of the total cross section for Λ production, in addition to phase space the $p\Lambda$ final state interaction has to be taken into account. In contrast, Σ^0 production is satisfactorily well described by phase space behaviour only [58]. This qualitatively different behaviour might be explained by the $\Sigma^0 p$ FSI being much weaker compared to the Λp system. However, the interpretation implies dominant S–wave production and reaction dynamics that can be regarded as energy independent. Within the present level of statistics, contributions from higher

---

[3] For further complementary theoretical approaches see references in [1, 58, 59].

partial waves can be neither ruled out nor confirmed at higher excess energies for Σ^0 production.

The energy dependence of the production ratio $\mathscr{R}_{\Lambda/\Sigma^0}$ is shown in Fig.4 in comparison with theoretical calculations obtained within the approach of [50] assuming a destructive interference of π and K exchange and employing different choices of the microscopic hyperon nucleon model to describe the interaction in the final state [61]. The result

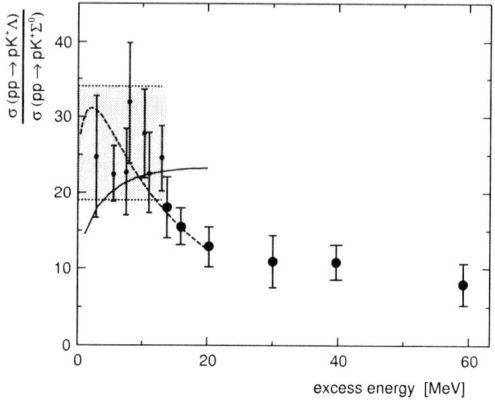

FIGURE 4. Λ/Σ^0 production ratio in proton–proton scattering as a function of the excess energy. Data are from [48] (shaded area) and [60]. Calculations [61] within the Jülich meson exchange model imply a destructive interference of K and π exchange using the microscopic Nijmegen NSC89 (dashed line [62]) and the new Jülich model (solid line [63]) for the YN final state interaction.

crucially depends on the details — especially the off–shell properties — of the hyperon–nucleon interaction employed. At the present stage both the good agreement found in [50] with the threshold enhancement (2) and for the Nijmegen model (dashed line in Fig.4) with the energy dependence of the cross section ratio should rather be regarded as accidental[4]. Calculations using the new Jülich model (solid line in Fig.4) do not reproduce the tendency of the experimental data. It is suggested in [61] that neglecting the energy dependence of the elementary amplitudes and higher partial waves might no longer be justified beyond excess energies of $Q = 20\,\text{MeV}$. However, once the reaction mechanism for close–to–threshold hyperon production is understood, exclusive data should provide a strong constraint on the details of hyperon–nucleon interaction models.

PRESENT AND FUTURE

Intermediate baryon resonances emerge as a common feature in the dynamics of the exemplary cases for threshold meson production in nucleon–nucleon scattering discussed in this article. However, this does not hold in general for meson production in the $1\,\text{GeV}/c^2$ mass range (for a discussion on η' production see [64]). Moreover, the

---

[4] In the latter case an SU(2) breaking in the $^3S_1$ ΣN channel had to be introduced [62] resulting in an ambiguity for the $\Sigma^0 p$ amplitude.

extent to which resonances are evident in the observables or actually govern the reaction mechanism depends on the specific channels, which differ in view of the level of present experimental and theoretical understanding.

The $N^*(1440)$ resonance dominates $\pi^+\pi^-$ production at threshold, and exclusive data allow to extract resonance decay properties in the low–energy tail of the Roper. Dynamical differences between the different isospin configurations of the $\pi\pi$ system and the behaviour at higher energies remains to be understood with first experimental clues appearing.

With three strongly interacting particles in the final state, a consistent description of η production close to threshold requires an accurate three–body approach taking into account the possible influence of higher partial waves. High statistics differential cross sections and polarization observables coming up should straighten out both the excitation mechanism of the $N^*(1535)$ and the admixture of higher partial waves.

At present, the available experimental data on the elementary strangeness production channels give evidence for both an important role of resonances coupling to the hyperon–kaon channels and on a dominant non–resonant kaon exchange mechanism. Experiments on different isospin configurations, high statistics and spin transfer measurements close to threshold should disentangle the situation in future.

From the cornerstone of total cross section measurements, it is apparent from the above examples to what extent our knowledge is presently enlarged by differential observables and what will be the impact of polarization experiments in future to get new perspectives in threshold meson production.

ACKNOWLEDGMENTS

The author gratefully acknowledges the pleasure to work with the CELSIUS/WASA and COSY–11 collaborations, and, in particular, thanks M. Bashkanov, H. Clement, R. Meier, P. Moskal and W. Oelert for helpful discussions . This work has been supported by The Swedish Foundation for International Cooperation in Research and Higher Education (STINT Kontrakt Dnr 02/192).

REFERENCES

1. Moskal, P., Wolke, M., Khoukaz, A., and Oelert, W., *Prog. Part. Nucl. Phys.*, **49**, 1–90 (2002).
2. Oset, E., and Vicente-Vacas, M. J., *Nucl. Phys.*, **A 446**, 584–612 (1985).
3. Bernard, V., Kaiser, N., and Meißner, U.-G., *Nucl. Phys.*, **B 457**, 147–174 (1995).
4. Gómez Tejedor, J. A., and Oset, E., *Nucl. Phys.*, **A 600**, 413–435 (1996).
5. Morsch, H. P., and Zupranski, P., *Phys. Rev.*, **C 61**, 024002 (2000).
6. Krehl, O., Hanhart, C., Krewald, S., and Speth, J., *Phys. Rev.*, **C 62**, 025207 (2000).
7. Hernández, E., Oset, E., and Vicente Vacas, M. J., *Phys. Rev.*, **C 66**, 065201 (2002).
8. Brodowski, W., et al., *Phys. Rev. Lett.*, **88**, 192301 (2002).
9. Johanson, J., et al., *Nucl. Phys.*, **A 712**, 75–94 (2002).
10. Pätzold, J., et al., *Phys. Rev.*, **C 67**, 052202 (2003).
11. Alvarez-Ruso, L., Oset, E., and Hernández, E., *Nucl. Phys.*, **A 633**, 519–546 (1998).
12. Alvarez-Ruso, L. (2002), private communications.
13. Hagiwara, K., et al., *Phys. Rev.*, **D 66**, 010001 (2002).

14. Bashkanov, M., et al., proceedings of this conference (2003).
15. Clement, H., et al., Annual Report 2002, Forschungszentrum Jülich (2003), Jül–4052.
16. Moalem, A., Gedalin, E., Razdolskaja, L., and Shorer, Z., *Nucl. Phys.*, **A 600**, 445–460 (1996).
17. Bernard, V., Kaiser, N., and Meißner, U.-G., *Eur. Phys. J.*, **A 4**, 259–275 (1999).
18. Gedalin, E., Moalem, A., and Razdolskaja, L., *Nucl. Phys.*, **A 650**, 471–482 (1999).
19. Moskal, P., et al., e–Print Archive: nucl–ex/0307005 (2003), Phys. Rev. **C**, in print.
20. Bergdolt, A. M., et al., *Phys. Rev.*, **D 48**, 2969–2973 (1993).
21. Chiavassa, E., et al., *Phys. Lett.*, **B 322**, 270–274 (1994).
22. Calén, H., et al., *Phys. Lett.*, **B 366**, 39–43 (1996).
23. Calén, H., et al., *Phys. Rev. Lett.*, **79**, 2642–2645 (1997).
24. Hibou, F., et al., *Phys. Lett.*, **B 438**, 41–46 (1998).
25. Smyrski, J., Wüstner, P., et al., *Phys. Lett.*, **B 474**, 182–187 (2000).
26. Moskal, P., et al., *Phys. Rev. Lett.*, **80**, 3202–3205 (1998).
27. Moskal, P., et al., *Phys. Lett.*, **B 474**, 416–422 (2000).
28. Balestra, F., et al., *Phys. Lett.*, **B 491**, 29–35 (2000).
29. Khoukaz, A., et al., Annual report 2000/01, Institute of Nuclear Physics, University of Münster (2001), URL http://www.uni-muenster.de/Physik/KP/anrep.
30. Nakayama, K., Haidenbauer, J., Hanhart, C., and Speth, J., *Phys. Rev.*, **C 68**, 045201 (2003).
31. Fix, A. (2003), private communications.
32. Fix, A., and Arenhövel, H., *Nucl. Phys.*, **A 697**, 277–302 (2002).
33. Schuberth, U., Phd thesis, Uppsala University (1995).
34. Moskal, P., et al., *Phys. Lett.*, **B 482**, 356–362 (2000), and references therein.
35. Calén, H., et al., *Phys. Lett.*, **B 458**, 190–196 (1999).
36. Abdel-Bary, M., et al., *Eur. Phys. J.*, **A 16**, 127–137 (2003).
37. Peña, M. T., Garcilazo, H., and Riska, D. O., *Nucl. Phys.*, **A 683**, 322–338 (2001).
38. Batinić, M., Švarc, A., and Lee, T.-S. H., *Phys. Scripta*, **56**, 321–324 (1997).
39. Santra, A. B., and Jain, B. K., *Nucl. Phys.*, **A 634**, 309–324 (1998).
40. Gedalin, E., Moalem, A., and Razdolskaja, L., *Nucl. Phys.*, **A 634**, 368–392 (1998).
41. Fäldt, G., and Wilkin, C., *Phys. Scripta*, **64**, 427–438 (2001).
42. Nakayama, K., Speth, J., and Lee, T.-S. H., *Phys. Rev.*, **C 65**, 045210 (2002).
43. Baru, V., et al., *Phys. Rev.*, **C 67**, 024002 (2003).
44. Zlomanczuk, J. (2003), private communications.
45. Winter, P., et al., *Phys. Lett.*, **B 544**, 251–258 (2002).
46. Czyżykiewicz, R., et al., proceedings of this conference (2003).
47. Balewski, J. T., et al., *Phys. Lett.*, **B 420**, 211–216 (1998).
48. Sewerin, S., Schepers, G., et al., *Phys. Rev. Lett.*, **83**, 682–685 (1999).
49. Baldini, A., Flaminio, V., Moorhead, W. G., and Morrison, D. R. O., *Total Cross-Sections for Reactions of High–Energy Particles*, vol. 12 B of *Landolt–Börnstein: New Series. Group 1*, edited by H. Schopper, Springer, Heidelberg, Germany, 1988, ISBN 3–540–18412–0.
50. Gasparian, A., et al., *Phys. Lett.*, **B 480**, 273–279 (2000).
51. Siebert, R., et al., *Nucl. Phys.*, **A 567**, 819–843 (1994).
52. Tan, T. H., *Phys. Rev. Lett.*, **23**, 395–398 (1969).
53. Balestra, F., et al., *Phys. Rev. Lett.*, **83**, 1534–1537 (1999).
54. Maggiora, M., *πN Newslett.*, **16**, 273–279 (2002).
55. Laget, J. M., *Phys. Lett.*, **B 259**, 24–28 (1991).
56. Rożek, T., and Grzonka, D., COSY Proposal 117, IKP, FZ Jülich, Germany (2002), URL http://www.fz-juelich.de/ikp/en/publications.shtml.
57. Shyam, R., Penner, G., and Mosel, U., *Phys. Rev.*, **C 63**, 022202 (2001).
58. Kowina, P., et al., proceedings of this conference (2003).
59. Wagner, M., et al., proceedings of this conference (2003).
60. Kowina, P., Wolke, M., et al., e–Print Archive: nucl–ex/0302014 (2003).
61. Gasparyan, A., *Symposium on Threshold Meson Production in pp and pd Interaction*, Forschungszentrum Jülich, Germany, 2002, vol. 11 of *Matter and Material*, pp. 205–211.
62. Maessen, P. M. M., Rijken, T. A., and de Swart, J. J., *Phys. Rev.*, **C 40**, 2226–2245 (1989).
63. Haidenbauer, J., Melnitchouk, W., and Speth, J., *AIP Conf. Proc.*, **603**, 421–424 (2001).
64. Moskal, P., et al., proceedings of this conference (2003).

Results from KLOE at DAΦNE

The KLOE collaboration[1]
Presented by C. Bloise

Laboratori Nazionali di Frascati, 00044 Frascati RM, Italy

Abstract.
   The KLOE experiment at the Frascati ϕ factory, DaΦne, has collected $\sim$500 pb^{-1}, i.e. 1.5×10^9 ϕ decays. At the ϕ factory it is possible to select pure K_L and K_S beams. Although the integrated luminosity is insufficient for precision tests of the CP, T symmetries in kaon decays, a wide number of topics in kaon and hadronic physics are accessible from the largest sample of Φ decays at rest collected so far. The cross section $\sigma(e^+e^- \to \pi^+\pi^-\gamma)$ below 1 GeV, relevant for the precise evaluation of the muon magnetic moment, has been measured with a statistical accuracy better than 1%. For the K_S, we obtained the ratio of the branching fractions $\Gamma(K_S \to \pi^+\pi^-(\gamma))/\Gamma(K_S \to \pi^0\pi^0) = (2.239 \pm 0.003_{stat} \pm 0.015_{syst})$, fully inclusive of the $\pi\pi\gamma$ final state. The analysis of the $\sim$20,000 K_S semileptonic decays $K_S \to \pi e\nu$ is being finalized providing precise measurements of both, the K_S semileptonic branching ratio, and Re x_+, i.e. the $\Delta S = \Delta Q$ rule violation parameter. For the K_L, we obtained the ratio $\Gamma(K_L \to \gamma\gamma)/\Gamma(K_L \to \pi^0\pi^0\pi^0) = (2.80 \pm 0.02_{stat} \pm 0.02_{syst}) \times 10^{-3}$, of interest to Chiral Perturbation Theory (ChPT), as well as preliminary results on the branching ratios to other decay modes. In particular, our measurements of the semileptonic decays of both, neutral, and charged kaons will improve the precision of the CKM matrix element $|V_{us}|$, clarifying the present disagreement between different experiments. The ϕ radiative decays, both in scalar and pseudo-scalar mesons, have been analyzed giving new measurements of the $\eta - \eta'$ mixing angle, and of the $\phi \to a_0(980)\gamma$, $\phi \to f_0(980)\gamma$ branching ratios.

KLOE began to take data for physics in year 2000. Since the first data taking campaign the machine luminosity increased countinously (Fig. 1) together with the data quality, and in year 2002 DaΦne has reached a peak luminosity of $8 \times 10^{31} cm^{-2}s^{-1}$, delivering to KLOE 4.2 pb^{-1} per day. The total integrated luminosity is at present $\sim$500 pb^{-1}, corresponding to a sample of 1.3×10^9 kaon pairs produced in a pure $J^{PC} = 1^{--}$ quantum state.

---

[1] The KLOE collaboration: A. Aloisio, F. Ambrosino, A. Antonelli, M. Antonelli, C. Bacci, G. Bencivenni, S. Bertolucci, C. Bini, C. Bloise, V. Bocci, F. Bossi, P. Branchini, S. A. Baulychjov, R. Caloi, P. Campana, G. Capon, T. Capussela, G. Carboni, G. Cataldi, F. Ceradini, F. Cervelli, F. Cevenini, G. Chiefari, P. Ciambrone, S. Conetti, E. De Lucia, P. De Simone, G. De Zorzi, S. Dell'Agnello, A. Denig, A. Di Domenico, C. Di Donato, S. Di Falco, B. Di Micco, A. Doria, M. Dreucci, O. Erriquez, A. Farilla, G. Felici, A. Ferrari, M. L. Ferrer, G. Finocchiaro, C. Forti, A. Franceschi, P. Franzini, C. Gatti, P. Gauzzi, S. Giovannella, E. Gorini, E. Graziani, M. Incagli, W. Kluge, V. Kulikov, F. Lacava, G. Lanfranchi, J. Lee-Franzini, D. Leone, F. Lu, M. Martemianov, M. Matsyuk, W. Mei, L. Merola, R. Messi, S. Miscetti, S. Moulson, S. Müller, F. Murtas, M. Napolitano, A. Nedosekin, F. Nguyen, M. Palutan, E. Pasqualucci, L. Passalacqua, A. Passeri, V. Patera, F. Perfetto, E. Petrolo, L. Pontecorvo, M. Primavera, F. Ruggieri, P. Santangelo, E. Santovetti, G. Saracino, R. D. Schamberger, B. Sciascia, A. Sciubba, F. Scuri, I. Sfiligoi, A. Sibidanov, T. Spadaro, E. Spiriti, M. Testa, L. Tortora, P. Valente, B. Valeriani, G. Venanzoni, S. Veneziano, A. Ventura, S. Ventura, R. Versaci, I. Villella, G. Xu

CP717, *Hadron Spectroscopy: Tenth International Conference,*
edited by E. Klempt, H. Koch, and H. Orth
© 2004 American Institute of Physics 0-7354-0197-7/04/\$22.00

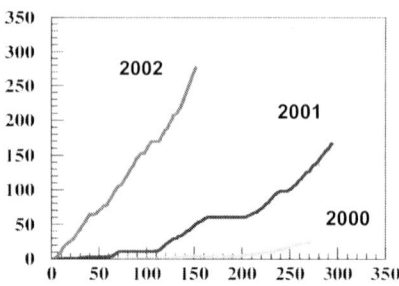

FIGURE 1. Integrated luminosity in KLOE in year 2000 ($\sim$25 pb^{-1}), 2001 ($\sim$170 pb^{-1}), and 2002 ($\sim$280 pb^{-1}).

In year 2003 there has been a major shutdown of DaΦne in order to:

- install a new interaction region (IP) for KLOE, designed to reduce the beam background inside the apparatus. The low-β quadrupoles have been mounted on a rotating structure which allows focusing optimization and operation at different KLOE soleinodal fields;
- rearrange the magnets in the straight sections: the new setup will increase the injection efficiency;
- modify the wigglers: the reshaping of the poles will improve the beam dynamical aperture and the lifetimes.

The DaΦne re-commissioning that started at the beginning of September should lead to operation with 110 bunches (the bunches were 50 in year 2002), at a peak luminosity of $2 \times 10^{32} cm^{-2} s^{-1}$, delivering to KLOE 10 pb^{-1} per day and 2 fb^{-1} per calendar year.

The detector. The design of the experiment has been optimized for the discrimination of the *CP*-violating decays $K_L \rightarrow \pi^+ \pi^-$ and $\pi^0 \pi^0$ from the much more abundant $K_L \rightarrow \pi \mu \nu$ and $K_L \rightarrow 3\pi^0$ decays. The detector must provide good momentum resolution for charged tracks, as well as full solid angle coverage and excellent energy and time resolution for photons. Moreover, given the rather long mean decay path of the K_L at DaΦne (3.4 m), a large detector is required in order to have a reasonable geometrical acceptance. The KLOE apparatus consists of a large drift chamber for the measurement of the charged particles, a sampling calorimeter made of lead and scintillating fibers, and a superconducting magnet providing the solenoidal field of 5.2 kGauss. The drift chamber [1], 2 m radius and 3.3 m long, is filled with a low-Z gas mixture of 90% Helium and 10% Isobutane, and enclosed by Carbon-Fiber/Epoxy walls. The light materials optimize the momentum resolution and reduce both, photon conversion and $K_L \rightarrow K_S$ regeneration. The transverse momentum resolution is $\sigma_{p_t}/p_t \sim 0.4\%$, and the kaon decay vertices are reconstructed with a precision of $\sim$3 mm. The electromagnetic calorimeter (ECAL) [2], 15 X_0 thick, is divided into a barrel and two C-shaped endcaps to optimize the hermecity. To complete the coverage of the solid angle, two small calorimeters

QCAL [3] are wrapped around the focusing quadrupoles. The energy resolution for photons is $\sigma_E/E = 5.7\%/\sqrt{E_{GeV}}$ and the time resolution is $\sigma_t = (54/\sqrt{E_{GeV}} \oplus 50)$ ps. The photon impact point is measured with a precision of ~ 1 cm$/\sqrt{E_{GeV}}$ along the fibers and ~ 1 cm in the transverse coordinate.

HADRONIC CROSS SECTION MEASUREMENT.

Present interest to improve the precision of the $\sigma(e^+e^- \to \pi^+\pi^-)$ below 1 GeV comes from the discrepancies found between different evaluations of the hadronic vacuum polarization contribution to the muon magnetic moment [4]. The anomalous magnetic moment of the muon receives contributions, in decreasing order, from QED processes, from the lowest order hadronic vacuum polarization $a_\mu^{\mathrm{had,LO}}$, from higher order hadronic loops, from light by light scattering, and from electroweak diagrams. The second largest contribution, $a_\mu^{\mathrm{had,LO}}$, is also affected by the largest error. $a_\mu^{\mathrm{had,LO}}$ is evaluated via the dispersion integral from the experimental data, and is affected by the uncertainties in the spectral function, especially those in the region of small Q^2, i.e. below 1 GeV^2 [5]. The data used for the spectral function in this energy range are the $e^+e^- \to \pi^+\pi^-$ annihilation cross section and the $\tau \to \pi\pi\nu$ decay, related to each other by the assumption of CVC and isospin conservation. Significant discrepancies remain between data from e^+e^- (dominated by CMD-2 results [6]) and τ (dominated by Aleph results [7]), especially in the energy region immediately above the ρ resonance (0.6:1 GeV^2).

Running at fixed e^+e^- center-of-mass energy, KLOE can determine the $\sigma(e^+e^- \to \pi^+\pi^-)$ by radiative return. The ISR processes reduce the effective energy for the $\pi\pi$ channel so that we can measure the $\sigma(e^+e^- \to \pi^+\pi^-)$ cross section from threshold to 1 GeV^2, provided that the knowledge of ISR, and ISR+FSR effects are well established. We are using the PHOKHARA generator [8] which is able to describe these processes with an accuracy of 5 per mil. A comparable precision (6 per mil) is achieved for the luminosity measurement, where the BABAYAGA [9] code has been interfaced with the KLOE MonteCarlo to provide the absolute luminosity scale by counting large-angle Bhabha's.

The analysis performed in KLOE to isolate the $\pi\pi\gamma$ events is reported in the A.Denig contribution to these proceedings. At present we have measured [10] the $\sigma(e^+e^- \to \pi^+\pi^-\gamma)$ on the basis of more than 1.5×10^6 large-angle $\pi\pi$ pairs, covering the kinematical region $(0.3 < s_\pi < 1)$ GeV^2. The KLOE preliminary result

$$a_\mu^{\mathrm{had,LO}} \times 10^{10} \,(0.37{:}0.93 \text{ GeV}^2) = 378.4 \pm 0.8_{stat} \pm 4.5_{syst} \pm 3.0_{theo} \pm 3.8_{FSR} \quad {}^2$$

is compatible with the CMD-2 measurement, confirming the discrepancy between e^+e^- and τ data. The analysis is being finalized: we expect to reach 1% precision on both experimental, and theoretical systematics. Also the $\pi\pi\gamma$ events with detectable, large-angle photons are being studied at KLOE to extend the hadronic cross section measurement

---

[2] The reported value is the outcome of an updated analysis presented at the Pisa Workshop SIGHAD03 - 8-10 October - and supersede the result $a_\mu^{\mathrm{had,LO}} \times 10^{10} \,(0.37{:}0.95 \text{ GeV}^2) = 374.1 \pm 1.1_{stat} \pm 5.2_{syst} \pm 3.0_{theo} \left({}^{+7.5}_{-0} FSR\right)$ presented at this Conference.

down to the $(2m_\pi)^2$ threshold, at $(0.08 < s_\pi < 0.4)$ GeV^2.

KAON PHYSICS : RECENT ACHIEVEMENTS.

With the present statistics KLOE is improving the precision on the kaon masses, life-times, branching ratios, and decay distributions, of interest for the study of a wide range of phenomena, including the test of the unitarity of the CKM matrix, the asymmetries in semileptonic decays, the processes described by ChPT, and the direct CP violation.

At the ϕ factory it is possible to tag $K_{S,L}$ and $K^{+,-}$ beams : the presence of one K_S (K_L) signals the K_L (K_S) on the other side, and the same for K^+ (K^-). In practice, the K_S are tagged searching for the interaction of the K_L in the calorimeter, i.e. an energetic cluster with typical delay of 30 ns due to the low momentum (110 MeV/c) of the kaons produced at DAΦNE . The excellent timing of the KLOE calorimeter provides in this case the K_S momentum with an accuracy of 2 MeV/c, the same obtained for the K_L beams that are selected using the charged decays $K_S \to \pi^+\pi^-$, i.e. requiring one vertex in the interaction region compatible with the hypothesis of two charged pions with invariant mass equal to the kaon mass. Using these simple criteria, the tagging efficiencies (70% for the K_L tagging, and 30% for the K_S) are suitable to collect huge samples of well defined particles. The charged kaons are tagged by searching for the 2-body decays, $K \to \mu\nu$ and $K \to \pi\pi^0$ that in the kaon rest frame have monochromatic charged secondaries.

The KLOE published results on kaon physics are those of references [11], [12], [13], [14], [15].

For the K_S, KLOE obtained on the basis of the data collected in year 2000, for the main decay modes, the ratio $\Gamma(K_S \to \pi^+\pi^-(\gamma))/\Gamma(K_S \to \pi^0\pi^0) = 2.236 \pm 0.003 \pm 0.015$ [12], consistent but slightly larger than the world average of $R_S{}^\pi = 2.197 \pm 0.026$. The result is, at the present accuracy level, fully inclusive of the contribution coming from the radiative $K_S \to \pi\pi\gamma$ decays. The phase shift difference in $K \to \pi\pi$ transitions with I=0 and I=2 amplitudes from our measurement is $(48 \pm 3)°$, in agreement with the strong $\pi\pi$ phase shift predicted by ChPT and with the value obtained from the $\pi\pi$ scattering data. In this anaysis the photon detection efficiency has been measured using $\Phi \to \pi^+\pi^-\pi^0$ events and the track reconstruction efficiency has been determined from the analysis in (p,θ) bins of the $K_S \to \pi^+\pi^-$ decays identified by looking for one pion only. Most of the systematics come from the statistical error on the control samples used to evaluate the selection efficiency from data and it is being improved with the analysis of larger data sets now available.

We are studying the largest sample, $\sim$20,000 events, of K_S semileptonic decays. They are selected [13] within the tagged K_S, looking for tracks of opposite charge with a vertex in the interaction region. In the hypothesis that both tracks are pions, their invariant mass must be smaller than 490 MeV/c to cut out the $K_S \to \pi^+\pi^-$ background. The background is furtherly reduced by the analysis of the time of flight of the two particles from the decay point to the calorimeter. The expected time of flight from the momentum and the length of the tracks is computed for each mass hypothesis, m=m$_e$ and m=m$_\pi$, and compared with the time measured by the calorimeter. This comparison

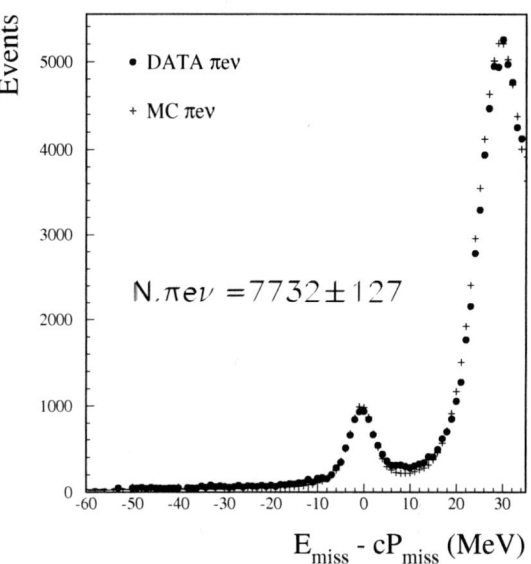

FIGURE 2. E_{miss}- c·P_{miss} distribution. The $K_S \to \pi e \nu$ events show up in the region around zero, the rest of the distribution is due to $K_S \to \pi^+ \pi^-$ decays. Solid markers are data, crosses are the result of the fit to the MC-expected shapes for the signal and the background.

is used also to isolate $\pi^+ e^-$ and $\pi^- e^+$ decays. The efficiencies of the analysis cuts have been controlled with a sample of $K_L \to \pi e \nu$ decays near the interaction region. The overall efficiency to detect $K_S \to \pi e \nu$ is 0.208 ± 0.004. The final event counting is done using the E_{miss}- c·P_{miss} discriminant variable, with a fit of the distribution where contributions from the signal and the background ($K_S \to \pi^+ \pi^-$) show up in separate regions (see Fig. 2). The preliminary result for the semileptonic branching ratio, coming from the analysis of 170 pb^{-1} and 7732±127 semileptonic decays, is BR($K_S \to \pi e \nu$) $= (6.81 \pm 0.12 \pm 0.10) \cdot 10^{-4}$. Assuming CPT conservation, the comparison of the K_S with the K_L semielectronic partial widths provides a probe of the $\Delta S = \Delta Q$ rule. Re x_+, the parameter measuring the $\Delta S = \Delta Q$ violation, from our value of BR($K_S \to \pi e \nu$) turns out to be Re $x_+ = (3.3 \pm 5.2 \pm 3.5) \cdot 10^{-3}$. This preliminary result has a precision comparable with the best available measurement [16]. We are analyzing the rest of the sample (280 pb^{-1}): the updated result, together with the expected improvement in the K_L semielectronic partial width from our data, will reduce the present error on Re x_+ by a factor of 1.5.

For the K_L, KLOE has recently published [14] a new measurement of the ratio $\Gamma(K_L \to \gamma\gamma) / \Gamma(K_L \to \pi^0\pi^0\pi^0)$, based on 362 pb^{-1} collected during 2001 and 2002. The $K_L \to \gamma\gamma$ decay rate is interesting for ChPT and is also connected with the $K_L \to \mu^+\mu^-$ decay, being the dominant contribution to the long-distance term of the process. The

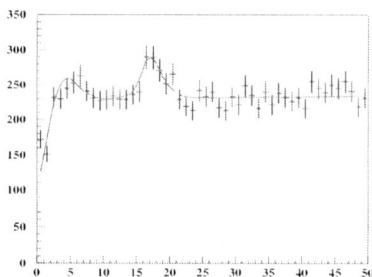

FIGURE 3. $K_L K_S \rightarrow \pi^+\pi^-\pi^+\pi^-$ events as a function of the difference in decay times ($\Delta T/\tau_S$). The interference pattern shows up at small ΔT ($\Delta T/\tau_S < 10$) and across the beam pipe wall location ($\Delta T/\tau_S \sim 15$), where the decays interfere with the coherent regeneration processes, $K_L \rightarrow K_S$. The curve is a fit to the interference function at fixed values of Γ_L and Γ_S.

$K_L \rightarrow \gamma\gamma$ events have a clear signature at the Φ-factory, being the only source of $\sim$250 MeV photon pairs that balance the momentum of tagging $K_S \rightarrow \pi^+\pi^-$ decay. The event selection can be highly inclusive so that the knowledge of the selection efficiency and the residual background can be very accurate. The final measurement, dominated by the statistical error on the $K_L \rightarrow \gamma\gamma$ events, is $\Gamma(K_L \rightarrow \gamma\gamma)/\Gamma(K_L \rightarrow \pi^0\pi^0\pi^0) = (2.79 \pm 0.02 \pm 0.02) \cdot 10^{-3}$, in good agreement with the recent result from NA48 [17].

The semileptonic decays of the kaons are used together with the hyperon semileptonic decays [18] and the $\tau \rightarrow K(n\pi)\nu$ [19] to derive the CKM matrix element $|V_{us}|$. Significant discrepancies remain between the different data sets and between the kaon semileptonic partial widths measured in years seventies (old data) and the recent results [20] obtained from experiment E865 at the BNL analyzing $\sim$70,000 K_{e3}^+ decays. Moreover, the old data show a 2.2 σ effect when are compared with the best determination of $|V_{ud}|$ imposing the unitarity of the CKM matrix, i.e. $|V_{ud}|^2 + |V_{us}|^2 \sim 1$. The preliminary results of KLOE from neutral kaon decays [21] are in agreement with the old measurements: work is in progress to improve the experimental precision of all the relevant quantities, including the charged kaon semileptonic decays, the Dalitz plot slope parameters, and the $K_L/K^\pm$ lifetimes.

At KLOE, because the initial state of the neutral kaon pairs is an antisymmetric superposition of K_S and K_L, the final state decay products show characteristic interference patterns. By studying the time dependence of these patterns for different final state combinations it is possible to measure CP and CPT violation parameters. This program can be attacked with few fb^{-1} of integrated luminosity. In Fig. 3 the first attempt to search for these interference patterns is presented: the $K_L K_S \rightarrow \pi^+\pi^-\pi^+\pi^-$ final states are selected and the difference in decay times reported. The interference pattern shows up at small ΔT ($\Delta T/\tau_S < 10$), and across the beam pipe wall location ($\Delta T/\tau_S \sim 15$) the interference with the coherent regeneration processes ($K_L \rightarrow K_S$) become evident.

The Dalitz plot of the decays of the kaons into three pions is sensitive to the $\Delta S = 1$ CP violating transitions, and is at the same time a good probe of the ChPT calculations. Experimentally, the $K^\pm \rightarrow \pi^\pm\pi^0\pi^0$ branching ratio has been measured thirty years ago

[22] with a statistical accuracy of 3%. Using the entire data set of tagged charged kaons, KLOE has measured a new value of the BR($K^{\pm} \rightarrow \pi^{\pm}\pi^0\pi^0$) [15], and the Dalitz plot analysis is in progress. The event selection requires a charged kaon track reconstructed in the drift chamber with a decay vertex associated to only one charged track, and at least four neutral clusters with the correct time of flight from the charged vertex to the calorimeter. Loose kinematical cuts have been applied to the charged kaon and to its decay products. Residual background comprises mainly $K^{\pm} \rightarrow \pi^{\pm}\pi^0$ and $K^{\pm} \rightarrow e^{\pm}\pi^0\nu$ events, and has been measured from a fit of the missing mass spectrum in the data to the sum of the signal and the background spectra independently obtained by MonteCarlo simulations. The total background fraction is $(0.75 \pm 0.11)\%$. Special sub-samples of the signal have been used to measure the selection efficiency. In general, control samples selected by cutting on calorimeter variables have been used to compute the efficiencies involving the tracking in the drift chamber, and vice-versa. The efficiencies have been evaluated on samples not used in the branching ratio measurement, and the main error source comes from the limited statistics of those samples. Our final results is BR($K^{\pm} \rightarrow \pi^{\pm}\pi^0\pi^0$) = $(1.781 \pm 0.013 \pm 0.016) \cdot 10^{-2}$, 3.5 times more precise than the world average published in the 2002 PDG [23].

RESULTS FROM Φ RADIATIVE DECAYS.

Φ radiative decays to pseudoscalars are studied in the context of the chiral Lagragians. In particular, the BR($\Phi \rightarrow \eta'\gamma$) provides information about SU(3)-symmetry breaking and the amount of gluonium content of the η' meson. In KLOE we measured [24] the ratio of branching fractions BR($\Phi \rightarrow \eta'\gamma$) / BR($\Phi \rightarrow \eta\gamma$) selecting $\eta' \rightarrow \pi^+\pi^-\eta$ ($\eta \rightarrow \gamma\gamma$), and $\eta \rightarrow \pi^+\pi^-\pi^0$ ($\pi^0 \rightarrow \gamma\gamma$) . The particles in the final state, i.e. $\pi^+\pi^-\gamma\gamma\gamma$, are the same so that most systematics uncertainties cancel in the ratio. The analysis [24] requires one charged vertex in the interaction region and three photons with the correct time of flight from the charged vertex to the calorimeter. A kinematic fit constraining the total energy, total momentum, and the photon time and positions is applied to the events to improve the photon energy determination and to reduce, by a loose cut on the χ^2 value, the contamination from $\Phi \rightarrow K_S K_L$ and $\Phi \rightarrow \pi^+\pi^-\pi^0$. To further reduce these sources of background additional cuts, on the energy of the photons, and on the pion energy endpoint, have been separately applied for the two channels. After these cuts the $\Phi \rightarrow \eta'\gamma$ sample is still dominated by $\Phi \rightarrow \eta\gamma$ events that are separated using the energy correlation of the two most energetic photons. These photons, of energy E1 and E2, come from η decay in the case of $\Phi \rightarrow \eta'\gamma$, so that the energies are strongly anticorrelated, while in the other case one of the photons is the radiative photon with energy of 363 MeV. The final η' event counting is done from a fit of the $\pi^+\pi^-\gamma\gamma$ invariant mass, assuming for the signal the shape obtained from MonteCarlo and for the background the shape obtained from the events selected just outside the signal region in the E1-E2 plane. The ratio of the branching fractions is R=$(4.70 \pm 0.47 \pm 0.31) \cdot 10^{-3}$ and the value of the pseudoscalar mixing angle in the flavor basis is $\varphi_P = (41.8^{+1.9}_{-1.6})°$. This value, taking into account of the constraints from other processes involving P$\rightarrow$Vγ transitions, indicates a gluonium fraction in η' below 15%, as opposed to the suggestion coming from the large measured value of the BR(B $\to$ $K\eta'$) [25].

At present the quark model can not clearly explain the phenomenology involving $f_0(980)$ and $a_0(980)$ scalar mesons. In particular, the decay rate of $\Phi \to \pi^0\pi^0\gamma$ is too large for the expected couplings of the f_0 to the KK and to the $\pi^0\pi^0$ mesons if the scalar (S) is an ordinary quark-antiquark state. These mesons may be 4-quarks structures or $K\bar{K}$ molecules: the $\Phi \to S\gamma$ branching ratio significantly change according to the different interpretations. KLOE performed the analysis of both the channels, $\pi^0\pi^0\gamma$ dominated by $\Phi \to f_0\gamma$ processes [26], and $\eta\pi^0\gamma$ coming mainly from $\Phi \to a_0\gamma$ decays [27]. We have measured the branching fractions and the differential $dN/dM_{\pi\pi}$, $dN/dM_{\eta\pi}$ spectra. The analyses include the study of the

- energy and momentum conservation to reduce various sources of background;
- $\pi\gamma$ and $\pi\pi$ invariant mass to cut the $\omega \to \pi^0\gamma$ contamination and to isolate $\pi^0\pi^0\gamma$ from $\eta\pi^0\gamma$ events;
- the photon pairing to identify $\pi^0\pi^0\gamma$ decays.

For the $\Phi \to \eta\pi^0\gamma$ analysis, the $\eta \to \gamma\gamma$ and the $\eta \to \pi^+\pi^-\pi^0$ decays have been considered. The angular distributions of the events show that the dominant contribution comes respectively from $\Phi \to f_0\gamma$ and $\Phi \to a_0\gamma$ decays. The final results, on the basis of data taken during 2000, are BR($\Phi \to \pi^0\pi^0\gamma$) = $(1.08 \pm 0.03 \pm 0.05) \cdot 10^{-4}$ and BR($\Phi \to \eta\pi^0\gamma$) = $(8.51 \pm 0.51 \pm 0.57) \cdot 10^{-5}$.

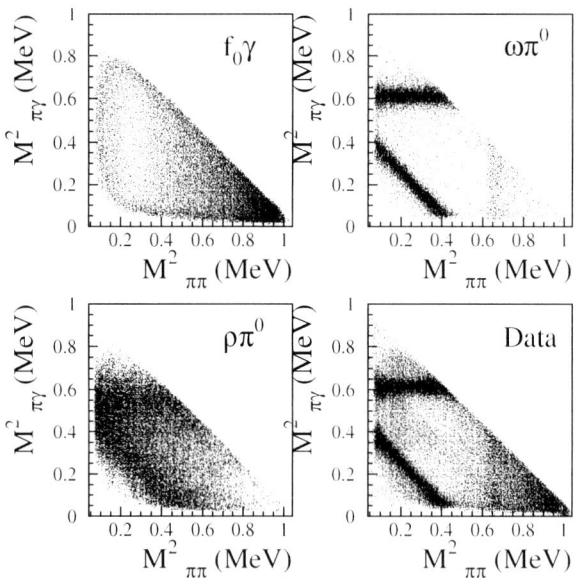

FIGURE 4. $M^2_{\eta\gamma}$ vs $M^2_{\pi^0\pi^0}$ distributions for : $\phi \to f_0\gamma$ MC simulation (top-left), $e^+e^- \to \omega\pi^0$ MC simulation (top-right), $\phi \to \rho\pi^0$ MC simulation (bottom-left) and real data (bottom-right). Data sample refer to the entire KLOE data set, i.e. $\sim$500 pb^{-1}.

The $dN/dM_{\pi\pi}$ from the selected $\pi^0\pi^0\gamma$ events and the $dN/dM_{\eta\pi}$ from the $\eta\pi^0\gamma$ sample have been used respectively to estimate the contributions coming from $\Phi \to f_0\gamma$ and $\Phi \to a_0\gamma$ processes. The terms from $S\gamma$ and $\rho\pi$ have been considered. The $S\gamma$ term has been treated by means of K^+K^- loop model . The coupling $g^2{}_{SKK}$ and the coupling ratio $g^2{}_{SKK}/g^2{}_{Sfinalstate}$ have been left as free parameters. The contribution from $\rho\pi$ turns out to be negligible for both the analyzed channels. For the $dN/dM_{\pi\pi}$ spectral function of the $\pi^0\pi^0\gamma$ sample, a better agreement with data is obtained including an additional term coming from $\Phi \to \sigma\gamma \to \pi^0\pi^0\gamma$ plus the interference between σ and f_0 channels. The σ contribution has been parameterized assuming $M_\sigma = 478$ MeV and $\Gamma_\sigma = 324$ MeV according to the measurement of reference [28]. The estimated branching fractions are $BR(\Phi \to f_0\gamma) = (4.47 \pm 0.21) \cdot 10^{-4}$ and $BR(\Phi \to a_0\gamma) = (7.4 \pm 0.7) \cdot 10^{-5}$, while the ratio of the coupling of the two scalars to the KK system is $R_{g^2} = g^2{}_{f_0KK} / g^2{}_{a_0KK} = 7.0 \pm 0.7$.

We are now re-evaluating with a model-independent analysis the contributions to the selected final states coming from different processes using the detailed description of the Dalitz plot provided by the new data available (see Fig. 4).

SUMMARY AND OUTLOOK

The KLOE measurement at 1% accuracy of the $\sigma(e^+e^- \to \text{hadrons})$ is forthcoming. The analysis of the collected sample of ~500 pb^{-1} will improve, among other topics, the precision of the kaon branching fractions, and the K_L, $K^\pm$ lifetimes. In particular the study of the semileptonic decays will provide a new measurement of the CKM matrix element $|V_{us}|$. By the end of year 2003 DaΦne will restart delivering data to KLOE at improved luminosity and data taking conditions, providing 2 fb^{-1} per calendar year. New data will allow us to continue and extend the physics program including also measurements of the parameters describing the shape of the interference patterns in the neutral kaon system [29].

REFERENCES

1. KLOE Collaboration, Adinolfi, M., et al., *Nucl. Instr. Meth.*, **A488**, 51 (2002).
2. KLOE Collaboration, Adinolfi, M., et al., *Nucl. Instr. Meth.*, **A482**, 364 (2002).
3. KLOE Collaboration, Adinolfi, M., et al., *Nucl. Instr. Meth.*, **A483**, 649 (2002).
4. Davier, M., Eidelman, S., Hocker, A., and Zhang, Z., Update estimate of the muon magnetic moment using revised results from e^+e^- annihilation (2003), hep-ph/0308213.
5. Davier, M., Eidelman, S., Hocker, A., and Zhang, Z., Confronting spectral functions from e^+e^- annihilation and τ decay: consequences for the muon magnetic moment (2002), hep-ph/0208177.
6. CMD-2 Collaboration, Akhmetshin, R., et al., *Phys. Lett.*, **B527**, 161 (2002).
7. ALEPH Collaboration, Barate, R., et al., Measurement of branching fractions in τ decays (2002), ALEPH 2002-030.
8. Czyz, H., Grzelinska, A., Kuhn, J., and Rodrigo, G., The radiative return at Φ- and B-factories: FSR at next-to-leading order (2003), hep-ph/0308312.
9. Carloni Calame, C., Lunardini, C., Montagna, G., Nicrosini, O., and Piccinini, F., Large-angle Bhabha scattering and luminosity at flavour factories (2000), hep-ph/0003268.

10. KLOE Collaboration, Aloisio, A., et al., Determination of $\sigma(e^+e^- \to \pi^+\pi^-)$ from radiative processes at DAΦNE (2003), hep-ex/0307051.
11. Antonelli, M., and Dreucci, M., Measurement of the K^0 mass from $\Phi \to K_S K_L, K_S \to \pi^+\pi^-$ (2002), http:://www.lnf.infn.it/kloe/pub/knote/kn181.ps.
12. KLOE Collaboration, Aloisio, A., et al., *Phys. Lett.*, **B538**, 21 (2002).
13. KLOE Collaboration, Aloisio, A., et al., *Phys. Lett.*, **B535**, 37 (2002).
14. KLOE Collaboration, Aloisio, A., et al., *Phys. Lett.*, **B566**, 61 (2003).
15. KLOE Collaboration, Aloisio, A., et al., Measurement of the branching ratio for the decay $K^\pm \to \pi^\pm\pi^0\pi^0$ with the KLOE detector (2003), http:://www.lnf.infn.it/kloe/pub/knote/kn190.ps.
16. CPLEAR Collaboration, Angelopoulos, A., et al., *Phys. Lett.*, **B444**, 38 (1998).
17. NA48 Collaboration, Lai, A., et al., *Phys. Lett.*, **B551**, 7 (2003).
18. Cabibbo, N., Swallow, E., and Winston, R., Semileptonic hyperon decays and CKM unitarity (2003), hep-ph/0307214.
19. Gamiz, E., et al., Determination of m_S and V_{us} from hadronic τ decays (2002), hep-ph/0212230.
20. E865 Collaboration, Thompson, J., et al., New, high statistics measurement of the $K^+ \to \pi^0 e^+\nu$ (K^+_{e3}) branching ratio (2003), hep-ex/0307053.
21. KLOE Collaboration, Aloisio, A., et al., KLOE prospects and preliminary results for K_{l3} decay measurements (2003), hep-ex/0307016.
22. Chiang, I., et al., *Phys. Rev.*, **D6**, 1254 (1972).
23. Particle Data Group, Hagiwara, K., et al., *Phys. Rev.*, **D66**, 010001 (2002).
24. KLOE Collaboration, Aloisio, A., et al., *Phys. Lett.*, **B541**, 45 (2002).
25. BABAR Collaboration, Aubert, B., et al., *Phys. Rev. Lett.*, **87**, 221802 (2001).
26. KLOE Collaboration, Aloisio, A., et al., *Phys. Lett.*, **B537**, 21 (2002).
27. KLOE Collaboration, Aloisio, A., et al., *Phys. Lett.*, **B536**, 209 (2002).
28. Aitala, E., et al., *Phys. Rev. Lett.*, **86**, 770 (2001).
29. Buchanan, C., et al., *Phys. Rev.*, **D45**, 4088 (1992).

Contradictions about Fine Structures in Meson Spectra and Proposed High-Resolution Hadron Spectrometer Using "Interactive" Solid-State Hydrogen Target

Bogdan C. Maglich

HiEnergy Technologies, Inc., 1601B Alton Parkway, Irvine, California 92606, USA and
Serbian Academy of Sciences and Arts, Belgrade, Yugoslavia

Abstract. High resolution has been discouraged in meson spectrometry for 4 decades by the *Doctrine of Experiments Incompatible with Theory (DEIT)*. DEIT *a priori* rejects narrow hadron resonances on the paradigm that <u>only broad</u> hadron peaks, $\Gamma \geq 100$ MeV, can exist - in spite of the accumulated evidence to the contrary. The facts are: Mesons 2 orders of magnitude narrower than 'allowed' for hadrons, have been confirmed; a new one was announced at this conference. Narrow meson structures have been repeatedly reported at high momentum transfer, $|t| > 0.2$, while they are absent at the low transfer, $|t| \sim 0.01$, where 99% of the experiments are performed. Modification of meson mass and width as a function of the density of nuclear matter in which they are produced, have been recently reported.

We postulate for meson spectra: (1) Intrinsic ('true') width, Γ, is different from the observable ('apparent') width, Γ': $\Gamma < \Gamma'$ (2) Γ of all meson states are narrow and can be observed only at or near the maximum $|t|$ reachable in the reaction, and (3) Γ of all meson resonances are subject to broadening as $|t|$ decreases. Since both Γ' and the production σ are inversely proportional to $|t|$, most of the observed spectra are produced at the lowest $|t| < 0.01$ and thus the peaks appear broad.

We have conceptually designed a novel type hadron spectrometer with an order of magnitude better resolution (0.1 MeV). It would operate at 2 orders of magnitude higher $|t|$ ($0.3 < |t| < 1$ (GeV/c)^2, than most experiments to date ($|t| < 0.01$). Mesons in the mass region $0.5 < M_x < 5$ GeV would be produced in $\pi P \rightarrow PX$ (baryons in $PP \rightarrow PP^*$) in a 'solid state hydrogen target' consisting of an array of plastic scintillator fibers, CH; collisions with C are electronically rejected. Missing mass of P is measured in the region of the maximum recoil angle.

Story of the suppression by DEIT for 17 years of the observation of a theoretically unexplained narrow peak, which turned out to be $\omega \rightarrow 2\pi$, and the related correspondence between Werner Heisenberg and this author, is narrated.

1. "Philosophy, When Applied a Priori, May Become Dogma" [1] - *Yuval Ne'eman (2001)*

Throughout the history of spectroscopy, better resolution had invariably led to the discoveries of finer and finer structures and new physics. In contrast, the attempts to build meson spectrometers with very fine mass resolution had put us repeatedly into the collision course with the *Doctrine of Experiments Incompatible with Theory* or DEIT (Sect. 3). DEIT discourages the experimentalists from improving their measurement accuracy on the paradigm that <u>only broad</u> hadron peaks can exist. Strong interaction coupling constant dictates that the physical widths cannot be substantially narrower

CP717, *Hadron Spectroscopy: Tenth International Conference,*
edited by E. Klempt, H. Koch, and H. Orth
© 2004 American Institute of Physics 0-7354-0197-7/04/$22.00

than Γ~ 100 MeV, unless the strongly produced state decays electromagnetically. Hence, a strongly decaying hadron with Γ<<100 MeV cannot be real. Such measurements are a priory rejected as statistical fluctuations or malfunctioning of the instrument.

When I first showed the ω° meson peak to a conference of international particle theorists at Berkeley a month before the discovery was announced [2], all of them said that hadrons cannot be narrow; you must have an error in your program. Two days later, the entire theory group reversed its position to "ω° must be narrow" because of the centrifugal barrier of a 3 π system, each pion pair with l=1.

The <u>a priori</u> rejection of narrow hadron resonances continued for 4 decades to this date in spite of the accumulated evidence to the contrary:

• Hadrons with <u>2 orders of magnitude</u> narrower Γ than "allowed" for hadrons have been experimentally confirmed from the beginning of the meson resonance era to date. E.g. in addition to ω°, we have φ (1040) with Γ= 4 MeV. A new narrow peak with Γ < 10 MeV has been reported in this meeting.

• A narrow 2 π peak exactly of the ω mass and width (Fig. 1A), observed a month before the ω discovery, had been officially declared a statistical fluctuation for 17 years until experimentally confirmed as G-parity violating ω→2π decay. The manipulations used to shield from the reality a theory-in-vogue that could not explain the effect, are described in Sect. 4.

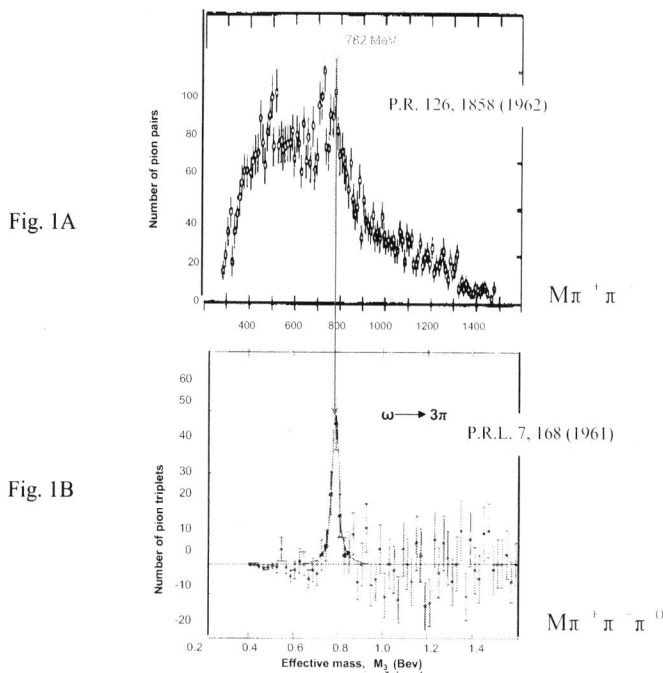

Fig. 1A

Fig. 1B

FIGURE 1. A. Effective mass of $\pi^+\pi^-$ from $\overline{P}P \rightarrow 5\pi$, observed 3 weeks before discovery of ω°, **(B)** at the very same mass as the second peak in A. Because no theoretical model could explain 2ρ's, publication of data in A was prevented for a year on the grounds that the equality of M and Γ with those of ω was a statistical fluctuation and the 2 peak structure is, in fact, 1-single peak, although the single-to-double peak likelihood ratio was ~10.

• Narrow meson structures have been repeatedly observed at high momentum transfer, |t| >0.2, while absent at the lowest transfer, |t| ~0.01, where 99% of experiments are performed. E.g. at |t| ~0.01, the peaks a_1(1260), $J^P=1^+$ and a_2 (1320) $J^P=2^+$, are broad and unresolved. But at 20 times greater |t|, a 2 peak structure was observed (Fig.2) in two independent Missing Mass Spectrometers both operating at |t| ~0.2, one at CERN [3], the other at BNL [4].

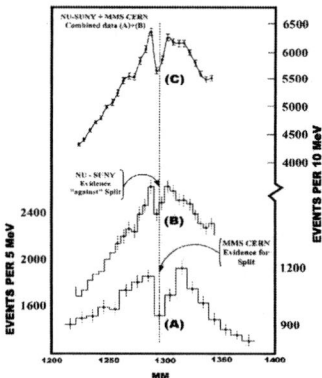

FIGURE 2. (A) split A_2 peak as measured by CERN Missing-Mass Spectrometer [3] at |t| ≥ 0.2 $(GeV/c)^2$. (B) Same measured at BNL [4]. Although resolution at BNL was 30% poorer than that at CERN, combined data (C)= (A) + (B) reject one-peak hypothesis. Nevertheless, a flawed analysis (Sect. 4) was used to prove that the 2-peak structure in (B) represented the evidence against 2-peak structure in 2(A).

• Variable mass? Measurement of the ω^o mass as a function of the density of hadronic matter which it was produced [5] appears to be compatible with the mass and width modification behavior expected from the Brown-Rho scaling [6].

•Variable Γ? There are rumors that most recent data from RHIC are indicative that even the narrowest hadron, φ, undergoes a width modification from 4 MeV to 2 MeV between low and high and centrality which is proportional to |t|).

•Fine structures keep appearing at high |t|. To this date, 46 reported measurements at low |t| identified at least 8 overlapping I=1 resonances in the ρ_3 mass region 1600-1750 MeV, also known as g-meson (g for Geneva). Referring to Fig. 3, this region was split into three peaks when |t| increased by a factor of 20 and the statistics and resolution improved [7]. An obvious feature of the high |t| spectrum is the fine structure. In contrast, the spectra obtained at low |t|<0.01, mostly in bubble chambers and in peripheral collisions, display broad peaks only.

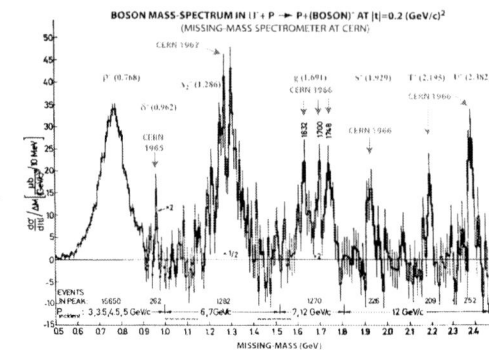

FIGURE 3. Combined charged meson spectrum from 9 Missing-Mass Spectrometry experiments at CERN, all at high $|t| \geq 0.2$. [P.R.L. <u>17</u>, 49 (1962)].

• Fine structures in annihilation meson spectra in the ρ_5 mass region 2000-2450 MeV. When the statistical sample in an entirely different hadron spectrometer, the Annihilation Spectrometer at BNL [8] reached 3×10^8 events in the mass spectrum and the significance of the 2 peaks, ρ_5 (2250) and ρ_5 (2350) T and U, reached 1,300σ, fine structures on top of the broad peaks began to appear as, at least, 6 little 10σ peaks (Fig. 4). In a number of independent low $|t|$ experiments in the very same mass region, at least 5 broad overlapping mesons ($\Gamma > 300$ MeV) have been observed, their J ranging from 1 to 6. How can these states be so broad in spite of the high J in view of the centrifugal barrier? We believe they are a superposition of many narrow unresolved peaks.

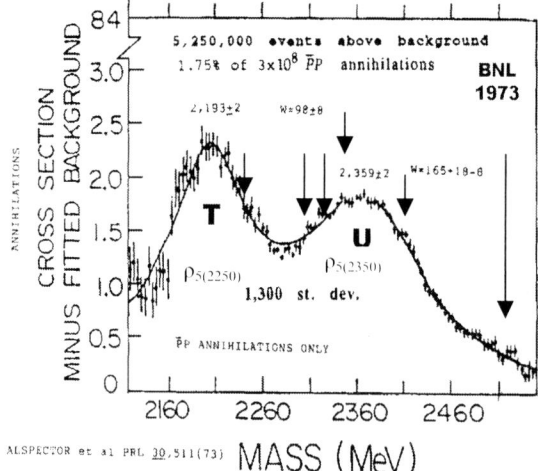

FIGURE 4. Neutral heavy meson spectrum measured by "annihilation spectrometer" at BNL [8]. Note: the two peaks from Fig. 3, ρ5 (2250) and ρ5 (2350), (originally named T and U) appear in the $\overline{P}P$ annihilation reaction at nearly the same masses, but with narrower widths than those observed in low $|t|$ πP collisions. The number of annihilation events in this experiment is 3×10^8, which equals roughly 10 times the total integrated world sample.

40

2. Momentum Transfer Broadening of "Observable Meson Widths"

The ever changing Γ's have been known to every meson experimentalist for decades but there has been no theoretical motivation to pursue the investigation of Γ vs. $|t|$ dependence. When the spectra are subdivided into $|t|$ bands, the statistics of the sub-samples were insufficient to prove it.

Our experience in measuring the particle spectra leads us to the following postulates:

1. There is a difference between the intrinsic ('true') width, Γ, and observable ('apparent') width, Γ', of meson resonances; always $\Gamma < \Gamma$'.

2. The intrinsic widths, Γ, of all meson states are narrow and can be observed only at or near the maximum value of the momentum transfer $|t|$ reachable in the reaction at which the resonance is produced.

3. The intrinsic widths, Γ, of all meson resonances are subject to broadening as the $|t|$ decreases. Since both the observable widths, Γ', and the production cross sections, σ , are inversely proportional to [t] most of the observed spectra are produced at lowest [t] <0.01 and thus appear broad.

3. Doctrine of 'Experiments Incompatible with Theory' ('DEIT')

The doctrine is based on the premise that theoretical understanding of particle physics has reached the degree of exactitude that equals that of the Newtonian mechanics. If a measurement is not compatible with theory, it is a violation of the Laws of Physics rather than a contradiction of the theory; hence such measurements must be incorrect and, therefore, are inadmissible.

For the purpose of proving that the effects not predicted by the theory were statistical flukes, DEIT has devised a radical revision of the information theory: Two-Tier 'Statistical' Test. This is done by attributing an enhanced statistical weight to the fact that the effect was predicted, while imposing an order of magnitude stricter acceptance test for unpredicted effects.

The acceptance threshold for the expected is that their relative likelihood of being real is simply greater than that of being a fluctuation, i.e. 2/3:1/3. In contrast, the acceptance threshold for an unpredicted effect is 100:1. E.g. even if an unpredicted effect has only 2% probability of being a statistical fluctuation, the 2% are considered the 'proof' that the effect is a statistical fluke, although its likelihood of being a real effect is 2/3 or better. The resistance to narrow peaks and the methodology for making them 'go away' is best illustrated by the following story.

4. How was a Real Effect Shoved Under the Rug for 17 Years?

In his story on the discovery of ω^o, Alvarez [2a], revealed an internal controversy that was rampant at Berkeley at the time: whether the 2 narrow π_+ π- peaks (Fig. 1A) was another discovery or an artifact. Alvarez's narrative makes no mention that the 2 peak structure was observed three weeks before ω^o, and that it had been accepted at

Berkeley as the first evidence for fine structure. When it turned out, however, that the M and Γ of the higher peak were exactly those of ω^0, the 2- peak structure became an embarrassment. No theoretical model was compatible with two ρ mesons! Hence, every attempt was then made by DIET to 'remove' the fine structure by insisting that it was a statistical fluctuation. We had been effectively prevented from publishing even a sanitized version of the data in Fig. 1A until a year later [10].

Heisenberg became visibly excited when I showed the 2 peak structure in Fig. 1A my lecture in Munich. He insisted that the higher peak was the G-parity violating $\pi+ \pi-$ decay of ω. I felt that Heisenberg could well be right. From my interactions with Gell-Mann back home, I learned that G-parity was not conserved in EM interactions, so ω could decay electromagnetically part of the time. The problem was that Heisenberg went on to insist that if you allowed for G violation, the observed Dalitz plot distribution of ω allows for an alternate J^P assignment for ω, that of a pseudoscalar, 0^-. Duerr-Heisenberg's Nonlinear-Spinor Theory needed a pseudoscalar meson and Heisenberg believed ω^0 was it. Soon thereafter, I received Heisenberg's letter (Fig. 7) explaining his point, followed by his paper [11]. It was perhaps for this reason that both, the 2 peaks in Fig. 1A and Heisenberg's position on G parity violation were ignored by the particle physics community. A new 'statistical analysis' was devised for the purpose of suppressing this real effect; it is described below.

Soon thereafter, the issue became moot with the discovery of η, $J^P=0^-$. Ten years later, the G parity violating decay of ω^0 was indicated [12] with 4 σ, and 7 years later - three years after Heisenberg's death - the $\omega \rightarrow 2\pi$ effect was definitely established [13] with 9 σ.

How to make a real physics effect disappear for 17 years? Fermi used to say that one cannot prove that something does not exist. You can prove only the limits of its existence, i.e. if it exists, it is smaller than …or it does not appear under certain circumstances…. etc. In our case (Fig. 1A), the 2-peak hypothesis gave 50% probability while 1-peak 5%. To claim that 5% represents the proof against 2 peaks, you have to obfuscate the Information Theory, which is exactly how the "statistical test" of the fine structure was conducted. A Monte Carlo code named GAME was devised by G. Lynch [9]. 100 histograms were generated with the assumption that there is a real 1 broad peak; the game was to see how many times it would 'look' like the 2 peaks like in Fig. 1A. As described by Alvarez, "The experimenter – who felt confident that his bump was significant - picked out 2 of the 100 computer generated histograms as looking significant"…The 'bump' was the 2 π peak at the ω mass and the physicist was me. The fact that one broad peak displayed a visual similarity to 2 peaks in 2% of cases has then declared "the evidence against 2 peaks," in spite of the fact that the data fitted the 2-peak hypothesis with 50% probability.

The test was flawed on 4 counts. (1) only one hypothesis was tested - 1 broad real peak looks like 2 narrow peaks, without testing the opposite assumption - 2 real narrow peaks that look as 1 broad peak, thus biasing the test up-front in favor of the 1-peak hypothesis; (2) qualitative visual inspection is scientifically meaningless; each of the 100 histograms should have been subjected to 2 quantitative χ^2- tests, one with the 1-peak and one with 2-peak hypothesis; (3) no χ^2- analysis was made to determine the probability for 1 narrow peak to have exactly the observed M and Γ of the ω; and (4) the erroneous premise that statistical analyses of the computer generated spectra can provide a more meaningful test than that of fitting the real data points with 1-peak and 2-peak hypotheses.

Similar obfuscations have been used over and over again to shove under the rug the resonant states not expected from the old Quark Model. A typical example was the campaign to prove that the 2 peak structure in the ρ-π resonance observed at CERN at the high [t]$\geq$0.2, known today as the a^1 (1260)-a^2 (1320) mass region is, in fact, 1 single peak. Since only one ρ-π resonance was compatible with the original simple Quark Model, this result triggered the "One-or-two- A$_2$ mesons controversy" between CERN [3] and BNL [4] that was rampant through the early 1970's.

L. Lederman, then chief spokesperson for the American Particle Physics community insisted that the success or failure of American Particle Physics depended as to whether or not there are one or two A$_2$ mesons, and declared, based on the data in Fig 2 B proves that there is only one A$_2$ meson.

To explain 2 peaks, one would need J=2, but the Old Quark model allowed no higher spin than J=1. Some theorists were fiercely defending the old Model on the grounds of its 'simplicity'. The exception was Gell-Mann, who told me at the time that 2 peaks can be easily accommodated by allowing J=2 but that he would prefer a simple model with J=1. The new Constituent Quark Model [15] allows all spin states, thus the issue is moot today. Since then, evidence for two A$_2$ mesons, renamed a$_1$ and a$_2$, has been reported in 97 measurements at low |t|; a very broad a$_1$ with J= 1 (28 papers), a$_2$ with J=2 (69 papers). The two mesons, a$_1$ and a$_2$ with J=1 and J=2, fit nicely the Model [15].

Nevertheless, the zealous defendants of the old Quark Model succeeded in burying the fine structure for 25 years.

Although the fit of the MM Spectrometer built at BNL was the only other instrument to operate at high |t|>0.2. BNL data to the 2-peak hypothesis (Fig. 2B) has 5 times greater probability than a 1-peak. Nevertheless, the fact that the 1- peak hypothesis had a non-zero probability was declared to represent the proof against 2 peaks! When the split in a$_1$-a$_2$ began to show up in the BNL Spectrometer data, the experiment was summarily stopped, instead of allocating more accelerator time needed to get a larger statistical sample.

In fact, the two experiments CERN [3] and BNL [4] resolved the contradiction about the widths: a$_1$ and a$_2$ peaks are broad and overlapping at a low momentum transfer |t|~0.01, where 99% of experiments operated; and narrow and separated when produced at an order of magnitude higher |t|~0.24, at which both MM Spectrometers operated.

In 1971, Paul Dirac approached me after my talk and asked about the fine structures, among other issues (Fig. 9). I told him about the resistance to my searches for fine structures and how DEIT insists that there is no need for improved resolutions for nothing narrow can exist. Dirac responded that, throughout the history of spectroscopy, better resolution always led to finer and finer structures and to new physics, and he would be surprised if this wouldn't happen in meson physics today, irrespective of the lack of theoretical predictions.

5. High Resolution Hadron Spectrometer with Solid State Hydrogen Target

We have designed a magnet-less Missing-Mass Hadron Spectrometer with a mass resolution of ±0.2 MeV that would be able to obtain, in 6 days, the spectra we collected at CERN (Fig. 3) in 6 months of continuous running. The instrument uses the reaction of the type $\pi + P \rightarrow P + X$ or $K+P \rightarrow P+X$, where X are the missing particle(s). Our mass resolution was ±15 MeV at that time; currently, typical resolutions are ± 5 MeV, although at RHIC ±1.2 MeV has been achieved. It will be a Maximum-Angle Jacobian-peak type missing-mass spectrometer[16] that would operate only at maximum momentum transfer, $|t|>0.25$ (GeV/c)^2.

The key element of the new spectrometer is the interactive solid state hydrogen target. It consists of an array of 100's of plastic scintillator (CH) filaments, about 5 cm long.

Each filament is viewed by 2 photomultipliers. Since this is a "maximum angle spectrometer" (Fig. 5), only the recoil protons emitted at an angle of 40° or higher are detected in a digital plane about 1 m away. Tight coincidence between the filament flashes and the plane, and the pulse height, will discriminate the interactions with P from those with C by rejecting: (1) high light output, (2) incorrect TOF and (3) incorrect angle. Since the mass resolution is determined mainly by the angular resolution, we project an overall mass resolution of ±0.2 MeV.

A California industrial fund has been established to support the development of this spectrometer and the associated theoretical physics studies.

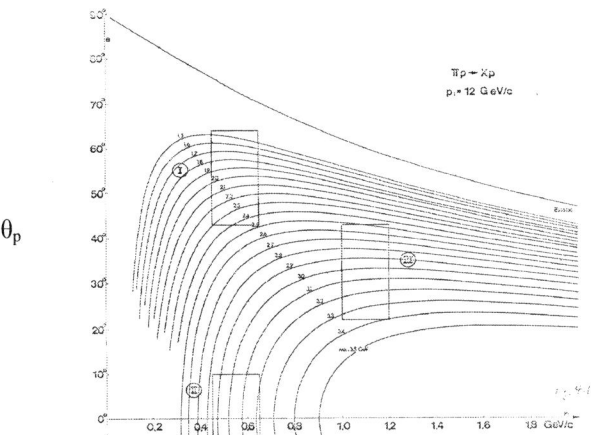

FIGURE 5. Proton Momentum vs. photon angle in LAB for $\pi\ P \rightarrow Px$ at $P_\pi=12$ GeV/c. Each line corresponds to a constant mass of X. Boxes I and III are the operating regions. See P.L. 18, 185 (1965).

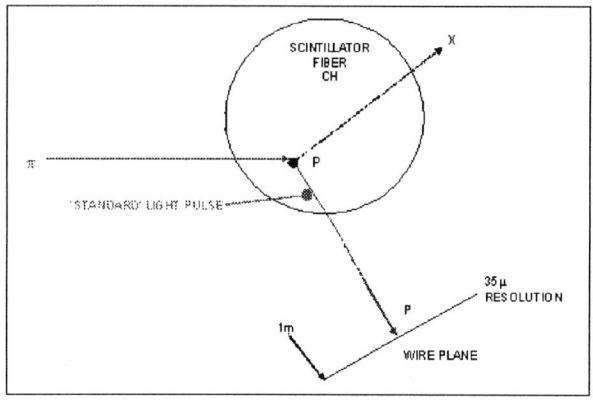

FIGURE 6. Interactive Solid State Hydrogen Target. Each CH fiber is viewed by 2 photomultipliers. P=proton; X= mesons.

FIGURE 7. Werner Heisenberg (right) and Bogdan Maglich (left) right after Maglich announced that neutral pseudo scalar meson, $J^P=0^-$, the cornerstone of Duerr-Heisenberg theory, had been found (η-meson) [1962 (German Phys. Soc. Annual Meeting, Stuttgart.] "Jetzt haben Sie Ihren Eigenmeson, einlassen Sie bitte dass ω^o Spin 1 hat?" –Now that you have your "Eigenmeson" – please admit that ω^o has spin one?" said Maglich, by coining the word Eigenmeson otherwise non-existent in German language - at which Heisenberg burst into laughter. - At the time, Heisenberg postulated that G-parity was violated in ω^o decay and that the narrow peak in $\pi+\pi-$ at the ω^o mass, observed by Maglich, was the G-parity violating 2π decay of ω^o. Neither this peak nor Heisenberg's interpretation of it were accepted by the particle physics community for 16 years. G-parity violating decay of ω was experimentally established in 1978, two years after Heisenberg's death. - In his letters to Maglich, Heisenberg argued that, if G parity was violated, ω^o could be a pseudoscalar, not vector. See Fig. 8 for Heisenberg-Maglich correspondence. [Photo Copyright 1976 by World Science Education, Princeton, New Jersey]

MAX-PLANCK-INSTITUT FÜR PHYSIK UND ASTROPHYSIK

INSTITUT FÜR PHYSIK
Prof. W. Heisenberg

MÜNCHEN 23 , den 13. Dez. 1961
AUMEISTERSTRASSE 8
TELEFON 363621

Herrn
Dr. Bogdan M a g l i ć
Lawrence Radiation Laboratory
University of California
B e r k e l e y 4, Calif., USA

Lieber Herr Maglić!

Haben Sie den besten Dank für Ihren Brief, dessen Inhalt uns alle hier sehr interessiert hat.

Vielleicht darf ich zuerst auf die (möglicherweise mißverstandene) Frage des Dalitz-Diagramms beim ω-Meson in Ihrer und Steinbergers Arbeit eingehen. Ich möchte zunächst ausdrücklich betonen, daß Dirr und mir irgendeine Kritik an Ihren schönen Messungen völlig ferngelegen hat. Vielmehr wollten wir nur auf eine, wie uns scheint, unvermeidliche Schwierigkeit bei der Interpretation der Experimente hinweisen, die ich etwas ausführlicher erklären möchte. Wenn man die Energien und Impulse der beim Zerfall entstehenden π-Mesonen gemessen hat, so pflegt man durch eine Ausgleichsrechnung dafür zu sorgen, dass Energie und Impuls im Ganzen beim Zerfall erhalten geblieben sind und dass daher der in das Dalitz-Diagramm einzutragende Punkt in das "erlaubte" Gebiet fällt. Wegen der unvermeidbaren Messfehler ist dieser Punkt aber etwas verschoben gegen den (nicht messbaren) richtigen Punkt. Denkt man sich diese Verschiebungen (vom richtigen zum "gemessenen" Punkt) als Pfeile in das Diagramm eingetragen, so werden im Inneren des erlaubten Gebietes alle Pfeilrichtungen statistisch verteilt vorkommen, am Rande aber werden die Pfeile bevorzugt nach innen oder parallel zum Rand, aber nicht nach aussen laufen, denn es darf ja kein Punkt nach aussen rücken. Die Folge ist, dass eine flache Verteilung im Dalitz-Diagramm, die sich etwa als Folge konstanter Matrixelemente ergibt (Fig. 1), je nach der Grösse der Messfehler in eine Verteilung vom Typ Fig. 2 oder Fig. 3 verwandelt wird.

Bei dem Dalitz-Diagramm für den Zerfall von τ-Mesonen, das Sie Ihrem Brief beigefügt haben, sind offenbar die Messfehler ausserordentlich gering, so dass Sie sich der Fig. 1 weitgehend genähert haben. Es scheint mir aber zweifelhaft – besonders da eines der drei π-Mesonen ja notwendig neutral ist – ob die unvermeidbaren Messfehler bei Ihren Messungen über das ω-Meson ähnlich klein sein können. Wenn man beim ω-Meson eine Verteilung von der Art der Fig. 1 erwarten könnte, so wäre, wie Sie mit Recht feststellen, die von Ihnen gemessene Verteilung nach den statistischen Fehlern allein nicht mit der gleichförmigen Verteilung vereinbar; wohl aber könnte eine durch unvermeidbare Messfehler verwaschene Verteilung nach der Art der Fig. 3 durchaus mit Ihren Messungen statistisch vereinbar sein. In Ihrer Arbeit haben Sie leider die Grösse dieses unvermeidlichen "Verwaschungseffektes" nicht angegeben. Ich vermute aber nach den Diskussionen mit unseren Experimentalphysikern, dass die Verwaschung doch nicht ganz klein sein kann. Daher kamen wir zu unserer Formulierung "The Dalitz plot favours slightly the symmetry ¹ ... ... but it is scarcely sufficient to decide between the three possibilities."

Jedenfalls glauben wir, dass es sehr gefährlich wäre, eine Analyse nur auf das Dalitz-Diagramm allein zu basieren und andere experimentelle Tatsachen, wie z.B. die sehr geringe Linienbreite, dabei ausser acht zu lassen.

Die Untersuchung des Verhältnisses $R = \dfrac{\omega \to 3\pi_0}{\omega \to \pi^+, \pi^0, \pi^-}$ scheint uns ausserordentlich wichtig. Sie schreiben leider in Ihrem Brief nur, dass Ihre Messungen gegen die Grössenordnung 1 dieses Verhältnisses sprechen, aber Sie sagen nicht, was Sie nun wirklich für dieses Verhältnis bekommen und mit welcher Methode Sie das Verhältnis bestimmen. Ich wäre Ihnen sehr dankbar, wenn Sie mir gerade über diesen Punkt noch ausführlichere Angaben schicken könnten.

Alles das, was ich bisher geschrieben habe, zielt natürlich in genau derselben Form auch für das sehr interessante η-Meson. Ich glaube es wäre zweckmässig, wenn man über alle derartigen Zustände zunächst eine Interpretation möglichst viele verschiedenartige Informationen sammeln könnte, z.B. über das Verhältnis verschiedener Zerfallsprozesse, über Linienbreite, Zerfall mit Lichtquantenemission usw., und dass man den Vergleich mit der Theorie erst am Ende vornehmen sollte.

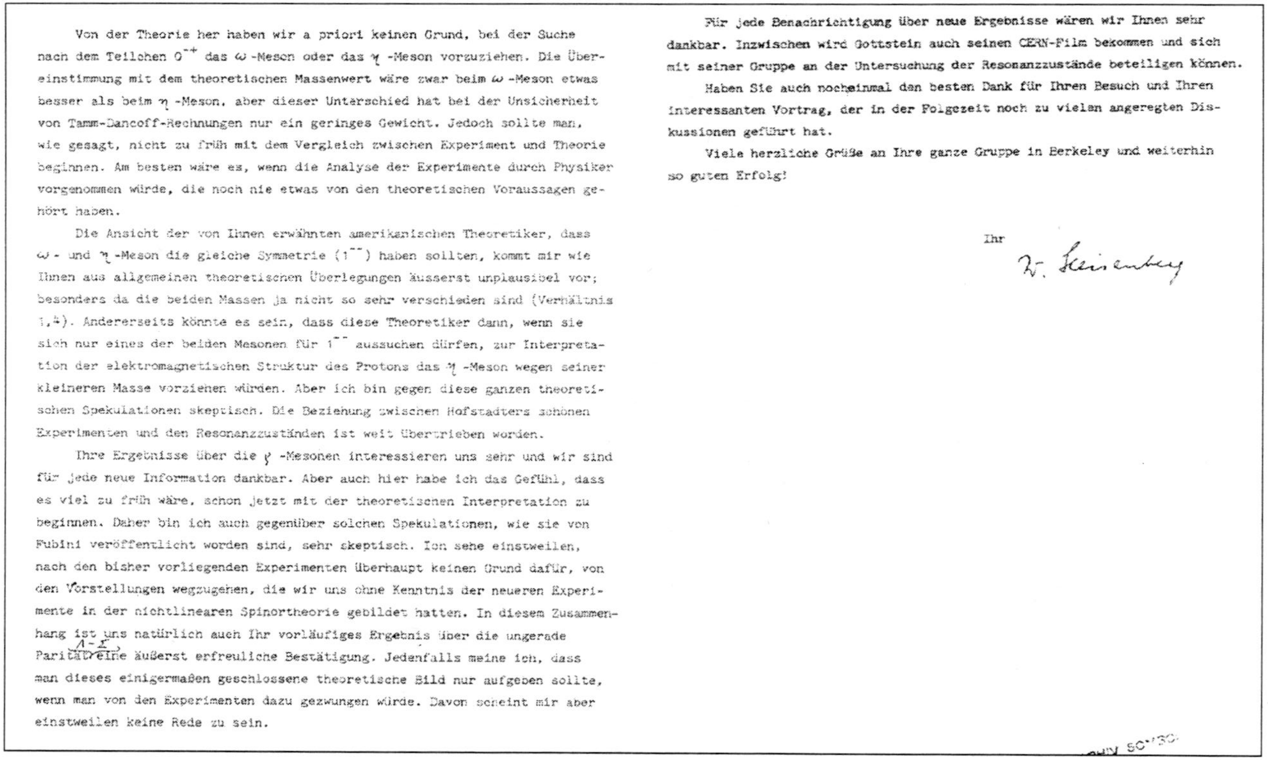

Von der Theorie her haben wir a priori keinen Grund, bei der Suche nach dem Teilchen 0^{-+} das ω-Meson oder das η-Meson vorzuziehen. Die Übereinstimmung mit dem theoretischen Massenwert wäre zwar beim ω-Meson etwas besser als beim η-Meson, aber dieser Unterschied hat bei der Unsicherheit von Tamm-Dancoff-Rechnungen nur ein geringes Gewicht. Jedoch sollte man, wie gesagt, nicht zu früh mit dem Vergleich zwischen Experiment und Theorie beginnen. Am besten wäre es, wenn die Analyse der Experimente durch Physiker vorgenommen würde, die noch nie etwas von den theoretischen Voraussagen gehört haben.

Die Ansicht der von Ihnen erwähnten amerikanischen Theoretiker, dass ω- und η-Meson die gleiche Symmetrie (1^{--}) haben sollten, kommt mir wie Ihnen aus allgemeinen theoretischen Überlegungen äusserst unplausibel vor; besonders da die beiden Massen ja nicht so sehr verschieden sind (Verhältnis 1,4). Andererseits könnte es sein, dass diese Theoretiker dann, wenn sie sich nur eines der beiden Mesonen für 1^{--} aussuchen dürfen, zur Interpretation der elektromagnetischen Struktur des Protons das η-Meson wegen seiner kleineren Masse vorziehen würden. Aber ich bin gegen diese ganzen theoretischen Spekulationen skeptisch. Die Beziehung zwischen Hofstadters schönen Experimenten und den Resonanzzuständen ist weit übertrieben worden.

Ihre Ergebnisse über die ζ-Mesonen interessieren uns sehr und wir sind für jede neue Information dankbar. Aber auch hier habe ich das Gefühl, dass es viel zu früh wäre, schon jetzt mit der theoretischen Interpretation zu beginnen. Daher bin ich auch gegenüber solchen Spekulationen, wie sie von Fubini veröffentlicht worden sind, sehr skeptisch. Ich sehe einstweilen, nach den bisher vorliegenden Experimenten überhaupt keinen Grund dafür, von den Vorstellungen wegzugehen, die wir uns ohne Kenntnis der neueren Experimente in der nichtlinearen Spinortheorie gebildet hatten. In diesem Zusammenhang ist uns natürlich auch Ihr vorläufiges Ergebnis über die ungerade Λ-Σ Parität eine äußerst erfreuliche Bestätigung. Jedenfalls meine ich, dass man dieses einigermaßen geschlossene theoretische Bild nur aufgeben sollte, wenn man von den Experimenten dazu gezwungen würde. Davon scheint mir aber einstweilen keine Rede zu sein.

Für jede Benachrichtigung über neue Ergebnisse wären wir Ihnen sehr dankbar. Inzwischen wird Gottstein auch seinen CERN-Film bekommen und sich mit seiner Gruppe an der Untersuchung der Resonanzzustände beteiligen können.

Haben Sie auch nocheinmal den besten Dank für Ihren Besuch und Ihren interessanten Vortrag, der in der Folgezeit noch zu vielen angeregten Diskussionen geführt hat.

Viele herzliche Grüße an Ihre ganze Gruppe in Berkeley und weiterhin so guten Erfolg!

Ihr

W. Heisenberg

FIGURE 8. Heisenberg to Maglich, Letter No. 1 (12/13/1961). Heisenberg points out that G-parity violation could make Dalitz plot for $J^P = 0^-$ (Fig. 1) look similar to that for $J^P = 1^-$ due to the statistical errors; hence, ω^0 could be the pseudo scalar needed for Duer-Heisenberg theory. For 11 additional Heisenberg to Maglich letters, see www.worldscienceinvestigation.org [Courtesy of Historical Archives, City of Sombor, Yugoslavia]

FIGURE 9. Paul Dirac and Bogdan Maglich at Everglades, Florida (January, 1971) during the Coral Gables Conference on Fundamental Interactions. Dirac was one of the few theorists who encouraged searches for fine structures. [Photo courtesy of Bogdan C. Maglich]

REFERENCES

1.   Yuval Ne'eman: HADRON 2001Ninth International Conference on Hadron Spectroscopy, p.259, AIP Conf. Proc. 619, 259(2002): Review of the history of Hadron Spectroscopy and Ne'eman's role in 1958-1964.

2. B. C. Maglich, L.W. Alvarez, L. Stevenson, A. Rosenfeld, Phys. Rev. 7, 168 (1961):

2a. "Bogdan Maglić, a visitor to our group, … made and important decision to concentrate on proton-antiproton annihilations into 5 pions … Although Bogdan Maglić originated the plan for this search, and pushed through the measurements by himself, he graciously insisted that the paper announcing his discovery of ω meson should be coauthored by three of us who had developed the chamber, the beam, and the analysis program that made it possible." SCIENCE 165 1088 (1969).

3. B. Levrat et al PL 22, 714(66); L. Dubal et al, Nucl. Phys. B3, 435(67). Seguinot et al, P.L. 19, 712(66). Now named $\pi_2(1670)$.

4. Bowen et al, PRL 26, 1663(71).

5. Ozawa et al, PRL 86, 5019(2001); see also PR B502, 59(2001).

6. Brown and Rho, PRL 66, 2720(91) and arXiv:nucl-th/0101015 v4 20 jul 2001

7. Seguinot et al, P.L. 19, 712 (66). Now named π_2 (1670).

8. Alspector et al, Phys. Rev. Lett. 30, 511 (1973).15

9. Ref. 2a p. 1082 – refers to program FAKE

10. J. Button et al, Phys. Rev.126, 1858 (1962).

11. H.-P. Duerr and W. Heisenberg, Nuovo Cimento 23, 807 (1962)

12. H.J. Behrend et al PRL 27,61 (71)

13 A.B. Wicklund et al, Phys. Rev. D17, 1179 (1978).

14. Leon Lederman, after dinner speech at Int'l. Symp. on Particle Physics, "Pions to quarks" (FERMILAB, May 1-4, 1985); Cambridge U. Press (1989). From my personal contacts with the creators of SU(3), I knew that they did not share such cataclysmic views.

15.   P.R. D32, 189(85). For an "easy" dynamic comparison of hadron data and model expectations, see Eric Swenson's (U. of Pittsburgh) web page: http://fafnir.phyast.pit.edu/exotica/mesons/SV.html

16. B. Maglich and G. Costa, Phys. Lett.18, 185 (1965)

Initial State Radiation Study at $\Upsilon(4S)$ in BaBar[1]

Evgeny Solodov

BudkerINP, Novosibirsk
on behalf of the BABAR Collaboration

Abstract. An analysis of the processes with initial state radiation (ISR) has been performed using 90 fb^{-1} of *BABAR* data. The selection of $\mu^+\mu^-$ and multi-hadron final states has been demonstrated accompanied with the detected ISR photon. The invariant mass of hadronic final state determines the virtual photon energy and data can be compared with direct e^+e^- cross sections. The present *BABAR* data are already competitive with e^+e^- machine data in 0.28-3.0 GeV energy range and demonstrate many interesting details usefull for low energy hadron spectroscopy.

In addition to light meson spectroscopy these data can be used for calculation of R - the ratio of $e^+e^- \rightarrow hadrons$ cross section to $e^+e^- \rightarrow \mu^+\mu^-$ - and thereby to impact the $(g-2)_\mu$ measurement.

The ISR technic gives an access to J/ψ production. The J/ψ decays to $\mu\mu$, 4π, $2K2\pi$ and 4K have been selected and new preliminary measurements of branching ratios performed with comparable or typically better accuracy than PDG.

INTRODUCTION

The possibility of using the initial state radiation (ISR) of hard photons at B-factories to study hadronic final state production at lower e^+e^- c.m. energies has been discussed previously [1, 2, 3]. The interest to this kind of study is rising up because of discrepancy between measured muon g-2 value and one predicted by Standard Model [4], where hadronic contribution is taken from e^+e^- experiments at low energies. The study of the ISR events at B-factories can give independent measurements of hadronic cross sections as well as contribute to low mass resonance spectroscopy.

The ISR cross section for a particular final state f depends on e^+e^- cross section $\sigma_f(s)$ and is obtained from:

$$\frac{d\sigma(s,x)}{dx} = W(s,x) \cdot \sigma_f(s(1-x)), \tag{1}$$

where $x = \frac{2E_\gamma}{\sqrt{s}}$; E_γ is the energy of the ISR photon in the nominal c.m. frame, and $\sqrt{s}$ is the nominal c.m. energy. The function W(s,x) describes the energy spectrum of the virtual photons and can be calculated with better than 1% accuracy [1, 2, 3]. ISR photons are produced at all angles relative to the collision axis, and it has been shown that the *BABAR* acceptance for such photons is around 10-15 % [3].

Events corresponding to $e^+e^- \rightarrow \mu^+\mu^-\gamma$ are providing ISR luminosity for the normalization of the hadronic cross section measurements. For a hadronic final state, f, the

---

[1] Work supported in part by Department of Energy contract DE-AC03-76SF00515.

CP717, *Hadron Spectroscopy: Tenth International Conference*,
edited by E. Klempt, H. Koch, and H. Orth
© 2004 American Institute of Physics 0-7354-0197-7/04/$22.00

normalized Born cross section at c.m. energy squared s', $\sigma_f(s')$, is obtained by relating the observed number of events in an interval ds' centered at s', $dN_{f\gamma}$, to the corresponding number of radiative di-muon events, $dN_{\mu\mu\gamma}$, by means of

$$\sigma_f(s') = \frac{dN_{f\gamma} \cdot \varepsilon_{\mu\mu} \cdot (1 + \delta^{\mu\mu}_{rad})}{dN_{\mu\mu\gamma} \cdot \varepsilon_f \cdot (1 + \delta^{f}_{rad})} \cdot \sigma_{e^+e^- \to \mu^+\mu^-}(s'), \tag{2}$$

where $s' = s(1-x)$; $\varepsilon_{\mu\mu}$ and ε_f are detection efficiencies, and $1 + \delta^{\mu\mu}_{rad}$, $1 + \delta^{f}_{rad}$ are the corrections excluding fraction of events when hard photon comes from final particles. This correction is important for di-muons and negligible for most of the hadronic final states. The Born cross section $\sigma_{e^+e^- \to \mu^+\mu^-}(s')$ is used. The radiative corrections to the initial state, acceptance for the ISR photon, and virtual photon properties are the same for $\mu^+\mu^-$ and f, and cancel in the ratio.

An advantage deriving from the use of ISR is that the entire range of effective collision energy is scanned in one experiment. This avoids the relative normalization uncertainties which can arise when data from different experiments are combined.

A disadvantage is that invariant mass resolution limits the width of the narrowest structure which can be measured via ISR production.

The resolution and absolute energy scale can be monitored directly using the width of the J/ψ resonance produced in the $e^+e^- \to J/\psi\gamma$ reaction. For a narrow resonance, such as the J/ψ the total production cross section can be calculated as

$$\sigma^{tot}_{J/\psi}(s) = \frac{12\pi^2\Gamma_{ee}}{m_{J/\psi} \cdot s} \cdot W(s,x); \; x = 1 - \frac{m^2_{J/\psi}}{s}, \tag{3}$$

where $m_{J/\psi}$ and Γ_{ee} are mass and electron partial width of J/ψ. For $s = m^2_{\Upsilon(4S)}$ the cross section is equal to 0.036 nb. With 130 fb^{-1} of $BABAR$ integrated luminosity about $4 \cdot 10^6$ of J/ψ's decay in the detector. The cross section for the final state f

$$\sigma^{f}_{J/\psi}(s) = \frac{12\pi^2\Gamma_{ee}B_f}{m_{J/\psi} \cdot s} \cdot W(s,x); \; x = 1 - \frac{m^2_{J/\psi}}{s}, \tag{4}$$

is propotional to the product $\Gamma_{ee} \cdot B_f$ or $\Gamma \cdot B_{ee} \cdot B_f$ where Γ and B_{ee}, B_f are the total width and branching fractions of J/ψ to e^+e^- and f.

The invariant mass of the final particles determines the position of J/ψ peak and detector mass resolution $\sim$8 MeV can be achieved by using a kinematic fit. Preliminary studies of some particular ISR processes have been performed with $BABAR$ data [5, 6] showing good detector efficiency and particle identification capability for this kind of events.

THE $\mu^+\mu^-\gamma$ FINAL STATE AND ISR LUMINOSITY

The data used in this analysis were collected with the $BABAR$ detector at the PEP-II asymmetric e^+e^- storage ring. The total integrated luminosity used in this analysis is $89.3 fb^{-1}$. Both data collected at $\Upsilon(4S)$ resonance and continum are used for this analysis.

The *BABAR* detector is described elsewhere [7]. The information from *BABAR* tracking system (Silicon Vertex Tracking - SVT and Drift Chamber - DCH) is used to measure angles and momenta of charged particles. Information from DIRC allows to identified kaons in final state. The IFR allows to identify muons and the photons are detected in the CsI calorimeter - EMC.

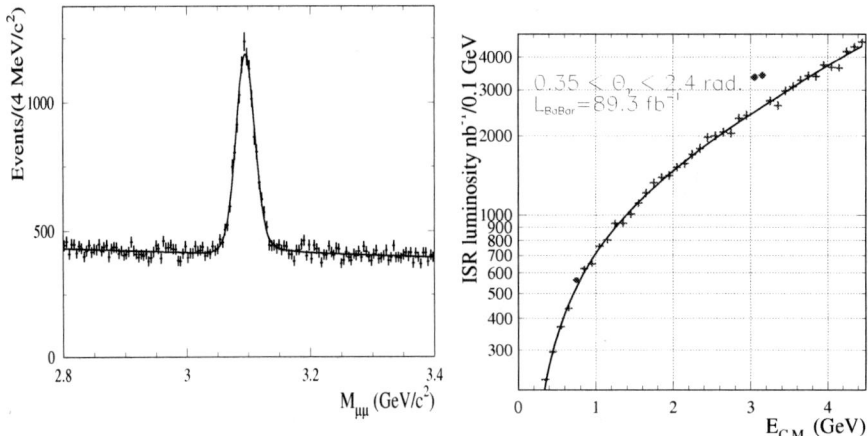

FIGURE 1. On the left: The $\mu^+\mu^-$ invariant mass distribution in J/ψ region. On the right: The calculated ISR luminosity integrated over 0.1 GeV for 89.3fb $^{-1}$ of integrated *BABAR* luminosity.

The pair of muons from $e^+e^- \rightarrow \mu^+\mu^-\gamma$ process are easily identified and the plot in fig. 1(left) shows invariant mass distribution of muon pairs in the J/ψ region. The invariant mass of the muon pair defines the effective collision energy, i.e. the c.m. energy of the virtual photon in wide range. The energy dependence of the ISR luminosity, dL, for the interval dE_{γ^*} centred at virtual photon energy E_{γ^*} is then obtained from

$$dL(E_{\gamma^*}) = \frac{dN_{\mu\mu\gamma}(E_{\gamma^*})}{\varepsilon_{\mu\mu}(E_{\gamma^*})\cdot(1+\delta^{\mu\mu}_{FSR})(E_{\gamma^*})\cdot\sigma_{e^+e^-\rightarrow\mu^+\mu^-}(E_{\gamma^*})}, \quad E_{\gamma^*} = m^{\mu\mu}_{inv},$$

where $dN_{\mu\mu\gamma}$ is the number of experimental di-muon events observed in this interval, $\varepsilon_{\mu\mu}$ - acceptance from simulation and $(1+\delta^{\mu\mu}_{FSR})$ - correction on final state radiation (FSR). The $\sigma_{e^+e^-\rightarrow\mu^+\mu^-}(E_{\gamma^*}$ Born cross section is used. The $\mu\mu\gamma$ events with hard photon emitted by final muons contribute to total number of observed events and should be removed for limunosity calculation.

The ISR luminosity vs. effective c.m. energy for $89.3 fb^{-1}$ integrated luminosity is shown in fig. 2(left) and is used for normalization of hadronic final states according to equation 1. The present *BABAR* data are equivalent to an e^+e^- machine scan in 0.1 GeV steps with a luminosity integral per point varying from 700 nb^{-1} at 1 GeV to 4 pb^{-1} at 4.5 GeV c.m. energy. This luminosity integral is already competitive with exising e^+e^- experimental data. The systematic error is estimated as 3% and should be increased to 5% for mass region below 1 GeV.

THE HADRONIC FINAL STATES

Currently the major hadronic final states $\pi^+\pi^-\gamma$, $K\bar{K}\gamma$, $p\bar{p}\gamma$, $\pi^+\pi^-\pi^0\gamma$, $4\pi\gamma$, $K\bar{K}\pi^0\gamma$, $5\pi\gamma$, $6\pi\gamma$ are under study in the *BABAR* Collaboration. The Monte Carlo generators for these processes are based on formulae from [8, 9, 10]. The analysis procedure for $\pi^+\pi^-\pi^+\pi^-$, $K^+K^-\pi^+\pi^-$ and $K^+K^-K^+K^-$ final states is described in [12]. Figure 2(left) presents the obtained $e^+e^- \to \pi^+\pi^-\pi^+\pi^-$ cross section in comparison with all existing e^+e^- data. The estimated systematic error is about 5%.

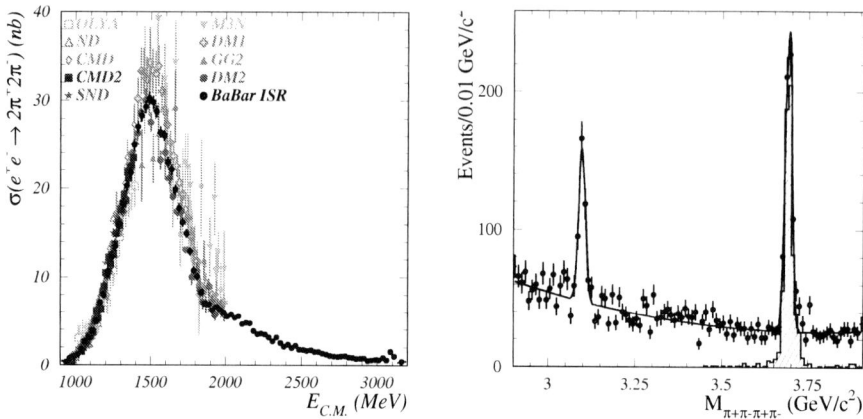

FIGURE 2. On the left: The $e^+e^- \to \pi^+\pi^-\pi^+\pi^-$ cross section obtaind from ISR at *BABAR* in comparison with all e^+e^- data. On the right: The signals from J/ψ and $\psi(2S)$ in 4π invariant mass. The shaded region at the latter corresponds to $\psi(2S) \to J/\psi\pi^+\pi^-$, with $J/\psi \to \mu^+\mu^-$.

The hadronic contribution to $(g-2)_\mu$ from this particular channel evaluated using all available e^+e^- data in 0.56-1.8 GeV range is $\alpha_\mu^{had}x10^{10} = 14.21 \pm 0.87_{exp} \pm 0.23_{rad}$. The τ decay data give $\alpha_\mu^{had}x10^{10} = 12.35 \pm 0.96_{exp} \pm 0.40_{SU2}$. The *BABAR* data in this energy range give $\alpha_\mu^{had}x10^{10} = 12.95 \pm 0.64_{exp} \pm 0.13_{rad}$ what shows a potential of the ISR measurements.

THE J/ψ DECAYS

The ratio of $\mu^+\mu^-$ events from J/ψ peak (see fig. 1(left)) to continum allows to calculate the product [6]

$$\Gamma(J/\psi \to e^+e^-) \cdot B(J/\psi \to \mu^+\mu^-) = 0.330 \pm 0.008 \pm 0.007 \text{ keV},$$

where first error is statistical from about 7800 observed events and the second one includes systematic errors from uncertainties in background estimation, line shape, radiative corrections and Monte Carlo statistic. Using the world averages for $B(J/\psi \to \mu^+\mu^-)$ and $B(J/\psi \to e^+e^-)$, we derive the J/ψ electronic and total widths: $\Gamma(J/\psi \to e^+e^-) = 5.61 \pm 0.20$ keV and $\Gamma = 94.7 \pm 4.4$ keV.

Figure 2(right) shows the J/ψ and $\psi(2S)$ signals in four charged tracks invariant mass. The later is seen due to the process $\psi(2S) \to J/\psi\pi^+\pi^- \to \mu\mu\pi^+\pi^-$ and can be easily isolated by requirement of J/ψ mass in one pair of charged particles (shaded histogram). By using 270±20 and 620±25 observed events respectively, detection efficiency from simulation and ISR luminosity the following products have been obtained:

$$B_{J/\psi\to4\pi} \cdot \Gamma_{J/\psi ee} = (1.95 \pm 0.14 \pm 0.13) \cdot 10^{-2} \text{ keV},$$

$$B_{\psi(2S)\to J/\psi\pi^+\pi^-} \cdot B_{J/\psi\to2\mu} \cdot \Gamma_{\psi(2S)ee} = (4.50 \pm 0.18 \pm 0.22) \cdot 10^{-2} \text{ keV}.$$

Using the world averages for $\Gamma_{J/\psi\to e^+e^-}$, $\Gamma_{\psi(2S)\to e^+e^-}$ and $B_{J/\psi\to2\mu}$ we derive the values $B_{J/\psi\to4\pi} = (3.70 \pm 0.27 \pm 0.36) \cdot 10^{-3}$ and $B_{\psi(2S)\to J/\psi\pi^+\pi^-} = 0.361 \pm 0.015 \pm 0.037$.

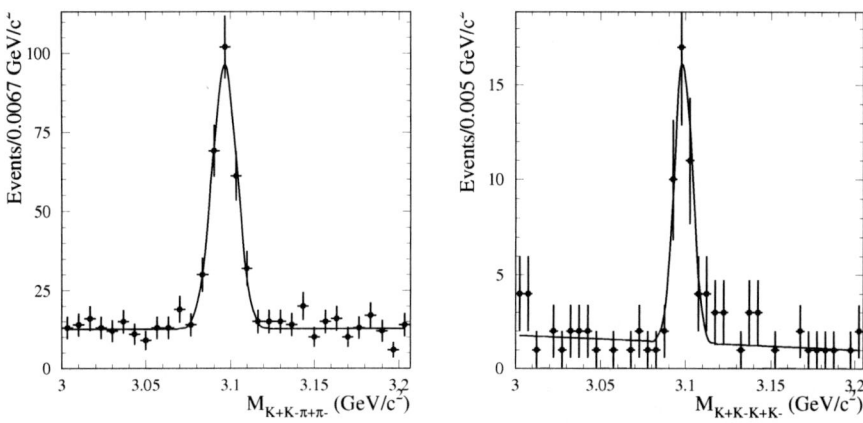

FIGURE 3. On the left: The signals from J/ψ in $2K2\pi$ invariant mass. On the right: The signals from J/ψ in $4K$ invariant mass.

Figures 3 show J/ψ signals in the $K^+K^-\pi^+\pi^-$ and $K^+K^-K^+K^-$ final states. 233±19 and 38.5±6.7 events have been observed respectively. Using Monte Carlo efficiency and ISR luminosity the following products have been obtained:

$$B_{J/\psi\to2K2\pi} \cdot \Gamma_{J/\psi ee} = (3.29 \pm 0.27 \pm 0.27) \cdot 10^{-2} \text{ keV},$$

$$B_{J/\psi\to4K} \cdot \Gamma_{J/\psi ee} = (3.6 \pm 0.6 \pm 0.5) \cdot 10^{-3} \text{ keV}.$$

Taking into accout world average value for $\Gamma_{J/\psi\to e^+e^-}$ we derive the relative decay rates

$$B_{J/\psi\to2K2\pi} = (6.25 \pm 0.50 \pm 0.62) \cdot 10^{-3},$$

$$B_{J/\psi\to4K} = (6.9 \pm 1.2 \pm 1.1) \cdot 10^{-4}.$$

CONCLUSION

The studies of some particular ISR processes have been performed with *BABAR* data showing good detector efficiency and particle identification capability for this kind of events. The preliminary $e^+e^- \rightarrow \pi^+\pi^-\pi^+\pi^-$ cross section with about 5% systematic error has been obtained from threshold to 4.5 GeV c.m. energy. The radiative return to J/ψ resonance allows to measure the relative brunching fractions with best to date accuracy.

ACKNOWLEDGMENTS

We are grateful for the extraordinary contributions of our PEP-II colleagues in achieving the excellent luminosity and machine conditions that have made this work possible. The success of this project also relies critically on the expertise and dedication of the computing organizations that support *BABAR*. The collaborating institutions wish to thank SLAC for its support and the kind hospitality extended to them. This work is supported by the US Department of Energy and National Science Foundation, the Natural Sciences and Engineering Research Council (Canada), Institute of High Energy Physics (China), the Commissariat à l'Energie Atomique and Institut National de Physique Nucléaire et de Physique des Particules (France), the Bundesministerium für Bildung und Forschung and Deutsche Forschungsgemeinschaft (Germany), the Istituto Nazionale di Fisica Nucleare (Italy), the Research Council of Norway, the Ministry of Science and Technology of the Russian Federation, and the Particle Physics and Astronomy Research Council (United Kingdom). Individuals have received support from the A. P. Sloan Foundation, the Research Corporation, and the Alexander von Humboldt Foundation.

REFERENCES

1. A.B. Arbuzov *et al.*, JHEP **9812**, 009 (1998).
2. S. Binner, J.H. Kuehn, K. Melnikov, Phys. Lett. **B459** 279 (1999).
3. M. Benayoun *et al.*, Mod. Phys. Lett. **A14**, 2605 (1999).
4. M.Davier *et al.*, LAL-02-81, Aug 2002. 44pp. e-Print Archive: hep-ph/0208177.
5. E.P.Solodov (for *BABAR* collaboration). "Study of e^+e^- collisions in the 1.5-3.0 GeV C.M. Energy Region Using ISR at *BABAR*", Invited talk at "International Workshop on e^+e^- Physics at Intermediate Energy", SLAC, Stanford, April 30-May 4, 2001, hep-ex/0107027.
6. "J/ψ production in $e^+e^- \rightarrow \mu^+\mu^-\gamma$ process", To be published in Phys.Rev.Lett.
7. *BABAR* Collaboration, B. Aubert *et al.*, "The *BABAR* Detector," hep-ex/0105044 (2001), submitted to Nucl. Instr. and Meth.
8. H.Czyz and J.H.Kuehn, Eur.Phys.J **C18**(2000)497-509 (hep-ph/0008262).
9. A.B.Arbuzov *et al.*, JHEP **9710**, 001 (1997), hep-ph/9702262.
10. M. Caffo, H. Czyz, E. Remiddi, Nuo. Cim. **110A**, 515 (1997); Phys. Lett. **B327**, 369 (1994).
11. E. Barberio, B. van Eijk and Z. Was. Comput. Phys. Commun. **66**, 115 (1991).
12. see talk by R.Stroili in this Proceedengs

Measurement of the
$e^+e^- \to \pi^+\pi^-\pi^+\pi^-, K^+K^-\pi^+\pi^-, K^+K^-K^+K^-$
Cross Sections Using Initial State Radiation at BABAR

R. Stroili for the BABAR Collaboration

Universitá di Padova & INFN, via Marzolo 8, 35131 Padova

Abstract. First results of a study of the $e^+e^- \to \pi^+\pi^-\pi^+\pi^-\gamma$ process with hard photon emitted from initial state are presented. About 60000 fully reconstructed events have been selected based on 89.3 fb^{-1} of BABAR data. The invariant mass of the hadronic final state defines the effective collision c.m. energy, and so BABAR ISR data can be compared to the relevant direct e^+e^- measurements. From obtained 4π mass spectrum we evaluate $e^+e^- \to \pi^+\pi^-\pi^+\pi^-$ cross section for the range of c.m.s. energy from 0.6 to 4.5 GeV. The systematic error of the cross section measurement is 5% and comparable with the best e^+e^- data. The cross sections for identified $2K2\pi$ and $4K$ final states also have been presented.

INTRODUCTION

The study of the ISR events at B-factories can give independent measurements of hadronic cross sections as well as contribute to low mass resonance spectroscopy.

The ISR cross section for a particular final state f depends on e^+e^- cross section $\sigma_f(s)$ and is obtained from:

$$\frac{d\sigma(s,x)}{dx} = W(s,x) \cdot \sigma_f(s(1-x))$$

where $x = \frac{2E_\gamma}{\sqrt{s}}$; E_γ is the energy of the ISR photon in the nominal c.m. frame, and $\sqrt{s}$ is the nominal c.m. energy. The function

$$W(s,x) = \beta \cdot ((1+\delta) \cdot x^{(\beta-1)} - 1 + \tfrac{x}{2})$$

describes the energy spectrum of the ISR photons. $\beta = \frac{2\alpha}{\pi} \cdot (2ln\frac{\sqrt{s}}{m_e} - 1)$, and δ takes into account vertex and self-energy corrections. At the $\Upsilon(4S)$ energy, $\beta = 0.088$ and $\delta = 0.067$.

Events corresponding to $e^+e^- \to \mu^+\mu^-\gamma$ are providing ISR luminosity for the normalization of the hadronic cross section measurements [1]. For a hadronic final state, f, the normalized Born cross section at c.m. energy squared s', $\sigma_f(s')$, is obtained by relating the observed number of events $dN_{f\gamma}$, to the corresponding number of radiative di-muon events, $dN_{\mu\mu\gamma}$, by means of

$$\sigma_f(s') = \frac{dN_{f\gamma} \cdot \varepsilon_{\mu\mu} \cdot (1+\delta_{rad}^{\mu\mu})}{dN_{\mu\mu\gamma} \cdot \varepsilon_f \cdot (1+\delta_{rad}^{f})} \cdot \sigma_{e^+e^- \to \mu^+\mu^-}(s')$$

CP717, *Hadron Spectroscopy: Tenth International Conference,*
edited by E. Klempt, H. Koch, and H. Orth
© 2004 American Institute of Physics 0-7354-0197-7/04/$22.00

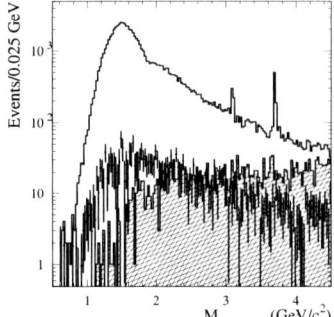

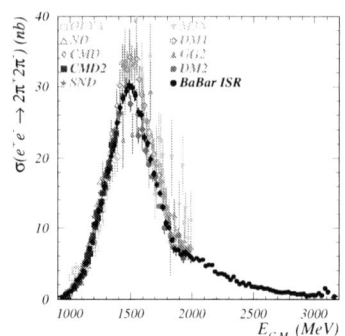

a) The four-pion invariant mass distribution for the signal region. The points indicate the estimated ISR-type background. The cross-hatched histogram corresponds to the non-ISR background.

b) The energy dependence of the $e^+e^- \rightarrow \pi^+\pi^-\pi^+\pi^-$ cross section obtained with *BABAR* ISR data (black points) in comparison with that resulting from individual e^+e^- production measurements.

FIGURE 1.

where $s' = s(1-x)$; $\varepsilon_{\mu\mu}$ and ε_f are detection efficiencies, and $1 + \delta_{rad}^{\mu\mu}$, $1 + \delta_{rad}^f$ are the corrections excluding fraction of events when hard photon comes from final particles. The Born cross section $\sigma_{e^+e^- \rightarrow \mu^+\mu^-}(s')$ is used. The radiative corrections to the initial state, acceptance for the ISR photon, and virtual photon properties are the same for $\mu^+\mu^-$ and f, and cancel in the ratio.

An advantage deriving from the use of ISR is that the entire range of effective collision energy is scanned in one experiment. This avoids the relative normalization uncertainties which can arise when data from different experiments are combined.

This paper reports the results from analysis of $\pi^+\pi^-\pi^+\pi^-$, $K^+K^-\pi^+\pi^-$ and $K^+K^-K^+K^-$ exclusive hadronic final states accompanied by a hard photon assumed to result from ISR.

DATA SELECTION AND ANALYSIS

The data used in this analysis were collected with the *BABAR* detector [2] at the PEP-II asymmetric e^+e^- storage ring. The total integrated luminosity used in this analysis is $89.3fb^{-1}$. Both data collected at $\Upsilon(4S)$ resonance and continuum are used for this analysis.

ISR events where selected requiring a high energy photon ($E_{\gamma CM} > 3$ GeV) opposite to momentum of charged tracks. A relatively clean sample of four-pion candidate events can be selected requiring that there be four charged tracks The number of detected photons (in addition to ISR photon) are not taken into account for four pion final state study. Electrons or positrons from beam losses are effectively removed from the charged track list.

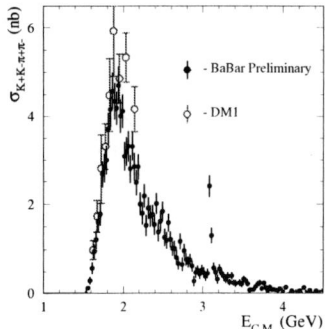
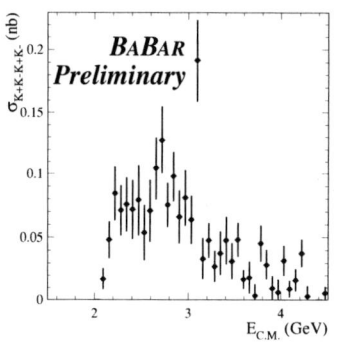

a) The c.m. energy dependence of the $e^+e^- \to K^+K^-\pi^+\pi^-$ cross section obtained from ISR events at BaBar.

b) The c.m. energy dependence of the $e^+e^- \to K^+K^-K^+K^-$ cross section obtained from ISR events at BaBar.

FIGURE 2.

To calculate acceptance and efficiencies a special package of Monte Carlo generators for radiative processes has been developed. The $4\pi\gamma$ final state simulation is based on the code developed by Kuehn and Czyz [3]. For the $2K2\pi$ and $4K$ final states a phase-space model has been used.

The radiative corrections - multiple (real) photon emission by initial and final particles have been added with the technique of structure functions [4, 5] and "PHOTOS" package. The accuracy of radiative corrections is better or about 1%.

For the background estimation a relatively big sample of the main ISR processes ($2\pi\gamma$, $3\pi\gamma$ ... $6\pi\gamma$, $2K2\pi\gamma$) has been simulated. General simulation (JETSET) with quark-antiquark and $\tau - \tau$ final states to estimate non-ISR type background were also used.

The constrained fit procedure uses the measured momenta and angles of charged particles and error matrix to solve four energy-momentum equations with the photon mass as the only constraint.

The four track sample can contain events with four pions, 4 kaons and events with 2 kaons plus 2 pions in the final state. To discriminate between these three possibilities, the three fits are performed for each event.

The main sources of background are other ISR multipion processes and non-ISR multipion production from e^+e^- collision. The background was studied using simulation and events from the χ^2 control sample.

The mass resolution obtained by simulation is $\sigma_{M4\pi}$=0.0062 GeV and $\sigma_{M4\pi}$=0.0075 for 1.5 GeV and 3 GeV regions respectively. This is confirmed by the data where the J/ψ width in 4π is 0.0080 GeV [1]. Figure 1a presents the invariant mass distribution.

Figure 1b presents the obtained cross section in 0.025 GeV step which is corresponding to about $3\sigma_{M4\pi}$ and bin-to-bin correlations are less than 1%.

The following corrections and systematic errors discussed above have to be added to the measured cross section determined from 4 track sample:

- Luminosity from $\mu\mu\gamma$ [1]: $\pm 3\%$; 5% for $m_{4\pi} < 1.0$ GeV.

- $\chi^2 < 30$ cut MC-DATA difference: $+3\pm2\%$
- Background subtraction: $\pm1\%$; $\pm10\%$ for $m_{4\pi} < 1.0$ GeV; $\pm3\%$ for $m_{4\pi} > 3.0$ GeV
- MC-DATA difference in track losses: $+3\pm2\%$
- Radiative (multiple FSR) correction accuracy: $\pm1\%$
- Acceptance from simulation (model dependent): $\pm2\%$ for <3 GeV; $\pm15\%$ for the rest

Assuming no correlation the total systematic error in the cross section is calculated to be 12% for $m_{4\pi} < 1.0$ GeV, 5.0% for 1-3 GeV mass region and 16% for higer masses.

$K^+K^-\pi^+\pi^-$ final state

The constrained fit in $2K2\pi$ hypothesis (kaon ID is required for one or two particles) allows to separate this final state with relatively low background. A χ^2 based selection is applied leaving negligible number of background events.

Using number of events, efficiency and ISR luminosity. the $e^+e^- \to K^+K^-\pi^+\pi^-$ cross section is calculated. Figure 2a shows the obtained cross section.

The systematic errors dominate by not correct acceptance simulation (10%) and a difference in kaon ID for data and MC (up to 5%/track) and estimated to be $\approx15\%$.

$K^+K^-K^+K^-$ final state

The constrained fit in 4K hypothesis allows to separate this final state. Main background for this process are $2K2\pi$ events. But requirements of 3 or 4 particles to have the kaon IDs and χ^2 cuts leave small number of these background events.

Using number of events, acceptance and ISR luminosity the $e^+e^- \to K^+K^-K^+K^-$ cross section is calculated. Figure 2b shows the obtained cross section. There are no e^+e^- data available for comparison.

Systematic errors dominate by not correct acceptance simulation and background subtraction procedure (10%) and difference in kaon ID for data and MC (up to 5% per track) and estimated to be $\approx25\%$.

Mass substructures

The different mass combinations were studied for data and MC to search for the states not included to the model. Figures 3a show the scatter plots of 3 pions and 2 pions vs. 4π mass for data and MC. Good agreement in general is seen except the narrow regions for J/ψ and $\psi(2S)$ whose decays are not simulated. From fig. 3a it's clearly visible a contribution from $a_1(1260)$ and $f_2(1270)$, a detailed PWA is needed to confirm it.

Figure 3b shows $\pi\pi$, K^+K^- and $K\pi$ mass combinations. The production seems to be dominated by $K^{*+}(892)K^{*-}(892)$ final state as it is seen in fig. 3b(a). If events in the $K^*(892)$ bands are removed the scatter plot $M_{\pi^+\pi^-}$ vs. $M_{K^+K^-}$ shows the presence of

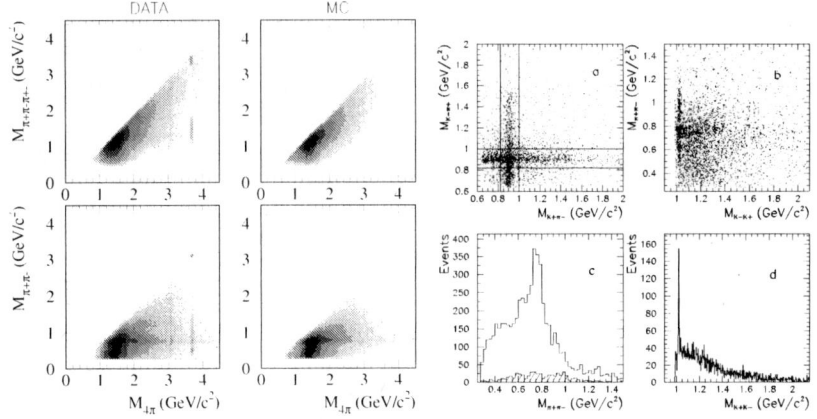

a) The $\pi^+\pi^-$ and three-pion vs. four-pion invariant mass distributions for data (left) and simulation (right) for the 4π data sample.

b) Invariant mass distributions for the $K^+K^-\pi^+\pi^-$ data sample.

FIGURE 3.

ρ^0 and ϕ resonancies - fig. 3b(b). One dimentional plot fig. 3b(c) shows $\pi^+\pi^-$ mass distribution for events not associated with ϕ and shaded hist for events from ϕ spike shown in fig. 3b(d). No evidence of $\phi f_0(980)$ has been observed so far.

CONCLUSIONS

The good detector resolution and PID capabilities of *BABAR* allow the measurement of the $e^+e^- \rightarrow \pi^+\pi^-\pi^+\pi^-$ cross section with ISR data in the energy range 0.6 to 4.5 GeV with a systematic error of about 5% in the central region. This can be compared with all e^+e^- data, which are available only up to 2.0 GeV - the maximum c.m. energy where exclusive studies of this channel have been performed. Figure 1b shows obtained cross section in comparison with all existing e^+e^-. It's the first time that the cross section for the process $e^+e^- \rightarrow K^+K^-K^+K^-$ has been measured.

REFERENCES

1. see talk by E. Solodov in this Proceedengs
2. *BABAR* Collaboration, B. Aubert *et al.*, "The *BABAR* Detector," hep-ex/0105044 (2001), submitted to Nucl. Instr. and Meth.
3. H.Czyz and J.H.Kuehn, Eur.Phys.J **C18**(2000)497-509 (hep-ph/0008262).
4. A.B.Arbuzov *et al.*, JHEP **9710**, 001 (1997), hep-ph/9702262.
5. M. Caffo, H. Czyz, E. Remiddi, Nuo. Cim. **110A**, 515 (1997); Phys. Lett. **B327**, 369 (1994).
6. E. Barberio, B. van Eijk and Z. Was. Comput. Phys. Commun. **66**, 115 (1991).

Experimental study of the $e^+e^- \to \pi^+\pi^-\pi^0$ reaction by SND detector in the energy range $\sqrt{s} = 0.42 - 1.38$ GeV[1]

M. N. Achasov*, K.I.Beloborodov*, A.V.Berdyugin*, A.G.Bogdanchikov*,
A.V.Bozhenok*, A.D.Bukin*, D.A.Bukin*, T.V.Dimova*, V.P.Druzhinin*,
V.B.Golubev*, A.A.Korol*, S.V.Koshuba*, E.V.Pakhtusova*,
E.A.Perevedentsev*, E.E.Pyata*, S.I.Serednyakov*, Yu.M.Shatunov*,
V.A.Sidorov*, Z.K.Silagadze*, A.A.Valishev* and A.V.Vasiljev*

**Budker Institute of Nuclear Physics, Siberian Branch of the Russian Academy of Sciences,
Laurentyev 11, Novosibirsk, 630090, Russia*

Abstract. The review of the SND results of the $e^+e^- \to \pi^+\pi^-\pi^0$ process study in the energy range $\sqrt{s} = 0.42 - 1.38$ GeV at VEPP-2M collider, based on about 2×10^6 selected events, is presented. The total cross section, parameters of the ρ, ω, ϕ resonances, and ω', ω'' states were obtained. It was found that $\rho\pi$ and $\omega\pi^0$ intermediate states describe the reaction dynamics. The experimental data cannot be described by a sum of only ω, ϕ, ω' and ω'' resonances contributions. This can be interpreted as a manifestation of the $\rho \to 3\pi$ decay, suppressed by G-parity, with relative probability $B(\rho \to 3\pi) = (1.01 {}^{+0.54}_{-0.36} \pm 0.034) \times 10^{-4}$.

INTRODUCTION

The $e^+e^- \to \pi^+\pi^-\pi^0$ cross section at low energies is determined by the transitions of light vector mesons V ($V = \omega, \phi, \omega', \omega''$) into the final state: $V \to \rho\pi \to 3\pi$. The mesons with zero isospin have large branching ratios: $B(\omega \to 3\pi) \simeq 0.9$, $B(\phi \to 3\pi) \simeq 0.15$ [1], $B(\omega' \to 3\pi) \sim 1$, $B(\omega'' \to 3\pi) \sim 0.5$ [2]. The process can also proceed via mechanism suppressed by the G-parity: $V \to \omega\pi^0 \to 3\pi$ [2, 3] or $V \to \rho\pi \to 3\pi$ ($V = \rho, \rho', \rho''$). The study of the reaction allows to determine the vector mesons parameters and provide information on the *OZI* rule violation in the $\phi \to 3\pi$ decay and on the *G*-parity violation in the processes $\rho \to 3\pi$. The process $e^+e^- \to 3\pi$ in the energy region $\sqrt{s}$ below 2.2 GeV was studied in several experiments during the last 30 years [4, 5, 6]. Recently the process $e^+e^- \to 3\pi$ was also studied by the Spherical Neutral Detector (SND)[2, 7, 8, 9, 10], the process dynamics was analyzed and the cross section was measured in the energy region $\sqrt{s}$ from 420 to 1380 MeV. This talk is a review of the SND results.

---

[1] Presented by M.N.Achasov, e-mail:achasov@inp.nsk.su

TABLE 1. The number of selected $e^+e^- \to 3\pi$ events.

$\sqrt{s}$	below 980 MeV	from 980 to 1060 MeV	from 1060 to 1380 MeV
N_{events}	1.2×10^6	5×10^5	6×10^3

DATA PROCESSING

The Spherical Neutral Detector (SND) [12] has operated since 1995 up to 2000 at VEPP-2M [11] e^+e^- collider in the energy range from 0.36 to 1.38 GeV. During six experimental years SND had collected data with integrated luminosity about 30 pb^{-1}. During the experimental runs, the first-level trigger [12] selects events with energy deposition in the calorimeter more than 180 MeV and with two or more charged particles. For analysis, events containing two charged and two or three neutral particles were selected. Extra photons in $e^+e^- \to 3\pi$ events can appear because of the overlap with the beam background or nuclear interactions of the charged pions in the calorimeter. Under these conditions the background sources are $e^+e^- \to e^+e^-\gamma\gamma, e^+e^-\gamma, \pi^+\pi^-(\gamma), \mu^+\mu^-(\gamma), 2\pi^{\pm}2\pi^0$, K^+K^-, $K_S K_L$ processes. To reject the collinear background, the cut on $\Delta\phi$ of the charged particles was imposed: $|\Delta\phi| > 5°$. To suppress the $e^+e^-\gamma\gamma$ events an energy deposition of the charged particles in the calorimeter was required to be small: $E_{cha} < 0.5\sqrt{s}$. The cut on dE/dx energy losses in the drift chamber rejected the K^+K^- events in the vicinity of the ϕ meson peak: $(dE/dx) < 3 \cdot (dE/dx)_{min}$. Then a kinematic fit was performed under the following constraints: the charged particles are assumed to be pions, the system has zero total momentum, the total energy is $\sqrt{s}$, and the photons originate from the $\pi^0 \to \gamma\gamma$ decays. The cut on the $\chi^2_{3\pi}$ was applied: $\chi^2_{3\pi} < 20$ at $\sqrt{s} < 1030$ MeV, and $\chi^2_{3\pi} < 5$ at $\sqrt{s} > 1030$ MeV. In the energy region above 900 MeV for additional suppression of the $e^+e^- \to 2\pi^{\pm}2\pi^0$ and $K_S K_L$ background, the events with exactly two photons were selected. The number of selected events is presented in the Table 1.

The cross section was calculated as the ratio of the number of selected events to integrated luminosity, detection efficiency obtained by Monte Carlo simulation, and radiative correction for the initial state calculated according to Ref.[13]. The obtained cross section is shown in Fig.1. The total systematic error $\sigma_{sys} \simeq 4 - 5\%$ includes the errors of detection efficiency, integrated luminosity, a model error, and an error due to background subtraction.

DATA ANALYSIS

In analysis of the $e^+e^- \to 3\pi$ reaction we took into account the $\rho\pi$ and $\omega\pi$ transition mechanisms, possible $\rho'\pi$ transition, and the interaction of the ρ and π mesons in the final state [14]:

$$\frac{d\sigma}{dm_0 dm_+} = \frac{4\pi\alpha}{s^{3/2}} \frac{|\vec{p}_+ \times \vec{p}_-|^2}{12\pi^2\sqrt{s}} m_0 m_+ \cdot |F|^2,$$

$$|F|^2 = \left| A_{\rho\pi}(s) \cdot \left(\sum_{i=+,0,-} \frac{g_{\rho\pi\pi}}{D_\rho(m_i)Z(m_i)} + a_{3\pi} \right) + A_{\omega\pi}(s) \frac{\Pi_{\rho\omega} g_{\rho\pi\pi}}{D_\rho(m_0)D_\omega(m_0)} \right|^2.$$

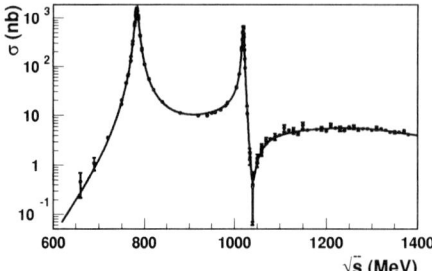 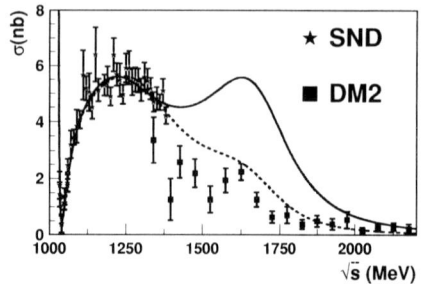

FIGURE 1. The $e^+e^- \to 3\pi$ cross section. The results of the SND [8, 2, 10] and DM2 [5] are shown. Dashed curve corresponds to the fit under assumption that a relative bias between the SND and DM2 data exists (DM2 data were scaled by a factor of 1.5. Solid curve is the result of the fitting to the SND data only.

A possible $\rho'\pi$ contribution was written as a constant term, because we searched it in the vicinity of the ϕ meson

$$a_{3\pi} = \frac{A_{\rho'\pi}(s)}{A_{\rho\pi}(s)} \sum_{i=+,0,-} \frac{g_{\rho'\pi\pi}}{D_{\rho'}(m_i)},$$

The $\rho\pi$ and $\omega\pi$ amplitudes are

$$A_{\rho\pi}(s) = \frac{1}{\sqrt{4\pi\alpha}} \sum_{V=\omega,\rho,\phi,\omega',\omega''} \frac{\Gamma_V m_V^2 \sqrt{m_V \sigma(V \to 3\pi)}}{D_V(s)\sqrt{W_{\rho\pi}(m_V)}} e^{i\phi_V}, \ A_{\omega\pi}(s) = \sum_{V=\rho,\rho',\rho''} \frac{g_{\gamma V} g_{\rho\omega\pi}}{D_V(s)}.$$

From the dipion mass spectra analysis in the ϕ-meson energy region [9], it was found that the experimental data can be described with $e^+e^- \to \rho\pi \to 3\pi$ transition only. The value of the constant term obtained by SND is consistent with zero and differs by 2σ from KLOE result [6]. The ρ meson mass and width were measured. The mass value agrees with the results obtained in other e^+e^- experiments. The main results of this analysis are presented in Table 2. The analysis of the dipion mass spectra in the energy region above 1.1 GeV [2] has shown that for their description the $e^+e^- \to \omega\pi \to 3\pi$ mechanism is required. The phase between $e^+e^- \to \omega\pi$ and $e^+e^- \to \rho\pi$ processes amplitudes was measured.

The $e^+e^- \to 3\pi$ cross section measured by SND [8, 2, 10] was analyzed together with the DM2 [5] results on the $e^+e^- \to 3\pi$ and $\omega\pi^+\pi^-$ processes. The SND and DM2 measurements agree poorly. So, to take into account possible relative systematic shift between experiments, the DM2 cross section was multiplied by a factor of 1.5 [2, 10]. It was found that for the good description of the data, the ω, ρ, ϕ, ω' and ω'' contributions should be taken into account [10]. The measured ω and ϕ mesons parameters are shown in Table 2 and ω' and ω'' parameters in Table 3.

The conventional view on the *OZI* suppressed $\phi \to 3\pi$ decay is that it proceeds through ϕ-ω mixing, i.e. in the wave function of the ϕ-meson which is dominated by

TABLE 2. The results of the dipion mass spectra analysis [9], results of $\omega \to 3\pi$ and $\phi \to 3\pi$ decays study [10, 8].

	SND	Other data	
m_ρ, MeV	775.0 ± 1.3	775.9 ± 0.5	(PDG-2002)
Γ_ρ, MeV	150.4 ± 3.0	147.9 ± 1.3	(PDG-2002)
$m_{\rho\pm} - m_{\rho 0}$, MeV	-1.3 ± 2.3	$0.4 \pm 0.7 \pm 0.06$	(KLOE[6])
$|a_{3\pi}| \times 10^5$, MeV^{-2}	0.01 ± 0.34	0.7 ± 0.1	(KLOE[6])
m_ω, MeV	$782.79 \pm 0.08 \pm 0.09$	782.57 ± 0.12	(PDG-2002)
Γ_ω, MeV	$8.68 \pm 0.04 \pm 0.24$	8.44 ± 0.09	(PDG-2002)
$\sigma(\omega \to 3\pi)$, nb	$1615 \pm 9 \pm 57$	636 ± 27	(PDG-2002)
$\sigma(\phi \to 3\pi)$, nb	$657 \pm 10 \pm 37$	1484 ± 29	(PDG-2000)
ϕ_ϕ, degree	$163 \pm 3 \pm 6$	$158 - 172$	[15]

TABLE 3. The ω' and ω'' parameters obtained from the fit [10, 2] of SND and DM2 data.

V	ω'	ω''
m_V, MeV	$1400 \pm 50 \pm 130$	$1770 \pm 50 \pm 60$
Γ_V, MeV	$870 \pm^{500}_{300} \pm 450$	$490 \pm^{200}_{150} \pm 130$
$\sigma(V \to 3\pi)$, nb	$4.9 \pm 1.0 \pm 1.6$	$5.4 \pm^{0.2}_{0.4} \pm 3.9$
$\sigma(V \to \omega\pi^+\pi^-)$, nb		$1.9 \pm 0.4 \pm 0.6$
$B(V \to e^+e^-)$	$\sim 6.5 \times 10^{-7}$	$\sim 1.6 \times 10^{-6}$
$\Gamma(V \to e^+e^-)$, eV	~ 570	~ 860
$B(V \to 3\pi)$	~ 1	~ 0.65
$B(V \to \omega 2\pi)$		~ 0.35

s quarks, there is an admixture of u and d quarks. An alternative to the ϕ-ω mixing is the direct decay [16]. Analysis of the $\Gamma(\phi \to e^+e^-)/\Gamma(\omega \to e^+e^-)$ ratio and $g_{\phi\rho\pi}$ and $g_{\omega\rho\pi}$ coupling constants obtained in SND experiments indicates that the direct transition is preferable to the ϕ-ω mixing as the main mechanism of the $\phi \to 3\pi$ decay [10].

The ω' and ω'' parameters obtained from the fits (Table 3) should be considered as rather approximate estimation of the ω' and ω'' resonances main parameters. To measure the parameters of these states precisely new data above 1.4 GeV required.

It was found that the experimental data cannot be described by a sum of ω, ϕ, ω' and ω'' resonances contributions. This can be interpreted as a manifestation of the $\rho \to 3\pi$ decay suppressed by G-parity. The obtained parameters of the decay $B(\rho \to 3\pi) = (1.01 \pm^{0.54}_{0.36} \pm 0.34) \times 10^{-4}$ and $\phi_\rho = -135 \pm^{17}_{13} \pm 9$ degree are in agreement with the theoretical values expected from the ρ-ω mixing $B(\rho \to 3\pi) = (0.4 - 0.6) \times 10^{-4}$ and $\phi_\rho \simeq -90$ degree.

Using the $e^+e^- \to 3\pi$ cross section obtained by SND detector, the contribution to the anomalous magnetic moment of the muon due to the $\pi^+\pi^-\pi^0$ intermediate state was calculated $a_\mu(3\pi, \sqrt{s} < 1.38 \text{GeV}) = (458 \pm 2 \pm 17) \times 10^{-11}$.

CONCLUSION

The $e^+e^- \to 3\pi$ cross section was measured in the SND experiment at the VEPP-2M collider in the energy region $\sqrt{s}$ below 1380 MeV. The experimental data were analyzed in the framework of the generalized vector meson dominance model. It was found that the $\omega\pi$ and $\rho\pi$ intermediate states describe the process dynamics. The ω and ϕ mesons parameters were obtained and parameters of the ω', ω'' resonances were estimated. Experimental data cannot be described by a sum of ω, ϕ, ω' and ω'' contributions. This can be interpreted as a manifestation of the $\rho \to 3\pi$ decay. The SND study of the $e^+e^- \to 3\pi$ process was reported in Ref.[8, 9, 2, 10]. Now the VEPP-2000 collider with the maximum center-of-mass energy 2 GeV is under construction [17]. The $e^+e^- \to 3\pi$ process study will be continued in future experiments with SND detector at VEPP-2000.

ACKNOWLEDGMENTS

This work was supported in part by Presidential Grant 1335.2003.2 for support of Leading Scientific Schools and by Russian Science Support Foundation.

REFERENCES

1. K. Hagiwara et al., Phys. Rev. D 66, 010001 (2002)
2. M.N. Achasov et al., Phys. Rev. D 66, 032001 (2002)
3. N.N. Achasov, A.A. Kozhevnikov, and G.N. Shestakov, Phys. Lett. 50B, 448 (1974); N.N. Achasov, N.M. Budnev, A.A. Kozhevnikov, and G.N. Shestakov, Yad. Fiz. 23, 610 (1976); N.N. Achasov and G.N. Shestakov, Fiz. Elem. Chastits. At. Yadra 9, 48 (1978)
4. J.E. Augustin et al., Phys. Lett. B 28, 513 (1969); D. Benaksas et al., Phys. Lett. B 42, 507 (1972); L.M. Kurdadze, et al, JETP Lett. 36, 274 (1982); S.I. Dolinsky et al., Phys. Rep. 202, 99 (1991); L.M. Barkov et al., JETP Lett. 46, 164 (1987); G.Parrour et al., Phys. Lett. B 63, 357 (1976), Phys. Lett. B 63, 362 (1976); A.D. Bukin et al., Yad. Fiz. 27, 976 (1978); R.R. Akhmetshin et al., Phys. Lett. B 476, 33 (2000), Phys. Lett. B 364, 199 (1995), Phys. Lett. B 434, 426 (1998); A. Cordier et al., Nucl. Phys. B 172, 13 (1980), G. Cosme et al., Phys. Lett. B 48, 155 (1974), Nuc. Phys. B 152, 215 (1979); B. Delcourt et al., Phys. Lett 113B, 93 (1982)
5. A. Antonelli et al., Z. Phys., C 56, 15 (1992)
6. A. Aloisio et al., Phys. Lett. B 561, 55 (2003)
7. M.N. Achasov et al., Phys. Lett. B 462, 265 (1999)
8. M.N. Achasov et al., Phys. Rev. D 63, 072002 (2001)
9. M.N. Achasov et al., Phys. Rev. D 65, 032002 (2002)
10. M.N. Achasov et al., hep-ex/0305049 (to be published in Phys. Rev. D)
11. A.N. Skrinsky, in Proc. of Workshop on physics and detectors for DAΦNE, Frascati, 1995, p.3
12. M.N. Achasov et al., Nucl. Instr. and Meth. A449, 125 (2000)
13. E.A. Kuraev, V.S. Fadin, Yad. Fiz. 41, 733 (1985)
14. N.N. Achasov and A.A. Kozhevnikov, Phys. Rev. D 49, 5773 (1994), Yad. Fiz. 56, 191 (1993), Int. J. Mod. Phys. A 9, 527 (1994)
15. N.N. Achasov, A.A. Kozhevnikov, Phys. Rev. D 61 054005 (2000), Yad. Fiz. 63, 2029 (2000)
16. N.N. Achasov et al., Yad. Fiz. 54, 1097 (1991), Int. J. of Mod. Phys. A vol.7 No.14 3187 (1992); N.N. Achasov, A.A. Kozhevnikov, Phys. Lett. B 233, 474 (1989), Yad. Fiz. 55, 809 (1992), Int. J. Mod. Phys. A 7, 4825 (1992), Yad. Fiz. 55, 3086 (1992), Part. World 3, 125 (1993), Phys. Rev. D 52 3119 (1995), Yad. Fiz. 59, 153 (1996)
17. Yu.M.Shatunov et al, in Proc. of the 2000 European Particle Acc. Conf., Vienna (2000), p.439

Antiproton-proton annihilation at rest into three pseudoscalar mesons

A. Sarantsev

Nusallee 14-16, 53115, Bonn, Germany and Orlova Rosha 1, 188300 Gatchina, Russia email

Abstract. We present the analysis of the data on reactions $p\bar{p}$ annihilation into three pseudoscalar mesons. The combined analysis of the data taken with liquid hydrogen, gaseous hydrogen and deuteron targets shows an excellent agreement for the reactions investigated and allows to control the fraction of the P-wave annihilation. The latter provides a better determination of tensor and vector states situated close to the edge of the phase volume. A possible presence of $a_0(1290)$ is investigated.

INTRODUCTION

Antiproton-proton annihilation is a rich source of mesons, and a large variety of meson resonances has been unrevealed in this process. Even though annihilation at rest is limited in phase space to production of resonances below 1.7 GeV, several resonances have been discovered by stopping antiprotons in H_2. Among them are $\eta(1420)$, $f_0(980)$, $K_1(1280)$, $f_2(1565)$, $f_0(1370)$, $f_0(1500)$ in various decay modes and $a_0(1490)$ (see [1] and references there in).

However to understand decay properties of resonances it is necessary to perform a combined analysis of a large set of reactions using unitarity and analiticity constraints. Such approach provides a unique information about branching ratios of the resonances and allows us to perform quark classification of the states. This is especially important in the case of strong final state interaction when properties of the physical resonances can deviate dramatically from $q\bar{q}$ or gluon-gluon states calculated in framework of quark models or lattice calculations.

In this paper we present a combined analysis of the reactions taken with the Crystal Barrel detector at LEAR by stopping antiproton in liquid, gaseous hydrogen and deuteron (see Table 1). We constrain our fit with the data from the CERN-Munich experiment [2] on the reaction $\pi^- p \to \pi^- \pi^+ p$, from the GAMS collaboration [3] on $\pi^- p \to \pi^0 \pi^0 p$, $\pi^- p \to \eta\eta\ p$, $\pi^- p \to \eta\eta' p$ and from BNL [4] on the reaction $\pi^- p \to K\bar{K}p$ where unitary $\pi\pi$ scattering amplitudes were used in the analysis.

The P-state fraction in the $\bar{p}p$ annihilation into mesons depends both from the production mechanism and the final state and it is difficult to determine it from the fit of one reaction. In reactions taken in liquid H_2, the P-state contribution is mostly small; including it in the analysis of one data set usually leads to a large number of ill-defined parameters. This problem can be solved in the combined analysis of the data taken by stopping antiprotons in gaseous and in liquid H_2. The increase of P-state annihilation fraction when going from liquid to 12 atm gaseous H_2 was calculated by Batty [5] and

CP717, *Hadron Spectroscopy: Tenth International Conference,*
edited by E. Klempt, H. Koch, and H. Orth
© 2004 American Institute of Physics 0-7354-0197-7/04/$22.00

TABLE 1. The list of the reaction analysed. Letters *L* and *G* denote liquid and gaseous targets.

Reaction	Target	Reaction	Target	Reaction	Target
$\bar{p}p \to \pi^0\pi^0\pi^0$	(L) H_2	$\bar{p}p \to \pi^+\pi^0\pi^-$	(L) H_2	$\bar{p}p \to K_S K_S \pi^0$	(L) H_2
$\bar{p}p \to \pi^0\eta\eta$	(L) H_2	$\bar{p}n \to \pi^0\pi^0\pi^-$	(L) D_2	$\bar{p}p \to K^+K^-\pi^0$	(L) H_2
$\bar{p}p \to \pi^0\pi^0\eta$	(L) H_2	$\bar{p}n \to \pi^-\pi^-\pi^+$	(L) D_2	$\bar{p}p \to K_L K^\pm \pi^\mp$	(L) H_2
$\bar{p}p \to \pi^0\pi^0\pi^0$	(G) H_2			$\bar{p}n \to K_S K_S \pi^-$	(L) D_2
$\bar{p}p \to \pi^0\eta\eta$	(G) H_2			$\bar{p}n \to K_S K^- \pi^0$	(L) D_2
$\bar{p}p \to \pi^0\pi^0\eta$	(G) H_2				

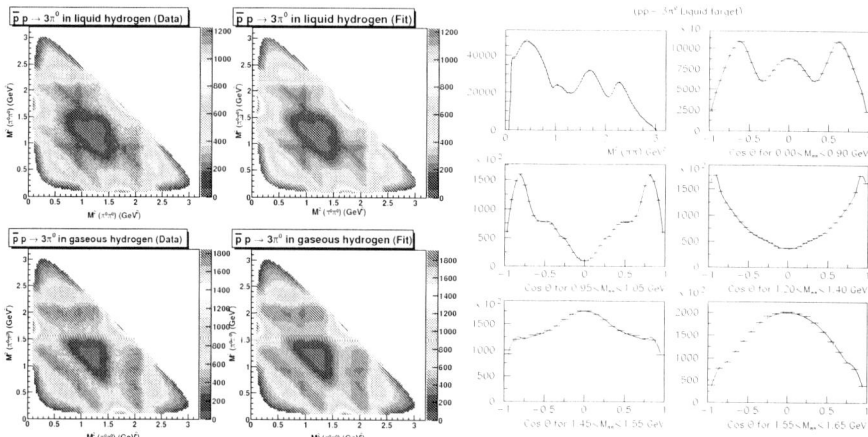

FIGURE 1. Experimental Dalitz plots and results of fit for reaction $p\bar{p} \to \pi^0\pi^0\pi^0$ in liquid and gaseous H_2. The mass and angular projections (data versus the fit) for the $p\bar{p} \to \pi^0\pi^0\pi^0$ annihilation in liquid hydrogen

we imposed this restriction in the fit.

FIT OF THE DATA.

In the present work the P-vector/K-matrix technique was used for the description of the meson scattering amplitudes in the 0^+0^{++}, 1^-1^{--} and $\frac{1}{2}0^{++}$ channels. Resonances in other channels were parameterised either with Flatté form ($a_0(980)$, $f_2(1560)$, $a_0(1450)$) or as T-matrix poles. The detailed description of the P-vector, K-matrix method can be found in [6] and references therein.

The previous analyses [8, 9] showed that for a successful description of the annihilation data and $\pi\pi$ scattering amplitudes the 5 pole parameterisation of the K-matrix in 0^+0^{++} channel is needed. Such parameterization leads to the 5 poles in the complex plane: a very broad one and 4 relatively narrow states: $f_0(975)$, $f_0(1370)$, $f_0(1500)$ and

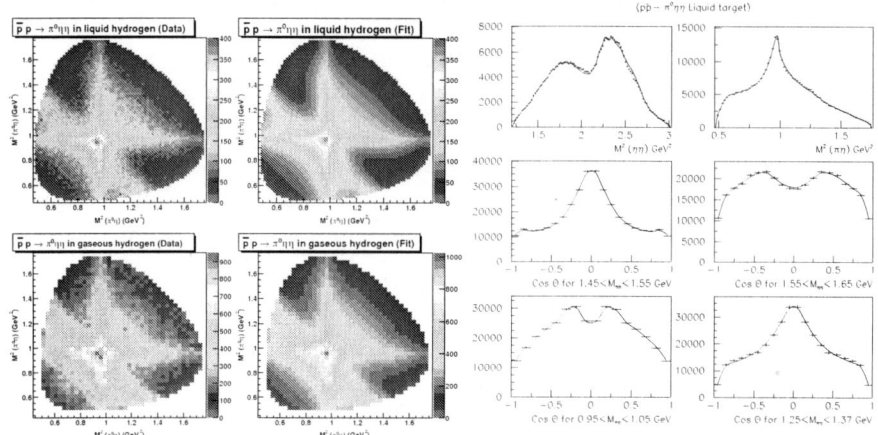

FIGURE 2. Experimental Dalitz plots and result of fit for reaction $p\bar{p} \rightarrow \pi^0\eta\eta$ in liquid and gaseous H_2. The description of mass and angular projections for the annihilation in liquid hydrogen.

$f_0(1750)$. The parameters for first four states are very stable in all solutions while parameters of fifth pole which is defined by BNL data only has an ambiguity in coupling to multimeson channel. As the result two solutions were found and detailed description of them is given in [6].

In $\pi^0\pi^0\pi^0$ and $\pi^0\eta\eta$ reactions (see Figs. 1,2) 0^{++} states, as well as $f_2(1275)$ and $f_2(1560)$ states give appreciable contributions. We found very strong production of $f_2(1560)$ from the P-wave annihilation (in agreement with analysis [7]) in $\bar{p}p \rightarrow 3\pi^0$ reaction (gaseous target) which allows us to define the parameters of this state with a high accuracy.

In $\pi^0\eta\eta$ reaction (Fig. 2) the only isovector resonance produced within the phase space is $a_0(980)$. The effective width, calculated at half height of the peak in $\pi\eta$ scattering amplitude, corresponds well to that given in PDG [1]. However, the simultaneous fit for decays of this state into $\pi\eta$ and $K\bar{K}$ channels reveal the Flatté structure with strong couplings to $\pi\eta$ and $K\bar{K}$.

For a successful description of the $\bar{p}p \rightarrow \pi^0\pi^0\eta$ reaction other four isovector states are needed: $a_0(1450)$, $a_2(1320)$, $a_2(1700)$ and $\pi_1(1400)$ (see Fig. 3).

We obtained the mass value for $a_0(1450)$ state a bit higher than in earlier analyses [10] and [8] but in the error region given there. The mass shift was mainly due to the latest high statistic data used in the fit.

One of the most intriguing states is $\pi_1(1400)$. This wave is an exotic one for the quark systematic and observation of such state is a very important issue. It shows up strongly in the reaction of antiproton annihilation with neutron [11], however in the present analysis of annihilation in hydrogen this wave showed up as a very broad contribution with width 600-1000 MeV and could not be distinguished from the non-resonant production. However if we impose the results obtained in [11] for this state the result changes only marginally and we conclude that the fit is compatible with the analysis [11].

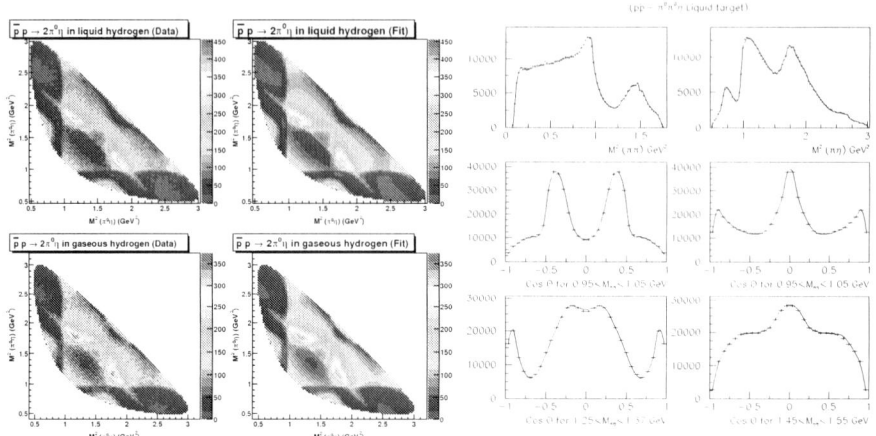

FIGURE 3. Experimental Dalitz plots and result of fit for reactions $p\bar{p} \to \pi^0\pi^0\eta$ in liquid and gaseous H_2. The description of mass and angular projections for the annihilation in liquid hydrogen.

The production of scalar and tensor mesons in $p\bar{p} \to \pi^+\pi^-\pi^0$ (see Fig. 4) is completely fixed from fit of neutral channels by isotopic relations. For fitting ρ meson sector 3×3 ($\pi\pi$, $\omega\pi$ and 4π) three pole K-matrix was used. The 4π channel was simulated as the $S(\pi\pi)\rho(770)$ decay mode. The K-matrix parameters are constrained from description of the CERN-München data fitted with the unitary amplitude.

Another source of the information about vector states are $\bar{p}n \to \pi^0\pi^0\pi^-$ and $\bar{p}n \to \pi^+\pi^-\pi^-$ data. The production of vector states from initial isospin 1 $\bar{p}p$ state in all reactions is related by isotopic coefficients only. The difference in P-wave fraction for annihilation in hydrogen and deuteron was described by real parameters and we found quite a large fraction (40-50%) for P-wave annihilation in deuteron.

The analysis of the production channels with kaons have only very few new parameters. All resonances in $K\bar{K}$ channel are fixed from the analysis of the annihilation data into pion and η-meson final states. The $IJ^P = \frac{1}{2}0^+$ wave was fitted as 3x3 ($K\pi$, $K\eta$ and $K\eta'$ channels) two pole K-matrix. The unitary amplitude was fixed from the fit of LASS data [12]. Including in the K-matrix expression Adler zero did not lead to any changes in the parameters or description. The result of the fit is shown in Figs. 5, 6.

In this framework the data can be described quite well: the χ^2 per point for fitted reactions is between 1.1 and 1.4 for reactions with π and η meson production and between 0.9 and 1.65 for final states with kaons. The description of the CERN-München data, GAMS and BNL data are on the same level as reported in [6] and we do not show them here. The parameters of the observed states are given in the Table 2.

The OBELIX collaboration reported an observation of the $a_0(1290)$ state [13]. Such state was not observed in the previous analysis of the Crystal Barrel data. Here we also found a good description of all reactions without including $a_0(1290)$. Moreover when this state was introduced in the fit, it either collapses with $a_0(980)$ or becomes very broad giving a marginal contribution to the cross section.

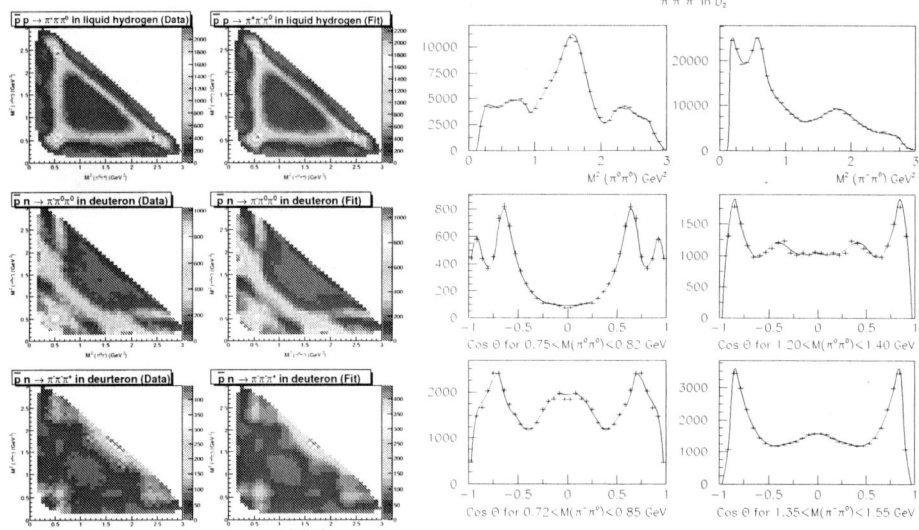

FIGURE 4. Experimental Dalitz plots and results of fit for reaction $p\bar{p} \to \pi^+\pi^-\pi^0$, $\bar{p}n \to \pi^-\pi^0\pi^0$ and $\bar{p}n \to \pi^+\pi^-\pi^-$. The mass and angular projections are gshown for $\bar{p}n \to \pi^-\pi^0\pi^0$ reaction.

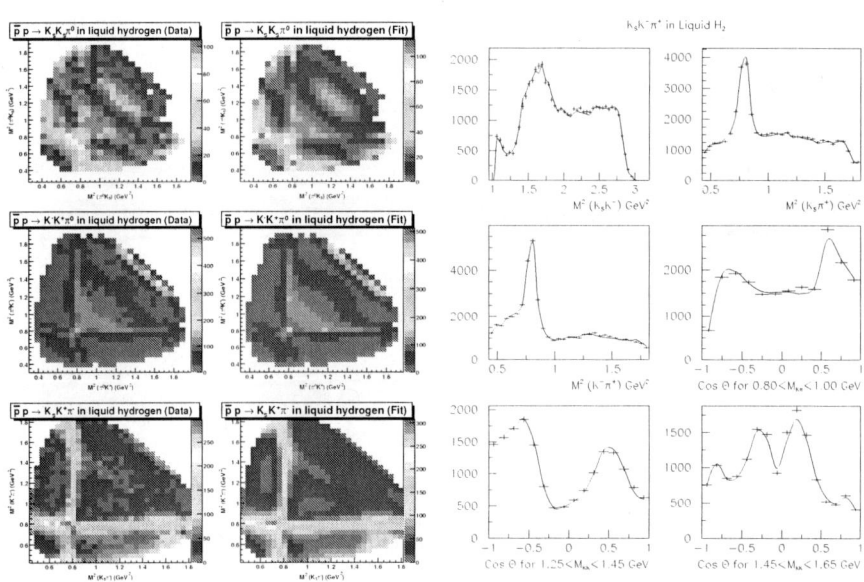

FIGURE 5. Experimental Dalitz plots and results of fit for reactions $p\bar{p} \to K_S K_S \pi^0$, $K^+ K^- \pi^0$ and $K^\pm K_S \pi^\mp$. The description of mass and angular projections are shown for the reaction $\bar{p}p \to K^\pm K_S \pi^\mp$.

TABLE 2. The characteristics of the mesons found in the combined analysis. "Pole" means the pole position in the K-matrix fit.

Resonance	Masses and widths (MeV)	Comments
$f_0(980)$	$(1015 \pm 15) - i(43 \pm 8)$	pole
$f_0(1300)$	$(1310 \pm 20) - i(160 \pm 20)$	pole
$f_0(1500)$	$(1496 \pm 8) - i(58 \pm 10)$	pole
$f_0(1530)$	$(1530^{+90}_{-250}) - i(560 \pm 140)$	pole
$f_0(1780)$	$(1780 \pm 30) - i(140 \pm 20)$ $(1780 \pm 50) - i(220 \pm 50)$	pole (sol. I) pole (sol. II)
$f_2(1560)$	$M = 1580 \pm 15 , \Gamma = 185 \pm 20 , \Gamma_{\pi\pi} = 25 \pm 15$	
$a_0(980)$	$M = 988 \pm 5 , g_{\pi\eta} = 0.436 \pm 0.02 , \left(\frac{g_{K\bar{K}}}{g_{\pi\eta}}\right)^2 = 1.23 \pm 0.10$	
$a_2(1320)$	$M = 1310 \pm 3 , \Gamma = 111 \pm 5$	
$\rho(1450)$	$M = 1490 \pm 40 , \Gamma = 400 \pm 50 , \Gamma_{\pi\pi} = 190 \pm 45$ $(1460 \pm 25) - i(180 \pm 15)$	K-matrix pole
$\rho(1700)$	$M = 1695 \pm 40 , \Gamma = 340 \pm 50 , \Gamma_{\pi\pi} = 210 \pm 45$ $(1650 \pm 40) - i(120 \pm 20)$	K-matrix pole
$K_0(1430)$	$(1415 \pm 25) - i(165 \pm 25)$ $M = 1220^{+50}_{-60}$	pole K-matrix
$K_0(1800)$	$(1820 \pm 40) - i(125 \pm 50)$ $M = 1885^{+50}_{-80}$	pole K-matrix
$K_1(1410)$	$M = 1415 \pm 15 , \Gamma = 195 \pm 15$	
$K_2(1430)$	$M = 1440^{+40}_{-20} , \Gamma = 130^{+40}_{-20}$	

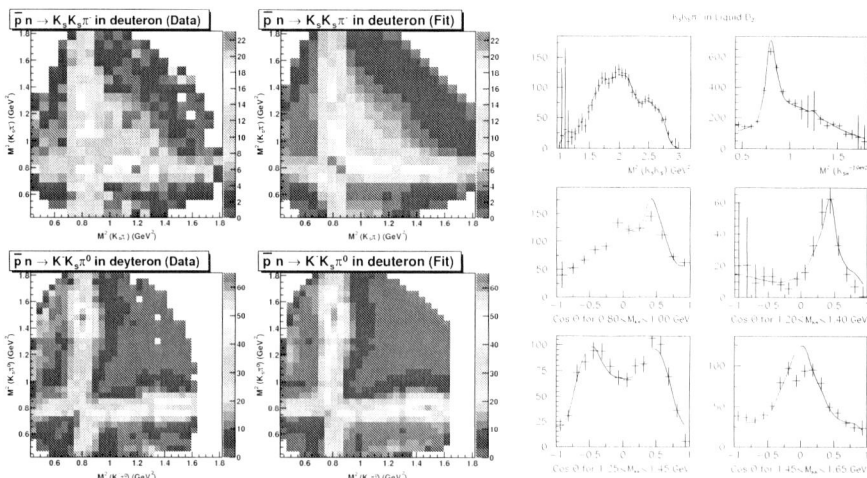

FIGURE 6. Experimental Dalitz plots and results of fit for reactions $p\bar{n} \to K_S K^- \pi^0$ and $K_S K_S \pi^-$. The mass and angular projections are shown for the reaction $\bar{p}n \to K_S K^- \pi^0$.

SUMMARY

We have performed the combined analysis of the large set of the data obtained by the Crystal Barrel collaboration on $\bar{p}p$ annihilation into three pseudoscalar mesons. The fit demonstrated an excellent compatibility of the data and allows us to define resonance parameters with a high accuracy. The properties of observed resonances are comprehensively investigated and contributions from different initial states in $\bar{p}p$ annihilation are calculated.

ACKNOWLEDGMENTS

I am very grateful to V. Anisovich, E. Klempt and D.V. Bugg for extremely useful discussions and comments. I acknowledge support from the Alexander von Humboldt Foundation.

REFERENCES

1. Review of Particle Physics, Phys. Rev. D 66 (2002)
2. B. Hyams et al., Nucl. Phys. **B64** (1973) 134.
3. D. Alde et al., Z. Phys.**C66** (1995) 375;
    A.A. Kondashov et al., Proc. 27th Intern. Conf. on High Energy Physics, Glasgow (1994) 1407;
    Yu.D. Prokoshkin et al., Physics-Doklady **342** (1995), 473;
    A.A. Kondashov et al, Preprint IHEP 95-137, Protvino (1995).
    F. Binon et al., Nuovo Cim. **A78** (1983) 313.
    F. Binon et al., Nuovo Cim. **A80** (1984) 363.
4. S.J. Lindenbaum and R.S. Longacre, Phys. Lett. **B274** (1992) 492;
    A. Etkin et al., Phys. Rev. **D25** (1982) 1786.
5. C.Batty, Nucl.Phys. **A601** (1996) 425.
6. V. V. Anisovich and A. V. Sarantsev, Eur. Phys. J. A **16** (2003) 229, hep-ph/0204328.
7. D.V.Bugg et al.,was published in Phys.Lett.
8. C. Amsler et al., Phys. Lett. **B342** (1995) 433,
    C. Amsler et al., Phys. Lett. **B353** (1995) 571.
    C. Amsler et al., Phys. Lett. **B355** (1995) 425.
9. V.V. Anisovich, Yu.D. Prokoshkin and A.V. Sarantsev, Phys.Lett. **B389** (1996) 388.
10. C. Amsler et al., Phys. Lett. **B333** (1994) 277.
11. Abele et al., Phys. Lett. **B423** (1998) 175.
12. D. Aston *et al.*, Nucl. Phys. B **296** (1988) 493.
13. A. Bertin *et al.* [OBELIX Collaboration], Phys. Lett. B **434** (1998) 180.

Analysis of Crystal Barrel data on the reaction $\bar{p}p$ annihilation at rest into $\pi^+\pi^-\pi^0\eta$

M. Matveev

Helmhotz-Institut für Strahlen- und Kernphysik der Universität Bonn , Nusallee 14-16, 53115, Bonn, Germany

Abstract.
    The preliminary results of partial wave analysis of the Crystal Barrel data on $\bar{p}p$ annihilation at rest into $\pi^+\pi^-\pi^0\eta$ are presented. The data provide the information about production of meson channels, such as $b_1\pi$, $h_1\eta$ from $^1S_0$ and $^3S_1$ $\bar{p}p$ state annihilation. The analysis shows the presence of the b_1^* resonance and a possible presence of ρ_2 in the mass region around 1600 MeV. The mass of b_1^* was found to be 1620 ± 15 MeV.

INTRODUCTION

One of the important step in understanding of the QCD structure at low energies is understanding of the meson spectrum and meson decay properties. A very exciting subject is an observation of exotic from point of view quark-antiquark classification states. However, unless such states have exotic quantum numbers, they are expected to be mixed strongly with nearby $q\bar{q}$ states. Therefore to locate exotic states the full picture of the spectrum is needed. Apart from search for exotics the spectrum of $q\bar{q}$ states by itself can reveal a very important information about confinement. The observation of the linear meson trajectories in the (n, M^2) plane [1] where n is the radial quantum number was not expected from classical nonrelativistic quark model calculations [2]. To explain such behaviour the new model had been put forward [3] which provide a new very exciting information about Dirac structure of the confinement potential.

However, some of the linear trajectories have states which had not been observed so far in the experiment. Many of this states either can not be produces from initial states available in most experiments or decay into many particle final states which are difficult to detect. An important source of the information are the data on the $p\bar{p}$ annihilation and during last years two large experiments: Obelix and Crystall Barrel had taken a high statistical data on these reactions. The analysis of three meson final state reveal a set of new states in the mass region below 1.7 GeV (see [4] and references therein) and provided a possibility to define properties of known mesons with a high accuracy.

However many reactions with four body final states are not analysed yet. There are both technical and theoretical problems here. In the four body final state the meson production has two mechanisms: direct production of two resonances decaying into two meson systems and production of a resonance and a spectator meson with consequent cascade decay of the resonance into three meson system. The theoretical problems in the investigation of such reactions are connected with understanding of contribution from

CP717, Hadron Spectroscopy: Tenth International Conference,
edited by E. Klempt, H. Koch, and H. Orth
© 2004 American Institute of Physics 0-7354-0197-7/04/$22.00

different singularities in the scattering amplitude. In four body final state amplitude the next to leading (pole) singularities will be triangle singularities and so called four body singularities. The triangle singularities have a logarithmic behavior [5] while four body singularities are square root singularities. The contribution of such processes can be comprehensively investigated in analysis of the high statistic data and it is one of the important item in the meson spectroscopy.

In this report we present a preliminary analysis of the data on the reaction $\bar{p}p \to \pi^+\pi^-\pi^0\eta$ at rest taken with liquid hydrogen target. The annihilation in liquid hydrogen is dominated by the S-wave annihilation and we expect to observe a production of known states $h_1(1170)$ and $b_1(1230)$ at low mass region. In $\pi\pi\eta$ channel we can reach energy up to 1740 MeV and search for b_1^* or ρ_2 resonances. The observation of the b_1^* state would help to establish b_1 trajectory and ρ_2 state is a kind of puzzle for quite a long time. From quark model calculations this state is predicted in the mass region around 1700 MeV but had not been observed up to now.

DETECTOR AND DATA SELECTION

The data were taken by the Crystal Barrel detector at the LEAR at CERN [6]. The detector has a nearly 4π acceptance for charge and neutral particles. The liquid-hydrogen target is surrounded by the Silicon Vertex detector and by the Jet Drift Chamber for charge particles detection. The Cesium Iodine calorimeter is used for detection of decay photons from neutral particles.

The trigger preselected online data with two charge particles in final state only. Data were selected by two long tracks in Drift Chamber for best momentum resolution. We use kinematical fit to satisfy energy and momentum conservation. Confidence level of the analyzed reaction was taken to be more then 10%. The main background for the selected $\pi^+\pi^-\pi^0\eta$ data comes from the reaction $\bar{p}p \to \pi^+\pi^-\pi^0\pi^0$. To supress the background we applied the cut for the confidence level of background hypothesis to be less then 0.01%. Events from process $\bar{p}p \to \omega\eta$ with ω decaying into three pions was suppressed by cutting events in the mass region of omega. These events are situated at the edge of available mass region and can be cut out without loosing the information about channel investigated. After all cuts, 77055 events were selected for the analysis. Phase-space Monte Carlo was used to simulate the detector acceptance and calculate the reconstruction efficiency which was found to be 11%.

PARTIAL WAVE ANALYSIS

To obtain reliably the characteristics of the mesons decaying into three meson channels it is necessary to control the production of two resonances decaying into two meson states. In $\pi\eta$ channel the lowest state is $a_0(980)$. A production of $a_2(1320)$, $a_0(1450)$ and $\pi_1(1400)$ is also possible, however the $\pi\eta$ mass distribution is already very small at the energy region higher 1300 MeV although some structure connected with $a_2(1320)$ shows up in the $\pi^0\eta$ channel (see Fig. 1). If we restrict the analysis to the S-wave annihilation the only channels which is produced within phase space $a_0(980)\rho(770)$

$$^3S_1 \rightarrow [\rho\, a_0]^s_{S,D} \; (\rho \rightarrow [\pi\pi]^{as}_P, \quad a_0 \rightarrow [\pi\eta]_S)$$

Other channels are either forbidden for annihilation from initial S-wave or (for example $a_2(1320)\rho(770)$) are suppressed by the phase space. In the present analysis we introduced this channel with masses and widths of $\rho(770)$ taken from PDG [4]. The parametrization of the $a_0(980)$ was taken in the Flatté form with parameters fixed from [7].

The second type of resonance production in this reaction is the resonances decaying by cascade mechanism into three meson system. The states produced with η-meson as a spectator decay into ρ and π system and in the $p\bar{p}$ annihilation at rest can only be reliably observed if their masses are below 1300 MeV. The most natural state to be produced in this channel is $h_1(1170)$. However we also included in the fit a possible production of $\omega\eta$ and $a_2(1320)\eta$ states:

$$\begin{aligned}
^1S_0 &\rightarrow [\eta\, a_2]_D \; (a_2 \rightarrow [\pi\rho]^{as}_D) \\
^3S_1 &\rightarrow [\eta\, h_1]_{S,D} \; (h_1 \rightarrow [\pi\rho]^s_{S,D}) \\
^3S_1 &\rightarrow [\eta\, \omega]_P \; (\omega \rightarrow [\pi\rho]^s_P)
\end{aligned} \tag{1}$$

The resonances produced with pion as a spectator is our main subject of interest in the present investigation. This is the only states which can be produced up to 1.74 Gev and can decay either into $\pi\pi$ system with η as the recoil particle or into $\pi\eta$-system with π meson as the recoil particle. Therefore the states produced here can be observed in different decay modes.

$$\begin{aligned}
^1S_0 &\rightarrow [\pi\rho]^{as}_P \\
^3S_1 &\rightarrow [\pi\rho]^s_P \; (\rho \rightarrow [\rho\eta]_P, \quad \rho \rightarrow [\pi a_2]^{as}_D) \\
^3S_1 &\rightarrow [\pi b_1]^s_{S,D} \; (b_1 \rightarrow [\rho\eta]_{S,D}, \quad b_1 \rightarrow [\pi a_0]^{as}_P, \quad b_1 \rightarrow [\pi a_2]^{as}_P) \\
^3S_1 &\rightarrow [\pi \rho_2]^s_P \; (\rho_2 \rightarrow [\rho\eta]_P, \quad \rho_2 \rightarrow [\pi a_0]^{as}_D, \quad \rho_2 \rightarrow [\pi a_2]^{as}_{S,D})
\end{aligned} \tag{2}$$

In presented analysis only S-wave annihilation of the $\bar{p}p$ -system was taken into account. The two body spectrum (Fig. 1) is dominated by the production of $\rho(770)$ and $a_0(980)$ resonances and can be described reasonably in this approach as well as angular distributions.

To describe the three body $\pi\pi\eta$ mass spectrum we have to introduce a new state with mass around 1600 MeV. This state is responsible for the peak in the mass region 1640 MeV (see Fig. 2). The fit favors 1^+1^{+-} quantum numbers for this new resonance. There is also a contribution from 2^{--} state. On the present state we can not say whether this is an indication of a ρ_2 resonance or a nonresonant contribution with 2^{--} quantum numbers.

The mass scan of the observed b_1^* states shows a good minimum in the region in the region 1620 MeV (see Fig. 3). As the result the mass was found to be 1620 ± 15 MeV. The contribution of other states in the fitted cross section is given in Table 1.

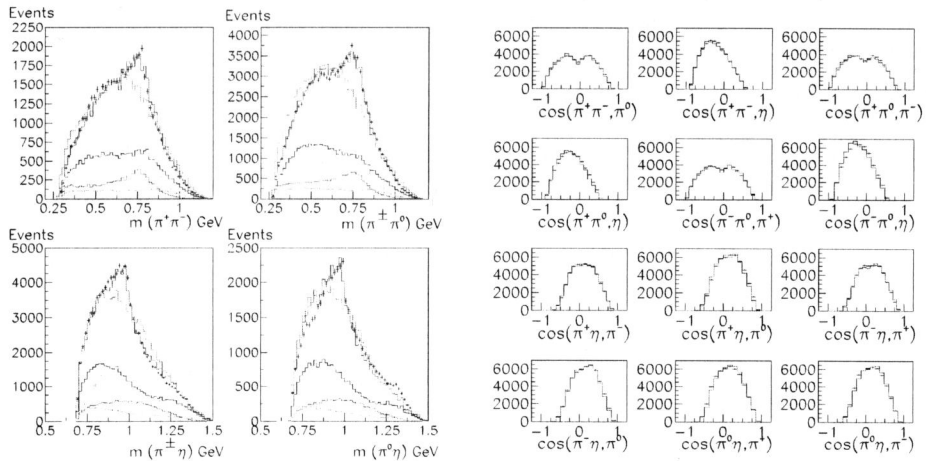

FIGURE 1. Invariant masses and angles for two particles states

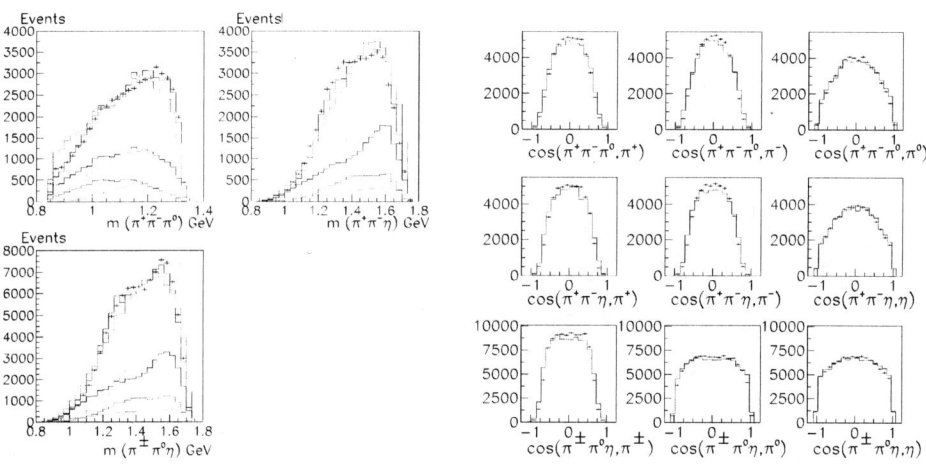

FIGURE 2. Invariant Masses and angles for three particles states

However one can see that fit still missing the perfect description of some mass and angular projections. The difference does not reveal any structure but some systematic deviations presents. A possible reason is a missing contribution from P-wave annihilation. Indeed including the P-wave annihilation improves the description of the data although do not provide any new structure to the fit. Another possibility is a contribution from four body singularities and is our main subject of further investigation.

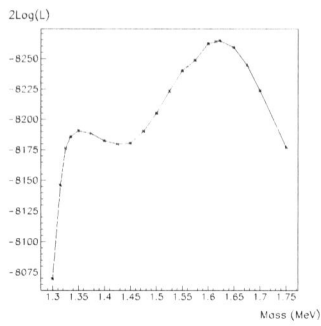

FIGURE 3. Mass scan for b_1^*

TABLE 1. Contributions

$^1S_0$	$\rightarrow$ $[\pi\rho(1450)]_{l=1}$	5%
$^1S_0$	$\rightarrow$ $[\pi\rho(1700)]_{l=1}$	5%
$^3S_1$	$\rightarrow$ $[\rho\, a_0]_{l=0,2}$	13%
$^3S_1$	$\rightarrow$ $[\eta\, h_1(1170)]_{l=0,2}$	9%
$^3S_1$	$\rightarrow$ $[\pi\, b_1(1230)]_{l=0,2}$	5%
$^3S_1$	$\rightarrow$ $[\pi\, b_1^*]_{l=0,2}$	42%
$^3S_1$	$\rightarrow$ $[\pi\rho(1450)]_{l=1}$	35%
$^3S_1$	$\rightarrow$ $[\pi\rho(1700)]_{l=1}$	16%
$^3S_1$	$\rightarrow$ $[\pi\,\rho_2]_{l=0,2}$	6%

ACKNOWLEDGMENTS

I would like to acknowledge the useful discussions and comments from E. Klempt, H. Kalinovski, A. Sarantsev and A. Anisovich.

REFERENCES

1. A.V. Anisovich, V.V. Anisovich, A.V. Sarantsev, Phys. Rev. **D62**, 051502, (2000)
2. S. Capstick and N. Isgur, Phys. Rev. **D34**, 2809, (1986)
3. R. Ricken, M. Koll, D. Merten, B.C. Metsch, H.R. Petry, Eur. Phys. J. **A9**, 221, (2000)
4. D.E. Groom *et al.* (Particle Data Group), Eur. Phys. J. **C15**, (2002)
5. A.V. Anisovich, V.V. Anisovich, Phys. Lett. **B345**, 321, (1995)
6. C. Amsler *et al.*, Nucl. Instrum. Methods. **A321**, 69 (1992)
7. V.V. Anisovich, D.V. Bugg, A.V. Sarantsev, B.S. Zou, Rev. **D50**, 1972, (1994)
8. A.V. Anisovich, V.V. Anisovich, V.N. Markov, M.A. Matveev, A.V. Sarantsev, J. Phys **G28**, 15, (2002)

Dalitz-plot analysis of D_S and D^+ to three pions in FOCUS

S.Malvezzi [1]

I.N.F.N, Milan, Via Celoria 16- 20133 Milan, Italy

Abstract. Preliminary FOCUS results of the Dalitz-plot analysis of D_s^+ and D^+ to three-pion decays are presented. The *K-matrix* formalism is applied for the first time to charm decays to fully exploit the already existing knowledge coming from the light-meson spectroscopy experiments. In particular, all measured dynamics of S-wave $\pi\pi$ scattering, characterized by broad/overlapping resonances and large non-resonant background, can be properly accounted for. The results are discussed, along with their possible implications on the controversial nature of the σ meson.

INTRODUCTION

Over the last decade Dalitz plot analysis has emerged as a powerful tool for investigating effects of resonant substructures, interference patterns and final-state interactions in the charm sector. The potentiality to investigate charm dynamics has revealed itself to be strongly connected with the knowledge of the light-meson sector. In particular, the need to model scalar particles populating our charm-meson Dalitz plots has led us to question the validity of the Breit–Wigner (BW) approximation for the description of broad resonances. A formalism for studying overlapping and multi-channel resonances was proposed long ago and is based on the *K-matrix* [1, 2] parametrization. This formalism, originating in the context of the two-body scattering, can be generalized to cover the case of resonance production in more complex reactions [3]. Its implementation allows us to include positions of the poles in the complex plane directly into our analysis and embed the results from spectroscopy experiments [4, 5] in our amplitudes.

THE DECAY AMPLITUDE

In this analysis the decay amplitude of the D meson into three-pion final state is written as:[2]

$$A(D) = a_0 \, e^{i\delta_0} + F_1 + \sum_i a_i \, e^{i\delta_i} B(abc|r_i), \tag{1}$$

where the first constant term represents the direct non-resonant three-body decay, F_1 the decay involving S-wave states and the complex functions $B(abc|r_i)$ the intermediate

---

[1] On behalf of the FOCUS collaboration, http://www-focus.fnal.gov/spires.html
[2] The coefficient and phase of the F_1 amplitude are fixed at 1 and 0 respectively.

CP717, *Hadron Spectroscopy: Tenth International Conference,*
edited by E. Klempt, H. Koch, and H. Orth
© 2004 American Institute of Physics 0-7354-0197-7/04/$22.00

two-body non-scalar resonances. $B(abc|r_i)$ are the usual Breit–Wigner terms of the traditional isobar model, whose explicit forms are given in [6]. F_1 is written in the context of the *K-matrix* approach and its expression is discussed below. The squared modulus of this amplitude gives the probability density function, which has to be multiplied by the three-pion phase-space to obtain the probability.

In general, the decay of a D meson into three pions involves the production of a $(IJ)^{PC}$ state with an accompanying pion. While the BW approximation is suitable for states with $J > 0$, since they are characterized by relatively narrow and isolated resonances, the treatment of S-wave states requires a more general formalism to account for the non-trivial dynamics due to the presence of broad and overlapping resonances [4, 7]. Furthermore the *K-matrix* formalism provides an elegant way of expressing the two-body unitarity constraint which, in the simple isobar model is not explicitly guaranteed. Minor violation effects are expected for narrow, isolated resonances but more severe ones for wide, overlapping states. For $J = 0$, only states with even isospin and positive P and C are allowed to couple strongly to $\pi^+\pi^-$. We limit ourselves to isoscalar S-wave states, $(00)^{++}$, since $I = 2$ must involve at least two $q\bar{q}$ pairs and no four-quark states with $I = 2$ are known. In the energy region relevant to this analysis the production of a $(00)^{++}$ state with an accompanying pion involves five channels $i = 1 \ldots 5$, namely, $1 = \pi\pi$, $2 = K\bar{K}$, $3 = \eta\eta$, $4 = \eta\eta'$ and $5 = $ multi-meson states (four-pion state mainly at $\sqrt{s} < 1.6$ GeV).

The amplitude for the particular channel $(00)^{++}_i\pi$ can be written in the context of the *K-matrix* formalism as

$$F_i = (I - iK\rho)^{-1}_{ij}P_j , \tag{2}$$

where I is the identity matrix, K is the *K-matrix* describing the isoscalar S-wave scattering process, ρ is the phase-space matrix for the five channels, and P is the 'initial' production vector into the five channels. In this picture the production process can be viewed as consisting of an initial preparation of several states, which are then propagated by the $(I - iK\rho)^{-1}$ term into the final state. The only amplitude that matters for the present analysis is F_1, since the S-wave in the final state is $\pi^+\pi^-$. In order to write the amplitude explicitly, we need an updated *K-matrix* parametrization of the $(00)^{++}$-wave scattering. To our knowledge, the only self-consistent description of S-wave isoscalar scattering is, at present, that given in the *K-matrix* representation by Anisovich and Sarantsev in [5] through a global fit of all available scattering data from the $\pi\pi$ threshold up to 1900 MeV. We use their *K-matrix* parametrization, namely:

$$K^{00}_{ij}(s) = \left\{ \sum_\alpha \frac{g^{(\alpha)}_i g^{(\alpha)}_j}{m^2_\alpha - s} + f^{\text{scatt}}_{ij} \frac{1\,\text{GeV}^2 - s^{\text{scatt}}_0}{s - s^{\text{scatt}}_0} \right\} \times \frac{s - s_A m^2_\pi/2}{(s - s_{A0})(1 - s_{A0})} , \tag{3}$$

where $g^{(\alpha)}_i$ is the coupling constant of the *K-matrix* pole m_α to the i meson channel; the parameters f^{scatt}_{ij} and s^{scatt}_0 describe a smooth part of the *K-matrix* elements; and the factor $\frac{s - s_A m^2_\pi/2}{(s - s_{A0})(1 - s_{A0})}$ suppresses a false kinematical singularity in the physical region near the $\pi\pi$ threshold (the Adler zero). The parameter values used in this paper are listed in Table 1, which was provided by the authors of [5]. Note that *K-matrix* and its parameters are by definition real.

TABLE 1. *K-matrix* parameters. Masses and coupling constants are in GeV. The f_{ij} terms are reported here just for $i = 1$ since these are the only values relevant to the three-pion decay.

m_α	$g_{\pi\pi}$	$g_{K\bar{K}}$	$g_{4\pi}$	$g_{\eta\eta}$	$g_{\eta\eta'}$
0.65100	0.24844	−0.52523	0.00000	−0.38878	−0.36397
1.20720	0.91779	0.55427	0.00000	0.38705	0.29448
1.56122	0.37024	0.23591	0.62605	0.18409	0.18923
1.21257	0.34501	0.39642	0.97644	0.19746	0.00357
1.81746	0.15770	−0.17915	−0.90100	−0.00931	0.20689

s_0^{scatt}	f_{11}^{scatt}	f_{12}^{scatt}	f_{13}^{scatt}	f_{14}^{scatt}	f_{15}^{scatt}
−3.30564	0.26681	0.16583	−0.19840	0.32808	0.31193

s_A	s_{A0}
1.0	−0.2

The *K-matrix* of Table 1 generates a physical *T-matrix*, $T = (I - i\rho \cdot K)^{-1}K$, which describes the scattering into the $\pi^+\pi^-$ final state with five poles, whose masses and half-widths are listed in Table 2.

TABLE 2. *T-matrix* poles. The $(m,-\Gamma/2)$ values are in GeV.

$f_0(980)$	$f_0(1300)$	$f_0(1200-1600)$	$f_0(1500)$	$f_0(1750)$
$(1.019, 0.038)$	$(1.306, 0.170)$	$(1.470, 0.960)$	$(1.489, 0.058)$	$1.746, 0.160$

The decay amplitude for the D meson into three-pion final state, where $\pi^+\pi^-$ are in a $(IJ^{PC} = 00^{++})$-wave, is thus

$$F_1 = (I - iK\rho)_{1j}^{-1} \left\{ \sum_\alpha \frac{\beta_\alpha g_j^{(\alpha)}}{m_\alpha^2 - s} + f_{1j}^{\text{prod}} \frac{1\,\text{GeV}^2 - s_0^{\text{prod}}}{s - s_0^{\text{prod}}} \right\} \frac{s - s_A m_\pi^2/2}{(s - s_{A0})(1 - s_{A0})}, \quad (4)$$

where β_α is the coupling to the m_α pole in the 'initial' production process, f_{1j}^{prod} and s_0^{prod} are the *P-vector* background parameters. The *P-vector* parameters are complex. The presence of the Adler-zero term, not required *a priori* in the F_1 amplitude numerator, should be investigated by studying its effect on the fit quality to the data.

THREE-PION PRELIMINARY RESULTS

The three-pion selected samples (Fig. 1) consist of 1527 ± 51 and 1475 ± 50 events for the D^+ and D_s respectively. The Dalitz plot analyses are performed on yields within $\pm 2\sigma$ of the fitted mass value. In our analysis the S-wave, isoscalar resonance parameters are fixed to those of Table 2 while the *P-vector* parameters (i.e., β_α, f_{1j}^{prod} and s_0^{prod}) are free in the fit. Vector and tensor resonances in the total D-meson decay amplitude, Eq. 1, can be reasonably well described, as mentioned above, in the context of the isobar model. Thus, their masses and widths are fixed to the PDG values and the coefficients and phases of their amplitudes ($a_0, a_i, \delta_0, \delta_i$ of Eq. 1) are additional free parameters.

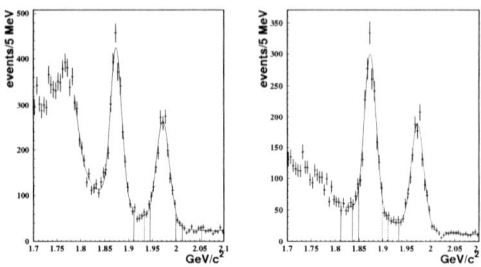

FIGURE 1. Signal and side-band regions of the three-pion invariant-mass distribution for D_s^+ and D^+ Dalitz-plot analysis respectively.

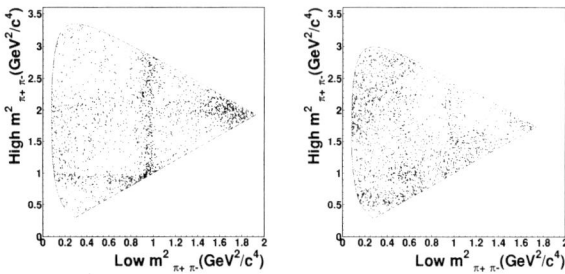

FIGURE 2. D_s^+ and D^+ Dalitz plots.

$$D_s^+ \rightarrow \pi^+ \pi^- \pi^+$$

The resulting fit fractions, as defined in [6], phases and amplitude coefficients for the D_s to three pions are quoted in Table 3. The contribution of the entire S-wave is computed as a single fit fraction since there is no possibility to distinguish on the real axis among the different terms, either resonance or background, contributing to the S-wave. On the other hand, the measurable quantities that carry all the physical information are the complex couplings at the production to the corresponding T-matrix physical poles. The resulting couplings, reported in Table 5, are computed by continuing the amplitude $F_1(s)$ into the complex s-plane to the position of the poles and evaluating the pole residues. The

TABLE 3. Fit results from the *K-matrix*-model for D_s^+.

decay channel	fit fraction (%)	phase (deg.)	amplitude coefficient
$(S-wave)\,\pi^+$	$87.04 \pm 5.60 \pm 4.17$	0 (fixed)	1 (fixed)
$f_2(1275)\,\pi^+$	$9.74 \pm 4.49 \pm 2.63$	$168.0 \pm 18.7 \pm 2.5$	$0.165 \pm 0.033 \pm 0.032$
$\rho^0(1450)\,\pi^+$	$6.56 \pm 3.43 \pm 3.31$	$234.9 \pm 19.5 \pm 13.3$	$0.136 \pm 0.030 \pm 0.035$

D_s^+ Dalitz projections of our data are shown in Fig. 3 with our final fit superimposed. The corresponding fit C.L., evaluated with a χ^2 estimator over a Dalitz plot with bin size chosen adaptively according to the statistics, is 3%. The $D_s^+ \rightarrow \pi^+ \pi^- \pi^+$ channel is known to be the best candidate for quantifying the role of the annihilation process in

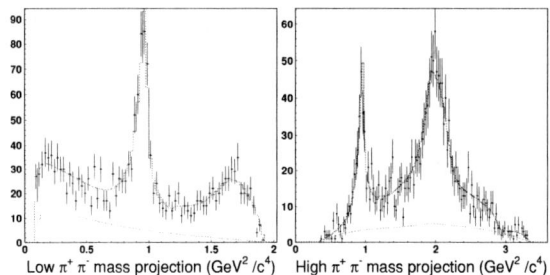

Low $\pi^+ \pi^-$ mass projection (GeV^2/c^4)   High $\pi^+ \pi^-$ mass projection (GeV^2/c^4)

FIGURE 3. D_s^+ Dalitz-plot projections with the *K-matrix* fit superimposed; the background shape under the signal is shown as well. The corresponding fit C.L. is 3%.

TABLE 4. Fit results from the *K-matrix*-model fit for D^+.

decay channel	fit fraction (%)	phase (deg.)	amplitude coefficient
$(S-wave)\,\pi^+$	$56.46 \pm 3.78 \pm 1.02$	0 (fixed)	1 (fixed)
$f_2(1270)\,\pi^+$	$12.26 \pm 1.73 \pm 0.21$	$-38.6 \pm 20.89 \pm 4.17$	$1.963 \pm 0.451 \pm 0.101$
$\rho^0(770)\,\pi^+$	$26.89 \pm 3.78 \pm 1.08$	$-128.9 \pm 18.54 \pm 3.75$	$2.908 \pm 0.762 \pm 0.227$

charm hadronic decays. It is interesting to note that neither the three-body non-resonant nor $\rho(770)\pi^+$ components are required by the fit, suggesting a marginal role of the annihilation diagram in the decay.

$$D^+ \to \pi^+ \pi^- \pi^+$$

The $D^+ \to \pi^+ \pi^- \pi^+$ Dalitz plot shows an excess of events at low $\pi^+\pi^-$ mass, which in the traditional isobar approach would require an *ad hoc* free-parameter Breit–Wigner having mass $m = 442.6 \pm 27.0$ and width $\Gamma = 340.4 \pm 65.4$ MeV/c^2. On the other hand we know that non-trivial structures can be generated by a complicate interplay among the S-wave resonances and the underlying non-resonant S-wave component. For this reason it is interesting to study this channel with a formalism that embeds the already measured S-wave $\pi^+\pi^-$ scattering dynamics. The complete fit results are reported in Table 4 and the resulting production coupling constants in Table 5.

The D^+ Dalitz projections are shown in Fig. 4; the fit C.L. is 6.8%. The most interesting feature of these results is the fact that a better treatment of the S-wave portion of the decay amplitude alone, such as that provided by the present *K-matrix* model, is enough to well reproduce the low-mass $\pi^+\pi^-$ structure of the D^+ Dalitz plot. This would suggest that any σ-like object in the D decay data is consistent with any σ-like object in $\pi^+\pi^-$ scattering. This observation certainly helps in better understanding the σ-puzzle; obviously, other studies and investigations with higher statistics are required to draw any final conclusions.

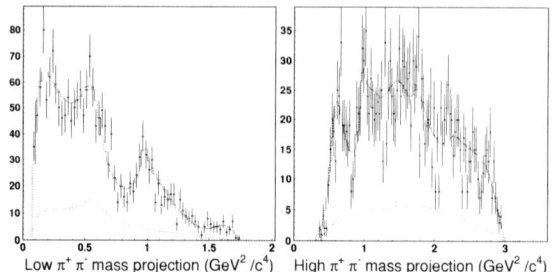

Low $\pi^+\pi^-$ mass projection (GeV^2/c^4)   High $\pi^+\pi^-$ mass projection (GeV^2/c^4)

FIGURE 4. D^+ Dalitz-plot projections with the *K-matrix* fit superimposed. The corresponding fit C.L. is 6.8%.

TABLE 5. D_s^+ and $D^+ \to \pi^+\pi^-\pi^+$ production coupling constants for the five *T-matrix* poles: they are referred to the $f_0(980)$ value. Phases are in degrees.

T-matrix pole	$(m, \Gamma/2)$ (GeV)	D_s^+ coupling constant	D^+ coupling constant *
$f_0(980)$	$(1.019, 0.038)$	$1 \cdot e^{i0}$ (fixed)	$1 \cdot e^{i0}$ (fixed)
$f_0(1300)$	$(1.306, 0.170)$	$(0.43\pm0.04)\,e^{i(-163.8\pm4.9)}$	$(0.96\pm0.03)\,e^{i(-164.3\pm2.3)}$
$f_0(1200-1600)$	$(1.470, 0.960)$	$(4.90\pm0.08)\,e^{i(80.9\pm1.06)}$	$(2.04\pm0.03)\,e^{i(2.9\pm1.9)}$
$f_0(1500)$	$(1.488, 0.058)$	$(0.51\pm0.02)\,e^{i(83.1\pm3.03)}$	$(0.71\pm0.03)\,e^{i(-188.3\pm2.3)}$
$f_0(1750)$	$(1.746, 0.160)$	$(0.82\pm0.02)\,e^{i(-127.9\pm2.25)}$	$(0.87\pm0.02)\,e^{i(62.5\pm1.4)}$

* The coupling to the $f_0(1200-1600)$ pole has to be looked at with a certain caution because of the intrinsic limitations of the approximation used for the 4-body phase-space when extrapolated very deeply into the complex plane.

CONCLUSIONS

The *K-matrix* formalism has been applied to perform a Dalitz plot analysis of the FOCUS D^+ and $D_s^+ \to \pi^+\pi^-\pi^+$ data. The results are extremely encouraging since the same parametrization of two-body resonances coming from light-quark experiments also works for charm decays; this result was not obvious beforehand. Furthermore, the same model is able to reproduce the non-trivial features of the $D^+ \to \pi^+\pi^-\pi^+$ Dalitz plot with a good C.L. The *K-matrix* method used here will find its full application to the forthcoming excellent statistics of charm experiments.

REFERENCES

1. E.P. Wigner, *Phys. Rev* **70** (1946) 15.
2. S.U. Chung *et al.*, *Ann. Physik* **4** (1995) 404.
3. I.J.R. Aitchison, *Nucl. Phys.* **A187** (1972) 417.
4. K.L. Au, D. Morgan, M.R. Pennington, *Phys. Rev.* **D35** (1987) 1633; M.R. Pennington, hep-ph/9905241.
5. V.V. Anisovich, A.V. Sarantsev, *Eur. Phys. J.* **A16** (2003) 229.
6. P.L. Frabetti *et al.*, *Phys. Lett.* **B407** (1997) 79.
7. S. Spanier and N.A. Törnqvist, Scalar Mesons (rev.), Particle Data Group, http://pdg.lbl.gov

MEASUREMENT OF THE HADRONIC CROSS SECTION AT DAΦNE WITH THE KLOE DETECTOR

The KLOE collaboration [1]
presented by Achim G. Denig

Universität Karlsruhe, IEKP, Postfach 3640, 76021 Karlsruhe, Germany

Abstract. We have measured the cross section $\sigma(e^+e^- \to \pi^+\pi^-\gamma)$ as a function of the $\pi^+\pi^-$ invariant mass, $M_{\pi\pi}$, with the KLOE detector at DAΦNE ($W = m_\phi = 1.02$ GeV). The photon in the above process is due to Initial State Radiation. Dividing by a theoretical radiator function, we obtain the cross section $\sigma(e^+e^- \to \pi^+\pi^-)$ for the mass range $0.37 < M_{\pi\pi}^2 < 0.93$ GeV^2. We extract the pion form factor and the hadronic contribution to the muon anomaly, a_μ.

HADRONIC CROSS SECTION AT DAΦNE

Motivation

Accurate measurements of the cross section for e^+e^- annihilation into hadrons are of importance for an interpretation of the recent new precision measurement of the anomalous magnetic moment of the muon [1]. Hadronic contributions to the photon spectral functions due to quark loops are not calculable in the framework of perturbative QCD. It is well known, however, that the hadronic piece of the spectral function is connected by unitarity to the cross section for $e^+e^- \to$ hadrons. A dispersion relation can thus be derived, giving the contribution to a_μ as an integral over the hadronic cross section, multiplied by an appropriate kernel. The process $e^+e^- \to \pi^+\pi^-$ below 1GeV is of special importance since it contributes to ∼60% to the total integral. The most recent evaluation of the dispersion integral [4] [5] gives the following values for the muon

[1] The KLOE Collaboration: A. Aloisio, F. Ambrosino, A. Antonelli, M. Antonelli, C. Bacci, G. Bencivenni, S. Bertolucci, C. Bini, C. Bloise, V. Bocci, F. Bossi, P. Branchini, S. A. Bulychjov, R. Caloi, P. Campana, G. Capon, T. Capussela, G. Carboni, G. Cataldi, F. Ceradini, F. Cervelli, F. Cevenini, G. Chiefari, P. Ciambrone, S. Conetti, E. De Lucia, P. De Simone, G. De Zorzi, S. Dell'Agnello, A. Denig, A. Di Domenico, C. Di Donato, S. Di Falco, B. Di Micco, A. Doria, M. Dreucci, O. Erriquez, A. Farilla, G. Felici, A. Ferrari, M. L. Ferrer, G. Finocchiaro, C. Forti, A. Franceschi, P. Franzini, C. Gatti, P. Gauzzi, S. Giovannella, E. Gorini, E. Graziani, M. Incagli, W. Kluge, V. Kulikov, F. Lacava, G. Lanfranchi, J. Lee-Franzini, D. Leone, F. Lu, M. Martemianov, M. Matsyuk, W. Mei, L. Merola, R. Messi, S. Miscetti, M. Moulson, S. Müller, F. Murtas, M. Napolitano, A. Nedosekin, F. Nguyen, M. Palutan, E. Pasqualucci, L. Passalacqua, A. Passeri, V. Patera, F. Perfetto, E. Petrolo, L. Pontecorvo, M. Primavera, F. Ruggieri, P. Santangelo, E. Santovetti, G. Saracino, R. D. Schamberger, B. Sciascia, A. Sciubba, F. Scuri, I. Sfiligoi, A. Sibidanov, T. Spadaro, E. Spiriti, M. Testa, L. Tortora, P. Valente, B. Valeriani, G. Venanzoni, S. Veneziano, A. Ventura, S. Ventura, R. Versaci, I. Villella, G. Xu.

CP717, *Hadron Spectroscopy: Tenth International Conference,*
edited by E. Klempt, H. Koch, and H. Orth
© 2004 American Institute of Physics 0-7354-0197-7/04/$22.00

anomaly, compared to the experimental value of the E821 collaboration:

Theory using τ data	$a_\mu^{theo} = (11\ 659\ 195.6 \pm 6.8) \times 10^{-10}$
Theory using e^+e^- data	$a_\mu^{theo} = (11\ 659\ 180.9 \pm 8.0) \times 10^{-10}$
Experiment BNL-E821	$a_\mu^{exp} = (11\ 659\ 203 \pm 8) \times 10^{-10}$

The first value is obtained including hadronic τ decay data, assuming conservation of the vector current (CVC) and correcting for isospin breaking effects. The second value uses e^+e^- data only (see also [6]), in particular the recent reanalysis [3] of the CMD-2 measurement [2] of the $\pi^+\pi^-$ channel (0.6% systematic error) in the energy range below 1 GeV. The $e^+e^- \to \pi^+\pi^-$ based result disagrees by $\sim$2 σ with the BNL measurement, while the value using τ decay data is in agreement with the experimental value. Further independent hadronic cross section measurements are needed to clarify the situation.

Initial State Radiation

Particle factories such as DAΦNE or the B-factories typically operate at fixed centre-of-mass energies: $W = m_\phi$ in the case of DAΦNE. Initial state radiation, ISR, is a complementary approach at particle factories which allows studying $e^+e^- \to$ hadrons over the entire energy range (from $2m_\pi$ to W). For a photon (energy E_γ) radiated before annihilation of the e^+e^- pair, the invariant mass of the $\pi^+\pi^-$ system is given by: $M_{\pi\pi}^2 = s_\pi = W^2 - 2WE_\gamma$. In general the $\pi^+\pi^-\gamma$ and $\pi^+\pi^-$ cross section are related through:

$$s_\pi \frac{d\sigma(\pi^+\pi^-\gamma)}{ds_\pi} = \sigma(\pi^+\pi^-, s_\pi) \times H(s_\pi) \tag{1}$$

Eq. 1 defines the radiator function $H(s_\pi)$ which we obtain from the Monte Carlo program PHOKHARA, a NLO generator for the $\pi^+\pi^-\gamma$ exclusive final state [7] [8] [9] [10] [11]. Our present analysis is based on the observation of ref.[7], that for small polar angles of the radiated photon, the ISR process dominates over the FSR process, which is an indistinguishable background to our ISR approach. In the following we present the cross section measurement of the reaction $e^+e^- \to \pi^+\pi^-\gamma$ with $\theta_\gamma < 15°$ or $\theta_\gamma > 165°$ [2]. No explicit photon detection is done since the KLOE electromagnetic calorimeter (EmC) does not cover angles smaller than 20°. We cut on the di-pion production angle, $\theta_{\pi\pi}$, which is calculated from the momenta of the two charged tracks. If only one photon is emitted, the following relation holds exactly: $\theta_\gamma = 180° - \theta_{\pi\pi}$. As we will show in the following, an efficient and almost background free signal selection can be obtained without photon tagging.

---

[2] For small s_π, the di-pion system is recoiling against the small angle photon, resulting in small angle pion tracks which cannot be detected in the KLOE drift chamber. We are therefore limited to measuring $\sigma(\pi^+\pi^-)$ for $s_\pi >$0.3 GeV^2. A complementary analysis in which photons are selected at large angles is in progress. In this case the photon can be tagged in the electromagnetic calorimeter and the kinematical acceptance allows us to measure events down to the 2π threshold.

ANALYSIS OF $\pi^+\pi^-\gamma$ EVENTS

We have analyzed KLOE data taken in 2001 with an integrated luminosity of 140pb^{-1}. After fiducial volume and selection cuts we collect $\sim 1.5 \times 10^6$ events. To obtain the cross section, we subtract the residual background from this spectrum and divide by the selection efficiency and the integrated luminosity.

We briefly comment in the following on the individual analysis items. Further details can be found in [12]. For a detailed description of the KLOE detector, which consists of a high resolution tracking detector ($\sigma_{p_T}/p_T \leq 0.4\%$) and an electromagnetic calorimeter ($\sigma_E/E = 5.7\%/\sqrt{E(\text{GeV})}$), we refer to ref. [13] [14].

- *Detection of two charged tracks*: with polar angle larger than $50°$, coming from a vertex in the fiducial volume $R < 8$ cm, $|z| < 7$ cm. The cuts on the transverse momentum $p_T > 160$ MeV or on the longitudinal momentum $|p_z| > 90$ MeV reject tracks spiralizing along the beam line, ensuring good reconstruction conditions. The probability to reconstruct a vertex in the drift chamber is $\sim 98\%$ and has been studied with $\pi^+\pi^-\pi^0$ and $\pi^+\pi^-$ data.

- *Identification of pion tracks*: A Likelihood Method (calibrated on real data), using the time of flight of the particle and the shape of the energy deposit in the electromagnetic calorimeter, has been developed to reject $e^+e^- \to e^+e^-\gamma$ background. Background from $e^+e^-\gamma$ events is drastically reduced like that. A control sample of $\pi^+\pi^-\pi^0$ has been used to study the behaviour of pions in the electromagnetic calorimeter and to evaluate the selection efficiency ($> 98\%$) for signal events.

- *Background subtraction*: $\mu^+\mu^-\gamma$ and $\pi^+\pi^-\pi^0$ events are rejected by a cut in a kinematic variable called track mass, m_{trk} This variable is calculated from the reconstructed pion momenta, $\vec{p}_+$, $\vec{p}_-$, applying 4-momentum conservation under the hypothesis that the final state consists of two particles with the same mass and one photon. For such $e^+e^- \to x^+x^-\gamma$ events, the value of m_{trk} peaks at m_π, m_μ, m_e for $x = \pi, \mu, e$ respectively, thus allowing a selection of signal events. The density distribution of the two track events in the $[s_\pi, m_{\text{trk}}]$ plane is very effective for separating signal from background. The final event selection is defined by: $(m_{\text{trk}} > 120) \cap (m_{\text{trk}} < 250 - 105\sqrt{1-(s_\pi/850000)^2}) \cap (m_{\text{trk}} < 220)$, all units in MeV.

- *Luminosity Measurement*: The integrated luminosity is measured with the KLOE detector using large angle Bhabha (LAB) events. The effective Bhabha cross section at large angles ($55° < \theta_{+,-} < 125°$) is about 430 nb. For the computation of the radiative corrections we use two independent Bhabha event generators: BHAGENF [15] [16]) and BABAYAGA ([17]). For each generator a systematic error of 0.5% is quoted by the authors. The two generators agree to better than 0.2% with each other. All selection efficiencies (trigger, EmC cluster, DC tracking) are $> 99\%$ and are well reproduced by the detector simulation program. We also obtain excellent agreement between the experimental distributions ($\theta_{+,-}, E_{+,-}$) and those obtained from Monte Carlo simulation. The experimental uncertainty in the acceptance due to all these effects is 0.4%. We assign a total systematic error for the luminosity of $\delta\mathscr{L} = 0.5\%_{\text{th}} \oplus 0.4\%_{\text{exp}}$.

TABLE 1. List of systematic uncertainties.

Acceptance	0.3%
Trigger + Offline Reconstruction Filter	0.6%
Tracking	0.3%
Vertex	0.7 %
Particle ID Estimator (Likelihood)	0.1 %
Track Mass	0.2 %
Background subtraction	0.5 %
Total experimental systematics	1.2 %

The contribution of the several analysis items to the total systematic error is shown in table 1. The value for the total systematic error of 1.2% is preliminary and expected to be reduced to ~1.0% soon. We have not unfolded data for mass resolution effects. MC studies have shown that the effect of detector smearing is small.

RESULTS

The result of our cross section measurement for $e^+e^- \to \pi^+\pi^-\gamma$ is shown in fig. 1. According to eq. 1 the radiator function $H(s_\pi)$ is needed in order to extract $\sigma(e^+e^- \to \pi^+\pi^-)$. We obtain $H(s_\pi)$ from PHOKHARA, setting $F_\pi(s_\pi) = 1$ and switching off the vacuum polarization of the intermediate photon in the generator. The radiator $H(s_\pi)$ is also shown in fig. 1, left. The H-function is theoretically known with a precision of ~0.5%. We take this value (together with the error on luminosity) as the theory error.

FSR and Vacuum Polarization Corrections
Two further radiative corrections have to be applied to our data before evaluating the hadronic contribution to the muon anomaly (see ref. [4] for details). The cross section has to be the *bare* cross section, i.e. vacuum polarization corrections must be subtracted. This can be done by correcting the cross section for the running of α.
The second correction deals with the treatment of FSR events. In our $\pi^+\pi^-\gamma$ cross section measurement a part of multiphoton events (more than 1 hard photon) are re-moved by the m_{trk} cut. The efficiency of this cut is evaluated by Monte Carlo using a recently published new version of PHOKHARA [11], in which simultaneously photons from ISR and FSR are simulated. A preliminary study shows, that the cross section is changed by less than 2% due to this next-to-leading-order FSR effect. We are working out the final corrections to be applied to data. For now, we apply an error of 1% as a conservative estimate of the uncertainty due to FSR corrections.

Hadronic contribution to the muon anomaly
We have evaluated the hadronic two-pion contribution to the muon anomaly, $a_\mu^{\pi\pi}$, by inserting our measured *bare* cross section $e^+e^- \to \pi^+\pi^-$ into the dispersion integral. In order to compare our result with the CMD-2 result we cover the same energy interval as CMD-2. The preliminary KLOE result (in 10^{-10} units) is in good agreement with the

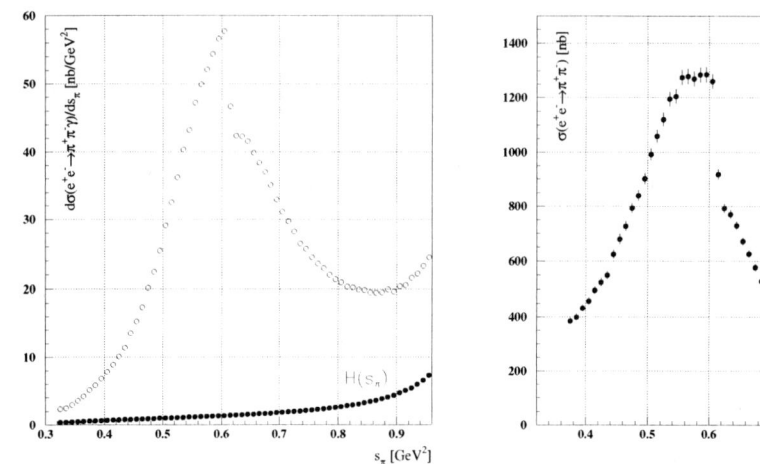

FIGURE 1. Left: Cross section for $e^+e^- \to \pi^+\pi^-\gamma$. The radiator $H(s_\pi)$, is also shown. Right: Bare cross section for $e^+e^- \to \pi^+\pi^-$.

CMD-2 value as can be seen in the following table [3] :

KLOE preliminary	$a_\mu^{\pi\pi} = 378.4 \pm 0.8_{stat} \pm 4.5_{syst} \pm 3.0_{theo} \pm 3.8_{FSR}$
CMD-2 reanalysis 2003 [3]	$a_\mu^{\pi\pi} = 378.6 \pm 2.7_{stat} \pm 2.3_{syst+theo}$

The actual systematic error of 1.2% will be reduced down to $\simeq 1\%$ in the very near future. Also the error on FSR will be further reduced. The statistical error is negligible. Of special interest is the energy region above the ρ peak because large discrepancies between evaluations based on τ data and $e^+e^- \to \pi^+\pi^-$ data are seen here. For this reason we have calculated $a_\mu^{\pi\pi}$ in the range below (<0.65 GeV^2) and above (>0.65 GeV^2) the ρ peak and compare the values with the ones from CMD-2[4]:

s_π / GeV^2	$a_\mu^{\pi\pi}$ [KLOE]	$a_\mu^{\pi\pi}$ [CMD-2]
0.37 – 0.65	309.4 ± 6.0	308.5 ± 2.8
0.65 – 0.93	68.8 ± 1.2	72.2 ± 0.7

Our data are in good agreement with CMD-2 below the ρ peak. For the energy range $0.65 < s_\pi < 0.93$GeV^2 our data are lower than CMD-2. Final corrections for FSR are still missing. We conclude that our data confirm the results from $e^+e^- \to \pi^+\pi^-$ and the discrepancy with respect to τ data.

---

[3] These numbers are the outcome of an updated analysis, presented at SIGHAD03 (October 8-10,2003) and supersede the result presented at HADRON2003: $a_\mu^{\pi\pi} = 374.1 \pm 1.1_{stat} \pm 5.2_{syst} \pm 2.6_{theo} {}^{+7.5}_{-0.0}|_{FSR}$

[4] This is our evaluation of $a_\mu^{\pi\pi}$ [CMD-2], based on the values tabulated in ref. [3]

REFERENCES

1. G.W. Bennet, et al., Phys. Rev. Lett. 89, (2002) 101. See also references 1-4 in this paper.
2. R.R. Akhmetshin et al., Phys. Let. B 527 (2002) 161-172
3. R.R. Akhmetshin et al., hep-ex/0308008
4. M. Davier, S. Eidelman, A. Höcker and Z. Zhang, Eur. Phys. J. C 27, (2003) 497, hep-ph/0208177
5. M. Davier, S. Eidelman, A. Höcker and Z. Zhang, hep-ph/0308213
6. S. Eidelman and F. Jegerlehner, Z.Phys. C **67** (1995) 585
7. S. Binner, J.H. Kühn, K. Melnikov, Phys.Lett. B **459** (1999) 279
8. G. Rodrigo, A. Gehrmann-De Ridder, M. Guilleaume, J.H. Kühn, Eur. Phys. J. C **22**, 81 (2001);
9. G. Rodrigo, H. Czyż, J.H. Kühn and M. Szopa, Eur. Phys. J C **24** (2002) 71
10. H. Czyż, A. Grzelińska, J. H. Kühn and G. Rodrigo, Eur. Phys. J. C **27**, 563 (2003).)
11. H. Czyż, A. Grzelińska, J. H. Kühn and G. Rodrigo, hep-ph/0308312
12. M. Incagli for the [KLOE collaboration], *Determination of* $\sigma(e^+e^- \to \pi^+\pi^-)$ *from radiative processes at DAΦNE*, hep-ex/0307051
13. M. Adinolfi *et al.* [KLOE collaboration], Nucl. Instrum. Meth. A **488** (2002) 51
14. M. Adinolfi *et al.* [KLOE collaboration], Nucl. Instrum. Meth. A **482** (2002) 364
15. F.A. Berends, R. Kleiss, Nucl.Phys. B **228** (1988) 537
16. E. Drago, G. Venanzoni, *A Bhabha Generator for DAΦNE including radiative corrections and φ resonance*, **INFN/AE-97/48** (1997)
17. C.M.C. Calame, C. Lunardini, G. Montagna, O. Nicrosini, F. Piccinini, Nucl.Phys. B **584** (2000) 459
18. A. Höfer, J. Gluza and F. Jegerlehner, Eur. Phys. J C **24** (2002) 51

Large N_c behavior of light resonances in meson-meson scattering

J.R.Peláez

Departamento de Física Teórica II, Universidad Complutense, 28040 Madrid, Spain

Abstract. By scaling the parameters of meson-meson unitarized Chiral Perturbation Theory amplitudes according to the QCD large N_c rules, one can study the spectroscopic nature of light meson resonances. The scalars σ, κ $f_0(980)$ and, possibly, the $a_0(980)$ do not seem to behave as $\bar{q}q$ states, in contrast to the vectors $\rho(770)$ and $K^*(892)$. The behavior shown by the scalars is naturally explained in terms of diagrams with intermediate $\bar{q}\bar{q}qq$-like states. Here we review our recent study and show how the results do not depend on the different fits to data.

INTRODUCTION

Although QCD is firmly established as the theory of strong interactions it becomes non-perturbative at low energies, and gives only little help to address the existence and nature of the lightest scalar mesons. Alternatively Chiral Perturbation Theory (ChPT) [1] has been devised as the QCD low energy Effective Lagrangian built as the most general derivative expansion respecting its symmetries, containing only π, K and η mesons. These particles are the QCD low energy degrees of freedom since they are Goldstone bosons of the QCD spontaneous chiral symmetry breaking. For meson-meson scattering ChPT is an expansion in even powers of momenta, $O(p^2), O(p^4)...$, over a scale $\Lambda_\chi \sim 4\pi f_0 \simeq 1\,\text{GeV}$. Since the u, d and s quark masses are so small compared with Λ_χ they are introduced as perturbations, giving rise to the π, K and η masses, counted as $O(p^2)$. At each order, ChPT is the sum of *all terms* compatible with the symmetries, multiplied by "chiral" parameters, that absorb loop divergences order by order, yielding finite results. The leading order is universal since there is only one parameter, f_0, that sets the scale of spontaneous chiral symmetry breaking. Different underlying dynamics manifest themselves with different higher order parameters. In Table I are listed the parameters that determine meson-meson scattering up to $O(p^4)$, called L_i. As usual after renormalization, they depend on an arbitrary regularization scale, as $L_i(\mu_2) = L_i(\mu_1) + \Gamma_i \log(\mu_1/\mu_2)/16\pi^2$, where Γ_i are constants given in [1]. In physical observables the μ dependence is canceled with that of the loop integrals.

The large N_c expansion [4] is the only analytic approximation to QCD in the whole energy region, also providing a clear definition of $\bar{q}q$ states, that become bound states when $N_c \to \infty$. The N_c scaling of the L_i parameters has been given in [1, 5], and is listed in Table I. In addition, the π, K, η masses scale as $O(1)$ and f_0 as $O(\sqrt{N_c})$. However, it is not known at what scale μ to apply the large N_c scaling, and it has been pointed out that the logarithmic terms can be rather large for $N_c = 3$ [6]. The scale dependence is certainly suppressed by $1/N_c$ for $L_i = L_2, L_3, L_5, L_8$, but not for $2L_1 - L_2, L_4, L_6$ and L_7.

CP717, *Hadron Spectroscopy: Tenth International Conference,*
edited by E. Klempt, H. Koch, and H. Orth
© 2004 American Institute of Physics 0-7354-0197-7/04/$22.00

Customarily, the uncertainty in the μ where the N_c scaling applies is estimated varying μ between 0.5 and 1 GeV [1]. We will check that this estimate is correct with the vector mesons, firmly established as $\bar{q}q$ states.

TABLE 1. $O(p^4)$ chiral parameters ($\times 10^3$) and their N_c scaling. In the ChPT column, L_1, L_2, L_3 come from [2] and the rest from [1]. The IAM columns correspond to different fits [3]

$O(p^4)$ Parameter	N_c scaling	ChPT $\mu = 770\,\text{MeV}$	IAM I $\mu = 770\,\text{MeV}$	IAM II $\mu = 770\,\text{MeV}$	IAM III $\mu = 770\,\text{MeV}$
L_1	$O(N_c)$	0.4 ± 0.3	0.56 ± 0.10	0.59 ± 0.08	0.60 ± 0.09
L_2	$O(N_c)$	1.35 ± 0.3	1.21 ± 0.10	1.18 ± 0.10	1.22 ± 0.08
L_3	$O(N_c)$	-3.5 ± 1.1	-2.79 ± 0.14	-2.93 ± 0.10	-3.02 ± 0.06
L_4	$O(1)$	-0.3 ± 0.5	-0.36 ± 0.17	0.2 ± 0.004	0 (fixed)
L_5	$O(N_c)$	1.4 ± 0.5	1.4 ± 0.5	1.8 ± 0.08	1.9 ± 0.03
L_6	$O(1)$	-0.2 ± 0.3	0.07 ± 0.08	0 ± 0.5	-0.07 ± 0.20
L_7	$O(1)$	-0.4 ± 0.2	-0.44 ± 0.15	-0.12 ± 0.16	-0.25 ± 0.18
L_8	$O(N_c)$	0.9 ± 0.3	0.78 ± 0.18	0.78 ± 0.7	0.84 ± 0.23
$2L_1 - L_2$	$O(1)$	-0.55 ± 0.7	0.09 ± 0.10	0.0 ± 0.1	-0.02 ± 0.10

ChPT is a low energy expansion, but in recent years it has been extended to higher energies by means of unitarization [3, 7, 8, 9, 10]. The main idea is that when projected into partial waves of definite angular momentum J and isospin I, physical amplitudes t should satisfy an elastic unitarity condition:

$$\text{Im}\, t = \sigma |t|^2 \quad \Rightarrow \quad \text{Im}\,\frac{1}{t} = -\sigma \quad \Rightarrow \quad t = \frac{1}{\text{Re}\, t^{-1} - i\sigma}, \tag{1}$$

where σ is the phase space of the two mesons, a well known function. However, from the right hand side we note that to have a unitary amplitude we only need $\text{Re}\, t^{-1}$, and for that we can use the ChPT expansion; this is the Inverse Amplitude Method (IAM) [7]. The IAM generates the ρ, K^*, σ and κ resonances not initially present in ChPT, ensures unitarity in the elastic region and respects the ChPT expansion. When inelastic two-meson processes are present the IAM generalizes to $T \simeq (\text{Re}\, T^{-1} - i\Sigma)^{-1}$ where T is a matrix containing all partial waves between all physically accessible states whereas Σ is a diagonal matrix with their phase spaces, again well known [3, 8, 9, 10]. With this generalization it was recently shown [3] that, using the one-loop ChPT calculations, the IAM generates the ρ, K^*, σ, κ, $a_0(980)$, $f_0(980)$ and the octet ϕ, describing two body π, K or η scattering up to 1.2 GeV. Furthermore, it has the correct low energy expansion, with chiral parameters compatible with standard ChPT, shown in Table I. Different IAM fits [3] are due to different ChPT truncation schemes equivalent up to $O(p^4)$ and the estimates of the data systematic error.

Those IAM results have been recently used [11] to study the large N_c behavior of the scattering amplitudes and the poles associated to resonances. The large N_c results are similar for all IAM sets, and to illustrate it, we show here the results of set III, whereas in [11] we used set II, reaching the same conclusions. Note that these ChPT amplitudes are fully renormalized, and therefore scale independent. Hence all the QCD N_c dependence appears correctly through the L_i and cannot hide in any spurious parameter.

RESULTS

Let us then scale $f_0 \to f_0 \sqrt{N_c/3}$ and $L_i(\mu) \to L_i(\mu)(N_c/3)$ for $i = 2,3,5,8$, keeping the masses and $2L_1 - L_2, L_4, L_6$ and L_7 constant. In Fig.1 we show, for increasing N_c, the modulus of the $(I,J) = (1,1)$ and $(1/2,1)$ amplitudes with the Breit-Wigner shape of the ρ and $K^*(892)$ vector resonances, respectively. There is always a peak at an almost constant position, becoming narrower as N_c increases. We also show the evolution of the ρ and K^* pole positions, related to their mass and width as $\sqrt{s_{pole}} \simeq M - i\Gamma/2$. We have normalized both M and Γ to their value at $N_c = 3$ in order to compare with the $\bar{q}q$ expected behavior: M_{N_c}/M_3 constant and $\Gamma_{N_c}/\Gamma_3 \sim 1/N_c$ The agreement is remarkable within the gray band that covers the uncertainty $\mu = 0.5 - 1\,\text{GeV}$ where to apply the large N_c scaling. We have checked that outside that band, the behavior starts deviating from that of $\bar{q}q$ states, which confirms that the expected scale range where the large N_c scaling applies is correct.

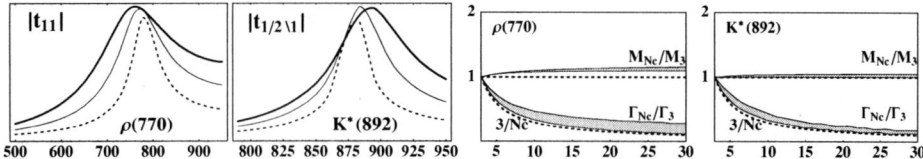

FIGURE 1. Left: Modulus of $\pi\pi$ and πK elastic amplitudes versus $\sqrt{s}$ for $(I,J) = (1,1),(1/2,1)$: $N_c = 3$ (thick line), $N_c = 5$ (thin line) and $N_c = 10$ (dotted line), scaled at $\mu = 770\,\text{MeV}$. Right: $\rho(770)$ and $K^*(892)$ pole positions: $\sqrt{s_{pole}} \equiv M - i\Gamma/2$ versus N_c. The gray areas cover the uncertainty $N_c = 0.5 - 1\,\text{GeV}$. The dotted lines show the expected $\bar{q}q$ large N_c scaling.

In Fig.2, in contrast, all over the σ and κ regions the $(0,0)$ and $(1/2,0)$ amplitudes decrease as $N_c \to \infty$. Their associated poles show a totally different behavior, since *their width grows with N_c*, in conflict with a $\bar{q}q$ interpretation. (We keep the M, Γ notation, but now as definitions). This is also suggested using the ChPT leading order unitarized amplitudes with a regularization scale [10, 12]. In order to determine their spectroscopic nature, we note that *in the whole σ and κ regions*, $\text{Im}\,t \sim O(1/N_c^2)$ and $\text{Re}\,t \sim O(1/N_c)$. Imaginary parts are generated from s-channel intermediate physical states. If it was a $\bar{q}q$ meson, with mass $M \sim O(1)$ and $\Gamma \sim 1/N_c$, we would expect $\text{Im}\,t \sim O(1)$ and a peak at $\sqrt{s} \simeq M$, as it is indeed the case of the ρ and K^*. Therefore, from $\bar{q}q$ states, the σ and κ can only get real contributions from ρ or K^* t-channel exchange, respectively. The leading s-channel contribution for the κ comes from $\bar{q}\bar{q}qq$ (or two meson) states, which are predicted to unbound and become the meson-meson continuum when $N_c \to \infty$ [13]. The same interpretation holds for the sigma, but in the large N_c limit $\bar{q}\bar{q}qq$ and glueball exchange count the same. Given the fact that glueballs are expected to have masses above 1 GeV, and that the κ is a natural $SU(3)$ partner of the σ, a dominant $\bar{q}\bar{q}qq$ component for the σ seems the most natural interpretation, although it could certainly have some glueball mixing.

The large N_c behavior of the $(0,0)$ amplitude in the vicinity of the $f_0(980)$ is shown in Fig.3. This resonance and the $a_0(980)$ are more complicated due to the distortions caused by the nearby $\bar{K}K$ threshold. We see that the characteristic sharp dip of the $f_0(980)$ vanishes when $N_c \to \infty$, at variance with a $\bar{q}q$ state. For $N_c > 5$ it follows again the $1/N_c^2$

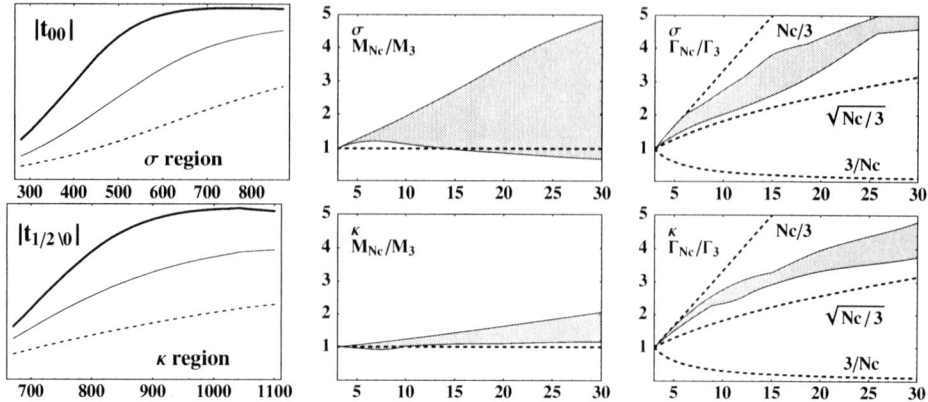

FIGURE 2. Top) Right: Modulus of the $(I,J) = (0,0)$ scattering amplitude, versus $\sqrt{s}$ for $N_c = 3$ (thick line), $N_c = 5$ (thin line) and $N_c = 10$ (dotted line), scaled at $\mu = 770\,\text{MeV}$. Center: N_c evolution of the σ mass. Left: N_c evolution of the σ width. Bottom: The same but for the $(1/2,0)$ amplitude and the κ.

scaling compatible with $\bar{q}\bar{q}qq$ states or glueballs. The $a_0(980)$ behavior, shown in Fig.4, is more complicated. When we apply the large N_c scaling at $\mu = 0.55 - 1$ GeV, its peak disappears, suggesting that this is not a $\bar{q}q$ state, and $\text{Im}\,t_{10}$ follows roughly the $1/N_c^2$ behavior in the whole region [1]. However, as shown in Fig.5, the peak does not vanish at large N_c if we take $\mu = 0.5\,\text{GeV}$. Thus we cannot rule out a possible $\bar{q}q$ nature, or a sizable mixing, although it shows up in an extreme corner of our uncertainty band. For other recent large N_c arguments in a chiral context see [14].

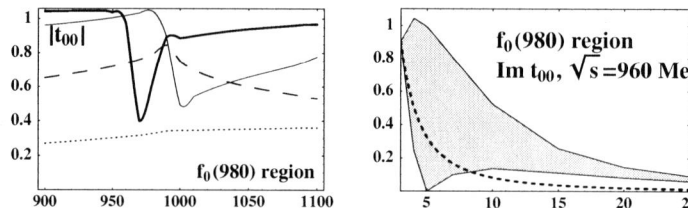

FIGURE 3. Right: Modulus of a $(I,J) = (0,0)$ scattering amplitude, versus $\sqrt{s}$, for $N_c = 3$ (thick), $N_c = 5$ (thin), $N_c = 10$ (dashed) and $N_c = 25$ (dotted), scaled at $\mu = 770\,\text{MeV}$. Left: $\text{Im}\,t_{00}$ versus N_c.

CONCLUSION

We have shown that the QCD large N_c scaling of the unitarized meson-meson amplitudes of Chiral Perturbation Theory is in conflict with a $\bar{q}q$ nature for the lightest scalars

---

[1] The idea of this work and the pole movements were presented by the author in two workshops [11]. While completing the calculations and the manuscript the results without the scale uncertainties have been confirmed [15] for all resonances, using the approximated IAM [8].

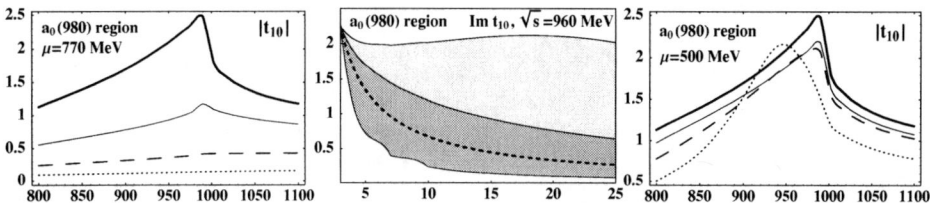

FIGURE 4. Right: Modulus of a $(I, J) = (1, 0)$ scattering amplitude, versus $\sqrt{s}$, for $N_c = 3$ (thick), $N_c = 5$ (thin), $N_c = 10$ (dashed) and $N_c = 25$ (dotted), scaled at $\mu = 770\,\text{MeV}$. Right: scaled at $\mu = 500\,\text{MeV}$. Center: Imt_{00} versus N_c. The dark gray area covers the uncertainty $\mu = 0.55 - 1\,\text{GeV}$, the light gray area from $\mu = 0.5$ to $0.55\,\text{GeV}$.

(not so conclusively for the $a_0(980)$), and strongly suggests a $\bar{q}\bar{q}qq$ or two meson main component, maybe with some mixing with glueballs, when possible.

ACKNOWLEDGMENTS

I thank A. Andrianov, D. Espriu, A. Gómez Nicola, F. Kleefeld, R. Jaffe, E. Oset, J. Soto and M. Uehara for their comments and support from the Spanish CICYT projects, BFM2000-1326, BFM2002-01003 and the E.U. EURIDICE network contract no. HPRN-CT-2002-00311.

REFERENCES

1. S. Weinberg, Physica **A96** (1979) 327. J. Gasser and H. Leutwyler, Annals Phys. **158** (1984) 142; Nucl. Phys. B **250** (1985) 465.
2. J. Bijnens, G. Colangelo and J. Gasser, Nucl. Phys. **B427** (1994) 427.
3. A. Gómez Nicola and J. R. Peláez, Phys. Rev. D **65** (2002) 054009 and AIP Conf. Proc. **660** (2003) 102 [hep-ph/0301049].
4. G. 't Hooft, Nucl. Phys. B **72** (1974) 461. E. Witten, Annals Phys. **128** (1980) 363.
5. A. A. Andrianov, Phys. Lett. B **157**, 425 (1985). A. A. Andrianov and L. Bonora, Nucl. Phys. B **233**, 232 (1984). D. Espriu, E. de Rafael and J. Taron, Nucl. Phys. B **345** (1990) 22 S. Peris and E. de Rafael, Phys. Lett. B **348** (1995) 539
6. A. Pich, hep-ph/0205030.
7. T. N. Truong, Phys. Rev. Lett. **61** (1988) 2526. Phys. Rev. Lett. **67**, (1991) 2260; A. Dobado, M.J.Herrero and T.N. Truong, Phys. Lett. **B235** (1990) 134. A. Dobado and J. R. Pelaez, Phys. Rev. D **47** (1993) 4883. Phys. Rev. D **56** (1997) 3057.
8. J. A. Oller, E. Oset and J. R. Pelaez, Phys. Rev. Lett. **80** (1998) 3452; Phys. Rev. D **59** (1999) 074001 and Phys. Rev. D **62** (2000) 114017. M. Uehara, hep-ph/0204020.
9. F. Guerrero and J. A. Oller, Nucl. Phys. B **537** (1999) 459 [Erratum-ibid. B **602** (2001) 641].
10. J. A. Oller and E. Oset, Nucl. Phys. A **620** (1997) 438; Phys. Rev. D **60** (1999) 074023.
11. J. R. Pelaez, hep-ph/0309292; hep-ph/0307018 and hep-ph/0306063.
12. M. Harada, F. Sannino and J. Schechter, hep-ph/0309206.
13. R. L. Jaffe, Proceedings of the Intl. Symposium on Lepton and Photon Interactions at High Energies. Physikalisches Institut, University of Bonn (1981) . ISBN: 3-9800625-0-3
14. V. Cirigliano *et al.* hep-ph/0305311. N. N. Achasov, hep-ph/0309118. T. Schaefer, hep-ph/0309158.
15. M. Uehara, hep-ph/0308241.

Observation of resonances in the reaction $\bar{p}p \to \eta\eta\pi^0$ at 5.2 GeV/c

Ismail Uman

Northwestern University, Evanston, IL 60208, USA
(Fermilab E835 Collaboration)

Abstract. Data from the Fermilab E835 experiment have been used to study the reaction $\bar{p}p \to \eta\eta\pi^0$ at 5.2 GeV/c. A sample of 22 million six photons events has been analyzed to construct the Dalitz plot containing $\sim 80k$ $\eta\eta\pi^0$ events. A partial wave analysis of the data has been done. At least five f_J-states decaying into $\eta\eta$ between ~ 1.4 and 2.3 GeV are observed. Two f_0 states are identified with the popular candidates for the lightest scalar glueball, $f_0(1500)$ and $f_0(1710)$. In addition, at least five a_J-states decaying into $\eta\pi^0$ up to 2.4 GeV are also observed. Masses, widths and spins of these resonances are determined by maximum likelihood analysis of the data and compared with the results in the literature.

INTRODUCTION

The analysis of antiproton-proton annihilation at antiproton momentum of 5.2 GeV/c into three pseudoscalar meson provides an attractive means of studying light mesons with mass in the range from 1.3 to 2.5 GeV/c^2. The annihilation into $\eta\eta\pi^0$ is a selective process in which only f_J-states ($I = 0$) decaying into $\eta\eta$ and a_J-states ($I = 1$) decaying into $\eta\pi^0$ with quantum numbers $J^{PC} = even^{++}$ can be formed. Theoretical calculations for states with angular momenta $L \leq 4$ predict five a_{even}^{++}-states and eleven f_{even}^{++}-states in this mass range [1]. Presently, only nine f_{even}^{++}-states and four a_{even}^{++}-states have been firmly identified [2].

In addition to the ordinary $q\bar{q}$ mesons *glueballs* (gg, ggg) and *hybrids* ($q\bar{q}g$) can be produced in antiproton-proton annihilations. Present lattice calculations predict the lowest scalar and tensor glueballs to have their masses in the same mass range as the light mesons [3]. The lowest glueball state with $J^{PC} \equiv 0^{++}$ is predicted at $M \simeq 1600$ MeV. It can mix with the $q\bar{q}$ members of isoscalar scalar mesons in the $1.3 \to 1.7$ GeV region. Thus $f_0(1370)$, $f_0(1500)$ and $f_0(1710)$ are considered as containing various level of glueball amplitudes [4].

The glueball candidate $f_0(1500)$ has been observed in J/ψ radiative decay, central production and proton-antiproton annihilation. The excitation of $f_0(1710)$ in proton-antiproton annihilation is still open to question.

A clear peak at $M(\eta\eta) = 1740$ MeV was observed by the Fermilab E760 experiment in the $\bar{p}p \to \pi^0\eta\eta$ reaction, but no J^{PC} determination was done [6]. Crystal Barrel $\bar{p}p$ annihilation measurements of $M(\eta\eta)$ at different $\bar{p}$ momenta have led to contradictory results [7],[8], [9].

Hybrid mesons ($q\bar{q}g$) can have exotic quantum numbers which are forbidden for $\bar{q}q$

CP717, *Hadron Spectroscopy: Tenth International Conference,*
edited by E. Klempt, H. Koch, and H. Orth
© 2004 American Institute of Physics 0-7354-0197-7/04/$22.00

mesons. States with $J^{PC} \equiv 1^{-+}$ can be produced inclusively in $\bar{p}p$ annihilation. The decay of a 1^{-+} hybrid candidate into $\eta\pi^0$ has been reported [10], [11].

DATA AND EVENT SELECTION

The E835 experiment is located in the Fermilab Antiproton Accumulator. The stored, cooled circulating $\bar{p}$ intersect a hydrogen cluster jet gas target, and the reaction products are detected in a detector system which surrounds the interaction region. Full details of detector performance can be found elsewhere [12]. Photons were detected in the central calorimeter which consists of 1280 lead glass blocks providing an angular coverage of $10.6^o < \theta < 70.0^o$ and 2π in ϕ. A forward calorimeter was used to veto events having clusters in the $3.3^o < \theta < 11.0^o$ region. A system of scintillator hodoscopes was used to reject charged particles in the angular range covered by the central calorimeter. The photon energy resolution was $\frac{\sigma_E(E)}{E} = \frac{0.06}{\sqrt{E(GeV)}} + 0.04$, and the angular resolution was $\sigma_\theta \sim 6$ mrad and $\sigma_\phi \sim 11$ mrad. Luminosity was measured by detecting recoil protons at $\sim 87.5^o$. The data were collected at $\bar{p}$ momenta of 5.2 GeV corresponding to center of mass energy range of $E_{cms} = 3.409 - 3.418$ GeV with a total luminosity of $\Sigma_L = 11.4 pb^{-1}$.

The reaction studied was $\bar{p}p \to \pi^0\eta\eta$, $\pi^0 \to 2\gamma$, $\eta_1 \to 2\gamma, \eta_2 \to 2\gamma$. Thus the final state consisted of six photons. Event selection was done with the following requirements.

Each $\pi^0\eta\eta$ event is required to satisfy a 7C kinematic fit with confidence level CL> 10%. Background events fitting $\pi^0\pi^0\pi^0$, $\eta\pi^0\pi^0$ and $\eta\eta\eta$ hypotheses were rejected if they have CL> 10^{-2}. These 'anticuts' have a non-negligible effect in the Dalitz plot. They reduced the statistics by about 32%, 7.2% and 1.5%, respectively. These events are mainly distributed in the upper edge of the Dalitz plot shown in Fig. 1. The effect of possible contamination due to these channels is negible for resonances having masses higher than 1.3 GeV.

A full Monte Carlo simulation of the detector using GEANT program has been used. About 5.5 million phase space $\bar{p}p \to \eta\eta\pi^0$ events were generated and submitted to the same cuts as the data events. The overall efficiency for event selection was found to be about 4.2%.

Additional background studies of events with 7 and 8 photon final states were performed. The worst background arises from $\omega\eta\pi^0$ and $\omega\pi^0\pi^0$ with one missing photon and from $4\pi^0$ with two photons lost. These backgrounds were estimated from the same sample of data using the same procedure that was used to select $\eta\eta\pi^0$ events. An additional MC simulation of 1 million $\omega\eta\pi^0$, $\omega\eta\pi^0$ and $4\pi^0$ events is used to estimate their relative feeddown in the $\eta\eta\pi^0$ channel and found to be 5.2%, 5.4% and 4.1%, respectively. These backgrounds can influence those resonances having masses lower than 1.3 GeV but are expected to have neglible effect at higher masses. After all cuts 83400 data events and 236275 MC events were left for the final analysis. Because of 2-fold symmetry the Dalitz plot and the projections shown in Figs. 1 and 2 have 166800 entries. A visual examination of Fig. 1 reveals the major characteristics of the f_J and a_J resonances. Complex structures are evident in the central part of the Dalitz Plot which may be due to $f_4(2050)$, $f_2(2150)$, $f_0(2100)$ and $f_4(2300)$. Diagonal bands presumably

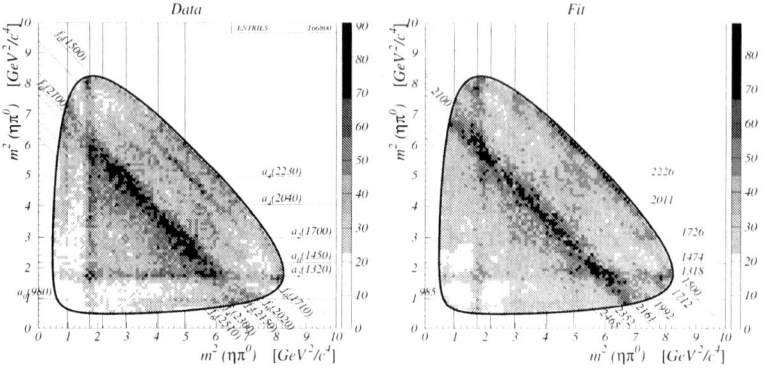

FIGURE 1. Dalitz plot (data and fit) of $\eta\eta\pi^0$ states at 3.42 MeV/c^2. The positions of quoted resonances are indicated in MeV.

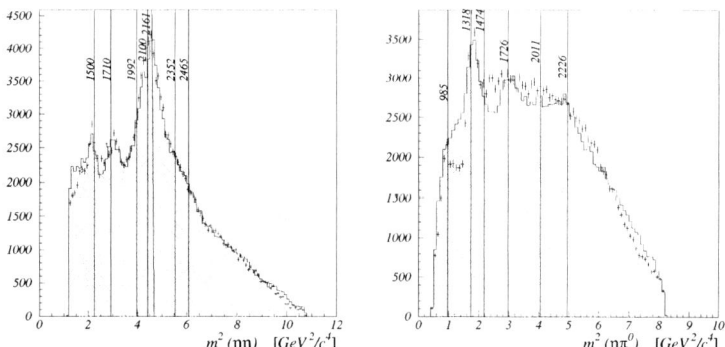

FIGURE 2. $\eta\eta$ (b) and $\eta\pi^0$ (c) mass projections. Curves are from the best fit. The positions of quoted resonances are indicated in MeV.

due to $f_0(1500)$ and/or $f'_2(1525)$ and $f_0(1710)$, and crossing bands corresponding to $a_2(1320)$ are clearly visible in the Dalitz Plot as well as in the projections (Fig. 2). Close to the lower edges horizontal and vertical bands are presumably due to $a_0(980)$. PDG also lists [2] a broad $a_4(2040)$ decaying into $\eta\pi^0$. Altough not visible by eye, the partial wave analysis described below also indicates the presence of $a_0(1450)$ and $a_2(1700)$.

PWA FORMALISM

The transition to the $\eta\eta\pi^0$ state in the isobar model is described in two steps. In the first step an f_J or a_J-state at an angle (θ, ϕ) with respect to the beam direction and a recoiling π or η (spectator) meson are produced. In the second step the f_J or a_J-state decays in two other mesons. A sequence of rotations and transformations to the rest frame of the resonance allows us to parametrise the decay of every resonance by helicity amplitudes

$$A^\lambda = G_\lambda \frac{Y_J^\lambda(\alpha,\beta)exp(i\delta_\lambda)}{M^2 - s - iM\Gamma} \tag{1}$$

where s is the invariant mass, M and Γ are the mass and width of the resonance. The spherical harmonics Y_J^λ for spin J and helicity λ depend exclusively on the decay angles α,β. In the denominator a relativistic Breit-Wigner amplitude with a constant width Γ is assumed. G_λ and the δ_λ are the magnitude and the phase of the complex coupling constant. Because of the many initial states a full description of the production process is not possible. However because of C-parity invariance the differential cross section is forward-backward symmetric. Therefore the production process of every resonance is approximated multiplying the relative amplitudes with polynomials in even powers of $cos\theta$. Since there are many initial partial waves two resonances, say f_J and a_J are expected to be only be partially coherent. In this case the intensity w is given by the equation

$$w = \sum_\lambda [|A_{f_J}^\lambda|^2 + |A_{a_J}^\lambda|^2 + +2c_\lambda \Re(A_{f_J}^\lambda A_{a_J}^{\lambda*})]$$

where the interference coefficient c_λ is constrained to lie within the range 0 (no coherence) and ± 1 (full coherence). The free parameters of the fit $(G_\lambda, \delta_\lambda, c_\lambda)$ are optimized in order to give the maximum value of the negative log-likelihood. A detailed description of this formalism may be found in [8].

RESULTS

To start with, we perform an initial fit including the channels $\bar{p}p \rightarrow f_0(1500)\pi^0$, $f_0(1710)\pi^0$, $f_4(2050)\pi^0$, $f_2(2150)\pi^0$, $f_4(2300)\pi^0$, $a_2(1700)\pi^0$, $a_2(1700)\pi^0$ and $a_4(2050)\eta$ with masses and widths fixed to PDG values. Amplitudes with $|\lambda| > 1$ are found to be negligible and the fit is obtained with the optimization of 13 parameters with a log-likelihood of $ln_L = -11661$.

An improved fit is then obtained by adding to the basic model space the $a_0(1450)$, $f_6(2510)$ and a new state $a_4(2230)$ and optimizing the production process of all states. Self-interference of $a_4(2040)$, $a_2(1320)$ and $a_4(2230)$ result in a large improvement in the log-likelihood by $\Delta ln_L / \Delta\#par. = 808/6$. Interference between $a_4(2040)/a_0(1450)$, $a_4(2040)/f_2(2150)$, $a_0(1450)/a_2(1320)$ are also found to be significant. Preliminary mass and width scans and spin test are then performed. The best fit (see Figs.1,2) is obtained by dropping $f_4(2050)$ from the fit (its decay branching ratio to $\eta\eta$ is only $\sim 0.1\%$) and adding $f_0(2020)$ and $f_0(2100)$.

Mass, width and spin tests are repeated for all resonances and the results are listed in Table 1. The optimization of the $f_0(1370)$ was not possible because it lies near the lower end of $\eta\eta$ phase space. The optimization results for a selection of resonances are shown in Figures 3, 4 and 5.

Additional resonances which were tried are $a_0(980)$, $f_2'(1525)$, $f_2(1640)$ $f_2(1910)$ and $\pi_1(1400)$. $a_0(980)$, $f_2'(1525)$, $f_2(1640)$, $f_2(1910)$ did not show significant contri-

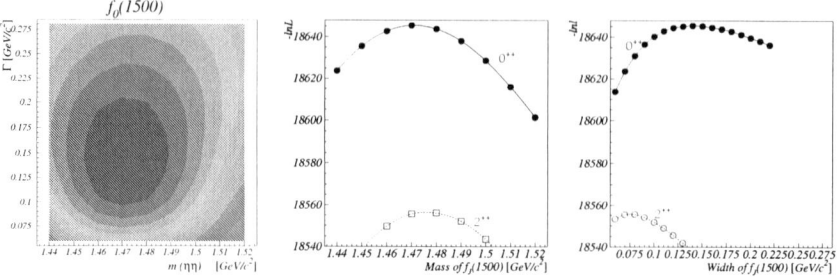

FIGURE 3. Optimization of $f_0(1500)$. The optimun is reached for $J = 0$ at a mass of $M = 1472 \pm 5$ MeV and width $\Gamma = 143 \pm 15$ MeV. $J = 2$ is excluded by a change of >89 in the log-likelihood.

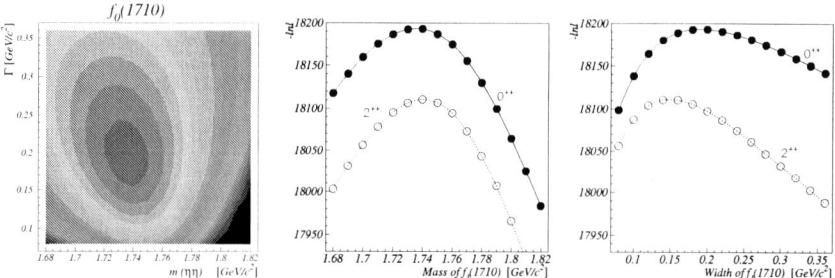

FIGURE 4. Optimization of $f_0(1740)$. The optimun is reached for $J = 0$ at a mass of $M = 1734 \pm 6$ MeV and width $\Gamma = 196 \pm 11$ MeV. $J = 2$ is excluded by a change of >83 in the log-likelihood.

bution ($< 1\%$). No convergence was obtained when the reported hybrid $\pi_1(1400)$ was included in the fits.

The 0^{++} state at $M = 1734 \pm 6$ MeV and $\Gamma = 196 \pm 11$ MeV can be identified with the $f_0(1710)$. We find two scalars which are presumably the $f_0(2020)$ and $f_0(2100)$, and a

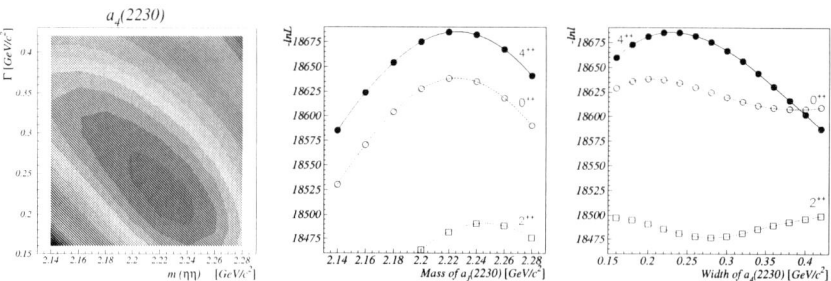

FIGURE 5. Optimization of $a_J(2230)$. The optimun is reached for $J = 4$ at a mass of $M = 2226 \pm 6$ MeV and width $\Gamma = 234 \pm 11$ MeV. $J = 0, 2$ are excluded by a change of $>47, 194$ in the log-likelihood, respectively.

TABLE 1. Preliminary results for masses, widths and J^{PC} of light quark resonances decaying into $\eta\eta$ and $\eta\pi^0$ as seen in $\bar{p}p \rightarrow \pi^0\eta\eta$ at 5.2 GeV/c. All errors are statistical only. Asterisks* mark states presently omitted from meson summary list of PDG'02. Daggers† mark f-states which have not been seen in $\eta\eta$ before, and a-states not seen in $\eta\pi$ before. Errors are statistical only.

PDG	Mass(MeV)			Width(MeV)		Intensity
Resonance	PDG'02	E835p	J^{PC}	PDG	E835p	%
$f_0(1500)$	1507 ± 5	1472 ± 5	0^{++}	109 ± 7	143 ± 15	2.8
$f_0(1710)$	1713 ± 6	1734 ± 4	0^{++}	125 ± 12	196 ± 11	6.2
$f_0(2020)^{*\dagger}$	1992 ± 16	2023 ± 7	0^{++}	442 ± 60	218 ± 18	5.9
$f_0(2100)^{*\dagger}$	$2060 - 2120$	2100 ± 6	0^{++}	$175 - 430$	308 ± 18	15.8
$f_2(2150)^*$	2159 ± 11	2138 ± 4	2^{++}	167 ± 30	161 ± 10	8.6
$f_4(2300)^{*\dagger}$	$2300 - 2340$	2352 ± 8	4^{++}	$150 - 280$	247 ± 17	7.4
$f_6(2510)^*$	2465 ± 50	2484 ± 14	6^{++}	255 ± 40	181 ± 25	0.8
$a_2(1320)$	$1318. \pm 1.$	1330 ± 2	2^{++}	$111. \pm 2.5$	71 ± 3	2.5
$a_0(1450)$	1474 ± 19	1350 ± 2	0^{++}	265 ± 13	135 ± 5	7.5
$a_2(1700)^*$	1726 ± 26	1740 ± 7	2^{++}	256 ± 40	345 ± 16	12.4
$a_4(2040)$	2011 ± 13	1986 ± 5	4^{++}	360 ± 40	302 ± 10	12.2
$a_4(2230)^{*\dagger}$	New	2226 ± 6	4^{++}	New	234 ± 11	18.4

$f_4(2300)$ decaying to $\eta\eta$. We confirm the $f_2(2150)$, $f_6(2510)$ and $a_2(1700)$ resonances. In addition a new $a_4(2230) \rightarrow \eta\pi^0$ state has been observed with $M = 2226 \pm 6$ MeV and $\Gamma = 234 \pm 11$ MeV. We note that this is the first time $f_0(1500)$ and $f_0(1710)$ are observed in the same $\bar{p}p$ experiment. For both of these our widths are larger than the averages reported in PDG [2]. We are continuing to investigate these and other features of our results. Among the improvement being investigated are the use of mass dependent widths.

REFERENCES

1. S. Godfrey and N. Isgur, Phys. Rev. D **32**, 189 (1985).
2. K. Hagiwara *et al.* [Particle Data Group Collaboration], Phys. Rev. D **66**, 010001 (2002).
3. C. J. Morningstar and M. J. Peardon, Phys. Rev. D **60**, 034509 (1999)
4. C. Amsler and F. E. Close, Phys. Rev. D **53**, 295 (1996); also
   F. E. Close and A. Kirk, Eur. Phys. J. C **21**, 531 (2001).
5. C. Amsler and F. E. Close, Phys. Lett. B **353**, 385 (1995)
6. T. A. Armstrong *et al.* [E760 Collaboration], Phys. Rev. D **48**, 3037 (1993).
7. A. V. Anisovich *et al.*, Phys. Lett. B **449**, 154 (1999).
8. A. Abele *et al.* [Crystal Barrel Collaboration], Eur. Phys. J. C **8**, 67 (1999).
9. C. Amsler *et al.* [Crystal Barrel Collaboration], Eur. Phys. J. C **23**, 29 (2002).
10. D. R. Thompson *et al.* [E852 Collaboration], Phys. Rev. Lett. **79**, 1630 (1997); also
    S. U. Chung *et al.* [E852 Collaboration], Phys. Rev. D **60**, 092001 (1999)
11. A. Abele *et al.* [Crystal Barrel Collaboration], Phys. Lett. B **423**, 175 (1998); also
    Phys. Lett. B **446**, 349 (1999).
12. G. Garzoglio *et al.*, Nucl. Instrum. Meth. A (in press).

An Analysis of the Low-lying States of Hadrons

H.Noya[*] and H.Nakamura[†]

[*]Institute of Physics, Faculty of Economics, Hosei University at Tama, Machida, Tokyo 194-0298, Japan, hnoya@mt.tama.hosei.ac.jp
[†]2-25-3, Moridai, Atsugi, Kanagawa 243-0037, Japan

Abstract. We developed the new mass formula to obtain the exotic baryon mass spectrum which was observed by Tatischeff et al.

In an earlier paper[1], we made an analysis based on the two center quark model (TQM) for the masses spectra of the low-lying states of the dibaryon and the exotic mesons, which have been found by Tatischeff et al. recently [2-4]. We assumed that they are the multiquark systems $q^k \bar{q}^h$ which consist of a quark q^k and an antiquark $\bar{q}^h$.

It is also assumed that the particles of those clusters are not in spatially excited state and then, by the virtue of the Pauli principle, the following mass formula with two parameters $M_0(k,h)$ and $M_1(k,h)$ holds as used in the Tatischeff's analysis. The formula is the following.

$$M = M_0(k,h) + M_1(k,h)[i_1(i_1+1) + 1/3 s_1(s_1+1) + i_2(i_2+1) + 1/3 s_2(s_2+1)] \quad (1)$$

where i_1 and s_1 (i_2 and s_2) represent the isospin and the spin of the quark cluster q^k (the antiquark cluster $\bar{q}^h$), respectively. The contribution from the color state of clusters was neglected. Assuming a very simple function for $M_0(k,h)$ and $M_1(k,h)$ to economize the free parameters, we calculated the mass spectra of $q^2\bar{q}^2$, $q^3 \bar{q}^3$, $q^4 \bar{q}^4$, $q^5 \bar{q}^5$, $q^7 \bar{q}$, $q^8 \bar{q}^2$, and $q^9 \bar{q}^3$ systems and compared with the data. The results were sufficiently good through many of the states with small i_1 and /or i_2 were still missing.

In this short article, we present a more reliable phenomenological analysis with the following assumptions.
(A) The color states of clusters are limited to 3 or 3^* state and then clusters q^3 and $\bar{q}^3$ must be eliminated.
(B) In general, the polarization of isospin holds and then the states with the maximum or nearly maximum $i_{1,2}$ are easily observable.
(C) The parameter $M_1(k,h)$ for the cluster q^k ($\bar{q}^h$) is the function of the particle number k (h) only.

With (C), the mass formula (1) is modified as

$$M = U(k,h) + M_1(k)f_1 + M_2(h)f_2 \quad (2)$$

CP717, Hadron Spectroscopy: Tenth International Conference,
edited by E. Klempt, H. Koch, and H. Orth

$$f_1 = i_1(i_1 + 1) + 1/3s_1(s_1 + 1) - k \tag{3}$$
$$f_2 = i_2(i_2 + 1) + 1/3s_2(s_2 + 1) - h \tag{4}$$

First, we analyze the mass spectra of the low-lying narrow exotic mesons below 800 MeV and low-lying dibaryon at 1880 - 1910 MeV assuming that they are $q^k \bar{q}^k$ (k = 2,4,5) and $q^7 \bar{q}$ states, respectively, using two parameters $U(k,h), M_1(k) f_1$ and $U(7,1), M_1(7)$.

TABLE 1. Table for $M_1(k)$ unit in MeV

k	2	4	5	7
$M_1(k)$	21	27	15	5.3

TABLE 2. Table for U(k,h) unit in MeV

(k,h)	(2,2)	(4,4)	(5,5)	(7,1)	(8,2)	(10,4)
U(k,h)	394	573	640	1917	1958	2050

The results for the optimum values shown in Table 1,2 are illustrated in Fig. 1,2. Applying (B) for $q_4\bar{q}^4$ and $q^5\bar{q}^5$, only the high-lying states with $i_1 = i_{max}$ and $i_2 = i_{max}, i_{max} - 1$ are illustrated by solid lines, where i_{max} represents the maximum isospin in the cluster.

As seen in Table 1, it seems that the function $M_1(k)$ decreases rapidly at around k = 7, indicating that the contribution from the color-magnetic vanishes for large particle number.

Therefore, hereafter we take

$$M_1(k) = 0 \quad \text{for} \quad k \geq 8. \tag{5}$$

In the second stage, we analyze the mass spectrum of the dibaryon at 1910-2130 MeV assuming they are $q^8\bar{q}^2$ and $q^{10}\bar{q}^4$. Note that the free-parameters in this analysis are U(8,2) and U(10,4). The results for the optimum values given in Table 2 are illustrated in Fig.3 together with the mass spectrum of the narrow dibaryon resonance. There is no adequate state for the dibaryon at 1941 MeV. We think that this dibaryon is a loose bound state of a nucleon and a narrow nucleon excited state $N^*(1004)$, which was found by Tatischeff et al.[5]. For reference, the results of the DCM analysis ($\Delta_{01} = 0$) [6] for the high-lying dibaryon states are also illustrated in the right-hand side of Fig.3.

The states at 2140 and 225) MeV correspond to a well-know two broad dibaryon resonance with $J^P = 2^+$ and 3^-, respectively. The single state at 2070 MeV with $I(J^P)$ = 0(0$^-$) configuration $[(1P_{1/2})(1S_{1/2})]^3$ and TS-combination [10][10][10] are assigned to a narrow dibaryon resonance d' at 2065 MeV.[7] The state with dashed line are the anomalous state in which two quarks in a diquark are excited to the $1P_{1/2}$ shell.

As one can see in Figures 1-3, a satisfactory good result has been obtained in the present analysis, which indicates that our assumption from (A) to (C) is correct. Further, the result for $M_1(k)$ suggests strongly that the color-magnetic interaction in a cluster vanishes for high particle number. We can estimate the energy E_q for a qq pair production in a multiquark system from the empirical value of the system $U(k,h)$. If the cluster is a simple assembly of quarks, the E_q must be 2m($\sim$ 600MeV), where m is the effective

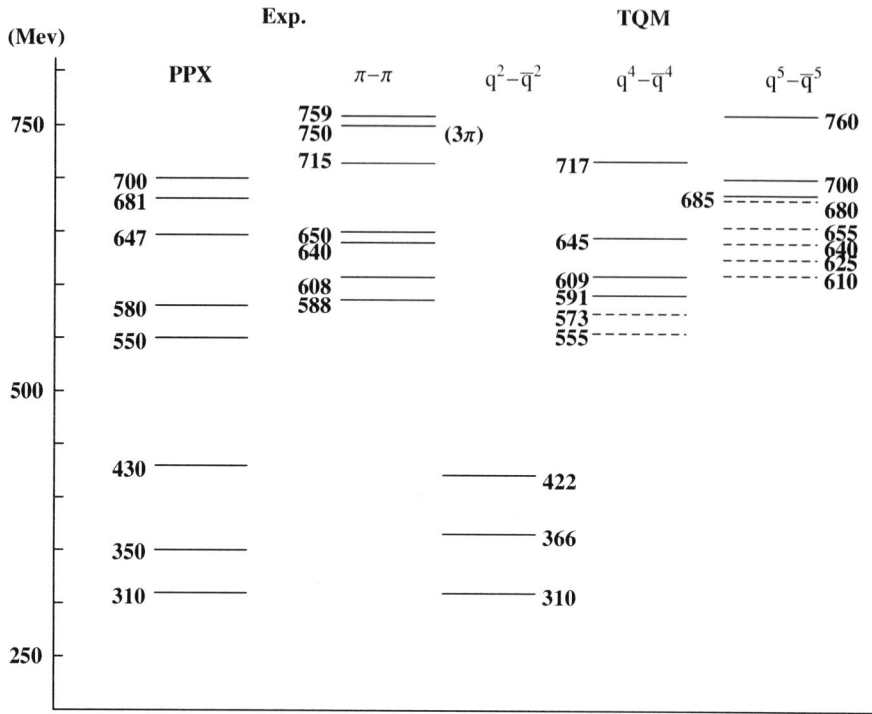

FIGURE 1. The mass spectrum of the narrow mesonic resonances below 1 Gev including the TQM results.

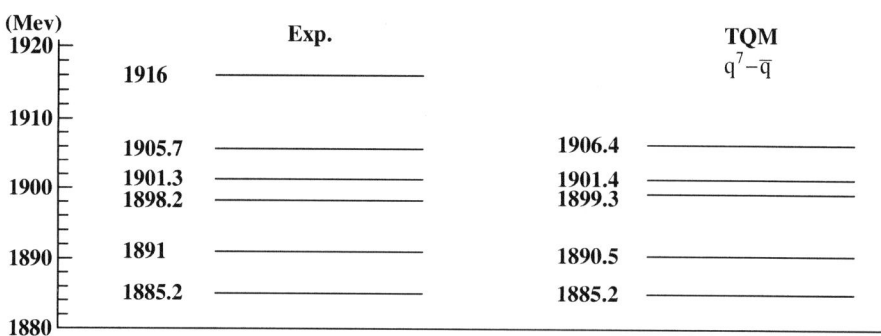

FIGURE 2. The mass spectrum of the narrow dibaryon resonances below 2 Gev including the TQM results.

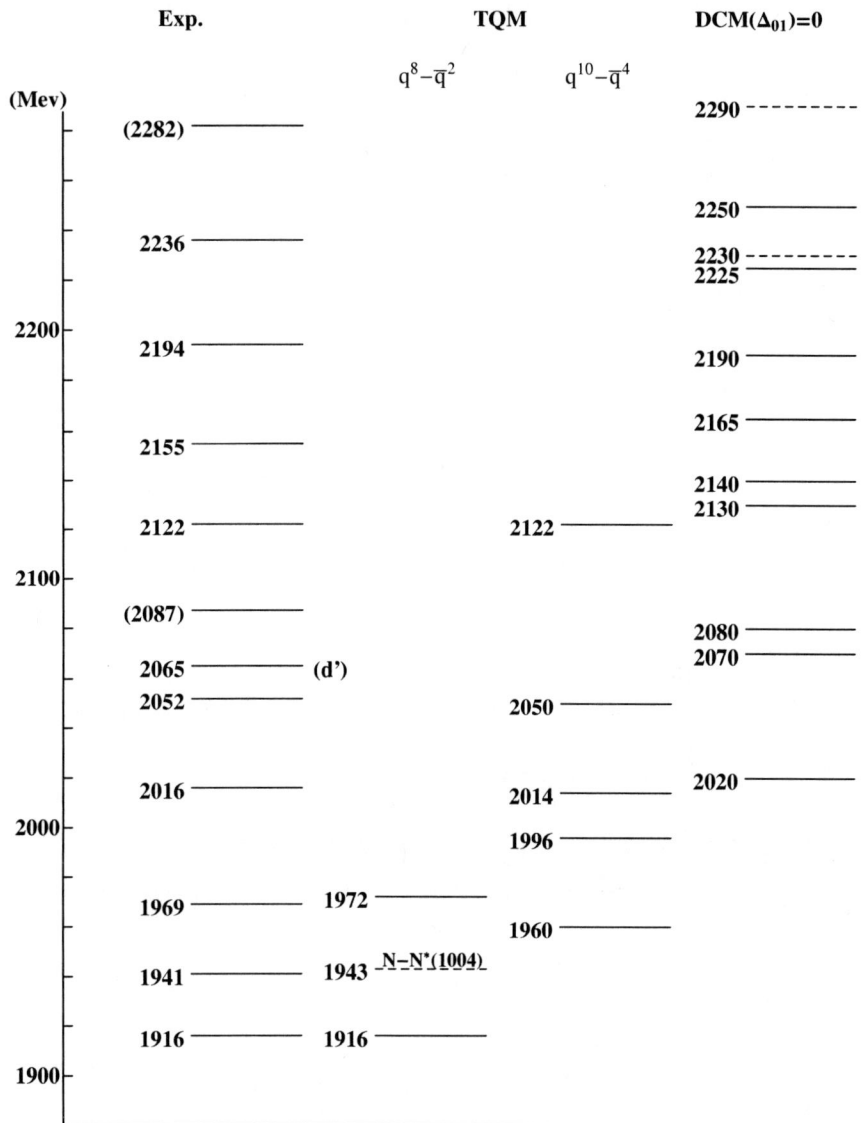

FIGURE 3. The mass spectrum of the narrow dibaryon resonances above 2 Gev including the TQM and the DCM results.

Exp.: B.Tatischeff et al. Phys.Rev.C59(1999)1878

DCM: Diquark Cluster Model Calculation

N.Konno.H.Nakamura and H.Noya Phys.Rev.D35(1987)239

mass of u and d quarks. But, strikingly, the estimated E_q-value is only 70-90 MeV for the exotic meson and 44 MeV for the dibaryon. Although we have treated only very specific small system in the present analysis, the obtained small Eq-value suggests strongly that, in the qq-pair production in a high energy multiquark system, a large reduction of the effective quark mass occurs. However we should note that this fact has very important meaning in the formation of our matter and antimatter world in the early universe.[8]

The good gained by results obtained in the present analysis strongly suggest that "the polarization of quanta"occurs in the high-energy reactions, not only the polarization of baryon (quark) number but also of isospin, though the firm evidence for the polarization of the spin was not obtained.

REFERENCES

1. H.Noya and H.Nakamura : 9th International Conference of the Hadron
Spectroscopy: Edt.D.Amelin and A.M.Zaitsev. AIP conference 619(2001) pp.533
2. J.Yonnet et al.:Phys.Rev.C63(2000)014001
3. B.Tatischeff et al.:Phys.Rev.C62(2000)054001
4. B.Tatischeff and J.Yonnet:Invited talk at the International Workshop
"Relativistic Nuclear Physics: From hundreds MeV to TeV"
Stara Lesna, Jun.26-July 1,2000
5. B.Taschiteff et al. Eur.Phys.J.A17(2003)245
6. H.Noya,N.Konno,Y.Uehara and H.Nakamura:Nucl.Phys.B(Proc.Suppl.)21(1991)335
7. R.Biliger,A.Clement and M.G.Schepkin:Phys.Rev.Lett.71(1993)42
8. H.Noya and H.Nakamura:Talk at the International Conference LEAP03

Observation of $K_s^0 K_s^0$ resonances in deep inelastic scattering at HERA

M. Barbi

McGill University, Physics Department
Montreal, Quebec
Canada
(on behalf of the ZEUS Collaboration)

Abstract. Inclusive $K_s^0 K_s^0$ production in deep inelastic ep scattering at HERA has been studied with the ZEUS detector using an integrated luminosity of 120 pb^{-1}. Two states are observed at masses of 1537^{+9}_{-8} MeV and 1726 ± 7 MeV, as well as an enhancement around 1300 MeV. The state at 1537 MeV is consistent with the well established $f_2'(1525)$. The state at 1726 MeV may be the glueball candidate $f_0(1710)$. However, it's width of 38^{+20}_{-14} MeV is narrower than the PDG value of 125 ± 10 MeV for the $f_0(1710)$.

INTRODUCTION

The $K_s^0 K_s^0$ system is expected to couple to scalar and tensor glueballs. This has motivated intense experimental and theoretical study during the past few years [1, 2]. Lattice QCD calculations [3, 4] predict the existence of a scalar glueball with a mass of 1730 ± 100 MeV and a tensor glueball at 2400 ± 120 MeV. The scalar glueball can mix with $q\bar{q}$ states with $I = 0$ from the scalar meson nonet, leading to three $J^{PC} = 0^{++}$ states whereas only two can fit into the nonet. Experimentally, four states with $J^{PC} = 0^{++}$ and $I = 0$ have been established [5]: $f_0(980)$, $f_0(1370)$, $f_0(1500)$ and $f_0(1710)$.

The state most frequently considered to be a glueball candidate is $f_0(1710)$ [5], but its gluon content has not yet been established. This state was first observed in radiative J/ψ decays [6] and its angular momentum $J = 0$ was established by the WA102 experiment using a partial-wave analysis in the K^+K^- and $K_s^0 K_s^0$ final states[7] . Observation of $f_0(1710)$ in $\gamma\gamma$ collisions may indicate a large quark content. A recent publication from L3 [8] reports the observation of two states in $\gamma\gamma$ collisions above 1500 MeV, the well-established $f_2'(1525)$ [5] and a broad resonance at 1760 MeV. It is not clear if the latter state is the $f_0(1710)$.

The ep collisions at HERA provide an opportunity to study resonance production in a new environment. Production of K_s^0 particles has been studied previously at HERA [9, 10, 11]. In this contribution, the first observation of resonances in the $K_s^0 K_s^0$ final state in inclusive deep inelastic ep scattering (DIS) is reported [12].

CP717, *Hadron Spectroscopy: Tenth International Conference,*
edited by E. Klempt, H. Koch, and H. Orth
© 2004 American Institute of Physics 0-7354-0197-7/04/$22.00

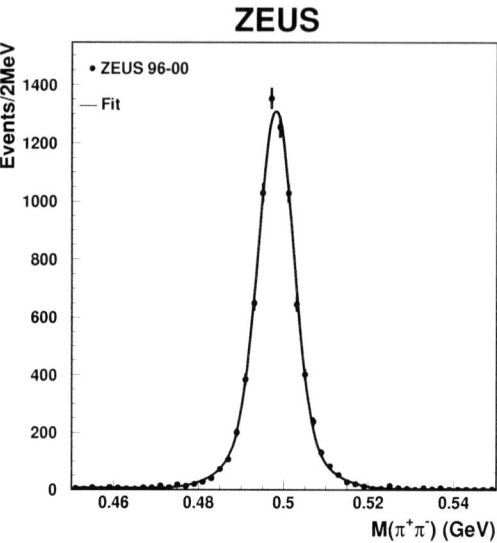

ZEUS

• ZEUS 96-00

— Fit

Events/2MeV

$M(\pi^+\pi^-)$ (GeV)

FIGURE 1. The distribution of $\pi^+\pi^-$ invariant mass for events with at least two K_s^0 candidates passing all selection cuts. The solid line shows the result of a fit using one linear and two Gaussian functions.

EVENT SELECTION AND K_S^0-PAIR CANDIDATES

A detailed description of the ZEUS detector can be found elsewhere [13].

The data used for this study correspond to a total integrated luminosity of 120 pb^{-1} collected in ZEUS during the 1996-2000 running period.

The inclusive neutral current DIS process $e(k) + p(P) \rightarrow e(k') + X$ can be described in terms of the following variables: the negative of the invariant-mass squared of the exchanged virtual photon, $Q^2 = -q^2 = -(k-k')^2$; the fraction of the lepton energy transferred to the proton in the proton rest frame, $y = (q \cdot P)/(k \cdot P)$; and the Bjorken scaling variable, $x = Q^2/(2P \cdot q)$.

A three-level trigger system was used to select events online [13]. The inclusive DIS selection was defined by requiring an electron found in the Uranium Calorimeter, and further requirements were applied to ensure a well defined data sample [12].

Oppositely charged track pairs reconstructed by the ZEUS central tracking detector (CTD) and assigned to a secondary vertex were selected and combined to form K_s^0 candidates. Both tracks were assigned the mass of a charged pion and the invariant-mass $M(\pi^+\pi^-)$ was calculated. Only events with at least one pair of K_s^0 candidates were selected. The invariant mass of the K_s^0 pair candidate $M(K_s^0, K_s^0)$ was reconstructed in the range $0.995 < M(K_s^0 K_s^0) < 2.795$ GeV. A detailed description of the K_s^0 pair candidate selection can be found in [12].

Figure 1 shows the distribution in x and Q^2 of selected events containing at least one pair of K_s^0 candidates.

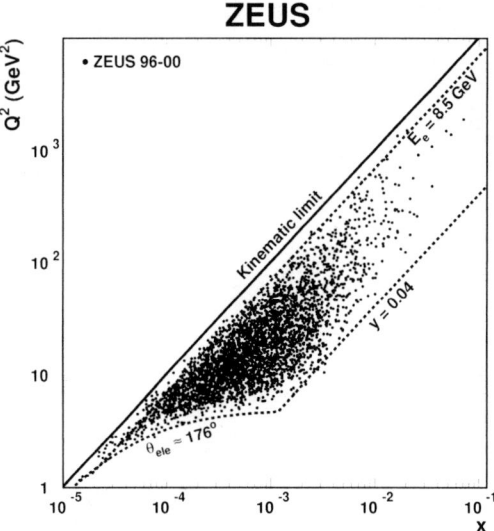

FIGURE 2. The distribution in x and Q^2 of events passing all selection cuts. The dashed lines delineate approximately the kinematic region selected. The solid line indicates the kinematic limit for HERA running with 920 GeV protons.

Figure 2 shows the $M(\pi^+\pi^-)$ distribution in the range $0.45 < M(\pi^+\pi^-) < 0.55$ GeV after the K_s^0 pair candidate selection.

RESULTS

The $K_s^0 K_s^0$ spectrum may have a strong enhancement near the $K_s^0 K_s^0$ threshold due to the $f_0(980)/a_0(980)$ state [14, 15, 16]. Since the high $K_s^0 K_s^0$ mass is the region of interest for this analysis, the complication due to the threshold region is avoided by imposing the cut $cos\theta_{K_s^0 K_s^0} < 0.92$, where $\theta_{K_s^0 K_s^0}$ is the opening angle between the two K_s^0 candidates in the laboratory frame.

After applying all selections, 2553 K_s^0-pair candidates were found in the range $0.995 < M(K_s^0 K_s^0) < 2.795$ GeV, where $M(K_s^0 K_s^0)$ was calculated using the K_s^0 mass of 497.672 MeV [5]. The momentum resolution of the CTD leads to an average $M(K_s^0 K_s^0)$ resolution which ranges from 7 MeV in the 1300 MeV mass region to 10 MeV in the 1700 MeV region. Figure 3 shows the measured $K_s^0 K_s^0$ invariant-mass spectrum. Two clear peaks are seen, one around 1500 MeV and the other around 1700 MeV, along with an enhancement around 1300 MeV. The data for $cos\theta_{K_s^0 K_s^0} > 0.92$ are also shown.

The distribution of Fig. 3 was fitted using three modified relativistic Breit-Wigner (MRBW) distributions and a background function $U(M)$;

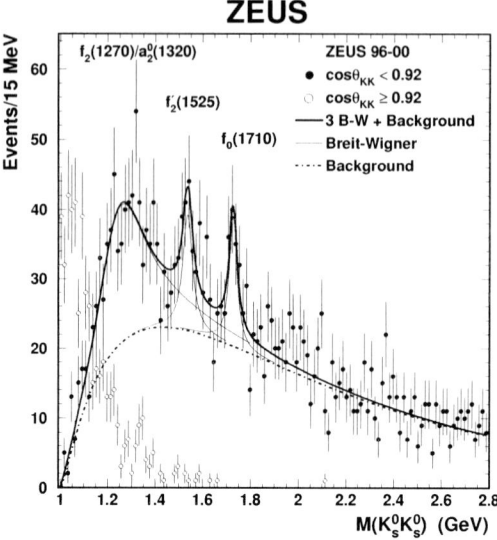

ZEUS

FIGURE 3. The $K_s^0 K_s^0$ invariant-mass spectrum for K_s^0 pair candidates with $cos\theta_{K_s^0 K_s^0} < 0.92$ (filled circles). The thick solid line is the result of a fit using three Breit-Wigners (thin solid lines) and a background function (dotted-dashed line). The K_s^0 pair candidates that fail the $cos\theta_{K_s^0 K_s^0} < 0.92$ cut are also shown (open circles).

$$F(M) = \sum_{i=1}^{3} \left(\frac{m_{*,i} \Gamma_{d,i}}{(m_{*,i}^2 - M^2)^2 + m_{*,i}^2 \Gamma_{d,i}^2} \right) + U(M) , \tag{1}$$

where $\Gamma_{d,i}$ is the effective resonance width, which takes into account spin and large width effects [17], $m_{*,i}$ is the resonance mass, and M is $K_s^0 K_s^0$ invariant mass. The background function is

$$U(M) = A \cdot (M - 2m_{K_s^0})^B \cdot e^{-C\sqrt{M - 2m_{K_s^0}}} , \tag{2}$$

where A, B and C are free parameters and $m_{K_s^0}$ is the K_s^0 mass defined by Hagiwara et al [5]. Monte Carlo studies showed that effects of the track-momentum resolution on the mass reconstruction were small compared to the measured widths of the states. Therefore the resolution effects were ignored in the fit.

Below 1500 MeV, a region strongly affected by the $\cos\theta_{K_s^0 K_s^0}$ cut, a peak is seen around 1300 MeV where a contribution from $f_2(1270)/a_2^0(1320)$ is expected. This mass region was fitted with a single Breit-Wigner.

Above 1500 MeV, the lower-mass state has a fitted mass of 1537^{+9}_{-8} MeV and a width of 50^{+34}_{-22} MeV, in good agreement with the well established $f_2'(1525)$. The higher-mass state has a fitted mass of 1726 ± 7 MeV and a width of 38^{+20}_{-14} MeV. The widths reported here were stable, within statistical errors, to a wide variation of fitting methods including those using 30 MeV bins rather than the default 15 MeV bins. The width is narrower than the PDG value (125 ± 10 MeV) [5] reported for $f_0(1710)$, but when it is fixed to this value, the fit is still acceptable.

It was found that 93% of the K_s^0-pair candidates selected within the detector and trigger acceptance are in the target region of the Breit frame [18, 19], the hemisphere containing the proton remnant. Of the K_s^0-pair candidates in the target region, 78% are in the region $x_p = 2p_B/Q > 1$, where p_B is the absolute momentum of the $K_s^0 K_s^0$ in the Breit frame. High x_p corresponds to production of the K_s^0-pair in a region where sizeable initial state gluon radiation may be expected. This is in contrast to the situation at $e^+ e^-$ colliders where the particles entering the hard scattering are colourless.

CONCLUSIONS

The first observation in ep deep inelastic scattering of a state at 1537 MeV, consistent with $f_2'(1525)$, and another at 1726 MeV, close to $f_0(1710)$, is reported. There is also an enhancement near 1300 MeV which may arise from the production of $f_2(1270)$ and/or $a_2^0(1320)$ states. The width of the state at 1537 MeV is consistent with the PDG value for the $f_2'(1525)$. The state at 1726 MeV has a mass consistent with the glueball candidate $f_0(1710)$, and is found in a gluon-rich region of phase space. However, it's width of 38^{+20}_{-14} MeV is narrower than the PDG value of 125 ± 10 MeV for the $f_0(1710)$.

REFERENCES

1. S. Godfrey, J. Napolitano, Rev. of Modern Phys. 71 (1999) 1411.
2. E. Klempt, Preprint hep-ex/0101031 (2000).
3. C.J. Morningstar, M. Peardon, Phys. Rev. D 60 (1999) 034509.
4. C. Michel and M. Teper, Nucl. Phys. B 314 (199) 347.
5. K. Hagiwara et al., Phys. Rev. D 66 (2002) 1.
6. BES Coll., J. Z. Bai et al., Phys. Rev. Lett. 77 (1996) 3959.
7. WA102 Coll., D. Barberis et al., Phys. Lett. B 453 (1999) 305.
8. L3 Coll., M. Acciari et al., Phys. Lett. B 501 (2001) 173.
9. ZEUS Coll., J. Breitweg et al., Eur. Phys. J. C 2 (1998) 77.
10. ZEUS Coll., M. Derrick et al., Z. Phys. C 68 (1995) 29.
11. H1 Coll., S. Aid et al., Nucl. Phys. B 480 (1996) 3.
12. ZEUS Coll., Preprint hep-ex/0308006 (2003).
13. ZEUS Coll., U. Holm (ed.), *The ZEUS Detector.* Status Report (unpublished), DESY (1993), available on http://www-zeus.desy.de/bluebook/bluebook.html
14. S. Godfrey and N. Isgur, Phys. Rev. **D32**, 189 (1985)
15. WA102 Collaboration, D. Barberis et al., Phys. Lett. **B 489**, 24 (2000)
16. T. Barnes, *IX International Conference on Hadron Spectroscopy*, AIP Conference Proc. Vol. 619, p. 447. Protvino, Russia (2002). Also in preprint hep-ph/0202157
17. J. Benecke and H. P. Dürr, Nuovo Cimento **56**, 269 (1968)
18. R. P. Feynman, *Photon-Hadron Interactions*, Benjamin, New York (1972)
19. K.H. Streng, T.F. Walsh and P.M. Zerwas, Z. Phys. **C 2**, 237 (1979)
20. R.P. Feynman, Photon-Hadron Interactions (1972).
21. K.H. Streng, T.F. Walsh and P.M. Zerwas, Z. Phys. C 2 (1979) 237.

Light Meson Radiative Decays

A. Donnachie

Department of Physics and Astronomy
University of Manchester
Manchester M13 9PL
England

Abstract. Radiative decays of excited vector mesons are shown to provide good discrimination between the 2^3S_1 and the 1^3D_1 excitations of the ρ and ω and between these and a possible vector hybrid. Radiative decays of the 1^3D_1 excitations of the ρ, ω and ϕ to the scalars $f_0(1370)$, $f_0(1500)$ and $f_0(1710)$ provide a flavour filter, clarifying the extent of glueball mixing in the scalar states. A complementary approach to the latter is provided by the radiative decays of the scalars to the ρ, ω and ϕ.

INTRODUCTION

Radiative transitions have proved their value in the baryon sector, successfully reproducing the magnitudes and relative phases of over 100 helicity amplitudes for photoexcitation of the proton and neutron[1]. In contrast, calculations of light-meson radiative decays have concentrated mainly on ground-state to ground-state decays. New and proposed facilities (Initial State Radiation and B-decays at BABAR and BELLE, J/ψ decay at CLEO-C, photo- and electroproduction at JLab, e^+e^- annihilation at Novosibirsk and Frascati, central production in pp collisions at COMPASS) promise greatly increased statistics and the possibility of studying radiative decays of excited light mesons.

In two recent papers[2, 3] the radiative decays of excited vector and scalar mesons have been studied with three objectives in mind: to discriminate between the radial (2^3S_1) and orbital (1^3D_1) excitations of the ρ and ω; to discriminate between these $n\bar{n}$ excitations and possible $J^{PC} = 1^{--}$ hybrids; to discriminate among different $q\bar{q}$ and glueball mixing scenarios in the scalars.

Several key radiative widths are found to be large, $\gtrsim 500$ keV, and offer strong discriminatory power.

THE MODEL

Wave functions are taken as Gaussian, that is of the form $\exp(-p^2/\beta^2)$ multiplied by the appropriate polynomial, and the parameter β found for each of the $1S$, $1P$, $2S$, $1D$ states by treating it as the variational parameter in the Hamiltonian

$$H = \frac{p^2}{m_q} + \sigma r - \frac{4}{3}\frac{\alpha_s}{r} + C$$

CP717, *Hadron Spectroscopy: Tenth International Conference,*
edited by E. Klempt, H. Koch, and H. Orth
© 2004 American Institute of Physics 0-7354-0197-7/04/$22.00

with standard quark-model parameters: $m_{u,d} = 0.33$ GeV, $m_s = 0.45$ GeV, $\sigma = 0.18$ GeV^2, $\alpha_s = 0.5$.

The pure electric-dipole ($E1$) transition is well-defined for heavy quarks, but is certainly a bad approximation for light quarks so we include the magnetic quadrupole ($M2$) transition as well.

This approach has a long history of success in the baryon sector even though the $M2$ terms are the same order in p^2 as $E1$ corrections such as anomalous magnetic moments of the constituents, spin-orbit terms, Thomas precession and binding effects. Some of these corrections can be calculated[4, 5]; some can only be estimated[4]. The success in the baryon sector suggests that the collective effect of these corrections is small.

Within this "leading multipole" hypothesis there are checks on our procedures. The ground-state to ground-state transitions, e.g. $\rho \to \eta\gamma$, $\omega \to \eta\gamma$, are given correctly by the model. The prediction for $\Gamma(f_1(1285) \to \gamma\rho)$, 1400 keV, is in good accord with experiment[6, 7, 8], 1320 ± 312 keV. The prediction that $\Gamma(f_2(1270) \to \gamma\rho)$ is appreciably smaller than $\Gamma(f_1(1285) \to \gamma\rho)$ is in qualitative accord with experiment as there is no evidence for the radiative decay of $f_2(1270)$ in either the MARK III or WA102 experiments[7, 8] and both have strong f_2 signals. From general considerations we can form a positivity constraint among a combination of widths which is satisfied by the model and allows a more general conclusion to be drawn, namely that $\Gamma(f_0 \to \gamma\rho) \sim \Gamma(f_1 \to \gamma\rho)$ in agreement with the calculation.

VECTOR MESON RADIATIVE DECAYS

The hadronic decays of the light $n\bar{n}$ vector mesons are in conflict[9] with the predictions[10] of the $^3P_0$ model. One solution is to invoke the presence of $J^{PC} = 1^{--}$ isovector and isoscalar hybrids suitably mixed with the 2^3S_1 and 1^3D_1 excitations of the ρ and ω. For numerical purposes we take the $\rho(1450)$ and $\omega(1420)$ to be the 2^3S_1 states and the $\rho(1700)$ and $\omega(1650)$ to be the 1^3D_1 states.

The first result is that radiative decays can distinguish between the 2^3S_1 and the 1^3D_1.

$$\Gamma(\rho(1450) \to f_2(1270)\gamma) \quad \sim 700\text{keV}$$
$$\Gamma(\rho(1700) \to f_2(1270)\gamma) \quad \sim 140\text{keV}$$
$$\Gamma(\rho(1450) \to f_1(1285)\gamma) \quad \sim 350\text{keV}$$
$$\Gamma(\rho(1700) \to f_1(1285)\gamma) \quad \sim 1100\text{keV}$$
$$\Gamma(\omega(1420) \to a_2(1270)\gamma) \quad \sim 420\text{keV}$$
$$\Gamma(\omega(1650) \to a_2(1270)\gamma) \quad \sim 90\text{keV}$$
$$\Gamma(\omega(1420) \to a_1(1285)\gamma) \quad \sim 340\text{keV}$$
$$\Gamma(\omega(1650) \to a_1(1285)\gamma) \quad \sim 1020\text{keV}$$

Radiative decays can also resolve the issue of the $J^{PC} = 1^{--}$ hybrid. As the $q\bar{q}$ pair in the hybrid is in a spin-singlet state, radiative decays to the spin-triplet $f_2(1270)$ and $f_1(1285)$ will be suppressed. For example, for the isovector hybrid, ρ_H, radiative decay widths to the spin-triplet $f_2(1270)$ and $f_1(1285)$ will be small, in contrast to the

TABLE 1. Radiative decay widths in keV for the $\rho(1690)$ and $\phi(1900)$ to the scalars $f_0(1370)$, $f_0(1500)$ and $f_0(1710)$ for the different mixing scenarios described in the text: light glueball (L), middleweight glueball (M) and heavy glueball (H).

	$\rho(1690)$			$\phi(1900)$		
	L	M	H	L	M	H
$f_0(1370)$	174	440	603	7	8	31
$f_0(1500)$	520	301	98	5	35	261
$f_0(1710)$				173	156	17

corresponding decays of the $\rho(1450)$ and $\rho(1700)$. The dominant radiative decay of ρ_H should be to the spin-singlet $b_1(1235)$, which is suppressed for the spin-triplet $\rho(1450)$ and $\rho(1700)$. Specific calculation [11] gives

$$\Gamma(\rho_H(1700) \to b_1(1235)\gamma) \sim 700 \text{keV}$$

So high-intensity e^+e^- annihilation to γX or J/ψ decay to $\gamma\gamma X$, with $X = f_2(1270)$, $f_1(1285)$ and $b_1(1235)$ would resolve the three isovector states and specify the mixing.

Radiative decays can also separate cleanly the $\phi(1690)$. For example the $f_2(1525)\gamma$ decay of the $\phi(1690)$ provides a unique signature for the $s\bar{s}$ state, albeit with a smallish width:

$$\Gamma(\phi(1690) \to f_2(1525)\gamma) \sim 200 \text{keV}$$

Radiative vector meson decays to the scalars can tell us something about the possible scalar glueball which is invoked to explain the existence of three scalar states when, if we have only $q\bar{q}$ states, there should be two. If there is no mixing among the scalars so that the $f_0(1370)$ is pure $n\bar{n}$ and the $f_0(1710)$ is pure $s\bar{s}$ then

$$\Gamma(\rho(1700) \to f_0(1370)\gamma) \sim 900 keV$$
$$\Gamma(\rho(1450) \to f_0(1370)\gamma) \sim 65 keV$$
$$\Gamma(\phi(1900) \to f_0(1710)\gamma) \sim 190 keV$$

Three different mixing scenarios have been proposed: the bare glueball is lighter than the bare $n\bar{n}$ state (the light glueball solution[12]); the mass of the bare glueball is between the bare $n\bar{n}$ state and the bare $s\bar{s}$ state (the middleweight glueball solution[12]); and the bare glueball is heavier than the bare $s\bar{s}$ state (the heavy glueball solution[13]). Each of the three mixing scenarios affects the radiative decays in a unique way, as can be seen in Table 1.

So the relative rates of the radiative decays of the $\rho(1700)$ to $f_0(1370)$ and $f_0(1500)$, and of the $\phi(1900)$ to $f_0(1500)$ and $f_0(1710)$ change radically according to the particular model for $q\bar{q}$-glueball mixing. In combination they can 'weigh' the bare glueball.

An important check on this phenomenology is provided by the decay $\omega(1650) \to a_0(1450)\gamma$, predicted width ~ 610 keV.

TABLE 2. Radiative decay widths in keV from the scalars $f_0(1370)$, $f_0(1500)$ and $f_0(1710)$ to the $\rho(770)$ and $\phi(1020)$ for the different mixing scenarios described in the text: light glueball (L), middleweight glueball (M) and heavy glueball (H).

	$\rho(770)$			$\phi(1020)$		
	L	M	H	L	M	H
$f_0(1370)$	443	1121	1540	8	9	32
$f_0(1500)$	2519	1458	476	9	60	454
$f_0(1710)$	42	94	705	800	718	78

SCALAR MESON RADIATIVE DECAYS

A complementary approach to flavour-filtering among the scalars is provided by the radiative decays of the scalars to the ground-state vectors ρ, ω and ϕ. The results for the three mixing scenarios are given in Table 2.

The width of the decay $f_1(1285) \to \rho\gamma$ is measured[6, 7, 8] and provides a good check on the model: 1320 ± 312 keV compared to a predicted value of ~ 1400 keV. The predicted width for the decay $f_2(1270) \to \rho\gamma$ is ~ 640 keV. Experimentally this width is small as neither MARK III nor WA102 see it although both have a large $f_2(1270)$ signal[7, 8]. The branching fractions for radiative decay of J/ψ to $f_1(1285)$ and $f_2(1270)$ are comparable[6] at $(6.1 \pm 0.9) \times 10^{-4}$ and $(1.30 \pm 0.14) \times 10^{-3}$, so the non-observation of any $f_2(1270)$ signal in the decay $J/\psi \to \gamma(\gamma\rho)$ is meaningful. A similar situation holds in central production in high-energy proton-proton interactions[8] and one can deduce[14] an upper limit on $\Gamma(f_2(1270) \to \rho\gamma)$ of 500 keV at 95% confidence level.

A further check on the phenomenology would be provided by the decay $a_0(1450) \to \omega\gamma$, predicted width ~ 2100 keV.

SINGLE QUARK TRANSITIONS

Independent of details of binding dynamics it is possible to obtain relations among helicity amplitudes, and hence widths, that depend only on the assumption that the mesons are $q\bar{q}$ P and S states. There are only three multipoles for $V \to \gamma P$ or $P \to \gamma V$, the leading electric dipole E_1, an "extra" electric dipole E_R and the magnetic quadrupole M_1. Thus it is possible to obtain relationships among the widths. For $n\bar{n}$ states, assuming equal phase space and equal form factors, for $V \to \gamma f_J$

$$\Gamma(\rho(2S) \to \gamma f_2) + 7\Gamma(\rho(2S) \to \gamma f_0) \geq 3\Gamma(\rho(2S) \to \gamma f_1)$$

and for $f_J \to \gamma V$

$$5\Gamma(f_2 \to \gamma\rho) + 7\Gamma(f_0 \to \gamma\rho) \geq 9\Gamma(f_1 \to \gamma\rho)$$

As $\Gamma(f_1 \to \gamma\rho) \sim 1300$ keV, this latter equation requires that one or other of f_0, f_2 must have a radiative width ~ 1000 keV. As there is no evidence for the f_2 decay it follows that $f_0 \to \gamma\rho$ should be large, in line with our specific calculations.

SUMMARY

In general, radiative decays are a better probe of meson structure than hadronic decays as the coupling to the charges and spins of constituents gives detailed information on wave functions. Specifically, light-quark radiative decays provide a strong discriminatory mechanism and act as a good flavour filter.

Some of the radiative decays may be detectable in present experiments. The E852 experiment[15] sees the $\omega\eta$ decay of the $\omega(1650)$. If the $\omega(1650)$ is the 1^3D_1 excitaion of the ω, the $^3P_0$ model predicts $\Gamma(\omega(1650) \to \omega\eta) \sim 13$ MeV and the radiative decay model finds $\Gamma(\omega(1650) \to a_1(1260)\gamma) \sim 1000$ keV $\sim 8\%$ of the $\omega\eta$ width. The E852 experiment has several thousand events in the $\omega\eta$ channel, so we may expect several hundred events in the $a_1\gamma$ channel.

The VES collaboration[16] sees the $\rho\eta$ decay of the $\rho(1450)$ and of the $\rho(1700)$, with several thousand events. Assuming that the $\rho(1450)$ and the $\rho(1700)$ are the 2^3S_1 and 1^3D_1 excitations of the ρ, the $^3P_0$ model predicts $\Gamma(\rho(1450) \to \rho\eta) \sim \Gamma(\rho(1700) \to \rho\eta) \sim 25$ MeV and the radiative decay model finds $\Gamma(\rho(1450) \to f_2(1270)\gamma) \sim 700$ keV, $\Gamma(\rho(1700) \to f_1(1285)\gamma) \sim 1100$ keV. So the radiative decays could again be present at the level of several hundred events.

REFERENCES

1. L. A. Copley, G. Karl and E. Obryk, *Nucl.Phys.* **B13**, 303 (1969); R. P. Feynman, M. Kislinger and F. Ravndal, *Phys.Rev.* **D3**, 2706 (1971); S. Capstick, *Phys.Rev.* **D46**, 2864 (1992)
2. F. E. Close, A. Donnachie and Yu. S. Kalashnikova, *Phys.Rev.* **D65**, 092003 (2002)
3. F. E. Close, A. Donnachie and Yu. S. Kalashnikova, *Phys.Rev.* **D67**, 074031 (2003)
4. A. le Yaouanc, L. Oliver, O. Pene and J. C. Raynal, *Z.Phys.* **C40**, 77 (1988)
5. F. E. Close and H. Osborn, *Phys.Lett.* **B34**, 400 (1971); F. J. Gilman and I. Karliner, *Phys.Rev.* **D10**, 2194 (1974) G. Karl, S. Meshkov and J. Rosner, *Phys.Rev.Lett.* **45**, 215 (1980); G. Hardekopf and J. Sucher, *Phys.Rev.* **D25**, 2938 (1982); F. Foster and G. Hughes, *Z.Phys* **C14**, 123 (1982); R. McClary and N. Byers, *Phys.Rev.* **D28**, 1692 (1983); S. Godfrey, G. Karl, and P. O'Donnell, *Z.Phys.* **C31**, 77 (1986) F. E. Close and Z. P. Li, *Phys.Rev.* **D42**, 2194 (1990)
6. Particle Data Group, *Eur.Phys.J.* **C15**, 1 (2000)
7. D. Coffman et al, *Phys.Rev.* **D41**, 1410 (1990)
8. D. Barberis et al, *Phys.Lett.* **B440**, 225 (1998)
9. A. Donnachie and Yu. S. Kalashnikova, *Phys.Rev.* **D60**, 114011 (1999)
10. T. Barnes, F. E. Close, P. R. Page and E. S. Swanson, *Phys.Rev.* **D55**, 4157 (1997)
11. F. E. Close and J. Dudek, private communication
12. F. E. Close and A. Kirk, *Eur.Phys.J.* **C21**, 531 (2001)
13. W. Lee and D. Weingarten, *Phys.Rev.* **D61**, 014015 (2000)
14. A. Kirk, private communication
15. P. Eugenio et al, *Phys.Lett.* **B497**, 190 (2001)
16. D. V. Amelin et al, *Nucl.Phys.* **A668**, 83 (2000)

Formation and interference of tensor mesons in the reaction $\gamma\gamma \to \pi^+\pi^-\pi^0$

S. Hou

Academia Sinica, Taipei, Taiwan; National Central University, Chungli, Taiwan

Abstract. The two-photon reaction $\gamma\gamma \to \pi^+\pi^-\pi^0$ was investigated using data collected with the Belle detector at KEKB. The spectrum was analyzed for resonance formation and interference between the dominant $a_2(1320)$ and higher mass radial excitation tensor states. The observation includes the newly reported $a_2'(1750)$, and evidences for $a_2''(1950)$ and $a_2'''(2190)$.

The Belle detector [1] is suitable for studies of resonance formation in two-photon collisions. We present a study of $a_2(1320)$, $a_2'(1750)$ and higher mass radial excitation states in $\pi^+\pi^-\pi^0$ final state using a data sample corresponding to a total integrated luminosity of 26.0 fb^{-1} at $\sqrt{s}$=10.58 GeV. The resonance parameters and spin-parity states were measured and compared to Monte Carlo for various spin-parity hypotheses.

The $\pi^+\pi^-\pi^0$ final state was analyzed for consecutive decays of $R \to I\pi$ and the di-pion isobar decay $I \to \pi\pi$. The di-pion isobars considered are ρ and f_2. The $a_2(1320)$ meson decays to $\rho^\pm\pi^\mp$. Higher mass resonances can decay to $\rho^\pm\pi^\mp$ and $f_2\pi^0$. The spin-parity is tested for $J^P = 2^-$ and 2^+. Interference occurs for $J^P = 2^+$, and the cross section is given by

$$d\sigma_{\gamma\gamma}^{J_z} \propto \left| \mathrm{BW}(a_2)\sum_I D^{J_z}(I) + \alpha'e^{i\phi'}\mathrm{BW}(a_2')\sum_I D'^{J_z}(I) + \alpha''e^{i\phi''}\mathrm{BW}(a_2'')\sum_I D''^{J_z}(I) +.. \right|^2 \quad (1)$$

$$\text{with} \quad \sum_I D^{J_z}(I) = \mathrm{BW}(\rho^+)T^{J_z}(\rho^+) + \mathrm{BW}(\rho^-)T^{J_z}(\rho^-) + \xi e^{i\psi}\mathrm{BW}(f_2)T^{J_z}(f_2), \quad (2)$$

where J_z is the spin polarization, BW denotes the Breit-Wigner resonance, and the interference is characterized by the amplitudes (α's) and the phase angles (ϕ's). The decay amplitude is a sum of di-pion Breit-Wigner terms multiplied by the transition amplitude (T^{J_z}) of orthogonal spherical harmonics for spin dependence. The interference between decay modes is given by the amplitude ξ and the phase angle ψ.

The data sample were selected for two-photon events with total energy below 5 GeV. Contamination from beam related background was eliminated by requiring final state photons with energy greater than 180 MeV. The mass resolution for π^0 is 6 MeV/c^2. The π^0 candidates were selected for photon pairs with invariant mass within three-sigma of the π^0 mass. A feed-forward neural network was developed for discrimination of different partial waves. The neural network has seven input nodes and multiple output nodes each corresponding to a partial wave. The input variables include di-pion invariant masses and $\cos\theta$ distributions of final state pions. To enhance discrimination against background, the set of partial waves used in training includes only $J^P = 2^+$ with helicity $\lambda = 0$ and 2, and three-pion phase space distribution. Background is presumably from

CP717, *Hadron Spectroscopy: Tenth International Conference*,
edited by E. Klempt, H. Koch, and H. Orth
© 2004 American Institute of Physics 0-7354-0197-7/04/$22.00

inclusive two-photon events resembled by the three-pion phase space distribution. The three output nodes, O_i, are combined into a single variable $NN_i = O_i \cdot (1 - O_j) \cdot (1 - O_k)$. The distributions obtained in two mass regions for the probability of 2^+ helicity 2 wave are shown in Fig. 1. The $a_2(1320)$ is dominant in the distribution with the three-pion invariant mass $m(3\pi)$ below 1.5 GeV/c^2 (Fig. 1.a), where the Monte Carlo is normalized to the PDG radiative width ($\Gamma_{\gamma\gamma} = 1$ keV) [2]. The Monte Carlo in the distribution for higher $m(3\pi)$ (Fig. 1.b) is arbitrarily normalized to $\Gamma_{\gamma\gamma} = 0.5$ keV. The data distributions follow well the helicity 2 wave predictions. The excess in low NN region is presumably from background. A selection cut on NN can improve the selection purity for resonance study.

The selection cuts were chosen on the most effective parameters, namely the three-pion transverse momentum squared p_t^2, final state photon energy E_γ, and neural network output NN. The cuts were optimized for selection efficiency and purity. The final values applied are $E_\gamma > 180$ MeV, $p_t^2 < 0.0005$ GeV^2/c^2 and NN > 0.2. The selection efficiency is shown in Fig. 2.a for Monte Carlo estimation made for the ratio of selected events to the generated. The Monte Carlo was generated for $J^P = 2^+$ helicity 2 wave containing resonances in $\rho\pi$ and $f_2\pi$ decay modes. The function shape has little dependence on the selection cuts, and therefore reduces the systematic uncertainty.

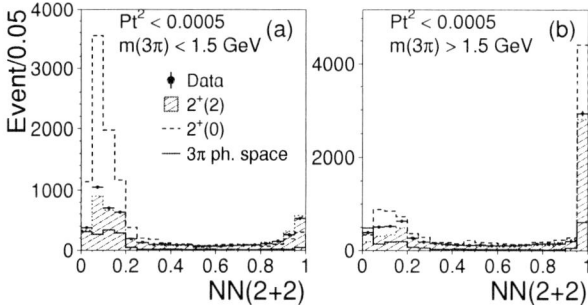

FIGURE 1. Neural network tests for events with $m(3\pi)$ (a) below and (b) above 1.5 GeV/c^2.

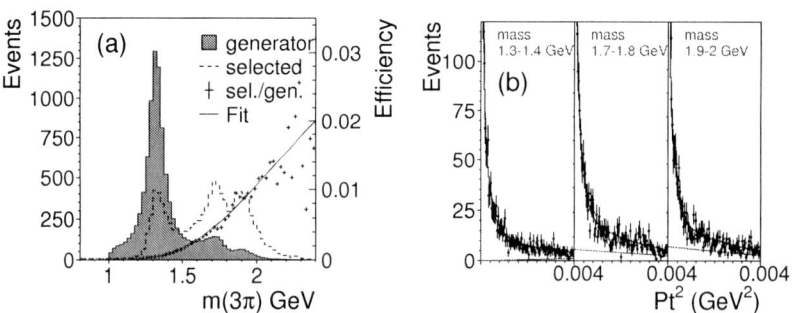

FIGURE 2. (a) Monte Carlo estimation for event selection efficiency; (b) p_t^2 distributions and the fits to Monte Carlo plus a linear background function.

The background estimation is made by comparing the difference in p_t^2 distribution to Monte Carlo prediction. Background in p_t^2 is approximately linear. It is estimated for p_t^2 distributions in 100 MeV/c^2 intervals of $m(3\pi)$. Each distribution was fitted to the Monte Carlo prediction with a linear background function. Examples are illustrated in Fig. 2.b. The background in each mass interval was integrated as a function of the three-pion mass for the mass spectrum analysis.

The spin-parity dependence is observed for the angular distribution of the three-pion decay plane. The norm, $\vec{N} = \vec{p}_{\pi^+} \times \vec{p}_{\pi^-}$, in the three-pion rest frame was measured for the squared magnitude normalized to the maximum kinetic energy,

$$\Lambda = \left| \vec{N}/Q \right|^2, \tag{3}$$

where $Q = E(3\pi) - 3m_\pi$. The Λ distributions, shown in Fig. 3, were made for two invariant mass intervals for $a_2(1320)$ and higher mass resonances. The χ^2 tests to Monte Carlo were made for spin-parity hypotheses and decay modes allowed. The distribution for $a_2(1320)$ selected with $m(3\pi) < 1.5$ GeV/c^2, is consistent with the $J^P = 2^+$ prediction with $\chi^2/\mathrm{ndf} = 1.8$ for helicity 2 state. The distribution in higher mass region is also in good agreement with the $J^P = 2^+$ prediction with $\chi^2/\mathrm{ndf} = 2.1$ for helicity 2 wave.

The spin-parity and helicity dependence is explicitly seen in the polar angle distribution of final state pions. The $\cos\theta$ distribution of π^0 and the difference in $\cos\theta$ for π^+ and π^- are shown in Fig. 4, for $m(3\pi)$ below and above 1.5 GeV/c^2. The asymmetric distribution seen for π^0 in low mass region is due to the boost and energy threshold for photons in the asymmetric beam configuration at Belle. Assuming the resonances are all of $J^P = 2^+$ states, the fractions of helicity states can be obtained by fitting the data distribution to the sum of Monte Carlo predictions and background. The data spectra are consistent with the predominant helicity 2 hypothesis. The helicity 2 fraction obtained are listed in Table 1.

The invariant mass spectrum of data, shown in Fig. 5.a, was tested to Monte Carlo prediction for having three tensor states at 1320, 1750 and 1950 MeV. The resonance

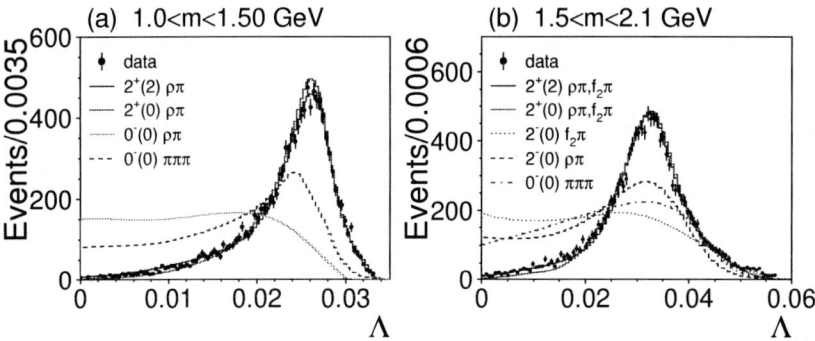

FIGURE 3. Distributions for the Λ parameter in two mass intervals for $a_2(1320)$ and higher mass resonances. Comparison with Monte Carlo was performed for possible spin-parity states.

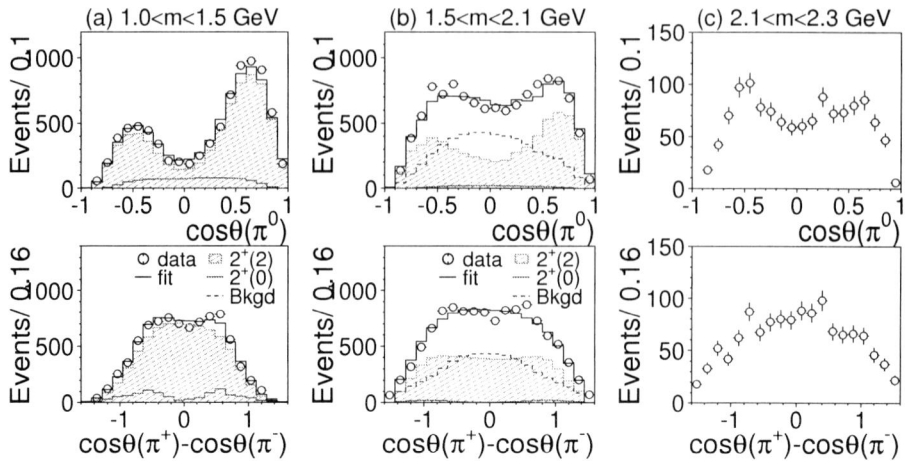

FIGURE 4. $\cos\theta$ distributions and the fits for the fractions of $J^P = 2^+$ helicity states.

TABLE 1. Helicity 2 fractions of the fits

	$a_2(1320)$	$a_2'(1750)$	$a_2''(1950)$
mass range (GeV/c^2)	$1.0 - 1.5$	$1.5 - 1.85$	$1.85 - 2.1$
fraction (%)	$100 \pm 2 \pm 5$	$92 \pm 2 \pm 5$	$96 \pm 2 \pm 5$

parameters were obtained by χ^2 minimization to Monte Carlo predictions, and the results are listed in Table 2. The systematic errors listed were estimated by varying the background fraction and the acceptance function applied.

Above 2.1 GeV/c^2 the invariant mass distribution of data is not interpreted by Monte Carlo. It is improved by adding a Breit-Wigner term in the fit. The Breit-Wigner term is accounted for 45% of the events above a background of 24%. These events were examined for spin-parity and decay modes. The di-pion mass spectra and $\cos\theta$ distributions (Fig. 4.c) are similar to those in lower mass regions. It is presumably a higher radial excitation state, $a'''(2190)$. The resonance parameters obtained are also listed in Table 2. The number of events obtained corresponds to a radiative width of $\Gamma_{\gamma\gamma}(2190) = 66 \pm 5 \pm 12$ eV.

Events for $a_2'(1750)$ and $a_2''(1950)$ with $1.5 < m(3\pi) < 2.1$ GeV/c^2 were analyzed for the decay interference of $\rho\pi$ and $f_2\pi$ intermediate states expressed by the decay amplitude D^{J_z} in Eq. 2. The di-pion invariant mass distributions are shown in Fig. 5.b for events selected in the corresponding $m(3\pi)$ intervals. The phase angle has the effect of shifting the di-pion mass peaks, and the amplitude enhances the mass peak of ρ or f_2. The histograms shown are the Monte Carlo of parameters close to the final results. Values obtained are listed in Table 2. The interference amplitudes and phase angles are compatible for the two radial excitation states.

In summary, the Monte Carlo used in this analysis has included the interference between tensor states. The invariant mass spectrum indicates the presence of multiple ra-

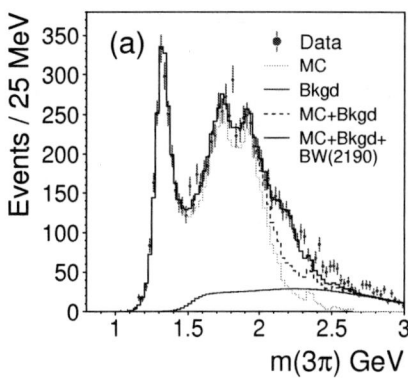
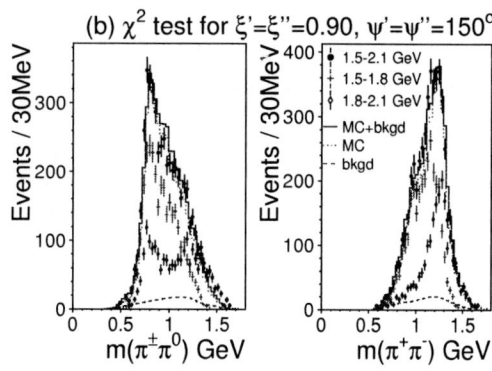

FIGURE 5. (a) $m(3\pi)$ distribution and the fit to Monte Carlo plus a Breit-Wigner resonance at 2190 MeV/c^2; (b) di-pion mass spectra and the Monte Carlo of interference parameters close to final fit.

TABLE 2. Resonance parameters

	$a_2'(1750)$	$a_2''(1950)$	$a_2'''(2190)$
mass range (GeV/c^2)	$1.0 - 1.5$	$1.5 - 1.85$	$1.85 - 2.1$
Tensor state parameters			
amplitude (α)	$0.47 \pm 0.01 \pm 0.04$	$0.58 \pm 0.02 \pm 0.05$	$-$
phase angle (ϕ deg.)	$149 \pm 2 \pm 12$	$149 \pm 2 \pm 12$	$-$
mass MeV/c^2	$1764 \pm 6 \pm 8$	$1949 \pm 6 \pm 8$	$2190 \pm 6 \pm 8$
width MeV	$237 \pm 4 \pm 10$	$229 \pm 8 \pm 10$	$185 \pm 10 \pm 10$
$\rho\pi, f_2\pi$ **decay interference**			
amplitude (ξ)	$0.91 \pm 0.10 \pm 0.07$	$0.91 \pm 0.07 \pm 0.07$	$-$
phase angle (ψ)	$151 \pm 4 \pm 12$	$149 \pm 4 \pm 12$	$-$

dial excitation states with the interference angles close to $180°$. The angular distributions of pions in $a_2(1320)$ decay, as well as those of higher mass resonances, are consistent with the prediction of $J^P = 2^+$ helicity 2 wave. The $a_2'(1750)$ observed is consistent with the L3 report [3]. It is accompanied by a higher radial excitation state $a_2''(1950)$. The enhancement observed above 2.1 GeV/c^2 indicates an additional radial excitation state $a_2'''(2190)$.

REFERENCES

1. Belle Collab., K. Abe et al., *Nucl. Inst. and Meth. A*, **479**, 117 (2002).
2. Particle Data Group, *Phys. Rev. D*, **66**, 010001 (2002).
3. L3 Collab., M. Acciarri et al., *Phys. Lett. B*, **413**, 147 (1997).

Recent Results of J/ψ at BES

Feng LU

Representing BES Collaboration
Institute of High Energy Physics,Academia Sinica – P. O. BOX 918-2,Beijing 100039,P. R. China

Abstract. Based on 5.8×10^7 J/ψ events collected by BESII, the mass and full width of η_c, as well as its decay branching ratios to $K^+K^-\pi^+\pi^-$, $\pi^+\pi^-\pi^+\pi^-$, $K^\pm K_S^0 \pi^\mp$, $\phi\phi$ and $p\bar{p}$ are measured. Partial Wave Analysis (PWA) of $J/\psi \to \gamma K\bar{K}$ and the study of $J/\psi \to \gamma\gamma V$ ($V = \rho, \phi$) are performed. We also reported the measurements of $J/\psi \to p\bar{p}$ and $J/\psi \to \pi^+\pi^-\pi^0$.

INTRODUCTION

With the upgraded Beijing Spectrometer (BESII)[1] at the Beijing Electron-Positron Collider (BEPC), which runs in 2-5GeV τ-charm energy region with the peak luminosity of $5 \times 10^{30} cm^{-2} s^{-1}$ at J/ψ energy, BES group has accumulated $(57.7 \pm 2.72) \times 10^6$ J/ψ events. J/ψ decays provide us a good laboratory in searching for glueballs, hybrids, exotics and the study of light hadron spectroscopy. All following results are based on above 58M BESII J/ψ sample.

THE MEASUREMENT OF η_C MASS, WIDTH AND BRANCHING RATIOS

The η_c samples are selected from the reactions $J/\psi \to \gamma\eta_c$, $\eta_c \to K^+K^-\pi^+\pi^-$, $\pi^+\pi^-\pi^+\pi^-$, $K^\pm K_S^0\pi^\mp$ (with $K_S^0 \to \pi^+\pi^-$), $\phi\phi$ (with $\phi \to K^+K^-$) and $p\bar{p}$. After event selection criteria[2], clear η_c signals appear in each decay mode. An unbinned maximum likelihood fit is performed for all five channels simultaneously. The mass, full width and production branching ratios of η_c are determined as: $M_{\eta_c} = 2977.5 \pm 1.0 \pm 1.2$ MeV/c^2, $\Gamma_{\eta_c} = 17.0 \pm 3.7 \pm 7.4$ MeV/c^2, $B(J/\psi \to \gamma\eta_c) \times Br(\eta_c \to K^+K^-\pi^+\pi^-) = (1.5 \pm 0.2 \pm 0.2) \times 10^{-4}$, $B(J/\psi \to \gamma\eta_c) \times B(\eta_c \to \pi^+\pi^-\pi^+\pi^-) = (1.3 \pm 0.2 \pm 0.4) \times 10^{-4}$, $B(J/\psi \to \gamma\eta_c) \times B(\eta_c \to K^\pm K_S^0\pi^\mp) = (2.2 \pm 0.3 \pm 0.5) \times 10^{-4}$, $B(J/\psi \to \gamma\eta_c) \times B(\eta_c \to \phi\phi) = (3.3 \pm 0.6 \pm 0.6) \times 10^{-5}$ and $B(J/\psi \to \gamma\eta_c) \times B(\eta_c \to p\bar{p}) = (1.9 \pm 0.3 \pm 0.3) \times 10^{-5}$. The results are consistent with PDG values within the errors, except $B(\eta_c \to \phi\phi)$. But it agrees with Belle's new measurement[3].

---

[1] BES Collaboration (J. Z. Bai *et al.*), Nucl. Instr. Meth. **A458**, 627 (2001).
[2] BES Collaboration (J. Z. Bai *et al.*), Phys. Lett. **B555**, 174 (2003).
[3] H.C. Huang *et al.*,hep-ex/0305068

CP717, *Hadron Spectroscopy: Tenth International Conference,*
edited by E. Klempt, H. Koch, and H. Orth
© 2004 American Institute of Physics 0-7354-0197-7/04/$22.00

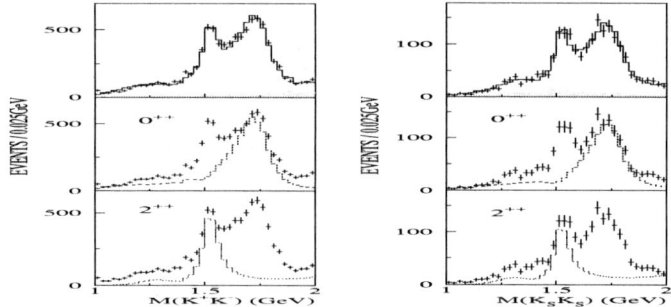

FIGURE 1. The global fit results of the $K\bar{K}$ invariant mass distributions from $J/\psi \to \gamma K^+ K^-$ and $J/\psi \to \gamma K_S^0 K_S^0$. The points are the data and the solid histograms in the top panels show the maximum likelihood fit. Histograms on subsequent panels show the complete 0^+ and 2^+ contributions including all interferences.

PARTIAL WAVE ANALYSIS OF $J/\psi \to \gamma K\bar{K}$

No clear $\xi(2230)$ signals are observed in either $K^+ K^-$ or $K_S^0 K_S^0$ invariant mass spectra of $J/\psi \to \gamma K\bar{K}$ process after final selection. A partial wave analysis is carried out using the relativistic covariant tensor amplitude method[4] for the $K\bar{K}$ invariant mass range of 1-2 GeV. The relative magnitudes and phases of the amplitudes are determined by a maximum likelihood fit. The estimated background events are included into the data samples with the opposite sign of log likelihood compared to data. These events cancel backgrounds in the data samples.

The global fit constrains phase variations as a function of mass to simple Breit-Wigner forms. It performs the optimum averaging of helicity amplitudes and their phases over resonances. The following channels are considered: $J/\psi \to \gamma f_2'(1525), \gamma f_0(1710), \gamma f_2(1270), \gamma f_0(1500), \gamma +$ broad 0^{++} and 2^{++} components. For the spin 0 amplitude, two interfering resonances ($f_0(1500), f_0(1710)$) and an interfering constant amplitude term, which is used to describe the broad S- wave contribution, are included. The mass and width of the $f_0(1500)$ are fixed to the PDG values. For the spin 2 amplitudes, the $f_2'(1525)$ and $f_2(1270)$ are included. We choose a resonance with the mass and width of 2250 MeV and 350 MeV to represent the structure in the high mass region. The tail of this resonance could contribute to the present fitted range. The mass and width of the $f_2(1270)$ are fixed to the PDG values. $J/\psi \to \gamma K^+ K^-$ and $J/\psi \to \gamma K_S^0 K_S^0$ samples are analyzed independently, and the fit results are obtained by their averaged values. Fig.1 and Table 1 show the global fit results where strong production of $f_2'(1525)$ and the S-wave resonance $f_0(1710)$ are clearly seen. This confirms earlier conclusions that 1.7 GeV mass region is dominated by 0^+. It is found that $f_0(1500)$ only has a statistical significance of 1.6σ. The helicity amplitude ratios of the $f_2'(1525)$ are consistent with theoretical predictions[5]. Another analysis of bin by bin

---

[4] B.S. Zou and D.V. Bugg, Eur. Phys. J. **A16** (2003) 537.
[5] M. Krammer, Phys. Lett. **B74** (1978) 361.

fit (40MeV bin width), which fixes $f_2'(1525)$ mass and width to the values in Table 1 from the global fit, gives well consistent results with the global fit values listed in Table 1.

TABLE 1. Measurements of the $f_2'(1525)$ and $f_0(1710)$ for the global fit. The first error is statistical, the second is systematic, and the third one in the branching fractions is for the model-dependence of the broad components.

	$f_2'(1525)$	$f_0(1710)$
M (MeV)	$1519 \pm 2^{+15}_{-5}$	$1740 \pm 4^{+10}_{-25}$
Γ (MeV)	$75 \pm 4^{+15}_{-5}$	166^{+5+15}_{-8-10}
$B(J/\psi \to \gamma X,$ $X \to K\bar{K})(\times 10^{-4})$	$3.42 \pm 0.15^{+0.69+1.55}_{-0.65-0.00}$	$9.62 \pm 0.29^{+2.11+2.81}_{-1.86-0.00}$
amp. ratios x^2	$1.00 \pm 0.28^{+1.06}_{-0.36}$	—
y^2	$0.44 \pm 0.08^{+0.10}_{-0.56}$	—

PRELIMINARY MEASUREMENT OF $J/\psi \to P\bar{P}$

In addition to the branching ratios of J/ψ decaying to baryon pairs, the information regarding the angular distribution of the decay products is especially interesting since theoretical models based on the first order QCD calculations give predictions for the angular distributions of two baryon final states. In general the angular distribution for J/ψ decaying to baryon pairs can be written as:

$$\frac{dN}{d\cos\theta_B} \propto 1 + \alpha \cos^2\theta_B, \tag{1}$$

where θ_B is the angle between the baryon direction and the positron beam direction.

63316 $J/\psi \to p\bar{p}$ events are selected from 58 million J/ψ events with a background level of about 1.5%. The fit of the angular distribution of data after efficiency correction gives $\alpha = 0.676 \pm 0.036 \pm 0.042$, and the branching ratio of $J/\psi \to p\bar{p}$ is determined to be $(2.26 \pm 0.01 \pm 0.12) \times 10^{-3}$. Here, the systematic errors in α come mainly from different Monte-Carlo simulation of MDC wire resolution, the influence of background angular distribution, while the systematic errors for the branching ratio has additional source which comes from the uncertainty of J/ψ total number. Table 2 shows the comparison of $J/\psi \to p\bar{p}$ results with those from MarkII and PDG.

TABLE 2. Comparison of the $J/\psi \to p\bar{p}$ results

	efficiency	Br($\times 10^{-3}$)	α
BESII	48.53%	$2.26 \pm 0.01 \pm 0.12$	$0.676 \pm 0.036 \pm 0.042$
MarkII	49.7%	$2.16 \pm 0.07 \pm 0.15$	0.61 ± 0.23
PDG	–	2.12 ± 0.10	0.63 ± 0.08

PRELIMINARY STUDY OF $J/\psi \to \gamma\gamma V (V = \rho, \phi)$

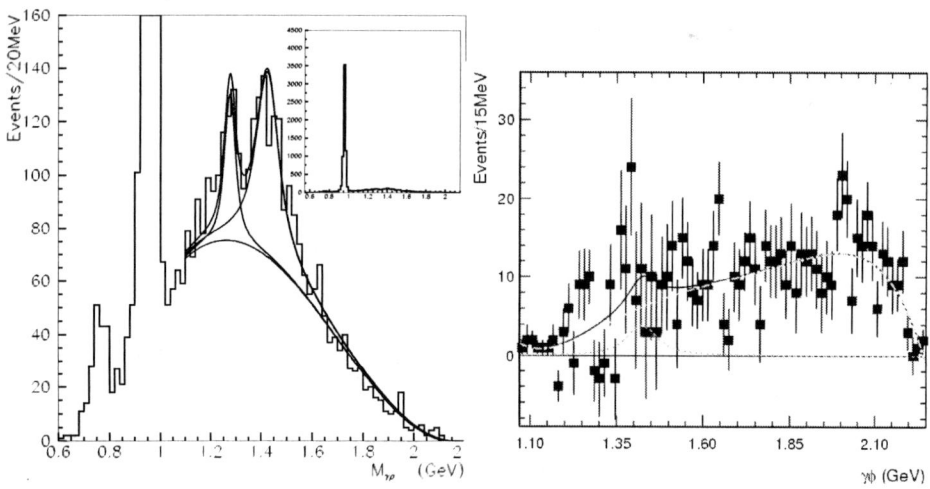

FIGURE 2. Left plot: The invariant mass of $\gamma\rho$. The inserted figure shows the full mass scale where $\eta'(958)$ is clearly observed. Right plot: The Breit-Wigner fit results of the invariant mass of $\gamma\phi$ after background subtraction.

The standard perturbative theory[6] predicts that the decay width ratio of $\eta(1440)$ to $\gamma\rho$ and $\gamma\phi$ should be 9:2 if it is a $q\bar{q}$ state, while should be 1:1 if it is a pure glueball state. A clear signal around 1420 MeV, which corresponds to $\eta(1440)$, is observed in $\gamma\rho$ invariant mass spectrum. The fitted result is showed in Fig2.(left) and Table 3. No significant signal of $\eta(1440) \to \gamma\phi$ is found in $J/\psi \to \gamma\phi$ channel. After subtracting backgrounds estimated from ϕ side bands (see Fig2.(right)), the upper limit of $B(J/\psi \to \gamma\eta(1420)) \times B(\eta(1420) \to \gamma\phi)$ is determined to be less than 0.82×10^{-4} at 95% C.L. .

TABLE 3. Summary on $J/\psi \to \gamma X (X \to \gamma\rho)$ Preliminary Study

Mass (MeV)	Width (MeV)	$B(J/\psi \to \gamma X)B(X \to \gamma\rho)$ $(\times 10^{-4})$	Events	Significance
$1276.1 \pm 8.1 \pm 8.0$	$40.0 \pm 8.6 \pm 9.3$	$0.38 \pm 0.09 \pm 0.06$	203 ± 49	6.3σ
$1424.4 \pm 9.8 \pm 11.0$	$101.0 \pm 8.8 \pm 8.8$	$1.07 \pm 0.17 \pm 0.11$	547 ± 86	9.3σ

PRELIMINARY STUDY OF $J/\psi \to \pi^+\pi^-\pi^0$

we present two independent measurements of the branching fraction of $J/\psi \to \pi^+\pi^-\pi^0$ on the study of the 58M J/ψ and 14M $\psi(2S)$ decays respectively. The absolute measurement is based on $J/\psi \to \pi^+\pi^-\pi^0$ directly, giving $Br(J/\psi \to \pi^+\pi^-\pi^0) = (21.35 \pm$

---

[6] M. S. Chanowitz, Phys. Lett. B164 (1985) 379.

$0.04 \pm 1.85) \times 10^{-3}$. The relative branching fraction is measured from a comparision of the rate for $\psi(2S) \to \pi^+\pi^- J/\psi$, $J/\psi \to \pi^+\pi^-\pi^0$(I) and $J/\psi \to \mu^+\mu^-$(II).

$$
\begin{aligned}
B(J/\psi \to \pi^+\pi^-\pi^0) &= \frac{N_I^{obs}/\varepsilon_I}{N_{II}^{obs}/\varepsilon_{II}/B(J/\psi \to \mu^+\mu^-)} \\
&= \frac{N_I^{obs}}{N_{II}^{obs}} \cdot \frac{\varepsilon_{II}}{\varepsilon_I} \cdot B(J/\psi \to \mu^+\mu^-)
\end{aligned}
\tag{2}
$$

$B(J/\psi \to \mu^+\mu^-)$ value is taken from PDG value, and $B(J/\psi \to \pi^+\pi^-\pi^0)$ is then $(20.9 \pm 0.2 \pm 1.1) \times 10^{-3}$, which is well consistent with the result of absolute measurement. The preliminary branching fraction is obtained by combining above two results as: $B(J/\psi \to \pi^+\pi^-\pi^0) = (2.10 \pm 0.11)\%$. It is inconsistent with PDG value.

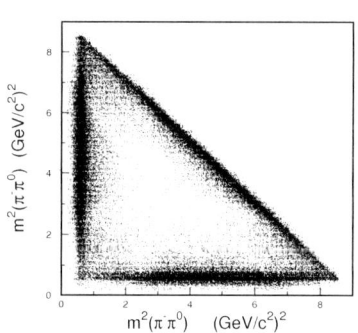
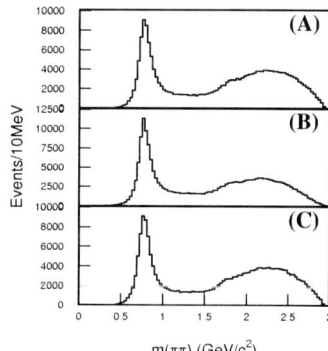

FIGURE 3. Left plot: the Dalitz plot for $J/\psi \to \pi^+\pi^-\pi^0$. Right plot: the distributions of invariant mass of two pions. A. $J/\psi \to \rho^+\pi^-$, B. $J/\psi \to \rho^0\pi^0$, C. $J/\psi \to \rho^-\pi^+$.

ACKNOWLEDGMENTS

The BES collaboration thanks the staff of the BEPC and the IHEP computing center for their hard efforts. This work is supported in part by the National Natural Science Foundation of China under contracts Nos. 19991480, 10225524, 10225525, the Chinese Academy of Sciences under contract No. KJ 95T-03, the 100 Talents Program of CAS under Contract Nos. U-11, U-24, U-25, and the Knowledge Innovation Project of CAS under Contract Nos. U-602, U-34 (IHEP); by the National Natural Science Foundation of China under Contract No.10175060 (USTC); and by the Department of Energy under Contract No. DE-FG03-94ER40833 (U Hawaii).

Recent Results of $\psi(2S)$ Decays at BES

XiaoHu Mo

Institute of High Energy Physics, CAS, Beijing 100039, China

Abstract. Using 14 million $\psi(2S)$ data sample collected with BES at BEPC, the recent results of $\psi(2S) \to VT$, $K_S^0 K_L^0$ (also $J/\psi \to K_S^0 K_L^0$), and $\chi_{cJ} \to B\bar{B}$ decays are presented.

Charmonium physics is always one of the interesting and intriguing field of particle physics. Charmonium provides us an excellent and simple system to study QCD, the production and decay mechanisms of heavy quarkonia and light hadron spectra from its decays, and can be treated non-relativistically and perturbatively. Using 14 M $\psi(2S)$ data sample collected with BEijing Spectrometer (BES) at BEPC, the recent results of $\psi(2S) \to VT$, $K_S^0 K_L^0$ (also $J/\psi \to K_S^0 K_L^0$), and $\chi_{cJ} \to B\bar{B}$ decays are presented. The BES detector is described in detail in Ref.[1].

STUDY OF VT CHANNEL IN $\psi(2S)$ DECAY

Both J/ψ and $\psi(2S)$ decays are expected to be dominated by annihilation into three gluons, with widths that are proportional to the square of the $c\bar{c}$ wave function at the origin [2]. This yields the pQCD expectation (so-called "12 %" rule) that

$$Q_h = \frac{\mathscr{B}_{\psi(2S)\to X_h}}{\mathscr{B}_{J/\psi \to X_h}} = \frac{\mathscr{B}_{\psi(2S)\to e^+e^-}}{\mathscr{B}_{J/\psi \to e^+e^-}} = (12.3 \pm 0.7)\% \ . \tag{1}$$

The violation of this rule was firstly revealed by MARK-I in VP channel (such as $\rho\pi$ and $K^*\bar{K}$ channel) [3], which leads to famous "$\rho\pi$ *puzzle*". This phenomenon was then confirmed by BES at higher sensitivity [4]. Afterwards, BES collaboration presented many other observations, one of them is about VT channel. Based on BES-I 4 M data, the upper-limits of four VT channels, $\omega f_2(1270)$, $\rho a_2(1320)$, $K^*(892)^0 \bar{K}_2^*(1430)^0 + c.c.$, and $\phi f_2'(1525)$, were given [5]. Now with BES-II 14 M $\psi(2S)$ date sample, all these upper-limits have been determined to be branching fractions. For these decay modes, studies focus on four-charged-track final states, such as $K^+K^-\pi^+\pi^-$ or $K^+K^-K^+K^-$, and those with additional two photons decayed from π^0, such as $\pi^+\pi^-\pi^+\pi^-\gamma\gamma$ or $K^+K^-\pi^+\pi^-\gamma\gamma$. After event selection, the invariant mass distributions for different channels are shown in Fig. 1. From the data fitting, the observed numbers of events are obtained, and M.C. simulation gives the corresponding efficiencies. The final results are shown in Table 1, together with statistic and systematic errors, and the resulting errors are at the level of 30 to 40 percent. Combining the corresponding results of J/ψ decay

CP717, *Hadron Spectroscopy: Tenth International Conference,*
edited by E. Klempt, H. Koch, and H. Orth
© 2004 American Institute of Physics 0-7354-0197-7/04/$22.00

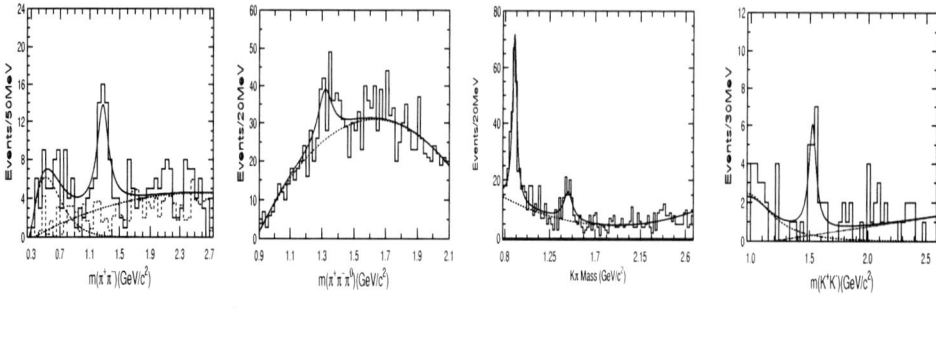

(a) ωf_2 final state   (b) ρa_2 final state   (c) $K^* \overline{K_2^*}$ final state   (d) ϕf_2 final state

FIGURE 1. Invariant mass distributions for VT channel(The dashed line indicates background and solid line indicates the synthetic fitting result).

from PDG2002 [6], the Q_h values were calculated as listed in Table 1. Comparing with 12 % rule, it is seen that the Q_h value of VT channel is greatly suppressed.

FIRST OBSERVATION OF $K_S^0 K_L^0$ IN $\psi(2S)$ DECAY

For PP channel study, the theoretical motivation, besides the pCQD rule test, involves the phase study which is very important in understanding the strong interaction mechanism of charmonium decay. A recent phenomenological analysis predicts a relation between the branching ratio of $K_S^0 K_L^0$ and the phase between the three-gluon and the one-photon annihilation amplitudes [7]. So the measurement of the $K_S^0 K_L^0$ branching ratio is important to determine the phase. From data analysis point of view, the event topology of $\psi(2S) \to K_S^0 K_L^0$ is fairly prominent: the neutral K_L almost leaves no information in Main Drift Chamber due to long decay lifetime, while the K_S swiftly decays into two pions. By the virtue of this characteristic topology of event, two good charged tracks are required with net charge zero; in addition, secondary vertex requirement is applied for K_S identification. With these requirements, the distribution of the momentum of K_S is obtained as shown in Fig. 2 (a). The different shaded histograms indicate different estimation and simulation of background, whose shape in the vicinity of signal region is described by exponential function. For signal events, the Gaussian function is used to fit the observed number of events. The final branching ratio is worked out to be $(5.25 \pm 0.47 \pm 0.63) \times 10^{-5}$. The similar study has also been made for $J/\psi \to K_S^0 K_L^0$ decay. The momentum distribution of K_S is shown in Fig. 2 (b) and the branching ratio is worked out to be $(1.86 \pm 0.47 \pm 0.63) \times 10^{-4}$. It is worth while to notice that the BES measurement result is considerably larger than that of the PDG value: $\mathscr{B}_{J/\psi \to K_S^0 K_L^0} = (1.08 \pm 0.47) \times 10^{-4}$.

In contrast with VT channel, the Q_h value for PP channel is enhanced greatly. Using BES measured branching ratios of $K_S^0 K_L^0$ decay from J/ψ and $\psi(2S)$, the Q_h value

TABLE 1. The results of $\psi(2S)$ and χ_{cJ} decays.

VT channel	$\mathscr{B}_{\psi(2S)}$ (10^{-4}) (from BES)	$\mathscr{B}_{J/\psi}$ (10^{-3}) (from PDG2002)	Q_h
ωf_2	$2.05 \pm 0.41 \pm 0.46$	4.3 ± 0.6	4.8 ± 1.5
ρa_2	$2.55 \pm 0.73 \pm 0.60$	10.9 ± 2.2	2.3 ± 1.1
$K^* \overline{K_2^*} + c.c.$	$1.64 \pm 0.33 \pm 0.41$	6.7 ± 2.6	2.4 ± 1.2
$\phi f_2'$	$0.48 \pm 0.14 \pm 0.12$	$1.23 \pm 0.06 \pm 0.20^*$	3.9 ± 1.6
PP channel	$\mathscr{B}_{\psi(2S)}$ (10^{-5}) (from BES)	$\mathscr{B}_{J/\psi}$ (10^{-4}) (from BES)	Q_h
$K_S^0 K_L^0$	$5.25 \pm 0.47 \pm 0.63$	$1.86 \pm 0.43 \pm 1.2$	28.2 ± 4.7
Decay mode	$\mathscr{B}_{Exp.}$ (10^{-4}) (from BES)	$\mathscr{B}_{The.}$ (10^{-4}) (by COM)	$R_{Exp./The.}$
$\chi_{c0} \to \Lambda\overline{\Lambda}$	$4.7^{+1.3}_{-1.2} \pm 1.0$	$-$	$-$
$\chi_{c1} \to \Lambda\overline{\Lambda}$	$2.6^{+1.0}_{-0.9} \pm 0.6$	0.366	7.1
$\chi_{c2} \to \Lambda\overline{\Lambda}$	$3.3^{+1.5}_{-1.3} \pm 0.7$	0.333	9.9

* This value from DM2 only.

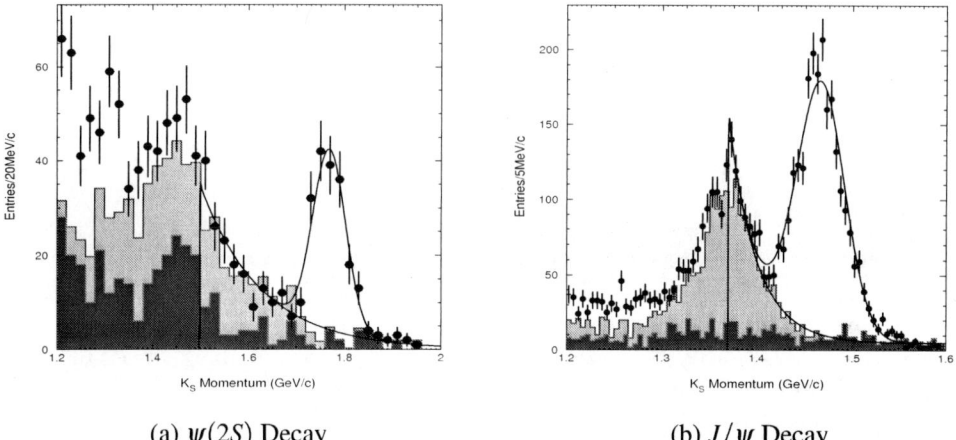

(a) $\psi(2S)$ Decay   (b) J/ψ Decay

FIGURE 2. The K_S^0 momentum distribution for (a) $\psi(2S)$ decay and (b) J/ψ decay. The dots with error bars are data, the dark shaded histogram is from K_S^0 mass sideband events, and the light shaded histogram is Monte Carlo simulated backgrounds. The curves shown in the plot are from a best fit of the distribution.

is calculated as $(28.2 \pm 4.7)\%$ with fairly high precision. Comparing to 12% rule, the deviation is greater than 3 σ. According to Ref. [7], the relation between branching ratio of $\psi(2S) \to K_S^0 K_L^0$ and the phase between the three-gluon and the one-photon annihilation amplitudes is shown in Fig. 3(a) , where three inputs correspond to three groups of branching ratios of $\psi(2S)$ to $\pi^+\pi^-$ or K^+K^-. Using the $K_S^0 K_L^0$ branching ratio,

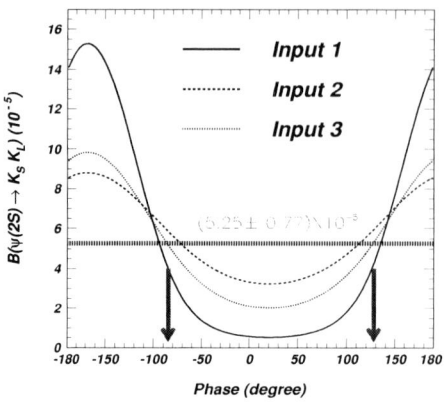

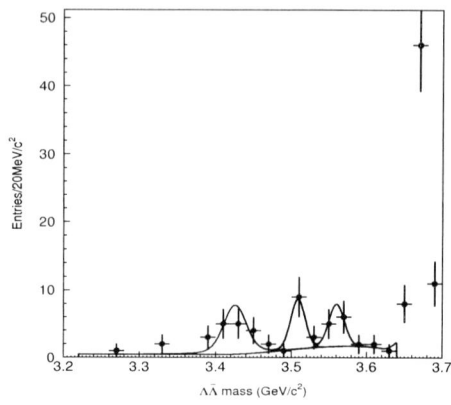

(a) Relation between $\mathcal{B}(K_S^0 K_L^0)$ and phase   (b) Mass distribution of $\gamma \Lambda \overline{\Lambda}$

FIGURE 3. (a) $\psi(2S) \to K_S^0 K_L^0$ branching ratio as a function of the relative phase for three different inputs: input one based on DASP results; input two on BES results; input three on K^+K^- result from BES and $\pi^+\pi^-$ result derived from pion form factor (for detail information, see Ref. [7]). (b) Mass distribution of $\gamma \Lambda \overline{\Lambda}$ candidates, fit with three mass resolution smeared Breit-Wigner functions and a background as estimated from data sideband and Monte Carlo simulation.

the phase is determined to be either $-85°$ or $130°$. Here the most interesting result is the large phase which supports the theoretically favored orthogonal phase assumption [8].

STUDY OF χ_{cJ} DECAYS

The large sample of $\psi(2S)$ decays permits studies of χ_{cJ} decays with high precision. Some theoretical papers of interest are given in Ref. [9], and experiment results from BES could refer to Refs. [10] and [11]. A recent analysis involving χ_{cJ} decay is the measurement of branching ratios of $\chi_{cJ} \to \Lambda\overline{\Lambda}$. The detailed information could be found in Ref. [12], where $\gamma\pi^+\pi^- p\overline{p}$ events with $\pi^+\pi^- p\overline{p}$ mass in the χ_{cJ} mass region are studied carefully. The background from non-$\Lambda\overline{\Lambda}$ event is estimated from the Λ mass sidebands of data distribution, while that from channels with $\Lambda\overline{\Lambda}$ production is estimated according to Monte Carlo simulation for the following decay modes: $\psi(2S) \to \Lambda\overline{\Lambda}, \Sigma\overline{\Sigma}^0, \Lambda\overline{\Sigma}^0 + c.c.$, $\Xi\overline{\Xi}^0 + c.c.$, and $\psi(2S) \to \gamma\chi_{cJ}$, $\chi_{cJ} \to \Sigma\overline{\Sigma}^0 \to \gamma\gamma\Lambda\overline{\Lambda}$. In addition, $\psi(2S) \to \pi^+\pi^- J/\psi \to \pi^+\pi^- p\overline{p}$ as background is also taken into consideration. The background shape is determined by combining two kinds of background estimation, and the observed numbers of events are obtained from fitting of the selected $\Lambda\overline{\Lambda}$ mass spectrum, as shown in Fig. 3(b), and the branching ratios could be found in Table 1.

For comparison, the relevant theoretical results are also listed in Table 1, where the theoretical calculation is based on Color Octet Mechanism (COM). According to the values listed in the table, it could be seen the results on χ_{c1} and χ_{c2} decays only agree marginally with model predictions.

SUMMARY

Based on 14 M $\psi(2S)$ data sample, the branching ratios of four VT channels, ωf_2, $\rho a_2, K^* \overline{K_2^*} + c.c.$, and $\phi f_2'$ are measured. The suppression of this decay mode with respect to "12%" rule is confirmed with better precision. The final state $K_S^0 K_L^0$ was firstly observed in $\psi(2S)$ decay. The Q_h of this final state is calculated with high precision, which is considerably enhanced comparing with "12%" rule. In addition, using the branching ratio of $\psi(2S) \rightarrow K_S^0 K_L^0$, the phase between the three-gluon and the one-photon annihilation amplitudes is determined to be either $-85°$ or $130°$. The branching ratios of $\chi_{cJ} \rightarrow \Lambda\overline{\Lambda}$ are measured, with these results, the effectiveness of the calculation based on Color Octet Mechanism was tested.

ACKNOWLEDGMENTS

I would like to acknowledge help from Prof. C. Z. Yuan and Dr. W. F. Wang, who provided corresponding results presented in this report. Thanks are also due to Prof. Y. S. Zhu, Prof. S. Jin and other colleagues of BES collaboration who gave many suggestions and comments for improvement of my report.

REFERENCES

1. J. Z. Bai, *et al.*, (BES Collab.), Nucl. Inst. Meth. **A344**, 319 (1994);**A458**, 627 (2001).
2. T. Appelquist and H. D. Politzer, Phys. Rev. Lett. **34**, 43 (1975);
   A. De Rújula and S. L. Glashow, Phys. Rev. Lett. **34**, 46 (1975).
3. M. E. B. Franklin *et al.*, (MARKII Collab.), Phys. Rev. Lett. **51**, 963 (1983).
4. Y. S. Zhu, Proc. of the 28th Intern. Conf. on High Energy Physics (Warsaw University, 1996), Eds. Z. Ajduk and A. K. Wroblewski (World Scientific, 1997) p.507.
5. J. Z. Bai, *et al.*, (BES Collab.), Phys. Rev. Lett. **81**, 5080 (1998).
6. Particle Data Group, K. Hagiwara *et al.*, Phys. Rev. **D66**, 01001 (2002).
7. C. Z. yuan, P. Wang and X. H. Mo, Phys. Lett. **B567**, 73 (2003).
8. J.-M. Gérard and J. Weyers, Phys. Lett. **B462**, 324 (1999).
9. J. Bolz, P. Kroll and G. A. Schuler, Phys. Lett. **B392**, 198 (1997); G. T. Bodwin, E. Braaten and G. P. Lepage, Phys. Rev. **D51**, 1125 (1995); P. Kroll and S. M. H. Wong, hep-ph/9710464; V. L. Chernyak and A. R. Zhitnitsky, Nucl. Phys. **B201**, 492 (1982); S. J. Brodsky and G. P. Lepage, Phys. Rev. **D24**, 2848 (1995); A. Duncan and A. H. Mueller, Phys. Lett. **B93**, 119 (1980).
10. J. Z. Bai, *et al.*, (BES Collab.), Phys. Rev. Lett. **81**, 3091 (1998).
11. F. Liu, Nucl. Phys. **A675**, 71c-75c (2000).
12. J. Z. Bai, *et al.*, (BES Collab.), Phys. Rev. **D67**, 112001 (2003).

Study of the $e^+e^- \to \pi^0\gamma$ process in the energy range $0.60 - 1.06$ GeV.[1]

M.N.Achasov*, K.I.Beloborodov*, A.V.Berdugin*, A.G.Bogdanchikov*,
A.V.Bozhenok*, A.D.Bukin*, D.A.Bukin*, A.V.Vasiljev*, T.V.Dimova*,
V.P.Druzhinin*, V.B.Golubev*, *A.A.Korol**, S.V.Koshuba*, A.P.Lysenko*,
E.V.Pakhtusova*, S.I.Serednyakov*, Yu.M.Shatunov*, V.A.Sidorov*,
Z.K.Silagadze*, A.N.Skrinsky* and Yu.V.Usov*

Budker Institute of Nuclear Physics
Siberian Branch of the Russian Academy of Sciences

Abstract. The process $e^+e^- \to \pi^0\gamma$ in the energy range $\sqrt{s} = 0.60 - 1.06$ GeV was studied at VEPP-2M collider with SND detector using $\sim 14\,pb^{-1}$ of integrated luminosity. Data were analyzed in the framework of the vector meson dominance model. Preliminary results on obtained cross-section and parameters of $\rho^0, \omega, \phi \to \pi^0\gamma$ radiative decays are presented.

INTRODUCTION

In the vector meson dominance (VMD) model the $e^+e^- \to \pi^0\gamma$ process is considered as a transition $e^+e^- \to \rho^0, \omega, \phi \to \pi^0\gamma$. Further the decay parameters can be related with $\omega, \phi \to \rho^0\pi^0$ and $\pi^0 \to 2\gamma$ [1]. Another approach to the decays is a non-relativistic quark model (NQM). In this model vector mesons considered as composite from valent quark with codirected spins, while pseudoscalar mesons with opposite directed spins. The decay may be explained as spin overturn with photon emission (magneto-dipole transition)[2]. The study of such decays is important for understanding light mesons structure and tests of strong interactions at low energies.

EXPERIMENT

The experiment was made with SND detector [3] at VEPP-2M collider[4]. The SND detector consists of tracking system, electromagnetic calorimeter and muon veto system. The principal part of the detector is an electromagnetic calorimeter. Full thickness of calorimeter for particles originating from the detector center is $13.4X_0$, total solid angle is $90\% \cdot 4\pi$. Calorimeter energy resolution for photons is $\frac{\sigma_E}{E} \approx \frac{4.2\%}{E(GeV)^{1/4}}$, its angle resolution is $\sigma_{\varphi,\theta} \approx \frac{0.82°}{\sqrt{E(GeV)}} \oplus 0.63°$ [5].

---

[1] Presented by A.A.Korol, *e-mail: korol@inp.nsk.su*

CP717, *Hadron Spectroscopy: Tenth International Conference,*
edited by E. Klempt, H. Koch, and H. Orth
© 2004 American Institute of Physics 0-7354-0197-7/04/$22.00

Presented work is an extension of the studies published in Ref.[6, 7] but using additional data. The data were collected in experiments in 1998 and 2000 [8]:

- PHI98: energy range $0.98 - 1.06$ GeV, integrated luminosity 7.83 pb^{-1} ($\sim 1.2 \cdot 10^7$ ϕ-mesons)
- OME00: energy range $0.60 - 0.97$ GeV, integrated luminosity 5.93 pb^{-1} ($\sim 2.5 \cdot 10^6$ ω-mesons)

ANALYSIS

The process $e^+e^- \to \pi^0\gamma$ was studied in the 3γ final state. Main background sources are $e^+e^- \to 3\gamma$ (QED origin); $e^+e^- \to 2\gamma$ misidentified because of additional clusters from machine background; $e^+e^- \to \eta\gamma$, and cosmic background. For luminosity calculation events of the $e^+e^- \to 2\gamma$ process were used.

The events were accepted by first level trigger which allows two or more clusters in calorimeter, no signal in either tracking or muon system, energy deposition in calorimeter greater then some threshold (which varied but never was greater than $0.4\sqrt{s}$).

Preliminary selection included following cuts: no charged tracks, number of clusters in calorimeter $N_{np} \geq 3$, calorimeter energy deposition $E_{cal} \geq 0.65\sqrt{s}$, calorimeter total momentum $P_{cal} \leq 0.3\sqrt{s}$, polar angles of two most energetic clusters $36° \leq \theta_{1,2} \leq 144°$, polar angle of the next by energy cluster $27° \leq \theta_3 \leq 153°$, its energy deposition $E_{cal\,3} \geq 0.1\sqrt{s}$.

Kinematic fit with energy and momentum conservation constraints was applied to selected events. It improved π^0 mass resolution from $\sigma_{m_{\gamma\gamma}} = 11.2$ MeV to $\sigma_{m_{\gamma\gamma}} = 8.6$ MeV. Cut on the fit quality parameter $\chi^2_{3\gamma} < 20$ was then applied. Selected events were subdivided into two classes: 108 MeV$\leq m_{\gamma\gamma} \leq 162$ MeV (class A) and the rest (class B). Class B events were used to cross-check our understanding of background contribution and estimate related systematic uncertainty. For class A $\sim 7 \cdot 10^4$ events were selected.

For luminosity calculation events were selected (class C) with no charged tracks, two or more clusters with calorimeter energy deposition $E_{cal\,1,2} \geq 0.3\sqrt{s}$, polar angles $36° \leq \theta_{1,2} \leq 144°$, acollinearity $\Delta\varphi_{12} \leq 10°$, $\Delta\theta_{12} \leq 25°$. Events satisfying preliminary selection cuts for A and B classes were excluded. It is necessary to note significant contribution of $e^+e^- \to \pi^0\gamma$ events (up to 10% at ω peak) to this class.

Cross section of the studied process was parametrized with usual variable width Breit-Wigner description [9]. More details published in Ref.[7]. Cross sections $\sigma_{V\pi^0\gamma}$ of the $e^+e^- \to V \to \pi^0\gamma$ transitions at resonances mass ($V = \rho^0, \omega, \phi$) were approximation parameters. Other parameters considered were relative phases $\varphi_{\rho\omega}$ and $\varphi_{\phi\omega}$, and constant real amplitude $a_{\pi^0\gamma}$ taking into account possible contribution from higher resonances decays. Masses and widths of ω, ρ^0, ϕ-mesons and branching ratios of ω and ϕ major decay modes were taken from previous SND measurements [12], other external parameters were taken from review of particle physics [13].

Visible cross section of $e^+e \to \pi^0\gamma$ was calculated taking into account radiative

TABLE 1. Approximation results

No.	1	2	3	4
χ^2/N	235/56	51/54	51/54	51/53
$\sigma_{\omega\pi^0\gamma}$, nb	174.1 ± 0.9	151.9 ± 1.5	152.1 ± 1.6	152.4 ± 1.6
$\sigma_{\rho^0\pi^0\gamma}$, nb	0	0.59 ± 0.07	0.58 ± 0.07	0.56 ± 0.07
$\sigma_{\phi\pi^0\gamma}$, nb	6.11 ± 0.17	5.58 ± 0.29	5.73 ± 0.31	5.64 ± 0.37
$\varphi_{\rho\omega}$,°		−9.3 ± 3.3	−9.3 ± 3.3*	−12.6 ± 1.1†
$\varphi_{\phi\omega}$,°	169 ± 8	151 ± 9	160 ± 9	160 ± 10
Re $a_{\pi^0\gamma}$,nb$^{\frac{1}{2}}$	0	0	0	−0.05 ± 0.07

* estimated from $\rho - \omega$ mixing
† estimated from $\rho - \omega$ mixing, $\Pi_{\rho\omega} = -3676 \pm 303$ MeV^2 fixed using $B_{\omega\to2\pi}$

correction [14] and dependency on registration efficiency from the energy of initial state radiation photons. Registration efficiency for all processes was calculated with Monte Carlo simulation.

Cross section of QED $e^+e^- \to 2\gamma$ was calculated using Ref.[15]. Cross section for $e^+e^- \to 3\gamma$ process was calculated from tree-level diagram corrected using Ref.[16]. Cosmic background events fraction was estimated from the data collected without beams.

Cross section approximation was done using maximum likelihood method. Classes A, B, C were approximated at the same fit. Integrated luminosity was recalculated at the every minimization step.

Four models were used for approximation: (1) no ρ meson decay contribution; (2) including ρ meson contribution, with fixed $\rho - \omega$ relative phase; (3) including ρ meson contribution, $\varphi_{\rho\omega}$ estimated from electromagnetic $\rho - \omega$ mixing [10, 11]; (4) including ρ meson contribution, with electromagnetic $\rho - \omega$ mixing fixed using $B_{\omega\to2\pi}$, and constant real amplitude.

RESULTS AND DISCUSSION

The approximation results are summarized in the Table 1. Model (1) contradicts to experimental data, models (2)–(4) describe data equally well. Obtained cross section is shown on Fig. 1. Main sources of systematic errors of cross section are uncertainty in luminosity determination (2%), in registration efficiency (2.5%), QED and $e^+e^- \to \eta\gamma$ background subtraction ($0.5 - 20\%$ depends on energy point). Cosmic background contribution was found negligible ($< 0.1\%$). Model and external parameters uncertainty also give contribution to systematic errors of cross section parameters.

Cross section parameters were found as:

$$\sigma_{\rho^0\pi^0\gamma} = (0.59 \pm 0.07 \pm 0.06)\,nb$$
$$\sigma_{\omega\pi^0\gamma} = (151.9 \pm 1.5 \pm 4.6)\,nb$$
$$\sigma_{\phi\pi^0\gamma} = (5.58 \pm 0.29 \pm 0.29)\,nb$$

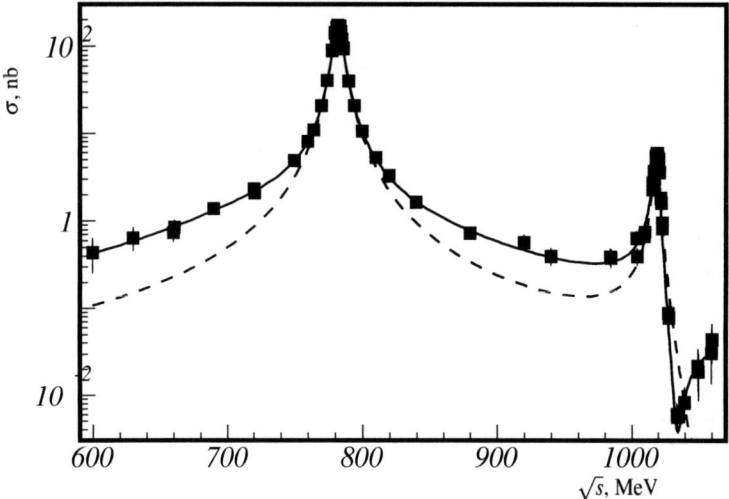

FIGURE 1. Cross section of the $e^+e^- \to \pi^0\gamma$ process. Dots represent experimental data, solid curve depicts approximated theoretical form with $\rho^0 \to \pi^0\gamma$ (model 2), dashed curve — without ρ^0 contribution (model 1).

TABLE 2. Comparison with previous experimental data.

Source	Γ_ω, keV	Γ_ρ, keV	Γ_ϕ, keV
This work	$733 \pm 7 \pm 22$	$80.4 \pm 9.7 \pm 7.6$	$5.98 \pm 0.31 \pm 0.31$
SND (2000,2003) [6, 7]	$788 \pm 12 \pm 27$	$77 \pm 17 \pm 11$	$5.40 \pm 0.16 {}^{+0.43}_{-0.40}$
PDG2002 [13]	747 ± 41	121 ± 31	5.28 ± 0.43
$\omega \to$*neutrals* (PDG2002 [13])	767 ± 79		
$\rho^\pm \to \pi^\pm\gamma$ (PDG2002[13])		68 ± 7	

$$\varphi_{\rho\omega} = -9.3° \pm 3.3° \pm 2.4°$$
$$\varphi_{\phi\omega} = 151° \pm 9° \pm 11°$$

and branching ratios of $\rho^0, \omega, \phi \to \pi^0\gamma$ decays:

$$B_{\omega\to\pi^0\gamma} = (8.45 \pm 0.09 \pm 0.25)\%$$
$$B_{\rho^0\to\pi^0\gamma} = (5.32 \pm 0.63 \pm 0.50) \cdot 10^{-4}$$
$$B_{\phi\to\pi^0\gamma} = (1.34 \pm 0.07 \pm 0.07) \cdot 10^{-3}$$

The measured decays partial widths and previous experimental data are summarized in the Table 2. Measured parameters agree with previous experimental data and mostly compatible with phenomenological expectations (see e.g. tables in Refs.[6, 7]). Relative phase of ρ and ω decays may be attributed to $\rho - \omega$ mixing ($-12.6° \pm 1.1°$ from

$B_{\omega \to 2\pi}$). Partial width ratios $\frac{\Gamma_{\omega \to \pi^0 \gamma}}{\Gamma_{\rho^0 \to \pi^0 \gamma}} = 9.11 \pm 1.10 \pm 0.90$ and $\frac{\Gamma_{\phi \to \pi^0 \gamma}}{\Gamma_{\omega \to \pi^0 \gamma}} = (8.16 \pm 0.43 \pm 0.48) \cdot 10^{-3}$ are compatible with SU(3) predictions.

CONCLUSION

The cross section of the $e^+ e^- \to \pi^0 \gamma$ process is measured in the c.m. energy region $\sqrt{s} = 0.60 - 1.06$ GeV using the data obtained by the SND detector at the VEPP-2M collider. Total integrated luminosity used is $\sim 14 \, pb^{-1}$, $7 \cdot 10^4$ events were selected for analysis.

Data were analyzed in the framework of VMD model, parameters of decays $\rho^0, \omega, \phi \to \pi^0 \gamma$ were obtained. Parameters of decays agree with previous measurements and phenomenological estimations. Partial width of the $\rho^0 \to \pi^0 \gamma$ decay is close to the $\rho^\pm \to \pi^\pm \gamma$ one. The relative phase for ρ^0, ω decays can be attributed to electromagnetic $\rho - \omega$ mixing.

ACKNOWLEDGMENTS

One of the authors thanks organizers of the conference for the invitation and support. This work was supported in part by Presidential Grant 1335.2003.2 for support of Leading Scientific Schools and by Russian Science Support Foundation.

REFERENCES

1. M.Gell-Mann, et al., *Phys. Rev. Lett.*, **8**, 261–262 (1962).
2. D.A.Geffen, and Wilson, W., *Phys. Rev. Lett.*, **44**, 370 (1980).
3. Achasov, M. N., et al., *Nucl. Instrum. Meth.*, **A449**, 125–139 (2000).
4. Skrinsky, A. N., "VEPP-2M status and prospects and ϕ-factory project at Novosibirsk," in *Proc. of Workshop on physics and detectors for DAΦNE 95*, 1995, vol. IV, p. 3.
5. Achasov, M. N., et al., *Nucl. Instrum. Meth.*, **A411**, 337–342 (1998).
6. M.N.Achasov, et al., *Eur. Phys. J.*, **C12**, 25–33 (2000).
7. Achasov, M. N., et al., *Phys. Lett.*, **B559**, 171–178 (2003).
8. Achasov, M. N., et al., *AIP Conf. Proc.*, **619**, 30–39 (2002).
9. N.N.Achasov, et al., *Sov. J. Nucl. Phys.*, **54**, 664–671 (1991).
10. Achasov, N. N., and Kozhevnikov, A. A., *Sov. J. Nucl. Phys.*, **55**, 449–459 (1992).
11. H.B.O'Connell, et al., *Prog. Part. Nucl. Phys.*, **39**, 201–252 (1997).
12. Achasov, M. N., et al., *Phys. Rev.*, **D63**, 072002 (2001), Achasov, M. N., et al., *Phys. Rev.*, **D65**, 032002 (2002), Achasov, M. N., et al., *Phys. Rev.*, **D66**, 032001 (2002), Achasov, M. N., et al. (2003), hep-ex/0305049 (to be published in Phys. Rev. D).
13. Hagiwara, K., et al., *Phys. Rev.*, **D66**, 010001 (2002).
14. E.A.Kuraev, and V.S.Fadin, *Sov. J. Nucl. Phys.*, **41**, 466–472 (1985).
15. Baier, V. N., et al., *Phys. Rept.*, **78**, 293–393 (1981).
16. Kuraev, E. A., and Silagadze, Z. K., *Jad. Fiz.*, **58**, 1843–1845 (1995).

Preliminary results of the analysis of the centrally produced $\phi\phi$ system

M.A.Reyes[a], M.C.Berisso[b], D.C.Christian[c], J.Felix[a], A.Gara[d],
E.E.Gottschalk[c], G.Gutierrez[c], E.P.Hartouni[e], B.C.Knapp[d],
M.N.Kreisler[b,e], S.Lee[b], K.Markianos[b], G.Moreno[a], M.H.L.S.Wang[b,c],
A.Wehman[c], D.Wesson[b]

[a] *Universidad de Guanajuato, León, Guanajuato, México*
[b] *University of Massachusetts, Amherst, Massachusetts, USA*
[c] *Fermilab, Batavia, Illinois, USA*
[d] *Columbia University, Nevis Laboratory, New York, USA*
[e] *Lawrence Livermore National Laboratory, Livermore, California, USA*

Abstract. We present preliminary results of the analysis of the centrally produced $\phi\phi$ system at 800 GeV/c in the reaction $pp \rightarrow p_{slow}(\phi\phi)p_{fast}$. A partial wave analysis of the data up to 2.6 GeV/c^2 shows that three waves with $J^{PC}LS = 2^{++}02$ and with different J_z and reflectivity are needed to describe the data. A fit to the wave amplitudes using Jost functions gives two resonances, one above and one below the $\phi\phi$ threshold. The resonance above threshold has a Breit-Wigner structure with $M_R = 2.243 \pm 0.015(stat) \pm 0.010(syst)$ GeV and $\Gamma_R = 0.368 \pm 0.033(stat) \pm 0.030(syst)$ GeV.

The first observation of $\phi\phi$ production was made using the BNL–MPS Spectrometer at 22.6 GeV/c [1] in the OZI [2] suppressed reaction,

$$\pi^- p \rightarrow \phi\phi n \tag{1}$$

A Partial Wave Analysis (PWA) showed that only three 2^{++} waves were necessary to fit this data [3]. The larger than expected cross section observed indicated that these states may not be considered conventional $q\bar{q}$ mesons [4].

We present here preliminary results of the analysis of the centrally produced $\phi\phi$ system in the 800 GeV/c doubly diffractive reaction,

$$pp \rightarrow p_{slow}(\phi\phi)p_{fast}, \quad \phi \rightarrow K^+ K^- \tag{2}$$

using events of this reaction selected from the 4×10^9 pp interaction data sample recorded by Fermilab E690 during the 1991 fixed target run.

The E690 apparatus consisted of a high rate, open geometry multiparticle spectrometer used to measure the target system (T) in $pp \rightarrow p_{fast}(T)$ reactions, and a beam spectrometer system used to measure the incident 800 GeV/c beam and

CP717, *Hadron Spectroscopy: Tenth International Conference,*
edited by E. Klempt, H. Koch, and H. Orth
© 2004 American Institute of Physics 0-7354-0197-7/04/$22.00

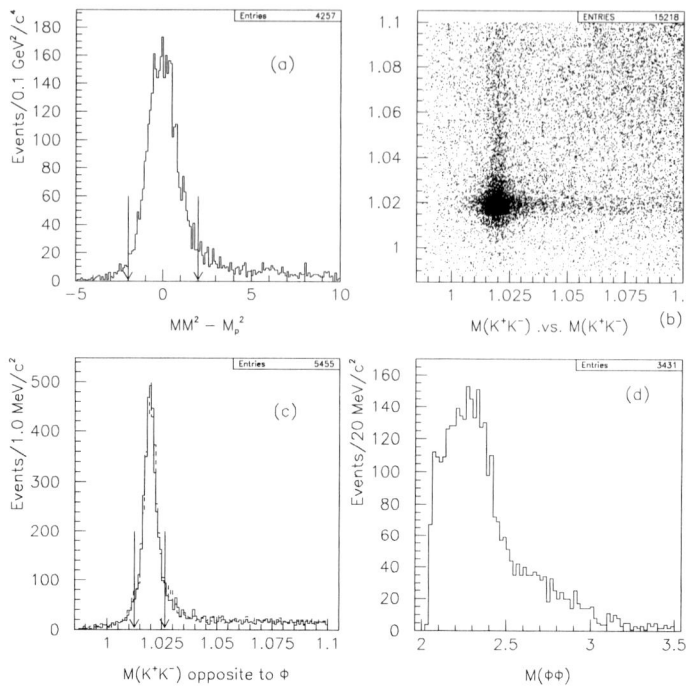

FIGURE 1. (a) Missing mass squared minus proton mass squared for events in reaction (2). (b) K^+K^- invariant mass, first vs. second pair. (c) K^+K^- invariant mass, when the other pair lies in the ϕ–mass band. (d) $\phi\phi$ invariant mass.

scattered proton. A liquid hydrogen target was located just upstream of the multiparticle spectrometer. The 96 cell Cherenkov counter located at the downstream end of the main spectrometer magnet used Freon 114 as a radiator and had a pion threshold of 2.57 GeV/c. The E690 apparatus has been described elsewhere [5].

After the track and vertex reconstruction stage of the data analysis, final state (2) was selected by requiring a primary vertex in the LH_2 target with two positive and two negative tracks, an incoming beam track and a fast forward proton. All four tracks were required to have Cherenkov information compatible with being charged kaons. The Cherenkov information for at least one track was required to be incompatible with the hypothesis that the track was a pion. A kinematical cut was made on the missing momentum of $p_z < 250\,\mathrm{MeV/c}$ or $\arctan(p_t/p_z) > 45$ degrees. This cut ensured that the missing target recoil proton was outside the geometrical acceptance of the spectrometer.

The missing mass squared (MM^2) minus proton mass squared shown in Fig.1a has a clear peak around zero for events in reaction (2). Fig.1b shows the scatter plot of the first versus second mass pair for the 7609 events selected with $m(K^+K^-) < 1.1$ GeV/c^2 and $-2 < MM^2 - m_p^2 < 2$ GeV^2/c^4. Fig.1c shows the K^+K^- invariant mass, when the other pair lies in the ϕ–mass band of $1.0124 < M(K^+K^-) < 1.0264$

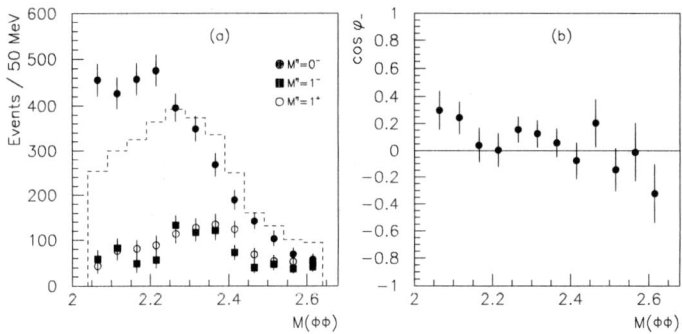

FIGURE 2. (a) $M(\phi\phi)$ data distribution (histogram) and acceptance corrected cross section (markers) from the PWA. (b) Phase difference between waves of negative reflectivity.

GeV/c². Fig.1d shows the $\phi\phi$ invariant mass after all selection cuts, showing a high bump between 2–2.5 GeV/c². 99.6% of events have only one combination per event into this plot. 3431 events with $M(\phi\phi) < 3$ GeV/c² remain.

Six angles are chosen to specify the spin and angular momentum of the $\phi\phi$ system. Two of them (γ, β) are defined as the Gottfried-Jackson (GJ) angles of one of the ϕ mesons, in the rest frame of the $\phi\phi$ system, with the z-axis in the direction of $\vec{p}_{fast} - \vec{p}_{beam}$, and in the y-axis in the direction of the $(\vec{p}_{fast} - \vec{p}_{beam}) \times (\vec{p}_{slow} - \vec{p}_{tgt})$ cross product, measured in the pp CM system. The rest are the two pairs of GJ angles $(\alpha_{1,2}, \theta_{1,2})$ for the K^+'s in their parent ϕ rest frames, with the z'-axis in the direction of $\vec{p}_\phi$, and with $y' = \hat{z} \times \hat{z}'$.

The allowable $\phi\phi$ basis vectors in terms of the total angular momentum J, orbital angular momentum L, parity P, and exchange reflectivity η, are given by [6]

$$G^{J^P L S M^\eta}(\gamma, \beta, \alpha_1, \alpha_2, \theta_1, \theta_2) = \text{Real}\left[\frac{(1-i) - \eta(1+i)}{2} \sum_{\mu, \lambda} C(1, 1, S | \mu, -\lambda) \times \right.$$

$$\left. C(L, S, J | 0, \mu - \lambda) e^{-iM\gamma} e^{i\mu\alpha_1} e^{i\lambda\alpha_2} d^J_{M, \mu-\lambda}(\beta) d^1_{\mu,0}(\theta_1) d^1_{\lambda,0}(\theta_2) \right] \quad (3)$$

where $M = |J_z|$. For this system $I = 0$, $C = +$, and $L + S =$ an even number.

We performed a PWA of our data divided in 12 bins of 50 MeV/c² beginning at 2.04 GeV/c², using the 54 waves with $0 \le J \le 4$, $0 \le L \le 3$, $0 \le S \le 2$, $0 \le M \le 2$; 14 of these have $M = 0$. For each mass bin, we selected the wave with best log-likelihood, and then added a second and third wave and selected the best fit with waves following from bin to bin. The results of this procedure are plotted in Fig.2. Only three waves are necessary to describe the data, all of them with $J^{PC} LS = 2^{++} 02$ (Fig.2a). One of the waves has $M = 0$ and $\eta = -1$, and the other two have $M = 1$ with reflectivities -1 and $+1$. The two waves with $\eta = -1$ interfere with a phase of about 90° (Fig.2b.) Including the two $2^{++} 02$ waves with $M = 2$ does not improve the significance of the PWA.

We fitted the wave amplitudes obtained from the PWA following Morgan and Pennington's approach [7]. We used Jost functions to parameterize the $\mathcal{S}$-matrix,

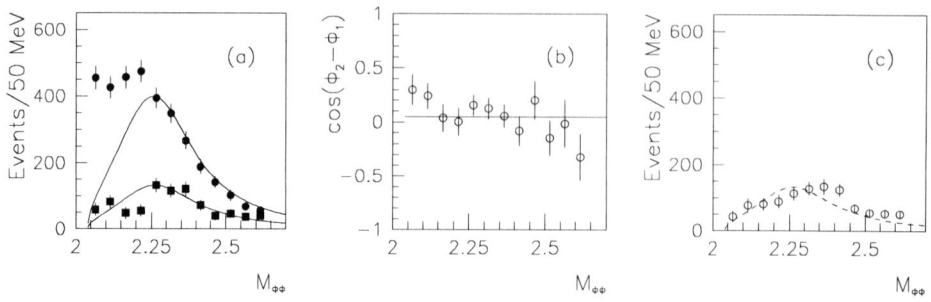

FIGURE 3. (a) Fit of negative reflectivity wave amplitudes with two k-poles (one BW.) (b) Phase between negative reflectivity waves. (c) Fit to the positive reflectivity wave amplitude.

and assumed two channel decays, XX and $\phi\phi$, where X is assumed to be a low mass scalar (π's, for example). Each Breit-Wigner (BW) structure introduced in the fit was parameterized using two k-poles in Riemann sheets II and III, and the production amplitudes are introduced as real numbers.

We first fitted the wave amplitudes using only two k-poles. In order to make this fit converge we excluded the first three 50 MeV bins. The BW parameters obtained from the fit [7] are $M_R = 2.249 \pm 0.015$ GeV and $\Gamma_R = 0.340 \pm 0.027$ GeV. This fit is shown in Fig.3.

Obviously, the data needs another set of k-poles below threshold to describe the $J_z = 0$ wave amplitude near threshold. For the four k-pole fit in all twelve bins, we assumed that for $|J_z| \neq 0$ production is only via the $\phi\phi$ channel, while for $|J_z| = 0$ production can be via the XX and $\phi\phi$ channels. From the four k-poles that resulted from the fit we extracted the following parameters for the BW structure above threshold:

$$M_R = 2.243 \pm 0.015(stat) \pm 0.010(syst) \text{ GeV}$$
$$\Gamma_R = 0.368 \pm 0.033(stat) \pm 0.030(syst) \text{GeV} \tag{4}$$

while the BW parameters extracted for the resonance below threshold are

$$M_R \sim 1.9 \text{ GeV}, \quad \Gamma_R \sim 0.3\text{GeV} \tag{5}$$

but its parameters can not be fixed by using only the $\phi\phi$ data. This fit is shown in Fig.4. For the first BW resonance, the statistical error in M_R and Γ_R comes from the fit (χ^2/d.o.f. of 1.4), while for the systematic error we only used the difference between the two k-pole fit and four k-pole fit.

CONCLUSIONS

We report preliminary results of the analysis of the centrally produced $\phi\phi$ system at 800 GeV/c. A PWA of the data shows that it can be described using three waves

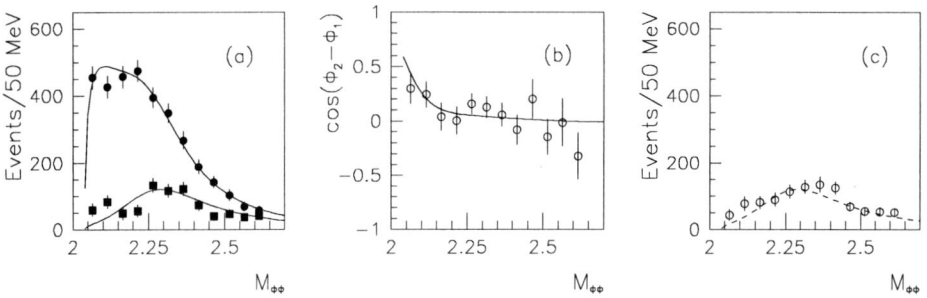

FIGURE 4. (a) Fit of negative reflectivity wave amplitudes with four k-poles (two BW's.) (b) Phase between negative reflectivity waves. (c) Fit to the positive reflectivity wave amplitude.

with $J^{PC}LS = 2^{++}02$, and with $M^{\eta} = 0^-$, 1^-, 1^+. The four k-pole fit to the wave amplitudes gives a BW resonance above threshold with the following parameters

$$M_R = 2.243 \pm 0.015(stat) \pm 0.010(syst) \text{ GeV}$$
$$\Gamma_R = 0.368 \pm 0.033(stat) \pm 0.030(syst)\text{GeV}$$

while the parameters for the resonance below threshold can not be accurately fixed using only the $\phi\phi$ channel.

Acknowledgements

This work was funded in part by the Department of Energy under Contracts No. DE-AC02-76CHO3000 and No. DE-AS05-87ER40356, the National Science Foundation under Grants No. PHY89-21320 and No. PHY90-14879, and CONACyT de México under Grants No. CONACyT-NSF/2001 and No. PROMEP/103.5/02/1436.

REFERENCES

1. A.Etkin *et al.*, *Phys.Rev.Lett.* **40**, 422 (1978).
2. S.Okubo, *Phys.Lett.* **5**, 165 (1963); G.Zweig, *CERN Report No. TH-401 and TH-412*, 1964 (unpublished); J.Iizuka, *Prog.Theor.Phys.*, Suppl. **37-38**, 21 (1966).
3. A.Etkin *et al.*, *Phys.Lett.* **B201**, 568 (1988).
4. K.Hagiwara *et al.*, *Euro.Phys.Jour.* **C3**, 1 (2002), pp.754-756.
5. E.P.Hartouni *et al.*, *Nucl.Instrum.Methods* **A 317**, 161 (1992); J.Uribe *et al.*, *Phys. Rev.* **D49**, 4373 (1994); D.C.Christian *et al.*, *Nucl.Instrum.Methods* **A345**, 62 (1994).
6. R.S.Longacre, *AIP Conf. Proc.***113**, 0051 (1984).
7. D.Morgan and M.R.Pennington, *Phys. Rev.* **D48**, 1185 (1993).

Vector Meson Property in Covariant Classification Scheme

Masuho Oda[1]

Faculty of Engineering, Kokushikan University, Tokyo 154-8515, Japan

Abstract. Recently our collaboration group has proposed the covariant classification shceme of hadrons, leading to possible existence of two ground state vector mesons. One is corresponding to ordinary ρ nonet and the other is extra ρ nonet. We investigate the decay property of $\omega(1250)$ and $\rho(1250)$ in the covariant classification scheme. And it is shown that $\omega(1250)$ is promising candidate of our extra ω meson.

INTRODUCTION

The existence of "extra" vector mesons $\rho(1250)$ and $\omega(1250)$ is pointed[1] out by recent and several old works. It has been reported that they have following properties. Concernig $\omega(1250)$: i) It was observed through $e^+e^- \to V \to \pi^+\pi^-\pi^0$ process and found that $\rho\pi$ intermediate state dominates there, ii) it was also found that the channel of $\omega(1250) \to \omega\pi\pi$ is strongly suppressed. Concerning $\rho(1250)$: It was seen in $e^+e^- \to V \to \pi^+\pi^-$ process. Although there is some wide enhancement in the cross section around 1.25(GeV), but peak is not clear. This indicates that $\rho(1250)$ has quite weak coupling to $\pi\pi$ channel and broad other decay channels. (i.e. $\Gamma_{tot}(\rho(1250) \to any)$ is large.)

The PDG[2] assignments of light-quark vector mesons are shown in Table.1. From this table we can see that these states, $\rho(1250)$ and $\omega(1250)$, are obviously out of PDG classification scheme based on non-relativistic qurak model. Recently our collaboration group has proposed[3, 4]the covariant classification scheme of hadrons, which is the extension of covariant oscillator qurak model (COQM) taking[2] into account all general Dirac-spinors of quarks inside hadron as physical components. It leads the existence of two ground state vector meson. One is ordinay ρ nonete and the other is extra ρ nonete. Extra vector mesons, $\omega(1250)$ and $\rho(1250)$, have the light mass to be assigned to our extra vector mesons predicted in the covariant classification sheme. The purpose of this talk is to point[5] out that this assignment is promising through the investigation of decay property of those mesons.

---

[1] Representing the collaboration group with M. Ishida, S. Ishida, T. Maede and K. Yamada
[2] In COQM, only boosted Pauli-spinors are taken as physical components.

CP717, *Hadron Spectroscopy: Tenth International Conference,*
edited by E. Klempt, H. Koch, and H. Orth
© 2004 American Institute of Physics 0-7354-0197-7/04/$22.00

TABLE 1. PDG assignments of light-quark vector mesons.

$N^{2S+1}L_J$	$(u\bar{u}-d\bar{d})/\sqrt{2}$	$(u\bar{u}+d\bar{d})/\sqrt{2}$	$s\bar{s}$
1^3S_1	$\rho(770)$	$\omega(783)$	$\phi(1020)$
2^3S_1	$\rho(1450)$	$\omega(1420)$	$\phi(1680)$
1^3D_1	$\rho(1700)$	$\omega(1650)$	
3^3S_1	$\rho(2150)$		

STRONG INTERACTION IN COVARIANT CLASSIFICATION SCHEME

Wave function of light-quark meson. In the covariant classification scheme mesons are described by the tensors in $\tilde{U}_{SF}(12)\otimes O(3,1)_L$-space,

$$\Phi_B^A(x_{1\mu},x_{2\mu}), \tag{1}$$

where $A=(\alpha,a)(B=(\beta,b))$ is the Drirac-spinor and flavor indices of quark (anti-quark), and $x_{1\mu}(x_{2\mu})$ is the space-time coordinate of quark (anti-quark). Our sheme is the boosted LS-coupling sheme. Concernig the interanl space-time part of Φ, we use[6] the definite metric type four-dimensional harmonic oscillator wave function. Concernig the spin- flavor part of Φ: Ground state mesons are assigned[3] to the representation $\mathbf{12}\times\mathbf{12^*}=\mathbf{144}$ of $\tilde{U}_{SF}(12)$ symmetry. The $\mathbf{144}$ is decomposed to four $((\mathbf{3}+\mathbf{1})_S,(\mathbf{3}+\mathbf{1})_F)$'s. Therefore, in the ground $q\bar{q}$ systems, there are the two sets (Non-reativistic and Super-reativistic) of pseudoscalar and vector nonets, and the two setes of scalar and axialvector nonets. In the following we will discuss only the spin wave function of pseudoscalar and vector mesons. The explicit form of them are given as follows:

Non-relativistic component

$$\chi_{Ps}^{(NR)}(p) = \frac{i\gamma_5}{2\sqrt{2}}(1+i\gamma v), \tag{2}$$

$$\chi_V^{(NR)}(p) = \frac{i\gamma_\mu}{2\sqrt{2}}(1+i\gamma v)\varepsilon_\mu(p), \tag{3}$$

Super-relativistic component

$$\chi_{Ps}^{(SR)}(p) = \frac{i\gamma_5}{2\sqrt{2}}(1-i\gamma v), \tag{4}$$

$$\chi_V^{(SR)}(p) = \frac{i\gamma_\mu}{2\sqrt{2}}(1-i\gamma v)\varepsilon_\mu(p), \tag{5}$$

where $\gamma v=\gamma_\mu p_\mu/M$ ($p_\mu(M)$ is the meoson momentum (mass)),and $\varepsilon_\mu(p)$ is the polarization vector of vector mesons. The Eqs.(3) and (4) are the usual Bargmann-Wigner (BW)[7] spin wave functions, and reduced to, at the rest frame of mesons, the spin wave functions of nonrelativistic quark model. The Eqs.(5) and (6), the chiral states, are the

141

extened BW spin wave functions. We assign the following two mixtures (Normal and Extra) to the physical states.

$$\chi_i^{(N/E)}(p) = \begin{pmatrix} cos\theta_i \\ -sin\theta_i \end{pmatrix} \chi_i^{(NR)}(p) + \begin{pmatrix} sin\theta_i \\ cos\theta_i \end{pmatrix} \chi_i^{(SR)}(p), \quad (i = P_S, V) \qquad (6)$$

where θ is a mixing angle. The $\chi_{P_S}^{(N)}$, putting[4] $\theta_{P_S} = 45°$, is assigned to the ordinary π-nonet. The $\chi_V^{(N)}$ ($\chi_V^{(E)}$) is assigned to the ordinary ρ nonet (the extra ρ nonet).

 Strong decay vertex. In our sheme, effective strong interactions which is corresponding to OZI allowed graph, for example 3-meson vertex is shownin Fig.1, is obtained directly through the overlappping of relevant constituent qurak line. The overlapping of quarks in the transition matrix elements must be chiral svmmetric, consistently

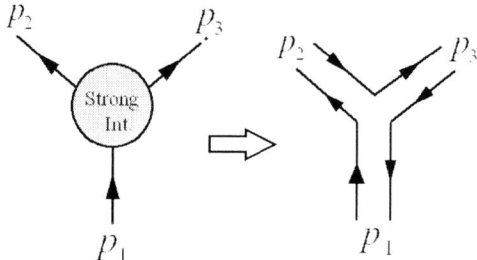

FIGURE 1. Mechanism of strong unteraction

with fundamental QCD. To guarantee this we introduce Φ_U and $\overline{\Phi}_U$ defined[8] as

$$\Phi_U(p) \equiv \Phi_U(p)(i\gamma v), \quad \overline{\Phi}_U(p) \equiv \gamma_4 \Phi^\dagger \gamma_4(-i\gamma v). \qquad (7)$$

It is worth noting that Φ_U and $\overline{\Phi}_U$ have an esseential role to give the right charge to chiral states as well as to ordinary sates. Thus the effective interaction, H, of 3-meson vertex is given as

$$H = g\left\{ \langle \overline{\Phi}_U(p_2)\overline{\Phi}_U(p_3)\Phi_U(p_1) \rangle + \langle \overline{\Phi}_U(p_2)\Phi_U(p_1)\overline{\Phi}_U(p_3) \rangle \right\}, \qquad (8)$$

where g is the coupling constant,and $\langle \quad \rangle$ represents to take trace over the Dirac-spinor and flavor indices. In Eq.(7) form factor effects, due to the overlapping integral of internal space-time wave function, is taken to be 1 because all mesons in interactions are in the ground state. From Eqs.(1)-(8), we obtain the matrix elements T (in momentum space) for relevant process asfollows:

$\rho^{(N/E)}(p_1) \to \pi^{(N)}(p_2)\pi^{(N)}(p_3)$

$$T = g\sqrt{2}A(\pm \cos\theta_V - \sin\theta_V)(p_2 - p_3)_\mu \frac{\varepsilon_\mu(p_1)}{\sqrt{2E_1}} \frac{1}{\sqrt{2E_2}} \frac{1}{\sqrt{2E_3}}, \qquad (9)$$

$\omega^{(N)}(p_1) \to \rho^{(N)}(p_2)\pi^{(N)}(p_3)$

$$T = g\sqrt{2}B \cos 2\theta_V \varepsilon_{\mu\kappa\lambda\sigma} p_{2\kappa} p_{1\mu} \frac{\varepsilon_\sigma(p_1)}{\sqrt{2E_1}} \frac{\varepsilon_\lambda(p_2)}{\sqrt{2E_2}} \frac{1}{\sqrt{2E_3}}, \qquad (10)$$

142

$$\omega^{(E)}(p_1) \to \rho^{(N)}(p_2)\pi^{(N)}(p_3) \quad and \quad \rho^{(E)}(p_2) \to \omega^{(N)}(p_1)\pi^{(N)}(p_3),$$

$$T = g\sqrt{2}(B\sin 2\theta_V + C)\varepsilon_{\mu\kappa\lambda\sigma}p_{2\kappa}p_{1\mu}\frac{\varepsilon_\sigma(p_1)}{\sqrt{2E_1}}\frac{\varepsilon_\lambda(p_2)}{\sqrt{2E_2}}\frac{1}{\sqrt{2E_3}}, \tag{11}$$

where

$$A = \frac{1}{2}\sqrt{\frac{M_1^3}{M_2^3}}, \quad B = \sqrt{\frac{M_1 M_2}{2M_3}}\left(\frac{M_1 + M_2}{M_1 M_2}\right), \quad C = \sqrt{\frac{M_1 M_2}{2M_3}}\left(\frac{M_1 - M_2}{M_1 M_2}\right). \tag{12}$$

Numerical result. From the matrix elements given in Eqs.(9)-(12) we can derive straightforwardly the formulas for strong decay width. In our scheme there exist two parameters ,the coupling constant g and the mixing angle θ_V. We use tentatively the following value to them.

$$g = 2.55 \, GeV^{-\frac{1}{2}}, \quad \theta_V = 13.0°. \tag{13}$$

We have given our predicted values in Table.2. From this table we see that: i) Concerning $\rho(770) \to \pi\pi$, $\omega(783) \to 3\pi$ and $\omega(1250) \to \rho(770)\pi$, taking into account the fact that we considered only chiral symmetric interaction, our predicted values seem to be consistent with experiments. ii) Concerning $\rho(1250)$, $\Gamma(\rho(1250) \to 2\pi)$ is very large, but it is diffidult to get definite conclusion because of very poor present experimental information.

TABLE 2. Predicted strong decay width of vector mesons.

Process	Decay width Γ(MeV)	
	Theory	Expriment
$\rho(770) \to 2\pi$	177	149.2±0.7[2]
$\omega(783) \to 3\pi$	3.5	7.5±0.1[2]
$\omega(1250) \to \rho(770)\pi$	357	426±135[9]
$\rho(1250) \to 2\pi$	35732	
$\rho(1250) \to \omega(783)\pi$	119	

SUMMARY AND REMARKS

We have studied the decay properties of extra vector mesons, $\omega(1250)$ and $\rho(1250)$, by the covariant classification scheme and obtained the result that $\omega(1250)$ is the promising candidate of our extra $\omega^{(E)}$.

There are the following future problems: i) At preset it is difficult to obtain the definite conclusion for $\rho(1250)$ because of poor experiments. However it is necessary to estimate phenomenologicaly the numerical value of $\Gamma_{tot}(\rho^{(E)} \to any)$. ii) In this talk, we considered only chiral symmetric limit. So, as a next step, we should consider the process in order of spontanueous chiral symmetry breaking. iii) It is important to search $\phi^{(E)}$ which is the member with $s\bar{s}$ componet of extra ρ nonet, phenomenologicaly.

REFERENCES

1. Typical references are given in the following.
   M. N. Achasov *et al.*, hep-ex/0109035.
   M. N. Achasov *et al.*, Phys. Lett. **B462** (1999), 365.
   V. Ivanchenko, hep-ph/0106041.
   A. Bertin *et al.*, Phys. Lett. **B408** (1997), 476.
   A. Bertin *et al.*, Phys. Lett. **B414** (1997), 220.
2. Particle Data Group, E. J. Weiberg et al., Phys. Rev. **D66** (2002), 1.
3. M. Ishida and S. Ishida, in this proceedings.
   S. Ishida, in this proceedings.
   S. Ishida and M. Ishida, Phys. Lett. **B539** (2002), 249.
4. M. Ishida and S. Ishida and T. Maeda, Prog. Theor. Phys. **104** (2000), 785.
5. T. Maeda, in proceedings of Nihon university and KEK symposium, 2003.
6. H. Yukawa, Phys. Rev. **91** (1953), 415, 416.
   T. Takabayashi, Nuovo. Cim. **33** (1964), 668
7. A. Salam, R. Delbourgo and J. Strathdee, Proc. R. Soc. London **A284** (1965), 146.
   B. Sakita and K. C. Wali, Phys. Rev. **B139** (1965), 1355.
8. S. Ishida (Ref.3).
9. M. N. Achasov *et al.*, hep-ex/0109035 (Ref.1).

Study of the Reaction $\pi^- p \to \eta' \pi^0 n$ at the VES Spectrometer

Yu. Gouz (VES collaboration[1])

IHEP, Protvino, Russia

Abstract. Here we present the study of the reaction $\pi^- p \to \eta' \pi^0 n$ at the VES spectrometer. The mass and angular distributions for the $\eta' \pi^0$ final state are compared with distributions for $\eta \pi^0$ events, selected from the same data sample. The ratio $R = BR(a_2^0(1320) \to \eta' \pi^0)/BR(a_2^0(1320) \to \eta \pi^0)$ was measured: $R = 0.047 \pm 0.019$, which is in agreement with PDG value [1]. It was shown also that in the $\eta' \pi^0$ final state the production of exotic wave with quantum numbers $J^{PC} = 1^{-+}$ proceeds mainly through unnatural parity exchange. This is an indirect evidence of the fact that the decay width of the exotic meson $\pi_1(1600)$ into $\rho\pi$ is much smaller than into $b_1(1235)\pi$.

INTRODUCTION

The $\eta(\eta')\pi$ systems produced in the pion beam are mainly interesting because the quantum numbers of such systems with orbital momentum of 1 are exotic: $J^{PC} = 1^{-+}$. The high statistics studies of the $\eta\pi^-$ [2, 3, 4], $\eta'\pi^-$ [4, 5] and $\eta\pi^0$ [6, 7] systems were performed in GAMS/NA12, VES and E852 experiments. In all the three final states the behavior of the exotic 1^{-+} waves is different.

In the $\eta'\pi^-$ system the 1^{-+} wave with positive exchange naturality is dominant, forming a bump at M≈1.6 GeV, which is now interpreted as a signal of an exotic meson $\pi_1(1600)$ [1, 5, 8]. The peak at 1.6 GeV is also seen in the 1^{-+} wave in $b_1(1235)\pi$ and possibly in $f_1(1285)\pi$ final states [4, 9, 10]; $\pi_1(1600)$ is a good candidate for a hybrid meson. In the $\eta\pi^-$ system the exotic wave with positive exchange naturality forms a broad bump at M≈1.4 GeV. This bump, together with its phase motion, allows interpretation as an exotic meson $\pi_1(1400)$ [1, 3]. It was noticed [11] that the P-wave states in the $\eta'\pi$ and $\eta\pi$ systems are essentially different. Assuming that η-meson is pure $SU_f(3)$ octet, one can show that the P-wave state in the $\eta\pi$ system cannot be a member of $SU_f(3)$ octet, but necessarily belongs to a $10 \oplus \overline{10}$ representation; the $\eta'\pi$ P-wave is, in contrary, pure $SU_f(3)$ octet. This may explain the difference between P-wave spectra in $\eta'\pi^-$ and $\eta\pi^-$.

In the $\eta\pi^0$ system the non-negligible P-wave is also present; however it does not

[1] VES collaboration: D.V. Amelin, V.A. Dorofeev, R.I. Dzhelyadin, A.B. Fenyuk, Yu.G. Gavrilov, Yu.P. Gouz, I.A. Kachaev, V.V. Kabachenko, A.N. Karyukhin, Yu.A. Khokhlov, A.N. Konoplyannikov, V.F. Konstantinov, V.V. Kostyukhin, V.D. Matveev, V.I. Nikolaenko, A.P. Ostankov, B.F. Polyakov, D.I. Ryabchikov, A.A. Solodkov, A.V. Solodkov, O.V. Solovianov, E.A. Starchenko, A.M. Zaitsev, A.V. Zenin.

CP717, *Hadron Spectroscopy: Tenth International Conference,*
edited by E. Klempt, H. Koch, and H. Orth
© 2004 American Institute of Physics 0-7354-0197-7/04/$22.00

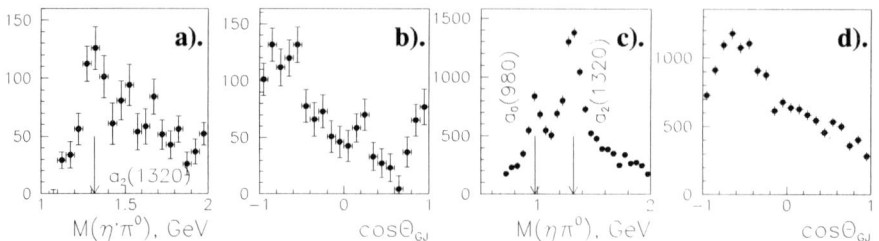

FIGURE 1. The effective mass and $\cos\theta_{GJ}$ distributions for $\eta'\pi^0$ (a,b,) and $\eta\pi^0$ (c,d) final states.

show proper phase motion with respect to the D-wave, so its resonance interpretation is problematic [6, 7]. The difference may be caused by the fact that $\eta\pi^0$ system is produced in charge exchange reaction.

Here we present the first study of the fourth final state of this type, $\eta'\pi^0$. The distributions in the $\eta'\pi^0$ system are compared to corresponding distributions for $\eta\pi^0$.

DATA ANALYSIS

The events of the reaction $\pi^- p \to \eta'\pi^0 n$ were selected from all available VES statistics in the decay mode $\eta' \to \pi^+\pi^-\eta$, $\eta \to \gamma\gamma$, $\pi^0 \to \gamma\gamma$. The events of the reaction $\pi^- p \to \eta\pi^0 n$ were selected from the same data sample with similar decay chain: $\eta \to \pi^+\pi^-\pi^0$, $\pi^0 \to \gamma\gamma$. Significant amount of background is present in both data sets. The events from side bands around η' peak in the $pi^+\pi^-\eta$ effective mass spectrum and η peak in the $\pi^+\pi^-\pi^0$ spectrum were used for the background evaluation and subtraction. Total of 2630 "signal" and 1415 "background" events were selected for the analysis of the $\eta'\pi^0$ system, 18547 "signal" and 8436 "background" events were selected for $\eta\pi^0$.

The mass spectra and distributions of cosine of the Gottfried-Jackson angle for $\eta'\pi^0$ and $\eta\pi^0$ systems are shown in Fig1. Clear peak of $a_2(1320)$ meson is seen in the $\eta'\pi^0$ mass spectrum. Peaks of $a_0(980)$ and $a_2(1320)$ mesons are seen in the $\eta\pi^0$ mass spectrum, in agreement with [6, 7]. The distributions of $\cos\theta_{GJ}$ show significant asymmetry in both systems.

Using the effective mass spectra, we determined the ratio of decay branchings of $a_2^0(1320)$ into $\eta'\pi^0$ and $\eta\pi^0$: $R \approx 0.047 \pm 0.019$, which agrees with the PDG value [1] obtained from $\eta'\pi^-$ and $\eta\pi^-$ final states [2, 5].

The conventional mass independent partial wave analysis [6, 7] of the $\eta\pi^0$ system was performed in 50 MeV bins with standard set of waves: $S0$, $P0$, $P-$, $D0$, $D-$ in the unnatural parity exchange sector and $P+$ and $D+$ in the natural parity exchange sector. The background subtraction was performed using the method of adding events from background sample into the minimization functional with opposite sign [2]. The results are shown in Fig.2. They agree with similar analyses with higher statistics done by GAMS/NA12 and E852 experiments [6, 7]: there are peaks of $a_0(980)$ meson in the

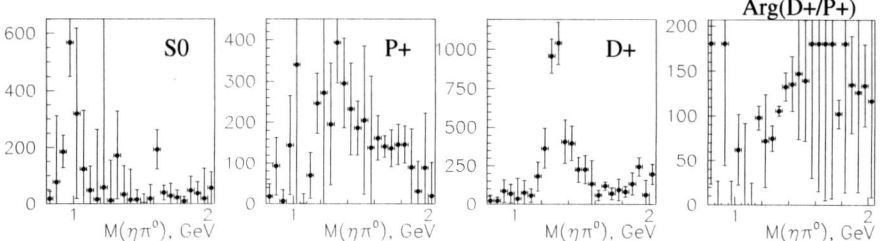

FIGURE 2. The results of Partial Wave Analysis of the $\eta\pi^0$ system: the spectra of $S0$, $P+$, $D+$ waves and phase difference between $P+$ and $D+$ waves.

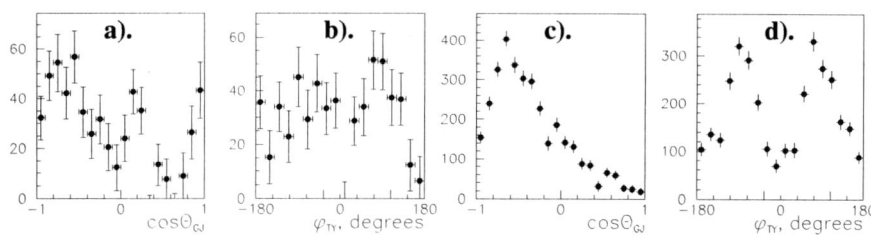

FIGURE 3. The $\cos\theta_{GJ}$ and ϕ_{TY} distributions for $\eta'\pi^0$ (a,b) and $\eta\pi^0$ (c,d) in the effective mass range of $1.45 \div 1.9$ GeV.

$S0$ and $a_2(1320)$ meson in $D+$ wave spectra, broad bump in the $P+$ wave spectrum; the phase difference between $P+$ and $D+$ waves shows rapid variation about 1.3 GeV caused by the $a_2(1320)$ meson.

It would be interesting to study the $\eta'\pi^0$ system at the effective mass of $\pi_1(1600)$. The $\cos\theta_{GJ}$ and ϕ_{TY} distributions for the $\eta'\pi^0$ system for the effective mass range $1.45 \div 1.9$ GeV are shown in Fig.3 a,b. For the comparison, the $\cos\theta_{GJ}$ and ϕ_{TY} distributions for the $\eta\pi^0$ system for the same effective mass range are shown in Fig.3 c,d. The $\cos\theta_{GJ}$ distribution shows significant asymmetry in both systems. The dominance of positive naturality exchange in $\eta\pi^0$ is clearly visible as big fraction of $\cos^2\phi_{TY}$ component in its ϕ_{TY} distribution. From the ϕ_{TY} distribution of $\eta'\pi^0$ one can conclude that the contribution of natural parity exchange, including $P+$ wave, is small.

The available $\eta'\pi^0$ statistics is insufficient for mass-independent partial wave analysis. Instead we performed partial wave analysis for the events in the whole effective mass range of $1.45 \div 1.9$ GeV, assuming that in this mass range there are no significant relative phase variations for any pair of partial waves (this is the case for $\eta\pi^-$, $\eta'\pi^-$ and $\eta\pi^0$ systems) and knowing that there is no significant acceptance variations. The results for $\eta'\pi^0$ and $\eta\pi^0$ are shown in Fig.4 a,b. Each bin represents intensity of one of partial waves. All possible solutions are shown as dotted bars in subbins of each bin; the solid

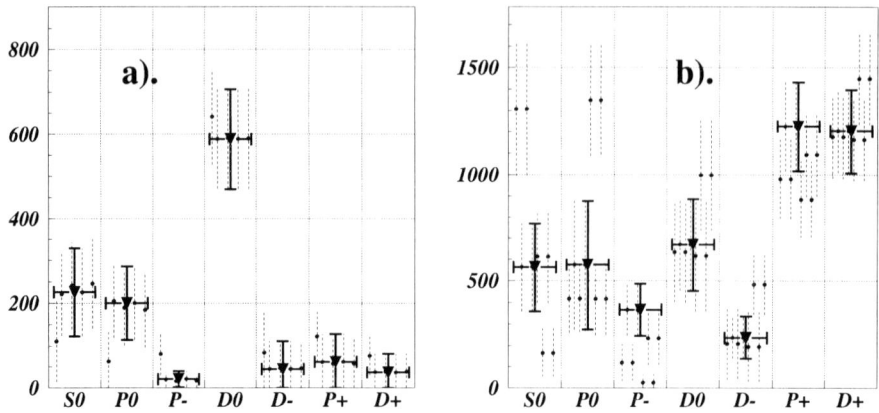

FIGURE 4. The results of partial wave analysis in the effective mass interval $1.45 \div 1.9$ GeV for $\eta' \pi^0$ (a) and $\eta \pi^0$ (b).

crosses correspond to preferred solutions[2].

The largest wave in the $\eta' \pi^0$ system is $D0$. Significant $P0$ wave is present, while $P+$ is compatible with zero. The $P+$ and $P0$ waves are produced via ρ- and $b_1(\rho_2)$-trajectory exchange, respectively. If we believe that P wave in $\eta' \pi^0$ system around 1.6 GeV is caused by exotic $\pi_1(1600)$-meson, then the smallness of $P+$ means that its coupling to the $\rho \pi$ channel is weaker than to $b_1 \pi$. This suppression is rather strong, because the b_1-trajectory lies lower than ρ-trajectory: $\alpha_\rho(0) \approx 0.5$, $\alpha_{b_1}(0) \approx -0.3$, and if the couplings to $\rho \pi$ and $b_1 \pi$ were equal, the ρ-exchange ($P+$) would dominate in the P-wave production. The P-waves in $\eta' \pi^0$ and $\eta \pi^0$ can be compared in the following way. We can evaluate the ratio of matrix elements squared for the production of $P+$ wave in $\eta' \pi^0$ and $\eta \pi^0$ systems: taking into account branching ratios of η and η' and phase space factors, we obtain: $|T_{P+}^{\eta' \pi^0}|^2 / |T_{P+}^{\eta \pi^0}|^2 \approx 0.1 \pm 0.1$ For the $P0$ wave the ratio of intensities in $\eta' \pi^0$ and $\eta \pi^0$ is higher, $\sim 0.7 \pm 0.3$. This means that the $P+$ wave in $\eta(\eta')\pi^0$ systems is produced in $SU_f(3)$ decuplet state, while $P0$ is produced equally well in both octet and decuplet states.

At present, the status of experimental observation of $\pi_1(1600)$ decay into $\rho^0 \pi$ is not clear. The observation of $\pi_1(1600)$ peak in the $J^{PC} = 1^{-+} \rho \pi$ wave with positive exchange naturality was reported in E852 [12] and early VES studies [10]; its intensity was such that it led to the ratio of decay widths of $\pi_1(1600)$ into $\rho \pi$ and $b_1 \pi$ of ~ 1.5. Later it was found [13, 14] that the shape of $1^{-+} \rho \pi$ wave is model dependent, and at

---

[2] The reasons for this choice of solution were the following: for $\eta' \pi^0$ - the $P0$ wave intensity should be higher than $P-$; for $\eta \pi^0$ - also that $D+$ intensity should be higher than both $S0$ and $D0$; actually the particular choice of solution does not affect the final conclusion.

some model settings the $\pi_1(1600)$ peak disappears. The study of the $\eta'\pi^0$ final state gives us an indirect evidence of the suppression of the $\rho\pi$ decay channel of $\pi_1(1600)$ and rules out $\rho\pi:b_1\pi \sim 1$

CONCLUSIONS

The reaction $\pi^- p \to \eta'\pi^0 n$ was studied at VES experiment. The distributions in the $\eta'\pi^0$ system were compared to distributions in the well studied $\eta\pi^0$ system. The ratio of branchings of $a_2(1320)$-meson decay into $\eta'\pi^0$ and $\eta\pi^0$ was determined: $R = 0.047 \pm 0.019$, which agrees with its PDG value. The P waves were studied in $\eta'\pi^0$ and $\eta\pi^0$ systems in the effective mass range of $1.45 \div 1.9$ GeV. It was found that in this mass interval in the $\eta'\pi^0$ system the P wave production proceeds mainly through unnatural parity exchange. This may mean that the coupling of exotic $J^{PC} = 1^{-+}$ state $\pi_1(1600)$ to the $\rho\pi$ channel is significantly weaker than to the $b_1(1235)\pi$ channel.

ACKNOWLEDGMENTS

This work was supported in part by grants RFBR 02-02-17479 and 00-15-96689 and NSh 1695.2003.2. The authors also express their thanks to the Organizational Committee of the HADRON 2003 conference for the financial support.

REFERENCES

1. K. Hagiwara et al, *Phys.Rev.*, **D66**, 010001 (2002).
2. G.M. Beladidze et al, *Phys.Lett.*, **B313**, 276–282 (1993).
3. S.U. Chung et al, *Phys.Rev.*, **D60**, 092001 (1999).
4. V. Dorofeev et al, "The $J^{PC} = 1^{-+}$ hunting season at VES," in *HADRON 2001*, edited by D. Amelin and A. Zaitsev, AIP Conference Proceedings 619, American Institute of Physics, Melville, New York, 2002, pp. 143–154.
5. E.I. Ivanov et al, *Phys.Rev.Lett.*, **86**, 3977–3980 (2001).
6. D. Alde et al, *Phys.Atom.Nucl.*, **62**, 421–434 (1999).
7. A.R. Dzierba et al, *Phys.Rev.*, **D67**, 094015 (2003).
8. Yu. Khokhlov et al, *Nucl.Phys.*, **A663**, 596–599 (2000).
9. D.V. Amelin et al, *Phys.Atom.Nucl.*, **62**, 445–453 (1999), hep-ex/9810013.
10. Yu. Gouz et al, "Study of the Wave with $J^{PC} = 1^{-+}$ in the Partial Wave Analysis of $\eta'\pi^-$, $f_1\pi^-$ and $\rho^0\pi^-$ Systems Produced in $\pi^- N$ Interactions at $p_\pi = 37$ GeV/c," in *ICHEP 1992*, edited by J. R. Sanford, AIP Conference Proceedings 272, American Institute of Physics, New York, 1993, pp. 572–576.
11. S.U. Chung, E. Klempt and J.G. Korner, *Eur.Phys.J.*, **A15**, 539–542 (2002), hep-ph/0211100.
12. G.S. Adams et al, *Phys.Rev.Lett.*, **81**, 5760–5763 (1998).
13. A. Zaitsev et al, "Search for Exotics in $I^G J^P = 1^- 1^-$, $1^- 0^+$ and $0^+ 2^+$ Waves," in *HADRON 1997*, edited by S. Chung and H. Willutzki, AIP Conference Proceedings 432, American Institute of Physics, Woodbury, New York, 1998, pp. 461–470.
14. I. Kachaev et al, "Study of reaction $\pi^- A \to \pi^+ \pi^- \pi^- A$ at VES setup," in *HADRON 2001*, edited by D. Amelin and A. Zaitsev, AIP Conference Proceedings 619, American Institute of Physics, Melville, New York, 2002, pp. 577–581.

Effects to Scalar Meson Decays of Strong Mixing between Low and High Mass Scalar Mesons

Tadayuki Teshima, Ichijiro Kitamura and Norikazu Morisita

Department of Applied Physics, Chubu University, Kasugai 487-8501, Japan

Abstract. We analyze the mass spectroscopy of low and high mass scalar mesons and get the result that the coupling strengths of the mixing between low and high mass scalar mesons are very strong and the strengths of mixing among $I = 1$, $1/2$ scalar mesons and those among $I = 0$ scalar mesons are almost same. Next, we analyze the decay widths and ratios of these mesons and get the results that the coupling constants A' for $I = 1$, $1/2$ which represents the coupling of high mass scalar meson $N' \to$ pseudoscalar mesons PP are almost same as the coupling A' for the $I = 0$. On the other hand, the coupling constant A for $I = 1$, $1/2$ which represents the low mass scalar meson $N \to PP$ are far from the coupling constant A for $I = 0$. We consider a resolution of this discrepancy.

INTRODUCTION

The $f_0(500)$ and $\kappa(900)$ confirmed recently, and $a(980)$ and $f_0(980)$ are considered to be a chiral partner of the pseudoscalar nonet as a Nambu-Goldstone boson [1] or $qq\bar{q}\bar{q}$ model [2]. On the other hand, high mass scalar mesons $a(1450)$, $K_0^*(1430)$, $f_0(1370)$ and $f_0(1710)$ are considered to construct $L = 1$ $q\bar{q}$ scalar nonet. We assume a strong mixing (inter-mixing) between low mass and high mass scalar nonets to explain the fact that the high mass $L = 1$ $q\bar{q}$ scalar nonet are so high compared to other $L = 1$ $q\bar{q}$ 1^{++} and 2^{++} mesons. Furthermore $f_0(1500)$ is considered to be a glueball candidate [3]. We analyzed the overall mixing among low mass scalar nonet which are assumed to be $qq\bar{q}\bar{q}$ and high mass scalar nonet $L = 1$ $q\bar{q}$ and glueball [3]. We have obtained the result that the inter-mixing is very strong and the mixing parameters λ_{01}^a, λ_{01}^K and λ_{01} producing the inter-mixing in $I = 1, 1/2$ and $I = 0$ mesons respectively are almost same [4].

We discuss the effects of the strong mixing between low and high mass scalar mesons to the decay processes of low and high mass scalar mesons. We estimated the decay coupling constants A for low mass scalar mesons (N)-pseudoscalar meson (P)-pseudoscalar meson (P) interaction and the coupling constant A' for high mass scalar (N')-PP interaction considering the mixing between $I = 1, 1/2$ low and high mass scalar mesons [5]. We also analyze A and A' and further the coupling constant A'' for glueball (G)-PP interaction considering the mixing among $I = 0$ low and high mass scalar mesons.

MIXING BETWEEN LOW AND HIGH MASS SCALAR MESONS

Inter-mixing between $I = 1$, $1/2$ low and high mass scalar mesons

CP717, *Hadron Spectroscopy: Tenth International Conference,*
edited by E. Klempt, H. Koch, and H. Orth
© 2004 American Institute of Physics 0-7354-0197-7/04/$22.00

Four quark light scalar meson states N_b^a are represented by quark fields q_a and anti-quark fields $\bar{q}^b$ as $N_b^a \sim \varepsilon_{bef}\bar{q}^e\bar{q}^f \varepsilon^{acd}q_cq_d$. The inter-mixing interactions are caused by the OZI rule allowed interaction and these are represented as the form;

$$L_{int} = -\lambda_{01}\varepsilon^{abc}\varepsilon_{def}N_a^d N_b^{\prime e} \delta_c^f = \lambda_{01}[a_0^+ a_0^{\prime -} + a_0^- a_0^{\prime +} + a_0^0 a_0^{\prime 0} + \kappa^+ K_0^{*-}$$
$$+\kappa^- K_0^{*+} + \bar{\kappa}^0 K_0^{*0} + \kappa^0 \bar{K}_0^{*0} - \sqrt{2}f_N f_N' - f_S f_N' - \sqrt{2}f_N f_S'], \qquad (1)$$

where $N_b^{\prime a}$ is the high mass scalar meson state, $N_b^{\prime a} \sim q_b\bar{q}^a$. The strength of this λ_{01} is considered to be very large because of the OZI rule allowed interaction.

We estimate the strength of the inter-mixing parameter λ_{01} for $I = 1$ $a_0(1450)$ and $a_0(980)$ mixing. We take the masses (unit:MeV) before mixing as $m_{\overline{a_0(980)}} = 1271 \pm 31$, $m_{\overline{a_0(1450)}} = 1236 \pm 20$ estimated from the relation $m^2(2^{++}) - m^2(1^{++})$

$= 2\left(m^2(1^{++}) - m^2(0^{++})\right)$. Diagonalising the mass matrix $\begin{pmatrix} m_{\overline{a_0(980)}} & \lambda_{01}^a \\ \lambda_{01}^a & m_{\overline{a_0(1450)}} \end{pmatrix}$

and taking the eigenvalues of masses $m_{a_0(980)} = 984.8 \pm 1.4$, $m_{a_0(1450)} = 1474 \pm 19$, we can get the result

$$\lambda_{01}^a = 0.600 \pm 0.028 \text{GeV}^2, \quad \text{mixing angle } \theta_a = 47.1 \pm 3.5°. \qquad (2)$$

Similarly, we estimate the strength λ_{01} for $I = 1/2$ $\kappa(900)$ and $K_0^*(1430)$ mixing. Using the masses before mixing and after mixing, $m_{\overline{\kappa(900)}} = 1047 \pm 62, m_{\overline{K_0^*(1430)}} = 1307 \pm 11, m_{\kappa(900)} = 900 \pm 70$, $m_{K_0^*(1430)} = 1412 \pm 6$, we get the results

$$\lambda_{01}^K = 0.507 \pm 84 \text{GeV}^2, \quad \text{mixing angle } \theta_K = 29.5 \pm 15.5°. \qquad (3)$$

It is confirmed that these coupling strengths are large and λ_{01}^K is as strong strength as λ_{01}^a.

Inter-mixing between $I = 0$ low and high mass scalar mesons

We consider the overall mixing between $I = 0$ low and $I = 0$ high mass scalar mesons and glueball. In these overall mixing, there are two types mixing; one is the intra-mixing among $I = 0, L = 1$ $q\bar{q}$ scalar mesons and glueball and among $I = 0$ $qq\bar{q}\bar{q}$ scalar mesons, the other is the inter-mixing between $I = 0$ low and $I = 0$ high mass scalar mesons and glueball. Overall mixing is expressed by the following mass matrix,

$$\begin{pmatrix} m_N^2 + 2\lambda_0 & \sqrt{2}\lambda_0 & \lambda_{01} & \sqrt{2}\lambda_{01} & 0 \\ \sqrt{2}\lambda_0 & m_S^2 + \lambda_0 & \sqrt{2}\lambda_{01} & 0 & 0 \\ \lambda_{01} & \sqrt{2}\lambda_{01} & m_{N'}^2 + 2\lambda_1 & \sqrt{2}\lambda_1 & \sqrt{2}\lambda_G \\ \sqrt{2}\lambda_{01} & 0 & \sqrt{2}\lambda_1 & m_{S'}^2 + \lambda_1 & \lambda_G \\ 0 & 0 & \sqrt{2}\lambda_G & \lambda_G & \lambda_{GG} \end{pmatrix}, \qquad (4)$$

where λ_0 and λ_1 are coupling strengths among isoscalar mesons in low mass and high mass scalar mesons with OZI suppression graphs, respectively, and λ_G is the transition strength between $q\bar{q}$ and glueball. λ_{GG} is the pure glueball mass. Using the input mass values (unit:GeV) $m_N = 1.271 \pm 0.031$, $m_S = 0.760 \pm 0.179$, $m_{N'}$

$= 1.236 \pm 0.02$, $m_{S'} = 1.374 \pm 0.003$, $m_{f_0(980)} = 0.980 \pm 0.010$, $m_{f_0(500)} = 0.500 \pm 0.100$, $m_{f_0(1370)} = 1.350 \pm 0.150$, $m_{f_0(1710)} = 1.715 \pm 0.007$, $m_{f_0(1500)} = 1.500 \pm 0.010$, we get the result for the case in which $f_0(1500)$ is assumed as glueball:

$$\lambda_{01} = 0.53 \pm 0.04 \text{GeV}^2, \quad \lambda_0 = 0.03 \pm 0.04 \text{GeV}^2, \quad \lambda_1 = 0.07 \pm 0.05 \text{GeV}^2,$$
$$\lambda_G = 0.23 \pm 0.06 \text{GeV}^2, \quad \lambda_{GG} = (1.53 \pm 0.03)^2 \text{GeV}^2,$$
$$^t(f_0(980), f_0(500), f_0(1370), f_0(1500), f_0(1710)) \tag{5}$$
$$= [R_{f_0(M)l}]^t (f_N, f_S, f_{N'}, f_{S'}, f_G),$$

$$[R_{f_0(M)l}] =$$
$$\begin{pmatrix} 0.72 \pm 0.06 & -0.39 \pm 0.10 & -0.15 \pm 0.11 & -0.56 \pm 0.04 & 0.15 \pm 0.05 \\ 0.23 \pm 0.09 & 0.79 \pm 0.08 & -0.53 \pm 0.08 & -0.10 \pm 0.06 & 0.11 \pm 0.04 \\ 0.05 \pm 0.08 & 0.43 \pm 0.06 & 0.68 \pm 0.0 & -0.48 \pm 0.04 & -0.28 \pm 0.05 \\ -0.42 \pm 0.12 & 0.01 \pm 0.04 & 0.06 \pm 0.06 & -0.36 \pm 0.12 & 0.81 \pm 0.10 \\ 0.53 \pm 0.10 & 0.17 \pm 0.04 & 0.46 \pm 0.05 & 0.51 \pm 0.04 & 0.45 \pm 0.17 \end{pmatrix}.$$

For the case in which $f_0(1710)$ is assumed as glueball,

$$\lambda_{01} = 0.44 \pm 0.04 \text{GeV}^2, \quad \lambda_0 = 0.02 \pm 0.05 \text{GeV}^2, \quad \lambda_1 = -0.08 \pm 0.05 \text{GeV}^2,$$
$$\lambda_G = 0.28 \pm 0.06 \text{GeV}^2, \quad \lambda_{GG} = (1.64 \pm 0.03)^2 \text{GeV}^2, \tag{6}$$
$$[R_{f_0(M)l}] =$$
$$\begin{pmatrix} 0.64 \pm 0.08 & -0.51 \pm 0.11 & -0.21 \pm 0.16 & -0.52 \pm 0.03 & 0.15 \pm 0.07 \\ 0.24 \pm 0.11 & 0.72 \pm 0.11 & -0.58 \pm 0.10 & -0.17 \pm 0.08 & 0.12 \pm 0.06 \\ 0.21 \pm 0.06 & 0.42 \pm 0.07 & 0.72 \pm 0.07 & -0.48 \pm 0.09 & -0.19 \pm 0.06 \\ 0.65 \pm 0.05 & 0.03 \pm 0.04 & 0.05 \pm 0.09 & 0.60 \pm 0.10 & -0.43 \pm 0.10 \\ 0.26 \pm 0.09 & 0.09 \pm 0.03 & 0.28 \pm 0.06 & 0.31 \pm 0.07 & 0.86 \pm 0.07 \end{pmatrix}.$$

DECAY PROCESSES OF SCALAR MESONS AND GLUEBALL

We analyze the decay processes of low mass scalar mesons (N) decaying to two pseudoscalar mesons (PP), high mass scalar mesons (N') decaying to PP and pure glueball (G) decaying to PP. We use the following interactions for NPP, $N'PP$ and GPP coupling with coupling constants A, A' and A'', respectively,

$$L_I = A\varepsilon^{abc}\varepsilon_{def}N_a^d\partial^\mu\phi_b^e\partial_\mu\phi_c^f + A'N_a'^b\{\partial^\mu\phi_b^c, \partial_\mu\phi_c^a\} + A''G\{\partial^\mu\phi_a^b, \partial_\mu\phi_b^a\}. \tag{7}$$

These interactions are represented graphically by the diagrams in fig. 1.

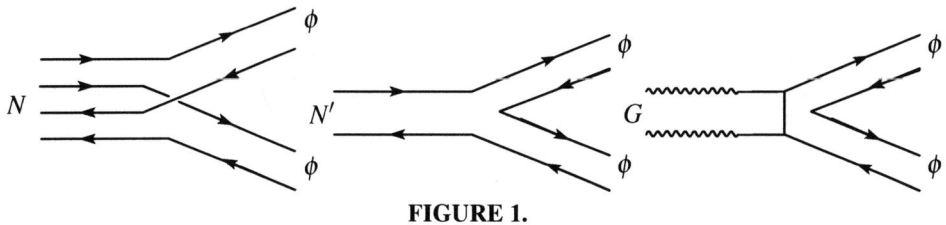

FIGURE 1.

$a_0(980)$, $a_0(1450)$ and $K_0^*(1450)$ meson decays

We analyzed the decay widths and ratios of $I = 1, 1/2$ scalar mesons. Using the PDG data [6], we estimated the allowed values for A and A' executing the χ^2 fit analysis. The results in taking maximum $\chi^2 \leq 5.348$ corresponding to the 50% C.L. on degree of freedom 6 are as follows;

$$A = 0.10 \pm 0.24 \text{GeV}^{-1}, \quad A' = -3.03 \pm 0.2 \text{GeV}^{-1}, \quad \theta_P = 49.0° \pm 3.0°. \tag{8}$$

We show the best-fit values for decay widths and ratios for these best fit A, A' in third column of Table I.

TABLE 1. Experimental data [6] and best-fit values for decay widths and ratios.

Decay width and ratio	Experimental data	Best fit value
$\Gamma(a_0(980) \rightarrow \text{all}(\pi\eta + K\bar{K}))$	75 ± 25MeV	36MeV
$\Gamma(a_0(980) \rightarrow K\bar{K})/\Gamma(a_0(980) \rightarrow \pi\eta)$	0.177 ± 0.024	0.156
$\Gamma(a_0(1450) \rightarrow \text{all}(\pi\eta + \pi\eta' + K\bar{K}))$	265 ± 13MeV	266MeV
$\Gamma(a_0(1450) \rightarrow K\bar{K})/\Gamma(a_0(1450) \rightarrow \pi\eta)$	0.88 ± 0.23	0.80
$\Gamma(a_0(1450) \rightarrow \pi\eta')/\Gamma(a_0(1450) \rightarrow \pi\eta)$	0.35 ± 0.16	0.49
$\Gamma(K_0^*(1430) \rightarrow \pi K)$	273 ± 44MeV	303MeV

$f_0(980)$, $f_0(1370)$, $f_0(1500)$ and $f_0(1710)$ meson decays

Isoscalar meson decay widths and decay ratios are analyzed using the data of PDG [6]. Analyzed processes are appeared in Table 2. Executing the χ^2 fit, we estimate the allowed values for A, A' and A'' in the $\chi^2 \leq 12.340$ corresponding to the 50% C.L. on degree of freedom 13. The allowed values for A, A', A'' and θ_P corresponding to the $f_0(1500)$ glueball case are as follows:

$$A = -2.88 \pm 0.16 \text{GeV}^{-1}, \quad A' = -2.28 \pm 0.08 \text{GeV}^{-1},$$
$$A'' = 0.305 \pm 0.034 \text{GeV}^{-1}, \quad \theta_P = (18.9 \pm 1.8)° \text{ or } (38.8 \pm 0.4)°, \tag{9}$$

and those corresponding to the $f_0(1710)$ glueball case are as follows:

$$A = -4.06 \pm 0.14 \text{GeV}^{-1}, \quad A' = -1.93 \pm 0.10 \text{GeV}^{-1},$$
$$A'' = 0.640 \pm 0.04 \text{GeV}^{-1}, \quad \theta_P = (50 \pm 2)°. \tag{10}$$

We showed the best fit values for decay widths and decay ratios on $A = -2.88 \text{GeV}^{-1}$, $A' = -2.28 \text{GeV}^{-1}$, $A'' = 0.305 \text{GeV}^{-1}$, $\theta_P = 18.9°$ for the $f_0(1500)$ glueball case and on

153

$A = -4.06\text{GeV}^{-1}$, $A' = -1.93\text{GeV}^{-1}$, $A'' = 0.640\text{GeV}^{-1}$, $\theta_P = 50°$ for the $f_0(1710)$ glueball case in third and forth column of Table 2.

TABLE 2. Experimental data [6] and best-fit values for decay widths and ratios.

Decay width and ratio	Exper. data	B.F.V.($f_0(1500)$)	B.F.V($f_0(1710)$)
$\Gamma_{f_0(980)\to \text{all}(\pi\pi + K\bar{K})}$	$70 \pm 30\text{MeV}$	48MeV	66MeV
$\Gamma_{f_0(980)\to\pi\pi}/\Gamma_{f_0(980)\to\pi\pi+K\bar{K}}$	0.74 ± 0.07	0.77	0.74
$\Gamma_{f_0(1370)\to\pi\pi}/\Gamma_{f_0(1370)\to\text{all}}$	0.26 ± 0.09	0.24	0.20
$\Gamma_{f_0(1370)\to K\bar{K}}/\Gamma_{f_0(1370)\to\text{all}}$	0.35 ± 0.13	0.02	0.02
$\Gamma_{f_0(1500)\to K\bar{K}}/\Gamma_{f_0(1500)\to\pi\pi}$	0.19 ± 0.07	0.15	0.09
$\Gamma_{f_0(1500)\to\eta\eta}/\Gamma_{f_0(1500)\to\pi\pi}$	0.18 ± 0.03	0.17	0.22
$\Gamma_{f_0(1500)\to\eta\eta'}/\Gamma_{f_0(1500)\to\pi\pi}$	0.095 ± 0.026	0.075	0.082
$\Gamma_{f_0(1500)\to\eta\eta'}/\Gamma_{f_0(1500)\to\eta\eta}$	0.29 ± 0.16	0.43	0.38
$\Gamma_{f_0(1710)\to\text{all}}$	$125 \pm 10\text{MeV}$	126MeV	123MeV
$\Gamma_{f_0(1710)\to\pi\pi}/\Gamma_{f_0(1710)\to K\bar{K}}$	0.39 ± 0.14	0.48	0.38
$\Gamma_{f_0(1710)\to K\bar{K}}/\Gamma_{f_0(1710)\to\text{all}}$	0.38 ± 0.14	0.51	0.58
$\Gamma_{f_0(1710)\to\eta\eta}/\Gamma_{f_0(1710)\to\text{all}}$	0.18 ± 0.08	0.17	0.19
$\Gamma_{f_0(1710)\to\eta\eta}/\Gamma_{f_0(1710)\to K\bar{K}}$	0.48 ± 0.15	0.33	0.33

These allowed regions are marked by ellipses in Fig. 2. There seems rather large discrepancy between these allowed regions. However, these allowed regions are spread out with the increases of maximum χ^2 in χ^2 fit. From 5.348 to 20 in χ^2 fit of $I = 1$, $1/2$ meson decays and from 12.34 to 30 in χ^2 fit of $I = 0$ meson decays, the allowed regions are spread as shown in fig. 2. The spread region for $I = 1, 1/2$ meson decay joins with the spread region for $I = 0$ meson decay $f_0(1710)$ glueball case at $(\sim -2.3\text{GeV}^{-1}, \sim -0.8\text{GeV}^{-1})$ in (A, A') plane. Also, the spread region for $I = 1, 1/2$ meson decay can join with the spread region for $I = 0$ meson decay $f_0(1500)$ glueball at $(\sim -1.7\text{GeV}^{-1}, \sim -1.4\text{GeV}^{-1})$ in (A, A') plane if maximum χ^2 is increased moreover in χ^2 fit.

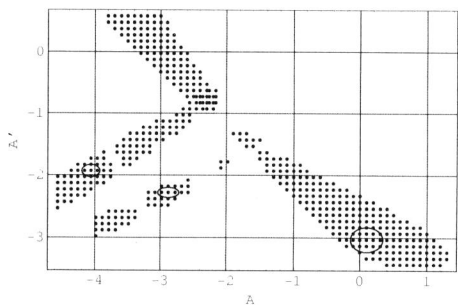

FIGURE 2.

REFERENCES

1. M. Ishida, *Prog. Theor. Phys.* **96**, 853(1996).
2. R. J. Jaffe, *Phys. Rev. D* **15**, 267(1977).
3. F. E. Close and A. Kirk, *Phys. Lett. B* **483**, 345(2000).
4. T. Teshima, I. Kitamura and N. Morisita, *J. Phys. G* **28**, 1391(2002).
5. D. Black, A. H. Fariborz and J. Schechter, *Phys. Rev. D* **61**, 074001 (2000).
6. K. Hagiwara et al. (Particle Data Group), *Phys. Rev. D* **66**, 010001(2002).

Dynamical selection rule in the decays of $\pi(1800)$.

V.Nikolaenko (on behalf of VES collaboration)[*] and VES collaboration:
D.Amelin, Yu.Gavrilov, Yu.Gouz, R.Dzheliadin, A.Fenyuk, I.Kachaev,
V.Kabachenko, A.Karyukhin, Yu.Khokhlov, A.Konopliannikov,
V.Konstantinov, V.Kostyuhin, V.Matveev, A.Ostankov, B.Polyakov,
D.Ryabchikov, A.A.Solodkov, A.V.Solodkov, O.Solovianov, E.Starchenko,
A.Zaitsev, A.Zenin

[*]*Department of Hadron Physics, IHEP, 142281, Protvino, Russia*

Abstract. Combined analysis of $\pi(1800)$ decays is performed using results of the VES experiment on its decays into $f_0(980)\pi$, $f_0(1500)\pi$, $f_0(600 \div 1300)\pi$, κK, $a_0(\eta\pi)\pi$, $\eta\eta'\pi$, $\rho\pi$ and K^*K channels. Relative branching ratios of those decay modes are measured. It is shown that the OZI selection rule is strongly violated in $\pi(1800)$ decays. Observed pattern of $\pi(1800)$ decays can be explained in terms of assumption, that $\pi(1800)$ decay proceeds via $SU(3)$-singlet and $SU(3)$-octet states.

INTRODUCTION.

The well established $\pi(1800)$ meson has been observed in three experiments [1],[2], [3], [4] in the following decay modes:

$$\pi(1800) \to \pi^-\pi^+\pi^-; \tag{1}$$

$$\pi(1800) \to K^-K^+\pi^-; \tag{2}$$

$$\pi(1800) \to \eta\eta\pi^-; \tag{3}$$

$$\pi(1800) \to \eta'\eta\pi^-; \tag{4}$$

and possibly in mode

$$\pi(1800) \to \omega\rho^-; \tag{5}$$

(see [5],[6]). [1] However, the information concerning the decay branching ratios remains scarce and fragmental. In this report we present results of simultaneous fits of $\pi(1800)$ resonance in the final states (1-4), which should decrease the experimental errors and transform the results for individual channels to a "common denominator".

---

[1] The situation with this decay mode is less clear in comparison with the previous channels. VES collaboration has reported that the peak position in this mode is $\sim$50 MeV lower than the PDG value (ref. [5]). This difference is less in E852 experiment (ref. [6])

CP717, *Hadron Spectroscopy: Tenth International Conference,*
edited by E. Klempt, H. Koch, and H. Orth

TABLE 1. Isobars included to the PWA.

Final state	2-meson isobars
$\pi^-\pi^+\pi^-$	$\varepsilon, \rho, f_0(980), f_0(1500) \rightarrow (\pi^+\pi^-)$
$K^-K^+\pi^-$	$\varepsilon + f_0(980), f_0(1500) \rightarrow (K^+K^-)$
	$K^*(890), K_0(1430) \rightarrow (K^+\pi^-)$
$\eta\eta\pi^-$	$a_0(980) \rightarrow (\eta\pi^-)$
	$f_0(1500) \rightarrow (\eta\eta)$
$\eta'\eta\pi^-$	$f_0(1500) \rightarrow (\eta'\eta)$
	$a_2(1320) \rightarrow (\eta'\pi)$

EXPERIMENTAL DATA AND PARTIAL WAVE ANALYSIS

Experimental data for final states (1), (2) and (3) correspond to one physics run at the VES setup, with π^- beam at 36.6 GeV/c and Be target. Estimated luminosity [2] for this run is $\sim 8\ nb^{-1}$. For the decay mode (4), with the lowest cross section, the experimental data from five physics runs at different beam momenta have been collected, with total luminosity $\sim 65\ nb^{-1}$. This analysis is based on the published results for channel (3), old data but new analysis for channels (1) and (2). New data are included for the channel(4). Decay mode (5) is not included to the combined fit because of the nature of the observed peak is not clear. The $(\omega\rho)$ peak is close to the $(\omega\rho)$ threshold and possibly contains a tail from the $\pi(1300)$, it is close also to the large $\pi_2(1670)$ signal, the peak position is ~ 50 MeV lower than the $\pi(1800)$ mass in other channels and, the last but not the least, there is no reference phase to verify the phase motion.

The reconstructed events for the decay modes (1) and (2) should have three charged tracks in the forward hemisphere. For the reaction (2), the positive track should be identified as the kaon, and at least one of the negative tracks should be identified too. Candidates for the decay modes (3) and (4) have been selected from events with 3 charged tracks and 4 γ in forward hemisphere. η-mesons in reaction (3) should have different decay modes: one in $\eta \rightarrow \gamma\gamma$ and another $\eta \rightarrow \pi^+\pi^-\pi^0$ mode. The η' in final state (4) was reconstructed via decay $\eta' \rightarrow \eta\pi^+\pi^-$, and both η are detected in 2γ mode.

A mass independent Partial Wave Analysis (PWA) has been performed for events in final states (1)-(4), after a subdivision for bins on the total mass. The $(\pi^-\pi^+\pi^-)$ sample ($\sim 8\,700\,000$ events) has been separated for two subsamples, at low $|t|<0.03\ GeV^2$ and at high $|t|$. A standard program for PWA of three pseudoscalar meson system has been applied [7],[8]. The isobars included for $J^P = 0^-$ waves are specified in Table 1. Details concerning the separation between $f_0(980)$ and ε parametrisation can be found in [3]. The parametrisation of $(K\pi)$ S-wave is taken from [9]. Results of mass-independent PWA for summary $J^P = 0^-$ waves, for all decay modes are shown in Fig. 1. Results for individual waves for the $\pi^-\pi^+\pi^-$ final state at low $|t|$ is shown at Fig. 2.

Results of mass-independent PWA are used as input for a combined fit. The fitted

---

[2] The luminosity for Be target was estimated from observed $a_2(1320)$ signal and a compilation of $a_2(1320)$ cross sections in π^-Be interactions.

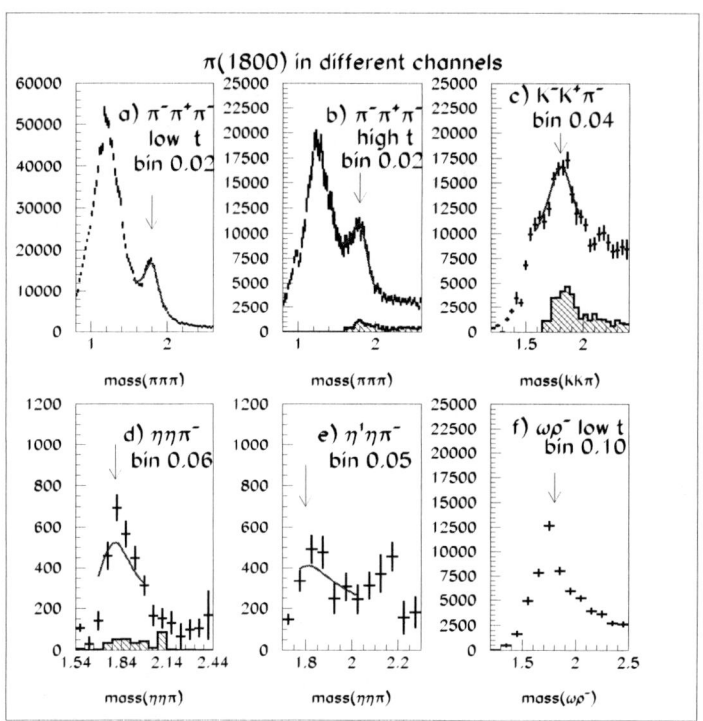

FIGURE 1. Mass spectra for summary 0^- waves after mass-independent PWA and combined fit results for the $\pi(1800)$ in different channels: a) $\pi^-\pi^+\pi^-$ at low $|t|$, $|t| < 0.03\ GeV^2$; b) $\pi^-\pi^+\pi^-$ at high $|t|$, $0.03 < |t| < 1.\ GeV^2$; c) $K^-K^+\pi^-$, $|t| < 0.20\ GeV^2$; d) $\eta\eta\pi^-$, $|t| < 0.19\ GeV^2$; e) $\eta'\eta\pi^-$, $|t| < 0.70\ GeV^2$; f) $\omega\rho^-$ at low $|t|$, $|t| < 0.08\ GeV^2$. Arrows indicate the mass of 1800 MeV. Curves show the fit results. Hatched histograms correspond to contributions from $f_0(1500)$ (not shown for plot e) because of other waves are very small below 1950 MeV).

model contains one common Breight-Wigner resonance and Background amplitudes (different for different decay modes). The modules of BG amplitudes were parametrised by polinomial functions of mass. The phases of BG amplitudes were taken as quadratic functions of mass. The resonance mass and width, the polinomial coefficients, the $\pi(1800)$ and BG branching ratios to different channels, and also phase offsets for different channels are taken as free parameters. The fitted $\pi(1800)$ mass is very close to the PDG mass, but the fitted width is ~ 270 MeV, i.e. greater than the PDG value. Results of the common fit are summarised in Table 2. One can notice a large negative interference between $f_0(600 \div 1300)\pi^-$ and $f_0(980)\pi^-$ subchannels.

A number of peculiarities is observed in the $\pi(1800)$ branching ratios:

- the branching into $(K^-K^+\pi^-)$ decay mode is large;
- in the $(\pi^-\pi^+\pi^-)$ final state, the ratio of branchings into $f_0(980)_{\pi^+\pi^-}\pi^-$ and into $f_0(600 \div 1300)_{\pi^+\pi^-}\pi^-$ is ~0.4, which is unusually large;
- decays into $(\rho\pi^-)$ and (K^*K^-) channels are suppressed;

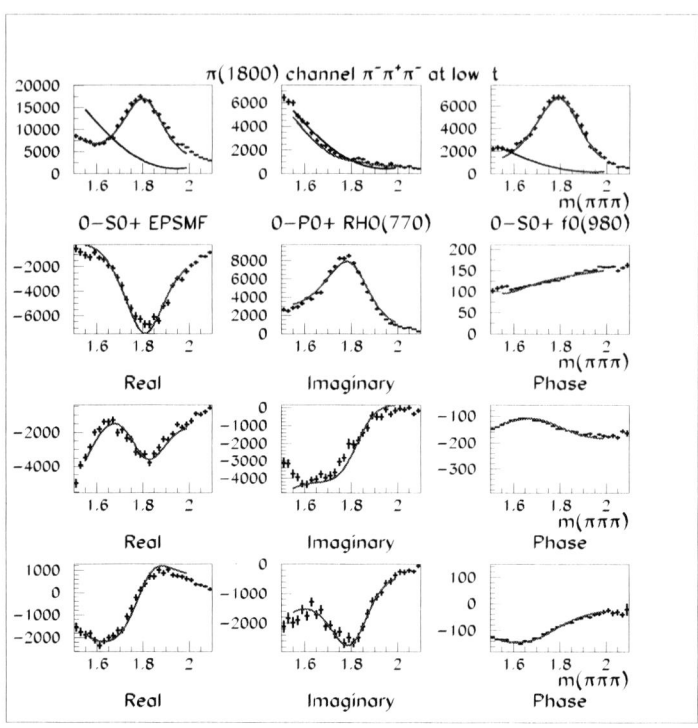

FIGURE 2. Fit results for $(\pi^-\pi^+\pi^-)$ final state at low $|t|$. Upper row: number of events in $(\varepsilon\pi)$, $(\rho\pi)$ and $(f_0(980)\pi)$ subchannels. Next rows: Real, Imaginary parts and Phases of interference terms between subchannels $(\varepsilon\pi)$-$(f_0(980)\pi)$, $(\varepsilon\pi)$-$(\rho\pi)$, and $(\rho\pi)$-$(f_0(980)\pi)$. Curves represent fit results (and fitted Backgrounds for upper row).

TABLE 2. Relative Branching Ratios for $\pi(1800)$ decays.

Final state	Subchannel	Relative width
$\pi^-\pi^+\pi^-$		1
	$f_0(600 \div 1300)_{\pi^+\pi^-}\pi^-$	1.1 ± 0.1
	$f_0(980)_{\pi^+\pi^-}\pi^-$	0.44 ± 0.15
	$f_0(1500)_{\pi^+\pi^-}\pi^-$	0.05 ± 0.03
	$\rho(770)\pi^-$	<0.03 at 90% C.L.
$K^-K^+\pi^-$		0.29 ± 0.10
	$K^*(892)K^-$	<0.03 at 90% C.L.
$\eta\eta\pi^-$		0.15 ± 0.06
	$a_0(980)_{\eta\pi^-}\eta$	0.13 ± 0.06
	$f_0(1500)_{\eta\eta}\pi^-$	0.012 ± 0.005
$\eta'\eta\pi^-$	$f_0(1500)_{\eta'\eta}\pi^-$	0.026 ± 0.010

- the branching into $(\eta'\eta\pi^-)$ is large.

The suppression of the decays $\pi(1800)$ into a pseudoscalar and a vector mesons is expected if $\pi(1800)$ represents the 2-nd radial exitation of π [10], as well as for the exotic treatment in the flux tube model [11]. However, the enhancement of the decay channels with (K^+K^-), $f_0(980)$ and η' needs a special explaination. The full set of unusual features can be explaned in the framework of one hypothesis: the $\pi(1800)$ decays mainly into $SU(3)$-singlet and $SU(3)$-octet.

CONCLUSIONS.

The simultaneous fit of $\pi(1800)$ decays into four final states, $(\pi^-\pi^+\pi^-)$, $(K^-K^+\pi^-)$, $(\eta\eta\pi^-)$ and $(\eta'\eta\pi^-)$ is performed.

- The decay of $\pi(1800)$ into $(f_0(1500)\pi)$ is seen in all four final states.
- A number of peculiarities in $\pi(1800)$ branching ratios is observed. They can be explained if one supposes that the $\pi(1800)$ decays mainly into $SU(3)$-singlet and $SU(3)$-octet (contrary to OZI predictions).

ACKNOWLEDGMENTS

This work is supported in part by the Russian Foundation of Basic Research grants RFBR 02-02-17479, RFBR 00-15-96689 and also NSh 1695.2003.2. Authors thank the organizers of the HADRON-2003 conference for hospitality and financial support.

REFERENCES

1. G. Bellini *et al., Phys. Rev. Lett.* 48(1982) p. 1697
2. S.Bityukov *et al., Phys. Lett.* B268(1991) p.137
   G.Beladidze *et al., Sov.J.Nucl.Phys*55(1992) p.1535, translated from YAF 55, p. 2748.
   E.Berdnikov *et al., Phys. Lett.* B337 p.219
   D.Amelin *et al., PAN* 59(1996), p. 976
   D.Amelin *et al., PAN* 62(1999) p.445,YAF 62(1999)487
   A.Zaitsev *"Searches for hybrid mesons"*, talk on International Symposium On Hadron Spectroscopy, Chiral Symmetry And Relativistic Description Of Bound Systems, Tokio, 2003
3. D.Amelin *et al., Phys. Lett.* B356(1995), p.595
4. S.Chung *et al., Phys. Rev.*. D65(2002), p.072001
5. D.Amelin *et al., "Study of the reaction $pi^-Be \rightarrow \omega\pi^-\pi^0 pBe$*, hep-ex/9810013v1
6. A.Popov *for E852 Collaboration, "A Study of the reaction $\pi p \rightarrow \omega\pi\pi^0 p$ at 18 GeV/c"*, AIP Conf.Proc.619:565-568,2002 (Protvino 2001)
7. G.Ascoli *et al., Phys.Rev.Lett* 25(1970)p.962;
   D.V.Brockway,University of Illinois report COO-1195-197(1970);
8. I.Kachaev *et al., "Study of reaction $\pi^-A \rightarrow \pi^+\pi^-\pi^-A$ at VES setup"*, AIP Conf.Proc.619:577-581,2002 (Protvino 2001)
9. D. Aston *et al., Nucl. Phys.* B296(1988)p.493
10. T.Barnes *et al., "Higher Quarkonia"*, hep-ph/9609339
11. F.E.Close, P.R.Page, *Nucl. Phys.*B443(1995)233

New contributions to $\phi \to \pi^0 \pi^0 \gamma$ and $\phi \to \pi^0 \eta \gamma$ decays [1]

J. Palomar*, **L. Roca***, E. Oset* and M. J. Vicente Vacas*

*Departamento de Física Teórica and IFIC Centro Mixto Universidad de Valencia-CSIC
Institutos de Investigación de Paterna, Apdo. correos 22085, 46071, Valencia, Spain*

Abstract. We study the radiative ϕ decay into $\pi^0 \pi^0 \gamma$ and $\pi^0 \eta \gamma$ taking into account mechanisms in which there are two sequential vector-vector-pseudoscalar or axial-vector–vector–pseudoscalar steps followed by the coupling of a vector meson to the photon, considering the final state interaction of the two mesons. There are other mechanisms in which two kaons are produced through the same sequential mechanisms or from ϕ decay into two kaons and then undergo final state interaction leading to the final pair of pions or $\pi^0 \eta$, this latter mechanism being the leading one.

INTRODUCTION

The radiative decays of the ϕ into $\pi^0 \pi^0 \gamma$ and $\pi^0 \eta \gamma$ have been the subject of intense study because one can get much information about the nature of the $f_0(980)$ and $a_0(980)$ resonances. The nature of the scalar meson resonances has generated a large debate, with new ideas brought by the claim that these resonances are dynamically generated from multiple scattering with the ordinary chiral Lagrangians [2]. These two reactions involving the decay of the ϕ are special. Indeed, the ϕ does not decay into two pions because of isospin symmetry it can decay into two charged kaons (with a photon attached to one of them) and the two kaons scatter giving rise to the two pions (or $\pi^0 \eta$). The radiative ϕ decay through this mechanism was studied in [3] and the results of lowest order chiral perturbation theory (χPT) were used to account for the $K^+ K^- \to \pi^0 \pi^0$ transition. Since the chiral perturbation theory $K^+ K^- \to \pi^0 \pi^0$ amplitude does not account for the $f_0(980)$, the excitation of this resonance has to be taken in addition, something that has been done more recently using a linear Sigma-model in [4].

The work of [5] leads to the excitation of the $f_0(980)$ in the $\pi^0 \pi^0$ production, or the $a_0(980)$ in $\pi^0 \eta$ production in a natural way, since the use of unitarized chiral perturbation theory ($U\chi PT$), as in [2], generates automatically those resonances in the meson meson scattering amplitudes and one does not have to introduce them by hand.

In addition to the mechanisms discussed before we have sequential $V \to VP \to PP\gamma$ process, which is known to provide the $\omega \to \pi^0 \pi^0 \gamma$ radiative width with accuracy and has been further extended to study $\rho \to \pi^0 \pi^0 \gamma$ and other radiative decays.

Another novelty of the present work is the consideration of sequential mechanisms

---

[1] Contribution to the X. International Conference On Hadron Spectroscopy, August 31 - September 6, 2003, Aschaffenburg, Germany. (Full article in [1])

CP717, *Hadron Spectroscopy: Tenth International Conference,*
edited by E. Klempt, H. Koch, and H. Orth
© 2004 American Institute of Physics 0-7354-0197-7/04/$22.00

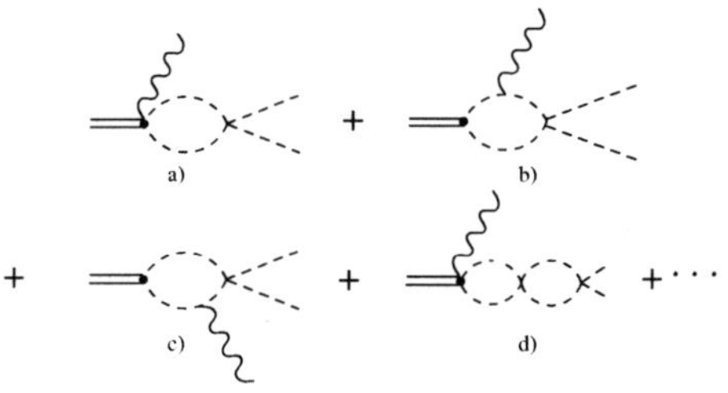

FIGURE 1. Loop diagrams included in the chiral loop contributions. The intermediate states in the loops are K^+K^-.

involving the exchange of an intermediate axial-vector meson ($J^{PC} = 1^{++}$ or 1^{+-}), both producing directly the final meson pair or through the intermediate production of kaons which undergo collisions and produce these mesons.

All the mechanisms considered here contribute moderately, but appreciably, to the ϕ radiative width. The good agreement with experiment is reached in spite of having in our approach a width for the $f_0(980)$ very small, of the order of 30 MeV, in apparent contradiction with the "visual" $f_0(980)$ width in the experiment, which looks much larger. The reason for this stems from the fact that, due to gauge invariance, the amplitude for the process contains as a factor the momentum of the photon, which grows fast as we move down to smaller invariant masses from the mass of the $f_0(980)$ where the photon momentum is very small. This distorts the shape of the resonance, making it appear wider.

MODEL

The mechanism for radiative decay using the tensor formulation for the vector mesons have been discussed in [5]. The diagrams considered are depicted in Fig. 1, where the loops contain K^+K^-. The vertices needed for the diagrams are obtained from the chiral Lagrangian for vector meson resonances [6], assuming ideal mixing between the ϕ and ω mesons.

In Fig. 1 the $MM \rightarrow MM$ transition amplitude implies the iterated loops implicit in the coupled channels Bethe Salpeter equation (BS) obtained in [2].

Following the lines of [7, 8] in the study of ρ and ω radiative decays, we also include sequential vector meson exchange here. They are depicted in Fig. 2, where we explicitly assume that the $\phi \rightarrow \rho^0\pi^0$ proceeds via the $\phi - \omega$ mixing through the Lagrangian
$$\mathscr{L}_{\phi\omega} = \Theta_{\phi\omega}\,\phi_\mu\,\omega^\mu.$$
It is worth mentioning that the theoretical expression for the $V \rightarrow P\gamma$ decay widths
$$\Gamma_{V\rightarrow P\gamma} = \tfrac{4}{3}\alpha C_i^2 \left(\frac{Gg f^2}{M_\rho M_V}\right)^2 k^3,$$ gives slightly different results to the experimental values from the PDG. For this reason the C_i coefficients were normalized so that the theoretical $V \rightarrow P\gamma$ decay widths agree with experiment.

Since the $\pi\pi$ interaction is strong in the region of invariant masses relevant in the

161

FIGURE 2. Diagrams for the tree level VMD mechanism.

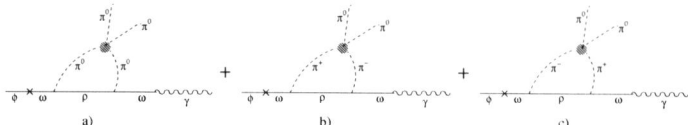

FIGURE 3. VMD diagrams with final state interaction of pions

present reaction we next consider the final state interaction of the pions in the sequential vector meson mechanism.

We must take into account the loop function of Fig. 3a, but on the same footing we must also consider those of Fig. 3b and 3c, where charged pions are produced and allowed to interact to produce the $\pi^0\pi^0$ final state. The thick dot in Fig. 3 means that one is considering the full $\pi\pi \to \pi\pi$ t-matrix, involving the loop resummation of the BS equation of ref. [2] and not just the lowest order $\pi\pi \to \pi\pi$ amplitude.

We have also to evaluate the diagrams analogous to those in Fig. 3 but with kaons and K^* in the intermediate states.

Since the mass of the ϕ is around 250 MeV higher than the ρ mass, and we are considering sequential vector meson mechanisms with ρ or K^* exchange, we should pay attention to the analogous mechanisms involving vector mesons with a similar mass difference with the ϕ on the upper side and these are the axial and vector mesons with $J^{PC} = 1^{+-}$ or 1^{++} (see Table 1). Therefore, the b_1 or a_1 axial vector mesons and the K_{1B}, K_{1A} strange axial vector mesons will play the role of the ρ or the K^* in former diagrams.

Because of the C parity of the states, the Lagrangians for the axial-vector–vector–pseudoscalar couplings have the structure of $\tilde{D} < B_{\mu\nu}\{V^{\mu\nu}, P\} >$ for the b_1 octet and $i\tilde{F} < A_{\mu\nu}[V^{\mu\nu}, P] >$ for the octet of the a_1, where the $<>$ means $SU(3)$ trace. In the last expressions V and P are the usual vector and pseudoscalar $SU(3)$ matrices respectively and B and A are axial vector $SU(3)$ matrices given in [9]. Note the use of the tensor formalism [6] in the previous Lagrangians. With the values for $\tilde{D} = -1000 \pm 120$ MeV and $\tilde{F} = 1550 \pm 150$ MeV, we are able to describe all the $A \to VP$ decays plus the radiative decays of the $a_1 \to \pi\gamma$.

The relevant mechanisms involving axial-vectors are those in which K, $\bar{K}$ are created and through scattering lead to the final $\pi^0\pi^0$ state, having also one of the K_1 resonances as intermediate state. These are not OZI forbidden and have a nonnegligible contribution.

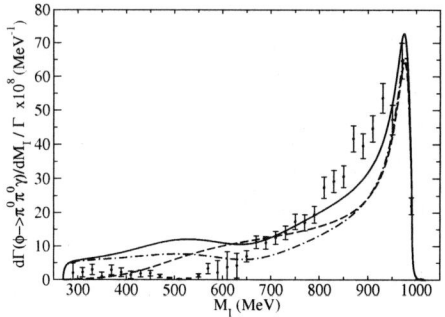

FIGURE 4. Different contributions to the two pion invariant mass distributions of the $\phi \rightarrow \pi^0\pi^0\gamma$ decay: Dashed line: chiral loops of Fig. 1. Dashed-dotted line: chiral loops of Fig. 1 + sequential VMD and its final state interaction. Solid line: idem plus the contribution of the mechanisms involving axial-vector mesons, (full model).

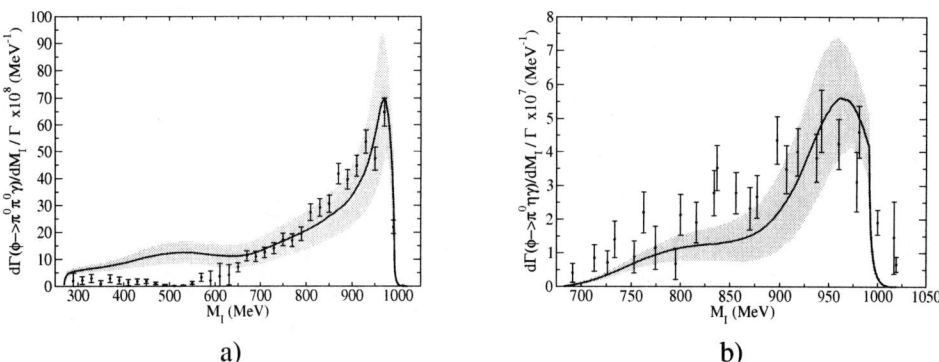

a)   b)

FIGURE 5. Final results for the $\pi^0\pi^0$ invariant mass distribution for the $\phi \rightarrow \pi^0\pi^0\gamma$ decay with the theoretical error band. Experimental data from [13].

RESULTS

In Fig. 4 we show the results of the different contributions. We should say that the loops of the sequential vector meson mechanisms involving kaons are relatively important but there is a strong cancellation between the mechanisms with a ϕ and an ω attached to the photon.

In Fig. 5a we show the final result but including an evaluation of the error band due to the uncertainties in the parameters of the model.

The branching ratio obtained in the present work is $BR(\phi \to \pi^0\pi^0\gamma) = (1.2 \pm 0.3) \times 10^{-4}$ to be compared with the experimental values
$$BR^{exp}(\phi \to \pi^0\pi^0\gamma) = (1.22 \pm 0.10 \pm 0.06) \times 10^{-4} \ [10], \ (0.92 \pm 0.08 \pm 0.06) \times 10^{-4}, [11], \ (1.09 \pm 0.03 \pm 0.05) \times 10^{-4} \ [13].$$

In Fig. 5a we can see that our results, considering the error band, fairly agrees with the experimental data except in the region around 500 MeV. Although the agreement with data at low masses is not very good, we must point out two sources of uncertainty in the experimental spectrum. First, the results in the low and intermediate mass region largely depend on the background subtraction dominated by the non-resonant $\omega\pi^0$ process. The size of this process is difficult to obtain because it has a strong background itself, mostly from the $\phi \to f_0\gamma$ process, as it is discussed in [15]. There, its magnitude has been obtained in a model dependent way assuming some a priori spectrum for the $\phi \to f_0\gamma$ process [15]. In fact, before the subtraction, the raw data resemble much more our calculated spectrum. Additionally, there is some uncertainty in the way the data are corrected to account for the experimental efficiency. This is done in [13] by dividing the observed spectrum by the effect of applying the experimental efficiency on some theoretical distribution. This unfolding procedure depends on the theoretical model used, which we think at low $\pi^0\pi^0$ masses is at least incomplete.

After the discussion of the former points the consideration of the $\phi \to \pi^0\eta\gamma$ decay requires only minimal technical details which one can see in ref. [1].In Fig. ??b we have plotted the full model performing the theoretical error analysis. We can see that when these uncertainties are considered we obtain a theoretical band in acceptable agreement with the experimental data.

The branching ratio obtained is $BR(\phi \to \pi^0\eta\gamma) = (0.6 \pm 0.2) \times 10^{-4}$ to be compared with the experimental values $BR^{exp}(\phi \to \pi^0\eta\gamma) = (0.88 \pm 0.14 \pm 0.09) \times 10^{-4}$ [12], $(0.90 \pm 0.24 \pm 0.10) \times 10^{-4}$ [11], $(0.85 \pm 0.05 \pm 0.06) \times 10^{-4}$ [14].

REFERENCES

1. J. E. Palomar, L. Roca, E. Oset and M. J. Vicente Vacas, arXiv:hep-ph/0306249.
2. J. A. Oller and E. Oset, Nucl. Phys. A **620** (1997) 438 [Erratum-ibid. A **652** (1999) 407] [arXiv:hep-ph/9702314].
3. A. Bramon, A. Grau and G. Pancheri, Phys. Lett. B 289 (1992) 97.
4. A. Bramon, R. Escribano, J. L. Lucio M, M. Napsuciale and G. Pancheri, Eur. Phys. J. C **26** (2002) 253 [arXiv:hep-ph/0204339].
5. E. Marco, S. Hirenzaki, E. Oset and H. Toki, Phys. Lett. B **470** (1999) 20 [arXiv:hep-ph/9903217].
6. G. Ecker, J. Gasser, A. Pich and E. de Rafael, Nucl. Phys. B **321** (1989) 311.
7. A. Bramon, R. Escribano, J. L. Lucio Martinez and M. Napsuciale, Phys. Lett. B **517** (2001) 345 [arXiv:hep-ph/0105179].
8. J. E. Palomar, S. Hirenzaki and E. Oset, Nucl. Phys. A **707** (2002) 161 [arXiv:hep-ph/0111308].
9. L. Roca, J. E. Palomar and E. Oset, arXiv:hep-ph/0306188.
10. M. N. Achasov *et al.*, Phys. Lett. B **485** (2000) 349 [arXiv:hep-ex/0005017].
11. R. R. Akhmetshin *et al.* [CMD-2 Collaboration], Phys. Lett. B **462** (1999) 380 [arXiv:hep-ex/9907006].
12. M. N. Achasov *et al.*, Phys. Lett. B **479** (2000) 53 [arXiv:hep-ex/0003031].
13. A. Aloisio *et al.* [KLOE Collaboration], Phys. Lett. B **537** (2002) 21 [arXiv:hep-ex/0204013].
14. A. Aloisio *et al.* [KLOE Collaboration], Phys. Lett. B **536** (2002) 209 [arXiv:hep-ex/0204012].
15. S. Giovanella ans S. Miscetti, KLOE note 178

The pion form factor from first principles

J. van der Heide

National Institute for Nuclear Physics and High-Energy Physics (NIKHEF), 1009 DB Amsterdam,
The Netherlands

Abstract. We calculate the electromagnetic form factor of the pion in quenched lattice QCD. The non-perturbatively improved Sheikoleslami-Wohlert lattice action is used together with the $\mathcal{O}(a)$ improved current. We calculate form factor for pion masses down to $m_\pi = 380\,MeV$. We compare the mean square radius for the pion extracted from our form factors to the value obtained from the 'Bethe Salpeter amplitude'. Using (quenched) chiral perturbation theory, we extrapolate our results towards the physical pion mass.

INTRODUCTION

The pion, being the lightest and simplest particle in the hadronic spectrum has been studied intensively in the past. Using a variety of effective and phenemenological models, the properties of the pion have been desribed with varying succes. However, these models make assumptions, for example, confinement is put in by hand in contrast to being the result of the underlying dynamics. Lattice QCD (LQCD) doesn't have this drawback since it is solves QCD directly.

Using LQCD, global properties of the pion such as the mass and the decay width have been calculated to satisfying accuracy. The form factor, which directly reflects the internal structure, is clearly an important challenge. The first lattice results were obtained by Martinelli and Sachrajda [1]. It was followed by a more detailed study by Draper *et al.* [2], who showed that the form factor could be described by a simple monopole form as suggested by vector meson dominance [3]. We extend [4] these studies by adopting improved lattice techniques [10–14], which means that we include extra operators in order to systematically eliminate all the $\mathcal{O}(a)$ discretisation errors.

Some aspects of the pion structure have been obtained [5–9] using 'the Bethe-Salpeter method'. We also use this approach and compare its predictions to the results of our direct calculation of the pion form factor.

Finally we study a chiral extrapolation to reach the physical limit.

THE METHOD

To extract the form factor we calculate the two- and three point functions of the pion, analogously to [2]. The two point function is given by

$$G_2(t,\mathbf{p}) = \sum_{\mathbf{x}} \left\langle \phi(t,\mathbf{x})\,\phi^\dagger(0,\mathbf{0}) \right\rangle e^{i\,\mathbf{p\cdot x}}, \tag{1}$$

CP717, *Hadron Spectroscopy: Tenth International Conference,*
edited by E. Klempt, H. Koch, and H. Orth
© 2004 American Institute of Physics 0-7354-0197-7/04/\$22.00

where $\phi^\dagger$ is an operator creating a state with the quantum numbers of the pion. By varying the interquark distance at the sink, t_f, we can improve the overlap with the physical pion and obtain information on the 'Bethe-Salpeter amplitudes'. The three point function is calculated as

$$G_3(t_f, t; \mathbf{p}_f, \mathbf{p}_i) = \sum_{\mathbf{x}_f, \mathbf{x}} \left\langle \phi(x_f) \, j_4(x) \, \phi^\dagger(0) \right\rangle e^{-i\,\mathbf{p}_f \cdot (\mathbf{x}_f - \mathbf{x}) \, -i\,\mathbf{p}_i \cdot \mathbf{x}} \qquad (2)$$

with j_4 the fourth component of the current, inserted at time t. Since the local current

$$j_\mu^L(x) = \bar{\psi}(x) \, \gamma_\mu \, \psi(x), \qquad (3)$$

is not conserved on the lattice, one can construct the Noether current belonging to our action

$$j_\mu^C = \kappa \left(\bar{\psi}(x)(1 - \gamma_\mu)U_\mu(x)\psi(x + \hat{\mu}) - \bar{\psi}(x + \hat{\mu})(1 + \gamma_\mu)U_\mu^\dagger(x)\psi(x) \right). \qquad (4)$$

This current however, still has $\mathcal{O}(a)$ corrections for $Q^2 > 0$. A conserved and improved current can be constructed [12–14]

$$j_\mu^I = Z_V \left\{ j_\mu^L(x) + a\, c_V\, \partial_\nu\, T_{\mu\nu} \right\}, \qquad (5)$$

with

$$\begin{aligned} T_{\mu\nu} &\quad - \quad \bar{\psi}(x)\, i\, \sigma_{\mu\nu}\, \psi(x)\,, \\ Z_V &\quad = \quad Z_V^0 \left(1 + a b_V\, m_q \right). \end{aligned} \qquad (6)$$

The bare-quark mass is defined as $m_q = \frac{1}{2a}(1/\kappa - 1/\kappa_c)$, where κ_c is the kappa value in the chiral limit and a denotes the lattice spacing. For our simulation, $\kappa_c = 0.13525$ [15]. Comparison of the currents will give us information on the importance of improvement.

DETAILS OF THE CALCULATION

On a $24^3 \times 32$ lattice, we generated $\mathcal{O}(100)$ quenched gluon configurations at $\beta = 6.0$. Subsequently, we calculated non-perturbatively improved ($c_{SW} = 1.796$ [10]) quark propagators for 5 different values of the hopping parameter, κ. These propagators were then combined to two- and three point functions for the pion with masses ranging from 360 to 970 MeV.[1] For the improved current, we use the parameters Z_V^0, b_V and c_V as determined by Bhattacharya et al. [17]. To extract the significant parameters from our numerical data, we use the following parametrisations. For the two point function, we have

$$G_2(t, \mathbf{p}) = \sum_{n=0}^{1} \sqrt{Z_R^n(\mathbf{p}) Z_0^n(\mathbf{p})} \left\{ e^{-E_{\mathbf{p}}^n t} + e^{-E_{\mathbf{p}}^n (N_\tau - t)} \right\}, \qquad (7)$$

---

[1] The lattice spacing $a = 0.105$ fm is taken from [16].

166

including the contribution of the ground state (n=0) and a first excited one (n=1). The Z_R^n denote the matrix elements,

$$Z_R^n(\mathbf{p}) \equiv |\langle \Omega | \phi_R | n, \mathbf{p} \rangle|^2 , \tag{8}$$

and are related to the 'Bethe-Salpeter amplitude', $\Phi(R) = \sqrt{Z_R^\pi(0)/Z_0^\pi(0)}$.

$E_\mathbf{p}^0$ and $E_\mathbf{p}^1$ are the energies of ground and excited state, respectively; R denotes the interquark distance. The three-point function is parametrised as

$$G_3(t_f, t; \mathbf{p}_f, \mathbf{p}_i) = F(Q^2) \sqrt{Z_R^0(\mathbf{p}_f) Z_0^0(\mathbf{p}_i)} e^{-E_{\mathbf{p}_f}^0 (t_f - t) - E_{\mathbf{p}_i}^0 t}$$

$$+ \left\{ \sqrt{Z_R^1(\mathbf{p}_f) Z_0^0(\mathbf{p}_i)} \langle 1, \mathbf{p}_f | j_\mu(0) | 0, \mathbf{p}_f \rangle e^{-E_{\mathbf{p}_f}^1 (t_f - t) - E_{\mathbf{p}_i}^0 t} + (1 \leftrightarrow 0) \right\} . \tag{9}$$

The production of pion pairs, 'wrap around effects' being due to the propagation of states beyond t_f and an elastic contribution from the excited state were ignored in our parametrisation after being estimated negligable.

All parameters in the two- and three-point functions - energies E, Z-factors and the form factor $F(Q^2)$ - were fit simultaneously to the data per configuration. For the three-point function, we let the current insertion time t vary from 0 to t_f. The value for the parameters and their error in these simultaneous fits are obtained through a single elimination jackknife procedure.

RESULTS

As a byproduct of our simulations we also obtain pion masses for the 5 different κ-values. They agree with the literature. We also checked the energie-momentum relation and up to the energies involved we found that a continuum relation provides the best description. These non-trivial tests indicate that our simulations are done correctly.

Using the different currents, we extract the form factor for our five κ-values. Comparison of the results for the Noether current and the improved current yields that the effect of improvement can be as large as 25% for the highest momenta considered. The improved results are shown in Fig.1. The high accuracy for the datapoint at $Q^2 = 0$ is due to the fact that we satisfy the Ward-Takahashi identity to 1 ppm.

As in the previous study [2] of the pion form factor we compare our results to a monopole form factor

$$F(Q^2) = \{1 + \frac{Q^2}{M_\rho^2}\}^{-1} , \tag{10}$$

which is suggested by the vector meson dominance ansatz. Fitting our data to this model, we extract a vector meson mass which is within 5% of the corresponding rho mass on the lattice [15]. From the behaviour of the form factor at low Q^2, we can extract the mean-square charge radius of the pion,

$$\left. \frac{dF(Q^2)}{dQ^2} \right|_{Q^2=0} = -\frac{1}{6} \langle r^2 \rangle_{FF} = -\frac{1}{M_V^2} \tag{11}$$

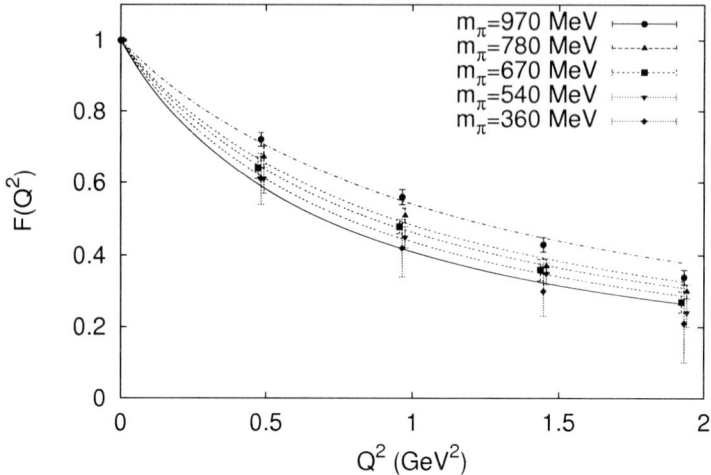

FIGURE 1. *Form factors as a function of Q^2 for the five pion masses. Curves: monopole fits to the data.*

where in the last step we assume Eq. 10 and use the fitted parameter M_V. The results are shown in Fig. 2 as a function of the pion mass. Previously, the 'Bethe-Salpeter-amplitude' $\Phi(R)$ has been used to obtain estimates of the charge radius,

$$\langle r^2 \rangle_{BS} := \frac{1}{4} \frac{\int d^3\vec{r}\, \vec{r}^2\, \Phi^2(|\vec{r}|)}{\int d^3\vec{r}\, \Phi^2(|\vec{r}|)} \,. \tag{12}$$

The results based on this procedure are also shown in Fig. 2. As can be seen these values are much lower than the actual values obtained from the form factor. Moreover, the Bethe-Salpeter results are almost mass independent, in accordance with the observations of Refs. [5–7, 9]. This is a known[7] deficit of the approach, which Fig. 2 makes quantitative. We extrapolate our results obtained with Eq. 11 using three different parametrisations. First, we try chiral perturbation theorie (χpt). At one-loop order the prediction for the radius [18] is

$$\langle r^2 \rangle_{\chi PT}^{one-loop} = c_1 + c_2 \log m_\pi^2 \tag{13}$$

In quenched χpt the rms is constant to this order. There are however indications that a mass dependence appears at higher order [19]. For our masses we restrict ourselves to a term linear in m_π^2 [20]. Lastly, we have also used the VMD prescription and assume a linear dependence of M_V on m_π^2. These three expectations are also plotted in Fig. 2.

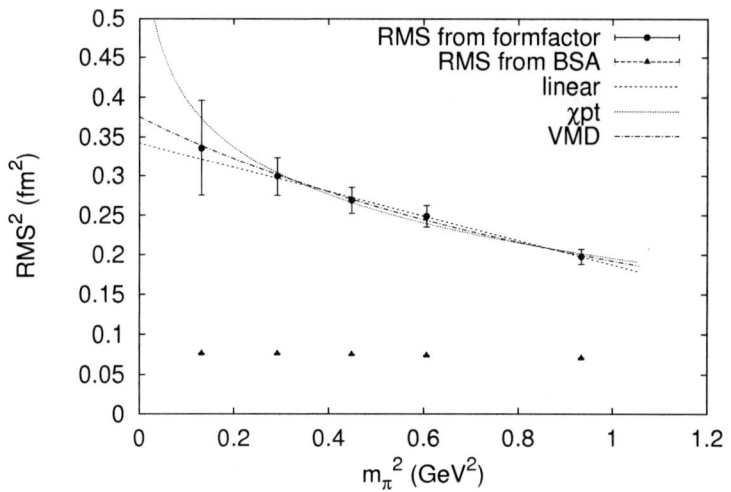

FIGURE 2. *Radius of the pion as obtained from Bethe-Salpeter amplitudes, Eq. 12, and from the form factor, Eq. 11. Also shown are three different parametrizations of* $\langle r^2 \rangle$ *(see text).*

ACKNOWLEDGEMENTS

This work has been done in collaboration with Justus Koch and Edwin Laermann.

REFERENCES

1. G. Martinelli and C.T. Sachrajda, Nucl. Phys. **B306** (1988) 865.
2. T. Draper, R.M. Woloshyn, W. Wilcox and K. Liu, Nucl. Phys. **B318** (1989) 319.
3. H.B. O'Connell, B.C. Pearce, A.W. Thomas and A.G. Williams, Prog. Part. Nucl. Phys. **39** (1997) 201.
4. J. van der Heide *et al.,,* Phys. Lett. **B566** (2003) 131.
5. M.-C. Chu, M. Lissia and J.W. Negele, Nucl. Phys. **B360** (1991) 31.
6. M.W. Hecht and T.A. DeGrand, Phys. Rev. **D46** (1992) 2155.
7. R. Gupta, D. Daniel and J. Grandy, Phys. Rev. **D48** (1994) 3330
8. P. Lacock *et al.*, Phys. Rev. **D51** (1995) 6403.
9. E. Laermann and P. Schmidt, Eur. Phys. J **C20** (2001) 541.
10. M. Lüscher *et al.*, Nucl. Phys. **B491** (1997) 323.
11. B. Sheikholeslami and R. Wohlert, Nucl. Phys. **B259** (1985) 572.
12. G. Martinelli, C.T. Sachrajda and A. Vladikas, Nucl. Phys. **B358** (1991) 212.
13. M. Lüscher, S. Sint, R. Sommer and H. Wittig, Nucl. Phys. **B491** (1997) 344.
14. M. Guagnelli and R. Sommer, Nucl. Phys. Proc. Suppl. **63** (1998) 886.
15. UKQCD Collaboration (K.C. Bowler *et al.*), Phys. Rev. **D62** (2000) 054506.
16. R.G. Edwards, U.M. Heller and T.R. Klassen, Nucl. Phys. **B517** (1998) 377.
17. T. Bhattacharya, R. Gupta, W. Lee and S. Sharpe, Phys. Rev. **63** (2001) 074505.
18. J. Gasser and H. Leutwyler, Ann. Phys. (N.Y.) **158** (1984) 142.
19. G. Colangelo and E. Pallante, Nucl. Phys. **B520** (1998) 433.
20. G. Colangelo, private communication.

The pion-pion scattering lengths from DIRAC

C. Santamarina

Basel University, Switzerland

on behalf of DIRAC Collaboration

Abstract. The scattering lengths of a two pion system are the *golden magnitudes* to test the QCD predictions in the low energy sector. The DIRAC (PS-212) experiment at CERN will obtain a particular combination of the S-wave isospin 0 and 2 scattering lengths by measuring the lifetime of pionium, the hydrogen-like $\pi^+\pi^-$ atom. This measurement tests the accurate predictions of the Chiral Perturbation Theory. The most recent experimental results are presented.

INTRODUCTION

DIRAC experiment, conducted at CERN, is measuring the lifetime of the ground state of pionium, the hydrogen-like $\pi^+\pi^-$ atom [1]. The decay of pionium is dominated by the strong channel $\pi^+\pi^- \rightarrow \pi^0\pi^0$ (BR=94%). The ground state lifetime is related to the S-wave isospin 0 and 2 scattering length difference by the Deser type formula known to NLO [2]:

$$\frac{1}{\tau} = \frac{2}{9}\alpha^3 \sqrt{M_\pi^2 - M_{\pi^0}^2 - \alpha^2 M_\pi^2/4} \left| a_0^0 - a_0^2 \right|^2 (1+\delta), \tag{1}$$

where α is the fine structure constant, M_π the mass of the charged pions, M_{π^0} the mass of the neutral pion and δ the correction to NLO ($\delta = 0.058$).

The Chiral Perturbation Theory has a precise prediction to $\circ$ (p^6) for this difference and hence for the lifetime of pionium [3]:

$$\left| a_0^0 - a_0^2 \right| = 0.265 \pm 0.004 \Rightarrow \tau = (2.9 \pm 0.1) \cdot 10^{-15} s,$$

which must be checked with similarly accurate experimental data. DIRAC goal is to achieve a 5% precision in the determination of the scattering length difference by measuring the lifetime with 10% precision.

EXPERIMENTAL METHOD

DIRAC measures the pionium lifetime analyzing the low relative momentum ($\vec{Q}$) spectrum of $\pi^+\pi^-$ pairs produced in the collisions of the $24 GeV/c$ proton beam with the target. The target consists of one or several metal layers of the overall thickness of $\sim 100\mu m$.

CP717, *Hadron Spectroscopy: Tenth International Conference,*
edited by E. Klempt, H. Koch, and H. Orth
© 2004 American Institute of Physics 0-7354-0197-7/04/$22.00

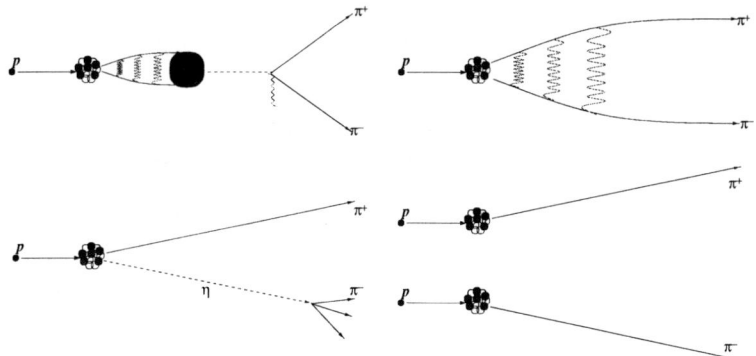

FIGURE 1. Pion pair production mechanisms: **Top left:**Atomic Pair. **Top right:**Coulomb Pair. **Bottom left:**Non Coulomb Pair. **Bottom right:** Accidental Pair.

This spectrum is formed of *Real* (time correlated) pairs, in which both pions come from a single proton-target interaction, and *Accidentals*, coming from two different interactions. Time correlated pairs are divided into:

- *Atomic Pairs (pairs resulting from atomic break-ups)*. Pionic atoms are created by the Final State Coulomb Interaction (FSCI) of low relative momentum $\pi^+\pi^-$ pairs. The sample of initial pionic atoms (N^A) evolves by colliding with the atoms of the metal target. The evolution terminates when the atoms break-up in one of the collisions or annihilate. If the number of broken atoms is given by n^A then, the shorter is the lifetime of pionium the smaller is the break-up probability $P_{br} = n^A/N^A$. The dependence of the break-up probability on the lifetime is accurately known [4], as shown in figure 3 for the Nickel target. *Atomic Pairs* are restricted to the $Q < 4MeV/c$ region.
- *Coulomb Pairs*. Their production mechanism is the same as for the atoms but leading to continuous spectrum eigenstates of the hydrogen-like hamiltonian of the $\pi^+\pi^-$ system. This is why the yield of atoms is proportional to the number of Coulomb pairs within a determined $\vec{Q} \in \Omega$ arbitrary region [1] [5]:

$$N^A = K \times \int_{\vec{Q} \in \Omega} \frac{dN^C}{dQ}. \tag{2}$$

The relative momentum distribution of *Coulomb Pairs* is given by:

$$\frac{dN^C}{dQ} \propto_T \left[\frac{2\pi M_\pi \alpha/Q}{1 - e^{-2\pi M_\pi \alpha/Q}} Q^2 \right], \tag{3}$$

where T stands for the transformation due to multiple scattering in the target and setup resolution of the *at production* distribution between brackets.

---

[1] In the experiment usually $\Omega = \{$Events with reconstructed $Q < 2MeV/c\}$.

- *Non Coulomb Pairs.* Non Correlated pairs in which one of the pions comes from the decay of a long-lived source [2]. Its relative momentum distribution is purely phase space driven and, hence, given by

$$\frac{dN^{NC}}{dQ} \propto_T [Q^2].$$ (4)

Apart from time correlated pairs also *Accidental* $\pi^+\pi^-$ *Pairs*, where each pion comes from a different proton-target interaction, are recorded. Their relative momentum distribution is also determined by phase space:

$$\frac{dN^{Acc}}{dQ} \propto_T [Q^2].$$ (5)

The *Accidental Pairs* sample is used to parameterize the *Real Pairs* pairs spectrum:

$$\frac{dN^{Real}}{dQ} = \frac{dn^A}{dQ} + \frac{dN^{NC}}{dQ} + \frac{dN^C}{dQ} = \frac{dn^A}{dQ} + (aR(Q)+b)\frac{dN^{Acc}}{dQ},$$ (6)

where $R(Q)$ is the ratio between Coulomb pairs and accidentals, obtained with Monte Carlo, and a and b are calculated by fitting the spectrum in the $Q > 4MeV/c$ region, free from atomic pairs.

The shape of the relative momentum distribution of *Atomic Pairs* can be obtained as:

$$\frac{dn^A}{dQ} = \frac{dN^{Real}}{dQ} - (aR(Q)+b)\frac{dN^{Acc}}{dQ}.$$ (7)

Moreover, the break-probability can then be experimentally determined:

$$P_{br} = \int \frac{dn^A}{dQ} \Big/ K \int_{\dot{Q} \in \Omega} aR(Q)\frac{dN^{Acc}}{dQ}.$$ (8)

An alternative method consists of using pure Monte Carlo distributions for *Coulomb*, *Non Coulomb* and *Atomic Pairs* to parameterize the *Real* pairs spectrum:

$$\frac{dN^{Real}}{dQ} = \alpha\frac{dn^A_{MC}}{dQ} + \beta\frac{dN^{NC}_{MC}}{dQ} + \gamma\frac{dN^C_{MC}}{dQ},$$ (9)

α, β and γ being parameters obtained from the fit to the whole spectrum [6].

SPECTROMETER

DIRAC spectrometer is a two arm telescope with a 1.65 T central magnet separating positive and negative charged particle channels [7]. It is located in the T8 channel of the

---

[2] A long lived source is a resonance with decay time large enough that it can fly a distance much larger than the pionium Bohr radius ($387\,fm$) from the primary interaction vertex.

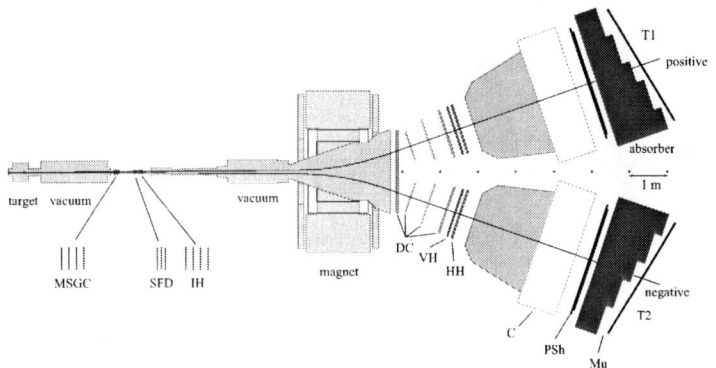

FIGURE 2. DIRAC spectrometer scheme. The top view shows the distribution of the detectors. MSGC=Micro Strip Gas Chambers, SFD=Scintillating Fiber Detector, IH=Ionization Hodoscopes detector, DC=Drift Chambers, VH=Vertical Hodoscopes, HH=Horizontal Hodospes, C=Čerenkov, PSh=Preshower, Mu=Muon counters.

PS East Hall B at CERN. The beam intensity is $\sim 10^{11}$ 24 GeV/c protons per spill (0.4 s).

The detectors are classified into *upstream* and *downstream* ones according to their position relative to the magnet. The *upstream* detectors consist of three sets of planes of Micro Strip Gas Chambers and Scintillating Fibers, used in tracking, and Ionization Hodoscopes, also used for triggering. *Downstream* we have two sets of Drift Chambers, the main tracking detector, Vertical and Horizontal hodoscope arrays for timing and triggering, Čerenkov detectors, to reject electrons at trigger level, Preshower, also used for triggering and electrons rejection offline, and muon detectors, for muons rejection.

The trigger system consists of three levels [8]. The T1 main trigger selects $\pi^+\pi^-$ events with coincidences upstream and the two arms downstream. The DNA+RNA makes a topological on-line analysis to perform the relative momentum cuts $Q_x < 3MeV/c$, $Q_y < 10MeV/c$ and $Q_l < 30MeV/c$. Finally, the T4 level selects events containing one track per arm satisfying $Q_x < 3MeV/c$ and $Q_l < 30MeV/c$ criteria.

The track resolution uses the standard Kalman filter procedure achieving an excellent relative momentum resolution of $\sigma_{Q_x} = \sigma_{Q_y} = 0.4MeV/c$ and $\sigma_{Q_L} = 0.6MeV/c$.

EXPERIMENTAL RESULTS

DIRAC started taking data in November 1998. Since that time we have collected thousands of $\pi^+\pi^-$ *Atomic Pairs*. For example, the accumulated Nickel target [3] statistics of 2000, 2001 and 75% of 2003 runs with tight cuts gives more than 12000 pairs.

However, the accuracy of the measurement, has been limited by systematic effects

---

[3] DIRAC main target is made of Nickel. The former $94\mu m$ thickness target was replaced in September 2001 by a $98\mu m$ one.

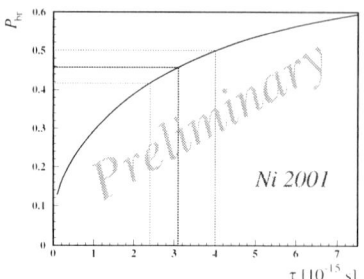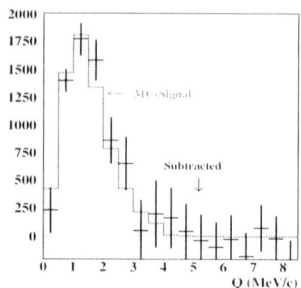

FIGURE 3. **Left:** Lifetime measurement. **Right:** Measured spectrum of $\pi^+\pi^-$ *Atomic Pairs* and comparison to the Monte Carlo corresponding distribution.

as the behavior of the electronic readout of some detectors or the parameterization of Multiple Scattering. In the last two years the major thrust of our research has been the study of systematics, the development of the Monte Carlo simulation of the experiment as well as dedicated measurements like the use of a multi-layer target [9] to deep in the knowledge of our data. These studies are foreseen to be finished early 2004. At the moment we have only a rough estimate of the systematic error of lifetime $\sim 1 \cdot 10^{-15}s$. Accounting for that, the result obtained with the use of Nickel target 2001 statistics, the best understood data, gives:

$$\tau = (3.1^{+0.9}_{-0.7}(stat.) \pm 1.(syst)) \cdot 10^{-15}s.$$

The pure Monte Carlo analysis leads to a compatible result of

$$\tau = (2.6 \pm 0.5(stat.) \pm 1.(syst)) \cdot 10^{-15}s.$$

In figure 3 we can also see the experimental shape of the spectrum of pairs from atomic break-up as defined in equation (7).

REFERENCES

1. B. Adeva et al. 1995 CERN/SPSLC 95-1 (Geneva: CERN); http://www.cern.ch/DIRAC
2. J. Gasser et al., Phys. Rev. D **64** 016008 (2001)
3. G. Colangelo et al., Phys.Lett. B **488**, 261 (2000)
4. C. Santamarina et al., J. Phys. B *in press.* http://arxiv.org/abs/physics/0306161
5. L.L. Nemenov, Sov. J. Nucl. Phys. **41**, 629 (1985)
6. C. Schuetz et al. (DIRAC Collaboration), proceedings of the XXVIIIth Rencontres de Moriond on QCD and Hadronic interactions. http://arxiv.org/abs/hep-ph/0305121
7. DIRAC collaboration: Nucl. Inst. and Meth. A *in press.* http://www.arxiv.org/abs/hep-ex/0305022
8. L. Afanasyev et al., Nucl. Instr. Meth. A **491**, 376 (2002)
9. D. Goldin et al. (DIRAC Collaboration), proceedings of the 8th Conference on the Intersections of Particle and Nuclear Physics. http://dirac.web.cern.ch/DIRAC/CIPANP2003.pdf

Last results from the DEAR experiment at DAΦNE

C. Curceanu (Petrascu)[a,e], G. Beer[i], A.M. Bragadireanu[a,e],
M. Cargnelli[d], J.-P. Egger[b,c], H. Fuhrman[d], C. Guaraldo[a], M. Iliescu[a,e],
T. Ishiwatari[d], K. Itahashi[g], M. Iwasaki[f], B. Lauss[h], V. Lucherini[a],
L. Ludhova[b], J. Marton[d], F. Mulhauser[b], T. Ponta[a,e], L.A. Schaller[b],
R. Seki[j,k], D. Sirghi[a,e], F. Sirghi[a], P. Strasser[f], J. Zmeskal[d]

[a] INFN - Laboratori Nazionali di Frascati, C.P. 13, Via E. Fermi 40, I-00044 Frascati, Italy;
[b] Universite de Fribourg, Institut de Physique, Bd. de Perolles, CH - 1700 Fribourg, Switzerland;
[c] Universite de Neuchatel, Institut de Physique, 1 rue A.- L. Breguet, CH-2000 Neuchatel,
Switzerland; [d] Institute for Medium Energy Physics, Boltzmanngasse 3, A-1090 Vienna, Austria;
[e] Institute of Physics and Nuclear Engineering "Horia Hulubei", P.O. Box MG - 6, R-76900,
Bucharest, Romania; [f] Institute of Physical and Chemical Research (RIKEN), 2-1 Hirosawa,
Wako, Saitama 351-01, Japan; [g] Tokyo Institute of Technology, 2-12-1 Ookoyana Meguro, Tokyo
152, Japan; [h] Physics Department, University of California and Berkeley, Berkeley CA 94720,
USA; [i] University of Victoria, Department of Physics and Astronomy, P.O. Box 3055 Victoria, BC,
Canada V8W3P6; [j] W.E. Kellogg Radiation Laboratory, California Institute of
Technology,Pasadena, CA 91125, USA; [k] Department of Physics and Astrophysics, California
State University, Northridge, CA 91330, USA

Abstract. The goal of the DEAR (DAΦNE Exotic Atom Research) experiment is the precision measurement of the K_α line shift and width, due to the strong interaction, in kaonic hydrogen and the first similar measurement in kaonic deuterium. The final aim is a precision determination of the antikaon-nucleon isospin dependent scattering lengths, in order to obtain the kaon nucleon sigma terms. In the first phase of the experiment, dedicated to the calibration and optimization of the setup, the experiment collected data on gaseous kaonic nitrogen: for the first time a pattern of three transitions ($7 \to 6$ at 4.6 keV, $6 \to 5$ at 7.6 keV and $5 \to 4$ at 14 keV) was measured, and the corresponding yields obtained. The kaonic hydrogen measurement followed, in the last months of 2002, for a total integrated luminosity of about 60 pb^{-1}. Data analyses are in progress, preliminary results being hereby presented, together with those on kaonic nitrogen. Future measurements: kaonic deuterium, kaonic helium and precise determination of the charged kaon mass are planned in the framework of the SIDDHARTA (SIlicon Drift Detector for Hadronic Atoms Research by Timing Application) experiment, which will replace DEAR on DAΦNE.

THE SCIENTIFIC PROGRAM OF DEAR

The objective of DEAR (DAΦNE Exotic Atom Research) [1] is the precise determination of the isospin dependent antikaon-nucleon scattering lengths, through an eV precision measurement of the K_α line shift and broadening, due to strong interaction, in kaonic hydrogen, and a similar (the first one) measurement in kaonic deuterium.

For kaonic hydrogen and kaonic deuterium the K-series transitions are of main experimental interest, since they are the only ones affected by the strong interaction. The K_α lines are clearly separated from the kigher K transitions. The shift ε and the width Γ of

CP717, *Hadron Spectroscopy: Tenth International Conference,*
edited by E. Klempt, H. Koch, and H. Orth
© 2004 American Institute of Physics 0-7354-0197-7/04/$22.00

the 1s state of kaonic hydrogen are related in a fairly model-independent way (Deser-Trueman formula [2]) to the real and imaginary part of the complex s-wave scattering length, a_{K^-p}:

$$\varepsilon + i\Gamma/2 = 412 \cdot a_{K^-p} \; eV \; fm^{-1} \tag{1}$$

A similar relation applies to the case of kaonic deuterium and to its corresponding scattering length, a_{K^-d}:

$$\varepsilon + i\Gamma/2 = 601 \cdot a_{K^-d} \; eV \; fm^{-1} \tag{2}$$

The measured scattering lenghts are then related to the isospin-dependent scattering lengths, a_0 and a_1:

$$a_{K^-p} = (a_0 + a_1)/2, \quad a_{K^-n} = a_1 \tag{3}$$

The extraction of a_{K^-n} from a_{K^-d} requires a more complicated analysis than the simple impulse approximation (K^- scattering from each free nucleon): higher order contributions associated with the K^-d three-body interaction have to be taken into account. This means to solve the three-body Faddeev equations by the use of potentials, taking into account the coupling among multichanneled interactions.

An accurate determination of the K^-N isospin dependent scattering lengths will place strong constraints on the low-energy K^-N dynamics, which in turn constraints the SU(3) description of chiral symmetry breaking [3]. Crucial information about the nature of chiral symmetry breaking, and to what extent the chiral symmetry must be broken, is provided by the calculation of the meson-nucleon sigma terms [4].

Presently only estimates, with 70% uncertainties, exist; a measurement of the K^-N scattering lengths at few percent level should allow the determination of sigma terms with a precision better than 20%.

The sigma terms are also important inputs for the determination of the strangeness content of the proton. The strangeness fraction depends of both kaon-nucleon and pion-nucleon sigma terms, being more sensitive to the first ones [5].

THE DEAR EXPERIMENTAL SETUP

The DEAR setup installed at the DAΦNE Accelerator at Laboratori Nazionali di Frascati contains a pressurized cryogenic target (23 K and 1.82 bar) for the kaonic hydrogen measurement, in order to optimize the distribution of the kaons stopped in the target, and to avoid an important loss of the signal due to the Stark effect.

The target cell was done in kapton 75μm thick, cylindrical shaped, with a diameter of 11 cm, reinforced with epoxy-fiberglass bars, in order to have as less material as possible in front of the CCDs, to avoid electronic transitions which could interfere with the line to be measured. The cryogenic setup was equipped with 16 CCD-55.

The setup was installed in one of the two interaction regions of DAΦNE and had periods of data taking starting from December 1999, when first collisions were achieved.

The first periods of data taking were dedicated to background understanding and reduction, by the use of appropriate shielding and machine optics solutions. The first measurement of an exotic atom at DAΦNE, namely kaonic nitrogen, followed in May 2001; this measurement was re-done in October 2002, when, for the first time, three transitions of kaonic nitrogen were clearly seen (7 → 6 at 4.6 keV, 6 → 5 at 7.6 keV and 5 → 4 at 14 keV) and the corresponding yields extracted. Results of this measurement are presented in Section 3.

The first measurement by DEAR of kaonic hydrogen was performed in the period November - December 2002. Preliminary results of this measurement are presented in Section 4.

KAONIC NITROGEN MEASUREMENT

The measurement of kaonic nitrogen (whose transitions have yields 20 times higher than kaonic hydrogen and therefore a fast feedback can be obtained) performed by DEAR had the following primary objectives:

- to prove the feasibility of the DEAR technique to produce and detect exotic atoms using the K^- beam from the ϕ-decay at DAΦNE;
- to perform, at the same time, the measurement of previously unobserved transitions of this exotic atom;
- to optimize the kaon stopping distribution inside the gaseous target.

These objectives were achieved in May 2001, when the first measurement of kaonic nitrogen 7 → 6 and 6 → 5 transitions, at 4.6 and 7.6 keV respectively, was performed. The results were published [6], suggesting a new method for a future precision measurement of the charged kaon mass.

In October 2002 the kaonic nitrogen spectrum was re-measured, just before the start of the kaonic hydrogen data taking.

The scientific aim of this run was the study of different degrader configurations to take into accout the effect of the boost in the ϕ-production.

A statistics corresponding to about 10.5 pb^{-1} integrated luminosity was collected.

A refined analysis of the spectrum was performed [7]. This analysis allowed, for the first time,to disentangle a complex of three kaonic nitrogen transition: the 7 → 6 transition, at 4.57 keV; the 6 → 5 transition, at 7.59 keV; the 5 → 4 transition, at 13.96 keV.

The continuous background subtracted spectrum is shown in Figure 1.

The number of events for each transition is reported in Table 1.

Apart of the kaonic nitrogen transitions, in the spectrum there are present lines corresponding to electronic transitions of materials present in the setup.

The transition yields, for the first time measured, were extracted by using Monte Carlo simulation of the setup and of the contributing physical processes. The preliminary results on the yields are reported in Table 1 and constitute important checks for cascade calculations in the field of exotic (kaonic) atom transitions.

The yields reported in Table 1 contain only the statistical errors; an accurate study on systematic errors is undergoing.

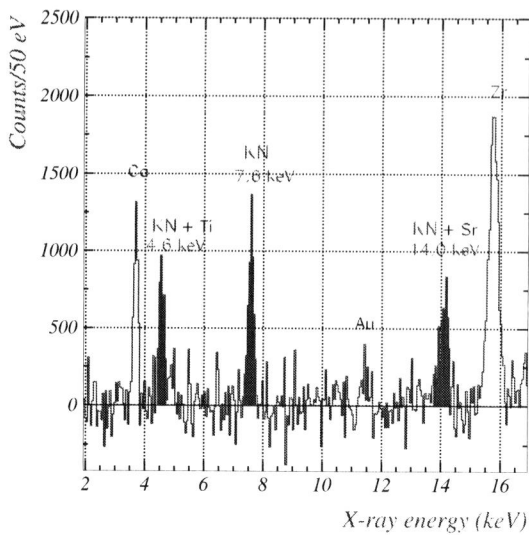

FIGURE 1. Background subtracted energy spectrum obtained for the run of October 2002 with kaonic nitrogen. Kaonic nitrogen transitions are clearly seen, as indicated in the figure.

TABLE 1. Kaonic nitrogen results.

Transition	Number of events	Yield of transition (%)
$7 \to 6$	2690 ± 650	33.7 ± 8.1
$6 \to 5$	5320 ± 395	55.5 ± 4.2
$5 \to 4$	1360 ± 330	66.4 ± 15.6

PRELIMINARY RESULTS FOR KAONIC HYDROGEN

In order to perform the first kaonic hydrogen measurement at DEAR the target was filled with hydrogen in cryogenic and pressurized conditions: 23 K and 1.82 bar. The kaonic hydrogen measurement lasted from 30 October to 22 December 2002. It was divided into two periods: from 30 October to 16 December a continuous run with kaonic hydrogen; the total integrated luminosity, measured by the DEAR kaon monitor, was about 60 pb^{-1}; from 16 December to 22 December a background run with no collisions in the DEAR Interaction Region - with a statistics equivalent to about half of the kaonic hydrogen one.

Two types of analyses are undergoing:
- a global fit of the kaonic hydrogen spectrum;
- a measured background subtracted spectrum analysis.

The two analyses give compatible preliminary results. The statistical evidence of the K-complex transition is (sum of individual contribution) better than 6σ. The very preliminary results on the shift and width for the K_α kaonic hydrogen like are:

$$\varepsilon = -(150 \pm 45) \; eV \tag{4}$$

$$\Gamma = (200 \pm 90) \; eV \tag{5}$$

Further analyses and checks are undergoing.

FUTURE PLANS

For the future, in order to achieve an eV level measurement of the shift and width for kaonic hydrogen and kaonic deuterium K_α lines, an upgrade of the setup is in progress. New fast triggerable devices, namely large area Silicon Drift Detectors (SDD), in the framework of the new SIDDHARTA (SIlicon Drift Detector for Hadronic Atoms Research by Timing Application) experiment - replacing the CCDs in DEAR, are planned to be used. The trigger is given by the entrance of the charged kaon in the target volume. The event can be identified and measured with high accuracy and low contamination by the use of a three-scintillator telescope, synchronized with the bunch frequency. The timing window (1 μs) to be used in trigger, together with the excellent energy resolution (140 eV at iron position) of the SDD will allow a dramatic decrease of the background so to obtain a signal/background ratio of about 1/1 for kaonic hydrogen and 1/5 for kaonic deuterium. First tests of a prototype SDD array, performed at the Beam Test Facility of LNF, gave excellent results.

Other exotic atoms (kaonic helium and other light atoms, as well as sigmonuim atom) and a high-precision measurement of the charged kaon mass will eventually become feasible.

ACKNOWLEDGMENTS

The DAΦNE group in warmly acknowledged for the very good cooperation and teamwork. Part of the work was supported by a "Transnational access to Research Infrastructure" (TARI) Contract No. HPRI-CT-1999-00088.

REFERENCES

. S. Bianco *et al.*, Rivista del Nuovo Cimento **22**, No. 11 (1999) 1
. S. Deser *et al.*, Phys. Rev. **96** (1954) 774; T.L. Truemann, Nucl. Phys. **26** (1961) 57;
. C. Guaraldo, Proceedings of III International Workshop on Physics and Detectors for DAΦNE, November 16-19, Frascati, Italy, Frascati Physics Series Volume XVI (1999) 642.
. E. Reya, Rev. Mod, Phys. **46** (1974) 545; H. Pagels, Phys. Rep. **16** (1975) 219
. R.L. Jaffe, C.L. Korpa, Comments Nucl. Part. Phys. **17** (1987) 163
. G. Beer *et al*, Phys. Lett **B535** (2002) 52
. T. Ishiwatari, Ph. D. Thesis, RIKEN (2003)

The New Pionic Hydrogen Experiment at PSI

Hermann Fuhrmann

Institute for Medium Energy Physics, Austrian Academy of Sciences, Vienna
for the Pionic Hydrogen Collaboration

Abstract. The hadronic shift ε_{1s} and broadening Γ_{1s} of the ground state in pionic hydrogen are measured with unprecedented precision. The expected accuracy is about 1% for broadening and 0.2% for shift. This will be an improvement by an order of magnitude for broadening as compared to previous experiments. The present status of the experiment, the status of data analysis and an outlook into the near future are presented.

INTRODUCTION

In a new high precision experiment at PSI (Paul Scherrer Institute, Villigen, Switzerland), the strong interaction shift ε_{1s} and broadening Γ_{1s} of the ground state in pionic hydrogen are measured [1]. The pion-nucleon isospin dependent scattering lengths can be derived from ε_{1s} and Γ_{1s} using Deser-type formulae. Thereby this measurement is a direct experimental test of this low energy approach to QCD such as the methods of Chiral Perturbation Theory.

Furthermore, the results can be compared to zero energy extrapolation from high energy scattering data and their validity checked.

The study of exotic atoms proves to be a very valuable method for investigating fundamental interactions. In particular, the strong interaction at zero energy can be experimentally studied with unprecedented precision [2, 3].

EXPERIMENTAL METHOD

X-rays resulting from $\pi H_{np \to 1s}$ transitions are measured. These transitions have a typical energy in the range of a few keV, whereas the expected hadronic effects are in the order of eV. Therefore the only way to obtain the required precision is the use of a high-resolution crystal spectrometer. Because of its small solid angle, a high pion flux and a high stopping rate in the target are necessary in order to keep beam time within realistic limits. The pion beam at PSI has the highest intensity in the world, making the PSI the best suited facility for such experiments. The high pion stopping rates are achieved with a superconducting cyclotron trap [2].

CP717, *Hadron Spectroscopy: Tenth International Conference,*
edited by E. Klempt, H. Koch, and H. Orth
© 2004 American Institute of Physics 0-7354-0197-7/04/$22.00

Experimental Setup

Figure 1 shows the experimental setup. Pions are focused on the target cell in the cyclotron trap. With our current setup a pion flux of $\sim 4\times10^9$ at a momentum of 110 MeV/c can be maintained.

The pions are decelerated first in a degrader and thereafter in the target gas, where finally pionic atoms are formed. X-rays emitted during cascade transitions are reflected by a spherically bent Bragg crystal and detected on a CCD (charge coupled device) detector array. Considering the rather low detection efficiency of the spectrometer ($\sim10^{-6}$), the lifetime of pions ($\sim10^{-8}$ s), and their typical stopping rate in gas targets ($\sim1\%$), the cyclotron trap and its target system had to be carefully optimized to achieve realistic event rates (~40 events per hour).

The cylindrically shaped target cell and its cryo-system allow measurements over a huge density range down to liquified hydrogen ($\sim 20K$). The exit window for the X-rays has a thickness of only 7.5μm. It is made of Kapton and supported by an aluminum structure [4].

The reflection angle depends on the energy of the X-rays thereby separating lines of different energy.

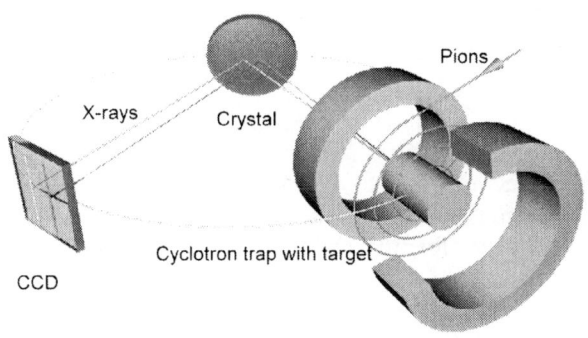

FIGURE 1. Scheme of the experimental set-up

The spectrometer is built in the so called Johann-geometry, where the bending radius of the crystal (in our case 3m) is twice the radius of the Rowland circle. The angle on the crystal between target cell and CCD can be maintained at an accuracy of some 0.1 arc s with piezo-actuators.

As detector, an array of six CCDs is used, each having 600×600 pixels with a size of 40×40μm^2. Its total area is 48×72mm^2.

Background stemming from the pions is suppressed by appropriate concrete shielding of the spectrometer. Cluster analysis of CCD data allows further suppression of background events. Therefore, we can achieve a typical signal to noise ratio of about 40:1.

Calibration

Figure 2 shows one method we use for calibrating the detector system for the 3p–1s transition in pionic hydrogen. Oxygen is mixed into hydrogen, because its 6h–5g transition line is known with sufficient high precision and close enough to the hydrogen line, which allows us to observe both lines simultaneously.

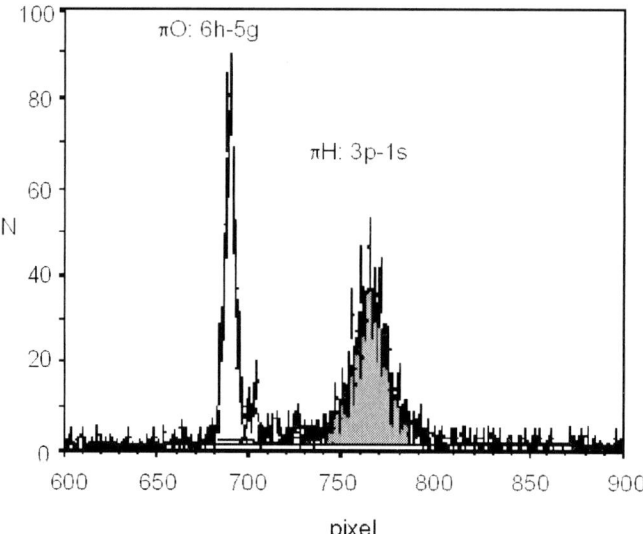

FIGURE 2. Energy calibration

As an alternative method, we can insert beryllium foils into the target cell and use certain pionic beryllium transition as calibration lines.

It should be noted that for the first time pionic lines were used for calibration, which reduce the systematic error stemming from the uncertainty in the pion mass (~3ppm) illustrating the advances in exotic atom physics research [2].

RESULTS

Shift

The weighted average from measurements from beam-times in 2000 to 2002 for the 3p – 1s transition is given as

$$E_{3p-1s} = 2885.928 \pm 0.008 \text{ (stat.)} \pm 0.007 \text{ (sys.) eV}$$

Actually the shift is obtained by subtracting the pure QED value for the transition in pionic hydrogen from the weighted average of measured energies. The QED value for the 3p – 1s transition used in this work is given as 2878.808 ± 0.006 eV. Hence we get the value for the shift as

$$\varepsilon_{1s} = 7.120 \pm 0.012 \text{ eV}$$

The convention about the sign of the shift is not consistent in literature, therefore it should be added, that the value obtained here corresponds to an attractive potential.

Since we used the same QED value as [8] our result is directly comparable, and, in fact, in agreement. However, our value already represents a reduction in the error by a factor ~3.

Width

The analysis of the width is more demanding, since various effects play a significant role in the broadening of the transition lines: Response function of the spectrometer, Doppler-broadening due to Coulomb-deexcitation, Coulomb explosion of molecules and density effects have to be carefully studied and disentangled. It implies, that the distribution of kinetic energy along the cascade processes must be determined [7].

Presently the analysis of our data is not finished. Hence only an upper limit for the value in the shift can be given at this point:
$$\Gamma_{1s} \leq 850 \text{ meV}$$

OUTLOOK

The first phase of this experiment is completed. We can now aim at our final goal, the determination of the hadronic shift and width in pionic hydrogen with unprecedented precision.

Besides high statistic measurements, it is planned to reduce systematic errors by several means. The response function of the crystals used will be studied with high precision by employing an ECRIT source. Extensive studies are planned in the near future, which will provide indispensable data input [5].

The systematic errors stemming from the apparatus will be improved significantly (mechanical stability, precise determination of the pixel size in the CCD, etc.). Pionic hydrogen cascade models and their resulting distribution in kinetic energy of pionic hydrogen will be improved with the help of the data we have obtained.

A measurement with muonic hydrogen, where hadronic broadening is absent, will be performed to determine the effects of Coulomb deexcitation. Advances in QED calculations and an improvement in the accuracy of the pion mass value will further add to the increase in precision [6].

Recent advances in the theoretical approach to low energy QCD will contribute to further clarify strong interaction at low energy [9].

ACKNOWLEDGMENTS

Among the many who deserve special thanks for their support of our experiment, I would like to single out this time Bruno Leoni from PSI. His tireless engagement at this experiment is great encouragement, and indispensable for its success.

REFERENCES

1. PSI experiment R-98.01; http://pihydrogen.web.psi.ch
2. Hennebach, M. "Precision Measurement of Ground State Transitions in Pionic Hydrogen,, thesis, Univ. Koln, FZ-Julich 2003.
3. Gotta D. et al., "Precision Spectroscopy of Pionic Hydrogen,, in *Proceedings of EXA02 International Workshop on Exotic Atoms*, edited by P. Kienle, H. Marton, J. Zmeskal, Austrian Academy of Sciences Press, Vienna 2003, pp. 71-79.
4. Marton, J., et al. "Target and Detector Systems for Exotic Atom Experiments,, in *ibid*, pp. 223-228.
5. Simons, L. et al. "Cyclotron Trap: Future Experiments,, in *ibid* pp. 197-204.
6. Indelicato, P. et al.: "Aspects of QED in the Framework of Exotic Atoms,, in *ibid* pp.61-70.
7. Siems, Th. "Kaskade und Coulomb-Explosion in leichten pionischen und myonischen Atomen,, thesis, Univ. Koln, FZ-Julich 1997.
8. H. Ch. Schroder et al., *Eur. Phys. J.* **C 21**, 473-488 (2001).
9. Ivanov, A. N. et al. nucl-th/0306047 to be published in *Eur. Phys. J.* **A**

$\eta \to \pi^0\gamma\gamma$ decay within a chiral unitary approach[1]

E. Oset[*], J. R. Peláez[†] and **L. Roca**[*]

[*]*Departamento de Física Teórica and IFIC Centro Mixto Universidad de Valencia-CSIC Institutos de Investigación de Paterna, Apdo. correos 22085, 46071, Valencia, Spain*
[†]*Dip. di Fisica. Universita' degli Studi, Firenze, and INFN, Sezione di Firenze, Italy Departamento de Física Teórica II, Universidad Complutense. 28040 Madrid, Spain.*

Abstract. We improve the calculations of the $\eta \to \pi^0\gamma\gamma$ decay within the context of meson chiral lagrangians. We use a chiral unitary approach for the meson-meson interaction, thus generating the $a_0(980)$ resonance and fixing the longstanding sign ambiguity on its contribution. This also allows us to calculate the loops with one vector meson exchange, thus removing a former source of uncertainty. In addition we ensure the consistency of the approach with other processes. First, by using vector meson dominance couplings normalized to agree with radiative vector meson decays. And, second, by checking the consistency of the calculations with the related $\gamma\gamma \to \pi^0\eta$ reaction. We find an $\eta \to \pi^0\gamma\gamma$ decay width of $0.47 \pm 0.10\,\text{eV}$, in clear disagreement with published data but in remarkable agreement with the most recent measurement.

INTRODUCTION

The $\eta \to \pi^0\gamma\gamma$ decay has attracted much theoretical attention, since Chiral Perturbation Theory (ChPT) calculations have sizable uncertainties and produce systematically rates about a factor of two smaller than experiment. Within ChPT, the problem stems from the fact that the tree level amplitudes, both at $O(p^2)$ and $O(p^4)$, vanish. The first non-vanishing contribution comes at $O(p^4)$, but either from loops involving kaons, largely suppressed due to the kaon masses, or from pion loops, again suppressed since they violate G parity and are thus proportional to $m_u - m_d$. The first sizable contribution comes at $O(p^6)$ but the coefficients involved are not precisely determined. The use of tree level VMD to obtain the $O(p^6)$ chiral coefficients by expanding the vector meson propagators, leads to results about a factor of two smaller than the "all order" VMD term, which means keeping the full vector meson propagator. All this said it has become clear that the strict chiral counting has to be abandoned since the $O(p^6)$ and higher orders involved in the full ("all order") VMD results are larger than those of $O(p^4)$.

Once the "all order" VMD results is accepted as the dominant mechanism, one cannot forget the tree level exchange of other resonances around the 1 GeV region. The $a_0(980)$ exchange, which was taken into account approximately in [2], was one of the main sources of uncertainty, since even the sign of its contribution was unknown.

After the tree level light resonance exchange have been taken into account, we should consider loop diagrams, since meson-meson interaction or rescattering can be rather

---

[1] Contribution to the X. International Conference On Hadron Spectroscopy, August 31 - September 6, 2003, Aschaffenburg, Germany. (Full article in [1])

CP717, *Hadron Spectroscopy: Tenth International Conference,*
edited by E. Klempt, H. Koch, and H. Orth
© 2004 American Institute of Physics 0-7354-0197-7/04/$22.00

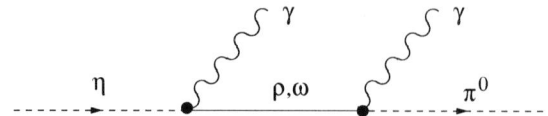

FIGURE 1. Diagrams for the VMD mechanism.

strong. First of all we find the already commented $O(p^4)$ kaon loops from ChPT, but also the meson loops from the terms involving the exchange of one resonance. The uncertainty from the latter was roughly expected [2] to be about 30% of the full width.

Another relevant question is that no attempts have been done to check the consistency of $\eta \to \pi^0 \gamma\gamma$ results withe the related channel $\gamma\gamma \to \pi^0 \eta$. The reason is not surprising since there are no hopes within ChPT to reach the $a_0(980)$ region where there are measurements of the $\gamma\gamma \to \pi^0 \eta$ cross section . On the other hand, the explicit SU(3) breaking already present in the radiative vector meson decays has not been taken into account when calculating the VMD tree level contributions.

The former discussion has set the stage of the problem and the remaining uncertainties that allow for further improvement. In recent years, with the advent of unitarization methods, it has been possible to extend the results of ChPT to higher energies where the perturbative expansion breaks down and to generate resonances up to 1.2 GeV. In particular these ideas were used to describe the $\gamma\gamma \to meson - meson$ reaction, with good results in all the channels up to energies of around 1.2 GeV [3]. With these techniques, and always within the context of meson chiral lagrangians, we will address three of the problems stated above: First, the $a_0(980)$ contribution, second, the evaluation of meson loops from VMD diagrams and, third, the consistency with the crossed channel $\gamma\gamma \to \pi^0 \eta$. In particular, we will make use of the results in [3], where the $\gamma\gamma \to \pi^0 \eta$ cross section around the $a_0(980)$ resonance was well reproduced using the same input as in meson meson scattering, without introducing any extra parameters.

With these improvements we are then left with a model that includes the "all order" VMD and resummed chiral loops.

MECHANISMS

Following [2] we consider the sequential VMD mechanism of Fig. 1 which can be easily derived from the VMD Lagrangians involving VVP and $V\gamma$ couplings

$$\mathcal{L}_{VVP} = \frac{G}{\sqrt{2}} \varepsilon^{\mu\nu\alpha\beta} \langle \partial_\mu V_\nu \partial_\alpha V_\beta P \rangle, \qquad \mathcal{L}_{V\gamma} = -4f^2 eg A_\mu \langle QV^\mu \rangle, \qquad (1)$$

where V_μ and P are standard $SU(3)$ matrices constructed with the nonet of vector mesons containing the ρ, and the nonet of pseudoscalar mesons containing the π, respectively. We also assume an ordinary mixing for the ϕ, the ω, the η and η'.

From Eq. (1) one can obtain the radiative widths for $V \to P\gamma$ obtaining a fair agreement with the experimental data in the PDG but the results can be improved by incorporating $SU(3)$ breaking mechanisms. For that purpose, we will normalize the couplings

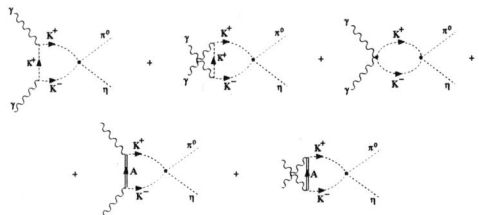

FIGURE 2. Diagrams for the chiral loop contribution.

so that the branching ratios agree with experiment. These will be called results with "normalized couplings". In this way we are taking into account phenomenologically the corrections to the $VP\gamma$ vertex from an underlying field theory.

The integrated width obtained using this sequential VMD contribution is $\Gamma = 0.57\,\text{eV}$ (universal couplings); $\Gamma = 0.30 \pm 0.06\,\text{eV}$ (normalized couplings), where the error has been calculated from a Monte Carlo Gaussian sampling of the normalization parameters within the errors of the experimental branching ratios.

Our VMD normalized result is within three standard deviations from the value presently given in the PDG: $\Gamma = 0.84 \pm 0.18\,\text{eV}$, but within one sigma of the more recent one presented in [4], $\Gamma = 0.42 \pm 0.14\,\text{eV}$. There are, however, other contributions that we consider next.

The contribution of pion loops to $\eta \to \pi^0\gamma\gamma$, evaluated in [2], proceeds, to begin with, through the G-parity violating $\eta \to \pi^0\pi^+\pi^-$ process but it is proportional to $m_u - m_d$, and we shall include it in the uncertainties of together with other isospin violating contributions.

The main meson loop contribution comes from the charged kaon loops at $O(p^4)$ and proceeds via $\eta \to \pi^0 K^+ K^- \to \pi^0\gamma\gamma$. Note that these loops are also suppressed due to the large kaon masses. That is why the $\eta \to \pi^0 a_0(980) \to \pi^0\gamma\gamma$ mechanism was included explicitly, with uncertainties in the size and sign of the $a_0(980)$ couplings. As commented in the introduction, the chiral unitary approach solves this problem by generating dynamically the $a_0(980)$ in the $K^+K^- \to \pi^0\eta$ amplitude.

We can illustrate this approach by revisiting the work done in [3] on the related process $\gamma\gamma \to \pi^0\eta$ where the chiral unitary approach was successfully applied around the $a_0(980)$ region. Since for the η decay the low energy region of $\gamma\gamma \to \pi^0\eta$ is also of interest, we will include next the VMD mechanisms also in this reaction. Once we check that we describe correctly $\gamma\gamma \to \pi^0\eta$, the results can be easily translated to the eta decay. We will finally add other anomalous meson loops that are numerically relevant for eta decay but not for $\gamma\gamma \to \pi^0\eta$.

In [3] it was shown that, within the unitary chiral approach, the $\gamma\gamma \to \pi^0\eta$ amplitude around the $a_0(980)$ region, diagrammatically represented at one loop in Fig. 2, factorizes as

$$-it = (\tilde{t}_{\chi K} + \tilde{t}_{AK^+K^-})t_{K^+K^-,\pi^0\eta} \tag{2}$$

with $t_{K^+K^-,\pi^0\eta}$ the full $K^+K^- \to \pi^0\eta$ transition amplitude.

The first three diagrams correspond to $\tilde{t}_{\chi K}t_{K^+K^-,\pi^0\eta}$ of Eq. (2). The meson meson scattering amplitude was evaluated in [5] by summing the Bethe Salpeter (BS) equation

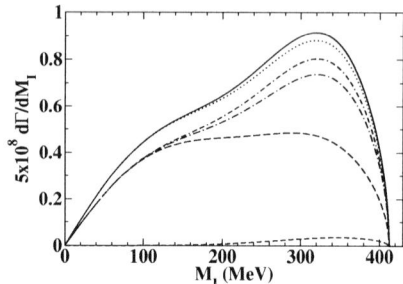

FIGURE 3. Contributions to the two photon invariant mass distribution. From bottom to top, short dashed line: chiral loops from Eq.(2); long dashed line: only tree level VMD; dashed-dotted line: coherent sum of the previous mechanisms; double dashed-dotted line: idem but adding the resummed VMD loops; continuous line: idem but adding the anomalous terms, which is the full model presented in this work (we are also showing as a dotted line the full model but substituting the full $t_{K^+K^-,\eta\pi^0}$ amplitude by its lowest order).

with a kernel formed from the lowest order meson chiral Lagrangian amplitude and regularizing the loop function with a three momentum cut off. Subsequently, other approaches like the inverse amplitude method or the N/D method were used and all of them gave the same results in the meson scalar sector. The BS equation with coupled channels can be solved algebraically, leading to the following solution in matrix form

$$t(s) = [1 - t_2(s)G(s)]^{-1}t_2(s), \qquad (3)$$

with s the invariant mass of the two mesons, t_2 the lowest order chiral amplitude and $G(s)$ a diagonal matrix, diag($G_{\bar{K}K}, G_{\eta\pi}$), accounting for the loop functions of two mesons. These G functions were regularized in [5] by means of a cut off.

In Eq. (2) there is another term, $\tilde{t}_{AK^+K^-}t_{K^+K^-,\pi^0\eta}$, which corresponds to the last two diagrams of Fig. 3 where the axial vector meson $K_1(1270)$ is exchanged.

In addition to the axial vector meson exchange in loops considered, we have to include the loops with vector meson exchange for completeness. In fact, some of the uncertainties estimated in [2] were linked to these loops. For consistency, once again we have to sum the series obtained by iterating the loops in the four meson vertex.

Of course, when introducing loops with vector meson exchange we have to consider loops involving a K^{*+} or a K^{*0} exchanged between the photons, which were not present at tree level.

In the $\eta \to \pi^0\gamma\gamma$ case, the meson loop diagrams correspond to those of $\pi^0\eta \to \gamma\gamma$ but considering the π^0 as an outgoing particle.

Since we are considering all the VMD diagrams and the chiral loops, we still have to take into account another kind of loop diagrams [2] which involve two anomalous $\gamma \to 3M$ vertices. Despite being $O(p^8)$ it has been found [2] that they can have a non negligible effect on the η decay.

RESULTS

Using the model described in the previous section, we plot in Fig. 4 the different contributions to $d\Gamma/dM_I$. We can see that the largest contribution is that of the tree level VMD (long dashed line). Let us recall that this is a new result as long as we are using the VMD couplings normalized to agree with the vector radiative decays. The resummation of the loops in Fig. 2 using Eq.(2), (short dashed line) gives a small contribution (0.011 eV in the total width), but when added coherently to the tree level VMD, leads to an increase of 30% in the η decay rate (dashed-dotted line). More interestingly, the shape of the $\gamma\gamma$ invariant mass distribution is appreciably changed with respect to the tree level VMD, developing a peak at high invariant masses. The resummed VMD loops leads, through interference, to a moderate increase of the η decay rate (double dashed dotted line), smaller than that of the chiral loops considered before. The last ingredient is the contribution of the anomalous mechanisms (continuous line), leading again to a moderate increase of the η decay rate, also smaller than the chiral loops from Eq.(2). These anomalous mechanisms have a very similar shape to the tree level VMD and interfere with it in the whole range of invariant masses.

$$\Gamma(\eta \to \pi^0 \gamma\gamma) = 0.47 \pm 0.10\, eV \tag{4}$$

where the theoretical error have been obtained considering the uncertainties from the vector meson radiative decays, the contribution of the 1^{+-} axial-vector mesons and the isospin violating terms.

Note that although we have considered a new error source from the uncertainties in the vector radiative decays, which turns out to be the largest one, we still have reduced the uncertainty from previous calculations.

The result of Eq. (4) is in remarkable agreement with the latest experimental number $\Gamma = 0.42 \pm 0.14\,eV$ [4], and lie within two sigmas from the earlier one in the PDG $\Gamma = 0.84 \pm 0.18\,eV$. Confirmation of those preliminary results would therefore be important to test the consistency of this new approach. Furthermore, precise measurements of the $\gamma\gamma$ invariant mass distributions would be of much help given the differences found with and without loop contributions.

REFERENCES

1. E. Oset, J. R. Pelaez and L. Roca, Phys. Rev. D **67** (2003) 073013 [arXiv:hep-ph/0210282].
2. L. Ametller, J. Bijnens, A. Bramon and F. Cornet, Phys. Lett. B **276** (1992) 185.
3. J. A. Oller and E. Oset, Nucl. Phys. A **629** (1998) 739 [arXiv:hep-ph/9706487].
4. S. Prakhov, in Proceedings of the International Conference of Non-Accelerator New Physics, Dubna, Russia; B. M. Nefkens and J. W. Price, in the Eta Physics Handbook, Proc. of the Uppsala Workshop, http://www.tsl.uu.se/ faldt/eta/Proceedings.html, Published in Phys. Scripta **T99** (2002) 114 [arXiv:nucl-ex/0202008].
5. J. A. Oller and E. Oset, Nucl. Phys. A **620**, 438 (1997) [Erratum-ibid. A **652**, 407 (1997)].

Molecule formation in the DK- and $K\bar{K}$-channel

Felix Philipp Sassen

Forschungszentrum Jülich, 52425 Jülich

Abstract. The possibility of molecule formation in the DK-channel coupled to the $D_s^+\pi$-channel has been investigated. With respect to the $K\bar{K}$-molecule known to be formed in the isoscalar channel we took a look at peripheral pion production in the reaction $\pi^- p \to \pi^0\pi^0 n$ and tried to fix the properties of a high lying scalar f_0' which mixes to the $K\bar{K}$-molecule by looking at its influence on the production data. Further more the changes in the properties of kaonium due to the $K\bar{K}$-molecule and due to a f_0'-state which originates from confinement are discussed.

THE $D_{sJ}^*(2317)$ A CANDIDATE FOR A DK-MOLECULE

Lately the BaBar collaboration has published the observation of a narrow state decaying to $D_s^+\pi^0$ at 2317 MeV[1]. Subsequently this $D_{sJ}(2317)$ has been confirmed by the Cleo[2] and the Belle collaboration[3]. Further more Cleo reports the observation of another narrow resonance $D_{sJ}(2463)$ at 2463 MeV and the absence of any signal in the decay channels $D_s^+\pi^+$ and $D_s^\dagger\pi$ at 2317 MeV[4]. The later result clearly shows that the resonance has isospin $I = 0$ and that the decay has to be isospin breaking. Inspired by the observation of molecule formation in the case of the $K\bar{K}$-channel[5] we want to check whether the observed resonance might be a mesonic molecule formed by attraction in the DK-channel, which stems from t-channel one boson exchanges.

We use the Blankenbecler-Sugar reductionof the Bethe-Salpeter equation to generate the T-matrix from a one boson exchange potential. Since the coupling $g_{DD\phi}$ and $g_{KKJ/\psi}$ are OZI suppressed we are left with three relevant diagrams. The potential in the DK-channel is generated by ρ and ω t-channel exchanges and the transition potential between the DK- and the $D_s^+\pi$-channel is generated by K^* exchange. Using the isospin bases and separating the isospin part of the potential from the momentum part one can directly tell from the isospin part, that there is strong attraction in the $I = 0$ channel while in the $I = 1$ channel repulsion and attraction nearly cancel, because the ρ-exchange switches from attractive to repulsive. (i.e. the isospin factor switches from -3 to 1 in the case of ρ-exchange while in the case of ω-exchange it stays -1 in both cases) Thus in agreement with experiment [4] it is only in the $I = 0$ case that we find enough attraction to form a DK-molecule. Since when working in the isospin basis all our potentials are isospin conserving we introduce isospin breaking by hand so that the formed molecule is able to decay despite being below the $D_s^+\eta$ threshold. (Later checks will show that we overestimated the effects of the splitting of the physical masses as well as the $\pi\eta$- and $\rho\omega$-mixing.) To fix the coupling constants of the different vertices we assumed $SU(4)$ symmetry to relate the the needed coupling constants to $g_{\pi\pi\rho}$ which was fixed from an analysis of $\pi\pi$ scattering[6]. A summary of the used values is given in ta-

CP717, *Hadron Spectroscopy: Tenth International Conference,*
edited by E. Klempt, H. Koch, and H. Orth

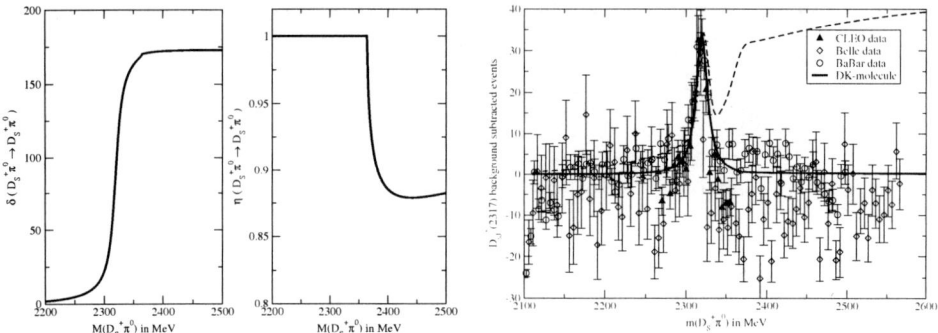

FIGURE 1. The two left panels show the phaseshifts and the inelasticities for $D_s^+ \pi^0$ scattering.(The magnitude of isospin breaking was fitted to the data.) The right panel shows a comparison of the calculated $D_s^+ \pi^0 \to D_s^+ \pi^0$ cross section (solid line) in comparison with the mass distributions measured by Babar[1], Cleo[2] and Belle[3]. The dashed line shows a calculation which uses only the physical masses and the $\pi\eta$-mixing to break isospin symmetry and which has the the experimental resolution folded to it. The curves as well as the data have been rescaled to coincide at 2317 MeV.

ble 1. Our results for $D_s^+ \pi^0$ phaseshifts and inelasticities are shown in figure 1 along with the cross section. As already mentioned the magnitude of isospin breaking was fitted. To improve on this shortcoming we calculate the magnitude of isospin breaking from the mass splitting of the D and the K mesons as well as from $\pi\eta$- and $\rho\omega$-mixing. To calculate the effects of the mass splitting between charged and neutral mesons we proceeded like in [7] and transform our potentials from the isospin basis to the particle basis to use the physical masses. The strength of the $\pi\eta$- and $\rho\omega$-mixing has been estimated from the results in[8] ($< \pi^0|H_{em}|\eta > \approx -4400$ MeV^2, $< \rho^0|H_{em}|\omega > \approx -4400$ MeV^2). The result of this calculation folded with the experimental resolution is shown as a dashed line in figure 1. The DK-molecule shows as a very sharp peak with width well below the experimental resolution and gets broadened by the experimental resolution. The effect of the DK-molecule is of the same order of magnitude as the influence of the opening $I = 1$ DK-channel, which contradicts the experimental findings. This shows that even though there is a DK-molecule formed in the $I = 0$ channel it is most probably not the origin of the peak observed by BaBar, Cleo and Belle. Since especially there is no $J = 1$ molecule formed, this finding makes scenarios like the "Chiral Multiplets of Heavy-Light Mesons" of Bardeen, Eichten and Hill[9] very probable.

THE $K\bar{K}$-MOLECULE IN $\pi^- P \to \pi^0 \pi^0 N$

Several experiments by the BNL E852[12] and the GAMS[13, 14] collaboration have been performed to measure the t-dependence of the pion production in the reaction $\pi^- p \to \pi^0 \pi^0 n$. All those experiments used peripheral reaction kinematics which means that we can model the production by the emission of a meson from the nucleon with the final state interaction given by pure meson-meson dynamics. Due to G-parity and the naturality of the exchanged particle we only need to consider the emission of either a

TABLE 1. The parameters for DK, $D_s^+\pi$ coupled channel scattering as fixed from $SU(4)$ relations. The phase convention for the mesons has been chosen in agreement with the convention of[10] taking $I_\pm, U_\pm, K_\pm$ to be positive. This convention differs from the one in [11, 5, 7] where $V_\pm$ instead of $U_\pm$ is chosen to be positive.

Potential	Iso-Factor $(I=0;I=1)$	Cutoff Λ	$g_1 * g_2$
K^* in t-channel	$(-;-\sqrt{2})$	2.2 GeV	$g_{\pi KK^*} * g_{D_s^+ K^* D} = -\frac{1}{2\sqrt{2}} g_{\pi\pi\rho}$
ρ in t-channel	$(-3;1)$	3.2 GeV	$g_{DD\rho} * g_{KK\rho} = \frac{1}{4} g_{\pi\pi\rho}$
ω in t-channel	$(-1;-1)$	3.2 GeV	$g_{DD\omega} * g_{KK\omega} = \frac{1}{4} g_{\pi\pi\rho}$

pion or an a_1. Since the emission of a pion mainly flips the helicity of the nucleon and the emission of an a_1 mainly conserves it, there is close to no interference between those two amplitudes. That is why we may model the production by equation (1), where the initial production is parameterized by Regge amplitudes very much like the ones found in[15].

$$N \propto \frac{1}{q_{beam}^2 s_{tot}} \int_{t_1}^{t_2} dt \left\{ A_{\pi\pi} |T_{\pi\pi\to\pi\pi}|^2 \frac{-t}{t-m_\pi^2} e^{b_\pi(t-m_\pi^2)} \right. \tag{1}$$

$$\left. + A_{\pi a_1} \left| T_{\pi a_1 \to \pi\pi} \right|^2 (1+tC) e^{b_{a_1} t} \right\} \times \theta(t_{min} - t)\theta(t - t_{max})$$

$$\left. \begin{matrix} t_{min} \\ t_{max} \end{matrix} \right\} = \frac{(m_\pi^2 - m_{\pi\pi}^2)^2 - (S_{12} \mp S_{34})^2}{4s} \tag{2}$$

$$S_{ij} = \sqrt{(s - (m_i + m_j)^2)(s - (m_i - m_j)^2)} \tag{3}$$

Here q_{beam} means the initial pion beam momentum, s_{tot} the invariant mass of the complete system and C, b_π and b_{a_1} should be considered free parameters. $A_{\pi\pi}$ and $A_{\pi a_1}$ do not directly depend on q_{beam} but they do depend on t as described in [16]. We chose $C = -4.4$ GeV^{-2} as in [15] and obtained the slope parameters b_π and b_{a_1} by fitting the t-dependence of the total cross section. Further more the off shell T-matrices have been calculated using the Jülich meson exchange model for meson-meson scattering. Equation (1) shows that the pion beam momentum at which the experiment is performed only alters the production in two ways. This is by an overall factor, which will not be observed since most experiments did not publish a normalization, and by the kinematical limits, which is of no importance for the beam momenta and invariant two pion masses $m_{\pi\pi}$ under consideration here. This means that the different experimental data shown in figure 2 should agree in shape. That is the case for the region of low momentum transfer to the nucleon, but at higher momentum transfer data starts to deviate above $m_{\pi\pi} = 1$ GeV. This region is most sensible to the admixture of a hard component f_0' to the $K\bar{K}$-molecule, because the contribution of the $K\bar{K}$-molecule diminishes at high momentum transfers. In figure 2 we also show two different calculations of cross sections corresponding to different admixtures of a f_0'. One calculation was fitted to the

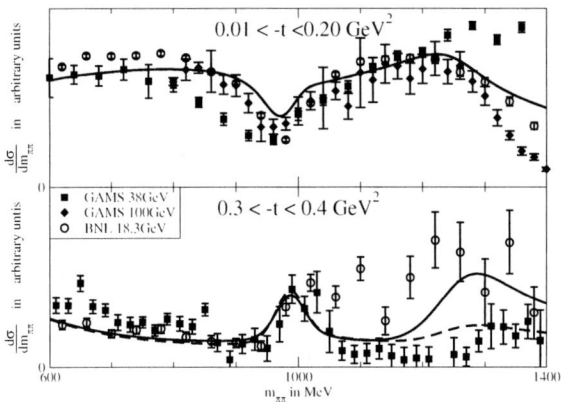

FIGURE 2. Cross section of $\pi^- p \to \pi^0 \pi^0 n$ as measured by experiments using different pion beam momenta. In the upper panel the momentum transfer t to the nucleon has been integrated over the region [0.01 GeV^2; 0.20 GeV^2] and in the lower panel over the region [0.3 GeV^2; 0.4 GeV^2]. (Circles: BNL at 18.3 GeV[12]; Squares: GAMS at 38 GeV[13]; Diamonds: GAMS at 100 GeV[14]) The solid line shows the results of our model with the coupling of the f_0' fitted to the BNL data. For the dashed line this coupling has been fitted to the GAMS data. In the upper panel these two curves coincide.

BNL data (solid) and the other was fitted to GAMS data (dashed). Both fits agree in the low momentum transfer region but deviate in the high momentum transfer case just as the data does. This means that we are not able to determine the admixture of the f_0' to the $K\bar{K}$ molecule from the available data. A more detailed discussion of this can be found in [17].

Further more the $\rho\rho$-channel coupled to the $\pi\pi$-, $K\bar{K}$- and πa_1- channel shows already sizable effects in the energy range sensitive to the parameters for the admixture of a bare state to the $K\bar{K}$-molecule. So one should also look at other observables sensitive to such an admixture.

In [18] the lifetime of kaonium has been shown to be such a quantity. We will shortly review the ideas and results of this paper here. As the $K\bar{K}$ binding energy is of the order of $10 - 20$ MeV which is much less then the reduced mass $m_{red} = 496$ MeV it is appropriate to work in the limit of non-relativistic effective field-theory. Then one uses the effective range expansion to construct a phase equivalent Bargmann potential. Since the Jost function can be constructed analytically from this potential the energy spectrum can be calculated from the Jost zeros λ. The zeros are quoted along with the lifetimes of the different energy levels in table 2 for two different effective ranges a_p^{eff}. To estimate the amount of distortion introduced by the strong interaction one should compare the values given for λ with $\lambda = 1/n$ in the pure coulomb case. The two values for the effective range a_p^{eff} quoted in table 2 have been chosen to represent two cases: a) the $f_0(980)$ is a pure $K\bar{K}$ molecule, the $a_0(980)$ is a pure cusp effect and the isovector $K\bar{K}$-channel does not couple to $\pi\eta$. b) the $f_0(980)$ has a hard component admixed to the $K\bar{K}$ molecule, the $a_0(980)$ is a pure cusp effect and the isovector $K\bar{K}$-channel couples to $\pi\eta$. Even so the lifetimes are very small one can see in table 2 the drastic influence of these changes on the lifetime of the different levels.

TABLE 2. Lifetime predictions for kaonium corresponding to the effective ranges of two different models for the strong $K\bar{K}$ interaction. The value of $a_p^{eff} = (1.940 - 1.199i) M_K^{-1}$ corresponds to the $f_0(980)$ being a pure strongly bound $K\bar{K}$ molecule and the $K\bar{K}$ channel not being coupled to the $\pi\eta$-channel, where as the value $a_p^{eff} = (1.156 - 3.268i) M_K^{-1}$ corresponds to a more loosely bound $K\bar{K}$-molecule which mixes with a hard f_0' and the $K\bar{K}$-channel being coupled to the $\pi\eta$-channel.

| Level | $a_p^{eff} = (1.940 - 1.199i) M_K^{-1}$ | | $a_p^{eff} = (1.156 - 3.268i) M_K^{-1}$ | |
	λ	Lifetime($\times 10^{-18}$sec)	λ	Lifetime($\times 10^{-18}$sec)
3rd	$0.2491 + 0.0005i$	199	$0.2494 + 0.0014i$	71
2nd	$0.3318 + 0.0009i$	84	$0.3322 + 0.0025i$	30
1st	$0.4965 + 0.0056i$	25	$0.4975 + 0.0056i$	8.9
Ground	$0.9863 + 0.0079i$	3.2	$0.9899 + 0.0223i$	1.1

CONCLUSION

Looking at the potentials arising from one boson t-channel exchanges in the DK-channel we find that in the $J = 0, I = 0$ channel the attraction is sufficient to form a molecule at the energy of the newly discovered $D_{sJ}^*(2317)$. Assuming isospin breaking by the physical masses and $\pi\eta$- as well as $\rho\omega$-mixing we find that the isospin breaking is to weak to account for the width and the strength of the observed resonance. We conclude that the observed resonance is most probable not a DK-molecule but some state arising from the confinement spectrum with a small admixture of the molecule. In the case of the $K\bar{K}$-molecule the production data is still not precise enough to pin down the amount of admixture from a compact state to the formed molecule within our model. We have shown that the lifetime of kaonium is very sensitive to such an admixture and a measurement of this quantity would be really helpful.

REFERENCES

1. Aubert, B., et al., *Phys. Rev. Lett.*, **90**, 242001 (2003).
2. Besson, D., et al., *Phys. Rev.*, **D68**, 032002 (2003).
3. Abe, K., et al., *hep-ex/0307052* (2003).
4. Stone, S., and Urheim, J., *hep-ph/0308166* (2003).
5. Lohse, D., Durso, J. W., Holinde, K., and Speth, J., *Nucl. Phys.*, **A516**, 513–548 (1990).
6. Janssen, G., Pearce, B. C., Holinde, K., and Speth, J., *Phys. Rev.*, **D52**, 2690–2700 (1995).
7. Krehl, O., Rapp, R., and Speth, J., *Phys. Lett.*, **B390**, 23–28 (1997).
8. Coon, S. A., and Scadron, M. D., *Phys. Rev.*, **C51**, 2923–2931 (1995).
9. Bardeen, W. A., Eichten, E. J., and Hill, C. T., *hep-ph/0305049* (2003).
10. Rabl, V., Campbell, J., George, and Wali, K. C., *J. Math. Phys.*, **16**, 2494 (1975).
11. de Swart, J. J., *Rev. Mod. Phys.*, **35**, 916–939 (1963).
12. Gunter, J., et al., *Phys. Rev.*, **D64**, 072003 (2001).
13. Alde, D., et al., *Z. Phys.*, **C66**, 375–378 (1995).
14. Alde, D., et al., *Eur. Phys. J.*, **A3**, 361–371 (1998).
15. Achasov, N. N., and Shestakov, G. N., *Phys. Rev.*, **D58**, 054011 (1998).
16. Petersen, J. L., *Phys. Rep.*, **2**, 155 (1971).
17. Sassen, F. P., Krewald, S., Speth, J., and Thomas, A. W., *Phys. Rev.*, **D68**, 036003 (2003).
18. Krewald, S., Lemmer, R. H., and Sassen, F. P., *hep-ph/0307288* (2003).

Roy's equations and the $\pi\pi$ experimental data

R. Kamiński, L. Leśniak* and B. Loiseau†

* Department of Theoretical Physics, H. Niewodniczański Institute of Nuclear Physics,
Polish Academy of Sciences, PL 31-342 Kraków, Poland
† Laboratoire de Physique Nucléaire et de Hautes Énergies,[1]
Groupe Théorie, Univ. P. & M. Curie, 4 Pl. Jussieu, F-75252 Paris, France[2]

Abstract.
Roy's equations are used to check if the scalar-isoscalar $\pi\pi$ scattering amplitudes fitted to experimental data fulfill crossing symmetry conditions. It is shown that the amplitudes describing the "down-flat" phase shift solution satisfy crossing symmetry below 1 GeV while the amplitudes fitted to the "up-flat" data do not. In this way the long standing "up-down" ambiguity in the phenomenological determination of the scalar-isoscalar $\pi\pi$ amplitudes has been resolved confirming the independent result of the recent joint analysis of the $\pi^+\pi^-$ and $\pi^0\pi^0$ data.

INTRODUCTION

In 1997 a new analysis of the $\pi^-p_\uparrow \to \pi^+\pi^-n$ reaction on a polarized target was performed in the $m_{\pi\pi}$ effective mass range from 600 to 1600 MeV [1]. For the first time the pseudoscalar (π-exchange) amplitude was separated from the pseudovector (a_1-exchange) amplitude in the region of the the the four-momentum transfer squared from -0.005 to -0.2 $(\text{GeV/c})^2$. Below 1000 MeV, where the S- and P-waves strongly interfere, the partial wave analysis of the $\pi^+\pi^-$ data provided us with two scalar-isoscalar solutions, called "up" and "down", which differ by their intensities. Lack of information on a sign difference between the phases of the S- and P-waves near the position of the ρ resonance led us to other two branches of the "up" and "down" amplitudes named "steep" and "flat". It was shown in [2] that both "up-steep" and "down-steep" S-wave isoscalar amplitudes significantly violate unitarity below 1 GeV and should be rejected as nonphysical. Two remaining "flat" amplitudes survived the unitarity check and other tests were needed to resolve the existing "up-down" ambiguity.

In 2001 new experimental data on the $\pi^0\pi^0$ production from the E852 collaboration appeared [3] and were used in a joint analysis of the $\pi^+\pi^-$ and $\pi^0\pi^0$ data [4]. The $\pi^0\pi^0$ data are very useful to compare with the $\pi^+\pi^-$ data due to an absence of the P-wave in the $\pi^0\pi^0$ channel and therefore a lack of the "up-down" ambiguity. The one-pion and a_1-exchange model described in [4] was used to calculate the S-wave intensities of the $\pi^0\pi^0$ production by choosing as an input the "up-flat" or "down-flat" phase shifts. The isospin relations between the $\pi^+\pi^- \to \pi^+\pi^-$ and $\pi^+\pi^- \to \pi^0\pi^0$

---

[1] Unité de Recherche des Universités Paris 6 et Paris 7, associée au CNRS
[2] This work has been performed in the framework of the IN2P3-Polish Laboratories Convention (project number 99-97).

amplitudes supplemented by the parameterization of the isotensor scalar amplitude taken from [1] were helpful in these calculations. It was shown that the $\pi^0\pi^0$ S-wave intensities determined for the "down-flat" phase shifts agree with the experimental values within the errors. However, for the "up-flat" phase shifts in the $m_{\pi\pi}$ range from 850 to 970 MeV important differences between the calculated $\pi^0\pi^0$ intensities and the corresponding experimental values occur. This fact led the authors to a conclusion that the "up-flat" data set should also be rejected.

ROY'S EQUATIONS AS A TEST FOR THE $\pi\pi$ AMPLITUDES

Another independent test of the "up-flat" and "down-flat" amplitudes consists in checking if they fulfill crossing symmetry conditions below 1 GeV. In order to achieve this task we have used Roy's equations [5] for the scalar-isoscalar, $\ell = 0$, $I = 0$, scalar-isotensor, $\ell = 0$, $I = 2$, and the vector-isovector, $\ell = 1$, $I = 1$, $\pi\pi$ partial waves determined in a wide $m_{\pi\pi}$ range. We were especially interested in the $m_{\pi\pi}$ region between 800 and 1000 MeV where differences between phase shifts of the "up-flat" and "down-flat" data sets are largest and reach about $45°$ (see Fig. 4 in [2]). In a recent analysis of Roy's equations [6] a special attention was put on the effective mass lower than 800 MeV.

As an input in Roy's equations we have used imaginary parts of the partial waves amplitudes $f_\ell^I(s)$ related to the $\pi\pi$ phase shifts δ_ℓ^I and inelasticities η_ℓ^I:

$$f_\ell^I(s) = \sqrt{\frac{s}{s-4\mu^2}}\frac{1}{2i}\left(\eta_\ell^I e^{2i\delta_\ell^I} - 1\right),\tag{1}$$

where μ is the charged pion mass and $s = m_{\pi\pi}^2$.

Below 970 MeV the following Padé representation of both the "up-flat" and "down-flat" phase shifts has been taken:

$$\tan\delta_0^0(s) = \frac{\sum_{i=0}^{4}\alpha_{2i+1}k^{2i+1}}{\Pi_{i=1}^{3}(k^2/\alpha_{2i}-1)},\tag{2}$$

where $k = \frac{1}{2}\sqrt{s-4\mu^2}$ is the pion momentum and $\alpha_j\,(j = 1,\ldots,7,9)$ are constant parameters. Above 970 MeV up to 2 GeV our coupled channel model [7] amplitude A, fitted to the "down-flat" data, and the amplitude C, constrained by the "up-flat" data, were applied. In the fits we also used the near threshold phase shifts calculated from the differences $\delta_0^0 - \delta_1^1$ obtained in the high statistics K_{e4} decay experiment [8]. The scattering length a_0^0 and the slope parameter b_0^0 are directly related to the constants α_j: $a_0^0 = -\alpha_1\mu$ and $b_0^0 = -\alpha_1\mu\left(0.5\mu^{-2} + \alpha_2^{-1} + \alpha_4^{-1} + \alpha_6^{-1} - \alpha_1^2\right) - \alpha_3\mu$.

The parameterization of isotensor wave using a rank-two separable potential model has been described in [9] where detailed analysis of the present study is presented.

For the P-wave, from the $\pi\pi$ threshold till 970 MeV, we have used an extended Schenk parameterization [6]:

$$\tan\delta_1^1(s) = \sqrt{1 - \frac{4\mu^2}{s}}\,k^2\left(A + Bk^2 + Ck^4 + Dk^6\right)\left(\frac{4\mu^2 - s_\rho}{s - s_\rho}\right),\tag{3}$$

where A is the P-wave scattering length and s_ρ is equal to the ρ-mass squared. Above 970 MeV we took the K-matrix parameterization of Hyams et al. [10]. The parameters

C and D were chosen to join smoothly the phase shifts given by both parameterizations around 970 MeV.

The contributions to Roy's equations from high energies ($m_{\pi\pi} > 2$ GeV) and from higher partial waves ($l > 1$) are called driving terms. They are composed of contributions from the $f_2(1270)$ and $\rho_3(1690)$ resonances and from the Regge amplitudes for the Pomeron, ρ- and f-exchanges. The Breit-Wigner parameterization with masses, widths and $\pi\pi$ branching ratios taken from [11] were used for $f_2(1270)$ and $\rho_3(1690)$. For the Regge parts we have used formulae of [6] without the u-crossed terms. We have found that the $f_2(1270)$ resonance dominates in the scalar isoscalar wave and that the introduction of the $\rho_3(1690)$ has a significant influence on the isotensor and isovector waves. In the isoscalar wave the Regge contributions are more than 10 times smaller than the resonance contributions but for the isospin 1 and 2 they are of the same order.

The thirteen constants, six for the scalar-isoscalar wave in (2), four for the isotensor wave and three for the isovector wave in (3), were calculated from the simultaneous fits to data and to Roy's equations separately for the "up-flat" and "down-flat" data. We have used the CERN MINUIT program with the χ^2 test function defined by

$$\chi^2 = \sum_{I=0,1,2} \left\{ \sum_{i=1}^{N_I} \left[\frac{\sin\left(\delta_\ell^I(s_i) - \varphi_\ell^I(s_i)\right)}{\Delta\varphi_\ell^I(s_i)} \right]^2 + \sum_{j=1}^{12} \left[\frac{\mathrm{Re}\, f_{out}^I(s_j) - \mathrm{Re}\, f_{in}^I(s_j)}{\Delta f} \right]^2 \right\}, \quad (4)$$

where $\varphi_\ell^I(s_i)$ and $\Delta\varphi_\ell^I(s_i)$ represent the experimental phase shifts and their errors, respectively, $s_j = [4j + 0.001]\mu^2$ for $j = 1,...,11$ and $s_{12} = 46.001\mu^2$. The real parts $\mathrm{Re}\, f_{in}^I$ have been calculated from (1) under an assumption that the inelasticity η_l^I is equal to 1 and the phase shifts δ_l^I are equal to $\phi_l^I(s_j)$. Other real parts, denoted by $\mathrm{Re}\, f_{out}^I$, constitute the output values calculated from Roy's equations. We take a Δf value of 0.5×10^{-2} to obtain acceptable fits to Roy's equations. 18 experimental values of the "up-flat" or "down-flat" data between 600 and 950 MeV were used in addition to six data taken from [8].

In Fig. 1a and 1b we present results of fits to the "up-flat" and "down-flat" phase shifts and to Roy's equations (solid lines). In both cases differences $|\,\mathrm{Re}\, f_{out}^I - \mathrm{Re}\, f_{in}^I\,|$ were of the order of 10^{-3} in all three partial waves. The χ^2 for 18 points between 600 and 970 MeV was 16.6 in the "down-flat" case and as large as 46.4 in the "up-flat" one. We see in Fig. 1a that the solid line lies distinctly below the "up-flat" data points between 800 and 970 MeV. In contrary, the corresponding line for the "down-flat" case in Fig. 1b is very close to experimental data in the same range of $m_{\pi\pi}$. In order to improve a fit to the "up-flat" data we have used constraints given by the good fit to the "down-flat" data. Two parameters were fixed by choosing the values of the scattering length and the slope parameter and two others by the values of phase shifts calculated from this fit at 500 and 550 MeV. A new fit with these constrains gave an improved value of $\chi^2 = 13$ for 18 "up-flat" data points, corresponding to the first part of χ^2 in (4) but provided us with an enormous value of $\chi^2 = 1.2 \times 10^4$ for the second part related to Roy's equations. The phase shifts for this amplitude are presented in Fig. 1a by the dotted line. It is clear that a simultaneous good fit to the "up-flat" data and to Roy's equations is impossible.

Apart of the fits to the "up-flat" and "down-flat" experimental points we have also performed fits to points shifted upwards and downwards by their errors. In these fits the

197

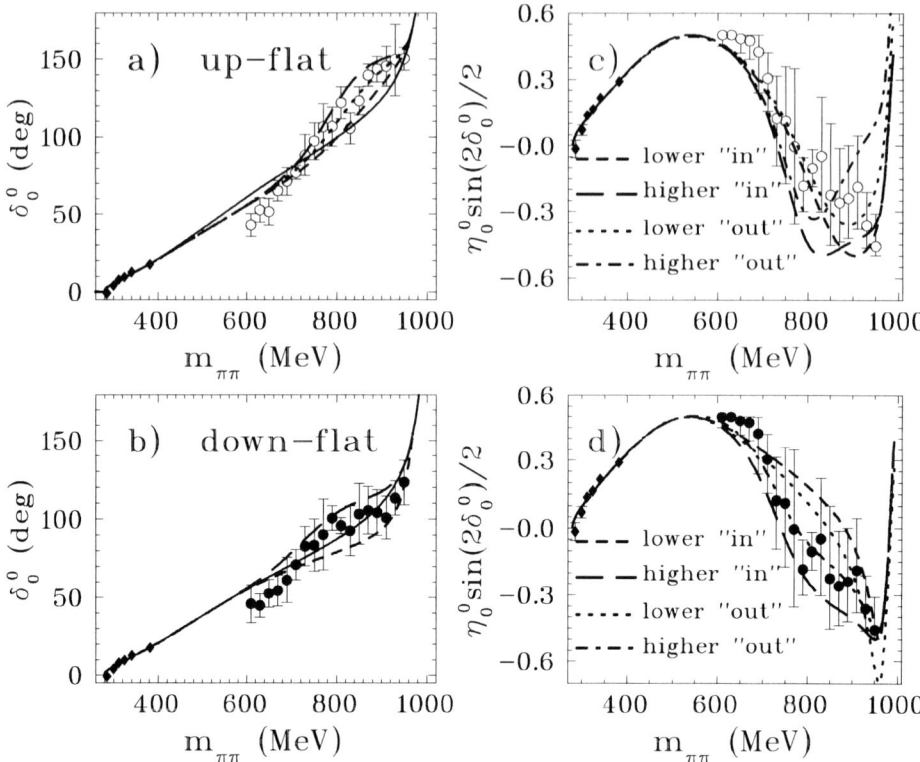

FIGURE 1. **a)** and **c)** correspond to the "up-flat" case, **b)** and **d)** correspond to the "down-flat" case. Fits to the scalar-isoscalar phase shifts of [1] and to Roy's equations are denoted by solid lines in **a)** and **b)**. Dotted line in **a)** between two dashed lines represents fit with constraints described in the text. Dashed lines in **a)** and **b)** represent fits to phase shifts moved upwards and downwards by their errors; the corresponding lines in **c)** and **d)** are called *higher* and *lower*, respectively. Lines in **c)** and **d)** correspond to real parts of input amplitudes *("in")* and real parts calculated from Roy's equations *("out")*, all multiplied by $2ks^{-1/2}$. Diamonds represent the K_{e4} data [8].

same four constraints described above were used below 600 MeV. Up to 937 MeV in the "down-flat" case in Fig. 1d the curves labeled *higher "in"* and *lower "in"* form a band including inside a band delimited by the lines *higher "out"* and *lower "out"*. All the curves lying inside these bands correspond to the amplitudes fulfilling the crossing symmetry so the "down-flat" data can be accepted as physical ones. In the "up-flat" case in Fig. 1c the output band lies outside of the input band from 840 to 970 MeV. It means that in this case the crossing symmetry is violated by the amplitudes fitted to the "up-flat" data.

In Fig. 2 we have presented the output results for the isotensor and isovector waves in the "down-flat" case only since in the "up-flat" case the curves are very similar. The "in" curves were not plotted because they are almost indistinguishable from the "out" ones.

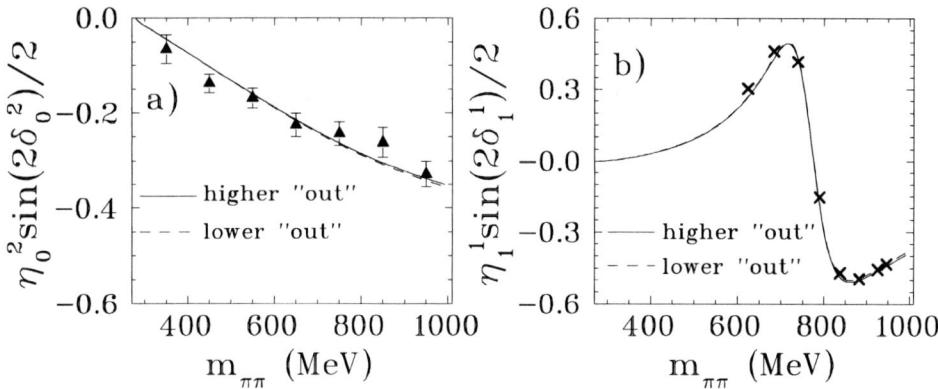

FIGURE 2. Real parts of isotensor **a)** and isovector **b)** $\pi\pi$ amplitudes (multiplied by $2ks^{-1/2}$) fitted to the "down-flat" data. Triangles in **a)** denote data of [12]. Crosses in **b)** are the pseudo-data calculated from the K-matrix fit of [10].

CONCLUSIONS

We have used Roy's equations as a tool to test if the amplitudes fitted to the "up-flat" and "down-flat" phase shifts extracted from the $\pi^- p_\uparrow \to \pi^+\pi^- n$ data fulfill crossing symmetry conditions. We have found that only the S-wave isoscalar amplitude corresponding to the "down-flat" data set can be accepted. The amplitude constrained to the "up-flat" data does not satisfy Roy's equations and should be rejected as nonphysical. This conclusion is in agreement with the independent results obtained from a joint analysis of the $\pi^+\pi^-$ and the $\pi^0\pi^0$ production data [4]. In this way a long standing "up-down" ambiguity in the $\pi\pi$ experimental data has been resolved in favour of the "down-flat" data set.

REFERENCES

1. Kamiński R., Leśniak L. and Rybicki K., Z. Phys. **C 74**, 79 (1997).
2. Kamiński R., Leśniak L. and Rybicki K., Acta Phys. Pol. **B31**, 895 (2000).
3. Gunter J. et al. (E852 Collaboration), Phys. Rev. **D 64**, 072003 (2001).
4. Kamiński R., Leśniak L. and Rybicki K., Eur. Phys. J. direct **C4**, 1 (2002).
5. Roy S. M., Phys. Lett. **B 36**, 353 (1971); Roy S. M., Helv. Phys. Acta **63**, 627 (1990).
6. Ananthanarayan B., Colangelo G., Gasser J. and Leutwyler H., Phys. Rep. **353**, 207 (2001).
7. Kamiński R., Leśniak L., B. Loiseau, Phys. Lett. **B 413**, 13 (1997).
8. Pislak S. et al., (E865 Coll.), Phys. Rev. Lett. **87**, 221801 (2001).
9. Kamiński R., Leśniak L. and B. Loiseau, Phys. Lett. **B 551**, 241 (2003).
10. Hyams B. et al., Nucl. Phys. **B 64**, 134 (1973).
11. K. Hagiwara et al. (Particle Data Group), Phys. Rev. **D66**, 010001 (2002).
12. Hoogland W. et al., Nucl. Phys. **B126**, 109 (1977).

Proposal to observe the strong Van der Waals force in $e + \bar{e} \to 2\pi$

Tetsuo Sawada[1]

Institute of Quantum Science, Nihon University, Tokyo, Japan 101-8308

Abstract

Large discrepancy of the p-wave phase shift data $\delta_1(v)$ of the π-π scattering from those of the dispersion calculation is pointed out. In order to determine which is correct, the pion form factor $F_\pi(v)$, which is the second source of information of the phase shift $\delta_1(v)$, is used. It is found that the phase shift obtained from the dispersion is not compatible with the data of the pion form factor. What is wrong with the dispersion calculation, is considered.

P-WAVE PHASE SHIFT $\delta_1(v)$ OF THE π-π SCATTERING

It is known that $\delta_1(v)$ is reproduced well by Wagner's straight line fit

$$\frac{s_\rho}{s}\left(\frac{v}{v_\rho}\right)^{3/2}\cot\delta_1^{(1)}(v) = \frac{s_\rho - s}{\sqrt{s_\rho}\Gamma_\rho} \tag{1}$$

in the low energy region, where v is the momentum squared in the center of mass system and $s = 4v + 4$ in which the unit $\mu^2 = 1$ is adopted. If we compare it with the effective range function $X_1(v)$, which is defined by

$$X_1(v) = \frac{v^{3/2}}{\sqrt{v+1}}\cot\delta_1^{(1)}(v) \quad, \tag{2}$$

we can rewrite Wagner's fit in terms of the effective range function

$$\frac{X_1(v)}{\sqrt{v+1}} = \tilde{c}(v - v_\rho) \quad. \tag{3}$$

>From the values of the mass and the width of the ρ-meson $m_\rho = 775.65$MeV. and $\Gamma_\rho = 143.85$MeV., the parameters of Eq.(3) are determined: $v_\rho = 6.721$ and $\tilde{c} = -1.576$ in the unit of $\mu = 1$. In figure 1, Wagner's fit and the data points are shown.

---

[1] Associate member of IQS for research.   e-mail address: t-sawada@fureai.or.jp

CP717, *Hadron Spectroscopy: Tenth International Conference,*
edited by E. Klempt, H. Koch, and H. Orth
© 2004 American Institute of Physics 0-7354-0197-7/04/$22.00

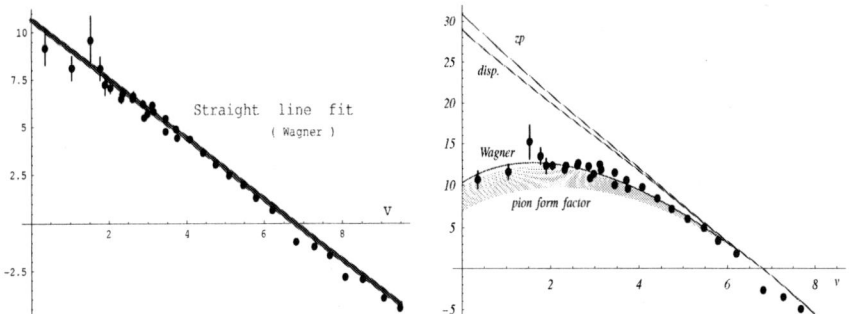

FIGURE 1. Effective range functions $X_1^{zp}(v)$, $X_1^{disp}(v)$ and $X_1^{wag}(v)$. The corridor is the result of the pion form factor.

Before the dispersion calculation, it is convenient to introduce the zero-potential amplitude $a_1^{zp}(v)$, which is characterized by vanishing of the left hand spectrum. Moreover it is expected to have the ρ-meson pole at right location which is specified by m_ρ and Γ_ρ. The following effective range function $X_1^{zp}(v)$ will do the job:

$$
X_1^{zp}(v) = c_0 + c_1 v + \frac{2}{\pi} \frac{v^{3/2}}{\sqrt{v+1}} \log(\sqrt{v} + \sqrt{v+1}) \tag{4}
$$

$$
\text{with} \quad c_0 = 31.72 \quad \text{and} \quad c_1 = -5.719 \quad .
$$

If we remember the relation between the amplitude $a_1(v)$ and $X_1(v)$

$$
\frac{a_1(v)}{v} = \frac{1}{X_1(v) - iv\sqrt{v/(v+1)}} = \frac{1}{X_1(v) - v(-v/(v+1))^{1/2}} \tag{5}
$$

the necessity of the logarithmic term in Eq.(4) is evident, because the term $-v(-v/(v+1))^{1/2}$ has cuts in $v < -1$ as well as in $v > 0$. We can numerically confirm the property that the amplitude does not have the left hand spectrum by computing

$$
\frac{K_1(v)}{v} \equiv \frac{a_1(v)}{v} - \frac{1}{\pi} \int_0^\infty dv' \frac{\text{Im} a_1(v')}{v'(v'-v)} \quad , \tag{6}
$$

which is sometimes called Kantor amplitude. If we use the zero-potential amplitude $a_1^{zp}(v)$ in evaluating Eq.(6), it must become identically zero, namely $K_1^{zp}(v) = 0$. In general, Kantor amplitude can be computed in principle from the experimental data, and can be used to explore the left hand spectrum namely to examine the force acting between the scattering particles.

In the π-π scattering the two-pion exchange spectrum is known to be computed from the crossing symmetry. The explicit form of the contribution of the two-pion exchange spectrum to the p-wave amplitude $a_1^{(1)}(v)/v$ is[1]

$$
\left(\frac{K_1^{2\pi}(v)}{v} \right) = \tag{7}
$$

201

$$= \frac{1}{\pi} \int_0^\infty 3\mathrm{Im}a_1^{(1)}(v'') \left(\frac{1 + 2(v+1)/v''}{v^2} 2Q_1(1 + 2(v''+1)/v) - \frac{1 + 2/v''}{6(v''+1)^2} \right) dv''$$

$$+ \frac{1}{\pi} \int_0^\infty [\frac{2}{3}\mathrm{Im}a_0^{(0)}(v'') - \frac{5}{3}\mathrm{Im}a_0^{(2)}(v'')] \left(\frac{2}{v^2}Q_1(1 + 2(v''+1)/v) - \frac{1}{6(v''+1)^2} \right) dv''$$

in which $\mathrm{Im}a_\ell^{(I)}(v'')$ are the imaginary part of the ℓ-th partial waves of isospin I. It turns out that $K_1^{2\pi}(v)/v$ is small.[2] The effective range function of the dispersion calculation $X_1^{disp}(v)$, which corresponds to the amplitude $(a_1^{zp}(v) + K_1^{2\pi}(v))/v$, stays close to $X_1^{zp}(v)$. In figure 2, the effective range curves $X_1^{zp}(v)$, $X_1^{disp}(v)$ and $X_1^{wag}(v)$, which is given in Eq.(3), are shown along with the data points. Although the locations of the ρ-meson and the slopes at $v = v_\rho$ are kept the same for three curves, $X_1^{wag}(v)$ deviates appreciably from other curves in the low energy region. The corridor just below Wagner's curve is the effective range function obtained from the pion form factor in the next section.

CROSS CHECK OF THE P-WAVE PHASE SHIFT $\delta_1(v)$

Because of the final state interaction, the phase of the pion form factor $F(v)$ coincides with the p-wave phase shift $\delta_1(v)$ at least in the elastic region of the corresponding π-π scattering. Let us introduce the phase function $\Delta(v)$ by $F_\pi(v) = |F_\pi(v)|e^{i\Delta(v)}$, which is expected to be equal to $\delta_1(v)$ in the low energy region. It is covenient to define a function

$$f(v) = \frac{1}{v+1} \log|F_\pi(v)| \quad . \tag{8}$$

If we remember $F_\pi(v)$ is normalized at $v = -1$, the denominator $(v+1)$ in Eq.(8) is necessary to remove zero at $v = -1$ and it also serve to make $f(v)$ to decrease at large $|v|$. The integral representation of $f(v)$ has the form of the Hilbert transformation:

$$\frac{\log|F_\pi(v)|}{v+1} = \frac{P}{\pi} \int_{-\infty}^\infty dv'\theta(v') \frac{\Delta(v')}{(v'+1)(v'-v)} \quad , \tag{9}$$

and whose inversion is

$$\theta(v)\frac{\Delta(v)}{v+1} = -\frac{P}{\pi} \int_{-\infty}^\infty dv' \frac{\log|F_\pi(v')|}{(v'+1)(v'-v)} \quad , \tag{10}$$

because the square of the Hilbert transformation is equal to minus identity.[1]

Equations (9) and (10) enable us to extract more complete information on the phase shift or on the pion form factor by analyzing the phase shift and the form factor data jointly. When the precise data of $|F_\pi(v)|$ were available in the space-like region ($v < -1$) as well as in the time-like region ($v > 0$), we could use Eq.(10) to evaluate the phase $\Delta(v)$. However even in such situation, we have to interpolate the data to the narrow unphysical region ($-1 < v < 0$), where experimental data are not available, although the interpolation function is strictly restricted by the condition that the integration of

Eq.(10) must vanish in $v < 0$. Since the precision of the data of the pion form factor in the threshold region is not sufficient, our program in this paper will become modest one to estimate the deviation of the phase shift in the sub-rho region, namely to determine the deviation coefficient a introduced in

$$\Delta(v) = a(\delta_1^{wag}(v) - \delta_1^{zp}(v)) + \delta_1^{zp}(v) \quad \text{in } 0 < v < v_\rho \tag{11}$$

by fitting the integration of Eq.(9) to the data of the pion form factor in the space-like region.

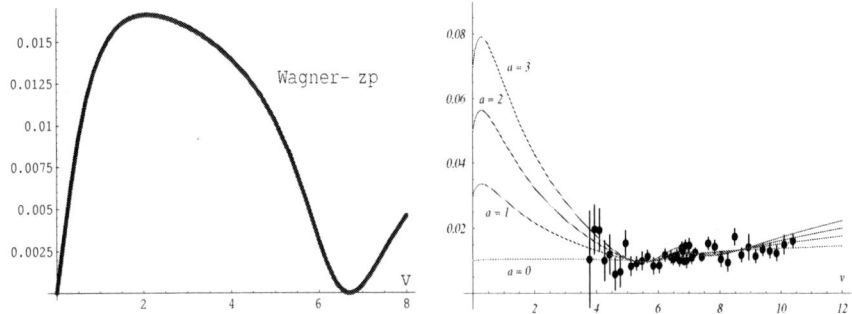

FIGURE 2. $(f(v) - f^{zp}(v))$ in the ρ-resonance region for $a = 0, 1, 2$ and 3.

In figure 3, the spectral function $(\delta_1^{wag}(v) - \delta_1^{zp}(v))/(v'+1)$ necessary to calculate $(f(v) - f^{zp}(v))$ is plotted against v. It is interesting to examine qualitatively what is the results of the Hilbert transformation of this spectral function in Eq.(9). Firstly $f(v)$ must shift upward in the space-like region, whereas in the ρ-meson and the higher energy region it must shift downward. Secondly because of the rapid raise of the curve of the spectral function at small v, $f(v)$ must have a narrow peak in the threshold region. Although the phase $\Delta(v)$ coincides with the π-π phase shift $\delta_1(v)$ in the ρ-resonance region, they deviate each other in the higher energy region where the inelasticity is not negligible. Therefore before we evaluate $f(v)$ in the space-like region, we must determine $\Delta(v)$ in the higher energy region, for various values of the deviation coefficient 'a' appeared in Eq.(11), in such a way that it reproduces the form factor $f(v)$ well in the ρ-resonance region. In figure 4, curves $(f(v) - f^{zp}(v))$ are plotted against v for $a = 0, 1, 2$ and 3, along with experimental data of CMD-2,[3] in which the ω-pole is removed. Although for $a = 0$ the curve continues monotonously to the space-like region, for $a > 0$ curves $(f(v) - f^{zp}(v))$ have narrow peaks in the threshold region. Our proposal is to observe such a narrow peak by measuring the cross section of $e + \bar{e} \to 2\pi$ precisely in the low energy region, and which serves to confirm that the long range force such as the strong Van der Waals force is acting between pions.

In figure 5 and 6, the same curves are plotted in the space-like region ($v < -1$) and in the unphysical region ($-1 < v < 0$). The data points in fig.5 are those of Amendolia et.al.in the region of small momentum transfer, whereas the points in fig.6 are other data in the space-like region.[4] The graphs indicate that the data points differ from the curve $a = 0$, which is obtained by choosing the zero-potential phase shift $\delta_1^{zp}(v')$ as $\Delta(v')$ in $0 < v' < v_\rho$ in the evaluation of $f(v)$. From the chi square search, $a = 1.1$ is the best fit

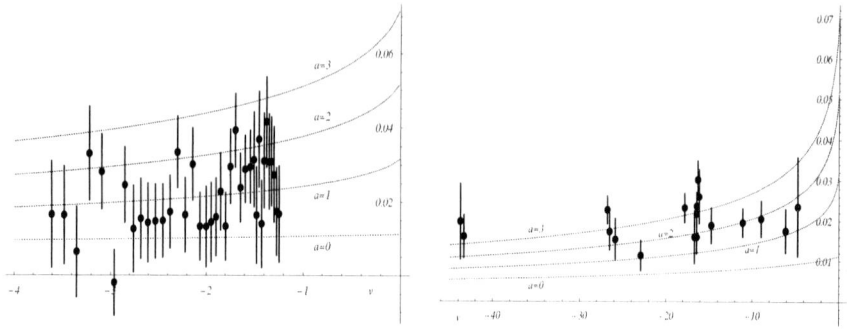

FIGURE 3. $(f(v) - f^{zp}(v))$ in the space-like region for $a = 0, 1, 2$ and 3.

for the data of low momentum transfer of fig.5. On the other hand, for the joint space-like data of fig.5 and 6, the minimum of chi square occurs at $a = 1.7$. Therefore the pion form factor data supports the Wagner's fit rather than the dispersin calculation or the zero-potential curve as shown in fig.2.

LONG RANGE INTERACTION IN THE π-π SCATTERING

In order to see what type of force is acting between pions, let us compute the contribution from the specrum on the left hand cut, namely the Kantor amplitude $K_1^{wag}(v)/v$ given in Eq.(6) by substituting $\Delta(v')$ by the phase shift $\delta_1^{wag}(v')$ which is close to the experimental data. In figure 7, $K_1^{wag}(v)/v$ is plotted against v.

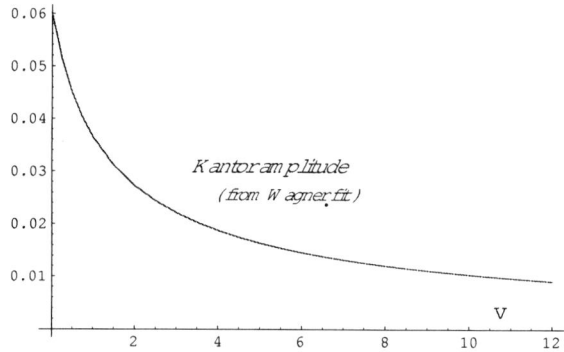

FIGURE 4. Kantor amplitude $K_1^{wag}(v)/v$ of π-π scattering. A cusp of the attractive sign appears at $v = 0$, which is characteristic to the long range force.

The curve $K_1^{wag}(v)/v$ is characterized by its large slope and very large curvature in the threshold region. On the other hand, since the spectrum of the short range force starts at far left, for example the 4-pion exchange spectrum starts at $v = -4$, $K_1(v)/v$ must be almost constant with small slope and extremely small curvature in the threshold region.

Therefore the curve indicates that strong force whose range is longer than that of the pion exchanges must acting.

In general the long range potential, whose asymptotic form is $V(r) \sim -C/r^\alpha$, induces a left hand spectrum in $a_1(v)$ starting from $v = 0$ and the threshold behavior of the spectrum is $\operatorname{Im} a_1(v') \approx C''(-v')^\gamma$. The powers α and γ are related by $\alpha = 2\gamma + 3$, and the coefficient C'' is proportional to C of the potential. In particular for $\alpha = 6$, which is the Van der Waals potential of the London type, the amplitude has the singular term $a_1(v) = -C''v^{3/2} + \cdots$, whereas for $\alpha = 7$, which is the Van der Waals potential of the Casimir-Polder type, the amplitude has the singular term $a_1(v') = C'''v^2 \log v + \cdots$. It is important that C'' and C''' are positive for the attractive potential. Figure 7 indicates that the behavior of the curve $K_1(v)/v$ is close to $c_0'' - C''\sqrt{v}$, namely case of $\alpha = 6$, althogh possibility of $\alpha = 7$, namely $c_0''' + C'''v \log v$, is not excluded. We can conclude that the attractive Van der Waals force dominates the pion-pion interaction rather than the short range force.

Finally we shall consider why the strong Van der Waals force appears in the hadron physics. When the hadron was regarded as an elementary particle, the interaction between hadrons must occur by the exchanges of mesons, and therefore it was inevitably short range. However after the introduction of the composite model of hadron, whose basic constructive force is strong or superstrong Coulomb type, because of the quantum fluctuation, we cannot avoid the strong Van der Waals force between the composite particles, namely between hadrons. Although the appearance of the Van der Waals force is simply a logical consequence of such composite model, what is important is its strength, which dominates the pion-pion interaction. It is known that the order of magnitude of the strength C of the Van der Waals potential $V(r) \sim -C/r^6$ is $C = (2/3)(^*e^2)^2 a_1^2 a_2^2 / \Delta E_1$, where $^*e^2$ is the "fine structure constant" of the basic Coulomb force whereas a_1 and a_2 are the radii of the composite particle 1 and 2 respectively. ΔE_1 is the first excitation energy. From the size of the cusp of figure 7, we can estimate the strength C, and which indicates that the fundamental Coulomb interaction is superstrong. Therefore the magnetic monopole model of hadron must be the favorite model, because from the charge quantization condition of Dirac $^*e^2$ is equal to $137/4$. If we remember that the Van der Waals interaction is universal, we can expect to observe the singular behavior also in other scatterings, whenever sufficiently precise data are available. In fact the attractive cusp is observed in the once subtracted S-wave amplitude $(a_0(v) - a_0(0))/v$ of the proton-proton scattering at $v = 0$, when the repulsive Coulomb singularity is properly removed.[5]

REFERENCES

1.   T.Sawada, Phys.Lett. **B225**, 291, (1989)
2.   T.Sawada, Nucl.Phys. **A675**, 375c, (2000)
3.   R.R.Akhmetshin et.al., arXiv:hep-ex/0308008, (2003)
4.   S.R.Amendolia et.al., Nucl.Phys. **B277**, 168, (1986)
        C.J. Bebek et.al., Phys.Rev. **D17**, 1693, (1978)
5.   T.Sawada, arXiv:nucl-th/0307023, (2003) and hep-ph/0004080, (2000)

Baryons

Hadron Spectroscopy at Jefferson Laboratory

Dennis P. Weygand

Thomas Jefferson National Accelerator Facility
Newport News, Virginia

Abstract. Recent results on hadron spectroscopy from Jefferson Laboratory's CEBAF Large Acceptance Spectrometer (CLAS) are presented. In particular we present results from the baryon resonance program for both electro- and photo- production. Also, we present very preliminary results on meson spectroscopy in γp interactions, and new results on the observation of the exotic baryon, the Θ^+.

INTRODUCTION

The Continuous Electron Beam Accelerator Facility (CEBAF) has been been operational since 1995 . The accelerator provides an electron beam up to 6 GeV which is separated and sent to three halls simultaneously. Halls A and C both are low acceptance double arm spectrometers restricted to an electron beam, while Hall B, which contains the CEBAF Large Acceptance Spectrometer (CLAS) [1], also has a bremsstrahlung induced photon tagging system [2] to permit both electron and photon beams. Hall A consists of two identical high resolution spectrometers with maximum momentum of 4 GeV/c , while Hall C consists of two symmetric focusing spectrometers, one with acceptance for high momentum particles, and the other for the detection of decay products. A schematic of the accelerator facility is shown in Fig. 1. Almost all of the hadron spectroscopy studies at CEBAF are performed at the CLAS facility, which began physics operations in 1997. CLAS is based on a six coil toroidal magnet which provides a primarily azimuthal field distribution, and is described in detail in Reference 1. Charged particle trajectory analysis from drift chambers [3] provides momentum resolution of 0.5% in the forward direction. A time-of-flight scintillator detector, and electromagnetic lead-scintillator sandwich calorimeters provide good particle identification, and a Cerenkov counter provides electron identification. Large acceptance is required to provide high reconstruction efficiency for multi-particle final states typical of hadronic reactions containing excited meson and baryon states. The luminosity of experiments using the tagged photon facility is limited to about 10^7 photons/second due to accidental coincidences with the tagger, while electron beam luminosities are limited by rates in drift chambers. A schematic of the detector is shown in Fig 2.

CP717, *Hadron Spectroscopy: Tenth International Conference,*
edited by E. Klempt, H. Koch, and H. Orth
© 2004 American Institute of Physics 0-7354-0197-7/04/$22.00

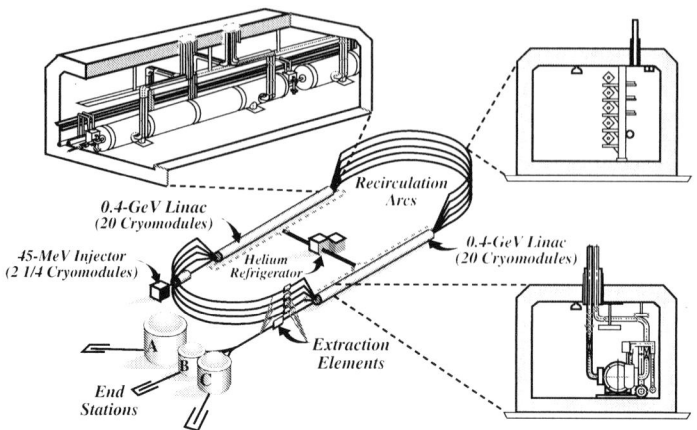

FIGURE 1.   The CEBAF accelerator site.

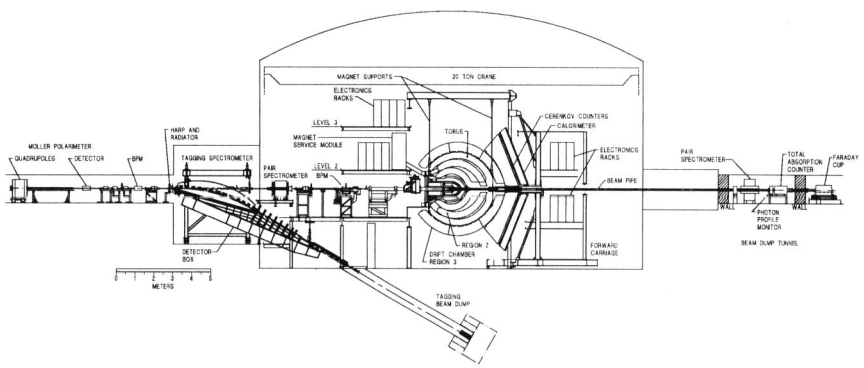

FIGURE 2.   Side view schematic of the CLAS detector.

THE FINAL STATE $\pi^+ \pi^- p$

Electromagnetic excitations of nucleon resonances, which are sensitive to the spin and spatial structure of the transition, give information on the properties of baryon structure. In the mass region above 1.6 GeV/c^2 there are many overlapping states, many of which are not well known. In addition, many of these states may decouple from single meson channels, and thus decay mainly to multi-pion final states, such as $\Delta \pi$ and $N\rho$, and therefore $N\pi\pi$. SU(6) X O(3) symmetry predicts many more states than have been observed experimentally: QCD mixing effects could decouple these states from the $N\pi$ channel, while strongly coupling them to the $N\pi\pi$ channel. CLAS has taken data on the reactions $ep \rightarrow e'p\pi^+\pi^-$ and $\gamma p \rightarrow p\pi^+\pi^-$. In the electroproduction reaction, the data was divided into three bins of Q^2: 0.5-0.8 (GeV/c)^2, 0.8-1.1 (GeV/c)^2, and (1.1-1.5) (GeV/c)^2. The data was corrected for acceptance, including reconstruction efficiency, and radiative effects, and binned in the following center-of-mass (CM) variables: $p\pi^+$

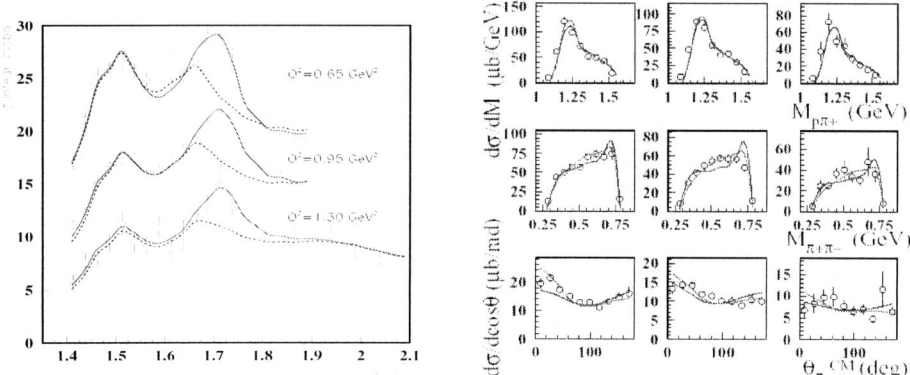

FIGURE 3. Left: Total cross section for $ep \to e'p\pi^+\pi^-$ for three different values of Q^2. Right: Differential cross sections from CLAS for the W bin 1.7 - 1.725 GeV/c^2 for the three different Q^2 intervals. The blue line corresponds to a fit where the hadronic parameters and position of the $D_{13}(1700)$ were allowed to vary. The red line corresponds to a fit where the photocouplings of all three states: $D_{13}(1700)$, $P_{13}(1720)$, and $P_{11}(1710)$ were allowed to vary, keeping the hadronic couplings fixed.

invariant mass, $\pi^+\pi^-$ invariant mass, π^- solid angle, , and y, the angle between the $p\pi^+$ and the hadronic plane. A model [5, 6] was used to describe the reaction in the kinematic range of interest as a sum of amplitudes of all possible production mechanisms of $\Delta\pi$ and $\rho\pi$. All other production mechanisms of $p\pi^+\pi^-$ were parameterized as phase space. A detailed treatment was developed for the non-resonant contributions to $\Delta\pi$ [5], while $\rho\pi$ non-resonant contributions were described through a diffractive ansatz. The model included twelve resonances listed in Table 1. The data was fit to this model in two steps, first fitting the non-resonant components keeping the resonant parameters within allowed values, and then fixing the non-resonant part and varying the resonant parameters. Generally the data fit the model very well, except for $W \sim 1700$ MeV. Here various fits were performed, described in Fig. 3 and Ref. [4]. We conclude that phenomenological calculations using existing PDG parameters provide poor agreement with our data. The solution is either to dismiss previously established hadronic parameters for the $P_{13}(1720)$, or introduce a new state with $J^P = \frac{3}{2}^+$ determined from the fit to the data, with a strong $\Delta\pi$ coupling, and small ρN coupling.

TABLE 1. Resonant states included in the fit of $ep \to e'p\pi^+\pi^-$.

State	J^P	State	J^P	State	J^P
$P_{11}(1440)$	$\frac{1}{2}^+$	$D_{13}(1520)$	$\frac{3}{2}^-$	$S_{11}(1535)$	$\frac{1}{2}^-$
$S_{11}(1650)$	$\frac{1}{2}^-$	$D_{15}(1675)$	$\frac{5}{2}^-$	$F_{15}(1680)$	$\frac{1}{2}^+$
$P_{13}(1720)$	$\frac{3}{2}^+$	$D_{13}(1700)$	$\frac{3}{2}^-$	$S_{31}(1620)$	$\frac{1}{2}^-$
$D_{33}(1700)$	$\frac{3}{2}^-$	$F_{35}(1905)$	$\frac{5}{2}^+$	$F_{37}(1950)$	$\frac{7}{2}^+$

CLAS has also investigated this final state through photoproduction. In this case the tagged photon beam was induced from a 2.4 GeV electron beam, which provided

TABLE 2. Baryon waves included in the fit of $\gamma p \rightarrow p\pi^+\pi^-$.

J^P	M	Isobars
$\frac{1}{2}^+$	$\frac{1}{2}$	$\Delta^{++}\pi^-, \Delta^0\pi+$
$\frac{1}{2}^-$	$\frac{1}{2}$	$\Delta\pi, pp$
$\frac{3}{2}^+$	$\frac{1}{2}, \frac{3}{2}$	$(\Delta\pi)_{l=1}, pp_{s=\frac{1}{2}}, (pp)_{s=\frac{3}{2}; l=1,3}, N^*(1440)\pi$

photon energies between 480 MeV to 2.28 GeV [2]. The analysis performed was a full partial wave analysis in the isobar model, and is described in detail in [7]. The observed intensity distribution is described as a square of appropriately summed amplitudes.

$$I(\tau) = \sum_\alpha \sum_\beta |^\alpha \psi_\beta(\tau)|^2 \qquad (1)$$

τ represents the set of variables necessary to define the configuration of the final state, and α represents non-interfering terms in the expansion, for example, the spins of initial state or final state particles, and β represents the partial wave decomposition. The processes considered in this case include s-channel baryon resonance production, with decay modes $\Delta\pi$ and ρp, as well as t-channel ρ production. Generally, the amplitudes $^\alpha\psi_\beta(\tau)$ can be factored into the product of a production amplitude and a decay amplitude for the given partial wave:

$$^\alpha\psi_\beta(\tau) = {}^\alpha V_\beta^\alpha A_\beta(\tau) \qquad (2)$$

Given the above description of the intensity distribution and an ensemble of experimental events, a likelihood function for the given experimental data set can be formed:

$$L = \left[\frac{\bar{n}^n}{n}e^{-\bar{n}}\right]\prod_i^n\left[\frac{I(\tau_i)}{\int I(\tau)\eta(\tau)d\tau}\right] \qquad (3)$$

In practice, the logarithm of the likelihood function is maximized by varying the production amplitudes. The integral in Eq. 3 is calculated by Monte Carlo methods, as the acceptance function $\eta(\tau)$ in general is not known in an analytic form. CLAS is simulated by a detailed GEANT model [1].

In the case of the reaction $\gamma p \rightarrow p\pi^+\pi^-$, separate, independent fits were done on the data in 20 MeV bins of photon beam energy. The waves included in the fit are listed in table 2. In Fig. 4 a comparison between the fit results and various differential cross sections for a sample bin, $1.72 < W < 1.75$ (GeV/c^2) is shown, and demonstrates that the fit provides a very good description of all of the relevant differential cross sections. Fig. 5 show the partial wave decomposition of the $J^P = \frac{5}{2}^+$ wave. One can clearly see the $\Delta\pi$ decay mode of the $N^*(1680)$. Fig. 6 shows the P_{13} and P_{33} waves: the $N^*(1720)$ and $\Delta(1600)$ can be seen in the partial wave decomposition.

The PWA fit can then be integrated over the entire phase space to determine the acceptance corrected total cross section. Figure 7 shows the total cross section measured in CLAS and compared to the ABBHHM [11] and CEA [10] results, as well as the various physics components.

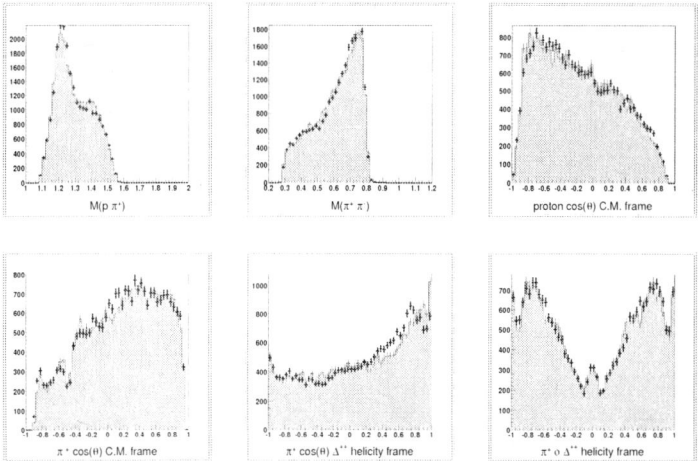

FIGURE 4. Comparison of the PWA fit to the data for the bin $1.72 < W < 1.75$ (GeV/c^2) for the reaction $\gamma p \to p\pi^+\pi^-$ The data is shaded, and the blue points are results of the fit.

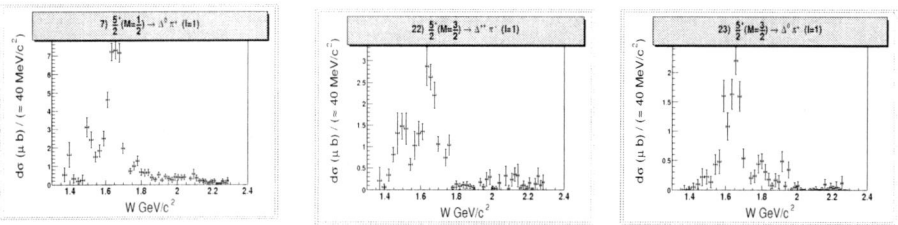

FIGURE 5. PWA results for the $J^P = \frac{5}{2}^+$ waves in the reaction $\gamma p \to p\pi^+\pi^-$.

$\gamma P \to \pi\pi\pi N$

The CLAS collaboration has also studied the reactions $\gamma p \to \pi\pi\pi N$ at 5 GeV. Photoproduction is expected to be an excellent way to produce the lowest lying $J^{PC} = 1^{-+}$ exotic mesons. BNL has observed the $\pi_1(1600)$ decaying to the final state $\pi^+\pi^-\pi^-$ [8][9].

In August and September in 2001 CLAS took data with a real photon beam with an energy range of 4.8 GeV to 5.47 GeV. The γ flux was $5 \times 10^6/sec$ into an 18 cm liquid hydrogen target. The total raw sensitivity was 2.7 events/pb.

The partial-wave analysis of these data was performed using the same program developed for the baryon analysis, but in this case including various meson waves produced in the t-channel and decaying to $\pi\pi\pi$. Each event is considered in the framework of an isobar model: an initial decay of a parent particle into a $\pi\pi$ isobar and an unpaired pion followed by the subsequent decay of the isobar. Each partial wave is characterized by the quantum numbers $J^{PC}[isobar]LM^\varepsilon$ — here J^{PC} are spin, parity and C-parity of the partial wave; M is the absolute value of the spin projection on the quantization axis; L is the orbital angular momentum between the isobar and the unpaired pion. In this case

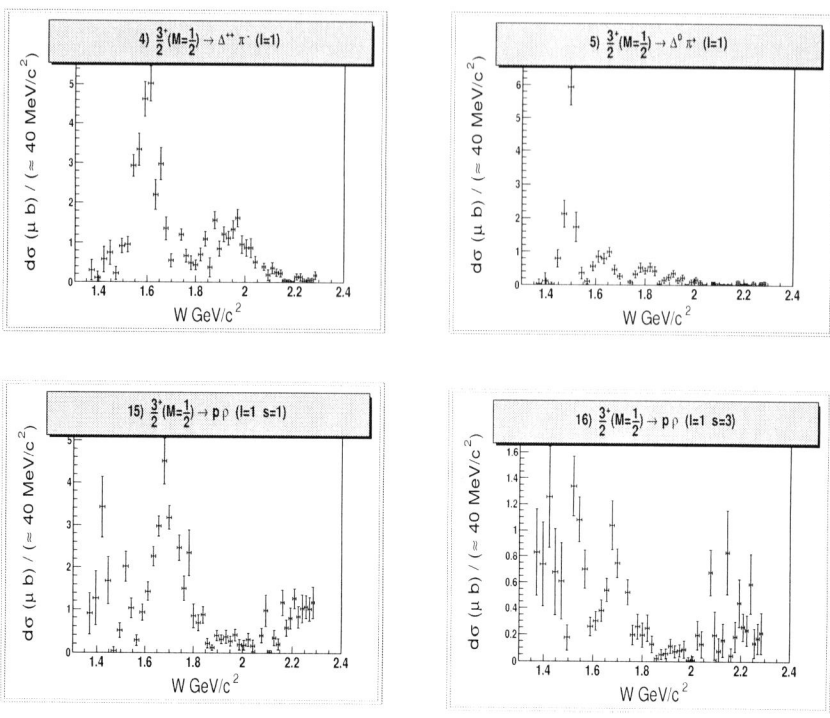

FIGURE 6. PWA results for the $J^P = \frac{3}{2}^+$ waves in the reaction $\gamma p \to p\pi^+\pi^-$.

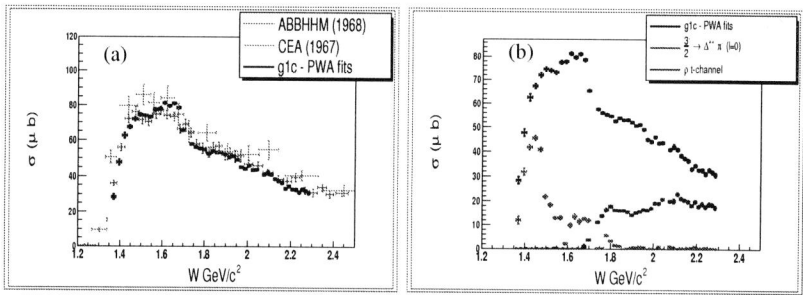

FIGURE 7. (a) Total cross section of the reaction $\gamma p \to \pi^+\pi^- p$ measured at CLAS, and compared to the ABBHHM and CEA results.(b) Various components of the total photon cross section from the PWA fits.

the spin-density matrix is expressed as eigenfunctions of the reflection operator through the reaction plane, ε [12]. In the most general case, the rank of the spin-density matrix can be four, in practice fits were limited to rank one. Relativistic Breit-Wigner functions with standard Blatt-Weisskopf factors were used in the description of the $\rho(770)$, and $f_2(1270)$ isobars; the $\pi\pi$ S-wave, σ, was parametrized according to Au, Morgan, and Pennington [13].

For the reaction $\gamma p \to \pi^+\pi^-\pi^0 p$ the three charged particles were measured in CLAS, while the π^0 was identified and measured from missing four-momentum. This reaction has considerable background from baryon resonances decaying to $p\pi$ through t-channel exchange processes. For the partial wave analysis, this background was included as an interfering background, with the baryon resonance decay modeled as an S-wave decay. More sophisticated descriptions of the background did not significantly improve the results. A total of 28 meson waves and 21 background (baryon) waves were included, and are listed in Table 3. Many other waves were tried in the partial wave fits, and were determined to be very small. The resulting partial wave analysis of the $\pi\pi\pi$ for the isoscalar 1^{--}, the exotic isovector 1^{-+}, isovector 1^{++}, and isovector 2^{++} waves are shown in Fig. 8. There is clear evidence for the photoproduction of the $a_1(1260)$ as well as the $\omega(1650)$. There is some evidence for production of the $a_2(1320)$ as well. In addition, there is a strong signal that corresponds in mass and width to the exotic $\pi_1(1600)$.

TABLE 3. Waves included in the fit of $\gamma p \to \pi^+\pi^-\pi^0 p$

J^{PC}	M^ε	L	Isobars
1^{++}	$0^+, 1^\pm$	0,2	$\rho(770)$
1^{--}	$0^-, 1^\pm$	1	$\rho(770)$
1^{-+}	$0^-, 1^+$	1	$\rho(770)$
2^{++}	$0^-, 1^+$	2	$\rho(770)$
2^{-+}	$0^+, 1^\pm$	0	$f_2(1270)$
2^{-+}	0^+	2	$f_2(1270)$
2^{-+}	$0^+, 1^\pm$	1,3	$\rho(770)$
2^{+-}	0^-	2	$\rho(770)$
3^{--}	$0^-, 1^+$	3	$\rho(770)$
3^{++}	$0^-, 1^\pm$	1	$f_2(1270)$
4^{++}	$0^-, 1^-$	3	$f_2(1270)$

From the same data run the reaction $\gamma p \to \pi^+\pi^+\pi^- n$ was studied. The three charged pions were measured in CLAS, while the neutron was identified and measured from missing four-momentum. Laboratory angle cuts on the pions, as well as selecting low t events, greatly reduces the baryon resonance background. Fig. 9 shows various distributions of the data: the t' distribution, the missing neutron mass, and di-pion effective masses.

The PWA results are shown in Fig. 10; waves included in the fit are listed in Table 4. While these results are very preliminary, there is a very clear signal for the $a_2(1320)$. There is some evidence for photoproduction of the $a_1(1260)$ and also the $\pi_2(1670)$. There is some strength in the exotic $J^{PC} = 1^{-+}$ partial wave near 1600 MeV/c^2, but it is not conclusive.

$$\Theta^+$$

The existence of an exotic S=+1 baryon has been suggested by several recent experiments [14]. CLAS has investigated the reaction $\gamma d \to K^+ K^- p(n)$ where the final state neutron is reconstructed from missing four-momentum. If the proton is a spectator in the

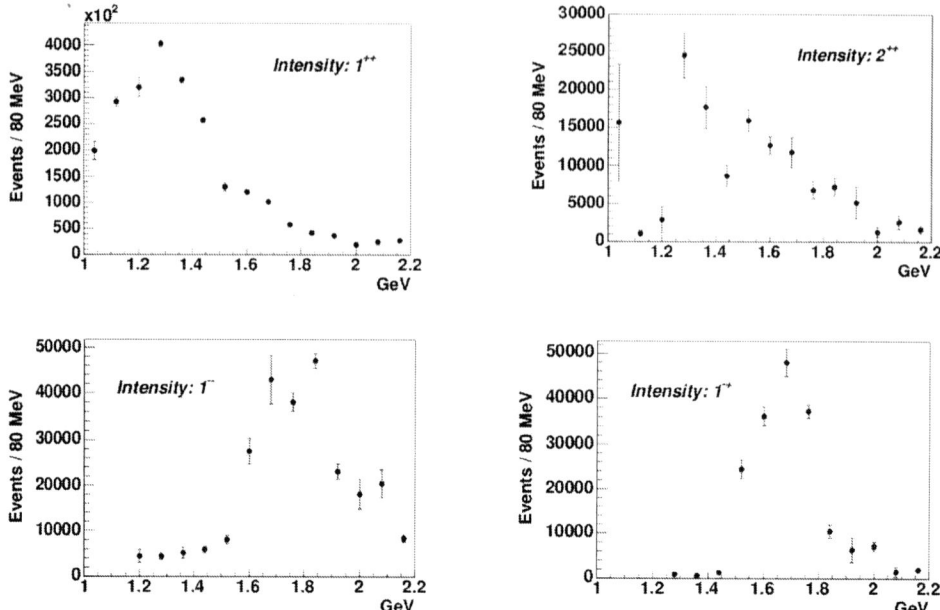

FIGURE 8. Partial wave decomposition of the reaction $\gamma p \rightarrow \pi^+\pi^-\pi^0 p$ (a) $J^{PC} = 1^{++}$ isovector, (b) $J^{PC} = 2^{++}$ isovector (c) $J^{PC} = 1^{--}$ isoscalar (d) $J^{PC} = 1^{-+}$ isovector

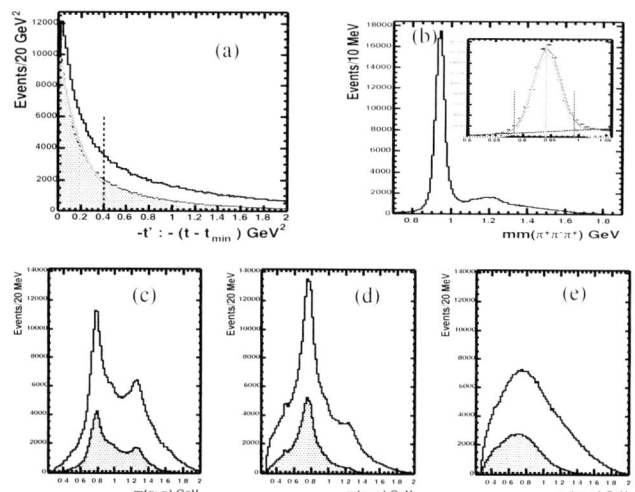

FIGURE 9. Various distributions of the reaction $\gamma p \rightarrow \pi^+\pi^+\pi^- n$ at 5.2 GeV. (a) t' distribution. Events to the left of the dotted line were selected for the PWA. The shaded area has cuts on the pion laboratory angles described in the text. (b) Missing mass off of $\pi^+\pi^+\pi^-$ showing the missing neutron. (c) Mass of the π^- and slow π^+ showing both the ρ and $f_2(1270)$ isobars. (d) Mass of the π^- and fast π^+. (e) Mass of $\pi^+\pi^+$. (c),(d), and (e) shaded areas are events selected for the final partial wave analysis.

TABLE 4. Partial waves included in the fit of $\gamma p \rightarrow \pi^+\pi^+\pi^- n$

J^{PC}	m^ε	L	Isobar	# Waves
0^{-+}	0^+	0	σ	1
0^{-+}	0^+	1	$\rho(770)$	1
1^{++}	$0^+, 1^\pm$	0,2	$\rho(770)$	6
1^{++}	$0^+, 1^\pm$	1	σ	3
1^{-+}	$0^-, 1^\pm$	1	$\rho(770)$	3
2^{++}	$0^-, 1^\pm, 2^\pm$	2	$\rho(770)$	5
2^{-+}	$0^+, 1^\pm$	1,3	$\rho(770)$	6
2^{-+}	$0^+, 1^\pm$	2	σ	3
2^{-+}	$0^+, 1^\pm$	0,2	$f_2(1270)$	6
Background				

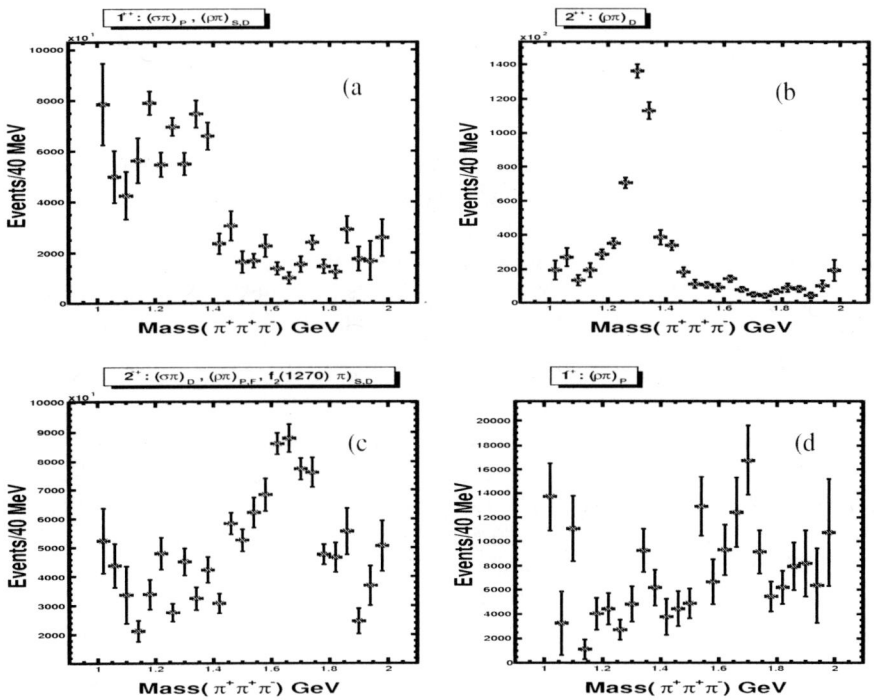

FIGURE 10. Partial wave decomposition of the reaction $\gamma p \rightarrow \pi^+\pi^+\pi^- n$. (a) $J^{PC} = 1^{++}$ partial wave. (b) 2^{++} partial wave. (c) 2^{-+} wave. (d) Exotic 1^{-+} wave.

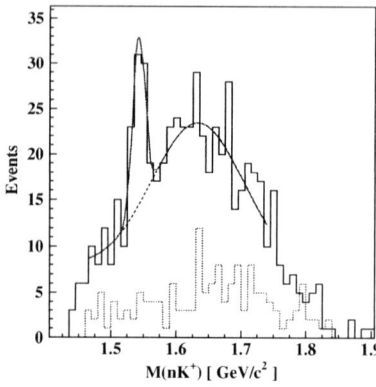

FIGURE 11. Invariant mass of the nK^+ showing a sharp peak at 1.542 GeV/c^2. A fit (solid line) to the peak with a smooth background (dashed line) gives a statistical significance of 5.8 σ.

interaction, its Fermi momentum is not sufficient to have it detected in CLAS. In some fraction of the events, however, both the neutron and proton partake in the interaction, perhaps through rescattering. The data presented here were taken with the CLAS tagged photon beam, with the incident electron at 2.474 and 3.115 GeV. The tagged photon flux was approximately $4 \times 10^6 \gamma$ per second. The integrated tagged photon flux above 1.51 GeV was 2.34×10^{12}. The beam was incident on a 10 cm liquid deuterium target.

Events were selected with the neutron momentum above 80 MeV/c. Known resonances, the $\phi(1020)$ and $\Lambda(1520)$ were removed from the analysis. Monte carlo simulations indicated that for Θ^+ events the K^+ momentum rarely exceeded 1.0 GeV/c, and events with $p_{K^+} > 1.0$ GeV/c were removed. The resulting invariant mass of the neutron and K^+ is shown in Fig. 11, and shows a 5.8 standard deviation peak over a smooth background at a mass of 1.542 GeV/c^2.

The CLAS collaboration has also searched for the Θ^+ baryon in γp interactions. In three separate data runs, with slightly different running conditions, CLAS took data with an 18 cm liquid hydrogen target and photon energies greater than 3.2 GeV. Two data runs (labeled a and b) had identical geometrical acceptance and trigger conditions, but different beam energies. Run a had a photon beam energy range 3.2–3.95 GeV, run b 3–5.25 GeV, and run c 4.8–5.47 GeV. Run c triggered on the events with at least 2 out of 6 CLAS sectors having signals, while runs a and b triggered on events with hits in opposite sectors. Runs a and b have the hydrogen target in the standard position, but in run c the target was moved upstream by 1 meter to improve CLAS acceptance in the forward direction. The estimated combined a and b integrated luminosity is about 2 events/pb, and run c has about 2.7 events/pb.

Events having a π^+, K^+, and K^- in the final state were selected for the analysis of the reaction $\gamma p \rightarrow \pi^+ K^- K^+ n$. The missing mass distribution off of the $\pi^+ K^+ K^-$ shows a very clear neutron peak in each data set; these events were selected for further analysis.

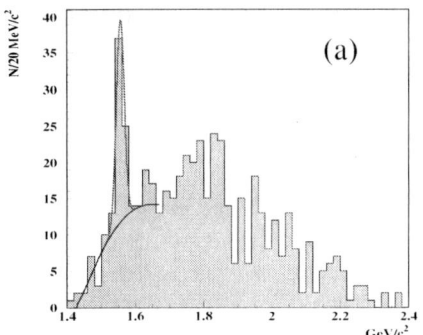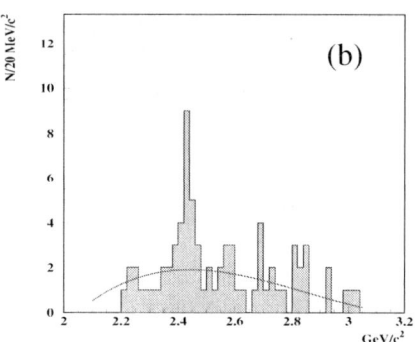

FIGURE 12. (a) The M_{nK^+} invariant mass spectrum in the reaction $\gamma p \to \pi^+ K^- K^+ (n)$ with the cut $cos\theta^*_{\pi^+} > 0.8$ and $cos\theta^*_{K^+} < 0.6$. $\theta^*_{\pi^+}$ and $\theta^*_{K^+}$ are angles between the π^+ and K^+ mesons and photon beam in the center of mass system. The background function we used in the fitting was obtained from the phase space simulation we completed. (b) The $M_{nK^+K^-}$ invariant mass spectrum calculated from the missing mass off of the π^+ in the reaction $\gamma p \to \pi^+ K^- K^+ (n)$ with the cuts $cos\theta^*_{\pi^+} > 0.8$ and $cos\theta^*_{K^+} < 0.6$. $\theta^*_{\pi^+}$ and $\theta^*_{K^+}$ are angles between the π^+ and K^+ mesons and photon beam in the center of mass system. The events in this plot have $M(K^+n)$ between 1.54 and 1.58 GeV/c^2. The shape of the background curve was obtained from our phase space simulation.

$\phi(1020)$ events were removed from the data sample by cutting events with $K^+ K^-$ effective mass less than 1.06 GeV/c^2.

Two angular cuts in the center of mass system were applied to extract the signal and suppress the background. Structure in the region of 1.55 GeV/c^2 appears after cutting $cos\theta^*_{\pi^+} > 0.8$, where $\theta^*_{\pi^+}$ is the center-of-mass angle between π^+ and the photon beam, and corresponds roughly to $-t < 0.28$ GeV/c^2 in our beam energy range, where $t = (k-p)^2$, k is the photon 4-momentum, and p is the pion 4-momentum. The reaction $\gamma p \to \pi^+ K^- K^+ n$ is dominated by meson resonance production decaying to $K^+ K^- \pi^+$ and small momentum transfer to the proton, and the excitation of baryon resonances decaying to $n\pi^+$. These processes have the K^+ moving forward in the center of mass system. To suppress such backgrounds, a cut was applied to select events having a positive kaon going in the backward direction with $cos\theta^*_K < 0.6$. The θ^+ peak was clearly observed in each of the three data sets; the resulting nK^+ mass spectrum were combined and are shown in Fig. 12(a).

The nK^+ effective mass distribution was fitted by the sum of a Gaussian function and a background function obtained from phase space simulation. The fit parameters are: $N_{events} = 41 \pm 10$, $M = 1555 \pm 1$ MeV/c^2, $\sigma = 11 \pm 3$ MeV/c^2. The mass scale uncertainty is estimated to be ± 10 MeV/c^2. This uncertainty is mainly due to the momentum calibration of the CLAS detector and the photon beam energy calibration. The statistical significance for the fit in Fig.12 calculated over a 40 MeV/c^2 mass window is 7.8 σ. The fact that the angular cuts we applied enhanced the Θ^+ signal suggests the possible production of a N^*/Δ^* that decays to Θ^+ and K^-; if the Θ^+ is an isoscalar, only an intermediate N^* ia possible. For the events with nK^+ effective mass between 1.54 and 1.58 GeV/c^2, the missing mass off of the π^+ is shown in Fig. 12(b).

SUMMARY

CLAS at Jefferson Laboratory has a continuing rich program in hadron spectroscopy. New high-quality data from the electroproduction and photoproduction of multi-meson decays of baryon resonances, eg. $p\pi^+\pi^-$ is currently being analyzed. Such studies will shed light on the missing baryon question. In addition, CLAS is currently analyzing data from the photoproduction of mesons. Initial partial wave analysis results are now becoming available. New proposals for high statistics real photon beam experiments, as well as near-real photons using small angle electron scattering, which produces linearly polarized photons, are in preparation. In addition, CLAS, in γd and γp interactions, has contributed significantly to the rapidly expanding study of exotic penta-quark states such as the $\Theta^+(1540)$.

ACKNOWLEDGMENTS

I wish to thank all of my collaborators at CLAS for their help with this talk, as well as the extraordinary efforts of the CEBAF staff which have made this program possible. This work is supported by the Department of Energy under contract DE-AC05-84ER40150.

REFERENCES

1. B. A. Mecking, *et al.* **Nucl. Instrum. Meth. A503**,513-553 (2003).
2. D. I. Sober, *et al.* **Nucl. Instrum. Meth. A440**, 263 (2000).
3. M. D. Mestayer, *et al.* **Nucl. Instrum. Meth. A449**, 81 (2000).
4. M. Ripani, *et al. Phys. Rev. Lett. 91*, No. 2:022002-1 (2003).
5. M. Ripani, *et al.* **Nucl. Phys. A672**, 220 (2000).
6. V. Mokeev,*et al.* **Phys. At. Nucl 64**, 1292 (2001).
7. John P. Cummings and Dennis P. Weygand, "An Object-Oriented Approach to Partial Wave Analysis", arXiv:physics/030952 (submitted to **Nucl. Instrum. Meth**) (2003).
8. G. S. Adams, *et al. Phys. Rev. Lett. 81*:5760-5763 (1998).
9. S. U. Chung, *et al.* **Phys. Rev. D65**:072001 (2002).
10. H. R. Crouch, Jr., *et al.*, **Phys. Rev.145**, 994-1000 (1966).
11. ABBHHM Collaboration, *Phys. Rev. Lett. 175*, 1669 (1968).
12. S. U. Chung and T. L. Trueman, **Phys. Rev. D11**, 633
13. K. L. Au, D. Morgan, and M. R. Pennington, **Phys. Rev. D35**, 1633 (1987).
14. T. Nakano, *et al.*, *Phys. Rev. Lett. 91*, 012002 (2003).
15. V. Barmin, *et al.* **Phys. At. Nucl 66**:1715-1718 (2003).

Spectroscopy experiments at ELSA

Christian Weinheimer

Helmholtz-Institut für Strahlen- und Kernphysik, University of Bonn, 53115 Bonn, Germany
Email: weinheimer@iskp.uni-bonn.de

Abstract. The spectroscopy of baryon resonances especially with high masses allows new insight into the properties of hadronic matter and to study the question of the relevant degrees of freedom in the region of non-perturbative QCD. The electron accelerator ELSA at Bonn provides electrons with energies up to 3.2 GeV which are used to produce unpolarized and polarized bremsstrahlung photons with energies up to 3.0 GeV for photoproduction experiments off the nucleon. The results of the 1997 and 1998 runs of the SAPHIR experiment are discussed. Preliminary results from the runs in 2000 and 2001 of the Crystal Barrel experiment at ELSA on single and double neutral meson production are presented. From the high statistics 2003 data of the Crystal Barrel-TAPS experiment asymmetries obtained with linearly polarized photons for the channel $\vec{\gamma}p \rightarrow p\eta\pi^0$ are shown.

INTRODUCTION

The spectrum and the properties of excited baryons reflect the behavior of QCD in the low energy regime. Unfortunately, for this low energy domain, the QCD Lagrangian cannot be solved by a perturbative expansion.

Except from some promising results from Lattice QCD calculations together with chiral perturbation theory extrapolations to real quark masses [1] constituent quark models have to be used to describe the baryon spectrum and properties. Common to these models is the use of a confining potential in combination with a short-range residual interaction. The latter differs between the various models, the most prominent examples are one-gluon exchange [2], Goldstone boson exchange [3] and instanton interactions [4]. Although these constituent quark models use somewhat different approaches, generally they are quite successful in describing the masses of low-lying resonances (see fig. 1). But fig. 1 also shows, that the constituent quark models exhibit in the low-mass region already a lot of short-comings: There are common difficulties to describe the low lying radial excitations of the nucleon or the $\Delta(1232)$, like the N(1440) P_{11} (not shown in fig. 1) or the $\Delta(1600)$ P_{33}, respectively. Another example are the negative parity delta resonances with masses around 1900 MeV $\Delta(1900)$ S_{31}, $\Delta(1940)$ D_{33}, $\Delta(1930)$ D_{35} which are predicted at significantly higher masses.

At even higher masses above 2 GeV, many of the predicted exited baryonic states seem not to be realized in nature: Have they not been observed in previous π scattering experiments or does this fact point to an underlying quark-diquark structure of the nucleon? In addition, there exist other poorly understood problems, like why exited states with similar quantum numbers do have so different decay channels, or why there are pairs of mass-degenerate resonances with opposite parity (*e.g.* see [8]).

CP717, *Hadron Spectroscopy: Tenth International Conference,*
edited by E. Klempt, H. Koch, and H. Orth
© 2004 American Institute of Physics 0-7354-0197-7/04/$22.00

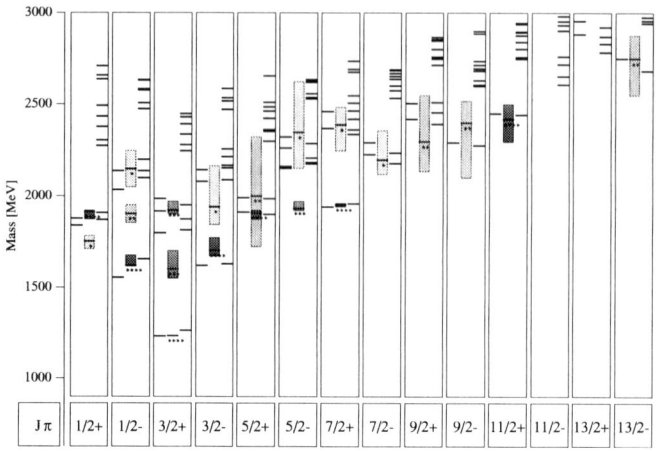

FIGURE 1. Comparison of established resonances according to the Particle Data Group [5] (middle lines, full boxes) with one-gluon exchange model (left lines) [2, 7] and the relativistic constituent quark model with instanton forces [4, 6] (right lines).

These open questions are being addressed at Bonn by the baryon spectroscopy experiments SAPHIR and Crystal Barrel at ELSA or CB-TAPS at ELSA, respectively. The former is optimized for charged decay products, whereas the latter experiments aim for the decays into multi-photon final states. Up to now the excitation of baryon resonances off the nucleon has been studied at Bonn by using bremsstrahlung photons produced by electrons from the accelerator ELSA.

This paper is structured as follows. In section 2 the electron accelerator ELSA and the photon tagging system are discussed briefly. The SAPHIR experiment and its recent results are reported in section 3. Section 4 is describing the Crystal Barrel at ELSA experiments and first preliminary results on single and double neutral meson production. First asymmetry data from the CB-TAPS at ELSA experiment are presented in section 5. The conclusions and an outlook are given in section 6.

THE ELECTRON ACCELERATOR ELSA AND THE PHOTON TAGGING SYSTEM

At the electron accelerator facility ELSA at Bonn unpolarized and polarized electrons from the two linacs are accelerated in a synchrotron to an energy of 1.6 GeV before they are extracted to the stretcher and synchrotron ring ELSA, in which the energy is further increased up to 3.5 GeV. By slow extraction a continuous beam of unpolarized or polarized electrons is provided for the experiments SAPHIR (up to 1998), Crystal Barrel at ELSA (2000-2001) and CB-TAPS at ELSA (2002-2003), respectively, and GDH (up to 2003).

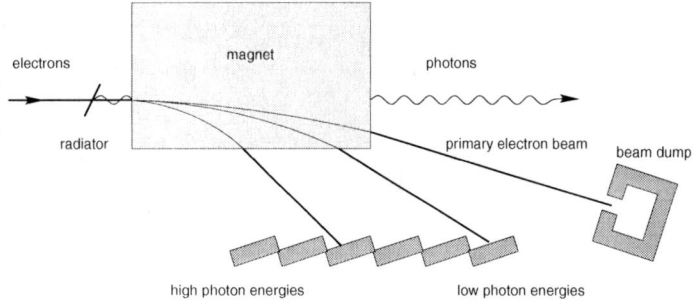

FIGURE 2. The energy-tagged real photon beam at ELSA consisting of a radiator target, a dipole magnet and position resolving detectors. Unscattered electrons are bend into the beam dump.

From these electrons real photons are created by bremsstrahlung in a thin radiator target. The momentum of each scattered electron is determined by a magnetic spectrometer consisting of a large dipole and position resolving detectors consisting of scintillating fiber hodoscopes and wire chambers (see figure 2). Knowing the primary electron beam energy E_0 and the energy of the scattered electron E_e the energy of each photon E_γ is determined by $E_\gamma = E_0 - E_e$. The large acceptance of the position resolving detectors provides a tagging range of $0.25E_0 < E_\gamma < 0.95E_0$ with a photon intensity of up to 10^7 1/s.

In addition to the amorphous radiator, from 2002 a well-oriented diamond crystal has been used as bremsstrahlung target. The crystal structure gives rise to coherent bremsstrahlung at certain photon energies. This coherent radiation is linearly polarized. The orientation of the crystal defines the maxima of polarization and intensity as function of photon energy [9].

THE SAPHIR EXPERIMENT AND RECENT RESULTS

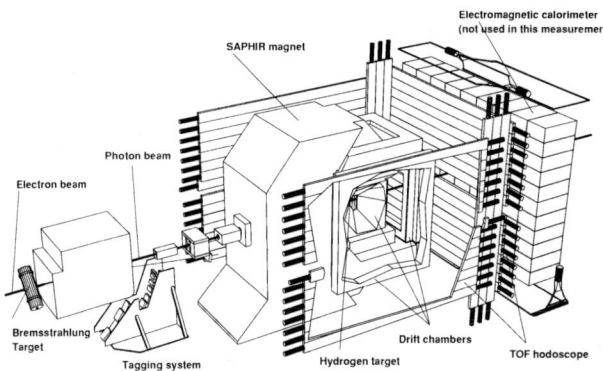

FIGURE 3. Schematic picture of the SAPHIR detector, for details please see text.

Fig. 3 shows the SAPHIR detector, which was operating from 1988 to 1998 at ELSA. It consists of a liquid hydrogen target surrounded by a cylindrical drift chamber in an open dipole magnet. In forward direction additional drift chambers improved the momentum resolution. Scintillator hodoscopes were used for triggering and timing purposes.

Recently the analysis of the last SAPHIR runs of 1997 and 1998 has been finished and submitted for publication. The omega photoproduction data of the channel $\gamma p \to p\omega \to \omega \pi^+ \pi^- \pi^0$ [10] were analyzed within a coupled-channel analysis [11], which describes the threshold region by a strong contribution from N(1710) P_{11}. If this could be confirmed, it would be of special interest, since this state – although having non-exotic quantum numbers – is being interpreted as a member of the antidecuplett of pentaquark states [12].

The analysis of the hyperon production data of the 1997 and 1998 runs show indications for new resonances [13]. As the previous SAPHIR data the new $\gamma p \to K^+ \Lambda$ data show indications for a new resonance around 1900 MeV. An analysis of the previous SAPHIR data suggested – not without controversal discussion – a state N(1895) D_{13} [14]. The $\gamma p \to K^+ \Sigma^0$ cross section exhibits a clear peak at a photon energy of 1450 MeV. The analysis of previous SAPHIR data suggested the existence of a resonance $\Delta(1900)$ S_{31} or a $\Delta(1910)$ P_{31}.

The SAPHIR collaboration has analyzed their data with respect to the exotic state Θ^+. Evidence for the reaction $\gamma p \to K_s^0 \Theta^+ \to \pi^+ \pi^- n K^+$ has been found [15] and reported at this conference [16].

THE CRYSTAL BARREL AT ELSA EXPERIMENT AND PRELIMINARY RESULTS

The Crystal Barrel at ELSA experiment (see figure 4) started in 2000. It consists of a a LH$_2$ cell of 5 cm length and 3 cm diameter as reaction target, which is surrounded

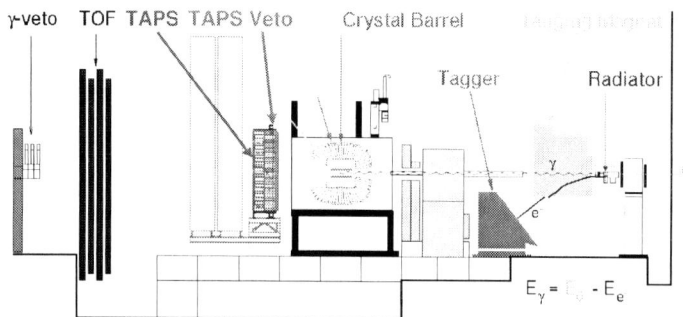

FIGURE 4. The Crystal Barrel-TAPS experiment at ELSA/Bonn. Before the TAPS detector was installed, in 2000 and 2001 the forward solid angle was closed by additional 90 CsI crystals ($12^o < \theta < 30^o$) and by a time-of-flight spectrometer ($\theta < 12^o$). Please see text for more details.

by three layers of scintillating fibers built to detect and to trigger on charged particles leaving the target (proton trigger). In addition, it provided an intersection point of a particle's trajectory with the detector and hence helped to identify clusters of charged particles in the barrel. The target region is surrounded by the Crystal Barrel calorimeter, which consists of 1380 CsI crystals covering 98 % of the full solid angle.

The large angular coverage together with the energy resolution of $\sigma(E_\gamma)/E_\gamma = 2.5$ % at 1 GeV and the angular resolution of 1.2^o for photons favors the CB-ELSA experiment for the detection of multi-photon final states with high efficiency and resolution. This allows not only to investigate single meson production like $\gamma p \to p\pi^0$ and $\gamma p \to p\eta$ but also to study subsequent decays of baryon resonances by single meson emission. Of special interest are the reactions $\gamma p \to p\pi^0\pi^0$ and $\gamma p \to p\eta\pi^0$. The latter has the advantage that a single isoscalar η meson connects only states of same isospin: *E.g.*, a baryon resonance which decays into η and $\Delta(1232)$ (the latter can be identified by its subsequent decay into $p\pi^0$) has to be a Δ^*.

In 2000 and 2001 data have been taken with electron energies of $E_0 = 1400$ MeV and $E_0 = 3200$ MeV of 3 week running time each. The good energy resolution and high background suppression for single neutral mesons decaying into photons is illustrated in fig. 5. Preliminary differential cross sections for $\gamma p \to p\pi^0$ and $\gamma p \to p\eta$ have been reported at this conference [17]. In the former case the data show nice agreement with the SAID predictions up to 2 GeV. At higher energies the resonance contributions decrease and a strong forward peaking indicates t-channel exchange. Fig. 6 shows the nice agreement of the Crystal Barrel at ELSA η photoproduction data with the two models SAID and MAID for photon energies up to 1.5 GeV and with other experiments for photon energies up to 2 GeV At energies above 2 GeV the data exhibit a strong forward peaking pointing to t-channel exchange, some enhancement in backward direction may indicate u-channel exchange. The high signal-to-background ratio (see fig. 5) did not require background subtraction in the case of π^0 photoproduction, for the η photoproduction data the background contribution has been determined from side-bins.

Fig. 7 shows the invariant $\gamma\gamma$ mass against the invariant mass of the 2 other photons in events with 4 photons in the final states. Clear $\gamma p \to p\pi^0\pi^0$ and $\gamma p \to p\eta\pi^0$ signals are visible.

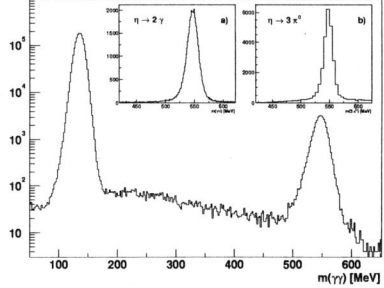

FIGURE 5. Invariant two photon masses after kinematical fit. The energy resolutions are $\sigma(\pi^0) = 8.5$ MeV and $\sigma(\eta) = 12$ MeV, respectively. The inserts show, that η mesons are reconstructed from $\eta \to 2\gamma$ (a) and $\eta \to \pi^0\pi^0\pi^0$ (b).

dσ/dΩ [μb/sr]

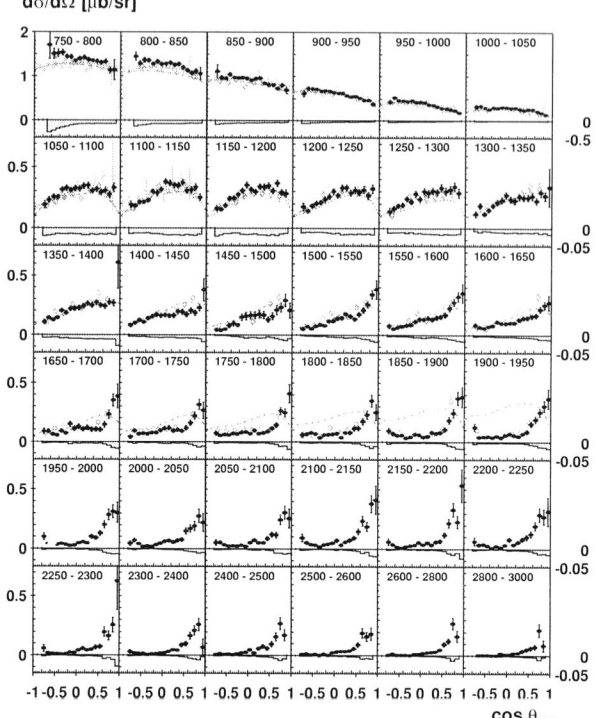

FIGURE 6. Preliminary differential cross sections for η photoproduction from CB at ELSA for photon energies from 750 MeV up to 3.0 GeV. For energies below 1.3 GeV, the photon flux was determined by a χ^2 fit to the π^0 photoproduction data to the SAID predictions. Above 1.3 GeV, normalization is taken from a measured photon flux scaled by a global factor of 0.75 in order to account for experimental uncertainties. Systematic errors are given as grey-shaded area at the bottom of each energy bin. Symbols indicate: CB-ELSA (■), TAPS (⋆), CLAS (○), GRAAL (△), MAID (dotted line), SAID (dashed line).

Fig. 8 shows the invariant mass of the $p\eta\pi^0$ system for $\gamma p \to p\eta\pi^0$ candidates (see also [18]). Selecting an energy range of 1800 MeV$< M(p\eta\pi^0)$ <2000 MeV a clear $\Delta(1232)$ production is seen in the invariant mass of the $p\pi^0$ subsystem (see fig. 9 left). Applying the same Δ-cut, the angular distribution of the η meson in the center of mass system (see fig. 9 right) shows some flat contribution, which may indicate the resonant production of a Δ^* state with a mass of about 1900 MeV. Requiring the invariant mass of the $p\eta\pi^0$ system somewhat higher (region iii of fig. 8 left) in addition to the $\Delta(1232)$ resonance higher mass contributions at about 1600 MeV become obvious in the $p\pi^0$ subsystem (see fig 8 right).

The application of an event-based partial wave analysis program [19, 20] to these data gave the following preliminary results: new resonances are required to describe the data. At least one new Δ state with a mass of about 2.2 GeV is needed. Unfortunately the solutions are ambiguous and therefore the open question of negative-parity Δ states around 1900 MeV cannot be answered yet. Polarization observables (s. next section) are

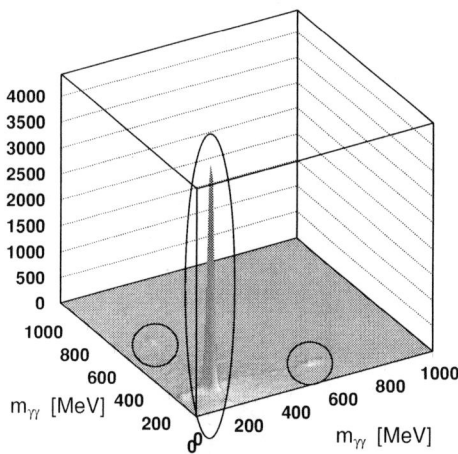

FIGURE 7. Invariant 2 photon masses in events with 4 photons in final states at $E_0 = 3.2$ GeV. The ellipses show the signal region for $\gamma p \to p\pi^0\pi^0$ (≈ 160000 events reconstructed) and $\gamma p \to p\eta\pi^0$ (≈ 22000 events reconstructed).

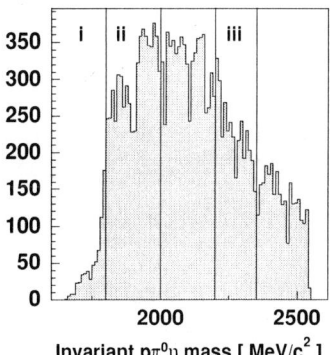 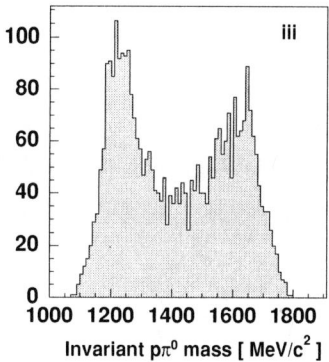

FIGURE 8. Invariant $p\eta\pi^0$ mass for $\gamma p \to p\eta\pi^0$ events at electron beam energy $E_0 = 3.2$ GeV (left) and invariant $p\pi^0$ mass for invariant $p\eta\pi^0$ masses of region iii of left plot (right).

needed to distinguish between the different contributing amplitudes.

The events with 4 photons in the final state from the run at $E_0 = 1.4$ GeV have been reconstructed. A preliminary partial wave analysis has been applied to the candidates for $\gamma p \to p\pi^0\pi^0$ [19]. The data were corrected for acceptance and normalized to the photon flux. Fig. 10 shows that the resulting preliminary total cross section is in reasonable agreement with data from other experiments. The preliminary result [19]

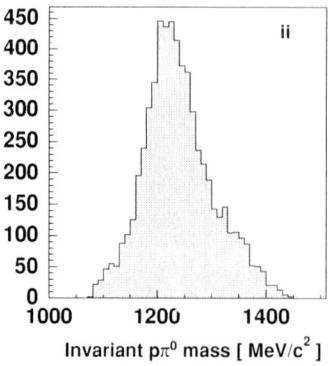

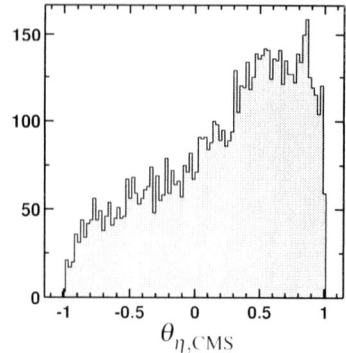

Invariant $p\pi^0$ mass [MeV/c^2]

$\theta_{\eta,CMS}$

FIGURE 9. Invariant $p\pi^0$ mass (left) and angular distribution of η meson in center of mass system (right) for $\gamma p \to p\eta\pi^0$ events at electron beam energy $E_0 = 3.2$ GeV and invariant $p\eta\pi^0$ masses of region ii of fig. 8 left.

of an event-based partial wave analysis confirms the parameters of known (***) and (****) resonances in the low-energy spectrum below 1.4 GeV. Dominant contributions are N(1520) D_{13}, Δ(1700) D_{33}, N(1680) F_{15} and N(1720) P_{13}. Contributions of the nucleon resonance N(1700) D_{13} are less dominant.

THE CB-TAPS AT ELSA EXPERIMENT AND FIRST RESULTS

End of 2001 the 90 most forward CsI crystals of the Crystal Barrel were removed and the TAPS detector [21] was installed (see fig. 4). The TAPS detector consists of 528

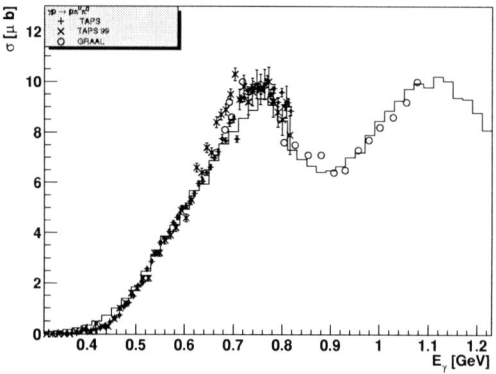

FIGURE 10. Preliminary total cross section (histogram) from partial wave analysis of $\gamma p \to p\pi^0\pi^0$ candidates at $E_0 = 1.4$ GeV [19] in comparison with results from the GRAAL and TAPS collaborations.

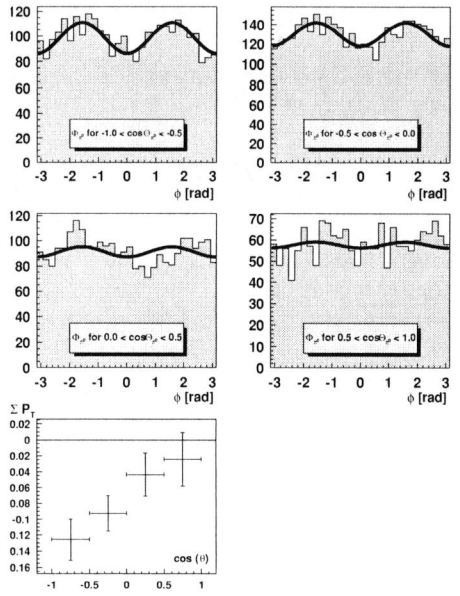

FIGURE 11. Preliminary Φ_{π^0} distributions for different Θ_{π^0} bins. The data are not acceptance corrected. The incoming photon energy is limited to 1440 MeV $\leq E_\gamma \leq$ 1640 MeV, i.e. the polarization maximum.

hexagonal BaF_2 crystals. The higher granularity and the faster timing improved the resolution, trigger and particle identification capabilities of the experiment. The data acquisition system was upgraded to allow for a higher photon flux at 8 times higher data taking rate. The freely oriented diamond crystal as radiator target allowed the use of linearly polarized photons. Longterm measurements with high statistics using liquid hydrogen, liquid deuterium and solid state targets started in autumn 2002.

In general, the use of linear polarization breaks the Φ symmetry. Thus, polarization allows a better determination of contributing amplitudes by adding further constraints in the PWA, e.g. small contributions may have large effects in certain polarization variables.

Fig. 11 shows the first photon asymmetries from the CB-TAPS period for $\vec{\gamma}p \to p\eta\pi^0$ candidate events prior to acceptance correction for energies around the maximum of the polarization P_T. In contrast to a two-body decay, in a three-body final state there is more than one asymmetry to observe. For illustration, from fig. 11 the asymmetry $P_T \cdot \Sigma(\pi^0)$ can be extracted for different $\Theta(\pi^0)$ bins. A clear angular depending asymmetry is visible (lower left), which will help to disentangle the ambiguous solution of the partial wave analysis.

CONCLUSION AND OUTLOOK

The experiments SAPHIR and Crystal Barrel (-TAPS) at the electron accelerator ELSA at Bonn have provided new data on photoproduction of baryon resonances.

The SAPHIR experiment has finished its analysis on ω and hyperon production. Indications for new resonances stemming from previous SAPHIR data seem to be confirmed. A signal for the exotic state Θ^+ has been reported. The Crystal Barrel experiment has preliminary results on photoproduction of single π^0 and η as well as of double π^0 and of $\pi^0\eta$. The former are compatible to previous experiments and models at low energy and expand the data base on photon energies up to 3 GeV. New data on $\gamma p \to p\pi^0\pi^0$ and $\gamma p \to p\pi^0\eta$ allowed preliminary partial wave analysis, confirming known resonances at low energies and requesting at least one new resonance at masses above 2 GeV. Linear polarization become available for the CB-TAPS experiment, first asymmetries look promising to disentangle the currently ambiguous partial wave solutions.

Beginning of 2004 TAPS will move to Mainz. The dismounted CsI crystals in forward direction will be reinstalled and upgraded by photomultiplier readout and fast plastic scintillator plates to detect charged particles in order to enhance the trigger and high-rate capabilities. The experiment will move to the former GDH position to allow for a polarized target and the use of polarized electrons to produce circular polarized photons as well. Data taking with single and double polarization observables will allow to further explore the high mass baryon resonance region. An upgrade with tracking of charged particles in a magnetic field as well as an electron scattering program are envisaged.

Acknowledgments: The detailed information and data obtained from the SAPHIR collaboration are acknowledged. The work of the author is supported by the grant WE 1843 of the Deutsche Forschungsgemeinschaft.

REFERENCES

1. Ch. Davies, *these proceedings*
2. S. Capstick and N. Isgur, Phys. Rev. **D34** (1986) 2809
3. L.Y. Glozman and D.O. Riska, Phys. Rept. **268** (1996) 263
4. U. Löring *et al.*, Eur. Phys. J. **A10** (2001) 309
5. K. Hagiwara *et al.* [Particle Data Group Collaboration], Phys. Rev. D **66** (2002) 010001.
6. U. Löring, B. C. Metsch and H. R. Petry, Eur. Phys. J. A **10** (2001) 395
7. S. Capstick and W. Roberts, Phys. Rev. **D47** (1993) 1994
8. E. Klempt, Phys. Lett. B **559** (2003) 144
9. D. Lohmann *et al.*, Nucl. Instr. Meth. **A343** (1994) 494
10. J. Barth *et al.*, Eur. Phys. J. **A18** (2003) 117
11. G. Penner and U. Mosel, Phys. Rev. **C66** (2002) 055212, nucl-th/0207069
12. Diakonov *et al.*, Z. Phys. **A359** (1997) 305
13. K. Glander *et al.*, Eur. Phys. J. **A**, *in print*, nucl-ex/0308025
14. C. Bennhold *et al.*, nucl-th/0008024
15. J. Barth *et al.*, Phys. Lett. **B572** (2003) 127, hep-ex/0307083
16. F. Klein, *talk at this conference Hadron 03*
17. O. Bartholomy, *these proceedings*
18. V. Credé, *these proceedings*
19. A. Anisovich, *these proceedings*
20. A. Sarantsev, *these proceedings*
21. B. Krusche and S. Schadmand, nucl-ex/03040800, *submitted to Prog. Part. Nucl. Phys.*

Status of Excited baryons at BES II

Xiaobin Ji

(Representing the BES Collaborations)

Institute of High Energy Physics, Beijing, China

Abstract. The πN system in $J/\psi \to \bar{N}N\pi$ is limited to be pure isospin 1/2 due to isospin conservation. This is a big advantage in studying $N^* \to \pi N$ from the J/ψ decays, compared with πN and γN experiments which mix isospin 1/2 and 3/2 for πN system. Based on the 58 million J/ψ data collected at the Beijing Electron Positron Collider, $J/\psi \to p\pi^-\bar{n}$ and other decay channels are analyzed. In the decay $J/\psi \to p\pi^-\bar{n}$, three peaks show up very clearly at 1400 MeV/c^2, 1500 MeV/c^2 and 1670 MeV/c^2 in the πN invariant mass spectrum with additional peak around 2030 MeV/c^2 which may be caused by a new N* resonance.

INTRODUCTION

BES is a large general purpose solenoid detector at the Beijing Electron Positron Collider(BEPC). BES II[2] is the upgrade of BES I[1]. At the end of April, 2001, BES had accumulated about 5.8×10^7 J/ψ data sample.

Baryon spectroscopy is important for understanding the internal structure of nucleon, but our understanding on baryon spectroscopy is still poor. A lot of theories[5] predict much more excited baryon states than we found. J/ψ decay has its own advantage to study baryon spectroscopy[4]. In the decay of $J/\psi \to N\bar{N}\pi$ or $N\bar{N}\pi\pi$, the isospin of $N\pi$ and $NN\pi$ system is limited to be 1/2, while in the πN scattering, the isospin may be 1/2 or 3/2. And many decay modes with a baryon-antibaryon pair proceed at a rate of order 10^{-3}.

N* IN THE DECAY $J/\psi \to P\pi^-\bar{N}$

Event selection

Anti-neutron cannot be detected directly in the BES detector. So in the events selection, two good charged tracks with opposite charge are required. The cosine of the polar angle of each charged track is limited within ± 0.8. For each charged track, the TOF and dE/dx measurements are used to calculate χ^2 values and the corresponding confidence levels($Prob_{pid}$) to the hypotheses that the particle is a pion, kaon and proton. The definition is:

$$\chi^2_{com} = \chi^2_{TOF} + \chi^2_{dE/dx}$$

$$Prob_{pid} = Prob(\chi^2_{com}, 2)$$

CP717, *Hadron Spectroscopy: Tenth International Conference*,
edited by E. Klempt, H. Koch, and H. Orth
© 2004 American Institute of Physics 0-7354-0197-7/04/$22.00

If only TOF or dE/dx has information, $Prob_{pid} = Prob(\chi^2, 1)$. The $p\pi$ invariant mass $m_{p\pi} > 1.15 \text{GeV}/c^2$ is required to remove the background from the $J/\psi \to \Lambda\bar{\Lambda}$. After these selection, the missing mass distribution is shown in the left figure of Fig. 1. Clear anti-neutron signal can be found and the background is relatively low.

$p\pi$ invariant mass spectrum

Since the background lies mainly at the high end of the missing mass distribution, so $m_{missing}$ is limited within 0.88 to 1.00 GeV/c^2 when we get $p\pi$ invariant mass spectrum. Fig. 1(right) shows the spectrum after the cut. Three clear peaks sitting around 1.50, 1.65 and 2.03 GeV/c^2, and the peak at 1.40 GeV/c^2 is also clear. According to PDG[3], the

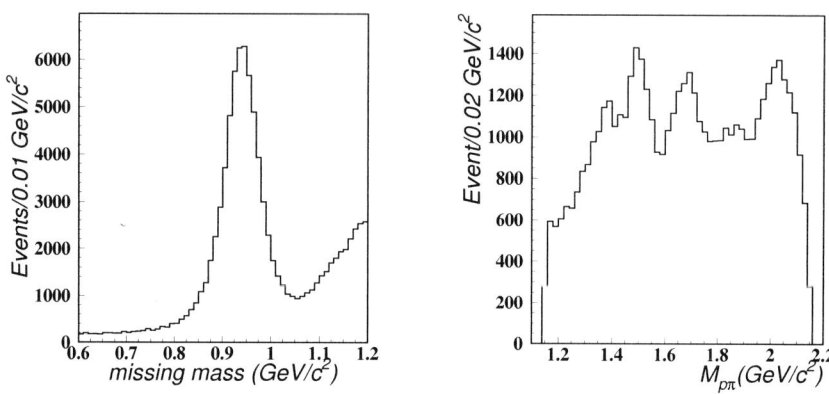

FIGURE 1. The missing mass(left) and invariant mass distribution of $p\pi^-$(right) in the decay $J/\psi \to p\pi^-\bar{n}$.

first peak may be caused by the known roper resonance N*(1440); two N* resonances, N*(1520) and N*(1535), may contribute to the second peak; three N* resonances lie around 1.67 GeV/c^2 or so. The peak around 2.03 GeV/c^2 may be caused by long-sought "missing" N* resonance(s), which has not be seen in both πN and γN data. Since the new peak's momentum is small and the orbital angular momentum of $L = 0$ is much preferred due to the suppression of the centrifugal barrier factor for $L \geq 1$. For $L = 0$, the spin-parity of $N^*(2030)$ is limited to be $1/2+$ and $3/2+$.

Background analysis

Background is estimated from the fitting of the missing mass distribution(Fig. 1). The Monte Carlo of $J/\psi \to anything$[6] is also used to estimate it. Both analyses show the background should less than 8% in the range 0.88 to 1.0 GeV/c^2 of the missing mass. And no peaks can be found in the $p\pi$ invariant mass distribution from the side band events of the missing mass.

Fitting of the $p\pi$ invariant mass spectrum

We try to fit the resonances in $p\pi$ invariant mass spectrum using relativistic Breit-Wigner formula

$$|BW|^2 \propto \frac{M_0\Gamma_0 q^{2l+1}k}{(M^2 - M_0^2)^2 + M_0^2\Gamma_0^2} \qquad (1)$$

where k is the momentum of $p\pi$ system, q is the momentum of one particle in the center-mass-system of p, π, l is the orbital angular momentum of $p\pi$ system. l is set to 1 for $N^*(1440)$ and 0 for other resonances. Efficiency variation is considered in the fitting. The fitting gives the mass and width of the possible new N^* resonance, which mass is $2065 \pm 3^{+15}_{-30}$ MeV/c^2 and width is $175 \pm 12 \pm 40$ MeV/c^2. The systematic error comes from the uncertain of the background mainly.

Here we just want to give some basic properties of the new possible N^* resonance around 2.03 GeV/c^2. The detailed information of these N^* resonances should be obtained from the partial wave analysis in the future.

Check from $J/\psi \to \bar{p}\pi^+n$

The charge conjugate channel is also analyzed. The event selection is almost the same as for $J/\psi \to p\pi^-\bar{n}$. The left figure of Fig. 2 shows the missing mass distribution of selected events. The signal is very clear. And the invariant mass distribution of $\pi^+\bar{p}$ (the right figure of Fig. 2) is almost identical to the π^-p (Fig. 1) in the decay $J/\psi \to p\pi^-\bar{n}$.

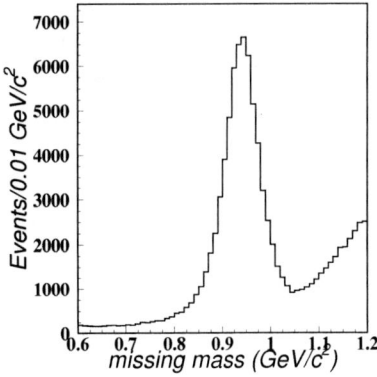
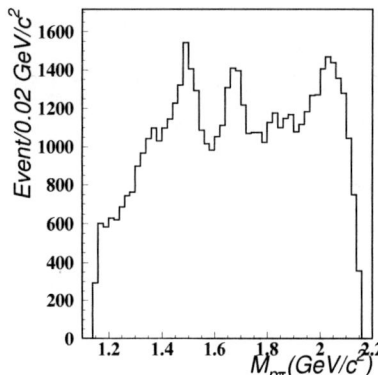

FIGURE 2. The missing mass(left) and invariant mass distribution of $\pi^+\bar{p}$(right) in the decay $J/\psi \to \bar{p}\pi^+n$.

THE ANALYSIS OF $J/\psi \to PK\Lambda$

In the event selection, four good charged tracks are required. Both proton and kaon should be identified by particle identification. And the χ^2 of the kinematic fit restricts within 20. The Λ signal is very clean and clear after the final selection. Fig. 3 shows the pK and $K\Lambda$ invariant mass distribution. In the pK invariant mass distribution, two clear

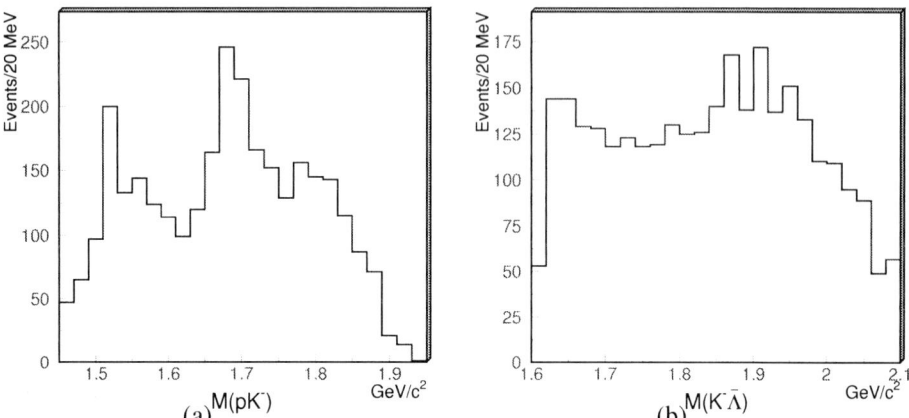

FIGURE 3. (a) pK invariant mass distribution; (b) $K\Lambda$ invariant mass distribution in the decay $J/\psi \to pK\Lambda$.

peaks may corresponding to $\Lambda(1520)$ and $\Lambda(1690)$; the bump at the threshold of ΛK invariant mass distribution may be caused by N*(1535) or N*(1650).

THE ANALYSIS OF $J/\psi \to P\bar{P}\pi^0$

Based on the 58 million J/ψ data, the events for $J/\psi \to p\bar{p}\pi^0$ have been selected and reconstructed. Fig. 4(a) is the invariant mass spectrum of the 2γ, the π^0 and η signals are clearly there. Fig. 4(b) is the $p(\bar{p})\pi^0$ invariant mass spectrum, which is similar with the $p\pi$ invariant mass spectrum in the decay $J/\psi \to p\pi^-\bar{n}$.

SUMMARY

In summary, the J/ψ decay at BEPC provides an excellent place to study the baryon excited states. Based on the 58 million J/ψ data sample, a possible new "missing" N* resonance with mass around 2.06 GeV/c^2 is found in the decay $J/\psi \to p\pi^-\bar{n}$.

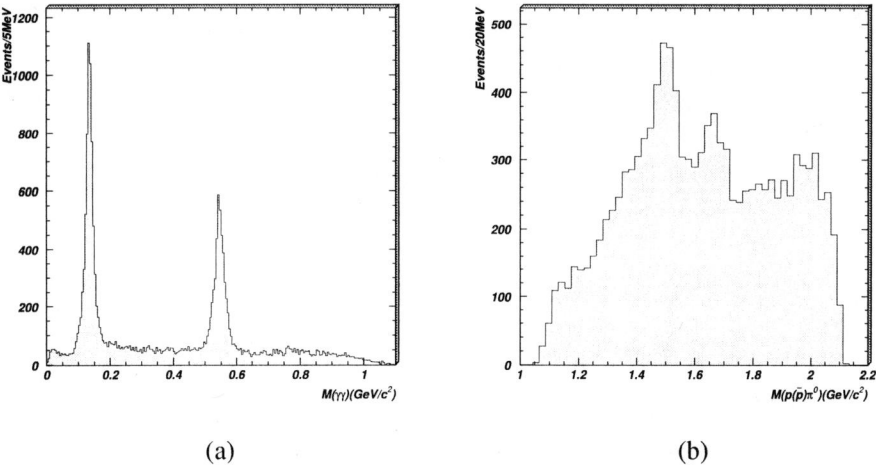

(a) (b)

FIGURE 4. $J/\psi \to p\bar{p}\pi^0$. (a) 2γ invariant mass spectrum; (b) $p(\bar{p})\pi^0$ invariant mass spectrum.

ACKNOWLEDGMENTS

The BES collaboration thanks the staff of BEPC for their hard efforts. This work is supported in part by the National Natural Science Foundation of China under contracts Nos. 19991480,10225524,10225525, the Chinese Academy of Sciences under contract No. KJ 95T-03, the 100 Talents Program of CAS under Contract Nos. U-11, U-24, U-25, and the Knowledge Innovation Project of CAS under Contract Nos. U-602, U-34 (IHEP); and by the National Natural Science Foundation of China under Contract No.10175060(USTC).

REFERENCES

1. BES Collaboration, J.Z. Bai *et al.*, *Nucl. Instr. Methods A*, **344**, 319 (1994).
2. BES Collaboration, J.Z. Bai *et al.*, *Nucl. Instr. Methods A*, **458**, 627 (2001).
3. K.Hagiwara *et al.*, *Phys. Rev. D*, **66**, 010001 (2002).
4. B.S. Zou, *Nucl. Phys. A*, **675**, 167(2000).
5. S. Capstick and N. Isgur, *Phys. Rev. D*, **34**, 2809 (1986); S. Capstick and W. Roberts, *Phys. Rev. D*, **47**, 1994 (1993); R. Bjiker, F. Iachello and A. Leviatan, *Ann. Phys.(N.Y.)*, **236**, 69 (1994).
6. Chen J.C. *et al.*, *Phys. Rev. D*, **62**, 034003 (2000).

Investigation of $\gamma p \rightarrow p \pi^0 \eta$ at ELSA in Bonn

Volker Credé for the CB-ELSA and CB/TAPS collaboration

Helmholtz-Institut für Strahlen- und Kernphysik, University of Bonn, Germany

Abstract. The Crystal-Barrel detector is the ideal instrument to study various multi-photon final states over the full dynamical range due to its almost 4π coverage of the solid angle and its high energy resolution. It allows to identify highly-excited baryon states by observing cascades of high-mass states to the ground state via the emission of pion and eta mesons. This could be observed in the 2000/2001 CB-ELSA data for the reaction $\gamma p \rightarrow p \pi^0 \eta$. Moreover, the investigation of $\gamma p \rightarrow \Delta(1232) \eta$ ($I = \frac{3}{2}$) allows to search for *missing* Δ states and will shed some light on the rather unknown Δ spectrum. Preliminary results of a partial wave analysis indicate the need for new resonances to describe the data. However, it could be shown that linearly polarised photons are very important in order to avoid ambiguities in determining the corresponding quantum numbers. In 2002/2003, polarised data have been taken off the proton as well as off the neutron with the Crystal-Barrel detector and TAPS in the forward direction. The latter has fast trigger capabilities and provides high granularity in the forward direction.

INTRODUCTION

Photon-induced reactions on the nucleon are a rich source of information for the baryon resonance spectrum. The full knowledge of possible baryon excitations and their properties would allow the extraction of the relevant degrees of freedom. Spectroscopic predictions are not possible in the non-perturbative regime of QCD. For this reason, effective theories and models are necessary in order to determine the masses, couplings and decay widths of resonances. Various constituent quark models are quite successful in describing the spectra. However, many open questions still remain. All models predict a series of hitherto unobserved states, for instance. The persistent non-observation would be a big problem as those models would have failed to describe physical reality. On the other hand, almost all existing data result from πN elastic scattering experiments and models focussing on baryon strong decays predict baryon states to be missing in πN analyses but to show up in electromagnetic production [3]. Thus, photoproduction experiments offer a large discovery potential.

The decay chain $\gamma p \rightarrow \Delta^* \rightarrow (\Delta \eta)(I = \frac{3}{2}) \rightarrow p \pi^0 \eta$ is a suitable reaction to study Δ states and to search for *missing* Δ^*. Additionally the region of Δ resonances with masses around 1950 MeV is of special interest in baryon spectroscopy. The PDG lists four well established states with positive parity in this mass region. In comparison, only three Δ^* with negative parity and poor experimental evidence are listed: $\Delta(1900)S_{31}$ (**), $\Delta(1940)D_{33}$ (*) and $\Delta(1930)D_{35}$ (***). A confirmation of those states with negative parity would be in contradiction with constituent quark models predicting the three states at masses clearly above 2 GeV [1, 4].

CP717, *Hadron Spectroscopy: Tenth International Conference,*
edited by E. Klempt, H. Koch, and H. Orth
© 2004 American Institute of Physics 0-7354-0197-7/04/$22.00

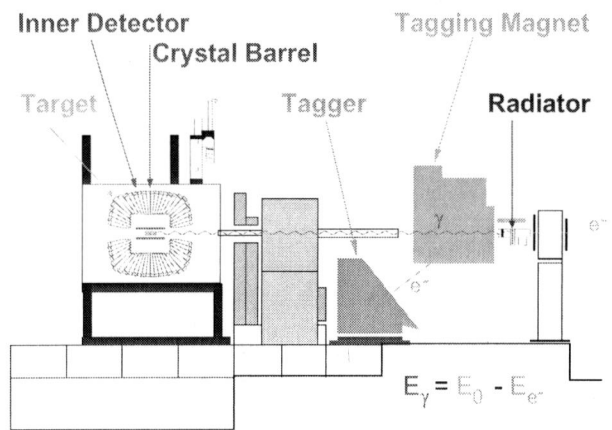

FIGURE 1. Setup of the CB-ELSA detector for a first series of measurements

THE CRYSTAL-BARREL EXPERIMENT AT ELSA

For the data presented here, electrons extracted from ELSA hit a primary radiation target with energy E_0 and produced bremsstrahlung. The corresponding energy of the photons ($E_\gamma = E_0 - E_{e^-}$) was determined in a tagging system by the deflection of the scattered electrons in a magnetic field. This detector provided a tagged beam in the photon energy range from 0.8 GeV up to 3.0 GeV. The setup of the CB-ELSA detector used for a first series of experiments is shown in Fig. 1. The calorimeter (Crystal-Barrel) consisting of 1380 CsI(Tl) crystals covering about 98 % of 4π solid angle is an ideal detector for photons. The photoproduction target in the center of the Crystal-Barrel (5 cm in length, 3 cm in diameter) was filled with liquid hydrogen. It was surrounded by a scintillating fibre detector built to detect and to trigger on charged particles leaving the target (proton trigger). In addition, it provided an intersection point of a particle's trajectory with the detector and hence helped to identify clusters of charged particles in the barrel. The general concept of the experiment is to combine the calorimeter with suitable forward detectors. Besides Time-Of-Flight walls in the start configuration, the TAPS detector (calorimeter consisting of 528 hexagonal BaF_2 crystals) was used in a second series of measurements. The latter is a fast trigger and provides an ideal granularity in the forward direction.

Investigation of the reaction $\gamma p \rightarrow p \pi^0 \eta$

Data was taken from December 2000 with the whole apparatus fully operational. Measurements at three different ELSA energies were performed: $E_0 = 1400$, 2600 and 3200 MeV. In the following, results are presented for a data run of $\approx 22\,000\ \pi^0\eta$ events at $E_0 = 3200$ MeV.

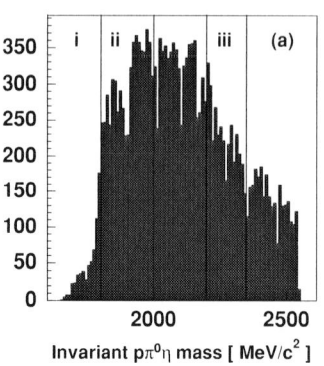

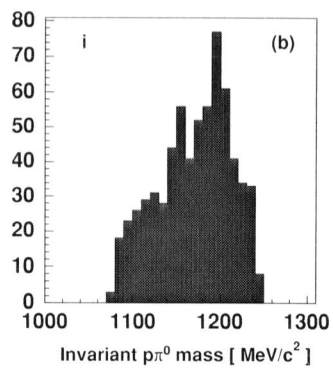

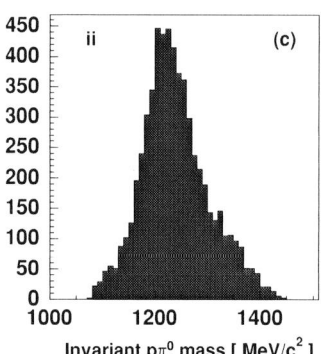

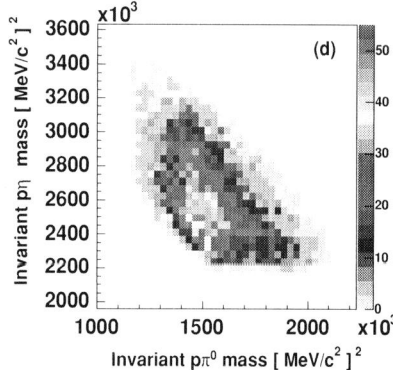

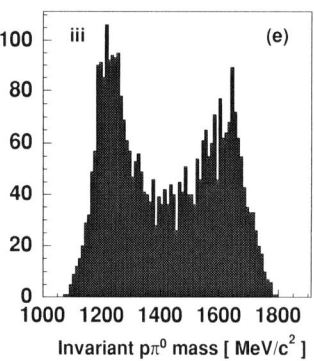

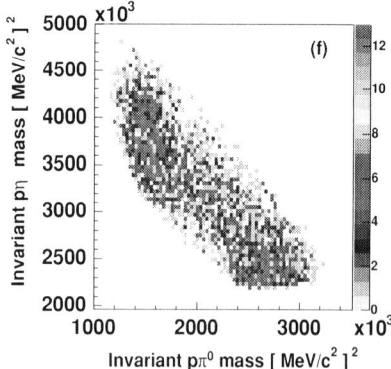

FIGURE 2. Different plots on the reaction $\gamma p \rightarrow p\pi^0\eta$. *(a) shows the total invariant $p\pi^0\eta$ mass. In (b),(c),(e), the $p\pi^0$ mass is plotted for the three different $p\pi^0\eta$ mass regions indicated in (a). Clear evidence for the $\Delta(1232)$ can be observed and, thus, hints for resonances decaying via $\Delta\eta$ become obvious. (d) and (f) show Dalitz plots for two different $p\pi^0\eta$ mass regions (ii and iii). Resonance structures become even more transparent.*

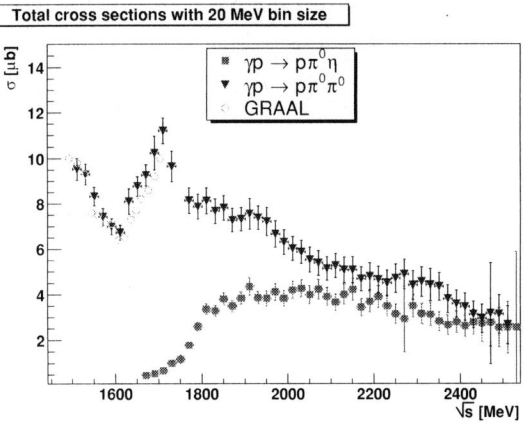

FIGURE 3. Preliminary total cross sections for the reaction $\gamma p \rightarrow p\pi^0\pi^0$ as well as $\gamma p \rightarrow p\pi^0\eta$. The low-energy part of the CB-ELSA double-pion cross section agrees well with the GRAAL data. It should be mentioned that no proper five-dimensional acceptance correction has been applied yet.

Figure 2 (a) shows the total invariant mass for the $p\pi^0\eta$ final state. No structures are visible at first sight. Different mass regions are indicated and the corresponding $p\pi^0$ mass spectra given. Hints for baryon resonances decaying into $\Delta\eta$ now become visible. In the total mass region around 1700 MeV, no structure can be seen, Fig. 2 (b). However, a clear peak at the Δ mass can be observed in the mass region around 1900 MeV, Fig. 2 (c). As a matter of fact, we expect a series of resonances in this mass region with positive as well as with negative parity. In principle, it would be very difficult to disentangle them. However, in the $\Delta\eta$ threshold region, we expect a small angular momentum between the emitted η meson and the $\Delta(1232)$. Hence, it should be possible to excite some resonances selectively. For orbital angular momenta $l = 0$ or 1, we should expect contributions from the $\Delta(1910)P_{31}$, $\Delta(1920)P_{33}$, $\Delta(1905)F_{35}$ ($l = 1$) and $\Delta(1940)D_{33}$ ($l = 0$).

For higher $p\pi^0$ masses, further resonance intensity may be hidden in a structure around 1600 MeV, Fig. 2 (e). One has to be careful interpreting structures in the mass projections as those are often reflections of the corresponding Dalitz plots (Fig. 2 (d) and (f)).

Figure 3 shows the total cross sections for the reactions $\gamma p \rightarrow p\pi^0\eta$ and $\gamma p \rightarrow p\pi^0\pi^0$. The latter agrees very well with the GRAAL data in the low energy region. Above 2 GeV both cross sections are almost equal in magnitude. However, it should be mentioned that no proper five-dimensional acceptance correction has been carried out yet.

Preliminary solutions of a partial wave analysis are based on an unbinned maximum likelihood fit taking all correlations among five independent variables properly into account (event-based fit) [5]. New resonances are needed to describe the data. There is evidence for a new Δ state at ≈ 2.2 GeV as well as hints for $\Delta^* \rightarrow a_0(980)p$ as the dominant contribution for $a_0(980)$ production. Solutions can be ambiguous and therefore

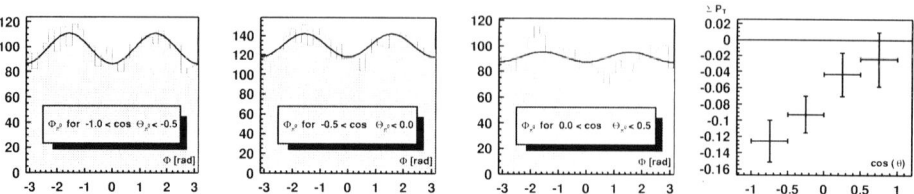

FIGURE 4. Φ_{π^0} distributions for different Θ_{π^0} bins. The data are not acceptance corrected. The incoming photon energy is limited to $1440\,\text{MeV} \leq E_\gamma \leq 1640\,\text{MeV}$, i.e. the polarisation maximum.

the question of negative-parity Δ states around 1950 MeV cannot be answered, yet. Polarisation data is needed to discriminate between different contributing amplitudes.

OUTLOOK

Simulations have shown that under certain conditions the mass and angular distributions of a Δ^* resonance ($J^P = 3/2^-$) cannot be distinguished from those of a Δ^* ($J^P = 1/2^+$), at least when interference effects are neglected. The CB/TAPS collaboration has taken data with linearly polarised photons created by coherent bremsstrahlung in a well-oriented diamond crystal. In general, the use of linear polarisation breaks the Φ symmetry. Thus, polarisation allows a better determination of contributing amplitudes by adding further constraints in the PWA, e.g. small contributions may have large effects in certain polarisation variables.

In a two-body decay, the use of linearly polarised photons (polarisation P_T) leads to a photon asymmetry Σ:

$$\sigma = \sigma_0 (1 + P_T \cdot \Sigma \cdot \cos(2\Phi))$$

In a three-body final state like $p\pi^0\eta$, there is more than one asymmetry depending on the choice of the corresponding Φ distribution. Fig. 4 shows the first results from a 2003 data-taking period. A photon asymmetry can clearly be extracted from the Φ_{π^0} distribution for different Θ_{π^0} bins, for instance. Even considering only 40 % polarisation and contributions from background processes, statistics should be sufficient to investigate the 1950 MeV/c^2 mass region and, therefore, contribute to the question of negative-parity states as well as to the problem of *missing* resonances.

REFERENCES

1. S. Capstick and N. Isgur, *Phys. Rev.* **D34**, 2809 (1986).
2. D. B. Lichtenberg, *Phys. Rev.* **178**, 2197 (1969).
3. E. S. Ackleh et al, *Phys. Rev.* **D54**, 6811 (1996).
4. U. Löhring et al, *EPJ* **A10**, 309 (2001).
5. A. Anisovich: Proceedings of this conference

Two-Pion Production in Proton-Proton Collisions

M. Bashkanov* for the CELSIUS-WASA and COSY-TOF Collaborations*,
S. Abdel-Samad[†], K.-Th. Brinkmann[**], H. Calén[‡], F. Cappellaro[§], H.
Clement*, L. Demiroers[¶], E. Doroshkevich[||], S. Dshemuchadse[**], C.
Ekström[‡], A. Erhardt*, W. Eyrich[††], K. Fransson[‡], H. Freiesleben[**], M.
Fritsch[‡‡], A. Gillitzer[†], L. Gustafsson[§], B. Höistad[§], M. Jacewicz[§], T.
Johansson[§], L. Karsch[**], S. Keleta[§], K. Kilian[†], I. Koch[§], J. Kress*, E.
Kuhlmann[**], S. Kullander[§], A. Kupsc[‡], S. Marcello[§§], P. Marciniewski[‡],
R. Meier*, K. Möller[¶¶], B. Morosov[***], H.P. Morsch[†], C. Pauly[¶], Y.
Petukhov[***], E. Roderburg[†], P. Schönmeier[**], W. Schröder[††], M.
Schulte-Wissermann[**], W. Scobel[¶], T. Sefzick[†], T. Skorodko*, J.
Stepaniak[†††], G.J. Wagner*, M. Wagner[††], U. Wiedner[§], P. Wintz[†], P.
Wüstner[†], J. Zabierowski[†††], J. Zlomanczuk[§] and P. Zupranski[‡‡‡]

*Physikalisches Institut, Universität Tübingen, D-72076 Tübingen
[†]Institut für Kernphysik, Forschungszentrum Jülich, D-52425 Jülich
[**]Institut für Kern- und Teilchenphysik, Technische Universität Dresden, D-01062 Dresden
[‡]The Svedberg Laboratory (TSL), Uppsala, Sweden
[§]Department of Radiation Sciences (ISV), Uppsala, Sweden
[¶]Hamburg University (HU), Hamburg, Germany
[||]Pysikalisches Institut, Universität Tübingen, D-72076 Tübingen
[††]Physikalisches Institut, Universität Erlangen-Nürnberg, D-91058 Erlangen
[‡‡]Institut für Experimentalphysik, Universität Bochum, D-44780 Bochum
[§§]INFN Torino, I-10125 Torino
[¶¶]Institut f'ur Kern- und Hadronenphysik, Forschungszentrum Rossendorf, D-01314 Dresden
[***]Joint Institute for Nuclear Research (JINR), Dubna, Russia
[†††]Soltan Institute of Nuclear Studies (INS), Lodz, Poland
[‡‡‡]Andrzej Soltan Institute for Nuclear Studies, PL-00681 Warsaw

Abstract. A program to measure the two-pion production in nucleon-nucleon collisions exclusively and in all channels from threshold up to $T_p = 1360$ MeV has been started at CELSIUS-WASA. First preliminary results for $pp\pi^+\pi^-$ and $pp\pi^0\pi^0$ channels are presented. In the near-threshold region they are consistent with a dominance of the Roper excitation and its successive decay into $N\sigma$ and $\Delta\pi$ channels. At energies $T_p > 1$ GeV we observe a very different behavior, which is at variance with theoretical predictions. At COSY-TOF exclusive measurements of the $pp \rightarrow pp\pi^+\pi^-$ reaction have been carried out with polarized beam. Preliminary results show non-zero analyzing powers which point to a sizeable admixture of $l \neq 0$ partial waves indicating possibly an influence of the ρ-channel.

INTRODUCTION

In nucleon-nucleon induced two-pion production the correlated two-pion exchange between the interacting nucleons is lifted onto the mass shell. This reaction gives rise to

CP717, *Hadron Spectroscopy: Tenth International Conference,*
edited by E. Klempt, H. Koch, and H. Orth
© 2004 American Institute of Physics 0-7354-0197-7/04/$22.00

a number of subsystems, where the different aspects of this process can be studied. Among these subsystems the $NN\pi$ system is special, since it facilitates the search for narrow, in particular NN-decoupled dibaryon resonances. First exclusive measurements at CELSIUS have provided [1] new and significant upper limits in the range of a few nb for the low-mass region $m_{dibaryon} < 2087$ MeV/c^2. For a review on the present status of dibaryon searches see [2].

In the $\pi\pi$ subsystem the dynamics in σ and ρ channels is of topical interest. In combination with the $N\pi$ and $N\pi\pi$ subsystems it gives access to the investigation of nucleon excitations and their decay properties. Particular emphasis is placed here on the investigation of the Roper resonance, the second excited state of the nucleon in the non-strange sector, since its nature and properties are still very poorly known.

EXPERIMENTAL RESULTS

Our first exclusive measurements of the $pp \rightarrow pp\pi^+\pi^-$ reaction near threshold [3, 4, 5] taken at CELSIUS reveal this reaction to be dominated by σ-exchange between the colliding nucleons followed by the Roper excitation in one of the nucleons with subsequent decay into $N\sigma$ or $\Delta\pi$ channels. From the observed interference of both decay routes into the final $N(\pi\pi)_{I=l=0}$ state their relative amplitudes and branchings have been determined [3, 5] in the low-energy tail of the Roper excitation. Though we observe this low-energy region to be strongly dominated by the $N\sigma$ channel, we also see a small but rapidly growing influence of the $\Delta\pi$ channel as the energy is increased. Due to its k^2 dependence the amplitude of the latter is even likely to take over at the position of the Roper resonance pole [3, 5].

These findings are basically in agreement with theoretical predictions of the Valencia group [6, 7]. However, there are also significant deviations in particular with respect to new measurements of this reaction in the threshold region with the polarized proton beam at COSY-TOF. There the preliminary data analysis yields partly non-zero analyzing powers, which point to perceptible admixtures of $l \neq 0$ partial waves in $pp \rightarrow pp\pi^+\pi^-$, in particular in the $\pi\pi$ system (Fig. 1).

With the new WASA 4π detector at CELSIUS a program has been started to measure the two-pion production exclusively in all channels over a wide energy range. First data for the $pp\pi^0\pi^0$ channel find the influence of the $N^* \rightarrow \Delta\pi \rightarrow N\pi\pi$ route to be considerably weaker than expected from the $pp\pi^+\pi^-$ channel assuming isospin invariance (Fig. 2). This finding possibly points to ρ channel admixtures in the latter channel — in accordance with the finding of non-zero analyzing powers there.

At incident energies above 1 GeV we observe drastic changes in the spectra. In $M_{p\pi}$ spectra we observe a clear peak which may be associated with Δ excitation, whereas the $M_{\pi\pi}$ spectra essentially fall back to phase space. The latter is very surprising and at variance with the theoretical prediction [6].

$$A(\theta)$$

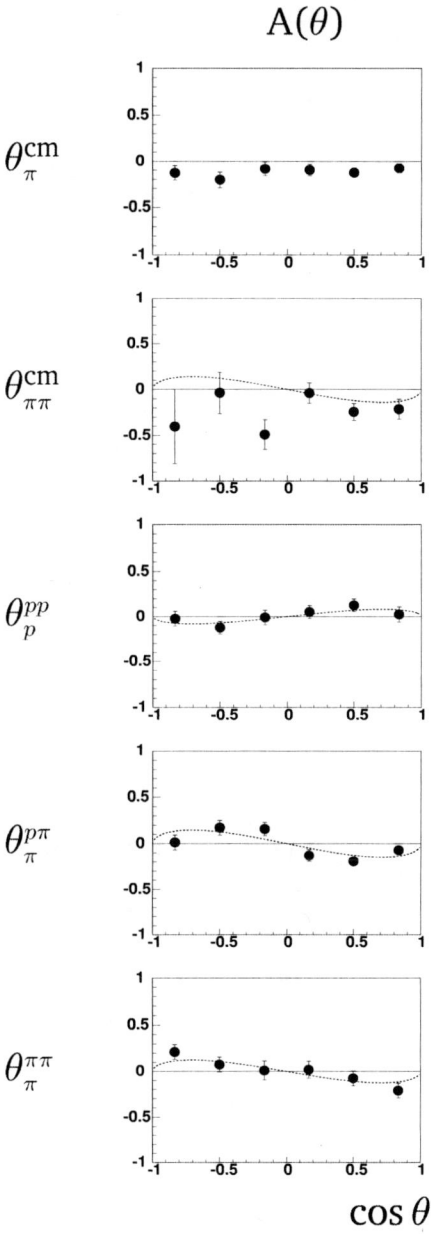

$$\cos\theta$$

FIGURE 1. Preliminary results for the analyzing powers of the reaction $\vec{p}p \rightarrow pp\pi^+\pi^-$ at $T_p = 750$ MeV measured at COSY with the TOF spectrometer [8]. Shown are the angular distributions for individual pions (Θ_π^{cm}) and for the pion pair ($\Theta_{\pi\pi}^{cm}$) in the overall center-of-mass system (cms), as well as for protons (Θ_p^{pp}) and pions ($\Theta_\pi^{p\pi}, \Theta_\pi^{\pi\pi}$) in the $pp, p\pi$ and $\pi\pi$ subsystems, respectively. The dotted lines are $\sin 2\Theta$ distributions fitted to the data. For the subsystems such a functional behaviour is a necessary, for the distributions in the overall cms an optional condition. For the latter also $\sin\Theta$ admixtures are allowed, see Ref. 9.

$$T_p = 775 \text{ MeV} \qquad T_p = 900 \text{ MeV}$$

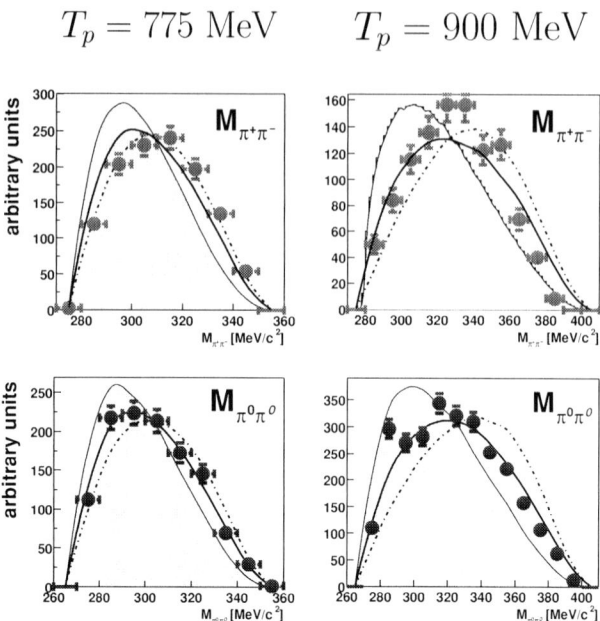

FIGURE 2. Preliminary results for $M_{\pi\pi}$ spectra of the reactions $pp \to pp\pi^+\pi^-$ (top) and $pp \to pp\pi^0\pi^0$ (bottom) at $T_p = 775$ (left) and 900 MeV (right) measured at CELSIUS with the WASA detector. The data are compared to phase space (shaded area) as well as to calculations assuming either a ratio of 3:1 (dotted) or 1:1 (solid) for the branching of the Roper resonance decay into $N^* \to \Delta\pi$ or $N^* \to N(\pi\pi)_{I=l=0}$. Note that the current PDG value [10] for this quantity is 4(2):1.

ACKNOWLEDGMENTS

We acknowledge the help of the CELSIUS and COSY crews during the course of these measurements. This work has been supported by BMBF (06 TU 201), FZ Jülich (FFE) and DFG (European Graduate School) and European Union programs for the access to research infrastructures.

REFERENCES

1.   W. Brodowski et al., Phys. Lett. **B550**, 147 (2002).
2.   E. Doroshkevich et al., Eur. Phys. J. **A17** (2003) 683
3.   W. Brodowski et al., Phys. Rev. Lett. **88**, 192301 (2002).
4.   J. Johanson et al., Nucl. Phys. **A712**, 75 (2002).
5.   J. Pätzold et al., Phys. Rev. **C67**, 052202(R) (2003).
6.   L. Alvarez-Ruso, E. Oset, E. Hernández, Nucl. Phys. **A633**, 519 (1998).
7.   E. Hernandéz, E. Oset, M.J. Vicente Vacas, Phys. Rev. **C66**, 065201 (2002).
8.   J. Kress, PhD thesis, Univ. Tübingen 2002
9.   H.O. Meyer et al., Phys. Rev. **C63**, 064002 (2001).
10. Particle Data Group, Phys. Rev. **D66**, 1 (2002).

Two-Pion Production On The Nucleon

S. Schneider[*] and S. Krewald[*]

[*]Institut für Kernphysik, Forschungszentrum Jülich, D-52425 Jülich, Germany

Abstract.
We have developed a microscopic model to study the two–Pion decay of the low lying N^* resonances, in particular the Roper Resonance $N^*(1440)$. We observe a sensitivity of the $m_{\pi\pi}$–distributions in $\pi N \to \pi\pi N$ to the decay parameters of the Roper Resonance.

INTRODUCTION

The understanding of QCD in the non-perturbative regime goes hand in hand with our understanding of the baryon spectrum. Experimental information on the masses and decays of excited baryons can be extracted from the analysis of πN scattering data and the transition to inelastic channels such as ηN and $\pi\pi N$. However, most of our knowledge of baryon resonances comes from πN scattering. In order to obtain more detailed information on the decay properties of resonances we study their two–Pion decay directly in the reaction $\pi N \to \pi\pi N$.

The most recent measurements of this reaction have produced detailed differential distributions for the reaction channels $\pi^\pm p \to \pi^\pm \pi^- n$ [1] and $\pi^- p \to \pi^0 \pi^0 n$ [2], the latter ones in an energy range from threshold up to 1.5 GeV.

On the theoretical side, there are predictions from Chiral Perturbation Theory [3] and from isobar models like that of the Valencia Group [4]. Chiral Perturbation Theory is the effective theory for the strong interactions at low energies, but resonances are not included as dynamical degrees of freedom. They only contribute to the size of the so-called low energy constants. In order to study resonances we develop a microscopic model by saturating the contact terms of Chiral Perturbation Theory with resonance exchanges.

OUR MODEL

Our Model contains resonance contributions from the excitation of Δ, $P_{11}(1440)$, $D_{13}(1520)$ and $S_{11}(1535)$ intermediate states. The exchanged mesons are the σ and the ρ meson, which have to be understood as a parametrisation of two-Pion exchange. In order to fulfil the requirements of chiral symmetry, we also include a $3\pi N$ and a 4π contact vertex that both appear at leading order in Chiral Perturbation Theory, and that we cannot saturate with resonance exchange. Some of the contributions are shown diagrammatically in Figure 1.

CP717, *Hadron Spectroscopy: Tenth International Conference*,
edited by E. Klempt, H. Koch, and H. Orth
© 2004 American Institute of Physics 0-7354-0197-7/04/$22.00

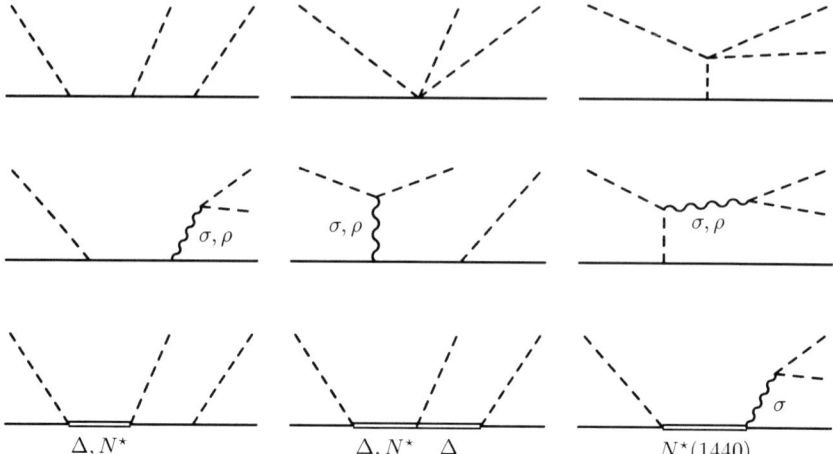

FIGURE 1. Some of the diagrams contributing to $\pi N \to \pi \pi N$.

The amplitudes are calculated in a relativistic framework, using the Rarita-Schwinger-Propagator for the spin-3/2 resonances $\Delta(1232)$ and $D_{13}(1520)$.

The parameters — coupling constants and the masses of the resonances — have been fitted to the $\pi\pi$ and πN phase shifts and inelasticities. Two-body unitarity is fulfilled by iterating the corresponding subgraphs of our model in a K-matrix approach.

For the σ, for example, we use the Lagrangian

$$\mathcal{L}_{\pi\pi\sigma} = -g_1 m_{\pi}^2 \vec{\pi}\vec{\pi}\sigma + \frac{g_2}{2} \partial_\mu \vec{\pi}\partial^\mu \vec{\pi}\sigma \tag{1}$$

allowing for scalar and gradient coupling at the $\pi\pi\sigma$-vertex. From a fit to the $I = J = 0$ partial wave of Pion-Pion scattering, we get the parameters $g_1^2/4\pi = 1.89\,m_{\pi}^{-2}$, $g_2^2/4\pi = 0.14\,m_{\pi}^{-2}$ and $m_\sigma = 835$ MeV. We have checked that we can saturate the Low Energy Constants of Chiral Perturbation Theory appearing at second order in πN-scattering with our parameters for the resonance exchanges.

RESULTS

The total cross sections in the five reaction channels $\pi^\pm p \to \pi^\pm \pi^+ n$, $\pi^- p \to \pi^0 \pi^0 n$ and $\pi^\pm p \to \pi^\pm \pi^0 p$ are shown in Fig. 2. For comparison, we also show the results of Chiral Perturbation Theory [3].

The data are reasonably well described by our model calculation from threshold up to $T_\pi \approx 0.4$ GeV, which corresponds to $\sqrt{s} = 1.38$ GeV and is below the second resonance region. Still we observe a sizable contribution from the Roper Resonance in the reaction channels $\pi^- p \to \pi^+ \pi^- n$ and $\pi^- p \to \pi^0 \pi^0 n$. We can ascribe this to the σN decay of the

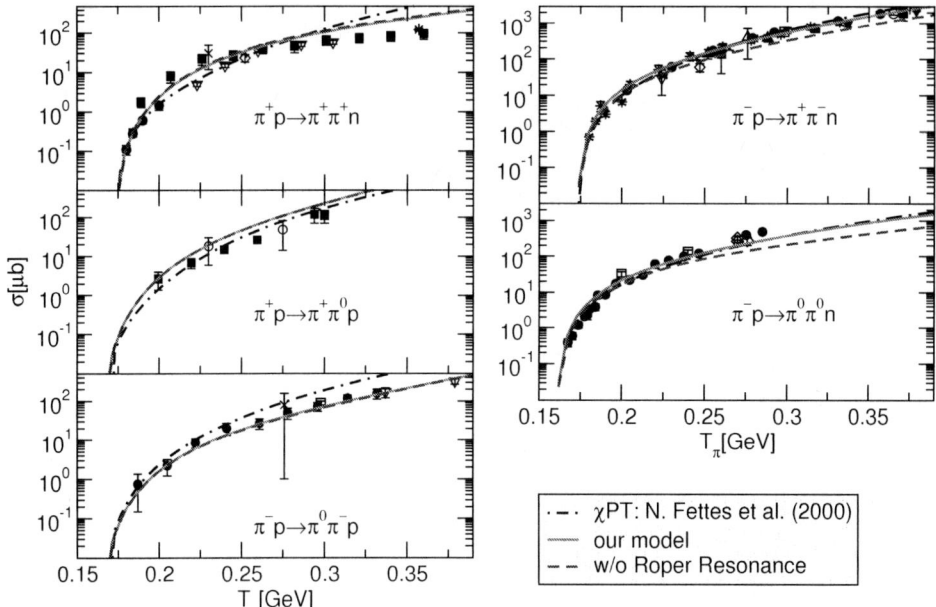

FIGURE 2. $\pi N \to \pi\pi N$ total cross sections. The dot-dashed lines show the results of Fettes et al. [3]. Our results are displayed as solid lines. Switching off all contributions from the Roper Resonance in our model, we get the total cross section shown by the dashed lines. The data are taken from [1, 5] and a compilation of older data in [6].

Roper Resonance. This already has been observed to be an important effect in the Oset and Vicente-Vacas Model [4].

The higher-lying $N^\star$ resonances contribute only marginally in this energy regime.

In Fig. 3 we show several differential distributions for $\pi^- p \to \pi^0\pi^0 n$ in the energy range $1.35\,\text{GeV} \leq \sqrt{s} \leq 1.50\,\text{GeV}$. We find good agreement between the experimental distributions and our model results for $d\sigma/dm_{\pi^0 n}$, $d\sigma/dm_{\pi^0\pi^0}$, $d\sigma/dt$ and $d\sigma/d\phi$.

The agreement for $d\sigma/d\cos\theta$ is still good at 1.35 GeV, but for the higher energies, the distributions even have the opposite curvature. Switching off the contributions from the Roper Resonance, the $d\sigma/d\cos\theta$ distribution becomes nearly flat. A change in the sign of the interference between the Resonance and the background would lead to a curvature similar to that in the data, but this would spoil the agreement for the other distributions, especially at the lower energies.

In order to study the interplay of the Roper $\to \sigma N$ and the Roper $\to \pi\Delta$ decay in more detail, we construct a variant of the model (Model B) in which the $\pi\Delta$ decay amplitude is enhanced by a factor of $\sqrt{8}$, while the σN decay is reduced by the same factor. Whereas we find a branching ratio $R = \frac{\Gamma(R\to\sigma N)}{\Gamma(R\to\pi\Delta)} = 0.544$ in the original model, we have almost only $\pi\Delta$ decay ($R = 0.008$) in Model B. We observe that this leads to an improvement for the lower energies. In particular, the $d\sigma/dm_{\pi^0\pi^0}$ distribution

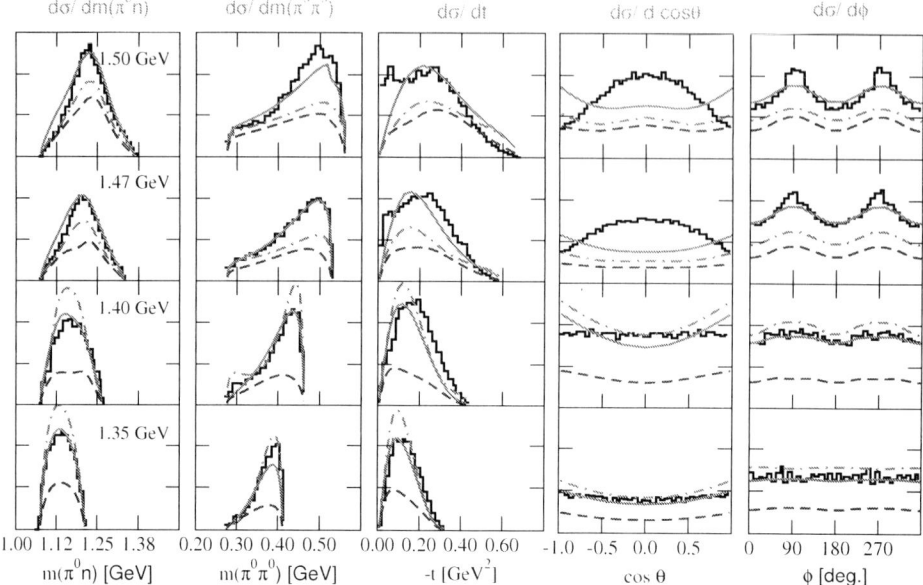

FIGURE 3. Differential distributions for $\pi^- p \to \pi^0 \pi^0 n$ from $\sqrt{s} = 1.35$ GeV (bottom) up to $\sqrt{s} = 1.50$ GeV (top). The solid lines show the results of our model, for the dashed lines, the contribution of the Roper Resonance has been switched off. The dot-dashed lines show the results we obtain, when we multiply the Roper $\pi \Delta$ coupling constant with a factor of $\sqrt{8}$ and at the same time decrease the Roper-σ-N coupling by the same factor. The experimental distributions are from [2] and have been scaled to the model results.

rises more steeply with the invariant pion mass, just as demanded by the data. The shape of the other distributions remains nearly unchanged. At 1.4 GeV we can see a shoulder developing at low $m_{\pi^0\pi^0}$. This is what remains of the double-peak structure of the Roper $\to \pi\Delta$ decay when it interferes with the σN decay and the background. At 1.47 GeV and 1.5 GeV, the $d\sigma/dm_{\pi^0\pi^0}$ and the $d\sigma/dt$ distributions in our Model B become too flat - here the decay parameters of the Roper Resonance from our original model are favoured. Furthermore, we need the strong σN decay in our original model in order to describe the early onset of the πN-inelasticities in the P_{11} partial wave. Still it is interesting to see that the data show some sensitivity to the decay properties of the Roper Resonance.

Altogether the agreement between the data and our model results is good with the exception of the $\cos\theta$-distribution. This observable seems to be sensitive to some parameters that do not influence the other distributions.

The experimental $m_{\pi\pi}$-distributions show a sensitivity to the decay parameters of the Roper Resonance. At energies below ≈ 1.4 GeV, they favour a strong Roper $\to \pi\Delta$ decay. This is somewhat counter-intuitive, because in πN scattering we need a strong σN decay in order to reproduce the inelasticities in the P_{11} partial wave directly above the $\pi\pi N$ threshold.

It will be interesting to apply our model also to the Photoproduction of two Pions in order to get a more complete picture of the decay properties of the low-lying $N^\star$ resonances.

REFERENCES

1. M. Kermani et al., *Phys. Rev. C*, **58**, 3419 (1998).
2. K. Craig et al., *Phys. Rev. Lett.*, **91**, 102301 (2003).
3. N. Fettes, V. Bernard, and Ulf-G. Meißner, *Nucl. Phys. A*, **669**, 269 (2000).
4. E. Oset and M.J. Vicente-Vacas, *Nucl. Phys. A*, **446**, 584 (1985).
5. J.B. Lange et al., *Phys. Rev. Lett.*, **80**, 1597 (1998).
6. V.V. Vereshagin et al., *Nucl. Phys.*, **A592**, 413–442 (1995).

Partial wave analysis of the two neutral pion photoproduction data.

A. Anisovich

*Helmhotz-Institut für Strahlen- und Kernphysik der Universität Bonn , Nusallee 14-16, 53115,
Bonn, Germany*

Abstract.
   Data on photoproduction of two neutral pions have been taken recently by the CB-ELSA experiment. The partial wave analysis of these data was performed using extended Rarita-Schwinger formalism. The data are dominated by production of resonances in γp channel with strong signal from D13(1520). The properties of this resonance like decay branching ratios and $A_{3/2}/A_{1/2}$ are determined with a good accuracy. The contribution from others baryons are discussed.

INTRODUCTION

The study of strong interactions at medium energies is one of the most challenging topics in nuclear and particle physics. Presently, we aim at an understanding of the effective degrees of freedom which govern the dynamics of quarks and gluons and are responsible for confinement.

   One approach to these questions is the investigation of baryon resonances. Baryons seem to be well described by the dynamics of constituent quarks bound by a confining potential. The fine structure splittings between different states then needs an additional interaction between these constituent quarks, and it is here where different quark models differ. A survey of presently known baryon resonances and their systematics can be found in [1]. The residual interaction can be chosen as due to an effective one-gluon exchange interaction [2], as exchange of Goldstone bosons [3] or as instanton-induced interactions [4]. All models get a reasonable description of the data for low-mass excitations but above 1.8 GeV they differ substantially. A particular striking example are the negative parity Δ^* resonances which quark models predict to have masses at about 2100 to 2200 MeV while the systematics of [1] demands a much lower mass, about 1950 MeV. In this mass range and above, the number of predicted states increases rapidly, but these states are so far not seen experimentally. This is the so-called missing resonance problem which has attracted considerable interest.

   There are several explanations for this phenomena:

1. Baryon resonances may have a quark-diquark structure [5].
   One can argue that scalar isoscalar diquarks are tightly bound and that one of the two oscillators in the three-body problem is frozen. Clearly, such a dynamic would reduce the number of states. However, in lattice gauge calculations of baryon ground states no such diquark clustering is observed. Also, the spectrum of baryon

CP717, *Hadron Spectroscopy: Tenth International Conference,*
edited by E. Klempt, H. Koch, and H. Orth
© 2004 American Institute of Physics 0-7354-0197-7/04/$22.00

resonances seems to be richer than diquark models would allow.

2. The missing baryon resonances may have only weak couplings to the Nπ channel. Most known resonances were discovered in πN elastic scattering, and the missing resonances may thus have escaped experimental discovery.

It is obvious that photoproduction experiments investigating these channels have a large potential for the discovery of new baryons. These reactions are measured by the CB-ELSA Collaboration in Bonn [6].

MESON PHOTOPRODUCTION AMPLITUDES IN THE DISPERSION TECHNIQUE USING THE MOMENT-OPERATOR EXPANSION

The moment–operator expansion is a powerful tool for the study of the analytical structure of amplitudes, in particular, for the determination of the resonance amplitudes. The advantages of this method are the following ones:

1. the technique is relativistic invariant;
2. all intermediate states are well under control allowing to take properly into account rescattering of final state particles. Rescattering processes are determined by the right-hand singularities of the amplitude;
3. in this technique there is no ambiguity related to the off-shell amplitudes.

Starting from early 1960s, the operator expansion was exploited for the amplitude analysis within the framework of dispersion–relation technique: in [7] the operator expansion was used for the calculation of rescattering processes in the reaction of three-particle production near threshold. A complete description of the method for non-relativistic kinematics was presented in [8].

The moment-operator expansion method is especially helpful for the investigation of particle spectra in the framework of multichannel K-matrix approach, or dispersion–relation N/D method. Within this method, the calculation of rescattering effects in multiparticle reactions is rather simple and straightforward.

For the two-fermion sector of nucleon-nucleon reactions, such as $NN \to NN$, $NN \to N\Delta$ and $\gamma d \to pn$, the relativistic operators have been constructed in [9].

The further development of this method can be found in [10] where the moment operator expansion for the two meson, two photon and fermion-antifermion states was presented. One of the application of this method is the analysis of the proton-antiproton annihilation in flight. The data obtained by Crystal Barrel Collaboration provide us with rich possibilities to study resonances in the mass region 1900-2400 MeV [11, 12]. The reconstruction of analytical meson amplitudes and definition of pole singularities are forced by the problem of classification of meson states, exotics included (glueballs, hybrids).

One of the advantages of the operator expansion method is a possibility to take promptly into account kinematical factors related to momenta of the incoming and outgoing particles. The account for kinematical factors is dictated by the necessity to investigate analytical structure of the amplitude near the production threshold of new particles. An alternative to the operator-expansion method consists in calculating kinematical factors in a standard expansion in spherical harmonics (see [13] as an example).

The moment-operator expansion has been recently used in the calculation of the radiative decays of scalar, tensor and vector mesons in the framework of double spectral integration technique [14]. The operators for $\gamma\gamma$-states are also used in analysis of rich experimental information on $\gamma\gamma \to$ *mesons* obtained at LEP (e.g. see [15]).

At present the further development of this technique was done which allow to use it for analysis of the photoproduction of nucleons. A complete set of formulae in terms of the moment–operator expansion has been calculated and coded. The formulae include photon-baryon and meson-baryon induced reactions and allow to include baryon resonances up to any spin; the formalism will be presented in a forthcoming publication [16]. Contributions from meson exchange diagrams can be added to the amplitude as background term or can be calculated on the basis of the Feynman rules.

Analysis of the data from $\gamma p \to p\pi^0\pi^0$

An analysis of the first so-far preliminary data obtained from the reaction $\gamma p \to p\pi^0\pi^0$ at beam energy 1.4 GeV shows that the following baryon resonances give the main contribution: $D_{13}(1520), P_{11}(1440), D_{13}(1700), P_{11}(1680)$. The total cross section $\gamma p \to p\pi^0\pi^0$ as determined in the partial wave analysis is shown in Fig. 1 and compared to data from the TAPS experiment at MAMI and the GRAAL experiment. The general agreement is rather good. The discrepancies between partial wave results and the other experiments might be due to a not yet perfect description of the data or inconsistencies between the experiments. Note that so far, the ELSA data are preliminary.

In any case, the results on the $N(1520)D_{13}$ are very stable. They are presented in Table 1 and compared to PDG values. The results demonstrate that with the techniques developed here parameters of baryon resonances can be determined faithfully. This is an important result since it provides solid foundations for the analysis of data at higher energies where new phenomena, in particular new baryon resonances, are expected.

The result shown above refer to what we call solution A. However spin-parity of the contribution at 1680 MeV is not well defined. The $P_{11}(1680)$ (solution A) can be replaced by the $F_{15}(1680)$ (solution B) or by the $D_{15}(1675)$ (solution C). Fig. 2 shows that these different solutions result in a similar description of the data.

ACKNOWLEDGMENTS

We wish to acknowledge support from the Alexander von Humboldt Foundation.

TABLE 1. Properties of the $N(1520)D_{13}$ derived from the partial wave analysis and comparison to PDG values.

	Our results	PDG
Mass	1525 ± 15 MeV	1515 to 1530 MeV
Width	115 ± 15 MeV	110 to 135 MeV
Ratio S/D waves in $\Delta(1232)\pi$	1.6 ± 0.4	0.7 ± 0.35
Helicity amplitude ratio $A_{3/2}/A_{1/2}$	-2.4	-6.9 ± 2.6

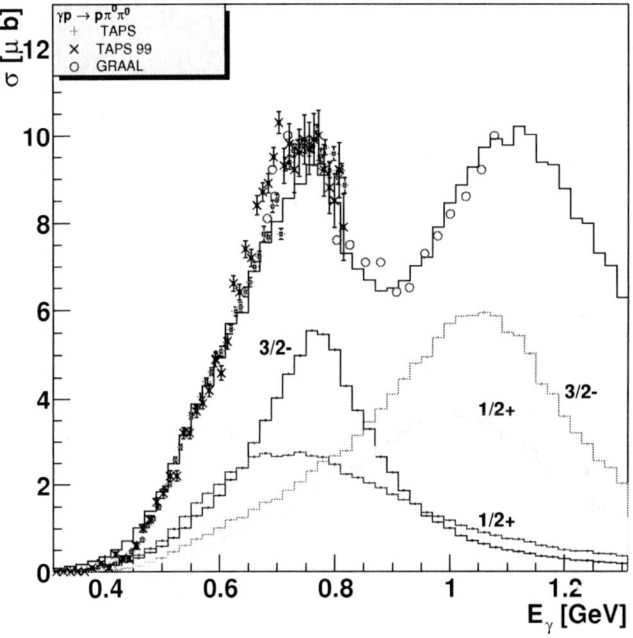

FIGURE 1. Total cross section from the partial wave analysis in comparison to TAPS and GRAAL data. The partial wave analysis result and its decomposition into resonance contributions are shown as solid lines.

REFERENCES

1. E. Klempt, Phys. Rev. **C66058201** (2002) 058201
2. S. Capstick and N. Isgur, Phys. Rev. D **34** (1986) 2809.
3. L. Y. Glozman, W. Plessas, K. Varga and R. F. Wagenbrunn, Phys. Rev. D **58** (1998) 094030.
4. U. Löring, K. Kretzschmar, B. C. Metsch and H. R. Petry, Eur. Phys. J. A **10** (2001) 309. U. Löring, B. C. Metsch and H. R. Petry, Eur. Phys. J. A **10** (2001) 395.

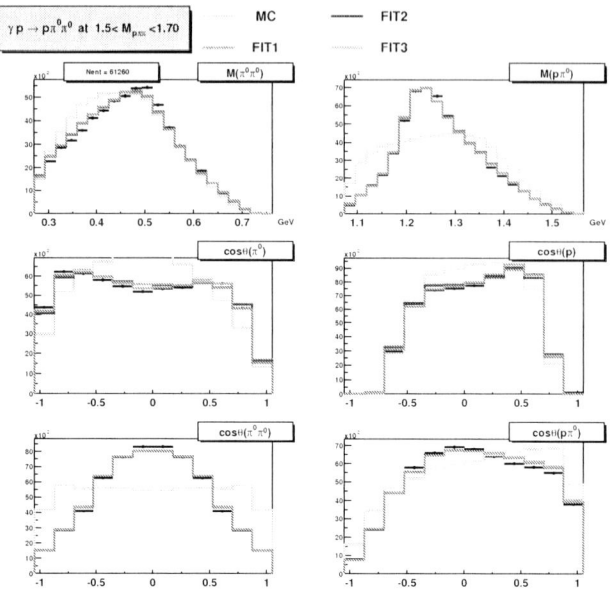

FIGURE 2. Two particle mass distributions (first line), angular distributions between final particles i and initial photon (second line) and angular distributions between decay and spectator particle in c.m.s. of two particles ij (third line). The points with error correspond to the data, solution A shown by the blue line, solution B by the pink line and solution C by the red line.

U. Löring, B. C. Metsch and H. R. Petry, Eur. Phys. J. A **10** (2001) 447.

5. D.B.Lichtenberg, Phys. Rev. **178** (1969) 2197

6. J.Junkersfeld for the CB-ELSA Collaboration, Acta Physica Polonica B33 (2002) 941

7. V.V. Anisovich and L.G. Dakhno, JETP **44**, 198 (1963);
    V.V. Anisovich and A.A. Anselm, UFN **88**, 287 (1965), [Sov. Phys. Uspekhi **88**, 177 (1966)].

8. C. Zemach, Phys. Rev. **97**, B97 (1965); **97**, B109 (1965).

9. A.V. Anisovich and A.V. Sarantsev, Sov. J. Nucl. Phys. **55**, 1200 (1992);
    V. V. Anisovich, M. N. Kobrinsky, D. I. Melikhov and A. V. Sarantsev, Nucl. Phys. A **544**, 747 (1992);
    V.V. Anisovich, D.V. Bugg and A.V. Sarantsev, Nucl. Phys. **A537**, 501 (1992);
    A.V. Anisovich and V.A. Sadovnikova, Eur. Phys. J. A2, 199 (1998).

10. A.V. Anisovich, V.V. Anisovich, V.N. Markov, M.A. Matveev, A.V. Sarantsev, J.Phys **G28**, 15, 2002

11. A.V. Anisovich, C.A. Baker, C.J. Batty et al., Phys. Lett. **B452**, 173 (1999); **B468**, 304 (1999); **B468**, 309 (1999).

12. A.V. Anisovich, C.A. Baker, C.J. Batty et al., Phys. Lett. **B452**, 187 (1999); **B452**, 173 (1999); **B452**, 180 (1999); **B500**, 222 (2001); **B507**, 23 (2001).

13. S.U. Chung, Phys. Rev. **D57**, 431 (1998). **A2**, 199 (1998).

14. A. V.Anisovich, V. V.Anisovich, D. V. Bugg and V.A. Nikonov, Phys. Lett. B **456**, 80 (1999);
    A. V. Anisovich and V. V. Anisovich, Phys. Lett. B **467**, 289 (1999);
    A. V. Anisovich, V.V. Anisovich, M.A. Nikonov, Phys.Atom.Nucl. **66**, 914, (2003).

15. A.V. Sarantsev et al., in preparation

16. A.V. Anisovich, E.Klempt, A.V. Sarantsev and U.Thoma "Amplitudes for photoproduction of baryon resonances in the operator expansion technique", in preparation

Single π^0 and η Photoproduction off the Proton at CB–ELSA

Olivia Bartholomy for the CB–ELSA Collaboration

Helmholtz–Institut für Strahlen– und Kernphysik,
University of Bonn, Germany

Abstract. Photoproduction of mesons provides an opportunity to access the properties of nucleon resonances. The study is complementary to πN scattering from which most known properties of resonances have been extracted. Many resonances predicted by quark model calculations have not been observed experimentally or are only weakly established. Photoproduction of mesons offers an additional tool to study the baryon spectrum, to gain information about masses, couplings, and decay widths of the contributing resonances.

In 2001, during its first period of taking data at the electron accelerator ELSA in Bonn, the CB–ELSA experiment gathered a large amount of high–quality data on meson photoproduction off the proton. The detector system is ideally suited for measuring photoproduction reactions with neutral mesons in the final state over the full angular range and at high energies. Differential cross sections of $\gamma p \rightarrow p\pi^0$ and $\gamma p \rightarrow p\eta$ have been extracted for incident photon energies up to $E_\gamma = 3$ GeV. At low energies, results of the TAPS, GRAAL, and CLAS experiments are well reproduced. New data points were added for forward angles of the meson and at energies above 2 GeV.

INTRODUCTION

When comparing theoretical predictions [1] with experimental findings [2], very good agreement is found at low masses, while experimental evidence for theoretically predicted states becomes sparse with growing M; more and more gaps appear on the experimental side. In order to account for these gaps, the so called *missing resonances*, several theoretical explanations have been suggested, e. g.

- Baryons are not formed by three individually interacting quarks, but have a quark–diquark structure [3]. The consequence would be one frozen degree of freedom inside the baryon resulting in less accessible states.
- The *missing resonances* might simply not couple to the so far mainly investigated channel πN, which has been studied in scattering experiments. Models predict that the missing states have a sufficiently strong coupling to photons and multi–particle final states, which makes photoproduction of multi–meson final states off the proton an ideal tool to study nucleon resonances [4, 5].

The goal of the CB–ELSA experiment is to contribute with its high–statistics and high–quality data to a clarification of the baryon spectrum and to help describe the internal structure of the nucleon.

CP717, *Hadron Spectroscopy: Tenth International Conference,*
edited by E. Klempt, H. Koch, and H. Orth
© 2004 American Institute of Physics 0-7354-0197-7/04/$22.00

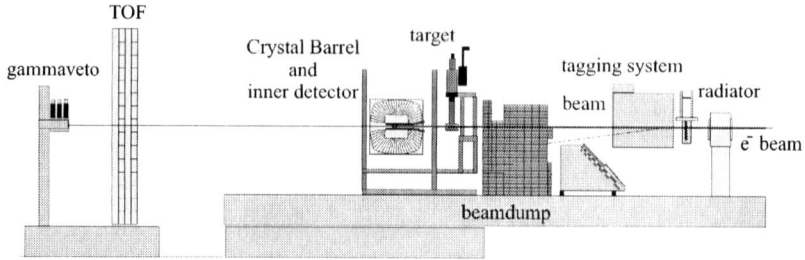

FIGURE 1. Experimental setup: Initial configuration of the CB–ELSA experiment

EXPERIMENTAL SETUP

The data presented here was taken in 2001 with the initial setup of the CB–ELSA experiment as shown in Fig. 1 at the electron accelerator facility ELSA in Bonn. The electron beam comes in from the right side and hits a radiator target where photons are produced via bremsstrahlung. In good approximation, the photon spectrum is proportional to $1/E_\gamma$. An energy is assigned to the photons by detecting the scattered electrons in a spatially and time resolving tagging detector. The trajectory of the electrons is bent in the field of a dipole magnet, so that a coordinate in the detector corresponds to an electron energy, and thus to the energy of the photon. The photons then are guided to a liquid hydrogen target. If no reaction takes place, the photons are detected by the γ–veto detector. If a proton is hit, the final state particles stemming from the reaction are detected in the main detector sytem. The heart of the CB–ELSA experiment, the Crystal Barrel detector, is a high–granularity calorimeter made of 1380 CsI crystals covering 98% of 4π. It is ideally suited for the detection of photons, and hence the investigation of neutral mesons in the final state which decay into photons. Charged particles are also detected in the calorimeter; they cannot be distinguished from photons by the Crystal Barrel alone, though. In order to identify them as charged particles, an inner detector consisting of three layers of scintillating fibres is used, which helps to determine the direction of charged particles.

DATA AND SELECTION

Data used for the analysis presented here was taken at incident electron energies of 1.4 and 3.2 GeV, resulting in photon energies from 0.3 to 3.0 GeV. A hit in the inner detector and the correct photon multiplicity in the Crystal Barrel were required, for the analyses of $\gamma p \to p\pi^0$ and $\gamma p \to p\eta$ two or six photons, respectively.

The event structure returned by tracking routines served as input for a kinematic fit with known four–vectors for the incident photon and final–state photons, but leaving the detected proton free of constraints (missing particle). Additional mass constraints were imposed by using the masses of mesons (the π^0 mass for $\gamma p \to p\pi^0$, the η mass for $\gamma p \to p\eta_{2\gamma}$, three π^0 masses for $\gamma p \to p\eta_{3\pi^0}$).

Background subtraction was applied by sidebin subtraction only for the η channel. The background underneath the π^0 signal was not explicitly treated, because it was

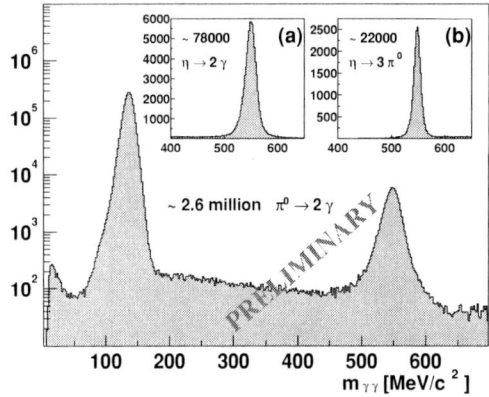

FIGURE 2. Invariant mass of two photons, (a) $\eta \to 2\gamma$, (b) $\eta \to 3\pi^0$

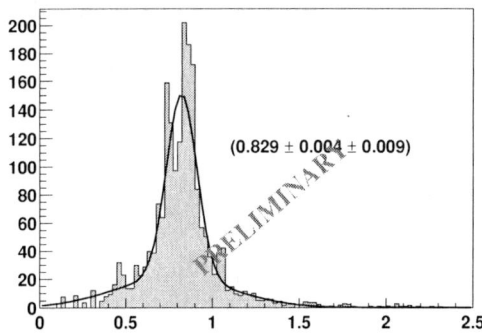

FIGURE 3. Branching ratio $\frac{\Gamma(\eta \to 3\pi^0)}{\Gamma(\eta \to 2\gamma)}$ as measured, PDG–Average: (0.832 ± 0.011)

of the order of only 10^{-3} compared to the signal. The mass spectra are shown in Fig. 2. A complete analysis of systematic errors was performed for both π^0 and η photoproduction.

In order to account for the acceptance of the detector system and reconstruction routines, a full GEANT–based Monte Carlo Simulation was carried out. The very good understanding of the CB–ELSA detection efficiency can be seen in the excellent agreement of the measured angular distributions with previously published data (Figs. 4 and 5) and the good match between the PDG value [2] for the branching ratio the two different analysed decay modes of the η meson into $3\pi^0$, a six photon final state, and into 2γ, as shown in Fig. 3.

The photon flux was determined by a χ^2 fit of the π^0 data to the SAID [6] prediction for energies up to 1.3 GeV. The cross sections for energies above 1.3 GeV were normalized to the measured photon flux, scaled by 0.75 to account for experimental deficiencies in the flux determination. The error of the normalization has been estimated to be in the order of 5% up to 1.3 GeV and 15% above.

dσ/dΩ [μb/sr]

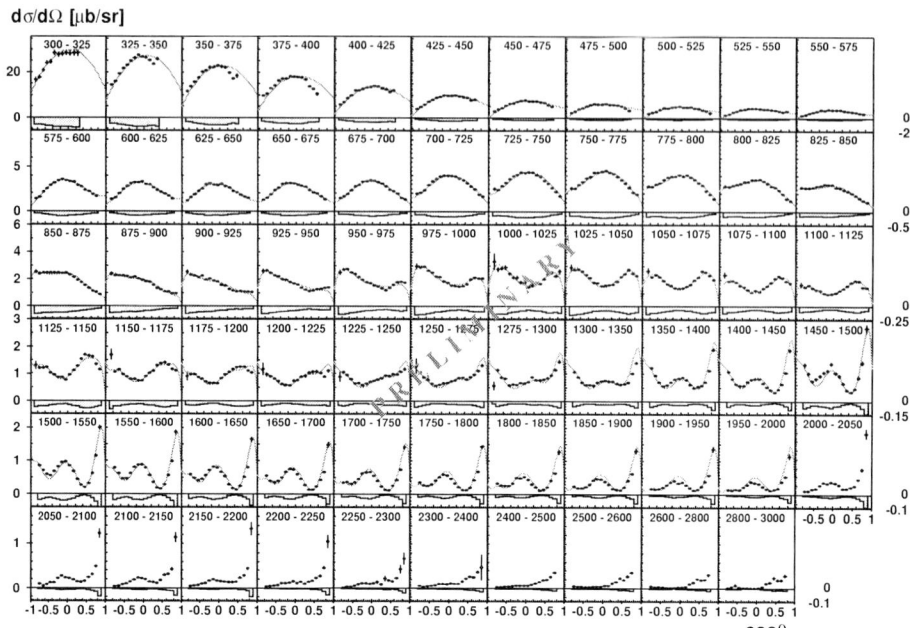

FIGURE 4. Differential cross sections of $\gamma p \rightarrow p\pi^0$ for photon energies between 0.3 and 3.0 GeV. ■: CB–ELSA, —— : SAID, ▨: systematic errors

PRELIMINARY RESULTS ON π^0 AND η PHOTOPRODUCTION

The preliminary results of CB–ELSA on the differential cross sections for neutral pion photoproduction as shown in Fig. 4 show a good agreement with the SAID model that had been fitted to previously measured data. Even the complicated structures indicating contributions from several nucleon and Δ resonances are well reproduced. Towards high photon energies, resonance contributions start to vanish around 2.2 GeV, a strong peak to forward angles of the meson develops, suggesting the production of pions via t–channel exchange.

The preliminary results of measured angular distributions for the η meson are shown in Fig. 5. Due to its isospin $I = 0$, only nucleon resonances can be excited in the intermediate state. The cross sections show fewer structures as in the π^0 case. Data is in good agreement with measurements from TAPS [8], GRAAL [9], and CLAS [10], and the SAID and MAID models [6, 7] below about 1.4 GeV. Close to threshold, the dominance of the $S_{11}(1535)$ is clearly seen as an isotropic distribution, while going to higher energies, an interference term of this resonance with a P wave appears, resulting in a distribution proportional to $\cos\theta_{cms}$, which changes sign around 1 GeV photon energy. At photon energies around 1.9 GeV the contribution of s–channel resonances disappears, a strong forward peaking (t–channel exchange) is observed. The rise of the cross sections at backward meson angles might indicate contributions in the u–channel.

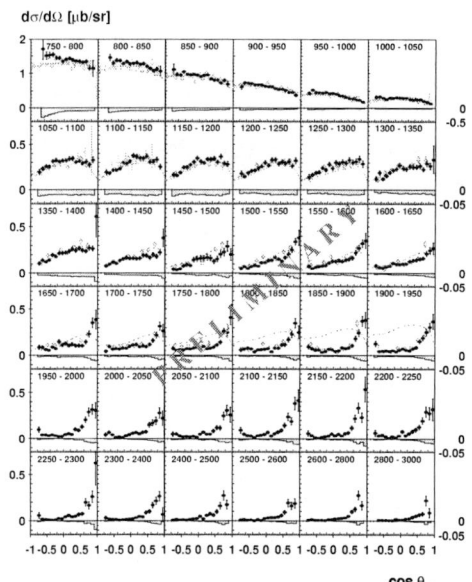

FIGURE 5. Differential cross sections of $\gamma p \to p\eta$ for photon energies between 0.75 and 3.0 GeV. ■: CB–ELSA, ○: CLAS, △: GRAAL, ⋆: TAPS, - - -: SAID, ⋯: MAID, ▨: systematic errors

SUMMARY AND CONCLUSION

The first experimental phase of the CB–ELSA experiment yielded high–statistics and high–quality data on photoproduction of neutral mesons. Angular distributions of $\gamma p \to p\pi^0$ and $\gamma p \to p\eta$ were derived for photon energies up to 3.0 GeV covering almost the full angular range. The preliminary results of the CB–ELSA experiment are compatible with previously published experimental findings. New data points are added at high energies and forward angles. An isobar model analysis is currently being performed, the results of which will contribute to a more detailed comprehension of resonance properties. Statistics is even improved for data taken more recently with the TAPS detector as a fast trigger covering the forward direction with high granularity.

REFERENCES

1. U. Löhring, K. Kretzschmar, B.Ch. Metsch and H.R. Petry: EPJ **A 10** (2001) 309
2. The Particle Data Group: Phys. Rev. **D 66** (2002) 1
3. D.B. Lichtenberg: Phys. Rev. **178** (1969) 2197
4. R. Koniuk and N. Isgur: Phys. Rev. **D 21** (1980) 1868
5. S. Capstick, W. Roberts: Phys. Rev. **D 49** (1994) 4570
6. http://gwdac.phys.gwu.edu/
7. http://www.kph.uni-mainz.de/maid/
8. B. Krusche et al.: Phys. Rev. Lett. **74** (1995) 3736
9. F. Renard et al.: Phys. Lett. **B 528** (2002) 215, J. Ajaka et al.: Phys. Rev. Lett. **81** (1998) 1797
10. M. Dugger et al.: Phys. Rev. Lett. **89** 2002 222002

Localizable Effective Theories, Bootstrap and the Parameters of Hadron Resonances

K. Semenov-Tian-Shanski[*], A. Vereshagin[†*] and V. Vereshagin[*]

[*]*St. Petersburg State University*
[†]*University of Bergen*

Abstract. We discuss the basic principles of constructing a meaningful perturbative scheme for effective theory. The main goal of this talk is to explain the approach and to present recent results [1] – [5] obtained with the help of the method of Cauchy forms in several complex variables.

Our work is aimed to develop a field-theoretic scheme providing the basis for dual models and, simultaneously, a link between two methods – the quantum field theory and the analytic theory of S-matrix. Besides, it concerns with the problem of renormalization of conventionally nonrenormalizable theories.

We rely on Weinberg's general scheme (see Chs. 2-5 of [6]) of constructing a quantum theory in terms of field operators that create and annihilate the *true* asymptotic states (corresponding to stable particles). Following [7], we call a theory effective if the interaction Hamiltonian (in the interaction picture) contains *all* terms consistent with the requirements of a given algebraic (linear) symmetry. The scheme is quantum *ab initio* and the problem of dynamical (nonlinear) symmetries requires special consideration. Here we do not discuss it.

We only consider a special class of *localizable* effective theories. An initial effective theory (constructed according to Weinberg's scheme solely from the fields of true asymptotic states) is called localizable if its tree level amplitudes can be reproduced in the framework of the well-defined tree level approximation based on the Hamiltonian of extended effective theory which – along with the fields of stable particles – also contains complementary auxiliary fields corresponding to fictitious unstable particles (resonances) of arbitrary high spin and mass. This implies that infinite sums of graphs providing formal expressions for tree level amplitudes of the initial theory converge, at least in certain small domains of the corresponding complex spaces. The existence of such domains is, in any case, necessary to assign meaning to the initial theory. It is precisely those domains where the tree level amplitudes of both theories must coincide with each other. Thus the extended theory just provides an analytic continuation of the tree level amplitudes constructed in the framework of the initial effective theory. In fact, our approach is just an attempt to extend Weinberg's quasiparticle method (see [8] and refs. therein) to the case of relativistic quantum theory.

It should be stressed that we only consider the case when the extended effective theory (as well as the initial one) does not contain massless particles of spin J > 1/2. Besides, we are only interested in constructing the effective *scattering* theory – the calculation of

CP717, *Hadron Spectroscopy: Tenth International Conference,*
edited by E. Klempt, H. Koch, and H. Orth
© 2004 American Institute of Physics 0-7354-0197-7/04/$22.00

Green functions is not implied. This means that the renormalization procedure possesses certain specific features allowing one to avoid attracting unnecessary renormalization prescriptions. The divergences which might occur in Green functions off the mass shell never bother us.

The last point deserves comment. In the case of customary renormalizable theories the number of renormalization prescriptions (RP) fixing the finite parts of counterterms is equal to that of coupling constants in the Hamiltonian (including mass and kinetic terms), *this latter one being finite.* "Hidden" couplings create no problem (see, e.g., [9]). In case of effective theories the situation looks quite different. They are renormalizable by construction because all possible local monomials are presented in the Hamiltonian. The problem is that the number of coupling constants is essentially infinite. This means that, to obtain finite results for all Green functions, one needs to point out a complete self-consistent infinite set of independent RP's. The structure of this set must provide a guarantee of convergence of infinite series appearing at every given order of loop expansion due to the presence of field derivatives of arbitrary high degree and order in the effective Hamiltonian. Otherwise, the resulting amplitudes would make no sense. The problem of constructing of such a set looks unsolvable until we have no regularity fixing its structure (possibly, up to a finite number of independent constants). The requirement of localizability is extremely useful in this very respect. As shown in [4], [5], in case of effective scattering theory it is possible to perform a detailed classification of parameters (combinations of coupling constants) appearing in expressions for the renormalized S-matrix elements of extended effective theory. This allows one to separate a group of *resultant* parameters. This group only contains the parameters which do contribute to renormalized S-matrix elements and thus require formulating the RP's. Finally, it is possible to show that crossing symmetry together with the requirement of convergence impose strong limitations on the allowed values of resultant parameters. In the case of tree level parameters describing the amplitudes of binary processes these limitations take a form of an infinite system of (algebraic) *bootstrap* equations connecting the values of the resultant coupling constants and the mass parameters appearing in the Hamiltonian. It is a direct consequence of the postulated properties of meromorphy and polynomial boundedness of the tree level amplitudes in every 3-dimensional band $B_x\{x \in \mathbf{R}, \ x \sim 0; \ v_x \in \mathbf{C}\}$, where x stands for (real) momentum transfer and v_x – for the corresponding (complex) energy-like variable.

We demand such properties for the following reasons. First, the polynomial boundedness property of the full (non-perturbative) amplitudes follows from the general axiomatic requirements. Hence it makes sense to construct the perturbation series in such a way that at every step the perturbative amplitude possesses the property of polynomial boundedness with the same degree of bounding polynomial as that of the full (non-perturbative) amplitude – this gives a chance to avoid strong corrections from the higher order terms. Besides, this requirement provides a guarantee that we deal with tempered distributions. Second, the property of meromorphy of tree level amplitudes follows from the *summability* requirement. The meaning of this term can be explained as follows. At every given order of loop expansion we have to take account of an infinite number of graphs. In absence of guiding principle fixing the summation order there is a danger either to get diverging series or to fall in contradiction with basic properties like crossing symmetry. That is why it looks reasonable to impose the summability requirement in the following form. *In every sufficiently small domain of the complex space of kinematical variables there must exist an appropriate order of summation of the formal sum of contributions coming from the graphs of a given loop order, such that the reorganized series happens convergent. Altogether, these series must define a unique analytic function with only those singularities which are presented in contributions of individual graphs or can be reproduced in the framework of the same loop order of extended theory containing auxiliary fields corresponding to unstable particles.* This definition can be considered as a generalization (or, better, detailing) of the localizability requirement. From this formulation it follows that to study the most general localizable effective theory it is quite sufficient to consider the case of extended theory with no *ad hoc* limitations on the structure of set of resonance parameters. In turn, this means that the tree level amplitudes of the extended theory must be meromorphic functions of kinematical variables. They must be polynomially bounded in certain energy-like variables at fixed values of the other ones,the degree of bounding polynomials being dictated by unitarity.

Here it is pertinent to stress one point. To present the results of resonance saturation of dispersion relations (DR) for the amplitudes with nondecreasing asymptotic behavior the following form is often used:

$$A(z,x) = P(z,x) + \sum_{k=1}^{\infty} \left[\frac{r_k(x)}{z - p_k(x)} \right]. \tag{1}$$

Here z stands for energy-like variable, x – for any real parameter (say, the momentum transfer) and $P(z,x)$ – for so-called subtracting polynomial in z with the coefficients depending on x. However, it should be noted that in the case of infinite number of resonances this form is to be taken just as a formal one. Usually, it is tacitly assumed that the series of pole contributions in the RHS converges. Unfortunately, this is not always true (see, e.g., [11]). This suggestion imposes too strong limitations on the values of resonance parameters $r_k(x)$ and $p_k(x)$. In fact, it states that resonances do not affect the asymptotic behavior of the amplitude in question. In case of strong interactions this looks too restrictive. For this reason in our work we use the method of Cauchy forms specially adjusted for the case of several variables (see [2], [4]); this allows us to avoid implicit postulating of the resonance spectrum properties. Using this method it is easy to write down the most general form of the result. In case when the amplitude has only simple poles at $z_k = p_k(x) \neq 0$ with residues $r_k(x)$ and behaves, say, like a constant (it only makes sense to speak about the contour asymptotics) for $x \in (a,b)$ and $|z| \to \infty$, it looks as follows

$$A(z,x) = A(0,x) + \sum_{k=1}^{\infty} \left[\frac{r_k(x)}{z - p_k(x)} + \frac{r_k(x)}{p_k(x)} \right], \qquad x \in (a,b). \tag{2}$$

It is possible to show (see [2]) that in this case the functional series $S_n = \sum_{k=1}^{\infty} \frac{r_k(x)}{p_k^n(x)}$ is certainly convergent for $n \geq 2$ and may happen divergent for $n < 2$. In case when $S_1 \leq M < \infty$ the only possibility to fulfil the postulated asymptotic behavior is to demand that $A(0,x) \neq 0$. This means that the asymptotics is completely formed by $A(0,x)$, i.e., the contribution of resonances is irrelevant. These problems disappear if one uses the Cauchy form (2).

The summability requirement imposes certain restrictions on the parameters of theory. As shown in cited above papers, those restrictions for the tree level parameters follow from the condition of identical coincidence of two Cauchy forms representing the amplitudes of cross-conjugated processes. Each one of these forms is only applicable in the corresponding 3-dimensional band (layer). In the intersection of those layers both forms are equally applicable and thus must coincide. This requirement leads to an infinite system of algebraic relations between the resultant parameters.

The corresponding mechanism may be illustrated by the following example. Consider a rational function of two complex variables $F(x,y)$. Let us suppose it has a single simple pole (in x) in the layer $B_y\{\ y \in \mathbf{R},\ y \in (-\eta,+\eta);\ x \in \mathbf{C}\}$ and also a single simple pole (in y) in the 'orthogonal' layer $B_x\{\ x \in \mathbf{R},\ x \in (-\xi,+\xi);\ y \in \mathbf{C}\}$. Let us also assume that $F(x,y)$ is decreasing at infinity in each layer and that it is regular at the origin $M(0,0)$. Now let us try to answer the following question: what is the structure of the set of numerical parameters providing a complete description of functions that possess these properties?

Every function regular at the origin is completely fixed by the coefficients f_{ij} of its series expansion $F(x,y) = \sum f_{ij}x^i y^j$. The above question can be rephrased in a more concrete way: how many independent combinations of these coefficients can be arbitrary and what are these combinations? Or, in terms of field theory: how many independent renormalization prescriptions is it necessary to impose in order to completely fix the amplitude $F(x,y)$ and what is the explicit form of those prescriptions?

In the layer B_y we have $F(x,y) = \frac{\rho(y)}{x - \pi(y)}$. By condition, the functions $\rho(y)$ and $\pi(y)$ are regular in the vicinity of the origin. Hence, $\pi(y) = \sum \pi_i y^i$, $\rho(y) = \sum\limits_{i=0} \rho_i y^i$. By analogy, in B_x: $F(x,y) = \frac{r(x)}{y - p(x)}$, where $p(x) = \sum p_i x^i$, $r(x) = \sum r_i x^i$. Hence in the intersection domain $B_x \cap B_y \equiv D_{xy}$:

$$\frac{r(x)}{y - p(x)} = \frac{\rho(y)}{x - \pi(y)}, \quad (x,y) \in D_{xy}\{x \in (-\xi,+\xi),\ y \in (-\eta,+\eta)\} . \qquad (3)$$

Substituting $\pi(y), \rho(y), p(x)$ and $r(x)$ into (3), we obtain an infinite system of conditions on the coefficients p_k, r_k, π_k, ρ_k:

$$r_{i+1}\pi_0 - p_{i+1}\rho_0 = r_i, \quad p_{i+1}p_0 - \pi_{i+1}r_0 = p_i, \quad r_{i+1}p_{j+1} = p_{i+1}\pi_{j+1} \quad i,j = 0,1,\dots \quad (4)$$

This system provides an example of what we call the bootstrap equations. Once solved, it permits to express the parameters p_i, r_i in terms of π_i, ρ_i. So it gives an answer to the question wether it is possible to carry out the analytic continuation from one layer to another. This is an infinite system of equations with respect to $2 \times \infty$ (formal notation!) unknown parameters, needed to reexpress the function $F(x,y)$ in the layer B_x in terms of the parameters defining it in the layer B_y. In general, it is very difficult to find solutions of such systems and even to show their solvability. Fortunately, in this simple example it turns out possible to write down the solution in explicit form. Separating the variables in (3), taking derivatives and solving the corresponding ordinary differential equations, one finds:

$$F(x,y) \doteq \frac{ad + bc}{-d + axy + bx + cy}. \qquad (5)$$

The important property of this result is that it contains only 4 arbitrary parameters! This means that the infinite system (4) only happens consistent if the function $F(x,y)$ defined in the layer B_y belongs to the four-parametric family (5). This is the only case when there exists the analytic continuation of this function from B_y into B_x with the desired properties. It is clear that in this case the continuation is unique.

This exercise gives an idea of the "power" of bootstrap restrictions. The direct analysis of the system (4) would lead to the same conclusion. In this simple example it happens possible. Unfortunately, the regular method of solving infinite algebraic systems is not known, except few trivial cases.

With the help of (5), one can express the parameters $f_{ij} = f_{ij}(a_1, a_2, a_3, a_4)$ in terms of "fundamental constants" a_i ($i = 1, \ldots, 4$). Then one can choose four arbitrary coefficients f_k ($k = 1, 2, 3, 4$) (or four arbitrary combinations) that allow the inversion $a_i = a_i(f_1, \ldots, f_4)$, and impose arbitrary "renormalization prescriptions" for these four quantities. The values of all other parameters should respect the conditions (4).

So, *to fix the amplitude $F(x,y)$ uniquely it is sufficient to impose four independent renormalization prescriptions defining the "fundamental" constants a, b, c, d.*

Precisely the same mechanism provides the system of bootstrap constraints for the parameters of pion-nucleon resonances (see [10]). The most important feature of this system is the renormalization invariance: *bootstrap equations are nothing but the restrictions for renormalization prescriptions.* This very property allows us to compare the bootstrap equations directly with known experimental data.

ACKNOWLEDGMENTS

The work was supported in part by INTAS (project 587, 2000), RFBR (grant 01-02-17152) and by Russian Ministry of Education (Programme Universities of Russia, project 02.01.001). The work by A. Vereshagin was supported by Meltzers Høyskole-fond (Studentprosjektstipend 2003).

REFERENCES

1. V. Vereshagin, *Phys. Rev. D* **55**, 5349 (1997).
2. A. Vereshagin and V. Vereshagin, *Phys. Rev. D* **59**, 016002 (2000).
3. A. Vereshagin, *πN Newsletter* **16**, 426 (2002).
4. A. Vereshagin, V. Vereshagin, and K. Semenov-Tian-Shanski, *Zap. Nauchn. Sem. POMI* **291**, Part 17, 78 (2002); English version is to appear in *J. Math. Sci. (NY)*, (2003).
5. A. Vereshagin and V. Vereshagin, hep-th/0307256, submitted to *Phys. Rev. D.*
6. S. Weinberg, *The Quantum Theory of Fields*, Cambridge University Press, Cambridge, 2000, vols. 1-3.
7. S. Weinberg, *Physica* **96A**, 327 (1979).
8. S. Weinberg, *Phys. Rev.* **133**, 1B232 (1964).
9. J. C. Collins, *Renormalization*, Cambridge University Press, Cambridge, 1984.
10. K. Semenov-Tian-Shanski, A. Vereshagin and V. Vereshagin, "Bootstrap and the parameters of pion-nucleon resonances" in these Proceedings.
11. B. V. Shabat, *An introduction to complex analysis*, Nauka, Moscow, 1969 (in Russian).

Bootstrap and the Parameters of Pion-Nucleon Resonances

K. Semenov-Tian-Shanski[*], A. Vereshagin[†*] and V. Vereshagin[*]

[*]St.Petersburg State University
[†]University of Bergen

Abstract. In this talk we demonstrate the results of application of the perturbative effective theory formalism developed in papers [1] – [6] to the calculation of πN elastic scattering amplitude. Restrictions on the contributing resonance parameters are obtained and the low energy coefficients are calculated.

INTRODUCTION

In [1] – [6] it is shown that when working in effective theory formalism (in the sense of Weinberg), the assumption that the perturbation theory (loop expansion for the scattering amplitudes) is self consistent, together with the general requirements of covariance, unitarity, causality and crossing, leads to certain restrictions for the effective Hamiltonian parameters. Moreover, using concrete renormalization scheme, it is also possible to obtain constraints (the *bootstrap equations*) for the *physical* parameters of the given amplitude. In other words: one can obtain restrictions for the particle spectrum and, thus, perform a comparison with the experiment.

We are going to discuss how to obtain those restrictions in case of πN-elastic scattering. As an example, we make the accurate estimate of the tensor-to-vector $\rho N N$ coupling ratio in complete agreement with the experimental data which has never been explained in model-independent way. Besides, we present the values of the first 48 coefficients in the expansion of the tree amplitude around the crossing symmetry point.

The mathematical background for these calculations and the formalism used is reviewed in more details in the talk [7].

πN ELASTIC SCATTERING

The amplitude $M_{a\alpha}^{b\beta}$ of the reaction $\pi_a(k) + N_\alpha(p,\lambda) \rightarrow \pi_b(k') + N_\beta(p',\lambda')$ can be written in the following form:

$$M_{a\alpha}^{b\beta} = i(2\pi)^4 \delta(k+p-k'-p')\left\{ \delta_{ba}\delta_\alpha^\beta M^+ + i\varepsilon_{bac}(\sigma_c)_{\cdot\alpha}^{\beta\cdot}M^- \right\} \ .$$

Here

$$M^\pm = \bar{u}(p',\lambda')\left\{ A^\pm + \left(\frac{k+k'}{2}\right)B^\pm \right\}u(p,\lambda) \ ,$$

CP717, *Hadron Spectroscopy: Tenth International Conference*,
edited by E. Klempt, H. Koch, and H. Orth
© 2004 American Institute of Physics 0-7354-0197-7/04/$22.00

$k \equiv k_\mu \gamma^\mu$, $a, b = 1, 2, 3$ and $\alpha, \beta = 1, 2$ stand for the isospin indices, λ, λ' — for polarizations of the initial and final nucleons, respectively, $\bar{u}(p', \lambda')$, $u(p, \lambda)$ — for Dirac spinors, and σ_c, $c = 1, 2, 3$ — for Pauli matrices. The invariant amplitudes $A^\pm$ and $B^\pm$ are the functions of an arbitrary pair of scalar kinematical variables $s \equiv (p + k)^2$, $t \equiv (k - k')^2$, and $u \equiv (p - k')^2$.

To construct the *tree amplitude* one needs to write down the contributions of all possible contact vertices and resonance exchange graphs.

We work in the framework of effective theory formalism. This means that, when constructing the Hamiltonian, we need to take account of *all* the terms consistent with (algebraic) symmetry properties of strong interactions; there are no limitations on the number and order of field derivatives. Besides, in order to avoid model dependence we reserve the possibility to work with infinite number of resonance fields and unbounded (though, of course, discrete) mass spectrum.

Altogether this means that the number of items contributing to the tree level amplitude is actually infinite. This creates a problem: we have no guiding principle allowing to fix the order of summation. The way out of this difficulty has been pointed out in [1] – [6]. It consists of switching to the *minimal* parametrization for the Hamiltonian and using the method of Cauchy forms. The important advantage of this approach is that it results in uniformly converging series of pole terms defining the amplitude as the polynomially bounded meromorphic function – no kind of singularities but simple poles can appear on this way. To construct the Cauchy form for the tree amplitude under consideration, one needs to establish the residues (which are the function of coupling constants and masses) at the corresponding pole terms (masses) and to fix the bounding polynomial degree — it happens quite sufficient for fixing the amplitude up to few unknown functions which, in turn, can be found from the *bootstrap equations*

The origin of bootstrap equations is quite natural. Using the technique of Cauchy forms, we can get well defined uniformly convergent expansions for the invariant amplitudes (we do not write them down here due to the lack of space) in three different bands on the Mandelstam plane: $B_s\{s \sim 0\}$, $B_t\{t \sim 0\}$ and $B_u\{u \sim 0\}$. This bands obviously has non-empty intersections (near the corners of Mandelstam's triangle), and the corresponding Cauchy forms are different in each band. Since we need the tree amplitude to posses crossing symmetry, each invariant amplitude should be a meromorphic function on all the Mandelstam plane. Thus the relevant Cauchy forms should coincide in the band intersection domains. This results to the set of functional equations (bootstrap equations) for the tree level invariant amplitudes, or, the same, to infinite set of numerical equations for Hamiltonian parameters[1].

If one uses the renormalized perturbation theory and imposes the physical renormalization prescriptions, in which the tree amplitude is expressed in terms of *physical* parameters, then the bootstrap equations becomes the restrictions for the physical (measurable) spectrum. In other words, the obtained bootstrap equations remains true after renormalization.

---

[1] It is interesting to note that in case of e.g. πN-elastic scattering some of those equations give explicit relations between bosonic and fermionic spectrum parameters, thus, demonstrating certain *supersymmetry* features.

It is these equations that can be tested substituting experimental data for resonance masses and widths. They also give a possibility to express one resonance parameter via the other, which, again, can be compared with the known data.

CALCULATION OF G_T/G_V

The quantities $G^T_{NN\rho}$ and $G^V_{NN\rho}$ (our minimal parametrization couplings can be related to them) were defined and fitted in [8] as couplings in the following effective Hamiltonian:

$$H^{NN\rho}_{eff} = -\bar{N}\left[G^V_{NN\rho}\gamma_\mu\vec{\rho}^\mu - G^T_{NN\rho}\frac{\sigma_{\mu\nu}}{4m}(\partial^\mu\vec{\rho}^\nu - \partial^\nu\vec{\rho}^\mu)\right]\frac{1}{2}\vec{\sigma}N,\qquad(1)$$

where σ_a are Pauli matrices and m is the proton mass.

The existing experimental data [8] give:

$$\frac{G^T_{NN\rho}}{G^V_{NN\rho}} \approx 6.1\ ,\quad \frac{G_{\pi\pi\rho}G^V_{NN\rho}}{4\pi} \approx 2.4\ ,\quad G_{\pi\pi\rho} \approx 6.0\ .\qquad(2)$$

Taking the relevant bootstrap equations (here - 2 of them) we treat the above couplings as unknown and express them via other resonance parameters[2], the resulting numerical equations being in complete agreement with (2) with 15% accuracy.

It should be noted, that the G_T/G_V ratio was recently calculated by the authors in the frame of KN-elastic scattering, again, in complete agreement with the experiment (to be published).

LOW-ENERGY COEFFICIENTS

Using the Cauchy forms technique, we have calculated the coefficients in the expansion of the *tree* amplitude around the crossing symmetry point ($t, \nu_t \equiv s - u = 0$). This coefficients certainly *will* be affected by loop corrections, however, as one can see from the Table 1, the tree level results are very close to the experimental values — this fact gives a hope that our way of constructing the tree amplitude [1]-[6] leads to nice convergence of loop expansion, at least, in low energy domain. In other words, the tree approximation gives nice description of the physical amplitude at low energies[3].

Introducing the new quantity

$$C^\pm = A^\pm + \frac{m\nu_t}{4m^2 - t}\tilde{B}^\pm,$$

---

[2] These particular equations seems to converge fast: among the known resonances only $N(0.94)$, $N(1.44)$, $\Delta(1.23)$ and one meson — $\rho(0.77)$, give significant contributions, other possible contributions are suppressed by the inverse squares of their mass.

[3] It should be noted that, in all the cases we checked, the bootstrap equations are consistent with the experimental data only if the tree amplitude asymptotic is taken in accordance with the corresponding Regge intersept. In other words, the tree amplitude *shall* have the asymptotic close to the physical one.

TABLE 1. Low energy coefficients (calculated at the tree level) and their experimental values (averaged). In the case of A^- it is meaningless to calculate errors: the corresponding quantities are too sensitive to the uncertainties in experimental data.

$\tilde{B}^+$	b_{00}^+	b_{01}^+	b_{02}^+	b_{03}^+	b_{10}^+	b_{11}^+	b_{20}^+	b_{21}^+
Experiment	-3.50 ±0.10	$+0.22$ ±0.10	-0.10 ±0.05	-0.00036 ±0.00004	-0.99 ±0.01	$+0.095$ ±0.015	-0.31 ±0.01	$+0.42$ ±0.08
Theory	-4.96	$+0.18$	-0.004	$+0.0001$	-1.00	$+0.07$	-0.19	$+0.02$

$\tilde{B}^-$	b_{00}^-	b_{01}^-	b_{02}^-	b_{03}^-	b_{10}^-	b_{11}^-	b_{20}^-	b_{21}^-
Experiment	$+8.37$ ±0.23	$+0.19$ ±0.07	$+0.019$ ±0.007	$+0.0021$ ±0.0002	$+1.08$ ±0.04	-0.063 ±0.011	$+0.30$ ±0.04	-0.32 ±0.07
Theory	$+8.56$	-0.071	$+0.002$	$+0.00003$	$+1.44$	-0.063	$+0.22$	-0.018

A^+	a_{00}^+	a_{01}^+	a_{02}^+	a_{03}^+	a_{10}^+	a_{11}^+	a_{20}^+	a_{21}^+
Experiment	$+25.5$ ±0.5	$+1.18$ ±0.05	$+0.035$ ±0.007	$+0.0060$ ±0.0005	$+4.60$ ±0.12	-0.051	$+1.19$ ±0.07	-0.056
Theory	$+30.2$	$+1.1$	$+0.04$	$+0.007$	$+6.28$	-0.25	$+1.23$	-0.087

C^+	c_{00}^+	c_{01}^+	c_{02}^+	c_{03}^+	c_{10}^+	c_{11}^+	c_{20}^+	c_{21}^+
Experiment	$+25.5$ ±0.5	$+1.18$ ±0.05	$+0.035$ ±0.007	$+0.0060$ ±0.0005	$+1.12$ ±0.02	$+0.15$ ±0.01	$+0.20$ ±0.01	$+0.034$ ±0.010
Theory	$+30.2$	$+1.1$	$+0.04$	$+0.007$	$+1.3$	-0.10	$+0.22$	-0.023

A^-	a_{00}^-	a_{01}^-	a_{02}^-	a_{03}^-	a_{10}^-	a_{11}^-	a_{20}^-	a_{21}^-
Experiment	-8.87	-0.34	$+0.1$	-0.0021	-1.25	$+0.023$	-0.338	$+0.305$
Theory	-9.85	$+0.2$	-0.004	$+0.00007$	-1.55	$+0.08$	-0.27	$+0.023$

C^-	c_{00}^-	c_{01}^-	c_{02}^-	c_{03}^-	c_{10}^-	c_{11}^-	c_{20}^-	c_{21}^-
Experiment	-0.50 ±0.05	-0.10 ±0.01	$+0.12$ ±0.04	$+0.00032$ ±0.00003	-0.17 ±0.01	-0.039 ±0.005	-0.038 ±0.004	-0.013 ±0.004
Theory	-0.6	$+0.09$	-0.0019	$+0.0001$	-0.18	$+0.026$	-0.035	$+0.006$

where $\tilde{B}^\pm$ is just $B^\pm$ with the nucleon pole subtracted[4], we define the low-energy coefficient $b_{mn}^\pm$, $a_{mn}^\pm$, and $c_{mn}^\pm$ in the following way:

$$\tilde{B}^+(t,v_t) = v_t \sum_{m,n} b_{mn}^+ (v_t^2)^m t^n, \quad \tilde{B}^-(t,v_t) = \sum_{m,n} b_{mn}^- (v_t^2)^m t^n,$$

$$\tilde{A}^+(t,v_t) = \sum_{m,n} a_{mn}^+ (v_t^2)^m t^n, \quad \tilde{A}^-(t,v_t) = v_t \sum_{m,n} a_{mn}^- (v_t^2)^m t^n,$$

---

[4] That is what can be compared with the experiment: the nucleon pole contribution is dominant in this momentum region but can be excluded in experimental data analysis. On the other hand, in our formulas we have this contribution explicitly and can simply remove it by hand.

$$\tilde{C}^{+}(t, v_t) = \sum_{m,n} c^{+}_{mn} (v_t^2)^m t^n, \quad \tilde{C}^{-}(t, v_t) = v_t \sum_{m,n} c^{-}_{mn} (v_t^2)^m t^n,$$

where all the expansions are around the point $t, v_t = 0$. Re-expanding corresponding Cauchy forms around this point in the above (Taylor) form, using experimental data for couplings and masses and neglecting all the contributions of the resonances with $M \geq 1.9$ GeV, we get numerical values for the coefficients[5] (see Table 1).

Actually, among baryons only $\Delta(1.23)$ and $N(1.44)$ give non-negligible contributions as well as σ among mesons, all other known resonances give less then 10%.

ACKNOWLEDGMENTS

The work was supported in part by INTAS (project 587, 2000), RFBR (grant 01-02-17152) and by Russian Ministry of Education (Programme Universities of Russia, project 02.01.001). The work by A. Vereshagin was supported by Meltzers Høyskole-fond (Studentprosjektstipend 2003).

REFERENCES

1. V. Vereshagin, *Phys. Rev. D* **55**, 5349 (1997).
2. A. Vereshagin and V. Vereshagin, *πN Newsletter* **15**, 288 (1999).
3. A. Vereshagin and V. Vereshagin, *Phys. Rev. D* **59**, 016002 (2000).
4. A. Vereshagin, *πN Newsletter* **16**, 426 (2002).
5. A. Vereshagin, V. Vereshagin, and K. Semenov-Tian-Shanski, *Zap. Nauchn. Sem. POMI* **291**, Part 17, 78 (2002); English version is to appear in *J. Math. Sci. (NY)*, (2003).
6. A. Vereshagin and V. Vereshagin, hep-th/0307256, submitted to *Phys. Rev. D*.
7. K. Semenov-Tian-Shanski, A. Vereshagin and V. Vereshagin, "Bootstrap and the parameters of pion-nucleon resonances" in these Proceedings.
8. M. M. Nagels et al., *Nucl. Phys.* **B109**, 1 (1976); O. Dumbrajs et al., *Nucl. Phys.* **B216** 277 (1983).

---

[5] The experimental data can be found in [8]: please note, that they use somewhat different definitions for low-energy coefficient, so one needs to perform certain recalculations to compare the results.

Study of π⁻p charge exchange scattering for forward angles

D.E.Bayadilov, Yu.A.Beloglazov, E.A.Filimonov, M.R.Kan,
N.G.Kozlenko, S.P.Kruglov, I.V.Lopatin, D.V.Novinsky, A.K.Rad'kov,
V.V.Sumachev

*Petersburg Nuclear Physics Institute, High Energy Physics Department, Meson Physics
Laboratory,188300, Gatchina, Leningrad distr., Russia*

Abstract. Differential cross sections of reaction π⁻p→π°n are measured for forward scattering angles. The experiment is carried out at the pion channel of the PNPI synchrocyclotron in Gatchina at ten momenta of incident pions in the range from 417 to 710 MeV/c. Measurements were made using the PNPI π°–spectrometer by detecting both photons from the decay π°→2γ. A brief description of the experimental setup is given, and results obtained are presented.

INTRODUCTION

Measurement of differential cross sections (DCS) of π⁻p charge exchange scattering π⁻p→π°n is a part of general program "Spectroscopy of nonstrange baryons using pion beams with energies from 300 to 2000 MeV", which is being performed by PNPI physicists since the 70ties. At present, an accuracy of determining characteristics of nonstrange excited baryons (i.e. π N resonances) is limited mainly by a lack of high quality experimental data on DCS of the π⁻p charge exchange scattering[1], especially in the region of low-lying pion-nucleon resonances. The only systematic set of such data was published in 1975 by Brown *et al.* [1] (Rutherford Appleton Laboratory, England) but among physicists involved in compiling πN data there exists an opinion that results of experiment [1] have rather big and hardly estimated systematic uncertainty due to an error in absolute calibration of the incident pions momenta. An attempt to measure CEX DCS was made at LAMPF in 1980, but results of this experiment have not been published - apparently because of some problem with systematic uncertainties. At last, CEX DCS were measured in the region of low-lying πN resonances at Brookhaven National Laboratory using the Crystal Ball, however till now only preliminary results in the restricted momentum range (from 147 to 322 MeV/c) are published [2]. And, again, the main problem is a reliable estimation of systematic uncertainties.

To fill in the existing gap in data-base and to solve ambiguities of experimental data on CEX DCS, two experiments were undertaken at PNPI. Measurements were made at the pion channel of the PNPI synchrocyclotron [3] and covered the energy range from

---

[1] Below we will note this reaction as CEX.

CP717, *Hadron Spectroscopy: Tenth International Conference,*
edited by E. Klempt, H. Koch, and H. Orth
© 2004 American Institute of Physics 0-7354-0197-7/04/$22.00

300 to 585 MeV (corresponding values of the incident pions momenta are from 417 to 710 MeV/c). The momentum spread of the pion channel is typically 6% (FWHM).

At the first stage, DCS of reaction $\pi^- p \to \pi^\circ n$ to the backward hemisphere were measured. The experiment was carried out by detecting the recoil neutron in coincidence with one of photons from the decay $\pi^\circ \to 2\gamma$.

Our further plans are to extend measurements to the region of smaller scattering angles. However, the method based on detecting the recoil neutrons has a limitation connected with the fact that for angles $\Theta^{c.m.} < 50°$ the energy of the recoil neutrons drops below 50 MeV and the efficiency of their detection becomes too small.

π° SPECTROMETER

To overcome this difficulty, we have designed and built at PNPI a new device - π°–spectrometer [4], which allowed us to extent the angular range of measuring CEX DCS up to forward angles.

The main principle of the π°–spectrometer is to determine E_{π° and Θ_{π° by measuring energies $E_{\gamma1}$, $E_{\gamma2}$ of two photons from the π° decay and angles $\Theta_{\gamma1}$, $\Theta_{\gamma2}$ of their emission and opening angle $\Psi_{\gamma\gamma}$ between these two photons. $E_{\gamma1}$, $E_{\gamma2}$, $\Theta_{\gamma1}$, $\Theta_{\gamma2}$ having been measured, the total energy E_{π° and emission angles Θ_{π° of π° meson (in the lab. system) can be evaluated using equations:

$$
E_{\pi^0} = \left[\frac{2M_{\pi_0}^2}{\left(1 - \cos\Psi_{\gamma\gamma}\right)\left(1 - X^2\right)} \right]^{1/2}
\tag{1}
$$

$$
\cos\Theta_{\pi^0} = \frac{E_{\gamma1}\cos\Theta_{\gamma1} + E_{\gamma2}\cos\Theta_{\gamma2}}{\sqrt{E_{\gamma1}^2 + E_{\gamma2}^2 + 2E_{\gamma1}E_{\gamma2}\cos\Psi_{\gamma\gamma}}}
\tag{2}
$$

where M_{π° is the mass of π° meson, $X=(E_{\gamma1}-E_{\gamma2})/(E_{\gamma1}+E_{\gamma2})$ characterizes the sharing of energy between two gammas and $\Psi_{\gamma\gamma}$ is the angle between the two gammas.

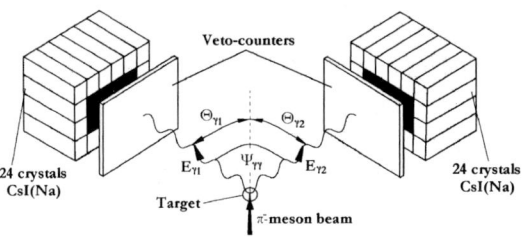

FIGURE 1. Schematic view of π^0 spectrometer.

A principal scheme of the π°–spectrometer is shown in Fig.1. The device consists of two total absorption electromagnetic calorimeters for detecting both gammas from the $\pi°$ decay. Each calorimeter is a 6×4 array of CsI(Na) crystals with an individual size of 6×6×30 cm³; the last figure is a thickness, it corresponds to 16.2 X_0.

The procedure of counter preparation included the following operations. First of all, the crystal surface was covered with transparent lacquer against air humidity. Then the crystal was wrapped by the white teflon tape of 70 μm thickness and by the aluminized mylar of 20 μm thickness, and placed into the container made of a black paper. The total package thickness was about 200 μm per one side. We used photomultiplier tubes (PMT) of FEU-97 type. Each PMT has an individual high voltage supplier mounted directly on a PMT base, which provides a good stabilization of a high voltage. This high voltage system was designed and manufactured at PNPI.

EXPERIMENT

The layout of the experimental setup is shown in Fig.2.

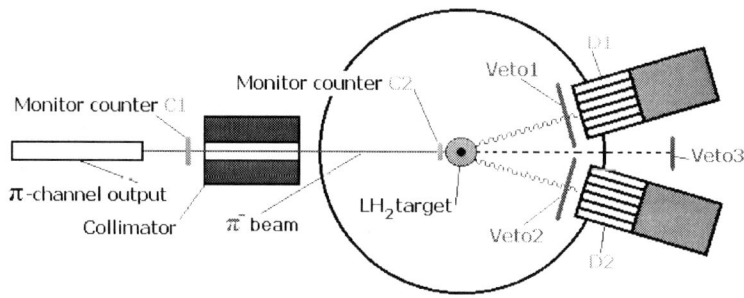

FIGURE 2. Experimental layout: LH$_2$ - liquid hydrogen target; D1 and D2 - right and left electromagnetic calorimeters of the $\pi°$-spectrometer; C1 and C2 - monitor counters; Veto3 - beam veto counter; Veto1 and Veto2 - veto counters placed in front of corresponding calorimeters.

Its basic elements are a liquid hydrogen target, a $\pi°$-spectrometer, and a set of monitor and veto counters. Since the experiment was planned to cover the range of incident pions momenta from 417 to 710 MeV/c, the angles between the axes of calorimeters and the π^--beam axis (in the symmetrical mode) were chosen to be equal 16° and the distance from the entrances of calorimeters to the target centre was 101 cm. In front of the calorimeters thin (5 mm) scintillation counters Veto1 and Veto2 were placed. The purpose of this veto counters was to ensure triggering the experimental setup only in cases when photons (but not charged particles emitted from the target) hit the calorimeter. The trigger was formed as the coincidence:

$$T = C1*C2*D1*D2*\underline{Veto3}*\underline{Veto1}*\underline{Veto2}$$

where C1 and C2 - the signals from monitor counters C1 and C2; D1 and D2 - the dynode signals from eight inner crystals of the right and left calorimeters, each of dynode signals was formed if the energy deposited in inner eight crystals of each calorimeter exceeds a definite threshold value; Veto3 - the signal from the veto counter placed at the beam axis just downstream of the calorimeters; Veto1 and Veto2 - the signals from the veto counters located in front of the calorimeters.

If the trigger T was formed, the following information was written on the PC hard disk: a charge collected from PMTs of each of 48 CsI(Na) crystals (readings of corresponding CDCs), time differences between passing the incident pion through the monitor counter C2 and entering photons from the $\pi^\circ \to 2\gamma$ decay to calorimeters D1 and D2 (readings of corresponding Time-to-Digital Converters TDCs), the number of monitor counts C1·C2 registered between this trigger and the previous one.

In the described experiment, a safety liquid hydrogen target (LHT) [5] with preliminary nitrogen cooling was used. An angle open for the detecting emitted particles was 270°. There is a window with an opening angle of 270° in the target housing, which is coated with a 200 μm-thick mylar film. A hydrogen container was a vertical cylinder 12 cm high and 10 cm in diameter, with walls made of 100 μm aluminum. The hydrogen liquefaction was performed by cooling the container with cold gaseous helium.

In order to evaluate a contribution of background processes – such as π^-p charge exchange scattering on protons of the scintillator C2[2] as well as π^-p quasifree charge exchange scattering on carbon nuclei of the scintillator C2 and aluminum nuclei of the walls of the LHT container - additional runs with an empty target were carried out at the same momenta at which measurements with a hydrogen-filled target were undertaken.

RESULTS

In Fig. 3 the obtained values $\dfrac{d\sigma^{cm}}{d\Omega}(0^0)$ are plotted as a function of the incident pion momentum. Shown by curves are momentum dependences predicted by two partial-wave analysis (PWAs) KH-80 and SM-02. The first PWA [6] was performed in 1980 by Prof.G.Hohler with collaborators and is used till now by the Particle Data Group for extracting characteristics of πN resonances presented in the Review of Particle Physics. The SM-02 PWA was made in 2002 by the group of physicists from the George Washington University headed by Prof.R.Arndt[3] (see http://gwdac.gwu.edu/analysis/pin_analysis.html); in contrast with the PWA KH-80, when performing SM-02 all πN experimental data published till the summer 2002 were included. One can see that there exists a marked disagreement between our new experimental results and the predictions of both PWAs - especially in the low-energy range.

---

[2] The counter C2 with the scintillator thickness of 2.5 mm was placed at a distance of 34 cm upstream of the LHT centre.
[3] Earlier the group of Prof. R.Arndt worked at Virginia Polytechnic Institute; the PWA procedure using by this group as well as characteristics of πN resonances extracted from results of this PWA are given in Ref. [7].

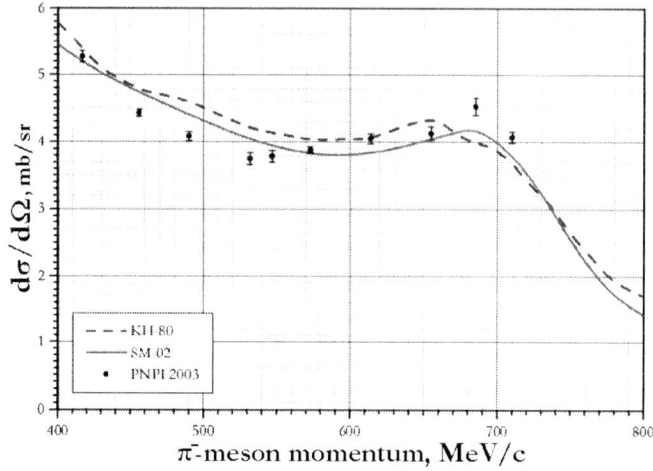

FIGURE 3. Momentum dependence of CEX DCS for $\Theta^{c.m.}=0°$ obtained in this experiment (•). Shown by curves are the predictions of different PWAs: solid curve - SM-02 (GWU), dashed curve - KH-80.

TABLE 1. Charge exchange DCS for $\Theta^{cm}=0°$

Momentun, MeV/c	Differential cross section (c.m.), mb/sr
417	5.27±0.09
456	4.43±0.06
490	4.08±0.07
532	3.75±0.09
547	3.79±0.09
573	3.88±0.05
614	4.05±0.07
655	4.13±0.10
685	4.53±0.13
710	4.07±0.08

ACKNOWLEDGMENTS

This work was made with financial support of the Russian Foundation for Basic Research and Russian Ministry of Industry, Science and Technology.

REFERENCES

1. R.M.Brown *et al.*, Nucl. Phys. B **117**, 12 (1976).
2. A.Koulbrdis for the Crystal Ball Collaboration, **AIP** Conference Proceedings, vol. 619, 2002, p. 701
3. V.A.Gordeev *et al.*, Prib. Tekhn. Exsp. **No 2**, 25 (1976)
4. D.E.Bayadilov *et al.*, PNPI Research Report 1998-1999, **Part 1**, Gatchina, 2000. p. 22.
5. S.P.Kruglov *et al.*, Prib. Tekh. Eksp. **No 6**, 213 (1997)
6. G.Hohler, "Pion Nucleon Scattering", in Landolt-Bornstein, **vol.9b** (Springer-Verlag, 1980).
7. R.A.Arndt *et al.*, Phys. Rev. **C 52**, 2120 (1995)

Baryon resonance analysis in a coupled-channel approach for energies up to $\sqrt{s} = 2$ GeV.

V. Shklyar* and U. Mosel*

*Heinrich-Buff-Ring 16 , D-35392 Giessen, Germany

Abstract. The pion-induced reactions are studied within a unitary coupled-channel effective Lagrangian model to obtain information on the baryon resonance spectra in up to 2 GeV energy region. We extend our previous calculations to include spin-$\frac{5}{2}$ resonance contributions. The effects of the spin-$\frac{5}{2}$ contributions to the different open channels are discussed.

INTRODUCTION

The analysis of the hadron spectra is an extremely important issue of the modern particle physics. On the one side, it is intended to provide an information of the different reaction dynamics in the resonance energy region where excitation of the nucleon is known to be important. On the other side, the precise knowledge of the resonance spectra is necessary to distinguish between various quark models. Some quark models (see [1] and Refs. therein) predict that the baryon resonance spectrum may be richer then discovered so far. This is the so-called problem of 'missing' nucleon resonances. One assumes that these states are weakly coupled to pion channels and are consequently not clearly seen in πN, $2\pi N$ and ηN reactions. The interest in study of the baryon resonances has also significantly grown last year since the pentaquark was discovered. One also assumes, that classification of the observed spectra can be different than it is thought before.

Despite on the great efforts made in the past, the situation with the identification of the resonance properties from experiment is not satisfactory. The Particle Data Group list a number of the resonances which properties differ from each other. Moreover, there are models which are not agree in the existence of some exited states. In view of these problems a unitary effective Lagrangian model has been developed which incorporates γN, πN, $2\pi N$, ηN, $K\Lambda$, $K\Sigma$ and ωN final states and deals with all available experimental data on pion- and photon-induced reactions [2]. A shortcoming of this study is the missing of higher-spin resonances with spin $J > \frac{3}{2}$. Since the spin-$\frac{5}{2}$ resonances have large electromagnetic couplings this limited the previous analysis of the Compton scattering data to the energy region $\sqrt{s} < 1.6$ GeV. Moreover, the extension to higher-spin baryon spectra becomes unavoidable for investigation of 'hidden' or 'missing' nucleon resonances. Particularly, study of the spin-$\frac{5}{2}$ part of the baryon spectra can shed light on the dynamics of the vector (ω and ρ) meson production which is itself a very intriguing question (see [3] and references therein).

CP717, *Hadron Spectroscopy: Tenth International Conference,*
edited by E. Klempt, H. Koch, and H. Orth

THE GIESSEN MODEL

In the present work we study the effects of spin-$\frac{5}{2}$ resonance contributions to πN, $2\pi N$, ηN, $K\Lambda$, $K\Sigma$, and ωN final states. To this end we extend the previous hadronic calculation [2] by including the D_{15}, F_{15}, D_{35}, F_{35} resonances and simultaneously analysing all available pion-induced reaction data in up to 2 GeV energy region. Due to the coupling-channel calculations this model provides a stringent test for the resonance contributions to the all open final states.

The Bethe-Salpeter coupled-channel equation is solved in the K-matrix approximation to obtain the scattering amplitudes for the reactions under consideration. The details of the model can be found in [2, 4]. In present calculations the following 19 resonances are included $P_{33}(1232)$, $P_{11}(1440)$, $D_{13}(1520)$, $S_{11}(1535)$, $P_{33}(1600)$, $S_{31}(1620)$, $S_{11}(1650)$, $D_{15}(1675)$, $F_{15}(1680)$, $D_{33}(1700)$, $P_{11}(1710)$, $P_{13}(1720)$, $P_{31}(1750)$, $P_{13}(1900)$, $P_{33}(1920)$, $F_{35}(1905)$, $D_{35}(1930)$, $F_{15}(2000)$, and $D_{13}(1950)$, which is denoted as $D_{13}(2080)$ by the PDG [5]. The obtained resonance parameters can be found in [4].

Similar to the spin-$\frac{3}{2}$ case in [2], the contributions from spin-$\frac{5}{2}$ states are investigated for two different types of the spin-$\frac{5}{2}$ projectors: for the 'conventional' (C) and Pascalutsa (P) prescriptions. While the first approach dates back to the original work of Rarita and Schwinger [6] and is widely used in the literature, the latter one assumes the gauge-invariant resonance coupling [7]. Although data quality is not good enough to distinguish between these two pictures now, this question is challenging for understanding of the baryon interactions.

RESULTS FOR THE PURELY HADRONIC CALCULATIONS

The results for the πN partial wave amplitudes are shown on the Figs. 1. We do not show here the corresponding P-p-π+ result since it almost coincides with the new P-calculations. The main differences are found for the conventional coupling calculations in comparison with the previous study. A substantially better description in the P_{13} partial wave is due to the additional off-shell background generated by spin-$\frac{5}{2}$ resonances. The same effect also improves the description of the real and imaginary high energy tails of the P_{31} and S_{31} amplitudes, respectively. The contribution from the spin-$\frac{5}{2}$ resonances can also be seen in the D_{33} amplitude which is also affected by spin-$\frac{5}{2}$ off-shell components. This leads to a worsening in the imaginary part of D_{33} above 1.8 GeV, giving however improvement in the corresponding real part. Due to the limit of the space we do not show the results for isospin-$\frac{3}{2}$ resonances. They can be found in [4].

The $D_{15}(1675)$, $F_{15}(1680)$, and $F_{35}(1905)$ resonances were included in our calculations. We have also found evidence for a second F_{15} state around 1.98 GeV which is rated two-star by [5].

Finally, we conclude that the main features of the considered spin-$\frac{5}{2}$ partial waves are well reproduced. From Figs. 1 one can see that there is no significant difference between the conventional and the Pascalutsa spin-$\frac{5}{2}$ couplings.

All considered resonances have a rather small decay ratios to the ηN, $K\Lambda$, $K\Sigma$, and

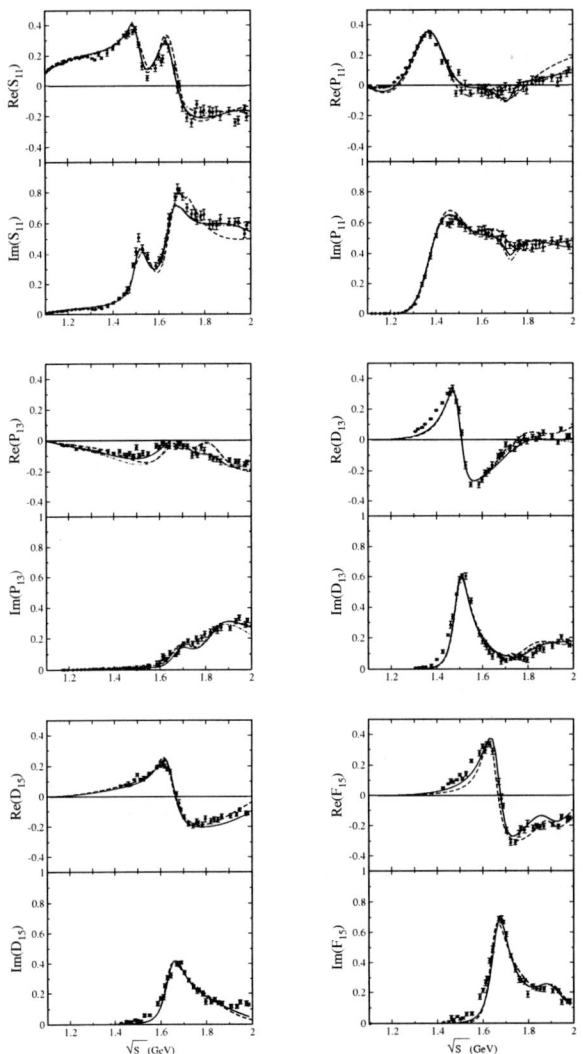

FIGURE 1. The $\pi N \rightarrow \pi N$ partial waves for $I = \frac{1}{2}$. The solid (dashed) line corresponds $C(P)$-calculations. The dash-dotted line is the best hadronic fit C-p-π+ from [2]. The data are taken from [8].

ωN channels. The only exception is the $F_{15}(2000)$ resonance where a small decay width to ηN has been found for the conventional coupling calculations [4].

In Fig. 2 the results for ηN, $K\Lambda$, $K\Sigma$, and ωN total cross sections are shown in comparison with best hadronic fit C-p-π+ from [2]. The main difference from the previous result is found in the ωN final state where a visible effect from the inclusion of

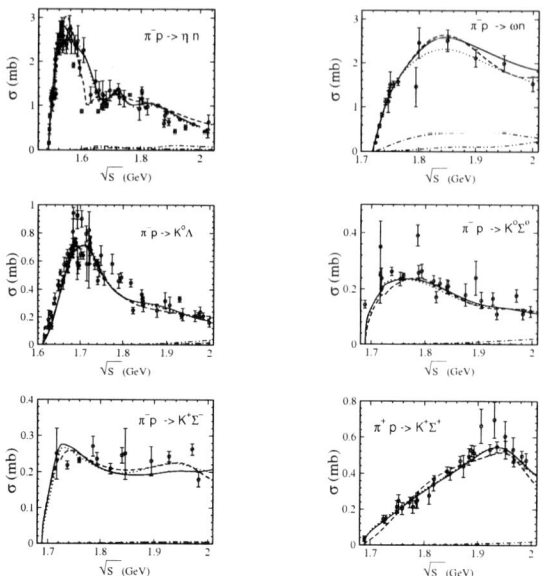

FIGURE 2. The total cross sections for the inelastic reactions. The solid (dashed) line corresponds to the C (P) result. The dotted line shows our previous results C-p-π+ [2]. The contributions from the spin-$\frac{5}{2}$ states are shown by dash-dotted (C) and dash-double-dotted (P) lines. For the data references see [2].

the spin-$\frac{5}{2}$ resonances is found in C-calculations. Although the $D_{15}(1675)$ and $F_{15}(1680)$ states are below the ω production threshold, they give noticeable contributions in the C-coupling calculations. This effect is, however, less pronounced in the P-calculations where the role of $D_{15}(1675)$ and $F_{15}(1680)$ are found to be not so important. The relative contributions from the different partial waves to the ωN reaction are shown in Fig. 3.

These results are similar to the pure hadronic calculations [2] where the main contributions are found to be from the P_{11}, P_{13} and D_{13} partial waves. Since the hadronic data include about 115 datapoints the couplings to the ωN channel are not enough constrained and inclusion of photoproduction data can change the situation [2]. Thus we conclude that more data on $\pi N \rightarrow \omega N$ reaction are needed to fix the contributions from the various partial waves to this channel in the purely hadronic fit. Looking to the ω-photoproduction reaction the new SAPHIR data give an opportunity to distinguish between various reaction mechanisms. This question is particularly addressed in [9].

SUMMARY

We have performed a first investigation of the pion-induced reactions on the nucleon within the effective Lagrangian coupled-channels approach including spin-$\frac{5}{2}$ resonances. To investigate the influence of additional background from the spin-$\frac{3}{2}$ and -$\frac{5}{2}$ resonances calculations using both the conventional and the Pascalutsa higher-spin couplings

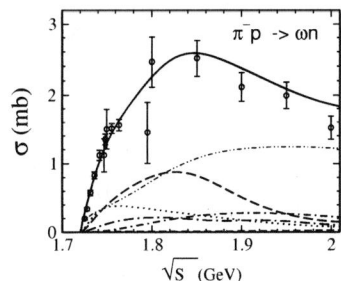

FIGURE 3. Contributions from S_{11}(dashed), P_{11}(dotted), P_{13}(long-dashed), D_{13}(dashed-double-dotted), D_{15}(dashed-dotted), D_{15}(long-dashed-dotted) partial waves to the $\pi N \rightarrow \omega N$ reaction. For the data references see [2].

have been carried out. A good description of the available experimental data has been achieved in all πN, $2\pi N$, ηN, $K\Lambda$, $K\Sigma$, and ωN final states within both frameworks. Apart from $2\pi N$ we find no significant contributions from other channels to the total πN inelasticities in the spin-$\frac{5}{2}$ waves. Nevertheless, the contributions from higher-spin resonances can be important in the ω-production channel. More data on this reaction are highly desirable to establish the role of different reaction mechanisms.

We have found evidence for $F_{15}(2000)$ resonance which is rated two-star by [5] and has not been included in the most recent resonance analysis by Vrana et al. [10].

We are proceeding with the extension of our model by performing a combined analysis of pion- and photon-induced reactions taking into account spin-$\frac{5}{2}$ states. Moreover, the decomposition of the $2\pi N$ channel into ρN, $\pi\Delta$ etc. states will be the subject of further investigations.

REFERENCES

1. S. Capstick and W. Roberts, nucl-th/0008028.
2. G. Penner and U. Mosel, Phys. Rev. C **66**, 055211 (2002). G. Penner and U. Mosel, Phys. Rev. C **66**, 055212 (2002).
3. A. I. Titov, T.-S. H. Lee, Phys. Rev. **66**, 015204 (2002); B. Kämpfer, A.I. Titov, B.L. Reznik, PANIC 02, Osaka, Japan (2002), nucl-th/0211078.
4. V. Shklyar, G. Penner and U. Mosel, submitted to Phys. Rev. C, also available via http://www.uni-giessen.de/~gd1267/.
5. K. Hagiwara et al., Phys. Rev. D **66**, 010001 (2002), http://pdg.lbl.gov.
6. W. Rarita and J. Schwinger, Phys. Rev. **60**, 61 (1941).
7. V. Pascalutsa, Phys. Rev. D **58**, 096002 (1998); V. Pascalutsa and R. Timmermans, Phys. Rev. C **60**, 042201 (1999).
8. M.M. Pavan, R.A. Arndt, I.I. Strakovsky, and R.L. Workman, Phys. Scr. T **T87**, 62 (2000); nucl-th/9807087, R.A. Arndt, I.I. Strakovsky, R.L. Workman, and M.M. Pavan, Phys. Rev. C **52**, 2120 (1995), updates available via: http://gwdac.phys.gwu.edu/.
9. V. Shklyar and U. Mosel, in preparation.
10. T.P. Vrana, S.A. Dytman, and T.-S.H. Lee, Phys. Rep **328**, 181 (2000).

Omega photoproduction as a probe of ωNN vertex.

A. Sibirtsev[*], K. Tsushima[†] and S. Krewald[**]

[*]Institut für Kernphysik, Forschungszentrum Jülich,D-52425 Jülich,
e-mail: a.sibirtsev@fz-juelich.de
[†]Department of Physics and Astronomy, University of Gergia, Athens, Georgia 30602, USA,
e-mail: tsushima@kn3.physast.uga.edu
[**]Institut für Kernphysik, Forschungszentrum Jülich,D-52425 Jülich,
e-mail: s.krewald@fz-juelich.de

Abstract. Within the one meson exchange and Regge models we study ω-meson photoproduction at energies above the s channel resonance region. Here the vector meson photoproduction is dominated by exchanges in the t channel and free parameters of the meson exchange contribution can be fixed explicitly by available data. Thus, at the resonance region the t channel background might be separated from the contribution which might come from baryonic resonances coupled to ω-meson. Furthermore, we investigate different model prescriptions for the ωNN vertex function by imposing gauge invariance as well as crossing symmetry. We show high precision measurements at small $|u|$ provide an access to ωNN vertex. Our predictions are in reasonable agreement with the results from CLAS Collaboration obtained recently at Jefferson Laboratory.

The ω-meson photoproduction off the nucleon at energies below $\simeq 5$ GeV is presently investigated experimentally at several laboratories. The experiments are conceptually designed for searching baryon resonances coupled to ω-meson. The most recent experimental results on angular spectra as well as future polarization mesurements are needed for an evaluation of the resonant contribution. The detailed knowledge of the nonresonant background is required for an extraction of the resonance properties from the data.

There has been extensive theoretical activity in developing one meson exchange model for the analysis of the ω-meson photoproduction [1, 2, 3]. The concept of these investigations is to fit the data by theoretical calculations which include both resonant and nonresonant contributions. Here the nonresonant contribution is dominated by the exchanges in the t channel. A long time ago, Berman and Drell [4] proposed that at small four momentum transfer squared, t, the vector meson photoproduction can be well understood in terms of exchanges of light mesons. The model analysis contains many free parameters, such as masses and widths of resonances, coupling constants and form factors at interaction vertices that appear both in s and t channels. These parameters are correlated and could not be fixed by experimental results in a model independent way.

We propose a different strategy in analysing the data on ω-meson photoproduction. The free parameters of the nonresonant contribution from t channel can be fixed at photon energies beyond the resonance region, i.e. at $2 \leq E_\gamma \leq 5$ GeV. Therefore, our approach requires experimental measurements of ω-photoproduction at energies up to $\simeq 5$ GeV, which are available at Jefferson Laboratory.

CP717, *Hadron Spectroscopy: Tenth International Conference,*
edited by E. Klempt, H. Koch, and H. Orth
© 2004 American Institute of Physics 0-7354-0197-7/04/$22.00

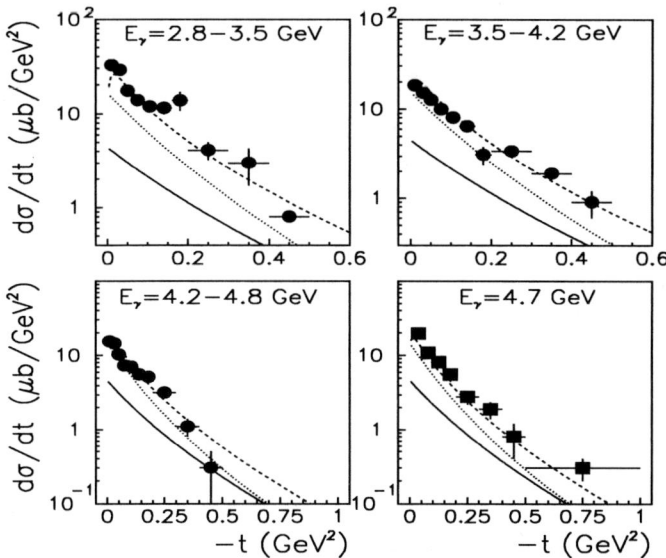

FIGURE 1. Differential $\gamma+p\rightarrow\omega+p$ cross section as a function of four momentum transfer squared t at different photon energies E_γ. The circles show Daresbury experimental results [8], while the squares are the SLAC data [7]. The solid lines indicate the calculations with Pomeron exchange, the dotted lines show the results with additional inclusion of f_2-meson exchange, while the dashed lines are the full model results with π, f_2 and Pomeron exchanges.

At energies beyond the s channel resonance region vector meson photoproduction off the nucleon is traditionally discussed in terms of exchanges in the t and u channels. At forward angles the vector meson photoproduction is dominated by exchanges in the t channel. Therefore the relevant data can fix the parameters of the one meson exchange model. However, the applicability of meson exchange model has to be investigated because the energy range considered starts to overlap with those energies where Regge theory is the adequate theoretical method [5].

Recently within the Regge model we perform [6] a systematical analysis of data on ω-meson photoproduction available from threshold up to HERA energy range and find that at photon energies above 20 GeV the reaction is entirely dominated by Pomeron exchange. At $E_\gamma<5$ GeV the dominant contribution to the $\gamma+p\rightarrow\omega+p$ reaction comes from the exchanges of π and f_2-meson trajectories.

Fig. 1 shows the Regge model calculations together with the experimental results [7, 8] collected at photon energies from 2.8 to 4.8 GeV. The contribution from the Pomeron exchange is indicated by the solid lines, while the sum of f_2 and Pomeron exchanges is shown by the dotted lines. The dashed lines in Fig. 1 show the full model results with inclusion of π, f_2 and Pomeron exchanges and describe the data well even at very low photon energies. Obviously, π-meson exchange becomes dominant with decreasing E_γ.

It is clear that an appropriate parameterization of the form factors and coupling constants at the interaction vertices allows us to reproduce the data even at very low

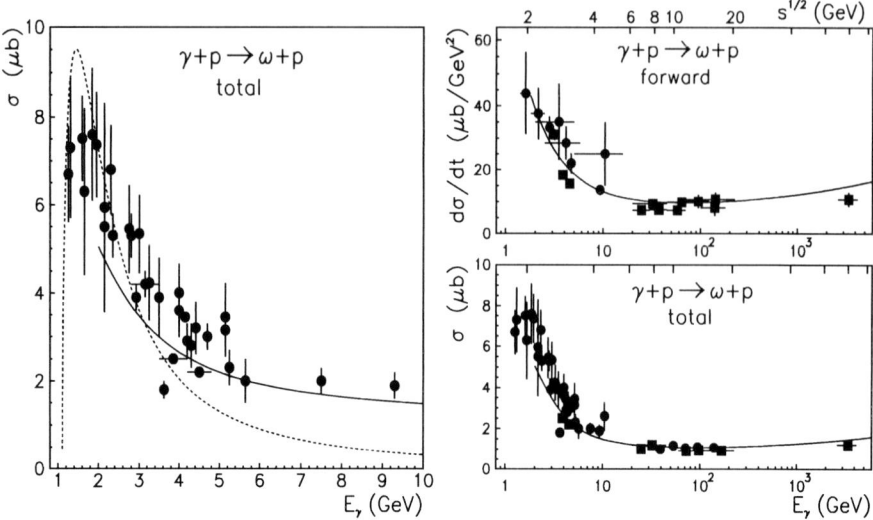

FIGURE 2. Total and forward $\gamma+p\rightarrow\omega+p$ cross section as a function of photon energy E_γ and invariant collision energy $\sqrt{s}$. The solid lines indicate the calculations with Regge theory, while the dashed line shows meson exchange model results.

energies. This parameterization can be used for substraction of the t channel background from the data collected at resonance region.

We found [6] that instead of Regge model parameterization of the π and f_2-meson exchange amplitudes it is possible to describe the ω-meson photoproduction at $E_\gamma<5$ GeV by one meson exchange model. Fig. 2 shows the data on total and forward $\gamma+p\rightarrow\omega+p$ cross section together with the Regge theory calculations and meson exchange model results. Here we include the π, η and σ-meson exchanges in the t channel as well as the nucleon exchange in the s and u channels. Within meson exchange models, the contribution from the nucleon exchange dominates at large angles.

Obviously both Regge and one meson exchange models contain free parameters, which can be readjusted in order to obtain reasonable agreement with the data. However, since these parameters can be fixed beyond the resonance region, the model finally might be used at low photon energies. At $E_\gamma<2$ GeV the contribution additional to t-channel might come from the baryonic resonances coupled to ω-meson. Therefore the discrepancy between our calculations and data at low energies should explicitly indicate the resonance contribution.

Another interest is ω-meson photoproduction at backward angles, i.e. where $|t|$ becomes large and $|u|$ still remains small. Here photoproduction is dominated by exchanges in the u channel. At low energies beyond the resonance region, the production amplitude is dominated by the nucleon exchange. At high energies it is due to the contribution from the nucleon Regge trajectory. Therefore, the data on ω photoproduction at small u should be sensitive to the prescription of the ωNN vertex function as well as to the methods generally used for restoration of gauge invariance of the nucleon exchange amplitude.

A naive evaluation of the amplitude and the introduction of phenomenological form

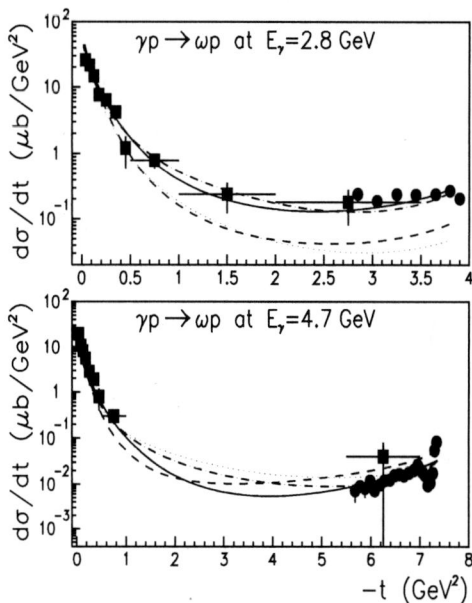

FIGURE 3. The data [14, 15] on the differential $\gamma p \rightarrow \omega p$ cross section as a function of t for different photon energies E_γ. The lines show our calculations with different prescriptions for the ωNN vertex functions given by model A (dashed), B (dotted), C (dashed-dotted), and D (solid).

factors in photoproduction reactions at the level of Born amplitudes violates gauge invariance. There are a few common recipes [9, 10, 11, 12, 13] proposed to restore gauge invariance and crossing symmetry. Our prelimenary study shows that agreement between the available data and the calculations substantially depends on recipe applied.

In order to understand whether it is necessary to use form factors in the ωNN vertices in s and u channels we first fit the data without ωNN form factors and vary the $g_{\omega NN}$ and $g_{\omega\gamma\sigma}$ coupling constants and cut-off parameters at the $\omega\gamma\pi$, $\omega\gamma\eta$ and $\omega\gamma\sigma$ vertices. We denote these calculations as Model A and show the results by the dashed lines in Fig. 3. The dotted lines show the results from Model B, which includes a form factor at the ωNN vertex in the u and s channels using the prescription from Refs. [3, 12]. The dashed-dotted lines indicate the calculations by Model C with recipe from Ref. [13], while the solid lines show the results obtained with a mild modification of the ansatz proposed in Ref. [13]. The lines in Fig. 3 show the results obtained by fitting the data and total χ^2 is 1796 for Model A, 1799 for B, 614 for C and 598 for D, respectively.

Therefore ω-meson photoproduction at backward angles can serve as a micro laboratory for investigation of different recipes for restoration of gauge invariance and crossing symmetry. However a possible acess to ωNN vertex function requires high precision measurements at small $|u|$. Very recently the CLAS Collaboration reported [16] new results on ω-meson photoproduction at $3.2 \leq E_\gamma \leq 3.92$ GeV. Fig.4 shows the CLAS data together with our calculations by Model D. We find a reasonable agreement between

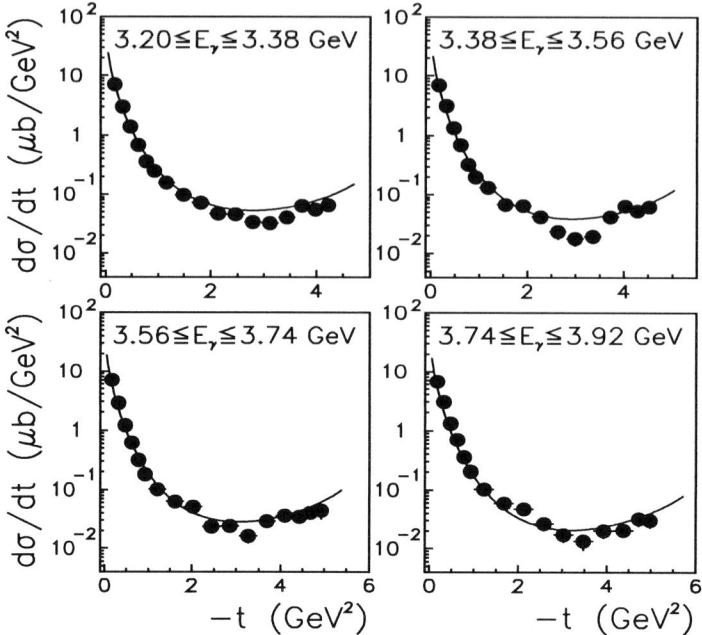

FIGURE 4. The recent CLAS data [16] on the differential $\gamma p \rightarrow \omega p$ cross section as a function of t for different photon energies E_γ. The lines show our calculations by model D.

new data and our calculations, although the parameters of Model D were fixed previously by experimental results from Refs. [14, 15]. The next step in data analysis is final evaluation of the ωNN vertex function and extraction of coupling constant.

REFERENCES

1. B. Friman and M. Soyeur, *Nucl. Phys.* **A 600**, 477 (1996).
2. Q. Zhao, Z. Li, and C.Bennhold, *Phys. Rev.* **C 58**, 2393 (1998).
3. Y. Oh, A.I. Titov and T.S.H. Lee, *Phys. Rev.* **C 63**, 025201 (2001).
4. S.M. Berman and S.D. Drell, *Phys. Rev.* **133**, 3791 (1964).
5. A. Donnachie and P.V. Landshoff, *Phys. Lett.* **B 296**, 227 (1992).
6. A. Sibirtsev, K. Tsushima and S. Krewald *Phys. Rev.* **C 67**, 055201,2003.
7. J. Ballam et al., *Phys. Rev.* **D 7**, 3150 (1973).
8. D.P. Barber et al., *Z. Phys.* **C 26**, 343 (1984).
9. K. Ohta, *Phys. Rev.* **C 40**, 1335 (1989).
10. R.L. Workman, H.W.L. Naus and S.J. Pollock, *Phys. Rev.* **C 45**, 2511 (1992).
11. H. Haberzettl, *Phys. Rev.* **C 56**, 2041 (1997)
12. H. Haberzettl, C. Bennhold, T. Mart and T. Feuster, *Phys. Rev.* **C 58**, 40 (1998)
13. R.M. Davidson and R. Workman, *Phys. Rev.* **C 63**, 025210 (2001).
14. R.W. Clifft et al., *Phys. Lett.* **B 72**, 144 (1977).
15. J. Ballam et al., *Phys. Rev.* **D 7**, 3150 (1973).
16. M. Battaglieri et al., *Phys. Rev. Lett.* **90**, 022002 (2003).

Hadron Spectra and Regge Trajectories

Gurjav GANBOLD

Bogoliubov Laboratory of Theoretical Physics, JINR, Dubna, Russia;
Institute of Physics and Technology, MAS, Ulaanbaatar, Mongolia;
Email: ganbold@thsun1.jinr.ru

Abstract. It is shown that the conception of analytic confinement may result in stable bound states of the constituent and carrier particles within simple relativistic quantum field models of the Yukawa interaction. Particularly, there exist bound states of massless carriers at relatively weak couplings and the excited bound states of massive quarks form asymptotically linear Regge trajectories. The obtained results satisfactorily explain the basic properties of the observed hadron spectra.

INTRODUCTION

The hadron spectroscopy and the phenomenon of the Regge trajectories (RTs) [1, 2, 3] as well many other subjects in the hadron physics are satisfactorily described by QCD, although it is a quite complicated theory because of its nonlinearity. Particularly, the propagators of quarks and gluons in the hadronization region are quite far from those determined by the Dirac and Klein-Gordon equations in QFT.

A theoretical approach made strictly within the quantum field theory is to introduce the analytic confinement (AC) of quarks and gluons into the consideration [4, 5, 6].

Recently, a detailed investigation on the hadron spectroscopy has been performed [8]. The RTs has been fitted by using a nonlinear formula $M^2(J) = a + b\,J + c\,J^2$, where $J = L + S$ and L, S are the orbital, spin quantum numbers. The conclusion is that the RTs of mesons and baryons are nonlinear [8] but the nonlinearity is relatively small: $|b| \gg |c|$. Besides, their slopes differ insignificantly for various hadron families. We may guess that the slope parameter does not depend on the details of the specific hadron bound state (BS) and may be considered a universal parameter governed only by the nature of quark-gluon interaction. Below, we will investigate this feature. For further consideration we take into account the following evidences:
- quarks and gluons are non-observable;
- the RTs of hadrons are almost linear and their slopes differ insignificantly.

These characteristics are hardly obtained within any local quantum field model. However, an introduction of the AC into consideration can fix this problem.

The AC assumes that the Fourier transforms of propagators of confined particles are entire analytic functions in the complex p^2-plane, so the equations for the free fields $S^{-1}(-\partial^2)\,\Psi(x) = 0$, $D^{-1}(-\partial^2)\,\varphi(x) = 0$ may result only in the trivial solutions $\Psi(x) \equiv 0$ and $\varphi(x) \equiv 0$. The AC can be achieved, particularly, by considering the background gluon self-dual homogenous vacuum field into consideration [7]. Further we postulate that the AC takes place.

CP717, *Hadron Spectroscopy: Tenth International Conference,*
edited by E. Klempt, H. Koch, and H. Orth
© 2004 American Institute of Physics 0-7354-0197-7/04/$22.00

A SIMPLE MODEL WITH AC

By investigating the "pure role" of the AC in formation of the two- and three-particle BS we simplify our task by omitting the spin, color and flavor. Later we will include these important quantum degrees of freedom. Consider a simple Yukawa model of two interacting scalar fields $\Phi(x)$ and $\varphi(x)$ (the prototypes of non-observable "quarks" and "gluons"). The partition function in the Euclidean space-time reads:

$$
\begin{aligned}
Z &= \iint \delta\Phi\delta\Phi^+ \int \delta\phi \; e^{-(\Phi^+S^{-1}\Phi)-\frac{1}{2}(\varphi D^{-1}\varphi)-g(\Phi^+\Phi\varphi)-g(\varphi\varphi\varphi)} \\
&= \iint \delta\Phi\delta\Phi^+ \; e^{-(\Phi^+S^{-1}\Phi)+\frac{g^2}{2}(\Phi^+\Phi\,D\,\Phi^+\Phi)-\frac{g^4}{6}(\Phi^+\Phi\,D\,\Phi^+\Phi\,D\,\Phi^+\Phi\,D)+\cdots}, \quad (1)
\end{aligned}
$$

where coupling constant g is supposed sufficiently small. Then, we introduce a complete orthonormal system $\{U_Q(y)\}$ with a set $Q = \{n, l, \{\mu\}\}$ of radial n, orbital l and magnetic $\{\mu\} = (\mu_1, ..., \mu_l)$ quantum numbers. Particularly (for mesons),

$$
(\Phi^+\Phi\,D\,\Phi^+\Phi) = \sum_Q \int dx \, J_Q(x)\, J_Q(x), \qquad J_Q(x) = \Phi^+(x)V_Q(\overset{\leftrightarrow}{\partial})\Phi(x),
$$

where $V_Q(\overset{\leftrightarrow}{\partial}) = (i)^l \int dy \, \sqrt{D(y_1)}U_Q(y)\, e^{\frac{y}{2}\overset{\leftrightarrow}{\partial}}$ is a nonlocal vertex and $\overset{\leftrightarrow}{\partial}=\overset{\rightarrow}{\partial}_x - \overset{\leftarrow}{\partial}_x$. By using the Gaussian functional representation we write

$$
e^{\frac{g^2}{2}\sum_Q \int dx J_Q(x) J_Q(x)} = \int \prod_Q \delta B_Q \; e^{-\frac{1}{2}\sum_Q (B_Q B_Q) + g\sum_Q (B_Q J_Q)}.
$$

Substituting this representation into (1) and by integrating over Φ we obtain

$$
Z = \int \prod_Q \delta B_Q \, \exp\left\{ -\frac{1}{2}\sum_{QQ'} (B_Q[\delta_{QQ'} - \alpha\Pi_{QQ'}]B_{Q'}) + W_I[gB] \right\}, \quad (2)
$$

where $\alpha = (g/4\pi)^2$ and a interaction functional of fields B_Q is introduced:

$$
W_I[gB] = -\mathrm{Tr}\left[\ln(1 - gB_Q V_Q S) + \frac{g^2}{2}B_Q V_Q S B_{Q'} V_{Q'} S \right].
$$

The polarization kernel reads

$$
\alpha\tilde{\Pi}_{QQ'}(p) = \iint dy dy' \, U_Q(y)\, \alpha\Pi_p(y,y')\, U_{Q'}(y'), \quad (3)
$$

$$
\alpha\Pi_p(y,y') = g^2 \, \sqrt{D(y)} \int \frac{dk}{(2\pi)^4} \, e^{-ik(y-y')}\tilde{S}\left(k+\frac{p}{2}\right)\tilde{S}\left(k-\frac{p}{2}\right) \sqrt{D(y')}.
$$

The kernel (3) should be diagonal on $\{U_Q(y)\}$, so we have the eigenvalue problem

$$
\int dy' \, \alpha\Pi_p(y,y')\, U_Q(y') = \lambda_Q(-p^2)\, U_Q(y) \quad \Rightarrow \quad \alpha\tilde{\Pi}_{QQ'}(p) = \lambda_Q(-p^2)\, \delta_{QQ'}. \quad (4)
$$

It is nothing else but the Bethe-Salpeter equation, written in the symmetric form.

Finally, by omitting the baryon part we rewrite the partition function for mesons:

$$Z = \int \prod_Q \delta \tilde{B}_Q \, e^{-\frac{1}{2}\sum_Q (B_Q G_Q^{-1} B_Q) + W_I[gB]}, \quad G_Q^{-1}(x) = [1 - \lambda_Q(-\partial^2)]\,\delta(x), \quad p^2 = -\partial^2.$$

This representation is completely equivalent to the initial one (1). It is a mathematical realization of the quark-hadron duality in the model under consideration. The variables $\{B_Q\}$ can be interpreted as fields of particles with quantum numbers $Q = \{nl\}$ and masses M_Q, if the Green function $\tilde{G}_Q(p^2) = (1 - \lambda_Q(-p^2))^{-1}$ has a simple pole in the Minkowski space ($p^2 = -M_Q^2$). Then, the mass of two-particle BS obeys the equation:

$$1 = \lambda_Q(M_Q^2). \tag{5}$$

1. The simplest "virton" approximation to the analytically confined quark and boson propagators (with $1/\Lambda$ - the average "radius" of AC region) allows us to solve explicitly the eigenvalues. Then, the mass spectra of "mesons" and "baryons" are [9]:

$$M_{q\bar{q}}^2 = 2\Lambda^2 \ln(\alpha_{c2}/\alpha) + 2(n + l/2) \cdot \Lambda^2 \ln(\alpha_{c2}), \quad \alpha_{c2} = \left(2 + \sqrt{3}\right)^2,$$

$$M_{qqq}^2 = 3\Lambda^2 \ln(\alpha_{c3}/\alpha) + 3(n + l/2) \cdot \Lambda^2 \ln(\alpha_{c3}), \quad \alpha_{c3} = \left((3 + \sqrt{5})/2\right)^2.$$

One can see that this specific form of AC leads to the linear and parallel RTs. The slope of RTs is defined only by Λ. Besides, BS exist for weak coupling $\alpha = (g/4\pi\Lambda)^2 < \alpha_{cN}$.

2. The next, more realistic approximation to the propagators is [6]:

$$\tilde{S}(p^2) = \frac{1}{p^2 + m^2}\left(1 - e^{-\frac{p^2 + m^2}{\Lambda^2}}\right), \quad \tilde{D}(p^2) = \frac{1}{p^2}\left(1 - e^{-\frac{p^2}{\Lambda^2}}\right).$$

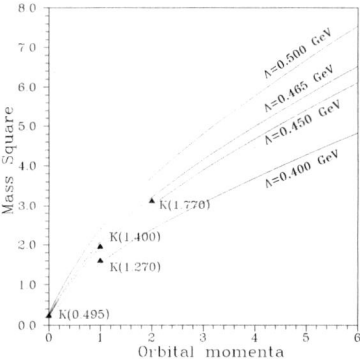

A variational estimate to the real and symmetric kernel $\Pi_p(y, y')$ results in the hadron mass spectra. An analysis shows that there exists a critical value α_c for the coupling constant $\alpha = (g/4\pi\Lambda)^2$. At $\alpha = \alpha_c$ there occurs a massless "glueball" built by massless "gluons". Generally, for two-particle BS:

1. If $\alpha < \alpha_c$, there exist stable BS with masses $M_l > m_{q1} + m_{q1}$.

2. If $\alpha > \alpha_c$, then meson BS can exist only for $m > m_c \neq 0$. Hereby, a massless meson arises as a BS of two massive quarks that contradicts the physical picture.

Therefore, we conclude that stable meson BS exist only for relatively weak coupling $\alpha < \alpha_c$. Our numerical estimation based on the experimental data for π, ρ, K, K^* mesons results in the optimal parameters $\Lambda = 520\,MeV$, $\alpha = 1.7$, $m_u = 50 MeV$, $m_s = 120 Mev$.

The meson RTs are plotted in Figure 1 in comparison with K-meson data.

We see that the RTs are **far not linear** for $l = 0 \div 4$, although the linearity occurs for larger l. Besides, the curvature of these RTs and their slopes depend on Λ considerably.

A similar behaviour and picture are obtained for the three-quark BS – baryons.

PARTICLES WITH SPIN, COLOR AND FLAVOUR

Now we consider a realistic model where quarks possess spin, color and flavor, and gluons are colored. The model Lagrangian reads:

$$\mathscr{L} = \left(\bar{\Psi}^i_\alpha [S^{-1}]^{ij}_{\alpha\beta} \Psi^j_\beta \right) + \frac{1}{2} \left(\phi^a \, [D^{-1}]^{ab} \, \phi^b \right) + g \left(\bar{\Psi}^i_\alpha \, (\Gamma^a)^{ij}_{\alpha\beta} \, \Psi^j_\beta \phi^a \right) + h \left(\phi^a \phi^b \phi^c \right) d^{abc},$$

where g and h are the quark-gluon and three-gluon coupling constants and the quark and gluon propagators in background of vacuum gluon field are the explicit entire analytic functions. Integrating out over the gluon variables we obtain the partition function

$$Z = \iint \delta\bar{\Psi}\delta\Psi \; e^{-\mathscr{L}_F[\bar{\Psi},\Psi] - \mathscr{L}_2[\bar{\Psi},\Psi] - \mathscr{L}_3[\bar{\Psi},\Psi] + \dots}$$

where $\mathscr{L}_2$ and $\mathscr{L}_3$ stand for the meson and baryon parts (see, (1)).

Mesons: For mesons we should extract a color singlet quark pair

$$\bar{\Psi}^i_\mu(x) \, \Psi^n_{\nu'}(y) = \frac{1}{3}\delta^{ni} \left(\bar{\Psi}_\mu(x) \, \Psi_{\nu'}(y) \right) + \dots, \quad (\bar{\Psi}\Psi) = \sum_{i=1}^{3}(\bar{\Psi}^i\Psi^i)$$

and make the Fierz transformation for the spin components:

$$\delta^{\mu\mu'}\delta^{\nu\nu'} = \frac{1}{4} \left(\delta^{\mu\nu'}\delta^{\nu\mu'} - (i\gamma_5)^{\mu\nu'}(i\gamma_5)^{\nu\mu'} - (i\gamma_\rho)^{\mu\nu'}(i\gamma_\rho)^{\nu\mu'} + (i\gamma_5\gamma_\rho)^{\mu\nu'}(i\gamma_5\gamma_\rho)^{\nu\mu'} \right)$$

We have estimated by variational means the mass spectrum of known mesons, the result is shown in Table 1. By omitting further details, we write the optimal parameters found

$$\Lambda = 600, \quad m_u = 120, \quad m_s = 180, \quad m_c = 1050, \quad \alpha_s = \frac{g^2}{4\pi} = 0.46.$$

Baryons: Color singlet combination of three color quarks (qqq) is $\varepsilon_{ijk}\varepsilon_{klm}/6$. The Fierz rule and CT invariance allow only the two-quark subparts:

$$i\varepsilon^{ijk} \left(\bar{q}^i_\mu \, [\gamma_\rho \, \mathscr{C}] \, \bar{q}^j_\alpha \right), \qquad \frac{i}{2}\varepsilon^{ijk} \left(\bar{q}^i_\mu \, [\gamma_\rho\gamma_\xi - \gamma_\xi\gamma_\rho] \, \mathscr{C} \, \bar{q}^j_\alpha \right)$$

TABLE 1. Estimated masses (in *MeV*) of some pseudoscalar and vector mesons $q\bar{q}$.

Quark pair	J^{PC}	Meson	M_0	Quark pair	J^{PC}	Meson	M_0
u,d	0^{-+}	$\pi(140)$	140	u,c	1^{--}	$D^*(2010)$	2110
u,d	1^{--}	$\rho(770)$	770	s,c	0^{-+}	$D_s(1970)$	2130
u,s	0^{-+}	$K(495)$	490	s,c	1^{--}	$D_s^*(2110)$	2170
u,s	1^{--}	$K^*(892)$	870	c,c	0^{-+}	$\eta(2980)$	2990
u,c	0^{-+}	$D(1870)$	2010	c,c	1^{--}	$J/\Psi(3090)$	3100

Therefore, the nucleon structure is ontained:

$$\mathcal{N}_\alpha(x,y,z) = \frac{1}{4}\varepsilon^{ijk}\left(u_\beta^i(x)\left[\mathscr{C}\,\gamma_\xi\right]^{\beta\nu} u_\nu^j(y)\right)\gamma_5\gamma_\xi^{\alpha\eta}d_\eta^k(z)$$

and the isobar is formed as follows:

$$\Delta_{\alpha\beta}(x,y,z) = \frac{1}{4}\varepsilon^{ijk}u_\alpha^i(x)\left[\mathscr{C}\,\gamma_\xi\right]^{\beta\nu} u_\nu^j(y)\left(\delta^{\alpha\beta}\,d_\xi^k(z)-(\gamma_\beta\gamma_\xi)^{\alpha\eta}\,d_\eta^k(z)\right).$$

Variational estimates for the masses of the Nucleon and Δ Isobar give satisfactory results provided the same input parameters of quark masses and $\Lambda = 600$ MeV.

We conclude that the AC provided with the weak coupling regime in QFT describes qualitatively the main features of meson spectra. We explain the formation of the stable two- and three-quark BS as well as glueballs - the scalar BS of massless gluons. By comparing our result with the recent experimental data we obtain stable and reasonable values for the confinement energy scale, the coupling constants and the masses of the constituent quarks of various flavours. The mechanism of the chiral symmetry breaking is naturally described in the last model. Of course, we cannot pretend to describe quantitavely all details of the hadron spectroscopy. Nevertheless, one can conclude that the slope of RTs weakly depends on specific details of hadron internal dynamics and may be considered a universal characteristic of quark-gluon interactions.

REFERENCES

1. G.F. Chew and S.C. Frautschi, Phys. Rev. Lett. **8** (1962) 41;
2. P.D.B. Collins, *An Introduction to Regge Theory and High Energy Physics*, Cambridge Univ. Press, Cambridge (1977);
3. T. Regge, Nouvo Cim. **14** (1959) 951; Nouvo Cim. **18** (1960) 947;
4. H. Leutwyler, Phys. Lett. **96B** (1980) 154; Nucl. Phys. **B179** (1981) 129;
5. G.V. Efimov and M.A. Ivanov, *The Quark Confinement Model of Hadrons* IOP Publishing Ltd, London, (1993);
6. G.V. Efimov and G. Ganbold, Phys. Rev. **D65** (2002) 054012,
7. G.V. Efimov, S.N.Nedelko, Phys. Rev.**D51** (1995) 176;
8. A. Tang and J.W. Norbury, Phys. Rev. **D62** (2000) 016006;
9. G. Ganbold, in *Sixth Workshop on Non-Perturbative QCD*, eds. H.M. Fried, Y. Gabellini and B. Muller, World Scientific, N.Y. (2002) 65-71;

Quark Spin Coupling in Light Baryons: A small review

Gabriel Karl

Department of Physics
University of Guelph
CANADA

Abstract. This is a very short review of quark spin couplings in light baryons. Most of the material comes from reference [6].

The subject I discuss has been rather controversial in the past, and to focus on issues rather than personalities, this small review will not mention any names, even in the references [1-7], not even for experimental papers. Nor is the review comprehensive, it focuses entirely on mistakes in the literature, including work by the speaker [7]. One should also not conclude from the statement that something is wrong in a reference that the reference is "completely wrong,,; the reference is fine except for the point which is being criticized.

The issues we focus on here are the mixing angles in P-wave baryons and the mass ordering among the lowest excited negative and positive parity baryons:

$$M(^2P\ 1520) > M(^2S\ 1470) > M(^2S\ 938).$$

The controversy is whether the coupling is "OGE,, or "OPE,, where the abbreviations stand for one-gluon-exchange-like, or one-pion-exchange-like. Neither the gluon nor the pion are true perturbative particles, but formulae which are postulated for the coupling of 'constituent quarks'. It would be nice to derive such formulae from a basic Lagrangian, like QCD, but this is not the case as far as I am aware- constituent quarks are difficult objects.

The conclusions we draw are, that mixing angles are independent of parameters for both OGE and OPE, but the data prefer the OGE fits. On the second issue, the mass ordering is parameter dependent for both OGE and OPE, and these three states can be fitted with both OPE and OGE couplings, but overall the OPE coupling gives a better fit. The two couplings have very similar form for a pair of fermions:

$$H_{OGE} = A[(8\pi/3)*\text{contact} + \text{tensor}]*\text{color coupling}$$

$$H_{OPE} = B[(-4\pi/3)*\text{contact} + \text{tensor}]*\text{flavor coupling}$$

CP717, *Hadron Spectroscopy: Tenth International Conference,*
edited by E. Klempt, H. Koch, and H. Orth
© 2004 American Institute of Physics 0-7354-0197-7/04/$22.00

where the term called contact is the same in OGE and OPE and similarly with the term called tensor. These are not written explicitly here. The main difference between OGE and OPE is in the relative weight in front of the contact term (8/3) versus (-4/3). The coefficient (8/3) is relevant to current-current (interaction of ampere loops) while the coefficient (-4/3) is relevant to electric dipole-electric dipole interaction which mimics the pion exchange (with zero-mass pions)because of symmetry properties. I am indebted to Prof. V.A.Novikov for explaining this last point to me. In any case these coefficients are discussed in detail in J.D.Jackson's well known textbook on Electromagnetism(Second or Third edition.),and are purely classical effects, although the contact term for hyperfine interaction is called "Fermi Contact Interaction,,. In the 1930's people thought that these terms could only arise from the Dirac equation because of Fermi's derivation.

In P-wave baryons, at angular momentum J=1/2 and 3/2 the total quark spin S could be either doublet or quartet. So the physical states are linear superposition of ^{2}P and ^{4}P components and there are mixing angles which describe the linear combination of doublet and quartet components. For both OPE and OGE interactions the linear superposition is independent of the strength of the interaction (B or A respectively) or of any spectroscopic data. There is an error in the literature for evaluating the OPE case[3] and this is corrected in [6]. The actual linear superposition differs in the OGE and OPE cases. The experimental data [5] agrees with the OGE but disagrees sharply with OPE linear combination especially as evaluated in [6]. So experiment supports the OGE Hamiltonian as far as mixing angles are concerned. More recently the OPE interaction was replaced by pseudoscalar plus vector meson exchange which can fix this problem. But it is hard to claim that OGE is "wrong,, and GBE(Goldstone boson exchange) is "right,,. All one can say is that the two Hamiltonians are equivalent phenomenologically. But "some Hamiltonians are more equivalent than others,, (with apologies to Orwell). The OGE coupling has a more glorious history and is relevant to atomic hyperfine structure, as well as to heavier quarks. Thus there is some bias in its favor, even though the application to constituent quarks is ad-hoc.

On the issue of the mass ordering of the nucleon, the S_{11} negative parity resonance and the Roper resonance, there is a general consensus in the literature (which includes a paper co-authored by the speaker) to the effect that OGE cannot put the Roper resonance below the S_{11} resonance[7]. But this is not the case. The Roper resonance can be put even below the nucleon- depending on the choice of parameters- the frequency of the oscillator and the strength of the hyperfine coupling. What is true is that OPE can fit more easily more states in this region, but that is a more subtle issue. It should be stressed that fitting states depends on parameters while finding mixing angles is parameter independent. So one should attach more weight to the description of mixing angles than the fitting of states.

REFERENCES:

1. Physics Letters **B73,**109(1977)
2. Physical Review**D18**,4187(1978)
3. Physics Reports **268**,263(1996)
4. Physical Review **D62**,054026(2000)
5. Nuclear Physics **B95**,516(1975),
   Eur.Phys.J.**A11**,217(2001)
   Physical Review**D68**,017502(2003)
6. Physical Review **D68,**054007(2003)
7. Physics Letters**124B**,520(1983)
   Physical Review**D51,**5068(1995)

Scalars

The Scalar Meson Sector and the σ, κ Problem

Wolfgang Ochs

Max-Planck-Institut für Physik, D-80805 München, Föhringer Ring 6, Germany

Abstract. In the light scalar meson sector ($M \lesssim 1.8$ GeV) one expects at least one $q\bar{q}$ nonet and a glueball, possibly also multi-quark states. We discuss the present phenomenological evidence for σ and κ particles; if real, they could be members of the lightest (quark or multi-quark) nonet together possibly with $a_0(980)$ and $f_0(980)$. Alternatively, the lightest nonet could include $f_0(980)$ but not σ and κ. Future decisive experimental studies, concerning tests of symmetry relations, especially in B-decays, are outlined.

INTRODUCTION

The light scalar meson sector represents still a big puzzle. There should be a nonet of P wave $q\bar{q}$ states with $J^{PC} = 0^{++}$ besides the other rather well known P wave nonets with 1^{++}, 2^{++} and 1^{+-}. In addition one expects the lightest glueball in the same channel. Furthermore, it is possible that more complex multi-quark bound states exist. In Table 1 we list the light scalar mesons according to the Particle Data Group [1].

Of particular recent interest are the σ and κ particles in the $\pi\pi$ I=0 and $K\pi$ I=$\frac{1}{2}$ channels, the lightest particles for the given isospin. Because of their large width as compared to their mass $\Gamma \gtrsim M$ these states are difficult to identify experimentally and therefore their status is still under debate, also at this conference. Besides the question of their existence one has to enlighten their role in spectroscopy, i.e. to determine their constituent structure and the multiplet they belong to. The high interest in these states also originates from their important role they play in meson theories based on chiral symmetry (reviews [2, 3, 4]).

In this talk the status of the phenomenology and possible future tests will be discussed.

TWO ROUTES FOR SCALAR SPECTROSCOPY

Among the various approaches to scalar spectroscopy we will contrast two routes which are essentially different in the classification of the states, although not unique among

TABLE 1. Scalar mesons below $\sim$2 GeV according to Particle Data Group [1], states with (?) not in the main listing

I = 0	$f_0(600)$ (or σ)	$f_0(980)$	$f_0(1370)$	$f_0(1500)$	$f_0(1710)$	$f_0(2020)$?
I = $\frac{1}{2}$		$\kappa(900)$?		$K_0^*(1430)$		$K^*(1950)$?
I = 1		$a_0(980)$		$a_0(1450)$		

CP717, *Hadron Spectroscopy: Tenth International Conference,*
edited by E. Klempt, H. Koch, and H. Orth
© 2004 American Institute of Physics 0-7354-0197-7/04/$22.00

themselves.

Route I: two multiplets below 1800 MeV

The upper multiplet includes the uncontroversial $q\bar{q}$ state $K^*(1430)$, then also the nearby $a_0(1450)$. In the isoscalar channel one observes $f_0(1370)$, $f_0(1500)$ and $f_0(1720)$. The 0^{++} glueball is assumed with mass around 1600 MeV as found in quenched lattice calculations.[1] Then the glueball and two members of the nonet can mix and generate the three observed f_0 states. After the original proposal [6] several such mixing schemes have been considered (review [7]) using different phenomenological constraints.

The lower mass states are now left over. The $f_0(980)$ and $a_0(980)$ could be $K\bar{K}$ molecules [8] or 4 quark states [9], recently proposed [10] to explain radiative ϕ decays. An interesting possibility would be to combine these two states with σ and κ into a second light nonet, either of $q\bar{q}$ or of $qq\bar{q}\bar{q}$ type. Such schemes appear in theories of meson meson scattering in a realization of chiral symmetry, for an outline, see [3].

Route II: one multiplet below 1800 MeV

In an alternative path one starts again with $K^*(1430)$ but takes $f_0(980)$ as the lightest isoscalar $q\bar{q}$ state [11, 12, 13, 14]. The identification of the other members of the nonet differs; in the phenomenological approach [12] the lightest isovector is $a_0(980)$ and the second isoscalar is $f_0(1500)$ (as in [11]) with a flavour mixing as in the pseudoscalar sector with $f_0(980)$ and η' near flavour singlet and $f_0(1500)$ and η near octet. The glueball is a rather broad state (width of order of mass) [12, 13, 14] centered around 1 GeV or a bit larger. In this case σ is the low mass component of the glueball whereas κ is not relevant for spectroscopy. Despite differences in detail σ and κ are not members of a nonet (σ possibly a mixture of glueball and $q\bar{q}$ [14]).

This leads to the important questions:

1. Are σ, κ poles in the amplitude, genuine resonances and members of a nonet?
2. Is $f_0(980)$ a member of a low mass or high mass multiplet?

We adress the second question below considering symmetry relations. The means to answer the first question are the detailed investigation of the relevant amplitudes (finding resonances, typically as circles in the complex amplitude plane ("Argand diagram") or from maximum phase variation), and the study of available production and decay channels to find out about the constituent structure.

In $2 \rightarrow 2$ scattering processes the standard method is the determination of the moments of the angular distribution from which the partial wave amplitudes can be determined ($\langle Y_L^0 \rangle \cong \sum c_{\ell m} \operatorname{Re}(A_\ell A_m^*)$) up to an overall phase and discrete ambiguities.

In Dalitz plot analyses of 3-body decays $R \rightarrow 1 + 2 + 3$ it will be useful to compare the fits with phase sensitive quantities. A staightforward generalization of the above [15] would be the study the moments in the relevant non-exotic channels (ij) which get the direct contribution from channel (ij) as above but an additional contribution from the crossed channel(s)

$$\langle Y_L^0 \rangle \cong \sum c_{\ell m} \operatorname{Re}(A_\ell A_m^*) + \text{crossed channel background.} \tag{1}$$

---

[1] The recent report by Bali [5] quotes the mass range 1.4 ... 1.8 GeV; the unquenched results are typically around 20% lower and decrease with decreasing quark mass.

This additional contibution is slowly moving if the spin of the crossed channel reso-nances is low. The comparison of the full resonance model (usually a Monte Carlo) with the channel moments (say, up to L=4) should reveal the fine structure from the interfer-ing partial wave amplitudes. Another possibility is the study of phases using a known resonance as analyser [16].

σ POLE

Summarizing a large variety of fits the PDG estimates the Breit-Wigner pole position as $M = (400 - 1200) - i(300 - 500)$ MeV reflecting a considerable fluctuation of the results. The width $\Gamma = 2\text{Im}\,M$ is of the order of the mass itself. Next, we will discuss the recent results which are based on phase determinations using angular distributions.

1. elastic $\pi\pi$ scattering

Data are obtained from single pion production using the One-Pion-Exchange model and from $K \to \pi\pi ev$ decays applying the Watson theorem. Recent studies using the Roy equations which implement analyticity, unitarity and crossing symmetry are found consistent with chiral symmetry constraints in the threshold region [17, 18]; also a unique phase shift solution "down flat" obtained from the polarized target data has been found [19], closely similar to the earlier results from unpolarized data [20]. In the analysis [17] also the σ pole is determined with remarkably small errors:

$$M = 470 \pm 30, \ \Gamma = 590 \pm 40 \text{ MeV}. \tag{2}$$

The mass is much below ~ 850 MeV where the phase shift passes $90°$. The origin of this pole and its connection to chiral symmetry can be understood by a qualitative argument (see [17]). Chiral symmetry leads to an amplitude zero in the isoscalar S wave amplitude T_0^0 near threshold (Adler zero) and results in a small scattering length. Unitarity then requires an imaginary part of the amplitude which rises more strongly than the real part, like $s^2 = E_{CM}^4$, in the chiral limit ($m_\pi = 0$)

$$T_0^0 = s/(16\pi F_\pi^2), \qquad \text{Im}\, T_0^0 = |T_0^0|^2 \sim s^2 \tag{3}$$

with the pion decay constant $F_\pi = 92.4$ MeV. Within a certain unitarization method one obtains the unitary amplitude

$$T_0^0 = s/(16\pi F_\pi^2 - is) \tag{4}$$

which has the correct threshold behaviour (3) and a pole at $\sqrt{s} = \sqrt{-i16\pi F_\pi} = 463 - i463$ MeV, not far away from the result of the full fit $\sqrt{s} = 470 - i295$ MeV in (2).

The errors in the pole determination take into account experimental errors and sys-tematic errors from the parametrization, including a single pole (σ) or two poles ($\sigma, f_0(980)$). From the above qualitative argument one may deduce that this distant pole is generated by the unitarization procedure to manage the rapid increase of the imaginary part near threshold, given the small scattering length.

Whether this distant pole so generated correspond to a short living particle to be classified into a flavour multiplet has to be thought of further.

The $\pi\pi$ S-wave amplitude in a larger mass range up to 1700 MeV (recent results [22]) can be viewed as broad object centered around 1000 MeV which interferes destructively with $f_0(980)$ and $f_0(1500)$ [21, 12]. Fits of the $\pi\pi$ elastic and various inelastic channels in a K-matrix formalism yield a mass $M \sim (1450 - i800)$ MeV without including a σ pole explicitly [13]. These K matrix results may not have satisfactory analytic properties at very low energies: if they are inserted into dispersion relations the σ pole reappears [23]. However, in the large energy range the S wave $\pi\pi$ amplitude (not counting the narrow resonances at 980 and 1500 MeV) describes only one circle in the Argand diagram corresponding to one broad state. This is most plausibly placed in the region 1.0-1.5 GeV; then the σ pole influences the low energy behaviour but does not generate an extra circle at low mass.

Investigating various production processes for this broad state it was concluded that it fulfills in all cases (except radiative J/ψ decays) the expectations for a glueball [12]. The recent observation of a broad peak in $K\bar{K}$ from $B \to K\bar{K}K$ by Belle [24] has also been taken as new evidence for this interpretation [25]. The glueball interpretation is also favoured by the K matrix analysis [13]. QCD sum rules require a broad glueball near 1000 MeV with the large decay into $\pi\pi$ [14].

2. Decay $D^+ \to \pi^+\pi^+\pi^-$

A promising source of information is provided by the 3-body decays of D and B mesons. In the isobar model one considers the final state to proceed through intermediate resonances in any non-exotic channel. In the simplest way one takes a sum of Breit Wigner resonances A_i, each multiplied by a constant amplitude and phase factor

$$T = a_0 e^{i\delta_0} + \sum_i a_i e^{i\delta_i} A_i(s_{12}) + \dots \tag{5}$$

where the dots refer to channels (13) and (23) if resonant.

Alternatively, one can express each channel (ij) by a multi-channel real K-matrix

$$K_{ij} = \sum_\alpha \frac{g_i^\alpha g_j^\alpha}{s_\alpha - s} + \dots \tag{6}$$

with poles s_α. The decay amplitude is then expressed by $F_i = (I - iK\rho)_{ij}^{-1} P_j$. The K matrix describes the propagation; with its poles it is universal in all processes and is determined from the multi-channel fit to 2-body collisions. The P vector contains the initial production amplitudes to be fitted for each process.

The decay $D^+ \to \pi^+\pi^+\pi^-$ has been analysed two years ago by the E791 Collaboration applying the isobar model [26]. They could not get a satisfactory fit without including the σ resonance in the fit to describe a peak at around 500 MeV. They obtained the parameters $M_\sigma = 478$ MeV and the width $\Gamma_\sigma = 324^{+24}_{-40} \pm 21$; this is considerably narrower than the elastic $\pi\pi$ scattering result (2). Clearly, the phases here are more rapidly varying than in elastic scattering which would imply strong rescattering effects.

A new result by the FOCUS collaboration presented at this conference [27] confirms the finding by E791 on the need for a σ contribution within the sum (5). An alternative fit has been carried out using the K matrix approach [13] which does not include σ explicitly. This fit contains five f_0 states found from 2 body collisions with the appropriate weights. The peak at low mass is then produced either by the S wave in this region itself or by the reflection from the crossed channel contributions. In any case, the phase variation over the peak region is smooth and does not cross 90° at the peak.

Further clarification should come from a careful study of phase sensitive quantities. As emphazised before [15] the angular moment $\langle Y_1^0 \rangle \propto \langle \cos\theta \rangle$ is proportional to the S-P interference and therefore a σ resonance should show a characteristic interference with the tail of the ρ – above the smooth background from the crossed channel.

An alternative possibility has been studied by the E791 collaboration [28] using the ΔA^2 method [16]. They selected the $f_2(1270)$ resonance in s_{12} as analyser and compared the difference of densities above and below the resonance mass $\Delta A^2 = A_+^2 - A_-^2$ as function of the conjugate mass s_{13} which is related to the amplitude phase

$$\Delta A^2(s_{13}) \propto \sin(2\delta(s_{13})). \tag{7}$$

They found a strong variation of the phase through the σ region by altogether 180° indicating a σ Breit Wigner resonance. As the f_2 is only rather weakly produced it will be important to confirm the effect with the clear ρ and $f_0(980)$ resonances. In these cases, it is more convenient to compare the two models with and without rapid phase variation directly to the two stripes A_+^2 and A_-^2 which contain the same phase sensitivity. Together with the $\langle \cos\theta \rangle$ moment variation it should be possible to get the wanted information on the behaviour of phases.

3. Central production $pp \to p(\pi\pi)p$

This process is assumed to be dominated at small momentum transfers between the protons by double Pomeron exchange. Recent measurements determined the angular distributions and the relative phases of amplitudes [29, 30].

The centrally produced $\pi\pi$ system peaks shortly above threshold below 400 MeV and this peak has been related to σ as well [29, 31]. In this case there is a simple dynamical explanation of the peak [15] in terms of the subprocess

$$\text{Pomeron Pomeron} \to \pi\pi \tag{8}$$

with one-pion-exchange. This interpretation is suggested by the close similarity of this process with $\gamma\gamma \to \pi\pi$, not only with respect to the S wave peak near 400 MeV but also to the very unusual peak in the D wave near 500 MeV.

Concerning the behaviour of the S wave phase we note that both experiments find the phase difference $\varphi_S - \varphi_{D_1^-}$ slowly rising from threshold to about 90° near 900 MeV very similar to elastic $\pi\pi$ scattering. On the other hand, the phase difference $\varphi_S - \varphi_{D_0^-}$ is rather energy independent below 1 GeV. This is difficult to explain assuming a common production mechanism. The presence of several production mechanisms with different spin couplings of the proton would invalidate the simple kind of analysis neglecting the spin effects. In any case, there is no rapid phase variation of the S wave near the peak of 400 MeV as expected from a simple Breit Wigner resonance. In this case a non-resonant mechanisme can be identified.

4. Decay $J/\psi \to \omega\pi\pi$

There is a sizable peak around 500 MeV in this process studied first by DM2 [32] and now by BES [33]. Again one may ask whether the peak can be represented by a normal

Breit Wigner resonance. Studying the $\pi\pi$ angular distribution as measured by DM2 we concluded [15] that the $\pi\pi$ phase shift has to increase slowly through the peak region, otherwise the interference with the nearly real D-wave (assumed to be the tail of f_2) would lead to a sign change of the interference term $\langle\cos^2\theta\rangle$.

Using the new high statistics data from BES on the $\pi\pi$ angular distributions and others it is actually possible to determine the S wave phase directly [33]; in this analysis the background D wave component is related not to f_2 but to the tail of $b_1(1235)$ in the crossed channel, albeit for a b_1 width considerably larger than given by the PDG. The resulting phase shifts behave smooth in the peak region as in elastic scattering but not like a local Breit-Wigner resonance. An explanation of such behaviour is suggested in terms of a σ pole with strongly energy dependent width (from Adler zero).

5. Other results

There are other channels where broad low mass peaks are observed, in particular decays of charmonia ψ', ψ'' and Y', Y'' into $\pi\pi$ and the respective ground state. These peaks may be related to σ as well [34]. Peaks are also seen in τ decays. As there are no phase studies available we do not discuss these further here.

6. Unsuccessful searches Finally we emphasize that in some reactions searches have been negative. In particular, CLEO [35] did not find any σ signal in the neutral D^0 decay $D^0 \to \pi^+\pi^-\pi^0$. Here the $\pi^+\pi^-$ mass specrum also shows a peak at small masses which is entirely explained by the crossed channel resonances.

κ POLE

At first sight, the low energy $K\pi$ scattering looks similar to $\pi\pi$: there is the possibility of a broad resonance close to threshold. However, there are some characteristic differences. In the following, we discuss the various observations.

1. elastic $K\pi$ scattering

The phase shifts of elastic scattering have been extracted from pion production experiments as in case of $\pi\pi$ scattering by the LASS collaboration [36]. The S wave in the region up to 1.6 GeV has been described by a superposition of a smooth background and the $K_0^*(1430)$ Breit-Wigner resonance

$$S = BG + BW\ e^{2i\delta_{BG}}, \qquad BG = \sin\delta_{BG}e^{i\delta_{BG}}, \qquad \cot\delta_{BG} = \frac{1}{aq} + \frac{bq}{2}. \qquad (9)$$

Another measurement is obtained from the semileptonic decays $D^+ \to K^-\pi^+\mu^+\nu$ by the FOCUS collaboration [37]. The Watson theorem relates the final state phase shifts to those of elastic scattering in the elastic region. Data are consistent with a constant $K\pi$ phase of $\varphi_{K\pi} = 45°$ in $800 < M_{K\pi} < 1000$ MeV. This is indeed in close agreement with the elastic scattering phase which varies between about 35° and 50° in this range [15].

The background phase in this parametrization rises up to about 50° at 1.5 GeV near the first resonance. Alternative parametrizations yield 70° [33]. This is quite different from $\pi\pi$ scattering, where the phase passes 90° already at 850 MeV, below the first resonance $f_0(980)$. The need for an extra state in $K\pi$ is therefore not evident, contrary to $\pi\pi$.

The question whether the elastic scattering data require a κ pole has been investigated by Cherry and Pennington [38]. They expand the scattering amplitude into a complete series of functions with correct branch cuts and truncate if no significant improvement is obtained. They find always the $K^*(1430)$ but not the κ. There is therefore no evidence for κ from the present data but the very low energy region below 800 MeV is not available yet in elastic scattering. Hopefully such data will be obtained from semileptonic D decays.

On the other hand, the data can also be described by models which include a κ pole. This is found in multi-channel fits using chiral symmetry constraints. Recent analyses in chiral perturbation theory yield acceptable fits to the $K\pi$ phase shifts [39, 40]. The position of the κ pole is found as $M_\kappa \sim 750 - i230$ MeV [40].

2. Decay $D^+ \rightarrow K^-\pi^+\pi^+$

The analysis by the E791 collaboration [41] proceeds similar at first to the corresponding one of the 3π final state: An isobar model fit including $K^*(1430)$ and constant background but no κ does not give a good fit. A satisfactory fit is found if the $K^*(1430)$ parameters are varied and a κ resonance is introduced with $M \sim 797 - i205$ MeV.

Further studies [42] have shown that the angular asymmetry $\langle \cos\theta \rangle$ in the $K^*(890)$ region between 800 and 1000 MeV is rather well fitted by both models with and without κ which implies that in this mass region both models have similar phases despite their different analytic expressions. At low masses $M < 800$ MeV a better description is obtained for angular distributions in a fit with κ. These results show the importance of the study of more details of the final state in the determination of the partial waves. It will be interesting to compare directly the total S wave phase in the full mass region of both models and compare with the elastic phase.

3. Decay $J/\psi \rightarrow K^*(890)K\pi$

This channel has been investigated by the BES collaboration [43] and there is some similarity to the corresponding channel with $\omega\pi\pi$. The K^* band, after subtraction of suitable side bands shows a κ peak of 3σ significance with mass $M \sim (771^{+164}_{-221} \pm 55) - i(110^{+112}_{-84} \pm 48)$. At this conference two analyses of BES data have been presented, both finding the κ albeit with different parameters: the first analysis with $M \sim (760 \pm 20 \pm 40) - i(420 \pm 45 \pm 60)$ MeV [33] also determines the $K\pi$ phase shifts and finds results consistent with elastic $K\pi$ scattering; the second one yields $M = (882 \pm 24) - i(167 \pm 41)$ MeV [44]. The considerable differences in the width results indicate the difficulty in the determination of this quantity.

4. Unsuccessful searches

Again, in some other D decay channels the κ has been searched for but could not be confirmed: In the channel $D^0 \rightarrow K^-\pi^+\pi^0$ the κ fraction was 0.4 ± 0.3 % [45] and in $D^0 \rightarrow K^0K^-\pi^+$ (15 ± 12) % [46].

TEST OF SYMMETRIES - σ, κ IN A SCALAR NONET?

As we have seen it is difficult to establish the σ, κ poles uniquely by fitting different parametrizations to the data, but in models respecting chiral symmetry these poles appear after unitarization. As they are quite far away from the real axis their interpretation in terms of real particles is not without doubt. Therefore we consider it a crucial test whether these particles also obey the relations following from the assumed underlying flavour symmetries. Here we consider the attractive possibility that $\sigma, \kappa, f_0(980)$ and $a_0(980)$ form a nonet, either built from $q\bar{q}$ or from $qq\bar{q}\bar{q}$. We consider here two such tests.

Tests in J/ψ decays

Symmetry relations involving the decays $J/\psi \to$ tensor + vector particles have been tested successfully by the DM2 Collaboration [47]. There are deviations from $SU(3)$ symmetry from electromagnetic interactions and quark mass effects which are taken from similar analyses of the vector + pseudoscalar final states. For example, using PGD results, one finds for

$$R^T = \frac{J/\psi \to K_2^{*0}(1430)\bar{K}^{*0}(890)}{J/\psi \to f_2(1270)\omega} = \frac{(3.4 \pm 1.3)\ 10^{-3}}{(4.3 \pm 0.6)\ 10^{-3}} = 0.8 \pm 0.3. \tag{10}$$

where $R^T = 1$ in the $U(3)$ symmetry limit with ideal mixing.

We assume now that the scalars fulfil the same relations as the tensors and assume the quark composition $\sigma \leftrightarrow f_2 = (u\bar{u} + d\bar{d})/\sqrt{2}$. Then the ratio R^T should equal the corresponding ratio for scalars R^S if we assume the same symmetry breaking in both multiplets. With the data from DM2 [32] for σ and the preliminary result from BES [43] for κ we find after correction for neutrals

$$R^S = \frac{J/\psi \to \kappa\bar{K}^{*0}(890)}{J/\psi \to \sigma\omega} = \frac{(0.19 \pm 0.18)\ 10^{-3}}{(2.4 \pm 0.45)\ 10^{-3}} = 0.08 \pm 0.08. \tag{11}$$

so the relation $R^T = R^S$ looks badly broken even if there are large errors. Adding an $s\bar{s}$ component to σ would make the agreement worse. At this state of the κ analysis we do not draw any final conclusion, but rather recommend the repetition of this exercise, once the data are considered firm. It would also be interesting to include results for $a_0(980)\rho$ and $f_0(980)(\phi/\omega)$ in the analysis.

Tests in charmless B-meson decays

Recently, a large production rate has been found for the decay [24, 48]

$$Br(B^+ \to K^+ f_0(980)) \sim 15 \times 10^{-6}, \tag{12}$$

comparable to $B \to K \pi$, also the decay $K_0^*(1430)\pi^-$, see Fig. 1 for a recent result by BELLE [24]. The $f_0(980)$ is a very interesting particle from our present viewpoint (double faced "Janus-particle"), as it could belong either to a multiplet of lower mass with σ, κ and $a_0(980)$ (Route I) or to one of higher mass with $a_0, K_0^*(1430)$ and f_0' (Route II); in the scheme [12] $a_0 \equiv a_0(980)$ and $f_0' \equiv f_0(1500)$ but other choices are possible.

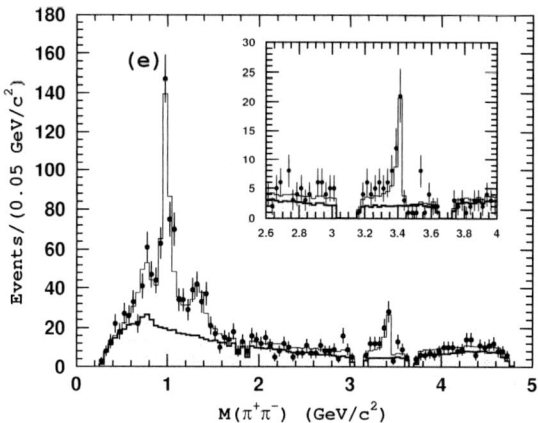

FIGURE 1. The $\pi\pi$ mass spectrum in $B^+ \to K^+\pi^+\pi^-$ with strong $f_0(980)$ production and little $f_0(1500)$ if any (from BELLE [24]).

Starting from this observation a strategy has been proposed [25] to find the members of the scalar nonet and to determine their flavour mixing. This involves the study of the decays

$$B \to K + \text{Scalars}, \qquad B \to K^* + \text{Scalars}. \qquad (13)$$

Some experience has been gained with the charmless decays as in (13) but with pseudoscalars instead of scalars. We follow here the phenomenological approach [49] which derives the 2 body decay rates from the processes $b \to su\bar{u}, b \to sd\bar{d}, b \to ss\bar{s}$ which get equal contributions from the QCD penguin diagrams and smaller contributions of order 20% from the CKM suppressed tree diagrams. Furthermore, for the flavour singlet mesons, there is a purely gluonic contribution from the penguin process $b \to sg$. In a simplified version [25] only the dominant penguin $sq\bar{q}$ and flavour singlet amplitudes are kept and this has allowed already acceptable fits to the available decay rates.

This simplified model is then taken over for the scalars and allows predictions for all members of the nonet in terms of only few parameters. There is the penguin amplitude P_{AB} for decay into hadrons from multiplets A, B ($s \to A$, spectator quark $\to B$), the gluonic amplitude $\gamma_{AB}P_{AB}$ and the amplitude with hadron A and B exchanged $\beta_{AB}P_{AB}$. For simplicity, as in the PP and PV case we assume $\gamma_S \equiv \gamma_{PS} \equiv \gamma_{VS}$ real, furthermore for the exchanged amplitude $\beta_{PS} = -\beta_{VS} = 1$. The $(-)$ sign comes from the fact that the VS state is in a P wave and this leads to alternating decay patterns in PV and PP final states [50] and the same happens with scalars.

For illustration we give in Table 2 some predictions for the two schemes with high or low mass nonet using the results of Table 2 in [15]. For the higher multiplet (Route II) we choose $f_0 \to f_0(980)$, $K^*(1430)$ and $f_0' \to f_0(1500)$, mixed like η', η as in the pseudoscalar sector [11, 12]. Taking the new data on $B^+ \to K^+ f_0(980)$ and $K_0^* \pi^+$ as input [24, 48] (neglecting $f_0 \to K\bar{K}$ decays) we determine $\gamma_s \approx -0.36$ and predict the rates for the other members of the nonet; in Table 2 we show the predictions for 3 more

TABLE 2. Predicted rates for scalars f_0' associated with K and $K^*(890)$, with rates $Kf_0(980)$ ($f_0 \to \pi\pi$ only) and $K_0^*(1430)\pi^+$ as input (underlined) [24] (see also text).

Route II:	f_0, f_0', K_0^*	$\gamma_s = -0.36$			
$B^+ \to$	$K^+ f_0$	$K^+ f_0'$	$K^{*+} f_0$	$K^{*+} f_0'$	$K_0^* \pi^+$
amplitude	$(3+4\gamma_s)/\sqrt{6}$	$\gamma_s/\sqrt{3}$	$(-1+4\gamma_s)/\sqrt{6}$	$(2+\gamma_s)/\sqrt{3}$	1
rate [10^{-6}]	<u>15</u>	1.6	38	34	<u>38</u>
Route Ia:	σ, f_0, κ	$\gamma_s = 0$			
$B^+ \to$	$K^+ f_0$	$K^+ \sigma$	$K^{*+} f_0$	$K^{*+} \sigma$	$\kappa\pi^+$
amplitude	$(1+\gamma_s)$	$(1+2\gamma_s)/\sqrt{2}$	$(-1+\gamma_s)$	$(1+2\gamma_s)/\sqrt{2}$	1
rate [10^{-6}]	<u>15</u>	7.5	15	7.5	15
Route Ib:	σ, f_0, κ	$\gamma_s = -0.5$			
rate [10^{-6}]	<u>15</u>	0	135	0	60

channels concerning f_0 and f_0'. Especially, f_0' has a small rate; this seems to be born out by the BELLE data [24] in the $\pi\pi$ channel (see Fig. 1), but there may be some signal in the $K\bar{K}$ channel which has to be analysed further. A clear prediction are the large rates with K^*. Other predictions follow easily.

For the lower multiplet (Route I) we choose $f_0 \to f_0(980) \sim s\bar{s}$ and $f_0' \to \sigma \sim (u\bar{u}+d\bar{d})/\sqrt{2}$. Now only the $f_0(980)$ rate is known. We first assume (Ia) the absence of the gluonic process ($\gamma_s = 0$). Then we obtain a σ rate half as big as the $f_0(980)$ rate whereas the data in Fig. 1 do not indicate any low mass effect at all. Next (Ib) we put $\gamma_s = -0.5$ to obtain a small σ rate, but then a large κ rate follows which is not easy to recognize in the data [24] but it should be looked for quantitatively.

This serves only as a simple exercise to demonstrate the relevance of the symmetry relations. As the dominant penguin processes are flavour symmetric all the nonet members should be produced in some K or K^* channel with rate comparable to $f_0(980)$. So it is interesting to find out which scalars are strongly produced and fulfill the approximate relations. In this way one may also learn whether $a_0(980)$ or $a_0(1430)$ is the isovector member.

CONCLUSIONS

1. **Poles** σ, κ are not necessarily required from the acceptable fits to data but they appear commonly in parametrizations with the small scattering length from chiral symmetry and with unitarity.
2. **Elastic scattering**: In $\pi\pi$ scattering besides $f_0(980)$ and $f_0(1500)$ there is apparently a slowly moving background amplitude describing an approximate circle in the complex plane; this could be a broad resonance with mass $\gtrsim 1$ GeV (a glueball?). In $K\pi$ scattering there is only one resonance, $K^*(1430)$, below 1800 MeV but no extra circle. The σ, κ represent distant poles in parametrizations ($\Gamma \gtrsim$ Mass); they modify the behaviour of amplitudes near threshold but do not generate circles. They are not necessarily real propagating particles.

3. **Low mass peaks**: A new development are the phase studies of the low mass effects in 3-body decays. There is still a controversy on whether the phase in the $\pi\pi$ and $K\pi$ channels move differently from elastic scattering. This question should be clarified by further studies of phase sensitive quantities, the ultimate goal being an energy independent phase shift analysis. In some cases the peaks with slow phase movement can be explained by non-resonant mechanisms.

4. **Symmetry relations for decay rates**: They represent the crucial test for the particle interpretation and the flavour properties of σ and κ poles and other scalars. One possibility is offered by J/ψ decays. A powerful approach is the study of decays $B \to K(K^*)+$ scalars which should allow to find the members of the lightest 0^{++} nonet and the flavour mixing of the isoscalars. In particular, if decay rates into σ and κ are measured, one may find out about their flavour symmetry and whether $f_0(980)$ belongs to a low mass or a high mass multiplet.

ACKNOWLEDGMENTS

This talk is based on common work with Peter Minkowski; I would like to thank him for the discussions and the collaboration and Heiri Leutwyler for correspondence on the sigma pole.

REFERENCES

1. Particle Data Group, K.Hagiwara et al., *Phys. Rev. D* **66**, 010001 (2002).
2. S. Spanier and N.A. Törnqvist, in PDG, Ref. 1.
3. F.E. Close and N.A. Törnqvist, *Phys. G* **28** R249 (2002).
4. S.F. Tuan, Tokyo Symposium on Hadron Spectroscopy, Feb. 2003, arXiv:hep-ph/0303248.
5. G.S. Bali, "Lattice calculations of hadron properties", arXiv:hep-lat/0308015.
6. C. Amsler and F.E. Close, *Phys. Rev.* **D53**, 295 (1996); *Phys. Lett.* **B353**, 385 (1995).
7. E. Klempt, *"Meson Spectroscopy"*, PSI Zuoz Summer School, Aug. 2000, arXiv:hep-ex/0101031.
8. J. Weinstein and N. Isgur, *Phys. Rev.* **D27** 588 (1983).
9. R.L. Jaffe, *Phys. Rev.* **D 15** 267, 281 (1977).
10. N.N. Achasov and V.V. Gubin, *Phys.Rev.* **D63**, 094007 (2001).
11. E. Klempt, B.C. Metsch, C.R. Münz and H.R. Petry, *Phys. Lett.* **B361**, 160 (1995).
12. P. Minkowski and W. Ochs, *Eur. Phys. J.* **C9**, 283 (1999).
13. V.V. Anisovich and A.V. Sarantsev, *Eur.Phys.J.* **A16**, 229 (2003).
14. S. Narison, *Nucl. Phys.* **B509** 312 (1998); *Nucl. Phys. B (Proc. Suppl.)* **64** 210 (1998).
15. P. Minkowski and W. Ochs, *Nucl. Phys. B (Proc. Suppl.)* **121** 123 (2003).
16. I. Bediaga and J.M. de Miranda, *Phys. Lett.* **B550**, 135 (2002).
17. G. Colangelo, J. Gasser and H. Leutwyler, *Nucl. Phys.* **B603**, 125 (2001).
18. F.J. Yndurain, "Low energy pion physics", arXiv:hep-ph/0212282.
19. R. Kaminski, L. Lesniak and B. Loiseau, *Phys. Lett.* **B551**, 241 (2003).
20. B. Hyams et al., *Nucl. Phys.* **B64**, 134 (1973).
21. D. Morgan and M.R. Pennington, *Phys. Rev.* **D48**, 1185 (1993).
22. Gunter et al. (E852 Collaboration), *Phys. Rev.* **D64**, 072003 (2001).
23. V.V. Anisovich and V.A. Nikonov, *Eur.Phys.J.* **A8**, 401 (2000).
24. A. Garmash et al. (Belle Collaboration) *Phys. Rev.* **D65** 092005 (2002); BELLE-CONF-0338, EPS Aachen (2003).
25. P. Minkowski and W. Ochs, arXiv:hep-ph/0304144.

26. E791 Collaboration, E.M. Aitala et al., *Phys. Rev. Lett.* **86**, 770 (2001).
27. S. Malvezzi for FOCUS collaboration, this conference; see also arXiv:hep-ex/0307055.
28. A. Reis, for E791 Collaboration, this conference; I. Bediaga, arXiv:hep-ex/0307008 (2003).
29. D. Alde et al. (GAMS Collaboration), *Phys. Lett.* **B397**, 350 (1997); R. Bellazzini et al., *Phys. Lett. B* **467**, 296 (1999).
30. D. Barberis et al. (WA102 Collaboration) *Phys. Lett. B* **453**, 325 (1999).
31. M. Ishida, *Prog.Theor.Phys.Suppl.* **149**, 190 (2003).
32. DM2 Collaboration, J.E. Augustin et al., *Nucl. Phys.* **B320**, 1 (1989).
33. D. Bugg, for BES collaboration, this conference.
34. T. Komada, M. Ishida and S. Ishida, *Phys. Lett.* **B508**, 31 (2001).
35. V.V. Frolov et al. (CLEO II Collaboration), arXiv:hep-ex/0306048 (2003).
36. D. Aston et al. (LASS Collaboration), *Nucl. Phys.* **B296**, 493 (1988).
37. J.M. Link et al. (FOCUS Collaboration), *Phys. Lett.* **B535**, 43 (2002).
38. S. Cherry and M.R. Pennington, *Nucl. Phys.* **A688**, 823 (2001).
39. J.A. Oller, E. Oset and J.R. Pelaez, *Phys.Rev.* **D59**, 074001 (1999); J.A. Oller and E. Oset, *Phys.Rev.* **D69**, 074023 (1999).
40. J.R. Pelaez and A. Gomez Nicola, *AIP Conf.Proc.* **660**, 102 (2003).
41. E.M. Aitala et al. (E791 Collaboration), *Phys. Rev. Lett.* **89**, 121801 (2002).
42. Carla Göbel, for E791 Collaboration, arXiv:hep-ex/0307003 (2003).
43. J.Z. Bai et al. (BES collaboration), arXiv:hep-ex/0304001.
44. T. Komada, for BES collaboration, this conference.
45. S. Kopp et al. (CLEO collaboration), *Phys.Rev.* **D63** 092001 (2001).
46. B. Aubert et al. (BaBar collaboration), ICHEP2002, arXiv:hep-ex/0207089.
47. A. Falvard et al. (DM2 Collaboration), *Phys. Rev.* **D38**, 2706 (1988).
48. B. Aubert et al. (BaBar Collaboration), BaBar-Conf-3/001, arXiv:arXiv:hep-ex/0303022.
49. A.S. Dighe, M. Gronau and J.L. Rosner, *Phys. Lett.* B367 357 (1996); *Phys. Rev. Lett.* **79** 4333 (1997); C.W. Chiang and J.L. Rosner, *Phys. Rev.* **D65** 074035 (2002).
50. H.J. Lipkin, *Phys. Lett.* **B415**,186 (1997).

Comments on the σ and κ

D.V. Bugg

Queen Mary, London E1 4NS, UK

Abstract.
The determination of σ and κ poles is discussed. Elastic scattering and production processes may be fitted simultaneously by including the Adler zero into the Breit-Wigner width $\Gamma(s)$. Data on $\pi\pi \to \pi\pi$ and K_{e4} decays give a σ pole at $(525 \pm 40) - i(247 \pm 25)$ MeV. A combined analysis with production data gives a better determination: $(542 \pm 25) - i(249 \pm 25)$ MeV. The Fourier transform of the matrix element for $\sigma \to \pi\pi$ reveals a compact interaction region with RMS radius ~ 0.4 fm. The analysis of LASS data for $K\pi$ elastic scattering, including the Adler zero, determines a κ pole at $(722 \pm 60) - i(386 \pm 50)$ MeV, consistent with BES data for $J/\Psi \to K^* \kappa$.

At this conference, Li presents BES data on $J/\Psi \to \omega\pi^+\pi^-$ [1], shown in Fig. 1. There is a clear σ peak. It also appeared in DM2 [2] and E791 data [3]. E791 claim a low mass $\kappa \to K\pi$ in $D^+ \to K^-\pi^+\pi^+$ [4]; preliminary BES data for $J/\Psi \to K^*(892)\kappa$ also require a κ peak [1]. Why do these peaks not appear in $\pi\pi$ and $K\pi$ elastic scattering?

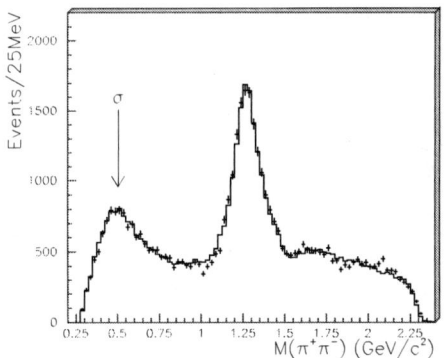

FIGURE 1. BES data for $J/\Psi \to \omega\pi^+\pi^-$.

The reason is simple. For a production process like $J/\Psi \to \omega\pi^+\pi^-$,

$$T_{prod} = \Lambda_\sigma / [M^2 - s - iM\Gamma(s)]. \tag{1}$$

Here M is the mass where the phase shift goes through $90°$ for real s; Λ_σ is a complex coupling constant $g\exp(i\phi)$. For elastic scattering there is an extra factor $M\Gamma(s)$ in the numerator, required to make the amplitude follow the unitary circle:

$$T_{el} = M\Gamma(s) / [M^2 - s - iM\Gamma(s)] = e^{i\delta} \sin\delta. \tag{2}$$

CP717, *Hadron Spectroscopy: Tenth International Conference,*
edited by E. Klempt, H. Koch, and H. Orth
© 2004 American Institute of Physics 0-7354-0197-7/04/\$22.00

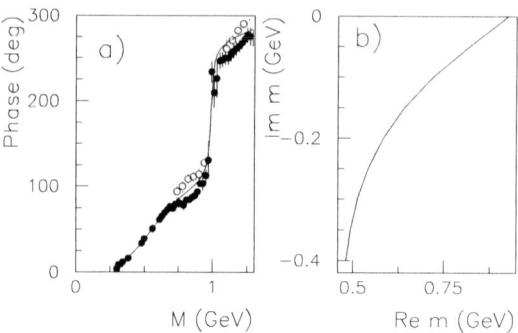

FIGURE 2. (a) Phase shifts for $\pi\pi \to \pi\pi$ and the fit; (b) the mass where the phase goes through $90°$ for complex s.

In $\pi\pi$ elastic scattering, there is a well-known Adler-Weinberg zero [5] at $s_A \simeq m_\pi^2/2$, just below the elastic theshold. This is a central feature of Chiral Perturbation Theory. Near threshold it cancels the pole, so $\delta(s)$ is rather featureless and shows little direct evidence for the pole. The same happens in $K\pi$ elastic scattering, where there is an Adler zero at $s_A \simeq M_K^2 - M_\pi^2/2$. The LASS data for $K\pi$ elastic scattering [6] display slowly rising phase shifts at low mass, again because the Adler zero suppresses the effect of the κ pole.

My approach is to include the Adler zero explicitly into the width:

$$\Gamma(s) = \rho(s)\frac{s - s_A}{M^2 - s_A}f(s)\exp\frac{s - M^2}{A}. \tag{3}$$

Several simple forms for $f(s)$ are tried, to test the stability of the extrapolation from the physical region to the pole at complex s. The best fit uses $f(s) = b_1 + b_2 s$. A combined fit is made to CERN-Munich phase shifts for $\pi^-\pi^+ \to \pi^-\pi^+$ [7], K_{e4} data [8], and $\pi^-\pi^+ \to \pi^0\pi^0$ [9,10]. The small phase shifts δ from $f_0(980)$, $f_0(1370)$ and $f_0(1500)$ are added to the σ phase (the Dalitz-Tuan prescription which preserves unitarity [11]). The σ pole lies at M $= (525 \pm 40) - i(247 \pm 25)$ MeV, i.e. a full width ~ 494 MeV. Systematic errors from the choice of $f(s)$ are included in the errors. The curve in Fig. 2(a) shows the fit. There are systematic discrepancies above 740 MeV between phase shifts of Hyams et al. (filled circles) and $\pi^0\pi^0$ data (open circles).

My pole agrees quite well with the result of Colangelo et al: M $= (490 \pm 30) - i(290 \pm 20)$ MeV [12]. Their determination has the virtue of fitting the left-hand cut consistently with the right-hand cut, using the Roy equations. They do not fit data above 800 MeV.

Mass and width are strongly correlated. Fortunately, BES data have an orthogonal correlation. Putting the two sets of data together, the pole is at $M = 542 \pm 25 - i(249 \pm 25)$ MeV. Parameter values are $M = 0.9264$ GeV, $A = 1.082$ GeV^2, $b_1 = 0.5843$ GeV, $b_2 = 1.6663$ GeV. For BES data alone, a Breit-Wigner amplitude of constant width or

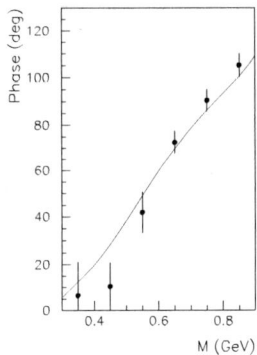

FIGURE 3. The phase of the $\pi\pi$ S-wave amplitude: the full curve fits BES data and elastic data; points with errors from BES data are fitted in slices of $\pi\pi$ mass 100 MeV wide.

the E791 formula give pole positions which agree within errors with that quoted above; this is an independent check on the stability of the pole. BES data determine the phase of the σ amplitude v. mass from interference between $\sigma \rightarrow \pi\pi$ and $b_1(1235) \rightarrow \omega\pi$ amplitudes. Results in Fig. 3 agree well with the phase of $\pi\pi$ elastic scattering.

A popular misconception is that the pole must be close to the mass M where the phase shift passes through 90° on the real axis. That is true only when Γ is constant. When Γ varies with s, the phase of the amplitude varies off the real s axis. Fig. 2(b) shows the position on the complex plane where the phase goes through 90°; there is a rapid variation with $Im\ m$.

A common assumption has been

$$\Gamma(s) = \Gamma_0\rho(s), \tag{4}$$

where $\rho(s)$ is phase space $2k/\sqrt{s}$; k is pion momentum in the σ rest frame. This formula has a problem: the s-dependence of the Breit-Wigner denominator creates a second pole below the $\pi\pi$ threshold at $M \simeq 210$ MeV. This pole cannot be avoided when using this formula. It is unphysical; a bound state decaying to $\gamma\gamma$ could not have escaped detection in extensive LEAR data.

Now we turn to the κ pole. LASS data for $K\pi \rightarrow K\pi$ are fitted with eqn. (3) setting $b_2 = 0$. Phases of the κ and $K_0(1430)$ are added. For $K_0(1430)$, a Flatté form including the Adler zero is used:

$$\Gamma(s) = \frac{s - s_A}{M^2 - s_A}[g_1\rho_{K\pi}(s) + g_2\rho_{K\eta'}(s)]. \tag{5}$$

There is freedom in fitting $K_0(1430)$ because it rides on top of the κ 'background'. It is constrained to fit the line-shape of the strong $K_0(1430)$ peak in BES data, which require $M = 1.513$ GeV, $g_1 = 0.304$ GeV, $g_2 = 0.380$ GeV and a pole at $1419 - i158$ MeV, close to the values of Aston et al: $M = 1412 \pm 6$ MeV, $\Gamma = 294 \pm 23$ MeV.

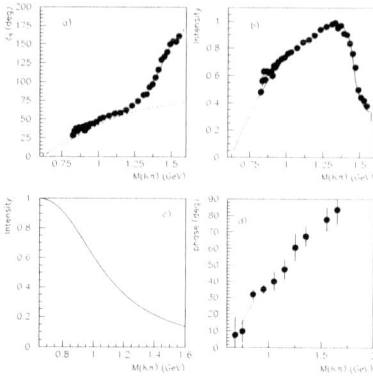

FIGURE 4. (a) $K\pi$ phase shifts from Lass [6]. The full curve shows the fit and the dashed curve the κ contribution. (b) The fit to the intensity in $K\pi \to K\pi$. (c) The intensity of the κ in production processes v. mass, normalised to 1 at its peak; dashed curve - E791 determination [4]. (d) The kappa phase v. $K\pi$ mass from BES data, compared with the fit.

The Adler zero is crucial for the κ pole. Without it, there is no pole. With it the pole appears. The fitted $K\pi$ scattering length is $0.19m_\pi^{-1}$, very close to Weinberg's prediction from current algebra of $0.18m_\pi^{-1}$ [5]. Fits are shown in Figs. 4(a) and (b). The pole position of the κ is at M $= 722 \pm 60 - i(386 \pm 50)$ MeV where errors are systematic. BES data for $J/\Psi \to K^*\kappa$ are consistent with this and give a pole with $Im\ M$ 300–400 MeV.

The full curve in Fig. 4(c) shows the intensity of the κ as it appears in *production processes*, (but without allowance for available phase space). Fig. 4(d) shows the phase of the κ amplitude from BES data in 100 MeV slices of mass.

We now return to the information contained in the matrix element $F(s)$ for $\pi\pi \to \pi\pi$. This matrix element is proportional to the square root of $\Gamma(s)$. Its Fourier transform reveals the radial dependence of the matrix element:

$$F(r) \propto (1/r) \int_0^\infty ds\ F(s) \sin kr. \tag{6}$$

Here r is the relative separation $r_1 - r_2$ between π^+ and π^-. Fig. 5 shows $F(r)$. The striking result is that $F(r)$ is quite compact, with an RMS radius of 0.39 fm. There is short range attraction and long-range repulsion; the net result is a small scattering length. The repulsive component is consistent with the usual confining potential. My result is quite different to that of the Ishida group [13], who have a short range repulsive interaction.

The curve on Fig. 5 measures the matrix element for the $\pi\pi$ interaction. The wave function of the σ could however extend to larger radius. This remark is relevant to the $f_0(980)$ and $a_0(980)$, whose masses are so close to the $K\bar{K}$ threshold that they inevitably have long-range $K\bar{K}$ components (like the deuteron), even though their formation depends on short range forces. Many authors suggest that σ, κ, $f_0(980)$ and $a0(980)$ make

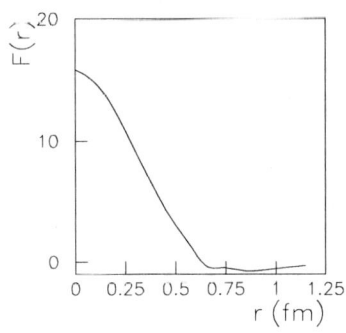

FIGURE 5. The radial dependence $F(r)$ of the matrix element for $\pi\pi \to \pi\pi$.

a nonet. For $a_0(980)$ and $f_0(980)$, the dominant exchange is K^* exchange in $\pi\pi \to K\bar{K}$. This may explain why $f_0(980)$ and $a_0(980)$ lie close to the KK threshold [15]. The recent discovery of the $D_s(2317)$ and $D_s(2460)$ [16, 17] suggests bound states of DK and D^*K; then $f_0(980)$ and $a_0(980)$ may be quasi-bound states of $K\bar{K}$. In all these cases, there is no Adler zero to inhibit quasi-bound states. Van Beveren and Rupp [18] remark that the Adler zero in $D\pi$ scattering leads to a broad 0^+ state at 2050–2150 MeV, above the $D\pi$ threshold, analogous to the σ. This may be the broad $D_0(2308)$ in $D\pi$ [19], where the analysis is presently done neglecting the Adler zero.

REFERENCES

1. W. Li, this conference.
2. J.E. Augustin et al., Nucl. Phys. B320 (1989) 1.
3. E.M. Aitala et al., Phys. Rev. Lett. 86 (2001) 770.
4. E.M. Aitala et al., Phys. Rev. Lett. 89 (2002) 121802.
5. S. Weinberg, Phys. Rev. Lett. 17 (1966) 616.
6. D. Aston et al., Nucl. Phys. B296 (1988) 253.
7. B. Hyams et al., Nucl. Phys. B64 (1973) 134.
8. S. Pislak et al., Phys. Rev. Lett. 87 (2001) 221801.
9. Particle Data Group, Phys. Rev. D66 (2002) 010001.
10. K. Takamatsu et al., Nucl. Phys. A675 (2000) 312c and Progr. Theor. Phys. 102 (2001) E52.
11. J. Gunter et al., Phys. Rev. D64 (2001) 072003.
12. R.H. Dalitz and S. Tuan, Ann. Phys. (N.Y.) 10 (1960) 307.
13. G. Colangelo, J. Gasser and H. Leutwyler, Nucl. Phys. B603 (2001) 125.
14. S. Ishida et al., Prog. Theor. Phys. 95 (1996) 745.
15. I am grateful to Bing Song Zou for this remark.
16. E.M. Aitala et al., hep-ex/0007028.
17. D. Besson et al., hep-ex/0305017.
18. E. van Beveren and G. Rupp, hep-ph/0305035.
19. D. Kassel, this conference.

Phase Motion in the Scalar Low-Mass $\pi^-\pi^+$ Amplitude in $D^+ \to \pi^-\pi^+\pi^+$

Alberto Reis, for the Fermilab E791 Collaboration

Centro Brasileiro de Pesquisas Físicas - CBPF - Brazil

Abstract.
   This work is a direct and model-independent measurement of the low mass $\pi^+\pi^-$ phase motion in the $D^+ \to \pi^-\pi^+\pi^+$ decay. The results show a strong phase variation, compatible with an isoscalar $\sigma(500)$ meson. This result confirms the previous Fermilab E791 result which found evidence for the existence of this scalar particle using a full Dalitz-plot analysis.

INTRODUCTION

Charm meson decays are a natural place for studying light scalar mesons. These decays are a clean environment: well defined initial state, very small non-resonant component and large coupling to scalars. In addition, D decays can provide insight into the quark content of these controversial particles, since the bulk of the hadronic width comes from modes for which there is a W-radiation amplitude.

Studies of light scalars using charm decays started a few years ago, when E791[1] presented results of Dalitz-plot analyses of the decays $D_s^+, D^+ \to \pi^-\pi^+\pi^+$[2, 3] and $D^+ \to K^-\pi^+\pi^+$ [4]. In particular, the analysis of the decays $D^+ \to \pi^-\pi^+\pi^+$ and $D^+ \to K^-\pi^+\pi^+$ showed evidence for the σ and κ mesons. In the case of $D^+ \to \pi^-\pi^+\pi^+$, the σ appears as an accumulation of signal events in low $\pi^+\pi^-$ mass. The E791 data could only be described by an amplitude having an s dependent phase (s being the $\pi^-\pi^+$ mass squared).

Several years later, the same effect - an accumulation of signal events in low $\pi^+\pi^-$ mass - was also observed in other decays from different experiments, for, instance, $D^0 \to \bar{K}^0\pi^+\pi^-$ from CLEO[5] and Belle[6], and $J/\psi \to \omega\pi^+\pi^-$ from BES[7]. Again, the description of these data requires amplitudes with s-dependent phase.

Since E791 publication of the $D_s^+, D^+ \to \pi^-\pi^+\pi^+$ results, two kinds of criticisms have been made: the quoted values of both σ and κ parameters are not correct because the simple Breit-Wigner formula is inadequate to describe broad scalars, especially when near the threshold; the other criticism is that one should not claim the existence of a resonance without showing the phase motion of the corresponding amplitude.

The Breit-Wigner formula, although being in this case only a naive approximation, has the key ingredient: an s-dependent phase. On the other hand, there is no single agreed way to treat broad scalars near threshold. The σ parameters depend strongly on the assumed functional form of its line shape.

In this work[8] the second criticism is addressed. The phase variation of the low-mass $\pi^+\pi^-$ amplitude is extracted in a model independent way.

CP717, *Hadron Spectroscopy: Tenth International Conference,*
edited by E. Klempt, H. Koch, and H. Orth
© 2004 American Institute of Physics 0-7354-0197-7/04/$22.00

THE AMPLITUDE DIFFERENCE METHOD

In Dalitz-plot analysis, resonant amplitudes are complex functions written in the general form $\mathscr{A} = f(s)e^{i\delta(s)}$. Non-resonant amplitudes, in contrast, have $\delta(s) = $ constant. If two resonant amplitudes cross in some region of the Dalitz-plot, they will interfere. The interference pattern in this crossing region depends on the s-dependent phases of both resonances. One can extract the phase motion of an unknown amplitude that crosses a well known resonance provided that:

- the contribution of other amplitudes is negligible in the crossing region between the amplitude under study and the known resonance;
- the integrated amplitude of the known resonance is symmetric with respect to an effective mass squared (m_{eff}^2).

The second condition ensures that, by comparing the amplitude below and above m_{eff}^2, we end up with an expression that involves only the desired phase $\delta(s_{13})$. Then, we can write the approximate amplitude of this phase space region in a simple way,

$$\mathscr{A}(s_{12},s_{13}) \simeq a_R \,\mathscr{BW}(s_{12})\,\mathscr{M}(s_{12},s_{13}) + a_s/(p^*/\sqrt{s}_{13})\,sin\delta(s_{13})\,e^{i(\delta(s_{13})+\gamma)} \quad (1)$$

where γ is the overall relative final state interaction (FSI) phase difference between the two amplitudes, a_R and a_s are respectively the real magnitudes of the known resonance and the under-study complex amplitude, $sin\delta(s_{13})e^{i\delta(s_{13})}$ represents the most general amplitude for a two-body elastic scattering; $p^*/\sqrt{s}_{13}$ is a phase space factor to make this description compatible with $\pi\pi$ scattering and $\mathscr{M}(s_{12},s_{13})$, $\mathscr{BW}(s_{12})$ are the angular function and Breit-Wigner for the known resonance, respectively.

The quantity $\Delta \mid \mathscr{A}(s_{13}) \mid^2 \equiv \mid \mathscr{A}(m_{eff}^2+\varepsilon,s_{13}) \mid^2 - \mid \mathscr{A}(m_{eff}^2-\varepsilon,s_{13}) \mid^2$, which is the difference of the amplitudes squared after integration over s_{12}, is computed in bins of s_{13}. It takes the form

$$\Delta \mid \mathscr{A}(s_{13}) \mid^2 = \frac{-4a_s a_R/(p^*/\sqrt{s}_{13})\varepsilon m_0\Gamma_0}{\varepsilon^2+m_0^2\Gamma_0^2}\,(sin(2\delta(s_{13})+\gamma)-sin\gamma)\mathscr{M}(s_{13})/(p^*/\sqrt{s}_{13})$$

$$(2)$$

If $\delta(s_{13})$ is an analytical function of s_{13}, then there will be maximum and minimum values of $\Delta \mid \mathscr{A}(s_{13}) \mid^2$, which we can use to determine both the constant term in the above equation and the phase γ,

$$\frac{-4a_s a_R/(p^*/\sqrt{s}_{13})\varepsilon m_0\Gamma_0}{\varepsilon^2+m_0^2\Gamma_0^2} \equiv \mathscr{C} = (\Delta \mid \mathscr{A}' \mid_{max}^2 - \Delta \mid \mathscr{A}' \mid_{min}^2)/2 \quad (3)$$

$$\gamma = sin^{-1}(\frac{\Delta \mid \mathscr{A}' \mid_{max}^2 + \Delta \mid \mathscr{A}' \mid_{min}^2}{\Delta \mid \mathscr{A}' \mid_{min}^2 - \Delta \mid \mathscr{A}' \mid_{max}^2}) \quad (4)$$

313

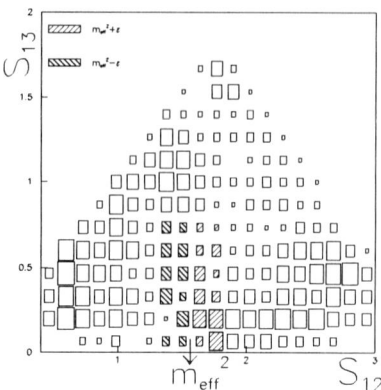

FIGURE 1. $D^+ \to \pi^- \pi^+ \pi^+$ Dalitz-plot folded distribution. The events used by the AD analysis are in the hatched region. The size of the area of each bin in the plot corresponds to the number of events in that bin.

where $\mathscr{A}' \equiv \mathscr{A}/(\mathscr{M}\sqrt{s_{13}}/p^*)$.

Finally, considering $\delta(s_{13})$ an increasing function of s_{13}, we have

$$\delta(s_{13}) = \frac{1}{2}\left(sin^{-1}\left(\frac{1}{\mathscr{C}}\Delta \mid \mathscr{A}'(s_{13}) \mid^2 + sin(\gamma)\right) - \gamma\right) \qquad (5)$$

This is, in essence, the idea of the amplitude difference method [9]. The method was shown to work in a "calibration" exercise using the $f_0(980)$ resonance in the $D_s^+ \to \pi^- \pi^+ \pi^+$ decay [10]. In this case we have the $f_0(980)$ contribution in both s_{12} and s_{13} axes. We were able to get a phase motion $\delta(s_{13})$ compatible with the $f_0(980)$ using the $f_0(980)$ in s_{12}.

PHASE MOTION OF SCALAR LOW-MASS $\pi^- \pi^+$ AMPLITUDE

The folded Dalitz-plot distribution of the $D^+ \to \pi^- \pi^+ \pi^+$ decay is shown Fig. 2. The horizontal and vertical axes are the squares of the $\pi^+ \pi^-$ invariant mass high (s_{12}) and low (s_{13}) combinations.

To study the low mass region in s_{13}, there are three possible well known resonances in s_{12} to act as a probe in this decay: $\rho(770)$, $f_0(980)$ and $f_2(1270)$. Figure 1 shows that the $\rho(770)$ and $f_0(980)$ are located in regions where other amplitudes can not be considered negligible. On the other hand, the tensor $f_2(1270)$, $m_0^2 = 1.61$ GeV$^2/c^4$, is placed where the $\rho(770)$, in the crossed channel reaches a minimum due to its decay angular distribution.

With the proper choice of $m^2_{eff} = 1.535 \text{ GeV}^2/c^4$, the integral over s_{12} of the $f_2(1270)$ amplitude squared is symmetrical: the number of events between m^2_{eff} and $m^2_{eff} + \varepsilon$ ($\varepsilon = 0.26 \text{ GeV}^2/c^4$) is equal to the number of events between m^2_{eff} and $m^2_{eff} - \varepsilon$. Moreover, in this region there is no significant contribution other than the $\pi\pi$ complex amplitude under study in s_{13} (the amount of $\rho(770)$ within this mass region was estimated to be $\sim 5\%$). The choice of the $f_2(1270)$ as the analyser amplitude satisfies the necessary conditions for the amplitude difference method.

The acceptance and the background must be similar between m^2_{eff} and $m^2_{eff} + \varepsilon$ and m^2_{eff} and $m^2_{eff} - \varepsilon$, otherwise there would be biases in $\delta(s_{13})$. Monte Carlo simulations show that the acceptance is nearly uniform in this region. The background in this region comes mostly from random combinations of three pions, and it is also uniformly distributed. Since we are subtracting two similar distributions, we considered the background only in the size of the statistical error.

The $f_2(1270)$ angular function has a zero at about $s_{13} \simeq 0.48 \text{ GeV}^2/c^4$. This means a singularity in $\mathscr{A}'$. This singularity is handled in the following way. The data is divided into ten s_{13} bins. The binning is such that the singularity is placed in the middle of one bin. Doing this, we isolate the singularity in a single bin (bin 6) and discard its further use in the analysis.

The values of γ and $\mathscr{C}$ were obtained solving Equations 3 and 4. The value of the phase γ from the amplitude difference method is in agreement with that of the full Dalitz-plot analysis, $\gamma_{AD} = 2.78 \pm 0.38 \pm 0.40$ and $\gamma_{Dalitz} = 2.59 \pm 0.19$[3].

Having γ and $\mathscr{C}$, the phase $\delta(s_{13})$ is obtained for each s_{13} bin using Equation 5. There are ambiguities arising from the sin^{-1} operation, so $\delta(s_{13})$ was determined with the assumption that the phase difference starts at zero at threshold and is an increasing, monotonic, smooth function of s_{13}.

The phase motion of the low $\pi^+\pi^-$ mass amplitude, including systematic and statistical errors, is shown in Fig. 2. In spite of the limited statistics, a strong phase variation is clearly observed. Starting from zero at the threshold the phase varies by about 180^0 and saturates at around $s_{13} = 0.6 \text{ GeV}^2/c^4$. This is the expected behavior of resonance. The observed phase motion supports the interpretation of the $\sigma(500)$ as a true resonance. A constant or slowly varying phase would disfavor this interpretation. With more statistics we could have more bins and the pole position could be inferred. Fig. 2 shows also the phase motion of the simple Breit-Wigner (solid line) used in [3]. Even considering the Breit-Wigner as a naive approximation, there is a qualitative agreement between its phase motion and the directly extracted $\delta(s_{13})$: the description of the data requires an amplitude with a strong phase variation.

CONCLUSIONS

A direct, model-independent measurement of the phase motion of the low mass $\pi^+\pi^-$ scalar amplitude was discussed. Using the well known $f_2(1270)$ tensor meson in the crossing channel as the base resonance, from the $D^+ \to \pi^-\pi^+\pi^+$ decay, the $\delta(s_{13})$ phase motion was extracted. We obtain a $\delta(s_{13})$ variation of about 180^0, which is the

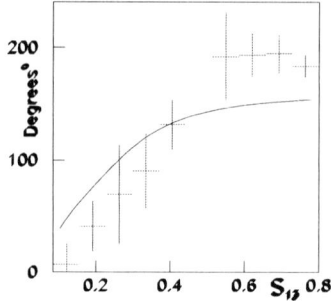

FIGURE 2. Phase motion vs s_{13}, with errors shown (systematic and statistical in quadrature). The continuous line is the Breit-Wigner phase motion, with the E791 parameters for the $\sigma(500)$.

expected behavior of a resonant amplitude. This result supports the interpretation of the $\sigma(500)$ as a true resonance, in agreement with the conclusions from our previous analysis of full $D^+ \rightarrow \pi^-\pi^+\pi^+$ Dalitz-plot. We could not extract the σ pole position with this method due to the limited statistics. The measurement of the correct σ pole position from a full Dalitz-plot analysis needs the correct functional form (theorists should agree on what it is). Whether the σ pole is the same in charm decay and scattering remains an open question.

REFERENCES

1. E791 Collaboration, E. M. Aitala et al., Eur. Phys. J. direct C1 (1999) 4; S. Amato et al., Nucl. Instrum. Meth. A324 (1993) 535; S. Bracker et al., IEEE Trans. Nucl. Sci. 43 (1996) 2457; S. Hanson et al., IEEE Trans. Nucl. Sci. 34 (1987) 1003; S. Bracker and S. Hansen, hep-ex/0210034.
2. E791 Collaboration, E.M. Aitala *et al.*, Phys. Rev. Lett. **86**(2001) 765.
3. E791 Collaboration, E.M. Aitala *et al.*, Phys. Rev. Lett. **86**(2001) 770.
4. E791 Collaboration, E.M. Aitala *et al.*, Phys. Rev. Lett. **89** (2002) 121801.
5. CLEO Collaboration, H. Muramatsu et al., Phys. Rev. Lett. **89** (2002) 251802.
6. Belle Collaboration, K. Abe et al., LP03, Batavia, hep-ex/0308043.
7. BES Collaboration, see W. Li talk in these proceedings.
8. E791 Collaboration, E. M. Aitala et al., Fermilab-Pub-03/296-E.
9. I. Bediaga and J. Miranda, Phys. Lett. **B550** (2002) 135.
10. I. Bediaga, hep-ex/0307008.

Generalized Weinberg's approach to the a_0/f_0 case

Yu.S.Kalashnikova

ITEP, 117218, B.Cheremushkinskaya 25, Moscow, Russia

Abstract. The problem of whether it is possible to distinguish composite from elementary particles is studied in the framework of Weinberg's approach. The possibility to extend this approach to the case of unstable particles in the presence of inelastic channels is considered. The interplay between the low-energy scattering data and the admixture of a bare state in the resonance is discussed, and the implications for the $a_0(980)/f_0(980)$ case are outlined.

The nature of $a_0(980)/f_0(980)$-mesons remains the most enigmatic question of meson spectrocsopy. Quark models [1] predict the 1^3P_0 $q\bar{q}$ states made of light quarks to exist at about 1 GeV. However, the vicinity of $K\bar{K}$ threshold suggests that significant $\frac{1}{\sqrt{2}}(u\bar{u} \pm d\bar{d})s\bar{s}$ admixture should be present in the wave functions of these mesons, either in the form of compact four-quark states [2] or in the form of $K\bar{K}$ molecules [3], [4]. The related question to be addressed in this regard is of the relative role of s- and t-channel force in the formation of such molecules [5].

The approaches of [3], [6] and [4] conclude that the t-channel force can be dominant in producing the attraction in $K\bar{K}$ channel necessary to form the a_0 and f_0 as hadronic molecules. On the other hand, as a_0/f_0 couple strongly to $K\bar{K}$ channel, one expects drastic unitarity effects, which are responsible for the dressing of the bare states seeded into mesonic continuum, the phenomenon described in the framework of coupled channel models [4], [7], [8].

Therefore the observed features of the $\pi\pi$, $\pi\eta$ and $K\bar{K}$ spectra could be explained both by potential-type interaction in these systems and by existence of "bare" confined states strongly coupled to mesonic channels, and the question persists whether it is possible to distinguish between different assignments for a_0/f_0.

Many years ago S.Weinberg [9] has considered a similar problem of "elementarity" of the deuteron, expressing the effective range n-p parameters in terms of field renormalization constant Z, which defines the admixture of a bare elementary-particle state in the deuteron. It was shown that the low-energy n-p data are consistent with small value of Z, so that the deuteron is indeed a molecular-type particle made of proton and neuteron.

To apply this approach, three requirements are needed. The particle should couple to a two-body channel with the threshold close to the nominal mass; this two-body channel should have zero orbital momentum; the particle must be stable, otherwise the factor Z is not defined. First two reqirements are met in the a_0/f_0 case, while the third one is not met: the decays $f_0 \rightarrow \pi\pi$ and $a_0 \rightarrow \pi\eta$ are known to be the main source of the width for these mesons. Nevertheless, it appears to be possible [10] to generalize Weinberg's

CP717, *Hadron Spectroscopy: Tenth International Conference,*
edited by E. Klempt, H. Koch, and H. Orth

approach to the case of unstable particles.

The starting point of such generalization is the dynamical scheme of the coupled channel model. It is assumed that the hadronic state is represented symbollically as

$$|\Psi\rangle = \begin{pmatrix} \sum_\alpha c_\alpha |\psi_\alpha\rangle \\ \sum_i \chi_i |M_1(i)M_2(i)\rangle \end{pmatrix}, \tag{1}$$

where the index α labels bare confined states $|\psi_\alpha\rangle$ with the probability amplitude c_α, and χ_i is the wave function in the i-th two-meson channel $|M_1(i)M_2(i)\rangle$. The wave function $|\Psi\rangle$ obeys the equation

$$\hat{\mathcal{H}}|\Psi\rangle = E|\Psi\rangle, \quad \hat{\mathcal{H}} = \begin{pmatrix} \hat{H}_c & \hat{V} \\ \hat{V} & \hat{H}_{MM} \end{pmatrix}, \tag{2}$$

where $\hat{H}_c$ defines the discrete spectrum of bare states, $\hat{H}_c|\psi_\alpha\rangle = E_\alpha|\psi_\alpha\rangle$, $\hat{H}_{MM}$ includes the free-meson part as well as direct meson-meson interaction (e.g., due to t- or u-channel exchange forces), and the term $\hat{V}$ is responsible for dressing the bare states. The latter is specified by the transition form factor $f^\alpha_{M_1(i)M_2(i)}(p)$,

$$\langle \psi_\alpha | \hat{V} | M_1(i)M_2(i)\rangle = f^\alpha_{M_1(i)M_2(i)M}(p), \tag{3}$$

where p is the relative momentum in the mesonic system $M_1(i)M_2(i)$. The function f decreases with p with some range β whose scale is set by the size scale of hadronic wave functions; the estimate for β is to be of order of a few hundred MeV.

In a simple case of only one bare state $|\psi_0\rangle$ and only one hadronic channel ($|K\bar{K}\rangle$) the system of equations (2) is easily solved, yielding for the $K\bar{K}$ scattering amplitude the form

$$F_{K\bar{K}}(k,k;E) = -\frac{2\pi^2 m f^2_{K\bar{K}}(k)}{E - E_0 + g_K(E)}, \quad k = \sqrt{mE}, \tag{4}$$

where

$$g_K(E) = \int \frac{f^2_{K\bar{K}}(p)}{\frac{p^2}{m} - E - i0} d^3p. \tag{5}$$

If the system possesses a bound state with the energy $-\varepsilon$, the admixture of a bare state in the bound state wave function, $|c_0|^2 = \cos^2\theta$, is defined from the expression

$$\tan^2\theta = \int \frac{f^2_{K\bar{K}}(p)d^3p}{(\frac{p^2}{m} + \varepsilon)^2}. \tag{6}$$

In the small binding limit $\sqrt{m\varepsilon} \ll \beta$ it is possible to express the effective range parameters in terms of the binding energy ε and angle θ in a model-independent way (for the details see [10]). The relations between scattering length a and effective range r_e and the binding energy ε and $Z = \cos^2\theta$ read

$$a = \frac{2(1-Z)}{2-Z}R + O(1/\beta), \quad r_e = -\frac{Z}{1-Z}R + O(1/\beta), \quad R = 1/\sqrt{m\varepsilon}, \tag{7}$$

318

coinciding with the ones obtained in [9].

In the case of unbound state one is to consider the continuum counterpart of Z, the spectral density of the bare state introduced in [11] and given by the expression

$$w(E) = 2\pi mk|c_0(E)|^2, \tag{8}$$

where $c_0(E)$ is found from the system of Eqs. (2) in the continuum. Due to the normalization condition

$$\int_0^\infty w(E)dE = 1 - Z \ or \ 1, \tag{9}$$

depending on whether there is a bound state or not, all the information on Z is encoded in the $w(E)$ too, and the generalization to the multichannel case is straightforward.

If one is interested only in the phenomena near $K\bar{K}$ threshold, it appears possible to express the spectral density $w(E)$ in terms of hadronic observables. Indeed, one can make use of the smooth dependence on energy of the intergal g_P for the light pseodoscalar channel, similar to (5). Then the near-threshold $K\bar{K}$ scattering amplitude is given by the Flattè-type expression

$$F_{K\bar{K}} = -\frac{1}{2k}\frac{\Gamma_K}{E - E_f + i\frac{\Gamma_K}{2} + i\frac{\Gamma_P}{2}}, \tag{10}$$

where

$$E_f = E_0 - \bar{E}_K - \bar{E}_P, \quad \Gamma_K = \bar{g}_{K\bar{K}}\sqrt{mE}, \quad \bar{g}_{K\bar{K}} = 4\pi^2 m f_{K0}^2,$$

$$\bar{E}_K = 4\pi m \int_0^\infty f_{K\bar{K}}^2(p)dp, \quad f_{K0} = f_{KK}(0),$$

and $\bar{E}_P$ and $\frac{1}{2}\Gamma_P$ are the real and imaginary parts of the integral g_P averaged over $K\bar{K}$ near-threshold region.

The spectral density can be written out as

$$w(E) = \frac{1}{2\pi}\frac{\Gamma_P + \bar{g}_{K\bar{K}}\sqrt{mE}\,\Theta(E)}{(E - E_f - \frac{1}{2}\bar{g}_{K\bar{K}}\sqrt{-mE}\,\Theta(-E))^2 + \frac{1}{4}(\Gamma_P + \bar{g}_{K\bar{K}}\sqrt{mE}\,\Theta(E))^2}. \tag{11}$$

Eq. (11) expresses the spectral density $w(E)$ in terms of hadronic observables (Flattè parameters), just in the same way as Weinberg's factor Z is expressed in terms of hadronic observables (effective range parameters) via Eqs. (7). Thus, Eq. (11) generalizes Weinberg's result to the case of unstable particles.

It is clear from the expression (11) that it is the singularity structure of the scattering amplitude which governs the behaviour of spectral density. In the elastic case in the presence of a bound state the pole positions in the k plane are given by

$$k_1 = i\sqrt{m\varepsilon}, \quad k_2 = -i\sqrt{m\varepsilon}\frac{2-Z}{Z}. \tag{12}$$

For a deuteron-like situation, i.e. for $Z \ll 1$, the second pole is far from the threshold and even moves to infinity in the limit $Z \to 0$. On the other hand, if Z is close to one, i.e. if there is considerable admixture of an elementary state in the wave function of

TABLE 1. Pole positions (in MeV/c) and W for various fits to a_0 (left) and f_0 (right) mesons.

Ref.	k_1	k_2	W_{a_0}	Ref.	k_1	k_2	W_{f_0}
[13]	-104+i55	104-i111	0.49	[17]	-58+i107	58-i729	0.17
[14]	-134+i71	134-i199	0.29	[18]	-65+i97	65-i477	0.23
[15]	-129+i44	129-i250	0.24	[16]	-69+i100	69-i804	0.14
[15]	-126+i73	126-i212	0.29	[19]	-84+i17	84-i351	0.21
[16]	-102+i97	102-i199	0.36				

the bound state, both poles are near threshold. In the limiting case $Z \to 1$ the poles are located equidistantly from the point $k = 0$. With inelasticity included, one expects the spectral density to be enhanced in the vicinity of the amplitude poles. If the poles are located in the near-threshold region, the spectral density in this region would be large. If, on the contrary, there is only one near-threshold pole, a considerable part of the spectral density is smeared over a much wider energy interval, which is a signal that the bare state admixture in the near-threshold resonance is small. So there is a one-to-one correspondence between the value of Z (or the behaviour of $w(E)$) and the pole counting scheme suggested by Morgan [12].

Several Flattè-like fits to $\pi\eta$ and $\pi\pi$ spectra were analysed in [10], and the pole positions and mear-threshold spectral densities for these fits were found. The results are given in Table 1 together with the integral

$$W_{a_0(f_0)} = \int_{-50MeV}^{50MeV} w_{a_0(f_0)}(E)dE . \tag{13}$$

of the spectral density over the region containing the $K\bar{K}$ threshold. The limit of integration is chosen to be twice as large as the peak width of the a_0/f_0 mesons. As the function $w(E)$ is normalized to unity, the integral (13) is a direct measure of bare state admixture in the a_0/f_0 mesons.

The spectral densities for various Flattè fits are shown at Fig.1. The a_0-meson looks like an above-threshold phenomenon, with considerable part of bare state spectral density peaked near $K\bar{K}$ threshold. The f_0 is a below-threshold resonance, with some small part of spectral density peaked below threshold. Obviously, this part can be viewed as Weinberg's "Z", smeared due to the presence of inelasticity, and this "Z" is definitely small.

The interelation between pole positions and near-threshold fraction of the bare state spectral is clearly seen from the Table 1. The case of a pair of pole singularities near the threshold corresponds to the bare state accidentally seeded into near-threshold region; the admixture of a bare state should be large in this case. In the opposite case of small bare state admixture one has only one stable pole position near threshold.

Indeed, in the a_0 case the fits lead to more equidistant positions of poles than in the f_0 case, and the near-threshold fraction of $w(E)$ is more sizable for the a_0 meson. Still, even for a_0-meson it is, averagely, about 30%, so the $a_0(980)$ contains a large admixture of mesonic components. As for the f_0 meson, the near-threshold fraction of spectral density is about 20% or less, and mesonic component in f_0 is large.

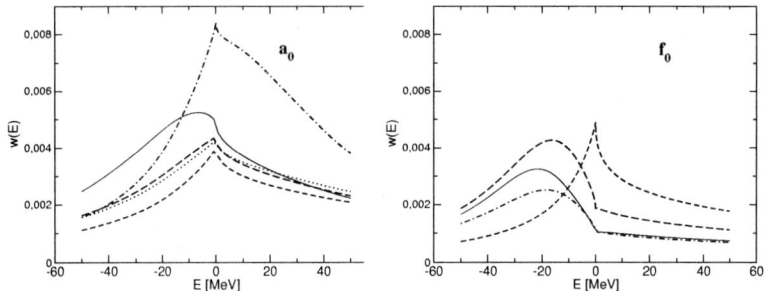

FIGURE 1. a) Spectral densities $w(E)$ for the a_0 meson based on the Flattè parameters taken from Ref. [13] (dashed-dotted line), Ref. [14] (dotted line), Ref. [15] (dashed line), Ref. [15] (long dashed line), and Ref. [16] (solid line). b) Spectral densities $w(E)$ for the f_0 meson based on the Flattè parameters taken from Ref. [17] (solid line), Ref. [18] (long-dashed line), Ref. [16] (dashed-dotted line), and Ref. [19] (dotted line).

We conclude, in such a way, that the simple $q\bar{q}$ assignement is unadequate for the $f_0(980)$-meson, and is not very appropriate for the $a_0(980)$-meson.

The author would like to thank V.Baru, J.Haidenbauer, C.Hanhart and A.Kudryavtsev for fruitful cooperation. Financial support of the grants 02-02-04001/436 RUS/13/652 and NSh-1774.2003.2 is gratefully acknowledged. This work is supported by the Federal Programme of the Russian Ministry of Industry, Science and Technology No 40.052.1.1.1112.

REFERENCES

1. Godfrey, S., and Isgur, N., Phys.Rev. **D32**, 189 (1985).
2. Jaffe, R.L., Phys. Rev. **D15**, 267, 281 (1977); Achasov, N.N., Devyanin, S.A., Shestakov, G.N., Phys. Lett. **B96**, 168, (1980).
3. Weinstein, J. and Isgur, N., Phys. Rev. **D27**, 588 (1979).
4. Locher, M.P., Markushin, V.E., Zheng, H.Q., Eur. Phys. J. **C4**, 317 (1998).
5. Isgur, N. and Speth, J., Phys. Rev. Lett, **77**, 2332 (1996).
6. Lohse, D., Durso, J.W., Holinde, K. and Speth, J., Phys. Lett. **B234**, 235 (1990); Janssen, G., Pearce, B.C., Holinde, K. and Speth, J., Phys. Rev. **D52**, 2690 (1995).
7. van Beveren, E., Dullemond, C. and Rupp, G., Phys. Rev. **D21**, 772 (1980); Rupp, G., van Beveren, E. and Scadron, M.D., Phys. Rev. **D65**, 078501 (2002); van Beveren, E. and Rupp, G., hep-ph/0304105.
8. Törnqvist, N.A., Z. Phys. **C68**, 674 (1995); Törnqvist, N.A. and Roos, M., Phys. Rev. Lett. **76**, 1575 (1996).
9. Weinberg, S., Phys. Rev. **137** B672 (1965).
10. Baru, V., Hanhart, C., Haidenbauer, J., Kalashnikova, Yu., Kudryavtsev, A., hep-p/0308129.
11. Bogdanova, L.N., Hale, G.M. and Markushin, V.E., Phys. Rev. **C44**, 1289 (1991).
12. Morgan, D., Nucl. Phys. **A543**, 632 (1992).
13. Teige, S. et al., Phys. Rev. **D59**, 012001 (2001).
14. Bugg, D.V., Anisovich, V.V., Sarantsev, A. and Zou, B.S., Phys. Rev. **D50**, 4412 (1994).
15. Achasov, N.N. and Kiselev, A.N., Phys. Rev. **D68**, 014006 (2003).
16. Antonelli, A., eConf. C020620 THAT06 (2002).
17. Achasov, M.N. et al., Phys. Lett. **B485**, 349 (2000).
18. Akhmetshin, R.R. et al., Phys. Lett. **B462**, 380 (1999).
19. Achasov, N.N. and Gubin, V.V., Phys. Rev. **D63**, 094007 (2001).

The properties of the σ and κ resonances in a new unitarization approach

H. Q. Zheng*, Z. Y. Zhou*, G. Y. Qin*† and Z. G. Xiao*

*Department of Physics, Peking University, Beijing 100871, China
†Present address: Department of Physics, McGill University, Montreal, Canada

Abstract. A new unitarization approach is discussed and applied to study the elastic πK scattering process. The existence of the light κ resonance is firmly established if the scattering length in the I=1/2 channel does not deviate too much from its value obtained from chiral perturbation theory, and a precise determination of the mass and width of the κ resonance requires a precise determination of the scattering length parameter.

In history the σ particle was firstly purposed by Gell-Mann and Levy in association with the linear σ model. Non-linear realization of chiral symmetry was later discovered [1] and since then there existed argument that the σ meson is unnecessary for chiral symmetry and even in contradiction to experiments. However there are also people, many of them are among the audience, insist on the existence of σ resonance which results in the return of σ in the Review of Particle Properties (named as $f_0(600)$) after disappearing for about 30 years. [2] The postulated κ resonance also has a rather long history [3] and the status is even more intriguing despite the recent experimental results from the E791 and the BES Collaborations. [4] The dynamics related to the σ and κ resonance is of highly non-perturbative, strong interaction nature. Since it is always not easy to separate a distant pole from the background contributions, conclusions found in the literature are very often model dependent.

However, in Ref. [5, 6], a model independent dispersive analysis has been proposed and it is demonstrated that the non-linear realization of chiral symmetry, or chiral perturbation theory (χPT) actually needs the σ resonance to accommodate for the experimental data. The crucial point to attack the problem is to consider the dispersion relations for the following two quantities,

$$F(s) = \frac{1}{2i\rho(s)}\left(S(s) - \frac{1}{S(s)}\right), \; \tilde{F}(s) = \frac{1}{2}\left(S(s) + \frac{1}{S(s)}\right), \tag{1}$$

where $\rho(s)$ is the kinematic factor. The functions $\rho(s)F(s)$ and $\tilde{F}(s)$ are respectively the analytic continuation of the imaginary and real part of the partial wave S-matrix defined in the elastic region and they satisfy the following dispersion relations:

$$F(s) = \alpha - \sum_j \frac{1/(2i\rho(z_j^{II}))}{S'(z_j^{II})(s - z_j^{II})} + \frac{1}{\pi}\int_L \frac{\mathrm{Im}_L F(s')}{s'-s}ds' + \frac{1}{\pi}\int_R \frac{\mathrm{Im}_R F(s')}{s'-s}ds',$$

CP717, Hadron Spectroscopy: Tenth International Conference,
edited by E. Klempt, H. Koch, and H. Orth

$$\tilde{F}(s) = \tilde{\alpha} + \sum_j \frac{1}{2S'(z_j^{II})(s - z_j^{II})} + \frac{1}{\pi} \int_L \frac{\mathrm{Im}_L \tilde{F}(s')}{s' - s} ds' + \frac{1}{\pi} \int_R \frac{\mathrm{Im}_R \tilde{F}(s')}{s' - s} ds' . \quad (2)$$

In the above expressions the subscript L represents dynamical cuts rather than the physical right hand cuts. R represents right hand cuts starting from the second physical threshold. The left hand cut is in general rather complicated but for equal mass scatterings like $\pi\pi$ scatterings the situation is much simplified: $L = (-\infty, 0]$ and $R = [4m_K^2, \infty)$. One subtraction to the dispersion integrals is understood. In Eq. (2) α and $\tilde{\alpha}$ are subtraction constants and z_j denotes pole positions on the second sheet. When z_j is real it represents a virtual state pole, when z_j is complex it must appear in one conjugate pair together with z_j^*, representing a resonance. The experimental curve of the function F is convex, yet chiral perturbation theory predicts a negative and concave left hand integral contribution. [5] This fact unambiguously establishes the existence of the σ resonance, if the chiral prediction to the cut integral is qualitatively correct. For more details we refer to Ref. [5].

>From Eq. (2) one obtains the generalized unitarity condition which holds on the entire complex s plane [6]:

$$\tilde{F}^2 + (\rho F)^2 = 1 . \quad (3)$$

This equation is used to obtain solutions of the simplest S matrices. Here 'simplest' means those solutions of unitary S matrices contain no cut integrals as appeared in Eq. (2) and contain minimal set of poles, i.e., one or two. The one pole solution represents a virtual/bound state whereas the two pole (on the second sheet) solution represents a resonance. The solution representing a resonance located at z_0 (having positive imaginary part) and z_0^* for un-equal mass scatterings is the following:

$$S(s) = \frac{M^2(z_0) - s + i\rho(s)sG[z_0]}{M^2(z_0) - s - i\rho(s)sG[z_0]} , \quad (4)$$

where

$$M^2(z_0) = \mathrm{Re}[z_0] + \frac{\mathrm{Im}[z_0]\,\mathrm{Im}[z_0\,\rho(z_0)]}{\mathrm{Re}[z_0\,\rho(z_0)]} , \quad G[z_0] = \frac{\mathrm{Im}[z_0]}{\mathrm{Re}[z_0\,\rho(z_0)]} . \quad (5)$$

The Eq. (4) is very interesting as it reveals the remarkable difference between a narrow resonance located far above the threshold and a light and broad resonance. In fact, $s = M^2(z_0)$ is the place where the resonance contribution to the phase shift passes $\pi/2$. However, a light and broad resonance corresponds to a very large $M(z_0)^2$. When $\mathrm{Re}[z_0] \leq (s_L + s_R)/2$ ($s_L = (m_K - m_\pi)^2$, $s_R = (m_K + m_\pi)^2$), the phase shift never reaches $\pi/2$! See fig. 1 for more illustrations.

It is worthwhile to make a pedagogical analysis to a widely used parameterization form found in the literature:

$$S = \frac{M^2 - s + i\rho(s)g}{M^2 - s - i\rho(s)g} . \quad (6)$$

For a sufficiently large M^2 and small g and for equal mass scatterings, such an S matrix contains a resonance and a *virtual* state [7]. However, the latter is not predicted by χPT

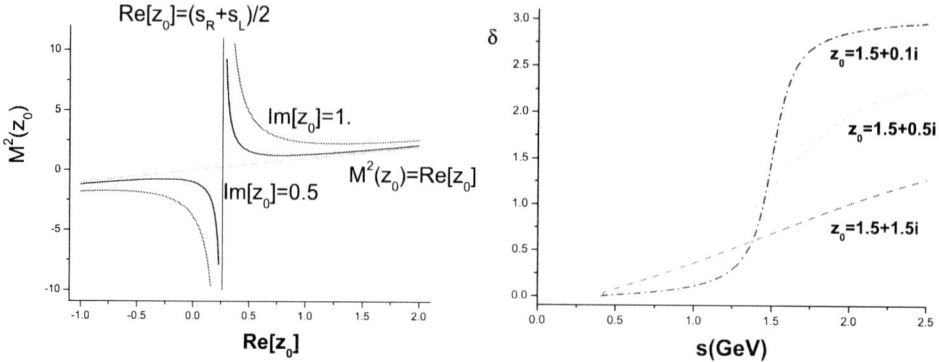

FIGURE 1. The left figure shows $M^2(z_0)$ as a function of $\mathrm{Re}[z_0]$, for fixed $\mathrm{Im}[z_0]$. The right figure give some examples of resonances and their contributions to the phase shift.

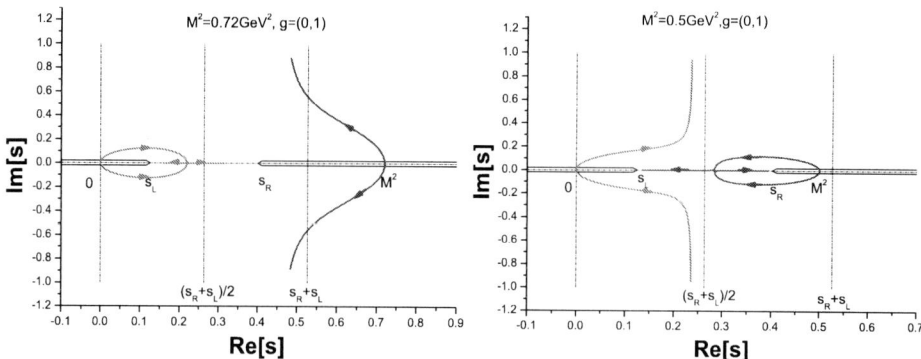

FIGURE 2. The traces of two pairs of resonances from Eq.(6) when increasing g for different M^2. We give two typical figures: left) $M^2 > (s_R + s_L)$; right) $M^2 < (s_R + s_L)$.

and violates the validity of chiral expansion at the pole position and therefore should be abandoned. For unequal mass scattering the S matrix Eq. (6) contains two resonance poles. The situation is depicted in fig. 2. The virtual state pole in $\pi\pi$ scattering and the resonance pole in πK scattering have a common origin: they are both generated from $s = 0$ due to the kinematical singularity of $\rho(s)$ at $s = 0$ and should therefore be removed on the same footing. Moreover, the physical pole contribution and the spurious pole contribution to the scattering length are additive and are both positive. The accompanied spurious pole can have a larger contribution than the resonance contribution itself if the resonance is light and broad! It is therefore incorrect to use Eq. (6) to discuss a light and broad resonance.

Since a unitary matrix divided by any unitary matrix is still unitary. If we single out all poles of an S matrix, we have without any loss of generality the following form:

$$S^{phy.} = \prod S^{poles} \cdot S^{cut} . \tag{7}$$

where S^{cut} no longer contains any pole and it can be parameterized as:

$$S^{cut} = e^{2i\rho f(s)},$$

$$f(s) = f_0 + \frac{s - s_0}{\pi}\int_L \frac{\text{Im}_L f(s')}{(s' - s_0)(s' - s)} + \frac{s - s_0}{\pi}\int_R \frac{\text{Im}_R f(s')}{(s' - s_0)(s' - s)}. \qquad (8)$$

It can be demonstrated that the discontinuity of f obeys the following simple relations: [8]

$$\text{disc} f = \text{disc}\{\frac{1}{2i\rho(s)}\log\left[S^{phy}(s)\right]\}. \qquad (9)$$

The equation (9) is useful when estimating the background contributions. For example we can simply approximate S^{phy} in Eq. (9) by $S^{\chi PT}$ on L,

$$\text{disc} f_L = \text{disc}\{\frac{1}{2i\rho(s)}\log\left[S^{\chi PT}(s)\right]\}, \qquad (10)$$

to estimate the background contributions.

Now it is at the stage to use the newly established unitarization scheme to study the πK scatterings. In here R in principle starts from $(m_K + m_\eta)^2$ to $+\infty$ but this cut is rather weak in practice and therefore it is appropriate to take $R = [(m_K + m'_\eta)^2, +\infty)$. We fit the LASS data [9] up to 1430MeV (about 20 MeV below the $K\eta'$ threshold). There are totally six parameters $a_0^{1/2}$, $a_0^{3/2}$, two pole parameters for κ and two for $K^*(1430)$. It is found appropriate to truncate the left hand integral at $\Lambda^2_{L,I=1/2} \simeq \Lambda^2_{L,I=3/2} \simeq 1.5\text{GeV}^2$ (since the once subtracted dispersion integral is convergent the numerical output is actually not very sensitive to the truncation) and we obtain:

$$\chi^2_{d.o.f.} = 38.35/(60 - 6);$$
$$M_\kappa = 594 \pm 79 MeV, \ \Gamma_\kappa = 724 \pm 332 MeV;$$
$$a_0^{1/2} = 0.284 \pm 0.089, \ a_0^{3/2} = -0.129 \pm 0.006;$$
$$M_{K^*} = 1456 \pm 8 MeV, \ \Gamma_{K^*} = 217 \pm 31 MeV. \qquad (11)$$

The fit quality of the above results are rather good though the width of the κ resonance is quite flexible. However the fit prefers a somewhat larger scattering length in the I=1/2 channel and much larger (in magnitude) scattering length parameter in the I=3/2 channel comparing with the χPT predictions to the πK scattering lengths [10]: $a_0^{1/2} = 0.18 \pm 0.02$, $a_0^{3/2} = -0.05 \pm 0.02$. It is necessary to further clarify the issue whether the κ resonance exist. According to Pennington, a resonance exist if without it the total χ^2 of the fit increase significantly. If we freeze the κ resonance, we get, $\chi^2_{d.o.f.} = 63.67/(60 - 4)$, $a_0^{1/2} = 0.446 \pm 0.006$ and other outputs are similar to Eq. (12). Comparing with the results in Eq. (12) the $\chi^2_{d.o.f.}$ given in this way is increased by a factor of 1.7. If this is still not enough to support the existence of κ, the value of $a_0^{1/2}$ given by freezing the κ degree of freedom is too large comparing with the χPT value: the fit value of $a_0^{3/2}$

is about 4 σ away from χPT result whereas $a_0^{1/2}$ is 14 σ away! The conclusion is that if $a_0^{1/2}$ does not deviate much from its value predicted by χPT then κ resonance must exist. On the contrary, it is estimated that if $a_0^{1/2}$ becomes larger than roughly 0.35 then the existence of the κ resonance becomes doubtful.

It is sometimes found in the literature the discussions on the constraint of Adler zero on the scattering amplitude. In our scheme, it is possible to embed this constraint into our parameterization form. In fact, the Adler zero automatically emerges in our approach if the subtraction constant f_0 is limited within certain range. This is because all the resonance(and virtual state) S matrices are real and less than 1 when $s_L < s < s_R$. On the contrary the cut integrals contribute a factor larger than 1, therefore the T matrix zero is obtainable in the right place when f_0 is confined to a certain range, which in turn put some constraints on the magnitude of the scattering length parameter itself. For example, within the range $0 < a_0^{1/2} < 0.26$ there exists a T matrix zero in the region $s_L < s < s_R$, and when $a_0^{1/2} \simeq 0.20$ the zero locates in the place close to the one loop χPT prediction. We make further fit by confining $a_0^{1/2}$ in the region 0.18 ± 0.02 (other conditions are the same as the fit described in above) and the results follow:

$$\chi^2_{d.o.f.} = 38.96/(60-6) ;$$
$$M_\kappa = 646 \pm 7 MeV , \ \Gamma_\kappa = 540 \pm 42 MeV ;$$
$$a_0^{1/2} = 0.2 , \ a_0^{3/2} = -0.128 \pm 0.006 ;$$
$$M_{K^*} = 1450 \pm 5 MeV , \ \Gamma_{K^*} = 232 \pm 25 MeV . \tag{12}$$

The Adler zero position is now at $s_A \simeq 0.245$GeV. If in this fit we further freeze the κ degrees of freedom we would obtain a total $\chi^2 \sim 750$. This clearly demonstrates the existence of the κ resonance. For more details of our analysis we refer to Ref. [8].

REFERENCES

1. S. R. Coleman, J. Wess and B. Zumino, Phys. Rev. **177** (1969)2239; C. G. Callan, S. R. Coleman, J. Wess and B. Zumino, Phys. Rev. **177**(1969)2247.
2. See for example, N. A. Tornqvist, Invited talk at YITP Workshop on Possible Existence of the sigma meson and its Implications to Hadron Physics, Kyoto, Japan, 12-14 Jun 2000, hep-ph/0008136.
3. R. Jaffe, Phys. Rev. **D15**(1977)267.
    Recent discussions may be found in, S. Ishida *et al.*, Prog. Theor. Phys.**98**(1997)621;
    S. N. Cherry and M. R. Pennington, Nucl. Phys. A688(2001)823.
4. E. M. Aitala *et al.*(E791 Collaboration), Phys. Rev. Lett.**89**(2002)121801;
    J. Z. Bai *et al.*(BES Collaboration), hep-ex/0304001.
5. Z. G. Xiao and H. Q. Zheng, Nucl. Phys. **A695**(2001)273.
6. J. Y. He, Z. G. Xiao and H. Q. Zheng, Phys. Lett. **B526**(2002)59; Erratum: *ibid.* **B549**(2002)362.
7. H. Q. Zheng, Talk given at International Symposium on Hadron Spectroscopy, Chiral Symmetry and Relativistic Description of Bound Systems, Tokyo, Japan, 24-26 Feb 2003, hep-ph/0304173.
8. H. Q. Zheng *et al.*, to appear.
9. D. Aston *et al.* (LASS Collaboration), Nucl. Phys. **B296**(1988)493.
10. Ulf-G.Meissner, Nucl. Phys. **B357**(1991)129.

Analysis of f_0 mesons with unitarized multi-state Breit-Wigner approach

V.Henner and T.Belozerova

Perm State University, 614990, Perm, Russia

Abstract. The properties of scalar $f_0(980)$, $f_0(1370)$, $f_0(1500)$ mesons are studied in $\pi\pi$ scattering and $p\bar{p}$ annihilation at rest with unitarized multi-state Breit-Wigner (BW) formulae described scattering amplitudes and production processes.

UNITARIZED MULTI-RESONANCE BW APPROACH

We consider a situation with several overlapping resonances with the same quantum numbers coupling to various hadron states and develop unitarized multi-resonance BW formula.

The use of the K-matrix rather than T-matrix is a simple way to satisfy unitarity. The actual physics states (the most often point of view is that they are closely related to BW description) and the K-matrix states are different. Thus, a writing down the S matrix as a sum of resonance terms and then to solve the constraints imposed by unitarity can give a direct way to extract resonances parameters from data.

Lets write the scattering matrix S in a resonance form:

$$S(E) = I - i\sqrt{\rho(E)}\,T(E)\,\sqrt{\rho(E)}, \quad T(E) = \sum_{r=1}^{N} \frac{\vec{g}_r \vec{g}_r}{E - \varepsilon_r(E)}. \tag{1}$$

Here $\varepsilon_r(E) \equiv \varepsilon_r^x(E) + i\varepsilon_r^y(E)$. Complex energy independent vectors of partial widths $\vec{g}_r \equiv \vec{g}_r^x + i\vec{g}_r^y$ have M-components (M is the number of channels). The scattering matrix T is free of threshold singularities that are included in the diagonal $M \times M$ phase states matrix $\rho(E)$. Below the k-th threshold $\rho_k(E) => i|(\rho_k(E)|$ (variable E is used instead of $s = E^2$ to simplify formulae). In further development of works [1] the resonances widths are energy dependent and we can study the case when resonances lie close to thresholds. A way to include background is described in [1].

A complexity of the vectors $\vec{g}_r$ is another way to take into account the relative phases between different BW terms, thus

$$T_{ij}(E) = \sum_{r=1}^{N} e^{i\varphi_{ij}^{(r)}} \frac{|g_{ir}| \cdot |g_{rj}|}{E - \varepsilon_r(E)}, \tag{2}$$

where φ_{ij}^r are real (constant) relative phases. Such phases are often introduced to take into account interference effects, but they should be determined from the unitarity constraints instead of considered as free parameters.

CP717, *Hadron Spectroscopy: Tenth International Conference,*
edited by E. Klempt, H. Koch, and H. Orth
© 2004 American Institute of Physics 0-7354-0197-7/04/$22.00

Find the conditions that should be imposed on vectors $\vec{g}_r$ to keep matrix S unitary and T invariant, i.e. $S^+(E)S(E) = I$ and $S_{ij}(E) = S_{ji}(E)$ identically on E. To satisfy these relations, vectors $\vec{g}_r$ are constructed in the way such that their imaginary parts $\vec{g}_r^y$ are combinations of the real parts $\vec{g}_r^x$:

$$\vec{g}_r^y = u_{r1}\vec{g}_1^x + u_{r2}\vec{g}_2^x + \ldots + u_{rN}\vec{g}_N^x, \qquad (r = 1, \ldots, N), \tag{3}$$

where matrix $U = \{u_{rk}\}_{r,k=1}^N$ is real antisymmetric matrix and vectors $\vec{g}_r$ satisfy the relations

$$\sum_{k=1}^M \rho_k(E)\theta_k(E)|g_{rk}|^2 = -\frac{2}{S}[S + 2Q_r]\varepsilon_r^y, \tag{4}$$

$$\sum_{k=1}^M \rho_k(E)\theta_k(E)\text{Re}(g_{qk}^*g_{rk}) = -\frac{2}{S}[F_{qr}(\varepsilon_q^x - \varepsilon_r^x) + G_{qr}(\varepsilon_q^y + \varepsilon_r^y)], \tag{5}$$

$$\sum_{k=1}^M \rho_k(E)\theta_k(E)\text{Im}(g_{qk}^*g_{rk}) = -\frac{2}{S}[G_{qr}(\varepsilon_q^x - \varepsilon_r^x) - F_{qr}(\varepsilon_q^y + \varepsilon_r^y)]. \tag{6}$$

Here $r = 1, \ldots, N$, $q = r+1, \ldots, N$. The constant coefficients S, Q_r, F_{qr} and G_{qr} are determined via the elements of matrix U.

Matrix U gives a measure of overlapping of resonances. If the resonances do not overlap, $|E_i - E_j| \gg \Gamma_i + \Gamma_j$, matrix elements $u_{rk} \to 0$, and vectors $\vec{g}_r$ are getting real and orthogonal: $g_r = g_r^x$, $(\vec{g}_r\vec{g}_q) = 0$. In this case $\varepsilon_r^x(E)$ can be arbitrary real functions of E (constants, if $\rho(E) \approx const$) and the resonances widths are

$$\Gamma_r(E) = -2\varepsilon_1^y(E) = \sum_{k=1}^M \rho_k(E)\,\theta_k(E)|g_{rk}|^2. \tag{7}$$

Taking, for instance, $\varepsilon_r(E) = m_r(E) - \frac{i}{2}\sum_{k=1}^M \rho_k(E)(g_{rk})^2$, where $m_r(E)$ are real functions (or constants), we obtain Flatte kind formula for several resonances:

$$T_{ij} = \sum_{r=1}^N \frac{g_{ir}g_{rj}}{E - m_r + \frac{i}{2}\sum_{k=1}^M \rho_k(g_{rk})^2}. \tag{8}$$

For overlapping resonances vectors $\vec{g}_r$ are complex and the expressions for masses and widths can be obtained from relations (4) - (6):

$$\varepsilon_r^y(E) = -\frac{S}{2(S + 2Q_r)}\sum_{k=1}^M \rho_k(E)\theta_k(E)|g_{rk}|^2, \qquad r = 1, \ldots, N.$$

If function $\varepsilon_1^x(E)$ is chosen as an arbitrary real function, then functions $\varepsilon_r^x(E)$ ($r = 2, \ldots, N$) are determined with

$$\varepsilon_r^x(E) = \varepsilon_1^x(E) + \frac{S}{2(F_{1r}^2 + G_{1r}^2)} \times$$

328

$$\times \sum_{k=1}^{M} \rho_k(E)\theta_k(E) \left[F_{1r}\mathrm{Re}(g_{1k}^* g_{rk}) + G_{1r}\mathrm{Im}(g_{1k}^* g_{rk})\right]. \tag{9}$$

The remaining $(N-1)(N-2)/2$ equations (5)-(6) imply the limitations on vectors $\vec{g}_r$ (when the number of resonances $N > 2$) [1]. For three resonances $(N = 3)$ these limitations are just the quadratic equations for each $k = 1,...,M$. When $N > 3$, the corresponding equations become nonlinear and practical usage of this method (if $\rho_k(E)$ are not constants) is not simple.

The solution of equation $E - \varepsilon_r^x(E) = 0$ gives the mass of the r-th resonance, $\hat{\varepsilon}_r$ (generally there might be several roots, but for f_0 resonances we obtained one root for each r).

Equation (7) gives the width of the r-th resonance:

$$\hat{\Gamma}_r = \Gamma_r(\hat{\varepsilon}_r) = \frac{S}{S+2Q_r} \sum_{k=1}^{M} \rho_k(\hat{\varepsilon}_r)\theta_k(\hat{\varepsilon}_r) |g_{rk}|^2. \tag{10}$$

Correspondingly, partial widths are given by

$$\hat{\Gamma}_{rk} = \frac{S}{S+2Q_r} \rho_k(\hat{\varepsilon}_r)\theta_k(\hat{\varepsilon}_r) |g_{rk}|^2. \tag{11}$$

Branching ratio of the r-th resonance into the k-th channel is

$$B_{rk} = \frac{\hat{\Gamma}_{rk}}{\hat{\Gamma}_r} = \frac{\rho_k(\hat{\varepsilon}_r)\theta_k(\hat{\varepsilon}_r) |g_{rk}|^2}{\sum_{k=1}^{M} \rho_k(\hat{\varepsilon}_r)\theta_k(\hat{\varepsilon}_r) |g_{rk}|^2}. \tag{12}$$

It is instructive to show two resonances case (for two overlapping resonances with constant widths the unitarized BW formula was derived first in [2]):

$$S_{ij} = \delta_{ij} - i\sqrt{\rho_i \rho_j} \left[\frac{g_{i1} g_{1j}}{E - \varepsilon_1(E)} + \frac{g_{i2} g_{2j}}{E - \varepsilon_2(E)}\right], \quad i,j = 1,...,M. \tag{13}$$

In this case $\mathbf{U} = \begin{pmatrix} 0 & -\alpha \\ \alpha & 0 \end{pmatrix}$, i.e. $\vec{g}_1^{\,y} = -\alpha \vec{g}_2^{\,x}$, $\vec{g}_2^{\,y} = \alpha \vec{g}_1^{\,x}$. Parameter α is not entirely free but limited by the condition $0 \leq \alpha < 1$. The coefficients in the unitary constraints (4) - (6) are: $S = 1 - \alpha^2$, $Q_1 = Q_2 = \alpha^2$, $F_{12} = -\alpha$, $G_{12} = 0$. According to the unitarity relation, $\varepsilon_1^y(E)$ and $\varepsilon_2^y(E)$ should obey the relation (4)

$$\varepsilon_r^y = -\frac{1-\alpha^2}{2(1+\alpha^2)} \sum_{k=1}^{M} \rho_k(E)\theta(E) |g_{rk}|^2, \quad r = 1,2.$$

Real part of "energy" of the first resonance $\varepsilon_1^x(E)$ is an arbitrary real function. Lets choose $\varepsilon_1(E)$ in a form similar to the Flatte formula:

$$\varepsilon_1(E) = m_1 - i\frac{1-\alpha^2}{2(1+\alpha^2)} \sum_{k=1}^{M} \rho_k |g_{1k}|^2. \tag{14}$$

329

Function $\varepsilon_2^x(E)$ is defined by relation (9):

$$\varepsilon_2^x(E) = \varepsilon_1^x - \frac{1-\alpha^2}{2\alpha} \sum_{k=1}^{M} \rho_k(E)\theta_k(E)\, g_{1k}^x g_{2k}^x.$$

Therefore, the complex "energy" of the 2-nd resonance can be given as

$$\varepsilon_2(E) = D(E) - i\frac{1-\alpha^2}{2(1+\alpha^2)} \sum_{k=1}^{M} \rho_k |g_{2k}|^2,$$

$$D(E) = m_1 + \frac{1-\alpha^2}{2(1+\alpha^2)} \sum_{k=1}^{M} \rho_k(1-\theta_k(E))(|g_{1k}|^2 - |g_{2k}|^2) -$$

$$- \frac{1-\alpha^2}{2\alpha} \sum_{k=1}^{M} \rho_k(E)\theta_k(E)\, g_{1k}^x g_{2k}^x.$$

The widths of the resonances are $\Gamma_r(\hat{\varepsilon}_r) = \frac{1-\alpha^2}{1+\alpha^2} \sum_{k=1}^{M} \rho_k(\hat{\varepsilon}_r)\,|g_{rk}|^2$. Free parameters is this scheme are m_1, α and coordinates of vectors $\vec{g}_1^x$, $\vec{g}_2^x$.

We have discussed so far the case of transitions between the same states i, j as occur in matrix T. Similarly to [3] it is simple to consider the slightly more general case in which g_{ir} is replaced by f_{pr} where p, the production channel, does not occur in the sums in $\Gamma_r(E)$ and $\varepsilon_r(E)$. For the transition from production state p to final state j we have

$$F_{pj} = \sum_{r=1}^{N} \frac{f_{pr} g_{rj}}{E - \varepsilon_r^x(E) + \frac{i}{2}\Gamma_r(E)}. \tag{15}$$

Vectors $\vec{f}_r$ like vectors $\vec{g}_r$ are complex. The amplitudes F and T have the same denominators (thus the same poles). The "running" masses and widths, $\varepsilon_r^x(E)$ and $\Gamma_r(E)$, are approximately constant only in case when all resonances are away from thresholds.

ANALYSIS OF $F_0(980), F_0(1370), F_0(1500)$ STATES

The model includes four channels: $\pi\pi$, $K\bar{K}$, $\eta\eta$, 4π. The phase factors $\rho_k(s)$ are:

$$\rho_k(s) = \sqrt{(s-s_k)/s}, \quad s_1 = 4m_\pi^2, \quad s_2 = 4m_K^2, \quad s_3 = 4m_\eta^2,$$

$$\rho_4(s) = \sqrt{(s-s_4)/s}/(1+\exp[\Lambda \cdot (s_0 - s)]),$$

$$\sqrt{s_4} = 4m_\pi, \quad s_0 = 1.9\,(GeV^2), \quad \Lambda = 6.0\,(GeV^{-2}).$$

This form of ρ_4 approximates either the $\rho\rho$ or $\sigma\sigma$ phase space ([4]). We describe well the data on the S-wave $\pi\pi$ scattering and available (see [5]) mass spectra of the processes

330

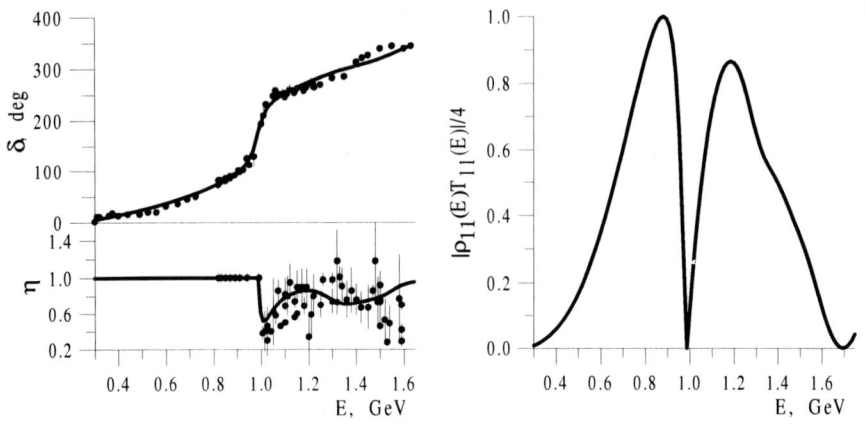

FIGURE 1. Phase δ and inelasticity η for $I = J = 0$ $\pi\pi$ scattering. Experimental points are from [6].

$p\bar{p} \to \pi^0(\eta\eta)$, $\pi^0(4\pi)$ (due to lack of space we do not present these figures). Tables show f_0 parameters.

Table 1. Parameters of f_0 mesons (in GeV)

Meson	Mass	Width
$f_0(980)$	0.987	0.117
$f_0(1370)$	1.356	0.269
$f_0(1500)$	1.549	0.151

Table 2. Branching ratios of f_0 mesons (in %) and relative (to $f_0(980)$) phases of $f_0(1370)$ and $f_0(1500)$ (in degrees).

State	$f_0(980)$	$f_0(1370)$	$\varphi_{f_0(1370)}$	$f_0(1500)$	$\varphi_{f_0(1500)}$
$\pi\pi$	77.29	8.19	-28.9	4.03	28.4
$K\bar{K}$	20.81	5.83	-38.0	3.37	18.2
$\eta\eta$	1.89	52.66	68.2	57.76	18.6
4π	0.01	33.31	88.7	34.84	26.5

REFERENCES

1. T.S.Belozerova, V.K.Henner, Phys. Part. Nucl. **29**, 63 (1998); Yad.Fizika **60**, 1998 (1997).
2. K.W.McVoy, Ann. Phys. **54**, 552 (1969).
3. I.J.R.Aitchison, Nuclear Physics **A189**, 417 (1972).
4. D.V.Bugg, A.V.Sarantsev, B.S.Zou, Nuclear Physics **B471**, 59 (1996).
5. N.N.Achasov, G.N.Shestakov, Phys.Rev, **53**, 3559, (1996).
6. B.Hyams, et al., Nucl.Phys. **B100**, 205, (1975).

The Light and Heavy Scalars in Unitarized Coupled Channel and Lagrangian Approaches

Frieder Kleefeld

Centro de Física das Interacções Fundamentais (CFIF), Instituto Superior Técnico,
Edifício Ciência, Piso 3, Av. Rovisco Pais, P-1049-001 LISBOA, Portugal
e-mail: kleefeld@cfif.ist.utl.pt

Abstract. Using ideas underlying the flavour-blind "Nijmegen Unitarised Meson Model" (NUMM) [1, 2, 3] we try to understand on the basis of a system of Schrödinger equations with one meson-meson and one (spinless) quark-antiquark channel coupled by a simple delta-shell transition potential the formation of (e.g. scalar) meson-meson scattering singularities in the complex momentum and energy plane[1]. Surprisingly we are able to describe without direct meson-meson interaction and without any need for glueballs the whole known scalar meson spectrum. "Light" scalar mesons (e.g. $f_0(600)$, $\kappa(800)$, $f_0(980)$, $a_0(985)$, $D_0^*(2290)$, ...) are identified to belong to the spectrum of the transition potential, while "heavy" scalars (e.g. $f_0(1370)$, $K_0^*(1430)$, $f_0(1500)$, $f_0(1710)$, $a_0(1450)$, $D_0^*(2621)$ (?), $D_0^*(2825)$ (?), $D_{sJ}^*(2928)$ (?), ...) are related to the confinement spectrum. Due to the particular value of the charm-strange reduced quark mass level-(anti)crossing in the complex momentum plane [4] occurs which relates the BABAR state $D_s(2317)$ [5] to the bare groundstate of the confinement spectrum, while the respective groundstate of the transition potential ends up as $D_{sJ}^*(2782)$ (?). We conclude with a short comment on (our) recent progress in the consistent quantum field theoretical effective description of resonances within a Lagrangian framework [6, 7, 8].

Hadronic excitations with scalar quantum numbers are a topic of heated dispute [9]. Unitarized coupled channel approaches as the one discussed here (see e.g. Ref. [10] and references therein) are particularly useful to understand the non-perturbative formation and nature[2] of (e.g. scalar) meson-meson scattering singularities in the complex momentum or energy plane, which then may enter as effective degrees of freedom with complex mass and coupling parameters effective Lagrangians describing — according to the "bootstrap" idea — meson-meson scattering already at tree level[3]. For a spherically symmetric situation we couple — in the simplest case — one Schrödinger equation describing a "bound" quark-antiquark system (confining potential $V_B(r)$) to one

---

[1] In early 1981 G. Rupp discovered the σ-pole of the NUMM when searching for poles "in his computer".

[2] "Heavy" scalars are — disregarding glueballs — mainly associated to the spectrum of the confining quark-antiquark interaction used (e.g. a harmonic oscillator potential), while "light" (dynamically generated [11]) scalars were identified by the author (see e.g. the comments in Refs. [10, 12]) to belong to the spectrum of the meson-(anti)quark transition potential (being to a good approximation of $^3P_0$-type).

[3] The Lagrangian of the Quark-Level Linear Sigma Model (QLLσM) [13, 11, 14] (and references therein) has not only been shown to be an excellent candidate to achieve this task as it reproduces with a minimum of parameters at tree-level a broad spectrum of experimental facts (including the correct prediction of the mass of the now experimentally confirmed $\kappa(800)$-meson) and allowed us to gain some insight in the quark-content of scalar mesons [15], yet could be also "derived" [6] from the Lagrangian of QCD. This "derivation" shows that "glueballs" and "quark-antiquark excitations" seem to be synonymous.

CP717, *Hadron Spectroscopy: Tenth International Conference,*
edited by E. Klempt, H. Koch, and H. Orth
© 2004 American Institute of Physics 0-7354-0197-7/04/$22.00

Schrödinger equation describing a meson-meson scattering continuum by a transition potential denoted by $V_T(r)$. I.e., we consider the the following coupled system of radial Schrödinger equations ($k := k_S := (2\mu_S(E - E_S^{(0)}))^{1/2}$, $k_B := (2\mu_B(E - E_B^{(0)}))^{1/2}$):

$$
\begin{aligned}
\left(d^2/dr^2 - L(L+1)/r^2 - 2\mu_S V_S(r) + k_S^2\right)\psi_S(r) &= 2\mu_S V_T(r)\,\psi_B(r),\\
\left(d^2/dr^2 - \ell(\ell+1)/r^2 - 2\mu_B V_B(r) + k_B^2\right)\psi_B(r) &= 2\mu_B V_T(r)\,\psi_S(r),
\end{aligned}
$$

with $\psi_S(0) = 0$ and $\psi_B(0) = 0$. Even though a majority of publications is trying to find the source of "light" scalars in the meson-meson scattering potential $V_S(r)$, we are disregarding this interaction in what follows completely (i.e. we set $V_S(r) = 0$). The conveniently normalized[4] eigensolutions $\phi_{n,\ell}(r)$ of the "bound" system for vanishing transition potential (i.e. $V_T(r)=0$) correspond to the respective eigenvalues $k_{B,n,\ell} = (2\mu_B(E_{B,n,\ell} - E_B^{(0)}))^{1/2}$. After integrating the "bound" problem using a Green function[5] $G_\ell(r,r';E - E_B^{(0)})$ and reinserting it into the "scattering" problem we arrive at the following generalized scattering problem (with $E = (k^2 + m_1^2)^{1/2} + (k^2 + m_2^2)^{1/2}$):

$$
(d^2/dr^2 - L(L+1)/r^2 + k^2)\psi_S(r) = -2\mu_S \Sigma_\ell V_T(r)\int_0^\infty dr' G_\ell(r,r';E - E_B^{(0)})V_T(r')\,\psi_S(r').
$$

Now we will approximate astonishingly well the $^3P_0$ transition potential by $V_T(r) = 2g_T(2\mu(E)/(2\mu_S))^{1/2}\,\delta(r - a)$ with $2\mu(E) = \partial k^2/\partial E = (E^4 - (m_1^2 - m_2^2)^2)/(2E^3)$ being the relativistic meson-meson phasespace, and hence reduce the generalized scattering problem to a (radial) scattering problem at an effective δ-shell described by the Schrödinger equation $K^2 \psi_L(\rho) = \left(-d^2/d\rho^2 + L(L+1)/\rho^2 + g\,\delta(\rho - 1)\right)\psi_L(\rho)$ (*) with $\rho := r/a$, $K := ak$, $\psi_L(\rho) := \psi_{S,L}(r)$. With $\lambda_\ell := 2g_T(a\,G_\ell(a,a;0))^{1/2}$ and $B_{n,\ell} := (a\phi_{n,\ell}(a)\phi_{n,\ell}^*(a))/(a\,G_\ell(a,a;0))$ the dimensionless coupling g displays the structure of the "Resonance Spectrum Expansion" (RSE) [16] of Rupp & van Beveren:

$$
g = 2\mu(E)\sum_\ell \lambda_\ell^2 \sum_{n=0}^\infty \frac{B_{n,\ell}}{E - E_{B,n,\ell}} \simeq 2\mu(E)\sum_\ell \lambda_\ell^2 \left(\sum_{n=0}^N \frac{B_{n,\ell}}{E - E_{B,n,\ell}} - 1\right).
$$

By construction there holds $\sum_{n=0}^\infty B_{n,\ell}(E_B^{(0)} - E_{B,n,\ell})^{-1} = -1$. The original idea of the RSE [16] was to consider the parameters $B_{n,\ell}$, $E_{B,n,\ell}$, λ_ℓ as free parameters to fit selective meson spectra conveniently. Empirically it has become clear [16, 17] that the parameters λ_ℓ and $B_{n,\ell}$ should be considered as "universal" for many different meson spectra provided the product $a\sqrt{\mu_B}$ is kept "universal"[6]. In the results displayed below we adopted the philosophy of the NUMM to describe all meson-spectra on the

---

[4] Orthonormality: $\int_0^\infty dr\,\phi_{m,\ell}^*(r)\,\phi_{n,\ell}(r) = \delta_{mn}$; completeness: $\sum_{n,\ell}\phi_{n,\ell}(r)\,\phi_{n,\ell}^*(r') = \delta(r - r')$.

[5] Green function: $G_\ell(r,r';E - E_B^{(0)}) = -2\mu_B\sum_{n,\ell}\phi_{n,\ell}(r)\,\phi_{n,\ell}^*(r')\,(k_B^2 - k_{B,n,\ell}^2 + i\varepsilon)^{-1}$.

[6] To keep $a\sqrt{\mu_B}$ constant became clear from calculations performed in e.g. Ref. [1, 2] based on a transition potential $V_T(r) = g\,\omega\rho_0^{-1}\,\delta(r\sqrt{\mu_B}\omega - \rho_0)\,\overline{V}_{int} = g(\mu_B a)^{-1}\,\delta(r - a)\,\overline{V}_{int}$ successfully applied to a wide range of vector meson spectra. This transition potential was a simplified version of the harmonic oscillator form of the $^3P_0$ transition potential inferred by G. Rupp [1] and successfully applied within the NUMM [1, 2, 3] in the representation $[V_T(r)]_{ij} = \bar{g}\,\omega\,c_{ij}\,(E/E_S^{(0)})^{1/2}\,(r/r_0)\exp(-(r/r_0)^2/2)$ with

basis of a harmonic oscillator potential[7] with an "universal" oscillator frequency $\omega = 190$ MeV, and constituent quark masses [14] $m_u = m_d = 337$ MeV, $m_s = 1.44 m_u$, $m_c \simeq m_D = 1865$ MeV. Hadronic resonances in meson-meson scattering are then determined for Eq. (∗) by solving the respective resonance condition $\cot \delta_L = i$ with[8] $\cot \delta_L = (n_L(K)/j_L(K)) - K/(g\,(j_L(K))^2)$. As in Ref. [16] we will choose for the description of scalar mesons ($L = 0$) a RSE with two P-wave ($\ell = 1$, $N = 1$) bare quark-antiquark ($q\bar{q}'$) states in each meson-meson (MM') scattering channel. Hence, the respective RSE resonance condition to be solved in each meson-meson scattering channel is[9]

$$\frac{2iK}{1 - \exp(2iK)} \simeq 2\mu_{MM'}^{q\bar{q}'}(K)\,\lambda^2 \left(\frac{B_{0,1}(\bar{\rho})}{E_{MM'}^{q\bar{q}'}(K) - E_{B,0,1}} + \frac{B_{1,1}(\bar{\rho})}{E_{MM'}^{q\bar{q}'}(K) - E_{B,0,1} - 2\omega} - 1 \right).$$

Up to now the bare groundstates of the harmonic oscillator have to be determined empirically. Here we choose[10] $E_{B,0,1} = 1310$ MeV [16] for S-wave $\pi\pi$-, $KK(I = 0)$-, πK-, $\pi\eta_{n\bar{n}}$-scattering, $E_{B,0,1} = 2440$ MeV for S-wave $D\pi$-scattering, and $E_{B,0,1} = 2545$ MeV [17] for S-wave DK-scattering. $a_{u\bar{s}}$ and $\bar{\rho}$ are then determined such that for given λ the mesons $\kappa(800)$ (pole-position $(714 - i228)$ MeV [16]) and $f_0(980)$ (pole-position 980 MeV) are reproduced simultaneously. In a good approximation $f_0(980)$ is here assumed to be purely strange [15]. In using $m_\pi = 140$ MeV and $m_K = 494$ MeV we obtain the approximate result $a_{u\bar{s}} \simeq 2.55357$ GeV^{-1} and $\bar{\rho} \simeq 1.45555$ yielding immediately $B_{0,1}(\bar{\rho}) \simeq 0.285546$ GeV, $B_{1,1}(\bar{\rho}) \simeq 0.0166127$ GeV, and $\lambda \simeq 1.11572$ GeV$^{-1/2}$. On the basis of these parameters and $m_{\eta_{n\bar{n}}} = 757.9$ MeV (for a mixing angle of $41.84°$ in $n\bar{n}$-$s\bar{s}$ basis [19]) we can determine the solution of further selective RSE resonance conditions of interest in choosing e.g. $M, M' \in \{\pi, \eta_{n\bar{n}}, K, D, \ldots\}$[11]. For S-wave πK-, $D\pi$-, and DK-scattering the results are illustrated graphically in Fig. 1. In all cases there

---

$r_0 := \rho_0\,(\mu_B\,\omega)^{-1/2}$. The flavour-blindness [10] of QCD is reflected here not only by the recoupling coefficients c_{ij} (or $\overline{V}_{int}$) [1, 2, 3, 18] but also by the "universal" values for ρ_0 and ω $((\mu_B\,\omega)^{-1/2}$ has the meaning of an oscillator length). In the RSE $(E/E_S^{(0)})^{1/2}$ finds the interpretation of the square root of the relativistic meson-meson phasespace, as $4\mu(E)/E_S^{(0)} = (E^4 - (m_1^2 - m_2^2)^2)/(E^3\,(m_1 + m_2)) \overset{m_1 = m_2}{\longrightarrow} E/E_S^{(0)}$.

[7] For the harmonic oscillator potential $V_B(r) = \frac{1}{2}\mu_B\,\omega^2\,r^2$ we have $E_{B,n,\ell} - E_B^{(0)} = \omega\,(2n + \ell + \frac{3}{2})$ and

$$B_{n,\ell}(\bar{\rho}) = \frac{2\omega\sqrt{\pi}}{\Gamma\left(\frac{1}{2}\left(\ell + \frac{3}{2}\right)\right)\Gamma\left(\frac{1}{2}\left(\ell + \frac{5}{2}\right)\right)\,I_{\frac{1}{2}(\ell + \frac{1}{2})}(\bar{\rho}^2/2)} \frac{(\bar{\rho}^2/2)^{\ell + \frac{1}{2}}\,e^{-\bar{\rho}^2}}{K_{\frac{1}{2}(\ell + \frac{1}{2})}(\bar{\rho}^2/2)}\,\frac{\Gamma\left(\ell + \frac{3}{2}\right)\,n!}{\Gamma\left(n + \ell + \frac{3}{2}\right)}\,\left|L_n^{(\ell + \frac{1}{2})}(\bar{\rho}^2)\right|^2.$$

Here we defined an "universal" parameter $\bar{\rho} := a\,(\mu_B\,\omega)^{1/2}$, while $I_\nu(z)$ and $K_\nu(z)$ are modified Bessel functions, and $L_n^{(\alpha)}(z)$ are standard generalized Laguerre polynomials.

[8] S-wave ($L = 0$) meson-meson scattering yields $i = -\cot K - K/(g\,\sin^2 K) \Leftrightarrow g = 2iK/(1 - \exp(2iK))$.

[9] Here we defined $\lambda := \lambda_1$, $E_{MM'}^{q\bar{q}'}(K) := ((K/a_{q\bar{q}'})^2 + m_M^2)^{1/2} + ((K/a_{q\bar{q}'})^2 + m_{M'}^2)^{1/2}$, and $\mu_{MM'}^{q\bar{q}'}(K) := ((K/a_{q\bar{q}'})^2 + m_M^2)^{1/2}\,((K/a_{q\bar{q}'})^2 + m_{M'}^2)^{1/2}/E_{MM'}^{q\bar{q}'}(K)$.

Note that $a_{n\bar{n}}\sqrt{\mu_{u\bar{u}}} = a_{s\bar{s}}\sqrt{\mu_{s\bar{s}}} = a_{u\bar{s}}\sqrt{\mu_{u\bar{s}}} = a_{c\bar{s}}\sqrt{\mu_{c\bar{s}}} = a_{c\bar{d}}\sqrt{\mu_{c\bar{d}}}$ with $\mu_{q\bar{q}'} := m_q\,m_{\bar{q}'}/(m_q + m_{\bar{q}'})$.

[10] Note that $k \simeq 766$ MeV $\simeq 4\omega$ resulting from $E_{B,1,1}$ seems to be to a good approximation "universal".

[11] For S-wave $\pi\pi$-, $KK(I = 0)$-, πK-, $\pi\eta_{n\bar{n}}$-, $D\pi$-, and DK-scattering we find the following pole positions in the complex energy plane: $\pi\pi$-scattering: $(516 - i412)$ MeV ($f_0(600)$), $(1385 - i81)$ MeV

334

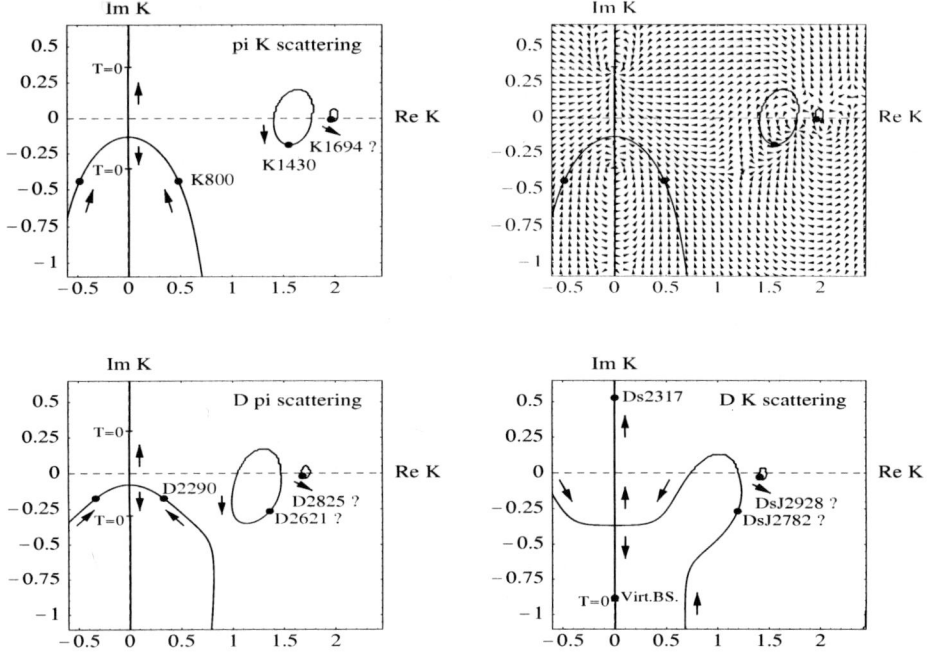

FIGURE 1. Propagation of poles for increasing λ^2 in the complex K-plane. Solid lines: curves with $\mathrm{Im}[\lambda^2] = 0$. "$T = 0$" indicates a zero of the amplitude due to vanishing phasespace ($\simeq$ "Adler-zero"). Resonances at $\lambda \simeq 1.11572$ GeV$^{-1/2}$ are denoted by •. S-wave scattering: $\underline{\pi K}$: $\kappa(800)$, $K_0^*(1430)$, $K_0^*(1694)$ (?); $\underline{D\pi}$: $D_0^*(2290)$, $D_0^*(2621)$ (?), $D_0^*(2825)$ (?); $\underline{DK}$: $D_s(2317)$, $D_{sJ}^*(2782)$ (?), $D_{sJ}^*(2928)$ (?).

starts for $\lambda = 0$ ($\Rightarrow g = 0$) a δ-shell pole trajectory at $K = \pi/2 - i\,\infty$. For $g \to \infty$ this pole should behave like a particle in a box and end up at $K \to \pi$. Instead, it collides for increasing λ with the pole stemming from the bare groundstate of the confinement problem and gets deflected either to the "left" (πK, $D\pi$) or to the "right" (DK)[12].

$(f_0(1370), E_{0,1}^{dressed})$, $(1694 - i\,4)$ MeV $(f_0(1710), E_{1,1}^{dressed})$; $KK(I = 0)$-scattering: 980 MeV $(f_0(980))$, 351 MeV (Virtual BS), $(1452 - i\,191)$ MeV $(f_0(1500), E_{0,1}^{dressed})$, $(1692 - i\,11)$ MeV $(f_0(1710), E_{1,1}^{dressed})$; πK-scattering: $(721 - i\,215)$ MeV $(\kappa(800))$, $(1404 - i\,130)$ MeV $(K_0^*(1430), E_{0,1}^{dressed})$, $(1694 - i\,7)$ MeV $(E_{1,1}^{dressed})$; $\pi\eta_{n\bar{n}}$-scattering: $(960 - i\,107)$ MeV $(a_0(985))$, $(1423 - i\,161)$ MeV $(a_0(1450), E_{0,1}^{dressed})$, $(1693 - i\,8)$ MeV $(E_{1,1}^{dressed})$; $D\pi$-scattering: $(2073 - i\,70)$ MeV $(D_0^*(2290))$, $(2621 - i\,163)$ MeV $(E_{0,1}^{dressed})$, $(2825 - i\,13)$ MeV $(E_{1,1}^{dressed})$; DK-scattering: $(2782 - i\,166)$ MeV $(D_{sJ}^*(???))$, 2244 MeV $(D_s(2317)$, $E_{0,1}^{dressed})$, 1907 MeV (Virtual BS, $E_{0,1}^{dressed})$, $(2928 - i\,20)$ MeV $(E_{1,1}^{dressed})$.

[12] Consequently, the origin of "light" scalar mesons, the strong distortions of the groundstates of the "observed" confinement spectrum, and the absence of "light" non-scalar mesons for realistic transition potentials (due to the centrifugal barrier) get a nice explanation. The flavour content for $f_0(600)$, $f_0(980)$, $f_0(1370)$, $f_0(1500)$ is well consistent with Ref. [15], while observed pole-positions correspond nicely to

Shortly we address effective Lagrangian approaches: scalar mesons are characterized within a Lagrangian close to bootstrap typically by complex mass and coupling parameters to be determined by a coupled channel approach. The formalism to describe fields within a non-Hermitian Lagrangian has been provided [6, 7, 8]. The QCD-Lagrangian has been mapped [6] into a QLLσM-Lagrangian replacing the gluon-quark interaction $g\,\overline{q_+^c}(x)\,\slashed{A}(x)\,q_-(x)$ by a *non-Hermitian* meson-quark interaction. The resulting scalar-meson-quark interaction $i\,\sqrt{2N_F/N_c}\,g\,\overline{q_+^c}(x)\,S(x)\,q_-(x)$ yields an *asymptotic free* theory which is PT-symmetric [21] admitting a *real spectrum* and a *probability interpretation*.

ACKNOWLEDGMENTS

This work has been supported by the *Fundação para a Ciência e a Tecnologia* of the *Ministério da Ciência e da Tecnologia (e do Ensinio Superior)* of Portugal, under Grants no. PRAXIS XXI/BPD/20186/99, SFRH/BDP/9480/2002, POCTI/FNU/49555/2002.

REFERENCES

1. G. Rupp, *Doctoral Thesis* (Catholic University of Nijmegen, 1982).
2. E. van Beveren et al., *Phys. Rev.* **D 21** (1980) 772; **D 22** (1980) 787; *Phys. Rev.* **D 27** (1983) 1527; J. E. Ribeiro, *Phys. Rev.* **D 25** (1982) 2406.
3. E. van Beveren et al., *Z. Phys.* **C 30** (1986) 615.
4. E. Hernández et al., *Rev. Mex. Fis.* **49 (S4)** (2003) 17; *Phys. Rev.* **A 67** (2003) 022721.
5. B. Aubert et al. [BABAR Collaboration], *Phys. Rev. Lett.* **90** (2003) 242001.
6. F. Kleefeld, *AIP Conf. Proc.* **660** (2003) 325 [hep-ph/0211460].
7. F. Kleefeld, hep-th/0310204; *Few-Body Systems Suppl.* **15** (2003) 201; *Acta Phys. Polon.* **B 30** (1999) 981; Vol. I of Proc. XIV ISHEPP 98, 17-22.8.1998, Dubna, Russia, pp. 69-77 (Eds. A.M. Baldin, V.V. Burov) [nucl-th/9811032]; F. Kleefeld, *Doctoral Thesis* (University of Erlangen-Nürnberg, 1999).
8. F. Kleefeld, E. van Beveren and G. Rupp, *Nucl. Phys.* **A 694** (2001) 470.
9. S. F. Tuan, hep-ph/0303248; D. V. Bugg, *Phys. Lett.* **B 572** (2003) 1.
10. E. van Beveren et al., *AIP Conf. Proc.* **660** (2003) 353 [hep-ph/0211411].
11. E. van Beveren, F. Kleefeld, G. Rupp, M. D. Scadron, *Mod. Phys. Lett.* **A 17** (2002) 1673.
12. E. van Beveren, G. Rupp, hep-ph/0306185.
13. R. Delbourgo, M. D. Scadron, *Mod. Phys. Lett.* **A 10** (1995) 251; *Int. J. Mod. Phys.* **A 13** (1998) 657; M. D. Scadron et al., *Nucl. Phys.* **A 724** (2003) 391; *AIP Conf. Proc.* **660** (2003) 311.
14. M. D. Scadron, G. Rupp, F. Kleefeld, E. van Beveren, hep-ph/0309109, hep-ph/0307003.
15. F. Kleefeld, E. van Beveren, G. Rupp, M. D. Scadron, *Phys. Rev.* **D 66** (2002) 034007.
16. E. van Beveren, G. Rupp, *Eur. Phys. J.* **C 22** (2001) 493.
17. E. van Beveren, G. Rupp, *Phys. Rev. Lett.* **91** (2003) 012003; hep-ph/0306051, hep-ph/0306155.
18. E. van Beveren, G. Rupp, *Eur. Phys. J.* **C 11** (1999) 717; *Phys. Lett.* **B 454** (1999) 165.
19. D. Kekez, D. Klabučar, M. D. Scadron, *J. Phys.* **G 27** (2001) 1775.
20. P. Krokovny [BELLE Collaboration], hep-ex/0210037.
21. C. M. Bender et al., hep-th/0303005; M. Znojil, quant-ph/0309100.

---

values obtained for a realistic transition potential [3]. The too large imaginary part of the $f_0(600)$ pole is an artefact of the used one-channel model. Furthermore, we observe a *twofold nature* of the $f_0(1710)$ being at the same time $n\bar{n}$ (relation to $f_0(600)$) and $s\bar{s}$ (relation to $f_0(980)$). The BABAR state $D_s(2317)$ [5] and BELLE state $D_0^*(2290)$ [20] are reproduced, while the respective next higherlying states are predicted to be $D_{sJ}^*(2782)$ (?) and $D_0^*(2621)$ (?). Note also the prediction of $K_0^*(1694)$ (?) and $a_0(1693\ldots1694)$ (?).

The κ Meson Production in
$J/\psi \to K^*(892)K\pi$ Process

T. Komada[1]

(for the BES Collaboration)

Department of Engineering Science, Junior College Funabashi Campus
Nihon University, Funabashi 274-8501, Japan

Abstract. An evidence for a low mass iso-spinor scalar meson, κ has been obtained in our analysis on the $K\pi$ system produced in the 58 million J/ψ decays into $K^*K\pi$. The data has been obtained by BES II at BEPC. The VMW method is used in the analysis. The analysis method is described with the background consideration, briefly.

INTRODUCTION

The existence of the low mass iso-scalar scalar meson σ has been widely accepted in recent years[1, 2]. It appeared newly as $f_0(600)$ or σ in the latest edition of the PDG table[3]. The existence of σ as the chiral partner of the Nambu-Goldstone pion demands an existence of κ in naturally. They are expected to form a chiral scalar nonet with $f_0(980)$ and $a_0(980)$, which is a chiral partner of the pseudoscalar π nonet.

The evidence for the existence of κ has been reported in the analyses of the $K\pi$ scattering phase shift data by several groups[5], including the sigma group. However, there appeared a criticism[6] similarly as that in the case of the σ meson. Meanwhile, the E791 collaboration has reported[7] the first evidence for κ in the production process through D^+ meson decays into $K^-\pi^+\pi^+$.

The purpose of the present report is to show the results of κ in our analysis on the BES II data with the VMW method[8], discussing on the contributions of relevant processes including backgrounds. Details of our works can be referred to the article which will be appeared in the Proceedings of the Nihon-U symposium in Tokyo[1].

METHOD OF ANALYSIS, VMW

The data for our analysis are those of the $K^*K\pi$ system in the 58 million J/ψ decays of BES II. The data selection and the acceptance correction by Monte Carlo simulation with the SIMBES code have been prepared[9] by N. Wu of IHEP(Beijing).

---

[1] The σ Collaboration : M. Y. Ishida(Tokyo Inst. Tech.), S. Ishida(Nihon U.), T. Komada(Nihon U.), T. Matsuda (Miyazaki U.), K. Takamatsu(KEK), T. Tsuru(KEK), K. Ukai(KEK), K. Yamada(Nihon U.) and I. Yamauchi(Tokyo Metro. Coll. Tech.)

CP717, *Hadron Spectroscopy: Tenth International Conference,*
edited by E. Klempt, H. Koch, and H. Orth
© 2004 American Institute of Physics 0-7354-0197-7/04/$22.00

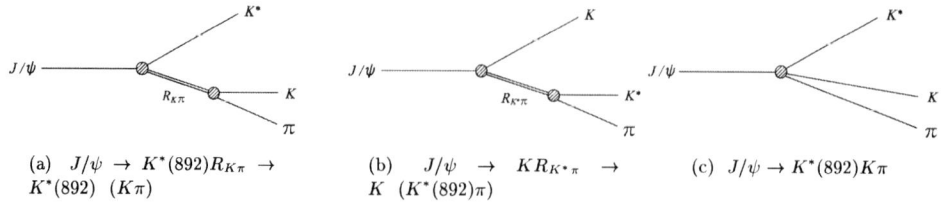

(a) $J/\psi \rightarrow K^*(892)R_{K\pi} \rightarrow$ (b) $J/\psi \rightarrow KR_{K^*\pi} \rightarrow$ (c) $J/\psi \rightarrow K^*(892)K\pi$
$K^*(892)$ $(K\pi)$ K $(K^*(892)\pi)$

FIGURE 1.   Three mechanisms of the decay process

We use in the analysis the VMW method which describes the process by the coherent sum of Breit-Wigner amplitudes of the relevant resonances with physically meaningful parameters. They are production couplings, masses, widths and relative phases. Non resonant direct production amplitude is included in the analysis. It interferes with resonant amplitudes.

In our analysis we consider the three mechanisms of relevant decay process; (a) $J/\psi \rightarrow K^*(892)R_{K\pi}$, (b) $J/\psi \rightarrow KR_{K^*\pi}$, (c) $J/\psi \rightarrow K^*K\pi$, which are depicted in Fig. 1 ($R_{K\pi}$ and $R_{K^*\pi}$ in Fig. 1 (a) and (b) are resonant particles decaying into $K\pi$ and $K^*\pi$, respectively, while the process (c) is the direct decay of J/ψ, interfering with the resonant states). In the case (a) of intermediate $K\pi$ resonances, we take into account the possible cases of $R_{K\pi}$ with, $J^P = 0^+$, 2^+, decaying into the $S-$ and $D-$wave states of $K\pi$ system, respectively. The most simple L agrangians relevant for production and decay of respective intermediate resonances, and the amplitudes in the respective cases of $S-$ and $D-$wave decays are given, respectively, by

Mechanism (a) (intermediate $K\pi$ resonance),

(S-wave)

$$L_S \sim \xi_\kappa \Psi_\mu K_\mu^* \kappa + g_\kappa \kappa K\pi + \cdots \,, \quad F_S = S_{h_\psi h_{K^*}}(r_\kappa e^{i\theta_\kappa} \Delta_\kappa(s_{K\pi}) + r_{K_0^*} e^{i\theta_{K_0^*}} \Delta_{K_0^*}(s_{K\pi}))$$

$$\Delta_\kappa(s_{K\pi}) = \frac{m_\kappa \Gamma_\kappa}{m_\kappa^2 - s_{K\pi} - i\sqrt{s_{K\pi}} \, \Gamma_\kappa(s_{K\pi})} \,, \quad \Gamma_\kappa(s_{K\pi}) = \frac{\mathbf{p} g_\kappa^2}{8\pi s_{K\pi}} \tag{1}$$

(D-wave)

$$L_D \sim \xi_{K_2^*} \Psi_\mu K_\nu^* K_{2\mu\nu}^* + g_{K_2^*} K_{2\mu\nu}^*(\partial_\mu K \partial_\nu \pi + \cdots) \,, \quad F_D = D_{h_\psi h_{K^*}} \, r_{K_2^*} e^{i\theta_{K_2^*}} \Delta_{K_2^*}^D(s_{K\pi})$$

$$\Delta_{K_2^*}^D(s_{K\pi}) = \frac{m_{K_2^*} \Gamma_{K_2^*} \left(\frac{F_D(s_{K\pi})}{F_D(m_{K_2^*}^2)} \right)}{m_{K_2^*}^2 - s_{K\pi} - i\sqrt{s_{K\pi}} \, \Gamma_{K_2^*}(s_{K\pi}) \left(\frac{F_D(s_{K\pi})}{F_D(m_{K_2^*}^2)} \right)^2} \,,$$

$$\Gamma_{K_2^*}(s_{K\pi}) = \frac{\mathbf{p}^5 g_{K_2^*}^2}{8\pi s_{K\pi}} \,, \quad F_D(s_{K\pi}) = \frac{1}{s_{K\pi} + m_{K_2^*}^2} \tag{2}$$

In the case (b) of intermediate $K^*\pi$ resonances with $J^P = 1^+$ decaying into the $S-$wave $K^*\pi$ system the corresponding formulas are given by

338

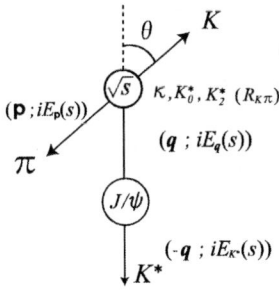

FIGURE 2. Kinematical variables

Mechanism (b) (intermediate $K^*\pi$-resonance),

$$L_{K_1} \sim \xi_{K_1}\Psi_\mu K_{1\mu}K + g_{K_1}K_{1\mu}K^*_\mu\pi \ , \quad F_{K_1} = B_{h_\psi h_{K^*}} \, r_{K_1}e^{i\theta_{K_1}}\Delta_{K_1}(s_{K^*\pi})$$

$$\Delta_{K_1}(s_{K^*\pi}) = \frac{m_{K_1}\Gamma_{K_1}}{m_{K_1}^2 - s_{K^*\pi} - i\sqrt{s_{K^*\pi}}\,\Gamma_{K_1}(s_{K^*\pi})} \ , \quad \Gamma_{K_1}(s_{K^*\pi}) = \frac{\mathbf{p}_{K^*}^2 g_{K_1}^2}{8\pi s_{K^*\pi}} \quad (3)$$

In the case (c) of the $K^*K\pi$ direct decay, the corresponding formulas are given by

Mechanism (c) (direct $K^*K\pi$),

$$L_{\text{direct } K\pi} \sim \xi_{K\pi}\Psi_\mu K^*_\mu K\pi \ , \quad F_{\text{direct } K\pi} = S_{h_\psi h_{K^*}} r_{K\pi}e^{i\theta_{K\pi}} \qquad (4)$$

In Eqs. (1), (2), (3) and (4) the S, D and B are the factors due to helicity-combinations among relevant particles given by $S_{h_\psi h_{K^*}} = \varepsilon_\mu^{h_\psi}\tilde{\varepsilon}_\mu^{h_{K^*}}$; $D_{h_\psi h_{K^*}} = \varepsilon_\mu^{h_\psi}\tilde{\varepsilon}_\nu^{h_{K^*}}\frac{1}{4}\{(r_\mu - \frac{q_\mu(m_K^2-m_\pi^2)}{s})(r_\nu - \frac{q_\nu(m_K^2-m_\pi^2)}{s}) - \frac{4\mathbf{p}_K^2}{3}(\delta_{\mu\nu} + \frac{q_\mu q_\nu}{s})\}$, $r_\mu = (p_K - p_\pi)_\lambda$; $B_{h_\psi h_{K^*}} = \varepsilon_\mu^{h_\psi}\tilde{\varepsilon}_\nu^{h_{K^*}}(\delta_{\mu\nu} + \frac{p_{K^*\pi\mu}p_{K^*\pi\nu}}{s_{K^*\pi}})$. Then, the total amplitude F squared for the process is given by

$$|F|^2 = \frac{1}{3}\sum_{h_\psi,h_{K^*}}|F_S + F_D + F_{K_1} + F_{\text{direct } K\pi}|^2$$

$$+\frac{1}{3}\left\{\frac{2}{3}\sum_{h_\psi}|\sum_{K_P^*}F_{K_P^*}|^2 + \sum_{h_\psi,h_{K^*}}|F_{K_S}|^2\right\} \qquad (5)$$

where summation is taken for $h_\psi, h_{K^*} = +,-,0$ and the factor $\frac{2}{3}$ comes from the fact that the initial J/ψ has only $\pm$ polarization.

In Eq. (5) we have taken into account the two backgrounds (the second term of the equation); the one is those from P-wave resonances $K_P^*(= K^*(892)$ and $K^*(1410))$ which are produced through $J/\psi \to (K\pi)_{\text{BG}}K_P^*$ with $m_{(K\pi)\text{BG}}$ being within the region of $K^*(892)$. $(K\pi)_{\text{BG}}$'s in the processes are considered to come from κ and non-resonant $K\pi$. The other does from $J/\psi \to K^*(892)K_S$ and $K_S \to \pi\pi$, where one π is misidentified with K. (As the values of mass and width for the background-K_S, we apply $m = 726$

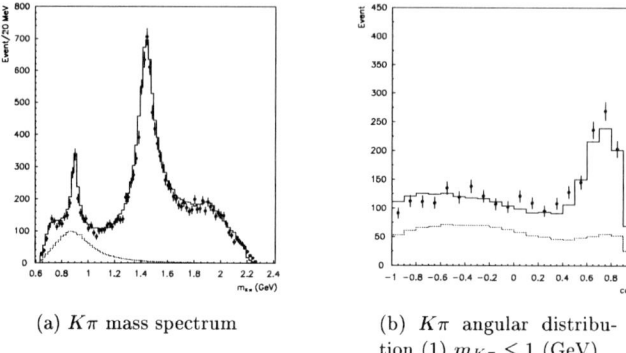

| (a) $K\pi$ mass spectrum | (b) $K\pi$ angular distribution (1) $m_{K\pi} \leq 1$ (GeV) |

FIGURE 3. Results of Analysis. The experimental data and fitted curves are given in a), where the contribution of κ resonance is also shown. The angular distributions below 1GeV are shown in b).

MeV and $\Gamma = 146$ MeV, which are estimated by Wu Ning with M. C. simulation). Corresponding formulas are given by

$$F_{K_p^*} = A_{h_\psi} r_{K_p^*} e^{i\theta_{K_p^*}} \Delta_{K_p^*}(s_{K\pi}), \quad A_{h_\psi} = \varepsilon_\mu^{h_\psi}(\delta_{\mu\nu} + \frac{q_\mu q_\nu}{s}) r_\nu$$

$$F_{K_S} = S_{h_\psi h_{K^*}} r_{K_S} \Delta_{K_S}(s_{K\pi}).$$

The differential decay width is given by

$$\Gamma = \frac{1}{64\pi^3 M_\psi^2} \int d\sqrt{s_{K\pi}} \int_{-1}^{1} d\cos\theta \quad \mathbf{q}(s_{K\pi}) \mathbf{p}(s_{K\pi}) \overline{|F|^2} \tag{6}$$

where the kinematical variables are defined in Fig. 3.

RESULTS OF ANALYSIS

Using the formulas given in §2 we have performed the fitting of the $K\pi$ mass spectrum, the $K^*\pi$ mass spectrum and the (K, π) angular distributions, simultaneously. The fitting parameters are production coupling r, initial phase θ, decay strength g(or Γ), mass m for respective (intermediate resonance and/or direct) amplitude. The experimental data with the fitted curves are given in Fig. 3, where the contributions of κ meson are also given. The angular distributions are shown below 1GeV of the $K\pi$ system. The values of χ^2, reduced χ^2 for the fitting, and respective contributions from $m_{K\pi}$ spectrum, $m_{K^*\pi}$ spectrum and angular distributions are given. $\chi^2/N_F = 375.8/(203 - 37) = 2.26$, $\chi^2_{(m_{K\pi})}/N_D(bin) = 2.27$ (82), $\chi^2_{(m_{K^*\pi})}/N_D(bin) = 1.40(81)$, $\chi^2_{(ang.\ dist.)}/N_D(bin) = 1.91$ (40).

The values of mass and width of κ meson obtained are

$$m_\kappa = 882 \pm 24 \text{ MeV}$$
$$\Gamma_\kappa = 335 \pm 82 \text{ MeV},$$

which are comparable with the values obtained through the analysis[5] of $K\pi$ scattering phase shifts.

$$m_\kappa = 905^{+65}_{-30} \text{ MeV} , \Gamma_\kappa = 545^{+235}_{-110} \text{ MeV} .$$

CONCLUDING REMARKS

We have analyzed the 58M J/ψ experimantal data (V103) of mass spectra and angular distributions using the VMW method, and obtained a clear evidence for the κ meson production.

We have tried also to fit the data without introducing the κ meson. The obtained values of reduced χ square are

with κ $\quad \chi^2/N_F = 375.8/(203-37) = 2.26$
without κ $\quad \chi^2/N_F = 882.0/(203-32) = 5.16$.

These show that the κ meson is indispensable to reproduce the experimental data. This is also clearly seen from the fit of $m_{K\pi}$ spectrum and angular distribution with the relevant $m_{K\pi}$ region.

Through the above investigation we may conclude that a clear evidence for existence of the κ meson has been given.

Finally we should like to add some remarks: Firstly the above obtained values of κ-mass and width depend slightly upon the parametrization of the background-K_S. So it is important to determine the magnitude of this (essentially experimental) background. Secondly in our method of analysis (VMW method), the effect of direct $(K^*K\pi)$-production amplitude, which are coherent to the other resonant amplitudes, are taken into account. This effect should be discriminated from the conventional, non-coherent backgrounds.

REFERENCES

1. T. Komada, in proceedings of " Hadron Spectroscopy, Chiral Symmetry and Relativistic Description of Bound Systems", KEK Proceedings 2003-7.
2. KEK-Proceedings 2000-4, Edited by S. Ishida, K. Takamatsu et al.
3. K. Hagiwara et al., Phys. Rev. **D66** (2002), 010001.
4. R. Delbourgo and M. D. Scadron, Phys. Rev. Lett. **48** (1982), 379.
    E. Beveren et al., Z. Phys. **C30** (1986), 651.
    M. Ishida, Prog. Theor. Phys. **101** (1999), 661.
5. S. Ishida et al., Prog. Theor. Phys. **98** (1997), 621; E. Beveren et al. in ref. [4]; D. Black, A. H. Fariborz, F. Sannino and J. Schechter, Phys. Rev. **D58** (1998), 054012.
6. S. N. Cherry and M. R. Pennington, Nucl. Phys. **A688** (2001), 823.
7. Fermilab E791 Collaboration, E. M. Aitala et al., Phys. Rev. Lett. **89** (2002), 121801.
8. S. Ishida et al., Prog. Theor. Phys. **88** (1992), 89; M. Ishida, S. Ishida and T. Ishida, Prog. Theor. Phys. **99** (1998), 1031; S. Ishida, Proc. Workshop on Hadron Spectroscopy, Frascati, 1999, ed. by T. Bressani et al., Frascati Phys. Series XV (1999), p. 85.
9. Wu Ning, in proceedings of " Hadron Spectroscopy, Chiral Symmetry and Relativistic Description of Bound Systems", KEK Proceedings 2003-7.

Sigma state and vacuum replica

Alexei V. Nefediev

ITEP, 117218, B.Cheremushkinskaya 25, Moscow, Russia

Abstract. The vacuum structure of a model for QCD with the four-quark interaction is studied and excited solutions — replicas — of the mass-gap equation are found, as a consequence of chiral symmetry breaking, which co-exists with the true vacuum of the theory. Quantum field theory is used to describe these new scalar states and the local operator creating the replica vacuum is constructed explicitly. A new interquark interaction due to these sigma-like states is identified and possible applications to hadronic processes are outlined.

In spite of progress in QCD in last decades the scalar hadronic sector of the theory still remains *terra incognita*, especially as far as states below 1 GeV scale are concerned. One example is given by the well-known puzzle of f_0/a_0 mesons, whose identification in terms of bare $q\bar{q}$ states, $K\bar{K}$ molecule, and so on is still an open problem. Another representative of the same family is the mysterious σ-meson — the lowest extremely broad scalar in the spectrum. In this contribution we address the physics of a new low-lying scalar state — the vacuum replica which was found recently to be very likely to exist in QCD [1] (a similar conclusion is made in a different approach in [2]). Whatever model for QCD is used, the chiral symmetry is broken spontaneously and the vacuum is full of strongly correlated quark-antiquark pairs with the $^3P_0$ coupling. Thus the basic conjecture of the approach is straightforward: the true vacuum of QCD is a scalar, and so are its "radial" excitations — radial excitations of the $^3P_0$ $q\bar{q}$ pairs in this vacuum [3] called vacuum replicas [1]. A quark propagating in the vacuum excites it, the excitation lives for a while and then it is absorbed by another quark, so that an extra interquark interaction arises as a consequence of the existence of the vacuum replicas. The communication of the quark with the vacuum, like with a medium, can be described in terms of effective particle-like exchanges, phonons in crystals being a textbook example of such a situation. As a result, a local field theory can be built to describe such replicas, as local in space-time objects, and to extract the aforementioned interquark interaction due to the exchange of the replica — a scalar particle-like object which competes, in the energy regions around $300 \div 400$ MeV, with the conventional σ-meson, a genuine $q\bar{q}$ bound state.

To exemplify this qualitative statement — which we believe to be general and model independent — we use a Hamiltonian model for QCD which can be traced back to QCD in the truncated Coulomb gauge [3, 4]:

$$H = \int d^3x \bar{\psi}(-i\vec{\gamma}\vec{\nabla})\psi + \int d^3x d^3y \left[\bar{\psi}\gamma_\mu \frac{\lambda^a}{2}\psi\right]_x K_{\mu\nu}^{ab}(\vec{x}-\vec{y}) \left[\bar{\psi}\gamma_\nu \frac{\lambda^b}{2}\psi\right]_y. \tag{1}$$

CP717, *Hadron Spectroscopy: Tenth International Conference,*
edited by E. Klempt, H. Koch, and H. Orth

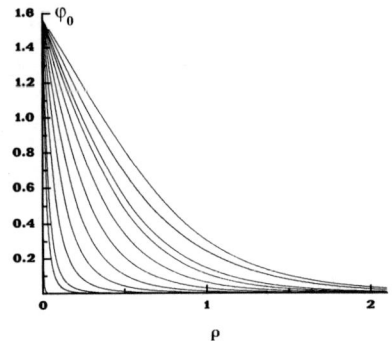
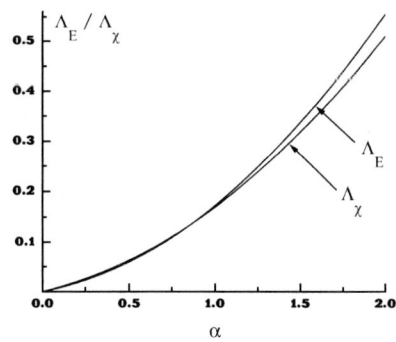

FIGURE 1. Solutions to the mass-gap equation (6) for $\alpha = 0.1, 0.3, 0.5, 0.7, 0.9, 1.0, 1.1, 1.3, 1.5,$ 1.7, 1.9, and 2.0 (plot (a); the curves localised closer to the origin correspond to smaller α's), and the mass parameters defining the chiral condensate, as $\langle \bar{q}q \rangle = -\Lambda_\chi^3$, and the excess of the vacuum energy density over the trivial vacuum, for two quark flavours and three colours, as $\Delta\mathcal{E}_{\text{vac}} = -\Lambda_E^4$, (plot (b)). All dimensional quantities are given in the units of K_0.

The explicit form of the kernel $K_{\mu\nu}^{ab}$ does not play an important role, provided it confines quarks and is defined by a potential,

$$K_{\mu\nu}^{ab}(\vec{x} - \vec{y}) = \delta^{ab} g_{\mu 0} g_{\nu 0} V(|\vec{x} - \vec{y}|). \tag{2}$$

Following the standard procedure we define dressed quarks [4],

$$q(\vec{x}, t) = \sum_{\xi=\uparrow,\downarrow} \int \frac{d^3 p}{(2\pi)^3} e^{i\vec{p}\vec{x}} [b_\xi(\vec{p}, t) u_\xi(\vec{p}) + d_\xi^\dagger(-\vec{p}, t) v_\xi(-\vec{p})], \tag{3}$$

$$
\begin{aligned}
u(\vec{p}) &= \frac{1}{\sqrt{2}} \left[\sqrt{1 + \sin\varphi} + \sqrt{1 - \sin\varphi}\, (\vec{\alpha}\hat{\vec{p}}) \right] u(0), \\
v(-\vec{p}) &= \frac{1}{\sqrt{2}} \left[\sqrt{1 + \sin\varphi} - \sqrt{1 - \sin\varphi}\, (\vec{\alpha}\hat{\vec{p}}) \right] v(0),
\end{aligned}
\tag{4}
$$

$\varphi(p)$ being the chiral (Bogoliubov) angle, and minimise the vacuum energy with respect to the latter:

$$\frac{\delta}{\delta\varphi} \langle 0|TH[\varphi]|0 \rangle = 0. \tag{5}$$

It was demonstrated in [5] that for an arbitrary power-like confining potential $V(r) = K_0^{\alpha+1} r^\alpha$ the mass-gap equation (5) takes the form:

$$p^3 S_p = \frac{4}{3} K_0^{\alpha+1} \Gamma(\alpha+1) \sin\frac{\pi\alpha}{2} \int_{-\infty}^{\infty} \frac{dk}{2\pi} \left\{ \frac{pk(S_k C_p - S_p C_k)}{|p-k|^{\alpha+1}} + \frac{C_k S_p}{(\alpha-1)|p-k|^{\alpha-1}} \right\}, \tag{6}$$

where $S_p = \sin\varphi(p)$, $C_p = \cos\varphi(p)$, and it is assumed that $\sin\varphi(-p) = \sin\varphi(p)$ and $\cos\varphi(-p) = -\cos\varphi(p)$. Then, for $0 < \alpha < 2$, this equation is convergent (the boundary

343

TABLE 1. Parameters of the kernel (8).

α_s	$\sqrt{\sigma_0}$, MeV	U, MeV	Λ, MeV
0.3	360	220	250

cases of $\alpha = 0$ and $\alpha = 2$ require special treatment [3, 5]) and possesses an infinite number of solutions $\varphi_n, n = 0, 1, \ldots$, as a consequence of a peculiar form of the dressed quark dispersive law E_p which becomes negative at small momenta. Treating the mass-gap equation (6) as a Schrödinger-like equation and using the quasiclassical quantisation procedure, one can estimate the behaviour of these solutions at the origin [5],

$$\varphi_n(p \sim 0) = \frac{\pi}{2} - \frac{p}{K_0} e^{C_\alpha(\pi n + \delta_\alpha)} + \ldots, \quad C_\alpha = \left[\frac{\sqrt{\pi}\Gamma\left(\frac{4-\alpha}{2}\right)}{2^\alpha \Gamma\left(\frac{1+\alpha}{2}\right)} \right]^{1/\alpha}, \tag{7}$$

δ_α being an n-independent correction to the leading regime. It is instructive to notice that the generalised formula for C_α in the D-dimensional space, similar to Eq. (7), contains $\Gamma\left(\frac{D-\alpha}{2}\right)$ in the numerator, so that for $\alpha = D$ only solution to the mass-gap equation with $n = 0$ survives and all replicas disappear. Such a strong confining potential, with $\alpha = 4$, is ruled out in our model since, as discussed above, the corresponding mass-gap eqution diverges. In the meantime, for $D = 2$, that is, for the 't Hooft model [6], the case $\alpha = D$ brings no divergences and can be studied. Indeed, the mass-gap equation possesses an unique solution in this case [7].

It was also argued in [5] that in real QCD the number of replicas should be finite, of order unity. Without loss of generality we consider only one replica, generalisation to the case of several replicas being trivial. As mentioned above, the explicit form of the kernel $K_{\mu\nu}^{ab}$ does not affect the qualitative results, leading, in the low-energy domain, only to numerical changes. We exemplify this by Fig. 1 where we give the profile of the ground-state solution φ_0 as well as the vacuum energy density and the chiral condensate for various powers α [5]. In what follows we consider a realistic form of the kernel $K_{\mu\nu}^{ab}$ which contains the linearly rising confining potential and, for phenomenological applications, also contains the short-range perturbative interquark interaction,

$$K_{\mu\nu}^{ab}(\vec{x} - \vec{y}) = \delta^{ab} g_{\mu\nu} \left[g_{\mu 0} \sigma_0 |\vec{x} - \vec{y}| - \frac{\alpha_s}{|\vec{x} - \vec{y}|} \left(1 - e^{-\Lambda |\vec{x} - \vec{y}|} \right) + U \right], \tag{8}$$

where $\alpha_s = 0.3$ and all dimensional scales are generated by the single nonperturbative scale fixed by the strength of the confining potential σ_0: $\Lambda \sim U \sim \sqrt{\sigma_0} \sim 300$ MeV (see Table 1). Such a kernel was considered in [1] and it was argued to be able to address light- and heavy-quark systems simultaneously. The corresponding mass-gap equation was solved numerically for the ground BCS vacuum, as well as for the replica (see Fig. 2).

The operator which relates the ground BCS vacuum, described by the chiral angle φ_0, and the replica, defined by φ_1, creates an infinite number of correlated $^3P_0$ $q\bar{q}$ pairs [4],

$$S = \exp\left(\frac{1}{2} \int \frac{d^3 p}{(2\pi)^3} \Delta\varphi C_p^\dagger \right), \quad C_p^\dagger = M_{\xi_1 \xi_2} b_{p\xi_1}^\dagger d_{-p\xi_2}^\dagger, \quad M = M_{3P_0} = (\vec{\sigma}\hat{\vec{p}}) i\sigma_2, \tag{9}$$

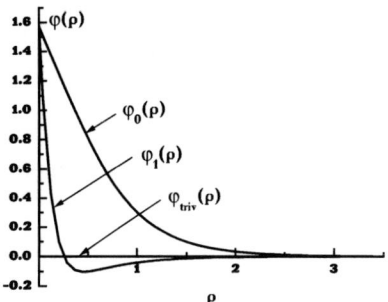
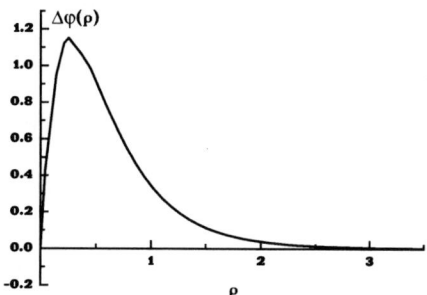

FIGURE 2. Solutions to the mass-gap equation with the kernel (8) and the difference $\Delta\varphi = \varphi_0 - \varphi_1$ [1].

the two vacua being orthogonal in the infinite volume,

$$\langle 0|1 \rangle = \langle 0|S|0 \rangle = \exp \left[N_C N_f V \int \frac{d^3 p}{(2\pi)^3} \ln \left(\cos^2 \frac{\Delta\varphi}{2} \right) \right] \underset{V \to \infty}{\longrightarrow} 0. \tag{10}$$

All quark creation/annihilation operators are defined in the BCS vacuum φ_0. From the point of view of the ground-state BCS vacuum, the replica is a coherent-like state belonging to the overcomplete set [8]

$$|\alpha, \theta\rangle = \prod_p \left(\cos^2 \frac{\alpha_p}{2} + \sin \frac{\alpha_p}{2} \cos \frac{\alpha_p}{2} C_p^\dagger[\theta_p] + \frac{1}{2} \sin^2 \frac{\alpha_p}{2} C_p^{\dagger 2}[\theta_p] \right) |0\rangle, \tag{11}$$

where $-\pi < \alpha_p \le \pi$, $0 \le \theta_p < 2\pi$, and the generalised operator $C_p^\dagger[\theta_p]$ is constructed similarly to Eq. (9), but with the help of the generalised $q\bar{q}$ coupling matrix $M[\theta_p] = e^{i(\vec{\sigma}\vec{p})\theta_p} M_{3P_0}$. Thus the replica is a coherent-like state, $|1\rangle = |\Delta\varphi, 0\rangle$, which minimises the vacuum energy locally and whose effect on quarks can be encoded in extra effective interquark interaction.

It was demonstrated in [8] that, in terms of quark fields, the operator S in (9) has the representation

$$S = T \exp \left[\int d^4 x f(x_0) \bar{q}(\vec{x}, x_0) \left(-i\vec{\alpha} \cdot \vec{\nabla} \right) q(\vec{x}, x_0) \right], \tag{12}$$

the function $f(t)$ being even with the Fourier transform such that $f(2E_p) = \frac{\Delta\varphi}{2p}$. The operator (12) describes the interaction of quarks with the vacuum through the so-called f-vertex [8] and it should be averaged over all possible moments of time $-T/2 < x_0^{(0)} < T/2$ when the replica was excited, T being the full time of the process (in Eq. (12) $x_0^{(0)} = 0$ but it can be easily restored, so that $f(x_0) \to f(x_0, x_0^{(0)})$ [8]). Besides that in order to consider reactions with multiple excitations of the replica, this operator should

345

be successively applied N times. The drawn picture corresponds to a one-dimensional diluted gas (f-chain) of noninteracting localised defects — excitations of the vacuum replica (f-vertices) — with the density $n = N/T$ which should tend to a finite limit as $T \to \infty$ [8].

Since $\Delta\varphi(p)$ is a small localised function (see Fig. 2), the operator (12) admits expansion in terms of f's, the lowest nontrivial contribution given by the second order. As a result, up to the order f^2, a new scalar particle-like object emerges, with the effective propagator

$$G(k) = nf(k_0)f(-k_0)(2\pi)^3\delta^{(3)}(\vec{k}) \tag{13}$$

attached to quark lines through the vertex $\vec{\alpha}\vec{p}$, $\vec{p}$ being the quark momentum. The propagator (13) possesses features standard for medium-related effects, reminiscent of those well known in the solid-state physics. Indeed, it is proportional to the density of the medium — the density of the defects in the chain, which corresponds to the linearised approximation to the medium. Besides, unlike the original interquark interaction given by the kernel $K_{\mu\nu}^{ab}(\vec{x} - \vec{y})$ in the Hamiltonian (1), the new interaction has a noninstantaneous form, which is also an inherent feature of the interaction with the medium. The rest of Feynman rules are left intact, and the energy-momentum conservation laws are preserved in the usual way. For example, quarks propagating in presence of the replica acquire a correction to their dispersive law, $\delta E_p = \frac{n}{8}(\Delta\varphi)^2$ [8].

Therefore we predict the existence of a scalar particle-like object with the energy distribution given by the function $f(\omega)$, peaked at $\omega \sim 400$ MeV, and strongly coupled to pions, with the dominating two-pion S-wave mode. We argue that the existence of such an object is not an artifact of a particular model for QCD, but is a consequence of chiral symmetry breaking. In hadronic processes this object reveals itself via "σ"-like exchange diagrams with and effective propagator (13). Leaving the low-energy limit of the theory intact, the vacuum replica may give a sizable effect at energies around 400 MeV, that is, in the region where the convention σ-meson plays dominating role, so that the two scalars co-exist. This work is in progress now and will be reported elsewhere.

The author would like to thank P. Bicudo and E. Ribeiro for fruitful cooperation. Financial support of INTAS grant YSF 2002-49, as well as the grant NS-1774.2003.2. is gratefully acknowledged. This work is supported by the Federal Programme of the Russian Ministry of Industry, Science and Technology No 40.052.1.1.1112.

REFERENCES

1. P. J. A. Bicudo, A. V. Nefediev, and J. E. F. T. Ribeiro, *Phys. Rev.* **D65**, 085026 (2002).
2. A. A. Osipov and B. Hiller, *Phys. Lett.* **B539**, 76 (2002).
3. A. Amer, A. Le Yaouanc, L. Oliver, O. Pene, and J.-C. Raynal, *Phys. Rev. Lett.* **50**, 87 (1983); A. Le Yaouanc, L. Oliver, O. Pene, and J.-C. Raynal, *Phys. Lett.* **B134**, 249 (1984); *Phys. Rev.* **D29**, 1233 (1984).
4. P. Bicudo and J. E. Ribeiro, *Phys. Rev.* **D42**, 1611 (1990).
5. P. J. A. Bicudo and A. V. Nefediev, hep-ph/0307302, *Phys. Rev. D*, in press.
6. G. 't Hooft, *Nucl. Phys.* **B75**, 461 (1974).
7. P. J. A. Bicudo and A. V. Nefediev, hep-ph/0308273, *Phys. Lett. B*, in press.
8. A. V. Nefediev and J. E. F. T. Ribeiro, *Phys. Rev.* **D67**, 034028 (2003).

Understanding $I = 2 \, \pi\pi$ Interaction

B.S.Zou[*], F.Q.Wu[*], L.Li[*†] and D.V.Bugg[**]

[*]Institute of High Energy Physics, CAS, Beijing 100039, China
[†]Peking University, Beijing 100087, China
[**]Queen Mary College, London, UK

Abstract. A correct understanding and description of the $I = 2 \, \pi\pi$ S-wave interaction is important for the extraction of the $I = 0 \, \pi\pi$ S-wave interaction from experimental data and for understanding the $I = 0 \, \pi\pi$ S-wave interaction theoretically. With t-channel ρ, $f_2(1270)$ exchange and the $\pi\pi \to \rho\rho \to \pi\pi$ box diagram contribution, we reproduce the $\pi\pi$ isotensor S-wave and D-wave scattering phase shifts and inelasticities up to 2.2 GeV quite well in a K-matrix formalism.

INTRODUCTION

Much attention has been paid to the isospin I=0 $\pi\pi$ S-wave interaction due to its direct relation to the σ particle and the scalar glueball candidates. However, to really understand the isoscalar $\pi\pi$ S-wave interaction, one must first understand the isospin I=2 $\pi\pi$ S-wave interaction due to the following two reasons: (1) There are no known s-channel resonances and less coupled channels in I=2 $\pi\pi$ system, so it is much simpler than the I=0 $\pi\pi$ S-wave interaction; (2) To extract I=0 $\pi\pi$ S-wave phase shifts from experimental data on $\pi^+\pi^- \to \pi^+\pi^-$ and $\pi^+\pi^- \to \pi^0\pi^0$ obtained by $\pi N \to \pi\pi N$ reactions, one needs an input of the I=2 $\pi\pi$ S-wave interaction. While the $I = 2 \, \pi\pi$ S-wave interaction can be extracted from the pure $I = 2 \, \pi^\pm\pi^\pm \to \pi^\pm\pi^\pm$ reactions, the $I = 0$ $\pi\pi$ S-wave interaction can only be extracted from $\pi^+\pi^- \to \pi^+\pi^-$ and $\pi^+\pi^- \to \pi^0\pi^0$ reactions which are mixture of $I = 0$ and $I = 2$ contributions. The relation between the $\pi^+\pi^- \to \pi^+\pi^-$, $\pi^0\pi^0$ S-wave amplitudes and the isospin decoupled amplitudes is as the following:

$$T_s(+-,+-) = T_s^{I=0}/3 + T_s^{I=2}/6, \tag{1}$$

$$T_s(+-,00) = T_s^{I=0}/3 - T_s^{I=2}/3. \tag{2}$$

The $T_s^{I=0}$ was usually extracted from $T_s(+-,+-)$ and $T_s(+-,00)$ information by assuming some kind of $T_s^{I=2}$ amplitude.

Up to now, experimental information on the I=2 $\pi\pi$ scattering mainly came from $\pi^+p \to \pi^+\pi^+n$ [1] and $\pi^-d \to \pi^-\pi^-pp$ [2, 3] reactions. The main features for the I=2 $\pi\pi$ S-wave phase shifts δ_0^2 and inelasticities η_0^2 are: (1) the δ_0^2 goes down more and more negative as the $\pi\pi$ invariant mass increases from $\pi\pi$ threshold up to 1.1 GeV; (2) the δ_0^2 starts to increase for energies above about 1.1 GeV; (3) the η_0^2 starts to deviate from 1 for energies above 1.1 GeV. The first feature can be well explained by the t-channel ρ exchange force [4, 5, 6, 7] while the effect of t-channel scalar exchange is extremely small and can be neglected [7]. Due to the relative poor quality of the I=2 $\pi\pi$ scattering

CP717, *Hadron Spectroscopy: Tenth International Conference*,
edited by E. Klempt, H. Koch, and H. Orth
© 2004 American Institute of Physics 0-7354-0197-7/04/$22.00

data above 1.1 GeV, the other two features are usually overlooked. In this work [8], we show in a K-matrix formalism [5, 9] that these two features can be well reproduced by the t-channel $f_2(1270)$ exchange and the $\pi\pi$-$\rho\rho$ coupled-channel effect, respectively.

Recently, the $\pi^+\pi^- \rightarrow \pi^+\pi^-$ scattering from the old πN scattering experiments with both unpolarized[10] and polarized targets[11] has been re-analyzed[12, 13] in combination with new information from $p\bar{p}$ and other experiments. The $\pi^+\pi^- \rightarrow \pi^0\pi^0$ scattering has also been studied by E852[14], GAMS[15] Collaborations and analyzed [16]. In Refs.[10, 12], a scattering length formula for I=2 $\pi\pi$ S-wave was used; in Ref.[16] another empirical parametrization was used. These parametrizations give similar phase shifts up to 1.1 GeV, but differ at higher energies. All these previous analyses have ignored the inelastic effects in the $I = 2$ channel. We will demonstrate that the correct description of the $I = 2$ S-wave interaction has significant impact on the extraction of the $I = 0$ $\pi\pi$ S-wave amplitude for energies above 1.1 GeV.

THEORETICAL FRAMEWORK

We follow the K-matrix formalism as in Refs.[5, 9]. For the two-channel case, the two-dimensional K matrix and phase space $\rho(s)$ matrix are

$$K = \begin{pmatrix} K_{11} & K_{12} \\ K_{12} & K_{22} \end{pmatrix}, \qquad \rho(s) = \begin{pmatrix} \rho_1(s) & 0 \\ 0 & \rho_2(s) \end{pmatrix}, \tag{3}$$

with i=1,2 representing $\pi\pi$ and $\rho\rho$ channel, respectively.

$$T_{11} = \frac{K_{11} - i\rho_2(K_{11}K_{22} - K_{12}K_{21})}{1 - i\rho_1 K_{11} - i\rho_2 K_{22} - \rho_1\rho_2(K_{11}K_{22} - K_{12}K_{21})}, \tag{4}$$

Ignoring the interaction between $\rho\rho$, we have $K_{22} = 0$; then

$$T_{11} = \frac{K_{11} + iK_{12}\rho_2 K_{21}}{1 - i\rho_1(K_{11} + iK_{12}\rho_2 K_{21})}, \tag{5}$$

where $iK_{12}\rho_2 K_{21}$ corresponds to the contribution of the $\pi\pi \rightarrow \rho\rho \rightarrow \pi\pi$ box diagram.

In order to obtain K_{11}, we incorporate the t-channel $f_2(1270)$ contribution into the t-channel ρ exchange term by the Dalitz-Tuan method [5, 12].

$$K_{11} = \frac{K_\rho(s) + K_{f_2}(s)}{1 - \rho_1^2(s)K_\rho(s)K_{f_2}(s)}. \tag{6}$$

The amplitudes for the t-channel ρ and $f_2(1270)$ meson exchange without considering the vertex form factor were given in Ref.[6] for studying $I = 0$ $\pi\pi$ scattering. In this work, we include a form factor of conventional monopole type to take into account the off-shell behavior of the exchanged mesons: $F(q^2) = (\Lambda^2 - m^2)/(\Lambda^2 - q^2)$ with (m, q) the mass and four-vector momentum, respectively, of exchanged mesons, and Λ the cutoff parameter to be determined by experimental data. The resulted K_ρ and K_{f_2} are given in our paper [8].

For energies above the $\rho\rho$ threshold, the inelastic effect should be taken into account in the $I = 2$ $\pi\pi$ channel. Note that unlike $I = 0$ $\pi\pi$ channel, the $I = 2$ $\pi\pi$ channel does not couple to the $K\bar{K}$ and $\omega\omega$ channels due to isospin conservation. The $\pi\pi$ channel couples to the $\rho\rho$ channel by the t-channel π exchange as shown in Fig.1.

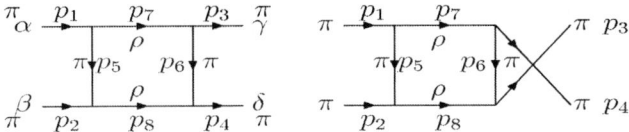

FIGURE 1. The $\pi\pi \rightarrow \rho\rho \rightarrow \pi\pi$ box diagrams

Assuming the on-shell approximation [18] for the $\rho\rho$ intermediate state, the amplitude for the box diagrams are calculated and projected to the S- and D-wave [8]. It is related to the K-matrix by $T_{box}^{I=2} = iK_{12}\rho_2 K_{12}$. For the phase space factor ρ_2, the width of the ρ meson is taken into account in our calculation.

NUMERICAL RESULTS AND DISCUSSION

From the formalism given above, we get I=2 $\pi\pi$ S-wave and D-wave phase shifts and inelasticities as shown in Fig.2. The t-channel ρ exchange alone (dashed lines) reproduces the phase shifts for energies up to 1.1 GeV very well with the form factor parameter $\Lambda_{\rho\pi\pi} = 1.5$ GeV [5], but underestimates the phase shifts at higher energies. The inclusion of the t-channel $f_2(1270)$ exchange (dot-dashed lines) increases the phase shifts especially for energies above 1 GeV and can reproduce the phase shift data very well with the form factor parameter $\Lambda_{f_2\pi\pi} = 1.7$ GeV. However they only contribute to the elastic scattering and cannot produce the inelasticities for energies above 1 GeV.

The experimental information on the inelasticities is scarce for the $I = 2$ $\pi\pi$ scattering. Two data points were given by Ref.[3] for energies $1 \sim 1.5$ GeV. For energies $1.5 \sim 2$ GeV, Ref.[2] estimated to be 0.5 ± 0.2 for the η_0^2 in one solution and assumed $\eta_0^2 = 1.53 - 0.475 m_{\pi\pi}$ (GeV/c^2) for another solution. The two solutions gave similar results for the $I = 2$ $\pi\pi$ S-wave phase shifts. For the $I = 2$ $\pi\pi$ D-wave scattering, the inelasticity could not be measured well and was assumed to be $\eta_2^2 = 1$ for the extraction of the δ_2^2.

Although there are only three data points with large error bars for the inelasticity parameter η_0^2 of the $I = 2$ $\pi\pi$ S-wave scattering, it is clear that the inelastic effect may be significant around 1.6 GeV. In order to reproduce this inelasticity, it is necessary to consider the $\pi\pi \leftrightarrow \rho\rho$ coupling channel effect. We find that including contribution from the $\pi\pi \rightarrow \rho\rho \rightarrow \pi\pi$ box diagram in our K-matrix formalism, the $I = 2$ S-wave inelasticity data can be very well reproduced without introducing any more free parameter as shown in Fig.2. The same diagram also predicts a broad shallow dip around 1.7 GeV for the inelasticity of the $I = 2$ D-wave scattering. Assuming $\eta_2^2 = 1$ for energies around 1.7 GeV may bias the δ_2^2 data around this energy. This may be the reason that the δ_2^2 data around 1.7 GeV has the largest discrepancy with our theoretical result. The box diagram has little influence to the phase shifts although it produces large inelasticity.

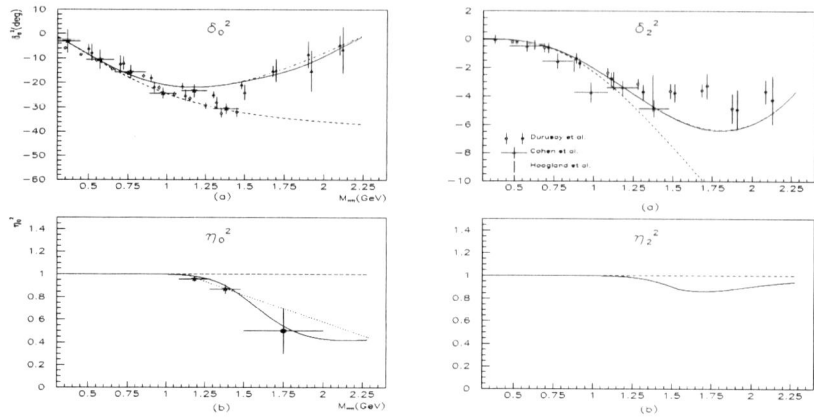

FIGURE 2. The $I = 2$ $\pi\pi$ S-wave (δ_0^2, η_0^2) and D-wave (δ_2^2, η_2^2) phase shifts and inelasticities. Data are from Ref.[1, 2, 3]. The solid curves represent the total contribution of ρ, f_2 exchange and the box diagram; dot-dashed curves from ρ and f_2 exchange; dashed curves from t-channel ρ exchange only.

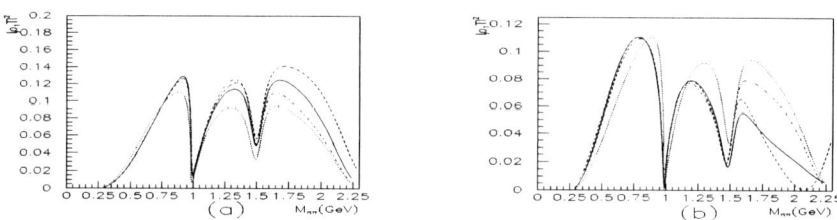

FIGURE 3. Full amplitudes squared of $\pi^+\pi^- \to \pi^+\pi^-$ (a) and $\pi^+\pi^- \to \pi^0\pi^0$ (b). The lines correspond to using $T_s^{I=0}$ from Ref.[5] plus various input of $T_s^{I=2}$: $T_s^{I=2} = 0$ (dotted lines); $T_s^{I=2}$ from Refs.[5, 10] (dot-dashed lines); $T_s^{I=2}$ from Ref.[16] (dashed lines); $T_s^{I=2}$ from this work (solid line).

Inspired by the new claim of a pentaquark state [19], we also explored the possibility of including an $I = 2$ s-channel resonance to reproduce the $I = 2$ $\pi\pi$ S-wave scattering data instead of using the t-channel f_2 exchange and the $\pi\pi \to \rho\rho \to \pi\pi$ box diagram, but failed to reproduce the data. While the η_0^2 needs the resonance with mass around 1.6GeV, the δ_0^2 needs the mass above 2.3 GeV with a much broader width.

To demonstrate the significance of possible impact of the $I = 2$ input for the extraction of $I = 0$ $\pi\pi$ amplitude, we calculate the full S-wave amplitudes for $\pi^+\pi^- \to \pi^+\pi^-$ and $\pi^+\pi^- \to \pi^0\pi^0$ according to Eqs.(1,2) with $T_s^{I=0}$ from Ref.[5] plus various $T_s^{I=2}$ inputs. The corresponding full S-wave amplitudes squared ($|\rho_1 T|^2$) are shown in Fig.3. The dotted lines are the results with $T_s^{I=2} = 0$. The dot-dashed lines correspond to the scattering length formula used in Refs.[5, 10, 12], which is similar to the result by considering only the t-channel ρ exchange contribution. The dashed lines use the new empirical $T_s^{I=2}$ formula of Ref.[16], which is similar to the result by considering t-channel ρ and f_2 exchange contributions, but ignoring the inelasticity caused by the $\pi\pi \leftrightarrow \rho\rho$ box diagram contribution. The solid lines are results with our $T_s^{I=2}$ including the t-channel ρ, f_2 exchange and the box diagram. It is clear that $T_s^{I=2}$ has significant contribution to the amplitudes for $\pi^+\pi^- \to \pi^+\pi^-$ and $\pi^+\pi^- \to \pi^0\pi^0$ processes, hence has significant impact on the extraction of the $T_s^{I=0}$ amplitude. Previous inputs of $T_s^{I=2}$ give similar results as our new $T_s^{I=2}$ for energies below 1.1 GeV, but differ from ours significantly for higher energies. The inclusion of both t-channel f_2 exchange and $\pi\pi \to \rho\rho \to \pi\pi$ box diagram contribution is important.

In summary, the basic features of $I = 2$ $\pi\pi$ scattering phase shifts and inelasticities can be well reproduced by the t-channel meson (ρ, f_2) exchange and the $\pi\pi \leftrightarrow \rho\rho$ coupled-channel effect in the K-matrix formalism. The t-channel ρ exchange provides repulsive negative phase shifts while the t-channel $f_2(1270)$ gives an attractive force to increase the phase shifts for $\pi\pi$ scattering above 1 GeV, and the coupled-channel box diagram causes the inelasticities. A correct description of the $I = 2$ $\pi\pi$ scattering has significant impact on the extraction of the $I = 0$ scattering amplitudes from $\pi^+\pi^- \to \pi^+\pi^-$ and $\pi^+\pi^- \to \pi^0\pi^0$ data, especially for energies above 1.2 GeV . A re-analysis of these data with our new description of the $I = 2$ $\pi\pi$ scattering will be carried out as our next step.

ACKNOWLEDGMENTS

The work is partly supported by CAS Knowledge Innovation Project (KJCX2-SW-N02) and the National Natural Science Foundation of China.

REFERENCES

1. W. Hoogland et al., Nucl. Phys. **B 126** (1977) 109.
2. N.B.Durusoy et al., Phys. lett. **45B** (1973) 517.
3. D.Cohen, T.Ferbel, P.Slattery, and B.Werner, Phys. Rev.**D7**(1973) 661.
4. J.Weinstein and N.Isgur, Phys. Rev. D **41** (1990) 2236; Phys. Rev. Lett. **48** (1982) 659.
5. L. Li, B.S. Zou, G.L. Li, Phy. Rev. **D 63** (2001) 074003.
6. B.S.Zou and D.V.Bugg, Phys. Rev. **D 50** (1994) 591.
7. D.Lohse et al., Phys. Lett. **B234** (1990) 235; Nucl .Phys. **A516** (1990) 513.
8. F.Q.Wu, B.S.Zou, L.Li and D.V.Bugg, hep-ph/0308152.
9. L. Li, B.S. Zou, G.L. Li, Phy. Rev. **D 67** (2003) 034025.
10. B. Hyams et al., Nucl. Phys. **B 64** (1973) 134.
11. H. Becker et al., Nucl. Phys. **B151** (1979) 46.
12. D.V. Bugg, A.V. Sarantsev, and B.S. Zou, Nucl. Phys. **B 471** (1996) 59.
13. R.Kaminski et al., Z.Phys. **C74** (1997) 79.
14. J.Gunter et al., Phys. Rev. **D64** (2001) 072003.
15. D.Alde et al., Z. Phys.**C66** (1995) 375; V.V.Anisovich et al., Phys. Lett.**B389** (1996) 388.
16. N.N.Achasov, G.N.Shestakov, Phys. Rev. **D67** (2003) 114018.
17. B.R. Martin et al., *Pion-Pion Interactions in Particle Physics*, Academic, New York, 1974.
18. Y.Lu, B.S.Zou and M.P.Locher, Z. Phys. **A345** (1993) 207; Z. Phys. **A347** (1994) 281.
19. T.Nakano et al., Phys. Rev. Lett. **91** (2003) 012002; V.V.Barmin et al., hep-ex/0304040; S.Stepanyan et al., hep-ex/0307018; J.Barth et al., hep-ex/0307083.

The $q\bar{q}$ spectra and the structure of the scalar mesons

J. Vijande*, F. Fernández* and A. Valcarce*

*Grupo de Física Nuclear, Universidad de Salamanca, E-37008 Salamanca, Spain

Abstract. The $q\bar{q}$ spectrum is studied within a chiral constituent quark model. It provides with a good fit of the available experimental data from light (vector and pseudoscalar) to heavy mesons including some recent results on charmonium. The 0^{++} light mesons and the new D states measured at different factories cannot be described as $q\bar{q}$ pairs and a tetraquark structure is suggested.

INTRODUCTION

Since Gell-Man conjecture, most of the meson experimental data were classified as $q\bar{q}$ states according to $SU(N)$ irreducible representations. Nevertheless a number of interesting issues remains still open as for example the understanding of the structure of the scalar mesons or the new D_s states measured on B factories.

The theoretical study of charmonium and bottomonium made clear that heavy-quark systems are properly described by nonrelativistic potential models reflecting the dynamics expected from QCD [1]. The light meson sector has been studied by means of constituent quark models, where quarks are dressed with a phenomenological mass and bound in a nonrelativistic potential, usually a harmonic oscillator [2]. Quite surprisingly a large number of properties of hadrons could be reproduced in this way [3]. In this talk we present the meson spectra obtained by means of a chiral constituent quark model in a trial to interpret some of the still unclear experimental data in the scalar sector.

SU(3) CHIRAL CONSTITUENT QUARK MODEL

Since the origin of the quark model hadrons have been considered to be built by constituent (massive) quarks. Nowadays it is widely recognized that because of the spontaneous breaking of chiral symmetry in the light quark sector at some momentum scale a constituent quark mass $M(q^2)$ appears. Once a constituent quark mass is generated such particles have to interact through $SU(3)$ Goldstone modes [pion, kaon, eta and sigma (which simulates the two-pion exchange)]. Explicit expressions of these potentials can be found elsewhere [4]. In the heavy quark sector, chiral symmetry is explicitly broken and therefore these interactions will not appear.

For higher momentum transfer quarks still interact through gluon exchanges. Following de Rújula *et al.* [5] the one-gluon-exchange (OGE) interaction is taken as a standard color Fermi-Breit potential. In order to obtain a unified description of light, strange

CP717, *Hadron Spectroscopy: Tenth International Conference,*
edited by E. Klempt, H. Koch, and H. Orth

and heavy mesons a scale dependent strong coupling constant has to be used [6]. We parametrize this scale dependence by

$$\alpha_s(\mu) = \frac{\alpha_0}{ln\left(\frac{\mu^2 + \mu_0^2}{\Lambda_0^2}\right)}, \tag{1}$$

where μ is the reduced mass of the $q\bar{q}$ system and α_0, μ_0 and Λ_0 are fitted parameters [4]. This equation gives rise to $\alpha_s \sim 0.54$ for the light quark sector, a value consistent with the one used in the study of the nonstrange hadron phenomenology [7], and it also has an appropriate high Q^2 behavior [8], given a value of $\alpha_s \sim 0.127$ at the Z_0 mass [9]. The δ function appearing in the OGE has to be regularized in order to avoid an unbound spectrum from below. To solve numerically the Schrödinger equation with this potential we use a flavor-dependent regularization [4].

The other nonperturbative property of QCD is confinement. Lattice QCD studies show that $q\bar{q}$ systems are well reproduced at short distances by a linear potential that is screened at large distances due to pair creation [10]. One important question which has not been properly answered is the covariance property of confinement. While the spin-orbit splittings in heavy quark systems suggest a scalar confining potential [11], a significant mixture of vector confinement has been used to explain the decay widths of P-wave D mesons [12]. Such property, being irrelevant for the central part of the interaction, determines the sign and strength of the spin-orbit Thomas precession term which is important for the scalar mesons. Therefore, we write the confining interaction as an arbitrary combination of scalar and vector terms $V_{CON}^{SO}(\vec{r}_{ij}) = (1 - a_s)V_V^{SO}(\vec{r}_{ij}) + a_s V_S^{SO}(\vec{r}_{ij})$ where $V_V^{SO}(\vec{r})[V_S^{SO}(\vec{r})]$ is the vector (scalar) spin-orbit contribution.

RESULTS

With the quark-quark interaction described above we have solved the Schrödinger equation for the different $q\bar{q}$ systems. Most of the parameters of the Goldstone boson fields are taken from the NN sector. The eta and kaon cutoff masses are related with the sigma and pion one as explained in Ref. [13]: $\Lambda[u(d)s] \simeq \Lambda(ud) + m_s$, m_s being the strange quark current mass. The confinement parameters a_c and μ_c are fitted to reproduce the energy difference between the ρ meson and its first radial excitation and the J/ψ and the $\psi(2S)$. The parameters involved in the OGE are obtained from a global fit to the hyperfine splittings well established in the Particle Data Group (PDG) [14]. Finally, the relative strength of the scalar and vector confinement is fitted to the energy of $a_1(1260)$ and $a_2(1320)$, ordering that cannot be reproduced with a pure scalar confinement. We obtain $a_s = 0.777$.

The spectra for the light pseudoscalar and vector mesons and for heavy mesons have been reported in Ref. [15]. The agreement with experimental data is remarkable. Let us emphasize that with only 11 parameters we are able to describe more than 110 states.

Recently Belle and BaBar [16, 17, 18] collaborations have reported new experimental measurements for the mass of the $\eta_c(2S)$. The average value from the different measurement is significantly larger than most predictions of constituent quark models and the previous experimental value of the PDG: $M[\eta_c(2S)] = 3594 \pm 5$ MeV. Such value cannot be easily explained in the framework of constituent quark models because the resulting $2S$ hyperfine splitting (HFS) would be smaller than the predicted for the $1S$ states [19, 20]. Based on this fact some authors have claimed for an α_s coupling constant depending on the radial excitation.

We predict a value $M[\eta_c(2S)] = 3627$ MeV, within the error bar of the last two Belle measurements, the ones obtained with higher statistics. Moreover the ratio 2S to 1S HFS is found to be 0.537, in good agreement with the experimental data, 0.479. The reason for this agreement can be found in the shape of the confining potential that also influences the HFS, the linear confinement being not enough flexible to accommodate both excitations [21].

Other puzzling state is a narrow resonance around 2317 MeV, known as $D_{S_J}^*(2317)$ reported by BaBar [22]. This state has been confirmed by CLEO [23] together with another possible resonance around 2460 MeV. Both experiments interpret these resonances as $J^P = 0^+$ and 1^+ states. This discovery has triggered a series of articles presenting alternative hypothesis [24]. The most striking aspect of these two resonances is that their masses are much lower than expected. We obtain a mass of 2470.9 MeV for the $J^P = 0^+$ and 2565.5 MeV for the $J^P = 1^+$. They are far from the experimental data although the rest of the states ($1^+, 2^+, 1^-$ and 0^-) agree reasonably well with the values of the PDG for both for the D's and D_s's states.

Concerning the scalar sector our results are shown in Table 1 (column three). We observe that for isovector states, there appears a candidate for the $a_0(980)$, the $^3P_0$ member of the lowest ^3P_J isovector multiplet. The other candidate, the $a_0(1450)$, is predicted to be the scalar member of a ^3P_J excited isovector multiplet. This reinforces the predictions of the naive quark model, where the LS force makes lighter the $J = 0$ states with respect to the $J = 2$. The assignment of the $a_0(1450)$ as the scalar member of the lowest ^3P_J multiplet would contradict this idea, because the $a_2(1312)$ is well established as a $q\bar{q}$ pair. The same behavior is evident in the $c\bar{c}$ and the $b\bar{b}$ spectra, making impossible to describe the $a_0(1450)$ as a member of the lowest ^3P_J isovector multiplet without spoiling the description of heavy-quark multiplets. However, in spite of the correct description of the mass of the $a_0(980)$, the model predicts a pure light-quark content, what seems to contradict some experimental evidences. The $a_0(1450)$ is predicted to be also a pure light quark structure obtaining a mass somewhat higher than the experiment.

In the case of the isoscalar states, one finds a candidate for the $f_0(600)$ with a mass of 402 MeV, in the lower limit of the experimental error bar and with a strangeness content around 8%. The $f_0(980)$ and $f_0(1500)$ cannot be found for any combination of the parameters of the model. It seems that a different structure rather than a naive $q\bar{q}$ pair is needed to describe these states. The $f_0(1500)$ is a clear candidate for the lightest glueball [25] and our results support this assumption. For the $f_0(1370)$ state (which may actually correspond to two different states [26]) we obtain two almost degenerate states around this energy, the lower one with a predominantly nonstrange content, and the other with a high $s\bar{s}$ content. Finally a state corresponding to the $f_0(1710)$ is obtained.

In the $I = 1/2$ sector, as a consequence of the larger mass of the strange quark as compared to the light ones, our model always predicts a mass for the lowest 0^{++} state

200 MeV greater than the $a_0(980)$ mass. Therefore, being the $a_0(980)$ the member of the lowest isovector scalar multiplet, the $\kappa(900)$ cannot be explained as a $q\bar{q}$ pair. We find a candidate for the $K_0^*(1430)$ although with a smaller mass.

THE SCALAR MESONS AS TETRAQUARK

Unlike the $q\bar{q}$ pairs tetraquark structures, suggested twenty years ago by Jaffe [27], can couple to 0^{++} without orbital excitation and therefore could be serious candidates to explain the structure of the scalar mesons.

In this section we study tetraquark bound states by solving the Schrödinger equation using a variational method where the spatial trial wave function is a linear combination of gaussians. The technical details are given in Ref. [28]. Due to the presence of the kaon-exchange there is a mixture among different configurations with the same isospin. In particular, in the isoscalar sector the configurations: $[(qq)(\bar{q}\bar{q})]$, $[(qs)(\bar{q}\bar{s})]$, and $[(ss)(\bar{s}\bar{s})]$ are mixed. The same happens in the isovector case for the configurations: $[(qq)(\bar{q}\bar{q})]$, and $[(qs)(\bar{q}\bar{s})]$, and in the $I = 1/2$ case for the configurations: $[(qq)(\bar{q}\bar{s})]$, and $[(qs)(\bar{s}\bar{s})]$. In all cases q stands for a u or d quark.

The results are shown in Table 1 (column four) where we present the lowest states for the three isospin sectors. As one can see, there appear two states, in the isoscalar and isovector sectors, with almost the same mass, although too high to be identified with the $f_0(980)$ and $a_0(980)$. In the $I = 1/2$ sector, there appears a candidate to be identified with the $\kappa(900)$. It has been recently argued the possible importance of three-body forces arising from the confining interaction for those systems containing at least three quarks [29]. We have performed a calculation including a three-body confining term as the one reported in Ref. [29] fixing its strength to reproduce the mass of the $f_0(980)$. The results are shown in Table 1 (column five) and as can be seen the degeneracy between the isoscalar and isovector states remains although their masses are now compatible with the experimental data. The lowest state of the isoscalar and $I = 1/2$ sectors are almost not affected.

TABLE 1. Light scalar meson masses in MeV

$(q\bar{q})$ state $(n^{2I+1,2S+1}L_J)$	Meson	$(q\bar{q})$	$(4q)$	$(4q)$+Three body	Experiment
$1^{3,3}P_0$	$a_0(980)$	983.5	1343	968	984.7±1.2
$2^{3,3}P_0$	$a_0(1450)$	1586.3			1474±19
$1^{1,3}P_0$	$f_0(600)$	402.7	604	644	400–1200
	$f_0(980)$		1325	1007	980±10
$1^{1,3}P_0$		1341.7			
$2^{1,3}P_0$	$f_0(1370)$	1391.2			1200–1500
$2^{1,3}P_0$	$f_0(1710)$	1751.8			1713±6
$3^{1,3}P_0$	$f_0(2020)$	1893.8			1992±16
	$\kappa(900)$		1026	922	797±43
$1^{2,3}P_0$	$K_0^*(1430)$	1213.5			1412±6
$2^{2,3}P_0$	$K_0^*(1950)$	1768.5			1945±30

Using the same interaction and formalism we have calculated the $D_{S_J}^*(2317)$ as a $[(uc)(\bar{u}\bar{s})]$ tetraquark. The result we obtain, M=2389 MeV, suggests that this state could also have a significant tetraquark component.

As a summary, we have found tetraquark bound states in the region of the light scalar mesons and in the $D_{S_J}^*(2317)$. Our results suggest that some states, as it is the case of the $a_0(980)$ and $f_0(600)$, could present a significant mixture of $q\bar{q}$ and tetraquark structures, but it assigns a clear tetraquark structure to the $f_0(980)$ and the $\kappa(900)$. However, more accurate calculations including the exchange terms in the variational wave function which are negligible for the heavy-light tetraquarks and the explicit coupling to $q\bar{q}$ channels should be done before drawing any definitive conclusion.

ACKNOWLEDGMENTS

This work has been partially funded by Ministerio de Ciencia y Tecnología under Contract No. BFM2001-3563 and by Junta de Castilla y León under Contract No. SA-109/01.

REFERENCES

1. E. Eichten *et al*, Phys. Rev. Lett. **34** 369 (1975).
2. N. Isgur, in *The new aspects of subnuclear physics*, (Pleenum Press, 1980) 107.
3. S. Godfrey and N. Isgur, Phys. Rev. D **32** 189 (1985).
4. J. Vijande and F. Fernández, in preparation.
5. A. de Rújula, H. Georgi, and S.L. Glashow, Phys. Rev. D **12** 147 (1975).
6. A.M. Badalian and V.L. Morgunov, Phys. Rev. D **60** 116008 (1999).
7. D.R. Entem, F. Fernández, and A. Valcarce, Phys. Rev. C **62** 034002 (2000).
8. S. Kluth. hep-ex/0309070
9. C.T.H. Davies *et al*, Phys. Rev. D **56** 2755 (1997).
10. G.S. Bali, Phys. Rep. **343** 1 (2001).
11. W. Lucha, F.F. Schöberl, and D. Gromes, Phys. Rep. **200** 127 (1991).
12. J. Sugiyama, S. Mashita, M. Ishida, and M. Oka, hep-ph/0306111.
13. M. Yan, Y. Huang, and X. Wang, hep-ph/0304188.
14. K. Hagiwara *et al*, Phys. Rev. D **66** 010001 (2002).
15. J. Vijande, F. Fernández, A. Valcarce, in Proc. of NAPP 2003 to be published in Fizaka B. hep-ph/0308318
16. S.K. Choi *et al*, Phys. Rev. Lett. **89** 102001 (2002).
17. K. Abe *et al*, Phys. Rev. Lett. **89** 142001 (2002).
18. R. Chistov, Talk given at the Flavor Physics and CP Violation Conference, Paris.
19. D. Ebert, R.N. Faustov, and V.O. Galkin, Phys. Rev. D **62** 034014 (2000).
20. E.J. Eichten and C. Quigg, Phys. Rev. D **49** 5845 (1994).
21. P. González, A. Valcarce, H. Garcilazo, and J. Vijande, Phys. Rev. D **68** 034007 (2003).
22. B. Aubert *et al*, Phys. Rev. Lett. **90** 242001 (2003).
23. D. Besson *et al*, hep-ex/0305017.
24. A. Dougall, R.D. Kenway, C.M. Maynard, and C. McNeile, hep-lat/0307001 and references therein.
25. C. Amsler, Phys. Lett. B **541** 22 (2002).
26. D. Black, A.H. Fariborz, and J. Schechter, Phys. Rev. D **61** 074001 (2000).
27. R.J. Jaffe, Phys. Rev. D **15** 267 (1977).
28. J. Vijande, F. Fernández, A. Valcarce, and B. Silvestre-Brac, in *Meson 2002*, (World Sci., 2003) 501.
29. V. Dmitrasinović, Phys. Lett. B **499** 135 (2001).

Combined analysis of the processes
$\pi\pi \to \pi\pi, K\overline{K}, \eta\eta$ and lower scalar nonets

Yu.S. Surovtsev[*], D. Krupa[†] and M. Nagy[†]

[*]*Bogoliubov Laboratory of Theoretical Physics, Joint Institute for Nuclear Research, Dubna 141
980, Russia Email: surovcev@thsun1.jinr.ru*
[†]*Institute of Physics, Slovak Academy of Sciences, Dúbravská cesta 9, 842 28 Bratislava, Slovakia*

Abstract. When simultaneously analysing experimental data on the processes $\pi\pi \to \pi\pi, K\overline{K}, \eta\eta$ in the channel with the vacuum quantum numbers, definite indications of the QCD nature of the f_0 resonances below 1.9 GeV are obtained. An assignment of the scalar mesons below 1.9 GeV to lower nonets is proposed. An approach, using only first principles (analyticity and unitarity) and the uniformizing variable, is applied.

INTRODUCTION

Already several decades a problem of scalar mesons draws permanently an attention of investigators. This is related to an important role played by these mesons in the hadronic dynamics. The f_0 mesons are most direct carriers of information about the QCD vacuum. Obviously, it is important to have a model-independent information on these states and on their QCD nature. It can be obtained only on the basis of the first principles (analyticity and unitarity) immediately applied to analysing experimental data. Earlier, we have proposed that method for 2- and 3-channel resonances [1] and applied it in the combined analysis of the processes $\pi\pi \to \pi\pi, K\overline{K}$ [2]. There were obtained a real evidence for the σ meson below 1 GeV and definite indications of the QCD nature of the f_0 states, which differ from the ones of many other authors (especially this concerns the $f_0(1370)$ and $f_0(1710)$ resonances). Therefore, a further combined analysis of the coupled processes (with adding other channels) is needed for studying scalar mesons.

THREE-COUPLED-CHANNEL FORMALISM

The processes $\pi\pi \to \pi\pi, K\overline{K}, \eta\eta$ are considered simultaneously in the 3-channel approach. The S-matrix is determined on the 8-sheeted Riemann surface. The elements $S_{\alpha\beta}$, where $\alpha, \beta = 1(\pi\pi), 2(K\overline{K}), 3(\eta\eta)$, have the right-hand cuts along the real axis of the s complex plane, starting with $4m_\pi^2$, $4m_K^2$, and $4m_\eta^2$, and the left-hand cuts. The Riemann-surface sheets are numbered according to the signs of analytic continuations of the channel momenta $k_1 = (s/4 - m_\pi^2)^{1/2}$, $k_2 = (s/4 - m_K^2)^{1/2}$, $k_1 = (s/4 - m_\eta^2)^{1/2}$ as follows: signs $(\mathrm{Im}\,k_1, \mathrm{Im}\,k_2, \mathrm{Im}\,k_3) = +++, -++, --+, +-+, +--, ---, -+ -, ++-$ correspond to sheets I, II, $\cdots$, VIII.

The resonance representations on the Riemann surface are obtained with the help of the formulae (see, ref. [1], Table 1), expressing analytic continuations of the matrix

CP717, *Hadron Spectroscopy: Tenth International Conference,*
edited by E. Klempt, H. Koch, and H. Orth

elements to unphysical sheets in terms of those on sheet I that have only zeros (beyond the real axis) corresponding to resonances. Starting from resonance zeros on sheet I, from these 3-channel formulae, we obtain 7 types of resonances corresponding to conjugate resonance zeros on sheet I of (a) S_{11}; (b) S_{22}; (c) S_{33}; (d) S_{11} and S_{22}; (e) S_{22} and S_{33}; (f) S_{11} and S_{33}; and (g) S_{11}, S_{22}, and S_{33}. For example, the arrangement of poles corresponding to a resonance of type (g) is: each sheet II, IV, and VIII contains a pair of conjugate poles at the points that are zeros on sheet I; each sheet III, V, and VII contains two pairs of conjugate poles; and sheet VI contains three pairs of poles. A resonance of every type is represented by a pair of complex-conjugate clusters (of poles and zeros on the Riemann surface). The cluster kind is related to the state nature. The resonance coupled relatively more strongly to the $\pi\pi$ channel than to the $K\overline{K}$ and $\eta\eta$ ones is described by the cluster of type (a); if the resonance is coupled more strongly to the $K\overline{K}$ and $\eta\eta$ channels than to the $\pi\pi$, it is represented by the cluster of type (e) (say, the state with dominant $s\bar{s}$ component).

We can distinguish, in a model-independent way, a bound state of colourless particles (e.g., $K\overline{K}$ molecule) and a $q\bar{q}$ bound state [1, 3].

For the simultaneous analysis of data on coupled processes we use the Le Couteur-Newton relations [4] that express the S-matrix elements of all coupled processes in terms of the Jost matrix determinant $d(k_1, k_2, k_3)$ that is the real analytic function with the only square-root branch-points at $k_i = 0$. To take into account these branch points, we must find proper uniformizing variable. However, it is impossible to map the 8-sheeted Riemann surface onto a plane with the help of a simple function. Therefore, we neglect the influence of the $\pi\pi$ threshold (however, unitarity on the $\pi\pi$ cut is taken into account). This approximation means the consideration of the nearest to the physical region semi-sheets of the Riemann surface. The uniformizing variable can be chosen as [1]

$$w = \frac{k_2 + k_3}{\sqrt{m_\eta^2 - m_K^2}}. \tag{1}$$

It maps our model of the 8-sheeted Riemann surface onto the w-plane divided into two parts by a unit circle centered at the origin. The sheets I (III), II (IV), V (VII) and VI (VIII) are mapped onto the exterior (interior) of the unit disk in the 1st, 2nd, 3rd and 4th quadrants, respectively. The physical region extends from the point w_π on the imaginary axis ($\pi\pi$ threshold, $|w_\pi| > 1$) down this axis to the point i on the unit circle ($K\overline{K}$ threshold), further along the unit circle clockwise in the 1st quadrant to point 1 on the real axis ($\eta\eta$ threshold) and then along the real axis to ∞. The type (a) resonance is represented in S_{11} by the pole on the image of the sheet II, of the sheet III, VI and VII and by zeros, symmetric to these poles with respect to the imaginary axis. Here the left-hand cuts are neglected in the Riemann-surface structure, and contributions on these cuts will be taken into account in the background.

On the w-plane, the Le Couteur–Newton relations are

$$S_{11} = \frac{d^*(-w^*)}{d(w)}, \quad S_{22} = \frac{d(-w^{-1})}{d(w)}, \quad S_{33} = \frac{d(w^{-1})}{d(w)}, \tag{2}$$

$$S_{11}S_{22} - S_{12}^2 = \frac{d^*(w^{*-1})}{d(w)}, \quad S_{11}S_{33} - S_{13}^2 = \frac{d^*(-w^{*-1})}{d(w)}.$$

The d-function is $d = d_B d_{res}$ with $d_B = \exp[-i\sum_{n=1}^{3} k_n(\alpha_n + i\beta_n)]$ describing the background, and

$$d_{res}(w) = w^{-\frac{M}{2}} \prod_{r=1}^{M} (w + w_r^*) \tag{3}$$

being the resonance part (M is the number of resonance zeros).

ANALYSIS OF EXPERIMENTAL DATA

For the $\pi\pi$-scattering, the data from the threshold to 1.89 GeV are taken from ref. [5]; below 1 GeV, from many works [2]. For $\pi\pi \to K\overline{K}$, practically all the accessible data are used [2]. The $|S_{13}|^2$ data for $\pi\pi \to \eta\eta$ from the threshold to 1.72 GeV are taken from ref. [6].

We obtain a satisfactory description: for the $\pi\pi$-scattering from ~ 0.4 GeV to 1.89 GeV ($\chi^2/\text{ndf} \approx 1.29$); for the process $\pi\pi \to K\overline{K}$ from the threshold to ~ 1.5 GeV ($\chi^2/\text{ndf} \approx 2.8$); for the $|S_{13}|^2$ data of the reaction $\pi\pi \to \eta\eta$ from the threshold to 1.5 GeV ($\chi^2/\text{ndf} \approx 0.95$). The total χ^2/ndf for all three processes is 1.95; the number of adjusted parameters is 29. The background parameters (in GeV^{-1} units) are $\alpha_1 = 1.51, \beta_1 = 0.0482, \alpha_2 = -0.93, \beta_2 = 0.139, \beta_3 = 0.89$.

Let us indicate the obtained poles of clusters for resonances on the complex energy plane $\sqrt{s}$, (in MeV units): for $f_0(600)$, type (a): $661 \pm 14 - i(595 \pm 22)$ on sheet II, $649 \pm 16 - i(595 \pm 9)$ on sheet III, $611 \pm 17 - i(595 \pm 28)$ on sheet VI, and $623 \pm 15 - i(595 \pm 26)$ on sheet VI; for $f_0(980)$: $1006 \pm 5 - i(34 \pm 8)$ on sheet II, and $962 \pm 18 - i(74 \pm 20)$ on sheet III; for $f_0(1370)$, type (b): $1384 \pm 21 - i(50 \pm 29)$ on sheet III, $1384 \pm 20 - i(78 \pm 26)$ on sheet IV, $1384 \pm 20 - i(182 \pm 23)$ on sheet V, $1384 \pm 20 - i(154 \pm 25)$ on sheet VI; for $f_0(1500)$, type (d): $1505 \pm 22 - i(320 \pm 25)$ on sheet II, $1500 \pm 30 - i(186 \pm 23)$ of the 2nd order on sheet III, $1505 \pm 20 - i(204 \pm 31)$ on sheet IV, $1505 \pm 21 - i(320 \pm 30)$ on sheet V, $1513 \pm 28 - i(318 \pm 27)$ of the 2nd order on sheet VI, and $1505 \pm 20 - i(204 \pm 35)$ on sheet VII; for $f_0(1710)$, type (c): $1699 \pm 22 - i(179 \pm 26)$ on sheet V, $1699 \pm 22 - i(168 \pm 23)$ on sheet VI, $1699 \pm 22 - i(71 \pm 19)$ on sheet VII, and $1699 \pm 26 - i(82 \pm 18)$ on sheet VIII. (To reduce the number of adjusted parameters, we supposed that simple poles of the resonance clusters arise from the simplest 3-channel Breit-Wigner form.)

Note a surprising result obtained for the $f_0(980)$ state. It turns out that this state lies slightly above the $K\overline{K}$ threshold and is described by a pole on sheet II and by a shifted pole on sheet III under the $\eta\eta$ threshold without an accompaniment of the corresponding poles on sheets VI and VII, as it was expected for standard clusters. This corresponds to the description of the $\eta\eta$ bound state.

For now, we did not calculate coupling constants in the 3-channel approach. Therefore, for subsequent conclusions, let us mention the results for coupling constants from our previous 2-channel analysis [2]: g_1 is the coupling constant with $\pi\pi$; g_2, with $K\overline{K}$. We have obtained (in GeV): for $f_0(665)$: $g_1 = 0.652 \pm 0.065$ and $g_2 = 0.724 \pm 0.1$, for $f_0(980)$: $g_1 = 0.167 \pm 0.05$ and $g_2 = 0.445 \pm 0.031$, for $f_0(1370)$: $g_1 = 0.116 \pm 0.03$ and $g_2 = 0.99 \pm 0.05$, for $f_0(1500)$: $g_1 = 0.657 \pm 0.113$ and $g_2 = 0.666 \pm 0.15$. From these values, we have concluded that the $f_0(980)$ and the $f_0(1370)$ are coupled essentially more strongly to the $K\overline{K}$ system than to the $\pi\pi$ one, $i.e.$, they have a dominant $s\overline{s}$

component. The $f_0(1500)$ has the approximately equal coupling constants with the $\pi\pi$ and $K\overline{K}$, which apparently could point up to its dominant glueball component. In the 2-channel case, $f_0(1710)$ is represented by the cluster corresponding to a state with the dominant $s\bar{s}$ component.

Our 3-channel conclusions on the basis of resonance cluster types generally confirm the ones drawn in the 2-channel analysis, besides the above surprising conclusion about the $f_0(980)$ nature.

Masses and widths of these states should be calculated from the pole positions. If we take the resonance part of amplitude as

$$T^{res} = \frac{\sqrt{s}\Gamma_{el}}{m_{res}^2 - s - i\sqrt{s}\Gamma_{tot}},$$

we obtain for masses and total widths the following values (in MeV): for $f_0(665)$, 889 and 1190; for $f_0(980)$, 1006 and 64; for $f_0(1370)$, 1386 and 156; for $f_0(1500)$, 1539 and 640; for $f_0(1710)$, 1701 and 164.

LOWER SCALAR 0^{++} NONETS

Now we can propose a following assignment of scalar mesons below 1.9 GeV to lower nonets. We exclude from this consideration the $f_0(980)$ as the $\eta\eta$ bound state. Then one can include to the lowest nonet the $f_0(600)$ as the eighth component of octet, the isodoublet $K_0^*(900)$ (or $\kappa(800)$), the isovector $a_0(980)$ and the $f_0(1370)$ as the SU(3) singlet. We consider the $K_0^*(900)$ (or κ), observed at analysing the $K-\pi$ scattering [7] and at studying the decay $D^+ \to K^-\pi^+\pi^+$ (Fermilab experiment E791) [8]. Then the Gell-Mann–Okubo formula

$$3m_{f_8}^2 = 4m_{K^*}^2 - m_{a_0}^2$$

gives $m_{f_8} = 0.87$ GeV. Our result for the σ-meson mass is $m_\sigma \approx 0.889 \pm 0.02$ GeV.

The second relation for masses of nonet, which is obtained only on basis of the quark contents of the nonet members and somehow restricts mass of the SU(3) singlet, is

$$m_\sigma + m_{f_0(1370)} = 2m_\kappa.$$

The left-hand side of this relation is $\sim 25\%$ bigger than the right-hand one if to take our mass values.

The next nonet could be formed of the $f_0(1500)$ (the eighth component of octet mixed with a glueball), the isodoublet $K_0^*(1430-1450)$, the isovector $a_0(1450)$ and the $f_0(1710)$ (SU(3) singlet). From the Gell-Mann–Okubo formula we obtain $m_{f_8} \approx 1.45$ GeV. In second formula

$$m_{f_0(1500)} + m_{f_0(1710)} = 2m_{K_0^*(1450)},$$

the left-hand side is $\sim 12\%$ bigger than the right-hand one.

Though the Gell-Mann–Okubo formula is fulfilled for both nonets rather satisfactorily, however, the breaking of the second relation (especially for the lowest nonet) tells us that the $\sigma - f_0(1370)$ and $f_0(1500) - f_0(1710)$ systems get additional contributions absent in the $K_0^*(900)$ and $K_0^*(1450)$, respectively.

CONCLUSIONS

1. In a combined model-independent analysis of data on the $\pi\pi$ scattering (from 0.4 to 1.89 GeV), processes $\pi\pi \to K\overline{K}$ (from the threshold to 1.5 GeV) and $\pi\pi \to \eta\eta$ (from the threshold to 1.5 GeV), a confirmation of the σ-meson with mass 0.889 GeV is obtained once more. This mass value rather accords with prediction ($m_\sigma = m_\rho$) on the basis of mended symmetry by Weinberg [9].

2. Consideration of the $\eta\eta$ channel is necessary for a consistent and reasonable representation of the obtained resonances.

3. The $f_0(980)$, $f_0(1370)$ and $f_0(1710)$ have the dominant $s\bar{s}$ component. Moreover, we obtain an additional indication for $f_0(980)$ to be the $\eta\eta$ bound state. Remembering a dispute [10] whether the $f_0(980)$ is narrow or not, we agree rather with the former.

4. The $f_0(1500)$ has the dominant flavour-singlet (e.g., glueball) component.

5. An assignment of the scalar mesons below 1.9 GeV to lower nonets is proposed. This assignment is not a solution of the scalar meson problem. However, it moves a number of questions and does not put the new ones. Therefore, this is probably a way to solve this problem.

ACKNOWLEDGMENTS

This work has been supported by the Grant Program of Plenipotentiary of Slovak Republic at JINR. M.N. were supported in part by the Slovak Scientific Grant Agency, Grant VEGA No. 2/3105/23; and D.K., by Grant VEGA No. 2/5085/99.

REFERENCES

1. D. Krupa, V.A. Meshcheryakov, and Yu.S. Surovtsev, *Nuovo Cim.* **A109**, 281 (1996).
2. Yu.S. Surovtsev, D. Krupa, and M. Nagy, *Eur. Phys. J.* **A15**, 409 (2002).
3. D. Morgan, and M.R. Pennington, *Phys. Rev.* **D48**, 1185 (1993).
4. K.J. Le Couteur, *Proc. Roy. Soc.* **A256**, 115 (1960); R.G. Newton, *J. Math. Phys.* **2**, 188 (1961); M. Kato, *Ann. Phys.* **31**, 130 (1965).
5. B. Hyams et al., *Nucl. Phys.* **B64**, 134 (1973); *ibid.* **B100**, 205 (1975).
6. F. Binon et al., *Nuovo Cim.* **A78**, 313 (1983).
7. S. Ishida et al., *Prog. Theor. Phys.* **98**, 621 (1997).
8. C. Gobel et al., (E791 Collaboration), hep-ex/0204018.
9. S. Weinberg, *Phys. Rev. Lett.* **65**, 1177 (1990).
10. B.S. Zou and D.V. Bugg, *Phys. Rev.* **D48**, R3948 (1993); D. Morgan and M.R. Pennington, *Phys. Rev. D48*, 5422 (1993).

Exotics

The Physics of the Newly Discovered Hadrons

D.O. Riska

Helsinki Institute of Physics, 00014 University of Helsinki, Finland

Abstract. An overview is given of the key physics issues that are associated with the recently discovered strange pentaquark $\theta(1540)^+$ and the low lying positive parity charm and charm-strange mesons as well as the narrow charmonium state with mass equal to $m(D) + m(D^*)$.

INTRODUCTION

The academic year 2002-2003 brought with it the first confirmed experimental discovery of an exotic pentaquark as well as of two low lying positive parity states in the spectra of both the charm and charm-strange mesons and finally a very narrow charmonium state with a mass very close to $m(D) + m(D^*)$. As all of these states have unexpected features they are all of exceptional theoretical interest.

THE ANTISTRANGE PENTAQUARK $\theta(1540)^+$

The state with strangeness +1, labeled $\theta(1540)^+$ was discovered by the LEPS collaboration at the SPRING 8 synchrotron [1] in the reaction $\gamma n \to K^+ Kn$. The state differs from the normal strange hyperons, which have strangeness -1. The default interpretation of the $\theta(1540)^+$ is therefore that it is an "antistrange" and therefore exotic $uudd\bar{s}$ pentaquark. The discovery has since been confirmed by the DIANA [2], CLAS [3] and SAPHIR [4] collaborations. Strange pentaquarks also admit an obvious interpretation as KN "molecular" states.

The main peculiarity of the $\theta(1540)^+$ is its remarkably narrow width. The 4 collaborations only report upper limits for the width: 25 MeV (LEPS and SAPHIR), 9 MeV (DIANA) and 21 MeV (CLAS). As the state lies 105 MeV above the threshold for KN decay it would a priori have been expected to have a width similar to those of the (non-exotic) strange hyperons in the same energy region. The width of the positive parity $\Lambda(1600)$ $1/2^+$ state is ~ 150 MeV, while that of the $\Lambda(1520)$ $3/2^-$ state is 15.6 MeV.

The $\Lambda(1405)$ $1/2^-$ state, which lies below the $\bar{K}p$ threshold, is however narrow. The low energy of this state is difficult to explain in constituent quark models, and indeed it is conventionally described as a molecular state. In the constituent quark model this state is interpreted as a flavor singlet combination of a u, a d and an s quark, which should be near degenerate with the $\Lambda(1520)$. In favor of the quark model interpretation one may invoke the charm analog of the $\Lambda(1450) - \Lambda(1520)$ strange doublet, the $\Lambda_c(2593) - \Lambda_c(2625)$, which most naturally is explained as an udc flavor singlet spin doublet.

CP717, *Hadron Spectroscopy: Tenth International Conference,*
edited by E. Klempt, H. Koch, and H. Orth
© 2004 American Institute of Physics 0-7354-0197-7/04/$22.00

Non-exotic pentaquarks is a very natural concept, and there is experimental evidence for the presence of such configurations in the proton. Another label for these are "sea-quark contributions". As an example Fermilab experiment E866 has revealed the presence and flavor asymmetry of the antiquark distributions in the proton sea $\bar{d}(x)/\bar{u}(x) > 1$ [5].

Charm pentaquarks

The original theoretical expectation was that exotic pentaquarks of the form $uudd\bar{c}$ might be stable, while antistrange pentaquarks $uudd\bar{s}$ were expected to be unstable and have large widths. Moreover on the basis of a schematic treatment of the colormagnetic hyperfine interaction, it was expected that the most stable charm pentaquark would be contain a strange quark: $uuds\bar{c}$, and have negative parity [6, 7]. An experimental search for such states in Fermilab experiment E791 revealed a few concentrated events at 2860 MeV in the $\phi\pi p$ channel, but without enough statistical significance for a claim of discovery [8].

The bound state version of the Skyrme model had in fact predicted that a spin $1/2$ state with $s = c = +1$ and positive parity should exist exist exactly at 2860 MeV [9]. In this model, which represents one version of the large N_C limit of QCD, chiral symmetry is spontaneously broken, and the key dynamics is due to the pions (the pseudo-Goldstone bosons of QCD). The model also predicts that a stable nonstrange charm pentaquark with $J = 1/2^+$ should exist below the charm strange pentaquark at ~ 2700 MeV. This state, which also appears in the chiral quark model [10], in which the hyperfine interaction is generated by pion and multipion exchange, is the charm analog of the $\theta(1540)^+$. This has been further elaborated in the direct flavor extension of the Skyrme model [11].

Parity

The search for the $\theta(1540)^+$ was driven by the theoretical prediction based on a non-topological chiral soliton model, that a narrow positive parity state with strangeness +1 might exist at 1530 MeV [12] - remarkably close to the energy of the discovered state. This energy was fixed by interpreting the state to be a member of a flavor antidecuplet, the nonstrange member of which was identified with the $N(1710)$ $1/2^+$ state, which conventionally is interpreted as non-exotic three-quark state.

A speed plot analysis of the extant data on K^+n scattering has revealed an interesting structure in the P_{01} partial wave in the region of 1.57 GeV, which is consistent with the $\theta(1540)^+$ [13]. The implication is that the parity of the $\theta(1540)^+$ is positive. Corresponding structure is seen only above 1.8 GeV in the S_{01} and D_{03} partial waves. A study of how the χ^2 of the partial wave fit to K^+n data depends on adding a narrow resonance in the interval 1520 - 1560 MeV shows that only if it occurs in the P_{01} wave is there a notable reduction of the value of χ^2, and that only if the width is exceedingly narrow (< 1 MeV) [14]. The present indications are therefore that the $\theta(1540)^+$ has positive parity.

The (tentative) positive parity of the $\theta(1540)^+$ is most readily understood in terms of the chiral quark model, in which pion and multipion exchange lead to a strong flavor and spin dependent interaction between the quarks, which can reverse the natural order of the $P-$ and $S-$states. The symmetry structure of the color-spin state of the 4 light quarks in the ground state of the $\theta(1540)^+$ is in this case $[31]_{spatial}[211]_{color}[1^4]_{space-color}$. This contrasts with the symmetry structure of the state in which all the light quarks are in the completely symmetric lowest $S-$state: $[4]_{spatial}[211]_{color}[211]_{space-color}$. This reversal of the natural ordering is also seen in the spectra of the nucleon and the $\Delta(1232)$, in which the lowest excited states, the $N(1440)$ and the $\Delta(1600)$, have positive parity and lie below the corresponding lowest states with negative parity state, the $N(1535)$ and the $\Delta(1700)$. While in these cases purely color and spin dependent interactions between the constituent quarks cannot bring about the required reversal, this is still possible in the case of the pentaquark [15].

The issue does not get clearer by the recent quenched lattice calculation of the energy of the $\theta(1540)^+$, which suggests that it has negative and not positive parity [16].

Isospin

The isospin of the $\theta(1540)^+$ is determined by its light flavor constituents $uudd$, and can be 0, 1 or 2. This may be determined by studying the decay patterns of the $\theta(1540)^+$ [17] and the reaction K^+p. A search for the θ^{++} in the latter reaction [4] saw no trace of such a particle, which should exist and be preferentially produced if the $\theta(1540)^+$ has isospin 1 or 2. This contraindicates the suggestion that the $\theta(1540)^+$ is an isotensor state with negative parity [18].

THE $D_S(2317)^+$ AND $D_S(2463)^+$

During the spring of 2003 the BABAR collaboration announced the discovery of an unexpectedly low lying positive parity narrow charm-strange meson, the $D_{s0}(2317)$ [19]. This discovery was rapidly confirmed by the CLEO [20] collaboration, which found another similar state 146 MeV above it, the $D_{s1}(2463)$. Because the $D_{s0}(2317)$ lies some 100 - 140 MeV below the prediction of most quark model calculations [23], the question arose as to whether the state is a $c\bar{s}$ state or a tetraquark.

Somewhat later the BELLE collaboration reported the discovery of the corresponding non-strange charm mesons with positive parity, the $D_0^*(2308)^0$ and the $D_1'(2427)^0$ [21]. The fact that these two states are very close in energy to the $D_{s0}(2317)$ and the $D_{s1}(2463)$ respectively, despite the presence of a strange quark in the latter, which a priori should have implied a mass difference of ~ 100 MeV between the corresponding non-strange and strange charm mesons, serves to emphasize the peculiarity of the $D_0^*(2308)^0$ and the $D_1'(2427)^0$ resonances.

TABLE 1. The $P-$ shell of the charm mesons and charmonium

Conventional notation	Heavy quark symmetry notation		$c\bar{q}$	$c\bar{s}$	$c\bar{c}$
$^3P_2$	$\frac{3}{2}$	2	$D_2^*(2460)$	$D_{s2}(2573)$	$\chi_{c2}(3556)$
$^3P_1$	$\frac{3}{2}$	1	$D_1(2420)$	$D_{s1}(2536)$	$\chi_{c1}(3510)$
$^1P_1$	$\frac{1}{2}$	1	$D_1'(2427)$	$D_{s1}(2463)$	?
$^3P_0$	$\frac{1}{2}$	0	$D_0^*(2308)$	$D_{s0}(2317)$	$\chi_{c0}(3415)$

Spin-orbit splitting

With the discovery of the two new positive parity charm and charm-strange mesons the $1P-$shells of the charm and charm strange mesons are complete. It is instructive to compare the spin-orbit splittings of these states to those of charmonium (Table 1). Because of the mass difference between the constituents of the charm-strange mesons, spin is not a good quantum number, and hence it is more natural to group the 4 $P-$states into two doublets, which are formed of the orbital angular momentum $L = 1$ and the spin of the light quarks (Table 1). The states within the doublets would be degenerate in the limit, where the mass of the heavy flavor constituent grows beyond bound, as the spin coupling of the heavy quark is $\sim 1/M$, where M is the mass of the heavy quark.

The ordering of the ^3P states of charmonium, in which the energies of the states (χ_{cJ}) increase with the value of the spin J, may be viewed as a consequence of the near balance between the spin-orbit interactions that are associated with the linear scalar confining interaction, and that associated with exchange of a single gluon:

$$V_{conf} = -\frac{c}{2M^2r}\vec{S}\cdot\vec{L}, \quad V_{gluon} = +\frac{2\alpha_s}{M^2r^3}\vec{S}\cdot\vec{L}. \tag{1}$$

Here $c \sim 1$ GeV/fm is the string tension and $\alpha_s \sim 0.3$ is the color coupling. Despite the simple form of this interaction model it matches the form of the heavy quark potential that is determined numerically by lattice methods remarkably well [22].

With the indicated conventional strengths of these interactions the gluon exchange spin-orbit interaction is the stronger one, and as a result the empirical ordering of the 3 χ_J states obtains. In a system, where one constituent is light and the other is heavy the ratio of the strengths of the scalar and vector components to the spin-orbit interaction changes, so that the relative strength of the gluon exchange spin-orbit interaction is reduced by a factor of 3:

$$V_{conf} \rightarrow_{m \ll M} -\frac{c}{4m^2r}\vec{S}\cdot\vec{L}, \quad V_{gluon} \rightarrow_{m \ll M} +\frac{\alpha_s}{3m^2r^3}\vec{S}\cdot\vec{L}. \tag{2}$$

Here m is the mass of the light constituent quark. As a consequence a reversal of the spin-orbit splitting in the charm mesons had been predicted [24]. The empirical charm and charm-strange meson spectra do not, however, bear out this out as they display the same ordering of the spin-orbit splittings as charmonium.

To obtain the empirical spin-orbit splitting of the positive parity D and D_s mesons, one may either increase the effective strength of the gluon exchange interaction by taking into account the running of the color coupling [25] or assume that the effective confining interaction is a mixture of scalar and vector coupling terms [26] in the heavy light mesons. Scalar confinement has the advantage that it leads to a simple explanation of the width of the J/ψ [27].

Structure of the positive parity D_s states

Quenched lattice calculations of hadron spectra tend to give results that are close to constituent quark model predictions. The low energy of the new positive parity D_s mesons therefore has been interpreted as indications that they are tetraquarks rather than $c\bar{s}$ states [28]. The continuum limit of a recent lattice calculation the $D_{s0}(2317)$ is barely consistent with the energy of the state that is below the threshold for strong decay [29].

There is however an intriguing way of understanding the low energy of the $D_{s0}(2317)$ and the $D_s(2463)$ as the parity partners of the corresponding ground states D_s and D_s^* [30, 31]. This is suggested by the almost equal ~ 350 MeV splitting between these two positive parity states and the corresponding ground states. If the mass of the strange quark in the charm-strange mesons were 0, these positive and negative parity states would be degenerate in energy by the chiral symmetry of QCD. The spontaneous breaking of the chiral symmetry leads to a splitting, which in the case of the heavy-light mesons should equal the value of the constituent quark mass, which indeed is of the order of ~ 350 MeV [32].

This suggestion implies that all strange-charm meson states should have parity partners. Specifically the prediction is that the other two $P-$state strange-charm mesons, the $D_{s1}(2536)$ and the $D_{s1}(2573)$ should have negative parity partners at 2721 MeV and 2758 MeV respectively [33].

THE $D_0^*(2308)^+$ AND $D_1'(2427)^+$

In contrast to the positive strange-charm mesons the two positive parity D meson states reported by the BELLE collaboration [21] have energies in good agreement with quark model predictions [25]. The empirical values for the widths of the two states $D_0^*(2308)^+$ and $D_1'(2427)^+$ are however much larger than what quark models would indicate. The chiral quark model [34] had predicted widths of the order 70 - 100 MeV, while the empirical values are $\Gamma = 276 \pm 21 \pm 18 \pm 60$ MeV and $\Gamma = 284^{+107}_{-75} \pm 24 \pm 70$ MeV respectively. As the uncertainty margin of the large widths of these two states is large it is likely that further experimental investigation may reduce the values.

The mass splittings between the $D_0^*(2308)^+$ and $D_1'(2427)^+$ and the corresponding ground states are 439 MeV and 417 MeV respectively. Given the wide uncertainty limits on the widths these splittings are sufficiently close to be viewed as equal and consistent with the parity doubling scenario [30], even though it is puzzling that these differences are larger than the corresponding splitting 350 MeV for the charm-strange mesons.

THE X(3872)

The BELLE collaboration very recently announced the discovery of an exceptionally narrow charmonium state at 3871.8±0.7±0.4 MeV [35]. This energy is clearly above the expected center of gravity of the ^3D charmonium states at ~ 3820 MeV. The energy is remarkably close to the sum of the D and D^* meson masses, which falls in the range 3871 - 3879 MeV, depending on the charge of the mesons.

The mass of the $X(3872)$ is also clearly larger than the twice the mass of the D meson, below which QCD inequalities would predict a narrow state, if the low lying positive parity D_s mesons are tetraquark states [36].

The most compelling suggestion for the nature of the $X(3872)$ is that it is a "deuson" state in the nomenclature of ref. [37], which is a narrow DD^* state that is bound by pion exchange. Because of the absence of three-pseudoscalar couplings pion exchange occurs in the DD^* system where the virtual pion is emitted and absorbed at the πDD^* vertices, but not in the DD system. Because the mass difference between the D^* and the D is very close to the pion mass, pion exchange interaction is in the case of the $X(3872)$ a very long range interaction, and hence the "deuson" will have a very spatially extended wave function [38].

The quantum number of the "deuson" is either 0^{-+} or 1^{++}. Such states would then also be expected near $2\,m(D^*)$ and $2\,m(B^*)$ [37].

OTHER NEW HADRON STATES

The overview above was restricted to the most recently discovered new hadron states, all of which have unexpected features, which call for further study. A number of other new hadrons have been reported within the last year. The SELEX collaboration has discovered the first doubly charmed hyperon - the Ξ_{cc}^+ with a mass of 3519 ± 1 MeV/c^2 [39]. The spectrum of the Ξ_{cc} states is expected to be organized in the same way as that of the Σ hyperons and the Σ_c hyperons. The flavor-symmetry of this state therefore should either be $[21]_F[21]_S[3]_{FS}$ (corresponding to the Σ and the Σ_c) or $[3]_F[3]_S[3]_{FS}$ (corresponding to the $\Sigma(1385)$) or the $\Sigma_c(2530)$). The splitting between these two states is expected to be of the order ~ 60 MeV [40].

Another intriguing narrow new state has been reported by the BES collaboration at 1859 MeV [41]. For the width an upper limit of 30 MeV has been set. This state is seen in $J/\psi \to \gamma p\bar{p}$ decays, and lies only ~ 20 MeV below the $p\bar{p}$ mass threshold. This may be the first compelling evidence for the existence of the long sought baryonium states. Several other experiments in fact also seen indications for structure, if less clear, near the $p\bar{p}$ threshold [42].

ACKNOWLEDGMENTS

I am much indebted to K. Hicks for information on the experimental work on the $\theta^+(1540)$. Research supported in part by the Academy of Finland through grant 54038.

REFERENCES

1. T. Nakano et al., Phys. Rev. Lett. **91**, 012002 (2003)
2. DIANA collaboration (V. V. Barmin et al.) hep-ex/0304040 (2003)
3. CLAS collaboration (S. Stepanyan et al.) hep-ex/0307018 (2003)
4. SAPHIR collaboration (J. Barth et al.) hep-ex/0307083 (2003)
5. R. S. Towell et al., Phys. Rev. **D64**, 052002 (2002)
6. C. Gignoux et al., Phys. Lett. **B193**, 323 (1987)
7. H. Lipkin , Phys. Lett. **B195**, 484 (1987)
8. E. Aitala et al., Phys. Rev. Lett. **81**, 44 (1998)
9. N. Scoccola et al., Phys. Lett. **B299**, 338 (1993)
10. F. Stancu, Phys. Rev. **D58**, 111501 (1998)
11. M. Praszalowicz, hep-ph/0308114 (2003)
12. D. Diakonov et al. Z. Phys. **A359**, 305 (1997)
13. N. G. Kelkar et al., J. Phys. **G29**, 1001 (2003)
14. R. A. Arndt et al., nucl-th/0308012 (2003)
15. B. K. Jennings et al., hep-ph/0308268 (2003)
16. S. Sasaki, hep-lat/0310014 (2003)
17. X. Chen et al., hep-ph/0307381 (2003)
18. S. Capstick et al., hep-ph/0307019 (2003)
19. BABAR collaboration (B. Aubert et al.), Phys. Rev. Lett. **90**, 242001 (2003)
20. CLEO collaboration hep-ex/0305100 (2003)
21. BELLE collaboration (K. Abe et al.) hep-ex/0307021 (2003)
22. G. Bali, Phys. Rept. **343**, 1 (2001)
23. S. Godfrey et al., Phys. Rev. **D43**, 1679 (1991)
24. H. Schnitzer, Phys. Lett. **B226**, 171 (1989)
25. T. A. Lähde et al., Nucl. Phys. **A674**, 141 (2000)
26. R. N. Cahn and J. D. Jackson, Phys. Rev. **D68**, 037502 (2003)
27. T. A. Lähde et al., Nucl. Phys. **...**, ... (1999)
28. G. Bali, hep-ph/0305209 (2003)
29. A. Dougall et al., hep-lat/0305209 (2003)
30. M. A. Nowak et al., Phys. Rev. **D48**, 4370 (1993)
31. W. A. Bardeen et al., Phys. Rev. **D49**, 409 (1993)
32. W. A. Bardeen, Phys. Rev. **D68**, 054024 (2003)
33. M. A. Nowak et al., hep-ph/0307302 (2003)
34. J. L. Goity et al., Phys. Rev. **D60**, 0034001 (1999)
35. BELLE collaboration(K. Abe et al.) hep-ex/0308029
36. S. Nussinov et al., hep-ph/0209095 (2002)
37. N. A. Törnqvist., Phys. Rev. Lett. **67**,556 (1992)
38. F. E. Close and P. R. Page, hep-ph/0309253 (2003)
39. SELEX collaboration (M. Mattson et al.) Phys. Rev. Lett. **89**,112001 (2002)
40. F. Coester et al., Nucl. Phys. **A634**, 335 (1998)
41. BES collaboration (J. Z. Bai et al.) Phys. Rev. Lett. **91**,022001 (2003)
42. S. L. Olsen, hep-ex/0305048 (2003)

$2\pi^0$ photoproduction from nuclei

F. Bloch, for the TAPS and A2 collaborations

Department of Physics and Astronomy University of Basel
Klingelbergstrasse 82 CH-4056 Basel (Switzerland)
E-mail: frederic.bloch@unibas.ch

Abstract. The photoproduction of π pairs ($\pi^0\pi^0$, $\pi^0\pi^\pm$) from ^{40}Ca has been measured with high statistical accuracy with the TAPS detector at the Mainz MAMI accelerator. Total cross sections and invariant mass distributions of the $\pi\pi$ pairs have been measured for incident photon energies from threshold up to 820 MeV. Evidence of a nuclear-mass dependence of the $\pi\pi$ invariant-mass distribution is found in the $\pi^0\pi^0$ channel while this dependence is not observed in the $\pi^0\pi^\pm$ channel. This indicates an in-medium modification of the $\pi\pi$ interaction in the I=J=0 channel. The total cross sections for both double pion channels agree with the elementary cross sections on the free nucleon, when the Ca cross section is scaled by $A^{2/3}$.

INTRODUCTION

The study of the in-medium properties of nucleons and mesons is one of the main challenging topics in nuclear physics. Although many changes of the fundamental properties in the nuclear medium are predicted, our knowledge of these in-medium effects is still poor. Measuring the reaction $A(\gamma,\pi^0\pi^0)$ allows to address simultaneously two different topics. The in-medium modifications of the meson-meson interaction and in-medium properties of nucleon resonances.

Correlated π pairs in the scalar-isoscalar J=I=0 channel are known as the sigma meson identified in the PDG as the $f_0(400\text{-}1200)$ [1]. Predictions of Roca et al. [2] study the meson-meson interaction in the scalar-isoscalar channel in the framework of a chiral-unitary approach at finite baryon density. The model dynamically generates the σ resonance, reproducing the meson-meson phase shifts in vacuum and accounts for the absorption of pions in the nucleus. A drop of the mass of the σ meson is also predicted for partial chiral symmetry restoration. This would occur at high densities and would induce a dropping of the σ mass down to the π mass (its chiral partner) [3]. The π, being a goldstone boson, is not expected to suffer from such mass modifications at such densities. This decrease of the σ mass should already be observable at normal nuclear densities, by measuring the σ through its 2π decay. The visible effect would there be a shift of the strength of the 2π mass distribution to smaller masses closer to the 2π production threshold with increasing baryons densities. A few years ago, the CHAOS collaboration [4] reported such an effect in pion induced reactions ($\pi^+A \to \pi^+\pi^\pm A'$ with $T_{\pi^+} = 283$ MeV). Photon induced reactions allow the study of such effects without the complications from initial state interactions which shadow the interior of the nuclei in π induced reactions. Consequently, larger effective densities can be reached. Data are presented for an incident-photon energy of $E_\gamma = 400\text{-}460$ MeV, which correspond to

CP717, Hadron Spectroscopy: Tenth International Conference,
edited by E. Klempt, H. Koch, and H. Orth

the same center-of-mass energy as was used in the pion induced experiment, enabling a direct comparison of the results. This energy range minimizes also effects of final states interactions with the medium as the outgoing π kinetic energy is such that the mean free path is maximized [5]. To study the nuclear mass dependence of the 2π mass in a different isospin channel than I=0, we measured simultaneously mass differential cross sections of the reaction $A(\gamma,\pi^0\pi^\pm)$ in the same energy range.

In-medium nucleon resonances are studied via the total $\pi\pi$ cross section. Total photoabsorption on the free proton shows a peak structure for incident-photon energies in the range 600-800 MeV. This region, called the second resonance region of the nucleon is made of three overlapping resonances ($P_{11}(1440)$, $D_{13}(1520)$, $S_{11}(1535)$). This peak structure is not seen on nuclei (from lithium up to heavy nuclei like uranium) [6]. Many explanations have been suggested to explain the suppression of the excitation strength in this region. A much discussed one is an in-medium broadening of the resonances (in particular the $D_{13}(1520)$) [7] [8]. The 2π channel, which is the predominant decay mode of the $D_{13}(1520)$, is then an obvious choice to study this region. Total cross section is measured in both $\pi^0\pi^0$ and $\pi^0\pi^\pm$ channels from threshold up to 820 MeV.

EXPERIMENTAL SETUP AND DATA ANALYSIS

The experiment was performed at the photon beam facility at MAMI. Tagged photons [9] with energies up to 820 MeV were produced via Bremsstrahlung with the MAMI electron beam [10]. Measurements were carried out using liquid hydrogen, carbon, calcium, and lead targets with thicknesses of 10 cm, 25 mm, 10 mm and 0.5 mm. Results for the carbon and lead targets have already been published in [11]. The angles and

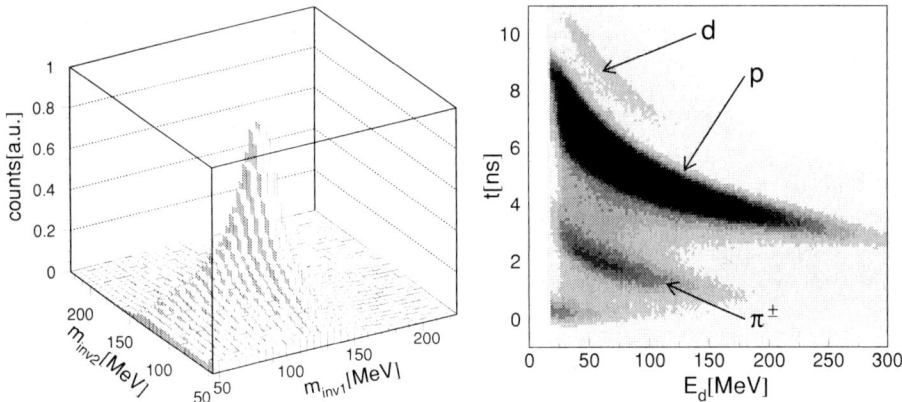

FIGURE 1. Left part: 2D plot of the two-photon invariant masses for events with four detected photons. Right part: time-of-flight versus energy deposited in the BaF$_2$. Three bands corresponding to deuterons, protons and charged π are visible.

energies of the pions were measured with the electromagnetic calorimeter TAPS [12]. In this experiment, TAPS consisted of 510 BaF$_2$ hexagonal scintillators [13]. Each of these modules has a length of 25 cm and an inner diameter of 5.9 cm. The crystals were arranged in 6 blocks containing 8×8 modules plus a rectangular forward wall of 138

BaF$_2$. A 5 mm thick plastic scintillator was placed in front of each crystal to differentiate between charged and neutral particles.

The discrimination of photons from particles was done with the hit pattern of the Veto detectors, a time-of-flight analysis and a pulse shape analysis of the BaF$_2$ signals. The π^0-mesons where then identified with a standard invariant mass analysis of coincident photon pairs (see fig.1 left). Further details are described in [14]. A time-of-flight versus energy analysis was used to separate particles with different masses, in particular in order to identify charged pions ($\pi^\pm$) (see fig.1 right and [14]). Due to the absence of a magnetic field, the two charged states, π^+ and π^-, can't be distinguished.

Above the η production threshold, the $\eta \rightarrow 3\pi$ decay is a potential background channel via events where only two of the three decay pions are detected. Most of this background can be suppressed with a missing mass analysis. The remaining background is subtracted using the η photoproduction cross section. This cross section is measured independently in the 2γ decay channel of the η with an invariant mass analysis.

Absolute cross sections are normalized using the thickness of the targets, the photon flux, detection efficiencies, geometrical acceptances and the branching ratio ($\pi^0 \rightarrow \gamma\gamma$). The geometrical acceptance and inefficiencies due to cuts and thresholds were deduced from a Monte-Carlo simulation based on GEANT3 [15] libraries and an event generator assuming a quasi-free production mechanism.

RESULTS

In fig.2 left, the measured invariant mass distributions of the π pairs are shown for incident-photon beam energy of E$_\gamma$ = 400-460 MeV. The dotted lines represent phase space distributions. In the $\pi^0\pi^0$ channel, we observe that the $\pi\pi$ mass follows the phase space distribution for carbon [11] and is shifted to a lower mass for calcium. For lead [11], despite the low statistics, a stronger shift than for calcium is seen. For the $\pi^0\pi^\pm$ channel, the mass distributions follow always phase space. This means that there is some evidence for an in-medium effect in the π-π interaction in the isospin I=0 channel that does not exist in the isospin I=1 channel. Final state interactions (FSI) of the pions are very unlikely responsible for that effect since they would also occur in the $\pi^0\pi^\pm$ channel. This finding is in line with the results from the pion induced reactions.

At higher incident photon beam energy (500-550 MeV) (see fig.2 right), no such difference can be clearly identified between the different isospin channels. This could be explained by stronger FSI which would shade effects like those seen closer to threshold (higher kinetic energy of the outgoing π results in a shorter mean free path).

The total cross section is shown in fig.3 for both channels $\pi^0\pi^0$ and $\pi^0\pi^\pm$. When scaled like A$^{2/3}$, the calcium cross section agrees with the proton and neutrons cross sections in $\pi^0\pi^\pm$ and with the deuteron cross section normalized to the mass number (which is the average of proton and neutron) in $\pi^0\pi^0$. This indicates that, above the immediate threshold region, due to strong final state interactions, only the low density surface region of the nuclei contributes to the measurement. On the surface of the nuclei, we do not observe an unexpected suppression of the resonance strength, as observed in total photoabsorption, which is not inconsistent with a possible broadening of the

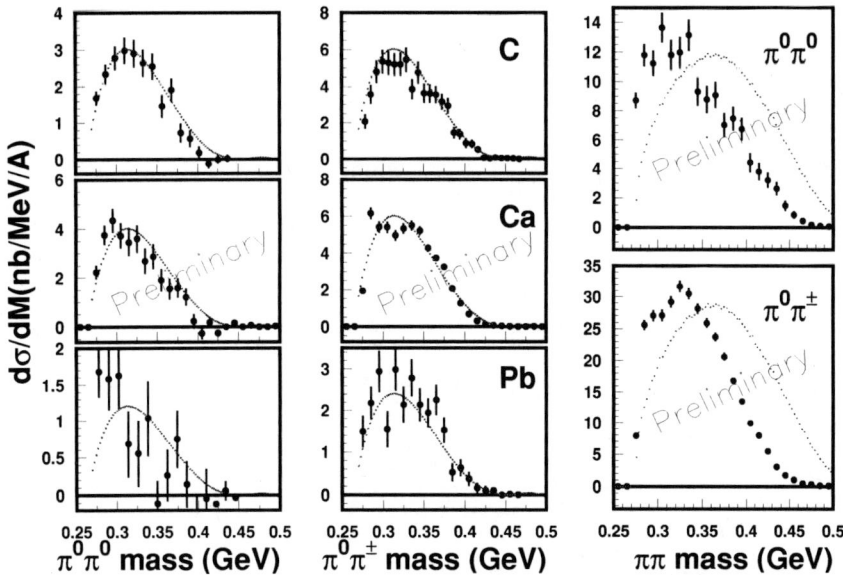

FIGURE 2. Differential cross sections of the reaction $A(\gamma, \pi^0\pi^0)$ and $A(\gamma, \pi^0\pi^\pm)$. Left part: with $A={}^{12}C, {}^{40}Ca, {}^{nat}Pb$ for incident photons in the energy range of 400-460 MeV. Right part: with $A={}^{40}Ca$ for incident photons in the energy range of 500-550 MeV. Error bars denote statistical uncertainties and the curves are phase space. The results for C and Pb are from [11].

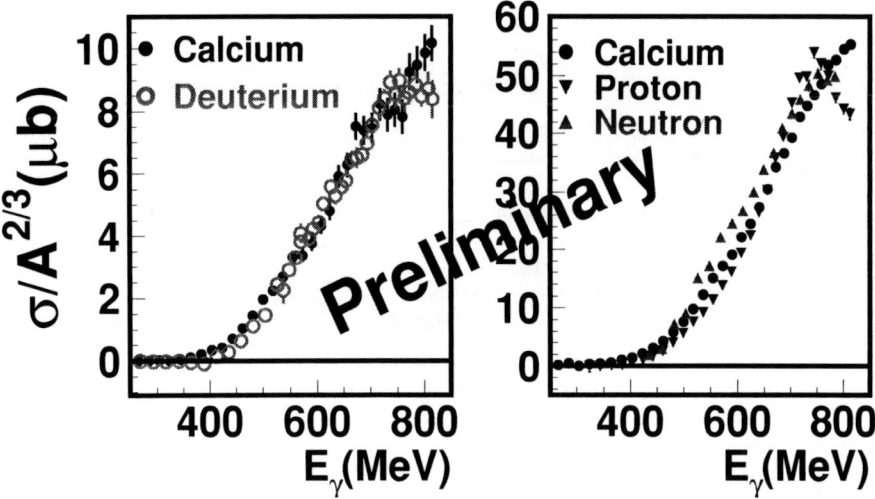

FIGURE 3. Total cross section of $\pi^0\pi^0$ (left part) and $\pi^0\pi^\pm$ (right part) photoproduction normalized to $A^{2/3}$, compared to the average nucleon cross section ($\sigma_d/2$).

375

resonances at normal nuclear density in the nuclear volume.

SUMMARY

We observed a significant in-medium effect on the $\pi\pi$-mass in the $A(\gamma,\pi^0\pi^0)$ (I=J=0) channel at E_γ=400-460 MeV. With increasing A, the strength of these distributions is shifting towards smaller invariant masses. No such effect has been observed in the $A(\gamma,\pi^0\pi^\pm)$ (I=1) channel. A large part of these modifications may be attributed to a change of the $\pi\pi$ interaction.

The total cross section scaled with the nuclear surface ($A^{2/3}$) agree with the deuteron normalized to the mass number or the proton and neutron cross section. This shows no significant in-medium effect on the surface of the nuclei contrarily to what had been observed in total photoabsorption.

ACKNOWLEDGMENTS

We thank the accelerator group of MAMI as well as many other technicians and scientists of the Institut für Kernphysik of the university of Mainz for their excellent support. We are indebted to all members of the TAPS/A2 collaborations who participated in the experiment. This work was supported by the Schweizerischer Nationalfond, the Deutsche Forschungsgemeinschaft (SFB 201), and the U.K. Engineering and Physical Sciences Research Council.

REFERENCES

1. D.E. Groom et al., Eur. Phys. J. C15 (2000) 1
2. Roca et al., nucl-th/020154
3. M. Lutz et al., Nucl. Phys. A542 (1992) 521
4. F. Bonutti et al., Nucl. Phys. A677 (2000) 213
5. W. Cassing et al., Phys. Report 188 (1990) 363
6. N. Bianchi et al., Phys. Lett. B325 (1994) 333
7. W.M. Alberico et al., Phys. Lett. B321 (1994) 177
8. L.A. Kondratyuk et al., Nucl. Phys. A579 (1994) 453
9. I. Anthony et al., NIM A301 (1991) 230
10. Th. Walcher, Prog. Part. Nucl. Phys. 24 (1990) 189
11. J.G. Messchendorp et al., Phys. Rev. Lett. 89 (2002) 222302
12. A.R. Gabler et al., Nucl. Instr. and Meth. A346 (1994) 168
13. M. Kotulla et al., JPPNP 50 (2003) 295
14. B. Krusche et al., Eur. Phys. J. A6 (1999) 309
15. R. Brun et al., GEANT, Cern/DD/ee/84-1, 1986

Evidence for Exotic Mesons

Joachim Kuhn

Carnegie Mellon University
Department of Physics
5000 Forbes Ave
Pittsburgh, PA 15217, USA

Abstract. States outside the constituent quark model have been hypothesized almost since the introduction of color. The possibility of self-interaction between gluons in QCD allows for a much richer spectrum than from $q\bar{q}$ mesons alone, namely hybrids ($q\bar{q}g$), glueballs (gg or ggg) and multiquarks ($qq\bar{q}\bar{q}$). In this context states with unusual (exotic) quantum numbers of spin, parity and charge-exchange, J^{PC}, that is combinations that cannot be achieved by a $q\bar{q}$ bound state are of particular interest. In this paper I will review evidence for several such states from experiments conducted at BNL (E852), CERN (Crystal Barrel) and IHEP (VES).

In addition I will also discuss several candidates for non-$q\bar{q}$ states with non-exotic quantum numbers.

INTRODUCTION

Normal $q\bar{q}$ mesons

In the constituent quark model mesons are bound states of a quark and an antiquark, $q\bar{q}$. For a fermion-antifermion pair with spin $\vec{S} = \vec{S}_1 + \vec{S}_2$, orbital angular momentum L, and a possible radial excitation, the total spin J, parity P and charge-conjugation C are uniquely determined:

Total spin	$\vec{J} = \vec{S} + \vec{L}$	$S = 0, 1$
Parity	$P = (-1)^{L+1}$	
Charge-conjugation	$C = (-1)^{L+S}$	

With the three lightest flavors of quarks (up, down, and strange) a total of nine states (a nonet) can be built combining a quark and an antiquark for each possible set of J^{PC}. Tab. 1 lists the possible meson states and their quantum numbers for $L \leq 3$. The first state listed in each row is the isospin $I = 1$ state, the next two are the two isoscalars and the last one denotes the four $I = \frac{1}{2}$ states that contain non-zero strangeness. It is evident that some combinations of J^{PC} are not allowed for regular $q\bar{q}$ states, namely $J^{PC} = 0^{+-}, 0^{--}, 1^{-+}, 2^{+-}, 3^{-+}, \ldots$ These combinations of J^{PC} are called *exotic* and experimental evidence of resonances with these quantum numbers (*exotic mesons*) is proof for states outside the normal $q\bar{q}$ picture of the quark model.

CP717, *Hadron Spectroscopy: Tenth International Conference*,
edited by E. Klempt, H. Koch, and H. Orth
© 2004 American Institute of Physics 0-7354-0197-7/04/$22.00

TABLE 1. Regular $q\bar{q}$ meson states for $L = 0, 1, 2$.

L	S	J^{PC}	**Mesons**
	1	4^{++}	$a_4\ f_4\ f_4'\ K_4$
$L = 3$	1	3^{++}	$a_3\ f_3\ f_3'\ K_3$
	1	2^{++}	$a_2\ f_2\ f_2'\ K_2$
	0	3^{+-}	$b_3\ h_3\ h_3'\ K_3$
	1	3^{--}	$\rho_3\ \omega_3\ \phi_3\ K_3$
$L = 2$	1	2^{--}	$\rho_2\ \omega_2\ \phi_2\ K_2$
	1	1^{--}	$\rho_1\ \omega_1\ \phi_1\ K_1$
	0	2^{-+}	$\pi_2\ \eta_2\ \eta_2'\ K_2$
	1	2^{++}	$a_2\ f_2\ f_2'\ K_2$
$L = 1$	1	1^{++}	$a_1\ f_1\ f_1'\ K_1$
	1	0^{++}	$a_0\ f_0\ f_0'\ K_0$
	0	1^{+-}	$b_1\ h_1\ h_1'\ K_1$
$L = 0$	1	1^{--}	$\rho\ \omega\ \phi\ K^*$
	0	0^{-+}	$\pi\ \eta\ \eta'\ K$

Hybrid mesons

Unlike Quantum Electrodynamics, where the photon is neutral and does therefore not (to first order) interact with charges, the gluon, the particle that transmits the strong force in Quantum Chromodynamics (QCD), necessarily carries the charge of the strong interaction, color. Therefore QCD allows the regular $q\bar{q}$ meson sector to be extended by states that carry constituent gluons in addition to the quark-antiquark pair ($q\bar{q}g^n$ - *hybrids*) or states that are entirely made out of gluons (gg^n - *glueballs*).

Regular and hybrid mesons can be described in the Flux Tube Model (FTM) [1, 2, 3], where the gluonic field is confined to a tube of flux connecting the quark-antiquark pair. An angular momentum excitation Λ of the flux tube around the $q\bar{q}$ axis gives rise to hybrid mesons, whereas the regular $q\bar{q}$ meson sector is recovered if the flux tube is in its ground state. By adding an angular momentum $\Lambda = 1$ to a $q\bar{q}$ state in a $L = 0$ configuration the following quantum numbers are possible

$$S = 0 \quad \Rightarrow \quad J^{PC} = 1^{++},\ 1^{--}$$

$$S = 1 \quad \Rightarrow \quad J^{PC} = (0, 1, 2)^{-+},\ (0, 1, 2)^{+-}$$

Note that the combinations $J^{PC} = 1^{-+}, 0^{+-}, 2^{+-}$ are *exotic* and should be clearly distinguishable from regular $q\bar{q}$ mesons. Mass predictions for these lightest hybrid states vary from $M = 1.8 - 2.0\ \text{GeV}/c^2$ [2].

In the FTM mesons decay when the flux tube breaks at any point along its length, creating a new $Q\bar{Q}$-pair from the vacuum [3]. Calculated decay ratios for regular $q\bar{q}$-mesons agree extremely well with the observed values [4]. An interesting selection rule arises for the decay of hybrids. The dominant decay of a $J^{PC} = 1^{-+}$ state should therefore be into an S- and a P-wave meson (e.g. $b_1\pi$), whereas breakup into two S wave mesons (e.g. $\eta\pi$) should be highly suppressed [3]

$$b_1(1235)\pi \,:\, f_1(1285)\pi \,:\, \rho(770)\pi \,:\, \eta'(958)\pi \,:\, \eta\pi =$$
$$170 \,:\, 60 \,:\, 5-10 \,:\, 0-10 \,:\, 0-10 \quad (1)$$

The masses of exotic mesons can also be calculated from first principles, using lattice gauge calculations [5]. In this case hybrid states are modeled as a $q\bar{q}$-pair joined by a non-straight path of color flux on a space-time lattice. [5] Calculations show that the $J^{PC} = 1^{-+}$ hybrid is the lightest one, with masses in the range of $M_H = 1800 - 2100 \text{ MeV}/c^2$ (see Tab 2). Decay properties of light hybrids are currently still beyond the possibilities of the very computing intensive calculations of the lattice gauge theories.

TABLE 2. Predicted masses for hybrid mesons from lattice gauge calculations.

Reference	M_H [MeV/c^2]
UKQCD [5]	1880 ± 200
MILC [6]	$1970 \pm 90 \pm 300$
MILC [7]	2110 ± 100
Lacock and Schilling [8]	1900 ± 200
Mei and Luo [9]	$2013 \pm 26 \pm 71$

HYBRIDS WITH EXOTIC QUANTUM NUMBERS

$\pi_1(1400)$

The first convincing evidence for a state with exotic quantum numbers was reported by the E852 collaboration from Brookhaven National Laboratory (BNL) in the reaction $\pi^- p \to \eta\pi^- p$ at $18\text{GeV}/c$ beam energy [10]. A partial wave analysis (PWA) shows a strong contribution from $a_2(1320)$ (Fig. 1a) and a smaller signal in the $J^{PC} = 1^{-+}$ wave (Fig 1b). The phase difference between these two waves (Fig. 1c) indicates interference in the region between $1.2 \text{ GeV}/c^2$ and $1.6 \text{ GeV}/c^2$. The intensities and phase difference can be fitted to two interfering Breit-Wigner resonances, resulting in mass $M = 1370 \pm 16^{+50}_{-30} \text{ MeV}/c^2$ and width $\Gamma = 385 \pm 40^{+65}_{-105} \text{ MeV}/c^2$ for the exotic state.

The result from E852 was confirmed by the Crystal Barrel experiment in antiproton-neutron annihilation to $\eta\pi^0\pi^-$ [11]. Fig.2a shows the Dalitzplot of this analysis, which could not be explained by the presence of $\rho(770)$ and $a_2(1320)$ alone. A χ^2-comparison gave agreement between the data and a fit (Fig. 2b bottom) if a $J^{PC} = 1^{-+}$ Breit-Wigner resonance with mass $M = 1400 \pm 20 \pm 20 \text{ MeV}/c^2$ and width $\Gamma = 385 \pm 50^{+50}_{-30} \text{ MeV}/c^2$ was included in the fit. Omission of this state resulted in significant structure in this comparison (Fig. 2b top).

The VES experiment analyzed the reaction $\pi^- Be \to \eta\pi^- Be$, with the η decaying into $\pi^+\pi^-\pi^\circ$ [12]. Although the P_+ wave shows a promising signal (Fig. 3b), a fit to non-resonant background or a broad resonance gave an equally satisfactory result.

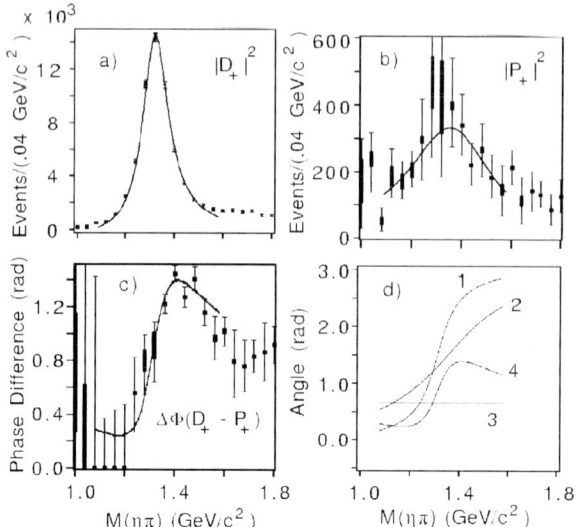

FIGURE 1. Results from E852 analysis of $\eta\pi^-$

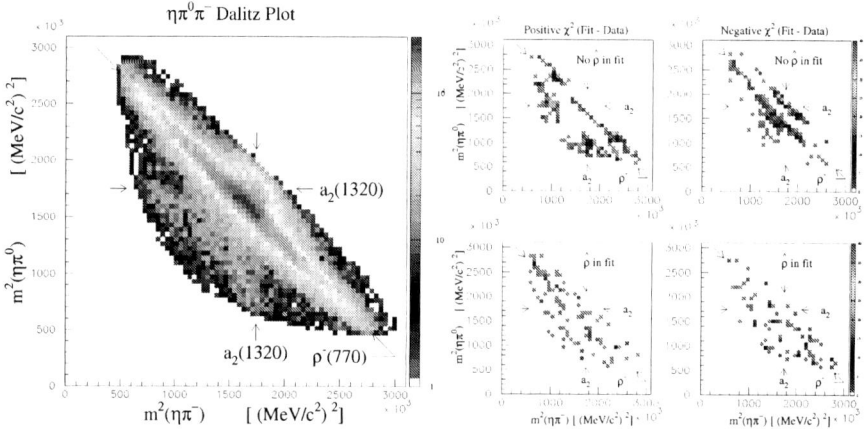

FIGURE 2. Dalitz plot for $\bar{p}d \rightarrow \eta\pi^-\pi^0 p$. The know resonances $a_2(3120)$ and $\rho(770)$ are indicated by the arrows.

$\pi_1(1600)$

A second exotic $J^{PC} = 1^{-+}$ state, the $\pi_1(1600)$ was observed in the reaction $\pi^- p \rightarrow \pi^+\pi^-\pi^- p$ [13] and later confirmed in the $\eta'(958)\pi^-$ final state [14]. In both cases a fit of the intensities and phase differences (Fig. 4) to Breit-Wigner shapes supported the existence of a resonance at 1600 MeV/c^2.

Again the VES collaboration confirmed the existence of an exotic signal in both, the

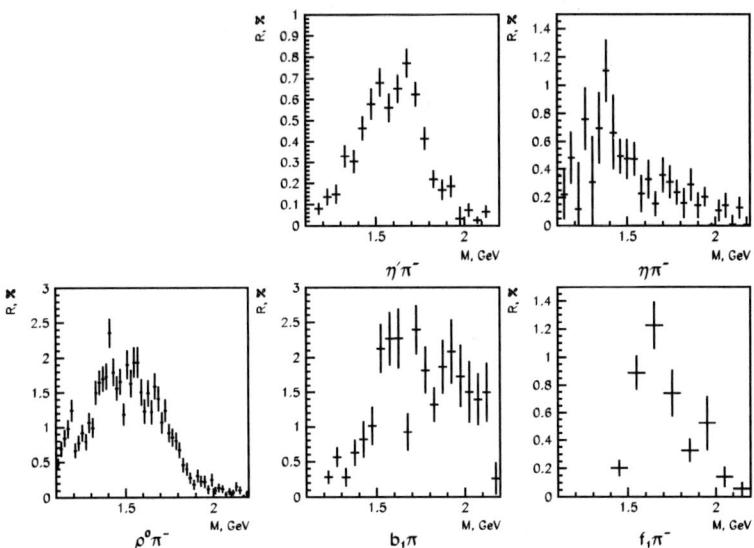

FIGURE 3. Intensities from the partial wave analysis of $\pi^- Be \rightarrow X\pi^- Be$ from the VES experiment. The final state particle X in the reaction is indicated below each figure.

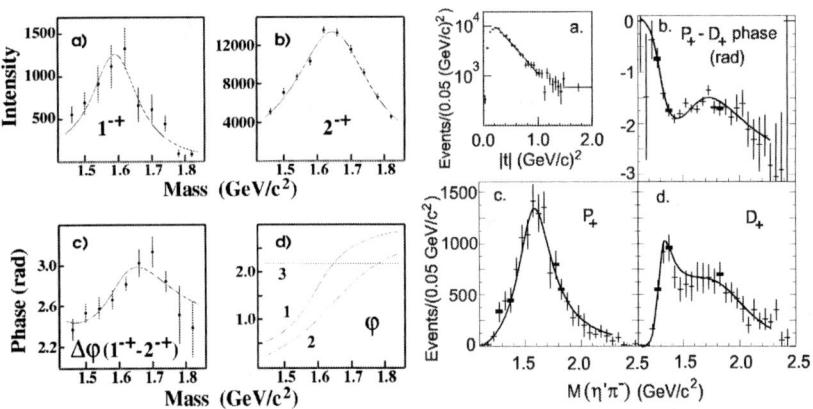

FIGURE 4. Intensities and phase differences from the PWA of $\rho\pi^-$ (left) and $\eta'\pi^-$ (right).

$\rho\pi^-$[15] and the $\eta'\pi^-$[12] final states, but was unable to claim resonant behavior. In the case of the $\rho\pi^-$ final state the small and rather broad exotic intensity (Fig. 3c) resulted in ambiguous solutions from fits to Breit-Wigner resonances. For the $\eta'\pi$ final state the phase was undetermined making it impossible to establish resonant behavior, although the shape of the intensity (Fig. 3a) indicates possible resonance dynamics.

Because of their rather complicated decay patterns the two strongest decay modes of $J^{PC} = 1^{-+}$ hybrids form the FTM predictions, $b_1(1235)\pi$ and $f_1(1285)\pi$, have only recently been analyzed by E852. The $f_1(1285)$ was detected in the $\pi^+\pi^-\eta$ decay mode from the reaction $\pi^-p \to \eta\pi^+\pi^-\pi^-p$ [16]. The three major $f_1\pi$ waves and the phase differences between them are shown in Fig. 5. The $J^{PC} = 1^{-+}$ wave shows considerable strength in the mass region from 1.5 to 1.9 GeV/c^2, indicating production of the $\pi_1(1600)$ in this channel. Again the distributions from the PWA fit (the points with the error bars) were fitted to Breit-Wigner shapes and the result is shown as the solid lines in Fig. 5, yielding mass $M = 1709 \pm 24 \pm 41$ MeV/c^2 and width $\Gamma = 403 \pm 80 \pm 115$ MeV/c^2. In addition a second pole at 2.0 GeV/c^2 was needed to describe the data. With only one pole in the exotic wave (dashed line in Fig. 5) the χ^2/DOF of the Breit-Wigner fit increased dramatically from $\frac{70.6}{47} = 1.5$ to $\frac{383.6}{46} = 8.3$.

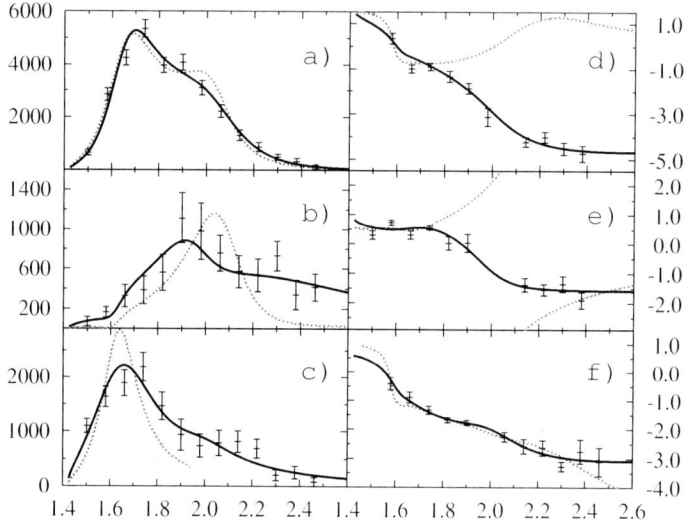

FIGURE 5. Results from the PWA (points with errorbars) and of the coupled-channel fit (solid line) of $1^{++}f_1\pi^-$ (a), $2^{-+}f_1\pi^-$ (b), and $1^{-+}f_1\pi^-$ (c) from the reaction $\pi^-p \to \eta\pi^+\pi^-\pi^-p$. The phase differences (d-f) are included in the fit. The dotted line indicates a fit with only one pole in the exotic wave.

The $b_1(1235)\pi$ final state was observed in the reaction $\pi^-p \to \omega\pi^\circ\pi^-p$ [17] and the intensity of the 1^{-+} $b_1\pi$ waves in negative and positive reflectivity from the PWA are shown in Fig. 6a and 6b, respectively. The intensities and phase differences of 4 waves were fitted simultaneously to Breit-Wigner distributions and the result of the fit is shown as the solid lines in Fig. 6[1]. Again the fit preferred a solution with two poles in the 1^{-+} wave. The masses and widths for the exotic states were found to be $M = 1664 \pm 8 \pm 4$ MeV/c^2, $\Gamma = 185 \pm 25 \pm 12$ MeV/c^2 and $M = 2000 \pm 20 \pm 10$ MeV/c^2, $\Gamma = 230 \pm 32 \pm 15$ MeV/c^2.

The VES collaboration again confirmed the existence of exotic signals in the $b_1\pi$ and

---

[1] Note that only the intensity of the $J^{PC} = 1^{-+}$ $b_1\pi$ waves are shown. The other waves from the Breit-Wigner fit were omitted here due to space limitations.

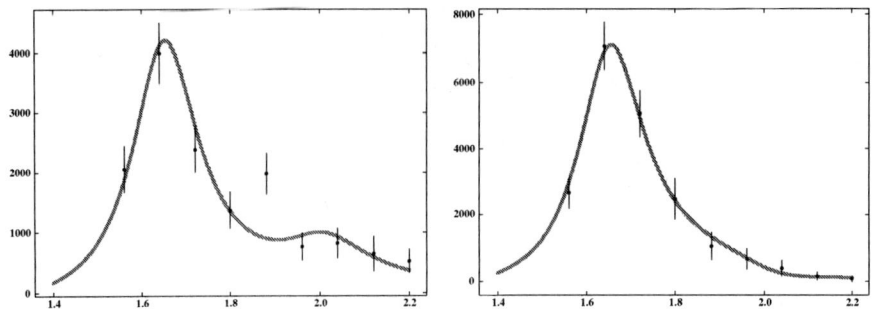

FIGURE 6. Results from the PWA (points with errorbars) and of a coupled-channel fit (solid line) of the $1^{-+} b_1 \pi$ wave in positive (left) and negative reflectivity (right).

$f_1 \pi$ waves in their analyses of $\omega \pi^\circ \pi^-$ and $\eta \pi^+ \pi^- \pi^-$, respectively (see Fig. 3d and 3e). In $f_1 \pi$ a Breit-Wigner fit was again unsuccessful due to model dependence of the phase of the reference wave [18]. The peak in the $b_1 \pi$ can be described by a Breit-Wigner resonance with a background term, but results are too preliminary to quote values for mass and width [19].

HYBRIDS WITH NON-EXOTIC QUANTUM NUMBERS

Hybrids with non-exotic quantum numbers are more difficult to detect since they look just like regular $q\bar{q}$ mesons. The main tools to distinguish them are therefore by observing an overpopulation of states with the same quantum numbers or by their decay properties. As quantum mechanical states these hybrids may however also mix significantly with regular meson states with the same quantum numbers and it may be very difficult to disentangle the spectrum.

$$J^{PC} = 0^{-+}$$

The $\pi(1800)$ is believed to be a candidate for a state with a large gluonic content, since it does not seem to decay to $\rho\pi$ and K^*K as expected for a radial excitation of the pion. Instead it has rather significant coupling to scaler mesons: $\sigma\pi$, $f_0(980)\pi$, $f_0(1500)\pi$ and $a_0(980)\eta$.

Recent observed decay modes of the $\pi(1800)$ from E852 include the $a_0(980)\eta$ and $f_0(1500)\pi$ from the analysis of the reaction $\pi^- p \to \eta\eta\pi^- p$ [20] and the $\omega\rho$ from $\pi^- p \to \omega\pi^\circ\pi^- p$ [17].

$$J^{PC} = 2^{-+}$$

There is an abundance of states in the $J^{PC} = 2^{-+}$ sector. While the $\pi_2(1670)$ and $\eta_2(1675)$ seem to be consistent with D-wave mesons, other observed states cannot be placed easily into the same nonet. The $\eta_2(1860)$ with its mass would be consistent with the singlet state of this nonet, but its large decay modes to $a_2\pi$ and $f_2\pi$ gives evidence that this state may be a hybrid [21]. An overpopulation of states is also observed in isovector channel. The strong $a_2\eta$ decay of the $\pi_2(1880)$ observed in $\eta\eta\pi^\circ$ makes it an ideal isospin partner for the $\eta_2(1860)$ [22].

The $\pi_2(1880)$ has also been observed by E852 in the $a_2\eta$ decay channel of the reaction $\pi^- p \to \eta\eta\pi^- p$ [20]. In addition it has also been seen in the $\eta\pi^+\pi^-\pi^-$ final state in the subsets $f_1\pi$ and $a_2\eta$ [16] (see also Fig. 7) and in the $\omega\pi^\circ\pi^-$ final state decaying to $b_1\pi$ [17] (Fig. 8).

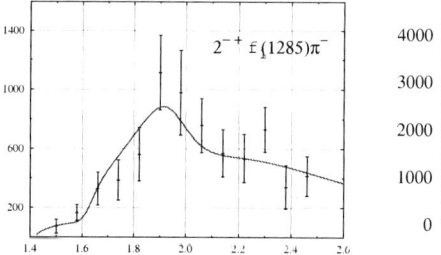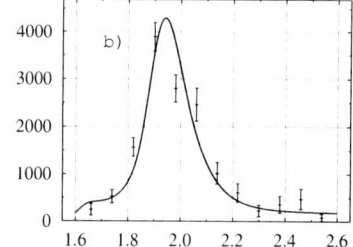

FIGURE 7. $\pi_2(1900)$ in the $f_1\pi$ (left) and $a_2\eta$ (right) decay mode from the E852 analysis of $\eta\pi^+\pi^-\pi^-$.

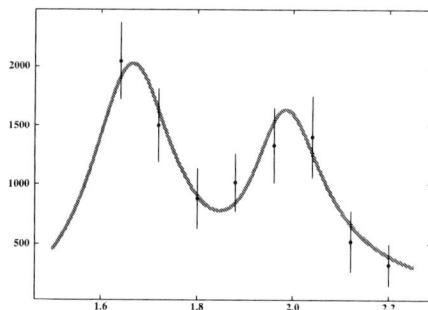

FIGURE 8. $\pi_2(1670)$ and $\pi_2(1900)$ in the $b_1\pi$ decay mode from the E852 analysis of $\omega\pi^\circ\pi^-$.

$$J^{PC} = 1^{++}$$

A slight overpopulation also exists in the isovector channel of $J^{PC} = 1^{++}$. The $a_1(1700)$ seems to be consistent with the first radial excitation of the $a_1(1230)$, making the second state at 2100 MeV/c^2, observed in decays into $f_1\pi$ and $a_1\eta$, a candidate for a hybrid meson [16] (Fig. 9).

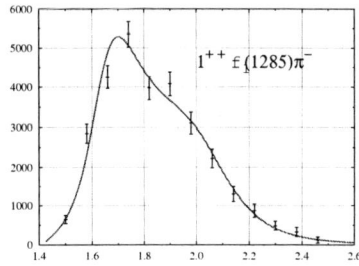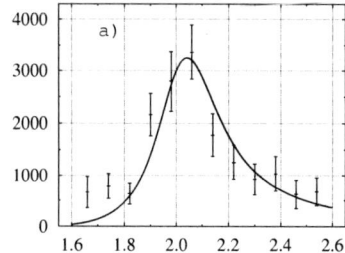

FIGURE 9. $a_1(2100)$ in the $a_1\eta$ (left) and $f_1\pi$ (right) decay mode from the E852 analysis of $\eta\pi^+\pi^-\pi^-$.

SUMMARY

Several hybrid mesons candidates exist now with both, exotic and non-exotic quantum numbers. Three states with the exotic combination $J^{PC} = 1^{-+}$ have been observed at 1400 MeV/c^2, 1600 MeV/c^2 and 2000 MeV/c^2. The interpretation as hybrids of the first two is problematic since the masses and some of its decays are inconsistent with predictions from the flux tube model and lattice gauge calculations. The $\pi_1(2000)$ has only been observed in the predicted strong decay channels and also agrees with the predictions for the mass, making it a very strong candidate for an exotic hybrid. Searching for other members of the 1^{-+} nonet, as well as members of other exotic nonets should provide clarification about the nature of these states.

Hybrid states with non-exotic quantum numbers are difficult to detect since they are entangled with the regular $q\bar{q}$ meson sector. Due to the observed overpopulation and the unusual decay properties of several observed states it may be possible to find evidence for hybrids also in these channels. The strongest candidates are the $\pi(1800)$, $\eta_2(1860)$, $\pi_2(1900)$ and $a_1(2100)$, which are now observed in numerous decay modes. Separating these states from regular mesons is difficult due to the interference effects (mixing) between states of equal quantum numbers.

ACKNOWLEDGMENTS

This research work was supported in part by the US Department of Energy, the US National Science Foundation and the Russian State Committee for Science and Technology. The authors also would like to thank the members of the Brookhaven MPS staff for their support in running the experiment and the members of the E852 collaboration for their efforts and contributions to this work.

REFERENCES

1. Isgur, N., and Paton, J., *Phys. Rev. D*, **31**, 2910 (1985).
2. Barnes, T., et al., *Phys. Rev. D*, **52**, 5242 (1995).
3. Close, F. E., and Page, P. R., *Nucl. Phys. B*, **443**, 233 (1995).
4. Kokoski, R., and Isgur, M., *Phys. Rev. D*, **35**, 907 (1987).
5. Lacock, P., et al., *Phys. Lett. B*, **401**, 309 (1997).
6. Bernard, C., et al., *Phys. Rev. D*, **56**, 7039 (1997).
7. Bernard, C., et al., *Nucl. Phys. B (Proc. Suppl.)*, **73**, 264 (1999).
8. Lacock, P., and Schilling, K., *Nucl. Phys. B (Proc. Suppl.)*, **73**, 261 (1999).
9. Mei, Z., and Luo, X., *hep-lat/0206012* (2002).
10. Thompson, D. R., et al., *Phys. Rev. Lett.*, **79**, 1630 (1997).
11. Abele, A., et al., *Phys. Lett. B*, **423**, 175 (1998).
12. Beladidze, G., et al., *Phys. Lett. B*, **313**, 276 (1993).
13. Adams, G. S., et al., *Phys. Rev. Lett.*, **81**, 5760 (1998).
14. Ivanov, E. I., et al., *Phys. Rev. Lett.*, **86**, 3977 (2001).
15. Amelin, D. V., et al., *Phys. Lett. B*, **356**, 595 (1995).
16. Kuhn, J., et al., *to be submitted to Phys. Lett. B* (2003).
17. Lu, M., Ph.D. thesis, Rensselaer Polytechnic Institute (2003).
18. Ryabchikov, D., et al., *Proceedings HADRON95 Manchester* (1995).
19. Amelin, D. V., et al., *Phys. of At. Nucl.*, **62**, 445 (1999).
20. Eugenio, P., et al., *Proceedings Gluonic Workshop, Jefferson Lab* (2003).
21. Anisovich, A. V., et al., *Phys. Lett. B*, **477**, 19 (2000).
22. Anisovich, A. V., et al., *Phys. Lett. B*, **500**, 222 (2001).

HERA-g, A NEW EXPERIMENT FOR GLUEBALL, HYBRID AND ODDERON STUDIES AT DESY

Corresponding Authors: M.Bruschi[*] and P.Schlein[†]

[*]I.N.F.N Sezione di Bologna, Bologna, Italy
e-mail address:bruschi@mail.desy.de
[†]University of California, Los Angeles, CA 90024, USA
e-mail address:schlein@ucla.edu

Abstract.
We propose a new, but relatively short, experimental program at HERA to use the existing HERA-B detector to run in the 920 GeV proton beam to study the production and decay properties of centrally-produced glueballs and hybrid mesons (the latter produced in P omeron-R eggeon collisions). A search for o dderon-Exchange will also be carried out by measuring isolated centrally-produced I=0, C=-1 states such as ω^0. A Level-1 trigger based on rapidity-gap vetoes at small and large angles outside the spectrometer aperture will efficiently select these events. We show the properties of such events extracted offline from $> 70 \cdot 10^6$ triggered events, corresponding to about 5 minutes of data taking with such a rapidity-gap trigger. For example, a 100 hour data-taking run in the manner described herein will already yield a factor of 1000 times more data than displayed in this Proposal. Such a data sample would allow fundamental contributions to be made to a number of important fields.The 100-hour run corresponds to about a factor 10 increase on WA-102 experiment data, for many channels.

Physicists and Institutes interested in participating in this project are asked to contact at the earliest possible time the corresponding authors.

The talk given by Peter Schlein on the plenary session of the HADRON 2003 conference was a summary of this new proposal submitted to the DESY PRC on 2 October 2003.

The full proposal, with the present list of authors, is accessible to all at the following web-site address:

http://www-hera-b.desy.de/subgroup/physics/glueball/doc/proposal.ps.

CP717, *Hadron Spectroscopy: Tenth International Conference*,
edited by E. Klempt, H. Koch, and H. Orth
© 2004 American Institute of Physics 0-7354-0197-7/04/$22.00

Exotic States in Crystal Barrel Analyses of Annihilation Channels

W. Dünnweber and F. Meyer-Wildhagen[1]

Sektion Physik, Universität München, D-85748 Garching

Abstract. Resonances with the exotic quantum numbers $J^{PC} = 1^{-+}$ are observed in $\rho\pi$ at m = 1400 - 1480 MeV/c^2 and in $b_1\pi$ at 1600 MeV/c^2. Comparison of the S- and P-wave $\bar{p}n$ branchings leads to the conclusion that the present $\rho\pi$ resonance is different from the exotic $\eta\pi$ resonance found in previous annihilation studies. Excited a_1 and π states are observed in the $\rho\pi, \rho'\pi$ and $\rho\omega$ systems. Further evidence for the 2-component character of the $E/\iota(1440)$ is found in its $\sigma\eta$ decay. A study of $\omega\pi$ decay of excited ρ states supports the existence of a ρ' about 250 MeV/c^2 below the $\rho'(1450)$.

INTRODUCTION

Recent analyses of Crystal Barrel data collected at LEAR focussed on π_1 resonances, carrying the exotic quantum numbers $J^{PC} = 1^{-+}$, and on supernumerary resonances in the mass region of radially excited mesons with $q\bar{q}$ quantum numbers. The clearest example, to date, of a π_1 produced in annihilation has been the $\eta\pi$ P-wave resonance at 1400 MeV/c^2 emerging in analyses of $\bar{p}n \rightarrow \eta\pi^-\pi^0$ (Refs. [1, 2, 3]). By comparison to the much weaker relative π_1 strength in $\bar{p}p \rightarrow \eta\pi^0\pi^0$ [4] , where the initial state is different, it became clear that a strong dynamic selectivity is at work. New results on exotic $\rho\pi$ and $b_1\pi$ resonances corroborate this finding.

Exotic intruders into the $q\bar{q}$ level scheme are difficult to recognize. Masses are shifted by configuration mixing. Level counting is not trivial, since assignments of peaks in different decay channels to a resonance may be confused by phase-space and dynamic effects leading to different peak positions. However, the decay branchings into modes that are characteristic of the model configurations may help to disentangle the level scheme. Examples are the decays of $qg\bar{q}$ hybrids into a P-wave and an S-wave meson [5, 6] or of molecular states into their constituent mesons [7, 8]. For radially excited $q\bar{q}$ states, specific decay modes should disappear as a result of the radial node of the wave function [9]. Analyses of Crystal Barrel data yield new decay branchings into $\rho\pi, \omega\rho, \omega\pi, \sigma\eta$ and K^+K^- for states in the m = 1300 - 1700 MeV/c^2 region where exotics compete with $q\bar{q}$ radial excitations.

---

[1] on behalf of the Crystal Barrel Collaboration, preliminary results

CP717, *Hadron Spectroscopy: Tenth International Conference,*
edited by E. Klempt, H. Koch, and H. Orth
© 2004 American Institute of Physics 0-7354-0197-7/04/$22.00

EXOTIC π_1 RESONANCES OF $\rho\pi$ AND $b_1\pi$

The annihilation reaction $\bar{p}d$ (at rest) $\to \pi^-3\pi^0 + p$ was studied with high statistics. About 200 000 fully reconstructed $\pi^-3\pi^0$ events are obtained when a cut of 100 MeV/c is imposed on missing momentum. This cut is sufficient to guaranty spectator status of the proton (see [3]). The channel is dominated by $\rho^-(\to \pi^-\pi^0)\pi^0\pi^0$. Although we are dealing with a 4-body channel, i.e. with 5 independent kinematic variables, the main features of the intermediate state are visible in a pseudo-Dalitz plot obtained by gating on ρ^- in the $\pi^-\pi^0$ invariant mass (Fig. 1, upper part). The prominent diagonal band is mainly from $\rho^-xf_2(1270)$ where f_2 decays to $\pi^0\pi^0$. Running parallel is the interference dip from $f_0(980) \to \pi^0\pi^0$ as part of the $\pi\pi$ s-wave (called σ below). The prominent vertical and horizontal structures are conglomerates of a few $\rho^-\pi$ resonances, predominantly a_1, π_2 and π_1. Partly these structures arise from π_0 symmetrization of the $\rho^-(\pi^-\pi^0)$ x $\sigma(\pi^0\pi^0)$ contribution.

A true representation of only measured quantities is the Goldhaber plot shown in the lower part of Fig. 1. The ρ^- band is a projection of the intermediate states discussed above. There is also $\pi^-\pi^0$ structure at 1200 MeV/c^2 indicating ρ^- excitation. Both representations of the data demonstrate the high selectivity of $\bar{p}n \to \pi^-3\pi^0$ which is an isospin 1 channel.

The partial-wave analysis assumes as intermediate configurations
 A) $(\pi^-\pi^0)$ x $(\pi^0\pi^0)$
 B) $((\pi^-\pi^0)\pi^0)$ x π^0 and charge permutations.
It is facilitated by conservation laws allowing only
 A) ρ x σ and ρ x f_2,
 B) a_1 x π^0, a_2 x π^0, π' x π^0, π_2 x π^0 and π_1 x π^0,
including possible excitations of these resonances and charge permutations. Allowed $\bar{p}n$ initial states are $^3S_1$ and $^1P_1$ with fitted contributions of 90% and 10% , respectively. Relativistic Breit-Wigner resonance amplitudes with mass dependent width and ℓ-dependent barrier penetration are used. The $\pi\pi$ s-wave is parametrized in the K-matrix formalism. A good description of the full experimental intensity distribution is achieved (see Fig. 1 and the corresponding invariant mass spectra in [10]).

The significance of the exotic π_1 contribution can be judged from a mass scan of the ℓn Likelihood of the fit to the data (Fig. 2). Clear evidence is obtained for a $\rho\pi$ resonance with the exotic quantum numbers $J^{PC} = 1^{-+}$. Modifications of the model space, concerning the resonance parameters of the excited ρ, π and a_1 states above 1200 MeV/c^2 do not obscure this evidence but lead to variations of the π_1 resonance parameters in the ranges $1400 < m$ (MeV) < 1480 and $250 < \Gamma(MeV) < 500$, as exemplified in Fig. 2. The indication of a second π_1 peak at 1600 MeV/c^2 depends on the choice of excited ρ and a_1 states. The relative $^1P_1$ and $^3S_1$ contributions of the observed π_1 intensity show a surprising selectivity. The $^1P_1$ contribution accounts for 5% of the total $\pi^-3\pi^0$ intensity, whereas the $^3S_1$ contribution is hardly significant ($< 1\%$).

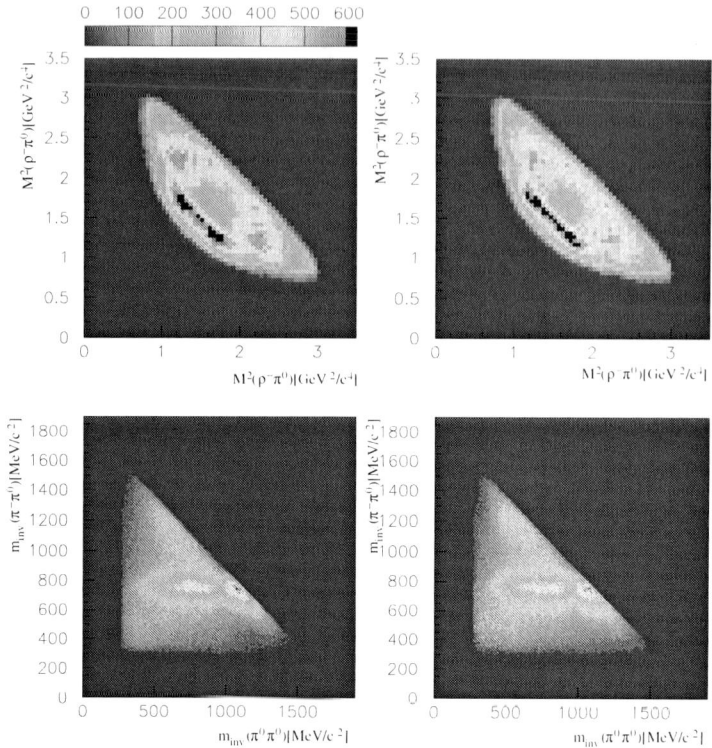

FIGURE 1. Intensity distributions of $\bar{p}n \rightarrow \pi^-\pi^0\pi^0\pi^0$ projected in two planes of different invariant mass combinations. The left and right hand sides show data and fit, respectively.

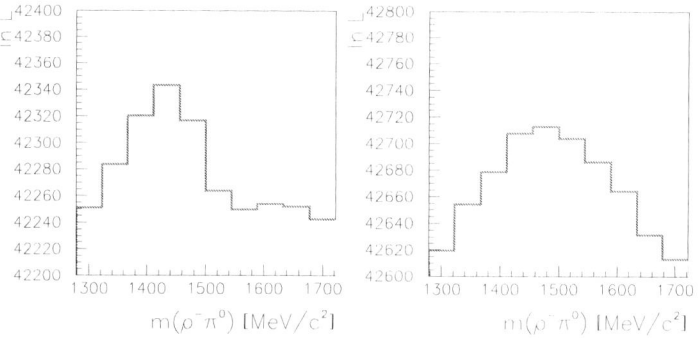

FIGURE 2. Mass scans for an exotic $\rho^-\pi^0$ resonance with quantum numbers $J^{PC} = 1^{-+}$. The fit models incorporate all allowed resonances (see text) with PDG parameters (left hand side) or with the masses of ρ' and a'_1 replaced by optimized values of 1280 and 1550 MeV/c^2, respectively (right hand side).

A recent analysis of Crystal Barrel data for $\bar{p}p \to \omega\pi^+\pi^-\pi^0$ [11] provides evidence for an exotic $\pi_1(1600)$ decaying to $b_1\pi$. As compared to a fit omitting this resonance, the gain in ℓn Likelihood is 271 for 7 parameters allowing for $\ell = 0,2$ decay and $^1S_0, ^3S_1$ initial $\bar{p}p$ states. According to experience with established resonances this change in Likelihood is sufficient to provide evidence for the $\pi_1(1600)$. The resonance parameters (m = 1590 ± 50, Γ = 280 ± 75 MeV/c^2) are in agreement with those from π-induced reactions [12]. A rather critical discussion of the evidence from those studies left the $b_1\pi$ resonance as the least contestable one, in contrast to $\eta'\pi$ and $\rho\pi$ [13]. There is no evidence for an additional π_1 around m = 1400 MeV/c^2 in the $b_1\pi$ system produced in annihilation.

The $\pi_1(1400)$ observed in $\eta\pi$ [3] is dominantly produced in the $\bar{p}n$ $^3S_1$ wave and, hence, cannot be identical to the m = 1400 - 1480 MeV/c^2 π_1 resonance in $\rho\pi$ which is mainly produced in the $^1P_1$ wave. For each of the two contributing $\bar{p}n$ states, X = $^3S_1$ or $^1P_1$, the relative Branching Ratio

$$R = \frac{BR(X \to \pi_1(\to \rho\pi)\pi)}{BR(X \to \pi_1(\to \eta\pi)\pi)}$$

must have the same value when π_1 is the same in both decay channels. Instead we find
$$R = 24.5 \pm 4 \text{ for } X = {}^1P_1,$$
$$R = 2 \pm 2 \text{ for } X = {}^3S_1.$$
These values were corrected for interferences due to exchange with the external π. Rescattering, e.g. from an intermediate $\sigma(\to \pi\pi)\rho$ state, estimated by using as an approximation [14] a smooth nonresonant term in the fit, modifies R by up to 30%, which would not resolve the above disagreement. To check this reasoning, the corresponding R values were also extracted for the $a_2(1320)$ decaying into $\rho\pi$ and $\eta\pi$ as well. In this case consistent values of R $\approx$ 5, which also agree with the known decay branching [12], are obtained for both X = $^3S_1$ and $^1P_1$.

It is concluded that two π_1 resonances are present in this mass region, one decaying preferentially into $\rho\pi$, the other into $\eta\pi$. The former one is probably different from the exotic $\rho\pi$ resonance candidate at m = 1600 MeV/c^2 seen in diffractive processes [12, 15]. In those studies the mass region below 1500 MeV/c^2 was not accessible due to background from a_1 feed-through [15]. It is noted that the $q\bar{q}q\bar{q}$ wave functions of $\rho\pi$ and $\eta\pi$ have different flavour symmetries (see [16, 17]).

RADIAL EXCITATIONS OF a_1, π, η, ρ AND EXOTIC INTRUDERS

A strong $a_1(J^{PC} = 1^{++})$ signal is observed in the $\pi^-3\pi^0$ data set introduced above. It accounts for 11% of the total intensity and shows up in the $\rho\pi$ and $\rho'\pi$ spectra and, with less significance, also in $\sigma\pi^-$ and $f_2\pi^-$. A mass scan (see [10]) yields m = 1550 MeV/c^2 which is hardly compatible with the $a_1(1700)$ found in diffractive production [18]. As a fingerprint of the first radial excitation of the $1^{++}q\bar{q}$ system, $\rho\pi$ decay with $\ell = 0$ is expected in the $^3P_0$ model calculations [9] to be strongly suppressed in comparison to $\ell = 2$. This is in agreement with the results for the VES resonance [18]. In contrast, $\Gamma(\ell = 0)/\Gamma(\ell = 2) \approx 4$ is found for the present $a_1(1550)$, which makes it a strong candidate for a non-$q\bar{q}$ 1^{++} state.

There are indications of two π states around m = 1800 MeV/c^2 in diffractive processes (see [19]). In $\bar{p}p$ annihilation into $\omega\pi^+\pi^-\pi^0$ a π state was observed at this mass, decaying into $\rho\omega$ [11]. In the $\bar{p}n \to \pi^- 3\pi^0$ data set, $\rho\pi$, $\rho'\pi$ and $\sigma\pi$ show $\pi(1800)$ signals of similar strength. Because of the vicinity to the phase space boundary it is difficult to decide whether the same resonance is dominant in these channels or whether splitting occurs.

The "E/i" is an old enigma of meson spectroscopy [12]. In recent studies [20] of $\bar{p}p \to \pi^+\pi^-\pi^+\pi^-\eta$ Crystal Barrel data, a 0^{-+} isoscalar resonance appears at 1405 MeV/c^2 in the systems $a_0(\to \pi\eta)\pi$ and $\sigma\eta$. Some evidence is also obtained for its radiative decay: $\Gamma(\pi^+\pi^-\gamma)/\Gamma(\pi^+\pi^-\eta) = 0.11 \pm 0.06$. Mass scans of an additional η show no signal in the region of the $\eta(1295)$ but give evidence for structure at 1490 MeV/c^2 (Fig. 3). This structure is consistent with the one seen so far only in K^*K[12]. The partial width for its decay into $\sigma\eta$ is about a factor of 4 larger than for $a_0\pi$. It is of high interest whether the E/i splitting can be reconciled with a single resonance, as a result of a radial node or of $K\bar{K}\pi$ rescattering or of interference with background, or whether a true dublet is present.

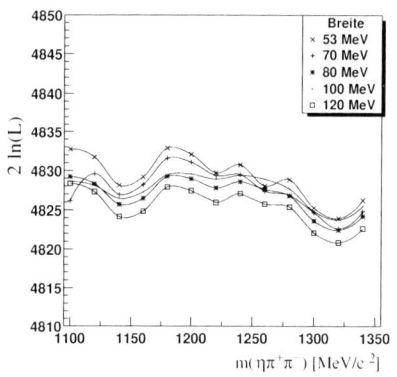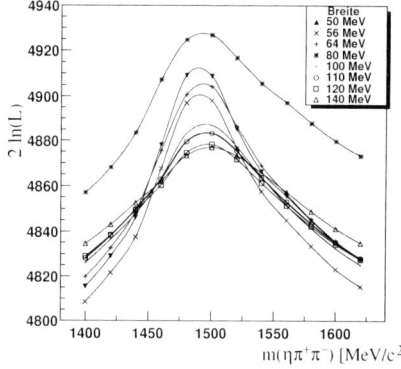

FIGURE 3. Mass scans of an η state with different given widths for $\bar{p}p$ annihilation at rest into $\pi^+\pi^-\pi^+\pi^-\eta$ [20], using a fit model that includes the $\eta(1405)$ with $\Gamma = 57$ MeV/c^2. An additional η appears at 1490 MeV/c^2 (right hand side). An $\eta(1295)$ is not observed (left hand side).

The number of ρ' states below 1.5 GeV/c^2 is debated since a long time [21]. New results on ρ' decays into $\omega\pi$ and $K\bar{K}$, observed in $\bar{p}n$ data [22], are in reasonable agreement with the 1 $^3D_1$ quark model assignment to the $\rho'(1700)$. If the experimental ρ' signals below 1.5 GeV/c^2 are ascribed to a single resonance, the branching ratios [22] would read $\Gamma(\pi\pi) : \Gamma(K\bar{K}) : \Gamma(4\pi) = 55 : 70 : 55 : 145$ (MeV/c^2), which is in dramatic disagreement with the $^3P_0$ quark model calculations for the first radial ($2\,^3S_1$). However, there is a strong indication in the $\omega\pi$ channel of a $\rho'(1200)$ (Fig.4), which is significantly below the $\rho'(1450)$ dominating in 4π. Close inspection of the phase motion relative to the $\rho'(1700)$ supports this evidence for two ρ' states below 1.5 GeV/c^2.

The quark model suggests to identify the $\rho'(1200)$ with a dominant $q\bar{q}$ configuration $2\,^3S_1$, since its mass is below the $K^*(1410)$, which is established as the strange member of the same nonet, and since $\Gamma(4\pi) \ll \Gamma(\omega\pi)$. This leaves the $\rho'(1450)$ as an intruder with dominantly exotic configuration.

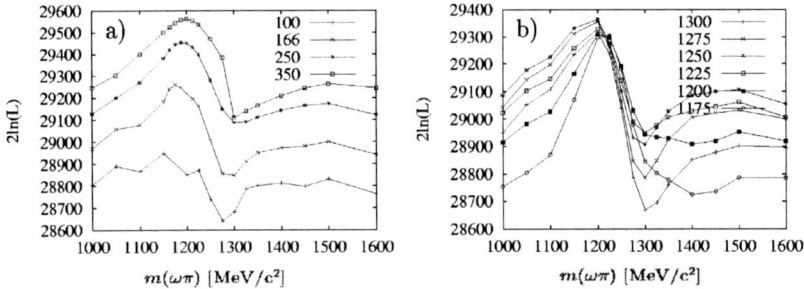

FIGURE 4. Scans for a second ρ' state in the $\bar{p}n \rightarrow \omega\pi^-\pi^0$ data [22] using a fit model that includes the additional channels $\rho^-(770)\omega, b_1(1232)\pi$ and $\rho(1700)\pi$. A resonance around 1200 MeV/c^2 is significant for different given values of the assumed ρ' width (a) and of the b_1 mass (b).

REFERENCES

1. W. Roethel, Diploma thesis (Munich 1995);
   K. Hüttmann, Diploma thesis (Munich 1997);
   K. Hüttmann and W. Dünnweber, Crystal Barrel Report No. 312 (1997)
2. W. Dünnweber, K. Hüttmann and W. Roethel, Proc. HADRON 97, p. 309
3. A. Abele et al., Crystal Barrel Collaboration, Phys. Lett. B**423** (1998) 175
4. A. Abele et al., Crystal Barrel Collaboration, Phys. Lett. B**446** (1999) 349
5. F.E. Close and P.R. Page, Nucl. Phys. B**443** (1995) 233;
   Phys. Rev. D**56** (1997) 1584
6. F. Iddir et al, Phys. Lett. B**205** (1988) 564
7. J. Weinstein and N. Isgur, Phys. Rev. D**41** (1990) 2236
8. T.H. Burnett and S.R. Sharpe, Annu. Rev. Nucl. Part. Sci 40 (1990) 327
9. T. Barnes, F.E. Close, P.R. Page, E.S. Swanson, Phys. Rev. D**55** (1997) 4157
10. F. Meyer-Wildhagen, Nucl. Phys. A**721** (2003) 605c
11. C.A. Baker et al., Phys. Lett. B**563** (2003) 140
12. Particle Data Group, Phys. Rev. D**66** (2002) 01000
13. D.V. Amelin et al., Proc. HADRON 01, p. 143 and p. 577
14. A.V. Anisovich et al., Nucl. Phys. A**690** (2001) 567; A.V. Sarantsev, private communication
15. G.S. Adams et al., Phys. Rev. Lett.**81** (1998) 5760
16. F.E. Close and H.J. Lipkin, Phys. Lett. B**196** (1987) 245
17. S.U. Chung and E. Klempt, Phys. Lett. B**563** (2003) 83
18. D.V. Amelin et al., Phys. Lett. B**356** (1995) 595
19. E. Klempt, Proc. HADRON 01, p. 463
20. C. Amsler et al., Crystal Barrel Collaboration, submitted to Eur. Phys. J.;
    J. Reinnarth, PhD thesis (Bonn 2003)
21. A. Donnachie and Yu.S. Kalashnikova, Proc. HADRON 01, p. 5
22. C. Amsler et al., Crystal Barrel Collaboration, submitted to Phys. Lett. B

B$_A$B$_{AR}$ Results on the D_s System

Robert N. Cahn

Lawrence Berkeley National Laboratory, 1 Cyclotron Rd., Berkeley, CA 94720

For the *BABAR* Collaboration

Abstract. The surprising discovery by the *BABAR* Collaboration of a narrow state with a mass of 2317 MeV decaying to $D_s\pi^0$ has been followed by other spectroscopic revelations. We focus on the *BABAR* results for the second state, at 2458 MeV, but mention related work from other experiments.

Introduction

The discovery [1] of a narrow state decaying into $D_s\pi^0$ was a surprise because it contradicted the expectations of apparently well established theory. The spectroscopy of systems of two heavy quarks is quite well understood after years of study in the ψ and Υ systems. Soon after the discovery of the charmonium system, DeRujula, Georgi, and Glashow proposed [2] that systems with one heavy quark and one light quark could be understood as analogs of the hydrogen atom, with spin-orbit couplings and hyperfine structure. This approach has been very successful in explaining the spectroscopy of heavy-quark light-quark systems.[3, 4, 5, 6].

A particularly impressive result is the understanding of p-wave states in the D (cs) system. By analogy with the hydrogen atom, we successively add interactions that break large symmetries down to smaller ones. Thus with only a spin-independent potential, all first level p-wave states are degenerate. If we write ℓ for the orbital angular momentum and s for the spin of the light quark, their sum $j = \ell + s$ is still a good quantum number when the spin-orbit interaction is included but hyperfine interactions are neglected. For p states we have $\ell = 1$ and thus $j = 1/2, 3/2$. The total angular momentum is obtained by adding the spin s' of the heavy quark: $J = j + s'$. Thus both the $j = 1/2$ and $j = 3/2$ states contribute to $J = 1$. In the limit of a very heavy quark (for which c is not an especially good candidate) the $j = 1/2$ and $j = 3/2$ states do not mix. In considering the emission of a pion by the $j = 3/2$ state with $J^P = 1^+$ to the s-wave ground states, $J^P = 0^-, 1^-$ we see that parity forbids decay to the D but allows decay to the D^*. The $D^*\pi$ can be in either s wave or d wave. However, since the s-wave ground states have $j = 1/2$, s-wave pion emission cannot reach them from the $j = 3/2$ p-wave state. The decay is instead by d-wave emission and is suppressed by the angular momentum barrier. As a result we predict a narrow $J = 1$ state in the D system. This is precisely what is seen. The other $J = 1$ state (with $j = 1/2$) should be broad and it is.

An entirely analogous prediction hold for the D_s system except that there the decay is by K emission in order to conserve both strangeness and isospin in the strong decay: $D_s(j = 3/2) \to D^*K$. This is indeed observed and is narrow. The second $J = 1$ state and

CP717, *Hadron Spectroscopy: Tenth International Conference,*
edited by E. Klempt, H. Koch, and H. Orth

the $J = 0$ state were expected to be broad as a consequence of s-wave K emission.

This last prediction failed because the the $J = 0$ state turned out to be 160 MeV lower in mass than models indicated and thus was below threshold for decay to DK. This striking failure of theoretical models has spurred both experimental and theoretical re-examination of the D_s system and heavy-quark light-quark systems, generally.

Discovery of the $D_{sJ}(2317)$

While *BABAR* detector[7] generally runs with PEP-II set at the $\Upsilon(4S)$ resonance to produce B-B pairs, there is necessarily a substantial production of cc continuum in the 91 fb^{-1} of integrated luminosity considered here.

Candidates for $D_s^+ \to K^+K^-\pi^+$ (throughout, we imply charge conjugate processes and states, as well) are isolated using particle identification information from dE/dx measurements in the drift chamber and silicon vertex tracker and especially from the DIRC, a Cerenkov detector that measures with precision the Cerenkov angle of emitted photons. Candidates for D_s are required to have masses within the range 1.955 GeV to 1.979 GeV and to form a vertex with a fit probability greater than 0.1%. The signal is further refined by requiring that the D_s decay be into one of the quasi-two-body modes, $\phi\pi$ or K^*K. We obtain additional discrimination by using the unique angular distribution for these decays by requiring that $|cos\theta_h| > 0.5$, where θ_h is the helicity angle of the vector particle's decay. Sideband regions, used for comparisons, are defined by the mass ranges (1.912 GeV, 1.934 GeV) and (1.998 GeV, 2.20 GeV).

The π^0 candidates are constructed from pairs of photons and are constrained to be consistent with the D_s trajectory and the beam envelope, with a fit probability greater than 1%. Again a signal region (122 MeV, 148 MeV) is defined, as well as sidebands, (90 MeV, 110 MeV) and (160 MeV, 180 MeV).

The D_s and π^0 candidates are combined to look for peaks in the invariant mass distribution. To eliminate backgrounds from B-meson decays and from continuum qq events, the $D_s\pi^0$ system is required to have momentum greater than 2.5 GeV in the CM. As seen in Fig. 1, a peak is apparent at a mass near 2.32 GeV.

Using data with $p(D_s\pi^0) > 3.5$ GeV in the CM, a fit with a Gaussian distribution for the signal and a third-order polynomial for the background finds 1267 ± 53 candidates, with a mass of 2316.8 ± 0.4 MeV and a standard deviation of 8.8 ± 0.4 MeV (statistical errors only). A analogous analysis using the decay channel $D_s^+ \to K^+K^-\pi^+\pi^0$ gives similar results: a mass of (2317.6 ± 1.3) MeV and a standard deviation of (8.8 ± 1.1) MeV. The measured width is consistent with the resolution and leads to the limit on the intrinsic width $\Gamma < 10$ MeV.

What is the $D_{sJ}(2317)$?

While more exotic explanation are possible, the simplest is that this is indeed the p-wave cs state with $J = 0$. It falls about 160 MeV lower than anticipated [3, 4, 5, 6]. Once that is accepted, the rest fits into place. The decay to $D_s\pi^0$ is isospin violating, but still

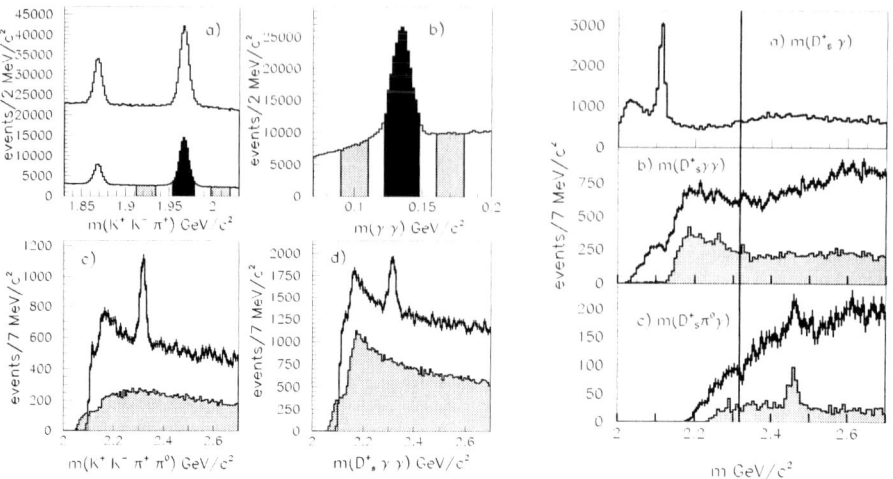

FIGURE 1. Left: Evidence for the $D_{sJ}(2317)$. (a) Distribution of the $K^+K^-\pi^+$ mass before and after cuts described in the text that isolate the signal. The signal region is dark, while the sidebands are shaded. The Cabibbo-suppressed decay of the D is visible. (b) The π^0 signal and sidebands. (c) The $D_s\pi^0$ mass distribution for signal π^0 with signal D_s (unshaded) and D_s sidebands (shaded). (d) The $D_s\gamma\gamma$ mass distribution for the D_s signal region and for the π^0 signal region (unshaded) and sideband region (shaded). Right: Searches for additional structure: (a) in $D_s^+\gamma$, (b) in $D_s^+\gamma\gamma$, (c) $D_s^+\pi^0\gamma$. The lower histograms of (b) and (c) correspond the $D_s^+\gamma$ masses that fall in the D_s^* signal region. The vertical line is at 2317 MeV. Taken from [1].

the best option available. It can be imagined to proceed through $\eta - \pi^0$ mixing [8].

If this surmise is correct, we should not find the radiative decay $D_{sJ}(2317) \to D_s\gamma$ (no $0 \to 0$ radiative transition), but could expect to find $D_{sJ}(2317) \to D_s^*\gamma$. In addition, we would expect to find a $J = 1$ state, which might be too light to decay to D^*K and which might thus be narrow. Among its decays then could be those to $D_s^*\pi^0$ and to $D_{sJ}(2317)\gamma$.

A Second State

Searches for structure in the channels $D_s\gamma$ and $D_s\gamma\gamma$ reveal nothing new[1]. See Figs.1a,b (Right). However, in the $D_s^+\pi^0\gamma$ channel, a small peak near 2460 MeV is evident, especially when the $D_s^+\gamma$ form a D_s^*. See Fig.1c (Right). The peak occurs near the crossing of the $D_{sJ}(2317)$ and D_s^* bands in the sense that if we add a random γ to the $D_{sJ}(2317)$ in such a way that the $D_s^+\gamma$ pair has a mass near that of the D_s^*, then the mass of the total $D_s^+\pi^0\gamma$ system must be close to 2460 MeV. A similar argument holds if we add a random π^0 to a D_s^*.

In this circumstance, the first priority is to make sure that the state at 2317 MeV was itself real and not simply an artifact of this possible higher state. Indeed the structure that would be produced from the decay of a state at 2460 MeV would be wider than that observed. Moreover, such decays could account for only one-sixth of the signal observed at the lower mass.

The study [1] of the 2460-MeV region in the first paper ended with the remark "...the complexity of the overlaping kinematics of the $D_s^*(2112)^+ \to D_s^+\gamma$ and $D_{sJ}(2317) \to D_s^+\pi^0$ decays requires more detailed study, currently underway, in order to arrive at a definite conclusion."

CLEO and BELLE

The CLEO and BELLE collaborations quickly confirmed the existence and narrowness of the $D_{sJ}(2317)$[9, 10, 11]. Moreover, CLEO identified the structure at 2460 MeV as a narrow state [9], as did BELLE, both in inclusive production through the cc continuum [10] and in exclusively reconstructed B decays[11].

BABAR Analysis of the Second State

BABAR did indeed carry out the detailed study. (The conference talk was given before complete results from BABAR on the $D_s(2458)$ were available. The BABAR paper on this was submitted Oct. 24, 2003. [12]) To address the question of whether there is really a signal at a mass near 2460 MeV or whether the structure shown above is just the result of superposing two known sources, the D_s^* and $D_{sJ}(2317)$ together with random π^0s or γs, we consider the variables

$$\Delta m_\gamma \equiv m(D_s^+\gamma) - m(D_s^+) \tag{1}$$

$$\Delta m_{\pi^0} \equiv m(D_s^+\gamma\pi^0) - m(D_s^+\gamma) . \tag{2}$$

A scatter plot of events in these two variables displays the crossing.

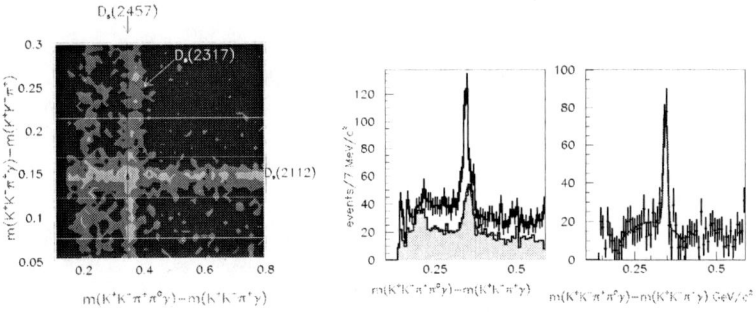

FIGURE 2. Left: Scatter plot of $D_s^+\pi^0\gamma$ events in the variables Δm_γ and Δm_{π^0}. Right: The distributions in the variable Δm_{π^0} for the (a) Δm_γ signal bands (white) and sidebands (shaded), (b) the difference between signal band and sidebands, with a Gaussian distribution plus polynomial fit.

Using the sidebands indicated in Fig. 2 (left), we can perform a subtraction to determine the distribution in the variable Δm_{π^0} This is shown in Fig. 2 (right).

The fit finds a yield of 140 ± 22 events with the peak at $\Delta m_{\pi^0} = 344.6 \pm 1.2$ MeV.

While this settles the question of whether there is a state near 2460 MeV, and confirms the results of CLEO and BELLE, it does not answer the question of whether its decay path is $D_s(2458) \to D_s^{*+} \pi^0$ or $D_s(2458) \to D_{sJ}(2317)\gamma$. Of course in the first instance the distribution of $D_s\gamma$ masses will be narrowly peaked near the D_s^{*+} while in the latter, there will be a narrow peak in the $D_s\pi^0$ channel.

To determine which is the case, we divide the region surrounding the signal area in the $m(D_s^+\gamma)$-$m(D_s^+\pi^0)$ plane into a three-by-three array of tiles, with the actual center coinciding with the peak intensity. Then by taking suitable linear combinations of tiles we can isolate a signal associated with just the $m(D_s^+\gamma)$ variable, and a signal associated with just the $m(D_s\pi^0)$ variable. We can then check whether these are broad or narrow, by comparing them with Monte Carlo generated to represent the two obvious possibilities for the decay path.

In Fig. 3 (left) we see the sideband-subtracted distributions in $m(D_s^+\gamma)$ and $m(D_s^+\pi^0)$ compared with Monte Carlo generated assuming the decay chain is $D_s(2458) \to D_s^{*+}\pi^0$. Figure 3 (right) shows the same data, but compared instead to Monte Carlo generated to represent the decay chain $D_s(2458) \to D_{sJ}(2317)\gamma$. It is apparent that the path $D_s(2458) \to D_s^{*+}\pi^0$ gives a far better description of the data.

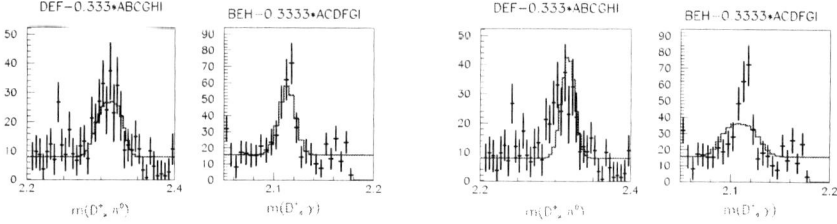

FIGURE 3. Left: Sideband-subtracted data for the distributions in Δm_{π^0} and Δm_γ compared to Monte Carlo (shown in the histogram) based on the decay path $D_s(2458) \to D_s^{*+}\pi^0$. Right: Sideband-subtracted data for the distributions in Δm_{π^0} and Δm_γ compared to Monte Carlo (shown in the histogram) based on the decay path $D_s(2458) \to D_{sJ}(2317)\gamma$.

The decay $D_s^{*+} \to D_s\gamma$ provides information on the polarization of the D_s^{*+}, and thus on the spin of its parent $D_s(2458) \to D_s^{*+}\pi^0$. If the $D_s(2458)$ is spin-zero, its parity must be odd. Otherwise, if the decay products have orbital angular momentum L, then the parity is $(-1)^L$. The total angular momentum is $J = L-1, L$, or $L+1$. In particular, if $D_s(2458)$ has natural spin-parity, then $J = L$ and there is a unique decay amplitude. For unnatural spin-parity (except $J^P = 0^-$) we have $J = L \pm 1$. With two decay amplitudes, the angular distribution of the photon relative to the line of flight of the D_s^{*+} is not uniquely determined. We find that for $J^P = 0^-$ the angular distribution is $\sin^2\theta$, while for natural spin-parity it is $1+\cos^2\theta$. Figure 4 shows our data compared to the expectations for various spin-parities. It is apparent that $J^P = 0^-$ is disfavored. This is in agreement with the BELLE result [11] obtained by studying $B \to DD_s(2357), D_s(2357) \to D_s\gamma$. The very existence of this decay excludes the assignment $J = 0$ for the $D_s(2357)$. The angular distribution excludes $J = 2$ and is consistent with $J = 1$.

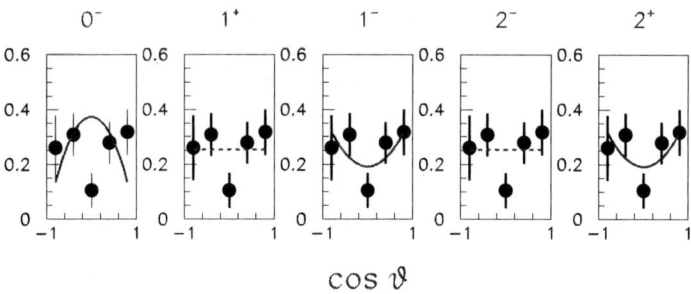

FIGURE 4. The observed angular distribution of the photon relative to the line of flight of the D_s^{*+} from the decay $D_s(2458) \to D_s^{*+}\pi^0$. The predictions for J^P of the $D_s(2458)$ are unique for 0^- and the natural spin-parity sequence $(1^-, 2^+, ...)$. Except for $J^P = 0^-$, for unnatural spin-parity the angular distribution is not determined a priori.

ACKNOWLEDGMENTS

The author was supported in part by the Director, Office of Science, Office of High Energy and Nuclear Physics, Division of High Energy Physics, of the U.S. DOE under Contract DE-AC03-76SF00098. The BABAR Collaboration is supported by the US DOE and NSF, the Natural Sciences and Engineering Research Council (Canada), Institute of High Energy Physics (China), the Commissariat à l'Energie Atomique and Institut National de Physique Nucléaire et de Physique des Particules (France), the Bundesministerium für Bildung und Forschung and Deutsche Forschungsgemeinschaft (Germany), the Istituto Nazionale di Fisica Nucleare (Italy), the Foundation for Fundamental Research on Matter (The Netherlands), the Research Council of Norway, the Ministry of Science and Technology of the Russian Federation, and the Particle Physics and Astronomy Research Council (United Kingdom).

REFERENCES

1. Aubert, B., et al., *Phys. Rev. Lett.*, **90**, 242001 (2003).
2. De Rujula, A., Georgi, H., and Glashow, S. L., *Phys. Rev. Lett.*, **37**, 785 (1976).
3. Godfrey, S., and Isgur, N., *Phys. Rev.*, **D32**, 189–231 (1985).
4. Godfrey, S., and Kokoski, R., *Phys. Rev.*, **D43**, 1679–1687 (1991).
5. Isgur, N., and Wise, M. B., *Phys. Rev. Lett.*, **66**, 1130–1133 (1991).
6. Di Pierro, M., and Eichten, E., *Phys. Rev.*, **D64**, 114004 (2001).
7. Aubert, B., et al., *Nucl. Instrum. Meth.*, **A479**, 1–116 (2002).
8. Cho, P. L., and Wise, M. B., *Phys. Rev.*, **D49**, 6228–6231 (1994).
9. Besson, D., et al., *Phys. Rev.*, **D68**, 032002 (2003).
10. Abe, K., et al. (2003), hep-ex/0307052, submitted to *Phys. Rev. Lett.*
11. Abe, K., et al. (2003), hep-ex/0308019, to appear in *Phys. Rev. Lett.*
12. Aubert, B., et al. (2003), hep-ex/0310050, submitted to *Phys. Rev. Lett.*

Evidence for the Strangeness S=+1 Pentaquark from LEPS and CLAS Experiments

K.H. Hicks[†]

*Department of Physics, Ohio University, Athens, OH 45701
and RCNP, Osaka University, Ibaraki, Osaka 567-0047, Japan*

Abstract. There are now several experimental collaborations that have seen evidence for a narrow state in the mass spectrum of the (nK^+) system. Two of these experiments, from the LEPS collaboration in Japan and the CLAS collaboration in the USA, are described briefly. Both use similar photoproduction reactions with a K^+K^- pair in the final state. In addition, data from the CLAS collaboration for the $\gamma p \to K_s K^+ n$ reaction are presented for the first time, which has no prominant peak in the (nK^+) mass spectrum when the K_s angle is limited to forward angles.

INTRODUCTION

Until recently, it was thought that the pentaquark, defined as a particle resonance with four quarks and one anti-quark, did not exist. This belief was based on exhaustive searches for a strangeness $S = +1$ resonance in the 1970's [1] yet only $S = -1$ particles were found. This result was a bit surprising because the rules of QCD do not forbid the existence of pentaquarks.

Nonetheless, progress was made in theoretical studies of the soliton model of the nucleon [2] which predicted, in addition to the usual octet and decuplet, an anti-decuplet ($\overline{10}$) of baryons that includes a $S = +1$ particle. Real progress was made when Diakonov, Petrov and Polyakov [3] predicted the mass of this particle using symmetries of the chiral soliton model and the key identification of the $S = 0$ baryon of the $\overline{10}$ with the spin $\frac{1}{2}$, P_{11} nucleon resonance at 1710 MeV. In this model, the mass of the pentaquark (called the Z^+ in Ref. [3] but since renamed the Θ^+ by the authors) was predicted to have a specific mass of 1530 MeV and a narrow width (< 15 MeV). This motivated experimters to look again for this $S = +1$ particle in already-existing data.

FIRST RESULTS

The first evidence for the Θ^+ pentaquark was reported in October, 2002 [4] by the LEPS collaboration in Japan. The reaction is $\gamma n \to K^- \Theta^+ \to K^- K^+ n$ where the neutron is bound inside a carbon nucleus, and only the K^+ and K^- were detected at forward angles ($\theta_{LAB}(K) < 30°$). The results are now published [5] and details of the measurement can be found there. The final spectrum is shown in Fig. 1 (bottom).

After the announcement by LEPS, the CLAS collaboration started to look for the Θ^+ in existing photoproduction data on a deuterium target. The advantage of a deuterium

CP717, *Hadron Spectroscopy: Tenth International Conference,*
edited by E. Klempt, H. Koch, and H. Orth
© 2004 American Institute of Physics 0-7354-0197-7/04/$22.00

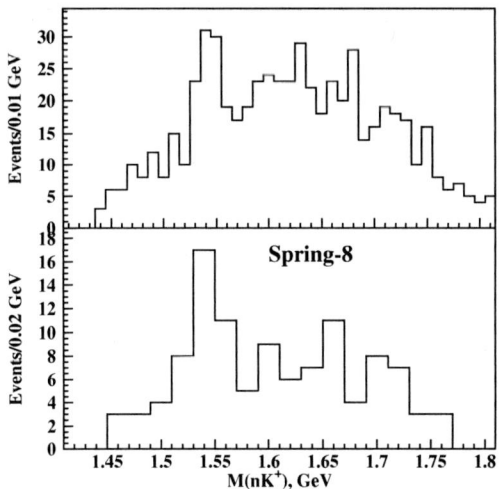

FIGURE 1. Comparison the invariant mass of the neutron-K^+ system, $M(nK^+)$ for the CLAS data (top) and LEPS data (bottom).

target is that a kinematically complete reaction can be measured, in contrast to the inclusive kinematics of the carbon target. The reaction $\gamma d \to K^- \Theta^+ p \to K^- K^+ p(n)$ was analyzed, where the neutron was deduced by the missing mass technique. Details of this result were first presented in February 2003 [6], followed by a more complete report in May [7]. Details can be found in Ref. [8]. The final spectrum is shown in Fig. 1 (top).

Comparing both figures, the prominent feature is a peak at the same mass, 1.54 GeV, with a width of < 25 MeV. The width of the peak is consistent with the known resolution of each experiment. This suggests that the intrinsic resolution of the resonance is smaller than the measured widths, although the small statistics prevents a definite conclusion. The shape of the background under the peaks is described in the respective references ([5, 8]). As long as the conservation laws of baryon number and strangeness hold, the resonance peak is a pentaquark made up of ($uudd\bar{s}$) construction. This is consistent with the Θ^+ prediction of Diakonov *et al.*, but until the spin and parity of this resonance is measured, we can not be sure that it is, indeed, the Θ^+ particle as predicted.

The CLAS data are shown again in Fig. 2 along with a fit to the peak. The background has been modeled by s-wave (non-resonant) photoproduction of $K^+ K^-$ pairs. The production cross section is just the phase space of 3-body ($K^+ K^-$ from a nucleon) and 4-body ($K^+ K^-$ from both nucleons in deuterium). Using this background shape, the fit gives a statistical significance of 4.7 σ, calculated as a fluctuation of the excess above the background shape (1 $\sigma = \sqrt{N_{Bg}}$, where N_{Bg} is the number of counts in the background within ± 20 MeV of the peak's central mass). The uncertainty in the statistical significance depends on the shape of the background, and different choices of background give values ranging from 4.6 to 5.8 σ.

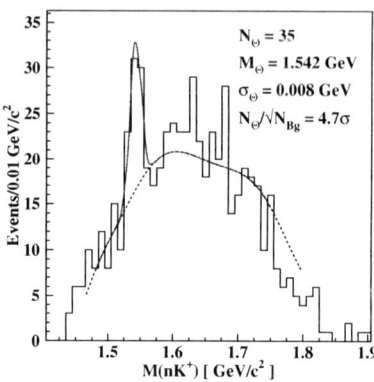

FIGURE 2. Peak fit using a MC background shape for the $M(nK^+)$ spectrum.

If the Θ^+ exists, then it should be produced in a variety of reactions. In addition to photoproduction from the neutron, which must be bound in a nucleus, it is natural to look for photoproduction from the proton. Preliminary results on the reaction $\gamma p \to K^- \pi^+ \Theta^+$ have already been presented [7]. Another reaction is $\gamma p \to \bar{K}^0 \Theta^+$ which was published by the SAPHIR collaboration [9] (which are also reported in these proceedings).

NEW EXPERIMENTAL RESULTS

We present here *preliminary* results of the CLAS collaboration [10] for analysis of the reaction $\gamma p \to \bar{K}^0 K^+ n$ where the neutron is measured by the missing mass technique. Fig. 3 shows the the mass calculated from the momentum and velocity of detected particles. The K^+ (top left) is detected directly. The K^0 mass (top right) is from K_s decay, made from the invariant mass of a detected $\pi^+ \pi^-$ pair. The n mass (bottom left) is from the missing momentum and energy. The $\Lambda(1520)$ mass (bottom right) is from the invariant mass of the deduced 4-vectors of the n and the K_s. All mass spectra show clear peaks with very little background, showing that particle identification is quite good. Several thousand good $(K_s K^+)$ events are identified for further analysis.

One must be careful in the analysis of the proton data because the K_s is a linear combination of K^0 and anti-K^0. Hence, the strangeness of the reaction is not uniquely identified by the K_s particle, and other reactions can produce the same final state. As shown above, the $\gamma p \to K^+ \Lambda^*$ reaction, where $\Lambda^* \to K^0 n$, has the same final state as $\gamma p \to \bar{K}^0 \Theta^+$ where the Θ^+ can decay to $K^+ n$. Also, since the K_s is determined from a $\pi^+ \pi^-$ pair, the reaction $\gamma p \to K^+ \Lambda^*$ followed by $\Lambda^* \to \Sigma \pi \to \pi^+ \pi^- n$ has the same particles in the final state.

After rejecting events where a πn invariant mass equals the Σ mass, and also events in the $\Lambda(1520)$ peak (see Fig. 3), the missing mass of the K_s^0 spectrum is shown in Fig. 4. This spectrum should show a peak at the mass of the Θ^+ if this state is produced with a

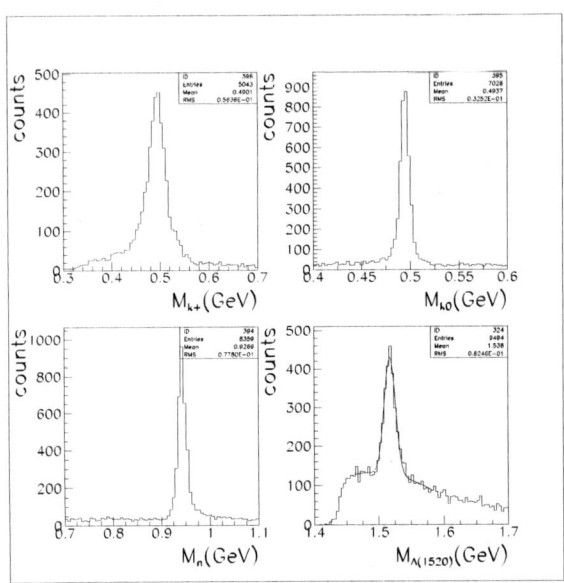

FIGURE 3. The masses calculated from coincident K^+, π^+ and π^- particles in photoproduction from a proton target at CLAS.

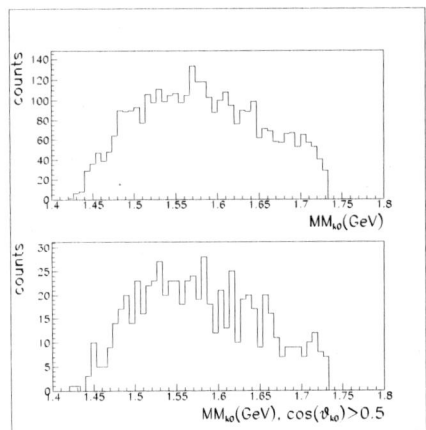

FIGURE 4. The missing mass spectrum for K_s^0 production from the reaction $\gamma p \to \bar{K}^0 n$ and event selection as described in the text[10].

cross section sufficiently larger than the non-resonant background. If the Θ^+ is produced in a t-channel process, then selecting events with forward angles of the K_s ($\cos \theta_{K_s} > 0.5$ where the angle is in the photon-proton center of mass system) could enhance the signal over the background. The bottom plot in Fig. 4 shows the spectrum after this event cut. In both cases, no prominent peak is observed.

SUMMARY AND CONCLUSIONS

The lack of a strong signal in the *preliminary* analysis of the $\gamma p \rightarrow K_s^0 K^+ n$ reaction at CLAS is surprising, considering that there are several measurements that have shown strong evidence for the Θ^+. One possible explanation is that the coupling constant at the $K^* p \Theta^+$ vertex is small, giving a small cross section in the t-channel. (The K^* vector meson is necessary as a virtual particle at forward angles since the neutral kaon couples to the photon primarily through a magnetic M1 transition). Of course, another possibility is that the Θ^+ does not exist, and that the other experiments are just unfortunate statistical fluctuations. (At the time of writing this paper, the latter possibility seems less likely, as more reports supporting the Θ^+ have been announced.)

A full understanding of the experimental situation for evidence of the Θ^+ must await future measurements. It is difficult to be patient at a time where excitement surrounds the announcement of a new particle, which could be the beginning of a new class of particles (pentaquarks). However, we must be cautious, and let the facts emerge. If the Θ^+ exists, then experiments with better statistics and more understanding of the background will find clear evidence for this particle. Until then, there are positive signs but no definite conclusions regarding the existence of pentaquarks.

ACKNOWLEDGMENTS

This work would not be possible without the hard work of many people in the LEPS and CLAS collaborations. The financial support of the U.S. Department of Energy (DOE), the National Science Foundation (NSF) and the Japanese Ministry of Education, Culture, Science and Technology (MEXT) is gratefuly acknowledged. The analysis in Figs. 3-4 was supported by the Italian Instituto Nazionale de Fisica Nucleare.

REFERENCES

1. Particle Data Group, Phys. Rev. **D66**, 010001 (2002).
2. M. Praszalowicz, *Workshop on Skyrmions and Anomalies*, M. Jezabek and M. Praszalowicz, eds., World Scientific, 1987, p. 164.
3. D.I. Diakonov, V. Yu. Petrov and M.V. Polyakov, Z. Phys. **A 359**, 305 (1997).
4. T. Nakano *et al.*, (PANIC Conference), Nucl. Phys. A721 (2003) 112c-117c.
5. T. Nakano *et al.*, Phys. Rev. Lett. **91**, 012002 (2003).
6. K. Hicks, U.S.-Japan Workshop on New Aspects of Quark Nuclear Physics, Honolulu, Feb. 19, 2003.
7. V. Kubarovsky, S. Stepanyan, Conference on the Intersections between Nuclear and Particle Physics (2003), *in press*.
8. S. Stepanyan, K. Hicks, *et al.*, hep-ex/0307018.
9. J. Barth *et al.*, (SAPHIR Collab.), hep-ex/0307083.
10. Analysis done by M. Battaglieri, R. Devita, M. Osipenko and M. Ripani of the INFN Genova group; parallel analysis by K. Hicks.

Penta-quark baryon: predictions from chiral solitons

Maxim V. Polyakov

St. Petersburg Nuclear Physics Institute, Russia; Institut de Physique, Université de Liège au Sart Tilman, B-4000 Liège 1 Belgium; Institut für Theor. Physik -II, Ruhr-Universität, D-44780 Bochum, Germany

Abstract. During the last year there has been mounting experimental evidence of the existence of a new type of baryon which cannot be made of three quarks but necessarily contains four quarks and one antiquark. It has been named Θ^+ and its minimal quark content is $uudd\bar{s}$. Its mass is about 1540 MeV and width less than about 10 MeV. This baryon has been predicted in 1997 with the correct mass and narrow width from the chiral soliton model of baryons [1]. In this contribution, we briefly summarize the main predictions of the soliton picture of baryons for exotic anti-decuplet.

All baryons observed until Autumn 2002 could be arranged into singlets, octets and decuplets of the Gell-Mann–Ne'eman $SU(3)$ flavour group and could therefore be considered as constructed of only three quarks. Higher $SU(3)$ multiplets ($\overline{10}$, 27, etc.) cannot be built of three quarks but require an addition of quark-antiquark pairs. Experimental searches of the states made of more than three quarks have been started already in sixties (see e.g. [2]). Clear "mark" of the baryon with minimal quark content $qqqq\bar{s}$ (pentaquark) is the positive strangeness of such states. Years of searches for such states have been summarized by the Particle Data Group in their 1986 edition [3]:

"However, the results permit no definite conclusion- the same story heard for 15 years.... The general prejudice against baryons not made of three quarks and the lack of any experimental activity in this area make it likely that it will be <u>another 15 years</u> before the issue is decided."

Unsuccessful searches lead to a widely spread belief that exotic baryons which cannot be constructed of three quarks and therefore belonging to higher $SU(3)$ multiplets, are not realized in nature. On the experimental side, the drive for hunting exotics weakened. On the theoretical side, the consequence was the appearance of a large class of quark models which assumed that baryons were made of precisely three quarks bound by some confining forces. In the situation when a) very few believed that exotic baryons could at all exist, b) previous specialized searches were unsuccessful and c) it was hard to imagine that a relatively light and narrow exotic baryon would not have been discovered merely by chance in the past, we have proposed [1] to make a fresh search of such penta-quark baryon, pointing out its mass, width and quantum numbers – isospin, spin and parity.

The soliton picture of baryons suggests a certain classification scheme for the low-lying baryons. In this scheme various baryons appear as rotational excitations of the same classical object – soliton. In the case of three light flavours, the first two low-lying

CP717, *Hadron Spectroscopy: Tenth International Conference*,
edited by E. Klempt, H. Koch, and H. Orth
© 2004 American Institute of Physics 0-7354-0197-7/04/$22.00

$SU(3)$ multiplets are the octet and the decuplet, just the same as in the quark model and in reality. The third rotational excitation is an *antidecuplet with spin* $1/2$ *and positive parity*. Probably the existence of the antidecuplet as the next $SU(3)$ rotational excitation has been first pointed out at the ITEP Winter School (February, 1984), see Ref. [4]. Other early references for the antidecuplet include Refs. [5, 6, 7]. However, the expected masses of its members had always been rather uncertain, at least up to ~ 100 MeV. Until Ref. [1], the question of width, also essential for experimental searches, had not been addressed at all. To make the Θ^+-mass prediction more definite, it was suggested in [1] to identify the non-strange member of the antidecuplet with the $N(1710)$, the only nucleon state listed in the PDG-tables having $J^P = 1/2^+$ and being in the expected mass interval.

In Fig. 1 we draw the $SU(3)$ diagram (from Ref. [1]) for the suggested antidecuplet in the (T_3, Y) axes, indicating its naive quark content as well as the (octet baryon + octet meson) content. In addition to the lightest Θ^+, there is an exotic quadruplet of $S = -2$ baryons (we call them $\Xi_{3/2}$). In Ref. [1] the following mass formula for the members of the antidecuplet was obtained [1]:

$$ M = \left[1890 - Y \times 180 \right] \text{ MeV}. \tag{1} $$

Note that this "soliton" mass formula is, to some extent, counterintuitive from the point of view of the naive quark model. For instance, strange baryon (Θ^+) appears to be lighter than the baryon with the nucleon quantum numbers. Up to now we were used to strange baryons being heavier than non-strange ones in a given multiplet. Also Θ^+ having 4 light+$\bar{s}$ quark content is about 540 MeV lighter than $\Xi_{3/2}^-$ with the quark content 3 light+2 s quarks. In the naive quark model one would expect the mass difference of about $\sim$150 MeV.

The observation of narrow peaks in the invariant mass spectra of nK^+ [9, 10, 11, 12, 13] and pK_S [14, 15, 16, 17, 18] events, verified by independent groups and laboratories worldwide, has solidified the evidence for an exotic baryon Θ^+ with strangeness $+1$, a mass of about 1540 MeV, and a narrow width. The existence of such a particle implies a whole new family of $SU(3)$ partners, beyond the familiar octets and decuplets.

What do we know about properties of Θ^+ penta-quark?

Its isospin mostly probably is zero [10, 12, 16] , in accordance with anti-decuplet interpretation of this state. Neither spin nor parity is measured.

Let us summarize our present knowledge of the Θ^+-width. The theoretical prediction, simultaneous with its mass, was $\Gamma_{\Theta^+} < 15$ MeV [1], unexpectedly narrow for strong decays. Existing measurements, instead of determining Γ_{Θ^+}, have only shown it to be smaller than experimental resolution (see Table 1). Most experimental publications have given an upper bound of about ~ 20 MeV. Xenon bubble chamber data, corresponding in essence to the charge exchange reaction $K^+ n \to K^0 p$, have provided the slightly lower bound of 9 MeV [14].

---

[1] Note that recent analysis of ref. [8] suggests a mass splitting of around 110 MeV.

TABLE 1. Properties of Θ^+

Collaboration	Mass (MeV)	Width (MeV)	Ref
DPP	1530	<15	[1]
LEPS	1540±10	<25	[9]
DIANA	1539±2	<9	[14]
CLAS/γn	1542±5	<21	[10]
CLAS/γp	1540±10	<32	[11]
ELSA	1540±4±2	<25	[12]
ITEP/ν	1533±5	<20	[15]
HERMES	1526±2.6±2.1	<19	[16]
CLAS/γp	1555±10	<26	[13]
ZEUS	1527±2	<24	[17]
NA49	1535		[18]
USC	1543	<6	[21]
GWU	1540–1550	≤1	[23]
Jülich	1545	<5	[22]
LBNL	1540	0.9±0.3	[25]

Less direct determinations [21], using previously measured K^+d total cross sections, have led to a stronger limitation $\Gamma_{\Theta^+} < 6$ MeV. A similar bound, < 5 MeV, was obtained in Ref. [22] within a more elaborated theoretical description. The partial-wave analysis (PWA) of available KN (elastic and charge exchange) scattering data similarly claims to exclude widths above 1–2 MeV [23]. A more detailed reexamination of the approach in Ref. [21] provides nearly the same result of $\Gamma_{\Theta^+} < 1.5$ MeV [24]. A similar method applied to the Xenon data [14] has allowed even the tentative claim of a lower limit $\Gamma_{\Theta^+} = 0.9 \pm 0.3$ MeV [25] (with additional assumptions and an unknown systematic uncertainty). One should emphasize, however, that all of these indirect treatments assume the existence of a Θ^+, which they can not confirm. Moreover, they are based mainly on rather old data, which may be shifted by the next generation of higher precision measurements.

Non-exotic members of the antidecuplet

We stress again that discovery of the Θ^+ penta-quark implies an existence of the whole new family of baryons– $SU(3)$ flavor partners. Their properties should be also very unusual. Identification of the non-exotic members of the anti-decuplet is very important.

Evidently, all of the above estimates for Γ_{Θ^+} are in sharp contrast with the width ~ 100 MeV ascribed [26] to the $N(1710)$, initially considered to be a unitary partner of the Θ^+ [1]. Of course, members of the same unitary multiplet can have different widths, but in the absence of a special reason (say, mixing with members of another multiplet) it would be more natural for them to have comparable widths.

The state $N(1710)$, though listed in the PDG Baryon Summary Table [26] as a 3-star resonance, is not seen in most recent analysis of pion-nucleon elastic scattering data [27]. Studies which have claimed to see this state have given widely varying estimates of its mass and width (from ~ 1680 MeV to ~ 1740 MeV for mass and from ~ 90 MeV to ~ 500 MeV for width). Branching ratios have also been given with large uncertainties

(10–20% for $N\pi$, 40–90% for $N\pi\pi$, and so on), apart from one which has been presented with greater precision ($6 \pm 1\%$ for $N\eta$).

Of course the non-observation of a broad $N(1710)$ state in pion-nucleon elastic analyses could be due to a very small πN branching ratio. Standard procedures used in partial-wave analysis (PWA) may also miss narrow resonances with $\Gamma < 30$ MeV. Therefore, the true unitary partner of the Θ^+ (if it is different from $N(1710)$ and sufficiently narrow) could have eluded detection.

The theoretical analysis, based on the soliton picture and assumption of $\Gamma_{\Theta^+} < 5$ MeV, shows that most probably $\Gamma_{N^*} < 30$ MeV. Having in mind all theoretical uncertainties, we can suggest several directions for experimental studies. First of all, one should search for possible new narrow nucleon state(s) in the mass region near ~ 1700 MeV. Searches may use various initial states (*e.g.*, πN collision or photoproduction). We expect the largest effect in the $\pi\pi N$ final state. Although $\pi\Delta$ is forbidden by $SU(3$, it was shown in [1] that the mixing of the antidecuplet with usual octet leads to strong enhancement of the $\pi\Delta$ decay mode. The final states ηN and $K\Lambda$ may also be interesting and useful, especially the ratio of ηN and πN partial widths as the latter is very sensitive to the value of the antidecuplet coupling constant and the octet–antidecuplet mixing angle. Another interesting possibility to separate antidecuplet and octet components of N^* is provided by comparison of photoexcitation amplitudes for neutral and charged isocomponents of this resonance [28], the point being that the antidecuplet component does not contribute to the photoexcitation of the charged component of N^* (see details in Ref. [28]).

Exotic $\Xi_{3/2}$

Recent experimental result [29] gives evidence for one further explicitly exotic particle $\Xi_{3/2}^{--}$ (see Fig. 1), with a mass 1862 ± 2 MeV and width < 18 MeV (*i.e.* less than resolution). Such a particle had been expected to exist as a member of the same antidecuplet containing the Θ^+, but its mass was predicted to be about 2070 MeV [1], essentially different from the experimental value. Let us note however that the soliton calculation of this mass difference requires some assumptions. In particular, it depends on the value of the σ-term, which is the subject of controversy. Its value, taken according to the latest data analysis [30], leads to an antidecuplet mass difference of about 110 MeV [8], what brings the mass of $\Xi_{3/2}$ to 1870 MeV. Also it is expected that the higher order symmetry breaking effects are larger for heavier members of the antidecuplet.

REFERENCES

1. D. Diakonov, V. Petrov and M. Polyakov, *Z. Phys.* **A359** (1997) 305, hep-ph/9703373
2. R.L. Cool et al., Phys. Rev. Lett. 17 (1966) 102.
3. M. Aguilar-Benitez *et al.*, *"Review Of Particle Properties. Particle Data Group,"* Phys. Lett. B **170** (1986) 1.
4. D.Diakonov and V.Petrov, *Baryons as solitons*, preprint LNPI-967 (1984), published in: *Elementary particles*, Moscow, Energoatomizdat (1985) vol.2, p.50 (*in Russian*).
5. M. Chemtob, Nucl. Phys. B **256** (1985) 600.
6. M.Praszalowicz, in: *Skyrmions and Anomalies*, M.Jezabek and M.Praszalowicz, eds., World Scientific (1987) p.112

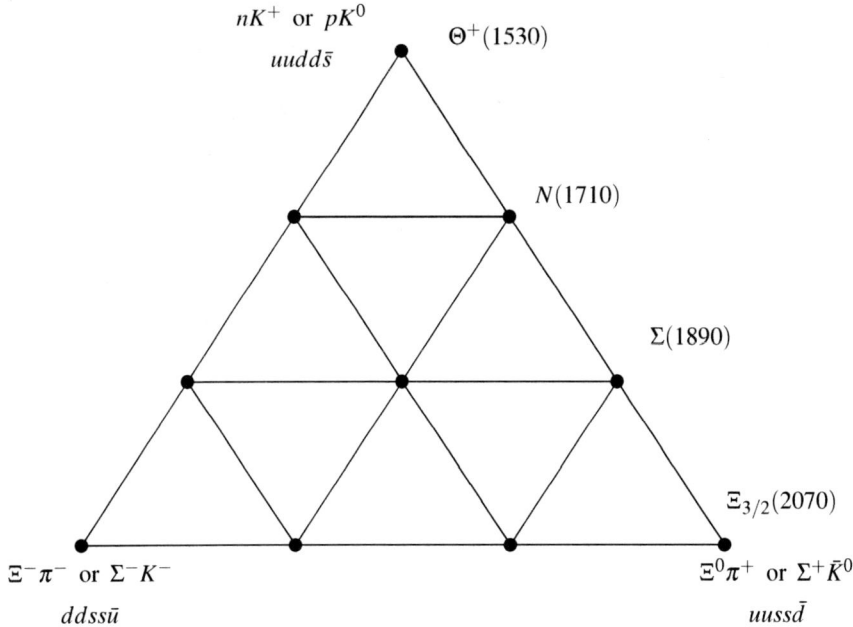

FIGURE 1. The suggested antidecuplet of baryons [1]. The corners of this (T_3, Y) diagram are exotic. We show their quark content together with their (octet baryon+octet meson) content, as well as the predicted masses.

7. H. Walliser, in *Baryon as Skyrme Soliton*, p. 247, ed. by G. Holzwarth, World Scientific, 1992; H. Walliser, Nucl. Phys. A **548** (1992) 649.
8. D. Diakonov and V. Petrov, hep-ph/0310212.
9. T. Nakano *et al.*, [LEPS Collaboration], Phys. Rev. Lett. **91**, 012002 (2003) [hep-ex/0301020].
10. S. Stepanyan *et al.*, [CLAS Collaboration], Phys. Rev. Lett. **91**, 252001 (2003) [hep-ex/0307018].
11. V. Koubarovsky and S. Stepanyan, [CLAS Collaboration], in *Proceedings of "Conference on the Intersections of Particle and Nuclear Physics (CIPANP2003), New York, NY, USA, May 19-24, 2003"*, to be published [hep-ex/0307088].
12. J. Barth *et al.*, [SAPHIR Collaboration], Phys. Lett. **B572**, 127 (2003) [hep-ex/0307083].
13. V. Koubarovsky *et al.*, [CLAS Collaboration], submitted to Phys. Rev. Lett., hep-ex/0311046.
14. V. V. Barmin *et al.*, [DIANA Collaboration], Phys. Atom. Nuclei **66**, 1715 (2003) [Yad. Fiz. **66**, 1763 (2003)] [hep-ex/0304040].
15. A. E. Asratyan *et al.*, to be published in Phys. Atom. Nuclei (2004) [Yad. Fiz. (2004)] [hep-ex/0309042].
16. A. Airapetian *et al.*, [HERMES Collaboration], submitted to Phys. Rev. Lett. hep-ex/0312044.
17. S. Chekanov [ZEUS Collaboration], http://http://www.desy.de/f/seminar/sem-schedule.html
18. D. Barna and F. Sikler [NA49 Collaboration], NA49 Note 294, http://na49info.cern.ch/cgi-bin/wwwd-util/NA49/NOTE?294
19. H. Walliser and V. Kopeliovich, J. Exp. Theor. Phys. **97**, 433 (2003) [Zh. Eksp. Teor. Fiz. **124**, 483 (2003)] [hep-ph/0304058].
20. M. Praszalowicz, Phys. Lett. **B575**, 234 (2003) [hep-ph/0308114].
21. S. Nussinov, hep-ph/0307357.

22. J. Haidenbauer and G. Krein, Phys. Rev. C **68**, 052201 (2003) [hep-ph/0309243].
23. R. A. Arndt, I. I. Strakovsky, and R. L. Workman, Phys. Rev. C **68**, 042201 (2003) [nucl-th/0308012].
24. R. A. Arndt, I. I. Strakovsky, and R. L. Workman, in *Proceedings of the VIII International Conference on Hypernuclear and Strange Particle Physics, Jefferson Lab, Newport News, VA, USA, October 14−18, 2003*, to be published [nucl-th/0311030].
25. R. N. Cahn and G. H. Trilling, to be published in Phys. Rev. D [hep-ph/0311245].
26. K. Hagiwara *et al.*, *Review of Particle Physics*, Phys. Rev. D **66**, 010001 (2002), http://pdg.lbl.gov.
27. R. A. Arndt, W. J. Briscoe, I. I. Strakovsky, R. L. Workman, and M. M. Pavan, submitted to Phys. Rev. C, nucl-th/0311089.
28. M. V. Polyakov and A. Rathke, Eur. Phys. J. A **18**, 691 (2003) [hep-ph/0303138].
29. C. Alt *et al.* [NA49 Collaboration], to be published in Phys. Rev. Lett., hep-ex/0310014.
30. M. M. Pavan, R. A. Arndt, I. I. Strakovsky, and R. L. Workman, in *Proceedings of 9th International Symposium on Meson-Nucleon Physics and the Structure of the Nucleon (MENU2001), Washington, DC, USA, July 26−31, 2001*, edited by H. Haberzettl and W. J. Briscoe, πN Newslett. **16**, 110 (2002) [Eprint hep-ph/0111066].

Investigation of Resonance Structures in the System of Two K_S-Mesons in the Mass Regions around 1070 and 2000 MeV

E.A. Fadeeva, SERP-E173 experiment [1]

ITEP, Moscow, Russia

Abstract. This report is devoted to previously unknown narrow resonances in the K_SK_S-system at masses 1070 MeV and 2000 MeV that are observed in experimental data coming from 6-m spectrometer (MIS ITEP). The experimental data on the production of K_S-pair were obtained in $\pi^- p$ interaction at 40 GeV by using a neutral trigger. The first state has the mass of about 1070 MeV and the width <3 MeV. Spin-parity is preferably $J^{PC} = 2^{++}$. The statistical confidence of the observed phenomenon is better than 7 standard deviations. The second one has the mass of about 2000 MeV and the width <7 MeV. Spin-parity is preferably $J^{PC} = 4^{++}$. The statistical confidence of the observed phenomenon is better than 6 standard deviations. Seeing their very small width these resonances are likely to be cryptoexotics (see [1],[2] for details).

INTRODUCTION

The experimental data we analyze had been accumulated by using 6-m spectrometer developed at the Institute for Theoretical and Experimental Physics (ITEP, Moscow) and installed at a 40GeV π^--meson beam from the accelerator at the Institute for High Energy Physics (IHEP, Serpukhov). A detailed description of the 6-m spectrometer was presented in [3]. The system of two K_S-mesons that was recorded under the experimental conditions of the 6-m spectrometer is produced in the following two reactions:

$$\pi^- p \to K_S K_S n, \tag{1}$$

$$\pi^- p \to K_S K_S + (n + m\pi^0, p + \pi^-, ...). \tag{2}$$

Reaction (1) is separated with a trigger facility based on veto counters surrounding the liquid-hydrogen target. Due to imperfect trigger operation, some fraction of events of the reaction (2) is recorded by the setup. The spectrometer records with high efficiency K_S-mesons going to the forward direction and decaying into to charged π-mesons. The precision of the measurement of the effective mass of the K_SK_S-system is better than 3 MeV in mass region around 1100 MeV, and better than 5-7 MeV in mass region around 2000 MeV. The recording efficiency is about 40% for the K_SK_S-system in the mass region around 2000 MeV and about 50% in the mass region 1070MeV. It depends on the K_S-meson momenta.

---

[1] I.A.Erofeev, O.N.Erofeeva, V.K.Grigor'ev, Yu.V.Katinov, V.I.Lisin, V.N.Luzin, V.N.Nozdrachev, Yu.P.Shkurenko, V.V.Sokolovsky, V.V.Vladimirsky. ITEP, Moscow.

CP717, *Hadron Spectroscopy: Tenth International Conference,*
edited by E. Klempt, H. Koch, and H. Orth
© 2004 American Institute of Physics 0-7354-0197-7/04/$22.00

In the analysis of the $K_S K_S$-system we used the following kinematical variables: the effective mass M_{KK} of the pair of K_S-mesons; the missing mass squared MM^2 defined as the squared mass of particles that are produced together with the $K_S K_S$-system and which are not recorded in the spectrometer; the 4-momentum transferred from the beam to the system being studied, t; the cosine of the Gottfried-Jackson angle, $cos\theta_{GJ}$; the Treiman-Yang angle, ϕ_{TY}.

The angles are calculated in the rest frame of the pair of K_S-mesons, the beam axis direction in this system being taken for the polar axis. The plane from which the Treiman-Yang angle is reckoned is spanned by the momenta of the beam and of the target proton in this reference frame.

RESONANCE STATES IN $K_S K_S$-SYSTEM

X(1070)

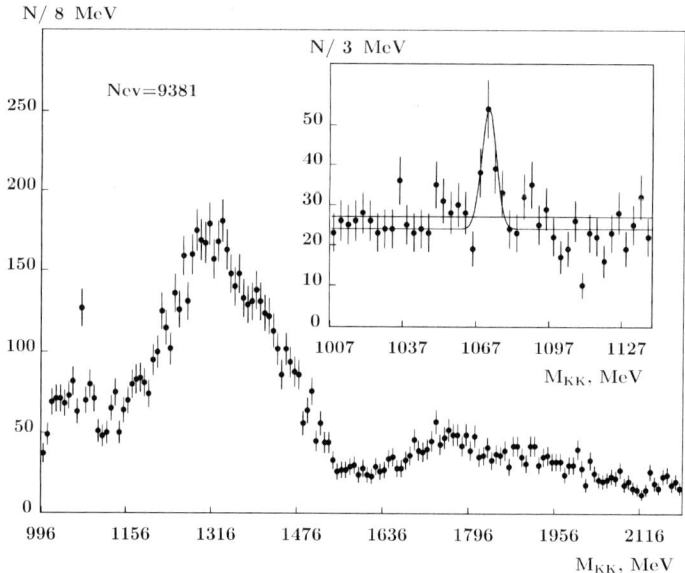

FIGURE 1. Effective-mass spectrum of two K_S-mesons.

The Figure (1) shows the mass spectrum of the $K_S K_S$-system from their production threshold with the bin width being 8 and 3 MeV, respectively. Here we use the following experimental cuts on transferred momentum ($0.0< -t <0.6$ GeV^2) and on missing mass squared ($-0.3< MM^2 <2.5$ GeV^2). The resonance feature manifests itself as a maximum in the vicinity of 1070 MeV. The figure (2) represents the distribution of number of events with respect to the transferred momentum. Momentum-transfer distribution of events from the effective mass region of the $K_S K_S$-system that are adjacent to the resonance region from the left and from the right ($1056< M_{KK} <1060$, $1084< M_{KK} <1088$ (MeV)) is represented by the curve and of events from the resonance region ($1068 < M_{KK} < 1076$ (MeV)) by the dots. It is seen that these distributions are quite different.

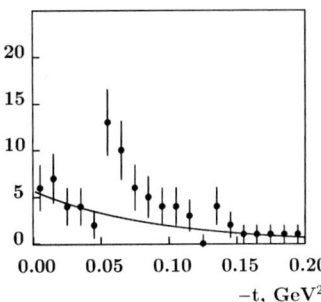

N/0.01 GeV²

FIGURE 2. Momentum-transfer distribution for X(1070).

In order to determine the parameters of the observed resonance feature and its statistical confidence, the experimental data were fitted by the Maximum-likelihood method (MLM). The main advantage of this method over the histogramming is that the mass and angles are not averaged over the bin width in the fitting procedure, so that the result does not depend on the choice of the reference point and the number of bins into which the mass range under study is divided. Describing the experimental data, we used the probability-density function $F(P;\Omega)$, where P is the set of the parameters (the amplitude, the mass M, the width σ appearing in the Gauss function and the coefficients of the squared amplitudes of the angular distributions). Elements of the phase space Ω are effective mass of two K_S-mesons, the cosine of the Gottfried-Jackson angle $cos\theta_{GJ}$, the Treiman-Yang angle ϕ_{TY}.

Fitting was performed for events falling within the range 1050-1120 MeV of K_SK_S masses. A first-degree polynomial proved to be sufficient for describing the mass dependence of the background. The resonance was approximated by a Gauss function. In order to obtain the most probable values of the parameters we minimized the functional:

$$\int_\Omega \varepsilon(\Omega)F(P;\Omega)d\Omega - \sum_{i=1}^{N} \ln F(P;\Omega_i). \qquad (3)$$

where $\varepsilon(\Omega)$ is the event-detection recording, N being the number of events. To compare the probabilities of experimental-data description with different parameter set, we calculated χ^2 by the formula:

$$\chi^2 = -2\ln L + const, \qquad (4)$$

where $L = \prod_{i=1}^{N} F(P;\Omega_i)$.

TABLE 1. Different sets of minimization for X(1070).

	Background N_{events}		Resonance N_{events}				Parameters MeV		$\delta\chi^2$	$N_{st.dev.}$
	S	D_{++}	S	D_0	D_+	D_{++}	M	σ		
1	876	178	–	–	–	–	–	–	0	–
2	824	172	48	–	–	10	1073.0	1.9	-38	5.8
3	781	182	–	19	32	–	1072.0	1.6	-50	7.0

The first line of the table 1 represents the result of fitting without allowing for the Gauss function and without discarding events falling within the resonance region. In the second line we tried to describe the resonance with the same set of the waves and the

413

same relations between them as those that observed for the background. In the third line is given the result of fitting in terms of the D_0- and D_+-waves.

Let us summarize results. We have obtained an indication of the existence of new resonance feature having a mass $1072\pm1.5\pm1.0$ MeV and a width less than 3 MeV. The spin-parity of this resonance is preferably $J^{PC} = 2^{++}$. The product of the cross section for X(1072) formation and the relevant branching ratio $\sigma BR(K_S K_S)$ is estimated at about 15 ± 7 nb. The statistical confidence of the observed phenomenon is better than 7 standard deviation.

Main feature of this resonance is its very small width (<3 MeV) (PDG (see [4]) in this mass region gives 50-150 MeV) and extraordinary cross-section dependence on the transferred momentum.

X(2000)

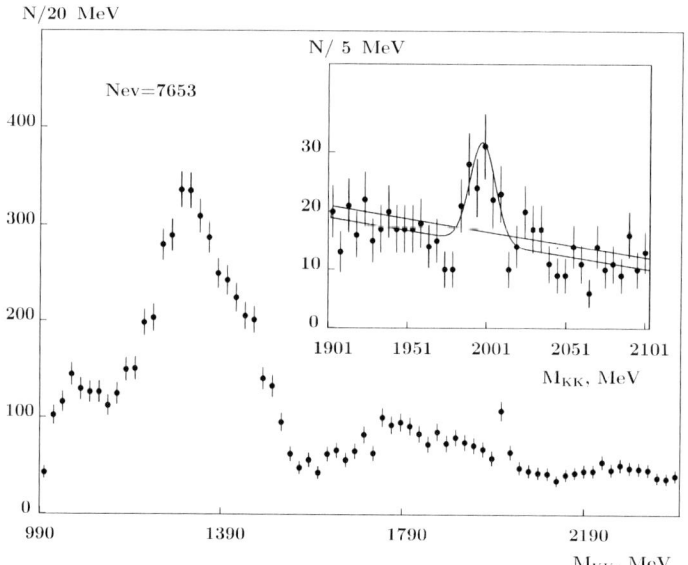

FIGURE 3.   Effective-mass spectrum of two K_S-mesons.

Figure (3) shows the effective-mass spectrum of system of two K_S-mesons from their production threshold to 2400 MeV with the bin width being 20 and 5 MeV, respectively. The resonance feature manifests itself as a maximum in the vicinity of 2000 MeV. We used the cut on transferred momentum $(0.0< -t <0.6$ GeV$^2)$. Versus X(1070), that is produced in both reaction (1) and (2), this resonance feature is produced only in reaction (1). To suppress the events going from reaction (2) the cut on missing mass squared $(0.3< MM^2 <1.5$ GeV$^2)$ is used.

The figure (4) represents the distribution of number of events with respect to the transferred momenta. Momentum-transfer distribution of events from the effective mass region of the $K_S K_S$-system that are adjacent to the resonance region from the left and from the right ($1950 < M_{KK} < 1985$, $2015 < M_{KK} < 2050$ (MeV)) is represented by the curve and of events from the resonance region ($1995 < M_{KK} < 2005$ (MeV)) by the dots.

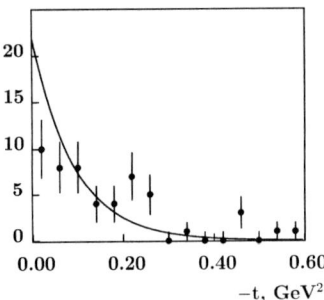

N/0.04 GeV^2

The data were fitted by the Maximum-likelihood method. Fitting was performed for events falling within the range 1850-2150 MeV of $K_S K_S$ masses. The table 2 gives the results of the fitting. The first line represents the result of fitting without allowing for the Gauss function and without discarding of events falling within the resonance region. In the second line we tried to describe the resonance with the same set of the waves and the same relations between them as those that observed for the background. In the third line we describe the resonance by G_+-wave. And in the fourth line is given the result of fitting in terms of the G_0- and G_+-waves.

FIGURE 4. Momentum-transfer distribution for X(2000).

TABLE 2. Different sets of minimization for X(2000).

	Background N_{events}				Resonance N_{events}					Parameters MeV		$\delta\chi^2$	$N_{st.dev.}$
	S	D_0	D_{++}	G_0	S	D_0	D_{++}	G_0	G_+	M	σ		
1	546	108	194	92	–	–	–	–	–	–	–	0	–
2	495	78	170	77	72	13	12	1	0	2000.5	6.6	-31	5.5
3	467	67	191	118	–	–	–	–	117	2001.0	6.5	-37	6.0
4	458	75	192	99	–	–	–	31	106	2000.8	6.9	-40	6.3

Now we summarize results. We have obtained an indication of the existence of new resonance feature having a mass $2000 \pm 3.0 \pm 3.0$ MeV and a width less than 7 MeV. The spin-parity of this resonance is preferably $J^{PC} = 4^{++}$. The product of the cross section for X(2000) formation and the relevant branching ratio $\sigma BR(K_S K_S)$ is estimated at about 9 ± 5 nb. The statistical confidence of the observed phenomenon is better than 6 standard deviation.

REFERENCES

1. Landsberg L.G., *Yad. Phys.* **57**, 47 (1994); *Phys. Atom. Nucl.* **57**, 42 (1994).
2. Landsberg L.G., *Usp. Phys. Nauk* **164**, 1129 (1994); *Phys. Uspekhi* **37**, 1043 (1994).
3. Nozdrachev V.N. et al., "The resonance structures of $K_S K_S$ and $\Lambda\bar{\Lambda}$ spectrum at MIS ITEP", in *CP619, Hadron Spectroscopy: Ninth International Conference*, edited by D. Amelin and A.M. Zaitsev, 2001, pp. 155-164.
4. Particle Data Group, *Eur. Phys. J.* (2002).
5. Vladimirsky V.V., Grigor'ev V.K., et al., *Yad. Phys.* **66**, 729 (2003); *Phys. Atom. Nucl.* **66**, 700 (2003).

Pentaquark in Lattice QCD

Shoichi Sasaki

Department of Physics, University of Tokyo, Tokyo 113-0033, Japan

Abstract. We present results of our numerical calculation of the mass spectrum for the pentaquark states in quenched lattice QCD. It is found that the spin-parity assignment of the lowest $\Theta(uudd\bar{s})$ state is most likely spin-1/2 and negative parity. We have also calculated the mass of the charm analog of the Θ and find that the $\Theta_c(uudd\bar{c})$ state lies much higher than the DN threshold, in contrast to several model predictions.

Recently, LEPS collaboration at Spring-8 has observed a very narrow resonance $\Theta^+(1540)$ in the K^- missing-mass spectrum of the $\gamma n \to nK^+K^-$ reaction on ^{12}C [1]. Remarkable observation is its positive strangeness ($S=+1$), which means that the observed resonance must contain a strange antiquark. Thus $\Theta^+(1540)$ cannot be a three quark state. It is an exotic baryon state with the minimal quark content $uudd\bar{s}$. This discovery is subsequently confirmed in different reactions by the DIANA Collaboration at ITEP [2], the CLAS Collaboration at JLab [3], and the SAPHIR Collaboration at ELSA [4].

Theoretically existence of such a state was predicted long time ago by the Skyrme model [5]. However, a prediction closest in mass and width with the experiment was made by Diakonov, Petrov and Polyakov using a chiral-soliton model [6]. They predicted that it should be a narrow resonance and stressed that it can be detected by experiment because of its narrow width. In a general group theoretical argument with flavor $SU(3)$, $S=+1$ pentaquark state should be a member of antidecuplet or higher dimensional representation such as 27-plet. Both the Skyrme model and the chiral-soliton model predict that the lowest $S=+1$ state appears in the antidecuplet, $I=0$, and its spin and parity should be $(1/2)^+$[5, 6]. Experimentally, spin, parity and isospin of $\Theta^+(1540)$ are not determined yet. After the discovery of the $\Theta^+(1540)$, many model studies for the pentaquark state are made with different spin, parity and isospin. Lattice QCD in principle can determine these quantum numbers of the $\Theta^+(1540)$, independent of such arbitrary model assumptions or the experiments. We stress that there is substantial progress in lattice study of excited baryons recently [7]. Especially, the negative parity nucleon $N^*(1535)$, which lies close to $\Theta^+(1540)$, has become an established state in quenched lattice QCD [7, 8]. Here we report that quenched lattice QCD is capable of studying the $\Theta^+(1540)$ as well.

Indeed, it is not so easy to deal with the $qqqq\bar{q}$ state rather than usual baryons (qqq) and mesons ($q\bar{q}$) in lattice QCD. The $qqqq\bar{q}$ state can be decomposed into a couple of color singlet states as qqq and $q\bar{q}$, in other words, can decay into two-hadron states even in the quenched approximation. For instance, one can start a study with a simple minded local operator for the $\Theta^+(1540)$, which is constructed from a neutron operator and a K^+ operator such as $\Theta = \varepsilon_{abc}(d_a^T C\gamma_5 u_b)d_c(\bar{s}_e\gamma_5 u_e)$ or its Fierz rearrangement. The two-

CP717, *Hadron Spectroscopy: Tenth International Conference,*
edited by E. Klempt, H. Koch, and H. Orth
© 2004 American Institute of Physics 0-7354-0197-7/04/$22.00

point correlation function composed of such an operator, in general, couples not only to the Θ state (single hadron) but also to the two-hadron states such as an interacting KN system [9, 10] Even worse, when the mass of the $qqqq\bar{q}$ state is higher than the threshold of the hadronic two-body system, the two-point function should be dominated by the two-hadron states. Thus, an operator with as little overlap with the hadronic two-body states is desired in order to identify the signal of the pentaquark state in lattice QCD.

For this purpose, we propose a local interpolating operator of antidecuplet baryons based on an exotic description as diquark-diquark-antiquark. There are basically two choices as $\bar{3}_c \otimes \bar{3}_c$ or $\bar{3}_c \otimes 6_c$ to construct a color triplet diquark-diquark cluster [11, 12]. The former which we adopt is rather plausible since the one-gluon exchange between quarks is attractive in the $\bar{3}_c$ channel in contrast with the 6_c. Therefore, we introduce the flavor antitriplet ($\bar{3}_f$) and color anti-triplet ($\bar{3}_c$) diquark field

$$\Phi_\Gamma^{i,a}(x) = \frac{1}{2}\varepsilon_{ijk}\varepsilon_{abc}q_{j,b}^T(x)C\Gamma q_{k,c}(x) \tag{1}$$

where C is the charge conjugation matrix, abc the color indices, and ijk the flavor indices. Γ is any of the 16 Dirac γ-matrices. Accounting for both color and flavor antisymmetries, possible Γs are restricted within 1, γ_5 and $\gamma_5\gamma_\mu$ which satisfy the relation $(C\Gamma)^T = -C\Gamma$. Otherwise, above defined diquark operator is identically zero. Hence, three types of flavor $\bar{3}_f$ and color $\bar{3}_c$ diquark; scalar (γ_5), pseudoscalar (1) and vector ($\gamma_5\gamma_\mu$) diquarks are allowed [13]. The color singlet $qqqq\bar{q}$ state can be constructed by the color antisymmetric parts of diquark-diquark with an antiquark as $(\bar{3}_c \otimes \bar{3}_c)_{\text{antisym}} \otimes \bar{3}_c = 3_c \otimes \bar{3}_c = 1_c$. In terms of flavor, $\bar{3}_f \otimes \bar{3}_f \otimes \bar{3}_f = 1_f \oplus 8_f \oplus 8_f \oplus \overline{10}_f$. Manifestly, in this description, the S=+1 state belongs to the flavor antidecuplet [14]. Automatically, the S=+1 state should have isospin zero. Then, the interpolating operator of the $\Theta(uudd\bar{s})$ is obtained as

$$\Theta(x) = \varepsilon_{abc}\Phi_\Gamma^{s,a}(x)\Phi_{\Gamma'}^{s,b}(x)C\bar{s}_c^T(x) \tag{2}$$

for $\Gamma \neq \Gamma'$. The form $C\bar{s}^T$ for the strange antiquark field is responsible for giving the proper transformation properties of the resulting Θ operator under parity and Lorentz transformations [15]. Remark that because of the color antisymmetry, the combination of the same types of diquark is not allowed. Consequently, we have three different types of exotic S=+1 baryon operators [13]:

$$\Theta_+^1 = \varepsilon_{abc}\varepsilon_{aef}\varepsilon_{bgh}(u_e^T Cd_f)(u_g^T C\gamma_5 d_h)C\bar{s}_c^T, \tag{3}$$

$$\Theta_{+,\mu}^2 = \varepsilon_{abc}\varepsilon_{aef}\varepsilon_{bgh}(u_e^T C\gamma_5 d_f)(u_g^T C\gamma_5\gamma_\mu d_h)C\bar{s}_c^T, \tag{4}$$

$$\Theta_{-,\mu}^3 = \varepsilon_{abc}\varepsilon_{aef}\varepsilon_{bgh}(u_e^T Cd_f)(u_g^T C\gamma_5\gamma_\mu d_h)C\bar{s}_c^T \tag{5}$$

where the subscript "$+(-)$" refers to positive (negative) parity since these operators transform as $_P\,\Theta_\pm(\vec{x},t)_P^{-1} = \pm\gamma_4\Theta_\pm(-\vec{x},t)$ (for $\mu = 1,2,3$) under parity. The first operator of Eq. (3) is proposed for QCD sum rules in a recent paper [15] independently. Recall that any of local type baryon operators can couple to both positive- and negative-parity states since the parity assignment of an operator is switched by multiplying the left hand side of the operator by γ_5. The desired parity state is obtained by choosing the appropriate projection operator, $1 \pm \gamma_4$, on the two-point function $G(t)$ and direction of propagation in time. Details of the parity projection are described in Ref. [8]. We emphasize that two of them can couple to both spin-1/2 and spin-3/2 states. Using them, it is possible to study the spin-orbit partner of the spin-1/2 Θ state, whose presence contradicts the Skyrme picture of the Θ [12]. However, we will not pursue this direction in this article. We utilize only the first operator of Eq. (3), which couples only to a spin-1/2 state, in the following numerical simulations.

It is worth mentioning that our proposed operators are slightly different from the diquark-diquark-antiquark suggested by Jaffe and Wilczek [16]. Their description strongly relies on the highly correlated diquarks, one of which should have angular momentum one. However, ours are combined with two different types of diquarks, which have different spin-parity. In addition, Glozman proposed different interpolating operators of the Θ as diquark-diquark-antiquark, which have an alternative color construction of the diquark-diquark cluster such as $\bar{3}_c \otimes 6_c$ [12].

We generate quenched QCD configurations on a lattice $L^3 \times T = 32^3 \times 48$ with the standard single-plaquette Wilson action at $\beta = 6/g^2 = 6.2$ ($a^{-1} = 2.9$ GeV). The spatial lattice size corresponds to $La \approx 2.2$fm, which may be marginal for treating the ground state of baryons without large finite volume effect. Our results are analyzed on 135 configurations. The light-quark propagators are computed using the Wilson fermions at four values of the hopping parameter $\kappa = \{0.1520, 0.1506, 0.1497, 0.1489\}$, which cover the range $M_\pi/M_\rho = 0.68-0.90$. $\kappa_s = 0.1515$ and $\kappa_c = 0.1360$ are reserved for the strange and charm masses, which are determined by approximately reproducing masses of $\phi(1020)$ and $J/\Psi(3097)$. We calculate a simple point-point quark propagator. To perform precise parity projection, we construct forward propagating quarks by taking the appropriate linear combination of propagators with periodic and anti-periodic boundary conditions in the time direction.

In this calculation, the strange (charm) quark mass is fixed at κ_s (κ_c) and the up and down quark masses are varied from $M_\pi \approx 1.0$ GeV ($\kappa = 0.1489$) to $M_\pi \approx 0.6$ GeV ($\kappa = 0.1520$). Then, we will perform the extrapolation to the chiral limit using five different κ values. Details of our analysis to distinguish between the pentaquark state and possible two-hadron states are described in Ref. [17].

In Fig. 1 we show the mass spectrum of the Θ states with the positive parity (open circles) and the negative parity (open squares) as functions of the pion mass squared. Mass estimates are obtained from covariant single exponential fits in appropriate fitting range where the pentaquark state dominates in the two-point function [17]. All fits have a confidence level larger than 0.3 and $\chi^2/N_{DF} < 1.2$. It is evident that the lowest state of the isosinglet S=+1 baryons has *the negative parity*[1]. We evaluate the mass of the Θ with both parities in the chiral limit. A simple linear fit for all five values in Fig. 1 yields $M_{\Theta(1/2^-)} = 1.76(9)$ GeV and $M_{\Theta(1/2^+)} = 2.62(9)$ GeV, if we use the scale set by r_0 from Ref. [18]. Those values should not be taken too seriously at this moment since they do not include any systematic errors. Such a precise quantitative prediction of

---

[1] The parity assignment in my talk at Hadron2003 was opposite. This flip stems from the consequence of correcting the form of the strange antiquark from $\bar{s}^T$ to $C\bar{s}^T$ in Eq.(2).

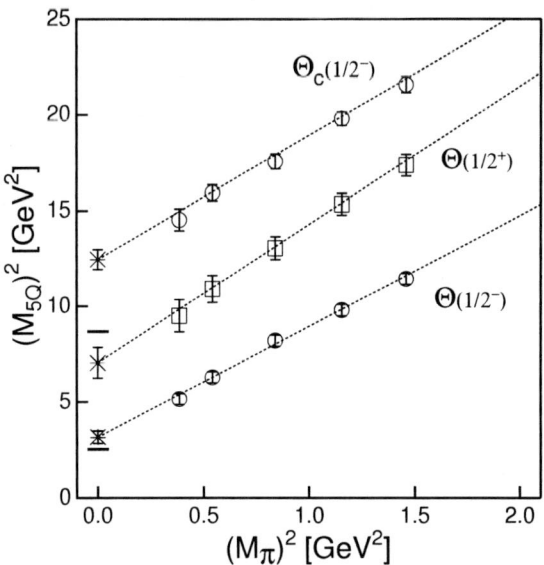

FIGURE 1. Masses of the spin-1/2 $\Theta(uudd\bar{s})$ states with both positive parity (open circles) and negative parity (open squares) as functions of pion mass squared. The charm analog of the isosinglet $\Theta_c(uudd\bar{c})$ state (open diamonds) is also plotted. Horizontal short bar represents the $KN(DN)$ threshold estimated by $M_N + M_K(M_N + M_D)$ in the chiral limit.

the pentaquark masses is not the purpose of the present paper. It is worth quoting other related hadron masses. The chiral extrapolated values for the nucleon, the kaon and the negative parity nucleon are $M_N = 1.05(3)$ GeV, $M_K = 0.52(1)$ GeV and $M_{N^*} = 1.73(5)$ GeV in this calculation. Comparing the negative parity Θ mass with the N^* mass, we obtain the remarkable observation $M_{\Theta(1/2^-)} \approx M_{N^*}$ in the chiral limit. It is in good agreement with the experimental fact that $\Theta^+(1540)$ is quite close to $N^*(1535)$. Thus, our results strongly indicate the J^P assignment of the $\Theta^+(1540)$ state is most likely $(1/2)^-$. This conclusion is consistent with that of a recent lattice study [19] if one corrects the parity assignment of their interpolating operator [20]. Note, however, that in contrast to Ref. [19] even the S-wave KN scattering state is also observed in our analysis [17], which strengthen the reliability of our result. Our conclusion is consistent with results obtained through a QCD sum rules approach [15] too.

Finally, we calculate the mass of other possible exotic states. The operator of exotic $\Xi_{3/2}(ssdd\bar{u}$ or $uuss\bar{d})$ can be treated by interchanging u and s or d and s in the operator of Eq. (3). As for the charm analog of the isosinglet $\Theta_c(uudd\bar{c})$, a strange antiquark is simply replaced by a charm antiquark. We remark that the negative parity state of the $\Xi_{3/2}$ can decay into either $\pi\Xi$ or $K\Sigma$ states. More complicated analysis is required than the case of the Θ. On the other hand, for the Θ_c, we take care of only the DN decay state for this one in similar to the Θ state. In this paper, we show results only for the lowest-lying spin-1/2 Θ_c state, which has the negative parity. As is evident in

Fig. 1, the Θ_c state lies much higher than the DN threshold in contradiction with several model predictions [16, 21]. The chiral extrapolated value of the Θ_c mass is about 640 MeV above the DN threshold in our calculation. This indicates that the anti-charmed Θ analog as either a very narrow resonance or a bound state is not to be expected.

We have calculated the mass spectrum of the $S=+1$ exotic baryon, $\Theta(uudd\bar{s})$, and the charm analog $\Theta_c(uudd\bar{c})$ in quenched lattice QCD. To circumvent the contamination from hadronic two-body states, we formulate the antidecuplet baryon interpolating operators using an exotic description like diquark-diquark-antiquark. We conclude that the J^P assignment of the $\Theta^+(1540)$ state is most likely $(1/2)^-$, which contradicts the Skyrme picture of the $\Theta^+(1540)$. Our results for the Θ_c state exclude the possibility of the charm analog of pentaquark like whichever a very narrow resonance or a bound state.

ACKNOWLEDGMENTS

It is a pleasure to acknowledge A. Hosaka, T. Nakano and L. Glozman for useful comments. I would also like to thank T. Doi, M. Oka and T. Hatsuda for fruitful discussions on the subject to determine the parity of the Θ state, and S. Ohta for helpful suggestions and his careful reading of the manuscript. This work is supported by the Supercomputer Project No.102 (FY2003) of High Energy Accelerator Research Organization (KEK) and JSPS Grant-in-Aid for Encouragement of Young Scientists (No. 15740137).

REFERENCES

1. T. Nakano *et al.*, Phys. Rev. Lett. **91**, 012002 (2003).
2. DIANA Collaboration, V. V. Barmin *et al.*, hep-ex/0304040.
3. CLAS Collaboration, S. Stepanyan *et al.*, hep-ex/0307018.
4. SAPHIR Collaboration, J. Barth *et al.*, hep-ex/0307083.
5. M. Chemtob, Nucl. Phys. B **256** (1985) 600.
6. D. Diakonov, V. Petrov and M. V. Polyakov, Z. Phys. A **359**, 305 (1997).
7. For recent reviews, see S. Sasaki, nucl-th/0305014 and C. Morningstar, nucl-th/0308026.
8. S. Sasaki, T. Blum and S. Ohta, Phys. Rev. D **65**, 074503 (2002).
9. M. Lüscher, Commun. Math. Phys. **105**, 153 (1986).
10. M. Fukugita *et al.*, Phys. Rev. D **52**, 3003 (1995).
11. C. E. Carlson, C. D. Carone, H. J. Kwee and V. Nazaryan, hep-ph/0307396.
12. L. Y. Glozman, hep-ph/0308232.
13. S. Sasaki, *Talk presented at the Int. Conf. on Hadron2003, Aschaffenburg, Germany, Sep. 1 - 6, 2003.*
14. A. Hosaka, hep-ph/0307232.
15. J. Sugiyama, T. Doi and M. Oka, hep-ph/0309271.
16. R. L. Jaffe and F. Wilczek, hep-ph/0307341.
17. S. Sasaki, hep-lat/0310014.
18. S. Necco and R. Sommer, Nucl. Phys. B **622**, 328 (2002).
19. F. Csikor, Z. Fodor, S. D. Katz and T. G. Kovacs, hep-lat/0309090.
20. Z. Fodor, in private communication.
21. M. Karliner and H. J. Lipkin, hep-ph/0307343.

Mass spectrum of heavy hybrid mesons in the QCD string model

D. S. Kuzmenko

Institute of Theoretical and Experimental Physics,
B.Cheremushkinskaya, 25, Moscow, Russia

Abstract. Using adiabatic hybrid meson potentials calculated in the QCD string model, the mass spectrum of heavy hybrid mesons is calculated in Born-Oppenheimer approximation within the accuracy of 100 MeV.

INTRODUCTION

A phenomenon of valence glue is very interesting from theoretical point of view, since it is directly related to the phenomenon of confinement. It is well known that states consisting of quark, antiquark, and valence gluon are present in QCD spectrum. In particular, the adiabatic spectrum of valence gluon excitations in the presence of static quark and antiquark was calculated with high accuracy in lattice QCD [1]. However, there is no established conception of valence gluon up to now. Several kinds of models with different treatments of excited glue exist, and precise lattice results [1] have to be used to distinguish the correct one. A QCD string approach, or QCD string model considered below treats hybrid meson as a system of point valence gluon joined by straight strings of confining field to quark and antiquark. It is based on the background field method and uses the Fock-Feynman- Schwinger proper time – path integral formalism [2] to derive the string Hamiltonian of hybrid meson [3]. In the recent paper [4] hybrid adiabatic potentials were calculated in the QCD string model and were shown to match well lattice data [1]. Since preliminary results have already been reported at this Conference held in Protvino two years ago [5], I shall only present final results and make estimations of an accuracy of the variational procedure been used. Relying on adiabatic potentials, I shall then calculate masses of heavy hybrid mesons in Born-Oppenheimer approximation.

STRING HAMILTONIAN OF HYBRID MESON

When the calculations of the spectrum of heavy hybrid mesons are concerned, one is interested first of all in dynamics of valence glue at small and intermediate distances between quark and antiquark. It is shown in the QCD string model, see [4] and refs. therein, that in this region the inertia J of the strings, joining the valence gluon with quark and antiquark, is much smaller than the "constituent mass" of valence gluon μ, appearing as *einbein field* in Fock-Feynman-Schwinger formalism. String Hamiltonian

CP717, Hadron Spectroscopy: Tenth International Conference,
edited by E. Klempt, H. Koch, and H. Orth
© 2004 American Institute of Physics 0-7354-0197-7/04/$22.00

of valence gluon takes the form [4]

$$H = \frac{\mu}{2} + \frac{p^2}{2\mu} + V^{\text{conf}} + V^{\text{OGE}} \qquad (1)$$

where first two terms describe the kinetic energy of valence gluon and contain the einbein μ which has to be eliminated through the stationary point condition, and last two terms are confining and one-gluon-exchange potentials,

$$V^{\text{conf}} = \sigma r_1 + \sigma r_2, \qquad V^{\text{OGE}} = -\frac{3\alpha_s}{2r_1} - \frac{3\alpha_s}{2r_2} + \frac{\alpha_s}{6R}, \qquad (2)$$

where r_1 and r_2 are distances from quark and antiquark to valence gluon, R is the quark-antiquark separation, and $\sigma \approx 0.18$ GeV^2 is the string tension. V^{conf} emerges from the area law for Wilson loop, while V^{OGE} represents the color-Coulomb perturbative OGE interaction. String inertia is taken into account in the order J/μ,

$$H^{\text{string}} = -\frac{\sigma}{6\mu^2}\left(\frac{1}{r_1}L_1^2 + \frac{1}{r_2}L_2^2\right), \qquad L_i = \mathbf{r}_i \times \mathbf{p}. \qquad (3)$$

Spin-orbit interaction of valence gluon emerges due to nonperturbative and perturbative potentials (2), which yield correspondingly

$$H^{\text{LS(np)}} = -\frac{\sigma}{2\mu^2}\left(\frac{\mathbf{L}_1\mathbf{S}}{r_1} + \frac{\mathbf{L}_2\mathbf{S}}{r_2}\right), \qquad H^{\text{LS(p)}} = \frac{3\alpha_s}{4\mu^2}\left(\frac{\mathbf{L}_1\mathbf{S}}{r_1^3} + \frac{\mathbf{L}_2\mathbf{S}}{r_2^3}\right). \qquad (4)$$

Hamiltonians (3), (4) treated as corrections lead to the energy decrease of the order of $100 \div 200$ MeV depending on the level.

VARIATION METHOD AND ITS ACCURACY

There is no exact solution of Schroedinger equation with Hamiltonian (1), (2). Therefore the variation method was used in [4] to calculate its spectrum. Having in mind a goal to study the stability of the variation procedure, let us consider the toy Hamiltonian

$$\tilde{H}(\mu) = \frac{\mu}{2} + \frac{p^2}{2\mu} + \sigma r \equiv \frac{\mu}{2} + h(\mu). \qquad (5)$$

An exact solution of the equation $h\Psi = \varepsilon\Psi$ in the case of S-wave is known to be the Airy function, $\Psi(r) = c/r \, \text{Ai}\left\{(2\mu\sigma)^{1/3}(r - \varepsilon/\sigma)\right\}$, where c is the constant; energy levels are defined by its zeros α_n: $\varepsilon_n(\mu) = (\sigma^2/2\mu)^{1/3}\alpha_{n+1}$. Taking the minimum, $E = E(\mu^*) = \min_\mu\left(\frac{\mu}{2} + \varepsilon(\mu)\right)$, one easily calculates the exact spectrum of toy Hamiltonian,

$$E_n = 2\mu^* = 2\left(\frac{\alpha_{n+1}}{3}\right)^{3/4}\sqrt{2\sigma}, \qquad E_0 = 2.346\sqrt{\sigma}. \qquad (6)$$

422

Consider now the variational method with probe wave functions $\Psi(\beta)$ depending on the variational parameter. After the calculation of averages $E(\mu,\beta) = \langle\Psi(\beta)|H(\mu)|\Psi(\beta)\rangle$ the variation over parameters μ and β has to be performed, $\min\limits_{\mu,\beta} E = E(\mu^*,\beta^*) = E^{\text{var}}$. Note that the choice of parameters order in variation is a question of convenience. Let us consider the oscillator probe w.f., $\Psi \sim \exp(-\beta r^2/2)$. Than the result for the ground level reads

$$E_0^{\text{var}} = 2\left(\frac{6}{\pi}\right)^{1/4}\sqrt{\sigma} = 2.351\sqrt{\sigma}. \tag{7}$$

Comparing to the exact answer (6) one can see that the accuracy of the variation procedure is high, $\delta E = (E_0^{\text{var}} - E_0)/E_0 = 0.2\%$, or $\Delta E = E_0^{\text{var}} - E_0 \approx 2$ MeV.

Another possible choice of probe wave function with one variation parameter is the w.f. of Coulomb-type, $\Psi \sim \exp(-\alpha r)$. Using this w.f. for toy Hamiltonian (5) one would get the ground level with the accuracy $\delta E = 4\%$, which is much worse than in the case of oscillator probe w.f.[1]. Therefore in what follows we proceed with probe functions of oscillator type.

In order to perform the evaluation of accuracy of variational procedure used in [4], we consider the string Hamiltonian (1), (2) in the case when positions of Q and $\bar{Q}$ coinside, i.e.

$$\hat{H}(\mu) = \frac{\mu}{2} + \frac{p^2}{2\mu} + 2\sigma r - \frac{3\alpha_s}{r}. \tag{8}$$

Minimization over μ is easily implemented: $\mu^* = \sqrt{\frac{2l+3}{2}}\beta$, where l is the angular momentum of valence gluon;

$$E(\beta) = \frac{2\sigma\langle r\rangle}{\beta} + \beta\left(\sqrt{\frac{2l+3}{2}} - 3\alpha_s\left\langle\frac{1}{r}\right\rangle\right), \tag{9}$$

where the dependence on β is written explicitly. Than the minimization over β leads to the energy levels

$$E_l^{\text{var}} = 2\sqrt{2\sigma\langle r\rangle\left(\sqrt{\frac{2l+3}{2}} - 3\alpha_s\left\langle\frac{1}{r}\right\rangle\right)} \tag{10}$$

One can see that at some value of α_s the expression (10) turns to zero, which means that the variational procedure becomes unapplicable. Critical values of constant coupling are $\alpha_s = 0.36$ for $l=0$ and $\alpha_s = 0.7$ for $l=1$, and grow further for higher levels. One can see that starting from $l=1$ they are far enough from the physical region of values, $\alpha_s = 0.2 \div 0.3$. Now we estimate the accuracy of the variation method calculating differences of levels (10) at two values of α_s: $\delta E^{\text{max}} = (E^{\text{var}}|_{\alpha_s=0.2} - E^{\text{var}}|_{\alpha_s=0.3})/E^{\text{var}}|_{\alpha_s=0.2} = 10\%$

---

[1] Note that calculations of ground level for H_2^+ ion with the probe w.f. of Coulomb type have also rather low accuracy, $(E_0^{\text{exp}} - E_0^{\text{var}})/E_0^{\text{exp}} \simeq 20\%$

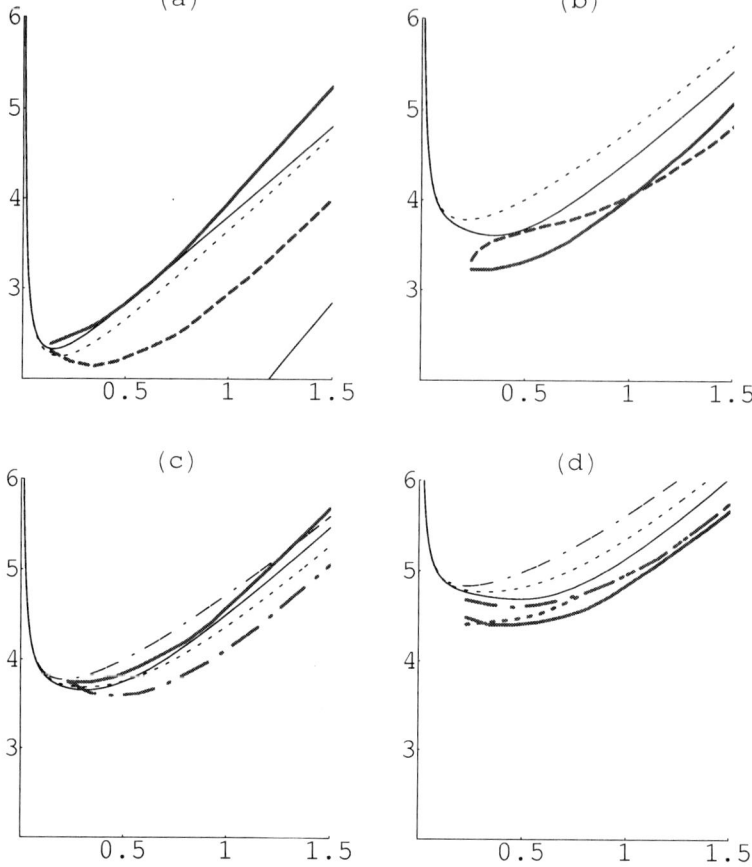

FIGURE 1. Adiabatic potentials in QCD string model from [4] (thin dense curves) compared to lattice curves from [1] in units $1/r_0 = 400$ MeV. Solid, dashed and dashed-dotted curves correspond to $\Lambda = 0, 1$ and 2. Quantum numbers of levels in atomic notations are: Σ_u^-, Π_u (a); Σ_g^+, Π_g (b); Σ_g^-, Π_g, Δ_g (c), and Σ_u^+, Π_u, Δ_u (d).

for first two levels, $l = 0, 1$. An absolute error, $\Delta E = (E^{\text{var}}|_{\alpha_s=0.2} - E^{\text{var}}|_{\alpha_s=0.3})$, equals 120 MeV for $l = 0$ and 160 MeV for $l = 1$.

We present results [4] of variational calculations of adiabatic potentials in QCD string model in comparison with lattice ones in Fig.1. One can see that, in average, descrepancy between corresponding analytic and lattice curves is indeed within 150 MeV.

SPECTRUM OF HEAVY HYBRID MESONS

We calculate spectra of heavy hybrid mesons masses in Born-Oppenheimer approximation assuming that the motion of valence quark and antiquark is slow compared to the motion of valence gluon, and using adiabatic potentials. One can see from Fig.1 that Q and $\bar{Q}$ are to be located around the potential minimum, where adiabatic potentials may be replaced by oscillatory ones. Therefore the problem reduces to the calculation of the three-dimentional oscillator spectrum, $M^{Q\bar{Q}g} = 2m_Q + E^{\text{osc}}$, where m_Q is the constituent mass of heavy quark. We choose standard values of parameters, $\alpha_s = 0.25$, $\sigma = 0.18$ GeV^2, $m_c = 1.29$ GeV, $m_b = 4.56$ GeV, and calculate lowest nonexotic states with quantum numbers $J^{PC} = 1^{+-}$: $M^{c\bar{c}g}(1^{+-}) = 4.61$ GeV, $M^{b\bar{b}g}(1^{+-}) = 10.53$ GeV. Then we can extract the mass of valence gluon M,

$$M \equiv M^{b\bar{b}g}(1^{+-}) - M^{\Upsilon(1S)} = 1.0 \text{ GeV}. \tag{11}$$

It is the mass that is related to correlation length of the gluon field, $\lambda \approx 1/M = 0.2$ fm. Note that direct calculations of the gluon correlation length from field correlators on the lattice [6] yield the same value, $\lambda = 0.2$ fm.

On can calculate the accuracy of Born-Oppenheimer approximation as $\Delta E^{Q\bar{Q}g} = M^2/2m_q$ to obtain $\Delta E^{c\bar{c}g} = 400$ MeV, $\Delta E^{b\bar{b}g} = 100$ MeV. These values indicate that the Born-Oppenheimer approximation does not work for $c\bar{c}g$ hybrid meson, while in the case of $b\bar{b}g$ hybrid its accuracy is of the same order as the accuracy of variation procedure considered above.

Most interesting from experimental point of view are exotic states with $J^{PC} = 1^{-+}$. There are two close states of hybrid mesons with these quantum numbers: $M_1^{b\bar{b}g}(1^{-+}) = 10.83$ GeV, where the valence gluon is in the state Π_u, and $M_2^{b\bar{b}g}(1^{-+}) = 10.92$ GeV, where it is in the state Σ_g^-. Note that the state Σ_g^- is missing on the lattice, and the state with Π_u has the mass $M_{1,\text{lat}}^{b\bar{b}g}(1^{-+}) = 10.8$ GeV [1].

To summarize, the uncertainty of adiabatic potentials in hybrid meson calculated in the QCD string model [4] is estimated as no greater than 150 MeV. An accuracy of Born-Oppenheimer adiabatic approximation is estimated for $b\bar{b}$-hybrid as ~ 100 MeV. A mass of valence gluon excitation $M = 1.0$ GeV is extracted from the lowest nonexotic $b\bar{b}$-hybrid meson state and the value of gluon correlation length $\lambda = 0.2$ fm is determined in agreement with direct lattice calculations. Masses of lowest exotic $b\bar{b}g$ states are calculated.

The author is grateful to Yu. S. Kalashnikova and Yu. A. Simonov for valuable comments. This work has been supported by the Federal Program of the Russian Ministry of Industry, Science and Technology No. 40.052.1.1.1112, and grant NSh-1774.2003.2.

REFERENCES

1. K. J. Juge, J. Kuti, C. Morningstar, in Proc. of the Third Int. Conf. on Quark Confinement and the Hadron Spectrum, Jefferson Lab, 1998, hep-lat/9809015; in Proc. of LATTICE98, Boulder, USA, 1998, Nucl. Phys. Proc. Suppl. **73**, 590 (1999).
2. Yu.A. Simonov, J.A. Tjon, Annals Phys. **300**, 54 (2002).
3. Yu.A.Simonov, Lectures at the XVII International School of Physics, Lisbon, 1999, hep-ph/9911237; Yu.S.Kalashnikova and Yu.B.Yufryakov, Phys.Lett. **B359**, 175 (1995); Phys.At.Nucl. **60**, 307 (1997).
4. Yu. S. Kalashnikova, D. S. Kuzmenko, Phys. Atom. Nucl. **66**, 955 (2003); hep-ph/0203128.
5. Yu. S. Kalashnikova, D. S. Kuzmenko, AIP Conf. Proc. **619**, 241 (2002).
6. A. Di Giacomo, H. Panagopoulos, Phys. Lett. B **285**, 133 (1992); A. Di Giacomo, E. Meggiolaro, H. Panagopoulos, Nucl. Phys. B **483**, 371 (1997).

Static potentials and confining field structure of baryons and $3g$-glueballs

D. S. Kuzmenko, Yu. A. Simonov

Institute of Theoretical and Experimental Physics,
B.Cheremushkinskaya, 25, Moscow, Russia

Abstract. Static potentials in baryons and $3g$-glueballs are calculated using the field correlator method. It is shown that the static baryon potential describes lattice data. For confining fields defined through the "connected probe" distributions in both types of hadrons are calculated. It is demonstrated that the structure of baryon is of Y type (but not the Δ-type one). A field distribution in Δ-type $3g$-glueballs is presented.

INTRODUCTION

Apart from the usual way to study the momentum structure of hadrons through their structure functions, the approach considered below deals with the space structure of hadrons with static sources. This approach is convenient from theoretical point of view, since Green functions of hadrons with static sources reduce to their Wilson loops. Recent computations in gluodynamics on the lattice of static potential [1, 2] and flux distribution in baryon [3] present a set of accurate data of numerical experiment. The same tasks were also considered at the same time analytically in the framework of the method of field correlators [4]-[9]. Using the nonperturbative exponential fall of gluon field correlators with the correlation length λ, the method of field correlators allows in particular to describe lattice simulations of the potential in baryon and explain its characteristic behavior [4], [5]. We shall demonstrate it in what follows, and present the static potential in the case of $3g$-glueball as well. Moreover, we shall consider flux distributions in terms of the "connected probe" [10].

$3G$-GLUEBALLS IN THE BACKGROUND FIELD METHOD

While the Wilson loop of baryon is well known, it is not so easy to handle $3g$-glueballs. One can construct the wave function in background field method, where the gluon field is splitted into two parts, "valence" g^a and "background" B^a. Valence gluons then acts as static sources bounded into gauge invariant physical states by confining background fields. Since the valence field transforms homogeniously, $G_i^j \rightarrow U_{j'}^{+j} G_{i'}^{j'} U_i^{i'}$, where $G_i^j(x) \equiv g_a(x) t_i^{(a)j}$, a is adjoint and $i, j, \ldots$ – fundamental color indexes, gauge-invariant wave functions of extended $3g$-glueballs may have structures of Y or of Δ type:

$$\Psi_Y^{(f)}(x,y,z,Y) = f^{abc} g_a(x,Y) g_b(y,Y) g_c(z,Y) \qquad (\text{and } f^{abc} \leftrightarrow g^{abc}), \qquad (1)$$

CP717, *Hadron Spectroscopy: Tenth International Conference,*
edited by E. Klempt, H. Koch, and H. Orth

$$\Psi_{\Delta}(x,y,z) = G_i^j(x)\Phi_j^k(x,y)G_k^l(y)\Phi_l^m(y,z)G_m^n(z)\Phi_n^i(z,x). \qquad (2)$$

Here $g_a(x,y) = g_a(x)\Phi^{ab}(x,y)$ denotes the extended gluon operator, and $\Phi(x,y) = \mathrm{Pexp}\, ig\int_y^x B_\mu dz_\mu$ is the parallel transporter or Schwinger line in the adjoint or fundamental representation. Coordinates x,y,z and Y in (1) apply to valence gluons and string junction positions respectively, f^{abc} and g^{abc} denote adjoint antisymmetric and symmetric symbols. Wilson loops are then given by vacuum averages $W = \langle \Psi^+(\bar{X})\Psi_i(X)\rangle$, where $X = x,y,z$ for Δ and x,y,z,Y for Y-states, and have a form

$$W^{Y,f} = \frac{1}{24}\left\langle f^{abc}f^{a'b'c'}\Phi^{aa'}(C_1)\Phi^{bb'}(C_2)\Phi^{cc'}(C_3)\right\rangle, \qquad (3)$$

$$W^{Y,d} = \frac{3}{40}\left\langle d^{abc}d^{a'b'c'}\Phi^{aa'}(C_1)\Phi^{bb'}(C_2)\Phi^{cc'}(C_3)\right\rangle, \qquad (4)$$

where C_i are trajectories of valence gluons. A Wilson loop of Δ-type glueball at large distances can be represented as a product of three rectangular quark-antiquark Wilson loops [5],

$$W^\Delta(\bar{X},X) = W(\bar{x},\bar{y}|x,y)W(\bar{y},\bar{z}|y,z)W(\bar{z},\bar{x}|z,x). \qquad (5)$$

STATIC POTENTIALS

Static potential in baryon calculated in bilocal approximation of the field correlator method [4] reads as

$$V_B(R_1,R_2,R_3) = \left(\sum_{a=b} - \sum_{a<b}\right) n_i^{(a)}n_j^{(b)}\int_0^{R_a}\int_0^{R_b} dl\, dl'\int_0^\infty dt\, \mathrm{D}_{i4,j4}(z_{ab}), \qquad (6)$$

where $\mathrm{n}^{(a)}$ is the unit vector directed from the string junction to the a-th quark, R_a the separation between this quark and the string junction, and $z_{ab} = (l\mathrm{n}^{(a)} - l'\mathrm{n}^{(b)}, t)$. It is expressed through the two-point correlation function $\mathrm{D}_{\mu_1 v_1, \mu_2 v_2}(x - x') \equiv (g^2/N_c)\,\mathrm{tr}\langle F_{\mu_1 v_1}(x)\Phi(x,x')F_{\mu_2 v_2}(x')\Phi(x',x)\rangle$ and minimal surfaces of Wilson loop. When shortest paths connecting x and x' are used, bilocal correlators are written in the general form containing two scalar formfactors $D(z)$ and $D_1(z)$ [11],

$$\mathrm{D}_{\mu_1 v_1, \mu_2 v_2}(z) = (\delta_{\mu_1\mu_2}\delta_{v_1 v_2} - \delta_{\mu_1 v_2}\delta_{\mu_2 v_1})D(z) +$$

$$\frac{1}{2}\left(\frac{\partial}{\partial z_{\mu_1}}(z_{\mu_2}\delta_{v_1 v_2} - z_{v_2}\delta_{v_1\mu_2}) + \frac{\partial}{\partial z_{v_1}}(z_{v_2}\delta_{\mu_1\mu_2} - z_{\mu_2}\delta_{\mu_1 v_2})\right)D_1(z), \qquad (7)$$

where $D(z) \sim \exp{-|z|/\lambda}$ falls with the correlation length $\lambda \approx 0.2$ fm, which is directly connected to the energy of valence gluon excitation, see e.g. [12]. Its contribution to the potential can be represented as a sum $V^d + V^{nd}$, where the first term corresponding to the first "diagonal" sum in (6) is just a sum of quark-antiquark potentials, $V^d(R_1,R_2,R_3) = \sum_a V^{Q\bar{Q}}(R_a)$. At large in comparison with λ distances $V^d \approx \sum_a \sigma R_a$, where $\sigma \approx 0.18$ GeV^2 is the string tension. The "nondiagonal" term V^{nd} corresponds to the second sum in (6) and saturates at large distances.

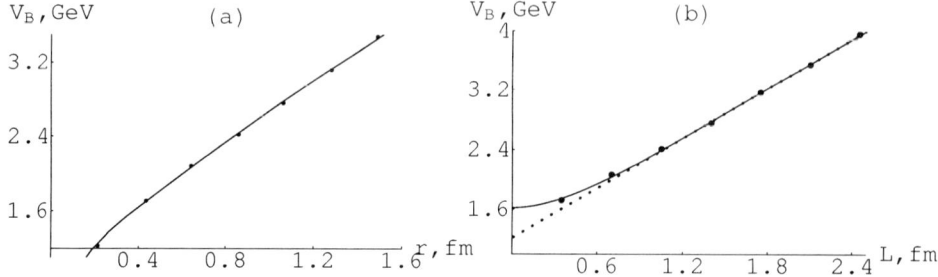

FIGURE 1. Baryon potential in equilateral triangle vs. quark separation r (solid curve) in comparison with the lattice data [2] (points) (a), and potential without the color-Coulomb part (solid curve) in comparison with the lattice data [1] (points) in dependence on the total length of the baryon string L (b).

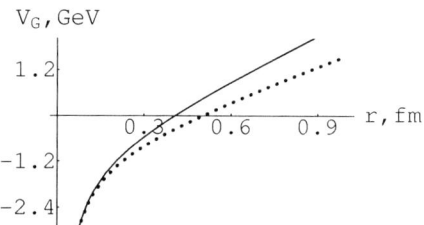

FIGURE 2. Potentials of three-gluon-glueballs V_G^Y (solid curve) and V_G^Δ (dotte curve) in equilateral triangle vs. sources separation r.

Function $D_1(z)$ in (7) has perturbative origin and corresponds to OGE color-Coulomb potential

$$V^{\text{Coul}} = -\frac{C_F \alpha_s}{2} \sum_{i<j} \frac{1}{r_{ij}}, \tag{8}$$

r_{ij} is the i-th and j-th quark separation. Moreover, in what follows we take into account the charge screening considered in [13].

A dependence of the baryon potential in equilateral triangle on the quark separation is given in comparison with the lattice data [2] in Fig. 1(a). The value of the string tension $\sigma = 0.17$ GeV^2 is fitted to obtain the lattice slope at large distances. The value of $\lambda = 0.21$ fm is calculated according to [13]. In Fig. 1(b) the behavior of the baryon potential with the color-Coulomb part subtracted is shown in comparison with the lattice data [1] as a function of the total length of baryon string $L = \sum_a R_a$ ($\sigma = 0.22$ GeV^2 is fitted, $\lambda = 0.18$ fm calculated according to [13]). A tangent (dots) with the slope σ demonstrates a decrease of the slope of potential at small distances. This effect is induced by the influence of the correlation length of confining fields [4].

For Y-type glueballs the Casimir scaling holds, $V_G^Y/V_B = C_A/C_F = 9/4$, where C_F and C_A are quadratic Casimir operators in fundamental and adjoint representations.

Potential in Δ-glueball in the case of equilateral triangle with the side r has a form $V_G^\Delta(r) = (C_A/C_F)V^{\text{Coul}}(r) + V^d(r) - 2V^{\text{nd}}(r)$ [5]. The behavior of V_G^Y and V_G^Δ in equilateral triangle in dependence on the source separation r is shown in Fig. 2. One can conclude that states of Δ-type are more favorite energetically than Y-type ones.

FIELDS DISTRIBUTIONS

Confining field in baryon calculated using the "connected probe" in the method of field correlators [8] reads as

$$\left(\varepsilon^{(B)}\right)^2 = \frac{2}{3}\left(\left(\varepsilon^{B}_{(1)}\right)^2 + \left(\varepsilon^{B}_{(2)}\right)^2 + \left(\varepsilon^{B}_{(3)}\right)^2\right), \tag{9}$$

where fields $\varepsilon^{B}_{(i)}$ calculated for the probe plaquette joint to the trajectory of i-th quark are expressed through confining fields of $Q\bar{Q}$ pairs $\varepsilon^{Q\bar{Q}}$ (which in turn are directly related to static potentials, [8]):

$$\varepsilon^{B}_{(1)}(\mathbf{x},\mathbf{R}^{(1)},\mathbf{R}^{(2)},\mathbf{R}^{(3)}) = \varepsilon^{Q\bar{Q}}(\mathbf{x},\mathbf{R}^{(1)}) - \frac{1}{2}\varepsilon^{Q\bar{Q}}(\mathbf{x},\mathbf{R}^{(2)}) - \frac{1}{2}\varepsilon^{Q\bar{Q}}(\mathbf{x},\mathbf{R}^{(3)}). \tag{10}$$

Normalizing coefficient 2/3 in (9) is chosen due to the condition that at large separations the field acting on quarks equals to σ.

A field calculated with connected probe in Δ-type glueball [8] is taken as

$$\varepsilon^{(G)}_{\Delta}(\mathbf{x},\mathbf{r}^{(1)},\mathbf{r}^{(2)},\mathbf{r}^{(3)}) = \sum_{i=1}^{3}\varepsilon^{Q\bar{Q}}(\mathbf{x}-\mathbf{r}^{(i)},\mathbf{r}^{(i+1)\mathrm{mod}3} - \mathbf{r}^{(i)}), \tag{11}$$

where $\mathbf{r}^{(i)}$ denotes the position of i-th valence gluon.

Distributions of confining field $\varepsilon^{(B)}$ and $\varepsilon^{(G)}_{\Delta}$ in the plane of static sources forming an equilateral triangle with the side 1 fm are shown in Fig. 3. Note that perturbative color-Coulomb field forming peaks around sources is not shown here. In Fig. 4 surfaces formed by the confining field with the value σ are shown for baryon and Δ-type glueball. Note that there is a maximum of the field in baryon around the string junction (in the centre of triangle).

SUMMARY

We have demonstrated how to construct Wilson loops of 3g-glueballs of Y- and Δ- types in the background field method. Static potential in baryon was calculated in bilocal approximation of the method of field correlator and was demonstrated to describe lattice data. Static potentials for 3g-glueballs were calculated and it was shown that Δ-type glueball is favourite energetically. Field distributions defined using the connected probe were plotted for baryon and Δ-type glueball.

This work has been supported by the Federal Program of the Russian Ministry of Industry, Science and Technology No. 40.052.1.1.1112 and grant NSh-1774.2003.2.

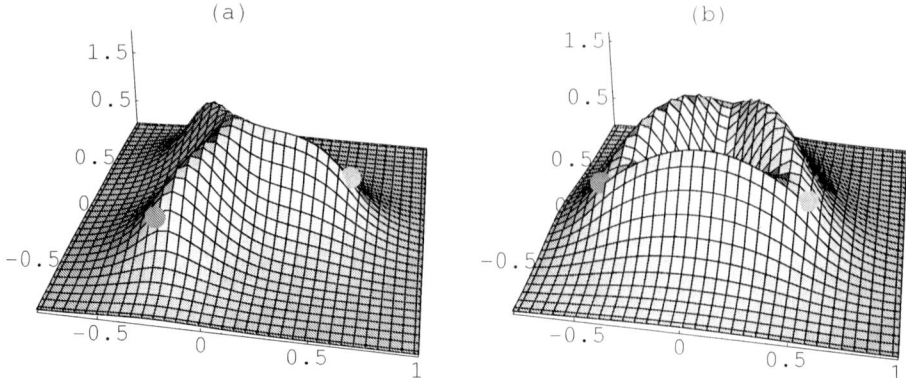

FIGURE 3. Distribution of the confining field $\varepsilon^{(B)}$ (a) $\varepsilon_\Delta^{(G)}$ (b) in the plane of sources for equilateral triangle with the side 1 fm. Positions of sources are marked by points.

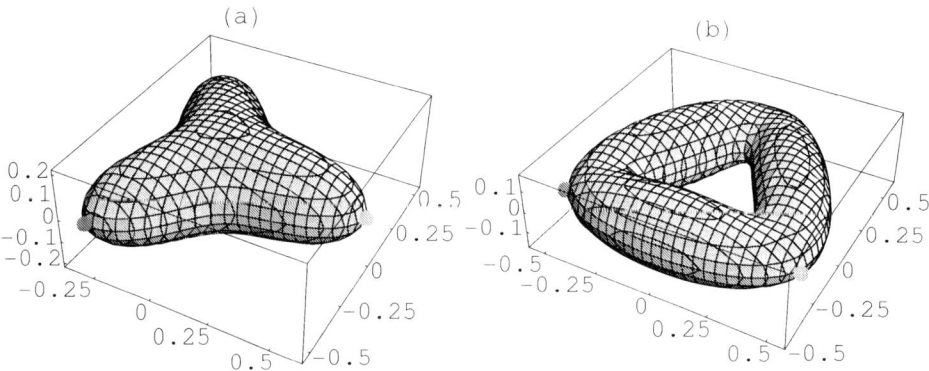

FIGURE 4. Surfaces $|\varepsilon^{(B)}(x)| = \sigma$ (a) and $|\varepsilon_\Delta^{(G)}(x)| = \sigma$ (b) at source separations 1 fm. Positions of sources are marked by points.

REFERENCES

1. T. T. Takahashi *et al.*, Phys. Rev. D **65**, 114509 (2002).
2. C. Alexandrou, Ph. de Forcrand, O. Jahn, hep-lat/0209062.
3. H. Ichie, V. Bornyakov, T. Streuer, G. Schierholz, Nucl. Phys. A **721**, 899 (2003).
4. D. S. Kuzmenko, hep-ph/0204250, Yad. Fiz. **66**, No.12 (2003) (in press)
5. D. S. Kuzmenko, Yu. A. Simonov, Yad. Fiz. **66**, 5 (2003).
6. D.S. Kuzmenko, Yu.A. Simonov, Phys. Lett. B **494**, 81 (2000).
7. D.S. Kuzmenko, Yu.A. Simonov, Yad. Fiz. **64**, 110 (2001).
8. D.S. Kuzmenko, Yu.A. Simonov, hep-ph/0302071, Yad. Fiz. **67**, No.2 (2004) (in press)
9. D.S. Kuzmenko, Yu.A. Simonov, V.I. Shevchenko, Usp. Fiz. Nauk (2003) (in press)
10. A.Di Giacomo *et. al.*, Phys.Lett. **B236** (1990) 199; Nucl.Phys. **B347** (1990) 441
11. H. G. Dosch, Phys. Lett. B **190**, 177 (1987); H. G. Dosch, Yu. A. Simonov, Phys. Lett. B **205**, 399 (1988); Yu. A. Simonov, Nucl. Phys. B **307**, 512 (1988);
    A. Di Giacomo, H. G. Dosch, V. I. Shevchenko and Yu.A. Simonov, Phys. Rept. **372**, 319 (2002).
12. D. S. Kuzmenko, this volume, hep-ph/0310033.
13. D. S. Kuzmenko, submitted to Phys. At. Nucl., hep-ph/0310017; this volume, hep-ph/0310035;
    D. S. Kuzmenko, V.I. Shevchenko, Yu. A. Simonov, Usp. Fiz. Nauk (2003) (in press).

Exotic charmonium hybrids at PANDA

Agnes Lundborg

Uppsala University

Abstract. Recent lattice-QCD calculations of the charmonium hybrid spectrum predict the ground state hybrid to be a spin-exotic with quantum number $J^{PC} = 1^{-+}$ at a mass of about 4.3 GeV/c^2. Such a low mass hybrid could be as narrow as $O(20 MeV/c^2)$ due to dynamical suppression of decay into open charm. The exotic quantum numbers prevent the state from mixing with conventional mesons and simplifies the identification of the state as a non-meson state. Lattice calculations name the most obvious hybrid charmonium decay channel to be a conventional charmonium and light hadrons. The detection of such a final state with seven photons and a lepton pair within the future PANDA detector at GSI is investigated with Monte Carlo methods at Uppsala University.

INTRODUCTION

After 25 years of high energy physics experiments, the strong force is to a large extent still unchartered territory. At small distances, where quarks experience asymptotic freedom, perturbative Quantum Chromo Dynamics has proven extremely successful. At larger distances gluon self-interaction leads to confinement of quarks into colorless hadrons. At this scale perturbative QCD no longer apply and phenomenological models struggle to clarify what really is going on. In this region the search is on for the the answer to the riddles of hadronic mass generation, confinement and chiral symmetry breaking. Knowledge of states with explicit gluonic content, so called glueballs and hybrids, can add important information on the processes involved.

QCD predicts the interaction between gluons and therefore hadronic states with excited glue, glueballs and hybrids, should exist. Still, not many hybrids or glueballs have been indisputably identified. In the region of light mesons the abundance of states is so large that there is almost always mixing between states with the same quantum numbers. To identify and classify heavily mixed states there is a need to understand the complete light-quark meson spectrum both experimentally and theoretically, a task which has proven to be very difficult.

Exotic quantum numbers. In the simple quark model a meson is described as a combination of a valence quark and an antiquark. This perspective restricts the accessible meson quantum numbers to a set of allowed J^{PC} combinations and most mesons can be described within this framework. States with different J^{PC} quantum numbers are referred to as spin-exotics, such states are pretty rare and therefore not likely to mix. They are golden channels: easier to isolate and identify as exotic. Examples of spin-exotics with $J^{PC} = 1^{-+}$ have been found in Crystal Barrel at CERN, VES at Serpukhov and E852 at Brookhaven.[1]

CP717, *Hadron Spectroscopy: Tenth International Conference,*
edited by E. Klempt, H. Koch, and H. Orth
© 2004 American Institute of Physics 0-7354-0197-7/04/$22.00

TABLE 1. D meson table (from PDG).

	Mass [MeV/c^2]	Width	J^P
$D^\pm$	1869	$10^{-4}\ eV$	0^-
D^0	1864	$10^{-3}\ eV$	0^-
D^{*0}	2007	$< 2.1\ MeV$	1^-
$D^{*\pm}$	2010	$96\ keV$	1^-
D_1^0	2422	$18.9\ MeV$	1^+
D_2^{*0}	2458	$234\ MeV$	2^+
$D_2^{*\pm}$	2459	$25\ MeV$	2^+

TABLE 2. A hybrid (H_c) production and decay channel with χ_{c_1} decay branching ratios.

$p\bar{p} \rightarrow H_c \pi^0$	
$H_c \rightarrow \chi_{c1}(\pi^0\pi^0)_{s-wave}$	
$\chi_{c_1} \rightarrow J/\psi\gamma$	$31.6 \pm 3.2\%$
$J/\psi \rightarrow e^+e^-$	$5.93 \pm 0.1\%$
$J/\psi \rightarrow \mu^+\mu^-$	$5.88 \pm 0.1\%$
$\pi^0 \rightarrow \gamma\gamma$	98.8%
Total χ_{c_1} decay BR	3.6%

CHARMONIUM HYBRID STATE

The charmonium region is populated by narrow well isolated states minimizing the probability of mixing. The charm quark mass, inbetween nonrelativistic and relativistic physics, adds further to the attractiveness of the charmonium system (often referred to as the positronium of QCD). Measurement of the lightest charmonium hybrid is one the PANDA physics goals.

The lowest energy excited gluonic field with $L = 1$ and negative CP couples to the heavy charm quark spins giving eight different hybrid states. The $J^{PC} = 1^{-+}$ is the lowest in mass[1], lattice-QCD predicts masses in the range 4.1-4.4 GeV/c^2. In this study the value 4.29 GeV/c^2 from a recent lattice-QCD calculation was used.[2]

Decay modes.   In the flux-tube model a symmetry argument[2] gives the selection rule that the $J^{PC} = 1^{-+}$ hybrid cannot decay into two states with the same nonrelativistic structure and zero internal angular momentum.[3] This rules out any combination of the first four states in table 1, giving as a first open charm channel $D^0\bar{D}_1^0$ ($\bar{D}^0D_1^0$) with a threshold at 4286 MeV[3]. If this picture is correct the width and the decay modes of the hybrid depends sensitively on the mass, for a low mass hybrid the width could be as narrow as 20 MeV[4].[4],[5]

With open charm decay forbidden or at least suppressed a slow decay into hidden charm is most probable. Assuming nonrelativistic decay (i.e. conservation of the heavy quark spin and angular momentum) the daughter charmonium has quantum numbers $J^{PC} = 1^{++}$, corresponding to the χ_{c_1}. This state is reached by emission of light hadrons where scalars are preferred to isoscalars[5].[6] The lightest scalar alternative is a two pion state in a relative s-wave. The χ_{c_1} can in turn decay electromagnetically to a J/ψ and a photon and the J/ψ to a lepton pair which can be used as a trigger. The branching ratios

---

[1] Same prediction from both lattice QCD and the flux-tube model.

[2] The phase of the hybrid state multiplied with the real-valued s-wave states integrates to zero over space.

[3] Note the width of the D_1^0 which softens the threshold.

[4] In reference [3] Page estimates the symmetry breaking in the case $\bar{D}D^{*0}$ to 4% giving a partial decay width into this channel of 1-4 MeV.

[5] Vacuum quantum number ($^3P_0$) quark pair emission.

for the decay chain is listed in table 2.

Production rates. The Crystal Barrel experiment at CERN has shown that an attractive prospect for the study of hybrid mesons is the gluon-rich proton-antiproton annihilations. In proton-antiproton annihilations in flight all nonexotic quantum numbers are directly accessed whereas spin-exotics can be produced only in association with other particles. The future PANDA experiment at GSI will let a antiproton beam ($p_{lab} = 1.5 - 15$ GeV/c) impinge on a dense proton target.[7] The charmonium hybrid could at $p_{lab} = 15$ GeV/c be produced in association with a π^0 or an η where the π^0 has to be produced with connected quark lines due to its isospin. For that process the π^0 production should be a factor of $\sqrt{2}$ larger from isospin coupling. Since there are no reliable predictions available both cases are studied.

A rough cross-section estimate is made based on the light hadron measurements. The suspected light hybrid and spin-exotic $\pi_1(1400)$ with $J^{PC} = 1^{-+}$ was, at Crystal Barrel, found to have a production intensity from nucleon-nucleon annihilation comparable to that of an ordinary meson.[1] Generalizing this to the hybrid charmonium case one can use the meson production channel $\bar{p}p \rightarrow J/\psi\pi$ measured at E760 with the cross section 130 pb[8], to set the scale[6]. With a similar cross section for our production channel and a total branching ratio of 4% into the final state of table 2 one can expect to have 70 events per day with the PANDA design luminosity of $L = 2 \times 10^{32}$ $cm^{-2}s^{-1}$.

THE PANDA SETUP

The $\backsim$ 100 hybrid events occuring each day within the detector has an overall hadronic background of 2×10^7 s^{-1}. The background giving fake signals should be significantly reduced since it has seven photons and a lepton pair in a very narrow energy range. To identify the hybrid and determine its quantum numbers a partial wave analysis on the full phase space is imperative requireing data with high resolution, high quality and large statistics. The PANDA multipurpose detector will cover 96% of the CMS phase space for formation states. It will be designed for charged particle tracking, photon calorimetry and identification of $\gamma, e^\pm, \mu^\pm, \pi^\pm, K^\pm, p$ and $\bar{p}$. For this a multipurpose detector is envisaged (fig. 1(a)).[7]

The photons will be detected in the Electromagnetic Calorimeter which has a central and a forward part. So far studies have been restricted to the central part (fig. 1(b)) giving fake acceptance losses for photons in the forward direction of about 6% per photon. The EMC parameters are listed in table 3.

---

[6]The cross section was measured at a $E_{CMS} = 3.52$ GeV and for PANDA the maximum energy available is $p_{lab} = 15$ GeV giving a slightly larger phase space factor.

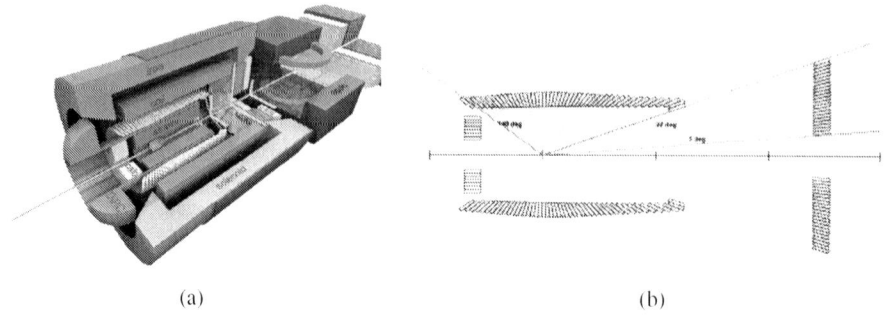

(a)   (b)

FIGURE 1. The PANDA multipurpose detector (a) and the central part of the Electromagnetic calorimeter (b).

TABLE 3. Design parameters of the PANDA EMC.

Material	$PbWO_4$
Thickness	$17X_0$
Crystalsize	$3.5cm \times 3.5cm \times 15cm$
Energy resolution	$\frac{1.54\%}{\sqrt{E[GeV]}} + 0.3\%$
Number of crystals in barrel detector	7150
Geometric coverage	$96\% \times 4\pi$
Ryman radius	$R_M = 2.2\ cm$
Time resolution	$< 20\ ns$

DETECTION

The final state particles are seven photons and a lepton pair. These are relatively energetic ($\backsim GeV/c^2$) and interact with the detector medium mainly by pair-production and brehmsstrahlung[7]. After Geant4 simulation of the detector-response only 37% of the events saw all photons detected and it was only in 28% of all events in which those could be combined into three pions (for hybrid production in association with a pion). In fig. 2(a) the three pions are shown in white on top of the combinatorics spectrum. At an antiproton beam momentum of 15 GeV/c the pion from the production reaction is produced with higher energy than the decay products of the hybrid. This separates true and false combinations $(\chi_{c1}\pi^0\pi^0)_{hybrid}$ and together with a cut in the missing mass (-0.05 $GeV/c^2 < E_{cand(H_c\pi^0)} - s < 0.4\ GeV/c^2$) and the momentum (-0.3 $GeV/c < p_{cand(H_c\pi^0)} < 0.4\ GeV/c$) in the CM-system a hybrid peak with 6.7% of the total events is obtained. In fig. 2(b) the identified hybrids are shown in black within a white combinatorics spectrum. The same cuts applied in the the case of production in

---

[7]For the 7 photons the distribution of pair production is approximately Poisson distributed with an expectation value of 13%.

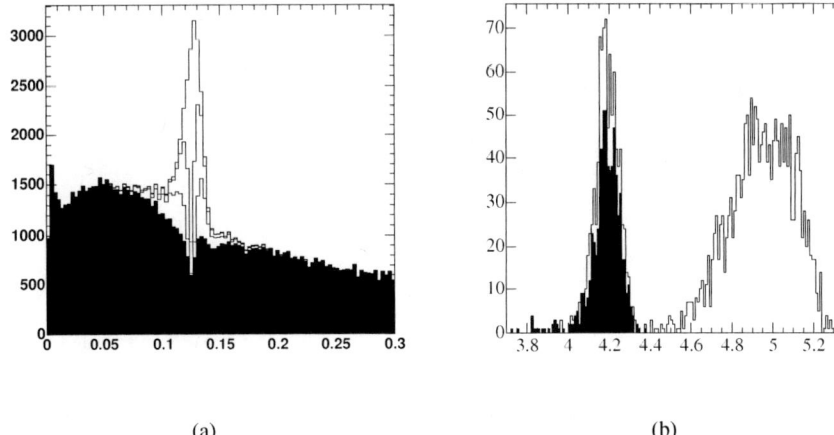

(a)   (b)

FIGURE 2. Geant4 spectra: (a) Three non-overlapping pion candidates in white on top of a black combinatorics spectrum. (b) Hybrid candidates in black within the combinatorics peaks, note the two-peak structure.

association with an η gives detection of 7.5% of the events.

CONCLUSION AND OUTLOOK

The channel $p\bar{p} \to H_c\pi^0, H_c \to \chi_{c1}(\pi^0\pi^0)_{s-wave}, \chi_{c1} \to J/\Psi\gamma, \pi^0 \to \gamma\gamma$ was estimated to be produced about 100 times a day at the HESR. A snap-shot of the Monte Carlo simulation of such events was given. Within the PANDA setup at least 6.7% of those events can be detected, for production in association with an η the number was 7.5% of the events. However much remains to be examined and optimized.

REFERENCES

1. Kuhn, J., "Beyond the quark model of hadrons from lattice QCD," Hadron spectroscopy - Tenth International Conference on Hadron Spectroscopy, 2003.
2. Bernard, C., and Hetrick, J., *Phys. Rev.*, **D56**, 7039 (1997).
3. Page, P., *Phys.Lett.*, **B402**, 183–188 (1997).
4. Close, F., *Phys. Rev.*, **D57** (1998).
5. Page, P., *Act aPhys.Polon.* (1998).
6. Michael, C., "Beyond the quark model of hadrons from lattice QCD," International Conference on Quark Nuclear Physics, 2002.
7. Gianotti, P., "Antiproton Physics at GSI," Hadron spectroscopy - Tenth International Conference on Hadron Spectroscopy, 2003.
8. Cester, R., "Formation of cc states from antiproton-proton annihilations in the Fermilab accumulator (E760)," in *Physics at SuperLEAR*, Inst.Phys.Conf.Ser. 124, 1992, pp. 91–103, paper presented at Workshop on Physics at SuperLEAR, Zurich, October 1991.

The isotensor pentaquark

Philip R. Page

Theoretical Division, MS B283, Los Alamos National Laboratory, Los Alamos, NM 87545, USA

Abstract. Further consequences of the 1540 MeV Θ^+ resonance as an isotensor pentaquark beyond Capstick *et al.* [1] are explored. It is argued that the SAPHIR data may not currently exclude the existence of the charged partner Θ^{++}. The usual prediction of the dominance of non-resonant $\Theta^+ K$, and $\Theta^+ K^*$, final states in photoproduction on the proton is argued not to obtain for an isotensor Θ^+. This enhances the importance of excited baryon final states, where the excited baryon decays to $\Theta^+ K$ or $\Theta^+ K^*$; as well as the non-resonant $\Theta^+ K\pi$ final state. The small width of the recently discovered Ξ^{--} cascade resonance to $\Xi^- \pi^-$ is easier to explain if Θ^+ is an isotensor pentaquark than if it is in the $\overline{10}$ representation, due to both an isospin and U-spin selection rule. A new production diagram for Θ^+ in the photoproduction on the deuteron is suggested.

AN ISOTENSOR PENTAQUARK EXPLAINS THE Θ^+ WIDTH

The consensus of various experiments is that the total width Γ of Θ^+ is less than 9 MeV [1]. More restrictive bounds on the width emerge from its non-observation in $K^+ d$ scattering ($\Gamma < 6$ MeV) [2] and K^+-nucleon scattering ($\Gamma \lesssim 1$ MeV) [3]. It was proposed that the narrowness of the Θ^+ can be explained if it is an isotensor state, in which case the decay to the kinematically allowed channels nK^+ and pK^0 is isospin violating [1]. Based on this hypothesis, an upper bound of roughly 0.45 MeV was put on the width [1], consistent with all experimental data above. If Θ^+ is isotensor, other charge states like Θ^{++} should exist.

SAPHIR MAY NOT HAVE EXCLUDED THE EXISTENCE OF Θ^{++}

In addition to the observation by SAPHIR of the Θ^+ with 63 ± 13 events, they also see a statistically insignificant Θ^{++} signal with 75 ± 35 events in the reaction $\gamma p \to \Theta^{++} K^- \to pK^+ K^-$ [4]. The SAPHIR detector appears to have an acceptance[1] that is about eight times higher in $K^+ K^- p$ than in $K^+ K_S^0 n$. An estimate for the ratio of cross-sections for Θ^{++} and Θ^+ production is then[2]

---

[1] SAPHIR estimates an acceptance ratio of 5000 events / 63 events / (3 to 4) $\times$ 1/2 $\times$ 2/3 = 7.6 [4].
[2] Noting that the observed K_S^0 is only produced half the time from $\bar{K}^0$ and that the detected $\pi^+ \pi^-$ mode of K_S^0 has a branching ratio of about 2/3. The branching ratio $Br(\Theta^{++} \to pK^+)$ is very close to unity if Θ^{++} is below the $NK\pi$ threshold [1].

CP717, *Hadron Spectroscopy: Tenth International Conference,*
edited by E. Klempt, H. Koch, and H. Orth
© 2004 American Institute of Physics 0-7354-0197-7/04/$22.00

$$\frac{\sigma(\gamma p \to \Theta^{++}K^-)}{\sigma(\gamma p \to \Theta^+\bar{K}^0)} = \frac{1}{2}\frac{2}{3}\frac{1}{8}\frac{N(\Theta^{++})}{N(\Theta^+)}\frac{Br(\Theta^+ \to nK^+)}{Br(\Theta^{++} \to pK^+)} \leq \frac{1}{24}\frac{75\pm35}{63\pm13} = \frac{1}{20\pm10} . \quad (1)$$

Thus the process $\gamma p \to \Theta^{++}K^-$ is at least 20 ± 10 times weaker than $\gamma p \to \Theta^+\bar{K}^0$. The cross-section $\sigma(\gamma p \to \Theta^+\bar{K}^0)$ was measured to be 200 nb [4]. A preliminary analysis by CLAS in the *same* reaction with the *same* photon energy range does not see the Θ^+ and gives a cross-section < 20 nb [5]. If the CLAS result is correct, then the serious discrepancy suggests that the SAPHIR analysis may well be in error, and the existence of the Θ^{++} is not excluded.

The photoproduction of an isotensor Θ through the process $\gamma p \to K\Theta$ cannot proceed via isospin conserving interactions through the isoscalar component of the photon, so that the process is taken to proceed through the isovector component (usually associated with the ρ^0 via vector meson dominance) [1] or the smaller isotensor component T^0 arising from four-quark Fock states. If $\gamma = c_0|I=0\rangle + c_1|I=1\rangle + c_2|I=2\rangle$ the scattering T-matrix element

$$\langle\gamma p|T|\Theta K\rangle = \sum_n \langle\gamma p|n\rangle\,\langle n|T|\Theta K\rangle =$$

$$c_1\left\langle\frac{3}{2}\frac{1}{2}\,\Big|\,2I_\Theta^z\,\frac{1}{2}I_K^z\right\rangle\sum_n\langle\rho^0 p|n\rangle\,\langle n|T|\Theta K\rangle_R + c_2\sum_n\langle T^0 p|n\rangle\,\langle n|T|\Theta K\rangle \quad (2)$$

where a formal sum over asymptotic states n has been inserted. The Clebsch-Gordon coefficient has been isolated explicitly from the $\langle n|T|\Theta K\rangle$ overlap corresponding to the $I=1$ photon component, noting that only intermediate states with $I=3/2$ can contribute in this case, and assuming that strong interactions conserve isospin. If the isotensor component of the photon is negligible, as is usually assumed, the amplitude ratio for scattering to Θ^+ ($I_\Theta^z = 0$) and Θ^{++} ($I_\Theta^z = 1$) is $\langle\frac{3}{2}\frac{1}{2}|20\frac{1}{2}\frac{1}{2}\rangle / \langle\frac{3}{2}\frac{1}{2}|21\frac{1}{2}-\frac{1}{2}\rangle = -\sqrt{\frac{2}{3}}$. The Θ^+ is produced with a cross-section $2/3$ that of Θ^{++} (the SAPHIR calculation obtained a factor of $1/3$ [4]). In Eq. 1 this ratio was estimated to be 20 ± 10, which led SAPHIR to conclude that Θ^{++} does not exist [4].

ISOTENSOR Θ^+ PHOTOPRODUCTION ON THE NUCLEON

Photoproduction of Θ^+ on the nucleon is typically calculated in hadronic models because perturbative QCD is not applicable [6]. However, in any of these models diagrams are missing, e.g. the *a priori* important proton sea diagram in Fig. 1, and there is a proliferation of unknown coupling constants. It is therefore instructive to return to a naïve quark level discussion understanding that the diagrams are not those of perturbative QCD, but represent Green's functions needed to evaluate the scattering T-matrix. The discussion will assume that there is a penalty for each time a quark-antiquark pair needs to be created from the vacuum.

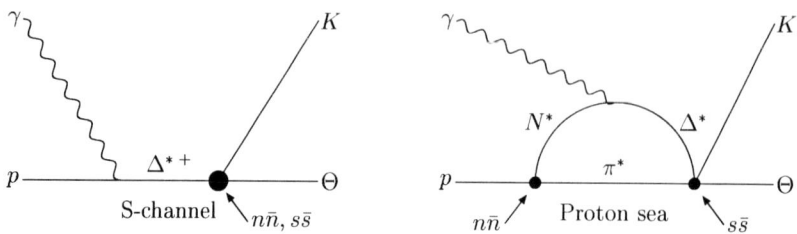

FIGURE 1. S-channel and proton sea production. An asterisk indicates an off-shell particle with the flavor structure of the corresponding on-shell particle. The site of pair creation is indicated by a filled circle, and the pair(s) that is created is shown. In the S-channel diagram the photon interacts first, and the two pairs are created later. In the proton sea diagram the light quark $n\bar{n}$ pair is created first, the photon subsequently interacts, and the strange quark $s\bar{s}$ pair is created last.

Consider the generic process $\gamma N \to K^\star \Theta^+$ for an isotensor Θ^+. Here $K^\star$ represents either the ground state K or an excited state with the flavor of K. In order for the process to happen, one light quark $n\bar{n} \equiv (u\bar{u} + d\bar{d})/\sqrt{2}$ pair and one $s\bar{s}$ pair must be created. The dominant production process is expected to be where one of the pairs that need to be created is created by the photon. (Such production processes can be found in the T-channel mechanisms suggested in Fig. 4 of Ref. [7], Fig. 1 of Ref. [4] and Refs. [6].) Because the isoscalar component of the photon does not contribute within isospin conserving interactions, this can only be the $n\bar{n}$ pair. The $s\bar{s}$ pair will then be created from gluons. The three light quarks in the incoming N will directly go into the outgoing Θ^+. However, this is isospin forbidden: the three light quarks in the Θ^+ is isospin $3/2$, and the N is isospin $1/2$. Since the would-be dominant production process vanishes within isospin symmetry, the production of the Θ^+ is more complicated. The photon interacts with a quark or antiquark, but the two $q\bar{q}$ pairs are both created from gluons.

$\gamma N \to K\Theta^+$ [4]: For this process intermediate resonance mechanisms of the type $\gamma N \to \Delta^\star \to K\Theta^+$ should be considered (similar to the S-channel diagram in Fig. 1). (An intermediate $N^\star$ is not allowed by isospin conservation.) The production process $\gamma N \to \Delta^\star$ requires no pair creation, while the decay $\Delta^\star \to K\Theta^+$ requires two pair creations, and has to compete with the various decay modes to a baryon and meson. There are also non-resonant mechanisms which require two pair creations, for example the proton sea diagram in Fig. 1. Hence the $\gamma N \to \Delta^\star$ and non-resonant $K\Theta^+$ mechanisms are competitive.

$\gamma N \to K\Theta^+\pi$ [7]: Since the $\gamma N \to K^\star \Theta^+$ process requires two pair creations from the isospin selection rule explicated above, other processes become competitive. (The branching ratio $K^\star \to K\pi$ is close to unity so that the pair creation needed for the decay incurs no penalty). Generally the process $\gamma N \to K\Theta^+\pi$ can happen by creating two $n\bar{n}$ pairs and one $s\bar{s}$ pair. This can only be competitive with the $\gamma N \to K^\star \Theta^+$ process if the photon creates one of the pairs. Processes with intermediate resonances of the type $\gamma N \to \Delta^\star\pi$ with $\Delta^\star \to K\Theta^+$ can happen via the photon creating an $n\bar{n}$ pair, with n going into the $\Delta^\star$, and $\bar{n}$ going into the π. Such a process involves two pair creations from the vacuum in the decay $\Delta^\star \to K\Theta^+$. Another possibility is $\gamma N \to \Delta^\star$ where $\Delta^\star \to K^*\Theta^+$ which again requires two pair creations for the decay. There are also non-resonant mechanisms which require two pair creations from the vacuum. (There are even non-resonant mechanisms coming from a four-quark Fock component in the photon, including a possible isotensor component, with *one* pair creation from the vacuum.) Hence the $\gamma N \to \Delta^\star \pi$, $\Delta^\star$, and non-resonant $K^*\Theta^+$ and $K\Theta^+\pi$ mechanisms are competitive.

THE Ξ^{--} AND ISOTENSOR Θ^+ CAN BE CONSISTENT

The recently discovered Ξ^{--} [8] can be put in the same $SU_F(3)$ multiplet as the Θ^+ if both are pentaquarks. An isotensor Θ^+ can only be put in a **35** representation of $SU_F(3)$ (mentioned in Ref. [9]), while an isoscalar Θ^+ can only be put in the $\overline{\mathbf{10}}$ representation. Both these representations also admit Ξ^{--}. In both representations the Θ^+ and Ξ^{--} are in the same V-spin multiplet. V-spin is an exact quantum number if $SU_F(3)$ is an exact symmetry of QCD. The p and Ξ^- are in the same V-spin multiplet and $SU_F(3)$ representation. Similarly for the K^0 and π^-. Hence the decay amplitude $\Theta^+ \to pK^0$ and that of $\Xi^{--} \to \Xi^-\pi^-$ are related by a V-spin Clebsch-Gordon relation. If the Θ^+ is isotensor the decay amplitude to pK^0 is zero within isospin symmetry. The V-spin relation implies that the decay amplitude $\Xi^{--} \to \Xi^-\pi^-$ is zero within $SU_F(3)$ symmetry. If the Θ^+ is isoscalar neither the Θ^+ nor the Ξ^{--} decay amplitudes are zero by these symmetry arguments.

The decay $\Xi^{--} \to \Xi^-\pi^-$ is also suppressed by a U-spin selection rule if Θ^+ is isotensor, but not if it is isoscalar. U-spin is an exact quantum number if $SU_F(3)$ is an exact symmetry of QCD. The U-spin of Ξ^{--} is 2 and 0 in the **35** and $\overline{\mathbf{10}}$ representations respectively. The U-spin of Ξ^- and π^- is $1/2$. The decay $\Xi^{--} \to \Xi^-\pi^-$ is hence U-spin forbidden only if Ξ^{--} is in the **35** representation, noting that this "fall-apart" decay proceeds without quark pair creation, i.e. the interaction is a U-spin singlet.

The fact that the decay $\Xi^{--} \to \Xi^-\pi^-$ is suppressed by two independent symmetry arguments if Θ^+ is isotensor goes a long way towards explaining the small < 18 MeV total width of Ξ^{--} [8]. The fly in the ointment is that Ξ^{--} can also decay to $\Xi^{*-}\pi^-$ and Σ^-K^- via fall-apart decay, and to $\Xi\pi\pi$ and $\Sigma K\pi$ via one vacuum $n\bar{n}$ pair creation. It remains to be explained why these decay widths are small.

PHOTOPRODUCTION THROUGH THE π IN THE DEUTERON

In the reaction $\gamma d \to p(n)K^+K^-$ studied at CLAS [10] the p and K^- must be detected in order to reconstruct the final state. If the p is a spectator, it will not be seen due to its small kinetic energy [10]. Only K^- which are not produced forward can be detected. Hence diagrams were suggested that are not of a spectator nature [10, 11]. The $KN \to \Theta^+$ fusion diagram of Ref. [11] is not allowed for an isotensor Θ^+ due to isospin conservation. The diagram originally suggested involved a $\gamma n \to \Theta^+K^-$ vertex [10], which was shown above to require two vacuum pair creations for an isotensor Θ^+, and

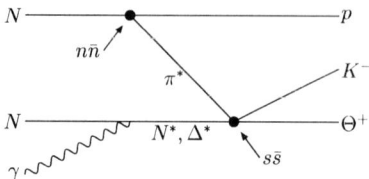

FIGURE 2. Photoproduction of Θ^+ on the deuteron. The neutron and proton in the deuteron are denoted by N. There are diagrams for each possible assignment of the neutron and proton to the label N. The other conventions are as is Fig. 1.

a K^- rescattering on the photon. In Fig. 2 a diagram is displayed which requires two vacuum pair creations and no rescattering, and should hence be dominant. The diagram is natural because the deuteron is an extended nucleus mainly bound by long-distance π-exchange.

ACKNOWLEDGMENTS

The help of both W. Roberts and S. Capstick (construction of the **27** and **35** multiplets), W. Roberts (noting the isotensor component of the photon), D. Weygand (noting the excited baryon production), and detailed discussions with L. Guo, K. Hicks and D. Weygand on their experimental data, are acknowledged. Additional helpful discussions with B. L. Berman, P. Bicudo, B. Cahn, A. Dolgolenko, T. Goldman, E. Klempt, H. J. Lipkin, C. A. Meyer, R. Schumacher, S. Sasaki, S. Stepanyan and M. Wagner are gratefully acknowledged. This research is supported by the U.S. Department of Energy under contract W-7405-ENG-36.

REFERENCES

1. S. Capstick, P.R. Page and W. Roberts, *Phys. Lett.* **B570**, 185 (2003).
2. S. Nussinov, hep-ph/0307357.
3. R. A. Arndt, I. I. Strakovsky and R. L. Workman, nucl-th/0308012.
4. J. Barth *et al.* (SAPHIR Collab.), hep-ex/0307083.
5. K. Hicks, *these proceedings*.
6. Y. Oh, H. Kim and S. H. Lee, hep-ph/0310019; W. Liu and C. M. Ko, nucl-th/0309023; *ibid.* nucl-th/0308034; M.V. Polyakov and A. Rathke, hep-ph/0303138.
7. V. Kubarovsky and S. Stepanyan for the CLAS Collab., Proc. of "Conf. on the Intersections of Particle and Nuclear Physics" (CIPANP 2003), 19-24 May 2003, New York, NY, hep-ex/0307088; R. A. Schumacher for the CLAS Collab., "Jefferson Lab User's Group Meeting Workshop", 11-13 June 2003, Newport News, VA.
8. C. Alt *et al.* (NA49 Collab.), hep-ex/0310014.
9. H. Harari and H. J. Lipkin, *Phys. Rev. Lett.* **13**, 345 (1964).
10. S. Stepanyan *et al.* (CLAS Collab.), hep-ex/0307018.
11. R. A. Schumacher, Proc. of "Electrophoto-production of Strangeness on Nucleons and Nuclei" (SENDAI03), June 2003, Sendai, Japan, nucl-ex/0309006.

Systematics of Quark–Antiquark States: Where are the Lightest Glueballs?

V.V. Anisovich

Petersburg Nuclear Physics Institute, 188300 Gatchina, Petersburg district, Russia

Abstract. The analysis of the experimental data of Crystal Barrel Collaboration on the $p\bar{p}$ annihilation in flight with the production of mesons in the final state resulted in a discovery of a large number of mesons over the region 1900–2400 MeV, thus allowing us to systematize quark-antiquark states in the (n,M^2) and (J,M^2) planes, where n and J are radial quantum number and spin of the meson with the mass M. The data point to linear meson trajectories, with a universal slope. Relying on these data and recent K-matrix analysis a nonet classification is performed. In the scalar-isoscalar sector, the broad resonance state $f_0(1200-1600)$ is superfluous for the $q\bar{q}$ classification, i.e. it is an exotic state. The ratios of coupling constants for the transitions $f_0 \to \pi\pi, K\bar{K}, \eta\eta, \eta\eta'$ point to the gluonium nature of the broad state $f_0(1200-1600)$. The lightest pseudoscalar glueball is also discussed.

The search for exotic mesons should be based on the classification of $q\bar{q}$-states. Exotic mesons are those which are superfluous for the $q\bar{q}$ systematics. The quark–antiquark systematics means:

(i) classification of $q\bar{q}$ states as states located on the (n,M^2) and (J,M^2) trajectories, and

(ii) determination of the quark–gluonium content of states from the analysis of the decay coupling constants, namely, hadronic and radiative decay couplings as well as weak ones.

For the hadronic decay couling constants, the most reliable information comes from the K-matrix analysis. In addition, the K-matrix analysis allows us to study bare states (the states before the onset of the decay processes).

SYSTEMATICS OF THE $q\bar{q}$-STATES ON THE (n,M^2) AND (J,M^2) PLANES.

The analysis of experimental data on the $p\bar{p}$ annihilation in flight with the production of mesons in the final state resulted in a discovery of the large number of mesons over the region 1900–2400 MeV [1]. This allowed us to systematize quark–antiquark states on the (n,M^2) and (J,M^2) planes. The data point to almost the linear meson trajectories on these planes, with a universal slope [2].

In Fig. 1, one can see the (n,M^2) trajectories for the $(I=1)$ states, which are drown for the a_1- and a_3-mesons (Fig. 1a), π-, π_2- and π_4-mesons (Fig. 1b), b_1- and b_3-mesons (Fig. 1c). All these trajectories reveal linear behaviour, such as

$$M^2 = M_0^2 + \mu^2(n-1),$$

CP717, *Hadron Spectroscopy: Tenth International Conference,*
edited by E. Klempt, H. Koch, and H. Orth
© 2004 American Institute of Physics 0-7354-0197-7/04/$22.00

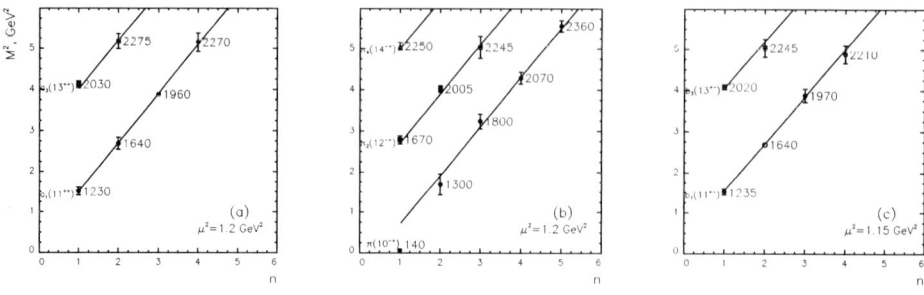

FIGURE 1. The (I=1)-mesons on the (n, M^2) planes: π, a_1, a_3, b_1, b_3 rajectories.

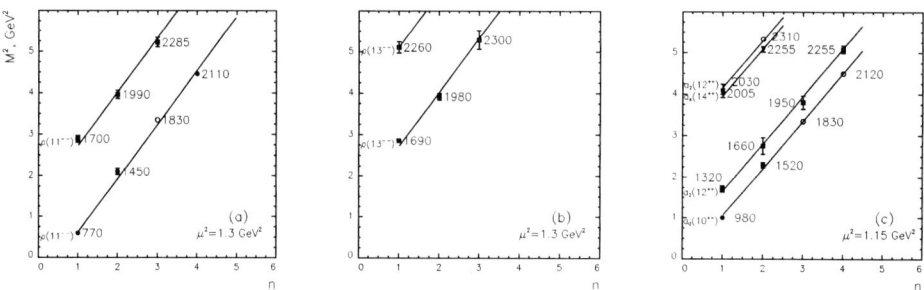

FIGURE 2. The (I=1)-mesons on the (n, M^2) planes: the ρ, ρ_3, a_0, a_2 trajectories

with $\mu^2 \simeq 1.2$ GeV^2; M_0 is the mass of the ground (basic) state, $n = 1$. The pion, being beyond the trajectory, is an exception, that is not a surprise, for the pion is a special particle in certain respect. In the classification, all these mesons should be treated as $q\bar{q}$-states. In Fig. 1c the state $b_1(1640)$ is shown which was not discovered in the experiment but is predicted by trajectories: the states we predict are denoted by open circles.

The trajectories $\rho, \rho_3, a_0, a_2, a_4$ shown in Fig. 2a demonstrate linear behaviour as well. The states $^3S_1$ $q\bar{q}$ and $^3D_1$ $q\bar{q}$ may mix with each other but considerable mass splitting of the $\rho(770)$ and $\rho(1700)$ states tells us that the mixing is not large.

The ρ_3 and a_0, a_2, a_4 trajectories are shown in Figs. 2b and 2c. The $a_0(980)$-resonance, which sometimes is discussed as a non-$q\bar{q}$ state, lays on the $q\bar{q}$-trajectory, that is an argument in favour of its quark–antiquark origin.

Figures 3 and 4 for the $I = 0$ states confirm the linearity of trajectories on the (n, M^2)-planes, with a universal slope $\mu^2 \simeq 1.2 - 1.3$ GeV^2. In this sector, we face a doubling of trajectories due to the existence of two flavour states $n\bar{n} = (u\bar{u} + d\bar{d})/\sqrt{2}$ and $s\bar{s}$.

In Fig. 3, one can see the f_J trajectories. Let me stress that $f_0(980)$, $f_0(1300)$ (the latter denoted as $f_0(1370)$ in the Particle Data compilation, 2000), $f_0(1500)$, $f_0(1750)$, which are sometimes discussed as candidates to exotics, lay quite comfortably on the linear $q\bar{q}$ trajectories. The K-matrix analysis [3] gives us one more state in the $(IJ^{++}00^{++})$-

442

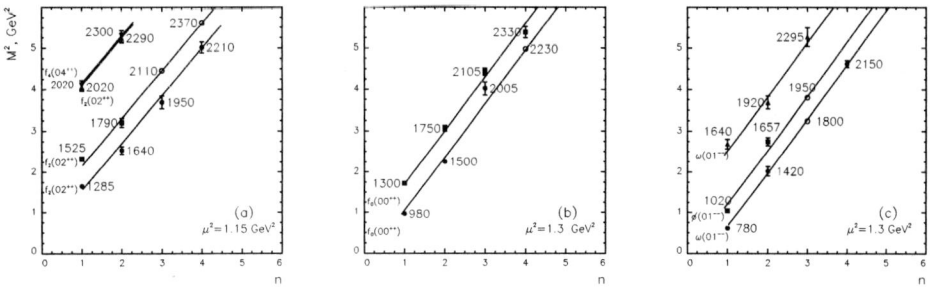

FIGURE 3. The (I=0)-states on the (n, M^2) planes: f_0, f_2, ω trajectories

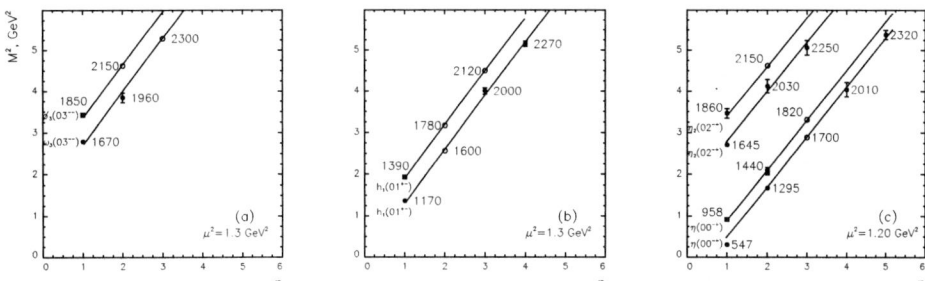

FIGURE 4. The (I=0)-states on the (n, M^2) planes: ω_3, h_1, η trajectories

sector, that is, a broad resonance $f_0(1200 - 1600)$: just this state may be considered as exotics, a descendant of the lightest scalar glueball. The light σ, if any, is also beyond the $q\bar{q}$ trajectories being in this way a candidate for exotics as well.

In Figs. 4, one can see the trajectoties ω_3, ϕ_3 and η_2. Let me emphasize that the situation in the 0^{-+} sector is more complicated than it is seen from Fig. 4c. On the one hand, the states $\eta(1440)$ and $\eta(1295)$ lay good enough on the linear (n, M^2) trajectory. On the other hand, experimental indications for $\eta(1295)$ are not convincing, and the resonance $\eta(1440)$ reveals itself in different reactions.

Now let us discuss the (J, M^2)-planes. The pion and a_1 trajectories and the daughter ones are given in Fig. 5a,b. The trajectories for different parities are degenerate — this is shown in Fig. 5c, where the combined presentation of π and a_1 trajectories is given.

The (J, M^2) trajectories for ρ and a_2 (Fig. 6) provide us unambigous information on the $a_0(980)$ resonance. It is clearly seen that $a_0(980)$ lays on the linear daughter trajectory — it is obvious from the combined picture (Fig. 6c). Supposing that $a_0(980)$ is non-$q\bar{q}$ meson, one should expect around 1 GeV another a_0 state, dominantly $q\bar{q}$. But additional a_0 state is definitely excluded by the experiment.

The (J, M^2) trajectories for the f_2 and ω resonances as well as for their daughter ones are shown in Fig. 7, and the combined presentation of Fig. 7c demonstrates the $f_0(980)$

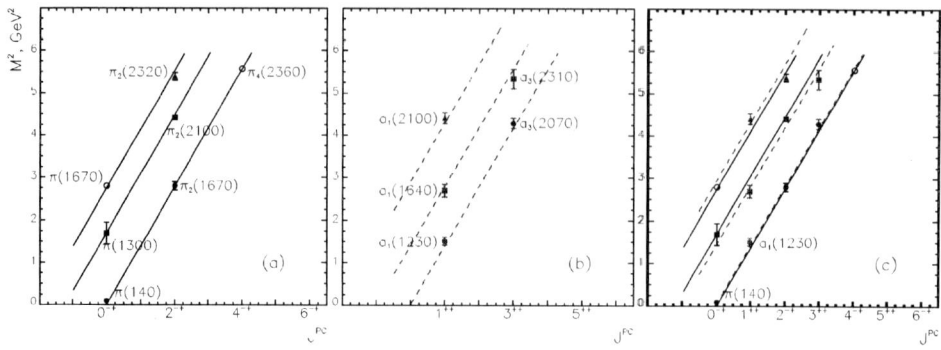

FIGURE 5. The π and a_1 trajectories on the (J, M^2) plane.

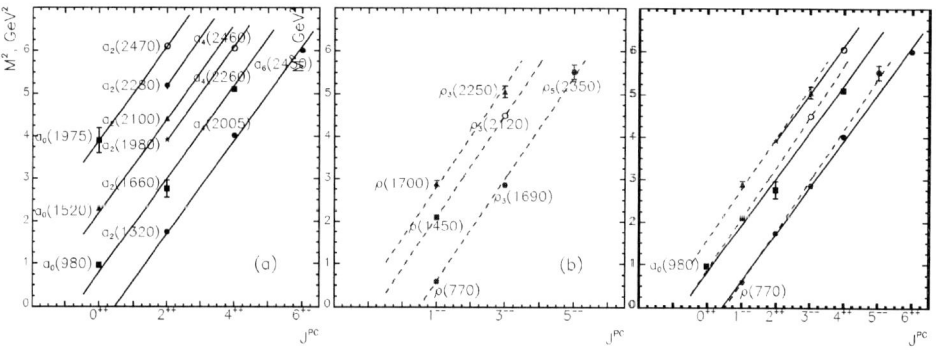

FIGURE 6. The ρ and a_2 trajectories on the (J, M^2) plane.

resonance laying on the daughter $q\bar{q}$ trajectory.

At last, figure 8 shows us the trajectories for the K-meson sector, the kaons with positive and negative parities lay on the degenerate trajectories. The κ meson discussed as a plausible 0^+ state with the mass ~ 900 MeV does not belong to linear trajectory, so it should be considered as an exotic state.

Summing up the results of meson systematization on the (n, M^2) and (J, M^2) trajectories, one can state that the resonances $f_0(980), f_0(1300), f_0(1500)$ and $f_0(1750)$ ($f_0(1710)$ in PDG 2000) are located on quark–antiquark trajectories. They originate from the standard quark–antiquark states.

One may conclude that there are extra 00^{++} states with respect to the $q\bar{q}$ systematics

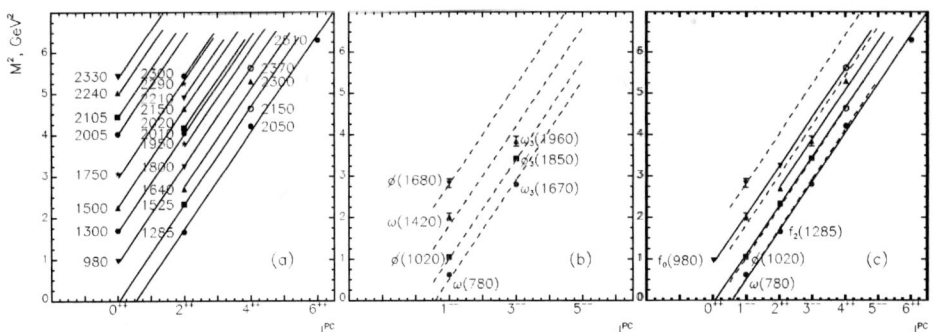

FIGURE 7. The f_2 and ω trajectories on the (J, M^2) plane.

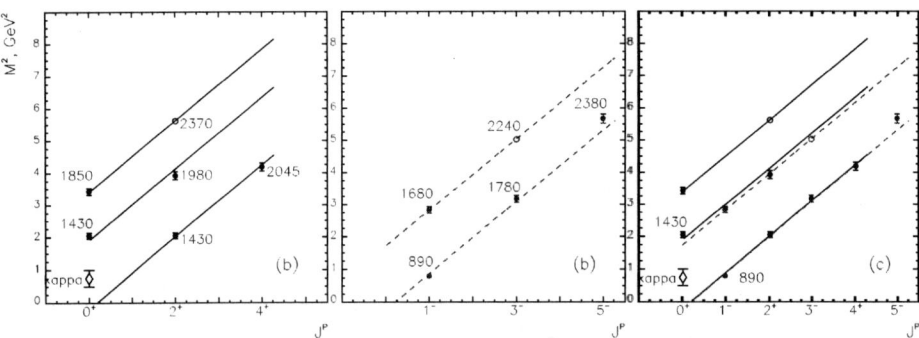

FIGURE 8. Trajectories on the (J, M^2) plane, K-meson sector.

on the (n, M^2) and (J, M^2) trajectories:

1) the broad state $f_0(1200 - 1600)$ (it follows from the K-matrix analysis that it is the descendant of the glueball),

2) the light σ-meson, $f_0(300 - 500)$ (if it exists),

3) κ-meson, $K_0(700 - 900)$ (if it exists).

In the pseudoscalar-isoscalar sector $(IJ^{PC} = 00^{-+})$, the situation in the mass region 1400-1800 MeV is rather uncertain, and one cannot exclude the states existing here, which are superfluous for the $q\bar{q}$ trajectories.

THE K-MATRIX ANALYSIS OF THE $(IJ^{PC} = 00^{++})$-WAVE.

The combined K-matrix analyses of the spectra have been performed for the mass interval $280 \leq M \leq 1950$ MeV by including the following final states:

$$I = 0: \quad \pi\pi, \ \eta\eta, \ K\bar{K}, \ \eta\eta', \ \pi\pi\pi\pi \ (\rho\rho, \ \sigma\sigma) \ [3],$$
$$I = 1: \quad \pi\eta, \ K\bar{K} \ [4],$$
$$I = \tfrac{1}{2}: \quad K\pi; \ [5].$$

Previous analysis [4] carried out in 1997–1998 was based on the experimental data as follows:

(1) GAMS data on the S-wave two-meson production in the reactions $\pi p \to \pi^0\pi^0 n$, $\eta\eta n$ and $\eta\eta' n$ at small nucleon momenta transferred, $|t| < 0.2$ $(\text{GeV}/c)^2$ [6, 7];

(2) GAMS data on the $\pi\pi$ S-wave production in the reaction $\pi p \to \pi^0\pi^0 n$ at large momentum transfers squared, $0.30 < |t| < 1.0$ $(\text{GeV}/c)^2$ [6];

(3) BNL data on the reaction $\pi^- p \to K\bar{K} n$ [8];

(4) CERN-Münich data on $\pi^+\pi^- \to \pi^+\pi^-$ [9];

(5) Crystal Barrel data on $p\bar{p}$ (at rest, from liquid H_2)$\to \pi^0\pi^0\pi^0, \ \pi^0\pi^0\eta, \ \pi^0\eta\eta$ [10, 11].

Now the experimental basis has much broadened, and additional samples of data are included into the analysis [3] of the 00^{++} wave as follows:

(6) Crystal Barrel data on proton-antiproton annihilation in gas: $p\bar{p}$ (at rest, from gaseous H_2) $\to \pi^0\pi^0\pi^0, \ \pi^0\pi^0\eta$ [12].

(7) Crystal Barrel data on proton-antiproton annihilation in liquid: $p\bar{p}$ (at rest, from liquid H_2)$\to \pi^+\pi^-\pi^0, \ K^+K^-\pi^0, \ K_S K_S \pi^0, \ K^+K_S\pi^-$ [12];

(8) Crystal Barrel data on neutron-antiproton annihilation in liquid deuterium: $n\bar{p}$(at rest, from liquid D_2)$\to \pi^0\pi^0\pi^-, \ \pi^-\pi^-\pi^+, \ K_S K^-\pi^0, \ K_S K_S \pi^-$ [12].

These data allowed us to perform more confident study of the two-kaon channels as compared to what had been done before. This is important for the conclusion about the quark-gluon content of scalar–isoscalar f_0-mesons under investigation.

(9) E852 Collaboration data on the $\pi\pi$ S-wave production in the reaction $\pi^- p \to \pi^0\pi^0 n$ at the nucleon momentum transfers squared $0 < |t| < 1.5$ $(\text{GeV}/c)^2$ [13].

Experimental data of the E852 Collaboration on the reaction $\pi^- p \to \pi^0\pi^0 n$ at $p_{lab} = 18$ GeV/c [13] together with the GAMS data on the reaction $\pi^- p \to \pi^0\pi^0 n$ at $p_{lab} = 38$ GeV/c [6] give us a solid ground for the study of the resonances $f_0(980)$ and $f_0(1300)$, for at large momenta transferred to the nucleon, $|t| \sim (0.5 - 1.5)$ $(\text{GeV}/c)^2$, the production of resonances is accompanied by a small background, thus allowing us to fix reliably their masses and widths.

The most important ingredients of the new analysis [3] are:

(i) the study of $K\bar{K}$-spectra that led to the determination of the flavour-octet and flavour-singlet components;

(ii) the study of $f_0(1300)$ produced without background: $\pi N \to \pi\pi N$ at large $|t|$.

Figure 9 taken from [3] demonstrates the complex-M plane for the 00^{++} sector. Here the masses and total widths of resonances are determined by the position of amplitude poles, $det|1 - i\hat{\rho}\hat{K}| = 0$, the decay couplings are determined by the pole residues.

Gradual transformation of bare states into real mesons occurs with the onset of decay channels as follows: $f_0^{bare}(700 \pm 100) \to f_0(980)$, $f_0^{bare}(1220 \pm 40) \to$

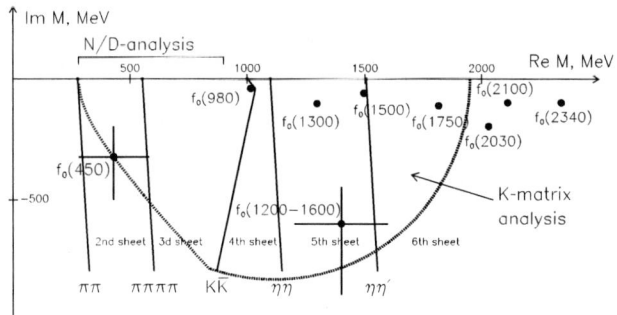

FIGURE 9. Complex-M plane in the $(IJ^{PC} = 00^{++})$ sector: masses and total widths of resonances are determined by the position of the amplitude poles $det|1 - i\hat\rho\hat K| = 0$, decay couplings are determined by the pole residues. Dashed line encircles the part of the plane where the K-matrix analysis reconstructs the analytical K-matrix amplitude: in this area the poles corresponding to resonances $f_0(980)$, $f_0(1300)$, $f_0(1500)$, $f_0(1750)$ and the broad state $f_0(1200 - 1600)$ are located. On the border of this area the light σ-meson denoted as $f_0(450)$ is shown (however, we have also rather good solution without σ-meson pole).

$f_0(1300)$, $f_0^{bare}(1230 \pm 40) \rightarrow f_0(1500)$, $f_0^{bare}(1580 \pm 40) \rightarrow f_0(1200 - 1600)$, $f_0^{bare}(1800 \pm 40) \rightarrow f_0(1750)$.

Although in the K-matrix analysis the pole position of the broad state $f_0(1200 - 1600)$ is defined with large errors, this pole is necessary in all the K-matrix solutions. Also in all solutions the requirement of the factorization of coupling constants is fulfilled: $g_{in} (M^2 - s - i\Gamma M)^{-1} g_{out}$.

The onset of the decay channels can be illustrated with an example of the f_0-levels in the potential well (Fig. 11): stable levels correspond to bare states (Fig. 11a), while overlapping resonances correspond to real mesons (Fig. 11b).

Figure 11a demonstrates that bare states also form linear trajectories. Here one can see two f_0^{bare} trajectories as well as a_0^{bare} and K_0^{bare} ones: all these trajectories have almost the same slopes as the trajectories of real resonances, which are also shown in Fig. 11b for the comparison.

To fix the nonet of bare states by using hadronic decay processes two parameters are needed only, namely, g, φ, where g is a universal decay coupling and φ the mixing angle for $n\bar n$ and $s\bar s$ components in $f_0^{bare}(1)$ and $f_0^{bare}(2)$. Decay couplings depend also on the suppression parameter for the strange quark production probability $\lambda = 0.6 \pm 0.2$, but its value is fixed by other reactions, for example, see [14].

Let us emphasize that these two parameters, g and φ, allowed us to describe ten decay reactions such as $f_0(1) \rightarrow \pi\pi, K\bar K, \eta\eta$, $f_0(2) \rightarrow \pi\pi, K\bar K, \eta\eta, \eta\eta'$, $K_0 \rightarrow \pi K$, $a_0 \rightarrow K\bar K, \eta\pi$. The constraints for decay couplings together with the placement of f_0^{bare} to linear (n, M^2) trajectories rigidly determine two nonets with $n = 1$ and $n = 2$. The extra state, which was found and investigated in [3, 4], namely, $f_0^{bare}(1580)$, should be identified as a scalar glueball, for its decay couplings satisfy the requirements inherent in the gluonium, see [15]. So, $f_0^{bare}(1580 \pm 40) \rightarrow glueball$.

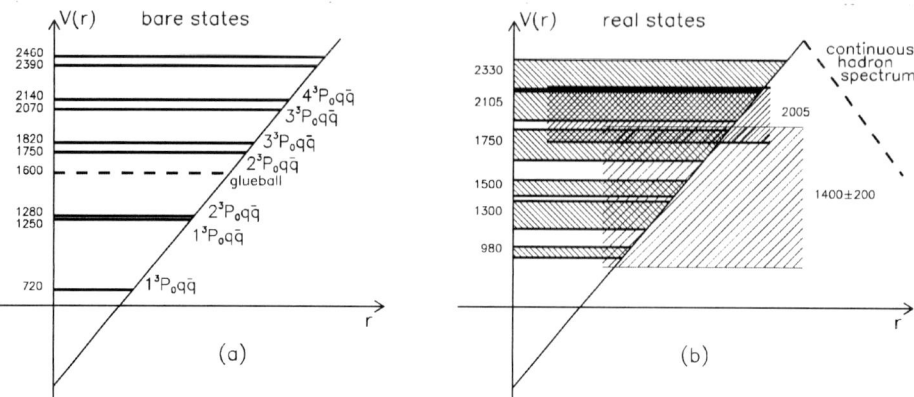

FIGURE 10. The f_0-levels in the potential well depending on the onset of the decay channels: bare states (a) and real resonances (b).

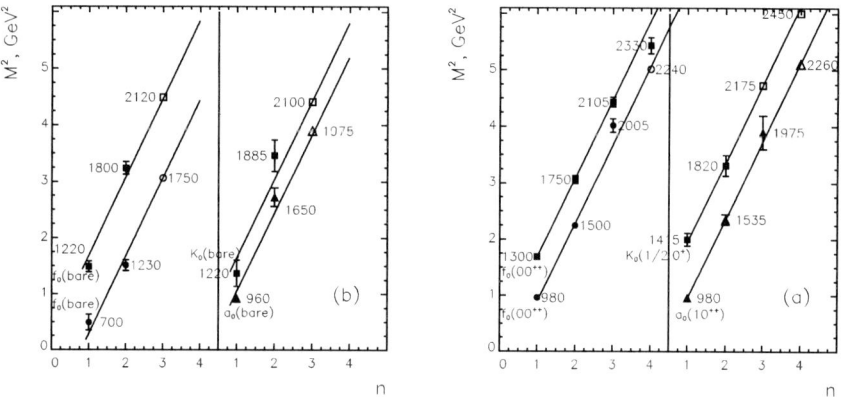

FIGURE 11. Linear trajectories on the (n, M^2) plane for bare states (a) and scalar resonances (b).

WHERE IS THE LIGHTEST PSEUDOSCALAR GLUEBALL?

In the pseudoscalar sector, in the mass region under discussion, one may expect the lightest pseudoscalar glueball, though the opinions about the mass of pseudoscalar glueball are rather different. According to lattice gluodynamic calculations, the mass of the lightest pseudoscalar glueball coincides with that of the tensor glueball, that is, it must be in the range 2100–2600 MeV [16], while, according to [17], its mass is close to that of the lightest scalar glueball: 1300–1700 MeV. The plausible existence of the light 0^{-+} glueball looks nice, for it might explain a considerable production of the 0^{-+} states in the radiative J/ψ decay, in particular, $J/\psi \to \gamma\eta'$ (according to [18], the admixture of the gluonium component in η' may be rather large, about 10%–20%). However, among narrow resonances one cannot see the candidates for pseudoscalar glueball: $\eta(1295)$ and $\eta(1440)$ lay on linear $q\bar{q}$ trajectories (Fig. 4c), though in [19] it was estimated that the value of gg component in the $\eta(1440)$ can be not small, $\sim 30\%$. Still, it is possible that pseudoscalar glueball, after the onset of decay channels, turned into the broad state in the region 1400–1500 MeV (as it occurred with scalar one) — the experimental data do not contradict this suggestion, see [20] and [21].

ACKNOWLEDGMENTS

The author thanks Organizing Committee for the invitation to attend HADRON 2003 and for the financial support. The RFFI grant N 01-02-17861 is also acknowledged.

REFERENCES

1. A.V. Anisovich, C.A. Baker, C.J. Batty *et al.*, Phys. Lett. B**449**, 114 (1999); B**452**, 173 (1999); B **452**, 180 (1999); B **452**, 187 (1999); B **472**, 168 (2000); B **476**, 15 (2000); B **477**, 19 (2000); B **491**, 40 (2000); B **491**, 47 (2000); B **496**, 145 (2000); B **507**, 23 (2001); B **508**, 6 (2001); B **513**, 281 (2001); B **517**, 261 (2001); B **517**, 273 (2001); Nucl. Phys. A **651**, 253 (1999); A **662**, 319 (2000); A **662**, 344 (2000).
2. A.V. Anisovich, V.V. Anisovich, and A.V. Sarantsev, Phys. Rev. D **62**:051502 (2000).
3. V.V.Anisovich, A.V. Sarantsev, Eur.Phys.J. A16, 229 (2003); Yad.Fiz. **66**, 690 (2003) [Phys.Atom.Nucl. **66**, 928 (2003)].
4. V.V. Anisovich, UFN **168** 481 (1998) [Physics-Uspekhi, **41** 419 (1998)]; V.V. Anisovich, A.A. Kondashov, Yu.D. Prokoshkin, S.A. Sadovsky, A.V. Sarantsev, Yad.Fiz. **60**, 1489 (2000) [Phys.Atom.Nucl. **60**, 1410 (2000)]; hep-ph/9711319 (1997).
5. A.V. Anisovich and A.V. Sarantsev, Phys. Lett. B**413**, 137 (1997).
6. D. Alde *et al.*, Zeit.Phys. C **66**, 375 (1995); Yu.D. Prokoshkin *et al.*, Physics – Doklady, **342**, 473 (1995).
7. F. Binon *et al.*, Nuovo Cim. A **78**, 313 (1983); **80**, 363 (1984).
8. S.J. Lindenbaum, R.S. Longacre, Phys.Lett. B **274**, 492 (1992); A. Etkin *et al.*, Phys.Rev. D **25**, 1786 (1982).
9. G. Grayer *et al.*, Nucl.Phys. B **75**, 189 (1974); W. Ochs, PhD Thesis, Münich University (1974).
10. V.V. Anisovich, D.S. Armstrong, I. Augustin *et al.*, Phys.Lett. B **323**, 233 (1994).
11. C. Amsler, V.V. Anisovich, D.S. Armstrong *et al.*, Phys.Lett. B **333**, 277 (1994); D.V. Bugg, V.V. Anisovich, A.V. Sarantsev, B.S. Zou, Phys.Rev. D **50**, 4412 (1994); C. Amsler *et al.*, Phys.Lett. B**342**, 433 (1995); B**355**, 425 (1995).
12. A. Abele *et al.*, Phys. Rev. D **57**, 3860 (1998); Phys. Lett. B **391**, 191 (1997); B **411**, 354 (1997); B **450**, 275 (1999); B **468**, 178 (1999); B **469**, 269 (1999); K. Wittmack, PhD Thesis, Bonn University, (2001); A.V.Sarantsev, talk at HADRON2003 Conference (2003).
13. J. Gunter *et al.*, (E852 Collaboration), Phys.Rev. D **64**:07003 (2001).
14. K. Peters and E. Klempt, Phys. Lett. B **352**, 467 (1995).
15. V.V. Anisovich, *Systematics of quark-antiquark states and scalar exotic mesons*, hep-ph/0208123.
16. G.S. Bali et al. Phys. Lett. B **309**, 378 (1993).
17. L. Faddeev, A.J. Niemi and U. Wiedner, *Glueballs, closed flux tubes and $\eta(1440)$*, hep-ph/0308240 (2003).

18.   V.V. Anisovich, D.V. Bugg, D.I Melikhov, and V.A. Nikonov, Phys. Lett. B **404**, 166 (1997).
19.   A.V. Anisovich, *Quark-gluon content of* $\eta(1295)$ *and* $\eta(1440)$, hep-ph/0104005 (2001).
20.   V.V. Anisovich, UFN **165**, 1225 (1995) [Physics Uspekhi **38**, 1179 (1995)].
21.   Z. Bai et al., Phys. Rev. Lett. **65**, 2507 (1990). Phys. Rev. Lett. **65**, 2507 (1990).

The gluon contents of the η and η' mesons

P. Kroll

Fachbereich Physik, Universität Wuppertal,
D-42097 Wuppertal, Germany
Email: kroll@physik.uni-wuppertal.de

Abstract. It is reported on a leading-twist analysis of the $\eta - \gamma$ and $\eta' - \gamma$ transition form factors. The analysis allows for an estimate of the lowest Gegenbauer coefficients of the quark and gluon distribution amplitudes.

One of the simplest exclusive observables is the form factor $F_{P\gamma^{(*)}}$ for the transitions from a real or virtual photon to a pseudoscalar meson P. Its behaviour at large momentum transfer is determined by the expansion of a product of two electromagnetic currents about light-like distances. The form factor then factorizes [1] into a hard scattering amplitude and a soft matrix element, parameterized by a process-independent meson distribution amplitude Φ_P. For space-like momentum transfer the form factor can be accessed in $e^+e^- \to e^+e^- P$. Such measurements have been carried through for quasi-real photons by CLEO [2] and L3 [3]. From the data on the form factors one may extract information about the meson distribution amplitudes by fitting the theoretical results to the experimental data. Here, in this talk, it is reported on recent attempts [4, 5] to perform such analyses to leading-twist NLO accuracy in the cases of the η and η' mesons.

As the valence Fock components of the η and η' mesons $SU(3)_F$ singlet and octet combinations of quark-antiquark parton states are chosen

$$|q\bar{q}_1\rangle = |u\bar{u} + d\bar{d} + s\bar{s}\rangle/\sqrt{3}, \qquad |q\bar{q}_8\rangle = |u\bar{u} + d\bar{d} - 2s\bar{s}\rangle/\sqrt{6}. \tag{1}$$

In addition the two-gluon Fock state, $|gg\rangle$, is to be taken into account which also possesses flavour-singlet quantum numbers and contributes to leading-twist order. Associated to each valence Fock component of the meson P is a distribution amplitude denoted by Φ_{Pi} ($i = 1, 8$) and Φ_{Pg}. The distribution amplitudes possess Gegenbauer expansions [1]

$$
\begin{aligned}
\Phi_{Pi}(\xi, \mu_F) &= \frac{3}{2}(1 - \xi^2)\left[1 + \sum_{n=2,4,\cdots} B_{Pn}^{(i)}(\mu_F)\, C_n^{3/2}(\xi)\right], \\
\Phi_{Pg}(\xi, \mu_F) &= \frac{1}{16}(1 - \xi^2)^2 \sum_{n=2,4,\cdots} B_{Pn}^{(g)}(\mu_F)\, C_{n-1}^{5/2}(\xi),
\end{aligned}
\tag{2}
$$

where $\xi = 2x - 1$, and x is the usual momentum fraction carried by the quark inside the meson. The Gegenbauer coefficients, B_{Pn}, which encode the soft physics, evolve with the factorization scale μ_F according to the relevant anomalous dimensions. The essential

CP717, *Hadron Spectroscopy: Tenth International Conference*,
edited by E. Klempt, H. Koch, and H. Orth
© 2004 American Institute of Physics 0-7354-0197-7/04/$22.00

point is that the singlet and gluon coefficients mix under evolution

$$B_{Pn}^{(1)}(\mu_F) \leftrightarrow B_{Pn}^{(g)}(\mu_F),$$
(3)

and that all coefficients evolve to zero for asymptotically large factorization scales. Hence

$$\Phi_{Pi} \to \Phi_{AS} = \frac{3}{2}(1-\xi^2), \qquad \Phi_{Pg} \to 0, \quad \text{for } \mu_F \to \infty.$$
(4)

It is important to note that the gluon distribution amplitude goes along with the following projector of a state of two incoming collinear gluons (colours a, b, Lorentz indices μ, ν and momentum fractions x, $1-x$) onto a pseudoscalar meson state

$$\mathscr{P}_{\mu\nu,ab}^g = \frac{i}{2}\sqrt{\frac{C_F}{n_f}}\frac{\delta_{ab}}{\sqrt{N_c^2-1}}\frac{\varepsilon_{\perp\mu\nu}}{x(1-x)}.$$
(5)

The anomalous dimensions have to be normalized accordingly [5]. The components of the transverse polarization tensor are $\varepsilon_{\perp 12} = -\varepsilon_{\perp 21} = 1$ and zero for all others.

The $\gamma^*(q,\mu)\,\gamma^{(*)}(q',\nu) \to P(p)$ vertex is parameterized as

$$\Gamma^{\mu\nu} = ie_0^2 F_{P\gamma}(\overline{Q},\omega)\varepsilon^{\mu\nu\alpha\beta}q_\alpha q'_\beta,$$
(6)

where $Q^2 = -q^2 \geq 0$, $Q'^2 = -q'^2 \geq 0$ and

$$\overline{Q}^2 = \frac{1}{2}(Q^2 + Q'^2), \qquad \omega = \frac{Q^2 - Q'^2}{Q^2 + Q'^2}.$$
(7)

Due to Bose symmetry the transition form factor is symmetric in ω. To leading-twist NLO accuracy the transition form factor reads $(P = \eta,\eta')$

$$
\begin{aligned}
F_{P\gamma^*} &= \frac{2}{3\sqrt{3}\,\overline{Q}^2}\int_{-1}^{1}\frac{d\xi}{1-\xi^2\omega^2}\left\{\left[\frac{f_P^{(8)}}{2\sqrt{2}}\Phi_{P8}(\xi,\mu_F) + f_P^{(1)}\Phi_{P1}(\xi,\mu_F)\right]\right.\\
&\quad \times \left[1 + \frac{\alpha_s(\mu_R)}{4\pi}\mathscr{K}_q(\omega,\xi,\overline{Q}^2)\right] + f_P^{(1)}\Phi_{Pg}(\xi,\mu_F)\frac{\alpha_s(\mu_R)}{4\pi}\mathscr{K}_g(\omega,\xi,\overline{Q}^2)\right\}.
\end{aligned}
$$
(8)

The NLO hard scattering kernels, $\mathscr{K}$, are calculated from the Feynman graphs shown in Fig. 1. The results - in the $\overline{\mathrm{MS}}$ scheme - can be found in the literature, see for instance [4, 5, 6]. The decay constants, $f_P^{(i)}$, are defined by matrix elements of $SU(3)_F$ singlet and octet axial vector currents:

$$\langle 0|J_{5\mu}^{(i)}|P(p)\rangle = if_P^{(i)}\,p_\mu.$$
(9)

The singlet decay constant $f_P^{(1)}$ depends on the scale [7] but the anomalous dimension controlling it is of order α_s^2. In a NLO calculation this effect is to be neglected for consistency. Note that the octet part of (8) also holds for the $\pi-\gamma$ form factor with the obvious replacement $\Phi_{P8} \to \Phi_\pi$, $f_P^{(8)} \to \sqrt{3}f_\pi$.

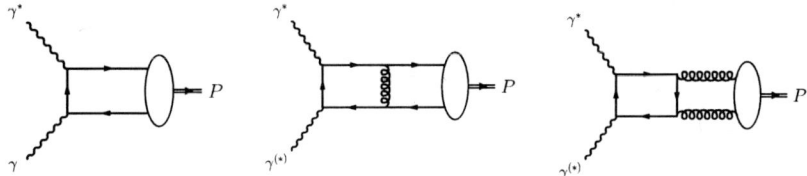

FIGURE 1. Sample Feynman graphs contributing to the transition form factors

Of particular interest is the limit $\omega \to 0$. Inserting the Gegenbauer expansion (2) into (8), one finds that the Gegenbauer coefficients of the quark and gluon distribution amplitudes first appears at order ω^n [4]. Hence, one obtains the prediction

$$F_{P\gamma^*}(\overline{Q}^2, \omega) = \frac{\sqrt{2}}{3\sqrt{3}} \frac{f_P^8 + 2\sqrt{2}f_P^1}{\overline{Q}^2} \left[1 - \frac{\alpha_s}{\pi}\right] + \mathcal{O}(\omega^2, \alpha_s^2). \tag{10}$$

Since the decay constants are known to amount to

$$f_\eta^{(8)} = 1.17 f_\pi, \quad f_\eta^{(1)} = 0.19 f_\pi, \quad f_{\eta'}^{(8)} = -0.46 f_\pi, \quad f_{\eta'}^{(1)} = 1.15 f_\pi, \tag{11}$$

with a accuracy of about 5% [8], (10) is a parameter-free prediction of QCD to leading-twist accuracy. Its theoretical status is comparable to that of the Bjorken sum rule [9]

$$\int_0^1 dx [g_1^p(x) - g_1^n(x)] = \frac{1}{6}\frac{G_A}{G_V}\left[1 - \frac{\alpha_s}{\pi} - 3.583(\frac{\alpha_s}{\pi})^2 - 20.215(\frac{\alpha_s}{\pi})^3 + \cdots\right], \tag{12}$$

and a few other observables among which is the famous result for the cross section ratio of e^+e^- annihilation into hadrons and into a pair of muons. It is known [10] that the perturbative series of the transition form factors are identical to that of the Bjorken sum rule. The prediction (10) well deserves experimental verification but there is no data as yet.

The real photon case, $\omega = 1$, is another interesting limit. Here data is available [2, 3] from which information about the distribution amplitudes can be extracted. For the case of the pion such analyses have been carried through immediately after the advent of the CLEO data in Ref. [11, 12] and, recently, in much greater detail in [4]. The η and η' data have been analyzed within the modified pertrubative approach in [13] and to leading twist NLO accuracy in [5]. Since the present quality of the data does not suffice to determine all six distribution amplitudes, one has to simplify matters and employ an $\eta - \eta'$ mixing scheme in order to reduce the number of free parameters. Since in hard processes only small spatial quark-antiquark separations are of relevance, it is sufficiently suggestive to embed the particle dependence and the mixing behaviour of the valence Fock components solely into the decay constants which play the role of wave functions at the origin. Following [8, 13], one may therefore take

$$\Phi_{Pi} = \Phi_i, \qquad \Phi_{Pg} = \Phi_g. \tag{13}$$

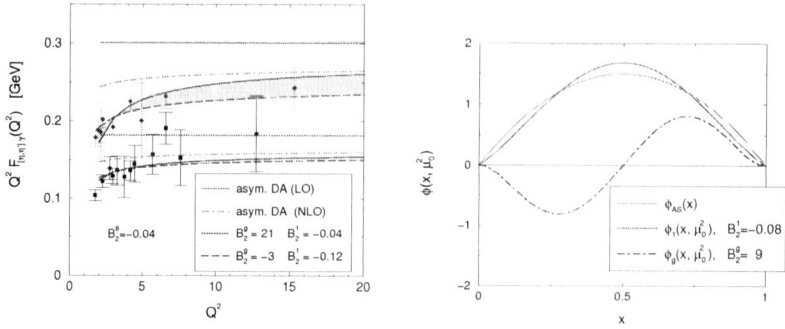

FIGURE 2. Left:The scaled $P\gamma$ transition form factor vs. Q^2. Data taken from [2, 3]; rombs represent the η' data, squares the η ones. Right: The flavour-singlet and gluon distribution amplitudes at the scale $\mu_0 = 1$ GeV^2

This assumption is further supported by the observations made in [13] that, as is the case for the pion [4, 11, 12] the quark distribution amplitudes are close to the asymptotic form, Φ_{AS}, for which the particle independence (13) holds trivially. The analysis is further simplified by truncating the Gegenbauer series in (2) at $n = 2$. The coefficients $B_2^{(i)}$, acting for all others, parameterize the deviations from the asymptotic form of the distribution amplitudes. Clearly, this is a serious assumption (note that to LO accuracy the transition form factors only fix the sum $1 + \sum B_n^{(i)}$) but in view of the large experimental errors as well of the limited range of momentum transfer in which data is available, one is forced to do so. Truncation at $n = 4$ does not lead to reliable results, all contributing Gegenbauer coefficients are highly correlated. A fit to the CLEO and L3 data provides

$$B_2^{(8)}(\mu_0) = -0.04 \pm 0.04, \quad B_2^{(1)}(\mu_0) = -0.08 \pm 0.04, \quad B_2^{(g)}(\mu_0) = 9 \pm 12, \quad (14)$$

where the following scales have been chosen: $\mu_0 = 1$ GeV, $\mu_F = Q$, $\mu_R = Q/\sqrt{2}$. The use of $\mu_F = Q/\sqrt{2}$ instead leads to values of the Gegenbauer coefficients which agree with those quoted in (14) almost perfectly. For comparison, B_2^π takes a value of -0.06 ± 0.03 as determined in [4].

The fit is compared to the data in Fig. 2. The insensitivity of the $\eta - \gamma$ transition form factor to the gluonic distribution amplitude is clearly seen which comes about as a consequence of the smallness of $f_\eta^{(1)}$, see (11). Although the present data are compatible with a leading-twist analysis as Fig. 2, the existence of power and/or higher-twist corrections cannot be excluded. This is a source of theoretical uncertainties in the results (14). Thus, for instance, the use of the modified perturbative approach in which quark transverse degrees of freedom and Sudakov suppressions are taken into account, leads to good agreement with experiment for the asymptotic distribution amplitudes [13].

Within errors the quark Gegenbauer coefficients for the octet and singlet case agree with each other and with the pion one. This implies not only approximate flavour symmetry but also the approximate validity of the OZI rule which is a prerequisite of the quark-flavour mixing scheme advocated for in [8]. Although the face value of $B_2^{(g)}$ is huge as compared to that of $B_2^{(1)}$ the gluonic distribution amplitude itself is not large as can be seen from Fig. 2, its $x \leftrightarrow 1 - x$ asymmetry and the numerical factors in (2) keep it small. Moreover since it only contributes to NLO its impact on the transition form factors is small resulting in large errors. In order to obtain more precise information on the gluonic distribution amplitude additional constraints from other reactions are required. The inclusive decay $\Upsilon(^1S) \to \eta'X$, discussed in [14], is one such possibility. Others are e.g. $B \to \pi\eta'$ or $\chi_{cJ} \to \eta'\eta'$. Finally it is to be emphasized that the approach presented in this article applies to all flavour-neutral pseudoscalar mesons, e.g. for the $\eta(1400)$. The properties of the valence distribution amplitudes (2) make it unlikely that a pseudoscalar meson possesses pure glueball properties. A substantial $q\bar{q}$ Fock component is always there. For flavour-neutral scalar mesons, on the other hand, the situation is different. The properties of the quark and gluon distribution amplitudes are reversed [15]. A strong gg Fock component is therefore not necessarily accompanied by strong $q\bar{q}$ one.

REFERENCES

1. G. P. Lepage and S. J. Brodsky, Phys. Rev. D **22**, 2157 (1980).
2. J. Gronberg *et al.* [CLEO Collaboration], Phys. Rev. D **57**, 33 (1998) [hep-ex/9707031].
3. M. Acciarri *et al.* [L3 Collaboration], Phys. Lett. B **418**, 399 (1998).
4. M. Diehl, P. Kroll and C. Vogt, Eur. Phys. J. C **22**, 439 (2001) [hep-ph/0108220].
5. P. Kroll and K. Passek-Kumericki, Phys. Rev. D **67**, 054017 (2003) [arXiv:hep-ph/0210045].
6. F. del Aguila and M. K. Chase, Nucl. Phys. B **193**, 517 (1981); E. Braaten, Phys. Rev. D **28**, 524 (1983); E. P. Kadantseva, S. V. Mikhailov and A. V. Radyushkin, Yad. Fiz. **44**, 507 (1986) [Sov. J. Nucl. Phys. **44**, 326 (1986)].
7. R. Kaiser and H. Leutwyler, Eur. Phys. J. C **17**, 623 (2000) [hep-ph/0007101].
8. T. Feldmann, P. Kroll and B. Stech, Phys. Rev. D **58**, 114006 (1998) [hep-ph/9802409] and Phys. Lett. B **449**, 339 (1999) [hep-ph/9812269].
9. D.J. Broadhurst and A.L. Kataev, Phys. Lett. B**544**, 154 (2002) [hep-ph/0207261].
10. B. Melic, D. Müller and K. Passek-Kumericki, Phys. Rev. D **68**, 014013 (2003) [hep-ph/0212346].
11. P. Kroll and M. Raulfs, Phys. Lett. B **387**, 848 (1996) [hep-ph/9605264].
12. I. V. Musatov and A. V. Radyushkin, Phys. Rev. D **56**, 2713 (1997) [hep-ph/9702443].
13. T. Feldmann and P. Kroll, Eur. Phys. J. C **5**, 327 (1998) [hep-ph/9711231].
14. A. Ali and A.Ya. Parkhomenko, Eur. Phys. J. C **30**, 367 (2003) [hep-ph/0307092].
15. M. K. Chase, Nucl. Phys. B **174**, 109 (1980); V. N. Baier and A. G. Grozin, Nucl. Phys. B **192**, 476 (1981).

A relativistic description of hadronic decays of the meson π_1

Nikodem Poplawski[†]

Nuclear Theory Center, Indiana University, Bloomington, IN 47405
This article is based on the author's Ph.D. thesis currently being written under the supervision of Prof. Adam Szczepaniak.

Abstract. The subject of this work is analysis of hadronic decays of exotic meson π_1 in a fully relativistic formalism, and comparison with the nonrelativistic results. The relativistic spin wave functions of mesons and hybrids are constructed based on unitary representations of the Lorentz group. The radial wave functions are obtained from phenomenological considerations of the mass operator. We find that decay channels $\pi_1 \to \pi b_1$ and $\pi_1 \to \pi f_1$ are favored, in agreement with results obtained using other models, thus indicating some model independence of the $S + P$ selection rules. We will also report on effects of meson final state interactions in exotic channels.

INTRODUCTION

In a region around 2 GeV a new form of hadronic matter is expected to exist in which the gluonic degrees of freedom are excited. In mesons these can result in resonances with exotic J^{PC} quantum numbers. The adiabatic potential calculations show π_1 (1^{-+}) as the lowest energy excited gluonic configuration [1]. The present models of hybrid decays (for instance [2, 3]) are nonrelativistic and therefore one should investigate corrections arising from fully relativistic treatment. The case of π_1 is of a special interest also because its evidence has been reported by the E852 collaboration and new experimental searches are planned for JLab and GSI.

RELATIVISTIC SPIN WAVE FUNCTION FOR MESONS AND HYBRIDS

For a system of non-interacting particles the spin wave function is constructed as an element of an irreducible representation of the Poincare group. We will assume $m_u = m_d = m$. In the rest frame of a quark-antiquark pair

$$l_q^\mu = (E(m_q, \mathbf{q}), \mathbf{q}), \ l_{\bar{q}}^\mu = (E(m_{\bar{q}}, -\mathbf{q}), -\mathbf{q}),$$

the normalized spin-1 wave function ($J^{PC} = 1^{--}$) is given by the Clebsch-Gordan coefficients and can be written in terms of Dirac spinors as

$$\Psi_{q\bar{q}}^{\lambda_{q\bar{q}}}(\mathbf{q}, \mathbf{l}_{q\bar{q}} = 0, \sigma_q, \sigma_{\bar{q}}) = \frac{1}{\sqrt{2}m_{q\bar{q}}} \bar{u}(\mathbf{q}, \sigma_q) \left[\gamma^i - \frac{2q^i}{m_{q\bar{q}} + 2m} \right] v(-\mathbf{q}, \sigma_{\bar{q}}) \varepsilon^i(\lambda_{q\bar{q}}), \quad (1)$$

CP717, *Hadron Spectroscopy: Tenth International Conference*,
edited by E. Klempt, H. Koch, and H. Orth
© 2004 American Institute of Physics 0-7354-0197-7/04/$22.00

where $m_{q\bar{q}}$ is the invariant mass and $\varepsilon^i(\lambda_{q\bar{q}})$ are polarization vectors corresponding to spin 1 quantized along the z-axis. The wave function of a $q\bar{q}$ system moving with a total momentum $\mathbf{l}_{q\bar{q}} = \mathbf{l}_q + \mathbf{l}_{\bar{q}}$ is given by

$$\Psi_{q\bar{q}}^{\lambda_{q\bar{q}}}(\mathbf{q}, \mathbf{l}_{q\bar{q}}, \lambda_q, \lambda_{\bar{q}}) = \sum_{\sigma_q, \sigma_{\bar{q}}} \Psi_{q\bar{q}}^{\lambda_{q\bar{q}}}(\mathbf{q}, \mathbf{l}_{\bar{q}} = 0, \sigma_q, \sigma_{\bar{q}}) D_{\lambda_q \sigma_q}^{*(1/2)}(\mathbf{q}, \mathbf{l}_{q\bar{q}}) D_{\lambda_{\bar{q}} \sigma_{\bar{q}}}^{(1/2)}(-\mathbf{q}, \mathbf{l}_{q\bar{q}}), \quad (2)$$

where the Wigner rotation matrix

$$D_{\lambda\lambda'}^{(1/2)}(\mathbf{q}, \mathbf{P}) = \left[\frac{(E(m,\mathbf{q})+m)(E(M,\mathbf{P})+M) + \mathbf{P}\cdot\mathbf{q} + i\boldsymbol{\sigma}\cdot(\mathbf{P}\times\mathbf{q})}{\sqrt{2(E(m,\mathbf{q})+m)(E(M,\mathbf{P})+M)(E(m,\mathbf{q})E(M,\mathbf{P})+\mathbf{P}\cdot\mathbf{q}+mM)}} \right]_{\lambda\lambda'}$$

corresponds to a boost with $\beta\gamma = \mathbf{P}/M$. One can show

$$\Psi_{q\bar{q}}^{\lambda_{q\bar{q}}}(\mathbf{q}, \mathbf{l}_{q\bar{q}}, \lambda_q, \lambda_{\bar{q}}) = -\frac{1}{\sqrt{2}m_{q\bar{q}}} \bar{u}(\mathbf{l}_q, \lambda_q) \left[\gamma^\mu - \frac{l_q^\mu - l_{\bar{q}}^\mu}{m_{q\bar{q}} + 2m} \right] v(\mathbf{l}_{\bar{q}}, \lambda_{\bar{q}}) \varepsilon_\mu(\mathbf{l}_{q\bar{q}}, \lambda_{q\bar{q}}), \quad (3)$$

where $\varepsilon^\mu(\mathbf{l}_{q\bar{q}}, \lambda_{q\bar{q}})$ are obtained from $(0, \varepsilon^i(\lambda_{q\bar{q}}))$ through a boost with $\beta\gamma = \mathbf{l}_{q\bar{q}}/m_{q\bar{q}}$. Similarly the normalized wave function for the spin-0 quark-antiquark pair ($J^{PC} = 0^{-+}$) is given by

$$\Psi_{q\bar{q}}(\mathbf{q}, \mathbf{l}_{q\bar{q}}, \lambda_q, \lambda_{\bar{q}}) = \Psi_{q\bar{q}}(\mathbf{l}_q, \mathbf{l}_{\bar{q}}, \lambda_q, \lambda_{\bar{q}}) = \frac{1}{\sqrt{2}m_{q\bar{q}}} \bar{u}(\mathbf{l}_q, \lambda_q) \gamma^5 v(\mathbf{l}_{\bar{q}}, \lambda_{\bar{q}}). \quad (4)$$

By coupling (3) or (4) for $\mathbf{l}_{q\bar{q}} = 0$ with one unit of the orbital angular momentum $L = 1$ and then making a boost (2), one obtains respectively the spin wave function for a quark-antiquark pair with quantum numbers $J^{PC} = 1^{+-}$

$$\Psi_{q\bar{q}}^{\lambda_{q\bar{q}}}(\mathbf{l}_q, \mathbf{l}_{\bar{q}}, \lambda_q, \lambda_{\bar{q}}) = \frac{1}{\sqrt{2}m_{q\bar{q}}(\mathbf{l}_q, \mathbf{l}_{\bar{q}})} \bar{u}(\mathbf{l}_q, \lambda_q) \gamma^5 v(\mathbf{l}_{\bar{q}}, \lambda_{\bar{q}}) Y_{1\lambda_{q\bar{q}}}(\bar{\mathbf{q}}), \quad (5)$$

or 0^{++}, 1^{++} and 2^{++}

$$\Psi_{q\bar{q}}^{\lambda_{q\bar{q}}}(\mathbf{l}_q, \mathbf{l}_{\bar{q}}, \lambda_q, \lambda_{\bar{q}}) = -\sum_{\lambda,l} \frac{1}{\sqrt{2}m_{q\bar{q}}} \bar{u}(\mathbf{l}_q, \lambda_q) \left[\gamma^\mu - \frac{l_q^\mu - l_{\bar{q}}^\mu}{m_{q\bar{q}} + 2m} \right] v(\mathbf{l}_{\bar{q}}, \lambda_{\bar{q}}) \varepsilon_\mu(\mathbf{l}_{q\bar{q}}, \lambda)$$

$$\cdot Y_{1l}(\bar{\mathbf{q}}) < 1, \lambda; 1, l | J, \lambda_{q\bar{q}} >, \quad (6)$$

with $\mathbf{q} = \Lambda(\mathbf{l}_{q\bar{q}} \to 0)\mathbf{l}_q$. In order to construct meson spin wave functions for higher orbital angular momenta L one need only to replace Y_{1l} with Y_{Ll}.

In the rest frame of the 3-body system corresponding to a $q\bar{q}$ pair with momentum $-\mathbf{Q}$ and transverse gluon with momentum $\mathbf{Q}$, the total spin wave function of the hybrid is obtained by coupling the $q\bar{q}$ spin-1 wave function (3) and the gluon wave function ($J^{PC} = 1^{--}$) to a total spin $S = 0, 1, 2$ and $J^{PC} = 0^{++}, 1^{++}, 2^{++}$ states respectively, and then with one unit of the orbital angular momentum to the exotic state 1^{-+}:

$$\Psi_{q\bar{q}g(S)}^{\lambda_{ex}}(\lambda_q, \lambda_{\bar{q}}, \lambda_g) = \sum_{\lambda_{q\bar{q}}, \sigma=\pm1, M, l} \Psi_{q\bar{q}}^{\lambda_{q\bar{q}}}(\mathbf{q}, \mathbf{l}_{q\bar{q}} = -\mathbf{Q}, \lambda_q, \lambda_{\bar{q}}) < 1, \lambda_{q\bar{q}}; 1, \sigma | S, M >$$

$$\cdot D^{(1)}_{\sigma\lambda_g}(\bar{\mathbf{Q}})Y_{1l}(\bar{\mathbf{Q}}) < S,M;1,l|1,\lambda_{ex} > .$$

The spin-1 Wigner rotation matrix $D^{(1)}$ relates the gluon helicity σ to its spin λ_g quantized along the z-axis. The corresponding normalized wave functions are then given by:

$$\Psi^{\lambda_{ex}}_{q\bar{q}g(S=0)} = \sqrt{\frac{3}{8\pi}} \sum_{\lambda_{q\bar{q}}} \Psi^{\lambda_{q\bar{q}}}_{q\bar{q}}(\mathbf{q},-\mathbf{Q},\lambda_q,\lambda_{\bar{q}})[\varepsilon^*(\lambda_{q\bar{q}}) \cdot \varepsilon^*_c(\mathbf{Q},\lambda_g)][\bar{\mathbf{Q}}\cdot\varepsilon(\lambda_{ex})],$$

$$\Psi^{\lambda_{ex}}_{q\bar{q}g(S=1)} = \sqrt{\frac{3}{8\pi}} \sum_{\lambda_{q\bar{q}}} \Psi^{\lambda_{q\bar{q}}}_{q\bar{q}}(\mathbf{q},-\mathbf{Q},\lambda_q,\lambda_{\bar{q}})[\varepsilon^*(\lambda_{q\bar{q}}) \times \varepsilon^*_c(\mathbf{Q},\lambda_g)] \cdot [\bar{\mathbf{Q}}\times\varepsilon(\lambda_{ex})],$$

$$\Psi^{\lambda_{ex}}_{q\bar{q}g(S=2)} = \sqrt{\frac{27}{104\pi}} \sum_{\lambda_{q\bar{q}}} \Psi^{\lambda_{q\bar{q}}}_{q\bar{q}}(-\mathbf{Q},\lambda_q,\lambda_{\bar{q}})\bar{\mathbf{Q}}\cdot[\varepsilon^*(\lambda_{q\bar{q}}) \otimes \varepsilon^*_c(\mathbf{Q},\lambda_g)] \cdot \varepsilon(\lambda_{ex}), \quad (7)$$

where $\bar{\mathbf{Q}} = \mathbf{Q}/|\mathbf{Q}|$, $(A \otimes B)_{ij} = 2A_{(i}B_{j)} - \frac{2}{3}\delta_{ij}(\mathbf{A}\cdot\mathbf{B})$ and $\varepsilon^i_c(\mathbf{Q},\lambda_g) = \varepsilon^j(\lambda_g)(\delta^{ij} - \bar{Q}^i\bar{Q}^j)$.

MESON AND HYBRID STATES

The π $(I = 1)$ and η $(I = 0)$ states $(J^{PC} = 0^{-+})$ are constructed in terms of annihilation and creation operators:

$$|M(\mathbf{P},I,I_3,\lambda) > = \sum_{all\ \lambda,c,f} \int \frac{d^3\mathbf{p}_q}{(2\pi)^3 2E(m,\mathbf{p}_q)} \frac{d^3\mathbf{p}_{\bar{q}}}{(2\pi)^3 2E(m,\mathbf{p}_{\bar{q}})} 2(E(m,\mathbf{p}_q) + E(m,\mathbf{p}_{\bar{q}}))$$

$$\cdot (2\pi)^3 \delta^3(\mathbf{p}_q + \mathbf{p}_{\bar{q}} - \mathbf{P}) \frac{\delta_{c_q c_{\bar{q}}}}{\sqrt{3}} \frac{F(I,I_3)_{f_q f_{\bar{q}}}}{\sqrt{2}} \Psi_{q\bar{q}}(\mathbf{p}_q,\mathbf{p}_{\bar{q}},\lambda_q,\lambda_{\bar{q}})\delta_{\lambda 0}\frac{1}{N(P)}$$

$$\cdot \psi_L(\frac{m_{q\bar{q}}(\mathbf{p}_q,\mathbf{p}_{\bar{q}})}{\mu}) b^\dagger_{\mathbf{p}_q\lambda_q f_q c_q} d^\dagger_{\mathbf{p}_{\bar{q}}\lambda_{\bar{q}} f_{\bar{q}} c_{\bar{q}}}|0 > . \quad (8)$$

In the above $\Psi_{q\bar{q}}$ is the spin-0 wave function (4), I denotes isospin and I_3 is its third component, c is color and f is flavor. ψ_L represents the orbital wave function resulting from the interaction between quarks that leads to a bound state (meson). Such a function depends (due to covariance) only on the invariant mass $m_{q\bar{q}}(\mathbf{p}_q,\mathbf{p}_{\bar{q}})$ [4]. Normalization constants are denoted by N (with $P = |\mathbf{P}|$) and μ's are free parameters, being scalar functions of meson quantum numbers. Finally, $F(I,I_3)$ is 2x2 isospin matrix ($f = 1$ for u and $f = 2$ for d)

$$F(0,0) = 1, \quad F(1,I_3) = \sigma^i \varepsilon^i(I_3).$$

The flavor structure of the η state (as well as of all meson states with isospin 0) was chosen as a linear combination $\frac{1}{\sqrt{2}}(|u\bar{u} > + |d\bar{d} >)$. Strictly speaking, those states are rather linear combinations $a|u\bar{u} > + b|d\bar{d} > + c|s\bar{s} >$. But $|s\bar{s} >$ does not contribute to

458

the amplitude of the decay of π_1 and therefore may be neglected in calculations, provided this amplitude is multiplied by a factor $\sqrt{1-|c|^2}$.

Similarly the $\rho\,(I=1)$ and $\omega,\phi\,(I=0)$ states ($J^{PC}=1^{--}$) are given by (8), but instead of $\Psi^{\lambda}_{q\bar{q}}\delta_{\lambda 0}$ one must use (3). The $b_1\,(I=1)$ and $h_1\,(I=0)$ states ($J^{PC}=1^{+-}$) contain the wave function (5) with $\mathbf{q}=\Lambda(\mathbf{P}\to 0)\mathbf{p}_q$. Finally, the $a\,(I=1)$ and $f\,(I=0)$ states ($J^{PC}=0,1,2^{++}$) correspond to (6).

The hybrid state in its rest frame is given by

$$|\pi_1(I_3,\lambda_{ex})\rangle = \sum_{\text{all }\lambda,c,f}\frac{1}{N_{ex}}\int \frac{d^3\mathbf{p}_q}{(2\pi)^3 2E(m,\mathbf{p}_q)}\frac{d^3\mathbf{p}_{\bar{q}}}{(2\pi)^3 2E(m,\mathbf{p}_{\bar{q}})}\frac{d^3\mathbf{Q}}{(2\pi)^3 2E(m_g,\mathbf{Q})}$$

$$\cdot(2\pi)^3 2(E(m,\mathbf{p}_q)+E(m,\mathbf{p}_{\bar{q}})+E(m_g,\mathbf{Q}))\delta^3(\mathbf{p}_q+\mathbf{p}_{\bar{q}}+\mathbf{Q})\frac{\lambda^{c_g}_{c_q c_{\bar{q}}}}{2}\frac{\sigma^i_{f_q f_{\bar{q}}}\varepsilon^i(I_3)}{\sqrt{2}}$$

$$\cdot\Psi^{\lambda_{ex}}_{q\bar{q}g}(\mathbf{p}_q,\mathbf{p}_{\bar{q}},\lambda_q,\lambda_{\bar{q}},\lambda_g)\psi'_L\left(\frac{m_{q\bar{q}}(\mathbf{p}_q,\mathbf{p}_{\bar{q}})}{\mu_{ex}},\frac{m_{q\bar{q}g}(\mathbf{p}_q,\mathbf{p}_{\bar{q}},\mathbf{Q})}{\mu_{ex'}}\right)b^{\dagger}_{\mathbf{p}_q\lambda_q f_q c_q}d^{\dagger}_{\mathbf{p}_{\bar{q}}\lambda_{\bar{q}} f_{\bar{q}} c_{\bar{q}}}a^{\dagger}_{\mathbf{Q}\lambda_g c_g}|0\rangle,$$

$$(9)$$

where the spin wave function $\Psi_{q\bar{q}g}$ was given in (7) for $S=0,1,2$ and the orbital wave function ψ'_L depends only on $m_{q\bar{q}}$ and the invariant mass of the three-body system. Here m_g denotes the effective mass of the gluon coming from its interaction with virtual partcles, and $\lambda^{c_g}_{c_q c_{\bar{q}}}$ are the Gell-Mann matrices. Constants N are fixed by normalization (m_M is mass of meson)

$$\langle \mathbf{P},\lambda,I_3|\mathbf{P}',\lambda',I'_3\rangle = (2\pi)^3 2E(m_M,\mathbf{P})\delta^3(\mathbf{P}-\mathbf{P}')\delta_{\lambda\lambda'}\delta_{I_3 I'_3}.$$

DECAYS OF π_1 AND NONRELATIVISTIC LIMIT

We will assume that a transverse gluon in π_1 creates a quark-antiquark pair and therefore the hybrid decays into two mesons. The Hamiltonian of this process in the Coulomb gauge is given by

$$H = \sum_{\text{all }c,f}\int d^3\mathbf{x}\,\bar{\psi}_{c_1 f_1}(\mathbf{x})(g\gamma\cdot\mathbf{A}^{c_g}(\mathbf{x}))\psi_{c_2 f_2}(\mathbf{x})\delta_{f_1 f_2}\frac{1}{2}\lambda^{c_g}_{c_1 c_2},$$

where

$$\psi_{cf}(\mathbf{x}) = \sum_{\lambda}\int \frac{d^3\mathbf{k}}{(2\pi)^3 2E(m,\mathbf{k})}[u(\mathbf{k},\lambda)b_{\mathbf{k}\lambda cf}+v(-\mathbf{k},\lambda)d^{\dagger}_{-\mathbf{k}\lambda cf}]e^{i\mathbf{k}\cdot\mathbf{x}}$$

and

$$\mathbf{A}^{c_g}(\mathbf{x}) = \sum_{\lambda}\int \frac{d^3\mathbf{k}}{(2\pi)^3 2E(m_g,\mathbf{k})}[\varepsilon_c(\mathbf{k},\lambda)a^{c_g}_{\mathbf{k}\lambda}+\varepsilon^*_c(-\mathbf{k},\lambda)a^{\dagger c_g}_{-\mathbf{k}\lambda}]e^{i\mathbf{k}\cdot\mathbf{x}}.\qquad(10)$$

Here g is the strong coupling constant. The matrix element $\langle M_1(\mathbf{P}_1)|\langle M_2(\mathbf{P}_2)|H|\pi_1\rangle$ (M stands for meson) will be a sum of two terms because pairs $b,b^{\dagger}$ and $d,d^{\dagger}$ appear twice and one can show they are equal. If $\mu_\eta = \mu_\pi$ then $A_{\pi\eta} = 0$ and the hybrid will not decay into π and η. The same occurs for $\rho + \omega$. Neither can it decay into two pions because of a relative minus sign from isospin that makes both terms cancel out. However, $\mu_{b_1} = \mu_\pi$ does not imply $A_{\pi b_1} = 0$ because the orbital wave functions of these mesons are different.

Since π and η have the same quantum numbers (except isospin) and therefore μ_π and μ_η should be almost equal (not exactly because $SU(3)_f$ is only an approximate symmetry and there is a contribution of $s\bar{s}$ in η), out of two channels $\pi\eta$, πb_1 the latter will be favored. However, free parameters μ need not to be close to each other for two mesons with different radial quantum numbers, making corresponding channels significant.

A nonrelativistic limit is obtained by ignoring Wigner rotation and using nonrelativistic phase space. For m large compared to μ's and P this limit should be approached by relativistic results. In this case for decays of π_1 into π and the S-meson $S_{q\bar{q}g} \neq 1$, whereas for those into π and the P-meson $S_{q\bar{q}g} \neq 0$. The difference in amplitudes coming from the spin wave function can be clearly seen, assuming $m_\eta = m_\rho$, $\mu_\eta = \mu_\rho$ and $m_{b_1} = m_{f_1} = m_{f_2}$, $\mu_{b_1} = \mu_{f_1} = \mu_{f_2}$ (the second condition is satisfied with a good approximation by masses of particles):

$$\Gamma_{\pi\rho} = \frac{1}{2}\Gamma_{\pi\eta}, \ \Gamma_{\pi f_1} = \frac{1}{8}\Gamma_{\pi b_1},$$

where $A_{\pi\eta}$, $A_{\pi b_1}$ are taken for $S = 0$ and $A_{\pi\rho}$, $A_{\pi f_{1,2}}$ for $S = 1$. Relations $A_{\pi\eta}$ vers. $A_{\pi b_1}$ and $A_{\pi\rho}$ vers. $A_{\pi f_{1,2}}$ depend on the orbital angular momentum wave functions ψ_L and ψ'_L. If $\mu_\rho = \mu_\pi$ then in a nonrelativistic limit π_1 would not decay into $\pi + \rho$. Therefore the width rate for this process is expected to be much smaller than that of πb_1.

ORBITAL WAVE FUNCTION

The orbital angular momentum wave function for a meson or a hybrid depends on the potential between quark and antiquark or for a $q\bar{q}g$ system. An explicit form of such a potential is not known exactly and such a function must be modeled. Because of the Lorentz invariance it may depend on momenta only through the invariant mass of particles. Moreover, it must tend to zero for large momenta fast enough to make the amplitude convergent. The most natural choice is the exponential function

$$\psi_L(m_{q\bar{q}}(\mathbf{p}_q, \mathbf{p}_{\bar{q}})/\mu) = e^{-m_{q\bar{q}}^2(\mathbf{p}_q \cdot \mathbf{p}_{\bar{q}})/8\mu^2}$$

for a meson, and

$$\psi'_L(m_{q\bar{q}}(\mathbf{p}_q, \mathbf{p}_{\bar{q}})/\mu_{ex}, m_{q\bar{q}g}(\mathbf{p}_q, \mathbf{p}_{\bar{q}}, \mathbf{Q})/\mu'_{ex}) = e^{-m_{q\bar{q}}^2(\mathbf{p}_q \cdot \mathbf{p}_{\bar{q}})/8\mu_{ex}^2} e^{-m_{q\bar{q}g}^2(\mathbf{p}_q \cdot \mathbf{p}_{\bar{q}} \cdot \mathbf{Q})/8\mu_{ex}'^2}$$

for a hybrid. The integrals for the decay amplitude are not elementary and must be computed numerically. In a nonrelativistic limit, however, they can be expressed in terms of the error function.

The free parameters of the presented model are m, m_g, μ's and g. The pion form factor constants f_π and F_π (whose behaviour is experimentally known) defined by

$$< 0|A^{\mu,i}(0)|\pi^k(\mathbf{p}) >= f_\pi p^\mu \delta_{ik}, \quad < \pi^i(\mathbf{p}')|V^{\mu,j}(0)|\pi^k(\mathbf{p}) >= F_\pi(p^\mu + p'^\mu)i\varepsilon_{ijk}, \quad (11)$$

allow us to fit m and μ_π (with the π state given by (8)). The axial and the vector currents are defined by

$$A^{\mu,i}(0) = \bar{\psi}_{cf}(0)\gamma^\mu \gamma_5 \frac{\sigma^i}{2}\psi_{cf}(0), \quad V^{\mu,j}(0) = \bar{\psi}_{cf}(0)\gamma^\mu \frac{\sigma^j}{2}\psi_{cf}(0), \quad (12)$$

with $\psi_{cf}(\mathbf{x})$ given in (10). By virtue of the Lorentz invariance f_π is a constant, whereas F_π is a function of $Q^2 = -(\mathbf{p} - \mathbf{p}')^2$. Impossibility of finding the generators H and M_{0i} of the Poincare group in presence of interaction together with normalization of states violate the Lorentz covariance between spatial and time components but do not break a rotational symmetry. The resulting form factors will depend on the frame of reference.

RESULTS AND SUMMARY

Taking $m_\pi = (3 * 140 + 770)/4 = 612MeV$ (in normalization) and $m = 306MeV$ gives $\mu_\pi = 220MeV$. Assuming also $g^2 = 10$, $m_g = 500MeV$ [5], $m_{ex} = 1.6GeV$ and equality of all parameters μ leads to the following values: $\Gamma_{\pi b_1} =150$MeV, $\Gamma_{\pi f_1} =20$MeV, $\Gamma_{\pi\rho} =3$MeV. In the nonrelativistic limit one obtains respectively: 230, 31 and 0 MeV.

Two important conclusions come from this work. Firstly, numerical results show that relativistic corrections arising from a Wigner rotation are significant. Therefore, models with no Wigner rotation are either nonrelativistic or inconsistent. These corrections decrease in general (for reasonable values of m) the width rates for π_1 or make them different from zero if they vanished for the NR case. Secondly, the π_1 prefers to decay into two mesons, one of which has no orbital angular momentum and the other has $L = 1$ ($S + P$ selection rule). This is in agreement with other models, for instance [6]. Calculations involving decay rates of π_1 into strange mesons and final state interactions ($b_1 \rightarrow \pi + \omega$, $\omega + \pi \rightarrow \rho$) are in preparation.

REFERENCES

1. K.J.Juge, J.Kuti, C.J.Morningstar, Nucl.Phys.Proc.Suppl. **63**, 326 (1998)
2. N.Isgur, R.Kokoski, J.Paton, Phys.Rev.Lett. **54**, 869 (1985)
3. F.E.Close, P.R.Page, Nucl.Phys. B**443**, 233 (1995)
4. B.Bakamjian, L.H.Thomas, Phys.Rev. **92**, 1300 (1953)
5. A.P.Szczepaniak, E.S.Swanson, arXiv:hep-ph/0308268 (2003)
6. P.R.Page, E.S.Swanson, A.P.Szczepaniak, Phys.Rev. D**59**, 034016 (1999)

Exotic meson searches - E852 data analysis

Maciej Swat

Physics Department,Indiana University,Bloomington,47405,USA

Abstract. I discuss the analysis of $\pi^- p \to \eta \pi^0 n$, $\eta \pi^- p$, $\eta' \pi^- p$ E852 data. I present a possible interpretation of the low mass exotic mesons reported earlier in $\eta \pi$ and $\eta' \pi$ channels.

INTRODUCTION

Strong interaction theory ,QCD, predicts that in addition to conventional mesons ($q\bar{q}$ - pair) one should observe mesons with excited gluonic degrees of freedom. Despite a lot of searchers for gluonic excitations, some of them reporting signals indicating presence of mesons with excited gluonic field, there is still controversy about the interpretation of these gluonic excitation candidates.

Mesons with excited gluons can be divided in two categories: glueballs - these particles are dominated by pure glue states and hybrids - objects composed of valence quark, anti quark and of excited gluonic field.

Observation of gluonic excitations may be problematic, because they can mix with ordinary mesons. Nevertheless some of the gluonic excitations are characterized by a set of J^{PC} quantum numbers that cannot be explained by a $q\bar{q}$ model thus making experimental identification of part of gluonic excitation spectrum unambiguous. The lowest lying J^{PC} combinations that cannot originate from valence quark model are $0^{--}, 0^{+-}, 1^{-+}, 2^{+-}, \ldots$. Mesons characterized by this set of J^{PC}'s are called exotic mesons.

I will discuss the analysis of $\eta \pi^0$, $\eta \pi^-$, $\eta' \pi^-$ data from BNL E852 experiment. Previous analysis of this data claimed the observation of signals consistent with the presence of exotic mesons in the $\eta \pi^-$ - so called $\pi_1(1400)$ [1]- and in $\eta' \pi^-$ - so called $\pi_1(1600)$ [2] - spectra. I will argue that signals previously interpreted as a signatures of exotic mesons can be explained by re-scattering effects.

THE $\eta \pi$ SPECTRUM

E852 used an 18 *GeV* π^- beam on hydrogen to produce different resonances which were seen through their decay into final state particles. The identification of resonances was done by means of partial wave analysis (PWA). From the J^{PC} quantum numbers of η and π one can conclude that $\eta \pi$ channels are well suited for exotic meson searches, because mesons decaying strongly into $\eta \pi$ will have J^{PC} quantum numbers equal to $0^{++}, 1^{-+}, 2^{++}$ for $L = 0$, $L = 1$, $L = 2$ respectively (L denotes relative orbital angular

CP717, *Hadron Spectroscopy: Tenth International Conference,*
edited by E. Klempt, H. Koch, and H. Orth
© 2004 American Institute of Physics 0-7354-0197-7/04/$22.00

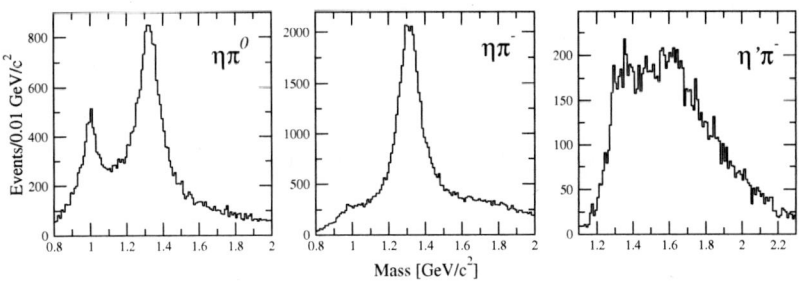

FIGURE 1. Mass spectra of $\eta\pi^0$, $\eta\pi^-$, $\eta'\pi^-$

momentum between η and π).Thus one can conclude that $\eta\pi$ spectrum should contain a_0 and a_2 resonances and possibly a component characterized by exotic quantum numbers 1^{-+}.

A partial wave analysis was performed for the $\eta\pi^0$, $\eta\pi^-$,$\eta'\pi^-$ channels as a function of t - the four momentum transfer and the invariant mass of the $\eta\pi^0$, $\eta\pi^-$,$\eta'\pi^-$ systems respectively. The acceptance corrected angular distribution of η or π (the intensity function $I(m_{\eta\pi},\Omega)$) - is decomposed into partial waves via the following relation:

$$I(m_{\eta\pi},\Omega) = \sum_{\varepsilon} |\sum_{L,M} a^{\varepsilon}_{LM}(m_{\eta\pi})Y^{\varepsilon}_{LM}(\Omega)|^2$$

where a_{LM} is referred to as a partial wave or production amplitude ,Y_{LM} denotes a decay amplitude ,$\varepsilon = +,-$ denotes naturality (reflectivity) of the production amplitude a^{ε}_{LM} , $L = J$ is the the value of orbital angular momentum of $\eta\pi$ system (also its total spin) and $M = 0,1$ is the projection of the $\eta\pi$ angular momentum on the z axis defined in the Gottfried-Jackson frame [3]. In the case of $\eta\pi$ systems it is sufficient to truncate above sum at L=2 implying that only S_0, P_0, P_-, D_0, D_-, P_+, D_+ partial waves should be considered. Waves with subscript '0' or '−' are due to negative naturality exchange e.g. b_1 and waves with subscript '+' are manifestation of natural exchange e.g. ρ .

It is well known, [3] that under the assumption that only S,P,D waves contribute to the spectrum, one can find up to 8 distinct sets of amplitudes in each invariant mass bin that give the same value of intensity function. It should be emphasized that these are only mathematical ambiguities. In other words if one knew exactly the interactions governing the $\eta\pi$ interactions one would be able to calculate production amplitudes uniquely. For this reason we know that only one of the ambiguous sets of production amplitudes (in a given mass bin) is the physical one. In order to identify the physical sets of production amplitudes we impose [4] physical constraints on possible mass bin to mass bin variation of amplitudes. In particular we have determined that for the $\eta\pi^0$ channel requiring S and D waves' intensities to be consistent with $a_0(980)$ and $a_2(1320)$, $S - D_0$ mass dependent phase to be consistent with two interfering Breit-Wigner resonances (again $a_0(980)$ and $a_2(1320)$) and requiring overall continuity of the phases and waves allowed us to uniquely select the physical amplitudes. In the case of $\eta\pi^-$ and $\eta'\pi^-$ channels ambiguous solutions cluster around each other and it is sufficient to impose continuity requirement to select the physical amplitudes.

In addition to PWA studies we also performed moment studies. Moments defined as

$$H(m_{\eta\pi}, LM) = \int d\Omega I(m_{\eta\pi}, \Omega) Y^{\star}_{LM}(\Omega)$$

are unambiguous functions of $m_{\eta\pi}$ and can be written in terms of partial wave amplitudes, e.g.

$$H(32) = \frac{2}{7}\sqrt{\frac{3}{2}}Re\{P_-D^{\star}_-\} - \frac{2}{7}\sqrt{\frac{3}{2}}Re\{P_+D^{\star}_+\}$$

One can also calculate moments without doing a PWA fit thus moments provide an important check on PWA results and a mass dependent fit to the moments can be compared to physical PWA solutions.

Fit to the data - $\eta\pi^0$ channel

Based on the assumption that there is an exotic meson present in $\eta\pi^-$ channel one should expect, due to isospin symmetry, to find its neutral partner in the $\eta\pi^0$ channel. To verify this hypothesis we have expressed the mass dependence of S, P, D waves of the $\eta\pi^0$ channel as a Breit-Wigner resonances and carried out mass dependent fits to positive naturality waves (P_+, D_+ and $P_+ - D_+$ phase difference) as well as a global fit to 12 non-vanishing moments. A Breit-Wigner parameterization of the S and D waves is fully justified as the shape of the the the intensity function unambiguously indicates presence of the $a_0(980)$ and $a_2(1320)$ mesons - see Fig.1. If indeed there is exotic meson present in the $\eta\pi^0$ spectrum one should be able to extract its mass and width by performing appropriate fit with the P waves given as a Breit-Wigner resonance. However, by fitting all the moments (global fit) or positive reflectivity waves only we were not able to find a self-consistent set of Breit-Wigner parameters that would properly describe the P-wave. In particular, depending on t-range and on type of fit, we were getting unphysical parameters of the P-wave resonance (either the decay width was very large or mass of the P-wave resonance was found to be anywhere between 1.0 and 10.0 GeV). At the same time the two well established mesons the $a_0(980)$ and $a_2(1320)$ were very well reproduced in our fits as Breit-Wigner resonances - see Fig.2. Thus we are lead to the conclusion that Breit-Wigner parameterization of the P wave is not appropriate. We do not see any reason in introducing so called $\pi_1(1400)$ resonance to explain the nature of the $\eta\pi^0$ P wave [4].

Final State Interactions (FSI) and the interpretation of $\eta\pi^-$ and $\eta'\pi^-$ P wave

By using a Breit-Wigner parameterization of the partial waves one implicitly assumes that η and π are produced via resonance decay and that this resonance comes from one meson exchange process. However it is entirely possible that double meson exchange

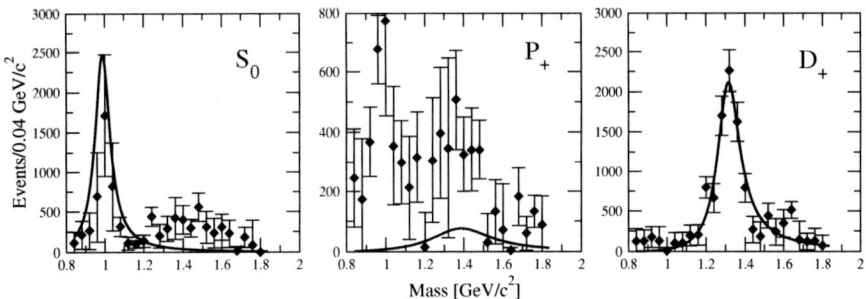

FIGURE 2. Physical PWA solutions for $\eta\pi^0$ channel -low-t- (filled diamonds) and fit to the moment results (continuous line). See also [4]

takes place and η and π are produced "separately" and then interact softly. Also one cannot exclude the possibility of channel mixing. Thus $\eta\pi^0$ analysis should take into account single and double meson exchange mechanisms and should be done in the coupled channel framework where, in addition to $\eta\pi^0$ channel, one considers $\eta'\pi^0$ channel as well. Unfortunately the experimental $\eta'\pi^0$ data sample had very low statistics so instead of studying neutral channels we used $\eta\pi^-$ and $\eta'\pi^-$ channels where $\eta \to \pi^+\pi^-\pi^0$ and $\eta' \to \pi^+\pi^-\eta$.

Our model of $\eta\pi$ interaction was inspired by the results of S-wave $\pi - \pi$ scattering where [5, 6] based on the chiral perturbation theory one can constrain $\pi\pi$ and KK final state interaction potentials so that when used in the coupled channel formalism they describe $\pi\pi$ S-matrix phase shift well above the threshold region. In particular one can reproduce the phase raise corresponding to $f_0(980)$ just by considering final state interactions in $\pi\pi$ and KK channels without the necessity of introducing explicit pole term.

Similarly for $\eta\pi^-$ and $\eta'\pi^-$ channels we can use low energy expansion methods to constrain soft interaction potentials [7] . We have parameterized partial waves in such a way that they were allowed to have contributions from the pole terms (in the absence of FSI they would be pure Breit-Wigners) and contributions from re-scattering (FSI) [8] . Similarly as for $\eta\pi^0$ channel we performed moment fits as well as fits to positive reflectivity waves only. In our fits we turned on and off the pole-term contributions to the P wave. We were able to conclude that fits with (continuous line) and without (dashed line) resonant contributions to the P wave are of the same quality with the only exception that in the case of the $\eta'\pi^-$ channel few mass bins around 1.6 GeV are better reproduced when we allow exotic resonance (i.e. with the pole term in the P wave included) to couple to $\eta\pi$ system -Fig.3. Nevertheless we find that bulk of the intensity of the $\eta'\pi^-$ P wave is due to FSI rather than to an underlying exotic resonance. We found also that the nature of the P wave of $\eta\pi^-$ channel can be entirely explained by FSI . This means that according to our studies it is unnecessary to introduce a low mass and large decay width exotic meson to explain the spectrum of the $\eta\pi^-$ and $\eta'\pi^-$ channels. It is also worth pointing out that even strong D_+ wave of the $\eta\pi^-$ channel has contributions from FSI

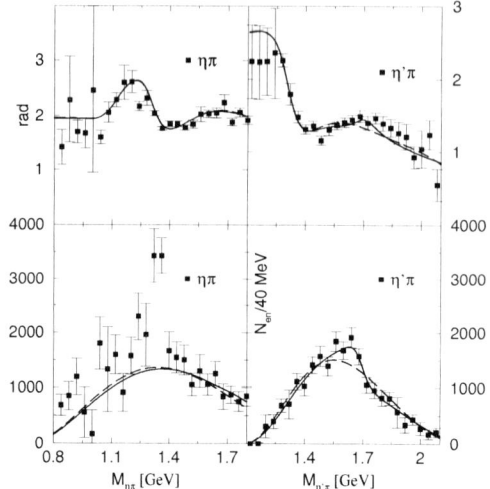

FIGURE 3. Exotic 1^{-+} wave coupled channel fit to the $\eta\pi^-$, $\eta'\pi^-$ data. Upper plots show phase difference in radians between P_+ and D_+ waves. The lower show the intensity of the P_+ wave. Continuous(dashed) line represents a fit with(without) pole term in the P wave

which are visible in the tail of D_+ spectrum. FSI provides also a natural interpretation of the wide structure seen in the $\eta'\pi^-$ D wave. This structure was previously parameterized as Breit-Wigner resonance [2].

SUMMARY AND ACKNOWLEDGMENTS

We have shown that final state interactions play an essential role in the understanding of the spectra of two pseudoscalar systems. In particular they provide us with natural interpretation of the exotic wave without necessity of introducing low mass exotic mesons.

This work was done in collaboration with Adam Szczepaniak, Alex Dzierba and Scott Teige and was supported by US Department of Energy.

REFERENCES

1. Chung, S. U., et al., *Phys. Rev.*, **D60**, 092001 (1999).
2. Ivanov, E. I., et al., *Phys. Rev. Lett.*, **86**, 3977–3980 (2001).
3. Chung, S. U., *Phys. Rev. D*, **56** (1997).
4. Dzierba, A. R., et al., *Phys. Rev.*, **D67**, 094015 (2003).
5. Oller, J. A., and Oset, E., *Phys. Rev.*, **D60**, 074023 (1999).
6. Oller, J. A., and Oset, E. (1997).
7. Bass, S. D., and Marco, E., *Phys. Rev.*, **D65**, 057503 (2002).
8. Szczepaniak, A. P., Swat, M., Dzierba, A. R., and Teige, S., *Phys. Rev. Lett.*, **91**, 092002 (2003).

The Θ^+ (1540) as a heptaquark with the overlap of a pion, a kaon and a nucleon

P. Bicudo* and G. M. Marques*

*Dep. Física and CFIF, Instituto Superior Técnico, Av. Rovisco Pais 1049-001 Lisboa, Portugal

Abstract. We study the very recently discovered Θ^+ (1540) at SPring-8, at ITEP and at CLAS-Thomas Jefferson Lab. We apply the same RGM techniques that already explained with success the repulsive hard core of nucleon-nucleon, kaon-nucleon exotic scattering, and the attractive hard core present in pion-nucleon and pion-pion non-exotic scattering. We find that the $K-N$ repulsion excludes the Θ^+ as a $K-N$ s-wave pentaquark. We explore the Θ^+ as heptaquark, equivalent to a $N+\pi+K$ borromean boundstate, with positive parity and total isospin $I=0$. We find that the kaon-nucleon repulsion is cancelled by the attraction existing both in the pion-nucleon and pion-kaon channels. Although we are not yet able to bind the total three body system, we find that the Θ^+ may still be a heptaquark state.

In this talk we study the exotic hadron Θ^+ [1] (narrow hadron resonance of 1540 MeV decaying into a nK^+) very recently discovered at Spring-8 [2], and confirmed by ITEP [3] and by CLAS at the TJNL [4]. This is an extremely exciting state because it may be the first exotic hadron to be discovered, with quantum numbers that cannot be interpreted as a quark and an anti-quark meson or as a three quark baryon. Very recent studies propose that the Θ^+ is a pentaquark state [5]. Exotic multiquarks are expected since the early works of Jaffe [6, 7], and some years ago Diakonov, Petrov and Polyakov [8] applied skyrmions to a precise prediction of the Θ^+. The nature of this particle, and its isospin, parity [9] and angular momentum are yet to be determined.

We start in this talk by reviewing the Quark Model (QM) and the Resonating Group Method (RGM) [10], which are adequate to study states where several quarks overlap. We show that the QM, together with chiral symmetry, produces hard core hadron-hadron potentials, which can be either repulsive or attractive. First we apply the RGM to show [11, 12] that the exotic $N-K$ hard core s-wave interaction is repulsive. This is consistent with the experimental data [13], see Fig. 1. We think that this excludes the Θ^+ as a bare s-wave pentaquark $udd\bar{u}s$ state or a tightly bound s-wave $N-K$ narrow resonance. The observed mass of the Θ^+ is larger than the sum of the K and N masses by $1540-939-493=118$ MeV, and this does not suggest a simple $K-N$ binding. However a π could also be present in this system, in which case the binding energy would be of the order of 20 MeV. Moreover this state of seven quarks would have a positive parity, and would have to decay to a p-wave $N-K$ system, which is suppressed by angular momentum, thus explaining the narrow width of the Θ^+. With this natural description in mind we then apply the RGM to show that the $\pi-N$ and $\pi-K$ hard core interactions are attractive. Finally we put together the $\pi-N$, $\pi-K$ and $N-K$ interactions to show that the Θ^+ is possibly a borromean [14] three body s-wave

CP717, *Hadron Spectroscopy: Tenth International Conference*,
edited by E. Klempt, H. Koch, and H. Orth
© 2004 American Institute of Physics 0-7354-0197-7/04/$22.00

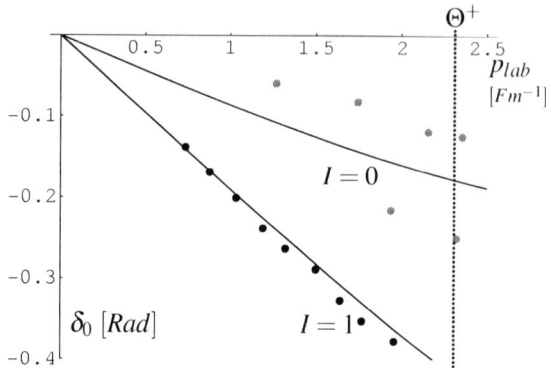

FIGURE 1. The $I = 0$ and $I = 1$ experimental [13] and theoretical (this talk and ref. [11]) s-wave phase shifts as a function of the kaon momentum in the laboratory frame.

boundstate of a π, a N and a K, with positive parity and total isospin $I = 0$.

The RGM was used by Ribeiro [15] to show that in exotic hadron-hadron scattering, the quark-quark potential together with Pauli repulsion of quarks produces a repulsive short range interaction. For instance this explains the $N - N$ hard core repulsion, preventing nuclear collapse. Deus and Ribeiro [16] used the same RGM to show that, in non-exotic channels, the quark-antiquark annihilation could produce a short core attraction. The RGM computes the effective hadron-hadron interaction using the matrix elements of the microscopic quark-quark interactions. The wave functions of quarks are arranged in anti-symmetrized overlaps of simple color singlet hadrons.

When spontaneous chiral symmetry breaking is included in the quark model [17, 18] the annihilation potential becomes crucial [19, 20] to understand the low mass of the π. The annihilation potential A is also present in the π Salpeter equation where it cancels most of the kinetic energy and confining potential $2T + V$. From the quark model with chiral symmetry breaking we get the matrix element [17],

$$\langle A \rangle_{S=0} = -\frac{2}{3}(2M_N - M_\Delta) . \tag{1}$$

While the standard quark-quark potential provides a repulsive (positive) overlap, it is clear in eq. (1) that the annihilation potential provides an attractive (negative) overlap. This confirms that the hard core can be attractive for non-exotic channels where annihilation occurs, while it is repulsive for exotic channels.

We summarize [11, 19, 20] the effective potentials for the different channels,

$$V_{K-N} = \frac{2 - \frac{4}{3}\vec{\tau}_A \cdot \vec{\tau}_B}{\frac{5}{4} + \frac{1}{3}\vec{\tau}_A \cdot \vec{\tau}_B} \frac{(M_\Delta - M_N)}{3} \left(\frac{2\sqrt{\pi}}{\alpha}\right)^3 e^{-\frac{p_\lambda^2}{2\beta^2}} \int \frac{d^3 p'_\lambda}{(2\pi)^3} e^{-\frac{p'^2_\lambda}{2\beta^2}}$$

$$V_{\pi-N} = \frac{2}{9}(2M_N - M_\Delta)\vec{\tau}_A \cdot \vec{\tau}_B \; \alpha^{-2} ,$$

$$V_{\pi-K} = \frac{8}{27}(2M_N - M_\Delta)\vec{\tau}_A \cdot \vec{\tau}_B \; \alpha^{-2} , \tag{2}$$

TABLE 1. This table summarizes the parameters μ, v, α, β (in Fm^{-1}) and scattering lengths a (in Fm) .

channel	μ	v_{th}	α	β	a_{th}	a_{exp}
$K - N_{I=0}$	1.65	0.50	3.2	3.2	-0.14	-0.13 ± 0.04 [13]
$K - N_{I=1}$	1.65	1.75	3.2	3.2	-0.30	-0.31 ± 0.01 [13]
$\pi - N_{I=\frac{1}{2}}$	0.61	-0.73	3.2	11.4	0.25	0.246 ± 0.007 [22]
$\pi - N_{I=\frac{3}{2}}$	0.61	0.36	3.2	3.2	-0.05	-0.127 ± 0.006 [22]
$\pi - K_{I=\frac{1}{2}}$	0.55	-0.97	3.2	10.3	0.35	0.27 ± 0.08 [23]
$\pi - K_{I=\frac{3}{2}}$	0.55	0.49	3.2	3.2	-0.06	-0.13 ± 0.06 [23]

where $\vec{\tau}$ are the isospin matrices. This parametrization in a separable potential enables us to use standard techniques [11] to exactly compute the scattering T matrix. The scattering length $d\delta_0/dk_{c.m.}$ is,

$$a = -\frac{\sqrt{\pi}}{\alpha} \frac{4\mu v}{\alpha^2 + \frac{\beta}{\alpha}4\mu v} . \tag{3}$$

The parameters and results for the relevant channels are summarized in Table 1. We have fitted α with the $I = 1$ $K - N$ scattering. We use $\beta = \alpha$ in the repulsive channels and fit it with the appropriate scattering lengths in the attractive $I = 1/2$ pionic channels.

Let us first apply our method to the $K - N$ exotic system, where the anti-quark $\bar{s}$ is present. In this case the $I = 0$ channel is less repulsive than the $I = 1$ channel. With our method we reproduce the $K - N$ exotic s-wave phase shifts – see Fig. 1 – where indeed there is no evidence for the Θ^+ state. In what concerns the $\pi - N$ system and the $\pi - K$ systems, the corresponding parameters in Table 1 are almost identical. There we find repulsion for $I = 3/2$ and attraction for $I = 1/2$. The repulsion in the $I = 3/2$ channel prevents a bound state in this channel. In what concerns the $I = 1/2$ channel, the attraction is not sufficient to provide for a bound state, because the π is quite light and the attractive potential is narrow. From eq. (3) we conclude that we have binding if the reduced mass is $\mu \geq -\frac{\alpha^3}{4\beta v}$. With the present parameters μ should be of the order of 1 Fm^{-1}. This is larger than the π mass, therefore it is not possible to bind the π to the K or to the N. All that we can get is a very broad resonance. For instance in the $\pi - K$ channel this is the kappa resonance [24], which has been recently confirmed by the scientific community. However, with a doubling of the interaction, produced by a K and a N, we expect the π to bind.

We now investigate the borromean [14] binding of the exotic Θ^+ constituted by a N, K and π triplet. In what concerns isospin, we need the π to couple to both N and K in $I = 1/2$ states for attraction, and the only candidate for binding is the total $I = 0$ state, see Fig. 2. Since the π is much lighter that the $N - K$ system, we study the borromean binding adiabatically. As a first step, we start by assuming that the K and N are essentially stopped and separated by $\vec{r}_N - \vec{r}_K = 2a\hat{e}_z$. This will be improved later. We also take advantage of the similarities in Table 1 to assume that the two heavier partners of the π have a similar mass of 3.64 Fm^{-1} and interact with the π with the same separable potential. Then we solve the bound state equation for a π in the potential

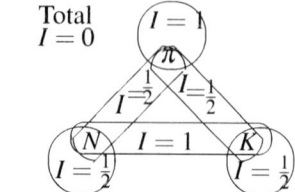

FIGURE 2. The isospin couplings in the Z/Θ.

$V_{\pi-N} + V_{\pi-K}$, where this potential is wider the direction of the $K - N$ z axis. The π energy is depicted if Fig. 3, and it is negative as expected. Once the π binding energy is determined, we include it in the potential energy of the $K - N$ system, which becomes the sum of the repulsive $K - N$ potential and the π energy. We find that for short distances the total potential is indeed attractive. Finally, using this $K - N$ potential energy we solve the schrödinger equation for the system, thus including the previously neglected K and N kinetic energies. However here we are not able to bind the $K - N$ system, because the total effective $K - N$ potential is not sufficiently attractive to cancel the positive $K - N$ kinetic energy. Nevertheless the K is heavier than the π, thus a small enhancement of the attraction would suffice to bind the heptaquark. We remark that existing examples of narrow resonances with a trapped K are the $f_0(980)$ [24], the $D_s(2320)$ [25] very recently discovered at BABAR [26], and possibly the $\Lambda(1405)$. Moreover we expect that meson exchange interactions and the irreducible three-body overlap of the three hadrons, that we did not include here, would further increase the attractive potential. Therefore it is plausible that a complete computation will eventually bind the $K - \pi - N$ system.

We conclude that the Θ^+ hadron very recently discovered cannot be an s-wave pentaquark. We also find that it may be a heptaquark state, composed by the overlap of a π, a K and a nucleon. This scenario has many interesting features. The Θ^+ would be, so far, the only hadron with a trapped π. Moreover the π would be trapped by a rare three body borromean effect. And the decay rate to a K and a N would be suppressed since the π needs to be absorbed with a derivative coupling, while the involved hadrons have a very low momentum in this state. Because the Θ^+ would be composed by a N and two pseudoscalar mesons, its parity would be positive, $J^P = \frac{1}{2}^+$, in agreement with

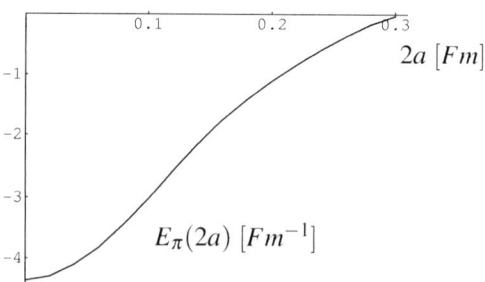

FIGURE 3. π energy as a function of the coordinate $|\vec{r}_N - \vec{r}_K| = 2a$.

the prediction of Diakonov, Petrov and Polyakov. The isospin of the Θ^+ would be $I = 0$ in order to ensure the attraction of the π both by the N and the K.

ACKNOWLEDGMENTS

We are very grateful to George Rupp for pointing our attention to the pentaquark state. The work of G. Marques is supported by Fundação para a Ciência e a Tecnologia under the grant SFRH/BD/984/2000.

REFERENCES

1. P. Schewe, J. Riordon, B. Stein; Physics News Update **644**, 1, June 30 (2003).
2. T. Nakano et al, arXiv:hep-ex/0301020, submitted to Phys. Rev. Lett.
3. V. Barmin et al, arXiv:hep-ex/0304040.
4. K. Hicks, to be published in the proceedings of the Conference on the Intersections of Particle and Nuclear Physics, New York may 2003, edited by Zohreh Parsa (2003).
5. H. Walliser and V.B. Kopeliovich arXiv:hep-ph/0304058; H. Gao and B.-Q. Ma, Mod. Phys. Lett. A **14** 2313 (1999) [arXiv:hep-ph/0305294]; F. Stancu and D. Riska arXiv:hep-ph/0307010; B. Wybourne arXiv:hep-ph/0307170; A. Hosaka arXiv:hep-ph/0307232; M. Karliner and H. Lipkin arXiv:hep-ph/0307243; M. Karliner and H. Lipkin arXiv:hep-ph/0307343; S. L. Zhu, arXiv:hep-ph/0307345; Xun Chen, Yajun Mao and Bo-Qiang Ma arXiv:hep-ph/0307381; C. E. Carlson, C. D. Carone, H. J. Kwee and V. Nazaryan, arXiv:hep-ph/0307396; L. Chen, V. Greco, C. Ko, S. Lee and W. Liu nucl-th/0308006.
6. R. L. Jaffe, Phys. Rev. D **15** 281 (1977).
7. D. Strottman, Phys. Rev. D **20** 748 (1979).
8. D. Diakonov, V. Petrov and M. V. Polyakov, Z. Phys. A **359** 305 (1997) [arXiv:hep-ph/9703373].
9. T. Hyodo, A. Hosaka and E. Oset, arXiv:nucl-th/0307105.
10. J. Wheeler, Phys. Rev. **52**, 1083 (1937); **52**, 1107 (1937).
11. P. Bicudo and J. E. Ribeiro, Z. Phys. C **38**, 453 (1988); P. Bicudo, J. E. Ribeiro and J. Rodrigues, Phys. Rev. C **52**, 2144 (1995).
12. I. Bender, H. G. Dosch, H. J. Pirner and H. G. Kruse, Nucl. Phys. A **414**, 359 (1984).
13. T. Barnes and E. Swanson, Phys. Rev. C **49** 1166 (1994); J. S. Hyslop, R. A. Arndt, L. D. Roper and R. L. Workman, Phys. Rev. D **46** 961 (1992).
14. M. V. Zhukov, B. V. Danilin, D. V. Fedorov, J. M. Bang, I. J. Thompson and J. S. Vaagen, Phys. Rept. **231**, 151 (1993); J. M. Richard, arXiv:nucl-th/0305076.
15. J. E. Ribeiro, Z. Phys. C **5**, 27 (1980).
16. J. Dias de Deus and J. Ribeiro Phys. Rev. D **21**, 1251 (1980).
17. P. J. Bicudo and J. E. Ribeiro, Phys. Rev. D **42**, 1611 (1990); 1625 (1990); 1635 (1990).
18. P. J. Bicudo, Phys. Rev. C **60**, 035209 (1999).
19. P. Bicudo, S. Cotanch, F. Llanes-Estrada, P. Maris, E. Ribeiro and A. Szczepaniak, Phys. Rev. D **65**, 076008 (2002) [arXiv:hep-ph/0112015]; P. Bicudo, Phys. Rev. C **67**, 035201 (2003).
20. P. Bicudo, M. Faria, G. M. Marques and J. E. Ribeiro, arXiv:nucl-th/0106071.
21. J. E. Ribeiro, Phys. Rev. D **25**, 2406 (1982); E. van Beveren, Zeit. Phys. C **17**, 135 (1982).
22. C. Itzykson and J. B. Zuber, 'Quantum Field Theory," published by Mcgraw-hill, New York, Usa (1980).
23. L. Nemenov, "Lifetime Measurement of $\pi^+\pi^-$ and $\pi^\pm K^\pm$ atoms to test low energy QCD" L23 Letter of Intent for Nuclear and Particle Physics Experiments at the J-PARC (2003).
24. E. van Beveren *et al.*, Z. Phys. C 30, 615 (1986).
25. G. Rupp and E. van Beveren, Phys.Rev.Lett.**91**, 012003(2003) [arXiv:hep-ph/0305035].
26. B. Aubert *et al.* [BABAR Collaboration], Phys. Rev. Lett. **90**, 242001 (2003) [arXiv:hep-ex/0304021].

Heavy Quarks

Recent results from Belle

P. Krokovny

Budker Institute of Nuclear Physics, Novosibirsk, Russia

Abstract. New results on hadron physics from the Belle experiment are presented.

INTRODUCTION

These results are obtained using various data samples from 80 fb^{-1} to 150 fb^{-1} taken with the Belle detector [1]. We identify B candidates by two kinematic variables: the energy difference, $\Delta E = (\sum_i E_i) - E_b$, and the beam constrained mass, $M_{bc} = \sqrt{E_b^2 - (\sum_i \vec{p}_i)^2}$, where $E_b = \sqrt{s}/2$ is the beam energy and $\vec{p}_i$ and E_i are the momenta and energies of the decay products of the B meson in the CM frame. The inclusion of charge conjugate modes is implicit throughout this report.

OBSERVATION OF 0^+ AND 1^+ BROAD $C\bar{U}$ STATES

A study of charmed meson production in B decays provides an opportunity to test predictions of Heavy Quark Effective Theory (HQET) and QCD sum rules. B decays to $D^{(*)}\pi$ final states are its dominant hadronic decay modes and are measured quite well [2]. The large data sample accumulated in the Belle experiment allows to study production of D meson excited states. D^{**}s are P-wave excitations of quark-antiquark systems that contain one charmed and one light (u,d) quark.

$B \to D^{**}\pi$ decays have been studied using the $D^+\pi^-\pi^-$ and $D^{*+}\pi^-\pi^-$ final states [3].

Figure 1 shows the ΔE distributions for the $B^- \to D^+\pi^-\pi^-$ and $B^- \to D^{*+}\pi^-\pi^-$ candidates. The following branching fractions are measured: $_{\mathrm{B}} (B^- \to D^+\pi^-\pi^-) = (1.02 \pm 0.04 \pm 0.15) \times 10^{-3}$ and $_{\mathrm{B}} (B^- \to D^{*+}\pi^-\pi^-) = (1.25 \pm 0.08 \pm 0.22) \times 10^{-3}$, without any assumption about the intermediate final states.

To study the dynamics of $B \to D^{(*)}\pi\pi$ decays, analyses of the Dalitz plots are performed. The description of the Dalitz plot $D^+\pi^-\pi^-$ includes amplitudes of the known $D_2^{*0}\pi^-$ mode, possible contributions of the processes with virtual $D^{*0}\pi^-$ and $B^{*0}\pi^-$ production and an intermediate $D^+\pi^-$ broad resonance structure with free mass and width. Figure 2(a) shows the $D^+\pi^-$ invariant mass distribution together with the resulting fit. A clear signal of the broad resonance with $J^P = 0^+$ is observed which can be identify as the scalar D_0^{*0} state. The results of the mass, width and branching fraction products are

CP717, *Hadron Spectroscopy: Tenth International Conference,*
edited by E. Klempt, H. Koch, and H. Orth

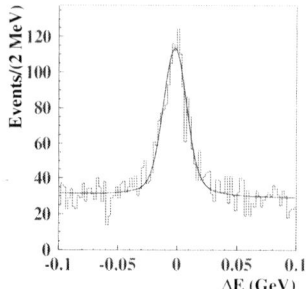 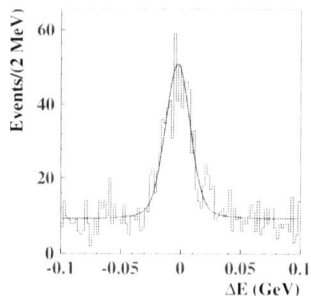

FIGURE 1. ΔE distributions for the $B^- \to D^+ \pi^- \pi^-$ (left) and $B^- \to D^{*+} \pi^- \pi^-$ (right) candidates.

TABLE 1. Branching fractions and resonance parameters for the $D^{(*)+} \pi^- \pi^-$ final states.

Mode	$\mathrm{B}\ (B^- \to D_X [D^{(*)+} \pi^-] \pi^-)$, 10^{-4}	$M(D_X)$, MeV	$\Gamma(D_X)$, MeV
$B^- \to D_2^{*0} [D^+ \pi^-] \pi^-$	$3.4 \pm 0.3 \pm 0.6 \pm 0.4$	$2462 \pm 2.1 \pm 0.5 \pm 3.3$	$45.6 \pm 4.4 \pm 6.5 \pm 1.6$
$B^- \to D_0^{*0} [D^+ \pi^-] \pi^-$	$6.1 \pm 0.6 \pm 0.9 \pm 1.6$	$2308 \pm 17 \pm 15 \pm 28$	$276 \pm 21 \pm 18 \pm 60$
$B^- \to D_1^0 [D^{*+} \pi^-] \pi^-$	$6.8 \pm 0.7 \pm 1.3 \pm 0.3$	$2421 \pm 1.5 \pm 0.4 \pm 0.8$	$23.7 \pm 2.7 \pm 0.2 \pm 4.0$
$B^- \to D_2^{*0} [D^{*+} \pi^-] \pi^-$	$1.8 \pm 0.3 \pm 0.3 \pm 0.2$	[5]	[5]
$B^- \to D_1^{\prime 0} [D^{*+} \pi^-] \pi^-$	$5.0 \pm 0.4 \pm 1.0 \pm 0.4$	$2427 \pm 26 \pm 20 \pm 15$	$384^{+107}_{-75} \pm 24 \pm 70$

presented in Table 1.

For the $D^{*+} \pi^- \pi^-$ final state the fit of the density distribution is performed in four dimensional phase space to take into account the angles of the pion from D^* decay. The fit function includes both known D_2^{*0}, D_1^0 intermediate state contributions and a broad $D^{*+} \pi^-$ resonance with free parameters. Figure 2(b) shows the $D^{*+} \pi^-$ invariant mass distribution as well as the resulting fit. Together with the narrow resonances a clear

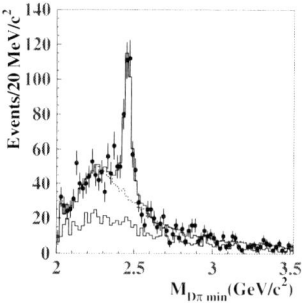 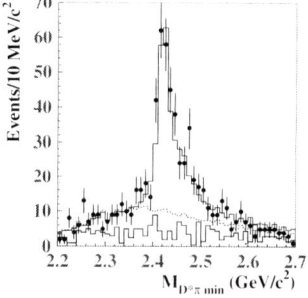

FIGURE 2. Minimal $D^+ \pi^-$ (a) and $D^{*+} \pi^-$ (b) invariant mass distributions. The points with error bars correspond to the B meson signal events, the hatched histogram shows the sidebands. The open histogram is the result of a fit while the dashed one shows the fit function without narrow resonance contribution.

signal of the broad state is observed. The angular distribution of $D^*\pi$ from this state is consistent with $J^P = 1^+$, $j_q = 1/2$. This state can be identified as a P-wave excitation of $c\bar{u} - D_1'^0$. The results of the mass, width and branching fraction products are presented in Table 1.

Together with observations of the broad resonances the branching ratios of B decay to the modes with known D^{**}: $D_1^0\pi^-$ and $D_2^{*0}\pi^-$ have been measured. Using these measurements the ratio of D_2^{*0} branching fractions $h = {}_B (D_2^{*0} \to D^+\pi^-)/{}_B (D_2^{*0} \to D^{*+}\pi^-) = 1.9 \pm 0.5$, consistent with the world average $h = 2.3 \pm 0.6$ [2], is obtained. The measured ratio $R = {}_B (B^- \to D_2^{*0}\pi^-)/{}_B (B^- \to D_1^0\pi^-) = 0.77 \pm 0.15$ is lower than the CLEO measurement 1.8 ± 0.8 [4] (although the results are consistent within errors) but is still a factor of two larger than the factorization prediction [6]. From our measurement it is impossible to determine whether the non-factorized part for tensor and axial mesons is large, or whether higher order corrections to the leading factorized terms should be taken into account.

Our measurements show that the narrow resonances compose $(36 \pm 6)\%$ of the $D\pi\pi$ decays and $(63 \pm 6)\%$ of the $D^*\pi\pi$ decays. This result is inconsistent with the QCD sum rule prediction and may indicate a large contribution from a color suppressed amplitude.

OBSERVATION OF NEW STATES $D_{sJ}^+(2317)$ AND $D_{sJ}^+(2457)$

The narrow $D_s\pi^0$ resonance at 2317 MeV, recently observed by the BaBar collaboration [7], is naturally interpreted as a P-wave excitation of the $c\bar{s}$ system. The observation of a nearby and narrow $D_s^*\pi^0$ resonance by the CLEO collaboration [8] supports this view, since the mass difference of the two observed states is consistent with the expected hyperfine splitting for a P-wave doublet with total light-quark angular momentum $j = 1/2$ [9, 10]. The observed masses are, however, considerably lower than potential model predictions [11], and similar to those of the $c\bar{u}$ $j = 1/2$ doublet states recently reported by Belle [3]. Measurements of the D_{sJ} quantum numbers and branching fractions (particularly those for radiative decays), will play an important role in determining the nature of these states.

We confirmed both resonances and measured masses for 0^+ and 1^+ states to be $(2317.2 \pm 0.5 \pm 0.9)$ MeV and $(2456.5 \pm 1.3 \pm 1.3)$ MeV respectively [12]. We also report the first observation of the radiative decay $D_{sJ}(2457) \to D_s\gamma$. Figure 3 shows the mass difference between the $D_s^{(*)}\pi^0$ and $D_s^{(*)}$ candidates. The ratio $\frac{{}_B (D_{sJ}(2457) \to D_s\gamma)}{{}_B (D_{sJ}(2457) \to D_s^*\pi^0)}$ is found to be $0.55 \pm 0.13 \pm 0.08$.

We also search for D_{sJ} production in $B \to DD_{sJ}$ decays [13]. We reconstruct $\bar{D}^0(D^-)$ mesons in the $K^+\pi^-$, $K^+\pi^-\pi^-\pi^+$ and $K^+\pi^-\pi^0$ $(K^+\pi^-\pi^-)$ decay channels. D_s^+ mesons are reconstructed in the $\phi\pi^+$, $\bar{K}^{*0}K^+$ and $K_S^0 K^+$ decay channels. D_{sJ} candidates are reconstructed from $D_s^{(*)}$ mesons and a π^0, γ, or $\pi^+\pi^-$ pair. The mass difference $M(D_{sJ}) - M(D_s^{(*)})$ is used to select D_{sJ} candidates. We use central mass values of 2317 MeV and 2460 MeV for $D_{sJ}(2317)$ and $D_{sJ}(2457)$ respectively and define signal regions within 12 MeV for the corresponding mass difference. We observe a clean signal for $B \to DD_{sJ}(2317)[D_s\pi^0]$ and $B \to DD_{sJ}(2457)[D_s^*\pi^0]$. We also observe for the first

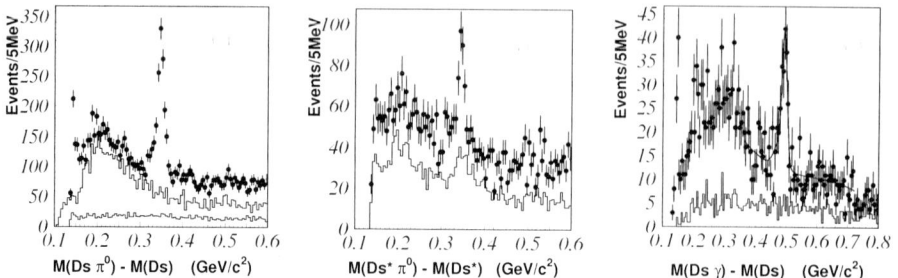

FIGURE 3. $M(D_s\pi^0) - M_{D_s}$ (a), $M(D_s^*\pi^0) - M_{D_s^*}$ (b) and $M(D_s\gamma) - M_{D_s}$ (c) mass-difference distributions. The signal is described using a double Gaussian and a third-order polynomial for the background. The histogram shows no structure for the $D_s^{(*)+}$ sidebands.

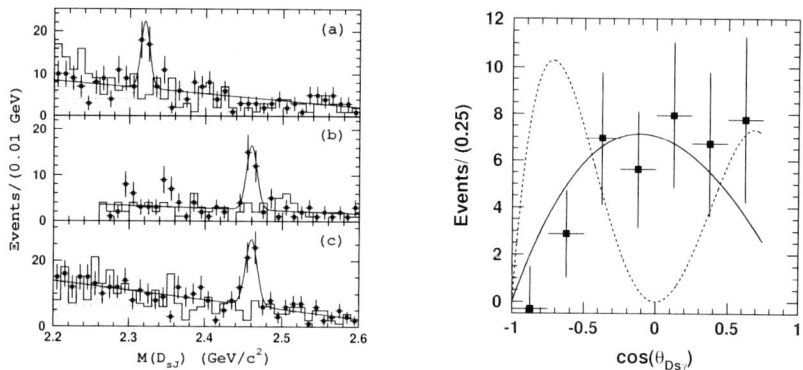

FIGURE 4. Left: $M(D_{sJ})$ distribution for the $B \to \bar{D}D_{sJ}$ candidates: (a) $D_{sJ}(2317) \to D_s\pi^0$, (b) $D_{sJ}(2457) \to D_s^*\pi^0$ and (c) $D_{sJ}(2457) \to D_s\gamma$. Right: the $D_{sJ}(2457) \to D_s\gamma$ helicity distribution. Points with errors represent the experimental data and curves are the results of the fits.

time the $D_{sJ}(2457) \to D_s\gamma$ decay. Figure 4(left) shows the invariant mass distributions for these decays. The measured branching fractions are presented in Table 2. We obtain the ratio $\frac{\text{B}\ (D_{sJ}(2457) \to D_s\gamma)}{\text{B}\ D_{sJ}(2457) \to D_s^*\pi^0)} = 0.38 \pm 0.11 \pm 0.04$, which is consistent with that from the continuum study.

We also study the helicity distribution for the $D_{sJ}(2457) \to D_s\gamma$ decay. The helicity angle $\theta_{D_s\gamma}$ is defined as the angle between the $D_{sJ}(2457)$ momentum in the B meson rest frame and the D_s momentum in the $D_{sJ}(2457)$ rest frame. The $\theta_{D_s\gamma}$ distribution in the data (Fig. 4(right)) is consistent with MC expectations for the $J = 1$ hypothesis for the $D_{sJ}(2457)$ (χ^2/n.d.f = 5/6), and contradicts the $J = 2$ hypothesis (χ^2/n.d.f.= 44/6). The $J = 0$ hypothesis is already ruled out by the conservation of angular momentum and parity in $D_{sJ}(2457) \to D_s\gamma$.

TABLE 2. $B \to DD_{sJ}$ branching fractions.

Decay channel	B, 10^{-4}	Signif.
$B \to \bar{D}D_{sJ}(2317) [D_s\pi^0]$	$8.5^{+2.1}_{-1.9} \pm 2.6$	6.1σ
$B \to \bar{D}D_{sJ}(2317) [D_s^*\gamma]$	$2.5^{+2.0}_{-1.8}(< 7.5)$	1.8σ
$B \to \bar{D}D_{sJ}(2457) [D_s^*\pi^0]$	$17.8^{+4.5}_{-3.9} \pm 5.3$	6.4σ
$B \to \bar{D}D_{sJ}(2457) [D_s\gamma]$	$6.7^{+1.3}_{-1.2} \pm 2.0$	7.4σ
$B \to \bar{D}D_{sJ}(2457) [D_s^*\gamma]$	$2.7^{+1.8}_{-1.5}(< 7.3)$	2.1σ
$B \to \bar{D}D_{sJ}(2457) [D_s\pi^+\pi^-]$	< 1.6	—
$B \to \bar{D}D_{sJ}(2457) [D_s\pi^0]$	< 1.8	—

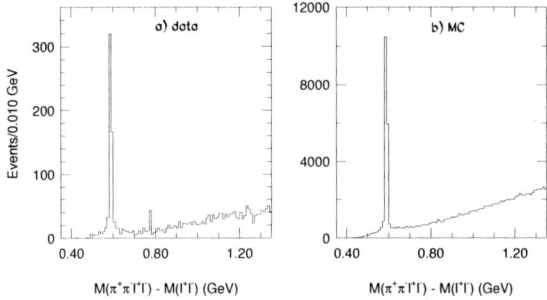

FIGURE 5. Distributions of $M(\pi^+\pi^-l^+l^-) - M(l^+l^-)$ for selected events in the ΔE-M_{bc} signal region for **(a)** Belle data and **(b)** generic $B\bar{B}$ MC events.

OBSERVATION OF A NEW NARROW CHARMONIUM STATE IN $B^\pm \to K^\pm\pi^+\pi^- J/\psi$ DECAY

A major experimental issue for the $c\bar{c}$ charmonium particle system is the existence of as yet unestablished charmonium states that are expected to be below threshold for decays to open charm and, thus, narrow. These include the $n = 1$ singlet P state, the $h_c(1P)$, and possibly the $n = 1$ singlet and triplet spin-2 D states, i.e. the $J^{PC} = 2^{-+}$ 1^1D_{c2} and $J^{PC} = 2^{--}$ 1^3D_{c2}, all of which are narrow if their masses are below the $D\bar{D}^*$ threshold. The observation of these states and the determination of their masses would provide useful information about the spin dependence of the charmonium potential.

We report on an experimental study of the $\pi^+\pi^- J/\psi$ and $\gamma\chi_{c0}$ mass spectra from exclusive $B^+ \to K^+\pi^+\pi^- J/\psi$ and $K^+\gamma\chi_{c0}$ decays [15] using a 152M $B\bar{B}$ event sample. For the $B \to K\pi^+\pi^- J/\psi$ study we use events that have a pair of well identified oppositely charged electrons or muons with an invariant mass in the range $3.077 < M_{l^+l^-} < 3.117$ GeV, a loosely identified charged kaon and a pair of oppositely charged pions.

Figure 5(a) shows the distribution of $\Delta M \equiv M(\pi^+\pi^-l^+l^-) - M(l^+l^-)$ for events in the ΔE-M_{bc} signal region. Here a large peak corresponding to $\psi(2S) \to \pi^+\pi^- J/\Psi$ is evident at 0.589 GeV. In addition, there is a significant spike in the distribution at 0.775 GeV. Figure 5(b) shows the same distribution for a large sample of generic $B\bar{B}$

TABLE 3. Results of the fits to the $\psi(2S)$ and $M = 3872$ MeV regions. The errors are statistical only.

Quantity	$\psi(2S)$ region	$M = 3872$ MeV region
Signal events	489 ± 23	35.7 ± 6.8
$M^{\mathrm{meas}}_{\pi^+\pi^- J/\psi}$ peak	3685.5 ± 0.2 MeV	3871.5 ± 0.6 MeV
$\sigma_{M\pi^+\pi^- J/\psi}$	3.3 ± 0.2 MeV	2.5 ± 0.5 MeV

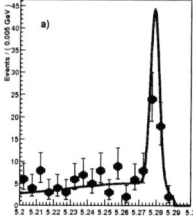 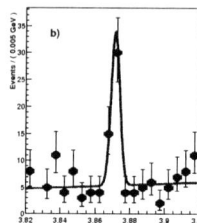 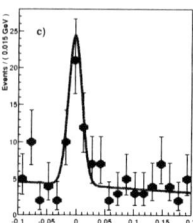

FIGURE 6. Signal-band projections of **(a)** M_{bc}, **(b)** $M_{\pi^+\pi^- J/\psi}$ and **(c)** ΔE for the $X(3872) \to \pi^+\pi^- J/\psi$ signal region with the results of the unbinned fit superimposed.

Monte Carlo (MC) events. Except for the prominent $\psi(2S)$ peak, the distribution is smooth and featureless.

We make separate fits to the data in the $\psi(2S)$ (3580 MeV $< M_{\pi^+\pi^- J/\psi} < 3780$ MeV) and the $M = 3872$ MeV (3770 MeV $< M_{\pi^+\pi^- J/\psi} < 3970$ MeV) regions using a simultaneous unbinned maximum likelihood fit to the M_{bc}, ΔE, and $M_{\pi^+\pi^- J/\psi}$ distributions. The results of the fits are presented in Table 3. Figures 6(a), (b) and (c) show the M_{bc}, $M_{\pi^+\pi^- J/\psi}$, and ΔE signal-band projections for the $M = 3872$ MeV signal region, respectively. The superimposed curves indicate the results of the fit. There are clear peaks with consistent yields in all three quantities. The signal yield of 35.7 ± 6.8 events has a statistical significance of 10.3σ. In the following we refer to this as the $X(3872)$.

We determine the mass of the signal peak relative to the well measured $\psi(2S)$ mass: $M_X = M^{\mathrm{meas}}_X - M^{\mathrm{meas}}_{\psi(2S)} + M^{\mathrm{PDG}}_{\psi(2S)} = 3872.0 \pm 0.6 \pm 0.5$ MeV. Since we use the precisely known value of the $\psi(2S)$ mass [2] as a reference, the systematic error is small. The measured width of the $X(3872)$ peak is $\sigma = 2.5 \pm 0.5$ MeV, which is consistent with the MC-determined resolution and the value obtained from the fit to the $\psi(2S)$ signal. From this we infer a 90% confidence level (CL) upper limit of $\Gamma < 2.3$ MeV.

We determine a ratio of product branching fractions for $B^+ \to K^+ X(3872)[\pi^+\pi^- J/\psi]$ and $B^+ \to K^+ \psi(2S)[\pi^+\pi^- J/\psi]$ to be $0.063 \pm 0.012 \pm 0.007$.

The decay of the $^3D_{c2}$ charmonium state to $\gamma\chi_{c0}$ is an allowed $E1$ transition with a partial width that is expected to be substantially larger than that for the $\pi^+\pi^- J/\psi$ final state; e.g. the authors of Ref. [16] predict $\Gamma(^3D_{c2} \to \gamma\chi_{c0}) > 5 \times \Gamma(^3D_{c2} \to \pi^+\pi^- J/\psi)$. Thus, a measurement of the width for this decay channel can provide important information about the nature of the observed state. We searched for an $X(3872)$ signal in the $\gamma\chi_{c0}$ decay channel, concentrating on the $\chi_{c0} \to \gamma J/\psi$ final state.

We select events with the same $J/\psi \to l^+l^-$ and charged kaon requirements plus two photons, each with energy more than 40 MeV. The signal-band projections of M_{bc}

480

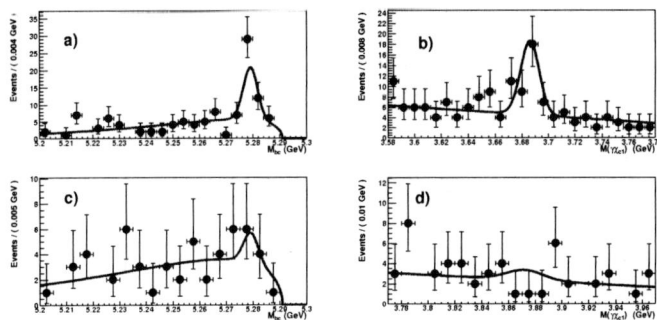

FIGURE 7. Signal-band projections of **(a)** M_{bc} and **(b)** $M_{\gamma\chi_{c0}}$ for the $\psi(2S)$ region with the results of the unbinned fit superimposed. **(c)** and **(d)** are the corresponding results for the $M = 3872$ MeV region.

and $M_{\gamma\chi_{c0}}$ for the $\psi(2S)$ region are shown in Figs. 7 (a) and (b), respectively, together with curves that represent the results of the fit. The fitted signal yield is $34.1 \pm 6.9 \pm 4.1$ events. The number of observed events is consistent with the expected yield of 26 ± 4 events based on the known $B \to K\psi(2S)$ and $\psi(2S) \to \gamma\chi_{c0}$ branching fractions [2] and the MC-determined acceptance.

The results of the application of the same procedure to the $M = 3872$ MeV region are shown in Figs. 7(c) and (d). Here, no signal is evident; the fitted signal yield is $3.7 \pm 3.7 \pm 2.2$. From these results, we determine a 90% CL upper limit on the ratio of partial widths of $\frac{\Gamma(X(3872) \to \gamma\chi_{c0})}{\Gamma(X(3872) \to \pi^+\pi^- J/\psi)} < 0.89$. This limit on the $\gamma\chi_{c0}$ decay width contradicts expectations for the $^3D_{c2}$ charmonium state.

The mass of the observed state is higher than potential model expectations for the center-of-gravity (cog) of the 1^3D_{cJ} states: $M_{cog}(1D) = 3810$ MeV [19, 18].

In summary, we have observed a strong signal for a state that decays to $\pi^+\pi^- J/\psi$ with $M = 3872.0 \pm 0.6 \pm 0.5$ MeV and $\Gamma < 2.3$ MeV (90% CL). This mass value and the absence of a strong signal in the $\gamma\chi_{c0}$ decay channel are in some disagreement with potential model expectations for the $^3D_{c2}$ charmonium state. The mass is within errors of the $D^0\bar{D}^{*0}$ mass threshold (3871.3 ± 1.0 MeV [2]), which is suggestive of a loosely bound $D\bar{D}^*$ multiquark "molecular state," as proposed by some authors [17].

MEASUREMENT OF THE $E^+E^- \to D^{(*)+}D^{(*)-}$ CROSS-SECTIONS

The processes $e^+e^- \to D^{(*)+}D^{(*)-}$ have not previously been observed at energies $\sqrt{s} \gg 2M_D$. A calculation in the HQET approach based on the heavy-quark spin symmetry [20], predicts cross-sections of about 5 pb^{-1} for $e^+e^- \to D\bar{D}^*$ and $e^+e^- \to D_T^*\bar{D}_L^*$ at $\sqrt{s} \sim 10.6$ GeV (the subscripts indicate transverse [T] and longitudinal [L] polarization of the D^*); the cross-section for $e^+e^- \to D\bar{D}$ is expected to be suppressed by a factor of $\sim 10^{-3}$.

This analysis [21] is based on 88.9 fb^{-1} of data taken at or near the $\Upsilon(4S)$ resonance. We reconstruct D^0 and D^+ mesons in the decay modes $D^0 \to K^-\pi^+$, $D^0 \to K^-\pi^+\pi^+\pi^-$ and $D^+ \to K^-\pi^+\pi^+$. D^{*+} mesons are reconstructed in the $D^0\pi^+$ decay mode.

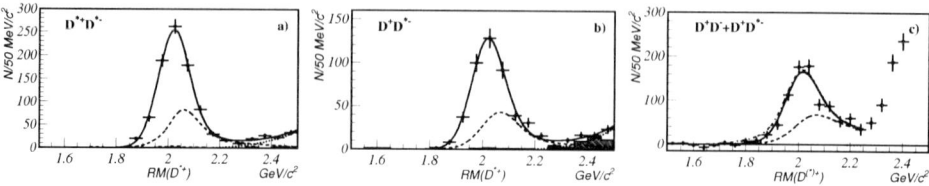

FIGURE 8. Distributions of the mass of the system recoiling against a) D^{*+}, and b) D^{+}. Points with error bars show the signal ΔM_{recoil} region; hatched histograms correspond to ΔM_{recoil} sidebands. The solid lines represent the fits described in the text; the dashed lines show the contribution due to events with ISR photons of significant energy. The dotted lines show the expected background contribution. c) The distribution of $M_{\text{recoil}}(D^{+})$ without any requirement on ΔM_{recoil}.

The processes $e^{+}e^{-} \rightarrow D^{(*)+}D^{(*)-}$ can be identified by energy-momentum balance in fully reconstructed events that contain only a pair of charm mesons. However, the reconstruction efficiency is small in this case. Taking into account two body kinematics, it is sufficient to reconstruct only one of the two charmed mesons in the event to identify the processes of interest. We choose the mass of the system recoiling against the reconstructed $D^{(*)}$ ($M_{\text{recoil}}(D^{(*)+})$) as a discriminating variable: $M_{\text{recoil}}(D^{(*)+}) = \sqrt{(\sqrt{s} - E_{D^{(*)+}})^2 - \vec{p}_{D^{(*)+}}^2}$, where $E_{D^{(*)+}}$ and $\vec{p}_{D^{(*)+}}$ are the CM energy and momentum of the reconstructed $D^{(*)+}$. For the signal a peak in the M_{recoil} distribution around the nominal mass of the recoiling D^{-} or D^{*-} is expected. This method provides a significantly higher efficiency, but also a higher background, in comparison to full event reconstruction. For the $e^{+}e^{-} \rightarrow D^{(*)+}D^{*-}$ processes we reconstruct in addition a slow pion from the $D^{*-} \rightarrow \bar{D}^{0}\pi_{slow}^{-}$ decay. This reduces the background to a negligible level.

We calculate the difference between the masses of the systems recoil mass against a $D^{(*)+}\pi_{slow}^{-}$ combination, and against the $D^{(*)+}$ alone, $\Delta M_{\text{recoil}} \equiv M_{\text{recoil}}(D^{(*)+} - M_{\text{recoil}}(D^{(*)+}\pi_{slow}^{-}))$. The variable ΔM_{recoil} peaks around the nominal $D^{*+} - D^{0}$ mass difference with a resolution of $\sigma_{\Delta M_{\text{recoil}}} \sim 1\,\text{MeV}$ as found by Monte Carlo simulation. For $e^{+}e^{-} \rightarrow D^{(*)+}D^{*-}$ we combine $D^{(*)+}$ candidates together with π_{slow}^{-} and require ΔM_{recoil} to be within a $\pm 2\,\text{MeV}$ interval around the nominal $D^{*+} - D^{0}$ mass difference.

The $M_{\text{recoil}}(D^{*+})$ and $M_{\text{recoil}}(D^{+})$ distributions are shown in Figs. 8(a) and 8(b), respectively. Clear signals are observed in both cases. The higher recoil mass tails in the signal distribution are due to initial state radiation (ISR). The hatched histograms show the M_{recoil} distributions for events in the ΔM_{recoil} sidebands.

Since the reconstruction efficiency depends on the production and $D^{*\pm}$ helicity angles, we perform angular analysis before computing cross-sections. A scatter plot of the helicity angles for the two D^{*}-mesons from $e^{+}e^{-} \rightarrow D^{*+}D^{*-}$ ($\cos\phi(D_{rec}^{*})$ vs $\cos\phi(D_{non-rec}^{*})$) for the recoil mass region $M_{\text{recoil}}(D^{*+}) < 2.1\,\text{GeV}$ is shown in Fig. 9(a). The distribution is fitted by a sum of three functions corresponding to the $D_{T}^{*}D_{T}^{*}$, $D_{T}^{*}D_{L}^{*}$ and $D_{L}^{*}D_{L}^{*}$ final states, obtained from Monte Carlo simulation. The fit finds the fractions of $D_{T}^{*}D_{T}^{*}$, $D_{T}^{*}D_{L}^{*}$ and $D_{L}^{*}D_{L}^{*}$ final states to be $(1.5 \pm 3.6)\%$, $(97.2 \pm 4.8)\%$ and $(1.3 \pm 4.7)\%$, respectively. Figure 9(b) shows the D^{*-} meson helicity distribution for $e^{+}e^{-} \rightarrow D^{+}D^{*-}$. The fraction of the $D^{+}D_{L}^{*-}$ final state is found from the fit to be equal

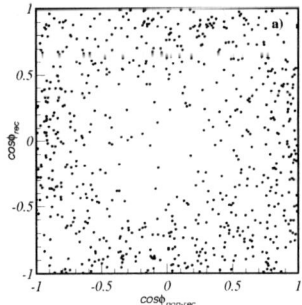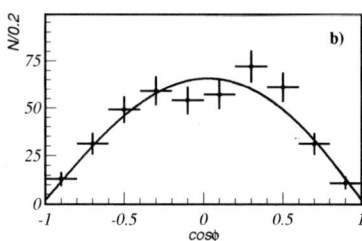

FIGURE 9. a) The scatter plot $\cos(\phi_{D^*_{rec}})$ vs $\cos(\phi_{D^*_{non-rec}})$ ($e^+e^- \to D^{*+}D^{*-}$). b) D^{*+} meson helicity angle distribution for ($e^+e^- \to D^+D^{*-}$) signal candidates.

to $(95.8 \pm 5.6)\%$.

We search for the process $e^+e^- \to D^+D^-$ by studying the recoiling against the reconstructed D^+ (M_{recoil}). Fig. 8(c) shows the distribution of $M_{recoil}(D^+)$ after D^+ mass sideband subtraction. To extract the $e^+e^- \to D^+D^-$ and $e^+e^- \to D^+D^{*-}$ yields we fit this distribution with the sum of two signal functions corresponding to D^- and D^{*-} peaks and a threshold function describing background events. The fit finds -13 ± 24 events in the D^- peak and 935 ± 42 in the D^{*-} peak. We obtains a $e^+e^- \to D^+D^{*-}$ cross-section of 0.61 ± 0.05 pb which agrees with the result using the ΔM_{recoil} method. For the $e^+e^- \to D^+D^-$ cross-section we set an upper limit of 0.04 pb at the 90% confidence level.

In summary, we report the first measurement of the cross-sections for the $e^+e^- \to D^{*+}D^{*-}$ and $e^+e^- \to D^+D^{*-}$ processes at $\sqrt{s} = 10.6$ GeV to be $0.65 \pm 0.04 \pm 0.07$ pb and $0.71 \pm 0.05 \pm 0.09$ pb, respectively, and set an upper limit on the $e^+e^- \to D^+D^-$ cross-section of 0.04 pb at 90% CL. The measured cross-sections are an order of magnitude lower than those predicted in Ref. [20], but their relative sizes are as predicted: the cross-sections for $e^+e^- \to D^{*+}D^{*-}$ and $e^+e^- \to D^+D^{*-}$ are found to be close each other, while the cross-section for $e^+e^- \to D^+D^-$ is much smaller. The helicity decomposition for $e^+e^- \to D^{*+}D^{*-}$ is found to be saturated by the $D^{*\pm}_T D^{*\mp}_L$ final state (the fraction is equal to $(97.2 \pm 4.8)\%$) and for $e^+e^- \to D^+D^{*-}$ — by the D^*_L final state $(95.8 \pm 5.6\%)$, in good agreement with the predictions of Ref. [20].

OBSERVATION OF $\eta_C(2S)$ PRODUCTION AND ITS MASS MEASUREMENT

Belle recently observed the $\eta_c(2S)$ production in exclusive B decays to $KKK^0_S\pi$, where the $\eta_c(2S)$ is reconstructed in the $K^\pm K^0_S\pi^\mp$ final state. The mass was measured to be $(3654 \pm 6 \pm 8)$ MeV [22] which is much larger than the previous Crystal Ball measurement of (3594 ± 5) MeV [23]. This year Belle also observed $\eta_c(2S)$ production $(108 \pm 24$ events) in double charmonia events $e^+e^- \to J/\psi\eta_c(2S)$ and confirmed a higher $\eta_c(2S)$ mass [24].

CONCLUSION

We have observed a strong signal for a new charmonium state that decays to $\pi^+\pi^- J/\psi$ with $M = 3872.0 \pm 0.6 \pm 0.5$ MeV, $\Gamma < 2.3$ MeV at 90% CL. We confirm the observation of $D_{sJ}(2317)$ and $D_{sJ}(2457)$ and report the first observation of the decay $D_{sJ}(2457) \to D_s\gamma$. We also observe D_{sJ} production in B decays. In $B^- \to D^{(*)+}\pi^-\pi^-$ decays all four P-wave D^{**} have been observed and their parameters have been measured. For the broad D_0^{*0} and D_1^{*0} states there are the first measurements.

REFERENCES

1. Belle Collaboration, A. Abashian *et al.*, Nucl. Inst. and Meth. **A 479**, 117 (2002).
2. K. Hagiwara *et al.* (Particle Data Group), Phys. Rev. **D 66**, 010001 (2002).
3. Belle Collaboration, K.Abe, *et al.* hep-ex/0307021, submitted to Phys. Rev. D.
4. CLEO Collaboration, G. Gronberg *et al.*, CLEO CONF 96-25, Proc. of the ICHEP 96, July 1996, Poland.
5. Fixed from the D_2^{*0} parameters obtained in the $B^- \to D^+\pi^-\pi^-$ analysis.
6. M. Neubert, Phys. Lett. **B 418**, 173 (1998).
7. BaBar Collaboration, B. Aubert *et al.*, Phys. Rev. Lett. **90**, (2003) 242001.
8. CLEO Collaboration, D. Besson *et al.*, hep/ex-0305017.
9. W. Bardeen, E Eichten and C. Hill, Phys. Rev. **D 68**, 054024 (2003).
10. In the heavy c-quark approximation, one expects two doublets of $c\bar{s}$ states with quantum numbers $J^P = 0^+, 1^+$ ($j = 1/2$) and $J^P = 1^+, 2^+$ ($j = 3/2$). The second doublet has already been observed in DK and D^*K decays.
11. J. Bartelt and S. Shukla, Ann. Rev. Nucl. Part. Sci. **45** 133, (1995).
12. Belle Collaboration, Y.Mikami *et al.*, BELLE-CONF-0340, hep-ex/0307052, submitted to Phys. Rev. Lett.
13. Belle Collaboration, P.Krokovny *et al.*, hep-ex/0308019 ,submitted to Phys. Rev. Lett.
14. A. Palano for the BaBar Collaboration, hep-ex/0309028.
15. Belle Collaboration, S.-K.Choi, S.L.Olsen *et al.*, hep-ex/0309032, submitted to Phys. Rev. Lett.
16. E.J. Eichten, K. Lane, and C. Quigg, Phys. Rev. Lett. **89**, 162002 (2002). See also P. Ko, J. Lee and H.S. Song, Phys. Lett. **B395**, 107 (1997).
17. See, for example, M.B. Voloshin and L.B. Okun, JETP Lett. **23**, 333 (1976); A. De Rujula, H. Georgi and S.L. Glashow, Phys. Rev. Lett. **38**, 317 (1977); and N. Tornqvist, Z. Phys. **C 61**, 525 (1994).
18. E. Eichten, K. Gottfried, T. Kinoshita, K.D. Lane and T.M. Yan, Phys. Rev. **D 21**, 203 (1980).
19. W. Buchmüller and S-H.H. Tye, Phys. Rev. **D 24**, 132 (1981).
20. A. G. Grozin, M. Neubert, Phys. Rev. **D 55**, 272 (1997).
21. Belle Collaboration, K.Abe *et al.* BELLE-CONF-0332, hep-ex/0307084.
22. Belle Collaboration, S.-K.Choi, S.L.Olsen *et al.*, Phys. Rev. Lett. **89**, 102001 (2002)
23. Crystal Ball Collaboration, C.Edwards *et al.*, Phys. Rev. Lett. **48**, 70 (1982).
24. Belle Collaboration, K.Abe *et al.*, BELLE-CONF-0331.

Hadron Physics with BABAR

Stefan M. Spanier

University of Tennessee, Knoxville
for the BABAR Collaboration

Abstract. The primary goal of BABAR is the measurement of *CP* violation in the *B* meson system. An integrated luminosity of greater than 120 fb^{-1} has been collected until Summer 2003 with an open trigger giving access to many non-*B* hadronic production modes. In this article, several new results will be presented.

THE BABAR EXPERIMENT

BABAR is a successful e^+e^- collision experiment at the asymmetric PEP II storage ring [1] of SLAC. The PEP II *B* factory is designed to operate at a luminosity of $3 \cdot 10^{33} cm^{-2} s^{-1}$ and a center-of-mass (CM) energy around 10.58 GeV. The energy of the electron beam is about 9.0 GeV, that of the positron beam 3.1 GeV, resulting in a Lorentz boost to the $\Upsilon(4S)$ resonance of $\beta\gamma = 0.55$. The BABAR detector [2] is optimized for the asymmetric configuration. Charged particle momenta are measured in a tracking system consisting of a 5-layer, double sided readout, silicon vertex tracker (SVT) ($15\mu m$ hit resolution for perpendicular tracks) and a 40-layer drift chamber (DCH) filled with a mixture of helium and isobutane, both operating in a 1.5 T superconducting solenoidal magnet. Both cover 92% of the solid angle in the laboratory and have an average tracking efficiency of 98%. The transverse momentum resolution at 1 GeV/*c* momentum is 1%. The electromagnetic calorimeter consists of 6580 CsI(Tl) crystals arranged in barrel and forward endcap subdetectors. The width of a π^0 is reconstructed to better than 7 MeV for photon energies above 30 MeV. Muons and long-lived neutral hadrons are identified in the instrumented flux return, composed of resistive plate chambers and layers of iron. Together with dE/dx information from SVT and DCH (mainly below 0.7 GeV/*c* track momentum) a detector of internally reflected Cherenkov light (DIRC) provides separation of kaons and pions (better than 2.5σ up to 4.3 GeV/*c* particle momentum). BABAR has collected an integrated luminosity of 125.4 fb^{-1} until Summer 2003 with an average logging efficiency of better than 96%.

THE CROSS SECTION

The resonance parameters of the $\Upsilon(4S)$ resonance, which is known for almost 20 years, have relative large uncertainties and the central mass values vary substantially from experiment to experiment [3]. The visible hadronic cross section has been scanned with the BABAR detector (see Fig. 1). The energy spread of the collider at this energy is

CP717, *Hadron Spectroscopy: Tenth International Conference,*
edited by E. Klempt, H. Koch, and H. Orth
© 2004 American Institute of Physics 0-7354-0197-7/04/$22.00

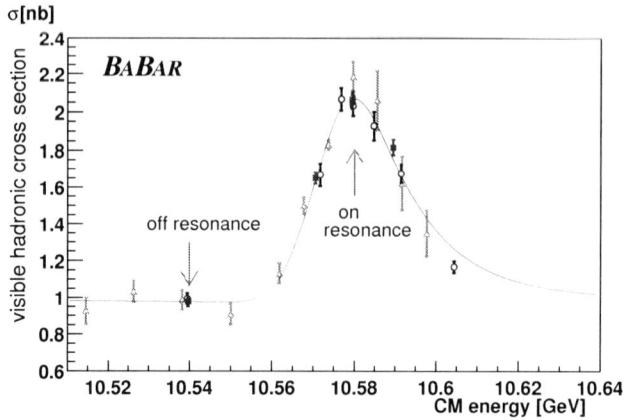

FIGURE 1. Data points and fit of the scan of the visible cross section of $e^+e^- \to hadrons$ near the $\Upsilon(4S)$. Different symbols represent different scans.

4.63 MeV. We measure:

$$M = (10.5793 \pm 0.0004 \pm 0.0012)\,\text{GeV}/c^2,$$
$$\Gamma_{tot} = (20.7 \pm 1.6 \pm 2.5)\,\text{MeV},$$
$$\Gamma_{ee} = (0.321 \pm 0.017 \pm 0.029)\,\text{keV},$$

with errors on mass and total width much lower than the present world average. The $\Upsilon(4S)$ lies only about 20 MeV above the kinematic threshold for open bottom production and decays nearly 100% into $\bar{B}B$ pairs. The maximum transverse momentum of the B mesons is 300 MeV/c. Therefore, B-decay events homogeneously occupy the transverse detector plane while in $e^+e^- \to \bar{q}q$ (continuum) events with light quarks $q = u, d, s, c$, charged and neutral particles appear lined up in jets. The $\bar{B}B$ events can be isolated based on the different event topology. In addition, one can sample the distributions of event variables for only the continuum events by reducing the CM energy below the $\Upsilon(4S)$ resonance (indicated 'off resonance' in Fig. 1).

On the other hand close to the $\Upsilon(4S)$ resonance the cross section for $e^+e^- \to (\bar{u}u, \bar{d}d, \bar{s}s)$ is 2.1 nb, for $e^+e^- \to \bar{c}c$ is 1.3 nb. Therefore, BABAR provides the opportunity to study light-quark and charm physics, and even free of $\bar{b}b$ background. Initial state radiation (ISR) provides access to CM energies below $\Upsilon(4S)$ with large statistics, e.g. J/psi and $\psi(2S)$. For precision tests of the Standard Model data for the R ratio is particularly needed in the CM energy range 1.4 - 3.4 GeV for which BABAR is well suited. Preliminary results of ISR measurements were presented during this conference [4]. Another production mode which is not discussed here is the τ-lepton decay. Since BABAR can afford to run an open trigger all these hadronic production processes including 2-photon ($\gamma\gamma$) fusion events are efficiently written to tape.

B MESON PHYSICS WITH BABAR

Already after two active years of the B factories the $B(\bar{b}d, \bar{b}u)$ meson is one of the if not the best studied meson. At BABAR B and $\bar{B}$ are produced in a coherent P-wave state (quantum entangled) so that time-dependent measurements depend on the relative separation between the two B-meson decay vertices. The boost, $\beta\gamma = 0.55$, dilates the typical B decay length of about 20μm to about 260μm straight down the beam (z) axis increasing the experimental precision. From the separation of the z position, Δz, one obtains the proper decay time difference $\Delta t \approx \Delta z/(\beta\gamma c)$.

Time Dependent Measurements

For charged Bs, evolution with time is just an exponential decay law (decay probability $f(\Delta t) = \frac{1}{\tau^+}e^{-\Delta t/\tau^+}$ with τ^+ the mean B^+ lifetime [1]). Neutral Bs can mix changing flavor $B^0 \rightarrow \bar{B}^0 + c.c.$ complicating their time evolution (time dependent probability to observe unmixed (+), $B^0\bar{B}^0$, or mixed (-), B^0B^0 and $\bar{B}^0\bar{B}^0$, events: $f_{\pm}(\Delta t) = \frac{1}{4\tau^0}e^{-\Delta t/\tau^0} \cdot (1 \pm \cos(\Delta m \Delta t))$, and τ^0 is the B^0 mean lifetime, Δm the $B^0 - \bar{B}^0$ oscillation frequency). In time-dependent CP–asymmetry measurements, each candidate event consists of a reconstructed neutral B meson, B_{CP}, and a partially reconstructed recoil B meson, B_{tag}. The B_{tag} is examined for evidence that it decayed either as B^0 or $\bar{B}^0$ (flavor tag). For decays of B_{CP} to a CP eigenstate the decay rate $f_+(f_-)$ when the tagging meson is a $B^0(\bar{B}^0)$ is given by

$$f_{\pm}(\Delta t) = \frac{e^{-|\Delta t|/\tau^0}}{4\tau^0}[1 \pm S\sin(\Delta m \Delta t) \mp C\cos(\Delta m \Delta t)], \tag{1}$$

where S and C are constants which depend on the final CP eigenstate. C is a measure of direct CP violation (CP violation in decay), and S is a measure of CP violation from the interference between mixing and decay. From the Standard Model one expects $S = -\eta_f \sin 2\beta$ and $C = 0$ with η_f the CP eigenvalue of B_{CP}. Results from BABAR as obtained from hadronic B decays [5, 6, 7]:

lifetime	$\tau^+ = 1.673 \pm 0.032 \pm 0.022$ ps
	$\tau^0 = 1.546 \pm 0.032 \pm 0.023$ ps
mixing	$\Delta m = (0.516 \pm 0.016 \pm 0.010)$ ps^{-1}
CP violation	$S = 0.741 \pm 0.067 \pm 0.033$
in $(\bar{c}c)K^0$	$C = 0.053 \pm 0.054 \pm 0.032$

The recent BABAR result for $\tau^0 = 1.523^{+0.024}_{-0.023} \pm 0.022$ ps [8]. The measurement of S in the charm modes is still dominated by statistical uncertainty. Figure 2 shows the reconstructed B^0 candidates for the Δm measurement.

---

[1] Charge conjugates are implied unless otherwise stated.

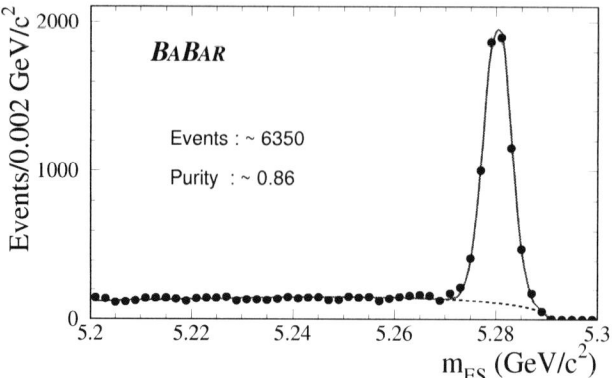

FIGURE 2. Distribution of the beam energy constraint mass for B^0 candidates decaying into multi-hadron final states which characterize the B flavor. The sample was used for the Δm measurement.

Rare B Decays

The decays of B mesons into rare hadronic final states also provide important information for the study of CP violation. A B decay mode accessing the angle α of the CKM unitarity triangle is $B^0 \to \pi^+\pi^-$ and has a branching fraction of $BF(B^0 \to \pi^+\pi^-) = (4.7 \pm 0.6 \pm 0.2) \cdot 10^{-6}$ [9]. Decays which proceed via loop diagrams (electroweak or gluonic penguins) are sensitive to new particles and interactions beyond the Standard Model (at the $\approx$ TeV scale). In particular time-dependent CP asymmetry measurements probe quite independent from hadronic decay models new weak phases that may change Standard Model predictions (see Fig. 3). Direct CP measurements in charmless hadronic B decays are also affected by new physics but depend on a good understanding of the long range hadronic physics for which on the other hand they provide an important input. BABAR has new results on several rare B decay modes, inlcuding

- $B^0 \to \pi^0\pi^0$, $BF = (2.1 \pm 0.6 \pm 0.3) \cdot 10^{-6}$ [10],
- $B^0 \to \eta'\phi$, $BF < 1.0 \cdot 10^{-6}$ at 90% C.L. [11],
- $BF(B^0 \to K_2^*(1430)^0\gamma) = (1.22 \pm 0.25 \pm 0.11) \cdot 10^{-5}$ and
  $B^+ \to K_2^*(1420)^=\gamma = (1.44 \pm 0.40 \pm 0.13) \cdot 10^{-5}$ [12],
- $B^0 \to Kl^+l^-, K^*l^+l^-, (l = e, \mu)$, $BF(B \to Kl^+l^-) = (0.65^{+0.14}_{-0.13} \pm 0.04) \cdot 10^{-6}$ and
  $BF(B \to K^*l^+l^-) = (0.88^{+0.33}_{-0.29} \pm 0.10) \cdot 10^{-6}$ [13],
- $BF(B^0 \to \rho^+\rho^-) = (27^{+7+5}_{-6-7}) \cdot 10^{-6}$, polarization $\Gamma_L/\Gamma = 0.99^{+0.01}_{-0.07} \pm 0.03$ [14],
- $B^+ \to \rho^+\pi^0$, $B^+ \to \rho^0\pi^+$, and $B^0 \to \rho^0\pi^0$ [15],
- $B^0 \to \bar{p}p$, $BF < 2.7 \cdot 10^{-7}$ at 90% C.L. [16],
- $B^0 \to K^+K^-K_S^0$, $BF = (23.8 \pm 2.3 \pm 2.2) \cdot 10^{-6}$ [17].

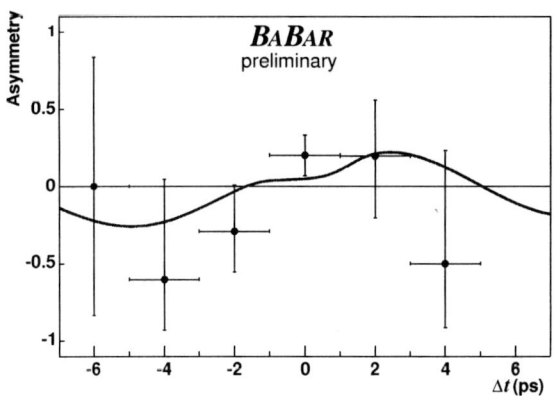

FIGURE 3. Time dependent asymmetry between B^0 and $\bar{B}^0$ tagged events for $B^0 \rightarrow \phi K_S^0$ candidate events. This decay is dominated by gluonic penguins. In the Standard Model the asymmetry is expected to be equal to the one in the $(\bar{c}c)K_S^0$ mode (up to 4%). The measured value of $S = 0.45 \pm 0.43 \pm 0.07$ [18].

CHARM PHYSICS WITH BABAR

The BABAR data sample corresponding to an integrated luminosity of 91 fb^{-1} contains about 220,000 $D^{*+} \rightarrow \pi^+ D^0 (\rightarrow K^- \pi^+) + c.c.$ events. Thus BABAR is competitive to other dedicated charm experiments.

Mixing and *CP* Violation

The observation of $D^0 - \bar{D}^0$ mixing is an indication of new physics since in the Standard Model it is predicted to be just below the sensitivity of current experiments (mixing rate $\approx 10^{-6}$). As new physics may introduce new weak phases *CP* violation in $D^0 - \bar{D}^0$ mixing needs to be considered. The dominant two-body decay of D^0 is the right-sign Cabibbo-favored decay $D^0 \rightarrow K^- \pi^+$. Evidence for mixing and *CP* violation will appear in the wrong-sign D^0 final state $K^+ \pi^-$, which is either reached via a doubly Cabibbo suppressed decay (tree-level amplitude) or via mixing followed by a Cabibbo-favored decay of the $\bar{D}^0$. BABAR has set improved limits on mixing and *CP* violation in wrong sign decays of the D^0 mesons [19] and in the measurements of lifetime ratios for decays to $K^- \pi^+$, $K^- K^+$, and $\pi^+ \pi^-$ [20], all in agreement with the Standard Model. Fig. 4 shows the reconstructed $D^0 \rightarrow K^+ \pi^-$ candidate events.

The New States

BABAR has observed a narrow state $D_{SJ}^*(2317)$ in the inclusive $D_S^+ \pi^0$ invariant mass distribution from $e^+ e^-$ data around the $\Upsilon(4S)$ (91 fb^{-1}). The observed width of about

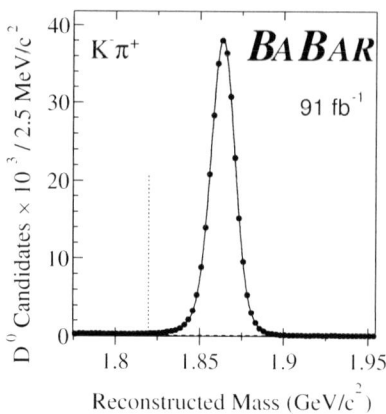

FIGURE 4. Invariant $K\pi$ mass for $D^0 \to K^-\pi^+$ decay events reconstructed from D^{*+} decays for the measurement of lifetime ratios in 91 fb^{-1}. The purity in the signal region is 99.4%.

8.6 MeV/c^2 is consistent with the experimental resolution. The D_S^+ is reconstructed from $K^+K^-\pi^+$ (or $K^+K^-\pi^+\pi^0$) and then combined with a π^0 reconstructed from a pair of photons. Each $K^+K^-\pi^+(\pi^0)\pi^0$ candidate must have a momentum in the e^+e^- CM frame greater than 2.5 GeV/c, which reduces combinatorial background from continuum and eliminates background from B-meson decays [21, 22]. The efficiency corrected $\cos\theta_h$ distribution, with θ_h being the helicity angle of the D_{SJ}^* decay with respect to its direction in the e^+e^- CM frame, is consistent with being flat. This is expected for a particle with spin-zero or for a particle of higher spin that is produced unpolarized. For a $(c\bar{s})$ state the decay into $D_S^+\pi^0$ violates isospin conservation thus explaining the small width. Parity conservation allows the spin-parity assignment of $\{0^+, 1^-, 2^+,..\}$. The low mass compared to those of the $D_{S1}(2536)^+$ and $D_{SJ}^*(2573)^+$ favors 0^+. In that case the decay to $D_S^+\gamma$ is excluded. At present there is no indication of such a decay at BABAR. The state has been confirmed by CLEO [23] and Belle [24, 25]. If the $D_{SJ}^*(2317)$ indeed is a $(c\bar{s})$ scalar meson, the low mass, small width, and decay mode are quite different from those predicted by potential models [26]. Since the grouping of scalar mesons is still unresolved [27] not much of a hint can be gained from possible scalar partner states.

In the inclusive $D_S^+\pi^0\gamma$ mass distribution of the same data sample a narrow state $D_{SJ}(2458)^+$ with a mass of $2458\pm1\pm1$ MeV/c^2 and a width which is consistent with the experimental resolution is observed [28]. It can be seen as an enhancement near in Fig. 6. Background under the peak originates from several sources. A peculiar background pattern emerges from the $D_S^*(2112)^+ \to D_S^+\gamma$ decay combined with an unassociated π^0 and the $D_{SJ}^*(2317)^+ \to D_S^+\gamma$ decay combined with an unassociated γ. Background peaks at a mass slightly higher than that of $D_{SJ}(2458)^+$ and is easily subtracted. The residual signal may decay to $D_S^+\pi^0\gamma$ through $D_S^*(2112)^+\pi^0$ or $D_{SJ}^*(2317)^+\gamma$. To disantangle these modes an unbinned maximum likelihood fit is applied. The fit uses the differences in the shapes of the $D_S^+\pi^0$ and $D_S^+\gamma$ mass distributions: the $D_{SJ}(2458)^+ \to D_S^*(2112)^+\pi^0$

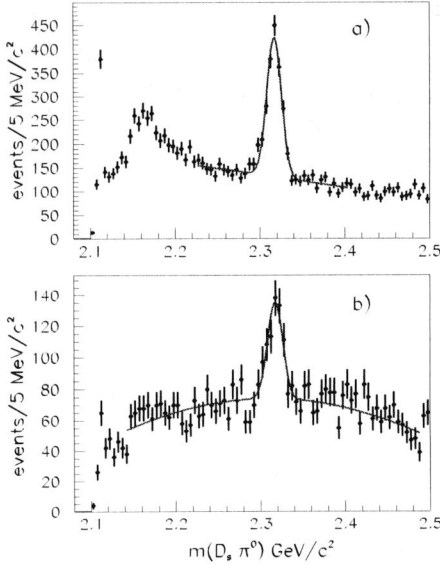

FIGURE 5. The $D_S^+ \pi^0$ mass distribution for (a) the decay $D_S^+ \to K^+ K^- \pi^+$ (1267 ± 53 signal candidates) and (b) the decay $D_S^+ \to K^+ K^- \pi^+ \pi^0$ (273 ± 33 signal candidates). The mass distributions are fit to a Gaussian plus a polynomial.

mode produces a narrow $D_S^+ \gamma$ and a wide $D_S^+ \pi^0$ mass distribution, opposite to the $D_{SJ}(2458)^+ \to D_{SJ}^*(2317)^+ \gamma$ mode. The data favor the decay through $D_S^*(2112)^+ \pi^0$:

$$\frac{BF(D_{SJ}(2458)^+ \to DSJ^*(2317)^+ \gamma)}{BF(D_{SJ}(2458)^+ \to D_S^*(2112)^+ \pi^0)} < 0.22 \qquad \text{at 95\% C.L..} \qquad (2)$$

An improved measurement of the mass of $D_{SJ}^*(2317)^+$ is obtained by taking into account background from the $D_{SJ}(2458)^+ \to D_S^*(2112)^+ \pi^0$ decay: $m(D_{SJ}^*(2317)^+) = 2317.3 \pm 0.4 \pm 0.8 c^2$ and rms width 7.3 ± 0.2 MeV/c^2. The distribution of the angle θ_h of the γ from the decay $D_S^*(2112)^+ \to D_S^+ \gamma$ in the $D_S^*(2112)$ rest frame relative to its direction of flight can be used to investigate the spin-parity of the $D_{SJ}(2458)^+$. The angular analysis of this decay mode disfavors $J^P = 0^-$. No conclusion can be drawn for $J^P = \{1^+, 2^-, 3^+, ..\}$ because the $D_S^*(2112)$ alignment cannot be predicted in these cases. The mass of the $D_{SJ}(2458)^+$ lies above the DK and below the D^*K thresholds. The narrow width and the isospin-violating decay to $D_S^*(2112)\pi^0$ indicate that the decay to DK is forbidden and suggest an unnatural spin-parity assignment. Belle has observed the decay $D_{SJ}(2458)^+ \to D_S^+ \gamma$ in production from both $\bar{c}c$ continuum [24] and B decay [25]. This rules out $J = 0$ and favors the $J^P = 1^+$ interpretation. The state is also observed by CLEO [23] at somewhat higher mass.

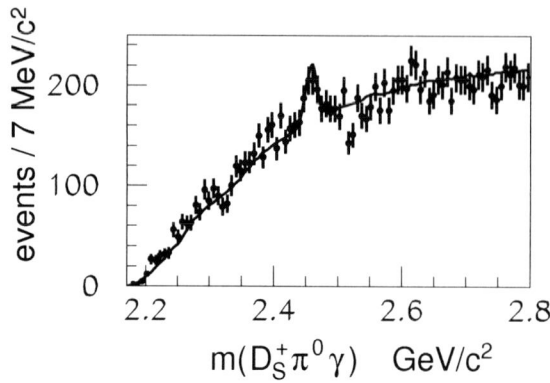

FIGURE 6. The mass distribution of selected $D_S^+\pi^0\gamma$ combinations with the result of a maximum likelihood fit (see text) superimposed.

Three-body D and B Decays

From a 22 fb^{-1} BABAR dataset, we measure [29] the branching fraction for each of the decays $D^0 \to K^0 K^- \pi^+$, $D^0 \to \bar{K}^0 K^+ \pi^-$ and $D^0 \to \bar{K}^0 K^+ K^-$ relative to that for the decay $D^0 \to \bar{K}^0 \pi^+ \pi^-$ as shown in Table 1. In these ratios, R_D, many systematic errors cancel. An amplitude analysis to these final states was performed [29] to

TABLE 1. Ratios of branching fractions, R_D, relative to the decay $D^0 \to \bar{K}^0 \pi^+ \pi^-$.

Channel	R_D /%, BABAR preliminary	R_D /%, PDG world average [3]
$D^0 \to K^0 K^- \pi^+$	$8.32 \pm 0.29 \pm 0.56$	11.7 ± 1.7
$D^0 \to \bar{K}^0 K^+ \pi^-$	$5.68 \pm 0.25 \pm 0.41$	8.9 ± 1.7
$D^0 \to \bar{K}^0 K^+ K^-$	$16.30 \pm 0.37 \pm 0.27$	17.2 ± 1.4

determine the relative fractions and phases of intermediate resonant and non-resonant amplitudes in D^0 decays. The analysis indicates, that the D^0 decays primarily to $K^0 K^- \pi^+$ via the $K^*(892)^+ K^-$ intermediate state. Only a small non-resonant contribution is required. The $D^0 \to \bar{K}^0 K^+ \pi^-$ final state contains several interfering amplitudes: $K^*(892)^- K^+$, $K^*(1430)^0 \bar{K}^0$, and $a_0(980)^+ \pi^-$ and a significant non-resonant. The $D^0 \to \bar{K}^0 K^+ K^-$ is dominated by $\phi(1020)\bar{K}^0$ which is interfering with a threshold scalar and $a_0(980)^+ K^-/a_0(980)^0 \bar{K}^0$. A small $f_0(980)\bar{K}^0$ amplitude is required to describe the Dalitz plot.

This demonstrates that the systematic study of three-body D decays can shed new light on light meson spectroscopy, in particular in the scalar meson sector. The knowledge gained about final state interaction in the analysis of high statistics D-meson decay Dalitz plots is important input for the analysis of three-body B-meson decays. Under the assumption that the three-body production is driven by 2-body final state interaction resonance parameters and phase shifts can be directly applied to B decays. The comparion between D and B decays will provide input for hadronic decay models. The analysis of charmless three-body B decays opens many new channels for CP studies.

BABAR has a first measurement of the branching fraction $BF(B^+ \rightarrow f_0(980)K^+; f_0(980) \rightarrow \pi^+\pi^-) = (9.2 \pm 1.2^{+2.1}_{-2.6}) \cdot 10^{-6}$ from the analysis of the $K^+\pi^+\pi^-$ Dalitz plot [30]. Resonances are searched for locally in the $K^+\pi^+\pi^-$ Dalitz plot with a likelihood fit to variables describing kinematics and event topology for signal and background. The systematic error of this measurement reflects the intensity variation due to interference with potential contributions near the $f_0(980)$ (the relative phase was sampled over the full range, but not yet fitted).

SUMMARY AND OUTLOOK

BABAR has proven not only to be a B meson factory, but also a charm-, ISR-, $\gamma\gamma$ and τ factory. BABAR enters now CP measurements in rare charmless B decays in the search for physics beyond the Standard Model, and its spectroscopy program has just started.

ACKNOWLEDGMENTS

I would like to thank the organizer of HADRON 2003 Conference who made it possible that this presentation could be given by phone for the first time during the Hadron Conference series to work around US travel visa problems.

REFERENCES

1. PEP II, SLAC-418, LBL-5379 (1993).
2. Aubert, B. (*BABAR Collaboration*), Nucl. Instr. and Mehtods **A479**, 1 (2002).
3. Hagiwara, K. et al., Particle Data Group, Phys. Rev. **D66**, 01001 (2002).
4. Solodov, E. and Stroili, R., *see BABAR presentations during this conference.*
5. Aubert, B. (*BABAR Collaboration*), Phys. Rev. Lett. **87**, 201803 (2001).
6. Aubert, B. (*BABAR Collaboration*), Phys. Rev. Lett. **88** 221802 (2002).
7. Aubert, B. (*BABAR Collaboration*), Phys. Rev. Lett. **89** 201802 (2002).
8. Aubert, B. (*BABAR Collaboration*), Phys. Rev. **D67** 072002 (2003).
9. Aubert, B. (*BABAR Collaboration*), Update of Time-Dependent CP-Violating Asymmetries in $B^0 \rightarrow h^+h^-$ Decays, Preliminary result presented at Lepton Photon Conference (2003).
10. Aubert, B. (*BABAR Collaboration*), Observation of the decay $B^0 \rightarrow \pi^0\pi^0$, *hep-ex/0308012* (2003).
11. Aubert, B. (*BABAR Collaboration*), Search for the B meson decay to $\eta'\phi$, *hep-ex/0309038* (2003).
12. Aubert, B. (*BABAR Collaboration*), Measurement of the $B^0 \rightarrow K_2^*(1430)\gamma$ and $B^+ \rightarrow K_2^{*+}(1430)\gamma$ branching fractions, *hep-ex/0308021* (2003).
13. Aubert, B. (*BABAR Collaboration*), Evidence for the rare decay $B^0 \rightarrow K^*l^+l^-$ and the measurement of the $B^0 \rightarrow Kl^+l^-$ branching fraction, *hep-ex/0308042* (2003).
14. Aubert, B. (*BABAR Collaboration*), Observation of the decay $B^0 \rightarrow \rho^+\rho^-$ and the measurement of the branching fraction and polarization, *hep-ex/0308024* (2003).
15. Aubert, B. (*BABAR Collaboration*), Measurement of Branching Fractions and CP-Violating Charge Asymmetries in $B^+ \rightarrow \rho^+\pi^0$ and $B^+ \rightarrow \rho^0\pi^+$ Decays, and Search for $B^0 \rightarrow \rho^0\pi^0$, *hep-ex/0307087* (2003).
16. Aubert, B. (*BABAR Collaboration*), A Search for $B^0 \rightarrow p\bar{p}$, preliminary result presented at Lepton Photon Conference (2003).

17. Aubert, B. (*BABAR Collaboration*), Measurement of the Branching Fraction for $B^0 \to K^+K^-K^0$, presented at Lepton Photon Conference (2003).

18. Aubert, B. (*BABAR Collaboration*), Time-dependent *CP*-Violating Asymmetries in $B^0 \to \phi K_S$, Preliminary result presented at Lepton Photon Conference (2003).

19. Aubert, B. (*BABAR Collaboration*), Search for D^0-$\bar{D}^0$ Mixing and a Measurement of the Doubly Cabibbo-suppressed Decay Rate in $D^0 \to K\pi$ Decays, *hep-ex/0304007* (2003).

20. Aubert, B. (*BABAR Collaboration*), Limits on $D^0 - \bar{D}^0$ Mixing and *CP* Violation from the Ratio of Lifetimes for Decay to $K^-\pi^+$, K^-K^+, and $\pi^-\pi^+$, Phys. Rev. Lett. **91**, 121801 (2003).

21. Aubert, B. (*BABAR Collaboration*), Phys. Rev. Lett. **90**, 242001 (2003).

22. Cahn, B., *see BABAR presentation during the conference*.

23. Besson, D. (*CLEO Collaboration*), Phys. Rev. **D68**, 032002 (2003).

24. Abe, K., (*Belle Collaboration*), *hep-ex/0307052* (2003).

25. Abe, K., (*Belle Collaboration*), *hep-ex/0308019* (2003).

26. Barnes, T., *see presentation during this conference*.

27. see many talks during dedicated sessions of this conference;
    Tornqvist, N., Spanier, S., The Scalar Mesons, Mini-Review in [3].

28. Aubert, B. (*BABAR Collaboration*), Observation of a Narrow Meson Decaying to $D_s^*(2112)^+\pi^0$ at a Mass of 2.458 GeV/c^2, BABAR-Pub 03/030 (2003), *submitted to Phys. Rev. Lett.*.

29. Aubert, B. (*BABAR Collaboration*), Dalitz Plot Analysis of D^0 Hadronic Decays $D^0->K^0K^-\pi^+$, $D^0->\bar{K}^0K^+\pi^-$ and $D^0->\bar{K}^0K^+K^-$, *hep-ex/0207089* (2003).

30. Aubert, B. (*BABAR Collaboration*), Measurements of the Branching Fractions of Charged B Decays to $K^\pm\pi^\mp\pi^\pm$ Final States, *hep-ex/0308065* (2003).

494

Recent Results from BES

Weiguo Li

Representing BES Collaboration, Institute of High Energy Physics, Chinese Academy of Sciences, Beijing 100039, China

Abstract. New results from BES on J/ψ, $\psi(2S)$ and $\psi(3770)$ decays are presented. Enhancement of the $p\bar{p}$ mass spectrum at the threshold is observed in the radiative J/ψ decays. Analyzes of sigma and kappa signals are reported in J/ψ to $\omega\pi^+\pi^-$ and J/ψ to $\bar{K}^{*0}K^+\pi^-/K^+K^-\pi^+\pi^-$ respectively. Results of J/ψ decaying to $\gamma K\bar{K}$, $p\bar{p}$, $\gamma\gamma V$ and 3π are presented. A search for new excited baryons in J/ψ decays is discussed. Results of four channels of $\psi(2S)$ to VT final states are reported and compared with 12% rule. First measurement of BF of $\psi(2S)$ to $K_s^0K_L^0$ is reported, and compared with that of J/ψ decays; results of χ_{cJ} decaying to $\Lambda\bar{\Lambda}$ are discussed. Evidence of $\psi(3770)$ to $J/\psi\pi^+\pi^-$ is presented. The status of BEPCII/BESIII project is briefly reported.

Beijing Electron Positron Collider (BEPC) and BEijing Spectroscopy (BES) had a major upgrade in 1996-1997, the upgraded BES is now called BESII [1]. After the successful run of data taking for R measurement in 1998, data had been taken at J/ψ, $\psi(2S)$ and $\psi(3770)$. 58M J/ψ and 14M $\psi(2S)$ hadronic events are produced at BESII, and about 20 pb^{-1} $\psi(3770)$ data is collected. The reported results here are based on these data samples. Most of the reported results are preliminary except the ones indicated to be already published.

To improve the consistency between data and MC simulations, extended works are done to improve the GEANT based simulation package called SIMBES. By careful tuning of the detector responses, the agreements of various distributions between data and MC simulations are much improved.

ENHANCEMENT OF $P\bar{P}$ MASS SPECTRUM AT THE THRESHOLD IN THE RADIATIVE J/ψ DECAY

Using 58M J/ψ data, radiative decay to $p\bar{p}$ is studied. Events with one gamma, one proton and one anti-proton are selected, the charged particle identification is mainly done by using dE/dx in the Main Drift Chamber and time of flight (TOF) information. The backgrounds are mainly from $\pi^0 p\bar{p}$, which is studied both by using data and MC simulations, and the shape of the $p\bar{p}$ mass spectrum agrees quite well between these two cases. it is found that the detector acceptance has some variation as a function of the invariant mass of the $p\bar{p}$ system. The fit of the mass spectrum can be seen in Fig. 1, an enhancement at the threshold is obvious. If treating it as a resonance, from the angle distributions, the state is preferred to be a 0^{-+} or a 0^{++}. The fitted mass is $1859^{+3+5}_{-10-25}MeV/c^2$ and its width is less than 30 MeV/c^2 at 90% C.L., when treating it

CP717, *Hadron Spectroscopy: Tenth International Conference,*
edited by E. Klempt, H. Koch, and H. Orth
© 2004 American Institute of Physics 0-7354-0197-7/04/\$22.00

as a s-wave(0^{-+}) resonance. The systematic error of the fitted mass is obtained by using MC to compare the fitted mass value with the input mass value. The state can also be fitted as a p-wave (0^{++}) resonance with its mass of $(1876.4 \pm 0.9)MeV/c^2$, and its width of $(4.6 \pm 1.8)MeV/c^2$. The fitting is somewhat better with the s-wave case. The detailed data selection and analysis can be referred to the already published paper[2].

It was studied to see if this possible new resonance be some known states, $\eta(1760)$ and $\pi(1800)$ were tried to fit the data, but the fitting quality was very bad, so it can be ruled out that it is a known state. This new resonance may be one of the long predicted baryonium states. BES is studying the other decay channels to understand more about the nature of this new finding.

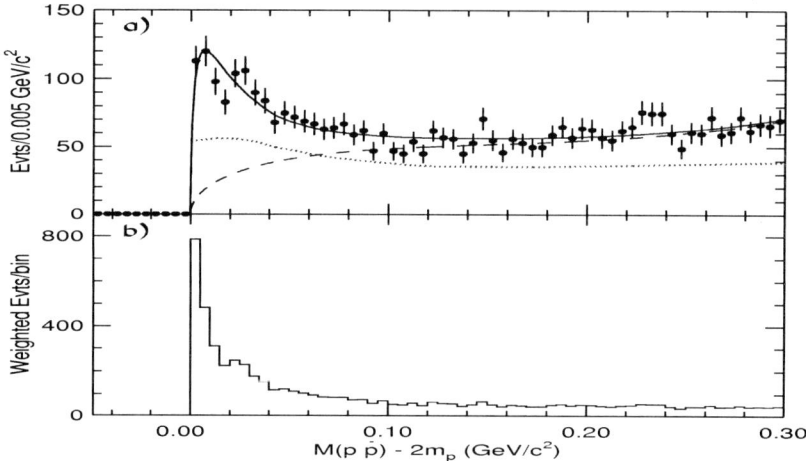

FIGURE 1. (a) The near threshold $M_{p\bar{p}} - 2m_p$ distribution for the $\gamma p\bar{p}$ event sample. The solid curve is the result of the fit described in the text; the dashed curve shows the fitted background function. The dotted curve indicates how the acceptance varies with $p\bar{p}$ invariant mass. (b) The $M_{p\bar{p}} - 2m_p$ distribution with events weighted by q^0/q.

SIGMA AND KAPPA ANALYSIS

Sigma is searched in the decay channel of $J/\psi \rightarrow \omega\pi^+\pi^-$. The requirements for event selection are as follows: four charged tracks with total Q of 0, each charged track should have good helix fitting quality, its polar angle $|cos\theta| < 0.8$, transverse momentum $p_{xy} > 60$ MeV/c, and with its vertex meeting the requirements of $V_{xy} < 2$ cm and $|V_z| < 20$ cm; two or more good photons with its energy deposit in shower counter > 30 MeV; dE/dx and TOF information are consistent with the particle assignments; each event should pass the four momenta constrained fit with fitting $\chi^2 < 40$; π^0 and ω are selected by requiring the invariant mass of two gammas or $\pi^+\pi^-\pi^0$ within the π^0 or ω mass by less than $40\ MeV/c^2$.

Fig. 2 (a) and (b) show the π^0 and ω signals; Fig. 2 (c) shows the $\pi^+\pi^-$ mass spectrum against the ω; Fig. 2 (d) shows the Dalitz plot of mass squared of $\omega\pi^-$ vs mass squared

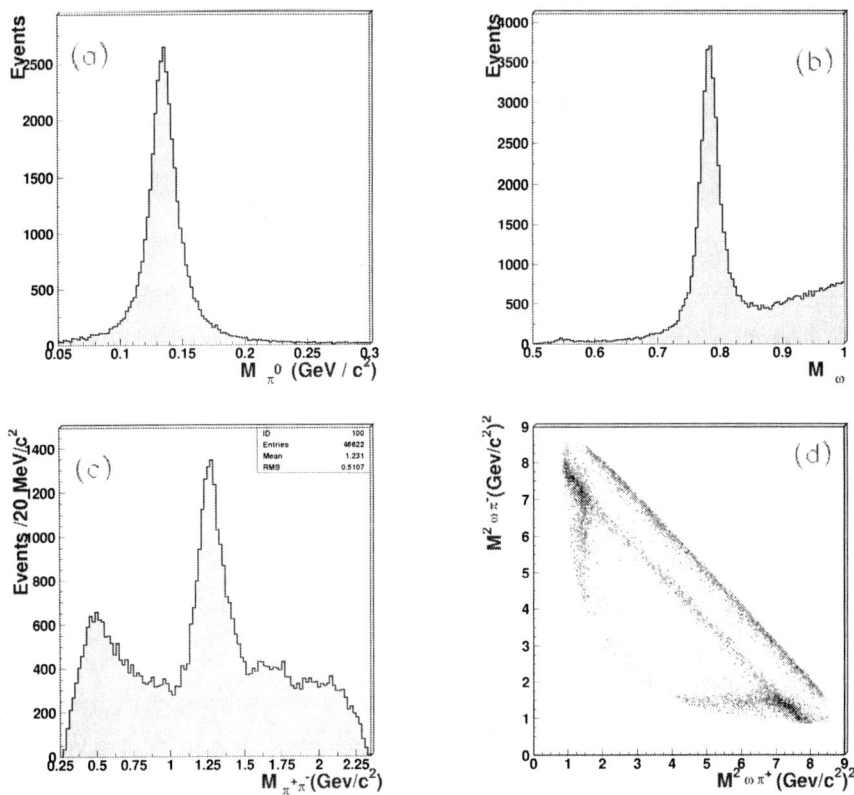

FIGURE 2. (a) and (b) are π^0 and ω signals. (c) The invariant mass spectrum of $\pi^+\pi^-$ against ω. (d) Dalitz plot of $M^2_{\omega\pi^-}$ vs $M^2_{\omega\pi^+}$.

of $\omega\pi^+$. From (c) and (d) some low mass enhancement with its mass peaking around $500\ MeV/c^2$ is evident, and the main contribution is regarded as the sigma signal.

From this stage, there are two different analysis approaches: method 1 fits only the $\pi^+\pi^-$ mass range <1.55 GeV, and treats ω as a particle; method 2 fits the whole data sample and uses the ω decay information. And the fitted functions for the sigma and the components in the fit are also different. In spite of these differences, the results of the fitted sigma parameters are quite similar from these two methods.

Here the results of method 2 are presented. Fig. 3 shows the data with fitted functions. The shaded areas in (a) and (b) represent the ω side band background contributions. The shaded area in Fig. 3 (c) represents the σ signal as part of the spin 0 contribution.

Three different fitting formulas for the sigma are used, The first uses the formula:

$$f = \frac{G_\sigma}{M^2 - s - iM\Gamma_{tot}(s)}$$

$$\Gamma_{tot}(s) = g_1 \frac{\rho_{\pi\pi}(s)}{\rho_{\pi\pi}(M^2)} + g_2 \frac{\rho_{4\pi}(s)}{\rho_{4\pi}(M^2)} \tag{1}$$

$$g_1 = f(s) \frac{s - m_\pi^2/2}{M^2 - m_\pi^2/2} exp[-(s - M^2)/\alpha]$$

Its pole position is $(542 \pm 10 \pm 40) - i(249 \pm 25 \pm 60)MeV/c^2$; the second formula treats $\Gamma(s)$ of simple Breit-Wigner as $\Gamma_0 \rho_{\pi\pi}(S)/\rho_{\pi\pi}(M^2)$, its pole position is $(566 \pm 15 \pm 50) - i(264 \pm 25 \pm 65)MeV/c^2$; The third formula treats Γ of simple Breit-Wigner as a constant, its pole position is $(540 \pm 10 \pm 40) - i(267 \pm 30 \pm 65)$ MeV, It can be seen that, the pole positions from three different fitting formulas are quite close to each other, although the mass and width values from these fitting are somewhat different. It is believed that the pole position is physically more meaningful.

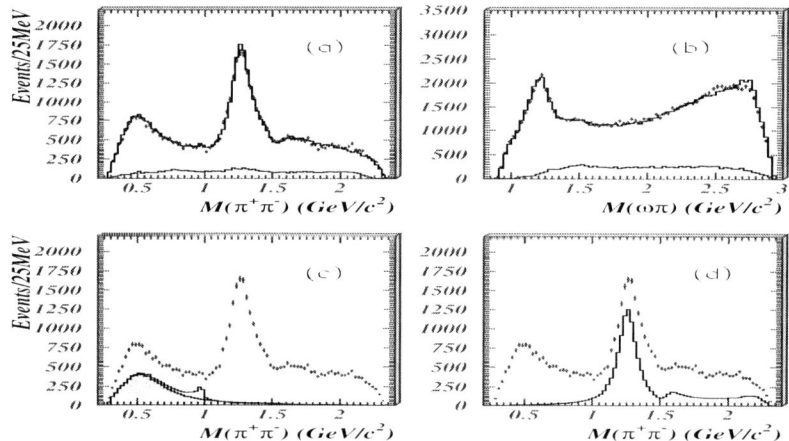

FIGURE 3. (a) The invariant mass spectrum of $\pi^+\pi^-$ against ω. (b) The invariant mass spectrum of $\omega\pi$. (c) The spin 0 contribution, the shaded area is the σ contribution. (d) The spin 2 contribution, $f_2(1270)$ is dominant.

Kappa is searched in the decay channel of J/ψ to $\bar{K}^*(892)^0 K^+\pi^-$ or $K^+K^-\pi^+\pi^-$. Firstly the events with $K^+K^-\pi^+\pi^-$ as the final decay products are selected, the requirements are as follows: four charged tracks with total Q of 0, the requirements for each track are the same as in the σ analysis described above; dE/dx and TOF information are consistent with the particle assignments; each event should pass the four momenta constrained fit with fitting $\chi^2 < 40$; the products of the probabilities, (the probability of four momenta constrained fit, the probability of each track assigned to a π or a K), should be smaller with the right $K^+K^-\pi^+\pi^-$ assignment compared with the other assignments as 4π, 4K and $K3\pi$. There are some K_S^0 and ϕ signals in the selected events, which are removed if the mass of $\pi^+\pi^-$ or K^+K^- is within the mass of K_S^0 or ϕ by less than 40 or 20 MeV/c^2 respectively.

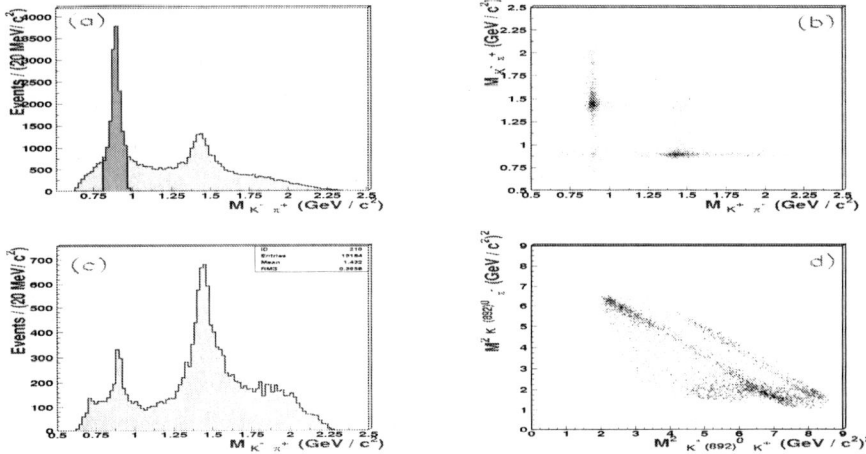

FIGURE 4. (a) The invariant mass of $K^-\pi^+$. (b) The scatter plot of $M^2_{K^-\pi^+}$ vs $M^2_{K^+\pi^-}$. (c) The invariant mass of $K^+\pi^-$ against $\bar{K}^*(892)^0$. (d) The Dalitz plot of $M^2_{K^*\pi^-}$ vs $M^2_{K^*K^+}$.

Fig. 4 (a) shows the mass spectrum of $K^-\pi^+$, the $\bar{K}^*(892)^0$ is visible, Fig. 4 (b) is the scatter plot of the mass spectrum $K^-\pi^+$ vs the mass spectrum of $K^+\pi^-$. Fig. 4 (c) shows the mass spectrum of $K^+\pi^-$ after $\bar{K}^*(892)^0$ is selected by requiring the mass of $K^+\pi^-$ within 80 MeV/c^2 of its mass. Fig. 4 (d) shows the Dalitz plot of mass squared of $\bar{K}^*(892)^0\pi^-$ vs mass squared of $\bar{K}^*(892)^0K^+$. From Fig. 4 (c) and (d), the broad enhancement at the mass around 800 MeV/c^2 is regarded as mainly the signal of the kappa particle.

After the $K^+K^-\pi^+\pi^-$ selection, there are several analysis approaches, one kind is to analyze the $K^+\pi^-$ system against the $\bar{K}^*(892)^0$, the other is to fit the $K^+K^-\pi^+\pi^-$ in the full phase space.

As mentioned above, the $\bar{K}^*(892)^0$ is selected by cutting the mass of $K^+\pi^-$ within 80 MeV/c^2 of $K^*(892)^0$ mass. Fig. 5 (a) shows the mass spectrum of $K^+\pi^-$ from the $\bar{K}^*(892)^0$ side band. Fig. 5 (b) shows the $K^+\pi^-$ mass spectrum after the side band background is subtracted. Fig. 5 (c) shows the kappa component in the $K^+\pi^-$ mass spectrum. Two methods are used in fitting the mass and angular distributions, both find it to be a 0^+ resonance. One uses a fitting formula as:

$$BW_\kappa = \frac{1}{m_\kappa^2 - s - i\sqrt{s}\Gamma_\kappa(s)}, \Gamma_\kappa(s) = \frac{g_\kappa^2 p_1(s)}{8\pi s} \qquad (2)$$

Its mass and width are (877 ± 85) and $(346 \pm 89)MeV/c^2$ respectively, the pole position is $[(851^{+43+107}_{-47-130}) - i(191^{+168+69}_{-53-32})]MeV/c^2$, consistent with the one using another fitting method, reported at the scaler session of this conference by Dr. Komada with the reported mass and width of $(882 \pm 24)MeV/c^2$ and $(335 \pm 82)MeV/c^2$ respectively.

The other kind of fitting uses the full phase space, as shown in Fig. 6, the fitted functions matched with data quite well.

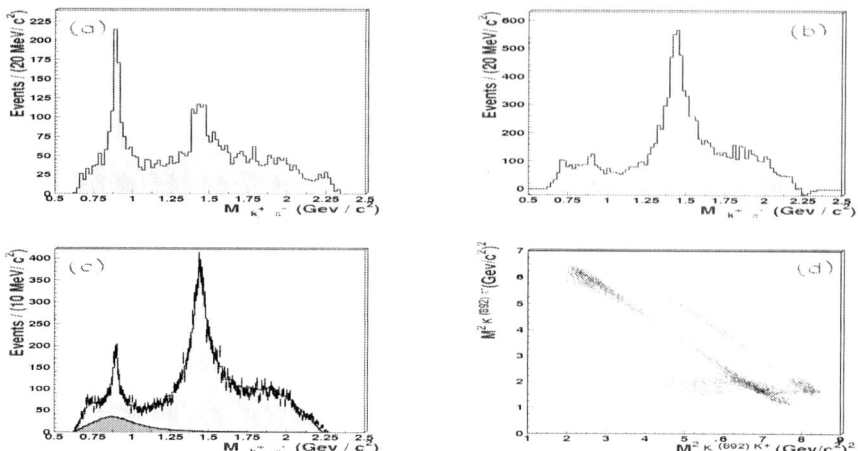

FIGURE 5. (a) The mass spectrum of $K^+\pi^-$ from the $\bar{K}^*(892)^0$ side band. (b) The $K^+\pi^-$ mass spectrum after the side band background is subtracted. (c) The kappa component in the $K^+\pi^-$ mass spectrum. (d) The Dalitz plot as in Fig. 4(d) using fitted functions.

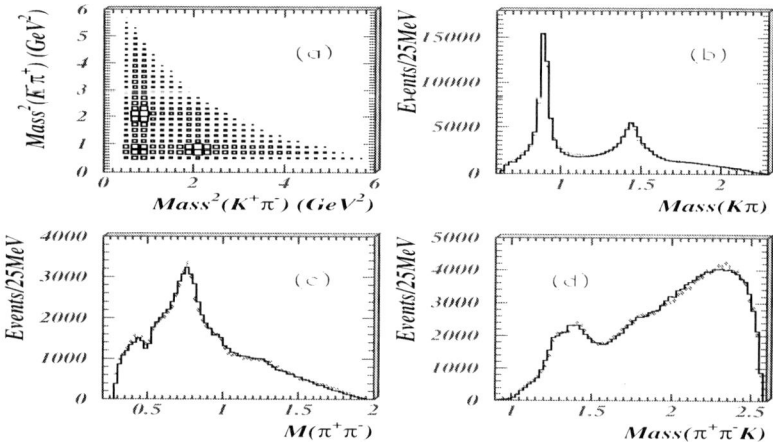

FIGURE 6. (a) The scatter plot of $M^2_{K^-\pi^+}$ vs $M^2_{K^+\pi^-}$. (b) The invariant mass of $K\pi$. (c) The invariant mass of $\pi^+\pi^-$. (d) The invariant mass of $K\pi^+\pi^-$, super-imposed on (b), (c) and (d) are fitted curves.

The kappa amplitude is written as:

$$
\begin{aligned}
f &= \frac{G_\kappa}{M_1^2 - s - iM_1\Gamma(s)} \\
\Gamma(s) &= \Gamma_0(s - s_A)\rho_{K\pi}(s)
\end{aligned}
\tag{3}
$$

The $K^0(1430)$ is fitted with a Flatte formula as:

500

$$f = \frac{M_2 g_1}{M_2^2 - s - i M_2 \Gamma(s)}$$

$$\Gamma(s) = \frac{s - s_A}{M_2^2 - s_A} [g_1 \rho_{K\pi}(s) + g_2 \rho_{\eta' K}(s)] \tag{4}$$

The pole position of kappa is obtained to be $(810 \pm 35(stat) - i(400 \pm 35(stat))MeV/c^2$.

We are checking these analyzes to see what is the source of the difference of the kappa parameters obtained from these different analyzes.

OTHER RESULTS FROM J/ψ DECAY

The detailed presentation was given by LU Feng at this conference. Here I summarize the main results, some of the figures are not repeated here.

The mass, width of η_c and its BF's of 5 decay channels

Using BESII 58 M J/ψ events, the radiative decays of J/ψ to $\gamma \eta_c$, and η_c to $p\bar{p}$, $K^+ K^- \pi^+ \pi^-$, $\pi^+ \pi^- \pi^+ \pi^-$, $K_S^0 K^\pm \pi^\mp$ and $\phi\phi$ are studied. The results of η_c mass and width are published[3]. From the combined fit, the fitted mass and width are $(2977.5 \pm 1.0 \pm 1.2)MeV/c^2$, and $(17.0 \pm 3.7 \pm 7.4)MeV/c^2$ respectively, with various systematic errors carefully studied. They are consistent with the PDG values of (2979.7 ± 1.5) and $(16.0^{+3.6}_{-3.2})MeV/c^2$.

The BFs of these five decay channels are listed in the Table 1.

TABLE 1. The Branching fractions of $BF(J/\psi \to \gamma\eta_c) \times BF(\eta_c \ decays)$.

process $J/\psi \to \gamma\eta_c$,	No. of events (fitted)	product of branching fractions
$\eta_c \to K^+ K^- \pi^+ \pi^-$	413 ± 54	$(1.5 \pm 0.2 \pm 0.2) \times 10^{-4}$
$\eta_c \to \pi^+ \pi^- \pi^+ \pi^-$	542 ± 75	$(1.3 \pm 0.2 \pm 0.4) \times 10^{-4}$
$\eta_c \to K^\pm K_S^0 \pi^\mp$	609 ± 71	$(2.2 \pm 0.3 \pm 0.5) \times 10^{-4}$
$\eta_c \to \phi\phi$	357 ± 64	$(3.3 \pm 0.6 \pm 0.6) \times 10^{-5}$
$\eta_c \to p\bar{p}$	213 ± 33	$(1.9 \pm 0.3 \pm 0.3) \times 10^{-5}$

Most of the BFs are consistent with PDG values, except that of $BF(\eta_c \to \phi\phi)$ which is smaller than that of PDG, but consistent with a recent Belle measurement[4].

The properties of $f(1710)$

It has been long debated if the $f(1710)$ is a 0^{++} or a 2^{++} state. From BESII data, two methods are used to study this state in J/ψ radiative decay to $K\bar{K}$[5]. One is the so called bin-by-bin fitting, the mass spectrum of $K\bar{K}$ is divided into small mass integrals, and each mass integral is subjected to the PWA (Partial Wave Analysis) to determine the contributions of different partial waves. From the fitting, it can be clearly seen that the main component at the mass of $1740\ MeV/c^2$ is 0^{++}, referring to the plots in [5].

Another method is the so called global fitting method. From Fig. 1 in Lu Feng's paper, it can be seen that the 2^{++} component is around $1525\ MeV/c^2$, consistent with $f_2'(1525)$ state. Its fitted mass and width are $(1519 \pm 2^{+15}_{-5})MeV/c^2$ and $(75 \pm 4^{+15}_{-5})$ MeV/c^2, and $BF(J/\psi \to \gamma f_2'(1525), f_2'(1525) \to K\bar{K}) = (3.42 \pm 0.15^{+0.69+1.55}_{-0.65-0}) \times 10^{-4}$. The 0^{++} component has its fitted mass and width of $(1740 \pm 4^{+10}_{-25})MeV/c^2$ and $(166^{+5+15}_{-8-10})MeV/c^2$, the $BF(J/\psi \to \gamma f_0(1710), f_0(1710) \to K\bar{K})$ is $(9.62 \pm 0.29^{+2.11+2.81}_{-1.86-0}) \times 10^{-4}$.

The bin-by-bin fitting gives results which are consistent with the global fit, where the parameters of $f_2'(1525)$ are fixed. The 0^{++} component has its fitted mass and width of $(1722 \pm 17)MeV/c^2$ and $(167^{+37}_{-29})MeV/c^2$, the $BF(J/\psi \to \gamma f_0(1710), f_0(1710) \to K\bar{K})$ is $(11.14^{+1.73}_{-1.20}) \times 10^{-4}$.

The global fit gives the prime results, and the bib-by-bin fit is a good cross-check.

It should be noted that, both in $J/\psi \to \gamma K\bar{K}$ and $\gamma p\bar{p}$, the $\xi(2230)$ is not seen in the invariant mass spectrum of $K\bar{K}$ and $p\bar{p}$ in the BESII data.

$J/\psi \to p\bar{p}$

The information regarding the angular distribution of the two baryon final states is especially interesting, for distinguishing various theoretical models. The angular distribution can be written as

$$\frac{dN}{d\cos\theta_B} \propto 1 + \alpha \cos^2 \theta_B \tag{5}$$

where θ_B is the angle between the baryon direction and the positron beam direction.

More than 60 K $J/\psi \to p\bar{p}$ events are selected and the background is small(at 1.5% level). Fitting the angular distribution with efficiency correction, referring to the relevant plot in LU Feng's paper at this conference, gives $\alpha = (0.676 \pm 0.036 \pm 0.042)$. And the $BF(J/\psi \to p\bar{p})$ is $(2.26 \pm 0.01 \pm 0.12) \times 10^{-3}$.

$J/\psi \to \gamma\gamma V (V = \rho, \phi)$

The standard perturbative theory [6] predicts that the partial width ratio between $\eta(1440)$ to $\gamma\rho$ and $\gamma\phi$ final states should be 9:2 if it is a $q\bar{q}$ state. Alternatively the ratio should be 1:1 if the $\eta(1440)$ is a pure glueball state. Figure 2 in LU Feng's paper gives the invariant mass spectrum of $\gamma\rho$ and $\gamma\phi$. The $\eta(1440)$ is clearly observed in $\gamma\rho$, with $BF(J/\psi \to \gamma\eta(1420)) \times BF(\eta(1420) \to \gamma\rho) = (1.07 \pm 0.17 \pm 0.11) \times 10^{-4}$. No significant signal of $\eta(1420) \to \gamma\phi$ is observed, the fitting gives $BF(J/\psi \to$

$\gamma \eta(1420)) \times BF(\eta(1420) \to \gamma\phi) = (0.31 \pm 0.30) \times 10^{-4}$, corresponding to 95% C.L. upper limit of $BF(J/\psi \to \gamma\eta(1420)) \times BF(\eta(1420) \to \gamma\phi) < 0.82 \times 10^{-4}$.

$$J/\psi \to \pi^+\pi^-\pi^0$$

Two independent measurements of the branching fraction of $J/\psi \to \pi^+\pi^-\pi^0$ are obtained based on 58M J/ψ and 14M $\psi(2S)$ respectively. One of the measurements uses J/ψ data to measure $J/\psi \to \pi^+\pi^-\pi^0$ directly, giving $BF(J/\psi \to \pi^+\pi^-\pi^0) = (21.35 \pm 0.04 \pm 1.85) \times 10^{-3}$. The other uses $\psi(2S) \to \pi^+\pi^-J/\psi$ decay, by measuring the relative branching fraction of $J/\psi \to \pi^+\pi^-\pi^0$ and $J/\psi \to \mu^+\mu^-$, and using $BF(J/\psi \to \mu^+\mu^-)$ from PDG2002, then $BF(J/\psi \to \pi^+\pi^-\pi^0) = (20.9 \pm 0.2 \pm 1.1) \times 10^{-3}$, which is well consistent with the result of the first measurement. The preliminary branching fraction is obtained by combining above two results: $BF(J/\psi \to \pi^+\pi^-\pi^0) = (2.10 \pm 0.11)\%$. This value is inconsistent with the previous measurements.

Search for excited baryon states in J/ψ decay

In the decay channel of $J/\psi \to p\pi^-\bar{n}$, a possible new N* resonance is observed with its mass and width of $(2065 \pm 3^{+15}_{-30})MeV/c^2$ and $(175 \pm 12 \pm 40)MeV/c^2$ respectively. The details may refer to the paper by JI Xiaobin in this conference proceeding.

RESULTS FROM $\psi(2S)$ DECAY

One important topic in $\psi(2S)$ decay is to study the so-called "12% " rule predicted by pQCD [7] that

$$Q_h = \frac{BF_{\psi(2S) \to X_h}}{BF_{J/\psi \to X_h}} = \frac{BF_{\psi(2S) \to e^+e^-}}{BF_{J/\psi \to e^+e^-}} = (12.3 \pm 0.7)\% \ . \tag{6}$$

Also through radiative $\psi(2S)$ decays, χ_{cJ} decays can be studied. Here some recent BES results in this area are summarized. The detailed information are given by MO Xiaohu in his talk given at this conference. the figures appeared in his paper are not repeated here to save space.

BFs of four VT decay channels

The decay of $\psi(2S)$ to four VT channels, $\omega f_2(1270)$, $\rho a_2(1320)$, $K^*(892)^0$ $\overline{K^*_2}(1430)^0$+c.c., and $\phi f'_2(1525)$, are studied based on 14M $\psi(2S)$ date sample. Signals are quite evident in Figure 1 of MO Xiaohu's paper and BFs are measured. The results are listed in Table 2, together with the corresponding results of J/ψ decay from

PDG2002 and the Q_h values. Comparing with 12% rule, the Q_h value of these four VT channels is suppressed.

TABLE 2. The results of $\psi(2S)$ and χ_{cJ} decays.

VT channel	$BF_{\psi p}$ (10^{-4}) (from BES)	$BF_{J/\psi}$ (10^{-3}) (from PDG2002)	$Q_h(\%)$
ωf_2	$2.05 \pm 0.41 \pm 0.46$	4.3 ± 0.6	4.8 ± 1.5
ρa_2	$2.55 \pm 0.73 \pm 0.60$	10.9 ± 2.2	2.3 ± 1.1
$K^* \overline{K_2^*} + c.c.$	$1.64 \pm 0.33 \pm 0.41$	6.7 ± 2.6	2.4 ± 1.2
$\phi f_2'$	$0.48 \pm 0.14 \pm 0.12$	$1.23 \pm 0.06 \pm 0.20$ (This value from DM2 only.)	3.9 ± 1.6
PP channel	$BF_{\psi'}$ (10^{-5}) (from BESII)	$BF_{J/\psi}$ (10^{-4}) (from BESII)	$Q_h(\%)$
$K_S K_L$	$5.25 \pm 0.45 \pm 0.63$	$1.86 \pm 0.04 \pm 0.14$	28.2 ± 4.7
Decay mode	$BF_{Exp.}$ (10^{-4}) (from BESII)	$BF_{The.}$ (10^{-4}) (by COM)	$R_{Exp./The.}$
$\chi_{c0} \to p\bar{p}$	$4.7^{+1.3}_{-1.2} \pm 1.0$	$-$	$-$
$\chi_{c1} \to p\bar{p}$	$2.6^{+1.0}_{-0.9} \pm 0.6$	0.366	7.1
$\chi_{c2} \to p\bar{p}$	$3.3^{+1.5}_{-1.3} \pm 0.7$	0.333	9.9

First Observation of $K_S^0 K_L^0$ in $\psi(2S)$ Decay

From data analysis point of view, the event topology of $\psi(2S) \to K_S^0 K_L^0$ is fairly simple: the neutral K_L^0 almost leaves no information in Main Drift Chamber due to its long decay lifetime, while the K_S^0 swiftly decays into two pions. By the virtue of this simple topology of event, two good charged tracks are required with a net charge of zero; in addition, secondary vertex requirement is applied for K_S^0 identification. With these requirements, the momentum distribution of K_S^0 is shown in Figure 2(a) of MO Xiaohu's paper. The backgrounds can be simulated quite well. A Gaussian function for signal events plus the backgrounds are used to fit the spectrum, resulting in $BF(\psi(2S) \to K_S^0 K_L^0) = (5.25 \pm 0.45 \pm 0.63) \times 10^{-5}$. The similar study has also been made for $J/\psi \to K_S^0 K_L^0$ decay. The momentum distribution of K_S^0 is shown in Figure 2 (b) of the same paper and the BF is worked out to be $(1.86 \pm 0.04 \pm 0.14) \times 10^{-4}$. It is worth while to notice that the BES measurement result is considerably larger than that of the PDG value: $BF(J/\psi \to K_S^0 K_L^0) = (1.08 \pm 0.14) \times 10^{-4}$.

In contrast to the VT channels, the Q_h value for $K_S^0 K_L^0$ channel is enhanced greatly to be $(28.2 \pm 4.7)\%$ with fairly high precision. Comparing to 12% rule, the deviation is greater than 3σ.

$$\chi_{cj} \to \Lambda\bar{\Lambda}$$

This result is published in [8]. By selecting $\gamma\pi^+\pi^-p\bar{p}$ events, $\Lambda\bar{\Lambda}$ signal is very clear. By fitting the $\Lambda\bar{\Lambda}$ mass spectrum, with the backgrounds obtained by MC from known decay channels, the BFs of three χ_{cJ} states to $\Lambda\bar{\Lambda}$ are obtained as listed in the Tab. 2.

For comparison, the relevant theoretical predictions are also listed in Tab. 2, where the theoretical calculation is based on Color Octet Mechanism (COM). It can be seen that the results on χ_{c1} and χ_{c2} decays only agree marginally with model predictions.

EVIDENCE OF $\psi(3770) \to J/\psi\pi^+\pi^-$

With about 8 pb^{-1} data sample, the non-$D\bar{D}$ events of $\psi(3770)$ decay to $J/\psi\pi^+\pi^-$ are searched. Events with four charged tracks are selected, with 2 leptons (e^+e^- or $\mu^+\mu^-$) and $\pi^+\pi^-$. To suppress gamma conversion background, the two π should have an opening angle between them $> 20^0$. The J/ψ signal is dominant in the invariant mass distribution of di-lepton pairs.

The searched signal can be seen in two different approaches. One is to further select the events with the conditions: $|E_{l^+l^-\pi^+\pi^-} - E_{cm}| < 2.5\sigma$, here σ is the detector energy resolution of the four tracks; and $|M_{ll} - M_{J/\psi}| < 150$ MeV. Then the recoil mass of $\pi^+\pi^-$ are examined. The other approach is to make a four momenta constrained fit with a fitting probability $> 1\%$, then the invariant mass of lepton pair is examined. The plots can be referred to the article by RONG Gang in this proceeding. From the Fig. 3 and 5 in that article, one peak corresponding to the $\psi(2S)$ decays and a smaller peak corresponding to the $\psi(3770)$ decays are clearly seen in both cases. Backgrounds come from the random combination as well as the initial state radiation, the latter may contribute some small enhancement under the $\psi(3770)$ peak, which is estimated by MC simulation.

The result from the second analysis method gives $BF(\psi(3770) \to J/\psi\pi^+\pi^-) = (0.59 \pm 0.26 \pm 0.16) \times 10^{-2}$, and the first method gives consistent result.

STATUS OF BEPCII/BESIII PROJECT

BEPC will have a major upgrade to change the collider from one ring machine to a two-rings machine. The expected luminosity will be $10^{33} \times cm^{-2} \times s^{-1}$ achieved by multiple bunches and micro-beta operation. The detector will almost entirely be rebuilt, it will mainly consist of: a main drift chamber with a small cell structure, He based working gas and Al field wires; a CsI crystal EM calorimeter; Time of Flight counters for charged particle identifications; Resistive Plate Chambers instrumented in the steel plates as a muon detector; a super-conducting magnet provides 1.0 tesla magnetic field.

The project is progressing quite well: most of the designs are finalized; a lot of the components have been ordered and/or in the stage of bidding. The project is scheduled to commission the machine and detector together at the end of 2006, and to take data sometimes in 2007. New groups are much welcome to join BESIII collaboration.

ACKNOWLEDGMENTS

The BES collaboration thanks the staff of the BEPC and the IHEP computing center for their hard efforts. This work is supported in part by the National Natural Science Foundation of China under contracts Nos. 19991480, 10225524, 10225525, the Chinese Academy of Sciences under contract No. KJ 95T-03, the 100 Talents Program of CAS under Contract Nos. U-11, U-24, U-25, and the Knowledge Innovation Project of CAS under Contract Nos. U-602, U-34 (IHEP); by the National Natural Science Foundation of China under Contract No.10175060 (USTC); and by the Department of Energy under Contract No. DE-FG03-94ER40833 (U Hawaii).

REFERENCES

1. J. Z. Bai, *et al.*, BES Collaboration, Nucl. Inst. Meth. **A344**, 319 (1994);**A458**, 627 (2001).
2. J. Z. Bai, *et al.*, BES Collaboration, Phys. Rev. Lett. 91, 022001 (2003).
3. J. Z. Bai, *et al.*, BES Collaboration, Phys. Lett. B555, 174 (2003)
4. BELLE Collaboration, hep-ex/0305068.
5. J. Z. Bai, *et al.*, BES Collaboration, Phys. Rev. **D68**, 052003 (2003)
6. M. S. Chanowitz, Phys. Lett. B164 (1985) 379.
7. T. Appelquist and H. D. Politzer, Phys. Rev. Lett. **34**, 43 (1975);
   A. De Rújula and S. L. Glashow, Phys. Rev. Lett. **34**, 46 (1975).
8. J. Z. Bai, *et al.*, BES Collaboration, Phys. Rev. **D67**, 112001 (2003).

Recent Hadron Physics Results from Fermilab

James S. Russ

Physics Department, Carnegie Mellon University, Pittsburgh, PA 15213 USA

Abstract. Hadron physics at Fermilab involves both new results from the collider experiments and continued analyses of the Fixed Target data sets. This summary highlights recent experimental and theoretical work on topics in hadron physics.

HADRON RESULTS FROM CDF AND D0

One of the major interests of both collider experiments is the study of B hadronic properties. Weak decay studies, e.g., CP violation in B decays, are only one aspect of the Run II heavy flavor program. The study of b- and c-hadron production in $\bar{p}p$ is an early highlight of the collider run. However, the Run I results have continued to be the subject of considerable theoretical activity, and we begin there.

The Run I B-production cross sections from CDF and D0 were about a factor of 3 larger than predicted by the NLO calculation of Frixione, Mangano, Nason and Ridolfi. [1]. Recently, Cacciari and Nason have re-examined the problem in light of new results on b-quark fragmentation. [2] Their analysis replaces the historic Peterson fragmentation function with a fragmentation analysis based on the moments of the LEP b-decay distributions. This leads to a change in slope of the yield with p_T and to much-improved agreement with the data, both in the central region (CDF) and at large rapidity $|y|$ (D0). Cacciari and Nason quote the former 3-fold data excess now as 1.7 ± 0.5(th) $\pm$ 0.5(expt), consistent with the upper limit of the theory uncertainty band due to (chiefly) to the scale uncertainty in the calculation.

CDF in Run II is able to exploit their new Two-Track Trigger to select hadronic charm decays. This has allowed them to measure for the first time the production distributions for a variety of charm hadrons from $\bar{p}p$ collisions. As an example, results for D^o production are shown in Fig. 1, along with a new theoretical treatment by Chicarelli and Nason. [3]. The experimental data have excellent statistical precision. The theory, including an uncertainty band from the scale choice, is also shown. The charm production results look very similar to the b-production seen above. The data lie just at the upper end of the scale-uncertainty band and follow the shape of the theory. This is suggestive of a common issue in production calculations, since fragmentation effects are quite different for b and c production. Also, the agreement improves at the largest p_T values, suggestive of reduced fragmentation effects as had been suggested by Run I measurements of b-jets. This will be pursued at the Tevatron. It will also be interesting to see how LHC measurements compare with theory at even larger p_T.

Another Run I surprise from the Collider experiments was the anomalously large production of J/ψ and Υ, especially at large p_T. The theory response was to add a color-

CP717, *Hadron Spectroscopy: Tenth International Conference*,
edited by E. Klempt, H. Koch, and H. Orth
© 2004 American Institute of Physics 0-7354-0197-7/04/$22.00

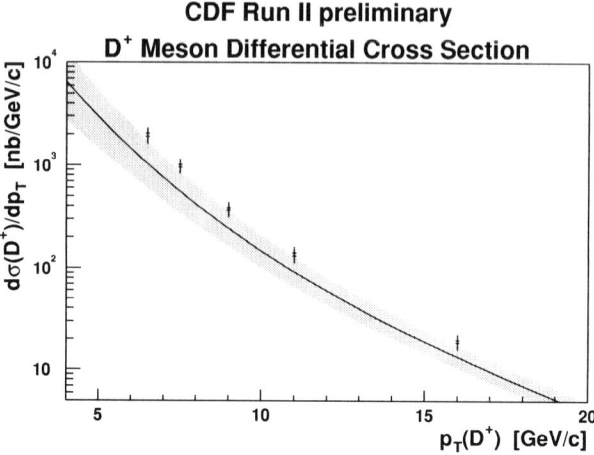

FIGURE 1. CDF differential cross section for inclusive D° production in p̄p collisions

octet component to the amplitude. Fitting the three new parameters of the model to the data produced a good representation of the yield. However, the theory predicted a large polarization of J/ψ at large p_T. This is not seen in either fixed target or collider data. This topic is naturally of great interest in Run II studies. Both experiments feature lepton triggers for these studies. The new CDF low-p_T di-lepton trigger allows measurement of the J/ψ production cross section down to $p_T = 0$ - a first for collider experiments. This differential cross section measurement will challenge theory to handle both large and small p_T regions consistently. No new results on the important polarization question have yet been released by either collaboration, but they are clearly in the near future.

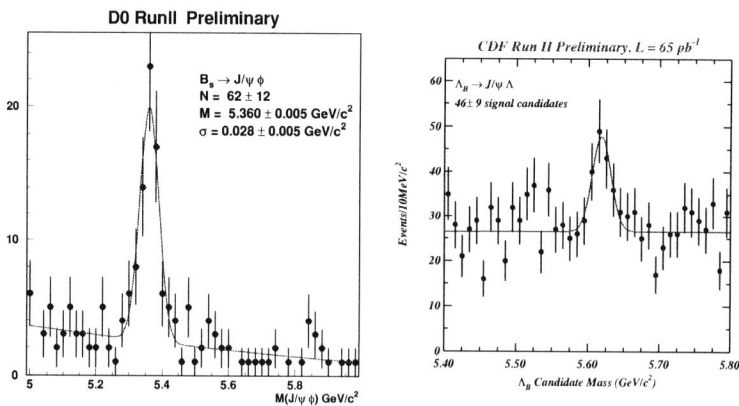

FIGURE 2. left: D0 mass plot for $B_s^0 \to J/\psi\phi$; right: CDF mass plot for $\Lambda_b^0 \to J/\psi\Lambda$

We close this section with examples of B-hadron exclusive final states as measured by CDF and D0. The quality of the data lead one to expect a series of B-hadron relative branching ratio measurements for the B_s and Λ_b systems, akin to the charm and beauty measurements from the e^+e^- machines. Fig. 2 shows good signal/noise quality for the D0 signal for the 4-charged-track decay $B_s \rightarrow J/\psi\phi$ on the left. On the right is the CDF signal for $\Lambda_b \rightarrow J/\psi\Lambda$. The Λ decays in the CDF drift chamber, and the background level is noticeably higher in this two-vertex reconstruction. CDF has improved signal/background in this channel already by recent analysis advances. These signals are from 20% of the current Run II data set. The early indications are that the colliders will indeed produce a rich set of B-hadron decays.

The last examples of B-hadrons that I will discuss from from CDF. These signals are relevant to CP-Violation studies that are not a part of this conference. However, since these are the first such hadrons studied at a hadron collider, let me show the signals. B_s mixing is a prime target for Run II studies. At left we see the first reconstruction of the exclusive decay $B_s \rightarrow D_s\pi$. It is clear that the Tevatron will produce a clean sample of these decays for mixing analysis, given enough integrated luminosity. At right, we see a sum plot of the overlapping decays B°, $B_s^\circ \rightarrow h^+h^-$, where h may be a kaon or pion. Kinematic analysis and limited particle ID capability allow CDF to separate the two meson signal into $\pi\pi$, πK, and KK final states. They have reported a preliminary result that that B_s decays are dominated by the KK channel, in contrast to B° decays, where the $K\pi$ decay mode dominates. The preliminary CDF B° branching ratio result $\pi\pi/ KK = 0.26 \pm .11 \pm .055$ agrees with the old Belle value in the PDG compilation. The most recent Belle data, presented at this conference, have much lower statistical error, but CDF and D0 ultimately may have more events. The CDF particle identification separates K and pi at $1.2\,\sigma$ over the momentum range relevant to these data. The Belle (or BaBar) particle identification is RICH-based and so reduces the mistags from about 30% to about 10%. Nevertheless, the CDF data will be an independent source of information about this very interesting example of B meson hadronic decays. These signals come from the first small sample of the ultimate luminosity. They suggest that the collider experiments can challenge the B-factories on certain CP asymmetry measurements.

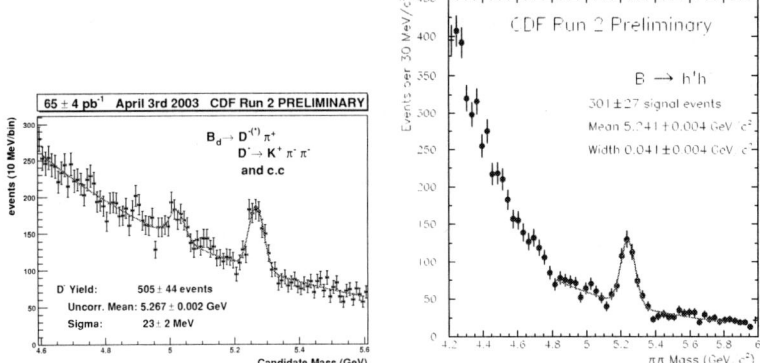

FIGURE 3. left: CDF mass plot for $B_s^\circ \rightarrow D_s\pi$; right: CDF mass plot for $B^\circ \rightarrow$ hadron hadron

FIXED TARGET D MESON DECAY STUDIES

The Fermilab experiments E791 (pion production) and FOCUS (photoproduction) have pioneered the study of light quark spectroscopy using Dalitz plot analyses of large samples of Cabibbo-favored and Cabibbo-suppressed final states of D-meson and D_s meson decays. One of the main points of discussion at this conference has been the role of scalar mesons in scattering and decay analyses. Especially controversial are the broad, low-mass enhancements $\sigma(550)$ and $\kappa(780)$. At this conference we have seen a major confrontation between K-matrix analyses (e.g., Malvezzi) and Breit-Wigner resonance amplitude analyses (e.g.,Reis). The reader can see their presentations in these proceedings for details. The issue is clearly one of choosing the correct phenomenology. Both approaches are self-consistent in that they both fit the Dalitz plots, but they give quite different interpretations of the resonant content of the decay amplitudes. Finding the right path through these analyses is important in view of the upcoming interest in analyzing B-meson Dalitz plots. In particular, to extract the CP-violating phase angle α from a Dalitz plot analysis of $B^o \to \pi\pi\pi$ decays requires us to have the $\sigma(550)$ problem under control.

I urge the experts on this problem to pool intellectual resources to form a Scalar Meson Working Group. At this conference David Bugg (Rutherford Appelton Laboratory) has organized a set of interested people. Collaboration by other interested physicists is welcome. Contact David for details (email: bugg@v2.rl.ac.uk).

PHOTOPRODUCTION

We heard discussion at this conference of the proliferation of states in the 1600-1800 MeV/c^2 mass region, some allegedly exotic. The exotic states are seen primarily in phase shift analyses of scattering data, with consequent ambiguities. Interpretations tend to differ between experiments. In a very different situation the FOCUS experiment reports the diffractive production of a narrow K$^+$K$^-$ state at 1750 MeV/c^2. [4] This state is seen clearly at small 4-momentum transfer t as shown in Fig. 4. The t-slope is consistent with experimental resolution. It does not decay to K^{*+}K$^-$, the dominant decay mode for the broad $\phi(1750)$ seen in other photoproduction experiments. Its quantum numbers and spectroscopic assignment are a puzzle to be resolved in the future. The state seems to be firmly established by these data. The question is about its quantum numbers and its place in the spectroscopy of unflavored mesons. If it is truly diffractive, with $J^P = 1^-$, then where does it fit into the set of ϕ-like objects? On the other hand, if it is made by the Primakoff effect (2 photon process), the quantum numbers can be anything in the natural sequence. Further work is clearly needed.

CHARM HADROPRODUCTION ASYMMETRIES

Charm hadroproduction studies in fixed target experiments have been carried out for the past 2 decades. The conventional picture of Heavy Quark production assumes factoriza-

510

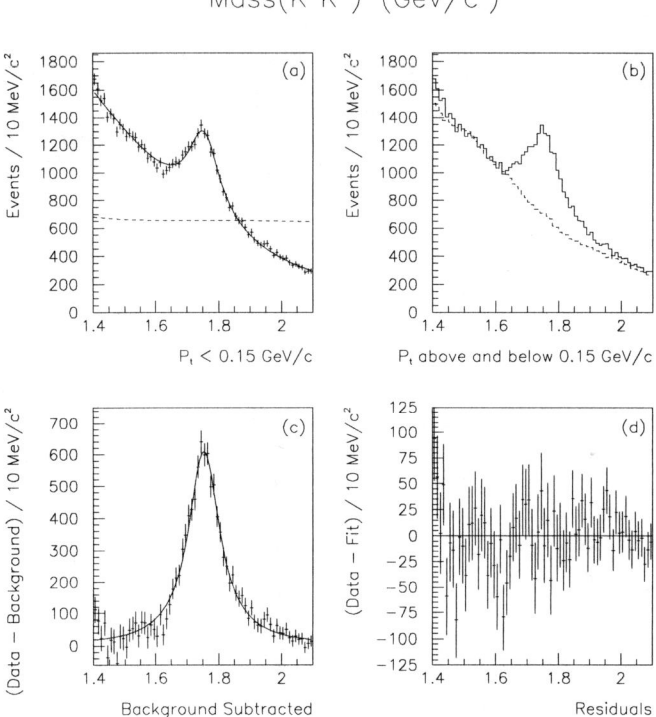

$$\text{Mass}(K^+K^-) \ (\text{GeV}/c^2)$$

FIGURE 4. K^+K^- mass distribution in different t regions, showing diffractive state at 1750 MeV/c^2

tion, i.e., a separation between the quarks produced in the hard scattering process and the hadron fragments that are left in the beam and target hadrons. Factorization allows one to use parton ideas from Deep Inelastic Scattering to analyze the quark production process at an unphysical high mass scale where the calculation is understood, then evolve the distributions to the m_Q-scale. The NLO parton calculation includes single gluon emission effects in the production process and was first analyzed by Nason, Dawson and Ellis. [5, 6] The NLO calculation leaves us with quark-level distributions. At this level there is almost complete symmetry between charm and anti-charm production. Observables are, however, related to hadrons. The data for charm and anti-charm hadrons shows striking species-dependent asymmetries that are not part of the underlying quark picture. [7, 8, 9]

Because these data cannot be explained by perturbative QCD, the questions raised focus on other physics: (a) why do some but not all charm hadrons show large asymmetry? (b) is there a beam-hadron dependence of the effect? and (c) why is it x_F dependent? These questions traditionally have been framed within two possible pictures of large non-perturbative effects: the Quark-Gluon-String Model (QGSM) originally included in the PYTHIA Monte Carlo [10] and the Intrinsic Charm Model (ICM) [11]. A different approach, using the Recombination Model originally developed by Hwa and Ochs for other production physics, has been recently applied to charm hadroproduction by Braaten, Mehen and collaborators. [12] This picture relates *all* production processes from all hadronic beams in one overall framework. There are few parameters, and the same parameter set works for multiple interactions at multiple energies. This approach seems to be an efficient analysis methodology to confront a wide range of single charm meson and baryon production data.

For details one should see the papers in Ref. [13]. The analysis adds to the basic tree-level scattering a recombination amplitude in which the produced charm quark combines with an antiquark from the beam hadron to produce the outgoing charm meson or with a quark from the beam hadron to produce an outgoing charm baryon. The model parametrizes the spin-color character of the charm state produced in the process with 4 parameters for each type - meson and baryon production. Preliminary results are shown in Fig. 5. The impressive feature is that the same parameter set covers different mesons and baryons from different beam hadrons and beam energies. A comprehensive analysis of all available data is underway. The figure shows the charm-anticharm asymmetry as a function of the Feynmann x parameter for the collision for D^* production by π^- at left and for Λ_c^+ production by protons at right. The fit parameters were set from *different* data that had very different asymmetries, yet the model gives a very nice representation of the data with no free parameters.

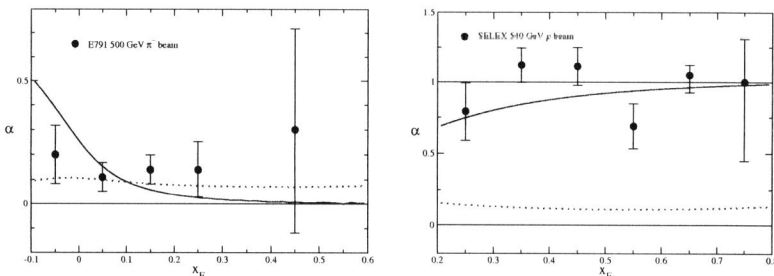

FIGURE 5. Recombination fits: (left) D^* production by π^- with parameters from D production; (right) Λ_c^+ production by protons with parameters from pion production

RECENT RESULTS ON DOUBLE CHARM BARYONS

The SELEX experiment (E781) at Fermilab recently published the first observation of the long-sought family of baryons with two charmed quarks, termed ccd(3520). [14]. Within the standard SU(4) representation of qqq states, one expects to see ground states as a Ξ_{cc} isodoublet and an Ω_{cc} singlet. The spectroscopy of excited states is more complicated than for the single charm baryons because of the degrees of freedom in the (cc) diquark. In a Born-Oppenheimer-like analysis of the QCD binding effects, one finds that the rotational degrees of freedom of the cc system may have splittings that are

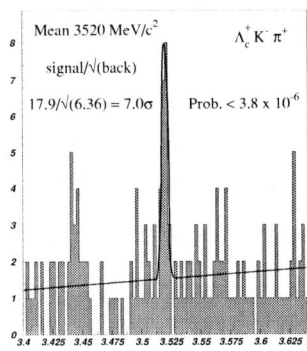

FIGURE 6. published SELEX plot for ccd(3520) discovery

not very different in size than the chromomagnetic hyperfine excitations seen in single-charm baryons.

SELEX studied the angular characteristics of the ccd(3520). A ground state is expected to decay isotropically, barring polarization effects. The SELEX data are summed over all azimuthal angles and are characterized by small p_T values, so no polarization asymmetries are expected. The isotropy condition was selected by studying phase space decay simulations. A cut that retained 80% of the simulation signal reduced the data yield from 16 events to 5 events - a clear indication of non-isotropic behavior. The mass distribution after the cut indicates a new significant mass enhancement at 3443 MeV/c^2. The lower-mass peak was visible in the original uncut mass distribution, shown in Fig. 6, but did not meet the 5σ discovery significance without the angular cut. The experimenters are continuing to explore these data and are preparing more publications on double charm baryons. They note several difficulties with the double charm interpretation. First, the weak lifetimes for the states are very short, much shorter than expected from theory. Why is this? Second, EM transitions between the excited state and ground state must be highly suppressed, even with a large weak decay rate. The symmetry of the cc diquark helps this, but a serious calculation is needed to understand a large suppression. Finally, the production mechanism is unusual, to say the least. A very large fraction of all Λ_c^+ seen in the SELEX data appear to arise from this double charm mechanism. The experimenters see these states only in baryon-nucleus collisions, not in meson-nucleus collisions. The FOCUS photoproduction experiment at Fermilab has 12 times more Λ_c^+ events from photon-nucleus interactions at about half the SELEX CM energy. They see no evidence for double charm in a wide variety of decay modes. Finding a way to confirm or refute the SELEX results is an important issue on the table for hadron physics.

CONCLUSIONS

The hadron physics program from Fermilab continues to emphasize hadrons containing heavy flavor, both in production and spectroscopy. Recent advances in theory have improved agreeeement with production data both for collider results and fixed target results. Spectroscopy of light mesons in charm decay has highlighted a conflict in the way to treat the broad scalar mesons - Breit-Wigner shapes or via the K-matrix formalism. This same issue is relevant to scattering experiments, and a common solution should be sought.

New particles are reported and invite more work at other machines. FOCUS has presented a new KK resonance at small t in photoproduction. It does not fit gracefully into the present vector meson spectroscopy. Questions still exist about its exact production mechanism, and the spin-parity analysis has not been done. SELEX has expanded its evidence for double charm production from baryon beams. Both the FOCUS and SELEX states require confirmation and offer future opportunities for hadron physics.

REFERENCES

1. Stefano Frixione, Michelangelo L. Mangano, Paolo Nason, and Giovanni Ridolfi. *Adv. Ser. Direct. High Energy Phys.*, 15:609–706, 1998.
2. M. Cacciari and P. Nason. *Phys. Rev. Lett.*, 89:122003, 2002.
3. M. Cacciari and P. Nason. hep-ph/0306212, 2003.
4. J. M. Link *et al. Phys. Lett.*, B545:50–56, 2002.
5. P. Nason, S. Dawson, and R. K. Ellis. *Nucl. Phys.*, B327:49–92, 1989.
6. P. Nason, S. Dawson, and R. K. Ellis. *Nucl. Phys.*, B335:260, 1990.
7. E. M. Aitala *et al. Phys. Lett.*, B371:157–162, 1996.
8. F. G. Garcia *et al. Phys. Lett.*, B528:49–57, 2002.
9. M. Kaya *et al. Phys. Lett.*, B558:34–40, 2003.
10. T. Sjostrand. *Comput. Phys. Commun.*, 82:74–90, 1994.
11. R. Vogt, S. Brodsky, and P. Hoyer. *Nucl. Phys.*, B383:643–684, 1992.
12. E. Braaten, Y. Jia, and T. Mehen. *Phys. Rev. Lett.*, 89:122002, 2002.
13. T. Mehen. hep-ph/0306178, 2003.
14. M. Mattson *et al. Phys. Rev. Lett.*, 89:112001, 2002.

The CLEO-c Research Program

Holger Stöck

University of Florida, Department of Physics, PO Box 118440, Gainesville, FL 32611-8440, USA

Abstract.
In spring 2003, the B physics era ended for the CLEO experiment with a final run at the $\Upsilon(5S)$ resonance. Over the summer the experiment and the CESR accelerator were modified to operate at lower center-of-mass energies between 3 and 5 GeV. In September 2003 the CLEO-c detector has begun to take its first data at the $\psi(3770)$ resonance, with which a new era for the exploration of the charmonium sector begins. The CLEO-c research program presented here will include studies of leptonic, semileptonic and hadronic charm decays, searches for exotic, gluonic matter and test for new physics beyond the Standard Model.

INTRODUCTION

The CLEO-c physics program includes a variety of measurements that will contribute to the understanding of important Standard Model processes as well as provide the opportunity to probe the physics that lies beyond the Standard Model. The dominant themes of this program are measurement of absolute branching ratios for charm mesons with the precision of the order of 1 - 2% (depending upon the mode), determination of charm meson decay constants and of the CKM matrix elements $|V_{cs}|$ and $|V_{cd}|$ at the 1 - 2% level and investigation of processes in charm and τ decays, that are expected to be highly suppressed within the Standard Model. Hence, a reconfigured CESR electron-positron collider operating at a center of mass energy range between 3 and 5 GeV together with the CLEO detector will give significant contributions to our understanding of fundamental Standard Model properties.

RUN PLAN AND DATA SETS

From the year 2003 to 2006 the CESR accelerator will be operated at center-of-mass energies corresponding to $\sqrt{s} \sim 4140 MeV$, $\sqrt{s} \sim 3770 MeV$ (ψ") and $\sqrt{s} \sim 3100 MeV$ (J/ψ). Taking into account the anticipated luminosity which will range from $5 \times 10^{32} cm^{-2} s^{-1}$ down to about $1 \times 10^{32} cm^{-2} s^{-1}$ over this energy range, the run plan will yield $3 fb^{-1}$ each at the ψ" and at $\sqrt{s} \sim 4140 MeV$ above $D_s \bar{D}_s$ threshold and $1 fb^{-1}$ at the J/ψ. These integrated luminosities correspond to samples of 1.5 million $D_s \bar{D}_s$ pairs, 30 million $D \bar{D}$ pairs and one billion J/ψ decays. As a point of reference, these datasets will exceed those of the Mark III experiment by factors of 480, 310 and 170, respectively. Table 1 summarizes the run plan.

CP717, *Hadron Spectroscopy: Tenth International Conference,*
edited by E. Klempt, H. Koch, and H. Orth
© 2004 American Institute of Physics 0-7354-0197-7/04/$22.00

TABLE 1. The 3-year CLEO-c run plan

Year	Resonance	Anticipated Luminosity (fb^{-1})	Reconstructed Events
1	$\psi(3770)$	~ 3	30M $D\bar{D}$, 6M tagged D
2	$\sqrt{s} \sim 4140 MeV$	~ 3	1.5M $D_s\bar{D}_s$, 0.3M tagged D_s
3	$\psi(3100)$	~ 1	60M radiative J/ψ

In addition, prior to the conversion to low energy a total amount of $4 fb^{-1}$ spread over the $\Upsilon(1S)$, $\Upsilon(2S)$, $\Upsilon(3S)$ and $\Upsilon(5S)$ resonances is taken to launch the QCD part of the program. These data sets will increase the available $b\bar{b}$ bound state data by more than an order of magnitude.

HARDWARE REQUIREMENTS

The conversion of the CESR accelerator for low energy operation requires the addition of 18 meters of wiggler magnets to enhance transverse cooling of the beam at low energies. 6 of 14 wigglers were installed in summer 2003 with additional 6 wigglers scheduled for installation in 2004. In the CLEO III detector the silicon vertex detector was replaced by a small, low mass inner drift chamber (Figure 1). In addition, the solenoidal field will be reduced to 1.0 T. No other requirements are necessary.

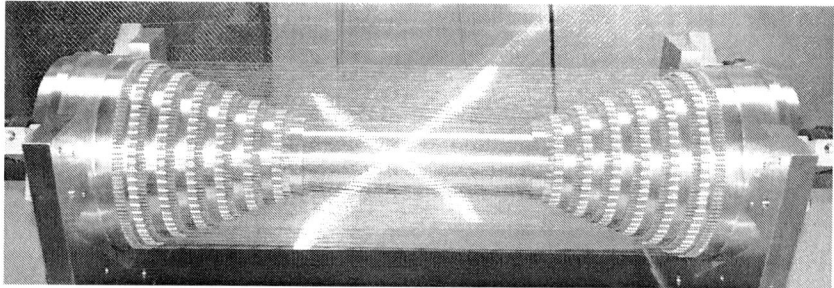

FIGURE 1. The new CLEO-c inner drift chamber, replacing the silicon vertex detector

PHYSICS PROGRAM

The following sections will outline the CLEO-c physics program. The first section will focus on the Ypsilon spectroscopy, the second section will describe the charm decay program, the third section will give an overview about the exotic, gluonic matter studies and the last section will descibe the oportunities for probing of new physics beyond the Standard Model.

Ypsilon Spectroscopy

From fall 2001 to spring 2003 CLEO has collected $4 fb^{-1}$ of data on the Υ resonances below the $\Upsilon(4S)$, as well as at the $\Upsilon(5S)$ resonance, which is currently beeing analyzed. So far, the only established states below $B\bar{B}$ threshold are the three vector singlet Υ resonances ($^3S_1$) and the six χ_b and χ'_b (two triplets of ^3P_J) that are accessible from these parent vectors via E1 radiative transitions (see Figure 2). By collecting substantial data samples at the $\Upsilon(1S)$, $\Upsilon(2S)$ and $\Upsilon(3S)$, CLEO will address a variety of outstanding physics issues.

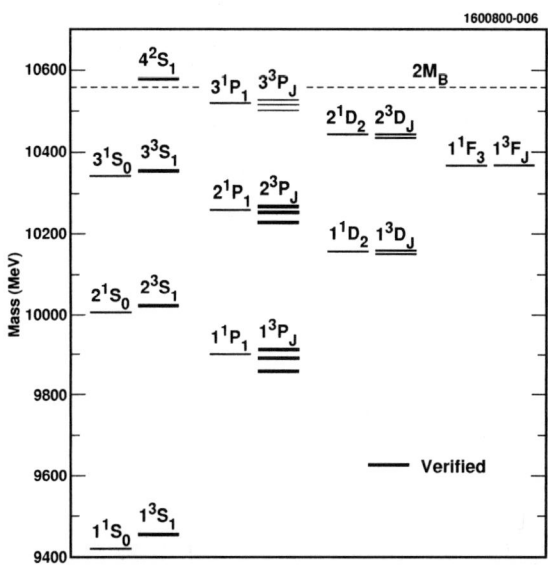

FIGURE 2. Approximate levels of the $b\bar{b}$ states. The name associated with the spin-parity assignments are $^1S_0 = \eta_b$, $^3S_1 = \Upsilon$, $^1P_1 = h_b$ and $^3P_J = \chi_b$ (triplets with J = 0,1,2).

- Discovery of η_b and Observation of h_b

  The η_b is the ground state of $b\bar{b}$. Most present theories [1] indicate the best approach would be the hindered M1 transition from the $\Upsilon(3S)$, with which CLEO might have a signal of 5σ significance in $1 fb^{-1}$ of data. In the case of the h_b, CLEO established an upper limit of $\mathcal{B}(\Upsilon(3S) \rightarrow \pi^+\pi^- h_b) < 0.18\%$ at 90% confidence level [2]. This result, based on $\sim 110 pb^{-1}$, already tests the theoretical predictions [3] for this transition which range from 0.1 - 1.0%. The resonance run program will measure the mass of the h_b, assuming the predictions are valid, to $\sim 5 MeV$.

- Observation of 1^3D_J states

  The $b\bar{b}$ system is unique as it has states with L = 2 that lie below the open-flavor threshold. These states have been of considerable theoretical interest, as indicated by many predictions of the center-of-gravity of the triplet and by a recent review [4]. In an analysis of the $\Upsilon(3S)$ CLEO data sample the $\Upsilon(1^3D_2)$ state could already be observed in the four-photon cascade $\Upsilon(3S) \rightarrow \gamma_1 \chi'_b \rightarrow \gamma_1 \gamma_2 \Upsilon(^3D_J) \rightarrow \gamma_1 \gamma_2 \gamma_3 \chi_b \rightarrow$

$\gamma_1\gamma_2\gamma_3\gamma_4\ell^+\ell^-$. The mass of the $\Upsilon(1^3D_2)$ state is determined to $10161.1 \pm 0.6 \pm 1.6 MeV/c^2$ [5].

- Search for glueball candidates in radiative $\Upsilon(1S)$ decays

  The BES collaboration has reported signals for a glueball candidate [6] in radiative J/ψ decay - a glue-rich environment. Naively one would expect the exclusive radiative decay to be suppressed in Υ decay by a factor of roughly 40, which implies product branching fractions for Υ radiative decay of $\sim 10^{-6}$. With $1 fb^{-1}$ of data and efficiencies of around 30% one can expect ~ 10 events in each of the exclusive channels, which would be an important confirmation of the J/ψ studies.

Charm Decays

The observable properties of the charm mesons are determined by the strong and weak interactions. As a result, charm mesons can be used as a laboratory for the studies of these two fundamental forces. Threshold charm experiments permit a series of measurements that enable direct study of the weak interactions of the charm quark, as well as tests of our theoretical technology for handling the strong interactions.

Leptonic Charm Decays

Measurements of leptonic decays in CLEO-c will benefit from the use of fully tagged D^+ and D_s decays available at the $\psi(3770)$ and at $\sqrt{s} \sim 4140 MeV$. The leptonic decays $D_s \rightarrow \mu\nu$ are detected in tagged events by observing a single charged track of the correct sign, missing energy, and a complete accounting of the residual energy in the calorimeter. The clear definition of the initial state, the cleanliness of the tag reconstruction, and the absence of additional fragmentation tracks make this measurement straightforward and essentially background-free. This will enable measurements of the yet barely known leptonic decay rates for D and D_s to a precision of 3 - 4% and will allow for incisive checks of theoretical calculations of the decay constants f_D and f_{D_s} at the 1 - 2 %. Table 2 summarizes the expected precision in the decay constant measurements.

TABLE 2. Expected decay constants errors for leptonic decay modes

Decay Mode	Decay Constant	Decay Constant Error %	
		PDG 2000	CLEO-c
$D^+ \rightarrow \mu^+\nu$	f_D	Upper Limit	2.3
$D_s^+ \rightarrow \mu^+\nu$	f_{D_s}	17	1.7
$D_s^+ \rightarrow \tau^+\nu$	f_{D_s}	33	1.6

Semileptonic Charm Decays

The CLEO-c program will provide a large set of precision measurements in the charm sector against which the theoretical tools needed to extract CKM matrix information precisely from heavy quark decay measurements will be tested and honed.

CLEO-c will measure the branching ratios of many exclusive semileptonic modes, including $D^0 \to K^- e^+ v$, $D^0 \to \pi^- e^+ v$, $D^0 \to K^- e^+ v$, $D^+ \to \bar{K}^0 e^+ v$, $D^+ \to \pi^0 e^+ v$, $D^+ \to \bar{K}^{0*} e^+ v$, $D_s^+ \to \phi e^+ v$ and $D_s^+ \to \bar{K}^{0*} e^+ v$. The measurement in each case is based on the use of tagged events where the cleanliness of the environment provides nearly background-free signal samples, and will lead to the determination of the CKM matrix elements $|V_{cs}|$ and $|V_{cd}|$ with a precision level of 1.6% and 1.7%, respectively. Measurements of the vector and axial vector form factors $V(q^2)$, $A_1(q^2)$ and $A_2(q^2)$ will also be possible at the $\sim 5\%$ level. Table 3 summarizes the proposed branching fractional errors.

TABLE 3. Expected branching fractional errors for semileptonic decay modes

Decay Mode	BR fractional error %	
	PDG 2000	CLEO-c
$D^0 \to K\ell v$	5	1.6
$D^0 \to \pi\ell v$	16	1.7
$D^+ \to \pi\ell v$	48	1.8
$D_s \to \phi\ell v$	25	2.8

HQET provides a successful description of the lifetimes of charm hadrons and of the absolute semileptonic branching ratios of the D^0 and D_s [7]. Isospin invariances of the strong forces lead to corrections of $\Gamma_{SL}(D^0) \simeq \Gamma_{SL}(D^+)$ in the order of $\mathcal{O}(tan^2\Theta_C) \simeq 0.05$. Likewise, $SU(3)_{Fl}$ symmetry relates $\Gamma_{SL}(D^0)$ and $\Gamma_{SL}(D^+)$, but a priori would allow them to differ by as much as 30%. However, HQET suggests that they should agree to within a few percent. A charm factory is the best place to measure absolute inclusive semileptonic charm branching ratios, in particular $\mathcal{B}(D_s \to X\ell v)$ and thus $\Gamma_{SL}(D_s)$.

Implications of the Leptonic and Semileptonic Measurements for CKM

Every weak decay involving leptons depends on both CKM elements and on hadronic matrix elements. As described in the sections above, CLEO-c data can be used for calibrating the theoretical tools that will determine the hadronic terms and for extracting the essential CKM elements.

Combining the leptonic and semileptonic measurements leads to "direct" determinations of the CKM elements $|V_{cd}|$ and $|V_{cs}|$. The results are shown in Table 4. For this table LatticeQCD is assumed and validated across a wide range of charm and onium decay measurements, to which CLEO-c will provide decay constants with 1% accuracy.

The impact of the entire suite of CLEO-c measurements on the current knowledge of the CKM matrix is summarized in the following paragraphs. Just to remind the reader,

TABLE 4. Collected results for $|V_{cd}|$ and $|V_{cs}|$

Decay Mode	CKM Element	CKM Precision
$D_s \to \mu^+ \nu$	$|V_{cs}|$	1.7
$D_s \to \tau^+ \nu$	$|V_{cs}|$	1.6
$D^0 \to K^- e^+ \nu$	$|V_{cs}|$	1.6
$D^+ \to \mu^+ \nu$	$|V_{cd}|$	2.3
$D^0 \to \pi^- e^+ \nu$	$|V_{cd}|$	1.7

the CLEO-c program of leptonic and semileptonic measurements has two components: one of calibrating and validating theoretical methods for calculating hadronic matrix elements, which can then be applied to all problems in CKM extraction in heavy quark physics; and one of extracting CKM elements directly from the CLEO-c data. The direct results of CLEO-c are the precise determination of $|V_{cd}|$, $|V_{cs}|$, f_D, f_{D_s}, and the semileptonic form factors. The precision knowledge of the decay constants f_D and f_{D_s}, together with the rigorous calibration of theoretical techniques for calculating heavy-to-light semileptonic form factors, are required for the direct extraction of CKM elements from CLEO-c. This also drives the indirect results, namely the precision extraction of CKM elements from experimental measurements of the B_d mixing frequency, the B_s mixing frequency, and the $B \to \pi \ell \nu$ decay rate measurements which will be done by a combination of efforts spread across BaBar, Belle, CDF, D0, BTeV, LHCb, ATLAS and CMS.

In Table 5 the combined projections are presented. In the determination of the CKM elements $|V_{cd}|$ and $|V_{cs}|$ from B and B_s mixing $|V_{tb}| = 1$ is used. The tabulation also includes improvement in the direct measurement of $|V_{tb}|$ itself which is expected from the Tevatron experiments [8].

TABLE 5. Combined projections for CKM elements at present and after CLEO-c

Present Knowledge		
$\delta V_{ud}/V_{ud} = 0.1\%$	$\delta V_{us}/V_{us} = 1\%$	$\delta V_{ub}/V_{ub} = 25\%$
$\delta V_{cd}/V_{cd} = 7\%$	$\delta V_{cs}/V_{cs} = 16\%$	$\delta V_{cb}/V_{cb} = 5\%$
$\delta V_{td}/V_{td} = 36\%$	$\delta V_{ts}/V_{ts} = 39\%$	$\delta V_{tb}/V_{tb} = 29\%$
After CLEO-c		
$\delta V_{ud}/V_{ud} = 0.1\%$	$\delta V_{us}/V_{us} = 1\%$	$\delta V_{ub}/V_{ub} = 5\%$
$\delta V_{cd}/V_{cd} = 1\%$	$\delta V_{cs}/V_{cs} = 1\%$	$\delta V_{cb}/V_{cb} = 3\%$
$\delta V_{td}/V_{td} = 5\%$	$\delta V_{ts}/V_{ts} = 5\%$	$\delta V_{tb}/V_{tb} = 15\%$

Hadronic Charm Decays

The D^0 is the best known of all the charm hadrons. The CLEO and ALEPH experiments by far provide the most precise measurements for the decay $D^0 \to K^- \pi^+$. They use the same technique by looking at $D^{*+} \to \pi^+ D^0$ decays and taking the ratio of the

D^0 decays into $K^- \pi^+$ to the number of decays with only the π^+ from the D^{*+} decay detected. The dominant systematic uncertainty is the background level in the latter sample. In both experiments, the systematic errors exceed the statistical errors. By using $D^0 \bar{D}^0$ decays, and tagging both D mesons, the background can be reduced to almost zero and the branching ratio fractional error can be improved significantly (see Table 6).

The D^+ absolute branching ratios are determined by using fully reconstructed D^{*+} decays, comparing $\pi^0 D^+$ with $\pi^+ D^0$ and using isotropic spin symmetry. Hence, this rate cannot be determined any better than the absolute D^0 decay rate using this technique. By using $D^+ D^-$ decays and a double tag technique the background can be reduced again to almost zero which leads to a significant improvement of the branching ratio fractional error (see Table 6).

TABLE 6.   Expected branching fractional errors for hadronic decay modes

	BR fractional error %	
Decay Mode	**PDG 2000**	**CLEO-c**
$D^0 \rightarrow K\pi$	2.4	0.5
$D^+ \rightarrow KK\pi$	7.2	1.5
$D_s \rightarrow \phi\pi$	25	1.9

Exotic, Gluonic Matter

With approximately one billion J/ψ produced, CLEO-c will be the natural glue factory to search for glueballs and other glue-rich states using $J/\psi \rightarrow gg \rightarrow \gamma X$ decays. The region of $1 < M_X < 3 GeV/c^2$ will be explored with partial wave analyses for evidence of scalar or tensor glueballs, glueball-$q\bar{q}$ mixtures, exotic quantum numbers, quark-glue hybrids and other new forms of matter predicted by QCD. This includes the establishment of masses, widths, spin-parity quantum numbers, decay modes and production mechanisms for any identified states, an in detail exploration of reported glueball candidates such as the tensor candidate $f_J(2220)$ and the scalar states $f_0(1370)$, $f_0(1500)$ and $f_0(1710)$, and the examination of the inclusive photon spectrum $J/\psi \rightarrow \gamma X$ with < 20 MeV photon resolution and identification of states with up to 100 MeV width and inclusive branching ratios above 1×10^{-4}. A Monte Carlo study of inclusive radiative J/ψ decays in CLEO-c is shown in Figure 3 based on a sample of 60 million J/ψ decays and assuming $\mathscr{B}(J/\psi \rightarrow \gamma f_J(2220)) = 8 \times 10^{-4}$. A monochromatic photon line from the $J/\psi \rightarrow \gamma f_J(2220)$ decay is clearly seen. The signal efficiency is 24%. With 10^9 J/ψ decays, CLEO-c will be able to discover any narrow resonance produced in radiative J/ψ decays with inclusive branching fractions of order 10^{-4} or greater.

In addition, spectroscopic searches for new states of the $b\bar{b}$ system and for exotic hybrid states such as $cg\bar{c}$ will be made using the $4fb^{-1}$ $\Upsilon(1S)$, $\Upsilon(2S)$, $\Upsilon(3S)$ and $\Upsilon(5S)$ data sets. Analysis of $\Upsilon(1S) \rightarrow \gamma X$ will play an important role in verifying any glueball candidates found in the J/ψ data.

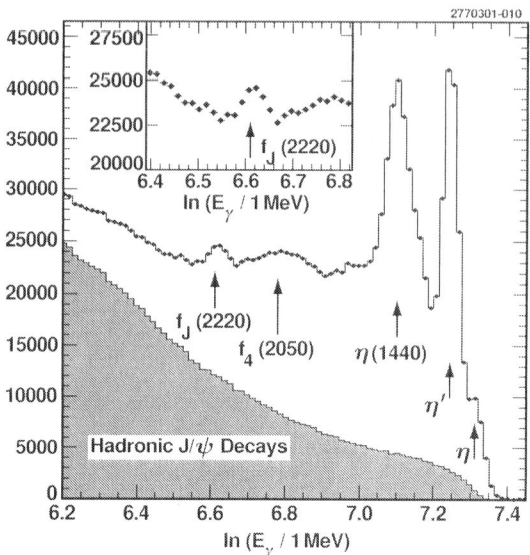

FIGURE 3. The inclusive photon spectrum from J/ψ decays from a Monte Carlo simulation in CLEO-c. Signals from η', $\eta(1440)$ and $f_J(2220)$ are clearly visible. A broad signal from $f_4(2050)$ production is also evident.

Charm Beyond the Standard Model

CLEO-c will have the opportunity to probe for new physics beyond the Standard Model. Three highlights - $D\bar{D}$-mixing, CP violation and rare charm decays - are discussed in the following sections.

$D\bar{D}$-Mixing

Within the Standard Model (SM), the processes which mediate the decays of charmed quarks and antiquarks can change the "charm" quantum number by one unit, $\Delta C = 1$. On the other hand, the mixing of D^0 and $\bar{D}^0$ necessitates changing a charm quark into an anti-charm quark, i.e. the "charm" quantum number must change by two units $\Delta C = 2$. This can be arranged in the SM only at one loop level and, therefore, is naturally suppressed. However, new physics (beyond the SM) contributions can generate $\Delta C = 2$ interactions as well. It is for this reason that neutral meson-antimeson mixing can provide important information about both the SM and new physics beyond the SM. The $D^0 - \bar{D}^0$ system is particularly interesting in this respect as it is the only system that is sensitive to the dynamics of the bottom-type quarks.

CLEO-c will have the important experimental advantage of operating at the $D^0\bar{D}^0$ threshold, where the D^0 and $\bar{D}^0$ are produced in the state that is quantum mechanically coherent. This allows new and simple methods to be used to measure the $D^0 - \bar{D}^0$

mixing parameters [9, 10]. Thus, in addition to the "standard" methods of searches for $D^0 - \bar{D}^0$ mixing, new tools and methods, unique to running at the $\psi(3770)$ (and higher resonances), become available for studies of these important parameters.

CP Violation

In addition to indirect CP violation, both SM and new physics effects can induce different contributions to the decay amplitudes of D mesons. This phenomenon can be traced back to the appearance of complex-valued couplings (CKM parameters) in the $\Delta C = 1$ Lagrangian that mediates D decays and leads to a CP-violating difference between decay rates of CP-conjugated states.

The production process

$$e^+ e^- \rightarrow \psi(3770) \rightarrow D^0 \bar{D}^0$$

produces an eigenstate of $CP+$, in the first step, since the $\psi(3770)$ has J^{PC} equal to 1^{--}. Now consider the case where both the D^0 and the $\bar{D}^0$ decay into CP eigenstates. Then the decays

$$\psi(3770) \rightarrow f_+^i f_+^j \ \text{or} \ f_-^i f_-^j$$

are forbidden, where f_+ denotes a $CP+$ eigenstate and f_- denotes a $CP-$ eigenstate. This is because

$$CP(f_\pm^i \ f_\pm^j) = (-1)^\ell = -1$$

for the $\ell = 1 \ \psi(3770)$

Hence, if a final state such as $(K^+ K^-)(\pi^+ \pi^-)$ is observed, one immediately has evidence of CP violation. Moreover, all $CP+$ and $CP-$ eigenstates can be summed over for this measurement. This measurement can also be performed at higher energies where the final state $D^{*0} \bar{D}^{*0}$ is produced. When either D^* decays into a π^0 and a D^0, the situation is the same as above. When the decay is $D^{*0} \rightarrow \gamma D^0$ the CP parity is changed by a multiplicative factor of -1 and all decays $f_+^i f_-^j$ violate CP [11]

Rare Charm Decays

Rare decays of charmed mesons and baryons provide "background-free" probes of new physics effects. In the framework of the Standard Model (SM) these processes occur only at one loop level. SM predicts vanishingly small branching ratios for these processes because of the absence in the SM of the super-heavy bottom-type quark supplemented by almost perfect GIM cancellation between the contributions of strange and down quarks. This is very different from the familiar case of bottom quark decays where the top quark contribution dominates the decay amplitude. It also makes the SM predictions for these transitions very uncertain, as the pertubative GIM cancellation mechanism is not effective for soft, long-distance contributions. In addition, in many cases annihilation topologies also give sizable contribution. At the end, any anomalous enhancement of a given branching ratio would have to be compared to the (dominant long-distance) SM amplitude. Fortunately, several model-dependent estimates exist indicating that the SM predictions for these processes are still far below current experimental sensitivities. Some examples are given in [12, 13]. From there is also follows that experiments which can measure rare D decay branching ratios at the level of 10^{-6}, such as CLEO-c, will start to confront models of new physics in an interesting way.

SUMMARY

The high-precision charm and quarkonium data will permit a broad suite of studies of weak and strong interaction physics. In the threshold charm sector measurements are uniquely clean and make possible the unambigous determinations of physical quantities discussed above. CLEO-c will utilize a variety of tools, namely J/ψ radiative decays, two-photon collisions (using almost real, as well as highly virtual space-like photons), deep inelastic Coulomb scattering and continuum production via $e^+ e^-$ annihilation to obtain significant new information on the spectrum of hadrons, both normal and exotic, and their decay channels. A quantitative improvement can be expected not only from the large accumulated statistics, but also from combining the results obtained using all these tools together with the results from the Υ resonance runs. The significance of this is better sensitivity, reduced systematics and a better chance to obtain a coherent picture of the hadron sector.

ACKNOWLEDGMENTS

I am delighted to acknowledge the invaluable contributions of many individuals to the development of the CLEO-c and CESR-c program and the outstanding contributions of my CLEO colleagues over the life of the experiment. The experimental aspects of this program are based on their effort and experience.

REFERENCES

1. Godfrey, S., and Rosner, J., *Phys. Rev. D*, **64**, 074011 (2001).
2. Butler, F., *Phys. Rev. D*, **49**, 40 (1994).
3. Kuang, Y.-P., and Yan, T.-M., *Phys. Rev. D*, **24**, 2874 (1981).
4. Godfrey, S., and Rosner, J., *Phys. Rev. D*, **64**, 097501 (2001).
5. Skwarnicki, T., "Heavy Quarkonia," in *Proceedings of Lepton-Photon 2003*, World Scientific, 2003.
6. BES Collaboration, J. B. e. a., *Phys. Rev. Lett.*, **76**, 3502 (1996).
7. Bigi, I., "Lifetimes of Heavy Hadrons," in *Proceedings of BCP3*, World Scientific, Singapore, 2000.
8. Swain, J., and Taylor, L., *Phys. Rev. D*, **58**, 093006 (1998).
9. Bigi, I., and Sanda, A., *Phys. Lett. B*, **171**, 320 (1986).
10. M. Gronau, Y. G., and Rosner, J., *Phys. Lett. B*, **508**, 37 (2001).
11. Bigi, I., and Sanda, A., *CP Violation*, Cambridge University Press, Cambridge, UK, 2000, p. 180.
12. S. Faijfer, P. S., S. Prelovsek, and Wyler, D., *Phys. Lett. B*, **487**, 81 (2000).
13. G. Burdman, J. H., E. Golowich, and Pakvasa, S., *Phys. Rev. D*, **52**, 6383 (1995).

Antiproton Physics at GSI

P. Gianotti

LNF - Via E. Fermi 40 - 00044 Frascati Italy

Abstract. A new facility for hadronic physics has been recently funded by the German Government. It is part of a major upgrade of the GSI accelerator complex presently running in Darmstadt. An intense, high momentum resolution antiproton beam, with momenta between 1.5 and 15 GeV/c, will be available at the High Energy Storage Ring (HESR), and the experimental activity will be carried out using a general purpose detector (PANDA), that will be built surrounding an internal target station installed at one of the two straight sections of the storage ring. The main characteristics of the new facility, the scientific program, and the design strategy of the PANDA detector are here presented.

INTRODUCTION

A major upgrade of the existing GSI accelerator facility in Darmstadt is planned and has been founded by the German Government. This is an ambitious project aiming to cover many aspects of modern experimental physics [1]:

- Nuclear structure physics: research with rare isotope beams to study nuclei far from the stability line.
- Nuclear matter physics: study of compressed and dense hadronic matter in nucleus-nucleus collisions.
- Plasma physics: high energy density matter produced using ion and laser beams.
- QED studies: ion-matter interactions in extremely strong electromagnetic fields.
- Antiproton physics: hadron spectroscopy studies to understand the hadron mass spectrum, and the fundamental properties of the strong interaction like the quark confinement, and the chiral symmetry breaking mechanism.

Experiments with antiprotons have demonstrated to be a rich source of high quality information for hadronic physics. With the new High Energy Storage Ring (HESR) for antiprotons at GSI, the physics of strange and charm quarks will be deeply explored. This is an energy region of transition between the perturbative regime and the low energy domain where phenomenological approaches are used to describe the strong interaction. With a high performance full solid angle magnetic spectrometer (the PANDA detector) some crucial points of this scientific field will be analyzed and hopefully clarified.

CP717, *Hadron Spectroscopy: Tenth International Conference,*
edited by E. Klempt, H. Koch, and H. Orth
© 2004 American Institute of Physics 0-7354-0197-7/04/$22.00

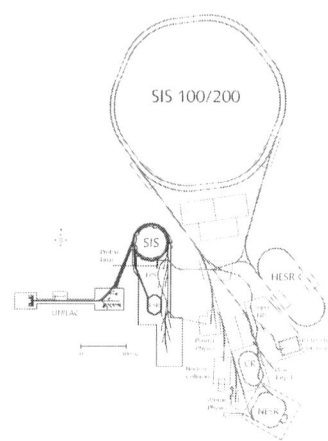

FIGURE 1. Layout of the present (blue) and future (red) GSI accelerator complex.

THE ANTIPROTON FACILITY AT GSI

The existing GSI accelerator complex will undergo a drastic change in the next years. The intensity and the energy of the ion beams available will be awfully increased and a new high quality antiproton beam will be provided. The heart of the new facility (see fig. 1) will consist of two separate synchrotron accelerator rings with 100 and 200 Tm maximum magnetic rigidity (SIS100/200). Both rings will have the same circumference, of about 1100 m, and will be installed in the same tunnel. The 30 GeV protons of the SIS100 could be used to produce antiprotons that will be subsequently collected, stored and cooled in two smaller storage rings. The pre-cooled antiprotons will be finally transfered, using the SIS100, into a dedicated High Energy Storage Ring (HESR) for the experimental activity. The HESR will provide a continuous antiproton beam of $5 \cdot 10^{10}$ particles. The momentum spread $\delta p/p$ will be 10^{-4} at the top machine luminosity $10^{32}\mathrm{cm}^{-2}\mathrm{s}^{-1}$, or 10^{-5} at $10^{31}\mathrm{cm}^{-2}\mathrm{s}^{-1}$. This high momentum resolution can be reached thanks to an high-energy electron cooler able to compensate the beam degradation introduced by antiproton target interactions and intra-beam scattering. The electron cooler and the target station will sit on the two opposite linear sections of the HESR machine.

THE ANTIPROTON EXPERIMENTAL PROGRAM

The physics of strong interaction is undoubtedly one of the most challenging areas of modern science. Quantum Chromodynamics (QCD) is reproducing the physics phenomena only at short distances, i.e. shorter than the size of the nucleon (10^{-15}m). Here perturbation theory is applicable yielding results of high precision and predictive power. At distance scales comparable to the nucleon radius, however, a different physics regime governs the phenomena and a different theoretical approach is required. This approach uses "effective" field theories which retain the fundamental symmetries of QCD, but uses baryons and mesons as the relevant degrees of freedom. The physics of strange and charmed quarks plays a crucial role in order to try to connect these two different energy domains. In fact, it interpolates between the limiting scales of QCD. The PANDA (antiProton ANnnihilation at DArmstadt) experiment, that will be installed on the High Energy Storage Ring, will have a scientific program devoted to these topics.

Experimentally, hadron structure can be investigated using different probes ($e^-, \pi, p, \bar{p},...$) each one with some specific advantage. Nevertheless, antiproton-proton annihilations copiously produce particles with gluon content, as well as particle-antiparticle pairs, allowing to access any quantum number for the final states. Therefore, HESR antiprotons are an excellent tool to perform spectroscopic studies of ordinary and exotic mesons in the energy range between 3 and 5 GeV. Furthermore, the use of antiprotons can address other open problems like in-medium modifications of hadrons properties, with the aim of checking the effects of chiral symmetry partial restoration on light mesons, and like the study of single and double hypernuclei. This field is a unique playground to get information on the hyperon-nucleon and hyperon-hyperon interactions. Finally, as soon as the luminosity of the HESR will reach the value of 10^{32} cm^{-2} s^{-1}, other more challenging topics could be accessed by the PANDA experiment: D-meson decay spectroscopy, the search for CP-violation signals in the charm and strange sector, the extraction of parton distribution functions from the inversed Deeply Virtual Compton scattering process.

In the following sub-sections the main topics of the PANDA scientific program will be illustrated.

Charmonium spectroscopy

Charmonium is considered the positronium spectrum of QCD. The spectrum (see fig. 2), including also spin and orbital effects, is well calculated, but the experimental knowledge in this sector has to be improved to verify the validity of the theoretical predictions. Recently, some new data are coming out from e^+e^- B-factories, but such experiments can directly access only 1^{--} states. Other quantum numbers can be obtained only by means of radiative transitions limiting the precision that can be achieved on the mass and width measurements. Furthermore, this reduces the cross-section of about 2 order of magnitude imposing long term stability problems and long dedicated data taking. On the contrary, $\bar{p}p$ annihilation proceeds via two or three gluons leading to the direct formation of all J^{PC} charmonium states. A second relevant advantage is the possibility to precisely tune the beam energy this directly translate to high resolution spectroscopy of the final states. LEAR and Fermilab experiments have successfully demonstrated these unique advantages, nevertheless a step forward will be reached at HESR by means of a high quality beam of higher energy and luminosity, together with a new hermetic detector with excellent detection capabilities both for charge and neutral particles.

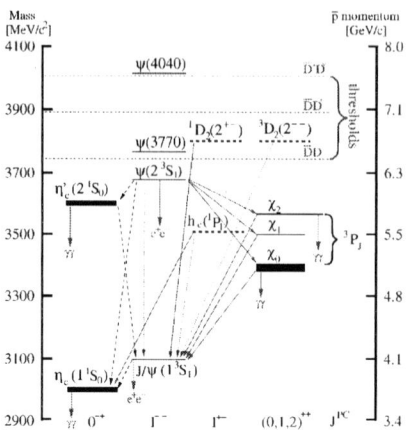

FIGURE 2. Spectrum of the charmonium states. Allowed decay modes are also shown.

Exotic states

The QCD spectrum is more rich than what is expected by the naive quark model. Extra states, containing explicitly gluons, must exist (hybrids and glueballs) together with molecular and multi-quark states. The additional degrees of freedom introduced by gluons allow hybrids and glueballs to have even spin-parity exotic quantum numbers. Hence exotic quantum numbers are an unambiguous signature to label gluonic hadrons. Two exotic states ($\pi_1(1440)$, $\pi_1(1600)$ [2, 3, 4]) with non conventional quantum numbers (1^{-+}) have been clearly seen in the data, and the best glueball candidates are probably the $f_0(1500)$ [6] and the $\eta_L(1440)$ [5]discovered in antiproton annihilations. The exotic states hunt has been carried on using different production techniques mainly in the light meson energy region, where a lot of ordinary resonances exist. This complicates the interpretation of the results because exotics tend to interfere with the overlapping conventional states.

From here the idea of moving in the charmonium energy region where the density of states, and therefore the interference, is smaller. Flux tube models, bag models, and LQCD agree that the mass of the lighter charmonium hybrid is laying between 3.9 and 4.5 GeV/c^2. They also attributed to it 1^{-+} quantum numbers. This mass value is above the $D\bar{D}$ threshold, and therefore this state is expected to decay into members of the D family. For this reason great attention in the D detection capability has been put into the PANDA detector design.

For what concern glueballs, LQCD in the quenched approximation [7] make rather detailed predictions for the mass spectrum (see fig. 3). These predictions gives for the lighter state (scalar 0^{++}) a mass value of 1500 MeV/c^2, in good agreement with the best experimental candidate $f_0(1500)$. On the contrary, for the pseudoscalar glueball (0^{-+}) a mass value above 2.5 GeV/c^2 is predicted, in disagreement with the best experimental candidate: $\eta_L(1440)$. An unambiguous way to check the model expectations could be that of looking to the so called "oddballs", glueballs with exotic quantum numbers.

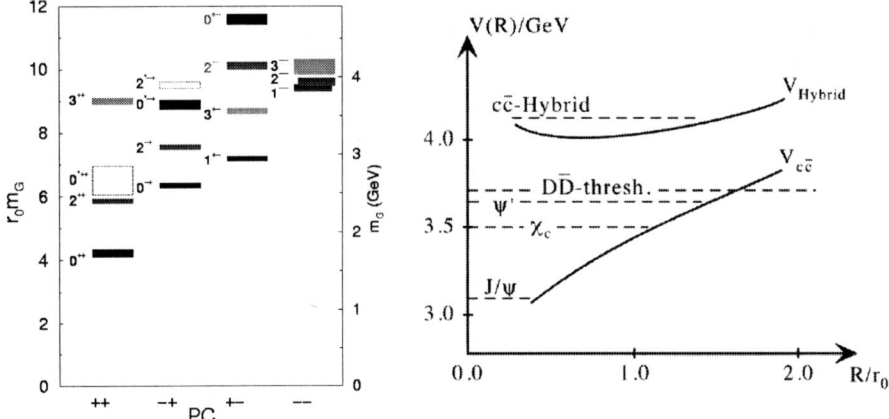

FIGURE 3. (left) LQCD glueballs spectrum (from ref[7]). (right) Potential between static quarks as a function of quark separation. $V_{c\bar{c}}$ represents the ground-state potential, V_{Hybrid} originates from the first excited state of gluonic flux giving rise to $c\bar{c}g$ hybrid states.

Oddballs cannot mix with normal mesons and consequently they are predicted to be rather narrow. Comparing oddball properties with those of non-exotic glueballs could be a good approach to understand glueball's structure. The lightest oddball, with $J^{PC} = 2^{+-}$, has a mass of 4.3 GeV/c^2.

The PANDA experimental program, for the exotic research, foresees two type of measurements: formation, and production. The former would generate states with non-exotic quantum numbers, while the second would yield an exotic particle together with an other meson (π, η,...) allowing to access even non-ordinary J^{PC}. This double possibility is a powerful experimental tool: the detection of a state in production and not in formation will be a clear, unique signature of its exotic behavior. The cross sections, expected for the exotic states are the same of ordinary mesons: ~ 100 pb for production experiments, ~ 1 μb for formation ones. In this kind of analysis the PANDA experiment will not have competitors. Experiments like BaBar or Belle will collect no more than 300 events for each hybrid or glueball state in around 5 years.

In medium modification of hadron properties

The investigation of hadron properties modification induced by nuclear matter is presently one of the main research activity at GSI. The goal of these studies is to understand the origin of hadron masses related to the spontaneous chiral symmetry breaking. This is done going into an environment, the nucleus, where the chiral symmetry is partially restored. Evidence for such an effect has been already seen for pions and kaons [8, 9], now the idea is to extend this research to the D meson family.

The J/ψ suppression in ultra-relativistic heavy-ion collisions is one of the parameters that indicate the formation of the quark-gluon plasma. Nevertheless, the J/ψ-N absorption cross section needs to be accurately measured in order to exclude other phenomena. The existing measurements give contradictory results hence the need of other precise data. The PANDA collaboration plans to measure J/ψ, Ψ', χ_J production in $\bar{p}$ annihilation on nucleons in nuclei on a series of nuclear targets.

Hypernuclear physics

The hypernuclear physics has a double valence. On one side it allows to study nuclear matter in presence of an explicit s quark, on the other it is a unique source of data on hyperon-nucleon interaction not accessible otherwise. Hypernuclear physics is not new, but in spite of its age it is experiencing a renewed interest thanks to the availability of better experimental conditions that are well suited to clarify some open problems of the field. As an example, at the Frascati ϕ-factory, the FINUDA experiment [10] is performing high precision measurements on some hypernuclei using the low energy kaon from the ϕ decay. This is the first time that an hypernuclear physics experiments is performed on a collider allowing to explore, at the same time, hypernuclear spectroscopy and hypernuclei decay modes. An other interesting topic of hypernuclear physics is the study of double hypernuclei. Up to now, only three candidates of double hypernuclei have been completely identified via their double pion decay, but if double hypernuclei could be produced at a reasonable rate, they could be a unique source of data on hyperon-hyperon interaction. Recent hypernuclear physics experiments, carried out at KEK and BNL, have demonstrated how powerful is the tool of performing spectroscopic studies using high resolution Ge γ-detectors. This allowed unprecedented measurements of the spin-orbit component of the hypernuclear levels. At PANDA, by slightly modifying the interaction-target region, this kind of measurements will be brought into the sector of double Λ hypernuclei and of Ω atoms.

The experimental idea foresees the production of baryons-antibaryons pairs inside a nuclear target. The antibaryon (e.g. a $\bar{\Xi}$) could be used as a trigger for the reaction, while the baryon (in this case Ξ^-) is slowed down and subsequently absorbed in a second active target. When a Ξ^- interacts with a proton, it produces two Λ with an energy release of only 28 MeV. Therefore, the probability that these two hyperons remain stuck to the nucleus is high. A new conception Ge-array detector will surround the second target allowing high precision spectroscopy of hypernuclear levels.

By using the same production technique other topics, like the possibility that the two Λs fuse to form a H-particle, could be explored.

THE PANDA EXPERIMENTAL APPARATUS

The experimental realization of the described physics program requires a dedicated high performance detector with a sophisticated trigger and a superior data acquisition system. In fact the expected rate of events will be $2 \cdot 10^7$ annihilations per second corresponding to a total $\bar{p}p$ cross-section of 100 mb. Simultaneously it must be capable to trigger on events with nb cross-section involving e^+e^-, $\mu^+\mu^-$, $\gamma\gamma$ and $\phi\phi$ pairs. In order to be able

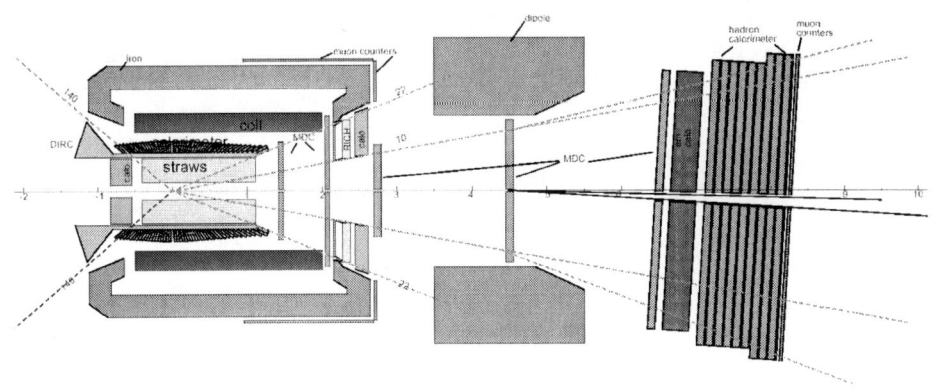

FIGURE 4. Side view of the proposed PANDA detector.

to efficiently detect D mesons excellent tracking capabilities in the vertex region, and a well-defined interaction point are also needed. On the other hand, to study in medium modifications of charmonium states, dilepton spectroscopy with mass resolution better than 1% is required. Finally the detector must be modular to allow element substitution for dedicated data taking (e.g. hypernuclear physics).

The detailed detector design is in progress, fig. 4 shows a first sketch of the PANDA apparatus. It consists of two distinct elements: a Target Spectrometer (TS) and a Foreword Spectrometer (FS). The TS surrounds the target station and consists of a radiation hard silicon pixel detector, for vertex reconstruction, a tracking system made of straw tubes and mini-drift chambers, for charge particle trajectory reconstructions, some ring imaging Cherenkov counters, using different radiator materials, to allow particle identification. The proposed electromagnetic calorimeter, for neutral particle detection, will consist of $PBWO_4$ crystals read-out by avalanche photo diodes. A superconducting solenoid of 2 T encloses the whole assembly and finally a muon detector is placed outside the iron yoke. The FS is placed downstream the interaction region after a dipole magnet. It houses micro drift chambers for the tracking, an electromagnetic and hadronic calorimeter, and a muon counter.

The international PANDA collaboration, 40 institutions of 9 different countries, is presently involved in the detailed drawing up of the scientific program and in the detector design. Following GSI future facility staging plan, PANDA will start the data taking during 2011.

CONCLUSIONS

After seven years from the LEAR accelerator shutdown, a new challenging project involving antiprotons is officially started in Europe. The characteristics of the new beam, together with the better performance of the detector involved, will determine a step forward in the hadronic physics studies allowing to clarify and to continue the analyses of many unclear hadronic structures suddenly interrupted by the unavailability of antiproton beams.

REFERENCES

1. "GSI Future Project, Conceptual Design Report", *see www.gsi.de/GSI_Future/cdr*.
2. for a recent review see H. Koch, LEAP03 conference proceeding, in press on *Nucl. Instr. Meth B.*
3. E. Abele *et al.*, *Phys. Lett.* **B446** (1999) 349; J. Reinhart, *Nucl. Phys.* **A692** (2001) 268c.
4. G. D. Adams *et al.*, *Phys. Rev. Lett.* **81** (1998) 5760; E. I. Ivanov *et al.*, *Phys. Rev. Lett.* **86** (2001) 3977.
5. F. Nichitiu *et al.*, *Phys. Lett.* **B 545** (2002) 261; C.Cicalò *et al.*, *Phys. Lett.* **B 462** (1999) 453 (*and ref. therein*).
6. A. Abele *et al.*, *Euro Phys. Jour.* **C 19** (2001) 667; D. barberis *et al.*, *Phys. Lett.* **B 471** (2000) 429.
7. C. J. Morningstar,M. Peardon, *Phys. Rev.* **D 60** (1999) 034509.
8. A. Gillitzer, *These proceedings*
9. S. H. Lee, *These proceedings*
10. M. Agnello *et al.*, Proceedng of the VIII Conference on Hypernuclear and Strange Particle Physics, in press on *Nucl. Phys.A.*

Short- and long-distance QCD effects in *B*-meson decays

Th. Feldmann

CERN Theory Division, CH-1211 Geneva 23, Switzerland

Abstract. Various exclusive and inclusive decays of *B* mesons are studied, at present, with dedicated experiments at "*B* factories". In order to compete with the experimental accuracy, we need a reliable theoretical framework to compute strong interaction effects in a hadronic environment. I discuss how the separation of (perturbatively calculable) short-distance QCD effects from (non-perturbative) long-distance phenomena helps to obtain precise theoretical predictions.

INTRODUCTION

The standard model (SM) for electroweak and strong interactions of elementary particles allows for precise theoretical predictions that are in remarkable agreement with present experimental results from high-energy experiments. However, an explanation of the SM and its parameters within a more fundamental theoretical framework remains an unresolved puzzle. Many questions in this respect (e.g. a grand unification of forces and matter, the origin of neutrino masses and its relation to the quark sector, the origin of *CP* violation, etc.) are related to the flavour sector of the SM, i.e. to the masses and coupling constants of the different quark and lepton species. *B*-meson decays are particularly useful to explore the parameters of the flavour sector in the SM or possible extensions of it. The *b* quark has a relatively long lifetime, and provides many rare decay modes that have small branching ratios and are sensitive to details of the interactions at small distances. In this way one hopes to reveal indirect effects from physics beyond the SM even before the direct detection of new particles.

From the QCD point of view, *b* quarks are interesting because they are the heaviest quarks that build pronounced hadronic bound states. The fact that the *b*-quark mass ($m_b \simeq 5$ GeV) is large with respect to typical hadronic scales leads to new approximate symmetries that can be observed in the *b*-hadron spectrum and decays. Furthermore, the heavy-quark mass provides a scale at which the strong coupling constant is still small, and perturbative computations of short-distance effects are possible. The short-distance dynamics has to be separated from (non-perturbative) long-distance physics related to the intrinsic QCD scale Λ_{QCD}. Technically this is achieved by an operator product expansion and from an effective field theory approach, respectively. In this way, *B*-meson decays can also be used to test and improve our theoretical understanding of QCD in different dynamical regimes.

CP717, *Hadron Spectroscopy: Tenth International Conference,*
edited by E. Klempt, H. Koch, and H. Orth

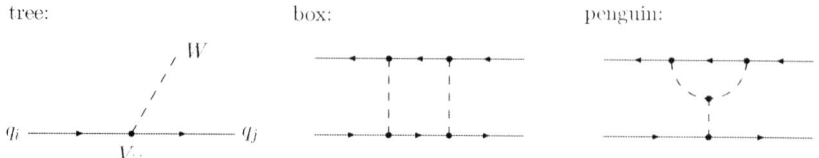

FIGURE 1. Weak flavour transitions in the standard model.

QCD AT THE ELECTROWEAK SCALE

In the SM, transitions between different quark flavours are mediated by charged W bosons. The relative strengths V_{ij} for flavour transitions $q_i \rightarrow q_j$ define the unitary Cabbibo–Kobayashi–Maskawa (CKM) matrix, which can be parametrized in terms of 3 real angles and one *CP*-violating phase. On the other hand, flavour-changing *neutral* currents (FCNCs) can only be induced via loop diagrams (box or penguin topologies, see Fig. 1). As a consequence, FCNCs are sensitive to the properties of virtual heavy particles in the loops, e.g. the top quark in the SM, or new particles in extensions of it.

Effective Hamiltonian

As in the case of the muon decay, we may "integrate out" the heavy particles (W, Z bosons, top quark, new physics) to arrive at an effective Hamiltonian (for a specific flavour transition), which has the schematic form

$$H_{\text{eff}} \propto G_F \sum_i C_i(\mu) \cdot \text{o}_i + \text{terms suppressed by } 1/m_W^2. \tag{1}$$

Here G_F is the Fermi constant, and $C_i(\mu)$ are effective coupling constants (Wilson coefficients) that encode the short-distance dynamics from physics *above* the scale $\mu = O(m_W)$. The dynamics of the remaining five quark flavours, leptons, gluons and photons at energy scales *below* μ is described by a set of operators o_i (four-fermion operators, and operators that couple fermions to the chromomagnetic or electromagnetic field strength). For more details, see the review in [1].

Matching

The Wilson coefficients can be calculated by "matching" scattering amplitudes in the SM and the effective theory. This requires computation of QCD (and QED) radiative corrections in both cases; see Fig. 2. Since the strong coupling constant is small at the matching scale, the Wilson coefficients have a perturbative expansion,

$$C_i(m_W) = c_i^{(0)} + \frac{\alpha_s(m_W)}{4\pi} c_i^{(1)} + \dots \tag{2}$$

full theory: effective theory:

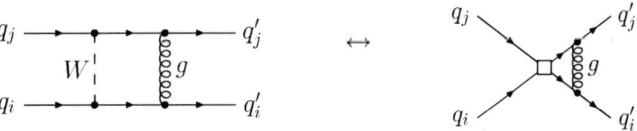

FIGURE 2. Example for radiative correction in the full and in the effective theory.

At present, the SM matching coefficients are known at order α_s^2 (i.e. at two loops). Different models for physics at and above the electroweak scale yield different matching coefficients:

$$c_i^{(n)} = c_i^{(n)}(m_t, m_W, \dots) + c_i^{(n)}(\text{new physics}) . \tag{3}$$

Therefore, experimental measurements of Wilson coefficients in weak decays test the SM and/or constrain new physics models.

Resummation of large logarithms

Radiative corrections to the matrix elements of the effective operators o_i in general would involve large logarithms $\ln m_b/\mu$, when calculated at a scale $\mu \sim m_W$. Since higher orders in perturbation theory also lead to higher powers of logarithms, the convergence of the perturbative series would be poor. This can be avoided by exploiting the fact that Wilson coefficients in the effective theory obey a renormalization-group equation, which allows the evolution of $C_i(m_W)$ to $C_i(m_b)$. At the low scale m_b, the large logarithms $(\ln m_b/m_W)^n$ are explicitly resummed in $C_i(m_b)$, and matrix elements of o_i only contain dynamics from energy scales below $\mu = m_b$. In order to derive the evolution equations, one has to calculate the anomalous-dimension matrix γ_{ij} that describes the scale dependence and mixing of operators in the effective theory. Currently, the calculation of three-loop anomalous dimensions is being completed (for a recent contribution, see [2]).

Example: $B \to X_s \gamma$. A prominent phenomenological example, where the theoretical machinery of the effective-Hamiltonian approach is relevant, is the inclusive rare radiative decay $B \to X_s \gamma$. It provides a stringent test of the contributions to the Wilson coefficient C_7, which is related to the effective $b \to s\gamma$ vertex. The comparison between experiment and theory reads

$$\text{Exp.:} \quad \text{BR}[B \to X_s \gamma] = (3.34 \pm 0.38) \times 10^{-4} \quad \text{[3]} \tag{4}$$

$$\text{SM:} \quad \text{BR}[B \to X_s \gamma] = (3.70 \pm 0.30) \times 10^{-4} \quad \text{[4, 5]} \tag{5}$$

The theoretical uncertainty is dominated by the renormalization-scheme dependence induced by the charm quark mass. The experimental and theoretical accuracy is already sufficient to put strong constraints on many new physics models (for a recent review on inclusive rare B decays, see [6]).

QCD AND HEAVY-QUARK EXPANSION

Because quarks are confined, experiments can only probe weak interactions in a hadronic environment. Technically, this amounts to considering hadronic matrix elements of the effective Hamiltonian (1). By construction, these matrix elements are sensitive to long-distance QCD dynamics, which is not accessible in perturbation theory. Nevertheless, some simplifications arise from the fact that the b quark mass is large compared to Λ_{QCD}. On the one hand, the strong coupling constant is small, $\alpha_s(m_b) \ll 1$, which implies that the dynamics at distances of order $1/m_b$ is still perturbative. On the other hand $\Lambda_{QCD}/m_b \ll 1$ provides a small expansion parameter, and in the heavy-quark limit ($m_b \to \infty$) the number of independent unknown hadronic quantities may be less than in the general case.

Heavy quark effective theory (HQET)

The above observations can be formalized in terms of an effective theory (HQET) where – to first approximation – heavy b and c quarks are replaced by static colour/flavour sources, moving with a fixed velocity v^μ,

$$h_v(x) = e^{im_Q v \cdot x} \frac{1 + \slashed{v}}{2} Q(x) . \tag{6}$$

The first few terms in the effective Lagrangian, resulting from integrating out "small" spinor components and "hard" quark and gluon modes (with virtualities of order m_Q^2), reads

$$L_{HQET} = \bar{h}_v \left\{ iv \cdot D + \frac{(i\vec{D})^2}{2m_Q} + C_m(\mu) \frac{g_s}{4m_Q} \sigma_{\mu\nu} G^{\mu\nu} + \dots \right\} h_v , \tag{7}$$

(where $v \cdot \vec{D} = 0$). The first term in this expansion is independent of the heavy-quark mass and diagonal in the heavy-quark spin. As a consequence, two new symmetries arise in the heavy-quark limit: heavy-flavour symmetry reflects the fact that the soft interactions of the b or c quark become the same, once the center-of-mass motion of the heavy quark is subtracted. Heavy-quark spin symmetry is related to the fact that soft interactions do not change the spin of the heavy quark. This has important implications for phenomenological observables such as the heavy hadron mass spectrum, or heavy meson transition form factors, to be discussed below.[1] The symmetries are broken by the subleading (kinetic and chromomagnetic) terms in the Lagrangian (7), as well as by perturbative matching coefficients for decay currents. For more details and references to the original literature, see for instance the review [8].

---

[1] HQET also applies to heavy-to-light decays as long as the energy transfer to the light quarks and gluons is small. In this kinematic regime, exclusive heavy-to-light processes are described by an effective low-energy Lagrangian for heavy and light mesons that combines HQET and chiral perturbation theory [7].

Example: Extracting $|V_{cb}|$ *and* $|V_{ub}|$. A well-known application of the heavy-quark mass expansion and HQET is the determination of the CKM elements $|V_{cb}|$ and $|V_{ub}|$ from $b \to c$ and $b \to u$ decays. At leading order in the $1/m_b$ expansion, the inclusive rates are just given by the partonic subprocess, which is calculable in perturbation theory (this has also been exploited in obtaining the theoretical prediction for $B \to X_s\gamma$ in (5)). Power corrections to the inclusive $B \to X$ rates start only at order $1/m_b^2$, and the corresponding non-perturbative parameters can be fitted to moments of experimentally measured inclusive decay spectra. As a result, one extracts [9]

$$
\begin{aligned}
|V_{cb}|_{\text{incl.}} &= 0.0421 \pm 0.0013 \,, \\
|V_{ub}|_{\text{incl.}} &= 0.00426 \pm 0.00013 \pm 0.00050 \,.
\end{aligned}
\tag{8}
$$

The theoretical accuracy is limited by the control on higher-order corrections in perturbation theory and in the heavy-quark expansion. For the extraction of $|V_{ub}|$ a subtle issue is how to separate the $b \to u\ell\nu$ spectrum from the $b \to c\ell\nu$ background.

Soft-collinear effective theory (SCET)

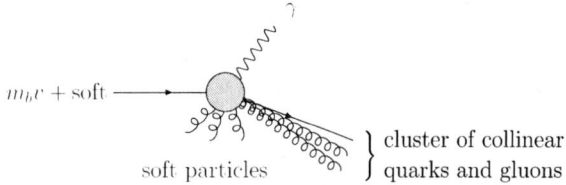

$m_b v + \text{soft}$

soft particles

$\left.\begin{array}{l}\end{array}\right\}$ cluster of collinear quarks and gluons

FIGURE 3. Kinematics for the decay $B \to X\gamma$ at large recoil energy.

The heavy-quark expansion can also be systematically applied to cases where the heavy quark decays into *energetic* light quarks and/or gluons, see Fig. 3. Examples are exclusive $B \to \pi\pi$ decays [10], or the endpoint spectrum in inclusive $B \to X_s\gamma$ decay [11]. In addition to the HQET field (6) and "soft" (i.e. low-energetic) light quark and gluon fields, the effective theory (SCET) for these cases includes "collinear" quark and gluon modes whose energy is proportional to the heavy-quark mass. In particular, collinear quarks in SCET are described in terms of "good" light-cone components ξ, in terms of which the Lagrangian that describes collinear quarks interacting with collinear gluons reads

$$
\mathcal{L}_{\text{coll}} = \bar{\xi} \left\{ in_- \cdot D + (i\slashed{D}_\perp - m) \frac{1}{in_+ \cdot D} (i\slashed{D}_\perp + m) \right\} \frac{\slashed{n}_+}{2} \xi \,.
\tag{9}
$$

Here n_-^μ is a light-like vector, which is determined by the jet axis or the momentum of the outgoing hadron(s), and n_+^μ is another light-like vector with $(n_+ \cdot n_-) = 2$. Contributions to heavy-to-light decays from the "bad" spinor components of collinear quarks, as well as from interactions between soft and collinear particles are suppressed by either $1/m_b$ or by α_s (see e.g. [12, 13]). In this way SCET provides an elegant alternative to understand factorization theorems for various QCD processes [14]. Although the physical motivation for separating short- and long-distance physics in HQET and SCET is similar, there are important differences to be noted:

- SCET gives no constraints for the light hadron spectrum.
- Interactions between energetic particles in the final state with soft spectators in the B meson are mediated by particles with virtualities of order $\mu^2 = m_b \Lambda_{QCD}$, which corresponds to a second short-distance scale in addition to m_b^2.
- The effective Lagrangian and currents in SCET are non-local.

Example: Forward–backward asymmetry zero in $B \to K^* \ell^+ \ell^-$. The suppression of the "bad" spinor components for light quarks in SCET and heavy quarks in HQET leads to the reduction of independent form factors for heavy-to-light transitions at large recoil [15]. As a consequence, to first approximation, the position of the forward–backward asymmetry zero in the decay spectrum of the exclusive $B \to K^* \ell^+ \ell^-$ can be predicted in a model-independent way [16, 17] in terms of the SM Wilson coefficients C_7 and C_9 for $b \to s\gamma$ and $b \to s\ell^+\ell^-$ transitions. First-order radiative corrections have been calculated in [18] and lead to the estimate $q_0^2 = 4.2 \pm 0.6$, where q_0^2 refers to the invariant mass of the lepton pair at which the forward–backward asymmetry vanishes.

QCD AND HADRONIC EFFECTS

After separating and calculating the short-distance QCD effects, we are left with matrix elements that encode the long-distance dynamics of quarks and gluons within hadrons. Since QCD perturbation theory cannot be applied in this case, we need alternative strategies to determine these hadronic parameters and obtain theoretical predictions for physical observables:

- extract hadronic parameters from one experimental observable, and insert them into the prediction for another;
- estimate them from non-perturbative calculations within lattice QCD;
- estimate them using QCD (light-cone) sum-rule techniques;
- tune phenomenological models to a subset of hadronic observables, and estimate other observables within that model.

In all these cases, the ultimate challenge is to give a reliable prediction of the systematic uncertainties. For the first option this apparently is a straightforward task, but only practicable if there are enough independent processes where one and the same hadronic quantity enters (which is the case, for instance, for parton distribution functions). In the lattice approach one has to deal with several (simultaneous) extrapolations: from finite lattice spacing to the continuum limit ($a \to 0$), from the "quenched approximation" ($n_f = 0$) to dynamical quarks, from quarks fitting on a finite lattice to realistic light- and heavy-quark masses. Here, progress is to be expected from improved lattice actions, increasing computer power and optimizing numerical algorithms, as well as from understanding the systematic effects for dynamical quark simulations better (see also C. Davies' talk at this conference). QCD sum rules are based on parton–hadron

TABLE 1. Estimates for B-meson decay constant and B_q^0–$\bar{B}_q^0$ mixing parameter from lattice QCD and QCD sum rules (numerical values taken from [20]).

	$f_{B_{u,d}}$	f_{B_s}	$\hat{B}_{B_d}$	B_{B_s}/B_{B_d}	$\xi = f_{B_s}\sqrt{\hat{B}_{B_s}}/f_{B_d}\sqrt{\hat{B}_{B_d}}$
Lattice QCD	203(27) MeV	238(31) MeV	1.34(12)	1.00(3)	$1.18(4)(^{12}_{0})$
QCD sum rules	208(27) MeV	242(29) MeV	1.67(23)	$\simeq 1$	

duality, which is applied to appropriate correlation functions, from which properties of the lowest contributing resonance can be extracted (for an introduction to QCD sum rules, see [19]). The suppression of contributions from higher resonances is achieved by a Borel transformation, and by relating the hadronic and the partonic representation of the spectral function above some threshold s_0. Among others, systematic uncertainties arise from varying the Borel parameter and the threshold parameter within a "stability window", and from power corrections parametrized in terms of quark and gluon condensates. Finally, concerning model estimates, the only way to quantify systematic uncertainties often is to compare the predictions of sufficiently many different models.

In the following I give some examples of hadronic parameters relevant to B physics. More details on the numerical estimates for various quantities can be found in [20].

f_B and B_B. Important phenomenological quantities that characterize the B meson are the decay constant and the B_q^0–$\bar{B}_q^0$ mixing parameter. They are defined as

$$\langle 0|\bar{q}\,\gamma_\mu\gamma_5\,b|B(p)\rangle = if_{B_q}\,p_\mu\,; \qquad \langle\bar{B}_q^0|_\circ\,(\Delta B = 2)|B_q^0\rangle\rangle = \frac{8}{3}\,B_{B_q}(\mu)\,f_{B_q}^2\,m_{B_q}^2\,, \qquad (10)$$

respectively, where $\circ$ $(\Delta B = 2)$ is a four-quark operator that induces $\bar{b}q \leftrightarrow \bar{q}b$. Numerical estimates for these quantities are summarized in Table 1 (where we considered the renormalization-group invariant quantity $\hat{B}_{B_q}$ for convenience). The ratio ξ defined in that table plays an important role in the analysis of the CKM triangle. The asymmetric error in the lattice estimate arises from the different ways of performing the chiral extrapolation to realistic light-quark masses, with or without including "chiral logs". We also remark that, in principle, B-meson and D-meson decay constants are related by HQET. However, at present, both the experimental accuracy in measuring f_D and the theoretical understanding of $1/m_c$ corrections are insufficient.

HQET parameters. In HQET the masses of the ground-state pseudoscalar (P) and vector (V) mesons to order $1/m_Q$ accuracy read

$$\begin{Bmatrix} m_P \\ m_V \end{Bmatrix} = m_Q + \bar{\Lambda}(m_q) + \frac{1}{2m_Q}\begin{Bmatrix} -\lambda_1(m_q) - 3\lambda_2(m_q) \\ -\lambda_1(m_q) + (1 - \delta_V(m_q))\lambda_2(m_q) \end{Bmatrix}. \qquad (11)$$

In the heavy-quark limit, pseudoscalar and vector meson belong to the same spin-symmetry multiplet characterized by a residual mass term $\bar{\Lambda}(m_q)$ that only depends on the flavour of the light degrees of freedom. The parameters λ_1 and λ_2 are related to the kinetic energy and chromomagnetic operator in (7), respectively.[2] The parameters in (11) can be determined from combing information from spectroscopy, lattice, QCD sum rules, and experimental data on inclusive B-meson decays (see [20] for a summary of quantitative results). Notice that the definition of the quantities on the right-hand side in (11) is renormalization-scheme and -scale dependent.[3]

Isgur–Wise form factor. In the heavy-quark limit, as a consequence of heavy-quark flavour and spin symmetry, all $B \to D^{(*)}$ form factors are given in terms of one Isgur–Wise function, which is normalized at zero recoil [30], e.g.

$$\langle D|\bar{h}_{v'}\Gamma h_v|B\rangle \propto \xi(v \cdot v')\,\mathrm{tr}\left[\frac{1+\not{v}'}{2}\Gamma\frac{1+\not{v}}{2}\right], \qquad \xi(1) = 1 . \tag{14}$$

Perturbative QCD corrections to $\xi(1)$ can be calculated in HQET. Non-perturbative corrections to (14) start only at order $1/m_Q^2$. The estimate thus obtained for the $B \to D^*$ form factor relevant to the analysis of $|V_{cb}|$ is [20]

$$h_{A1}^{B \to D^*}(1) = \eta_A\left[1 + \delta_{1/m^2} + \delta_{1/m^3} + \dots\right] = 0.91^{+0.03}_{-0.04} .$$

The knowledge of the form factor enables one to extract the matrix element $|V_{cb}|$ from exclusive decay modes. In practice, one also needs an estimate of the slope $\xi'(1)$ to extrapolate the experimental data to zero recoil. In this way one obtains $|V_{cb}|_{\mathrm{excl.}} =$

---

[2] We included a correction δ_V arising from the mixing between different vector mesons obtained from light degrees of freedom with spin-parity $j^P = 1/2^-$ and $j^P = 3/2^-$, respectively. The mixing angle between members of different spin-symmetry multiplets vanishes in the heavy-quark limit, but may be of the order of $10°$ for the two lightest axial-vector D mesons [21].

[3] The recent measurement of new excited D and D_s resonances [22] has renewed the interest in theoretical predictions for heavy-meson spectra [23]. The analogous formula to (11) for the lowest-lying heavy scalar (S), and axial-vector (A) states reads

$$\left\{\begin{array}{c} m_S \\ m_A \end{array}\right\} = m_Q + \bar{\Lambda}'(m_q) + \frac{1}{2m_Q}\left\{\begin{array}{c} -\lambda_1'(m_q) - 3\lambda_2'(m_q) \\ -\lambda_1'(m_q) + (1 - \delta_A(m_q))\lambda_2'(m_q) \end{array}\right\} . \tag{12}$$

In general, the HQET parameters for the two different (would-be) spin multiplets in (11) and (12) are independent of each other. Additional constraints are obtained if one Taylor-expands all quantities around the chirally symmetric limit (i.e. vanishing quark condensate, $\langle \bar{q}q \rangle \to 0$), and if one *assumes* that higher-order terms in that expansion are suppressed [24, 25]. In this case the mass splitting between chiral multiplets is expected to be $\bar{\Lambda}'(m_q) - \bar{\Lambda}(m_q) \propto |\langle \bar{q}q \rangle|$. The empirical values for this quantity are in fair agreement with phenomenological models based on spontaneous chiral symmetry breaking [26, 27, 28, 29]. Furthermore, neglecting contributions proportional to $\langle \bar{q}q \rangle$ or m_q in terms that are already suppressed by $1/m_Q$, one expects

$$\lambda_1(m_q) \simeq \lambda_1'(m_q) \equiv \lambda_1 , \quad \lambda_2(m_q) \simeq \lambda_2'(m_q) \equiv \lambda_2 , \quad \delta_V(m_q) \simeq \delta_A(m_q) \equiv \delta . \tag{13}$$

In other words, to first approximation $1/m_Q$ corrections should not depend on the flavour and parity of the light degrees of freedom, which appears to be in good agreement with experimental data. Nevertheless, one has to keep in mind that neither $1/m_c$ nor $\langle \bar{q}q \rangle$ are very small on hadronic scales, and higher-order terms in (11–13) may be non-negligible.

0.0402 ± 0.0020 [9], which is in good agreement with the number obtained from inclusive modes (8).

Heavy-to-light decays. In order to extract information on heavy-to-light decays from *exclusive* channels, one needs (among others) the corresponding transition form factors, for instance to obtain C_7 from $B \to K^*\gamma$ or $|V_{ub}|$ from $B \to \pi(\rho)\ell v$. In contrast to the heavy-to-heavy case, the values of heavy-to-light form factors themselves are not restricted by a simple symmetry principle. Often, one uses a simple parametrization, e.g. for the relevant form factor for $B \to \pi\ell v$ decays:

$$f_+^{B\to\pi}(q^2) = \frac{f_+^{B\to\pi}(0)}{(1 - q^2/m_{B^*}^2)(1 - \alpha q^2/m_{B^*}^2)} \,, \qquad (15)$$

where the first term in the denominator reflects the fact that for $q^2 \simeq m_b^2$ the form factor is dominated by the nearest vector-meson pole, and $f_+^{B\to\pi}(0) = 0.2\text{--}0.3$ and $\alpha = 0.3\text{--}0.6$ are fitted to lattice, QCD sum rules, or model estimates [20]. With increasing experimental data on the exclusive decay spectra, which would also help to test and refine different model estimates, significant improvement is expected for the determination of $|V_{ub}|$ from $B \to \pi\ell v$ and $B \to \rho\ell v$ decays. Currently, the result for $|V_{ub}|$ from exclusive decays is somewhat smaller than the result from the inclusive analysis (but compatible within the rather large uncertainties) [9].

Other exclusive B decays into light mesons require additional non-perturbative input, as soon as radiative QCD corrections are considered. The first systematic treatment of this issue has been discussed in [10] for the case of non-leptonic $B \to \pi\pi$ decays. It has been shown that (to leading order in the $1/m_b$ expansion and including first-order α_s corrections) the decay amplitude can be factorized into perturbative short-distance coefficient functions and non-perturbative matrix elements that define heavy-to-light transition form factors, or light-cone distribution amplitudes for B mesons and pions. Comparison with experimental data from $B \to \pi\pi$ and $B \to \pi K$ decays gives additional constraints on the CKM triangle (with some uncertainties coming from the parametrization of the B-meson distribution amplitude and from estimating $1/m_b$ corrections).

SUMMARY

The different energy scales relevant to B-meson decays imply that different dynamical aspects of QCD are probed: at the electroweak scale amplitudes are sensitive to the parameters that describe flavour transitions in the Standard Model or its possible extensions. Perturbative QCD corrections are included by matching the full theory onto an effective Hamiltonian and evolving the corresponding Wilson coefficients to scales of the order of the heavy-quark mass. At this scale, strong interactions are still perturbative. Furthermore, hadronic matrix elements can be expanded in inverse powers of the heavy-quark mass. This is formally described in terms of heavy quark effective theory (for b decays into charm quarks or low-energetic light quarks and gluons), of soft-collinear effective theory (for decays into energetic light quarks and gluons). Finally, the non-perturbative dynamics related to the intrinsic QCD scale is described in terms of hadronic parameters that have to be extracted from experiment or estimated from lattice, QCD sum rules or models. Reducing the theoretical uncertainties from each of the above dynamical regimes, and comparing the results with experimental data from *B* factories, enables us to test the flavour sector of the Standard Model, to find or constrain indirect new physics contributions, and to improve our understanding of perturbative and non-perturbative QCD.

ACKNOWLEDGMENTS

I would like to thank M. Nowak and A. Polosa for interesting discussions about the D_s resonances, and T. Hurth for a careful reading of the manuscript and helpful comments.

REFERENCES

1. Buchalla, G., Buras, A. J., and Lautenbacher, M. E., *Rev. Mod. Phys.*, **68**, 1125–1144 (1996).
2. Gambino, P., Gorbahn, M., and Haisch, U., hep-ph/0306079 (2003).
3. Jessop, C., SLAC-PUB-9610 (2002).
4. Gambino, P., and Misiak, M., *Nucl. Phys.*, **B611**, 338–366 (2001).
5. Buras, A. J., Czarnecki, A., Misiak, M., and Urban, J., *Nucl. Phys.*, **B631**, 219–238 (2002).
6. Hurth, T., hep-ph/0212304 (2003).
7. Wise, M. B., *Phys. Rev.*, **D45**, 2188–2191 (1992); Burdman, G., and Donoghue, J. F., *Phys. Lett.*, **B280**, 287–291 (1992); Yan, T.-M., et al., *Phys. Rev.*, **D46**, 1148–1164 (1992).
8. Neubert, M., *Phys. Rep.*, **245**, 259–396 (1994).
9. Schubert, K. R., talk presented at *Lepton Photon 2003*, Fermilab, 11-16 August 2003 [http://conferences.fnal.gov/lp2003/program/S6/schubert_s06.pdf].
10. Beneke, M., Buchalla, G., Neubert, M., and Sachrajda, C. T., *Phys. Rev. Lett.*, **83**, 1914–1917 (1999).
11. Bauer, C. W., Fleming, S., and Luke, M. E., *Phys. Rev.*, **D63**, 014006 (2001).
12. Bauer, C. W., Fleming, S., Pirjol, D., and Stewart, I. W., *Phys. Rev.*, **D63**, 114020 (2001).
13. Beneke, M., Chapovsky, A. P., Diehl, M., and Feldmann, T., *Nucl. Phys.*, **B643**, 431–476 (2002).
14. Bauer, C. W., et al., *Phys. Rev.*, **D66**, 014017 (2002).
15. Charles, J., Le Yaouanc, A., Oliver, L., Pène, O., and Raynal, J. C., *Phys. Rev.*, **D60**, 014001 (1999).
16. Burdman, G., *Phys. Rev.*, **D57**, 4254–4257 (1998).
17. Ali, A., Ball, P., Handoko, L. T., and Hiller, G., *Phys. Rev.*, **D61**, 074024 (2000).
18. Beneke, M., Feldmann, T., and Seidel, D., *Nucl. Phys.*, **B612**, 25–58 (2001).
19. Colangelo, P., and Khodjamirian, A., hep-ph/0010175 (2000).
20. Battaglia, M., et al., hep-ph/0304132 (2003).
21. Kilian, U., Körner, J. G., and Pirjol, D., *Phys. Lett.*, **B288**, 360–366 (1992); Di Pierro, M., and Eichten, E., *Phys. Rev.*, **D64**, 114004 (2001); Abe, K., et al., hep-ex/0307021 (2003); Cheng, H.-Y., hep-ph/0307168 (2003).
22. Aubert, B., et al., *Phys. Rev. Lett.*, **90**, 242001 (2003); Besson, D., et al., hep-ex/0305017 (2003); Abe, K., et al., hep-ex/0307041 (2003).
23. See also the numerous contributions to the proceedings of this conference.
24. Bardeen, W. A., Eichten, E. J., and Hill, C. T., hep-ph/0305049 (2003).
25. Nowak, M. A., Rho, M., and Zahed, I., hep-ph/0307102 (2003).
26. Nowak, M. A., Rho, M., and Zahed, I., *Phys. Rev.*, **D48**, 4370–4374 (1993).
27. Bardeen, W. A., and Hill, C. T., *Phys. Rev.*, **D49**, 409–425 (1994).
28. Ebert, D., et al., *Nucl. Phys.*, **B434**, 619–646 (1995), *Phys. Lett.*, **B388**, 154–160 (1996).
29. Deandrea, A., et al., *Phys. Rev.*, **D58**, 034004 (1998), and hep-ph/0307069.
30. Isgur, N., and Wise, M. B., *Phys. Lett.*, **B232**, 113 (1989) and **B237**, 527 (1990).

Observation of the $\eta'_c(2^1S_0)$ Resonance in Photon-Photon Fusion at CLEO

Kamal K. Seth

Northwestern University, Evanston, IL 60208, USA
(CLEO Collaboration)

Abstract. We report on the observation of $\eta'_c(2^1S_0)$, the radial excitation of $\eta_c(1^1S_0)$, the ground state of charmonium in the two-photon fusion reaction $\gamma\gamma \to \eta'_c \to K^0_S K^{\pm}\pi^{\mp}$ in 13.4 fb^{-1} of CLEO II/II.V data and 9.2 fb^{-1} of CLEO III data. The data have been analyzed to extract the η'_c resonance parameters. The results from recent observations of η'_c by Belle and BaBar are also presented.

INTRODUCTION

After nearly 75 years of purely phenomenological study of the strong interaction, we now have a fundamental theory of it in Quantum Chromodynamics, QCD. Although the QCD interaction between quarks and gluons is flavor-independent, it is best studied in heavy quark (c, b) spectroscopy, because the quarks are not too relativistic and the strong coupling constant is small enough to allow use of perturbative methods. The study is further facilitated because, in contrast to light quark hadrons, $c\bar{c}$ and $b\bar{b}$ states are narrow and widely separated.

The central part of the $q\bar{q}$ interaction has been successfully modeled in terms of a Coulombic potential, $\propto 1/r$, and a confinement potential, $\propto r$. However, the spin-dependence of the $q\bar{q}$ potential is not well understood. In particular, the spin-spin hyperfine interaction, which results into the splitting between the spin-singlet (^1L_J) and spin-triplet (^3L_J) states of a $q\bar{q}$ meson, is very poorly known, primarily because the experimental identification of spin-singlet states is very difficult. No spin singlet states have been identified in $b\bar{b}$ bottomonium, and h_c, the singlet P wave ($^1P_1$) state has not been firmly identified in charmonium. The only singlet state of quarkonium firmly identified before now is the ground state of charmonium, $\eta_c(1^1S_0)$, and the only hyperfine splitting we know is:

$$M(J/\psi, 1^3S_1) - M(\eta_c, 1^1S_0) = 115 \pm 2 \text{ MeV}.$$

This splitting is entirely attributed to the vector Coulombic interaction in which the $\vec{s}_1 \cdot \vec{s}_2$ interaction is a contact interaction, i.e., finite only for $L = 0$. This is perhaps all right because the $1S$ states, J/ψ and η_c are bound at a radius, $r \approx 0.4f$ where the $q\bar{q}$ potential is dominated by its Coulombic part. Nothing is really known about the spin-spin contribution of the confinement potential, which would be sampled by higher lying $2S(\psi', \eta'_c)$ and $1P(h_c, \chi_J)$ states, which reside at $r \approx 0.8f$. It is 'assumed' that the confinement potential is Lorentz scalar, and therefore makes no contribution to the $\vec{s}_1 \cdot \vec{s}_2$ interaction. The experimental evidence for this strong 'conjecture' is scant. Since

CP717, Hadron Spectroscopy: Tenth International Conference,
edited by E. Klempt, H. Koch, and H. Orth

h_c has not been firmly identified, the only hope of shedding light on the true nature of the long-range hyperfine interaction is through the identification of $\eta'_c(2^1S_0)$.

History of Searches of $\eta'_c(2^1S_0)$

1982: Edwards *et al.* (Crystal Ball, SLAC) [1] reported a ~ 90 MeV peak in the inclusive photon spectrum, $e^+e^- \to \psi' \to \gamma X$, and interpreted it as due to η'_c, with $M(\eta'_c) = 3594 \pm 5$ MeV, $\quad \Gamma(\eta_c) < 8.0$ MeV

1983-1995: No confirmation of the CB result followed, and PDG dropped η'_c from its charmonium states summary.

Non-Observations

Since 1995 there have been several unsuccessful attempts to identify η'_c. Only upper limits could be established for product branching ratios.

1995: Fermilab E760 [2], $p\bar{p} \to (\eta'_c) \to \gamma\gamma$
$B(p\bar{p} \to \eta'_c) \times \Gamma(\eta'_c \to \gamma\gamma) \leq 0.7$ eV, (90% CL), $\quad M = 3584 - 3624$ MeV

1998: DELPHI(LEP) [3], $\gamma\gamma \to (\eta'_c) \to hadrons$
$R \equiv \frac{\Gamma(\gamma\gamma \to \eta'_c) \times B(\eta'_c \to hadrons)}{\Gamma(\gamma\gamma \to \eta_c) \times B(\eta_c \to hadrons)} \leq 0.34$, (90% CL), $\quad M = 3500 - 3800$ MeV

1999: L3(LEP) [4], $\gamma\gamma \to (\eta'_c) \to hadrons$
$R < 2/(6.9 \pm 2.7) = 0.21 - 0.48$, (95% CL), $\quad M = 3500 - 4000$ MeV

2001: Fermilab E835[5], $p\bar{p} \to (\eta'_c) \to \gamma\gamma$
$B(p\bar{p} \to \eta'_c) \times \Gamma(\eta'_c \to \gamma\gamma) \leq 0.7$ eV, (90% CL), $\quad M = 3575 - 3660$ MeV

Reported Observations

Recently, Belle[6,7], BaBar[8], and CLEO[9] have reported observations of η'_c.

2002: Belle[6], $B \to K(K_sK\pi)$, and Belle[7] $\Upsilon(4S) \to J/\psi X$

2003: BaBar[8], CLEO[9], both in $\gamma\gamma \to K_sK\pi$

We describe these results in the following.

Belle Results

η'_c **in B-decay**: Belle[6] studied 41.8 fb^{-1} of e^+e^- data at $\Upsilon(4s)$, $\sqrt{s} = 10.58$ GeV, with 45 million $B\bar{B}$ pairs. They selected $K^+(K_sK^{\pm}\pi^{\mp})$ events which could be attributed to B decays, and found B peaks in $M(K_sK\pi)$ bins containing η_c and η'_c. Their results are given in Table 1. We note that they find an unexpectedly large $M(\eta'_c)$. It is 60 MeV larger than that claimed by the Crystal Ball [1].

η'_c **in double $c\bar{c}$ production**: Belle[7] also studied 46.2 fb^{-1} of e^+e^- at $\sqrt{s} \approx 10.6$ GeV. They studied $e^+e^-(continuum) \to J/\psi X$, and found recoils against J/ψ for $X = \eta_c, D^{*+}, D^0, \chi_{c0}, \eta'_c$. A disquieting feature of the results presented in Table 2 is that $M(\eta'_c)$ is 32 MeV smaller than that found from B-decay (Table 1).

TABLE 1. Belle results from B-decays.

Peak	# Events	Mass(MeV)	Γ(MeV)	$\frac{Br(B \to K\eta_c') \times Br(\eta_c' \to K_s K\pi)}{Br(B \to K\eta_c) \times Br(\eta_c \to K_s K\pi)}$
$\eta_c(1^1S_0)$	104 ± 14	2979 ± 2	11 ± 11	
$\eta_c'(2^1S_0)$	39 ± 11	$3654 \pm 6 \pm 8$	15^{+24}_{-15} or < 55	$0.38 \pm 0.12 \pm 0.05$ (90% CL)

TABLE 2. Belle results for double charmonium production.

Peak	# Events	Mass(MeV)	Γ(MeV)	significance
$\eta_c(1^1S_0)$	67^{+13}_{-12}	$2962(13)$		$6.7(5.9)\sigma$
$\eta_c'(2^1S_0)$	42^{+15}_{-13}	$3622(12)$		3.4σ

The announcements by Belle[6,7] that they had identified $\eta_c'(2S)$, albeit at two different masses 3654/3622 MeV, rekindled our interest in trying to find η_c' in two photon fusion at CLEO, since we had earlier successfully indentified η_c[10] in the analysis of 13.4 fb^{-1} of CLEO II/II.V data at $\Upsilon(4S)$ in the reaction

$$e^+e^- \to e^+e^-(\gamma\gamma), \gamma\gamma \to \eta_c \to K_s K\pi$$

We have now extended our analysis of 13.4 fb^{-1} of CLEO II/II.V data to the η_c' region, and also analyzed 9.2 fb^{-1} of CLEO III data. Preliminary results were presented at several conferences[9]. These are the measurements we present here in some detail. For completeness, we also present the nearly contemporaneous BaBar results[8] for the same reaction.

Observation of $\eta_c'(2^1S_0)$ in two-photon fusion at CLEO

Originally, 13.4 fb^{-1} of CLEO II/II.V e^+e^- data taken at $\Upsilon(4S)$ and vicinity was examined for $e^+e^- \to e^+e^-(\gamma\gamma), \gamma\gamma \to \eta_c' \to K_s^0 K^\pm \pi^\mp$. The event selection required:
- Exactly four reconstructed charged tracks in the detector
- One $K_s \to \pi^+\pi^-$ candidate (standard K_s selection criteria)
- P_T, transverse momentum of the system ≤ 600 MeV
- Amount of neutral unmatched energy, ≤ 200 MeV
- $|cos\theta| < 0.91$ for charged π and K
- Charged particle identification using dE/dx and TOF information

The $M(K_s K\pi)$ data were fitted with two Breit Wigner resonances, and background curves of several different parametrizations. In addition to the strong signal for η_c, a clear signal was observed for η_c' (Fig. 1). Two dimensional analysis, fitting $M(K_s K\pi)$ and P_T distributions simultaneously, confirmed that the observed η_c' signal is of two-photon origin.

In order to obtain an independent confirmation of the η_c' observation, 9.2 fb^{-1} of CLEO III data was examined. The CLEO III detector contains a new Drift Chamber

TABLE 3. Results for η_c and η'_c as obtained from two-photon fusion data from CLEO II/II.V and CLEO III. All errors are statistical only.

Data Set	Events	Mass (MeV)	significance	Efficiency
CLEO II/II.V - η_c	287 ± 28	2984.7 ± 2.1	15.9σ	10.0%
CLEO III - η_c	203 ± 22	2982 ± 2	14.2σ	9.8%
CLEO II/II.V + III - η_c	490 ± 36	2983.3 ± 1.4		
CLEO II/II.V - η'_c	36^{+15}_{-11}	3642.7 ± 4.1	5.0σ	13.8%
CLEO III - η'_c	29 ± 8	3642.5 ± 3.6	5.7σ	11.6%
CLEO II/II.V + III - η'_c	65^{+17}_{-14}	3642.6 ± 2.7		

TABLE 4. Two-photon fusion results from BaBar[8].

Resonance	# Events	Mass(MeV)	Width (MeV)
η_c	1715 ± 70	$2983.3 \pm 1.2 \pm 1.8$	$33.3 \pm 2.5 \pm 0.8$
η'_c	86 ± 23	$3632.2 \pm 5.0 \pm 1.8$	$20 \pm 10 \pm 4$

and a RICH detector, and much superior particle identification (π/K separation) was achieved. Event selection criteria were essentially the same as those for CLEO II/II.V data, except for π/K identification by means of the RICH detector. The spectra and the results for CLEO II/II.V and CLEO III, which are mutually consistent, are shown in Fig. 1 and Table 3.

It is of interest to compare η_c and η'_c excitations. We do so in terms of the ratio:
$R(\eta'_c/\eta_c) \equiv (B(\gamma\gamma \to \eta'_c) \times B(\eta'_c \to K_s K\pi))/(B(\gamma\gamma \to \eta_c) \times B(\eta_c \to K_s K\pi))$.
The preliminary results are:
CLEO II : $R(\eta'_c/\eta_c) = 0.17^{+0.07}_{-0.06}$,
CLEO III : $R(\eta'_c/\eta_c) = 0.29 \pm 0.09$.

$\eta'_c(2^1S_0)$ in $\gamma\gamma$ fusion from BaBar

As mentioned earlier, BaBar has also made an observation of η'_c in 88 fb^{-1} of e^+e^- data, $e^+e^- \to e^+e^-(\gamma\gamma)$, $\gamma\gamma \to \eta'_c \to K_s K\pi$, taken at $\Upsilon(4S)$[8]. Their results are presented in Table 4.

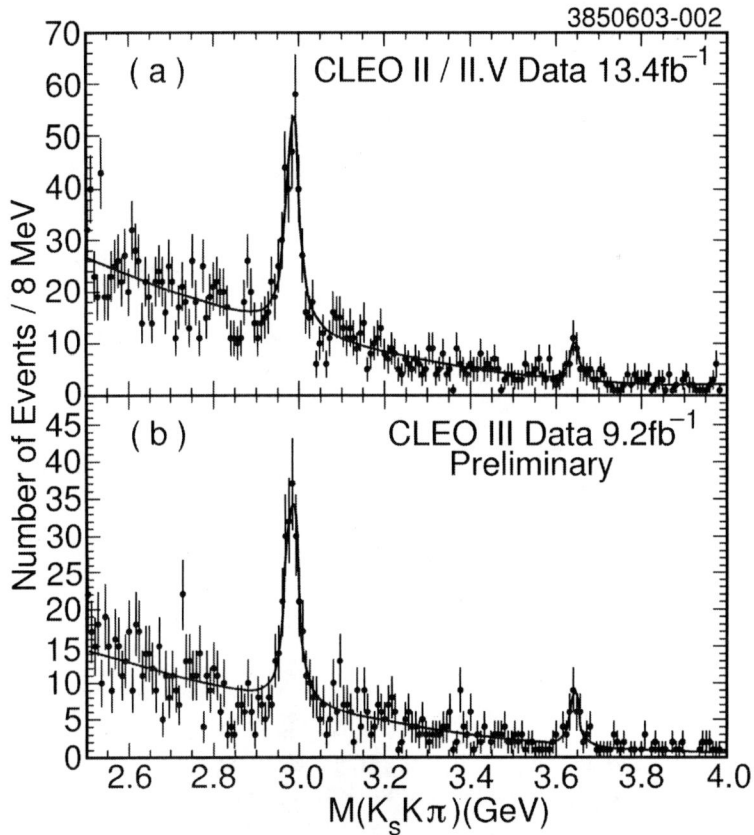

FIGURE 1. Invariant mass spectra for $M(K_s K\pi)$ from CLEO II/II.V and CLEO III two-photon fusion data. η_c and η_c' enhancements are clearly visible. The lines show the fits described in the text.

In a recent update, Belle has reported[11] new results from 101 fb^{-1} of data for $e^+e^- \rightarrow J/\psi X$ Their new results are presented in Table 5. We note that they now report $M(\eta_c)$ and $M(\eta_c')$ which are 8-10 MeV larger than those reported earlier (Table 2).

The present result for $M(\eta_c')$ and the earlier result from B-decay have been summarized by Seuster[12] to give Belle's weighted average as $M(\eta_c') = 3639 \pm 6$ MeV

TABLE 5.   Recent results from Belle for double charmonium production[11].

Peak	# Events	Mass(MeV)	significance	$\frac{\sigma(J/\psi+\eta_c')}{\sigma(J/\psi+\eta_c)}$
$\eta_c(1^1S_0)$	175 ± 28	2972 ± 7	9.9σ	
$\eta_c'(2^1S_0)$	108 ± 24	3630 ± 8	4.4σ	0.54 ± 0.15

TABLE 6.   Summary of η_c' results from CLEO, BaBar, and Belle.

	Mass, $M(\eta_c')$ (MeV)	Width, $\Gamma(\eta_c')$ (MeV)
CLEO[9]	3642.6 ± 2.7	10^{+22}_{-10} MeV < 46 MeV (90% CL)
BaBar[8]	3632.2 ± 5.0	$20 \pm 10 \pm 4$
Belle[6,7,11,12]	3639 ± 6	15^{+24}_{-15} MeV < 55 MeV (90% CL)

Summary of results for η_c'

From Table 6 it is clear that η_c' has been firmly identified with $< M(\eta_c') >= 3640.1 \pm 2.2$ MeV. This leads to the $2S$ hyperfine splitting $\Delta M_{hf}(2S) \equiv M(\psi') - M(\eta_c') = 45.9 \pm 2.3$ MeV. The width of η_c' has not been firmly established.

Theoretical predictions for ΔM_{hf}

It is instructive to compare the experimental hyperfine splittings for 1S and 2S charmonia with theoretical predictions for the same. The post-1978 theoretical predictions based on potential models and lattice calculations are listed in Table 7. In addition, we have the perturbative prediction:

$$\frac{M(\psi')-M(\eta_c')}{M(J/\psi)-M(\eta_c)} = \frac{|\psi_{2S}(0)|^2}{|\psi_{1S}(0)|^2} = \left(\frac{M(\psi')^2}{M(J/\psi)^2}\right)\frac{\Gamma(\psi' \to e^+e^-)}{\Gamma(J/\psi \to e^+e^-)},$$

which gives, $\Delta M_{hf}(2S) = 68 \pm 7$ MeV

We notice that the experimental value of $\Delta M_{hf}(2S)$ is considerably smaller than most of the theoretical predictions. No theoretical predictions for $\Gamma(\eta_c')$ are available.

548

TABLE 7. Theoretical predictions for hyperfine splittings of $1S$ and $2S$ states of charmonium. All predictions are based on potential model calculations, except the two marked with asterisks, which are based on quenched lattice calculations.

Year	Authors	$\Delta M_{hf}(1S)$	$\Delta M_{hf}(2S)$
1981	Eichten, Fienberg	115	83
1985	Godfrey, Isgur	130	60
1987	Olsson, Suchyta	97-136	35-41
1994	Gupta *et al.*	105-109	61-64
1994	Eichen, Quigg	117 (fix)	78(8)
1996	Chen, Oakes	106-146	54-93
2000	Ebert *et al.*	117 (fix)	103
* 1997	Bali *et al.*	124-192	94-106
* 2002	Okamoto *et al.*	54-85	25-43
2002	Lahde, Riska	102	38
2003	Badalian, Bakker	117 (fix)	57(8)
2003	Ebert *et al.*	117 (fix)	98
2003	EXPERIMENT	115 ± 2	46 ± 2

ACKNOWLEDGMENTS

We gratefully acknowledge the effort of the CESR staff in providing us with excellent luminosity and running conditions. This work was supported by the National Science Foundation, the U.S. Department of Energy, the Research Corporation, and the Texas Advanced Research Program. The author also wishes to thank D. Joffe for the preparation of this manuscript.

REFERENCES

1. C. Edwards *et al.*, *Phys. Rev. Lett.* **48** (1982) 70.
2. T. A. Armstrong *et al.*, E760 Collaboration, *Phys. Rev. D* **52** (1995) 4839.
3. P. Abreu *et al.*, DELPHI Collaboration, *Phys. Lett. B* **441** (1998) 479.
4. M. Acciari *et al.*, L3 Collaboration, *Phys. Lett. B* **461** (1999) 155.
5. M. Ambrogiani *et al.*, E835 Collaboration, *Phys. Rev. D* **64** (2001) 052003.
6. S. K. Choi *et al.*, Belle Collaboration, *Phys. Rev. Lett.* **89** (2002) 102001.
7. K. Abe *et al.*, Belle Collaboration, *Phys. Rev. Lett.* **89** (2002) 142001.
8. G. Wagner, BaBar Collaboration, QCD-Moriond (Les Arcs) 2003; hep-ex/0305083; J. Boyd, BaBar Collaboration, Photon-2003 (Frascati); E. Robutti, BaBar Collaboration, Photon-2003 (Frascati).
9. Z. Metreveli, CLEO Collaboration, APS/DNP (Philadelphia) 2003, *Bull. Amer. Phys. Soc.* **48 (2)** (2003) 12-17; D. Urner, CLEO Collaboration, CIPANP 2003 (New York) 2003; J. Ernst *et al.*, CLEO Collaboration, HEP 2003 Europhysics Conf. (Aachen) 2003, EPS-253; hep-ex/0306060.
10. G. Brandenburg *et al.*, *Phys. Rev. Lett.* **85** (2000) 3095.
11. T. Uglov, Belle Collaboration, HEP 2003 Europhysics Conf. (Aachen) 2003.
12. R. Seuster, Belle Collaboration, HEP 2003, Europhysics Conf. (Aachen) 2003.

Indication for Existence of Chiral-Axial Vector Meson in ($c\bar{n}$)-System

I.Yamauchi[1], M.Ishida[2], S.Ishida[3], T.Komada[4], K.Yamada[4]

1) Department of Mechanical Engineering, Tokyo Metropolitan College of Technology,
Tokyo 140-001, Japan
2) Department of Physics, Meisei University, Tokyo 191-8506, Japan
3) Research Institute of Quantum Science, College of Science and Technology,
Nihon University, Tokyo 101-0062, Japan
4) Department of Engineering Science, Junior College Funabashi Campus,
Nihon University, Funabashi 274-8501,Japan

Abstract. In the covariant level-classification scheme of hadrons proposed recently the existence of chiral scalar and axial-vector mesons, as respective chiral partners of the pseudo-scalar and vector mesons, were predicted in the ground states of Heavy-Light meson systems. The narrow resonances $Ds_J(2317)$ and $Ds_J(2463)$, observed recently in the final states $Ds \pi^0$ and $Ds*^+ \pi^0$ are to be assigned, respectively, as the scalar and axial-vector chiral particles in the ground-state ($c\bar{s}$) system. In this work we investigate the possibility of existence of the chiral axial-vector ($c\bar{n}$) meson, D_1^x, by re-analyzing the experimental data presented recently by the Belle Collaboration.

1. INTRODUCTION

The level classification of hadrons has been carried out successfully on the basis of the non-relativistic quark model. However, this non-relativistic (NR) scheme seems to be now confronted with a serious difficulty [1]: The existence of the light-mass iso-spinor scalar meson σ to be identified with the chiral partner of the π meson as a Nambu-Goldstone boson, has become widely accepted[2]. Subsequently, the possible existence of κ(900) meson, iso-spinor scalar meson, has been pointed out. Thus we are able to naturally identify the σ-meson nonet [2] with the members {σ(600), κ(900), $a_0(980)$, $f_0(980)$}. However, there are no appropriate seats for these members of the σ-nonet. Corresponding to this situation we have proposed a few years ago a new level-classification scheme, which has a relativistically covariant framework and is in conformity with the approximate chiral symmetry. In this new classification scheme "chiral states,,, which are out of the NR scheme, are expected to exist in the lower mass region, and the σ-nonet is naturally assigned as the ($q\bar{q}$) S-wave chiral state[3].

In the heavy-light (HL) quark meson system, it is expected that the approximate chiral symmetry of constituent light quark is valid. Correspondingly in the covariant classification scheme the chiral states[3,4], the scalar and axial-vector mesons, are predicted to exist, in addition to the conventional pseud-scalar and vector mesons, in the ground states.

CP717, Hadron Spectroscopy: Tenth International Conference,
edited by E. Klempt, H. Koch, and H. Orth
© 2004 American Institute of Physics 0-7354-0197-7/04/$22.00

The new narrow resonances Ds(2317) and Ds(2463), observed recently by BABAR Collaboration and CLEO Collaboration, successively, are naturally assigned as the chiral scalar and axial-vector mesons, respectively, in the ($c\bar{s}$) system [4].

A few years ago we had shown already [5] some indications for the chiral axial-vector D_1^x in the ($c\bar{n}$) system and the scalar B_0^x in the ($b\bar{n}$) system by reanalyzing the experimental data of CLEO and of L3 Collaboration, respectively [6]. In this work we investigate the possibility of existence of the chiral axial-vector ($c\bar{n}$) meson, D_1^x, by reanalyzing the experimental data [7] presented recently by the Belle Collaboration.

2. ANALYSIS OF D*π MASS SPECTRA IN SEARCH FOR D_1^x

2.1 Experimental data and project for reanalysis

As the experimental data to be reanalyzed, we choose the mass spectra of D*π system obtained through the process.

$$e^+ + e^- \rightarrow Y(4S) \rightarrow B\bar{B} \rightarrow D^*\pi + \cdots\cdots \qquad (2.1)$$

We consider the inclusive mass distribution of $D^*\pi$ events. As a description of the irrelevant background given by Belle Collaboration, which comes from the sideband MC simulation, we apply the formula

$$\text{B.G.} = \alpha(\Delta M)^\beta \times \exp(-\gamma_1(\Delta M) - \gamma_2(\Delta M)^2 - \gamma_3(\Delta M)^3)\ (\Delta M \equiv M(D^*\pi) - (m_{D^*} + m_\pi)) \quad (2.2)$$

with the fitting parameters α, β, γ_1, γ_2, γ_3.

The conventional three ($c\bar{n}$) P wave mesons may contribute, as the intermediate states, to the final $D^*\pi$ system as

$$D_1^* \rightarrow D^* + \pi \text{ (S-wave)} \qquad (2.3)$$
$$D_1 \rightarrow D^* + \pi \text{ (D-wave)} \qquad (2.4)$$
$$D_2^* \rightarrow D^* + \pi \text{ (D-wave)}, \qquad (2.5)$$

where D_1^*, D_1, D_2^* have, respectively, $^{jq}L_J = {}^{1/2}P_1, {}^{3/2}P_1$ and $^{3/2}P_2$ (here, $j_q = s_q + L$ is the total angular momentum of light quark). The respective partial wave states denoted in Eqs. (2.3)-(2.5) are deduced from the heavy quark symmetry (HQS).

In this work, the possible contribution from, in addition to the above conventional resonances, the chiral axial-vector meson D_1^x is taken into account as

$$D_1 \rightarrow D^* + \pi \text{ (S-wave)}, \qquad (2.6)$$

where we have inferred that the S-wave decay is dominant because of the small Q value.

Out of the three resonances in Eqs. (2.3)-(2.5), D_1 and D_2^*, decaying into the D-wave, have comparatively small widths ~20MeV, while D_1^*, decaying into the S-wave, has a large width, ~350MeV, reflecting the physical situation that all three resonances have small Q values and that the widths contain the kinematical factor $\Gamma \propto P^{2l+1}$(P being the relative momentum of the final system).

We apply the VMW method, where the absolute amplitude squared is given by

$$|M(s)| = |\Sigma\ r^i\ e^{i\theta i}\Delta_i(s)|^2 + \Sigma\ |r^j\ \Delta_j(s)|^2 + \text{B.G.}\ \ (\Delta_i(s) = -m_i\Gamma_i/(s - m_i^2 + im_i\Gamma_i)), \qquad (2.7)$$

here the summation with respect to the suffix i (j) is over the resonant particles (D_1^* and D_1^x (D_1 and D_2^*)) and $m_{i(j)}(\Gamma_{i(j)})$ represents their masses (widths), and $r_i(\vartheta_i)$ represents their production strength (phase). The relevant mass spectrum is given by

$$\Gamma(s) = \frac{1}{2\sqrt{s}} \int \frac{d^3 p_{D^*} \cdot d^3 p_\pi}{(2\pi)^3 2E_{D^*} \cdot (2\pi)^3 2E_\pi} \times (2\pi)^4 \delta^{(4)}(P - P_{D^*} - P_\pi)|M(s)|^2, \tag{2.8}$$

where $S = -(p_\mu^2)$, and p_μ is the total 4-momentum of the system.

In order to obtain the least-χ^2 solution, we restrict the values of our parameters, resonance mass and widths, as given in Table 1. The regions for D_1, D_2^* and D_1^* are determined from the center values and error of the respective resonances reported by Belle Collaboration.

TABLE. 1 Regions of the value of the mass and width(in MeV)

Resonances	D_1	D_2^*	D_1^*
Mass(m)	2419-2461	2456-2466	2350-2450
Width(Γ)	21-32	39-54	180-580

Regarding the chiral particle D_1^x, we have investigated the following three cases of its width, (1) narrow width case, (2) medium width case, and (3) broad width case. In case (1) no restrictions are imposed on the parameters. In case (2) the width of D_1^x is fixed to be $\Gamma(D_1^x)=150$MeV, while in case (3) is fixed as $\Gamma(D_1^x)=300$MeV. These two cases (2) and (3) were set up to see qualitatively the validity of the two theoretical predictions[4], $\Gamma'=133$MeV ($M(D_1^x)=2,250$MeV) and $\Gamma=530$MeV ($M(D_1^x)=2,360$MeV) corresponding to taking the mass splitting between chiral partners $\Delta M^x \equiv M(D_1^x) - M(D^*)=240$MeV and 350 MeV, respectively.

Here it is to be noted on the general situation of our analysis reflecting the low statistics of the data, as: Firstly the solution with least χ^2 gives $\Gamma(D_1^x)=22$MeV (case(1)) almost the same value as the bin-width of data, due to the large effect of one special point with $M(D^*\pi) \approx 2,25$ GeV. Secondly the contribution of D_1^x interferes with that of D_1^* with large width, and it is impossible to fix separately the parameters of two broad resonances by fitting of low statistics data. The cases (2) and (3) with the fixed values of $\Gamma(D_1^x)$ were set corresponding to this situation. We have carried out the χ^2–fitting of mass spectra, in all the above three cases both with and without D_1^x.

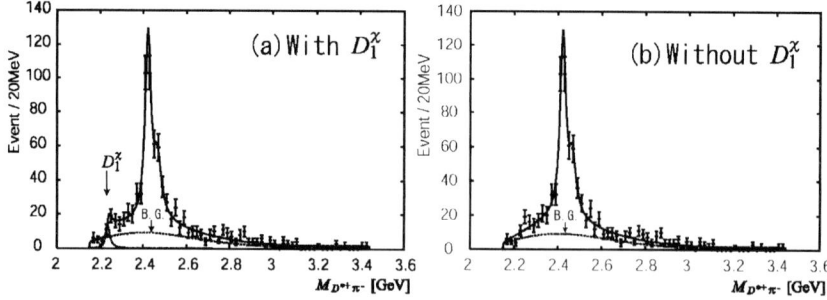

Fig. 1 Fitting in case (1) ;(a) with D_1^x and (b) without D_1^x

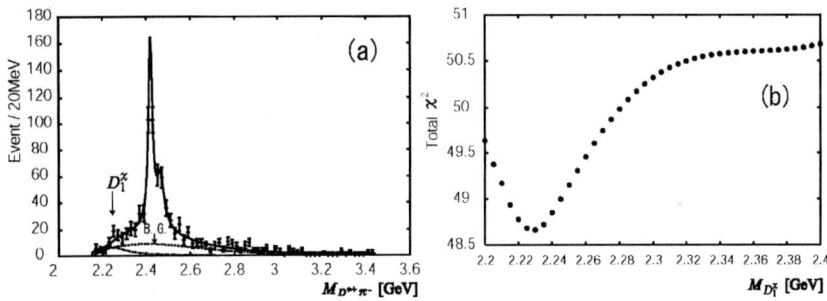

Fig.2 Results of analysis in case (2) with fixed $\Gamma(D_1^x)=150\text{MeV}$: (a) fitting with the mass of D_1^x $M(D_1) = 2,229$ MeV at the minimum χ^2 point; (b) Variation of χ^2 versus $M(D_1^x)$.

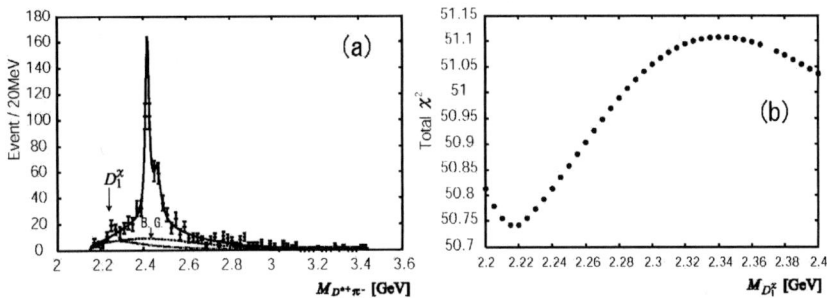

Fig.3 Results of analysis in case (3) with fixed $\Gamma(D_1^x)=300\text{MeV}$: (a) fitting with the mass of D_1^x $M(D_1^x) = 2,218$ MeV at the minimum χ^2 point; (b) Variation of χ^2 versus $M(D_1^x)$.

2.2 Result of analysis

The results of our analysis in case (1) both with and without D_1^x, are displayed in Fig.1 (a) and (b) respectively. Here first, we fit B.G. events given by Belle Collaboration. Then with this fixed values of B.G. we fit the experimental data on $D^*\pi$ mass spectra. In Fig.1 the respective contributions from D_1^x and B.G. are also shown.

Next we try to fit the experimental data with the same B.G. in the case(2) (case(3)) with the fixed value of $\Gamma(D_1^x)=150\text{MeV}$ (300MeV). In these cases of medium and broad width we have made the fitting as follows: First we investigate the variation of χ^2 versus $M(D_1^x)$. Then, with the value of $M(D_1^x)$ at this minimum χ^2 point, we perform our fitting. The results are given in Fig.2(a) and Fig3(a), respectively, for case (2) and case (3). The variation of χ^2 versus $M(D_1^x)$ are shown in Fig.2(b) and Fig.3(b), respectively, where we see the minimum χ^2 surely exist around the point $M(D_1) \approx 2,22$ GeV. The obtained values of mass and width of relevant resonances are collected in Table 2. In Table 3 the values of χ^2 and $\bar{\chi}^2$ ($=\chi^2/(\text{No. of data points} - \text{No. of parameters})$) in the respective cases of the fitting are given.

TABLE 2. Fitted value of mass and width of resonances (in MeV)

Case	D_1^x			D_1	D_2^*	D_1^*
	(1) Narrow-Width	(2) Medium-Width	(3) Broad-Width	In case (1)		
Mass(m)	2236	2229	2218	2433	2466	2350
Width( )	22.3	150	300	32	46.3	474

TABLE 3. Value of reduced χ^2

Case	With D_1^x			Without D_1^x
	(1)	(2)	(3)	
Reduced χ^2	46.4/(64-13) =0.911	48.7/(64-13) =0.954	50.7/(64-13) =0.995	54.91/(64-9) =0.998

CONCLUDING REMARKS

In this work we have investigated the possibility of existence of the chiral axial-vector ($c\bar{n}$) meson, D_1^x, by re-analyzing the experimental data presented by Belle Collaboration. In §2 we have carried out our reanalysis by applying the formulas Eqs. (2.7) and (2.8). There as the intermediate resonant particles, we have taken into account D_1^x, a new chiral axial-vector meson not included in the original analysis. As a result, we obtained some indication for existence of D_1^x with m≈2240MeV and Γ≈22 MeV. These results are similar to our previous results [5] obtained by reanalysis of the data by CLEO Collaboration [6]. However, the results seem to be affected largely from the one-special point, since of low-statistics of the data. The obtained values of the reduced $\bar{\chi}^2 = \chi^2/$(No. of data points-No. of param.) are $\bar{\chi}^2$=46.4/(64-13)=0.911 for the case with D_1^x and 54.9/(64-9)=0.998 for the case without D_1^x. The $\bar{\chi}^2$-values are not so much different between the cases with and without D_1^x. Accordingly we have investigated the possibility of broader D_1^x width cases. For the tentatively-fixed widths, $\Gamma(D_1^x)$=150 and 300 MeV, the result of minimum χ^2 search shows m(D_1^x)=2227 and 2218 MeV, respectively(*). These results of the broader width cases seem to support the prediction of the SU(3) LσM, m(D_1^x)=2250 MeV and $\Gamma(D_1^x)$=133 MeV[4]. However, the statistical accuracy of the data is very poor so it is necessary to have the more accurate data in order to obtain a definite conclusion.

ACKNOWLEDGMENTS

The authors would like to express their deep gratitude to Professors K.Takamatsu, T. Tsuru and K. Ukai for their encouragement.

REFERENCES

1. Proceedings on "Hadron Spectroscopy, Chiral Symmetry and Relativistic Description of Bound Systems,,, Nihon University, Tokyo, KEK Proceedings 2003-7.
2. Proceedings on Possible Existence of _-Meson and Its Implications to Hadron Physics, YITP, Kyoto, 2000("_-Meson 2000,,), Soryushiron Kenkyu (Kyoto) 102 (2001), E1;KEK-proceedings/2000-4.
3. M. Ishida and S. Ishida, in this proceedings.

(*)We have investigated the minimum-χ^2 search also in the case of $\Gamma(D_1^-)$=500 MeV. The result shows that there is no minimum points and the χ^2 decreases up to the m(D_1^-)=2400 MeV

4. S. Ishida, in this proceedings.
5. D. Ito et al., Prog. Theor. Phys. **108** (2002), 953.
6. CLEO Collaboration, P. Avery et al., Phys. Lett. **B331** (1994), 236.
    L3 Collaboration, S. Goldfarb et al., Phys. Lett. **B465** (1999), 323.
7. P. Krokovny, in this proceedings.
    K. Abe et al. (Belle Collaboration), in proc. of 31st International Conferences on High Energy Physics, Amsterdam, 24-31 July 2002 (ICHEP 2002).

The BABAR Resonance as a Four-quark Meson

K. Terasaki

Yukawa Institute for Theoretical Physics, Kyoto University, Kyoto 606-8502, JAPAN

Abstract.
A possible assignment of the new resonance observed at the B factories to a four-quark meson, $\hat{F}_I^+ \sim \{[cu][\bar{s}\bar{u}] - [cd][\bar{s}\bar{d}]\}$, that some of them can be observed as narrow resonances. Implication of existence of four-quark mesons in hadronic weak interactions is also discussed.

Recently a scalar $D_s^+\pi 0$ resonance with a mass $\simeq 2.32$ GeV and a width ~ 10 MeV has been observed at the B factories [1]. While it has been interpleted as an ordinary $c\bar{s}$ state such as the chiral partner of D_s^+ [5] or the excited $c\bar{s}$ state [6], their predicted mass values are still in controversy [7]. of iso-triplet scalar four-quark mesons in contrast with the above proposals. Four-quark mesons, $\{qq\bar{q}\bar{q}\}$, can be classified into four types [18],

$$\{qq\bar{q}\bar{q}\} = [qq][\bar{q}\bar{q}] \oplus (qq)(\bar{q}\bar{q}) \oplus \{[qq](\bar{q}\bar{q}) \pm (qq)[\bar{q}\bar{q}]\}, \tag{1}$$

where parentheses and square brackets denote symmetry and anti-symmetry, respectively. within the framework of $q = u$, d and s. It can be extended straightforwardly to the one including $q = u$, d, s and c [20]. In this talk, we concentrate on the $[cq][\bar{q}\bar{q}]$ mesons (with $q = u$, d, s). $\hat{F}_I^+ \sim \{[cu][\bar{s}\bar{u}] - [cd][\bar{s}\bar{d}]\}$ and estimate the masses of the lighter class of the $[cq][\bar{q}\bar{q}]$ mesons using the simple quark counting and taking $\Delta m_s = m_s - m_n \simeq 0.1$ GeV, $(n = u, d)$, at ~ 2 GeV scale and the measured $m_{\hat{F}_I} = 2.32$ GeV as the input data [24]. To estimate the mass values of the heavier class of $[cq][\bar{q}\bar{q}]$ mesons, we use $\Delta m_c = m_c - m_n \simeq 1.5$ GeV and $m_{\hat{\kappa}^*} \simeq 1.6$ GeV [18] as the input data. Their estimated mass values are listed in **TABLE 1**.

TABLE 1. Ideally mixed scalar $[cq][\bar{q}\bar{q}]$ mesons (with $q = u$, d, s), where S and I denote the strangeness and the isospin. ($\dagger$) : Input data

S	$I = 1$	$I = \frac{1}{2}$	$I = 0$	Mass(GeV)
1	$\hat{F}_I$		$\hat{F}_0$	2.32($\dagger$)
	$\hat{F}_I^*$		$\hat{F}_0^*$	(3.1)
0		$\hat{D}$		2.22
		$\hat{D}^*$		(3.0)
		$\hat{D}^s$		2.42
		$\hat{D}^{s*}$		(3.2)
-1		$\hat{E}0$		2.32
		$\hat{E}^{0*}$		(3.1)

CP717, *Hadron Spectroscopy: Tenth International Conference,*
edited by E. Klempt, H. Koch, and H. Orth

As seen in **TABLE 1**, the four-quark mesons with $*$ have large masses enough to decay into two vector mesons in addition to two pseudoscalar mesons, so that they will be broad. On the contrary, the estimated masses of $[cq][\bar{q}\bar{q}]$ without $*$ are near (or lower than) the thresholds of two body decays through I-spin conserving strong interactions, so that their phase space volumes are small even if kinematically allowed. Besides, it is seen [25], from the crossing matrices for color and spin in Ref. [18], that the wavefunction overlapping between the scalar $[qq][\bar{q}\bar{q}]$ meson and the two pseudoscalar $q\bar{q}$ meson state is small since the former is dominated by the $\bar{\mathbf{3}} \times \mathbf{3}$ of color $SU_c(3)$ and the $\mathbf{1} \times \mathbf{1}$ of spin $SU(2)$. Therefore, some of them can decay through I-spin conserving interactions but their rates will be small and they can be observed as narrow resonances such as the new resonance. Since some of them are not massive enough to decay into two pseudoscalar mesons through I-spin conserving interactions, their dominant decays may be I-spin non-conserving ones (unless their masses are higher than the expected ones). The $\hat{F}_I$ mesons form an iso-triplet, $\hat{F}_I^{++}$, $\hat{F}_I^+$ and $\hat{F}_I 0$, where the I-spin symmetry is always assumed in this talk unless we note. Then all of them can have the same type of kinematically allowed decays, $\hat{F}_I \to D_s^+ \pi$, with different charge states. The $\hat{D}$ and $\hat{D}^s$ form two independent iso-doublets. The $\hat{D}$ can decay into $D\pi$ final states and the kinematical condition is similar to the one in the decay, $\hat{F}_I \to D_s^+ \pi$, as long as the mass value of $\hat{D}$ in **TABLE 1** is taken. The dominant decay of $\hat{D}^s$ which contains an $s\bar{s}$ pair would be $\hat{D}^s \to D\eta^s \to D\eta$. Because of $m_{\hat{D}^s} \simeq m_D + m_\eta$, however, it is not clear if such a decay is allowed kinematically, as long as the value of $m_{\hat{D}^s}$ in **TABLE 1** is taken. Even if allowed, the rate would be much smaller than the ones for the above decays because of smaller phase space volume. The $\hat{F}_0^+$ is an iso-singlet counterpart of the $\hat{F}_I$ mesons. It cannot decay into the $D_s^+ \pi 0$ as long as the I-spin is conserved, so that it will decay dominantly through I-spin non-conserving interactions. In this case, the width of the $\hat{F}_0^+$ will be extremely narrow (much narrower than the $\hat{F}_I$ and $\hat{D}$ mesons). If its mass should be higher (by $\sim 50\,\mathrm{MeV}$ or more) because of some I-spin dependent force, it could decay dominantly into the DK final states (but could be still narrow.) The $\hat{E}0$ is an iso-singlet scalar meson with charm $C = 1$ and strangeness $S = -1$, i.e., $\hat{E}0 \sim [cs][\bar{u}\bar{d}]$. It cannot decay into $D\bar{K}$ final states unless it is massive enough. If its mass is almost the same as the $\hat{F}_0^+$, it cannot decay through strong interactions or electromagnetic interactions [27] as there are no ordinary mesons with $C = 1$ and $S = -1$. If it can be created, therefore, it will have a very long life. Now we study numerically decays of the $[cq][\bar{q}\bar{q}]$ mesons. Consider a decay, $A(\mathbf{p}) \to B(\mathbf{p}') + \pi(\mathbf{q})$,

$$\Gamma(A \to B\pi) = \frac{1}{2J_A + 1} \frac{q_c}{8\pi m_A 2} \sum_{spins} |M(A \to B\pi)|^2, \qquad (2)$$

where J_A, q_c and $M(A \to B\pi)$ denote the spin of A, the center-of-mass momentum of the final mesons and the decay amplitude, respectively. To calculate the amplitude, we here use the PCAC (partially conserved axial-vector current) hypothesis and a hard pion approximation in the infinite momentum frame (IMF), i.e., $\mathbf{p} \to \infty$ [29, 30]. In this approximation, the amplitude is evaluated at a slightly unphysical point, i.e., $m_\pi 2 \to 0$, and is given approximately by the asymptotic matrix element of A_π, $\langle B|A_{\bar\pi}|A\rangle$, as

$$M(A \to B\pi) \simeq \left(\frac{m_A 2 - m_B 2}{f_\pi}\right)\langle B|A_{\bar\pi}|A\rangle, \qquad (3)$$

where A_π is the axial counterpart of the *I*-spin. *Asymptotic matrix elements* (matrix elements taken between single hadron states with infinite momentum) of A_π can be parameterized by using *asymptotic flavor symmetry* (flavor symmetry of the asymptotic matrix elements). For asymptotic symmetry and its fruitful results, see Ref. [30] and references therein. We here list only the related asymptotic matrix elements [20],

$$\langle D_s^+|A_{\pi^-}|\hat{F}_I^{++}\rangle = \sqrt{2}\langle D_s^+|A_{\pi 0}|\hat{F}_I^+\rangle = \langle D_s^+|A_{\pi^+}|\hat{F}_I^0\rangle$$
$$= -\langle D0|A_{\pi^-}|\hat{D}^+\rangle = 2\langle D^+|A_{\pi 0}|\hat{D}^+\rangle$$
$$= -2\langle D0|A_{\pi 0}|\hat{D}^0\rangle = -\langle D^+|A_{\pi^+}|\hat{D}^0\rangle. \qquad (4)$$

$\Gamma(\hat{F}_I^+ \to D_s^+\pi 0) \sim 10$ MeV, as an example, since we do not find any other decays which can have large rates, and use it as the input data when we estimate the rates for the other decays. [However, the number ~ 10 MeV should not be taken too literally since it is still tentative, i.e., a possibility to take $\Gamma(\hat{F}_I^+ \to D_s^+\pi 0) \sim$ *a few* or *several* MeV is not excluded.] The results are listed in **TABLE 2**. All the calculated rates of the $\hat{F}_I$ and $\hat{D}$ mesons are lying in the region near the input data, so that they will be observed as narrow resonances in the $D_s^+\pi$ and $D\pi$ channels, respectively. As for the $\hat{D}^s \to D\eta$ decays, because of $m_{\hat{D}^s} \simeq m_D + m_\eta$, it is not clear if they are kinematically allowed. Besides, the decay is sensitive to the η η' mixing scheme which is still model dependent [31]. Therefore, we need more precise and reliable values of $m_{\hat{D}^s}$, η-η' mixing parameters and decay constants in the η-η' system to obtain a definite result.

TABLE 2. Dominant decays of scalar $[cq][\bar{q}\bar{q}]$ mesons and their estimated widths. $\Gamma(\hat{F}_I^+ \to D_s^+\pi 0) \sim 10$ MeV is used as the input data. The decays into the final states between angular brackets are not allowed kinematically as long as the parent mass values in the parentheses are taken.

Parent (Mass in GeV)	Final State	Width (MeV)
$\hat{F}_I^{++}(2.32)$	$D_s^+\pi^+$	
$\hat{F}_I^+(2.32)$	$D_s^+\pi 0$	~ 10
$\hat{F}_I 0(2.32)$	$D_s^+\pi^-$	
$\hat{D}^+(2.22)$	$D0\pi^+$	~ 10
	$D^+\pi 0$	~ 5
	$D^+\pi^-$	~ 10
	$D0\pi 0$	~ 5
$\hat{D}^s(2.42)$	$D\eta$	$-$
$\hat{F}_0^+(2.32)$	$D_s^+\pi 0$	$-$ (*I*-spin viol.)
$\hat{E}0(2.32)$		$-$

In summary we have studied the decays of the scalar $[cq][\bar{q}\bar{q}]$ mesons into two pseudoscalar mesons by assigning the new resonance to the $\hat{F}_I^+$ and assuming the I-spin conservation. All the allowed decays are not very far from the corresponding thresholds, so that their rates have been expected to saturate approximately their total widths. Therefore, we have used the measured width as the input data. the other models and to confirm it, therefore, it is important to observe these narrow resonances. Although we have not studied numerically, we can qualitatively expect that the $\hat{D}^s$ will be much narrower than the $\hat{F}_I$ and $\hat{D}$ mesons. The $\hat{F}_0^+$ decays through I-spin non-conserving interactions, so that it should be extremely narrow. The $\hat{E}0$ will decay through weak interactions if it is created as long as its mass is below the $\hat{E}0 \rightarrow D\bar{K}$ threshold. We have studied, so far, the strong decay properties of a group of the four-quark mesons. If their existence is confirmed, it will be very helpful in understanding of hadronic weak decays of K and charm mesons. The heavier class of $[qq][\bar{q}\bar{q}]$ and $(qq)(\bar{q}\bar{q})$ mesons can play an important role in hadronic weak decays of charm mesons, since the masses of some of the related members are expected to be close to the ones of the parent charm mesons. For example, the expected mass, $m_{\hat{\sigma}^{s*}} \simeq 1.8$ GeV, of the $\hat{\sigma}^{s*} \sim [us][\bar{s}\bar{u}] + [ds][\bar{s}\bar{d}]$ is close to the m_{D0} but the one, $m_{\hat{\sigma}^*} \simeq 1.45$ GeV, of the $\hat{\sigma}^* \sim [ud][\bar{d}\bar{u}]$ is much lower than the m_{D0} as seen in Ref. [18]. Therefore, the former can contribute to the intermediate state of the $D0 \rightarrow K^+K^-$ decay and enhance strongly the decay while the latter can contribute to the $D0 \rightarrow \pi^+\pi^-$ decay but cannot so strongly enhance it. In this way, we can obtain a solution to the long standing puzzle [32],

$$\frac{\Gamma(D0 \rightarrow K^+K^-)}{\Gamma(D0 \rightarrow \pi^+\pi^-)} \simeq 3, \tag{5}$$

consistently with the other two-body decays of charm mesons [20]. Confirmation of the existence of four-quark mesons is very much important not only in hadron spectroscopy but also in hadronic weak interactions of K and charm mesons.

ACKNOWLEDGMENTS

The author would like to thank Professor T. Onogi for providing information of the new resonance, discussions and encouragements, and Professor T. Kunihiro for discussions and encouragements.

REFERENCES

1. B. Aubert et al., the BABAR Collaboration, Phys. Rev. Lett. **90**, 242001 (2003); D. Besson et al., the CLEO Collaboration, hep-ex/0305017; K. Abe et al., the BELLE Collaboration, hep-ex/0307041.
2. B. Aubert et al., the BABAR Collaboration, Phys. Rev. Lett. **90**, 242001 (2003). hep-ex/0304021.
3. D. Besson et al., the CLEO Collaboration, hep-ex/0305017.
4. K. Abe et al., the BELLE Collaboration, hep-ex/0307041.
5. M. A. Nowak, M. Rho and I. Zahed, Phys. Rev. D **48**, 4370 (1993); W. A. Bardeen and C. T. Hill, Phys. Rev. D **49**, 409 (1994).
6. R. N. Cahn and J. D. Jackson, hep-ph/0305012.

7. For example, S. Godfrey, hep-ph/0305122; G. S. Bali, hep-ph/0305209; R. D. Kenway, C. M. Maynard and C. MacNeile, the UKQCD Collaboration, hep-lat/0307001; A. Deandra, G. Nardulli and A. D. Polosa, hep-ph/0307069; J. Hofmann and M. F. Lutz, hep-ph/0308263; Y. -B. Dai, C. -S. Huang, C. Liu and S. L. Zhu, hep-ph/0306274.

8. S. Godfrey, hep-ph/0305122.

9. G. S. Bali, hep-ph/0305209.

10. R. D. Kenway, C. M. Maynard and C. MacNeile, the UKQCD Collaboration, hep-lat/0307001.

11. A. Deandra, G. Nardulli and A. D. Polosa, hep-ph/0307069.

12. J. Hofmann and M. F. Lutz, hep-ph/0308263.

13. Y. -B. Dai, C. -S. Huang, C. Liu and S. L. Zhu, hep-ph/0306274.

14. T. Barnes, F. E. Close, and H. J. Lipkin, hep-ph/0305025.

15. A. P. Szczepaniak, hep-ph/0305060.

16. H.-Y. Chen and W.-S. Hou, hep-ph/0305038.

17. T. E. Browder, S. Pakvasa and A. A. Petrov, hep-ph/0307054.

18. R. L. Jaffe, Phys Rev. D**15**, 267 (1977); **15**, 281 (1977).

19. S. Hori, Prog. Theor. Phys. **36**, 131 (1966).

20. K. Terasaki and S. Oneda, Phys. Rev. D **38**, 132 (1988); **47**, 199 (1993); K. Terasaki, *ibid* **59**, 114001 (1999).

21. K. Terasaki and S. Oneda, Phys. Rev. D **38**, 132 (1988).

22. K. Terasaki and S. Oneda, Phys. Rev. D **47**, 199 (1993).

23. K. Terasaki, Phys, Rev. D **59**, 114001 (1999).

24. K. Terasaki, Phys. Rev. D **68**, 011501 (2003).

25. K. Terasaki, hep-ph/0309119.

26. S. Okubo, Phys. Lett. **5**,165 (1963); G. Zweig, CERN Report No. TH401 (1964); J. Iizuka, K. Okada and O. Shito, Prog. Theor. Phys. **35**,1061 (1965).

27. H. J. Lipkin, Phys. Lett. **B70**,113 (1977); M. Suzuki and S. F. Tuan, *ibid* **B133**, 125 (1983). H. J. Lipkin, Phys. Lett. **B70**,113 (1977).

28. M. Suzuki and S. F. Tuan, Phys. Lett. **B133**, 125 (1983).

29. K. Terasaki, S. Oneda and T. Tanuma, Phys. Rev. D **29**, 456 (1984).

30. S. Oneda and K. Terasaki, Prog. Theor. Phys. Suppl. **82**, 1 (1985).

31. T. Feldmann, Int. J. Mod. Phys. A **15**, 159 (2000).

32. Particle Data Group, K. Hagiwara et al, Phys. Rev. D **66**, 1001 (2002).

33. K. Terasaki, Int. J. Mod. Phys. A **16**, 1605 (2001) and references quoted therein.

560

Hard core attraction in hadron scattering and the family of the Ds meson molecule

Pedro Bicudo* and Gonçalo M. Marques*

*Dep. Física and CFIF, Instituto Superior Técnico, Av. Rovisco Pais 1049-001 Lisboa, Portugal

Abstract. We study the discovered Ds(2317) at BABAR, CLEO and BELLE, and find that it belongs to a class of strange multiquarks, which is equivalent to the class of kaonic molecules bound by hard core attraction. In this class of hadrons a kaon is trapped by a s-wave meson or baryon. To describe this class of multiquarks we apply the Resonating Group Method, and extract the hard core kaon-meson(baryon)interactions. We derive a criterion to classify the attractive channels. We find that the mesons f0(980), Ds(2457), Bs scalar and axial, and also the baryons with the quantum numbers of Λ, Ξ_c, Ξ_b and also Ω_{cc}, Ω_{cb} and Ω_{bb} belong to the new hadronic class of the Ds(2317).

Recently new narrow scalar resonances $D_s(2317)$ and $D_s(2457)$ were discovered at BABAR [1], CLEO [2] and BELLE [3]. These positive parity resonances [4] were also predicted by Nowak, Ro and Zahed [5, 6] as chiral partners of the well known negative parity $D_s(1968)$ and $D_s(2112)$. Here we find that the new D_s resonances can be undesrtood as tetraquarks, or equivalently as D-K molecules [7] bound by the hardcore attraction. These are not standard hadrons because they are neither quark-antiquark mesons nor three quark baryons. The experimental discovery of these hadrons also prompts us to study the new hadronic class which includes the $D_s(2317)$. In our framework the masses and couplings of hadrons are microscopically computed at the quark level and in a chiral invariant framework, [8, 9, 10].

In this talk we study the family of all possible narrow tetraquark and pentaquark resonances [11] where the quark s, or the Kaon play a crucial role. We start by reviewing the Resonating Group Method (RGM) [12]. The RGM, together with chiral symmetry, produces hard core hadron-hadron potentials, which can be either repulsive or attractive. We derive a criterion to discriminate which systems bind and which are unbound. We apply this criterion to find, among the s-wave hadrons, the candidates to trap a kaon. Finally we compute the binding energy of the selected hadrons, the positive parity mesons $f_0(980), D_s^{(0+)}, D_s^{(1+)}, B_s^{(0+)}, B_s^{(1+)}$ and the negative parity baryons $\Lambda(1405), \Xi_c^{(-)}, \Xi_b^{(-)}$ and also $\Omega_{cc}^{(-)}, \Omega_{cb}^{(-)}$ and $\Omega_{bb}^{(-)}$.

The RGM computes the effective multiquark energy using the matrix elements of the microscopic quark-quark interactions. The multiquark state is decomposed in anti-symmetrized combinations of simpler colour singlets, the baryons and mesons. The RGM was first used in hadronic physics by Ribeiro [13] to show that in exotic hadron-hadron scattering, the quark-quark potential together with the Pauli repulsion of quarks produces a repulsive short range interaction. For instance this explains the $N - N$ hard core repulsion, preventing nuclear collapse. Deus and Ribeiro [14] used the same RGM

CP717, *Hadron Spectroscopy: Tenth International Conference,*
edited by E. Klempt, H. Koch, and H. Orth
© 2004 American Institute of Physics 0-7354-0197-7/04/$22.00

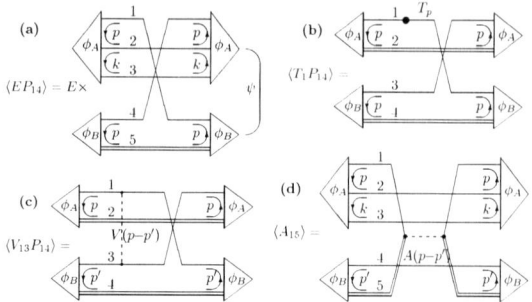

FIGURE 1. We show examples of RGM overlaps, in (a) the norm overlap for the meson-baryon interaction, in (b) a kinetic overlap the meson-meson interaction, in (c) an interaction overlap the meson-meson interaction, in (d) the annihilation overlap for the meson-baryon interaction.

to show that, in non-exotic channels, the quark-antiquark annihilation could produce a short core attraction. This is confirmed in several studies [8, 9, 10, 13, 14, 15].

The energy of the multiquark is computed with the matrix elements of the hamiltonian,

$$H = \sum_i T_i + \sum_{i<j} V_{ij} + \sum_{i\bar{j}} A_{i\bar{j}}, \tag{1}$$

which includes the quark(antiquark) kinetic energies, the quark(antiquark)-quark(antiquark) interaction proportional to the colour dependent $\frac{\vec{\lambda}_i}{2} \cdot \frac{\vec{\lambda}_j}{2}$, and the quark-antiquark annihilation. Once the internal energy of each cluster is accounted, and we also use the antisymmetry of the baryon wave-function, the remaining energy of the meson-meson or baryon-meson system is computed with the overlap,

$$
\begin{aligned}
_{mesB}^{barA} &= 3\langle \phi_B \, \phi_A | - (V_{14} + V_{15} + 2V_{24} + 2V_{25}) P_{14} \\
&\quad + A_{15} | \phi_A \phi_B \rangle \\
_{mesB}^{mesA} &= \langle \phi_B \, \phi_A | (1 + P_{AB}) [-(V_{13} + V_{23} + V_{14} + V_{24}) \\
&\quad \times P_{13} + A_{23} + A_{14}] | \phi_A \phi_B \rangle,
\end{aligned} \tag{2}
$$

where we use an obvious notation (see Fig. 1). The colour, spin and flavour contributions to the overlaps simply provide an algebraic factor. The momentum (or space) integrations, is estimate them with the variational method to produce a geometrical overlap,

$$\sum_{sfc} \int_p \phi_A^\dagger \phi_B^\dagger H \phi_B \phi_A \simeq M \sum_{sfc} \int_p \phi_A^\dagger \phi_B^\dagger \phi_B \phi_A, \tag{3}$$

where M is the sum of hadronic masses corresponding to the Hamiltonian H.

When spontaneous chiral symmetry breaking is included in the quark model [8, 16] the annihilation potential turns ou to be crucial [9, 10] to understand the low mass of the π. The annihilation potential A is also present in the π Salpeter equation where it cancels most of the kinetic energy and confining potential $2T + V$. From the quark model with chiral symmetry breaking we get the sum rules [8],

$$\langle 2T + V + A \rangle_{S=0} = M_\pi \simeq 0$$

$$\langle 2T + V \rangle_{S=0} = \frac{2}{3}(2M_N - M_\Delta)$$

$$\Rightarrow \langle A \rangle_{S=0} \simeq -\frac{2}{3}(2M_N - M_\Delta). \tag{4}$$

In eq. (4) the matrix elements are evaluated for s-wave and spin 0 wave-functions of light quarks. Similar sum rules also exist for spin 1 wave-functions, involving only the mass of the Δ baryon. The annihilation potential for the strange quark mass is smaller by a factor of σ which is a power of the constituent quark mass ratio $M_{u,d}/M_s$. From eqs. (4) it is clear that the annihilation potential provides an attractive (negative) overlap.

In what concerns the quark-quark(antiquark) potential, in the present case of s-wave states, with a $S = 0$ kaon, the only potential which may contribute is the hyperfine potential, proportional to $\frac{\vec{\lambda}_i}{2} \cdot \frac{\vec{\lambda}_j}{2} \vec{S}_i \cdot \vec{S}_j$. We find that the total matrix element is a hyperfine splitting,

$$\langle P_{13}(V_{13} + V_{23} + V_{14} + V_{24}) \rangle = 4\frac{2}{3}(M_\Delta - M_N) \tag{5}$$

and it is repulsive (positive).

These results are independent of the particular quark model that we choose to consider, providing it is chiral invariant. We therefore arrive at the criterion
- *whenever the two interacting hadrons have a common flavour, the repulsion is increased,*
- *when the two interacting hadrons have a matching quark and antiquark the attraction is enhanced .*

In this paper we are interested in the class of resonances which can be understood as a S=-1 kaon $s\bar{u}$ or $s\bar{d}$ trapped by a s-wave hadron. With our criterion we can exclude all hadrons with an antiquark $\bar{u}$ or $\bar{d}$ or with a quark s because the exchange overlap $\langle P_{13} \rangle$ would be allowed, and this certainly contributes to repulsion. We assume that the attraction is not sufficient to overcome this repulsion. In what concerns attraction we need a quark u or d in the s-wave partner of the kaon, in order to produce annihilation. This excludes the mesons $\eta, \eta' \omega, \phi$. Moreover we specialize in systems which are possible to study experimentally, where the kaon partner is a hadronic resonance with a very narrow width. This restricts the kaon partner to s-wave mesons and baryons, and excludes wider resonances like the meson ρ and the baryon Δ. The pion is also excluded because it is too light to bind to the kaon , all that we can get is a very broad resonance, the kappa resonance [17]. Therefore the hadrons which are best candidates to strongly bind the Kaons $s\bar{l}$ are the s-wave hadrons with flavour $l\bar{s}, l\bar{c}, l\bar{b}, lll, llc, llb, lcc, lcb, lbb$. This is expected to result in the $f_0(980)$ [17], the $D_s(2320$ [4] the $D_s(2463)$ [1], the $\Lambda(1405)$ and several other predicted resonances.

A convenient basis for the meson and baryon wave-functions is the harmonic oscillator basis, parametrized by the inverse radius α. We summarize [9, 10, 15] the effective potentials computed for the different channels,

$$V_{K-K} = 2(1+\sigma)^2 \frac{1}{6} \langle A \rangle |\phi^{\alpha}_{000}\rangle \langle \phi^{\alpha}_{000}| \,,$$

$$V_{K-D,D^*,B,B^*} = 2\frac{1}{6} \langle A \rangle |\phi^{\alpha}_{000}\rangle \langle \phi^{\alpha}_{000}| \,,$$

$$V_{K-N} = 4\frac{1}{6} \langle A \rangle |\phi^{\alpha}_{000}\rangle \langle \phi^{\alpha}_{000}| \,,$$

$$V_{K-\Sigma_c,\Sigma_b} = \frac{7}{3}\frac{1}{6} \langle A \rangle |\phi^{\alpha}_{000}\rangle \langle \phi^{\alpha}_{000}| \,,$$

$$V_{K-\Xi_{cc},\Xi_{cb},\Xi_{bb}} = 2\frac{1}{6} \langle A \rangle |\phi^{\alpha}_{000}\rangle \langle \phi^{\alpha}_{000}| \,, \tag{6}$$

where the colour and spin factors contribute respectively with $1/3$ and $1/2$, $\langle A \rangle$ is of the order of 430 MeV and the geometrical factor provides the separable potential $|\phi^{\alpha}_{000}\rangle \langle \phi^{\alpha}_{000}|$. The remaining factor is the flavour factor, the only one that turns out to differ in the s-wave kaon-hadron annihilation. The parameter α is are the only model one, and from the $D_s(2320)$ channel we α=285 MeV.

This parametrization hard core interaction in a separable potential, enables us to use standard techniques [15] to exactly compute the scattering T matrix. The binding occurs when the T matrix has a pole for a negative relative energy, and this happens when,

$$-4\mu v \geq \frac{\alpha^3}{\beta} \,. \tag{7}$$

The results are displayed in Tables 1 and 2. We conclude that the $f_0(980$ and the $D_s^{(0+)}, D_s^{(1+)}, B_s^{(0+)}, B_s^{(1+)}$ belong to the same class of tetraquark hadronic resonances. This class is consistent with the picture of a kaon trapped by a s-wave meson. In what concerns pentaquarks, where the kaon is trapped for instance by a nucleon to produce a Λ, or by other hadrons with u or d light quarks, our results predict that we have also have binding with the quantum numbers of the Λ and Ξ_c^+, Ξ_c^0, Ξ_b^0, Ξ_b^- and also of the Ω_{cc}, Ω_{cb}, and Ω_{bb}.

Here we neglected the coupling to the pion-hadron channels. This is correct for the multiquarks of Table 1 where this coupling is isospin violating. However in the systems of Table 2, where we found a deep binding, we plan to include the coupling to the pion-hadron channels. This coupling is expected to decrease substantially the binding energy. We also expect that our method addresses the protonomium recently discovered at BES [18] and the deuson X(3872) recently discovered at BELLE [19].

ACKNOWLEDGMENTS

We are very grateful to Emilio Ribeiro for discussions on the RGM and to George Rupp for discussions on hadronic resonances. The work of G. Marques is supported by Fundação para a Ciência e a Tecnologia under the grant SFRH/BD/984/2000.

TABLE 1. This table summarizes the parameters μ, v, α, β and binding energies B (in MeV). The italic binding energy B_{th} of the $D_s(1327)$ is fitted from experiment

channel	μ_{exp}	v_{th}	$\alpha = \beta$	B_{th}	B_{exp}
$D_s(2317) = \frac{K^-\bar{D}^0 + \bar{K}^0 D^-}{\sqrt{2}}$	392	-143	285	46	46
$D_s(2457) = \frac{K^-\bar{D}^{*0} + \bar{K}^0 D^{*-}}{\sqrt{2}}$	398	-143	285	47	46
$\frac{K^-\bar{B}^0 + \bar{K}^0 B^-}{\sqrt{2}}$	453	-143	285	55	-
$\frac{K^-\bar{B}^{*0} + \bar{K}^0 B^{*-}}{\sqrt{2}}$	454	-143	285	55	-
$\frac{K^-\Xi_{cc}^{++} + \bar{K}^0\Xi_{cc}^+}{\sqrt{2}}$	442	-143	285	53	-
$\frac{K^-\Xi_{cb}^+ + \bar{K}^0\Xi_{cb}^0}{\sqrt{2}}$	466	-143	285	56	-
$\frac{K^-\Xi_{bb}^0 + \bar{K}^0\Xi_{bb}^-}{\sqrt{2}}$	475	-143	285	58	-

TABLE 2. Results for channels open to pion decay.

channel	μ_{exp}	v_{th}	$\alpha = \beta$	B_{th}	B_{exp}
$f_0(980) = \frac{K^-K^+ + \bar{K}^0 K^0}{\sqrt{2}}$	248	-207	285	59	12 ± 10
$\Lambda(1405) = \frac{K^- p + \bar{K}^0 n}{\sqrt{2}}$	325	-286	285	149	30 ± 4
$\frac{K^-\Sigma_c^{++} + \bar{K}^0\Sigma_c^+}{\sqrt{2}}$	412	-167	285	68	-
$\frac{K^-\Sigma_b^+ + \bar{K}^0\Sigma_b^0}{\sqrt{2}}$	456	-167	285	75	-

REFERENCES

1. B. Aubert *et al.* [BABAR Collaboration], Phys. Rev. Lett. **90**, 242001 (2003) [arXiv:hep-ex/0304021].
2. D. Besson *et al.* [CLEO Collaboration], Phys. Rev. D **68**, 032002 (2003) [arXiv:hep-ex/0305100].
3. P. Krokovny *et al.* [Belle Collaboration], arXiv:hep-ex/0308019.
4. G. Rupp and E. van Beveren, Phys.Rev.Lett. **91**, 012003(2003) [arXiv:hep-ph/0305035].
5. M.A. Nowak, M. Rho and I. Zahed, Phys. Rev. D**48**, 4370 (1993) hep-ph/9209272.
6. W.A. Bardeen and C. T. Hill, Phys. Rev. D**49**, 409 (1994) hep-ph/9304265.
7. N. Tornqvist, Phys. Rev. Lett. **67**, 556 (1991).
8. P. J. Bicudo and J. E. Ribeiro, Phys. Rev. D **42**, 1611 (1990); 1625 (1990); 1635 (1990).
9. P. Bicudo, S. Cotanch, F. Llanes-Estrada, P. Maris, E. Ribeiro and A. Szczepaniak, Phys. Rev. D **65**, 076008 (2002) [arXiv:hep-ph/0112015]; P. Bicudo, Phys. Rev. C **67**, 035201 (2003).
10. P. Bicudo, M. Faria, G. M. Marques and J. E. Ribeiro, arXiv:nucl-th/0106071.
11. R. L. Jaffe, Phys. Rev. D **15** 281 (1977); D. Strottman, Phys. Rev. D **20** 748 (1979).
12. J. Wheeler, Phys. Rev. **52**, 1083 (1937); **52**, 1107 (1937).
13. J. E. Ribeiro, Z. Phys. C **5**, 27 (1980).
14. J. Dias de Deus and J. Ribeiro Phys. Rev. D **21**, 1251 (1980).
15. P. Bicudo and J. E. Ribeiro, Z. Phys. C **38**, 453 (1988); P. Bicudo, J. E. Ribeiro and J. Rodrigues, Phys. Rev. C **52**, 2144 (1995).
16. P. J. Bicudo, Phys. Rev. C **60**, 035209 (1999).
17. E. van Beveren *et al.*, Z. Phys. C 30, 615 (1986).
18. J. Z. Bai *et al.* [BES Collaboration], Phys. Rev. Lett. **91**, 022001 (2003) [arXiv:hep-ex/0303006].
19. K. Abe *et al.* [Belle Collaboration], arXiv:hep-ex/0308029.

Charm- and Bottom- Baryons: A Variational Approach Using Heavy Quark Symmetry

C. Albertus*, J. E. Amaro*, E. Hernández[†] and J. Nieves*

*Dpto. de Física Moderna, U. Granada, Spain.
†Grupo de Física Nuclear, Facultad de Ciencias, U. Salamanca, Spain.

Abstract. We evaluate masses of bottom and charmed baryons using several non-relativistic quark potentials which parameters have been adjusted to the meson spectra. Heavy Quark Symmetry leads to important simplifications of the three body problem, which turns out to be easily solved by a simple variational ansatz. The wave functions obtained can be readily used to compute further observables as mass densities or form factors. The quark-quark potentials explored so far, show an overall good agreement with the experimental masses.

INTRODUCTION

The non-relativistic constituent quark model (NRCQM), using QCD-inspired potentials, has proved to be an excellent tool to predict properties of hadrons.

In the case of baryons including one heavy flavour (c, b) and two light ones (u, d, s), it is possible to take advantage of yet another property of QCD: Heavy Quark Symmetry (HQS) Ref. [1, 2]. In the limit in which the mass of the heavy quark is infinity, the quantum numbers of the light degrees of freedom are well defined always. Furthermore, in this limit, the masses of the baryons depend only on the quark content and on the light-light quantum numbers of the baryon. All of this is a clear simplification for solving the three body problem. Thus, for bottom- and charm-baryons we can consider the quantum numbers of the two light quark system to be fixed, and neglect corrections terms in the wave function that scale as $\mathcal{O}\left(\Lambda_{QCD}/m_{c,b}\right)$.

The aim of this work is to determine masses and other properties like mass densities and electromagnetic form factors for baryons containing a heavy quark and two light ones. This study includes all baryons compiled in Table 1 and some more details will be given elsewhere [3].

Table 1. Baryons considered in this work. The information enclosed in the different columns is strangeness, spin-parity, isospin, spin-parity of the light degrees of freedom and quark content. The spin-parity of the light quarks, fifth and eleventh columns, in some cases are determined thanks to HQS.

Baryon	(S)	J^P	(I)	$s_l^{\pi_l}$	Quark content	Baryon	(S)	J^P	(I)	$s_l^{\pi_l}$	Quark content
$\Lambda_{c,b}$	(0)	$\frac{1}{2}^+$	(0)	0^+	$(u,d)c,b$	$\Xi'_{c,b}$	(-1)	$\frac{1}{2}^+$	$(\frac{1}{2})$	1^+	$(u,s)c,b$
$\Sigma_{c,b}$	(0)	$\frac{1}{2}^+$	(1)	1^+	$(u,u)c,b$	$\Xi^*_{c,b}$	(-1)	$\frac{3}{2}^+$	$(\frac{1}{2})$	1^+	$(u,s)c,b$
$\Sigma^*_{c,b}$	(0)	$\frac{3}{2}^+$	(1)	1^+	$(u,u)c,b$	$\Omega_{c,b}$	(-2)	$\frac{1}{2}^+$	(0)	1^+	$(s,s)c,b$
$\Xi_{c,b}$	(-1)	$\frac{1}{2}^+$	$(\frac{1}{2})$	0^+	$(u,s)c,b$	$\Omega^*_{c,b}$	(-2)	$\frac{3}{2}^+$	(0)	1^+	$(s,s)c,b$

CP717, *Hadron Spectroscopy: Tenth International Conference,*
edited by E. Klempt, H. Koch, and H. Orth
© 2004 American Institute of Physics 0-7354-0197-7/04/$22.00

THE THREE-BODY PROBLEM

The intrinsic hamiltonian that describes the dynamics of the baryon is given by[1]

$$H = -\frac{\vec{\nabla}_1^2}{2\mu_1} - \frac{\vec{\nabla}_2^2}{2\mu_2} + \frac{\vec{\nabla}_1 \cdot \vec{\nabla}_2}{m_h} + V_{l_1 h}(\vec{r}_1, spin) + V_{l_2 h}(\vec{r}_2, spin) + V_{l_1 l_2}(\vec{r}_1, \vec{r}_2, spin), \quad (1)$$

where $\vec{r}_i$ is the position of the i-th light quark with respect to the heavy one, m_h stands for the mass of the heavy quark, while μ_i accounts the reduced mass of the heavy and the i-th light quark system, $V_{l_i h}$ and $V_{l_1 l_2}$ are the light–heavy and light–light interaction potentials, and the words *spin* stands for possible spin dependence of the potentials. Note the presence of the Hughes-Eckart term $\vec{\nabla}_1 \cdot \vec{\nabla}_2 / m_h$ that results from the separation of the CM movement.

The potentials used in this work are the one proposed by R.K. Bhaduri et al. in Ref. [4], and the set of potentials proposed by B. Silvestre-Brac and C. Semay that can be found in Ref. [5]. The parameters of those potentials have been adjusted in the meson sector. For their use in the qq sector they have to be adequately transformed. We use the prescription $V_{ij}^{qq} = \frac{1}{2} V_{ij}^{q\bar{q}}$ that assumes a $\vec{\lambda}_i \cdot \vec{\lambda}_j$ color dependence in all terms of the potential [5].

To solve the three-body problem one can use Faddeev equations [5]. This is a non-trivial task from the computational point of view, and leads to wave functions that are difficult to use in other contexts. Here we propose an extremely simple variational scheme. As it is usual, we assume an antisymmetric wave-function for the color degrees of freedom and the spin-flavour wave function is determined by the quantum numbers specified in Table 1. Finally for the spatial wave function, we propose the ansatz

$$\psi(r_1, r_2, r_{12}) = NF(r_{12})\phi_1(r_1)\phi_2(r_2) \quad (2)$$

where N is a normalization factor, $\phi_i(r_i)$ is the ground state wave function for the $V_{l_i h}^{qq}$ potential, and $F(r_{12})$ is a Jastrow correlation function in the relative distance of the two light quarks r_{12}^2. For F we take

$$F(r_{12}) = \left(1 - e^{-c_1 r_{12}}\right) \sum_{j=2}^{4} a_j e^{-b_j^2 (r_{12} - d_j)^2} \quad (3)$$

where the term $e^{-c_1 r_{12}}$ would be excluded in those cases where the potential V_{ll}^{qq} does not show a repulsive hard core at the origin. Taking into account that the color wave function is antisymmetric we use symmetrized wave functions in the spin-isospin and orbital degrees of freedom of the two light quarks.

This variational scheme shows clear resemblances to that succesfully used in the study of double Λ hypernuclei [6].

---

[1] In this hamiltonian the motion of the center of mass (CM) of the baryon has been taken out.

[2] We have assumed that the relative orbital angular momentum between the light quarks is zero. Thus the spatial wave function can only depend on r_1, r_2 and r_{12}

PRELIMINARY RESULTS

Our results for the masses obtained with the AL1 potential of Ref. [5] are given in Tables 2 and 3. We find good agreement with experimental data [7], when available, and with previous results from lattice [8] and Faddeev calculations [5].

In Table 4 we give the mass radii obtained also for the AL1 potential. Our results agree with the ones obtained in Ref. [5] using a Faddeev approach. Conclusions are similar when the potential of [4] or potentials AL2, AP1 or AP2 of Ref. [5] are used.

Table 2. Masses for the bottom- baryons considered. The spin-parity of the light degrees of freedom is shown in the second column. Results with our variational approach and with a Faddeev calculation from Ref. [5] are included. Lattice QCD [8] and experimental values [7], when available, are also given.

B	s^π	content	$M_{exp.}$ [MeV]	$M_{Latt.}$ [MeV]	M_{Var} [MeV]	$M_{Fad.}$ [MeV]
Λ_b	0^+	udb	5624 ± 9	5640 ± 60	5640	5638
Σ_b	1^+	llb		5770 ± 70	5846	5845
Σ_b^*	1^+	llb		5780 ± 70	5877	
Ξ_b	0^+	lsb		5760 ± 60	5805	5806
Ξ_b'	1^+	lsb		5900 ± 70	5941	
Ξ_b^*	1^+	lsb		5900 ± 80	5972	
Ω_b	1^+	ssb		5990 ± 70	6034	6034
Ω_b^*	1^+	ssb		6000 ± 70	6065	

Table 3. As in Table 2 for the charm sector.

B	s^π	content	$M_{exp.}$ [MeV]	$M_{Latt.}$ [MeV]	M_{Var} [MeV]	$M_{Fad.}$ [MeV]
Λ_c	0^+	udc	2285 ± 1	2270 ± 50	2291	2285
Σ_c	1^+	llc	2452 ± 1	2460 ± 80	2453	2455
Σ_c^*	1^+	llc	2518 ± 2	2440 ± 70	2542	
Ξ_c	0^+	lsc	2469 ± 3	2410 ± 50	2476	2467
Ξ_c'	1^+	lsc	2576 ± 2	2570 ± 80	2571	
Ξ_c^*	1^+	lsc	2646 ± 2	2550 ± 80	2657	
Ω_c	1^+	ssc	2698 ± 3	2680 ± 70	2677	2675
Ω_c^*	1^+	ssc		2660 ± 80	2761	

Table 4. Results for mass radii using this variational here and those from the Faddeev calculation of Ref. [5].

B	$\langle r^2 \rangle$ $[fm^2]$(Var)	$\langle r^2 \rangle [fm^2]$ (Fad.)
Λ_b	0.045	0.045
Σ_b	0.054	0.054
Ξ_b	0.048	0.048
Ω_b	0.054	0.054
Λ_c	0.095	0.104
Σ_c	0.117	0.121
Ξ_c	0.096	0.104
Ω_c	0.102	0.108

Finally in Figs. 1 and 2 we give our results for the charge density and electric form factor of the Λ_b and Ω_b^- baryons.

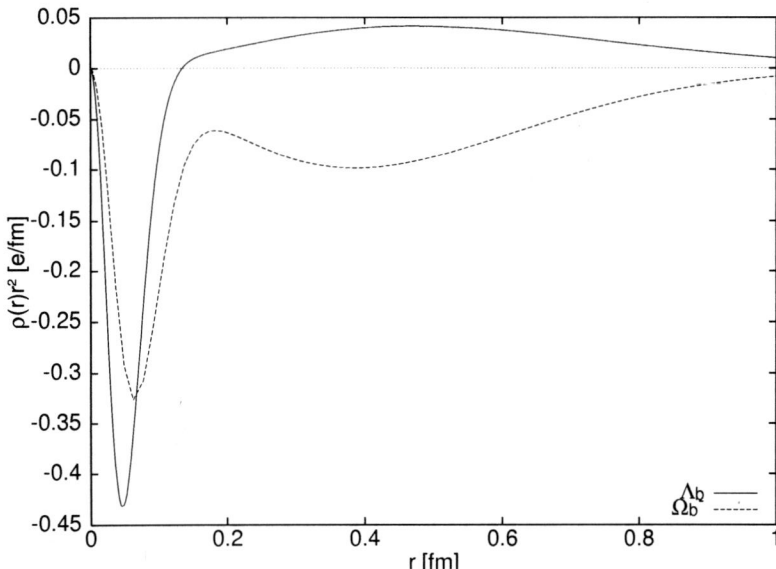

Figure 1. Charge density times r^2 for Λ_b (solid) and Ω_b^- (dashed).

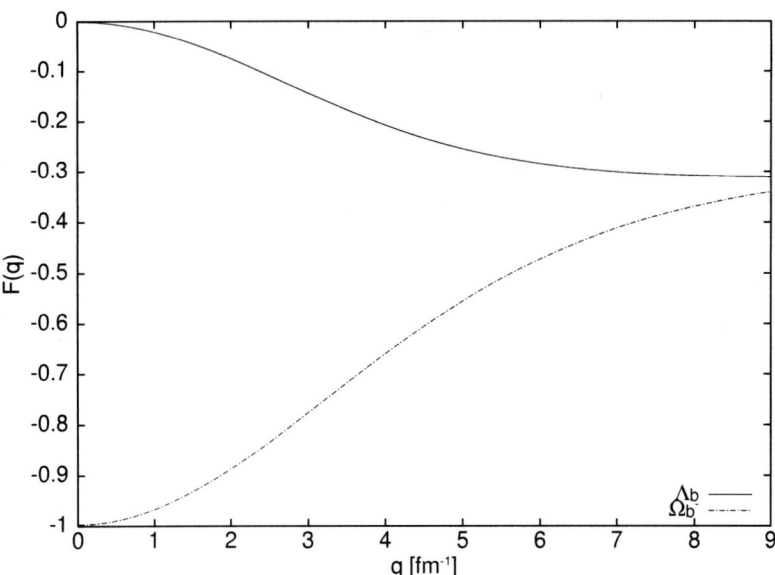

Figure 2. Electric form factors for Λ_b (solid) and Ω_b^- (dot-dashed). The value at the origin is the charge of the baryon, 0 for Λ_b and -1 for Ω_b^-

CONCLUDING REMARKS

The use of HQS simplifies considerably the solution of the three body problem in baryons with a heavy quark. Here we propose a method based on a simple variational approach that provides us with simple and portable wave functions that can be used in other contexts. Our results agree with previous ones obtained in the lattice or using a more complicate Faddeev approach. Calculations with potentials obtained from chiral quark models [9] and the study of the semileptonic decay of bottom baryons into charmed ones are under consideration.

ACKNOWLEDGMENTS

This work is supported by Spanish DGICYT and FEDER funds, under contracts no. BFM2000-1326 and BFM2002-03218 and by Junta de Andalucía and Junta de Castilla y León under contracts no. FQM225 and no. SA109/01. C. A. wishes to acknowledge a PhD. grant from Junta de Andalucía.

REFERENCES

1. Isgur, N., and Wise, M. B., *Phys. Lett.*, **B232**, 113 (1989).
2. Neuber, M., *Phys. Rep.*, **245**, 259 (1994).
3. Albertus, C., Amaro, J. E., Hernández, E., and Nieves, J. (2003), in preparation.
4. Bhaduri, R. K., Cohler, L. E., and Nogami, Y., *Nuovo Cimento*, **A65**, 58 (1981).
5. Silvestre-Brac, B., *Few-Body Systems*, **20**, 1–25 (1996).
6. Albertus, C., Amaro, J. E., and Nieves, J., *Phys. Rev. Lett.*, **89**, 032501 (2002).
7. K. Hagiwara, *et al.*, *Phys. Rev.*, **D66**, 010001 (2002).
8. K. C. Bowler, *et al.*, *Phys. Rev.*, **D54**, 3619 (1996).
9. Blanco, L., Fernández, F., and Valcarce, A., *Phys. Rev.*, **C59**, 428 (1999).

The phase between the three gluon and one photon amplitudes in quarkonium decays

Ping Wang

Institute of High Energy Physics, CAS, Beijing 100039, China

Abstract. The phase between three-gluon and one-photon amplitudes in $\psi(2S)$ and $\psi(3770)$ decays is analyzed.

MOTIVATIONS

It has been known that in J/ψ decays, the three gluon amplitude a_{3g} and one-photon amplitude a_γ are orthogonal for the decay modes 1^+0^- (90°) [1], 1^-0^- $(106 \pm 10)°$ [2], 0^-0^- $(89.6 \pm 9.9)°$ [3], 1^-1^- $(138 \pm 37)°$ [4] and $N\bar{N}$ $(89 \pm 15)°$ [5].

J. M. Gérard and J. Weyers [6] augued that this large phase follows from the orthogonality of three-gluon and one-photon virtual processes. The question arises: is this phase universal for quarkonium decays? How about $\psi(2S)$, $\psi(3770)$ and $\Upsilon(nS)$ decays?

QUARKONIUM PRODUCED IN ELECTRON-POSITRON COLLIDING EXPERIMENTS

Recently, more $\psi(2S)$ data has been available. Most of the branching ratios are measured in e^+e^- colliding experiments. For these experiments, there are three Feynman diagrams [7], as shown in Fig. 1, which contribute to the processes. Although such formulas were

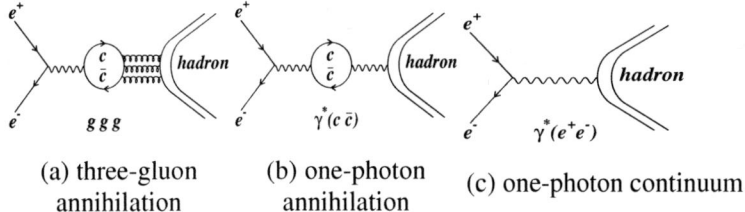

(a) three-gluon annihilation (b) one-photon annihilation (c) one-photon continuum

FIGURE 1. The Feynman diagrams of $e^+e^- \to$ *light hadrons* at charmonium resonance.

written in the early years after J/ψ was discovered, but the diagram in Fig. 1(c) is usually neglected. This reflects a big gap between theory and the actual experiments.

How important is this ampitude? For $\psi(2S)$, at first glance, $\sigma_{Born} = 7887$nb; while $\sigma_c \approx 14$nb. But for e^+e^- processes, initial state radiation modifies the Breit-Wigner cross section. With radiative correction, $\sigma_{r.c.} = 4046$nb; more important, the e^+e^- colliders

CP717, *Hadron Spectroscopy: Tenth International Conference,*
edited by E. Klempt, H. Koch, and H. Orth
© 2004 American Institute of Physics 0-7354-0197-7/04/$22.00

have finite beam energy resolution, with Δ at the order of magnitude of MeV; while the width of $\psi(2S)$ is only 300KeV. Here Δ is the standard deviation of the guassian function which describes the C.M. energy distribution of the electron-positron. This reduces the observed cross section by an order of magnitude. For example, with $\Delta = 1.3$MeV (parameter of BES/BEPC at the energy of $\psi(2S)$ mass), $\sigma_{obs} = 640$nb. If $\Delta = 2.0$MeV (paramters of DM2/DCI experiment at the same energy), $\sigma_{obs} = 442$nb.

The contribution from direct one-photon annihilation is most important for pure electromagnetic process, like $\mu^+\mu^-$, where the continuum cross section is as large as the resonance itself and the interference is apparent. This is seen in the $\mu^+\mu^-$ cross section curve in the experimental scan of $\psi(2S)$ resonance, as shown in Fig. 2.

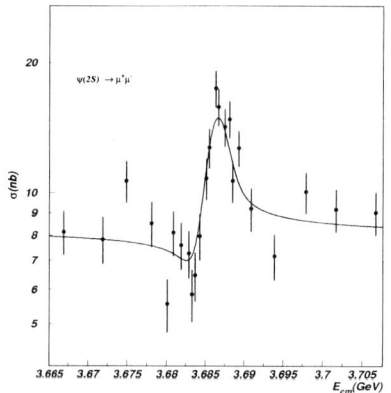

The observed cross section depends on experimental details: s_m, Δ, etc. [8]. The resonance cross section depends on the beam energy resolution of the e^+e^- collider; on the other hand, the continuum cross section depends on the invariant mass cut s_m in the selection criteria. This is seen from the treatment of the radiative correction [9]:

$$\sigma_{r.c.}(s) = \int_0^{1-\frac{s_m}{s}} dx F(x,s) \frac{\sigma_0(s(1-x))}{|1-\Pi(s(1-x))|^2} .$$

FIGURE 2. $\mu^+\mu^-$ curve at $\psi(2S)$ resonance scaned by BES

PURE ELECTROMAGNETIC DECAY

BES reports $\mathcal{B}(\psi(2S) \to \omega\pi^0) = (3.8 \pm 1.7 \pm 1.1) \times 10^{-5}$. What it means is the cross section of $e^+e^- \to \omega\pi^0$ at $\psi(2S)$ mass is measured to be $(2.4 \pm 1.3) \times 10^{-2}$ nb. About 60% of this cross section is due to continuum [10]. This gives the form factor $\mathcal{F}_{\omega\pi^0}(M_{\psi(2S)}^2)/\mathcal{F}_{\omega\pi^0}(0) = (1.6 \pm 0.4) \times 10^{-2}$. It agrees well with the calculation by J.-M. Gérard and G.López Castro [11] which predicts it to be $(2\pi f_\pi)^2/3s = 1.66 \times 10^{-2}$ with f_π the pion decay constant. Similarly π form factor at $\psi(2S)$ is revised [10].

$\psi(2S) \to 1^-0^-$ AND 0^-0^- DECAYS

The $\psi(2S) \to 1^-0^-$ decays are due to three-gluon amplitude a_{3g} and one-photon amplitude a_γ. With these two amplitudes, a previous analysis [12] yielded $a_{3g} \approx -a_\gamma$, i.e. the phase ϕ between a_{3g} and a_γ is 180° and $\phi = 90°$ is ruled out. Here the SU(3) breaking amplitude ε is small compared with a_{3g}. But these branching ratios so far are all measured by e^+e^- experiments. So actually we have three Feynman diagrams and three amplitudes. The analysis should be based on Table 1:

TABLE 1. $e^+e^- \to \psi(2S) \to 1^-0^-$ process

modes	amplitude	B.R.(in 10^{-4})
$\rho^+\pi^-$ ($\rho^0\pi^0$)	$a_{3g}+a_\gamma+a_c$	< 0.09
$K^{*+}K^-$	$a_{3g}+\varepsilon+a_\gamma+a_c$	< 0.15
$K^{*0}\overline{K^0}$	$a_{3g}+\varepsilon-2(a_\gamma+a_c)$	$0.41\pm0.12\pm0.08$
$\omega\pi^0$	$3(a_\gamma+a_c)$	$0.38\pm0.17\pm0.11$

In Table 1, a_{3g} interferes with $a_\gamma+a_c$, destructively for $\rho\pi$ and $K^{*+}K^-$, but constructively for $K^{*0}\overline{K^0}$ (ε is a fraction of a_{3g}). Fitting measured $K^{*+}K^-$ and $\rho\pi$ modes with different ϕ's are listed in Table 2.

It shows that a $-90°$ phase between a_{3g} and a_γ is still consistant with the data within one standard deviation of the experimental errors [13].

TABLE 2. Calculated results for $\psi(2S) \to K^{*+}K^-$ and $\rho^0\pi^0$ with different ϕ.

ϕ	$\mathscr{C} = \left\|\dfrac{a_{3g}}{a_\gamma}\right\|$	$\sigma_{pre}(K^{*+}K^-)$(pb)	$\mathscr{B}^0_{K^{*+}K^-}(\times10^{-5})^*$	$\sigma_{pre}(\rho^0\pi^0)$(pb)	$\mathscr{B}^0_{\rho^0\pi^0}(\times10^{-5})$
$+76.8°$	$7.0^{+3.1}_{-2.2}$	37^{+24}_{-23}	$5.0^{+3.2}_{-3.1}$	64^{+43}_{-41}	$9.0^{+6.1}_{-6.0}$
$-72.0°$	$5.3^{+3.1}_{-2.6}$	19^{+14}_{-14}	$3.1^{+2.3}_{-2.3}$	33^{+25}_{-24}	$5.5^{+4.1}_{-4.0}$
$-90°$	$4.5^{+3.1}_{-2.6}$	12^{+9}_{-9}	$2.0^{+1.5}_{-1.5}$	22^{+17}_{-17}	$3.7^{+2.9}_{-2.9}$
$180°$	$3.4^{+3.0}_{-2.2}$	$4.0^{+4.3}_{-3.2}$	$0.39^{+0.42}_{-0.31}$	$7.8^{+8.6}_{-6.7}$	$1.0^{+1.1}_{-0.8}$
BES observed		< 9.6		< 5.8	

$_*$ The superscript 0 indicates that the continuum contribution in cross section has been subtracted.

The newly measured $\psi(2S) \to K_S K_L$ from BES-II [15], together with previous results on $\pi^+\pi^-$ and K^+K^-, is also consistant with a $-90°$ phase between a_{3g} and a_γ. This is discussed in more detail by X.H. Mo in this conference.

$\psi(3770) \to \rho\pi$

J.L.Rosner [16] proposed that the $\rho\pi$ puzzle is due to the the mixing of $\psi(2S)$ and $\psi(1D)$ states, with the mixing angle $\theta = 12°$. In this scenario, the missing $\rho\pi$ decay mode of $\psi(2S)$ shows up instead as decay mode of $\psi(3770)$, enhanced by the factor $1/sin^2\theta$. He predicts $\mathscr{B}_{\psi(3770)\to\rho\pi} = (4.1\pm1.4) \times 10^{-4}$. With the total cross section of $\psi(3770)$ at Born order to be (11.6 ± 1.8) nb , $\sigma^{Born}_{e^+e^-\to\psi(3770)\to\rho\pi} = (4.8\pm1.9)$ pb .

But one should be reminded that for $\psi(3770)$, the resonance cross section, with radiative correction is only 8.17nb, while the continuum is 13nb. So to measure it in e^+e^- experiments, we must know the cross section $e^+e^- \to \gamma^* \to \rho\pi$. The cross section $\sigma_{e^+e^-\to\gamma^*\to\rho\pi}(s)$ can be estimated by the electromagnetic form factor of $\omega\pi^0$, since from SU(3) symmetry, the coupling of $\omega\pi^0$ to γ^* is three times of $\rho\pi$ [17]. The $\omega\pi^0$ form factor measured at $\psi(2S)$ is extrapolated to $\sqrt{s} = M_\psi(3770)$ by $|\mathcal{F}_{\omega\pi^0}(s)| = 0.531$ GeV/s . With this, the continuum cross section of $\rho\pi$ production at $\psi(3770)$ $\sigma^{Born}_{e^+e^- \to \gamma^* \to \rho\pi} = 4.4$ pb . Compare the two cross sections, the problem arises : how do these two interfere with each other?

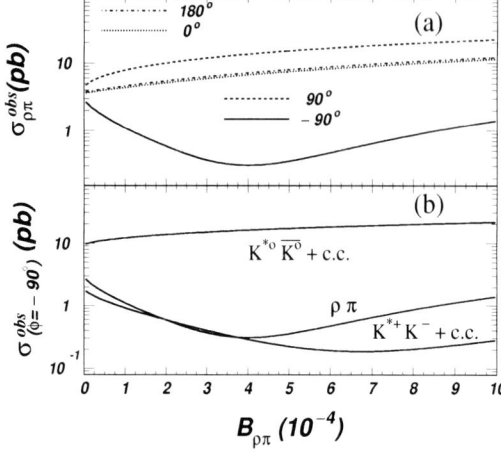

If the phase between a_{3g} and a_γ is $-90°$, then as in the case of $\psi(2S)$, the interference between a_{3g} and a_c is destructive in $\rho\pi$ and $K^{*+}K^-$ modes, but constructive in $K^{*0}\overline{K^0}$ mode [18]. The $e^+e^- \to \rho\pi$ cross section at $\psi(3770)$, as a function of $\mathcal{B}_{\psi(3770)\to\rho\pi}$ for different ϕ's are shown in Fig. 3(a); while the $e^+e^- \to \rho\pi, K^{*+}K^-, K^{*0}\overline{K^0}$ cross sections as functions of $\mathcal{B}_{\psi(3770)\to\rho\pi}$ for $\phi = -90°$ are shown in Fig. 3(b). To measure $\psi(3770) \to \rho\pi$ in e^+e^- collision, we must scan the $\psi(3770)$ peak (as we measure $\Gamma_{ee}, \Gamma_{total}$ and $M_{\psi(3770)}$).

FIGURE 3. (a) $e^+e^- \to \rho\pi$ cross section as a function of $\mathcal{B}_{\psi(3770)\to\rho\pi}$ for different phases, and (b) $e^+e^- \to K^{*0}\overline{K^0}+c.c., K^{*+}K^-+c.c.,$ and $\rho\pi$ cross sections as functions of $\mathcal{B}_{\psi(3770)\to\rho\pi}$.

Fig. 4(a) shows the $e^+e^- \to \rho\pi$ cross section vs C.M. energy for different ϕ's. Fig. 4(b) shows the $e^+e^- \to K^{*0}\overline{K^0}$ cross section with $\phi = -90°$.

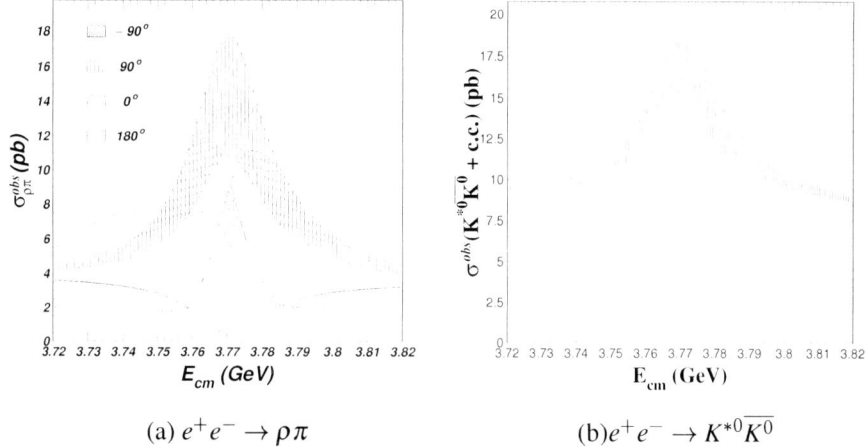

(a) $e^+e^- \to \rho\pi$ (b) $e^+e^- \to K^{*0}\overline{K^0}$

FIGURE 4. (a) $e^+e^- \to \rho\pi$ cross section vs C.M. energy for different phases: $\phi = -90°, +90°, 0°,$ and $180°$ respectively. (b) $e^+e^- \to K^{*0}\overline{K^0}$ cross section vs C.M. energy with $\phi = -90°$.

574

MARK-III gives $\sigma_{e^+e^- \to \rho\pi}(\sqrt{s} = M_{\psi(3770)}) < 6.3$ pb, at 90% C.L. [19]. It favors $-90°$.

$\psi(3770) \to 1^-0^-$ modes test the universal orthogonal phase between a_{3g} and a_γ in quarkonium decays as well as Rosner's scenario. A small cross section of $e^+e^- \to \rho\pi$ at $\psi(3770)$ peak means $\mathscr{B}(\psi(3770) \to \rho\pi) \approx 4 \times 10^{-4}$. (With radiative correction, the cancellation between a_{3g} and a_c cannot be complete. With a practical cut on the $\rho\pi$ invariant mass, the cross section is a fraction of 1pb.) It also implies the phase of the three gluon amplitude relative to one-photon decay amplitude is around $-90°$. These will be tested by the $20pb^{-1}$ of $\psi(3770)$ data by BES-II, or $5pb^{-1}$ of $\psi(3770)$ data by CLEO-c.

SUMMARY

- The universal orthogonality between a_{3g} and a_γ found in various decay modes of J/ψ can be generalized to $\psi(2S)$ and $\psi(3770)$ decays. A $-90°$ phase between a_{3g} and a_γ is consistant with the data on $\psi(2S) \to 1^-0^-$ and 0^-0^- modes.

- The $\psi(3770) \to \rho\pi, K^{*+}K^-, K^{*0}\overline{K^0}$ test the universal $-90°$ phase, as well as Rosner's scenario on $\rho\pi$ puzzle. This should be pursued by BES-II and CLEO-c.

- The exisiting $\Upsilon(nS)$ data should be used to test the phase in bottomonium states.

REFERENCES

1. M. Suzuki, Phys. Rev. **D63**, 054021 (2001).
2. J. Jousset *et al.*, Phys. Rev. **D41**, 1389 (1990); D. Coffman *et al.*, Phys. Rev. **D38**, 2695 (1988); N.N.Achasov, Talk at Hadron2001.
3. M. Suzuki, Phys. Rev. **D60**, 051501(1999).
4. L. Köpke and N. Wermes, Phys. Rep. **74**, 67 (1989).
5. R. Baldini, *et al.* Phys. Lett.**B444**, 111 (1998).
6. J. M. Gérard and J. Weyers, Phys. Lett. **B462**, 324 (1999).
7. S.Rudaz, Phys.Rev.D14,298(1976); P. Wang, C. Z. Yuan, X. H. Mo and D. H. Zhang, hep-ex/0210063, submitted to Phys.Rev.Lett.
8. P. Wang, C. Z. Yuan, X. H. Mo and D. H. Zhang, hep-ex/0210063, submitted to Phys. Rev. Lett.
9. E. A. Kuraev and V. S. Fadin, Sov. J. Nucl. Phys. **41**, 466 (1985); G. Altarelli and G. Martinelli, CERN **86-02**, 47 (1986); O. Nicrosini and L. Trentadue, Phys. Lett. **B196**, 551 (1987); F. A. Berends, G. Burgers and W. L. Neerven, Nucl. Phys. **B297**, 429 (1988); *ibid.* **304**, 921 (1988).
10. P.Wang, X.H.Mo and C.Z.Yuan, Phys.Lett. B557(2003)192.
11. J.-M. Gérard and G.López Castro, Phys. Lett. **B425**, 365 (1998).
12. M.Suzuki, Phys. Rev. **D63**, 054021 (2001).
13. P. Wang, C. Z. Yuan, X. H. Mo, hep-ex/0303144, submittd to Phys.Rev.Lett.
14. P.Wang, C.Z.Yuan and X.H.Mo, hep/0303144, submittd to Phys.Rev.Lett.
15. X. H. Mo's talk in this proceeding.
16. J.L.Rosner, Phys. Rev. **D64** (2001) 094002.
17. H. E. Haber and J. Perrier, Phys. Rev. **D32** (1985) 2961.
18. P.Wang, C.Z.Yuan, X.H.Mo, hep-ph/0308041, to be published in Phys.Lett.B
19. Yanong Zhu, Ph. D. thesis California Institute of Technology, 1988, Caltech Report No. CALT-68-1513; W. A. Majid, Ph. D. thesis) University of Illinois, 1993, UMI-94-11071-mc.

Results of experiment Focus on direct CP violation with charm D mesons, D^0 mixing and lifetime difference

Gianluigi Boca[1]

Dipartimento di Fisica Nucleare e Teorica
via Bassi 6
27100 Pavia

Abstract. A summary of some of the results of the experiment Focus, a high statistics charm photoproduction experiment performed at Fermilab, will be presented. Measurements will be shown of limits on the direct CP violation of the charmed D^+ and D^0 and on the mixing of the D^0 meson (lifetime difference between the CP-even and CP-odd states of the D^0 and 'wrong sign' decays).

The experiment Focus at Fermilab

The experiment used a γ beam (with typically 180 GeV mean energy) on a 4-segment (total 10% interaction lengths) beryllium oxide target. More than 6.3×10^9 triggers were collected during the 1996–97 run from which more than 1 million charmed particles were reconstructed. The particles from the interaction were detected in a large-aperture magnetic spectrometer with excellent vertex measurement (typical resolution of 30 fs for 2-track vertices), Cerenkov particle identification with π/K separation from 4.5 GeV/c to 61 GeV/c, two electromagnetic and one hadronic calorimeter, a μ detection system.

Search for direct CP violation in charm mesons

Direct CP violation occurs when the decay rate for a particle differs from the decay rate

---

[1] On th behalf of the Focus collaboration : **UC Davis**: J.M. Link, M. Reyes, P.M. Yager; **Rio de Janeiro, CBPF**: J.C. Anjos, I. Bediaga, C. Gobel, J. Magnin, A. Massafferri, J.M. de Miranda, I.M. Pepe, A.C. dos Reis; **Mexico, IPN**: S. Carrillo, E. Casimiro, E. Cuautle, A. Sanchez-Hernandez, C. Uribe, F. Vazquez; **Colorado U.**: L. Agostino, L. Cinquini, J.P. Cumalat, B. O'Reilly, J.E. Ramirez, I. Segoni, M. Wahl; **Fermilab**: J.N. Butler, H.W.K. Cheung, G. Chiodini, I. Gaines, P.H. Garbincius, L.A. Garren, E. Gottschalk, P.H. Kasper, A.E. Kreymer, R. K. Kutschke; **Frascati**: L. Benussi, M. Bertani, S. Bianco, F.L. Fabbri, A. Zallo; **Illinois U., Urbana**: C. Cawlfield, D.Y. Kim, K.S. Park, A. Rahimi, J. Wiss; **Indiana U.**: R. Gardner, A. Kryemadhi; **Korea U.**: C.H. Chang, Y.S. Chung, J.S. Kang, B.R. Ko, J.W. Kwak, K.B. Lee; **Kyungpook Natl. U.**: K. Cho, H. Park; **INFN, Milan & Milan U.**: G. Alimonti, S. Barberis, M. Boschini, A. Cerutti, P. D'Angelo, M. DiCorato, P. Dini, L. Edera, S. Erba, M. Giammarchi, P. Inzani, F. Leveraro, S. Malvezzi, D. Menasce, M. Mezzadri, L. Moroni, D. Pedrini, C. Pontoglio, F. Prelz, M. Rovere, S. Sala, E. Simili; **North Carolina U.**: T.F. Davenport, III; **Pavia U. & INFN, Pavia**: V. Arena, G. Boca, G. Bonomi, G. Gianini, G. Liguori, M.M. Merlo, D. Pantea, S.P. Ratti, C. Riccardi, P. Vitulo; **Puerto Rico U., Mayaguez**: H. Hernandez, A.M. Lopez, H. Mendez, A. Paris, J. Quinones, W. Xiong, Y. Zhang; **South Carolina U.**: J.R. Wilson; **Tennessee U.**: T. Handler, R. Mitchell; **Vanderbilt U.**: D. Engh, M. Hosack, W.E. Johns, M. Nehring, P.D. Sheldon, K. Stenson, E.W. Vaandering, M. Webster; **Wisconsin U., Madison**: M. Sheaff.

CP717, *Hadron Spectroscopy: Tenth International Conference,*
edited by E. Klempt, H. Koch, and H. Orth
© 2004 American Institute of Physics 0-7354-0197-7/04/$22.00

of its CP-conjugate particle. In the case of the charm D meson that means $A_{cp} \neq 0$, where $A_{cp} \equiv \frac{\Gamma(D \to f) - \Gamma(\bar{D} \to \bar{f})}{\Gamma(D \to f) + \Gamma(\bar{D} \to \bar{f})}$, D is any charm meson, f is the decay final state, $|\bar{D}> \equiv CP|D>$, $|\bar{f}> \equiv CP|f>$. A general description of the cases in which CP violation may happen can be found for instance in [1]. For charm, Standard Model theory predicts that the SCS decay channels have the largest CP asymmetry, with typical $A_{cp} \leq 10^{-3}$[2, 3, 4, 5]. The CF decays shouldn't show any CP asymmetry except for $D \to K_s + pions$ due to the fact that the K_s and K_L^0 in reality are not exact CP eigenstates and they have different lifetime. In particular for the decay $D^+ \to K_s \pi^+$ theory predicts $A_{cp} = -2\Re e(\varepsilon_K) \sim -0.33\%$[6]. The experiment Focus published[8] A_{cp} for the following decays :

- $D^+ \to K^+ K^- \pi^+$; $\quad A_{CP} = \frac{\eta(D^+ \to K^+ K^- \pi^+) - \eta(D^- \to K^+ K^- \pi^-)}{\eta(D^+ \to K^+ K^- \pi^+) + \eta(D^- \to K^+ K^- \pi^-)} = (0.6 \pm 1.1 \pm 0.5)\%$

$$\text{with } \eta(D^+ \to K^+ K^- \pi^+) \equiv \frac{\Gamma(D^+ \to K^+ K^- \pi^+)}{\Gamma(D^+ \to K^- \pi^+ \pi^+)}$$

- $D^0 \to K^+ K^-$; $\quad A_{CP} = \frac{\eta(D^0 \to K^+ K^-) - \eta(\bar{D}^0 \to K^+ K^-)}{\eta(D^0 \to K^+ K^-) + \eta(\bar{D}^0 \to K^+ K^-)} = (-0.1 \pm 2.2 \pm 1.5)\%$

$$\text{with } \eta(D^0 \to K^- \pi^+) \equiv \frac{\Gamma(D^0 \to K^+ K^-)}{\Gamma(D^0 \to K^- \pi^+)}$$

- $D^0 \to \pi^+ \pi^-$ $\quad A_{CP} = \frac{\eta(D^0 \to \pi^+ \pi^-) - \eta(\bar{D}^0 \to \pi^+ \pi^-)}{\eta(D^0 \to \pi^+ \pi^-) + \eta(\bar{D}^0 \to \pi^+ \pi^-)} = (4.8 \pm 3.9 \pm 2.5)\%$

$$\text{with } \eta(D^0 \to \pi^+ \pi^-) \equiv \frac{\Gamma(D^0 \to \pi^+ \pi^-)}{\Gamma(D^0 \to K^- \pi^+)}$$

where the first error is statistic, the second systematic. The use of a normalizing channel (use of η instead of Γ) is necessary in Focus to account for the different production rate of D and $\bar{D}$. Under the (very good) assumption that the CF normalizing decay channels don't exhibit CP violation, that is equivalent to the definition of A_{CP} given at the beginning. The flavour of the D^0 was established using the D^* tagging.

All these asymmetries are consistent with zero and in agreement, within errors, with the existing measurements (see Fig.1 and 2). Presently the precision of the world average is at the percent level. Only next generation experiments at the B or C factories will be able to achieve the one order of magnitude better precision necessary to challenge the theory predictions
. Focus measured also for the first time in the world two A_{CP} involving K_s^0 [10] :

- $D^+ \to K_s^0 \pi^+$; $\quad A_{CP} = \frac{\eta(D^+ \to K_s^0 \pi^+) - \eta(D^- \to K_s^0 \pi^-)}{\eta(D^+ \to K_s^0 \pi^+) + \eta(D^- \to K_s^0 \pi^-)} = (-1.6 \pm 1.5 \pm 0.9)\%$

$$\text{with } \eta(D^+ \to K_s^0 \pi^+) \equiv \frac{\Gamma(D^+ \to K_s^0 \pi^+)}{\Gamma(D^+ \to K^- \pi^+ \pi^+)}$$

- $D^+ \to K_s^0 K^+$; $\quad A_{CP} = \frac{\eta(D^+ \to K_s^0 K^+) - \eta(D^- \to K_s^0 K^-)}{\eta(D^+ \to K_s^0 K^+) + \eta(D^- \to K_s^0 K^-)} = (6.9 \pm 6.0 \pm 1.5)\%$

$$\text{with } \eta(D^+ \to K_s^0 K^+) \equiv \frac{\Gamma(D^+ \to K_s^0 K^+)}{\Gamma(D^+ \to K^- \pi^+ \pi^+)}$$

Again, the results are consistent, within errors, with no CP violation.

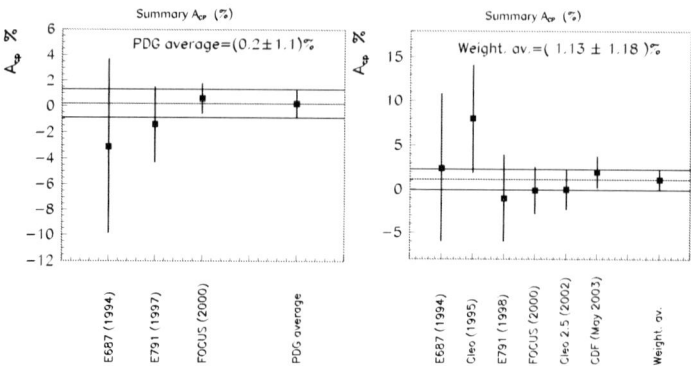

FIGURE 1. Summary of the latest results for the A_{CP} of the $D^+ \to K^+K^-\pi^+$ (left) and the $D^0 \to K^+K^-$ decay channels.

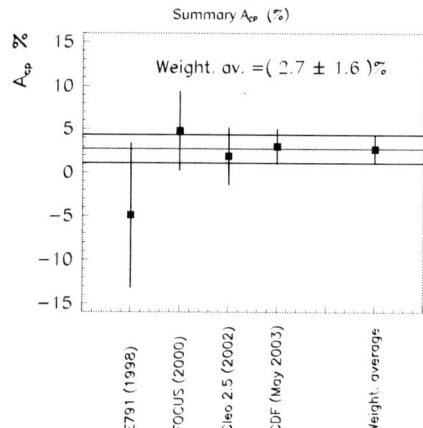

FIGURE 2. Summary of the latest results for the A_{CP} of the $D^0 \to K^+K^-$ decay channel

Mixing of the D^0 meson : lifetime difference and 'wrong sign' decays

The phenomenon of mixing in the $D^0\overline{D}^0$ system has been described in many articles and books and lately in [6].The variables governing mixing are $x \equiv \frac{m_1 - m_2}{\gamma}$ and $y \equiv \frac{\gamma_1 - \gamma_2}{\gamma_1 + \gamma_2} = \frac{\gamma_1 - \gamma_2}{2\gamma}$ where m_1 and γ_1 are mass and width of the D_1 state (usually defined as the more massive) analogous of the K_S^0 state, and $\gamma = \frac{\gamma_1 + \gamma_2}{2}$. Theoretical predictions[7] cover a range many orders of magnitude wide, with $|x|, |y|$ always less than 10^{-3}.

Focus published two measurements : the ratio of the lifetimes of the D^0 mesons decaying via $D^0 \to K^-\pi^+$ and K^-K^+ to measure the lifetime differences between CP even and CP odd final states, and the 'wrong sign' decays $D^0 \to K^+\pi^-$ that may occur either via a doubly Cabibbo suppressed (DCS) decay or through the mixing of the D^0 into $\overline{D}^0$ followed by the Cabibbo favored (CF) decay $\overline{D}^0 \to K^+\pi^-$.

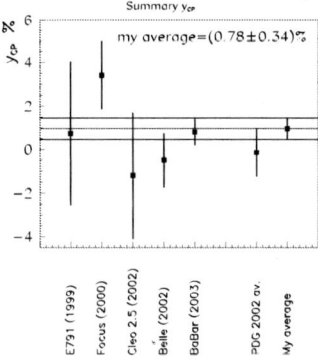

FIGURE 3. Summary of the latest results for the y_{CP}

Lifetime difference between the CP-even and CP-odd states of the D^0

Under the assumption that : 1) CP violation in the $D^0 \bar{D}^0$ system is negligible; 2) the lifetime difference $\frac{\gamma_1 - \gamma_2}{\gamma} \ll 1$, then it follows : a) the decays $D^0 \to K^- \pi^+$, $\bar{D}^0 \to K^+ \pi^-$, $D^0 \to K^- K^+$ and $\bar{D}^0 \to K^+ K^-$ are all exponential; b) the decay constants of $D^0 \to K^- \pi^+$ and $\bar{D}^0 \to K^+ \pi^-$ are the same and so are the decay constants of $D^0 \to K^- K^+$ and $\bar{D}^0 \to K^+ K^-$; c) the experimental lifetime ratio defined as: $y_{CP} \equiv \frac{\tau(D^0 \to K^- \pi^+)}{\tau(D^0 \to K^- K^+)} - 1$ is equal to $\frac{\Gamma(\text{CP even}) - \Gamma(\text{CP odd})}{\Gamma(\text{CP even}) + \Gamma(\text{CP odd})} \equiv y$ (where there is no need to distinguish between D^0 and $\bar{D}^0$ channels).

Focus measured[9] $y_{CP} = (3.42 \pm 1.39 \ stat \pm 0.74 \ sys)\%$ and in Fig. 3 a comparison with the world measurements up to now is shown. The agreement is good within errors and consistent with no lifetime difference at the 0.5 % level.

D^0 'wrong sign' decays

The observed ratio of $D^0 \to K^+ \pi^-$ to $D^0 \to K^- \pi^+$ can be used to obtain a relationship between the D^0 mixing and doubly Cabibbo suppressed decay parameters. Again under the assumption that CP violation in the $D^0 \bar{D}^0$ system is negligible, this ratio is given by

$$R_{WS} \equiv \frac{N(D^0 \to K^+ \pi^-)}{N(D^0 \to K^- \pi^+)} = \frac{N(\bar{D}^0 \to K^- \pi^+)}{N(\bar{D}^0 \to K^+ \pi^-)} = R_{DCS} + \sqrt{R_{DCS}} y' + R_M \qquad (1)$$

where $R_{DCS} \equiv \frac{\Gamma(D^0 \to K^+ \pi^-)}{\Gamma(D^0 \to K^- \pi^+)} = \frac{\Gamma(\bar{D}^0 \to K^- \pi^+)}{\Gamma(\bar{D}^0 \to K^+ \pi^-)}$; $R_M \equiv \frac{1}{2}(x^2 + y^2) = \frac{1}{2}(x'^2 + y'^2)$; $x' = x \cos \delta + y \sin \delta$; $y' = y \cos \delta - x \sin \delta$ and δ is the 'strong phase' defined as $\delta \equiv arg\left(-\frac{<K^+\pi^-|H_i|\bar{D}^0>}{<K^+\pi^-|H_i|D^0>}\right) = arg\left(\frac{<K^-\pi^+|H_i|D^0>}{<K^-\pi^+|H_i|\bar{D}^0>}\right)$. Focus measured[11] $R_{WS} = (0.404 \pm 0.085 \ stat \pm 0.025 \ sys)\%$. Using the measured value of R_{WS} and (1) one obtains an expression for R_{DCS} as a function of the mixing parameters x' and y'. In Fig. 4 R_{DCS} is plotted as a function of y' for two values of x' that cover the CLEO [12] 95% CL of $|x'| < 0.028$. For comparison, the mixing measurements of CLEO and FOCUS [9] are also included. If one further assumes that charm mixing is sufficiently small, the doubly Cabibbo suppressed branching R_{DCS} ratio is simply equal to R_{WS}. For comparison, Fig.

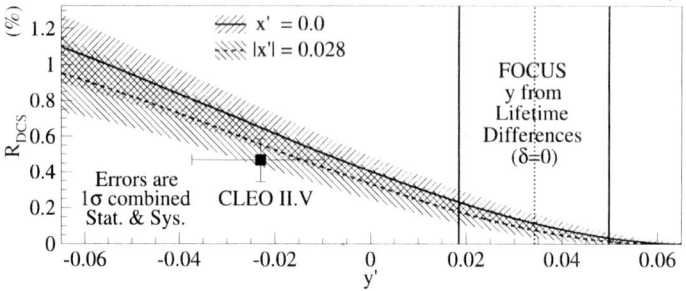

FIGURE 4. R_{DCS} plotted as a function of y'. Contours are given for two values of x' covering the 95% CL of the CLEO II.V result.

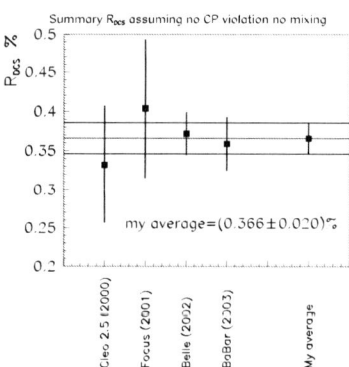

FIGURE 5. Summary of the most recent results for the R_{DCS} under the assumption of no CP violation, no mixing in the $D^0\overline{D}^0$ system.

5 shows the most recent measurements of this branching ratio, made with the assumption of no mixing.

REFERENCES

1. I.I. Bigi and A.I. Sanda, *CP Violation*, Cambridge University Press (2000).
2. F. Buccella *et al.*, *Phys.Rev.* **D51** (1995) 3478.
3. I.I. Bigi, *Proceedings of the Tau-Charm Factory Workshop*, (SLAC 1989), SLAC-Report-343.
4. M. Golden and B. Grinstein, *Phys. Lett.* **B222** (1989) 501.
5. F. Close and H. Lipkin, *Phys. Lett.* **B372** (1996) 306.
6. D.Benson, S.Bianco, I.Bigi,F.L.Fabbri, 'A Cicerone for the physics of charm', hep-ex/0309021, (2003).
7. H.Nelson, hep-ex/9908021 (1999)
8. J. M. Link *et al.*, Phys. Lett. **B491**, 232 (2000)
9. J. M. Link *et al.*, Phys. Lett. **B485**, 62 (2000).
10. J. M. Link *et al.*, Phys. Rev. Lett. **87**, 162001 (2001).
11. J. M. Link *et al.*, Phys. Rev. Lett. **86**, 2955 (2001).
12. R. Godang *et al.*, Phys. Rev. Lett. **84**, 5038 (2000).

E835 at FNAL: Charmonium Spectroscopy in $\bar{p}p$ Annihilations

C. Patrignani[*†], M. Ambrogiani[**], M. Andreotti[**], S. Bagnasco[†‡],
W. Baldini[**], D. Bettoni[**], G. Borreani[‡], A. Buzzo[†], R. Calabrese[**],
R. Cester[‡], G. Cibinetto[**], P. Dalpiaz[**], G. Garzoglio[§], K. E. Gollwitzer[§],
M. Graham[¶], A. Hahn[§], D. Joffe[‖], J. Kasper[‖], G. Lasio[††], M. Lo Vetere[†],
E. Luppi[**], M. Macrí[†], M. Mandelkern[††], F. Marchetto[‡], M. Marinelli[†],
W. Marsh[§], E. Menichetti[‡], Z. Metreveli[‖], R. Mussa[‡], M. Negrini[**],
M. Obertino[‡¶], M. Pallavicini[†], N. Pastrone[‡], T. Pedlar[‖], S. Pordes[§],
E. Robutti[†], J. Rosen[‖], P. Rumerio[‖], R. Rusack[¶], A. Santroni[†], J. Schultz[††],
K. K. Seth[‖], M. Stancari[††**], G. Stancari[§**], S. Seon-Hee[¶], A. Tomaradze[‖],
I. Uman[‖], T. Vidnovic III[¶] and S. Werkema[§]

*representing the FNAL-E835 collaboration
†I.N.F.N. and Universitá di Genova, Italy
**I.N.F.N. and Universitá di Ferrara, Italy
‡I.N.F.N. and Universitá di Torino, Italy
§Fermi National Accelerator Laboratory, U.S.A
¶University of Minnesota, U.S.A.
‖Northwestern University, U.S.A
††University of California at Irvine, U.S.A.

Abstract. E835 studied the properties of the charmonium states formed in $p\bar{p}$ annihilations. Recent and preliminary results are presented, including new preliminary measurements of χ_{c_J} masses and widths (improving the knowledge of fine splitting of P states), a preliminary measurement of $\psi(2S)$ branching ratios, and of the two photon partial width of the χ_{c_0}.

INTRODUCTION

The gross features of charmonium spectrum are in general reasonably well described by non relativistic potential models.

More sophisticated models, including non relativistic lattice QCD calculations, predict non negligible and sometimes relevant discrepancies from the naïve potential model for the energy splittings and the partial widths of charmonium states.

Precision measurements on the charmonium spectrum can provide an important test ground to assess the validity of methods devised to incorporate non perturbative, relativistic effects.

E835 has used the same experimental technique pioneered by experiment R704[1] at CERN's ISR and succesfully adopted later by experiment E760[2] at Fermilab.

Charmonium states are studied in $\bar{p}p$ annihilations where all charmonium states can be formed. The signal is extracted from the huge non resonant hadronic background by

CP717, *Hadron Spectroscopy: Tenth International Conference,*
edited by E. Klempt, H. Koch, and H. Orth
© 2004 American Institute of Physics 0-7354-0197-7/04/$22.00

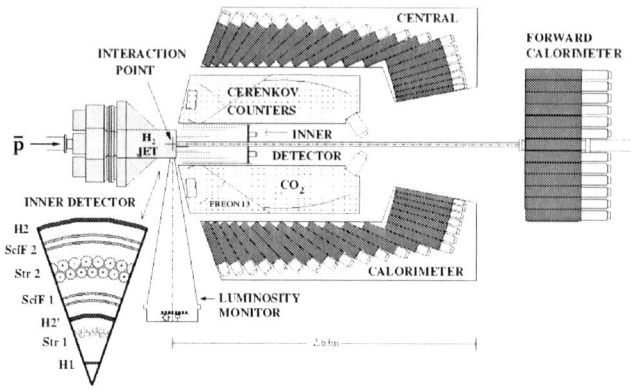

FIGURE 1. The E835-II detector layout

selecting charmonium decay modes to electromagnetic final states, or (as first proved in this experiment) by studying the interference with resonant continuum processes[15]. The required high luminosity is achieved by using a H_2 molecular cluster jet target[5] intersecting the coasting antiproton beam in the Fermilab Antiproton Accumulator[6].

Resonance parameters are determined from the excitation curve

$$\sigma(E_{CM}) = \int G_{beam}(E) \cdot \sigma_{BW}(E) dE; \quad \text{where}: \ \sigma_{BW}(E) \propto \frac{\Gamma_R^2 \mathscr{B}_{p\bar{p}} \mathscr{B}_{fin}}{(E - M_R)^2 + \Gamma_R^2/4}$$

. The center of mass energy and its distribution (G_{beam}) are determined from the beam momentum distribution, allowing the measurement of mass and width with precision < 100 keV[4].

EXPERIMENT E835 AT FERMILAB

Experiment E835 at FNAL took data in 1996/1997 (E835-I) for an overall integrated luminosity of ≈ 140 pb^{-1}, and again in 2000 (E835-II) for an integrated luminosity of ≈ 110 pb^{-1}.

The E835 detector is a major upgrade of the E760 detector[3]. It is a non-magnetic, large acceptance, cylindrical spectrometer, covering the complete azimuth (ϕ) and from $2°$ to $70°$ in polar angle (θ). It consists of a lead glass electromagnetic calorimeter divided into a barrel and a forward sections; the inner part of the barrel section is instrumented with a multicell threshold Čerenkov counter, triggering hodoscopes and a tracking system. In order to withstand the ~ 5 MHz asynchronous interaction rate, all read out channels are instrumented with multi-hit TDCs. The tracking system and the forward calorimeter have been further upgraded in year 2000 (Fig. 1).

The integrated luminosity for each energy setting was obtained by counting the number of recoil protons from $\bar{p}p$ elastic scattering in a silicon detector located at $\theta = 86.5°$ from the beam direction [11].

A typical stack of 5×10^{11} antiprotons was accumulated in 18 hours, the beam was decelerated from the accumulation energy (8.9 GeV) to the value appropriate of each resonance, then data were taken for approximately 48 h until the whole stack was used.

The density of the jet target[5] was continuously adjusted (10^{13}–10^{14} atoms/cm^3) to compensate for the beam loss keeping the instantaneous luminosity constant at $\sim 2 \times 10^{31}$ cm^{-2}sec^{-1}. The stochastic cooling kept the beam momentum constant with a $(\Delta p/p)_{rms} \approx 2 \times 10^{-4}$, compensating for energy losses in the jet target.

EVENT SELECTION

Candidate events for $e^+e^- X$ final states are selected online by the trigger if there are at least two charged tracks originating from the interaction point associated to a signal in the threshold Čerenkov counter and there are at least 2 energy deposits in the Central Calorimeter (CCAL) with large invariant mass ($\geq 2.2 GeV/c^2$).

The offline "inclusive" selection requires two identified electrons, based on a likelihood ratio test evaluated from the dE/dX in the scintillators, the number of ph.e. in Čerenkov and the shower lateral shape in the CCAL.

Exclusive final states are selected requiring then to be kinematically and topologically compatible with the final state under study (e^+e^-, $J/\psi\gamma$, $J/\psi\pi^+\pi^-$, etc.).

Candidate events for $\gamma\gamma$ final states are selected online by the trigger if there are no charged tracks originating from the interaction point and there are at least 2 energy deposits in the Central Calorimeter (CCAL) with large invariant mass ($\geq 2.2 GeV/c^2$), or if the total energy detected in CCAL exceeds 80% of the total energy,

The offline selection requires that the events be kinematically and topologically compatible with the final state under study ($\gamma\gamma$, $\pi^0\pi^0(\eta\eta) \rightarrow 4\gamma$, etc.). The background from non resonant neutral few body reactions ($\pi^0\gamma$, $\pi^0\pi^0$, etc.) populates mostly the backward-forward region in the center of mass, and the analysis is generally restricted to central values of $\cos\theta^*$ to increase the signal to background ratio.

PRELIMINARY RESULTS ON χ_{c_J} MASSES, WIDTHS AND $\mathcal{B}(p\bar{p})$

The masses and widths of χ_{c_J} states are studied in the reaction $\bar{p}p \rightarrow \chi_{c_J} \rightarrow J/\psi\gamma$.

E835 performed two scans at the χ_{c_0}: one in 1997 across the unstable transition energy of the accumulator[7], and a second, more accurate one, in 2000[8] after a major upgrade of the accumulator[6], when we also did scan $\chi_{c_{1,2}}$ to directly compare with the previous measurements for these states performed by E760[9]. Fig. 2 shows the results for the scans performed in E835-II, and measurements of resonance parameters are summarized in tab. 1, where the results are averaged taking into account common systematics.

The agreement on $\chi_{c_{1,2}}$ masses and widths measured ten years apart, after major upgrades in the detector and with a completely different setting of the antiproton accumulator is remarkable, and consistent with our estimate of the systematic error of less than 100 keV.

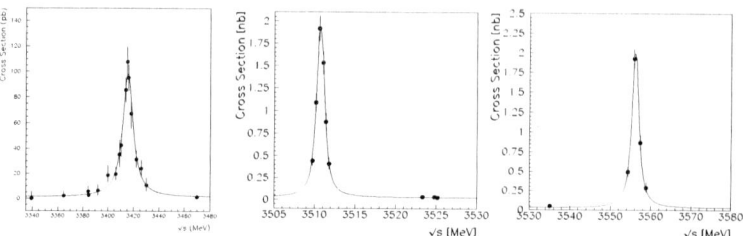

FIGURE 2. Resonance scans performed at the χ_{c_0} (left), χ_{c_1} (middle) and χ_{c_2} (right) by E835-II. charmonium experiments at FNAL.

TABLE 1. χ_{c_J} resonance parameters measured in $\bar{p}p \to \chi_{c_J} \to J/\psi\gamma$ by FNAL charmonium formation experiments

χ_{c0}	E835-II	E835-I	Average
M [MeV/c^2]	3515.4±0.4±0.2	3417.4$^{+1.8}_{-1.9}$±0.2	3415.5±0.4± 0.07
Γ_{TOT} [MeV]	9.8±1.0±0.1	16.7$^{+5.2}_{-3.7}$± 0.1	10.1±1.0
$\mathcal{B}(p\bar{p})\mathcal{B}(J/\psi+\gamma)[\times 10^7]$	27.2±1.9±1.3	29.3$^{+5.7}_{-4.7}$±1.5	27.6±2.5± 0.7
χ_{c1}	E835-II (prelim)	E760	Average
M [MeV/c^2]	3510.64±0.10±0.07	3510.53±0.10± 0.07	3510.59±0.07± 0.07
Γ_{TOT} [MeV]	0.88 ± 0.09	0.88± 0.14	0.88±0.08
$\mathcal{B}(p\bar{p})\Gamma(J/\psi+\gamma)[eV]$	18.8±0.7±0.6	21.8±2.7± 1.2	19.0±0.7± 1.2
χ_{c2}	E835-II (prelim)	E760	Average
M [MeV/c^2]	3556.10±0.15±0.07	3556.15±0.11±0.07	3556.13±0.09±0.07
Γ_{TOT} [MeV]	1.93±0.22	1.98±0.18	1.96±0.14
$\mathcal{B}(p\bar{p})\Gamma(J/\psi+\gamma)[eV]$	25.8±1.9±0.8	28.2±2.9±1.5	26.5±1.6±1.4

From these values we can directly determine the hyperfine splitting of P states (tab. 2) to better than 0.5%. These quantities pose severe constraints to any model, and can be used to test the accuracy of NRQCD lattice calculations. It must be noticed that the systematic error on the mass is mostly due to the absolute error on the mass scale, calibrated using the ψ' mass value from [10]. We expect this systematic error to be reduced to ≈ 20 keV thanks to the improved ψ' mass determination reported by [21].

TABLE 2. Hyperfine splitting of P charmonium states as measured by FNAL charmonium experiments

	Measured (preliminary)
$\Delta M_{12} = M(\chi_{c2}) - M(\chi_{c1})$ (MeV/c^2)	45.55±0.11
$\Delta M_{10} = M(\chi_{c1}) - M(\chi_{c0})$ (MeV/c^2)	95.2±0.4
$\rho = \dfrac{\Delta M_{21}}{\Delta M_{10}}$	0.478±0.002
$M_{c.o.g} = \dfrac{M(\chi_0) + 3M(\chi_1) + 5M(\chi_2)}{9}$ (MeV/c^2)	3525.31±0.07

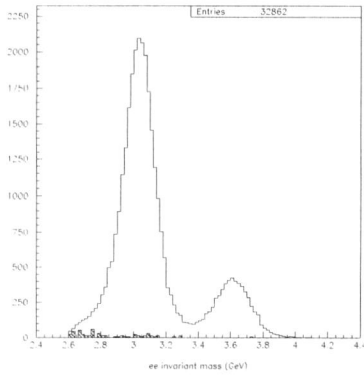

FIGURE 3. e^+e^- invariant mass of events used for $\psi(2S)$ branching ratios measurement (E835-II). The background level measured on a data sample outside the resonace region is shown as shaded area

RESULTS ON $\psi(2S)$ DECAYS TO e^+e^- AND $J/\psi X$

E835 collected (in the two runs) an integrated luminosity of ≈ 25 pb^{-1} at the $\psi(2S)$ formation energy, collecting an almost background free sample of 7500 $\psi(2S) \rightarrow e^+e^-$ and 36000 $\psi(2S) \rightarrow J/\psi X \rightarrow e^+e^- X$ with purity $> 98\%$ (fig. 3) from the formation of $\approx 3M$ $\psi(2S)$. From the subsample of events which can be fully reconstructed into exclusive final states, we can provide precise measurements of

$$\frac{\mathscr{B}(\psi(2S) \rightarrow \text{exclusive final state})}{\mathscr{B}(\psi(2S) \rightarrow J/\psi X)}$$

for the electronic decay (e^+e^-) and all final states involving a J/ψ.

The candidates for e^+e^-, $J/\psi \pi\pi$ (which have a distinctive final state topology), and $J/\psi\eta$ can be identified without relevant ambiguities from other $J/\psi X$ decays by kinematic fitting to the final state under study. The "internal background" contamination in each channel due to other $J\psi X$ modes is determined by MC simulation.

$\psi(2S) \rightarrow J/\psi\eta$, $J/\psi\pi^0$ and $\chi_{cJ}\gamma \rightarrow J/\psi\gamma\gamma$ are all studied in the final state $e^+e^-\gamma\gamma$, and the number of candidates for each decay mode can be determined by a fit to the Dalitz plot $J/\psi\gamma\gamma$ (fig. 4), taking into account the internal background contribution from $J/\psi\pi^0\pi^0$ and the non resonant $J/\psi\pi^0$ cross section (measured on 30 pb^{-1} taken in the range 3570 MeV $< \sqrt{s} < 3660$ MeV).

The angular distribution measured on the $\psi(2S) \rightarrow e^+e^-$ sample [20] is used to determine the $\psi(2S)$ helicity formation amplitudes, required to calculate the acceptances for all final states, since the detector covers only $\approx 50\%$ of the solid angle in the center of mass frame. The multipole amplitudes for the radiative decays $\psi(2S) \rightarrow \gamma\chi_{cJ}$ and $\chi_{cJ} \rightarrow \gamma J/\psi$ are taken from [18] and our own measurement [22]. The results are summarized in table 3. The number of exclusive events in all final states, corrected by acceptance and efficiency, fully accounts for the number of inclusive $\psi(2S) \rightarrow J/\psi X$ in our sample.

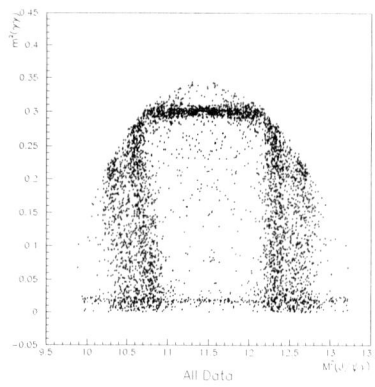

All Data

FIGURE 4. Dalitz plot for $J/\psi\gamma\gamma$ candidates (E835-I and E835-II samples)

TABLE 3. Preliminary results on $\psi(2S)$ branching ratios

Decay Mode	$\mathcal{B}(mode)/\mathcal{B}(J/\psi X)$ (%)		
	E760	E835-I	E835-II (PREL)
e^+e^-	$1.44\pm0.08\pm0.02$	$1.28\pm0.03\pm0.02$	$1.21\pm0.06\pm0.02$
$J/\psi\pi^+\pi^-$	49.6 ± 3.7	–	51.8 ± 2.7
$J/\psi\pi^0\pi^0$	32.3 ± 3.3	$32.8\pm1.3\pm0.8$	29.9 ± 3.3
$J/\psi\eta$	6.1 ± 1.5	7.2 ± 0.9	5.3 ± 0.8
		E835-I and E835-II (PREL)	
$J/\psi\eta$		5.6 ± 0.5	
$\chi_0\gamma \to J/\psi\gamma\gamma$		0.23 ± 0.08	
$\chi_1\gamma \to J/\psi\gamma\gamma$		5.40 ± 0.38	
$\chi_2\gamma \to J/\psi\gamma\gamma$		3.30 ± 0.28	
$J/\psi\pi^0$		0.41 ± 0.09	
$\sum J/\psi X$ excl. modes		99 ± 3	

Using $\mathcal{B}(\psi' \to J/\psi X) = 0.579 \pm 0.019$ [17] our measurements translates into $\mathcal{B}(\psi(2S) \to J/\psi\pi^0) = (0.24\pm0.05)\%$, $\mathcal{B}(\psi(2S) \to \chi_2\gamma \to J/\psi\gamma\gamma) = (1.91\pm0.16)\%$, $\mathcal{B}(\psi(2S) \to \chi_1\gamma \to J/\psi\gamma\gamma) = (3.13 \pm 0.22)\%$, and $\mathcal{B}(\psi(2S) \to \chi_0\gamma \to J/\psi\gamma\gamma) = (0.13 \pm 0.05)\%$ that are significantly larger than the most precise previous measurements from Crystal Ball [18], but are not incompatible with more recent determinations of $\mathcal{B}(\chi_{cJ} \to J/\psi\gamma)$ [17] and $\mathcal{B}(\psi(2S) \to \chi_{cJ}\gamma)$[19].

RESULTS ON $\eta_C(1^1S_0)$

The $\eta_c(1^1S_0)$ state was observed in the reaction $\bar{p}p \to \eta_c \to \gamma\gamma$[14]. The measured cross section for $\bar{p}p \to \gamma\gamma$ is shown in Figure 5 as a function of the center-of-mass energy. The events were selected for $|\cos\theta_\gamma^*| \leq 0.2$ to optimize the signal-to-background ratio. The data were fitted with the maximum likelihood method to a Breit-Wigner shape plus a power law background. As shown in E760 [13] this background is largely due to misidentified $\pi^0\pi^0$ and $\pi^0\gamma$ events, whose feed-down contribution can be precisely determined from the differential cross section measured by us on a data sample taken by a dedicated trigger. The values of the resonance parameters are $M_{\eta_c} = 2984.1 \pm 2.1 \pm 1.0$ MeV/c^2, total width $\Gamma = 20.4^{+7.7}_{-6.7} \pm 2.0$ MeV and $\Gamma(\eta_c \to \gamma\gamma)\mathscr{B}(\eta_c \to p\bar{p}) = (4.6^{+1.3}_{-1.1} \pm 0.4)$ eV, from which, using the value $\mathscr{B}(\eta_c \to p\bar{p}) = (1.2 \pm 0.4) \times 10^{-3}$[17] we derive $\Gamma(\eta_c \to \gamma\gamma) = 3.8^{+1.1+1.9}_{-1.0-1.0}$ keV. Figure 6 shows a comparison of the mass and total width

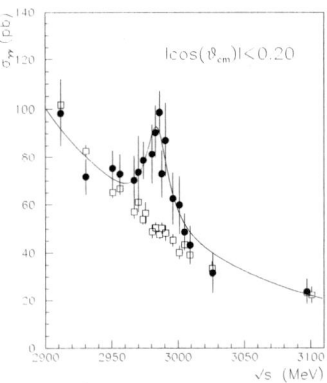

FIGURE 5. Measured cross section for $\bar{p}p \to \gamma\gamma$ vs center-of-mass energy at η_c. The events were selected for $|\cos\theta_\gamma^*| \le 0.2$. Open squares show feed-down from $\pi^0\pi^0$ and $\pi^0\gamma$ reactions, whose differential cross section is determined from samples simultaneously selected by a dedicated trigger.

of the η_c measured by E835 to other measurements. As it can be seen from these plots, the consistency between the different measurements is rather poor. While for the total width the most recent measurements are compatible to each other and seem to disagree with earlier determinations, for the mass the situation is less clear.

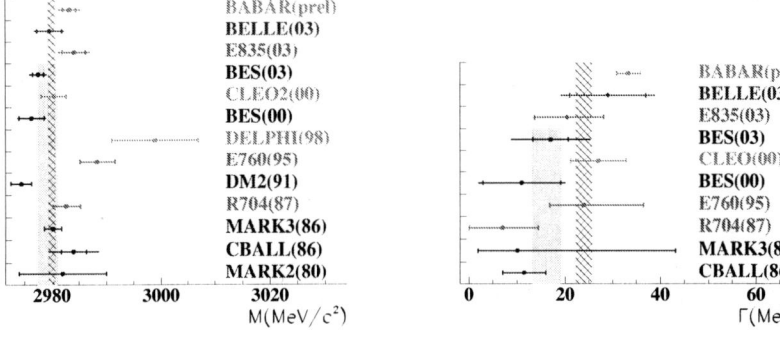

FIGURE 6. Comparison of the η_c mass and total width measured by E835 to other measurements. The light shaded area denotes the PDG average, the darker shade shows the average including more recent measurements.

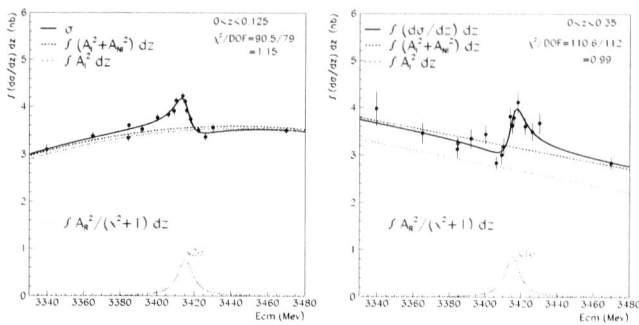

FIGURE 7. Integrated cross section as a function of E_{CM} (square resonant amplitude enlarged for plotting purposes)

$\bar{p}p \to \chi_{c0} \to \pi^0\pi^0$ AND $\eta\eta$

E835 has succesfully studied the χ_{c_0} resonance formation by measuring the interference with hadronic continuum processes $\pi^0\pi^0$[15] and $\eta\eta$, which have cross sections at least one order of magnitude larger.

The differential cross section $\bar{p}p \to \chi_{c_0} \to \pi^0\pi^0(\eta\eta)$ as a function of E_{CM} and $z \equiv |\cos\theta^*|$, is fitted to the sum of an interfering and a non-interfering part (Fig. 7):

$$\frac{d\sigma}{dz}(x,z) = \left| \frac{-A_R}{x+i} + A_I\, e^{i\delta_I} \right|^2 + \left| A_{NI}\, e^{i\delta_{NI}} \right|^2$$

where $x = \frac{E_{CM}-M_{\chi_0}}{\Gamma_{\chi_0}/2}$; $z \equiv |\cos\theta^*|$, A_R resonant amplitude; $A_I(x,z)$ interfering part of non resonant amplitude (λ=0); $A_{NI}(x,z)$ non interfering part of non resonant amplitude (λ=1), and λ initial state helicity of $\bar{p}p$.

The mass and width obtained by the fit ($M = 3514.7^{+0.7}_{-0.6} \pm 0.2$ MeV/c^2, $\Gamma = 8.4^{+1.7}_{-1.6} \pm 0.1$ GeV) are consistent with the (more precise) values determined in this experiment studying the $J/\psi\gamma$ final state (tab. 1). We then fix their values to determine $\mathcal{B}(\chi_{c0} \to p\bar{p}) \times \mathcal{B}(\chi_{c0} \to \pi^0\pi^0) = (5.09 \pm 0.81(\text{sta}) \pm 0.25(\text{sys})) \times 10^{-7}$, and $\mathcal{B}(\chi_{c0} \to p\bar{p}) \times \mathcal{B}(\chi_{c0} \to \eta\eta) = (4.0 \pm 1.2) \times 10^{-7}$ (preliminary). From our own measurements (taking into account common systematics) we obtain:

$$\frac{\mathcal{B}(\chi_{c0} \to J/\psi\gamma)}{\mathcal{B}(\chi_{c0} \to \pi^0\pi^0)} = 5.34 \pm 0.93 \pm 0.34$$

Taking our measurements of Γ[8] and $\mathcal{B}(\chi_{c0} \to \pi^0\pi^0)=\frac{1}{2}\mathcal{B}(\chi_c \to \pi^+\pi^-) = (2.5 \pm 0.35)10^{-3}$ from the literature[16], our measurement translates into $\Gamma(\chi_{c0} \to J/\psi\gamma) = 131 \pm 26 \pm 8 \pm 18$ keV that is in excellent agreement with the value of 137 ± 20 keV, predicted by the E_γ^3 scaling of $\Gamma(^3P_{1,2} \to J/\psi\gamma)$.

Our derived value for $\mathcal{B}(\chi_{c0} \to J/\psi\gamma) = 1.34 \pm 0.23 \pm 0.09 \pm 0.19\%$ is almost a factor 2 larger than the value measured by Crystal Ball[18], and indirectly supports our preliminary value for $\mathcal{B}(\psi(2S) \to \chi_{c0}\gamma \to J/\psi\gamma\gamma)$

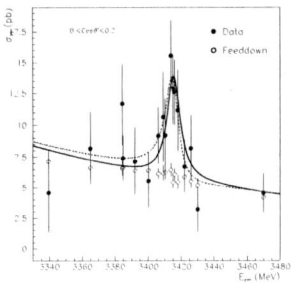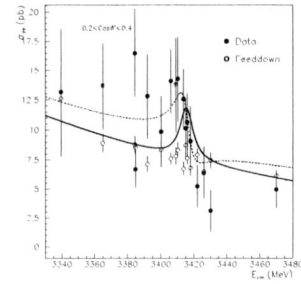

FIGURE 8. Measured cross section for $\bar{p}p \to \gamma\gamma$ at the χ_{c0} formation energy for: (left) $|\cos\theta^*| < 0.2$; (rigth) $0.2 < |\cos\theta^*| < 0.4$

$$\bar{p}p \to \chi_{c0} \to \gamma\gamma$$

We have measured the cross section for $\bar{p}p \to \gamma\gamma$ at the formation energy of the χ_{c0}.

The incoherent background from $\pi^0\pi^0$ and $\pi^0\gamma$ (calculated by MC based on the differential cross section for these reaction measured on data taken by a dedicated trigger) is sufficiently low up to $|\cos\theta^*| < 0.4$, but in this case the resonant $\bar{p}p \to \chi_{c_0} \to \gamma\gamma$ can interfere with non resonant $\bar{p}p \to \gamma\gamma$ which is expected to have a cross section of ≈ 10 pb^{-1} in the acceptance of our detector, mostly at large values of $|\cos\theta^*|$.

Because of uncertainties in the non resonant differential cross section, we restrict our measurement to $|\cos\theta^*| < 0.2$ (see fig. 8), where the fits with and without interference yield compatible results, and the possible effect of interference is included in the systematic error. We obtain a preliminary measurement of $\mathcal{B}(\chi_{c0} \to p\bar{p}) \times \mathcal{B}(\chi_{c0} \to \gamma\gamma) = (6.52 \pm 1.18(\text{stat}) \pm 0.55(\text{sys})) \times 10^{-8}$.

From the ratio between this value and our own measurement of $\mathcal{B}(\chi_{c0} \to p\bar{p}) \times \mathcal{B}(\chi_{c0} \to \pi^0\pi^0)$, using again our measurements for Γ [8] and $\mathcal{B}(\chi_c \to \pi^0\pi^0)$ from the literature[16] we can derive

$$\Gamma(\chi_{c0} \to \gamma\gamma) = 3.1 \pm 0.8 \pm 0.5 \; keV$$

which is in reasonable agreement with PQCD (with first order radiative correction)

$$\frac{\Gamma_{\gamma\gamma}}{\Gamma_{gg}} = \frac{8\alpha^2}{9\alpha_S^2} \times \frac{(1 + 0.2\alpha_S/\pi)}{(1 + 9.5\alpha_S/\pi)}$$

that would predict, for $\Gamma(\chi_{c0} \to gg) \simeq \Gamma_{TOT}(\chi_{c0})$ and $\alpha_S = 0.28$:

$$\Gamma_{\gamma\gamma,PQCD} = 3.26 \pm 0.33 \, keV$$

even if the theoretical reliability of this prediction has been questioned [23].

CONCLUSIONS

Experiment E835 took its last data in november 2000, concluding the fruitful experience at the Fermilab antiproton accumulator started 15 years earlier by E760, which significantly improved our knowledge of total and partial widths and splittings for the charmonium system.

The ability to identify high energy electrons and the excellent angular resolution for photons of our detector, have been exploited for measurements not related to charmonium spectroscopy, such as the proton electromagnetic form factors in the time-like region[24] at large momentum transfer, and in the study of light quarks resonances and search for glueball and hybrids[25].

Many results have been already published, but the analysis is still ongoing on part of the data, including those taken at the h_c, where we aim at a study in more than one decay channel, including the difficult but unambiguous radiative decay mode $\eta_c \gamma$.

REFERENCES

1. C. Baglin *et al.*, CERN Proposal CERN/ISRC/80-14, 1980 (unpublished).
2. V. Bharadwaj *et al.*, Fermilab Proposal P760, 1985 (unpublished).
3. T.A. Armstrong *et al.*, Fermilab Proposal P835(revised), 1992 (unpublished).
4. T. A. Armstrong *et al.* [E760 Collaboration], Phys. Rev. D **47** (1993) 772.
5. D. Allspach *et al.*, Nucl. Inst. Meth. **A410**(1998) 195.
6. D.P. McGinnis *et al.*, Nucl. Inst. Meth. **A506**(2003) 205.
7. M. Ambrogiani *et al.* [E835 Collaboration], Phys. Rev. Lett. **83** (1999) 2902.
8. S. Bagnasco *et al.* [Fermilab E835 Collaboration], Phys. Lett. **B533** (2002) 237.
9. T.A. Armstrong *et al.*, Nucl. Phys. **B373** (1992) 35.
10. A.A. Zholentz *et al.* Phys. Lett. **B96** (1980) 214.
11. S. Trokenheim *et al.*, Nucl. Inst. Meth. **355** (1995) 380.
12. T.A. Armstrong *et al.*, Phys. Rev. Lett. **69** (1992) 2337.
13. T.A. Armstrong *et al.*, Phys. Rev. **D52** (1995) 4839.
14. M. Ambrogiani *et al.*, Phys. Lett. **B566** (2003) 45.
15. M. Andreotti *et al.* Phys. Rev. Lett. **91** (2003) 091081; hep-ex/0308055.
16. J.Z. Bai *et al.* Phys. Rev. Lett. **81** (1998) 3091.
17. K. Hagiwara *et al.* [Particle Data Group Collaboration], Phys. Rev. D **66** (2002) 010001 and 2003 web update.
18. M.J. Oreglia *et al.* Phys. Rev. **D25** (1982) 2259; J. Gaiser *et al.* Phys. Rev. **D34** (1986) 711.
19. T. Skwarniki; talk at Lepton-Photon 03, Fermilab.
20. Seon-Hee Seo, to be published in the proceedings of CIPANP 03
21. V.M. Aulchenko *et al.*, Phys. Lett. **B573** (2003)63.
22. M. Ambrogiani *et al.* Phys. Rev. D **65** (2002) 052002.
23. J.P.Ma and Wang (hep-ph/0203082)
24. M. Andreotti *et al.* Phys. Lett B **559** (2003) 20.
25. Ismail Uman, this proceedings.

New precision measurement of the J/ψ- and ψ'-meson masses

V.M. Aulchenko[*], S.A. Balashov[*], E.M. Baldin[*], M.Yu. Barnyakov[*],
S.E. Baru[*], I.V. Bedny[*], O.L. Beloborodova[*], A.E. Blinov[*], V.E. Blinov[*],
A.V. Bogomyagkov[*], A.E. Bondar[*], D.V. Bondarev[*], A.R. Buzykaev[*],
S.I. Eidelman[*], V.R. Groshev[*], S.E. Karnaev[*], V.A. Kiselev[*],
S.A. Kononov[*], K.A. Kotov[*], E.A. Kravchenko[*], E.V. Kremyanskaya[*],
E.B. Levichev[*], V.M. Malyshev[*], A.L. Maslennikov[*], O.I. Meshkov[*],
S.E. Mishnev[*], N.Yu. Muchnoi[*], A.I. Naumenkov[*], S.A. Nikitin[*],
I.B. Nikolaev[*], A.P. Onuchin[*], S.B. Oreshkin[*], Yu.A. Pakhotin[*],
S.V. Peleganchuk[*], S.S. Petrosyan[*], V.V. Petrov[*], A.O. Poluektov[*],
A.A. Polunin[*], G.E. Pospelov[*], I.Ya. Protopopov[*], G.A. Savinov[*],
A.G. Shamov[*], D.N. Shatilov[*], A.I. Shusharo[*], B.A. Shwartz[*],
V.A. Sidorov[*], E.A. Simonov[*], Yu.I. Skovpen[*], A.N. Skrinsky[*],
A.M. Soukharev[*], A.A. Talyshev[*], V.A. Tayursky[*], V.I. Telnov[*],
Yu.A. Tikhonov[*], K.Yu. Todyshev[*], G.M. Tumaikin[*], Yu.V. Usov[*],
A.I. Vorobiov[*], A.N. Yushkov[*], A.V. Zatsepin[*] and V.N. Zhilich[*]

Budker Institute of Nuclear Physics, 630090 Novosibirsk, Russia

Abstract. A new high precision measurement of the J/ψ- and ψ'-meson masses has been performed at the VEPP-4M collider using the KEDR detector. The resonant depolarization method has been employed for the absolute calibration of the beam energy. The following mass values have been obtained:

$$M_{J/\psi} = 3096.917 \pm 0.010 \pm 0.007 \text{ MeV},$$
$$M_{\psi'} = 3686.111 \pm 0.025 \pm 0.009 \text{ MeV}.$$

The relative measurement accuracy has reached $4 \cdot 10^{-6}$ for J/ψ and $7 \cdot 10^{-6}$ for ψ', approximately 3 times better than in the previous precise experiments. The description of the experiment and the error analysis can be found in hep-ex/0306050, Phys. Lett. B573 (2003) 63-79.

CP717, *Hadron Spectroscopy: Tenth International Conference,*
edited by E. Klempt, H. Koch, and H. Orth
© 2004 American Institute of Physics 0-7354-0197-7/04/$22.00

Some Results on $\psi(3770)$ **Production and Decay from BES-II**

Gang RONG (for BES Collaboration)

Institute of High Energy of Physics, CAS, Beijing, 100039

Abstract. The $\psi(3770)$ resonance production is studied based on the data taken from 3.666 GeV to 3.897 GeV using the BES-II detector at the BEPC. The evidence of $\psi(3770)$ decays to a non-$D\bar{D}$ final state is observed. A total of 6.8 ± 3.0 $\psi(3770) \to J/\psi\pi^+\pi^-$ events are obtained from a data sample of 8.0 ± 0.5 pb^{-1} taken around 3.773 GeV. The branching fraction is determined to be $BF(\psi(3770) \to J/\psi\pi^+\pi^-) = (0.59 \pm 0.26 \pm 0.16)\%$, corresponding to the partial width of $\Gamma(\psi(3770) \to J/\psi\pi^+\pi^-) = (139 \pm 61 \pm 41)$ keV.

INTRODUCTION

The $\psi(3770)$ (called ψ'' for short) resonance was first observed by MARK-I [1]. Later on, DELCO [2] and MARK-II [3] made the cross section scan over the ψ'' resonance to measure the resonance parameters. Using the world averaged values of the ψ'' mass and the branching fraction for $\psi'' \to e^+e^-$, and the Breit-Wigner formula for ψ'' production at the peak, we obtain the Born order cross section for ψ'' production at the peak to be $\sigma_{\psi''} = 12.4 \pm 1.9$ nb. MARK-III measured the observed cross section [4] for $D\bar{D}$ production at 3.768 GeV to be $\sigma_{D\bar{D}}^{obs} = 5.0 \pm 0.5$ nb. At this c.m. (center-of-mass) energy, the cross section is reduced by a factor of 0.71 due to initial state radiation (ISR) [5]. After correcting the observed cross section $\sigma_{D\bar{D}}^{obs}$ for the effects of ISR, we obtain the Born order cross section for $D\bar{D}$ production to be $\sigma_{D\bar{D}} = 7.1 \pm 0.7$ nb.

The large discrepancy between $\sigma_{\psi''}$ and $\sigma_{D\bar{D}}$ indicates that either, contrary to what is generally expected, ψ'' could substantially decay into non$-D\bar{D}$ final states or the measured cross sections for $D\bar{D}$ and ψ'' productions suffer from large systematic shifts. Otherwise, there may exist some other effects which are responsible for the discrepancy.

Since the mass of ψ'' is above open charm-pair threshold and its width is two orders of magnitude larger than that of the $\psi(2S)$, it is thought to decay almost entirely to pure $D\bar{D}$ [2]. However, Lipkin pointed out that the ψ'' could decay to non$-D\bar{D}$ final states with a large branching fraction [6]. There are theoretical calculations [7, 8, 9, 10] that estimate the partial width for $\psi'' \to J/\psi\pi^+\pi^-$ based on the multipole expansion in QCD. Recently Kuang [10] used the Chen-Kuang potential model to obtain a partial width for $\psi'' \to J/\psi\pi\pi$ in the range from 37 to 170 keV, corresponding to 25 to 113 keV for $\psi'' \to J/\psi\pi^+\pi^-$ from isospin symmetry.

To understand what is for the large discrepancy between the $\sigma_{\psi''}$ and $\sigma_{D\bar{D}}$, one has

CP717, *Hadron Spectroscopy: Tenth International Conference,*
edited by E. Klempt, H. Koch, and H. Orth
© 2004 American Institute of Physics 0-7354-0197-7/04/$22.00

to more precisely measure the resonance parameters and the $D\overline{D}$ cross section. To these end, BES-II Collaboration operated a cross section scan experiment over the ψ'' resonance in 2003, and collected about 8 pb^{-1} data around 3.773 GeV in the time period from 2001 to 2002. Using these data samples, we studied ψ'' production and searched for the non-$D\overline{D}$ decays of ψ''.

BES-II DETECTOR

BES-II is a conventional cylindrical magnetic detector that is described in detail in Ref. [11]. A 12-layer Vertex Chamber (VC) surrounding the beryllium beam pipe provides input to the event trigger, as well as coordinate information. A forty-layer main drift chamber (MDC) located just outside the VC yields precise measurements of charged particle trajectories with a solid angle coverage of 85% of 4π; it also provides ionization energy loss (dE/dx) measurements which are used for particle identification. Momentum resolution of $1.7\%\sqrt{1+p^2}$ (p in GeV/c) and dE/dx resolution of 8.5% for Bhabha scattering electrons are obtained for the data taken at $\sqrt{s} = 3.773$ GeV. An array of 48 scintillation counters surrounding the MDC measures the time of flight (TOF) of charged particles with a resolution of about 180 ps for electrons. Outside the TOF, a 12 radiation length, lead-gas barrel shower counter (BSC), operating in limited streamer mode, measures the energies of electrons and photons over 80% of the total solid angle with an energy resolution of $\sigma_E/E = 0.22/\sqrt{E}$ (E in GeV) and spatial resolutions of $\sigma_\phi = 7.9$ mrad and $\sigma_Z = 2.3$ cm for electrons. A solenoidal magnet outside the BSC provides a 0.4 T magnetic field in the central tracking region of the detector. Three double-layer muon counters instrument the magnet flux return and serve to identify muons with momentum greater than 500 MeV/c. They cover 68% of the total solid angle.

ψ'' PRODUCTION

To obtain the resonance parameters of the ψ'', we performed a fine inclusive hadronic cross section scan experiment from 3.66 to 3.89 GeV, which cover $\psi(2S)$ and ψ'' resonances. The resonance parameters can be extracted from a fit to the observed inclusive hadronic cross sections.

For a set of experimental data taken at c.m. energy $E_{cm,i}$, the observed hadronic event production cross section is obtained by

$$\sigma^{obs}(E_{cm,i}) = \frac{N_{hadron}(E_{cm,i})}{L(E_{cm,i})\,\varepsilon_{hadron}(E_{cm,i})},$$

where $N_{hadron}(E_{cm,i})$ is the number of hadronic events observed at the c.m. energy $E_{cm,i}$; $L(E_{cm,i})$ is the integrated luminosity of the data collected at the $E_{cm,i}$, $\varepsilon_{hadron}(E_{cm,i})$ is the efficiency for detection of the hadronic events at the $E_{cm,i}$, and $i = 1, 2, ..., n$ are the energy points at which the data were collected.

593

The hadronic events are accepted if the following selection criteria is satisfied:

- the charged tracks must be with a good helix fit and the number of dE/dx hits is required to be greater than 14 per charged track;
- the distance of closest approach to the nominal interaction point is less than 2.0 cm;
- $|\cos(\theta)| < 0.84$, where θ is the polar angle of the charged track.
- $2.0 \ ns \ < T_{TOF} < T_p + 2.0 \ ns$, where T_{TOF} is the time-of-flight of the charged particle, and T_p is the expected time-of-flight of proton with a given momentum.
- $p < E_b + 0.1 \times E_b \times \sqrt{(1 + E_b^2)}$, where p is the charged track momentum and E_b is the beam energy in GeV;
- For each event, the total energy deposited in BSC should be greater than 28% of the beam energy.

Some beam-gas associated background events can also pass the above selection criteria. However, since the beam-gas associated background events are produced at random Z positions in the beam pipe, while the hadronic events are produced around $Z = 0$. To remove the beam-gas associated background events from the selected hadronic event sample, the averaged Z of the charged tracks in each of the events is calculated. By fitting to the averaged Z distribution with a Gaussian function plus a polynomial, the number of hadronic events can be obtained. At this step, the main sources of background are $e^+e^- \to \tau^+\tau^-$ and the final states from two photon processes. The number of background events due to these two processess can be determined by means of Monte Carlo simulation. The systematic uncertainty in the total number of the selected hadronic events due to the hadronic selection criteria is estimated to be about 2%.

The integrated luminosities of the data sets are determined by using the large-angle Bhabha scattering events. The selection and analysis of the events are based on the reconstructed momentum in MDC, the reconstructed energy and position of the showers in BSC. The numbers of the Bhabha scattering events are obtained by fitting to the $\delta\phi = |\phi_1 - \phi_2| - 180^o$ distributions with two Gaussian functions plus a polynomial, where the ϕ_1 and ϕ_2 are the azimuthal angles of the two energy clusters in the BSC. The systematic uncertainty in the measured values of the luminosities mainly arise from the difference between the real data and Monte Carlo sample. This uncertainty is estimated to be about 2.1%.

Due to ISR, the actual c.m. energies are distributed from the nominal collision c.m. energy E_{cm} to the energy of 0.28 GeV. In order to simulate the ISR and inclusive hadronic event productions, we developed a Monte Carlo package to simulate the inclusive hadronic event productions and decays in the all energy region. At an actual c.m. energy, final hadronic states are produced by sub-generators such as LUND model and some resonances generators, such as ψ'', $\psi(2S)$, J/ψ, ϕ, ρ and so on according to the corresponding Born Cross Sections of these sub-processes, respectively. The sub-processes decay into all possible final states according to the known decay modes and branching fractions. Figure 1 shows the Monte Carlo efficiencies for detection of the hadronic events produced at the different nominal c.m. energies. The systematic uncertainty in the efficiencies due to the generators is estimated to be about 2.5%.

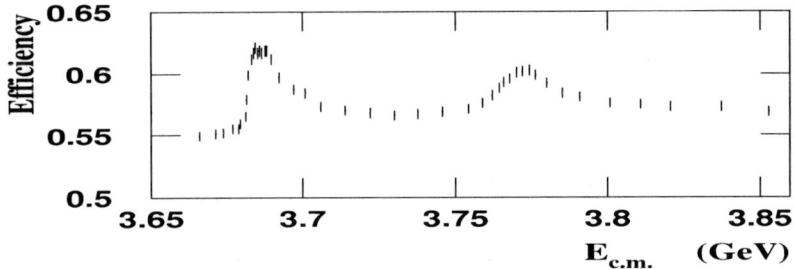

FIGURE 1. The Monte Carlo efficiency for detection of inclusive hadronic events vs the c.m. energies.

To determine the resonance parameters (mass $M_{\psi''}$, the total width Γ_{tot}, and leptonic width Γ_{ee}), the observed hadronic cross sections are fitted to a function that described the resonance shape, the nonresonance continuum backgrounds, the tail of the J/ψ, and the resonance of $\psi(2S)$. Since the contribution of $\psi(2S)$ to the total cross sections is relatively large in the ψ'' resonance region, the parameters of the two resonances are fitted simultaneously. The BEPC energy spread is in float in the fit. The shapes of the J/ψ tail and $\psi(2S)$ are given by the standard formula [12]. For the ψ'' resonance, we use a p-wave Breit-Wigner shape with an energy-dependent total width [3] to fit the observed hadronic cross sections. The ISR correction is made to the theoretical function based on Kuraev and Fadin [13].

Figure 2 shows the observed cross sections with the best fit. The fit[1] gives the mass difference of the ψ'' and $\psi(2S)$ is

$$\delta M = 86.9 \pm 1.0 \pm 0.5 \quad \text{MeV} \quad (\text{BES} - \text{II Preliminary !}).$$

EVIDENCE FOR $\psi'' \to J/\psi\pi^+\pi^-$

To search for the decay of $\psi'' \to J/\psi\pi^+\pi^-$, $J/\psi \to e^+e^-$ or $\mu^+\mu^-$, $\mu^+\mu^-\pi^+\pi^-$ and $e^+e^-\pi^+\pi^-$ candidate events are selected. These are required to have four charged tracks with zero total charge. Each track is required to have a good helix fit, to be consistent with originating from the primary event vertex, and to satisfy $|\cos\theta| < 0.85$, where θ is the polar angle.

Pions and leptons must satisfy particle identification requirements. For pions, the combined confidence level (CL), calculated for the π hypothesis using the dE/dx and TOF measurements, is required to be greater than 0.1%. In order to reduce γ conversion

---

[1] The error analysis on the resonance parameters is still in progress, the final results of some parameters are not ready to be reported at the conference yet.

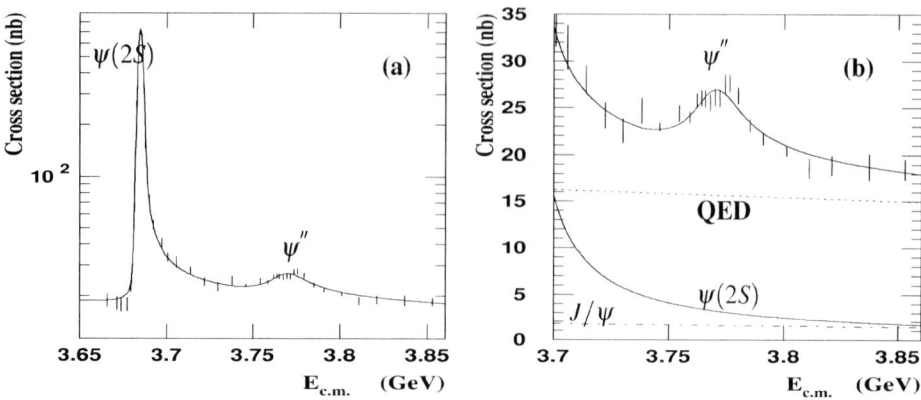

FIGURE 2. The observed inclusive hadronic cross sections vs the nominal c.m. energies. (a) the error bar represent the observed cross sections and the curve is the best fit; (b) shows the observed cross sections with the fit and the contributions from continuum background, $\psi(2S)$ tail and J/ψ tail.

background, in which the e^+ and e^- from a converted γ are misidentified as π^+ and π^-, an opening angle cut, $\theta_{\pi^+\pi^-} > 20°$, is imposed. For electron identification, the combined confidence level, calculated for the e hypothesis using the dE/dx, TOF and BSC measurements, is required to be greater than 1%, and the ratio $CL_e/(CL_e + CL_\mu + CL_\pi + CL_K)$ is required to be greater than 0.7. If a charged track hits the muon counter, and the z and $r\phi$ positions of the hit match with the extrapolated positions of the reconstructed MDC track, the charged track is identified as a muon. The data passed the selection criteria and particle identification are called pass-1 data.

To search for the decay of $\psi'' \to J/\psi\pi^+\pi^-$, the recoil masses of the $\pi^+\pi^-$ from the pass-1 data are calculated by

$$M_{\pi^+\pi^-}^{\mathrm{REC}} = \sqrt{(E_{cm} - E_{\pi^+\pi^-})^2 - |\vec{P}_{\pi^+\pi^-}|^2},$$

where E_{cm} is the c.m. energy, $E_{\pi^+\pi^-}$ and $\vec{P}_{\pi^+\pi^-}$ are the total energy and momentum of the $\pi^+\pi^-$ system, respectively. Figure 3(a) shows the recoil mass distribution for the events whose total energies of the 4 charged particles are within $\pm 2.5\sigma_{E_{\pi^+\pi^-l^+l^-}}$ window of the nominal c.m. energy, and the invariant masses of the Di-lepton are within ±150 MeV of the nominal J/ψ mass. Two peaks are observed. The higher one is from $\psi(2S)$ events produced via radiative return to the peak of the $\psi(2S)$, while the small enhancement at lower mass around 3.1 GeV is mainly from ψ'' decays. This is confirmed by a Monte Carlo simulation of ψ'' and $\psi(2S)$ production at the nominal c.m. energies around 3.770 GeV with subsequent decays to $J/\psi\pi^+\pi^-$ as shown in figure 3(b), where the two peaks are also clearly seen. In the figure 3(b), the solid histogram is for $\psi'' \to J/\psi\pi^+\pi^-$, while the dashed one is for $\psi(2S) \to J/\psi\pi^+\pi^-$ due to ISR. The Monte Carlo simulation includes leading-log-order radiative return, in which

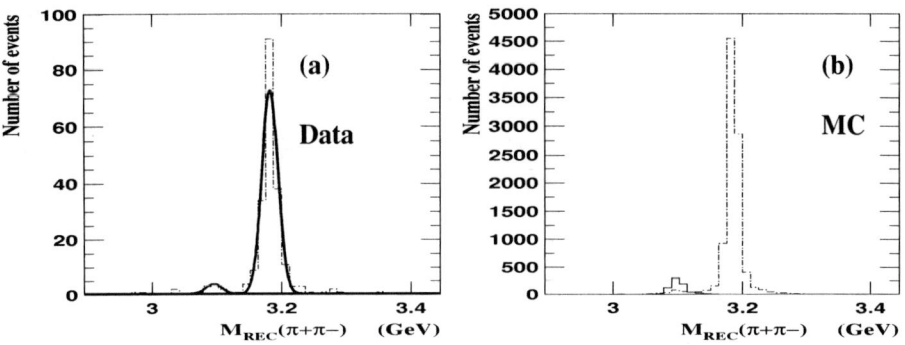

FIGURE 3. Distribution of the recoil masses of the $\pi^+\pi^-$ systems coming from (a) $l^+l^-\pi^+\pi^-$ events from the pass-1 data sample, (b) Monte Carlo $\psi(2S) \to J/\psi\pi^+\pi^-$ and $\psi'' \to J/\psi\pi^+\pi^-$ events.

the effective center-of-mass energies after ISR are generated according to Ref. [13]. The $\psi(2S)$ and ψ'' are generated with energy dependent Breit-Wigner functions, and the Lorentz boost of the hadron system recoiled by ISR photons and the beam energy spread are taken into account. The differential cross sections for $\psi(2S)$ and ψ'' production with subsequent decays to $J/\psi\pi^+\pi^-$ versus effective (or actual) energy determined by our Monte Carlo simulation are shown in figure 4, where the branching fraction for $\psi'' \to J/\psi\pi^+\pi^-$ is set to be 0.6%, while the branching fractions for $\psi(2S) \to J/\psi\pi^+\pi^-$ and $J/\psi \to l^+l^-$ ($l = e$ or μ) are taken from the Particle Data Group (PDG) [14].

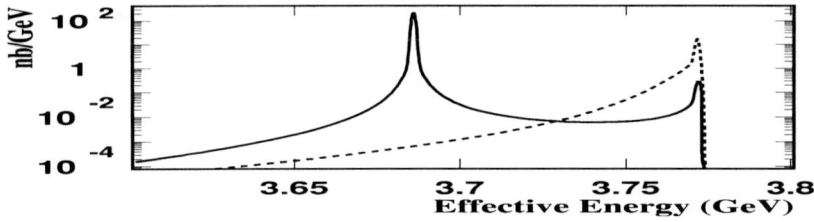

FIGURE 4. Differential cross sections for $\psi(2S)$ (solid curve) and ψ'' (dashed curve) productions and subsequent decays to $J/\psi\pi^+\pi^-$ with $J/\psi \to l^+l^-$ versus effective energy determined from Monte Carlo simulation at a nominal c.m. energy of 3.773 GeV.

In order to reduce background and improve momentum resolution, the events from the pass-1 data are subjected to four-constraint kinematic fits to either the $e^+e^- \to \mu^+\mu^-\pi^+\pi^-$ or the $e^+e^- \to e^+e^-\pi^+\pi^-$ hypothesis. Events with a confidence level greater than 1% are accepted.

Figure 5(a) shows the fitted Di-lepton masses of the accepted events. There are clearly two peaks. The lower mass peak is mainly due to $\psi'' \to J/\psi\pi^+\pi^-$, while the higher one is due to $\psi(2S) \to J/\psi\pi^+\pi^-$. The events from the higher peak is produced by radiative return to the $\psi(2S)$ peak, its energy will be approximately 3.686 GeV,

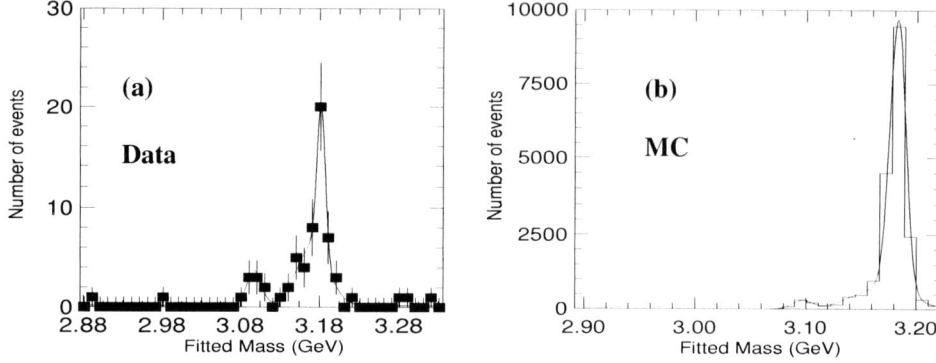

FIGURE 5. Momentum distribution of the $K^\mp \pi^\pm$ combinations for events for which the charged tracks passed the selection criteria.

while the c.m. energy is set to 3.773 GeV in the kinematic fitting. Therefore, the Di-lepton masses calculated based on the fitted momenta from the J/ψ coming from $\psi(2S) \to J/\psi \pi^+ \pi^-$ are shifted upward to about 3.18 GeV. Fitting to the mass spectrum with a Gaussian function to describe the $\psi'' \to J/\psi \pi^+ \pi^-$ signal, two Gaussian functions to represent the backgrounds of $\psi(2S) \to J/\psi \pi^+ \pi^{-2}$ and a Chebyshev polynomial for combinatorial background, we obtain 8.2 ± 2.9 signal events including the backgrounds of $\psi(2S) \to J/\psi \pi^+ \pi^-$ decays which is due to $\psi(2S)$ tail around the nominal c.m. energies. The 9 events in the J/ψ signal region in the figure 5(a) are all from the lower mass peaks in the figure 3(a).

Backgrounds from QED radiative processes with γ conversion, two-photon backgrounds, such as $e^+e^- \to e^+e^- \mu^+\mu^-$ (where the slow muons are misidentified as pions) and $e^+e^- \to e^+e^-\pi^+\pi^-$, and $e^+e^- \to \tau^+\tau^-$, are negligibly small. Candidate $J/\psi \pi^+\pi^-$ events could also be produced in the continuum process, $e^+e^- \to l^+l^-\pi^+\pi^-$. From a sample of 5.1 pb^{-1} taken in the energy region from 3.544 to 3.600 GeV with the BES-I detector and a sample of 6.0 pb^{-1} taken at 3.65 GeV with BES-II detector, no $J/\psi \pi^+\pi^-$, $J/\psi \to l^+l^-$ events are observed. Hence the continuum background is also negligible.

However, there is $\psi(2S) \to J/\psi \pi^+\pi^-$ background produced from the tail of the $\psi(2S)$ that can pass the event selection criteria and yield fitted Di-lepton masses to be around 3.097 GeV. This background, the main background to the $\psi'' \to J/\psi \pi^+\pi^-$ signal, is estimated by Monte Carlo simulation. Figure 5(b) shows the fitted mass distribution of the Di-lepton coming from the $\psi(2S) \to J/\psi \pi^+\pi^-$ due to the radiative return, in which a tail enhancement of the $\psi(2S)$ under the ψ'' is seen. It is induced by ISR return effects and raises the contribution of $\psi(2S)$ tail. From this 400 times large Monte Carlo sample, we estimate that $1.4 \pm 0.1 \pm 0.1$ out off 8.2 ± 2.9 events are due to

---

[2] One Gaussian function is for $\psi(2S)$ production due to ISR return for the nominal c.m. energy of 3.773 GeV, and the other one is for the nominal c.m. energies around 3.773 GeV.

$\psi(2S) \to J/\psi\pi^+\pi^-$, where the first error is statistical and the second systematic arising from the uncertainty in $\psi(2S)$ resonance parameters. There are total of 2.2 ± 0.4 background events in the signal region from 3.071 to 3.123 GeV in the figure 5(a), among which 0.80 ± 0.36 is from combinatorial background and $1.4 \pm 0.1 \pm 0.1$ is from $\psi(2S) \to J/\psi\pi^+\pi^-$. The probability of that the 9 events observed in the signal region is due to the fluctuation of 2.2 ± 0.4 background events is 8.2×10^{-4}. After background subtraction, 6.8 ± 3.0 signal events remain. In this analysis, possible interference between the $\psi(2S)$ background and the ψ'' is neglected since the interference term cancels after integrating over the pion momenta [9].

The total number of ψ'' events is obtained from our measured luminosities and the cross sections for ψ'' production at the center-of-mass energies E. The Born level cross section at the energy E is given by

$$\sigma_{\psi''}(E) = \frac{12\pi\Gamma_{ee}\Gamma_{\text{tot}}(E)}{(E^2 - M^2)^2 + M^2\Gamma_{\text{tot}}^2(E)},$$

where the ψ'' resonance parameters, Γ_{ee} and M are taken from the PDG [14], $\Gamma_{\text{tot}}(E)$ is chosen to be energy dependent and normalized to the total width Γ_{tot} at the peak of the resonance [14][3]. In order to obtain the observed cross section, it is necessary to take into account the ISR correction. The observed ψ'' cross section, $\sigma_{\psi''}^{obs}$, is reduced by a factor $g(s_{nom}) = \sigma_{\psi''}^{obs}(s_{nom})/\sigma_{\psi''}^{B}(s_{nom})$, where s_{nom} is the nominal c.m. energy squared. The ISR correction for ψ'' production is calculated using a Breit-Wigner function and the radiative photon energy spectrum [13]. With the calculated cross sections for ψ'' production at each energy point around 3.773 GeV and the corresponding luminosities, the total number of ψ'' events in the data sample is determined to be $N_{\psi''}^{prod} = (5.7 \pm 1.3) \times 10^4$, where the error is mainly due to the uncertainty in the observed cross section for ψ'' production.

The detection efficiency for the decay channel is determined to be $\varepsilon_{\psi'' \to J/\psi\pi^+\pi^-} = 0.171 \pm 0.002$, where the error is statistical. Using these numbers and the known branching fractions for $J/\psi \to e^+e^-$ and $\mu^+\mu^-$ [14], the branching fraction for the non-$D\bar{D}$ decay $\psi'' \to J/\psi\pi^+\pi^-$ is measured to be

$$BF(\psi'' \to J/\psi\pi^+\pi^-) = (0.59 \pm 0.26 \pm 0.16)\%,$$

where the first error is statistical and the second systematic. Using Γ_{tot} from the PDG [14], this branching fraction corresponds to a partial width of

$$\Gamma(\psi'' \to J/\psi\pi^+\pi^-) = (139 \pm 61 \pm 41) \text{ keV}.$$

The dominant systematic uncertainty is due to the uncertainty in the total number of ψ'' produced ($\pm 24\%$). Other systematic uncertainties are due to the efficiency ($\pm 10\%$) and the background shape ($\pm 6\%$). All systematic uncertainties are added in quadrature.

SUMMARY

In summary, the ψ'' production has been studied with the BES-II detector at the BEPC, and the branching fraction for $\psi'' \to J/\psi\pi^+\pi^-$ has been measured. From a total of $(5.7 \pm 1.3) \times 10^4$ ψ'' events, 6.8 ± 3.0 non$-D\bar{D}$ decays of $\psi'' \to J/\psi\pi^+\pi^-$ events are observed, leading to a branching fraction of $BF(\psi'' \to J/\psi\pi^+\pi^-) = (0.59 \pm 0.26 \pm 0.16)\%$, and a partial width $\Gamma(\psi'' \to J/\psi\pi^+\pi^-) = (139 \pm 61 \pm 41)$ keV.

ACKNOWLEDGMENTS

We would like to thank the BEPC staff for their strong efforts and the members of the IHEP computing center for their helpful assistance. We acknowledge Professor Yu-Ping Kuang and Professor Kuang-Ta Chao for many helpful discussions on the non$-D\bar{D}$ decay of $\psi(3770)$. This work is supported in part by the National Natural Science Foundation of China under contracts Nos. 19991480, 10225524, 10225525, the Chinese Academy of Sciences under contract No. KJ 95T-03, the 100 Talents Program of CAS under Contract Nos. U-24, U-25, and the Knowledge Innovation Project of CAS under Contract Nos. U-602, U-34(IHEP); by the National Natural Science Foundation of China under Contract No. 10175060(USTC); and by the Department of Energy under Contract No. DE-FG03-94ER40833 (U Hawaii).

REFERENCES

1. P. A. Rapidis *et al.*, (MARK-I Collaboration), Phys. Rev. Lett. **39**, 526 (1978).
2. W. Bacino *et al.*, Phys. Rev. Lett. **40**, 671 (1978).
3. R.H. Schindler *et al.*, Phys. Rev. **D21** 2716(1980).
4. J. Adler et al., Phys. Rev. Lett. **40**, 89 (1988)
5. J.Z. Bai *et al.*, (BES Collaboration), Phys. Rev. **D62** 012002(2000).
6. H. J. Lipkin, Phys. Lett. **B179**, 278 (1986).
7. K. Lane, Harvard Report No. HUTP-86/A045, (1986).
8. Y. P. Kuang and T. M. Yan, Phys. Rev. **D24**, 2874 (1981).
9. Y. P. Kuang and T. M. Yan, Phys. Rev. **D41**, 155 (1990).
10. Y. P. Kuang, Phys. Rev. **D65**, 094024 (2002).
11. J. Z. Bai *et al.* (BES Collaboration), Nucl. Instr. Meth. **A458**, 627 (2001).
12. G. Bonneau and F. Martin, Nucl. Phys. B27, 381(1971)
13. E. A. Kuraev and V. S. Fadin, Yad Fiz. **41**, 377 (1985); [Sov. J. Nucl. Phys. **41**, 466(1985)].
14. K. Hagiwara *et al.* (Particle Data Group), Phys. Rev. **D66** 010001-741(2002).

Observation of a Narrow $D_{sJ}^+\pi^0$ Resonance by CLEO

David G. Cassel

Laboratory for Elementary-Particle Physics, Cornell University, Ithaca, NY 14053-5001, USA

Abstract. CLEO discovered a new narrow particle tentatively named the $D_{sJ}^+(2463)$ decaying to $D_s^{*+}\pi^0$ and confirmed the existence of the $D_{sJ}^{*+}(2317)$ decaying to $D_s^+\pi^0$ discovered by BaBar.

INTRODUCTION

CLEO discovered a new narrow particle tentatively named the $D_{sJ}^+(2463)$ decaying to $D_s^{*+}\pi^0$ [1]. The search that led to this discovery was motivated by the BaBar discovery of an unexpected new narrow state called $D_{sJ}^{*+}(2317)$ which decays to $D_s^+\pi^0$ [2]. In the same analysis, CLEO confirmed the existence of the D_{sJ}^{*+}. The final states for the two new particle decays differ only by a soft photon, and it is very easy to miss this photon or to associate a spurious photon with the decay. Hence, very careful analyses are required to separate decays of the two particles. Belle confirmed both states in B decays [3, 4] as well as in continuum $e^+e^- \to c\bar{c}$ [4, 5]. BaBar also observed the $D_{sJ}^+(2463)$ state [6, 7].

Heavy quark symmetry illuminates the spectroscopy of D_s^+ ($c\bar{s}$) mesons [6, 8]. In the limit $m_c \to \infty$, the angular momentum ($j = l+s_s$) of the light $\bar{s}$ quark decouples from the spin of the c quark s_c. For $l > 0$, spin-orbit forces split different j values, and for large finite m_c, spin-spin and tensor forces remove further degeneracies. This leads to the following spectroscopy:

- The $l = 0$, $j = \frac{1}{2}$ states combine with s_c to produce the 0^- (D_s) and the 1^- (D_s^*), the two lowest lying $c\bar{s}$ mesons, both of which are well known [9].
- The $l = 1$, $j = \frac{3}{2}$ states combine with s_c to produce 1^+ and 2^+ states. These states are above the DK and D^*K thresholds but are relatively narrow because they decay via D waves. They have been observed [9].
- The $l = 1$, $j = \frac{1}{2}$ states combine with s_c to produce 0^+ and 1^+ states, and this 1^+ state will mix with the 1^+ state from $j = \frac{3}{2}$. According to conventional wisdom, they would also be above the DK and D^*K thresholds, but they could decay via S waves. Hence, these states would be wide and hard to detect, so it was not considered surprising that they had not been observed.

Adherents to this conventional wisdom concerning the $l = 1$, $j = \frac{1}{2}$ states were shocked when BaBar reported a new narrow resonance, $D_{sJ}^*(2317)^+ \to D_s^+\pi^0$. The signal is very large and robust. This mass is below most (but not all [10,

CP717, *Hadron Spectroscopy: Tenth International Conference,*
edited by E. Klempt, H. Koch, and H. Orth
© 2004 American Institute of Physics 0-7354-0197-7/04/$22.00

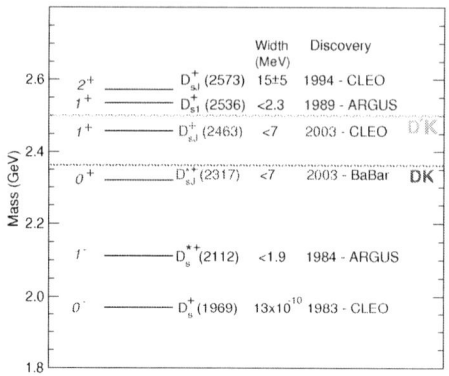

FIGURE 1. Spectroscopy of the $c\bar{s}$ mesons after discovery of the D_{sJ}^{*+} and D_{sJ}^{*}.

11]) theoretical predictions for the missing $c\bar{s}$ states and it is also below the DK threshold. Hence, the state should be narrow and the isospin violating strong decay $D_{sJ}^{*} \to D_s^{+}\pi^0$ could be favored. Other measurements support a 0^+ assignment for the D_{sJ}^{*}. BaBar data also show a peak in $D_s^{*+}\pi^0$ at $M \approx 2460$ MeV, but BaBar did not claim it as a signal. The low value of the D_{sJ}^{*} mass led to a flurry of theoretical activity [12] including exotic, as well as relatively conventional, interpretations of the result.

OBSERVATION OF TWO SIGNALS

We studied $D_{sJ}^{*+} \to D_s^{+}\pi^0$ and $D_{sJ}^{+} \to D_s^{*+}\pi^0$ decays in 13.5 fb^{-1} of CLEO II and CLEO II.V data, about $\sim \frac{1}{6}$ of the BaBar integrated luminosity. The decay chains studied and the invariant mass distributions observed are illustrated in Fig. 2. Details of the event selection and reconstruction are described elsewhere [1]. In this report, a quantity such as $M(D_s^{+})$ is the invariant mass of the particles in the D_s^{+} candidate. We observe significant peaks near 350 MeV in the mass difference distributions, $\Delta M(D_{sJ}^{*}) \equiv M(D_s\pi^0) - M(D_s)$ and $\Delta M(D_{sJ}) \equiv M(D_s^{*}\pi^0) - M(D_s^{*})$. We tentatively identify the peaks as $D_{sJ}^{*}(2317)^+$ (0^+) and $D_{sJ}(2463)^+$ (1^+).

The two decay chains differ only in the low-energy (~ 145 MeV) γ from D_s^{*+} decay, and the peaks in the two mass difference distributions $\Delta M(D_{sJ}^{*})$ and $\Delta M(D_{sJ})$ are both near 350 MeV. Therefore it is possible for a D_{sJ} decay to appear in the D_{sJ}^{*} signal if the low energy γ from a D_s^{*} decay is not detected. We call this process Feed Down. Alternately, if a random low energy photon and the D_s from D_{sJ} decay reconstruct to the D_s^{*} mass, the combination will appear in the D_{sJ} signal. We call this process Feed Up. Separating the two signals requires careful analyses and cross checks. We used three methods to establish the existence of the two states in our data and/or separate their contributions. Our excellent understanding of the CLEO CsI calorimeter and its high resolution are the crucial keys to success.

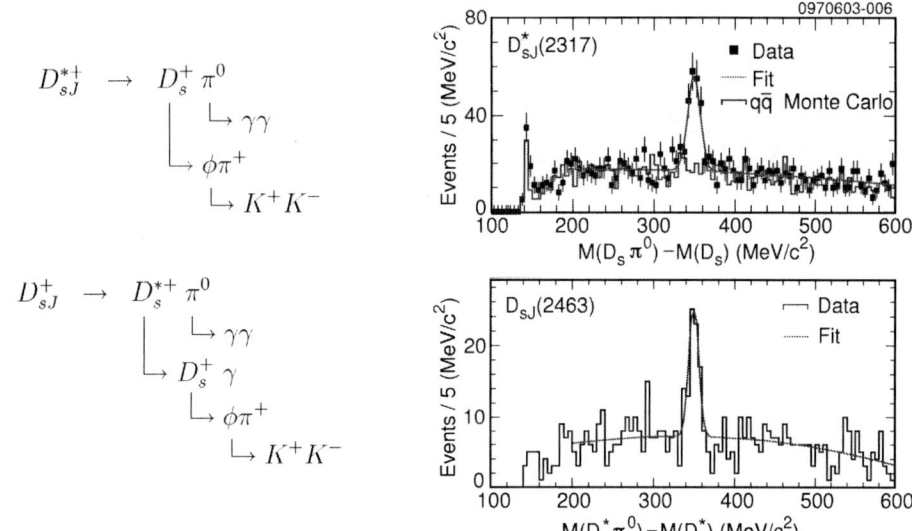

$$D_{sJ}^{*+} \rightarrow D_s^+ \pi^0$$
$$\quad\quad \hookrightarrow \gamma\gamma$$
$$\quad \hookrightarrow \phi\pi^+$$
$$\quad\quad \hookrightarrow K^+K^-$$

$$D_{sJ}^+ \rightarrow D_s^{*+} \pi^0$$
$$\quad\quad \hookrightarrow \gamma\gamma$$
$$\quad \hookrightarrow D_s^+ \gamma$$
$$\quad\quad \hookrightarrow \phi\pi^+$$
$$\quad\quad\quad \hookrightarrow K^+K^-$$

FIGURE 2. The decay chains studied by CLEO and the invariant mass distributions for $D_s^+ \pi^0$ (top) and $D_s^{*+}\pi^0$ (bottom) obtained from CLEO data. The normalization of the $e^+e^- \rightarrow q\bar{q}$ Monte Carlo histogram is absolute. The MC simulation does not include either D_{sJ}^{*+} or D_{sJ}^+, but does include the isospin violating decay $D_s^{*+} \rightarrow D_s^+ \pi^0$, which is responsible for the sharp peak near 145 MeV.

SEPARATING THE D_{sJ}^* AND D_{sJ}

From Monte Carlo simulations we determined that the resolution of the $\Delta M(D_{sJ}^*)$ peak would be $\sigma_{sig} = 6.0 \pm 0.3$ MeV for D_{sJ}^* decays and $\sigma_{FD} = 14.9 \pm 0.4$ MeV for candidates arising from the Feed Down process. The resolution that we obtain from a fit to the data illustrated in Fig. 2 is $\sigma = 8.0^{+1.3}_{-1.1}$ MeV. Hence the signal in the $\Delta M(D_{sJ}^*)$ peak is not entirely due to Feed Down. Similarly, we determined that the resolution of the $\Delta M(D_{sJ})$ peak would be $\sigma_{sig} = 6.0 \pm 0.3$ MeV for D_{sJ} decays and $\sigma_{FU} = 14.9 \pm 0.6$ MeV for candidates arising from the Feed Up process. The resolution that we obtain from a fit to the data illustrated in Fig. 2 is $\sigma = 6.1 \pm 1.0$ MeV, demonstrating that the $\Delta M(D_{sJ})$ peak is not entirely from the Feed Up process.

The distributions of $\Delta M(D_s\gamma\pi^0) \equiv M(D_s\gamma\pi^0) - M(D_s\gamma)$ for combinations with $M(D_s\gamma)$ in the D_s^* signal region and $M(D_s\gamma)$ in a D_s^* sideband are illustrated in Fig. 3. The D_{sJ} signal in the sideband distribution is very small, so Feed Up from $D_{sJ}^*(2317)$ decays is not significant. Furthermore, a fit of the D_s^* signal distribution to the sideband distribution and a Gaussian signal gives $N_{sig} = 45.7 \pm 11.7$ events. The signal in the D_s^* distribution is about 5.7σ above the number of combinations in the sideband distribution.

We can determine the actual yields of D_{sJ}^* and D_{sJ} decays in our data by taking cross feed into account using the following quantities: N_0 and N_1, the numbers of

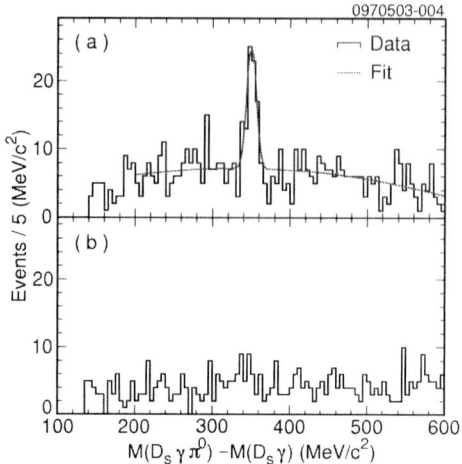

FIGURE 3. The $M(D_s\gamma\pi^0) - M(D_s\gamma)$ distribution for (a) candidates with $M(D_s\gamma)$ in the D_s^* signal region and (b) candidates with $M(D_s\gamma)$ in a D_s^* sideband.

D_{sJ}^* and D_{sJ} observed, R_0 and R_1, the real yields of D_{sJ}^* and D_{sJ} within our cuts (*i.e.*, produced × efficiencies), and f_0 and f_1, cross feed coefficients determined from Monte Carlo simulation. Then to first order: $N_0 = R_0 + f_1 R_1$ and $N_1 = R_1 + f_0 R_0$. By identifying D_{sJ}^* and D_{sJ} candidates with the same (cuts except for the additional D_s^* mass cut in the case of the D_{sJ}), we obtain from data (N's) and Monte Carlo simulations (f's)

$$D_{sJ}^* : \quad N_0 = 190 \pm 19 \quad \text{and} \quad f_0 = (9.1 \pm 0.7 \pm 1.5)\% \tag{1}$$

$$D_{sJ} : \quad N_1 = 55 \pm 10 \quad \text{and} \quad f_1 = (84 \pm 4 \pm 10)\%. \tag{2}$$

Then inversion using the numbers in Eqs. (1) and (2) yields:

$$D_{sJ}^* : R_0 = 155 \pm 23 \quad \text{and} \quad D_{sJ} : R_1 = 41 \pm 12 \tag{3}$$

This D_{sJ} yield is consistent with $N_{sig} = 45.7 \pm 11.7$ obtained from the fit to a signal plus sideband, and is $> 5\sigma$ above the sum of the combinatorial background plus the Feed Up yield.

D_{sJ}^* AND D_{sJ} MASSES AND WIDTHS

Our understanding of the CLEO II and II.V detectors and their excellent mass resolutions allow us to fit the D_{sJ}^* signal to two Gaussians with floating means and standard deviations to account for Signal (S) and Feed Down (F). The results are:

- $\mu_S = 350.0 \pm 1.2$ MeV $\qquad \sigma_S = 6.0 \pm 1.2$ MeV
- $\mu_{FD} = 344.9 \pm 6.1$ MeV $\qquad \sigma_{FD} = 16.5 \pm 6.3$ MeV

The widths obtained from the fit are consistent with the resolutions obtained from Monte Carlo simulations. After including systematic uncertainties we find

$$\Delta M(D_{sJ}^*) = 350.0 \pm 1.2 \pm 1.0 \text{ MeV and } \Gamma(D_{sJ}^*) < 7 \text{ MeV (90\% CL)}. \tag{4}$$

Table 1. Search for other D^*_{sJ} decay modes. Ratios are 90% CL ULs for $\mathcal{B}(\text{Mode})/\mathcal{B}(D_s\pi^0)$. Predictions are from Ref. [13].

Mode	Yield	Efficiency	Ratio	Prediction
$D_s^+\pi^0$	135 ± 23	$(9.7 \pm 0.6)\,\%$	—	
$D_s^+\gamma$	-19 ± 13	$(18.5 \pm 0.1)\,\%$	< 0.052	0
$D_s^{*+}\gamma$	-6.5 ± 5.2	$(7.0 \pm 0.5)\,\%$	< 0.059	0.08
$D_s^+\pi^+\pi^-$	2.0 ± 2.3	$(19.8 \pm 0.8)\,\%$	< 0.019	0
$D_s^{*+}\pi^0$	-1.7 ± 3.9	$(3.6 \pm 0.3)\,\%$	< 0.11	0

Table 2. Search for other D_{sJ} decay modes. Ratios are 90% CL ULs for $\mathcal{B}(\text{Mode})/\mathcal{B}(D_s^*\pi^0)$. Predictions are from Ref. [13].

Mode	Yield	Efficiency	Ratio	Prediction
$D_s^{*+}\pi^0$	41 ± 12	$(6.0 \pm 0.2)\,\%$	—	
$D_s^+\gamma$	40 ± 17	$(19.8 \pm 0.4)\,\%$	< 0.49	0.24
$D_s^{*+}\gamma$	-5.1 ± 7.7	$(9.1 \pm 0.3)\,\%$	< 0.16	0.22
$D_s^+\pi^+\pi^-$	2.5 ± 5.4	$(19.5 \pm 1.5)\,\%$	< 0.08	0.20
$D_{sJ}^*(2317)^+\gamma$	3.6 ± 3.0	$(2.0 \pm 0.1)\,\%$	< 0.58	0.13

We determined the mass and width of the D_{sJ} by subtracting the sideband distribution in Fig. 3(b) from the signal distribution in Fig. 3(a). Fitting the resulting distribution to a single Gaussian and a parabolic background yields

$$\Delta M(D_{sJ}) = 351.2 \pm 1.7 \pm 1.0 \text{ MeV and } \Gamma(D_{sJ}) < 7 \text{ MeV (90\% CL)} \quad (5)$$

after including systematic uncertainties.

Chiral multiplet models [10, 11] predict $M(1^+) - M(1^-) = M(0^+) - M(0^-)$ or $M(D_{sJ}^+) - M(D_s^{*+}) = M(D_{sJ}^{*+}) - M(D_s^+)$. CLEO finds,

$$\delta(\Delta M) = (351.2 \pm 1.7) - (350.0 \pm 1.2) = 1.2 \pm 2.1 \text{MeV}, \quad (6)$$

in excellent agreement with these predictions.

SEARCHES FOR OTHER DECAYS AND CHARGES

We searched for a number of plausible other decay modes of the D^*_{sJ} and D_{sJ}, but found no signals. The 90% CL ULs from these searches are given in Tables 1 and 2. The absence of some of these modes encourage the assumed spin-parity assignments for these particles. If the D^*_{sJ} is 0^+, $D_{sJ}^{*+} \to D_s^+\gamma$ and $D_{sJ}^{*+} \to D_s^+\pi^+\pi^-$ decays would be forbidden (by angular momentum and by angular momentum and parity, respectively). We do not observe either decay at the level of a few per cent of the $D_s^+\pi^0$ yield, supporting the 0^+ assignment. If the D_{sJ} is 1^+, angular momentum and parity forbid $D_{sJ} \to DK$. The D_{sJ} lies above the DK threshold but it is narrow suggesting that $D_{sJ} \not\to DK$. Naive $D\pi$ molecule models suggest that D^*_{sJ} and D_{sJ} should exist in other charge states. We searched for signals in neutral and doubly charged states decaying to $D^\pm_s \pi^\mp$, $D^{*\pm}_s \pi^\mp$, $D^\pm_s \pi^\pm$, and $D^{*\pm}_s \pi^\pm$. We see no signals in any of these modes and find that $\sigma(\text{mode})\, \mathcal{B}(\text{mode}) \lesssim 10\%$ of the corresponding $D^+_s \pi^0$ or $D^{*+}_s \pi^0$ rate (90% CL ULs).

SUMMARY AND CONCLUSIONS

CLEO observed a new narrow state tentatively identified as $D_{sJ}(2463)^+$ and confirmed the existent of the $D^*_{sJ}(2317)^+$ state observed by BaBar. CLEO results for the masses and widths of these states are given in Eqs. (4) and (5). The measured decay widths are consistent with 0. The mass differences $\Delta M(D^+_{sJ})$ and $\Delta M(D^{*+}_{sJ})$ are equal (Eq. (6)) within quite small uncertainties, in accord with chiral multiplet models [10, 11].

CLEO results also support J^P assignments of 0^+ and 1^+, respectively, to the D^{*+}_{sJ} and D^+_{sJ} states. However, high statistics angular analyses and searchers for decays that can bear on J^P are required to firmly establish J^P values for these states. Belle [3, 4, 5] and Babar [6, 7] have already inaugurated this program.

Finally, CLEO found no significant signals in searches for other $D^{(*)}_{sJ}$ charge states.

ACKNOWLEDGEMENTS

I am delighted to acknowledge the contributions of my CLEO and CESR colleagues, whose outstanding effort resulted in the results presented here. This work was supported by the US National Science Foundation and US Department of Energy. I appreciate the invitation to speak at this conference and the successful effort of the organizers and staff to make the conference so pleasant.

REFERENCES

[1]   CLEO Collaboration, D. Besson et al. Phys. Rev. D **68**, 032002 (2003).

[2]   BaBar Collaboration, B. Aubert et al., Phys. Rev. Lett. **90**, 242001 (2003).

[3]   Belle Collaboration, P. Krokovny et al., hep-ex/0308019, (submitted to Phys.Rev.Lett.).

[4]   P. Krokovny in these proceedings.

[5]   Belle Collaboration, Y. Mikami et al., hep-ex/0307052, (submitted to Phys.Rev.Lett.).

[6]   R.N. Cahn in these proceedings.

[7]   S. Spanier in these proceedings.

[8]   R.N. Cahn, and J.D. Jackson, Phys. Rev. D **68**, 037502 (2003).

[9]   Particle Data Group, K.Hagiwara et al., Phys. Rev. D **66**, 010001 (2002).

[10]   M.A. Nowak, M. Rho, and I. Zahed, Phys. Rev. D **48**, 4370 (1993).

[11]   W.A. Bardeen and C.T. Hill, Phys. Rev. D **49**, 409 (1994).

[12]   See Ref. [1] for further references to theory papers.

[13]   W.A. Bardeen, E. Eichten, and C.T. Hill, Phys. Rev. D **68**, 054024 (2003).

Heavy Flavour Physics at HERA

J. Wagner, representing the ZEUS and H1 Collaborations

DESY FH1, Notkestrasse 85, D 22606 Hamburg, Germany

Abstract. Recent results on the production of open charm and beauty in electron proton scattering at HERA are reviewed, focussing on results obtained with the full HERA-I dataset. These results are compared with perturbative QCD calculations at NLO as well as Monte Carlo predictions at LO.

INTRODUCTION

At HERA electrons (or positrons) of energy 27.5 GeV are collided with 920 GeV protons providing a center of mass energy $\sqrt{s} \approx 318$ GeV. During the HERA-I run the HERA experiments H1 and ZEUS each collected data corresponding to an integrated luminosity of about $110 \, \text{pb}^{-1}$, thus allowing significant tests of perturbative QCD in the area of heavy quarks.

In electron proton collisions heavy quarks are predominantly produced via the photon gluon fusion (PGF) mechanism, where a photon emitted by the incoming electron interacts with a gluon in the proton forming a quark-anti-quark pair. The major contribution is due to the exchange of almost real photons corresponding to a photon vituality $Q^2 \approx 0$ (photoproduction). In deep inelastic scattering (DIS) Q^2 is large, often defined experimentally by $Q^2 > 2 \, \text{GeV}^2$.

Heavy quark production can be described by a factorization ansatz consisting of four parts: proton structure, perturbatively calculable hard cross section, photon structure and fragmentation. Under the assumption that any three of the four parts are understood by theory, it is possible to study the fourth part using heavy quarks.

In the heavy quark area, the investigation of the proton structure means to determine for example the structure function F_2^c and the parton densities, especially the gluon density. Further, different evolution models for the parton densities like DGLAP or CCFM can be tested.

The hard cross section has to be calculated at a scale μ which may be choosen to be the heavy quark mass m_Q if other scales such as Q^2 are small. The transverse momentum p_T of a heavy flavour hadron, the transverse energy E_T of a heavy flavour jet or the virtuality are also possible scales leading to a multi scale problem. At large scales the heavy quarks have to be treated in the same way as the light quarks u, d, s.

At low Q^2 the photon exhibits hadronic behaviour, leading to reactions in which a parton inside the photon scatters off a parton in the proton. If the photon fluctuates into a $c\bar{c}$ pair before the hard interaction the process is called charm excitation. In these resolved processes the relative momentum fraction x_γ of the interacting parton to the photon momentum is $x_\gamma < 1$ while it is $x_\gamma = 1$ if the photon interacts directly as pointlike object.

The fourth part of the factorization ansatz - the fragmentation - describes the transition

CP717, *Hadron Spectroscopy: Tenth International Conference,*
edited by E. Klempt, H. Koch, and H. Orth

of coloured quarks into colourless hadrons. Here the question of the universality of fragmentation is important.

The data will be compared with leading order (LO) MC predictions as well as with next to leading order (NLO) calculations. Three different schemes are distinguished for the NLO calculations. The fixed order massive scheme ([1] for photoproduction and [2] for DIS) should give the best description near to threshold and thus for example at low p_T and low Q^2. The four flavour massless scheme [3] should hold at large scales for example at large p_T. Apart from the light quarks the heavy quarks are considered in this scheme also as active flavours in the proton or photon. The third scheme involves the matched calculations [4], which are an unification of both schemes and should provide a good description in the whole kinematic region. For the LO MC generators different evolution models like DGLAP or CCFM can be used while in the NLO calculations always DGLAP evolution is applied. The MC generators AROMA [5], RAPGAP [6] and PYTHIA [7] use DGLAP evolution, while CASCADE [8] is based on CCFM evolution. AROMA and CASCADE simulate only the direct photon contribution, but CASCADE contains implicitly some contributions from charm excitation. In contrast to the DGLAP evolution model CCFM requires no ordering of the transverse gluon momentum k_T along the gluon ladder. This leads to a k_T dependent gluon density and hard scattering matrix element.

CHARM PRODUCTION

Most measurements of charm production at HERA use the reconstruction of D^* mesons, exploiting the well known mass difference method. But to determine the relative fragmentation fractions of charm hadrons, H1 and ZEUS performed measurements of all charm ground states (D^+, D^0, D_s, D^{*+}, Λ_c (ZEUS)) in DIS or photoproduction respectively [9, 10]. The fragmentation fractions obtained at HERA are in good agreement with those from e^+e^- annihilations. ZEUS has measured the fragmentation function for D^* mesons in addition using D^* events associated with a jet in photoproduction [11]. Again good agreement with the fragmentation functions determined from the e^+e^- experiments OPAL and ARGUS is found. Altogether this indicates fragmentation universality.

Differential D^* coss section in DIS are measured [12, 13]. The data of the two experiments are consistent, but small differences of the theoretical interpretation appear. As an example the pseudorapidity η_{D^*} is shown in figure 1a) and b). At H1 the NLO calculations in the figure 1a), using peterson fagmentation, CTEQ5F3 as proton parton density and $m_c = 1.4\,\text{GeV}$ as the central value for the charm mass, is systematically below the data especially in the forward region (large η_{D^*}). However, the CASCADE MC leads to a good description of the data. At ZEUS the NLO calculation is doing a better job especially if the ZEUS NLO QCD fit to the inclusive DIS data from ZEUS and the fixed target experiments is used as a parameterization of the parton densities. The best description is obtained if the Lund string fragmentation instead of the Peterson fragmentation is used.

In the same analysis ZEUS has determined the charm contribution F_2^c to the structure function F_2 of the proton shown in figure 1c) [12]. Nice agreement between the ZEUS

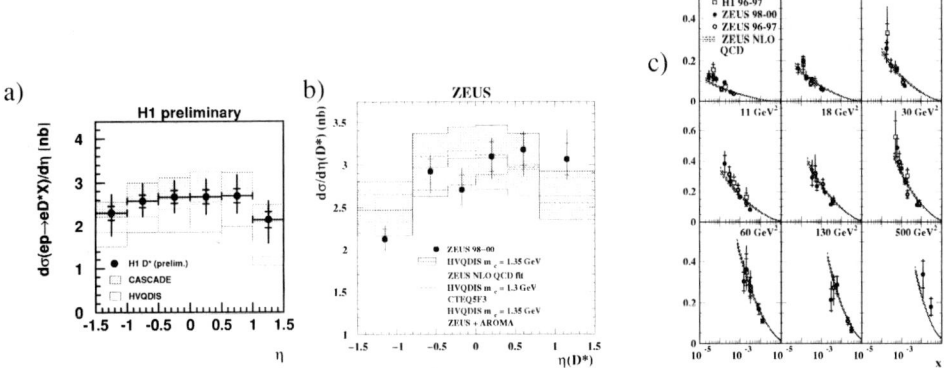

FIGURE 1. a) + b) Psedorapidity of the D^* in DIS measured by H1 (a) or by ZEUS (b) respectively. c) Charm contribution F_2^c to the structure function F_2 compared with the NLO QCD fit.

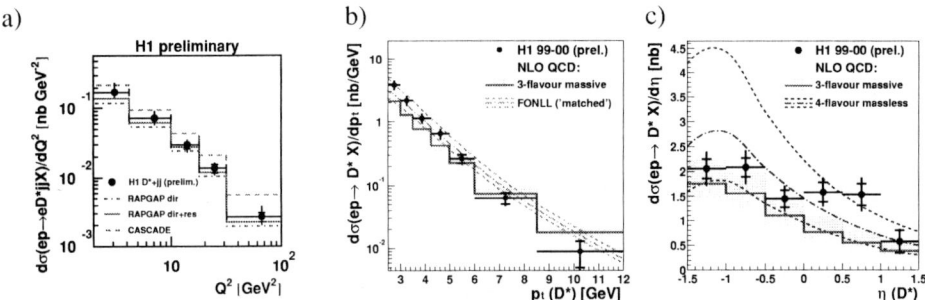

FIGURE 2. a) Q^2 distribution for D^* events associated with two jets. b) + c) Differential D^* cross sections measured in photoproduction by H1. Shown are the transverse momentum $p_T(D^*)$ and the pseudorapidity η_{D^*} of the D^*.

and H1 data and the NLO QCD fit is obtained, meaning that the prediction of the charm contribution to F_2 from scaling violations is consistent with the F_2^c measurement.

To study the production process in more detail H1 has selected events with two jets in addition to a D^* (see figure 2a) [13]. The CASCADE MC which describes the inclusive D^* data at H1 well, lies now above the data. The direct RAPGAP MC is below the data but the description of the data improves with the resolved contribution. However, the sum of direct and resolved contributions does not give a good description of the H1 inclusive D^* data.

The D^* measurements in photoproduction are compared to the three different schemes of the NLO calculations. As an example the $p_T(D^*)$ and η_{D^*} distributions obtained by H1 are presented in figures 2b) and c) [14]. The massive NLO calculations are below the data especially at low $p_T(D^*)$, while the massless and matched calculations give a reasonable agreement. ZEUS has measured in addition double differential cross sections with a larger Q^2 range [15] compared to the H1 measurement. The matched calculations

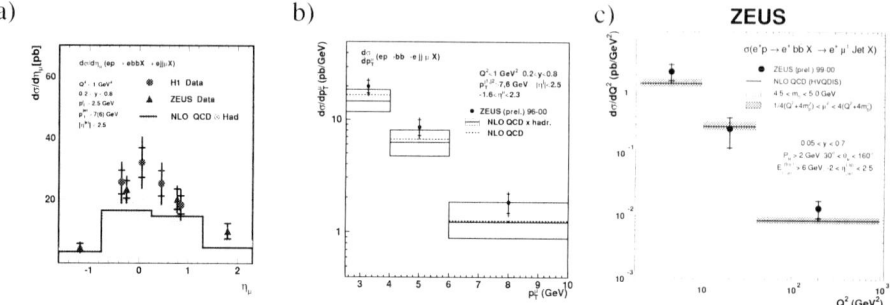

FIGURE 3. Visible cross section $\sigma(ep \to eb\bar{b}X \to e\mu\ dijet(jetDIS)X)$ measured in photoproduction (a+b) and in DIS (c).

are only at low $p_T(D^*)$ close to the data and in general the NLO calculations are below the data at medium $p_T(D^*)$ and large η_{D^*}.

To investigate the photon structure ZEUS has studied D^* mesons associated with two jets [16]. x_γ is reconstructed with the jets and then the direct and resolved data set are separated via a cut on x_γ at 0.75. From the $\cos\theta^*$ distribution, the angle between the proton and the D^* jet in the rest frame of the two jets, a large contribution from charm out of the photon is confirmed in the LO picture.

BEAUTY PRODUCTION

At HERA the total $b\bar{b}$ cross section is about two orders of magnitude smaller than the $c\bar{c}$ cross section. Nevertheless the theoretical prediction, should be more reliable for beauty. Both experiments use semileptonic decays of beauty hadrons to tag beauty. Both the relatively long lifetime of b hadrons (H1, using their silicon vertex detector) and the high masss of the b hadrons (H1, ZEUS) are exploited. Due to the high b mass the decay particles emerge at larger angles relative to the outgoing heavy flavour meson or jet. Measurements have been performed in DIS and photoproduction [17, 18, 19]. Figures 3a) and b) show the transverse momentum of the muon $p_T(\mu)$ and the pseudorapidity of the muon η_μ in photoproduction. The data from both experiments are consistent. The massive NLO QCD prediction is somewhat below the data but the shape is reasonably well described. At H1 it seems that the discrepancies between NLO predictions and data are larger at small $p_T(\mu)$, while at ZEUS this effect is not observed. In DIS the NLO QCD calculations describe apart from the shape also the normalization. As an example the Q^2 distribution obtained by ZEUS is shown in figure 3c). The LO MC generators are all below the data though CSCADE is closest.

If in addition to the muon a D^* is reconstructed to get more information about the heavy quark final state, it is possible to separate charm and beauty via the charge and angle correlations between the two particles. In the case of charm only one correlation is possible. Here the angle $\Delta\Phi$ between the D^* and the muon is large and the two particles have opposite charges. For beauty three correlations are possible due to the fact that here the muon can come from the semileptonic decay of a beauty or charm hadron. The $\Delta\Phi$

610

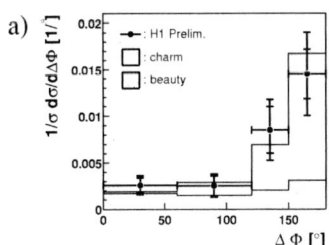
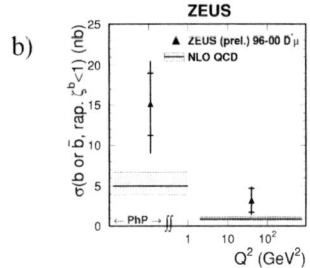

FIGURE 4. $D^*\mu$ correlations. a) $\Delta\Phi$ distribution obtained by H1, where $\Delta\Phi$ denotes the azimuthal angle difference between the D^* and the muon. b) Comparison between the ZEUS b or $\bar{b}$ cross section and the NLO QCD prediction.

distribution obtained by H1 is shown in figure 4a) [20]. The AROMA MC describes the shape of the data quite well. In figure 4b) [21] a comparison between the ZEUS b or $\bar{b}$ cross section and the NLO QCD predictions is presented. The data are all above the NLO QCD calculations but the errors are still large.

CONCLUSIONS AND OUTLOOK

Charm production at HERA has been extensively studied. The QCD calculations - LO predictions or NLO calculations - still have some problems to describe all aspects observed in the data. New precise differential beauty measurements have been shown. The data points are in photoproduction all above the NLO QCD predictions with discrepancies of about $\lesssim 1.5\,\sigma$ for each measurement.

The work is ongoing to finalize analyses on HERA-I data and first HERA-II data will come in near future. With the improved detectors (silicon vertex detectors, forward tracking, trigger,..) the investigation of new kinematic regions like the forward region is possible. Further the statistics will increase and with that new measurements like F_2^b or double tagging should become possible.

REFERENCES

1. S. Frixione, M.L. Mangano, P. Nason, G. Ridolfi, Phys. Lett., **B348** (1995), 633.
   S. Frixione, P. Nason, G. Ridolfi, Nucl.Phys., **B454** (1995), 3
2. E. Laenen, S. Riemersma, J. Smith, W.L. van Neerven, Phys.Lett., **B291** (1992), 325.
   E. Laenen, S. Riemersma, J. Smith, W.L. van Neerven, Nucl.Phys., **B392** (1993), 162 and 229.
   E. Laenen, S. Riemersma, J. Smith, W.L. van Neerven, Nucl.Phys., **B347** (1995), 143.
   B.W. Harris, J. Smith, Nucl.Phys., **B452** (1995), 109. and **B353** (1995), 535.
3. B.A. Kniehl, hep-ph/0211008, 2002.
4. M. Cacciari, S, Frixione, P. Nason, hep-ph/0102134, 2001.
5. G. Ingelman et al., hep-ph/9605285, 1996.
6. H. Jung, Comp.Phys.Comn. **86** (1995) 147.
7. T. Sjostrand, Comp.Phys.Comm. **82** (1994), 474.
8. H. Jung and G. Salam, Eur.Phys.J. **19** (2001), 351.

9.   H1 Collaboration, http://www-h1.desy.de/h1/www/publications/htmlsplit/H1prelim-02-076.long.html.
10. ZEUS Collaboration, http://www-zeus.desy.de/physics/phch/conf/eps03/564/paper.pdf.
11. ZEUS Collaboration, http://www-zeus.desy.de/physics/phch/conf/amsterdam_paper/HFL/cfrag.ps.gz.
12. ZEUS Collaboration, DESY-03-115, submitted to Phys.Rev.D
13. H1 Collaboration, http://www-h1.desy.de/h1/www/publications/htmlsplit/H1prelim-03-074.long.html.
14. H1 Collaboration, http://www-h1.desy.de/h1/www/publications/htmlsplit/H1prelim-03-071.long.html.
15. ZEUS Collaboration, http://www-zeus.desy.de/physics/phch/conf/amsterdam_paper/HFL/dstargamma.ps.gz.
16. ZEUS Collaboration, DESY-03-015, submitted to Physics Letters B.
17. H1 Collaboration, http://www-h1.desy.de/h1/www/publications/htmlsplit/H1prelim-03-072.long.html.
18. ZEUS Collaboration, http://www-zeus.desy.de/physics/phch/conf/amsterdam_paper/HFL/disB.ps.gz.
19. ZEUS Collaboration, http://www-zeus.desy.de/physics/phch/conf/amsterdam_paper/HFL/b_php.ps.gz.
20. H1 Collaboration, http://www-h1.desy.de/h1/www/publications/htmlsplit/H1prelim-02-071.long.html.
21. ZEUS Collaboration, http://www-zeus.desy.de/physics/phch/conf/eps03/575/paper.pdf.

Theoretical Concepts

Lattice QCD meets experiment in hadron physics

Christine Davies[*] and Peter Lepage[†]

[*]Dept. of Physics and Astronomy, University of Glasgow, Glasgow, G12 8QQ, UK
[†]Laboratory for Elementary-Particle Physics, Cornell University, Ithaca, NY 14853, USA

Abstract. We review recent results in lattice QCD from numerical simulations that allow for a much more realistic QCD vacuum than has been possible before. Comparison with experiment for a variety of hadronic quantities gives agreement to within statistical and systematic errors of 3%. We discuss the implications of this for future calculations in lattice QCD, particularly those which will provide input for *B* factory experiments.

INTRODUCTION

QCD is a key component of the Standard Model of particle physics. It gives us a rich spectrum of bound states of quarks and gluons whose properties are predictable from QCD if we can solve the theory. QCD is strongly coupled in this regime, however, and we need the non-perturbative techniques of lattice QCD to do this from first principles.

Most (but not all) questions which lattice QCD can address require calculations with a precision of a few percent to answer them. These include the spectrum of hadrons, their internal structure and decay rates. In particular, the hunt for internal inconsistencies in the Standard Model which could lead to new physics requires calculations of hadronic weak matrix elements to 2-3% to match the experimental errors that will become possible.

Figure 1 shows the recent status of the combined experimental and theoretical efforts to pin down the vertex of the Cabibbo-Kobayashi-Maskawa unitarity triangle [1]. The different circular constraints on the vertex come from different decay rates of *B* and *K* mesons and are found by dividing experimental rates by theoretical results obtained in lattice QCD. The current constraints are strongly limited by current lattice QCD errors of around 20%.

Lattice QCD is hard and numerically very expensive. Recent progress [2] has at last made precision calculations look possible and we will concentrate on that work and its implications in this review.

LATTICE QCD CALCULATIONS

Lattice QCD calculations proceed by the discretisation of a 4-d box of space-time into a lattice. The QCD Lagrangian is then discretised onto that lattice. The spacing between the points of the lattice, a, is ≈ 0.1fm in current calculations and the length of a side of the box is $L \approx 3.0$fm. Thus our simulations can cover energy scales from ≈ 2 GeV down

CP717, *Hadron Spectroscopy: Tenth International Conference*,
edited by E. Klempt, H. Koch, and H. Orth

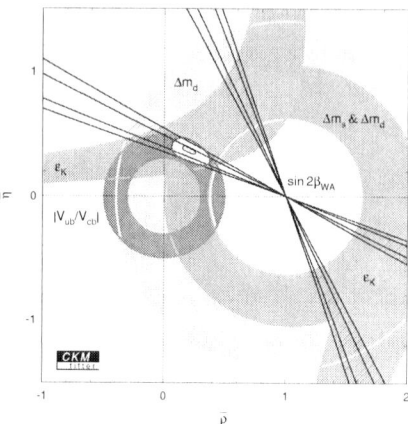

FIGURE 1. Recent constraints on the unitarity triangle from CKMfitter [1]

to ≈ 100 MeV.

The Feyman Path Integral is evaluated numerically in a two-stage process. In the first stage sets of gluon fields ('configurations') are created which are representative 'vacuum snapshots'. In the second stage, quarks are allowed to propagate on these background gluon field and hadron correlators are calculated. The dependence of the correlators on lattice time is exponential. From the exponent the masses of hadrons of a particular J^{PC} can be extracted, and from the amplitude, simple matrix elements.

QCD as a theory has a number of unknown parameters, the overall dimensionful scale of QCD ($\equiv$ the bare coupling constant) and the bare quark masses. To make predictions, these parameters must be fixed from experiment. In lattice QCD we do this by using one hadron mass for each parameter. The quantity which is equivalent to the overall scale of QCD on the lattice is the lattice spacing.

Lattice calculations are hard and time-consuming. Progress has occurred in the last thirty years through gains in computer power but also, often more importantly, through gains in calculational efficiency and physical understanding. One particular area which revolutionised the field from the mid-1980s was the understanding of the origin of discretisation errors and their removal by improving the lattice QCD Lagrangian. Discretisation errors appear whenever equations are discretised and solved numerically. They manifest themselves as a dependence of the physical result on the unphysical lattice spacing. In lattice QCD, as elsewhere, they are corrected by the adoption of a higher order discretisation scheme. The complication in a quantum field theory like QCD is the presence of radiative corrections to the coefficients in the higher order scheme which must be determined (using perturbation theory, for example [3]).

Physical understanding of heavy quark physics on the lattice has also made a huge difference to the feasibility of calculating matrix elements relevant to the B factory programme on the lattice. The use of non-relativistic effective theories requires the lattice to handle only scales appropriate to the physics of the non-relativistic bound states and not the (large) scale associated with the b quark mass. B physics is now one of the areas where lattice QCD can make most impact.

One area which has remained problematic, but which this year's results have addressed successfully, is the handling of light quarks on the lattice. In particular the problem is that of how to include the dynamical (sea) $u/d/s$ quark pairs that appear as a result of energy fluctuations in the vacuum. We can safely ignore $b/c/t$ quarks in the vacuum because they are so heavy, but we know that light quark pairs have significant effects, for example in screening the running of the coupling constant and in generating Zweig-allowed decay modes for unstable mesons.

Because quarks are fermions, they cannot be simulated directly on the computer, but must be 'integrated out' of the Feynman Path Integral. This leaves an QCD Lagrangian in terms of gluon fields which includes $\ln(\det(M))$ where M is an enormous ($10^7 \times 10^7$) sparse matrix. The inclusion of dynamical quarks is then numerically very expensive, particularly as the quark mass is reduced towards the small values which we know the u and d quarks have.

Many calculations even today use the 'quenched approximation' in which the light quark pairs are ignored. Results then suffer from a systematic error of $\mathcal{O}(20\%)$. A serious problem with the quenched approximation is the lack of internal consistency which means that the results depend on the hadrons that were used to fix the parameters of QCD. This ambiguity plagues the lattice literature.

Other calculations have included 2 flavours of degenerate dynamical quarks, i.e. u and d, but with masses 10-20$\times$ the physical ones. This approximation is better than the quenched approximation but large uncertainties remain because the s quark is omitted. Results must also be extrapolated to the physical u/d quark mass and chiral perturbation theory is a good tool for this. However, chiral perturbation theory only works well if the u/d quark mass is light enough and, for errors at the few percent level, this means less than $m_s/2$. This has been impossible to achieve in most calculations.

New results this year [2] have included u, d and s quarks in the vacuum, with light enough u/d masses to perform accurate chiral extrapolations. The results use a new discretisation of the quark action - the numerically fast improved staggered formalism. This formalism is well-matched to the supercomputing power of a few Tflops that is currently achievable.

Improved staggered quarks

The starting point for the staggered quark formalism is the naive discretisation of the Dirac quark action onto a lattice. This action has good features: chiral symmetry and discretisation errors that appear only as the square and higher powers of the lattice spacing. The naive discretisation suffers from the notorious doubling problem, however. A single quark species on the lattice gives rise to 16 quark species, or tastes, on a 4-d lattice. The additional tastes appear around the edges of the Brillioun zone, where $p \approx \pi/a$, as copies of a $p \approx 0$ quark. This would not be a problem if there were no interaction between the different tastes since the quark action would then fall apart into 16 different pieces in an appropriate basis and we could take $\det(M)^{(1/16)}$ in simulations to give the effect of 1 quark flavour.

There is interaction between the different tastes, however. It is mediated by highly virtual gluons, with momenta around π/a. A quark of one taste can absorb or emit such a high momentum gluon and turn into a quark of another taste. The effects of this taste-changing interaction are quite severe for the naive action, giving rise to large discretisation errors (even though formally of $\mathcal{O}(a^2)$) and large perturbative renormalisation factors, e.g. for the quark mass, when translating from the lattice scheme to the continuum. The degeneracy in mass of mesons made from quarks of different taste is lost. This is most noticeable for the pions because there is a light Goldstone boson.

Because the taste-changing interaction is a high momentum one it can be understood in lattice perturbation theory. In particular, the effects can be significantly improved by suppressing the coupling of quarks to gluons of momenta π/a in any direction. This is achieved by 'smearing' the gluon field in the action in a particular way [4, 5], and can be thought of as part of the standard Symanzik programme for systematically removing discretisation errors from lattice actions.

It is simple to 'stagger' the naive action and its improved variant to remove an exact degeneracy of a factor of 4 in tastes which arises from the spin degree of freedom. This results in an action with 4 doublers which can be simulated on the lattice using $\det(M)^{(1/4)}$ per flavour. It is very fast numerically because there is only one spin degree of freedom per site and the eigenvalues of M are well behaved. This is what has allowed the MILC collaboration to generate ensembles of configurations which include u, d, and s quarks in the vacuum with much more realistic masses than before [6].

Some worries remain about potential non-locality in the action as the result of taking the fourth root. However, this causes no problem in perturbative QCD where a simple power series in x is obtained for an action with $\det(M)^x$. Stringent non-perturbative tests are also then needed. Luckily these tests are possible in this formalism with present day computers because of its speed, and are exactly the calculations required to test (lattice) QCD. The results, shown in the next section, speak for themselves.

RECENT RESULTS

The MILC collaboration have made sets of ensembles of gluon field configurations which include 2 degenerate light dynamical quarks (u,d) and 1 heavier one (s) [6]. Taking the u and d masses the same makes the lattice calculation much faster and leads to negligible errors in isospin-averaged quantities. The dynamical s quark mass is chosen to be approximately correct based on earlier studies (in fact the subsequent analysis shows that it was slightly high and further ensembles are now being made with a lower value). The dynamical u and d quarks take a range of masses down as low as a sixth of the (real) m_s. The sets of ensembles divide into two different values of the lattice spacing, 0.13fm and 0.09fm, and the spatial lattice volume is $(2.5\text{fm})^3$, reasonably large. Analysis of hadronic quantities on these ensembles has been done by the MILC and HPQCD collaborations [2].

There are 5 bare parameters of QCD relevant to this analysis: $\alpha_s, m_{u/d}, m_s, m_c$ and m_b. The lattice spacing takes the place of α_s in lattice QCD. It is important that these parameters are fixed using the masses of 'gold-plated' hadrons, i.e. hadrons which are

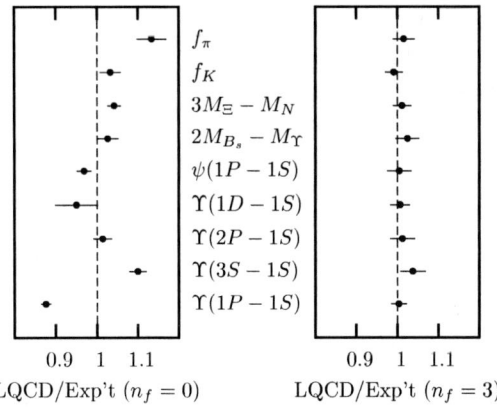

FIGURE 2. Lattice QCD results divided by experiment for a range of 'gold-plated' quantities which cover the full range of hadronic physics [2]. The unquenched calculations on the right show agreement with experiment across the board, whereas the quenched approximation on the left give systematic errors of $\mathcal{O}$(10-20%).

well below their strong decay thresholds. Such hadrons are well-defined experimentally and theoretically and should be accurately calculable in lattice QCD. Using them to fix parameters will not then introduce unnecessary additional systematic errors into lattice results for other quantities. This has not always been done in past lattice calculations, particularly in the quenched approximation. It becomes an important issue when lattice QCD is to be used as a precision calculational tool. We use the radial excitation energy in the Υ system (i.e. the mass splitting between the Υ' and the Υ) to fix the lattice spacing and m_π, m_K, m_{D_s} and m_Υ to fix the quark masses.

We can then focus on the calculation of other gold-plated masses and decay constants. If QCD is correct and lattice QCD is to work it must reproduce the experimental results for these quantities precisely. Figure 2 shows that this indeed works for the unquenched calculations with u, d and s quarks in the vacuum. A range of gold-plated hadrons are chosen which range from decay constants for light hadrons through heavy-light masses to heavyonium. This tests QCD in different regimes in which the sources of systematic error are very different and stresses the point that QCD predicts a huge range of physics with a small set of parameters.

References [7, 8, 9, 10] give more details on the quantities shown in Figure 2. Here we will discuss some of these. Figure 3 shows the radial and orbital splittings in the $b\bar{b}$ (Υ) system for the quenched approximation ($n_f = 0$) and with the dynamical MILC configurations with 3 flavours of dynamical quarks. Our physical understanding of the Υ system is very good and there are a lot of gold-plated states well below decay thresholds, which makes it a valuable system for lattice QCD tests. We use the standard lattice NRQCD effective theory for the valence b quarks, which takes advantage of the non-relativistic nature of the bound states. The lattice NRQCD action is accurate through v^4 where v is the velocity of the b quark in its bound state. This means that spin-independent splittings, such as radial and orbital excitations, are simulated through next-to-leading-

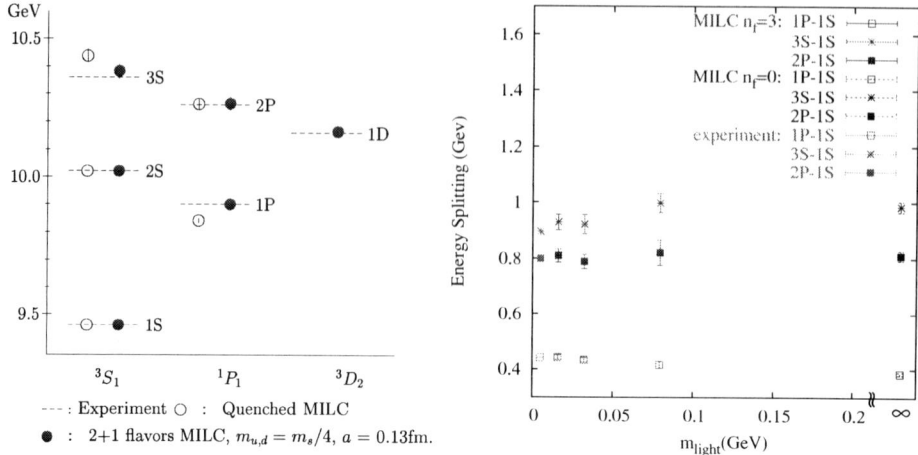

FIGURE 3. Radial and orbital splittings in the Υ system from lattice QCD in the quenched approxima- tion and including u, d and s dynamical quarks. In this plot the lattice spacing was fixed from the radial excitation energy i.e. the splitting between the Υ' and the Υ and the b quark mass was tuned to get the Υ mass correct. The right hand plot shows these splittings plotted as a function of the bare dynamical u/d quark mass for several ensembles of MILC configurations. The leftmost lattice points are the ones used in the left-hand plot and Figure 2.

order and should be accurate to $\approx 1\%$. Thus the test of QCD using these splittings is a very accurate one. The fine structure in the spectrum is only correct through leading- order at present and more work must be done to bring this to the same level and allow tests against, for example, the splittings between the different χ_b states [8].

The Υ system is a good one for looking at the effects of dynamical quarks because we do not expect it to be very sensitive to dynamical quark masses. The momentum transfer inside an Υ is larger than any of the u, d or s masses and so we expect the radial and orbital splittings to simply 'count' the number of dynamical quarks once we have reasonably light dynamical quark masses. The righthand plot of Figure 3 shows this to be true - the splittings are independent of the dynamical u/d quark mass in the region we are working in (and therefore for the points plotted in the left hand figure of Figure 3 and in Figure 2).

The π and K decay constants are important light hadron matrix elements, related to the purely leptonic decay rate via a W, and experimentally well-known. These are very sensitive to light quark masses and require a well-controlled extrapolation in the u/d quark mass and interpolation in the s quark mass to get accurate results to compare to experiment. Chiral perturbation theory can be used to perform the u/d quark mass extrapolation provided the masses used on the lattice are small enough for the expansion in powers of quark mass ($\equiv m_\pi^2/(1\text{GeV}^2)$) and its logarithms to work well. In practise this means that second order chiral perturbation theory should work at the 2% level for $m_{u/d} < m_s/2$. Note that the error is set by the largest quark mass used in the chiral fits, *not* the smallest.

Figure 4 shows the results and chiral extrapolation for the decay constants on the

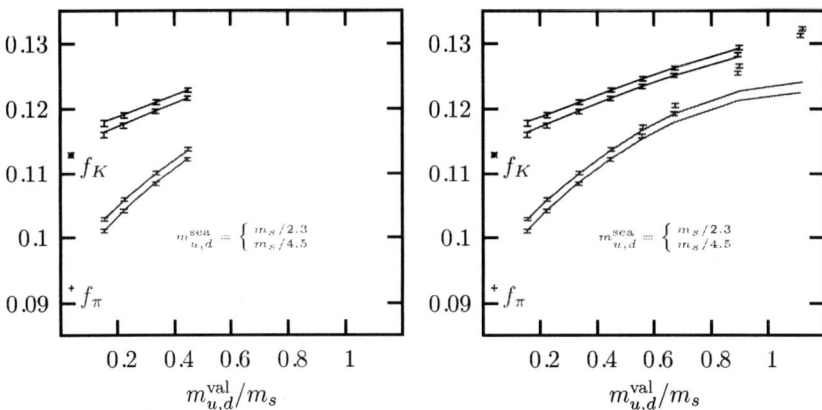

FIGURE 4. Results for the π and K decay constants as a function of light quark mass for two dynamical MILC ensembles at a lattice spacing of 0.09fm. The plot on the left shows the chiral extrapolation using only results with valence u/d quark masses $< m_s/2$ [2]. The chiral extrapolation must subsequently be corrected for the incorrect valence and sea s quark mass to give the results in Figure 2. The plot on the right shows that this chiral fit from light u/d quark masses does not agree well with the data for $m_{u/d} > m_s/2$.

ensembles of MILC configurations with $m_{u/d}^{sea} = m_s/2.3$ and $m_s/4.5$ at a lattice spacing of 0.09 fm. The curves in the left plot show the chiral extrapolation using only results with $m_{u/d}^{valence} < m_s/2$. This extrapolation has to be corrected, using the lattice results, to interpolate to the physical s quark mass for both sea and valence s quarks. This then gives the results shown in Figure 2 which agree with experiment. The plot on the right shows what happens when the chiral extrapolation fit obtained in the left plot is evaluated for larger valence $m_{u/d}$. The f_π results start to show clear disagreement for $m_{u/d} > m_s/2$, which makes the problem of performing accurate chiral extrapolations using results with $m_{u/d} > m_s/2$ obvious. Previous lattice calculations have been forced by computing cost to work only in this regime, with the added problem that the sea $m_{u/d}$ is also large [11].

Another gold-plated hadron mass is that of the nucleon. A full chiral extrapolation of results for this on the MILC configurations has not yet been done. The left-hand plot of Figure 5 shows very encouraging signs that an answer in agreement with experiment will be found [7]. There is a clear sign of dependence on the lattice spacing, however, which will have to be taken into account. Combinations of baryon masses can be made which are relatively insensitive to u/d quark masses and other effects and it is one of these, $3m_\Xi - m_N$, which is plotted in Figure 2.

It is important to realise that accurate lattice QCD results are not going to be obtainable in the near future for every hadronic quantity of interest. What these results show is that 'gold-plated' quantities should now work. Gold-plated hadrons are those well below decay threshold for strong decays. Unstable hadrons, or even those within 100 MeV or so of Zweig-allowed decay modes, have a strong coupling to their real or virtual decay channel which is not correctly simulated on the lattice. The problem is that, with the lattice volumes being used, the allowed non-zero momenta are typically greater

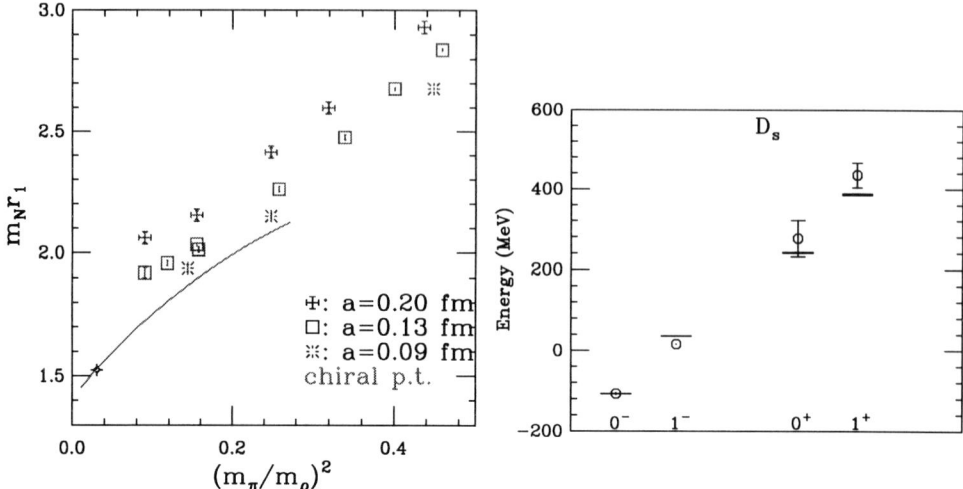

FIGURE 5. The left-hand plot shows results for the nucleon mass on MILC ensembles for different lattice spacings and dynamical quark masses. The nucleon mass is given in units of r_1, a parameter from the heavy quark potential whose physical value is 0.32fm. The dynamical quark mass is indicated by the variable m_π^2/m_ρ^2. The curve roughly indicates chiral perturbation theory [7]. The right-hand plot shows the spectrum of D_s states obtained from the MILC dynamical configurations with $m_{u/d} = m_s/4$ and lattice spacing 0.13fm [12].

than 400 MeV and this significantly distorts the decay channel contribution. Much larger simulations will be necessary to handle these hadrons.

Gold-plated hadrons include: $\pi, K, D, D_s, J/\psi, \Upsilon, B, B_s, p, n, \Lambda, \Omega$ etc. The following are *not* gold-plated: $\rho, \phi, D^*, D_{sJ}, \Delta, N^*$, pentaquarks, glueballs and hybrids in general. Lattice calculations will not get the masses right for non-gold-plated hadrons even when light dynamical quarks are included. This does not preclude lattice calculations giving useful qualitative results and insight but these points should be borne in mind for any quantitative comparison.

Figure 5 also shows the spectrum of D_s states obtained on the dynamical MILC configurations [12]. The valence c quarks are simulated using an effective theory which, in a similar way to the Υ above, should be accurate for spin-independent splittings and not quite so accurate for fine structure in the spectrum. The hyperfine splitting between the D_s and D_s^*, for example is currently missing a radiative correction to the term in the action proportional to the spin coupling to the chromo-magnetic field. This is being calculated in lattice perturbation theory [3]. Also shown are the scalar and axial vector orbital excitations compared to the recent experimental results for these mesons. The lattice calculation is giving a high result, albeit with large statistical errors at present. However, a high result is consistent with the fact that these mesons are not gold-plated and the lattice calculation does not currently include correctly the coupling to their decay modes.

Decay rates which can be accurately calculated for gold-plated hadrons are those in which there is at most one (gold-plated) hadron in the final state. This therefore includes

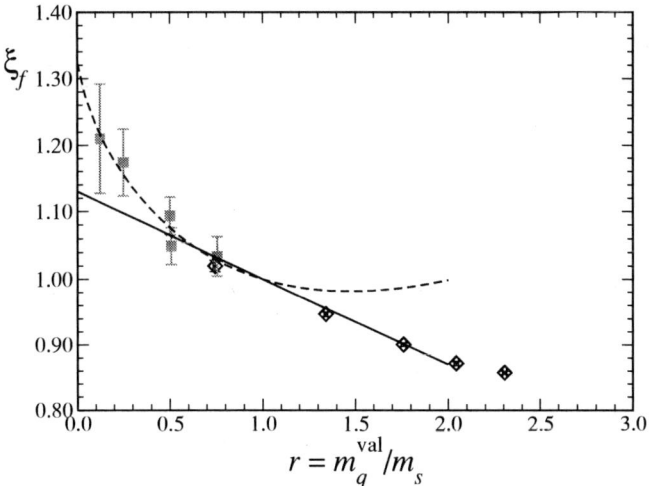

$$r = m_q^{val}/m_s$$

FIGURE 6. Results for the ratio of f_{B_s}/f_{B_d}, as a function of valence u/d quark mass in units of m_s [14]. The grey squares are from the dynamical MILC ensembles including u, d and s dynamical quarks [13]. The black diamonds are from the previous best calculation which included 2 flavours of dynamical quarks with masses $> m_s2$ [11]. The straight line is a linear extrapolation for the 2 flavour results, the curve includes the possibility of logarithms from chiral perturbation theory. This ratio, for physical u/d masses, appears in the ratio of oscillation frequencies for B_s and B_d mesons, which it is hoped to measure experimentally.

leptonic and semi-leptonic decays and the mixing of neutral B and K mesons. Luckily there is a gold-plated decay mode available to extract each element (except V_{tb}) of the CKM matrix which mixes quark flavours under the weak interactions in the Standard Model:

$$
\begin{pmatrix}
\begin{array}{ccc}
\mathbf{V_{ud}} & \mathbf{V_{us}} & \mathbf{V_{ub}} \\
\pi \to l\nu & K \to l\nu & B \to \pi l\nu \\
 & K \to \pi l\nu & \\
\mathbf{V_{cd}} & \mathbf{V_{cs}} & \mathbf{V_{cb}} \\
D \to l\nu & D_s \to l\nu & B \to Dl\nu \\
D \to \pi l\nu & D \to Kl\nu & \\
\mathbf{V_{td}} & \mathbf{V_{ts}} & \mathbf{V_{tb}} \\
\langle B_d|B_d\rangle & \langle B_s|B_s\rangle &
\end{array}
\end{pmatrix}
$$

As described earlier, the determination of the CKM elements and tests of the self-consistency of the CKM matrix are the current focus for the search for Beyond the Standard Model physics and lattice calculations of these decay rates will be a key factor in the precision with which this can be done.

First calculations on the dynamical MILC configurations have concentrated on the B and B_s leptonic decay rates [13, 12], because these are simplest. They are parameterised by the decay constants, f_B and f_{B_s}, and these are an important component of the mixing rate for these mesons, which constrains V_{ts} and V_{td}. Again one issue in extracting reliable lattice results for f_B and f_{B_s} is the chiral extrapolation in the u/d quark mass. Figure 6 shows results on the MILC configurations for the ratio of f_{B_s}/f_{B_d} plotted against the valence u/d quark mass [14]. The data extend into the region $m_{u/d} < m_s/2$ which will allow an accurate chiral extrapolation for the first time. Although the statistical errors are currently rather large, it seems likely that the result for this ratio will be larger than previous estimates based on extrapolations from larger u/d masses, and including only two flavours of dynamical quarks [11]. Further calculations of gold-plated matrix elements are in progress [15].

CONCLUSIONS

The impact of lattice QCD calculations has been hindered by the difficulty of including a realistic QCD vacuum. This has led to a level of systematic error far greater than the few percent needed to provide input to tests of the Standard Model, particularly those testing the CKM matrix at B factories. New results this year look set to herald a brighter future in which accurate calculations, at least for gold-plated quantities, are available from the lattice at last.

ACKNOWLEDGMENTS

We are grateful to PPARC, NSF, DoE and the EU for funding this work, and to all our collaborators on [2] for many useful discussions. A version of this talk was also presented at the 2003 Lepton Photon Conference and is published in the proceedings of that meeting.

REFERENCES

1. A. Höcker, H. Lacker, S. Laplace, F. Le Diberder, Eur. Phys. J. C**21** (2001) 225, hep-ph/0104062, http://www.slac.stanford.edu/~laplace/ckmfitter.html.
2. C. Davies *et al*, MILC/HPQCD/FNAL/UKQCD collaborations, hep-lat/0304004.
3. H. Trottier, Proceedings of LAT03, hep-lat/0310044.
4. G. P. Lepage, Phys. Rev. D**59**, 074502 (1999), hep-lat/9809157.
5. K.Orginos, D. Toussaint and R. L. Sugar, Phys. Rev D**60**, 054503 (1999), hep-lat/9903032.
6. C. Bernard *et al*, Phys. Rev. D**64**, 054506 (2001), hep-lat/0104002.
7. S. Gottlieb, Proceedings of LAT03, hep-lat/0310041.
8. A. Gray *et al*, Nucl. Phys. B (Proc. Suppl. **119**), 592 (2003), hep-lat/0209022; C. Davies *et al*, *ibid* 595, hep-lat/0209122.
9. M. di Pierro *et al*, Proceedings of LAT03, hep-lat/0310042.
10. C. Aubin *et al*, Proceedings of LAT03, hep-lat/0309088.
11. S. Aoki *et al*, hep-ph/0307039.
12. M. di Pierro *et al*, Proceedings of LAT03, hep-lat/0310045.
13. M. Wingate, Proceedings of LAT03, hep-lat/0309092.
14. A. Kronfeld, Proceedings of LAT03.
15. J. Shigemitsu *et al*, Proceedings of LAT03, hep-lat/0309039; M. Okamato *et al*, Proceedings of LAT03, hep-lat/0309107.

Strong Decays: Past, Present and Future

T.Barnes

Department of Physics, University of Tennessee, Knoxville, TN 37996, USA
Physics Division, Oak Ridge National Laboratory, Oak Ridge, TN 37831, USA

Abstract. In this talk I review the history of models of strong decays, from the original model through applications to charmonium, light and charmed mesons, glueballs and hybrids. Our current rather limited understanding of the QCD mechanism of strong decays is stressed. Regarding current and future applications of strong decay models, we note that in certain channels the very strong coupling predicted between $q\bar{q}$ basis states and the two-meson continuum may lead to strongly mixed states and perhaps molecular two-meson bound states. The relevance to the $D_{sJ}^*(2317)$ is discussed.

HISTORICAL INTRODUCTION

Origins of the $^3P_0$ model

Micu

The earliest reference in which the currently widely accepted microscopic physics of strong decays appears is due to Micu [1], in the paper "Decay Rates of Meson Resonances in a Quark Model". Micu was concerned with understanding the light P-wave mesons in the quark model, especially their widths. In the quark model, describing decays required the production of a $q\bar{q}$ pair; in lieu of a microscopic model of interactions between quarks she made the plausible assumption that the pair was produced with vacuum (0^{++}) quantum numbers, therefore in a $^3P_0$ state. No explicit quark model wavefunctions were assumed, so the implicit overlap integrals were described by two free parameters, taken from data. Micu applied this simple model to approximately 30 known light meson decays, and found that it was reasonably successful in explaining the observed partial widths. She concluded that "... this decay model proves once more the surprising viability of the quark model."

The ORSAY group

This work was follows by a series of decay model calculations by the ORSAY group (LeYaouanc *et al.*), who introduced explicit nonrelativistic quark model wavefunctions in the calculations of the decay amplitudes and a fundamental pair production amplitude γ. This cast the $^3P_0$ decay model in essentially the form in which it is used today.

CP717, *Hadron Spectroscopy: Tenth International Conference,*
edited by E. Klempt, H. Koch, and H. Orth
© 2004 American Institute of Physics 0-7354-0197-7/04/$22.00

In their initial 1973 paper [2] the ORSAY group set up the formalism of $^3P_0$ decays in the quark model and applied it to light baryon and meson decays and couplings. The importance of the D/S amplitude ratios in the decays $b_1 \to \omega\pi$ and $a_1 \to \rho\pi$ as crucial tests of the assumed $^3P_0$ quantum numbers of the $q\bar{q}$ pair was first stressed in this paper; this early and striking success of the $^3P_0$ model was strong evidence in favor of this model of strong decays. This paper also notes that the model predicts strong three-meson effective couplings and form factors, which is an application that has not yet been widely exploited, but is of great importance for the currently fashionable meson effective lagrangians and meson exchange models of hadronic reactions.

This introductory paper was followed by detailed studies of baryon decays [3, 4], which considered ≈ 100 baryon decay amplitudes. Finally, with the discovery of charmonium and states above open-charm threshold, the open-charm decays of charmonia were considered [5], and it was noted that the $\psi(4040)$ (elsewhere suggested as a D^*D^* molecule candidate due to its anomalous strong branching fractions) could be accepted as a conventional 3^3S_1 $c\bar{c}$ state, since the nodes of the radial wavefunction could plausibly weaken the DD and DD* modes.

It is straightforward to give Feynman rules for this $^3P_0$ model, since as noted by Ackleh *et al.* [6] it corresponds to the nonrelativistic limit of a $\bar{\psi}\psi$ decay interaction. The two diagrams that describe $q\bar{q}$ decays are shown in Fig.1 below.

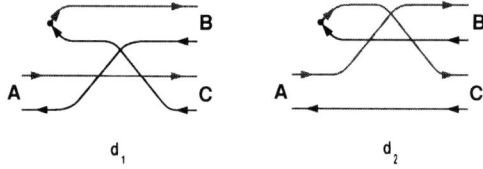

FIGURE 1.   The two independent meson decay diagrams in the $^3P_0$ model.

The Cornell charmonium decay model

The discovery of charmonium motivated the next important development in decay models. This was the introduction of a new microscopic model for the QCD mechanism underlying strong decays, as well as studies of coupled-channels effects, in a series of papers by the Cornell group [7, 8, 9].

Eichten *et al.* assumed that strong decays were driven by $q\bar{q}$ pair production from the linear confining interaction. This implied that the size of the phenomenological $q\bar{q}$ pair production amplitude γ in the $^3P_0$ model was proportional to the $q\bar{q}$ string tension, to the extent that these models could be compared.

A crucial difference between the Cornell model and most other decay and confinement models was their assumption of *timelike vector confinement*. This was subsequently rejected due to its inaccuracy in describing the splittings of the P-wave χ_J multiplet; vector confinement gives a positive spin-orbit term that does not agree with the data,

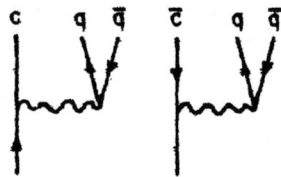

FIGURE 2. The Cornell decay model: pair production from linear vector confinement.

whereas scalar confinement has negative spin-orbit that leads to a good description of the χ_J masses. This issue is certainly not settled, and recent theoretical work by Swanson and Szczepaniak again argues in favor of vector confinement models [10].

The Cornell group did not give individual decay amplitudes and partial widths for resonances above open-charm threshold, but instead used a "resolvant" formalism to infer their combined contribution to R. This gave results that are quite similar to the current experimental R, and (even if this model of confinement proves incorrect) shows that this decay model merits further study.

Finally, the very topical issue of the level of mixing between "bare" $c\bar{c}$ basis states and the two-meson continuum was also considered by the Cornell group. They found that this is a significant effect, with (in their final version of this model [9]) the 1P χ_J states being $\approx 10\%$ meson-meson and the 2S $\psi' \approx 20\%$ meson-meson. They warn however that their two-meson intermediate states are truncated at DD^*, so the actual two-meson component may be much larger.

MODERN "SURVEYS" OF DECAY MODES

$\xi(2230)$ brou ha ha

Most of the detailed "survey" studies of strong decays that have appeared since 1980 have assumed the $^3P_0$ model, either as originally formulated or in one of several variants that modify the phase space, kinematics or the spatial dependence of the pair production amplitude. These survey papers typically attempt to consider as many open-flavor decay modes as possible, due to the "brou ha ha" associated with the question of the $\xi(2230)$.

The $\xi(2230)$ was originally reported in J/ψ radiative decays by MarkIII at SLAC [11] in KK final states (apparently both charged and neutral kaons). The state appeared remarkably narrow, hence it was considered a possible glueball candidate. However there were problems with confirmation of the $\xi(2230)$; DM2 for example did not see it, and they had slightly better statistics than Mark III. For those with long memories, the reports of the $\Theta(1542)$ at this meeting are disturbingly reminiscent of the $\xi(2230)$.

Although the narrow width of the $\xi(2230)$ made it a plausible tensor glueball candidate, assuming that it did exist, this exciting possibility was dampened by a "spoiler" paper by Godfrey *et al.* [12], who noted that L=3 $s\bar{s}$ states were expected near this mass, and they should "accidentally" be quite narrow (due to limited phase space and centrifical barriers), provided that their decays were dominated by decays to two S-wave mesons (KK, KK* and K*K*). The $^3F_2$ $s\bar{s}$ assignment was preferred because gg intermediate states in J/ψ radiative decays were predicted to populate $J^{PC} = 0^{\pm+}$ and 2^{++} final states preferentially.

This *a priori* plausible assumption of dominant decays to S-wave meson pairs was subsequently tested by Godfrey *et al.* [13], who carried out a more complete set of decay calculations. They were surprised to find that the decays of a $^3F_2$ $s\bar{s}$ meson were *not* dominated by S-wave meson pair final states. Instead the dominant mode was the S+P final state $KK_1(1270)$ by a considerable margin, so this $s\bar{s}$ state was instead predicted to be rather broad, once all the higher modes were included. Thus the $\xi(2230)$ could no longer be claimed to be an accidentally narrow $^3F_2$ $s\bar{s}$ state.

This specific result has broad implications. Evidently one should be cautious in using strong decay calculations to estimate total widths; calculations of *all* open-flavor decay amplitudes and partial widths are prudent. The assumption of dominance by a specific set of low-lying S-wave final mesons may well be inaccurate. For this reason, several recent papers on strong decays of light mesons have given results for all allowed open-flavor modes, and this approach should probably be followed in future studies of baryon and heavy meson strong decays as well.

Recent decay surveys

The past two decades have seen a series of detailed calculations of strong decay amplitudes and partial widths using the $^3P_0$ model or a variant. For mesons the best known is the encyclopaedic work by Godfrey and Isgur [14], and for baryons the corresponding paper by Capstick and Isgur [15]. These studies have been followed by more complete studies of decay modes. The paper of Barnes, Close, Page and Swanson [16] evaluates *all* open two-body modes of all nonstrange mesons expected in the quark model to 2.1 GeV; this was recently extended to all open modes of strange mesons in the same multiplets by Barnes, Black and Page [17]. Calculations of nonstrange baryon decays were extended to include modes with vector mesons by Capstick and Roberts [18]; an extensive review of baryon strong decays was recently published by the same authors [19].

New results for decays of strange mesons

As this is a conference that is largely devoted to the physics of mesons, it may be of interest to review some of the recent theoretical results on strange meson decays [17]. First, regarding the axial vectors, from the very broad width predicted for a 1^{++} state at the mass of the $f_2'(1525)$ (ca. 400 MeV) it is clear that there is no "hidden" axial vector state with the same mass and width as the f_2'. The near-threshold $f_1(1420)$ does appear consistent with the predicted width, although the nearby KK^* threshold will modify the shape of this resonance. Second, one often hears the speculation that $s\bar{s}$ mesons might be rather narrow, and hence offer attractive experimental targets. In our survey the five narrowest $s\bar{s}$ states we found without widely accepted experimental candidates were

$^1D_2$ $\eta_2(1850)$, $\Gamma = 129$ MeV, dominant mode KK^*

$^3F_4$ $f_4(2200)$, $\Gamma = 156$ MeV, dominant modes KK, KK^*, K^*K^*

3^1S_0 $\eta(1950)$, $\Gamma = 175$ MeV, dominant modes KK^*, K^*K^*

2^1P_1 $h_1(1850)$, $\Gamma = 193$ MeV, dominant modes KK^*, K^*K^*, $\eta\phi$

$^3D_2$ $\phi_2(1850)$, $\Gamma = 214$ MeV, dominant modes KK^*, $\eta\phi$

Evidently these are only moderately narrow states. As one might expect the final states KK, KK^* and K^*K^* are important for $s\bar{s}$. The mode $\eta\phi$ is much more attractive, however, as we do not expect $n\bar{n}$ ($n = u, d$) mesons to couple significantly to this channel. This mode is in effect an "$s\bar{s}$ filter", and should be much more attractive for identifying $s\bar{s}$ states than the open-strange modes involving kaons.

Turning to open-strange mesons, we find that the "strangest" state in the known strange meson spectrum is the $K^*(1414)$. First, the mass of this state appears much too light for a 2S radial vector kaon, given the nonstrange candidates $\rho(1465)$ and $\omega(1419)$. It is also very surprising that it would have a lower mass than the radial pseudoscalar kaon $K(1460)$. The decays of the $K^*(1414)$ are also a problem; in the $^3P_0$ model the dominant mode is predicted to be πK, with a branching fraction of about 30%. Experimentally, the branching fraction observed by LASS was only 6%. These discrepancies in mass and decays suggest a problem with a simple 2S radial K^* assignment. One exciting possibility is that we may be seeing the effect of mixing with exotic vector hybrid states; since there is no C-parity in kaons, the 1^{--} and 1^{-+} basis states mix, giving an overpopulation of vector states and a different mass matrix for 1^- kaons than for 1^{--} ρ and ω excited vectors. A simple comparison of strange and nonstrange excited vectors may therefore show the presence of the additional 1^{-+} hybrid basis states, and with significant hybrid mixing we would expect the excited kaons to have rather different masses and decay amplitudes than their nonstrange partners.

There are also very interesting issues regarding the singlet-triplet mixing angles of excited kaons (analogous to the $K_1(1273)$-$K_1(1402)$ mixing angle), which can be determined from decay amplitudes of these excited states. Another interesting aspect of kaon strong decays is the relative strength of η and η' modes, which depends on the angular quantum numbers in a complicated manner due to an interference between the $n\bar{n}$ and $s\bar{s}$ components of the η and η' [17]. These selection rules also have applications to B decays involving an η or η' [20].

STRONG DECAYS OF EXOTICA

Hybrids

The most interesting and influential predictions of meson decay models in recent years have been the predictions for the decay modes of exotica, specifically hybrid (excited glue) mesons. In the flux tube model of Isgur and Paton, hybrids are treated as states of quark, antiquark and flux-tube, in which the flux-tube is spatially excited about the quark-antiquark axis. When this simple quantum mechanical picture is combined with a $^3P_0$ model for pair production, which is assumed to take place along the path of the flux tube, one obtains a simple, intuitive picture of the open-flavor decay amplitudes of gluonic hybrid mesons.

Isgur, Paton and Kokoski [21] found that this simple picture gave a plausible explanation as to why the predicted rich spectrum of hybrids had not been clearly identified; the preferred decays to conventional two-meson final states showed a strong preference for the so-called "S+P modes", in which one of the final mesons had an internal orbital excitation. Since the P-wave mesons have secondary decays, a search for hybrids would require a study of complicated multimeson final states, which had not previously been considered systematically. In addition the numerical scales of the strong widths of hybrids were in many cases rather large, so that the hybrids would be difficult to identify. Isgur *et al.* cited the I=1 exotic with $J^{PC} = 1^{-+}$ as a case of special interest, since the total width of this state was found to be rather narrow, $\Gamma \approx 150\,\text{MeV}$, for an initial hybrid mass of 1.9 GeV. The preferred decay modes of this hybrid in the flux tube decay model were $b_1\pi$ and $f_1\pi$; this observation has stimulated several experimental studies of these rather complicated final states.

Isgur *et al.* only considered the decays of the J^{PC}-exotic hybrids. These flux tube hybrid decay calculations were extended to nonexotic hybrids by Close and Page [22], who found that some very interesting and relatively narrow nonexotics should appear in the spectrum as an overpopulation relative to the naive quark model, provided that the flux tube picture of hybrids and their decays is reasonably accurate. Two notable cases are an extra ω (predicted to be only 100 MeV wide, and to decay mainly to the KK_1 channels) and an extra π_2. This nonexotic hybrid π_2 is predicted to have a very characteristic $b_1\pi$ mode; this mode is forbidden to the conventional $q\bar{q}$ quark model π_2 (presumably the $\pi_2(1670)$), because it is an $(S_{q\bar{q}}=0) \rightarrow (S_{q\bar{q}}=0)+(S_{q\bar{q}}=0)$ transition. These are forbidden for $(q\bar{q}) \rightarrow (q\bar{q})+(q\bar{q})$ in the $^3P_0$ model as well as in OGE pair production and linear scalar pair production models. In contrast they are allowed for decays of flux-tube hybrids, since the π_2 hybrid actually has the quarks in an $S_{q\bar{q}}=1$ configuration. The very strong experimental limit of the $b_1\pi$ mode of the $\pi_2(1670)$ reported by VES is implicitly a constraint on the size of the hybrid component of the $\pi_2(1670)$ state vector.

As a final topic in hybrid decays, there have been very interesting recent results from LGT on the *closed-flavor* decays of heavy hybrids. The UKQCD collaboration [23] has found that some of these modes, notably to a χ state and a scalar (presumably an effectively $\pi\pi$) are remarkably large. This result is very surprising in view of the weakness of the known closed-flavor $c\bar{c}$ and $b\bar{b}$ dipion decays, but if correct suggests very attractive modes for heavy hybrid searches, such as $(\gamma \, \mathrm{J}/\psi) + (\pi\pi)_S$ for charmonium hybrids.

Glueballs

The search for glueballs is a central component of the more general search for exotica in hadron spectroscopy. The well known LGT study of the glueball spectrum by Morningstar and Peardon [24] shows that the currently experimentally accessible glueballs have non-exotic quantum numbers, since the lightest predicted glueball exotic is a 2^{+-} state at a very high mass of ~ 4 GeV. At the presently accessible masses of up to ca. 2.5 GeV just three glueballs are anticipated, a 0^{++} near 1.6 GeV and a 0^{-+} and 2^{++} near 2.3 GeV. The scalar channel is most interesting at present because we have two experimental candidates for this lightest glueball, the $f_0(1500)$ and $f_0(1710)$.

Although width and decay mode predictions for glueballs are obviously of paramount importance, there has been little work in this area. The naive expectation that glueballs should have flavor-blind decays is clearly strongly violated by both experimental candidates; the $f_0(1500)$ shows a strong preference for $\pi\pi$ modes over KK, and the $f_0(1710)$ shows the inverse pattern. This has been attributed to strong mixing between the pure glue basis state and the scalar quarkonium $|n\bar{n}\rangle$ and $|s\bar{s}\rangle$ basis states, analogous to η-η' mixing in the 0^{-+} sector (see for example the work of Amsler and Close [25]).

An alternative explanation for the violation of flavor-singlet decay symmetry in the scalar glueball candidates has been suggested by Sexton et al. [26, 27], based on a LGT study of the glueball-Ps-Ps three-point function. In this early lattice study they found a strong dependence of this glueball decay coupling on the mass assumed for the final pseudoscalar mesons, which if correct would skew mixing angle determinations using flavor-symmetric couplings, as assumed by Amsler and Close. The Sexton et al. LGT couplings favored the $f_0(1710)$ over the $f_0(1500)$ as a scalar glueball candidate.

The very important topic of glueball decay amplitudes merits more careful consideration in future LGT studies. The decay couplings the scalar glueball to vector meson pairs (which may be important for the $f_0(1500)$) and the couplings of the lightest 0^{-+} and 2^{++} excited glueballs (as regards favored modes and total widths) are also important topics for future lattice studies.

STUDIES OF THE UNDERLYING QCD DECAY MECHANISMS

Strong decay amplitudes are crucial properties of hadrons, which if understood even at a phenomenological level can be used to identify plausible candidates for the various conventional quark model hadrons as well as the exotica predicted by model studies and lattice QCD.

Of course there is a deeper question, which is the problem of what fundamental interaction in QCD is responsible for hadron strong decays at the quark-gluon level. Although we can develop phenomenological models such as the $^3P_0$ model without understanding the decay mechanism, these models are simply approximate descriptions of decays with unknown and perhaps large systematic approximations. The QCD mechanism that drives the open-flavor strong decays we have discussed here is not well established, and has been studied in surprisingly few reference. Here we will discuss the results of two of these references, which reach rather similar general conclusions.

The original Cornell charmonium group perhaps surprisingly did not assume the $^3P_0$ decay model. Instead they assumed a nonperturbative microscopic model for strong decays, in which $q\bar{q}$ pair production took place through the linear confining interaction, treated as an exchange between the constituent c and $\bar{c}$ quarks and the produced light $q\bar{q}$ pair (see Fig.2). This model is interesting in part because it is so highly constrained; the numerical value of the string tension is reasonably well known from spectroscopy, and therefore gives decay predictions that have no free parameters. This model gives rather good predictions for the behavior of R (specifically charmonium production above open charm threshold), which implicitly involves width calculations for the higher vector resonances. (These resonances were not treated individually in the Cornell studies, instead the contributions to R were determined implicitly using an effecting interaction containing decay loops.) Nonetheless the assumption of a vector confining interaction is controversial, and would presumably fail the D/S ratio test in decays such as $b_1 \to \omega\pi$.

A more recent paper by Ackleh *et al.* [6] investigates various possible QCD mechanisms for strong decays. This reference concludes that the naive OGE pair production diagram in most cases gives a rather small contribution to the decay amplitude, which (as was assumed by the Cornell group) is instead dominated by pair production from the linear confining interaction. Unlike the Cornell model however, Ackleh *et al.* assume a scalar confining interaction, which gives a D/S ratio for $b_1 \to \omega\pi$ that is close to experiment. Again the decay rates are known in this type of model in terms of the string tension; the simple linear scalar potential gives somewhat larger decay amplitudes than are observed experimentally.

A NOVEL APPLICATION: V-PS SCATTERING FROM FSI

Watson's theorem implies that final state interactions induce a phase factor of $e^{i\delta_f}$ in the decay amplitude of a resonance into the channel f. If there is only a single distinguishable channel, as in $\rho \to \pi\pi$, this phase is not observable. However in many decays a given final state spans several different channels, which are distinguished for example by internal angular momenta.

An example of this FSI effect was recently exploited by Nozar *et al.* [28] in a very interesting new "application" of strong decays. The decay $b_1 \to \omega\pi$ is well known as a textbook example of a meson decay to more than one channel, since the $\omega\pi$ system can be in both S-wave ($^3S_1$) and D-wave ($^3D_1$) states. The ratio of D-wave to S-wave amplitudes was very important historically in selecting the $^3P_0$ model as a realistic description of pair production quantum numbers in meson decays.

Since the S- and D-wave states have different scattering phase shifts, they will have different FSI phases; after rescattering the final $\omega\pi$ state will be of the form

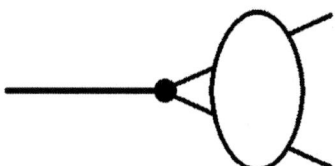

FIGURE 3.   The meson decay FSI effect exploited by Nozar *et al.* [28] to study $\omega\pi$ scattering.

$$|\omega\pi\rangle = a_S\, e^{i\delta_S}\, |\omega\pi(^3S_1)\rangle + a_D\, e^{i\delta_D}\, |\omega\pi(^3D_1)\rangle\,, \tag{1}$$

so the famous D/S ratio is actually a complex number. The relative phase $\delta_S - \delta_D$ appears in the $a_D a_S$ cross term in the squared amplitude, and thus can be inferred directly from the observed $\omega\pi$ angular distribution. The study of b_1 decay can thus give us very exciting information on the scattering amplitudes of strongly unstable resonances, which would not otherwise be accessible.

This measurement was carried out by Nozar *et al.* [28], who found an $\omega\pi$ relative scattering phase at the b_1 mass of

$$\left[\delta_S - \delta_D\right]_{\omega\pi} = -10.54° \pm 2.4°\,. \tag{2}$$

This is numerically rather similar to a quark model prediction of $\delta_S(\omega\pi) = -14°$ by Barnes, Black and Swanson [29]. (One expects the $\omega\pi$ D-wave phase shift to be small at the b_1 mass.)

Clearly this is just one example of a large number of scattering amplitudes of strongly unstable resonances that can be inferred from high-statistics studies of strong decays, and future measurements should provide phase shifts that will be very interesting for theorists attempting to understand hadron scattering and FSI effects.

A caution is appropriate; we have assumed that the underlying decay amplitudes themselves are relatively real in different channels, and also that FSI effects are diagonal. Of course both these assumptions are suspect, and should be tested by comparing scattering phase shifts inferred from the decays of different initial mesons.

FUTURE DECAYS: UNQUENCHING THE QUARK MODEL

One of the exciting new discoveries [30] discussed at this meeting was the observation of the charm-strange mesons $D_{sJ}^*(2317)$ and $D_{sJ}^*(2357)$, which are presumably 0^+ and 1^+ states respectively. The masses of these states are quite surprising, since the normally accurate potential model of Godfrey and Isgur [14] predicts much higher masses of 2.48 GeV and 2.55 GeV respectively.

Since these states are predicted by the $^3P_0$ model to be very broad [31] and are quite close to DK threshold, one possibility is that the usual neglect of decay couplings in quark model calculations is inaccurate here, and the states have been displaced downwards by ca. 150 MeV due to decay loops. This may imply that the states have large DK and DK* molecular components respectively [32], rather like the K$\bar{\text{K}}$ molecules, instead of being simple $c\bar{s}$ quark model states.

This topic of the contribution of virtual decay loop diagrams to hadron properties, known as "unquenching the quark model", is an important but rather obscure issue. Explicit evaluation typically finds that individual loop diagrams are large, but that there may be significant cancellations (see for example [33]). Future studies of strong decays will undoubtedly include investigations of the effects of these virtual decay loops, since the $D_{sJ}^*(2317)$ and $D_{sJ}^*(2357)$ may have "announced" to us that in some strongly-coupled channels these effects cannot be ignored.

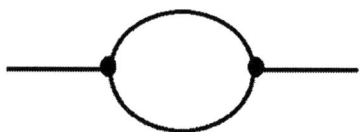

FIGURE 4.   The generic loop diagram representing the second-order effects of virtual decay of a meson into the two-meson continuum.

ACKNOWLEDGMENTS

I would like to thank the organizers of HADRON03 for their kind invitation to present this summary of the history, current status and possible future of research in hadron strong decays. This research was supported in part by the U.S. National Science Foundation through grant NSF-PHY-0244786 at the University of Tennessee, and by the U.S. Department of Energy under contract DE-AC05-00OR22725 at Oak Ridge National Laboratory (ORNL).

REFERENCES

1.   L.Micu, *Nucl. Phys. B*, **10**, 521 (1969).
2.   A.LeYaouanc, L.Oliver, O.Péne and J.C.Raynal, *Phys. Rev. D*, **8**, 2223 (1973).
3.   *ibid.*, *Phys. Rev. D*, **9**, 1415 (1974).
4.   *ibid.*, *Phys. Rev. D*, **11**, 1272 (1975).
5.   *ibid.*, *Phys. Lett. B*, **71**, 397 (1977).
6.   E.S.Ackleh, T.Barnes and E.S.Swanson, *Phys. Rev. D*, **54**, 6811 (1996).
7.   E.Eichten, K.Gottfried, T.Kinoshita, K.D.Lane and T.-M.Yan, *Phys. Rev. Lett.*, **36**, 500 (1976).
8.   *ibid.*, *Phys. Rev. D*, **i7**, 3090 (1978).
9.   *ibid.*, *Phys. Rev. D*, **21**, 203 (1980).

10. E.S.Swanson, *Nucl. Phys. Proc. Suppl.*, **64**, 312 (1998).
11. R.M.Baltrusaitis et al., *Phys. Rev. Lett.*, **56**, 107 (1986).
12. S.Godfrey, R.Kokoski and N.Isgur, *Phys. Lett. B*, **141**, 439 (1984).
13. H.G.Blundell and S.Godfrey, *Phys. Rev. D*, **53**, 3700 (1996).
14. S.Godfrey and N.Isgur, *Phys. Rev. D*, **32**, 189 (1985).
15. S.Capstick and N.Isgur, *Phys. Rev. D*, **34**, 2809 (1986).
16. T.Barnes, F.E.Close, P.R.Page and E.S.Swanson, *Phys. Rev. D*, **55**, 4157 (1997).
17. T.Barnes, N.Black and P.R.Page, *Phys. Rev. D*, **68**, 054014 (2003).
18. S.Capstick and W.Roberts, *Phys. Rev. D*, **49**, 4570 (1994).
19. *ibid.*, *Prog. Part. Nucl. Phys.*, **45**, S241 (2000).
20. H.J.Lipkin, *Phys. Lett. B*, **494**, 248 (2000).
21. N.Isgur, R.Kokoski and J.Paton, *Phys. Rev. Lett.*, **54**, 869 (1985).
22. F.E.Close and P.R.Page, *Nucl. Phys. B*, **443**, 233 (1995).
23. C.McNeile, C.Michael and P.Pennanen (UKQCD Collaboration), *Phys. Rev. D*, **65**, 094505 (2002).
24. C.J.Morningstar and M.Peardon, *Phys. Rev. D*, **60**, 034509 (1999).
25. C.Amsler and F.E.Close, *Phys. Rev. D*, **53**, 295 (1996).
26. J.Sexton, A.Vaccarino and D.Weingarten, *Nucl. Phys. B*, **42**, 279 (1995).
27. J.Sexton, A.Vaccarino and D.Weingarten, *Phys. Rev. Lett.*, **75**, 4563 (1995).
28. M.Nozar *et al.* (E852 Collaboration), *Phys. Lett. B*, **541**, 35 (2002).
29. T.Barnes, N.Black and E.S.Swanson, *Phys. Rev. C*, **63**, 025204 (2001).
30. B.Aubert *et al.* (BABAR Collaboration), *Phys. Rev. Lett.*, **90**, 242001 (2003).
31. S.Godfrey and R.Kokoski, *Phys. Rev. D*, **43**, 1679 (1991).
32. T.Barnes, F.E.Close and H.J.Lipkin, *Phys. Rev. D*, **68**, 054006 (2003).
33. P.Geiger and N.Isgur, *Phys. Rev. Lett.*, **67**, 1066 (1991).

Aspects of Confinement: a Brief Review

Eric S. Swanson

Department of Physics and Astronomy, University of Pittsburgh, Pittsburgh PA, 15260

Abstract. A brief and biased overview of the phenomenon of confinement in QCD is presented in three parts: (1) the definition of confinement, (2) properties of confinement, (3) ideas of confinement. The second part chiefly consists of a brief review of recent lattice computations related to confinement while the third summarizes some of the current analytical approaches to understanding confinement. These include the Dyson-Schwinger formalism in Landau gauge, Hamiltonian QCD in Coulomb gauge, and the vortex picture of confinement.

DEFINITIONS OF CONFINEMENT

Confinement is the poster boy of nonperturbative physics for good reason: it is associated with a linear potential with a string tension,

$$\sigma \propto \Lambda^2 e^{-\int \frac{dg}{\beta(g)}}$$

which is nonperturbative in the coupling.

Although this statement has intuitive appeal, one must be careful in defining confinement. For example, it is often loosely defined as the absence of free quarks in nature. But it is conceivable, and even possible, that there exists a coloured scalar particle which can form bound states with quarks. The resultant particles would then carry flavour and fractional electric charge[1], which is likely not the intent of the definition.

Similarly, requiring that all observable particles be colour singlets encompasses gauge theories in both the confinement and the Higgs phases. The latter also manifests colour singlet states because colour charge is completely screened.

One may hope to improve the situation by focussing on the expected physical properties of confinement. For example, the appearance of a long range linear potential between quarks is a reasonable requirement. Of course, the problem here is that string breaking will occur once the potential energy approaches the quark pair creation threshold. The conventional way out of this difficulty is to consider the work to separate two quarks as the quark masses approach infinity. Thus one is led to the strange position of defining quark confinement in a limit which removes quarks from the theory.

Centre Vortices

It is traditional to implement the last definition in lattice gauge theory with the aid of the Wilson loop:

CP717, *Hadron Spectroscopy: Tenth International Conference,*
edited by E. Klempt, H. Koch, and H. Orth
© 2004 American Institute of Physics 0-7354-0197-7/04/$22.00

$$\langle WL \rangle = \int DU_\ell \text{tr}[\prod_{i \in C} U_i] e^{-S_{YM}} \tag{1}$$

where C is a large planar loop and the pure gauge action is given by

$$S_{YM} = \frac{6}{g^2} \sum_P \left(1 - \frac{1}{N_c} \Re\text{tr}[U_P]\right). \tag{2}$$

Here the gluonic degrees of freedom are represented by link variables $U_\ell = \exp(igT^a A_\mu^a(x))$ where ℓ represents a link on a spacetime lattice which starts at the point x and points in the $\hat{\mu}$ direction and the lattice spacing has been set to unity. The sum in the action is over plaquettes, P, which are the smallest closed loops permitted on the lattice. U_P is a product of link variables around a plaquette and is thus a lattice implementation of the two-forms of gauge theories. It is measurements of the Wilson loop which have provided the most compelling demonstrations of the confinement phenomenon.

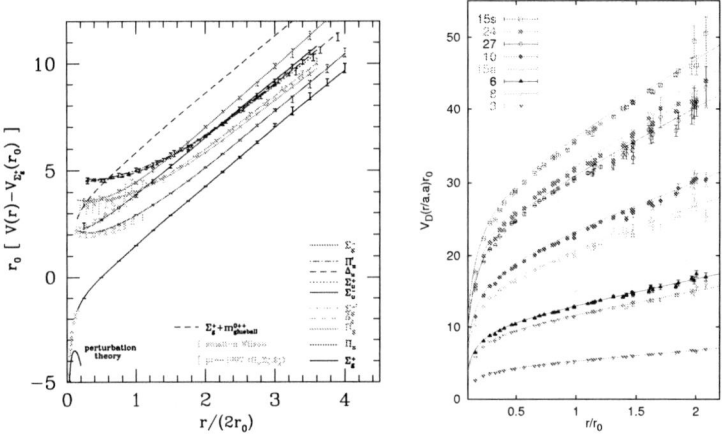

FIGURE 1. (left) Wilson Loop Measurements of Various Static Quark Potentials[2]. (right) Casimir Scaling of Confinement[9].

The lattice definition of QCD and confinement makes it especially easy to see the significance of the centre group of SU(3). The centre is defined as the set of all elements of the gauge group which commute with all other elements. Evidently these are given by the set $\{z = \exp(2\pi n/N_c)\text{I}\}$ where $n = 0, \ldots, N_c - 1$. This set forms the group Z_{N_c}. It is clear that multiplying all temporal link variables on a particular time slice by an element of the group, $U_t(t_0, \mathbf{x}) \to zU_t(t_0, \mathbf{x})$ does not change the value of a plaquette, implying that QCD is invariant under global Z_{N_c} transformations.

We now consider a Wilson loop which extends in the temporal direction for the entire length of the lattice (and, because of the imposition of boundary conditions, encircles the lattice), called a Polyakov line. If one makes the transformation discussed above it is clear that the Polyakov line picks up a phase: $PL(\mathbf{x}) \to zPL(\mathbf{x})$. Since the theory is invariant under such a transformation one must have either $\langle PL(\mathbf{x}) \rangle = 0$ if the vacuum of QCD shares the symmetry of the action or $\langle PL(\mathbf{x})\rangle \neq 0$ if the theory is in the Z_{N_c} broken phase.

The connection with confinement arises once it is realized that the Polyakov line measures the free energy of a static quark at the position $\mathbf{x}$: $\langle PL(\mathbf{x})\rangle = \exp(-ET)$ where T is the temporal extent of the lattice. But the free energy of an isolated confined quark is infinite, thus *QCD must be in the unbroken Z_{N_c} symmetry phase if it is confining.*

It is useful to note that Z_{N_c} transformations are an example of "singular gauge transformations" (which are *not* gauge transformations!), namely they may be generated by performing a gauge transformation which is periodic modulo a Z_{N_c} phase factor:

$$U_t(t,\mathbf{x}) \rightarrow g(t,\mathbf{x})U_t(t,\mathbf{x})g^{\dagger}(t+1,\mathbf{x}) \tag{3}$$

with $g(T+1,\mathbf{x}) = z^*g(1,\mathbf{x})$. For example the symmetry transformation $U_t(t_0,\mathbf{x}) \rightarrow zU_t(t_0,\mathbf{x})$ may be achieved by setting $g(t,\mathbf{x}) = 1$ for $t \leq t_0$ and $g(t,\mathbf{x}) = z^*$ for $t > t_0$. Singular gauge transformations are of central interest to the study of confinement because the structures associated with them are often postulated to cause confinement. We shall return to this in section III below.

Kugo-Ojima Criteria

The final definition of confinement to be considered here was first investigated by Kugo and Ojima shortly after the invention of QCD[3]. Their starting assumption was that only BRST singlets may be allowed as physical states if confinement is to hold. A sufficient condition for this is that the ghost propagator in Landau gauge is enhanced in the infrared:

$$D_G(k) = -\frac{1}{k^2}\frac{1}{1+u(k^2)} \equiv -\frac{G(k^2)}{k^2} \tag{4}$$

with $u(0) = -1$. This may be related to the gluon propagator via Dyson-Schwinger equations and implies that there is an infrared suppression of the transverse gluon correlator:

$$D_{\mu\nu}(k) = \frac{Z(k^2)}{k^2}\left(\delta_{\mu\nu} - \frac{k_\mu k_\nu}{k^2}\right) \tag{5}$$

with $Z(k^2)/k^2 \rightarrow 0$ as $k \rightarrow 0$[4].

It is interesting to note that these criteria are consistent with results derived by Zwanziger in noncovariant gauges[6].

PROPERTIES OF CONFINEMENT

At present the only reliable method for determining properties of confinement is with lattice computations. Indeed, the detection of confinement (with the Wilson loop) was one of the first lattice computations and it continues as the standard bearer of lattice gauge theory.

Recent investigations have revealed much beyond the static quark interaction. Perhaps the most famous examples are plots of action or field density which clearly show the formation of tubes of gluonic flux forming between the colour source and sink[7]. Of course, such tubes fit in very well with the naive notion of linear quark confinement. More interesting is recent investigations of the *dynamics* of flux tubes, as represented by the higher surfaces in Fig. 1(left). These may be interpreted as adiabatic energy surfaces describing hybrid mesons.

Another type of flux tube investigation places small kinks in Wilson loops to study the spin-dependence of the long range confining force[8]. The results are consistent with the supposition that confinement is of a Lorentz scalar nature (it is important to note, however, that mapping the lattice results to this type of interaction is an effective, low energy approximation only).

Casimir Scaling

Figure 1(right) shows in a compelling way the property of *Casimir scaling* of confinement. The figure was obtained by measuring the Wilson loop for sources in various representations of SU(3). The interaction between colour triplets is the lowest surface in the figure and forms the template for the others. In the figure one sees higher surfaces with sources in the 8, 6, 15_A, 10, 27, 24, and 15_S representations. The curves are obtained by multiplying a fit to the lowest (fundamental representation) surface by the quadratic Casimir, $C_R^2 = \langle R|T^a T^a|R \rangle$ divided by C_F^2. The quadratic Casimir is given by $(p^2 + q^2 + pq)/3 + p + q$ where (p,q) is the Dynkin index of the representation. The agreement is remarkable and is a strong indication that the colour structure of confinement may be written as

$$\int \bar{\psi} T^a \psi \ldots \bar{\psi} T^a \psi$$

where the ellipsis represents Lorentz and spatial dependence.

String Behaviour

In pre-QCD days strings were invoked as a fundamental theory of QCD. With the advent of QCD, strings have survived as an effective description of flux tubes. In particular one expects that

(i) transverse flux tube profiles should be logarithmically divergent with the tube length (due to string 'roughening');

(ii) the ground state potential should exhibit a universal 'Luescher term', $-\pi/(12R)$;

(iii) the string excitation spectrum should be that of bosonic string modes, π/R.

Juge, Kuti, and Morningstar have carried out a detailed analysis of the relationship of the hybrid surfaces of Fig. 1(left) to string excitations[10]. They have found that surface excitations only have π/R splittings for very large source separation (roughly 4 fermi or greater); see Figure 2(left). Furthermore, there is a cross over region at about 1 fermi where the surfaces move from a perturbative behaviour (characterized by the 'gluelump' spectrum) to a string-like behaviour.

Point (ii) has been examined with a new algorithm by Luescher and Weisz[11]; their main results are shown in Figure 2(right). One sees that the expected behaviour of the confinement potential is achieved at a source separation of roughly 1/2 fm or less. String-

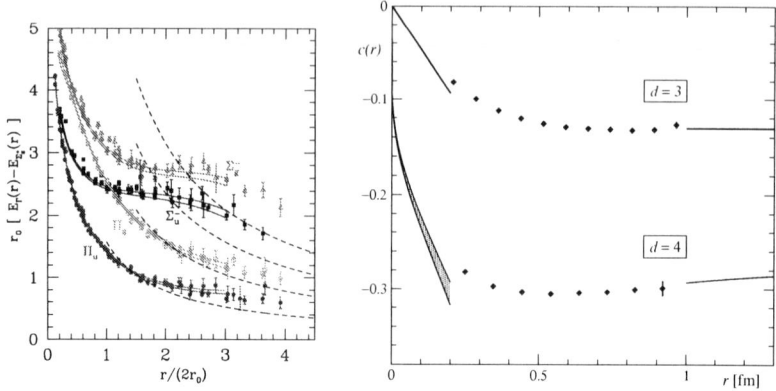

FIGURE 2. (left) Hybrid Surface Energy Differences. $2r_0$ is roughly 1 fermi[10]. (right) The Luescher Term in 4-d ($-\pi/12$) and 3-d ($-\pi/24$)[11].

like behaviour at such short distances may be called precocious and is very difficult to understand in light of the conclusion of Juge, Kuti, and Morningstar. It appears that the existence of a $1/r$ interaction with the expected Luescher coefficient is not necessarily indicative of string dynamics.

Baryonic Flux Tubes

Finally, investigations of the static baryon interaction have begun. The chief point of interest is whether the expected flux tubes form into a 'Y' shape or a 'Δ' shape (ie., is the effective baryonic quark interaction two-body or three-body?). This may be addressed by carefully examining the baryonic energy under a variety of quark configurations. Figure 3 shows the flux tube which arises in one such configuration (which seems to be an interpolation of Y and Δ). Current results are mixed, with some groups claiming support for the two-body hypothesis [13] and some for the three-body hypothesis[14]. Finally, a strong operator dependence in the flux tube profiles has been observed[15], which clearly needs to be settled before definitive conclusions can be reached.

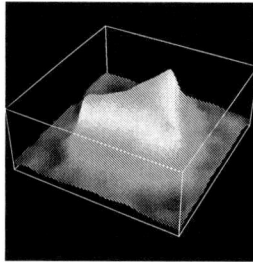

FIGURE 3. The Baryonic Flux Tube. The distance between the tube junction and a quark is approximately 1/2 fm[12].

IDEAS OF CONFINEMENT

The final portion of this report deals with analytic attempts to describe confinement. Unfortunately, there is a great deal of prejudice concerning this issue, with many people believing that any such attempts cannot succeed. This is not a useful point of view to take and is, in fact, at odds with generations of successes in field theory, many-body physics, and condensed matter physics:

nonperturbative does not mean intractable!

A trivial example is provided by elementary calculus. As is well known the function $f(x) = e^{-\frac{1}{x}}$ has a zero radius of convergence and hence no sensible 'perturbation theory'. However, if a sufficiently clever theorist were able to derive the 'Dyson-Schwinger' differential equation $f'(x) = \frac{1}{x^2} f(x)$, almost anyone could then derive any desired property of f.

More substantial examples are provided by the Gell-Mann – Brueckner resolution to the infrared divergence problem in the degenerate electron gas, the Galitskii method for dealing with strong short range repulsive interactions in nuclear physics, and Laughlin's explanation of the fractional quantum Hall effect.

The point is this: QCD is described by a Lagrangian and therefore all of the properties of QCD are carried by its diagrams. To make progress nonperturbatively one must find an infinite set of diagrams to 'sum' which capture the majority of the physics one is interested in. It would be a completely new occurrence in physics if this approach should not work for confinement, requiring the introduction of a concept such as 'intrinsic incomputability' (with commensurate implications for attempts at deriving the 'theory of everything').

In the following a very brief summary of Dyson-Schwinger and Green's function approaches to confinement are presented. I shall also recall some salient features of the vortex picture of confinement. Space constraints forbid discussing many other popular ideas of confinement such as the dual superconductor picture, merons, and monopoles.

The Dyson-Schwinger Formalism

There has been much progress in the Dyson-Schwinger approach to confinement in recent years. The idea is to truncate the infinite series of Dyson-Schwinger equations (which are equivalent to the field theory) in such a way that the leading infrared behaviour of the ghost and gluon propagators may be reliably extracted. This endeavour is assisted by appealing to known properties of nonperturbative vertices, either through Slavnov-Taylor identities, the application of BRST symmetries, or properties of special gauges (in particular, the ghost-gluon vertex is not renormalized in Landau gauge). It has been found that including the ghost propagator in the coupled set of D-S equations is important. One may then solve for the infrared behaviour of the ghost and gluon propagators under various approximations or solve the coupled integral equations numerically[16].

A convincing demonstration of confinement results. Analytic work finds that the infrared propagators behave as

$$Z(k) \to \left(\frac{k^2}{\sigma}\right)^{2\kappa}, \; k \to 0 \tag{6}$$

and

$$G(k) \to \left(\frac{\sigma}{k^2}\right)^{\kappa}, \; k \to 0. \tag{7}$$

with $\kappa \approx 0.6$. This value is very stable under modification of the truncations or Ansätze. Equations 6 and 7 demonstrate that the Kugo-Ojima confinement criteria are met. Furthermore, numerical solutions are in remarkable agreement with lattice results[5], indicating the utility of the truncations made.

Green's Function Approach in Coulomb Gauge

It has been 25 years since Gribov first noted that the imposition of Coulomb gauge in non-Abelian theories is beset with ambiguities[17]. The ambiguity arises because more than one solution to the gauge constraint, $\nabla \cdot \mathbf{A} = 0$ may exist. Gribov showed that the condition for this to occur was that nontrivial solutions to the equation $\nabla \cdot \mathbf{D} = 0$ had to exist. Here $\mathbf{D}$ is the covariant derivative in adjoint representation. Gribov proposed to resolve the ambiguity by restricting the gauge configurations to those with a positive definite value of $\det(\nabla \cdot \mathbf{D})$, called the Gribov region. He also noted that such a constraint imposes a boundary in field space, and this boundary can affect the gluon dispersion relation, causing confinement.

Much progress has occurred since Gribov framed his conjectures. It is now known that the Gribov region does not resolve the Gribov ambiguity, rather a smaller region called the fundamental modular region (FMR) is required. Furthermore, the FMR is contained in the Gribov region, the FMR incloses the origin, it is convex, and its boundary sometimes coincides with the boundary of the Gribov region[18]. Topological field configurations are realized through the imposition of nontrivial boundary conditions at the boundary of the FMR. Finally, Zwanziger has argued that physical matrix elements draw their support solely from the intersection of the Gribov and fundamental modular boundaries[19].

It is suggestive that the operator $\nabla \cdot \mathbf{D}$ appears in the instantaneous portion of the interaction in the QCD Hamiltonian in Coulomb gauge[21]:

$$V_{Coul} = -\frac{1}{2} \int d^3x d^3y \, \mathrm{J}^{-1} \rho^a(\mathbf{x}) \langle \mathbf{x}a | \frac{g}{\nabla \cdot \mathbf{D}} \nabla^2 \frac{g}{\nabla \cdot \mathbf{D}} | \mathbf{y}b \rangle \mathrm{J} \; \rho^b(\mathbf{y}) \tag{8}$$

where $\mathrm{J} = \det(\nabla \cdot \mathbf{D})$ is the Faddeev-Popov determinant and the colour density is given by $\rho^a = -f^{abc} \mathbf{A}^b(\mathbf{x}) \cdot \mathbf{E}^c(\mathbf{x})$. If vacuum field configurations are dominated by those near the boundary of the Gribov region there will be a strong infrared enhancement of the instantaneous interaction, which may cause confinement. This old observation has recently received support from several lattice computations[22, 23].

It is known that the non-Abelian Coulomb interaction is renormalization group invariant and that it generates the complete running coupling of QCD[24]. Furthermore, quarks decouple from transverse gluons in the static limit and therefore this interaction must generate confinement. Indeed, Zwanziger has recently shown that the non-Abelian Coulomb interaction provides an upper bound to the Wilson loop interaction[20] and has conjectured that this bound is saturated (numerical evidence in favour of this conjecture is provided in Ref. [22]).

It is clear that the instantaneous interaction is an important element of QCD. Recently an analytical attempt to understand this operator has been made in the Greens function approach[25]. The central idea was to compute the kernel appearing in Eq. 8 by summing all diagrams which contribute at leading order in the infrared[26]. The resulting Dyson equation was solved with the gluon propagator equation (which was obtained with the aid of a Gaussian Ansatz for the pure gauge vacuum). Numerical solution of the coupled nonlinear integral equations yielded a nontrivial gluonic quasiparticle dispersion relation and a linearly confining instantaneous interaction in remarkable agreement with lattice Wilson loop results[25]. These results are in agreement with the Kugo-Ojima confinement criteria, namely the inverse Faddeev-Popov operator (which is the Coulomb gauge analogue of the ghost propagator) is infrared enhanced. The infrared behaviour of the transverse gluon propagator is more problematic, but it is now understood to arise from strong modifications of the vacuum Ansatz near the Gribov boundary[27].

Vortices

Singular gauge transformations generate local objects of great interest. For example, zero-dimensional gauge dislocations are associated with instantons, one-dimensional with monopoles, and two-dimensional with vortices. A vortex is a sheet (in 4-d) of infinite field strength which is associated with a Z_{N_c} singular gauge transformation (see Ref. [28] for more details). It has been postulated that vortices drive confinement[29] because they are localized field configurations which percolate the lattice. The argument is quite general and relies on the fact that localized field distributions contribute independently to the expectation value of the Wilson loop operator, and hence yield an area law interaction.

There are many other attractive features of the vortex confinement picture, for example, deconfinement may be viewed as a vortex de-percolation phase transition[31] and the vortex density has been shown to scale in maximal centre gauge, thereby establishing their physical nature[32]. Lastly, it has recently been shown that monopoles may be understood in terms of vortex self intersections, thereby introducing topological field configurations into the discussion[33].

This compelling picture is strongly supported by lattice computations permitted by recent advances in lattice gauge fixing algorithms. In particular figure 4 shows that the Wilson loop string tension is maintained if one projects onto vortex gauge configurations (upper green line) *and* that linear confinement disappears if vortices are removed (lower red line).

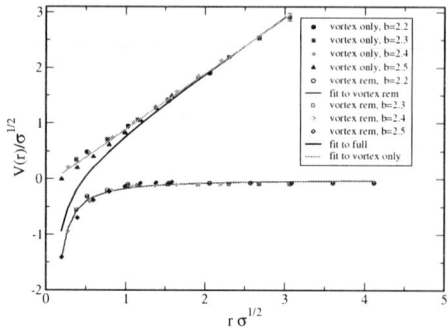

FIGURE 4. The SU(2) Wilson Loop Potential with and without Vortices[28].

CONCLUSIONS

Great strides have been made recently in the effort to understand confinement. New lattice algorithms allow a detailed study of the importance of different model gauge configurations and of the onset of string-like behaviour in flux tubes. Multiquark states are also beginning to be probed. At the same time, analytical methods are rapidly approaching the time when quantitatively accurate computations of nonperturbative phenomena are possible. And we are already seeing convincing demonstrations of the general features of confinement.

Of course much remains to be done. The analytical formalisms need to examine the robustness of the assumed truncations and the effects of topology and boundary conditions. At the same time, these approaches can check and compare with a new breed of fixed-gauge lattice results. There are many open issues in the lattice approach a well. For example, too many models of the QCD vacuum seem to be supported by the lattice and one needs to find some commonality among the different mechanisms and gauges. It is hoped and likely that issues such as the nature of baryonic flux tubes, the infrared behaviour of the Landau gauge gluon propagator, the onset of stringiness in flux tubes (is it at 1/2 or 4 fermi?), and the nature of the non-Abelian Coulomb interaction will all be resolved in the short term.

ACKNOWLEDGMENTS

I am grateful to the organizers of Hadron03 for the invitation to speak on such a fascinating topic and for providing a wonderful venue for discussing hadronic physics. I wish to thank Reinhard Alkofer, Pierre van Baal, Jeff Greensite, Hugo Reinhardt, Adam Szczepaniak, and Daniel Zwanziger for many illuminating discussions on this topic. This work was supported by the DOE under contracts DE-FG02-00ER41135 and DE-AC05-84ER40150.

REFERENCES

1. J. Greensite, "The confinement problem in lattice gauge theory", Prog. Part. Nucl. Phys. **51**, 1 (2003).
2. K.J. Juge, J. Kuti, and C.J. Morningstar, Nucl. Phys. Proc. Suppl. **63**, 326 (1998).
3. T. Kugo and I. Ojima, Prog. Theor. Phys. Suppl. **66**, 1 (1979).
4. T. Kugo, arXiv:hep-th/9511033.
5. R. Alkofer and C. S. Fischer, "The Kugo-Ojima confinement criterion and the infrared behavior of Landau gauge QCD", arXiv:hep-ph/0309089.
6. D. Zwanziger, Nucl. Phys. B **485**, 185 (1997).
7. G. S. Bali, K. Schilling and C. Schlichter, Phys. Rev. D **51**, 5165 (1995).
8. G. S. Bali, K. Schilling and A. Wachter, Phys. Rev. D **56**, 2566 (1997).
9. G. S. Bali, Phys. Rev. D **62**, 114503 (2000).
10. C. J. Morningstar, K. J. Juge and J. Kuti, Nucl. Phys. Proc. Suppl. **73**, 590 (1999); K. J. Juge, J. Kuti and C. Morningstar, Phys. Rev. Lett. **90**, 161601 (2003).
11. M. Luscher and P. Weisz, JHEP **0207**, 049 (2002).
12. H. Ichie, V. Bornyakov, T. Streuer and G. Schierholz, "The flux distribution of the three quark system in SU(3)", arXiv:hep-lat/0212024.
13. C. Alexandrou, P. De Forcrand and A. Tsapalis, Phys. Rev. D **65**, 054503 (2002).
14. T. T. Takahashi, H. Suganuma, Y. Nemoto and H. Matsufuru, Phys. Rev. D **65**, 114509 (2002).
15. F. Okiharu and R. M. Woloshyn, "A study of colour field distributions in the baryon", arXiv:hep-lat/0310007.
16. P. Watson and R. Alkofer, Phys. Rev. Lett. **86**, 5239 (2001); C.S. Fischer, R. Alkofer, and H. Reinhardt, arXiv:hep-ph/0202195; C. Lerche and L. von Smekal, arXiv:hep-ph/0202194; D. Atkinson and J.C. Bloch, Phys. Rev. **D58**, 094036 (1998); D. Zwanziger, arXiv:hep-th/9410019.
17. V.N. Gribov, Nucl. Phys. **B139**, 1 (1978).
18. P. Van Baal, "The QCD Vacuum", arXiv:hep-lat/9709066; "Nonperturbative Analysis, Gribov Horizons, and the Boundary of the Fundamental Domain", arXiv:hep-lat/9208027.
19. D. Zwanziger, "Non-perturbative Faddeev-Popov formula and infrared limit of QCD", arXiv:hep-ph/0303028.
20. D. Zwanziger, Phys. Rev. Lett. **90**, 102001 (2003).
21. N. H. Christ and T. D. Lee, Phys. Rev. D **22**, 939 (1980).
22. A. Cucchieri and D. Zwanziger, Phys. Rev. D **65**, 014001 (2002).
23. J. Greensite and S. Olejnik, Phys. Rev. D **67**, 094503 (2003).
24. D. Zwanziger, Nucl. Phys. B **518**, 237 (1998).
25. A. P. Szczepaniak and E. S. Swanson, Phys. Rev. D **65**, 025012 (2002).
26. E. S. Swanson and A. P. Szczepaniak, "Constructing confinement", arXiv:hep-ph/0205079.
27. A. P. Szczepaniak,"Confinement and gluon propagator in Coulomb gauge QCD", arXiv:hep-ph/0306030.
28. K. Langfeld, "Vortex Matter in SU(3) Lattice Gauge Theory", arXiv:hep-lat/0307030.
29. G. 't Hooft, Nucl. Phys. B **153**, 141 (1979); J.M. Cornwall, Nucl. Phys. B **157**, 392 (1979).
30. K. Langfeld, "Vortex induced confinement and the Kugo-Ojima confinement criterion", arXiv:hep-lat/0204025.
31. K. Langfeld, Phys. Rev. D **67**, 111501 (2003).
32. See K. Langfeld, H. Reinhardt, and O. Tennert, Phys. Lett. **B419**, 317 (1999). We note that the story is not simple since density scaling is not seen in laplacian gauge. Furthermore, only 62% of the SU(3) string tension (as opposed to all of the SU(2) string tension) is recovered upon vortex projection in the maximal centre gauge[28] .
33. H. Reinhardt, Nucl. Phys. B **628**, 133 (2002).

Quark model description of hadrons

Bernard Metsch

Helmholtz-Institut für Strahlen– und Kernphysik
Rheinische Friedrich–Wilhelms–Universität Bonn
Nußallee 14-16, D-53115 Bonn, Germany
E-mail: `metsch@itkp.uni-bonn.de`

Abstract. In this contribution I will try to give an overview of what has been achieved in constituent quark models of mesons and baryons by a comparison of some selected results from various *ansätze* with experimental data. In particular I will address the role of relativistic covariance, the nature of the effective quark forces, the status of results on electromagnetic and strong-decay observables beyond the mere mass spectra, as well as some unresolved issues in hadron spectroscopy.

INTRODUCTION

Although some appreciable progress has been made in *ab initio* calculations of low-lying baryon resonances within the lattice gauge approach, still the only comprehensive description of the complete known spectrum of hadrons (focussing on light quark flavours) with masses up to 3 GeV, which addresses such issues as linear Regge-trajectories, parity doublets in the baryon spectrum, the conspicuous structure of scalar excitations of hadrons, is in fact the constituent quark model, which assumes that the majority of meson and baryon excitations can be effectively described as $q\bar{q}$– and q^3– bound states of (constituent) quarks and that the coupling to more complicated configurations (such as strong decay channels) can be treated perturbatively. Although recent experimental findings hint at the existence of exotic meson and baryon resonances, this scheme at least constitutes a framework to judge what is to be considered as exotic.

Since quarks, even when adopting constituent, effective quark masses, move in hadrons with velocities which are a significant fraction of the velocity of light and most non-static observables involve processes at rather large momentum transfers, the quark model description should be based on the usual concepts of quantum field theories. In spite of this, traditionally the quark dynamics in quantitative constituent quark models has been formulated on the basis of the non-relativistic Schroedinger equation and relativistic corrections have at best been parameterized. Recent calculations on electromagnetic form factors elucidated the role of Poincare invariance in calculating electromagnetic currents.

The ultimate goal of any hadron model is to obtain a unified description of

- Mass spectra of (*e.g.* light-flavoured) hadrons from the ground states up to the highest masses < 3 GeV and highest angular momenta $J < 8$ observed, addressing such isuues as: Regge-trajectories, scalar excitations, (pseudo)scalar mixings (for mesons), parity doublets (for baryons), undetected resonances, etc.

CP717, *Hadron Spectroscopy: Tenth International Conference,*
edited by E. Klempt, H. Koch, and H. Orth
© 2004 American Institute of Physics 0-7354-0197-7/04/$22.00

- electroweak properties, such as electroweak form factors, radiative decays and transitions, semi-leptonic weak decays, etc.
- strong (two-body) decays and interactions.

Even within the framework of the constituent quark model, the various approaches found in the literature do not only differ appreciably with respect to there scope, but also in the modelling of the effective quark interactions used and in the assumptions concerning the dynamical equations. Here we can distinguish between (a) field theoretical approaches, which implement relativistic covariance in the basic set-up, such as: Lattice-gauge theory of QCD, Dyson-Schwinger/Bethe-Salpeter approaches relying on a parametrization of the infrared gluon propagator, see *e.g.* [1], instantaneous approximations to this, on the basis of a parametrization of confinement and using instanton-induced interactions, which allows for addressing the complete light-flavoured hadron spectrum and not merely the ground and some lower excited states and (b) quantum mechanical approaches on the basis of the Schrödinger equation with relativistic corrections using confinement potentials and effective quark interactions based (alternatively) on O(ne) G(luon) E(xchange), see *e.g.* [4] or G(oldstone) B(oson) E(xhange), see *e.g.* [2]. Here Dirac's instant–, point– or front– formulation of relativistic quantum mechanics is invoked to subsequently calculate various currents.

MESONS

In the following we will sketch the various assumptions and approximations made in constituent quark models by focussing on mesons: Adopting the framework of quantum field theory mesons are described as bound $q\bar{q}$ states with $M^2 = \bar{P}^2$, described by the Bethe-Salpeter amplitude $\chi_{\alpha\beta}(x_1, x_2) := \langle 0|T\left[\psi_\alpha(x_1)\bar{\psi}_\beta(x_2)\right]|\bar{P}\rangle$, which enters in a set of coupled equations which mutually determine the full propagators for the fermions and exchange bosons and the dressed vertex functions involved. In practise one truncates this set of equations by making an *Ansatz* for some n-point function and solving the equations (Bethe-Salpeter-Equation (BSE) for two particles or the Dyson-Schwinger-equation (DSE) for the self-energy) of lower order. In particular, based on an effective gluon propagator with a specific infrared behaviour this leads to the renormalization-group-improved rainbow-ladder approach [1], of which we will quote some interesting results. In a simplified *Ansatz* one can refrain from solving the DSE and assume that the fermion propagator has the free form $S(p) \approx i\left[\gamma^\mu p_\mu - m + i\varepsilon\right]^{-1}$ and to account for the self-energy contributions by introducing a constituent mass m. Furthermore one could assume that the irreducible interaction kernel is given by a single gluon exchange (OGE) in Coulomb gauge, possibly with a running coupling, where a Coulomb part of the interaction is instantaneous, and thus in the no-retardation limit $k^2 \rightarrow -|\vec{k}|^2$ arrive at an instantaneous OGE–potential. Such instantaneous interaction kernels allow for a parametrization of confinement by a string-like potential and, defining the Salpeter-Amplitude as $\Phi(\vec{p}) = \int\frac{dp^0}{2\pi}\chi(p^0, \vec{p})\Big|_{(P=M,\vec{0})}$, one then arrives at the Salpeter-Equation

(instantaneous Bethe-Salpeter Equation)

$$\Phi(\vec{p}) = \int \frac{d^3p}{(2\pi)^3} \frac{\Lambda_1^-(\vec{p})\gamma^0[V(\vec{p},\vec{p}')\Phi(\vec{p}')]\gamma^0\Lambda_2^+(-\vec{p})}{M+\omega_1+\omega_2}$$
$$- \int \frac{d^3p}{(2\pi)^3} \frac{\Lambda_1^+(\vec{p})\gamma^0[V(\vec{p},\vec{p}')\Phi(\vec{p}')]\gamma^0\Lambda_2^-(-\vec{p})}{M-\omega_1-\omega_2}, \tag{1}$$

with the projectors $\Lambda_i^\pm(\vec{p}) = (\omega_i(\vec{p}) \pm H_i(\vec{p}))/2\omega_i(\vec{p})$, the Dirac Hamiltonian $H_i(\vec{p}) = \gamma^0(\vec{\gamma}\cdot\vec{p}+m_i)$ and where $\omega_i(\vec{p}) = \sqrt{m_i^2 + \vec{p}^2}$. If one now drops the first term on the r.h.s. of Eq.(1) one arrives at the reduced Salpeter-equation, which then has the from of a Schrödinger equation with relativistic kinetic energy and relativistic corrections to the potential (contained in $\Lambda^\pm$). This can be considered the starting point of virtually all "relativized" constituent quark models.

Pioneering work in this spirit was performed already almost two decades ago by the group around Nathan Isgur both for mesons, see [3], and later on for baryons, see [4] and references therein. Here it was assumed that the quark interactions in hadrons can effectively be described by a linear confinement potential and spin-dependent parts of one gluon exchange; relativistic effects in the interactions were accounted for by parametrizations. This also holds for the description of annihilation contributions to pseudoscalar mixings. The scope of the calculation e.g. for mesons is a unified description of all resonances, both with light and with heavy flavours, and also includes a calculation of a multitude of electroweak and strong decay observables, which in spite of the more than a dozen model parameters can still be considered as rather efficient.

On the other hand one can also take the full Salpeter equation as a starting point for constituent quark model calculations: here the instantaneous interaction kernel consists of the Fourier transform of a string-like linearly rising confinement potential with an appropriate Dirac structure which avoids large spin-orbit splittings, supplemented by a spin-flavour dependent interaction motivated by instanton effects, see [6]. The latter has the decisive advantageous property to incorporate the $U_A(1)$ anomaly quantitatively and thus to account immediately for the splitting and mixing of (pseudo)scalar mesons. The total number of parameters in this approach amount to seven. As an example a comparison of the isoscalar mass spectrum for two versions of the confinement potential (Model $\mathscr{A}$ and Model $\mathscr{B}$ employing confinement Dirac structures $(\frac{1}{2}(\mathbf{I}\otimes\mathbf{I} - \gamma_0\otimes\gamma_0)$ and $\frac{1}{2}(\mathbf{I}\otimes\mathbf{I} - \gamma_5\otimes\gamma_5 - \gamma^\mu\otimes\gamma_\mu)$, respectively) with experimental data and the results from the calculation of Godfrey and Isgur is given in Fig. 1. Apart from the scalar sector the results are rather similar. While the 'relativized' quark model calculation resort to the rather *ad hoc* 'mock-meson' method, in the field theoretical approaches based on the Bethe-Salpeter equation the calculation of decay amplitudes in the Mandelstam-formalism is straightforward and parameter free, albeit numerically tedious. A comparison of the results for pseudoscalar decay constants is given in Table 1, for some radiative transitions in 2 and of the $\omega \to \pi\gamma$ and $K^* \to K\gamma$ transition form factors in Fig. 2. The Dyson-Schwinger approach leads to an excellent description of some observables, but at the time is unfortunately limited to calculations on properties of the lowest pseudoscalar and vector mesons mainly. This restriction does not applies to the instantaneous Bethe-Salpeter approach, which simultaneously describes the whole mass spectrum and

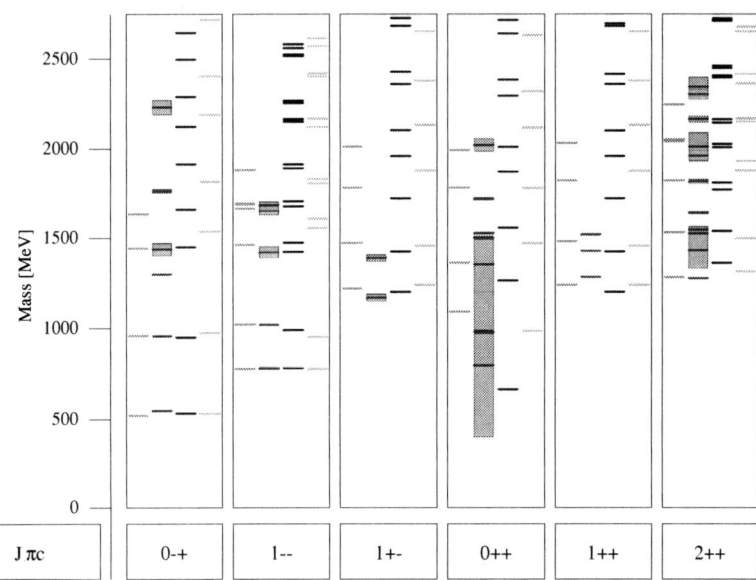

FIGURE 1. Spectrum of S- and P wave isoscalar mesons. From left to right each column (of fixed spin j, parity π and charge parity c) displays the results from the Godfrey-Isgur 'relativized' calculation [3], the experimental resonance position with a box indicating the error, and two versions of the relativistic calculation on the basis of the Salpeter equation with instanton-induced forces [6].

TABLE 1. Pseudoscalar decay constants in [MeV]

	Model $\mathscr{A}$	Model $\mathscr{B}$	DSE	Exp.	OGE
f_π	212	219	132	130.7 ± 0.46	184
f_K	248	238	154	159.8 ± 1.84	235

without introducing new parameters does fairly well also for observables at higher momentum transfer, see [6]. Such observables were not calculated in the relativized quark model, nevertheless the *ad hoc* "mock-meson method" gives a remarkable description of a multitude of other experimental data.

BARYONS

Although some pilot studies on (ground states of) baryons as q^3-systems have been done in the Dyson-Schwinger approach within a diquark-quark picture [1], the majority of constituent quark models of baryons still rely on the non-relativistic treatment with (some) relativistic corrections. If one insists on a description of the whole mass spectrum implementing relativistic covariance both in the quark dynamics and in the calculation of currents needed for decay observables, again, as for mesons, the instantaneous Bethe-Salpeter equation seems to be an appropriate starting point. Baryons are thus described

TABLE 2. Decay widths in [keV] of radiative meson transitions

Decay	Model $\mathcal{A}$	Model $\mathcal{B}$	DSE	Exp.	OGE
$\rho^{\pm} \to \pi^{\pm}\gamma$	35	21	53	67 ± 9	67^{*}
$\rho^{0} \to \pi^{0}\gamma$	35	21		117 ± 30	67
$\rho^{0} \to \eta\gamma$	50	40		57 ± 11	51
$\omega \to \pi^{0}\gamma$	315	185	479	717 ± 42	642
$\omega \to \eta\gamma$	5.5	4.4		5.5 ± 0.8	5.4
$K^{*\pm} \to K^{\pm}\gamma$	48	29	90	50 ± 5	67
$K^{*0} \to K^{0}\gamma$	102	70	130	117 ± 10	118
$\eta' \to \rho^{0}\gamma$	87	28		60 ± 5	135
$\eta' \to \omega\gamma$	9.7	3.1		6.1 ± 0.8	13.4
$\phi \to \eta\gamma$	58	35		58 ± 2	66
$\phi \to \eta'\gamma$	0.01	0.08		0.30 ± 0.16	0.26

* fitted

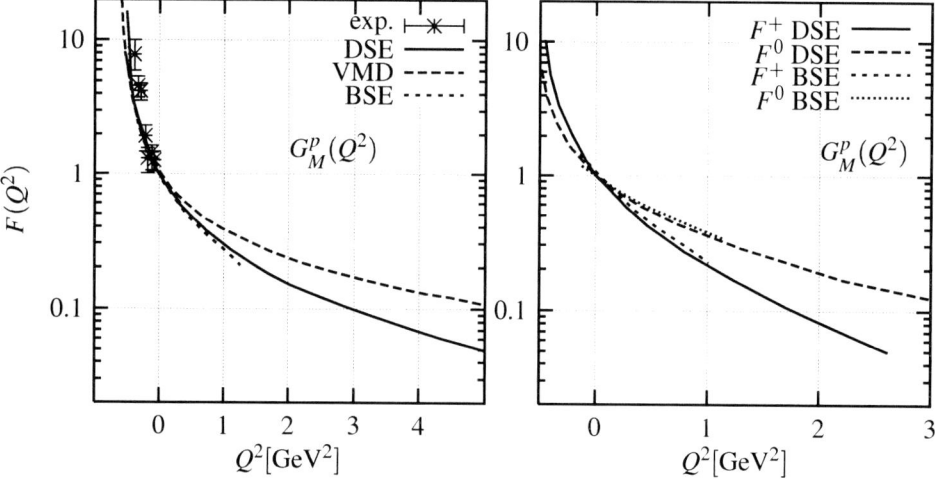

FIGURE 2. Left: comparison of $\omega - \pi - \gamma$-transition form factor calculated in the Dyson-Schwinger approach (DSE) of [1] with experimental data in the time-like region, the results from the instantaneous Bethe-Salpeter approach (BSE) and simple ω-vector meson dominance; Right: predictions for the charged (F^{+}) and neutral (F^{0}) $K^{*} - K - \gamma$-transition form factors

by the homogeneous Bethe-Salpeter equation with three-particle and two-particle in-stantaneous interaction kernels, which implement confinement and a spin-dependent in-teraction to account for the major mass splittings. Again the assumption of effective constituent quark propagators of the free form together with an approximate treatment of the two-body interactions [7] allows to formulate the dynamics in terms of Salpeter amplitudes (e.g. in the rest frame of the baryons): $\Phi_{M}(\vec{p}_{\xi}, \vec{p}_{\eta}) := \int \frac{dp_{\xi}^{0}}{2\pi} \frac{dp_{\eta}^{0}}{2\pi} \chi_{M}(p_{\xi}, p_{\eta})$

which then fulfills the Salpeter equation:

$$(\mathscr{H}\Phi_M)(\vec{p}_\xi,\vec{p}_\eta) = \sum_{i=1}^{3} H_i \,\Phi_M(\vec{p}_\xi,\vec{p}_\eta)$$

$$+ \left(\Lambda_1^+ \otimes \Lambda_2^+ \otimes \Lambda_3^+ + \Lambda_1^- \otimes \Lambda_2^- \otimes \Lambda_3^-\right)$$

$$\gamma^0 \otimes \gamma^0 \otimes \gamma^0 \int \frac{d^3 p'_\xi}{(2\pi)^3} \frac{d^3 p'_\eta}{(2\pi)^3} \, V^{(3)}(\vec{p}_\xi,\vec{p}_\eta,\vec{p}'_\xi,\vec{p}'_\eta) \,\Phi_M(\vec{p}'_\xi,\vec{p}'_\eta)$$

$$+ \left(\Lambda_1^+ \otimes \Lambda_2^+ \otimes \Lambda_3^+ - \Lambda_1^- \otimes \Lambda_2^- \otimes \Lambda_3^-\right)$$

$$\gamma^0 \otimes \gamma^0 \otimes \mathbf{1} \int \frac{d^3 p'_\xi}{(2\pi)^3} \left[V^{(2)}(\vec{p}_\xi,\vec{p}'_\xi) \otimes \mathbf{1} \right] \Phi_M(\vec{p}'_\xi,\vec{p}_\eta)$$

$$+ \text{ cycl. perm. (123).} \tag{2}$$

Again, if one would drop all terms involving the negative energy projectors Λ^- one arrives at a Schrödinger-type equation with relativistic corrections. Although the full Salpeter hamiltonian (2) is not positive definite with respect to the scalar product of the Salpeter amplitudes and thus positive and negative energy solutions occur, the negative energy solutions can (via the CPT-transformation) be mapped to positive energy solutions of opposite parity and consequently this approach leads to the same number of states as the non-relativistic quark model. Again, adopting a linear three-body confinement potential with a suitable spin dependence avoiding large spin-orbit splittings and the instanton-induced interaction to account for the major spin-dependent splittings with only seven parameters an excellent description has been obtained for all light-flavoured baryons, see [8, 9], including selective parity doubling and the Regge-trajectories up to the highest measured masses and total angular momenta. In 3 the results for N- and

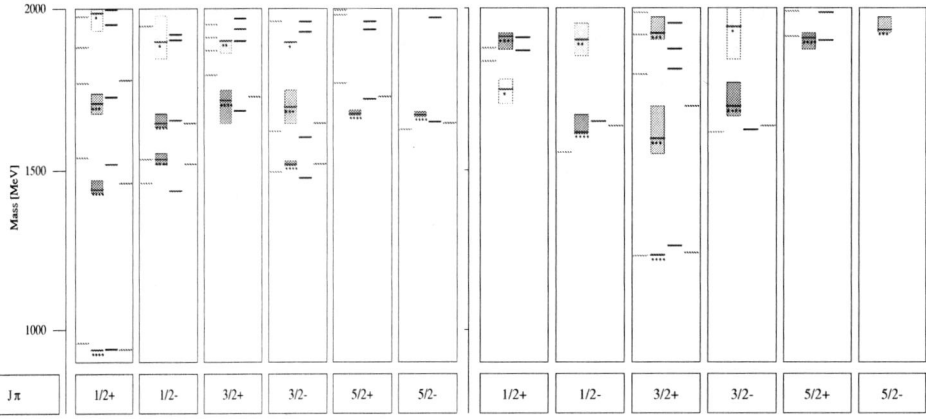

FIGURE 3. Low lying N- (left) and Δ-resonances (right). In each column from left to right the calculated result from the relativized quark model with a OGE-based quark interaction [4], the experimental data, the result from the instantaneous Bethe-Salpeter Equation with instanton-induced interactions and the results from a quark model calculation with Goldstone boson exchange [2] is displayed.

Δ-resonances of low-mass and low angular momenta is compared to experimental data as well as to the results from the relativized constituent quark model using parts of the OGE as a residual interaction, see [4] and the results from a constituent quark model developed by the Graz-group, which employs (flavour dependent) modified Yukawa-type potentials based on Goldstone-Boson-Exchange, see e.g. [2]. The latter treatment has the obviously satisfactory feature to be able to reproduce the first excited states of positive parity below the negative parity states, whereas the other treatments do yield a low lying Roper-like resonance but slightly above the lowest negative parity states. All calculations can not account for some negative parity Δ-resonances at approximately 1.9 GeV, see also the contribution of Ch. Weinheimer to this conference.

As for the mesons, in the Bethe-Salpeter approach electroweak currents can be calculated covariantly and (in lowest order) parameter free within the Mandelstam formalism [10]. The results for the magnetic moments of octet and decuplet baryons are given in Table 3 together with experimental data and the results which the Graz-group obtained employing the point-form of Dirac's relativistic quantum mechanics. Although the cal-

TABLE 3. Magnetic moments in μ_N of octet and decuplet baryons

Baryon	BSE	Exp.	GBE	Baryon	BSE	Exp.	GBE
p	2.77	2.793	2.70	Ξ^0	-1.33	-1.250	-1.27
n	-1.71	-1.913	-1.70	Ξ^-	-0.56	-0.6507	-0.67
Λ	-0.61	-0.613	-0.59	Δ^+	2.07	$2.7 \pm 1.5 \pm 1.3$	2.08
Σ^+	2.51	2.458	2.34	Δ^{++}	4.14	3.7-7.5	4.17
Σ^-	-1.02	-1.160	-0.94	Ω^-	-1.66	-2.0200	-1.59
Σ^0	0.75	–	0.70				

culational frameworks and the quark dynamics differ substantially in both approaches the results are remarkably similar and stress the importance of a relativistically covariant calculation of electromagnetic currents. This holds *a forteriori* for the calculation of electromagnetic (transition) form factors, see *e.g.* the comparison in Fig. 4. For more results on electroweak transition form factors we refer to [10]. Some new, representative results for semi-leptonic decays, calculated from the weak baryonic currents in the Mandelstam formalism, are listed in Table 4. Electroweak currents, provided that they are calculated in a relativistically covariant framework, can thus be satisfactory calculated in lowest order.

This is no longer holds *a priori* for the calculation of strong two-body decays, where channel couplings and mixing can be important, and in principle resonances could be even generated dynamically through such effects. Nevertheless it seems interesting to investigate to what extent a lowest order calculation, without any introduction of new parameters can describe some experimental features.

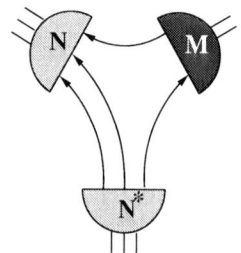

In the framework of the Mandelstam formalism the amplitude for the strong mesonic decay of excited baryons can be obtained in lowest order by evaluating the simple quark loop diagram displayed on the left, which involves the vertex functions (amputated Bethe-Salpeter-amplitudes) of the participating meson, obtained from the calculation on mesons [6], and of the initial and final baryon.

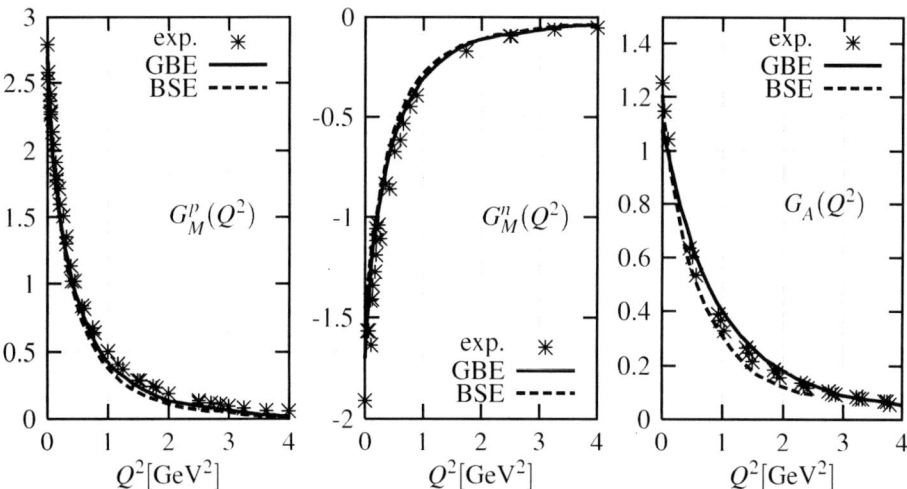

FIGURE 4. Comparison of the form factors calculated in the point-from approach in the constituent quark model with Goldstone Boson Exchange (GBE) of [2] and in the Mandelstam formalism on the basis of the Bethe-Salpeter Equation (BSE) [10] with experimental data; Left: Magnetic form factor of the proton; middle: Magnetic form factor of the neutron and right: axial form factor (adapted from [2])

TABLE 4. Decay rates and axial vector couplings of semi-leptonic decays of baryons.

Decay		$\Gamma\,[10^6\text{s}^{-1}]$ Exp.	Calc.	g_A/g_V Exp.	Calc.
n	$\to pe^-\bar{\nu}_e$			1.2670 ± 0.0035	1.21
Λ	$\to pe^-\bar{\nu}_e$	3.16 ± 0.06	3.10	-0.718 ± 0.015	-0.82
Σ^+	$\to \Lambda e^+\nu_e$	0.25 ± 0.06	0.20		
Σ^-	$\to \Lambda e^-\bar{\nu}_e$	0.38 ± 0.02	0.34		
Σ^-	$\to ne^-\bar{\nu}_e$	6.9 ± 0.2	4.91	0.340 ± 0.017	0.25
Ξ^0	$\to \Sigma^+e^-\bar{\nu}_e$	0.93 ± 0.14	0.91	$1.32^{+0.21}_{-0.17}\pm0.05$	1.38
Ξ^-	$\to \Sigma^0e^-\bar{\nu}_e$	0.5 ± 0.1	0.51		
Ξ^-	$\to \Lambda e^-\bar{\nu}_e$	3.3 ± 0.2	2.30	-0.25 ± 0.05	-0.27
Ω^-	$\to \Xi^0e^-\bar{\nu}_e$	68 ± 34	46		
Λ	$\to p\mu^-\bar{\nu}_\mu$	0.60 ± 0.13	0.47		
Σ^-	$\to n\mu^-\bar{\nu}_\mu$	3.04 ± 0.27	1.60		
Ξ^-	$\to \Lambda\mu^-\bar{\nu}_\mu$	2.1 ± 1.3	1.04		

Although in general the calculated partial widths are too small to account for the experimental values quantitatively, appreciable decay widths are found only for the well established resonances, the predicted values for higher lying resonances being in general smaller by at least an order of magnitude, see also Fig. 5, thus explaining why these have not been observed so far in elastic pion-nucleon scattering. This observation is in accordance with previous findings cited in [11, 4]. In Table 5 the calculated

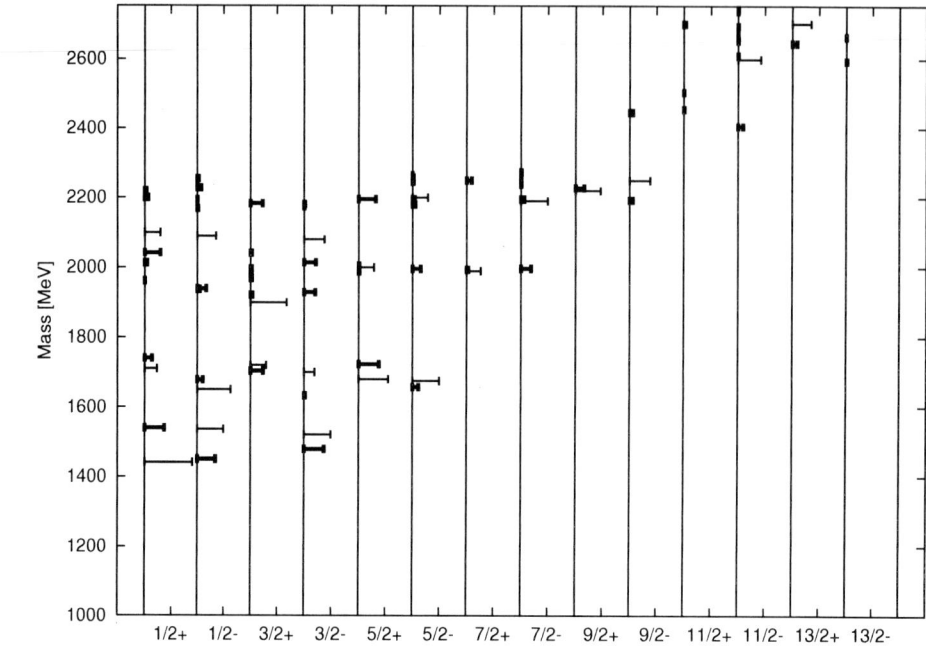

FIGURE 5. Decay amplitudes (proportional to the square root of the partial decay width) of strong $N^* \to N\pi$ decays. In each column (*i.e.* for each spin and parity $J\pi$) the experimental value (thin horizontal bars at the experimental resonance position) is compared to the calculated value (thick horizontal bars at the calculated resonance position).

partial decay widths of some selected low lying N- and Δ-resonances are compared to experimental data and recent results from the Graz group with a relativistic elementary meson emission [2] (with no extra parameters) as well as the results from the relativized quark model invoking the $^3P_0$-model [11].

TABLE 5. Partial decay widths in MeV of some strong two-body decays of N- and Δ-resonances.

Decay	BSE	GBE	$^3P_0$	Exp.	Decay	BSE	$^3P_0$	Exp.
$S_{11}(1535) \to N\pi$	33	93	216	$(68 \pm 15)^{+45}_{-23}$	$\to \Delta\pi$	1	2	< 2
$S_{11}(1650) \to N\pi$	3	29	149	$(109 \pm 26)^{+29}_{-4}$	$\to \Delta\pi$	5	13	$(6 \pm 5)^{+2}_{0}$
$D_{13}(1520) \to N\pi$	38	17	74	$(66 \pm 6)^{+8}_{-5}$	$\to \Delta\pi$	35	35	$(24 \pm 6)^{+3}_{-2}$
$D_{13}(1700) \to N\pi$	0.1	1	34	$(10 \pm 5)^{+5}_{-5}$	$\to \Delta\pi$	88	778	seen
$D_{15}(1675) \to N\pi$	4	6	28	$(68 \pm 7)^{+14}_{-5}$	$\to \Delta\pi$	30	32	$(83 \pm 7)^{+17}_{-6}$
$P_{11}(1440) \to N\pi$	38	30	412	$(228 \pm 18)^{+65}_{-65}$	$\to \Delta\pi$	35	11	$(88 \pm 18)^{+25}_{-25}$
$P_{33}(1232) \to N\pi$	62	34	108	$(119 \pm 0)^{+5}_{-5}$				
$S_{31}(1620) \to N\pi$	4	10	26	$(38 \pm 7)^{+8}_{-8}$	$\to \Delta\pi$	72	18	$(68 \pm 23)^{+14}_{-14}$
$D_{33}(1700) \to N\pi$	2	3	24	$(45 \pm 15)^{+15}_{-15}$	$\to \Delta\pi$	52	262	$(135 \pm 45)^{+45}_{-45}$

CONCLUSION

In conclusion we think that we have demonstrated, that constituent quark models provide a very useful tool in understanding hadron properties in a unified way: This not only involves a description of the mere mass spectrum, but also numerous decay amplitudes and electroweak (transition) form factors. In particular the field theoretical approaches which rely on the description of bound states of quarks through coupled Bethe-Salpeter/Dyson-Schwinger equations have provided very interesting results, unfortunately so far only for the ground states and some low-lying excited states. In this respect the approach based on the instantaneous Bethe-Salpeter equation, using free-form fermion propagators with constituent masses, implementing confinement by a string-like linearly rising potential and with instanton-induced interactions to explain the spin-dependent mass splittings seems to be a very efficient compromise combining the advantages of a relativistically covariant field theoretical treatment with the successful concepts of the (non-relativistic) constituent quark model. Adopting the point form of Dirac's Relativistic Quantum Mechanics does improve the description of observables within the latter category drastically, and supports the main findings of our treatment that a relativistic treatment of decay amplitudes, especially for processes at higher momentum transfers is absolutely imperative.

ACKNOWLEDGMENTS

I like to acknowledge the longstanding fruitful collaboration with Herbert Petry and the contributions by Matthias Koll, Ulrich Löring, Dirk Merten, Christian Haupt and Sascha Migura who did most of the calculations.

REFERENCES

1. Maris, P., Roberts, C.D., *Int. J. Mod. Phys.*, E**12**, 297-365 (2003); e-Print Archive: nucl-th/0301049.
2. Plessas, W., "Baryons as relativistic three quark systems", in *Lisbon 2002, Nuclear dynamics*, 139-150 (2002); e-Print Archive: nucl-th/0306021.
3. Godfrey, St., Isgur, N., *Phys. Rev.*, **32**, 189 (1985).
4. Capstick, S., Roberts, W., *Prog. Part. Nucl. Phys.*, **45**, 241 (2000).
5. Ricken, R., Koll, M., Merten, D., Metsch, B.Ch., Petry, H.R., *Eur. Phys. J.*, A**9**, 221 (2000).
6. Koll, M., Ricken, R., Merten, D., Metsch, B.Ch., Petry, H.R., *Eur. Phys. J.*, A**9**, 73 (2000).
7. Löring, U., Kretzschmar, K., Metsch, B.Ch., Petry, H.R., *Eur. Phys. J.*, A**10**, 309-346 (2001).
8. Löring, U., Metsch, B.Ch., Petry, H.R., *Eur. Phys. J.*, A**10**, 395-446 (2001).
9. Löring, U., Metsch, B.Ch., Petry, H.R., *Eur. Phys. J.*, A**10**, 447-486 (2001).
10. Löring, U., Kretzschmar, K., Merten, D., Metsch, B.Ch., Petry, H.R., *Eur. Phys. J.*, A**14**, 477-489 (2002).
11. Capstick, S., Roberts, W., *Phys. Rev.* D**49**, 4570-4586 (1994).

Chiral dynamics with strange quarks

Ulf-G. Meißner

Universität Bonn, HISKP (Th), D-53115 Bonn, Germany

Abstract. In the first part of the talk, I review what we know (or rather do not know) about the structure of the QCD vacuum in the presence of strange quarks. Chiral perturbation theory allows to study reactions of pions and kaons and to further sharpen our understanding of symmetry violation in QCD. I review recent progress on the description of pion-kaon scattering, in particular concerning isospin violation and the extraction of threshold and resonance parameters from Roy-Steiner equations. In the third part, it is shown how a unitary extension of chiral perturbation theory leads to novel insight into the structure of the $\Lambda(1405)$.

INTRODUCTION: S QUARK MYSTERIES

The strange quark plays a special role in the QCD dynamics at the confinement scale. Here, I will discuss some open questions surrounding chiral dynamics with strange quarks, pertinent to the structure of the strong interaction vacuum as well as to the structure of light mesons and baryons. Some of these issues are: Since $m_s \sim \Lambda_{QCD}$, is it appropriate to treat the strange quark as light or should it be considered heavy, as in the so–called heavy kaon effective field theory, see [1, 2, 3] ? Why is the OZI rule so badly violated in the scalar sector with vacuum quantum numbers? One example is the reaction $J/\Psi \to \phi\pi\pi/\bar{K}K$, which is OZI suppressed to leading order, but even has an additional doubly OZI suppressed contribution. The $\pi^+\pi^-$ event distribution shows a clear peak at the energy of 980 MeV, which is due to the f_0 scalar meson. This lets one anticipate that the dynamics of the low-lying scalar mesons and the mechanism of OZI violation are in some way related. More generally, it is of interest to learn about the phase structure of $SU(N_c)$ gauge theory at large number of flavors N_f. In QCD, we know that asymptotic freedom is lost for $N_f \geq 17$ but from the study of the two-loop β function one expects that there is a conformal window around $N_f \simeq 6$ [4]. This lets one contemplate the question whether there is already a rich phase structure even for the transition from $N_f = 2$ to $N_f = 3$? Some lattice studies seem to indicate a strong flavor sensitivity when going from $N_f = 2$ to $N_f = 4$ [5, 6]. As discussed by many speakers at this conference, the nature of the low–lying scalar mesons is still very much under debate (a topic I will not entertain in detail). In the baryon sector, there are also some "strange" states with non-vanishing strangeness. More precisely, what is the nature of some strange baryons like the $\Lambda(1405)$ or the $S_{11}(1535)$, are these three quarks states or meson-baryon bound states ? The latter scenario was already contemplated many years ago by Dalitz and collaborators [7] and has been rejuvenated with the advent of coupled channel calculations using chiral Lagrangians to specify the driving interaction. In the following, I address some of these issues.

CP717, *Hadron Spectroscopy: Tenth International Conference,*
edited by E. Klempt, H. Koch, and H. Orth
© 2004 American Institute of Physics 0-7354-0197-7/04/$22.00

THE VACUUM IN THE PRESENCE OF S QUARKS

There are many phenomenological as well as theoretical indications that the chiral symmetry (χS) of three–flavor QCD is spontaneously broken, abbreviated as SχSB. Now the question arises what are the order parameters of the SχSB ? Consider the current-current correlator between vector and axial currents,

$$\Pi_{\mu\nu}^{ab}(q) = i \int d^4 x e^{iqx} \langle 0|T\{V_\mu^a(x)V_\nu^b(0) - A_\mu^a(x)A_\nu^b(0)\}|0\rangle . \tag{1}$$

In the three flavor chiral limit, it can be written in terms of meson and continuum contributions and worked out explicitly,

$$\Pi_{\mu\nu}^{ab}(0) = -\frac{1}{4}g_{\mu\nu}\delta^{ab}F^2(3) . \tag{2}$$

If $\Pi_{\mu\nu}^{ab}(0) \neq 0$, then we have S$\chi$SB. We have thus identified an order parameter of spontaneous chiral symmetry breaking, namely the pion decay constant in the chiral limit,

$$\lim_{m_u,m_d,m_s\to 0} F_\pi = F(3) . \tag{3}$$

Its non-vanishing is a sufficient and necessary condition for SχSB,

$$\Pi_{\mu\nu}^{ab}(0) \neq 0 \leftrightarrow F(3) \neq 0 \leftrightarrow \text{S}\chi\text{SB} . \tag{4}$$

Naturally, there are many other possible order parameters. Often considered is the light quark condensate,

$$\langle 0|\bar{q}q|0\rangle = \langle 0|\bar{u}u|0\rangle^{(3)} = \langle 0|\bar{d}d|0\rangle^{(3)} = \langle 0|\bar{s}s|0\rangle^{(3)} \equiv -\Sigma(3) , \tag{5}$$

because the scalar-isoscalar operator $\bar{q}q$ mixes right- and left-handed quark fields. As will be discussed below, the quark condensate plays a different role than the pion decay constant. Other possible color-neutral order parameters of higher dimension are e.g. the mixed quark–gluon condensate $\langle 0|\bar{q}^i\sigma^{\mu\nu}G_{\mu\nu}^\alpha T_{ij}^\alpha q_j|0\rangle^{(3)}$ or certain four-quark condensates $\langle 0|(\bar{q}\Gamma_1 q)(\bar{q}\Gamma_2 q)|0\rangle$ with Γ_i some Dirac operator. It goes without saying that the spontaneously and explicitly broken chiral symmetry can be systematically analyzed in terms of an effective field theory - chiral perturbation theory (CHPT) (or some variant thereof). We now turn to the flavor dependence of these various order parameters. For this, consider QCD on a torus, or in an Euclidean ($t \to -ix^0$) box of size $L \times L \times L \times L$, understanding of course that we have to take the infinite volume limit at its appropriate place. Quark and gluon fields are then subject to certain boundary conditions, which are anti-periodic and periodic, in order. Analyzing the spectrum of the QCD Dirac operator, one arrives e.g. at the Banks–Casher relation. Also, the order parameters F^2 and Σ are dominated by the IR end of the Dirac spectrum [8], and one therefore expects a paramagnetic effect,

$$\Sigma(N_f+1) < \Sigma(N_f) \sim 1/L^4 , \quad F^2(N_f+1) < F^2(N_f) \sim 1/L^2 , \tag{6}$$

indicating a suppression of the chiral order parameters with increasing number of flavors. We note that the condensate is most IR sensitive. These results are exact, the question is now how strong this flavor dependence is or how it can be tested or extracted from some observables.

In the standard scenario of SχSB, terms quadratic in the quark masses are small, as has been recently confirmed for the *two flavor* case from the analysis of the BNL E865 K_{e4} data [9]. If these terms are also small in the three flavor case, the so-called Gell-Mann–Oakes–Renner ratio $X(3)$ stays close to one,

$$X(3) \equiv \frac{2\hat{m}\Sigma(3)}{F_\pi^2 M_\pi^2} \sim 1 \,, \tag{7}$$

with $\hat{m} = (m_u + m_d)/2$ the average light quark mass. There are many successes supporting this scenario, as one example I will discuss pion–kaon scattering in the next section. However, there is also some information pointing towards a more complicated phase structure (suppression of $\Sigma(3)$), as discussed next.

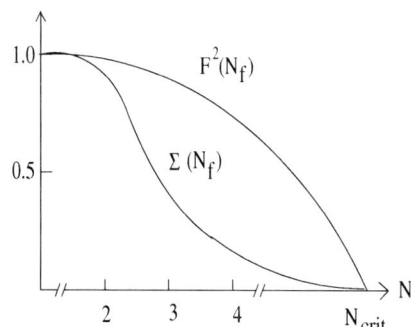

Figure 1. Flavor dependence of chiral symmetry breaking order parameters. A speculative scenario with a strongly suppressed three flavor condensate is depicted. In the standard scenario, the two lines for the pion decay constant and the condensate would be very close to each other.

Moussallam [10, 11] investigated a sum rule for the OZI violating correlator $\Pi_z \sim \langle \bar{u}u(x)\bar{s}s(0)\rangle_c$, which has the form

$$\Pi_z(m_s) = \frac{1}{\pi} \int_0^\infty \frac{ds}{s} \sigma(s) \,, \tag{8}$$

and allows to relate $\Sigma(3)$ with $\Sigma(2)$. This sum rule is super-convergent. In the approximation that the spectral function can be saturated by two–particle intermediate $\pi\pi$ and $K\bar{K}$ states, it can be expressed entirely in terms of the (non)strange scalar form factors of the pion and the kaon. These form factors can be calculated within CHPT, but are needed at higher energies here (for one particular calculation in a unitarized version of CHPT, see e.g. [12]). This gives the spectral function for energies below $\simeq 1.6\,$GeV. One can use various T-matrices for the $\pi\pi \to \pi\pi/K\bar{K}$ system to get an idea of the uncertainty in this energy domain. Above that energy, one can use pQCD. Putting all the various pieces together, one obtains

$$\Sigma(3) = \Sigma(2)\,[1 - 0.54 \pm 0.27] \,, \tag{9}$$

where the central value indicates a large suppression of the three flavor condensate but the uncertainties are large enough to give marginal consistency with the standard scenario. For a discussion of the stability of this result against some higher order corrections, see [11]. For more investigations of such a scenario see [13, 14]. Clearly, more work is needed to further quantify such results and to reduce the uncertainties.

PION-KAON SCATTERING

Pion–kaon scattering is the simplest scattering process involving strange quarks. Furthermore, since to one–loop accuracy all low–energy constants (LECs) are known from other processes, one can predict e.g. the S–wave scattering lengths (given here in the basis of total isospin 1/2 and 3/2). This has been done long time ago [15, 16] (in units of M_π^{-1}),

$$a_0^{1/2} = 0.18 \pm 0.03 \, [0.22 \pm 0.02] \, , \quad a_0^{3/2} = -0.05 \pm 0.02 \, [-0.045 \pm 0.008] \, . \quad (10)$$

The CHPT predictions are compared to the then existing data/Roy equation analysis in Fig. 2. Obviously, no firm conclusion could be drawn (the dark hatched ellipse comes in later).

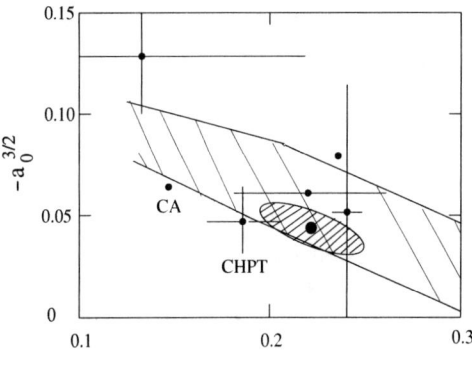

Figure 2. S-wave scattering lengths for πK scattering. The CHPT (CA) predictions are shown by the cross (black dot). The older data/Roy equation analysis can be traced back from [15, 16]. The dark hatched ellipse refers to the new dispersive analysis of [17].

In the light of more recent and more precise data from the eighties, that never were analyzed using dispersive methods, a novel evaluation of the Roy-Steiner equations was called for. This was recently achieved by Büttiker et al. [17]. They solved the Roy–Steiner equations for the S– and P–waves using all available input from $\pi K \to \pi K$ and $\pi\pi \to K\bar{K}$ and employing Regge theory for large energies. The outcome of this analysis are the S– and P–wave phase shifts below the matching energy of 1 GeV and the amplitude in the interior of the Mandelstam plane, in particular (sub)threshold parameters. It turns out that the resulting phase shifts are mostly in poor agreement with the existing low energy data, e.g. the mass of the spin-1 K^* mesons from the crossing of the P-wave isospin 1/2 phase through $\pi/2$ happens at 905 ± 3 MeV, visibly different from the PDG value of 891.7 ± 0.3 MeV. This needs further investigation. The resulting S–wave scattering lengths are given in the square brackets in Eq. (10), they come out consistent with the CHPT predictions, pointing toward the validity of the standard scenario. Similarly, the LECs extracted in [17] agree well with earlier determinations based on $X(3) \simeq 1$.

Another method of extracting the S-wave scattering lengths is the precise measurements of the characteristics of pion-kaon bound states, so-called πK atoms. In order to relate the lifetime and the energy shift to the scattering lengths, one has to make use of modified Deser formulae that include NLO effects in isospin breaking,

$$\Gamma_{\pi^0 K^0} \propto \left(a_0^{3/2} - a_0^{1/2} + \varepsilon\right)^2 (1 + \kappa), \tag{11}$$

$$\Delta E_{2S-2P}^{\text{str}} \propto \left(a_0^{3/2} + 2a_0^{1/2} + \varepsilon'\right)^2 (1 + \kappa'), \tag{12}$$

where ε and ε' represent, respectively, the isospin violating corrections in the regular part of the scattering amplitudes $\pi^- K^+ \to \pi^0 K^0$ and $\pi^- K^+ \to \pi^- K^+$ at threshold, while $\kappa(\kappa')$ is an additional contribution only calculable within the bound state formalism. There are two sources of isospin violation, the strong contributions $\propto m_u - m_d$ and electromagnetic contributions $\propto \alpha = e^2/4\pi$. It is most efficient to collect these two small parameters as $\delta \in \{m_u - m_d, \alpha\}$ and expand the corrections to order δ in all channels. To one-loop accuracy, ε and ε' have been calculated in [18, 19] (see also [20, 21])

$$a_0(\pi^- K^+ \to \pi^0 K^0) = -\sqrt{2}a_0^- \left\{ (1. \pm 0.8\%) + \underbrace{(1.3 \pm 0.1)\%}_{O(m_u - m_d)} + \underbrace{(0. \pm 1.1)\%}_{O(\alpha)} \right\} \tag{13}$$

$$a_0(\pi^- K^+ \to \pi^- K^+) = (a_0^- + a_0^+) \left\{ (1. \pm 16.1\%) + \underbrace{0.2\%}_{O(m_u - m_d)} + \underbrace{(0.9 \pm 3.2)\%}_{O(\alpha)} \right\} \tag{14}$$

where we have switched to the isospin basis for the scattering amplitudes, $T^+ = (T^{1/2} + 2T^{3/2})/3$ and $T^- = (T^{1/2} - T^{3/2})/3$. All quoted errors for the different contributions are due to the uncertainties in the respective strong and electromagnetic LECs. Note further that to leading order, the isospin violating corrections to the elastic scattering length are entirely given in terms of the pion mass difference, and thus are of electromagnetic origin. From the above equations, we can give rather precise predictions for ε and ε',

$$\varepsilon = 1.3 \pm 1.2 \%, \quad \varepsilon' = 1.1 \pm 3.2 \%. \tag{15}$$

We can thus conclude that the extraction of the strong scattering amplitudes at threshold from the lifetime and level shift of the πK atoms is sufficiently well under control. The isospin breaking effects in both cases are only of the order of 1% with an uncertainty of 1% and 3%, respectively. What remains to be done is a equally precise calculation of the bound state corrections κ and κ' [22].

THE NATURE OF THE $\Lambda(1405)$

In this section, I will discuss some issues in the framework of SU(3) baryon chiral perturbation theory and extensions thereof. First, it is often stated that three flavor baryon CHPT does not converge due to the large kaon mass and/or unitarity corrections.

While that is true in certain cases, there are many examples where indeed one can make precise predictions. As one particular example, let me consider the charge radii of the ground state baryon octet. To fourth order (complete one–loop calculation), the charge radii can be can given in terms of two LECs. These parameters can be fixed from the well measured proton and neutron electric radii, so that predictions for the other members of the octet emerge. On the other hand, the radius of the Σ^- can be obtained by scattering a highly boosted hyperon beam off the electronic cloud of a heavy atom (elastic hadron–electron scattering). Such an experiment has been first carried out at CERN, demonstrating the feasibility of the method and later repeated with much better accuracy at FNAL. The theoretical prediction (published before the data came out) compares well with the result from the SELEX collaboration,

$$\langle r_{\Sigma^-}^2 \rangle_{\text{th}} = 0.67 \pm 0.03 \text{ fm}^2 \text{ [23]}, \quad \langle r_{\Sigma^-}^2 \rangle_{\text{exp}} = 0.61 \pm 0.12 \pm 0.09 \text{ fm}^2 \text{ [24]}. \quad (16)$$

For a more detailed discussion of the status of SU(3) baryon CHPT, see e.g. [25].

Next, I discuss $K^- p$ scattering. For this process, a purely perturbative treatment is not possible (for an explicit demonstration, see [26]) due to the strong channel couplings and the appearance of a subthreshold resonance, the $\Lambda(1405)$, which is supposed to be a meson-baryon bound state rather than a genuine 3-quark state. First speculations about its possible unconventional nature date back to [7]. Since then many (QCD-inspired) models have been considered, but the first work of supplementing coupled channel dynamics with chiral Lagrangians which allows to dynamically generate the $\Lambda(1405)$ was reported in [27], see also [28] and the review [29]. A non-perturbative resummation scheme is mandatory to generate a bound state or a resonance. There exist many such approaches, but it is possible and mandatory to link such a scheme tightly to the chiral QCD dynamics. Such an improved approach was developed for pion–nucleon [30] and later applied to $\bar{K}N$ scattering [31]. The starting point is the T–matrix for any partial wave, which can be represented in closed form if one neglects for the moment the crossed channel (left-hand) cuts (for more explicit details, see [30])

$$T = \left[\tilde{T}^{-1}(W) + g(s) \right]^{-1}, \quad (17)$$

with $W = \sqrt{s}$ the cm energy (note that the analytical structure is much simpler when using W instead of s). $\tilde{T}$ collects all local terms and poles (which can be most easily interpreted in the large N_c world) and $g(s)$ is the meson-baryon loop function (the fundamental bubble) that is resummed by e.g. dispersion relations in a way to exactly recover the right-hand (unitarity) cut contributions. The function $g(s)$ needs regularization, this can be best done in terms of a subtracted dispersion relation and using dimensional regularization. It is important to ensure that in the low-energy region, the so constructed T–matrix agrees with the one of CHPT (matching). In addition, one has to recover the contributions from the left-hand cut. This can be achieved by a hierarchy of matching conditions,

$$\circ \ (p) \ : \ \tilde{T}_1(W) = T_1^\chi(W),$$
$$\circ \ (p^2) \ : \ \tilde{T}_1(W) + \tilde{T}_2(W) = T_1^\chi(W) + T_2^\chi(W),$$
$$\circ \ (p^3) \ : \ \tilde{T}_1(W) + \tilde{T}_2(W) + \tilde{T}_3(W) = T_1^\chi(W) + T_2^\chi(W)$$

$$+ T_3^{\chi}(W) + \tilde{T}_1(W)g(s)\tilde{T}_1(W) , \quad (18)$$

and so on. Here, T_n^{χ} is the T–matrix calculated within CHPT to $\circ$ (p^n). Of course, one has to avoid double counting as soon as one includes pion loops, this is achieved by the last term in the third equation (loops only start at third order in this case). In addition, one can also include resonance fields by saturating the local contact terms in the effective Lagrangian through explicit meson and baryon resonances (for details, see [30]). In particular, in this framework one can cleanly separate genuine quark resonances from dynamically generated resonance–like states. The former require the inclusion of an explicit field in the underlying Lagrangian, whereas in the latter case the fit will arrange itself so that the couplings to such an explicit field will vanish. It was observed in â Ǎĺ[31] that there are indeed two poles close to the nominal $\Lambda(1405)$ resonance, as earlier found in the cloudy bag model [32], and later confirmed in [33, 34]. The physics behind these two poles was recently revealed in [35]. Starting from an SU(3) symmetric Lagrangian to couple the meson octet to the baryon octet (in that limit, all octet Goldstone boson masses and all octet baryon masses are equal), one could in principle generate a variety of resonances according to the SU(3) decomposition,

$$8 \otimes 8 = 1 \oplus 8_s \oplus 8_a \oplus 10 \oplus \overline{10} \oplus 27 . \quad (19)$$

As it turns out, the leading order transition potential is attractive only in the singlet and the two octet channels, so that one a priori expects a singlet and two octets of bound states. However, the two octets come out degenerate, see Fig. 3. This has no particular dynamical origin but rather is a consequence of the actual values of the SU(3) structure constants. In the real world, there is of course SU(3) breaking of various origins. This was parameterized in [35] in terms of a symmetry breaking parameter x in the expressions for the meson M_i and baryon masses m_i as well as the subtraction constants a_i via $M_i^2(x) = M_0^2 + x(M_i^2 - M_0^2)$, $m_i(x) = m_0 + x(m_i - m_0)$ and $a_i(x) = a_0 + x(a_i - a_0)$, with $M_0 = 368\,\mathrm{MeV}$, $m_0 = 1151\,\mathrm{MeV}$ and $a_0 = -2.148$, where $0 \leq x \leq 1$. The motion of the various poles in the complex energy plane as a function of x is shown in Fig. 3. We note that the two octets split, in particular, one moves to lower energy ($I = 0, 1426\,\mathrm{MeV}$) close to the position of the singlet ($I = 0, 1390\,\mathrm{MeV}$). These are the two poles which combine to give the $\Lambda(1405)$ as it appears in various reactions.

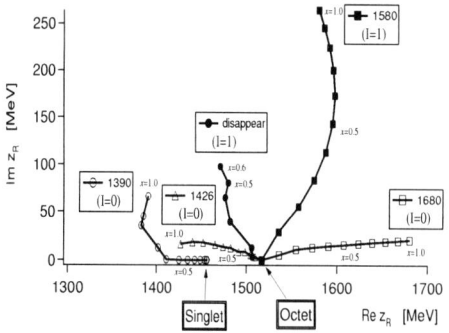

Figure 3. Trajectories of the poles in the scattering amplitudes obtained by changing the SU(3) breaking parameter x gradually. In the SU(3) limit ($x = 0$), only two poles appear, one corresponding to the singlet and the other to the two degenerate octets. The symbols correspond to the step size $\delta x = 0.1$.

The question is now how these two different poles can actually be disentangled in experiments? For that, one has to determine the couplings of these resonances to the physical states by studying the amplitudes close to the pole and identifying them with $T_{ij} = g_i g_j/(z - z_R)$ where z_R is the pole position and the g_i are in general complex numbers. As shown in [35], in the $I = 0$ channel the first resonance couples more strongly to $\pi\Sigma$ while the second one has a stronger coupling to the $\bar{K}N$ channel. We thus conclude that there is not just one single $\Lambda(1405)$ resonance, but *two*, and that what one sees in experiments is a *superposition* of these two states. Then, in the case that the $\Lambda(1405)$ is produced from the $\bar{K}N$ initial state, the peak is narrower as if it were produced from an $\pi\Sigma$ initial state. Therefore it is clear that, should there be a reaction which forces the initial channels to be $\bar{K}N$, then this would give more weight to the second resonance and hence produce a distribution with a shape corresponding to an effective resonance narrower than the nominal one and at higher energy. Such a case indeed occurs in the reaction $K^- p \rightarrow \Lambda(1405)\gamma$ studied theoretically in Ref. [36]. It was shown there that since the $K^- p$ system has a larger energy than the resonance, one has to lose energy emitting a photon prior to the creation of the resonance and this is effectively done by the Bremsstrahlung from the original K^- or the proton. Hence the resonance is initiated from the $K^- p$ channel and leads to a peak structure in the invariant mass distribution which is narrower and appears at higher energies than the experimental $\Lambda(1405)$ peaks observed in hadronic experiments performed so far. Experiments of producing the $\Lambda(1405)$ with (real or virtual) photons have been performed or are underway or will be done at SPRING-8, JLab and ELSA. Clearly, these should be able to verify (or falsify) the two pole nature of this particular baryon resonance. In the coupled channel approach matched to CHPT, there is also an interesting enhancement of the $I = 1$ amplitudes in the vicinity of the $\Lambda(1405)$. Independently of whether this can be interpreted as a resonance or as a cusp, the fact that the strength of the $I = 1$ amplitude around the $\Lambda(1405)$ region is not negligible should have consequences for reactions producing $\pi\Sigma$ pairs in that region. This has been illustrated for instance in [37], where the photoproduction of the $\Lambda(1405)$ via the reaction $\gamma p \rightarrow K^+ \Lambda(1405)$ was studied. It was shown there that the different sign in the $I = 1$ component of the $|\pi^+\Sigma^-\rangle$, $|\pi^-\Sigma^+\rangle$ states leads, through interference between the $I = 1$ and the dominant $I = 0$ amplitudes, to different cross sections in the various charge channels, a fact that has been confirmed experimentally very recently [38].

CONCLUDING REMARKS

There are many fascinating open problems in the large field of chiral dynamics with strange quarks. I have addressed three particular recent issues here and refer the reader to [39] for a much broader exposition. Certainly, one of the most important projects in the near future is to combine the chiral coupled channel dynamics with covariant quark models such as the Bonn one (see [40] and references therein) to solve the outstanding problem of the strong decay widths in such type of models and to get a better handle on the true nature of a variety of meson and baryon resonances, which have been one of the central issues of this conference.

ACKNOWLEDGEMENTS

I thank Véronique Bernard, Daisuke Jido, Bastian Kubis, José Antonio Oller, Eulogio Oset and Angels Ramos for enjoyable collaborations, Paul Büttiker and Jan Stern for useful communications and the organizers for their hospitality.

REFERENCES

1. A. Roessl, Nucl. Phys. B **555**, 507 (1999) [arXiv:hep-ph/9904230].
2. S.M. Ouellette, [arXiv:hep-ph/0101055].
3. M. Frink, B. Kubis and U.-G. Meißner, Eur. Phys. J. C **25**, 259 (2002) [arXiv:hep-ph/0203193].
4. T. Banks and A. Zaks, Nucl. Phys. B **196**, 189 (1982).
5. C. Sui, Nucl. Phys. Proc. Suppl. **73**, 228 (1999) [arXiv:hep-lat/9811011]; R.D. Mawhinney, Nucl. Phys. Proc. Suppl. **83**, 57 (2000) [arXiv:hep-lat/0001032].
6. Y. Iwasaki, K. Kanaya, S. Kaya, S. Sakai and T. Yoshié, Prog. Theor. Phys. Suppl. **131**, 415 (1998) [arXiv:hep-lat/9804005].
7. R.H. Dalitz and S.F. Tuan, Ann. Phys. (NY) **10**, 307 (1960).
8. S. Descotes, L. Girlanda and J. Stern, JHEP **0001**, 041 (2000) [arXiv:hep-ph/0010537].
9. S. Pislak, *et al.* (BNL-E865 Collaboration), Phys. Rev. Lett. **87**, 221801 (2001) [arXiv:hep-ex/0106071].
10. B. Moussallam, Eur. Phys. J. C **14**, 111 (2000) [arXiv:hep-ph/9909292].
11. B. Moussallam, JHEP **0008**, 005 (2000) [arXiv:hep-ph/0005245].
12. U.-G. Meißner and J.A. Oller, Nucl. Phys. A **679**, 671 (2001) [arXiv:hep-ph/0005253].
13. S. Descotes, JHEP **0103**, 002 (2001) [arXiv:hep-ph/0012221].
14. S. Descotes and J. Stern, Phys. Rev. D **62**, 054011 (2000) [arXiv:hep-ph/9912234].
15. V. Bernard, N. Kaiser and U.-G. Meißner, Phys. Rev. D **43**, 2757 (1991).
16. V. Bernard, N. Kaiser and U.-G. Meißner, Nucl. Phys. B **357**, 129 (1991).
17. P. Büttiker, S. Descotes-Genon and B. Moussallam, preprint HISKP-TH-03/18.
18. B. Kubis and U.-G. Meißner, Nucl. Phys. A **699**, 709 (2002) [arXiv:hep-ph/0107199].
19. B. Kubis and U.-G. Meißner, Phys. Lett. B **529**, 69 (2002) [arXiv:hep-ph/0112154].
20. A. Nehme and P. Talavera, Phys. Rev. D **65**, 054023 (2002) [arXiv:hep-ph/0107299].
21. A. Nehme, Eur. Phys. J. C **23**, 707 (2002) [arXiv:hep-ph/0111212].
22. J. Gasser and J. Schweizer, work in progress.
23. B. Kubis and U.-G. Meißner, Eur. Phys. J. C **18**, 747 (2001) [arXiv:hep-ph/0010283].
24. I. Eschrich *et al.* [SELEX Collaboration], Phys. Lett. B **522**, 233 (2001) [arXiv:hep-ex/0106053].
25. U.-G. Meißner, in M. Shifman (ed.), " At the frontier of particle physics", vol. 1, pp. 417-505, [arXiv:hep-ph/0007092].
26. N. Kaiser, Phys. Rev. C **64**, 045204 (2001) [arXiv:nucl-th/0107006].
27. N. Kaiser, P.B. Siegel, and W. Weise, Nucl. Phys. A **594**, 325 (1995) [arXiv:nucl-th/9505043].
28. E. Oset and A. Ramos, Nucl. Phys. A **635**, 99 (1998) [arXiv:nucl-th/9711022].
29. J. A. Oller, E. Oset and A. Ramos, Prog. Part. Nucl. Phys. **45**, 157 (2000) [arXiv:hep-ph/0002193].
30. U.-G. Meißner and J.A. Oller, Nucl. Phys. A **673**, 311 (2000) [arXiv:nucl-th/9912026].
31. U.-G. Meißner and J.A. Oller, Phys. Lett. B **500**, 263 (2001) [arXiv:hep-ph/0011146].
32. P. J. Fink, G. He, R. H. Landau and J. W. Schnick, Phys. Rev. C **41**, 2720 (1990).
33. D. Jido, A. Hosaka, J. C. Nacher, E. Oset and A. Ramos, Phys. Rev. C **66** (2002) 025203 [arXiv:hep-ph/0203248].
34. C. Garcia-Recio, J. Nieves, E. Ruiz Arriola and M. J. Vicente Vacas, [arXiv:hep-ph/0210311].
35. D. Jido, J. A. Oller, E. Oset, A. Ramos and U.-G. Meißner, Nucl. Phys. A **725**, 181 (2003) [arXiv:nucl-th/0303062].
36. J. C. Nacher, E. Oset, H. Toki and A. Ramos, Phys. Lett. B **461** (1999) 299 [arXiv:nucl-th/9902071].
37. J. C. Nacher, E. Oset, H. Toki and A. Ramos, Phys. Lett. B **455**, 55 (1999) [arXiv:nucl-th/9812055].
38. J. K. Ahn *et al.* [LEPS Collaboration], Nucl. Phys. A **721**, 715 (2003).
39. U.-G. Meißner, Phys. Scripta **T99**, 68 (2002) [arXiv:hep-ph/0201078].
40. D. Merten, U. Löring, K. Kretzschmar, B. Metsch and H. R. Petry, Eur. Phys. J. A **14**, 477 (2002) [arXiv:hep-ph/0204024].

Charmed meson resonances from chiral coupled-channel dynamics

E.E. Kolomeitsev[*] and M.F.M. Lutz[†]

*The Niels Bohr Institute
Blegdamsvej 17, DK-2100 Copenhagen, Denmark
†Gesellschaft für Schwerionenforschung (GSI)
Planck Str. 1, D-64291 Darmstadt, Germany

Abstract. Charmed meson resonances with quantum numbers $J^P = 0^+$ and $J^P = 1^+$ are generated in terms of chiral coupled-channel dynamics. At leading order in the chiral expansion a parameter-free prediction is obtained for the scattering of Goldstone bosons off charmed pseudo-scalar and vector mesons. The recently announced narrow open charm states observed by the BABAR and CLEO collaborations are reproduced. We suggest the existence of states that form an anti-triplet and a sextet representation of the SU(3) group. In particular, so far unobserved narrow isospin-singlet states with negative strangeness are predicted.

Recently a new narrow state of mass 2.317 GeV that decays into $D_s^+ \pi^0$ was announced [1]. This result was confirmed [2] and a second narrow state of mass 2.463 GeV decaying into $D_s^* \pi^0$ was observed. Such states were first predicted in [3, 4] based on the spontaneous breaking of chiral symmetry. The theoretical predictions [3, 4, 5, 6] rely on the chiral quark model which predicts the heavy-light $0^+, 1^+$ resonance states to form an anti-triplet representation of the SU(3) group. If one insists on a non-linear realization of the chiral SU(3) group and excludes any further model assumptions, no a priori prediction can be made for the existence of chiral partners of any given state.

Recently it was shown [7] that solving the coupled-channel Bethe-Salpeter equation with the interaction kernel following from a non-linear chiral SU(3) Lagrangian one is able to predict two octets and a singlet multiplets of the light 1^+ mesons consistent with the empirical spectrum. Similar results were obtained for light meson resonances with $J^P = 0^+$ quantum numbers [8, 9, 10, 11, 12, 13]. In view of the evident success of the chiral coupled-channel dynamics to predict the existence of a wealth of meson and baryon resonances in the (u, d, s)-sector of QCD [7, 14, 15] it is expectable that the chiral SU(3) symmetry should also predict spectra of hadrons with open charm or beauty (see also [16]). In this talk we review the χ-BS(3) approach [17, 18, 19, 20, 7, 14, 15] as applied to open-charm meson resonances [21, 22]. We will not have space to discuss further exciting results concerning open-beauty meson resonances [21], open-charm baryon resonances [23] or results in the (u, d, s)-sector of QCD [7, 14, 15].

Heavy-light meson states with quantum numbers $J^P = 0^+$ and $J^P = 1^+$ may be studied by considering the s-wave scattering of Goldstone bosons off the heavy-light ground state mesons with $J^P = 0^-$ and $J^P = 1^-$. If scalar or axial vector resonances exist they should manifest themselves as poles in the corresponding scattering amplitudes. The starting point to describe low-energy scattering processes is the chiral SU(3) Lagrangian

CP717, Hadron Spectroscopy: Tenth International Conference,
edited by E. Klempt, H. Koch, and H. Orth
© 2004 American Institute of Physics 0-7354-0197-7/04/$22.00

[24, 25, 26, 27] including heavy-light 0^- and 1^- fields. A systematic approximation scheme arises due to a successful scale separation justifying chiral power counting rules [28]. Our effective field theory for the scattering of Goldstone bosons off any heavy field is based on the assumption that the scattering amplitudes are perturbative at subthreshold energies with the expansion parameter Q/Λ_χ. The small scale Q is to be identified with any small momentum of the system. The chiral symmetry-breaking scale is

$$\Lambda_\chi \simeq 4\pi f \simeq 1.13 \text{ GeV},$$

with the parameter $f \simeq 90$ MeV determined by the pion decay process. Once the available energy is sufficiently high to permit elastic two-body scattering a further typical dimensionless parameter $m_K^2/(8\pi f^2) \sim 1$ arises [17, 18, 19] if strangeness is considered explicitly. This extra parameter invalidates any perturbative calculation within chiral SU(3) effective theory. Since this ratio is uniquely linked to two-particle reducible diagrams it is sufficient to sum those diagrams keeping the perturbative expansion of all irreducible diagrams, i.e. the coupled-channel Bethe-Salpeter equation has to be solved. This is the basis of the χ-BS(3) approach developed in [17, 18, 19, 7].

We identify the leading-order Lagrangian density [24, 25, 26, 27] describing the interaction of Goldstone bosons with pseudo-scalar and vector mesons,

$$\mathcal{L}(x) = \frac{1}{8f^2} \text{tr} \left[P(x)(\partial^\nu P^\dagger(x)) - (\partial^\nu P(x)) P^\dagger(x) \right] [\Phi(x), (\partial_\nu \Phi(x))]_-$$

$$- \frac{1}{8f^2} \text{tr} \left[P^\mu(x)(\partial^\nu P_\mu^\dagger(x)) - (\partial^\nu P^\mu(x)) P_\mu^\dagger(x) \right] \left[\Phi(x), (\partial_\nu \Phi(x)) \right]_- , \quad (1)$$

where Φ is the octet of Goldstone boson fields and P and P_μ are the triplets of massive pseudo-scalar and vector-meson fields in the matrix representation.

Within the χ–BS(3) approach [19, 7] the s-wave scattering amplitudes, $M_{jP}^{(I,S)}(\sqrt{s})$ take the simple form

$$M_{j\cdot}^{(I,S)}(\sqrt{s}) = \left[1 - V^{(I,S)}(\sqrt{s}) J_{jP}^{(I,S)}(\sqrt{s}) \right]^{-1} V^{(I,S)}(\sqrt{s}). \quad (2)$$

The effective interaction kernel $V^{(I,S)}(\sqrt{s})$ in (2) is determined by the leading order chiral SU(3) Lagrangian (1),

$$V^{(I,S)}(\sqrt{s}) = \frac{C^{(I,S)}}{8f^2} \left(3s - M^2 - \bar{M}^2 - m^2 - \bar{m}^2 - \frac{M^2 - m^2}{s} (\bar{M}^2 - \bar{m}^2) \right), \quad (3)$$

where (m, M) and $(\bar{m}, \bar{M})$ are the masses of initial and final mesons. We use capital M for the masses of heavy-light mesons and small m for the masses of the Goldstone bosons. The matrix of coefficients $C^{(I,S)}$, that characterize the interaction strength in a given channel, and the loop functions $J_{jP}^{(I,S)}(\sqrt{s})$ are given [7]. As expected from heavy-quark symmetry the interaction kernels as well as the loop functions are identical for the 0^- and 1^- sectors in the limit $M \to \infty$.

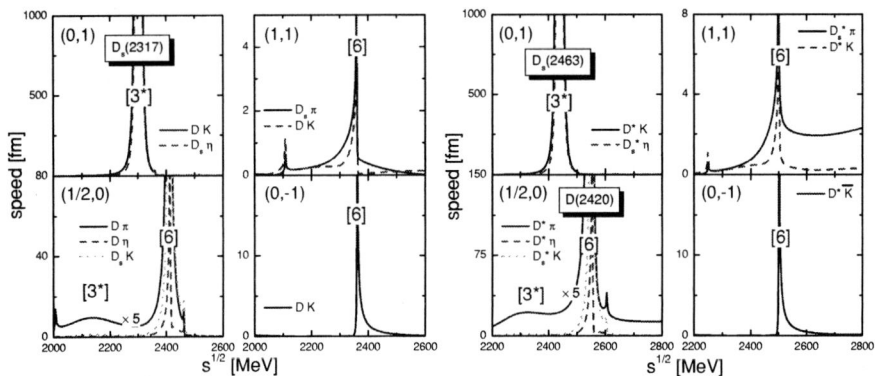

FIGURE 1. Speed plots for heavy-light scalar (left panel) and axial-vector (right panel) mesons with isospin (I) and strangeness (S).

In order to guarantee the perturbative nature of the scattering amplitude at subthreshold energies the $\chi-\mathrm{BS}(3)$ approach insists on a renormalization condition of the form

$$M^{(I,S)}(\sqrt{s} = \mu^{(I,S)}) = V^{(I,S)}(\sqrt{s} = \mu^{(I,S)}) \tag{4}$$

with the natural subtraction scales

$$\mu_{0^+}^{(I,0)} = M_{D(1867)}, \quad \mu_{0^+}^{(I,\pm1)} = M_{D_s(1969)}, \quad \mu_{0^+}^{(I,2)} = M_{D(1867)},$$
$$\mu_{1^+}^{(I,0)} = M_{D(2008)}, \quad \mu_{1^+}^{(I,\pm1)} = M_{D_s(2110)}, \quad \mu_{1^+}^{(I,2)} = M_{D(2008)}. \tag{5}$$

A crucial ingredient of the χ-BS(3) approach is a matching of s- and u-channel unitarized scattering amplitudes at subthreshold energies [19, 7]. This construction reflects our basic assumption that diagrams showing an s-channel or u-channel unitarity cut need to be summed to all orders at least at energies close to where the diagrams develop their imaginary part. By construction, a matched scattering amplitude satisfies crossing symmetry exactly at energies where the scattering process takes place. At subthreshold energies crossing symmetry is implemented approximatively only, however, to higher and higher accuracy when more chiral correction terms are considered. Insisting on the renormalization condition (4,5) guarantees that subthreshold amplitudes match smoothly and therefore the final 'matched' amplitudes comply with the crossing-symmetry constraint to high accuracy. A conceivable small variation of the subtraction scales around their natural values (5) has very little effect on the results. In fact chiral correction terms modify the effective interaction $V(\sqrt{s})$ rather than giving rise to a modification of the subtraction scale [19, 7, 22]. Changing the optimal subtraction scale (5) would deteriorate the quality of the matching of u- and s-channel unitarized amplitudes [19, 7].

We turn to the results of the χ-BS(3) approach for charmed mesons. It is instructive to explore first the SU(3) multiplet structure of the resonance states formed by the chiral coupled-channel dynamics. First the 0^+ sector is discussed in a 'heavy' SU(3) limit [15, 7] with $m_{\pi,K,\eta} = 500$ MeV and $M_D = 1800$ MeV. In this case we obtain an anti-triplet of mass 2204 MeV with poles in the $(0,+1),(1/2,0)$ amplitudes. The sextet channel does not show a bound-state signal in this case. However if the attraction is increased slightly by using $f = 80$ MeV rather than the canonical value 90 MeV, poles at mass 2298 MeV arise in the $(1, +1), (1/2, 0), (0, -1)$ amplitudes. This finding reflects that the Weinberg-Tomozawa interaction,

$$\bar{3} \otimes 8 = \bar{3} \oplus 6 \oplus \overline{15} \tag{6}$$

predicts attraction in the anti-triplet and sextet channel but repulsion for the anti-15-plet. In contrast performing a 'light' SU(3) limit [15, 7] with $m_{\pi,K,\eta} \sim 140$ MeV together with $M_D = 1800$ MeV we do not find any signal of a resonance in both anti-triplet and sextet channels. Analogous results are found in the 1^+ sector.

To study the formation of meson resonances we generate speed plots [29] as suggested by Höhler [30] (for definitions cf. [21]). Fig. 1 shows the spectra of 0^+ (left panel) and 1^+ (right panel) as they arise in calculations with physical masses. We predict a bound state of mass 2303 MeV in the $(0, 1)$-sector of 0^+ mesons (see Fig. 1, left panel). According to [5, 6] this state can be identified with a narrow resonance of mass 2317 MeV recently observed by the BABAR collaboration [1]. Since we do not consider isospin violating processes like $\eta \to \pi_0$ the latter state is a true bound state in our present scheme. Given the fact that our computation is parameter-free this is a remarkable result. In the $(1, +1)$-speeds where we expect a signal from the sextet a strong cusp effect at the $KD(1867)$-threshold is seen. The large coupling constant to the $\pi D_s(1969)$ channel leads to the broad structure seen in the figure. Fig. 1 (left panel) illustrates that in the $(\frac{1}{2}, 0)$-sector we predict a narrow 0^+ state of mass 2413 MeV just below the $\eta D(1867)$-threshold and a broad state of mass 2138 MeV. Modulo some mixing effects the heavier of the two is part of the sextet the lighter a member of the anti-triplet. The latter $(\frac{1}{2}, 0)$-state was expected to have a large branching ratio into the $\pi D(1867)$-channel [16, 6]. This is confirmed by our analysis. Finally in the $(0, -1)$-speed a pronounced cusp effect at the $\bar{K} D(1867)$-threshold is seen.

The spectrum predicted for the 1^+ states is very similar to the spectrum of the 0^+ states. Fig. 1 (right panel) demonstrates that it is shifted up by approximatively 140 MeV with respect to the 0^+ spectrum. The bound state in the $(0, 1)$-sector comes at 2440 MeV. Thus the mass splitting of the 1^+ and 0^+ states in this channel agrees very well with the empirical value of about 140 MeV measured by the BABAR and CLEO collaborations [1, 2]. A narrow structure at 2552 MeV is predicted in the $(\frac{1}{2}, 0)$-channel which may be identified with the $D(2420)$-resonance [31]. Even though the resonance mass is overestimated by about 130 MeV our result is consistent with its small width of about 20 MeV. The triplet state in this sector of mass 2325 MeV has again a quite large width reflecting the strong coupling to the $\pi D(2008)$-channel. Finally we obtain strong cusp effects at the $\bar{K} D(2008)$- and $KD(2008)$-thresholds in the $(0, -1)$- and $(1, +1)$-sectors. It is interesting to speculate whether chiral correction terms conspire to slightly increase the net attraction in these sectors. This would lead to a $(0, -1)$-bound state. The fact that we overestimate the mass of the sextet state $D(2420)$ by about 130 MeV we take as a prediction that this should indeed be the case [22]. An analogous statement holds for the 0^+ sector since due to heavy-quark symmetry chiral correction effects in the 0^+ and 1^+ are identical at leading order.

To summarize: We presented a coupled-channel description of the meson-meson scattering in the open charm sector using the chiral SU(3) Lagrangian involving light-heavy $J^P = 0^-$ and $J^P = 1^-$ fields that transform non-linearly under the chiral SU(3) group. The major result of our study is the prediction of the charmed mesonic states with $J^P = 0^+, 1^+$ quantum numbers forming anti-triplet and sextet representations of the SU(3) group. This differs from the results implied by the chiral quark model leading to anti-triplet states only. Our result suggests the existence of $J^P = 0^+, 1^+$ states with unconventional quantum numbers $(I, S) = (1, 1)$ and $(I, S) = (0, -1)$.

REFERENCES

1. BABAR Collaboration, B. Aubert et al., hep-ex/0304021.
2. CLEO Collaboration, D. Besson et al., hep-ex/0305017, hep-ex/0305100.
3. M.A. Nowak, M. Rho and I. Zahed, Phys. Rev. D **48** (1993) 4370
4. W.A. Bardeen and C.T. Hill, Phys. Rev. D **49** (1994) 409.
5. W.A. Bardeen, E.J. Eichten and Ch.T. Hill, hep-ph/0305049.
6. M.A. Nowak, M. Rho and I. Zahed, hep-ph/0307102.
7. M.F.M. Lutz and E.E. Kolomeitsev, nucl-th/0307039.
8. E. Van Beveren et al., Z. Phys. C **30** (1986) 615.
9. J.D. Weinstein and N. Isgur, Phys. Rev. D **41** (1990) 2236.
10. G. Janssen, B.C. Pearce, K. Holinde, J. Speth, Phys. Rev. D **52**, 2690 (1995).
11. J.A. Oller, E. Oset and J.R. Pelaez, Phys. Rev. D **59**, 074001 (1999); Erratum-ibid. **60**, 099906 (1999).
12. J. Nieves, M.P. Valderrama and E. Ruiz Arriola, Phys. Rev. D **65**, 036002 (2002).
13. A.G. Nicola and J.R. Pelaez, Phys. Rev. D **65**, 054009 (2002).
14. C. García-Recio, M.F.M. Lutz and J. Nieves, nucl-th/0305100.
15. E.E. Kolomeitsev and M.F.M. Lutz, nucl-th/0305101.
16. E. van Beveren and G. Rupp, Phys. Rev. Lett. **91**, 012003 (2003); help-ph/0306051;
17. M.F.M. Lutz and E.E. Kolomeitsev, Proc. of Int. Workshop XXVIII on Gross Properties of Nuclei and Nuclear Excitations, Hirschegg, Austria, January 16-22, 2000. nucl-th/0004021
18. M.F.M. Lutz und E.E. Kolomeitsev, Found. Phys. **31**, 1671 (2001).
19. M.F.M. Lutz and E.E. Kolomeitsev, Nucl. Phys. A **700**, 193 (2002).
20. M.F.M. Lutz, GSI-Habil-2002-1.
21. E.E. Kolomeitsev and M.F.M. Lutz, hep-ph/0307133, Phys. Lett. B (in print).
22. J. Hofmann and M.F.M. Lutz, hep-ph/0308263.
23. M.F.M. Lutz and E.E. Kolomeitsev, hep-ph/0307233, Nucl. Phys. A (in print).
24. S. Weinberg, Phys. Rev. Lett. **17**, 616 (1966); Y. Tomozawa, Nuov. Cim. A **46**, 707 (1966).
25. M.B. Wise, Phys. Rev. D **45**, 2188 (1992).
26. T.-M. Yan et al., Phys. Rev. **46**, 1148 (1992).
27. G. Burdman and J. Donoghue, Phys. Lett. B **280**, 287 (1992).
28. S. Weinberg, *The quantum theory of fields*, Vol. II, University Press, Cambridge (1996).
29. F.T. Smith, Phys. Rev. **118**, 349 (1960).
30. G. Höhler, πN NewsLetter **9**, 1 (1993).
31. K. Hagiwara et al., Phys. Rev. D **66**, 010001 (2002).

Chiral doublings of heavy-light hadrons: New charmed mesons discovered by BABAR, CLEO and BELLE

Maciej A. Nowak

M. Smoluchowski Institute of Physics, Jagiellonian University,
30-059 Kraków, Reymonta 4, Poland
e-mail: nowak@th.if.uj.edu.pl

Abstract. We remind the chiral doubling scenario[1, 2] for hadrons built of heavy *and* light quarks. Then we recall the arguments why new states $D_s(2317)$, $D_s(2460)$, $D_0(2308)$ and $D'_1(2427)$ should be viewed as chiral partners of D_s, D_s^*, D and D^*, respectively. We summarize with the list of predictions based of chiral doubling scenario for other heavy-light hadrons.

Recently, experimental physics of hadrons with open charm has provided several spectacular discoveries:

– First, BaBar [3] has announced new, narrow meson $D_{sJ}^*(2317)^+$, decaying into D_s^+ and π^0. This observation was then confirmed by CLEO [4], which also noticed another narrow state, $D_{sJ}(2463)^+$, decaying into D_s^* and π^0. Both states were confirmed by Belle [5], and finally, the CLEO observation was also confirmed by BaBar [6]. From the moment of discovery, both states triggered a flurry of activity among the theorists. Experimental results were surprising, since such states were neither expected below the DK and $D*K$ thresholds nor were expected to be that narrow. Till today, several theoretical constructions were proposed to explain the masses, quantum numbers and decay patterns, most of them discussed during this conference.

– Second, Belle has not only measured the narrow excited states D_1, D_2 with foreseen quantum numbers $(1+, 2+)$, but provided also first evidence for two new, broad states $D_0*(2308 \pm 17 \pm 15 \pm 28)$ and $D'_1(2427 \pm 26 \pm 20 \pm 17)$. Both of those are *ca* $350 - 400$ MeV higher above the usual D_0, D^* states and seem to have opposite parity.

– Third, Selex has provided preliminary data for doubly charmed baryons [7]. On top of known since December *ccd* state (3520), four other cascade $j = 1/2$ states are visible. Their masses are a challenge for standard estimations based on potential models [8]. This calls for alternative predictions, based either on chiral solitonic models or diquark scenarios.

In this talk, we recall that actually the presence of this type of states was predicted by theoretical arguments already in 1992 and 1993, and is in fact required from the point of view of symmetries of the QCD interactions. The two, mentioned above particles observed by BaBar and CLEO are the first, theoretically anticipated [1, 2] *chiral partners* of hadrons built out of light and heavy quarks. As such, they should represent rather a *pattern* of spontaneous breakdown of chiral symmetry than isolated events.

CP717, *Hadron Spectroscopy: Tenth International Conference,*
edited by E. Klempt, H. Koch, and H. Orth
© 2004 American Institute of Physics 0-7354-0197-7/04/$22.00

Strong interactions involve three light flavors (u, d, s) and three heavy flavors (c, b, t) with respect to the QCD infrared scale. The light sector (l) is characterized by the spontaneous breaking of chiral symmetry, while the heavy sector (h) exhibits heavy-quark (Isgur-Wise) symmetry [9]. In our original work [1] we addressed the question of the form of the heavy-light effective action in the limit where light flavors are massless, while the heavy flavors are infinitely massive. The novel aspect of our original derivation was that consistency with the general principles of spontaneously broken chiral symmetry requires the introduction of *chiral partners* in the form of a $(0^+, 1^+)$ multiplet of pseudoscalars and transverse vectors [1]. In the heavy-quark limit, the splitting between the chiral partners is small and of the order of the "constituent quark mass". The chiral corrections to the splitting were recently shown to be of order $m_\pi^2 / 4 m_h$, and therefore small irrespective of an effective Lagrangian analysis [10].

In brief, to leading order in the heavy-quark mass, the one-loop effective action for the chiral doubler $(0^+, 1^+)$

$$G = \frac{1 + v\!\!\!/}{2} (\gamma^\mu \gamma_5 \tilde{D}_\mu^* + \tilde{D}) . \tag{1}$$

duplicates the known action [11] for the standard $(0^-, 1^-)$ multiplet

$$H = \frac{1 + v\!\!\!/}{2} (\gamma^\mu D_\mu^* + i \gamma_5 D) . \tag{2}$$

i.e.

$$\mathscr{L}^G = -\frac{i}{2} \text{Tr}(\bar{G} v^\mu \partial_\mu G - v^\mu \partial_\mu \bar{G} G) + \text{Tr} V_\mu \bar{G} G v^\mu - \mathbf{g}_G \text{Tr} A_\mu \gamma^\mu \gamma_5 \bar{G} G - \mathbf{m}_G(\Sigma) \text{Tr} \bar{G} G \tag{3}$$

The axial A_μ and vector V_μ currents are light currents, contributing to transitions in odd or even number of pions, respectively (or generically, $SU(N_l)$ Goldstone bosons). The key difference is the opposite sign in the sign of the constituent mass contribution in (3), with respect to similar term for H multiplet. The sign flip follows from the γ_5 difference in the definition of the fields H and G. In other words: it is sensitive to the parity content of the heavy-light field since $H v\!\!\!/ = -H$ and $G v\!\!\!/ = +G$. The result is a split between the heavy-light mesons of opposite chirality. This unusual contribution of the chiral quark mass stems from the fact that it tags to the *velocity* $H v\!\!\!/ \bar{H}$ of the heavy field and is therefore sensitive to *parity*. The reparametrization invariance (invariance under velocity shifts of the heavy quark to order one) introduces mass shifts that are parity insensitive to leading order in $1/m_h$ [12]. Chiral partners communicate with each other via light axial currents

$$\mathscr{L}_{HG} = \sqrt{\frac{\mathbf{g}_G}{\mathbf{g}_H}} \text{Tr}(\gamma_5 \bar{G} H \gamma^\mu A_\mu) - \sqrt{\frac{\mathbf{g}_H}{\mathbf{g}_G}} \text{Tr}(\gamma_5 \bar{H} G \gamma^\mu A_\mu) \tag{4}$$

with no vector mixing because of the parity.

We visualize chiral doublers scheme for mesons in the form of cartoon, see Fig. 1. The three-dimensional "cube" is aligned along three "directions":
- chiral symmetry breaking (horizontal, green)

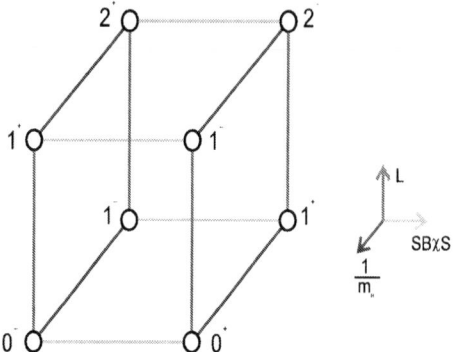

FIGURE 1. Cartoon representing *schematic* classification of chiral doublers.

- Isgur-Wise symmetry breaking (skew, red)
- total *light* angular momentum (vertical, blue).

The corners of the cube represent generic $h\bar{l}$ mesons, i.e. we expect similar "cubic" patterns for $c\bar{s}$, $c\bar{u}$, $c\bar{d}$, $b\bar{s}$, $b\bar{u}$, $b\bar{d}$ mesons. Let us focus on $c\bar{s}$ states, i.e D_s-cube. Lower left rung represents known pseudoscalar 0^- $D_s(1969)$ and vector 1^- $D_s^*(2112)$, belonging to $j_l = 1/2$ light angular momentum representation. The splitting between them is an $1/m_c$ effect and is expected to vanish in infinitely heavy charm quark limit, i.e. both particles would have form the H multiplet. The upper left rung corresponds to $j_l = 3/2$ representation, i.e. 1^+ and 2^+ *excited* multiplet. Here $D_{s1}(2536)$ and $D_{sJ}^*(2573)$ are the candidates, separated by (smaller for excited states) $1/m_c$ origin mass splitting. Similar pattern applies for the non-strange charmed mesons (D-cube), i.e. $D(1865)$, $D^*(2010)$, $D_1(2420)$ and $D_2(2460)$. This "left plaquette" of the cube completes the standard, "pre-BaBarian" charmed meson spectroscopy.

The novel aspect is the existence of the chiral doublers, i.e. the appearance of the *right plaquette*. First, we expect two chiral partners for D_s and D_s^*, representing right lower rung. Here newly discovered $D_{sJ}^*(2317)$ and $D_{sJ}(2463)$ are the candidates for the $(0^+, 1^+)$ scalar-axial G multiplet. The averaged splitting for $(0^+, 0^-)$ and the averaged splitting for $(1^+, 1^-)$ are 349.2 ± 0.8 and $346.8 + 1.1$, respectively, i.e. almost identical, as predicted a decade ago [1, 2]. Naturally, the splitting within the G multiplet, i.e. between the masses of the new BaBar state and CLEO state, is identical to the splitting between the $(1^-, 0^-)$ pair. The narrowness of the new states is basically the consequence of the kinematic constraints, as pointed by [13]: since the chiral split is smaller than the mass of the kaon, these states live longer. On top of this effect, the isospin conservation most probably forces the pionic decay via virtual η decay, suppressing the rate even further [13]. Electromagnetic transitions, estimated on the basis of chirally doubled lagrangians in [13] are also in agreement with the experimental data. It is noteworthy to stress, that chiral Ward identities additionally constraint the amplitudes of the pionic decays for the H multiplets, G *and* for $G - H$ pionic transitions [10].

Let us move now towards the excited states. On the basis of the chiral doublers scenario, we would also expect the chiral partners for the excited $j_l = 3/2$ multiplet, i.e. *new* chiral pair $(1^-, 2^-)$ [12]. Alternatively, this pair could be also viewed as the

$j_l = 3/2$ excitation of the BaBar-Cleo $(0^+, 1^+)$ multiplet. Our prediction for the masses of this *new* pair reads:

$$m(\tilde{D}_{s1}) = 2721 \pm 10 \text{MeV}$$
$$m(\tilde{D}_{s2}) = 2758 \pm 10 \text{MeV} \tag{5}$$

where we used [10] as an input the observed BaBar and CLEO splitting for the chiral multiplet $(0^+, 1^+)$ and the mass formulae obtained in [12]. Note that the chiral splitting for excited states is approximately half of the chiral splitting for the ground pair.

This completes the identification of corners of the "cube". Left and right plaquettes are chiral copies, front and back plaquettes become degenerate in infinite mass of the heavy quark and the lower and upper plaquettes are separated by the excitation of total *light* angular momentum j_l. We do not discuss here the possibility of even higher angular excitations, i.e. additional $j_l = 5/2$ plateau, pointing only that first such states $(2^-, 3^-)$ may naturally appear above 3 GeV.

Let us move now towards non-strange charmed mesons. Here two states from Belle, $D_0^*(2308)$ and $D_1'(2427)$ are natural candidates for lower right rung of the D-cube, i.e. for the chiral doublers of $D(1825)$ and $D^*(2010)$. There are however broad, since neither kinematic nor isospin restrictions apply here, contrary to their strange cousins.

One of the arguments against the interpretation that the above pair might be a chiral doubler is based on the values of the "chiral" mass shift, which (modulo experimental errors) seems to be equal of even larger for the non-strange mesons than for the strange ones. Actually, this argument acts rather in favor for the chiral doublers scheme [10]. A simple parametrization for the constituent quark mass (with good comparison to instanton liquid model and lattice data) was quoted in [14]

$$\Sigma(m_l) \approx m_l + \Sigma(0)(\sqrt{1 + (m_l/d)^2} - m_l/d) \tag{6}$$

with $\Sigma \approx 345 \text{MeV}/c^2$, $d \approx 198 \text{MeV}$. For a strange quark mass $m_s \approx 150 \text{MeV}$, the second term is reduced to $\Sigma(0)/2$, making the combination (6) weakly dependent on m_l and of order Σ all the way up to the strange quark mass. Thus, both mass splittings are about the same for (u, d, s) heavy-light mesons. Taking e.g. the value $\Sigma(0) = 400$ MeV one could easily estimate the effects of chiral splitting for strange quark to be of order 350 MeV, so even smaller than for non-strange quarks. We would like to mention, that several simplified models just add the *constant* chiral dressing to the current mass, ignoring the feedback of the explicit breaking of the chiral symmetry on the vacuum effects.

Let us mention for completeness, that the chiral doubling should be even more pronounced for bottom mesons, since the $1/m_h$ corrections are three times smaller. For $m_s = 150$ MeV, we expect [10] the chiral partners of B_s and B_s^* to be 323 MeV heavier, while the chiral partners of B and B^* to be 345 MeV heavier, i.e. close to predictions in [13]. We note that any observation of chiral doubling for B mesons would be a strong validation for chiral doublers proposal. Indeed, in the recently proposed alternative scenarios discussed during this conference (multiquark states, hadronic molecules, modifications of quark potential, unitarization) a repeating pattern from charm to bottom calls for additional assumptions.

Finally, let us mention very briefly the consequences of the chiral doublers scheme for the heavy light baryons, i.e. *hhl* and *hll* type. Already in [1] we pointed, that the opposite parity baryons could be described as chiral solitons of the action of (3) type. In the light of presented here preliminary Selex data [7] and planned COMPASS experiment, the issue of doubly heavy baryons is no longer academic. It is tempting to speculate, that the observed splitting between the signals for the *ccu* states of opposite parity, 3780 MeV and 3460 MeV [7] is of chiral origin (here 340 MeV). Indeed, double heavy baryons, with the heavy diquark in spin 1 state behaving alike the heavy color source $\bar{3}_c$ resemble mesonic configurations and are subjected to chiral doubling due to the light quark physics, a point also noted recently by [13].

We do not discuss here to what extent the newly discovered chiral partners can shed more light on effects of charmonium absorption/regeneration in thermal models with medium effects. We stress that this is an important issue in light of the current and future experiments at RHIC and LHC as well as at GSI, probing the restoration of the chiral symmetry.

ACKNOWLEDGMENTS

This talk is based on work done in collaboration with Mannque Rho and Ismail Zahed [1, 10, 12]. This work was partially supported by the Polish State Committee for Scientific Research (KBN) grant 2P03B 09622 (2002-2004). I am very grateful to David Cassel, Murray Moinester, Jim Russ, Bob Cahn, Ken Hicks, Lonya Glozman, Pavel Krokovny, Kamal Seth, Ted Barnes, Dan-Olof Riska, Su Houng Lee, Maxim Polyakov, Zhenya Kolomeitsev, Thorsten Feldmann, Klaus Goeke, Frank Close and many others for passionate discussions on new particles during this conference.

REFERENCES

1. M.A. Nowak, M. Rho, I. Zahed, Phys. Rev. **D48** (1993) 4370.
2. W. Bardeen and C. Hill, Phys. Rev. **D49** (1994) 409.
3. BABAR Coll., B. Aubert et al., hep-ex/0304021.
4. CLEO Coll., D. Besson et al., hep-ex/0305017, hep-ex/0305100.
5. BELLE Coll., P. Krokovny et al., hep-ex/0308019.
6. BABAR Coll., B. Aubert et al., hep-ex/0310050.
7. James Russ and Murray Moinester, talks at Hadron03, Aschaffenburg, 31-08-06.09. 2003, Germany.
8. V.V. Kiselev, A. K. Likhoded, hep-ph/0103169.
9. N. Isgur and M. B. Wise, Phys. Rev. Lett. **66** (1991) 1130. For early suggestions, see also E.V. Shuryak, Phys. Lett.**93B** (1980) 134, Nucl. Phys. **B198** (1982) 83.
10. M.A. Nowak, M. Rho and I. Zahed, hep-ph/0307102.
11. M.B. Wise, Phys. Rev. **D45** (1992) R2118;. T.-M. Yan et al, Phys. Rev. **D46** (1992) 1148; G. Burdman and J.F. Donoghue. Phys. Lett. **B280** (1992) 287.
12. M. A. Nowak and I. Zahed, Phys. Rev. **D48** (1993) 356.
13. W.A. Bardeen, E.J. Eichten, C.T. Hill, Phys. Rev. **D68** (2003) 054024.
14. M. Musakhanov, hep-ph/0104163 and reference to unpublished results of P. Pobylitsa therein.

The $K \to \pi\pi$ Electroweak Penguin Matrix Elements in the Chiral Limit

K. Maltman[*†], V. Cirigliano[**] and J.F. Donoghue and E. Golowich[‡]

[*]Dept. Math and Stats, York University, Toronto, ON CANADA M3J 1P3
[†]CSSM, Univ. Adelaide, Adelaide,SA, Australia 5005
[**]Dept. Fisica Teorica, IFIC, Univ. de Valencia - CSIC, E-46071 Valencia, Spain
[‡]Physics Dept., University of Massachusetts, Amherst, MA 01003 USA

Abstract. We present a model-independent determination of the $SU(3)_F$ chiral limit values of the $K \to \pi\pi$ matrix elements of the electroweak penguin operators $Q_{7,8}$ in the Standard Model. This determination is accomplished using a combination of dispersive and finite energy sum rule techniques, with hadronic τ decay data as input. The consistency between independent determinations is shown to be excellent. Implications for the value of ε'/ε in the Standard Model are also discussed.

In the Standard Model, ε'/ε is known to be dominated by contributions from the gluonic penguin operator, Q_6, and electroweak penguin (EWP) operator, Q_8[1]. For example, in the NDR scheme for γ_5, with central values for the CKM matrix elements, and the known Wilson coefficients of the effective strangeness changing non-leptonic weak decay Hamiltonian[2], one has [1]

$$\frac{\varepsilon'}{\varepsilon} \simeq \frac{25 \cdot 10^{-4}}{\text{GeV}^3} \left[B_6 \left(1 - \Omega_{IB} \right) - 0.4 B_8 + 0.1 \right]_{\mu = 2 \text{ GeV}} , \tag{1}$$

where $\Omega_{IB} = 0.06 \pm 0.08$ is an isospin-breaking correction [3], μ is the renormalization scale, and B_6, B_8 are the deviations of the $I_{\pi\pi} = 0$ Q_6 and $I_{\pi\pi} = 2$ Q_8 matrix elements from their large N_c (factorization) values. We describe recent progress in determining the Q_8 matrix element (ME).

The $8_L \times 8_R$ chiral structure of the EWP operators means that their $K \to \pi\pi$ ME's, $M_{7,8}$, survive in the $SU(3)$ chiral limit. A soft π, K evaluation then yields [4]

$$M_8 \equiv \langle (\pi\pi)_2 | Q_8 | K^0 \rangle_\mu = -2 \left[\langle O_1 \rangle + \frac{3}{2} \langle O_8 \rangle \right]_\mu / 3F_\pi^{(0)^3}$$

$$M_7 \equiv \langle (\pi\pi)_2 | Q_7 | K^0 \rangle_\mu = -2 \langle O_1 \rangle_\mu / F_\pi^{(0)^3} , \tag{2}$$

where $F_\pi^{(0)}$ is the π decay constant in the chiral limit and $\langle O_{1,8} \rangle$ are the VEV's of the 4-quark operators

$$O_1 \equiv \bar{q} \gamma_\mu \frac{\tau_3}{2} q \, \bar{q} \gamma^\mu \frac{\tau_3}{2} q - \bar{q} \gamma_\mu \gamma_5 \frac{\tau_3}{2} q \, \bar{q} \gamma^\mu \gamma_5 \frac{\tau_3}{2} q ,$$

$$O_8 \equiv \bar{q} \gamma_\mu \lambda^a \frac{\tau_3}{2} q \, \bar{q} \gamma^\mu \lambda^a \frac{\tau_3}{2} q \quad - \bar{q} \gamma_\mu \gamma_5 \lambda^a \frac{\tau_3}{2} q \bar{q} \gamma^\mu \gamma_5 \lambda^a \frac{\tau_3}{2} q . \tag{3}$$

CP717, Hadron Spectroscopy: Tenth International Conference,
edited by E. Klempt, H. Koch, and H. Orth

which also determine the $D = 6$ term in the OPE of the flavor ud V-A correlator $\Delta\Pi_{ud}$ [4]. Writing

$$[\Delta\Pi_{ud}]_{OPE} \equiv \sum_{D=2,4,6,\cdots} \left[a_D(\mu) + b_D(\mu) \log(Q^2/\mu^2) \right] , \qquad (4)$$

the $D = 2$ OPE term is proportional to $m_{u,d}^2$ (and hence numerically negligible), while the $D = 4$ term is dominated by the light quark condensate contribution, which is proportional to $\langle (m_u + m_d)\bar{u}u \rangle$ (and hence also chirally suppressed). The $D = 6$ OPE term is thus numerically dominant. The coefficients a_6 and b_6 are given by expressions involving $\langle O_{1,8} \rangle$ and scheme-dependent coefficients whose values, in the same scheme as used for the calculation of the coefficients of effective non-leptonic strangeness-changing Hamiltonian, may be found in Ref. [5]. It turns out that a_6 is numerically dominated by its $\langle O_8 \rangle$ contribution, and b_6/a_6 is small [5]. $\langle O_{1,8} \rangle$ can then be determined by writing down dispersion relations and/or finite energy sum rules (FESR's) for $\Delta\Pi_{ud}$. These sum rules involve various weight functions, and employ experimental data on the spectral function of $\Delta\Pi_{ud}$, $\Delta\rho(s)$ [5, 6], as input. $\Delta\rho(s)$ is measured in hadronic τ decay [7, 8].

The dispersive sum rules are of the form [5]

$$\langle O_1 \rangle_\mu - \frac{3C_8}{8\pi} \langle \alpha_s O_8 \rangle_\mu = \bar{I}_1(\mu) \qquad (5)$$

$$\langle (2\pi\alpha_s + \alpha_s^2) O_8 \rangle_\mu + A_1 \langle \alpha_s^2 O_1 \rangle_\mu = 2\pi\alpha_s(\mu) \bar{I}_8(\mu) \qquad (6)$$

where C_8 and A_1 are scheme-dependents constant, whose values in those schemes previous employed in calculating the coefficients of the effective weak Hamiltonian are known [5], and

$$\bar{I}_1(\mu) = \frac{3}{(4\pi)^2} [I_1(\mu) + H_1(\mu)],$$

$$\bar{I}_8(\mu) = \frac{1}{2\pi\alpha_s(\mu)} \left[I_8(\mu) - H_8(\mu) \right] , \qquad (7)$$

with

$$I_1(\mu) = \int_0^\infty ds \, s^2 \ln\left(\frac{s + \mu^2}{s} \right) \Delta\rho(s) ,$$

$$I_8(\mu) = \int_0^\infty ds \, \frac{s^2 \mu^2}{s + \mu^2} \Delta\rho(s), \qquad (8)$$

and

$$H_1(\mu) = \int_{\mu^2}^\infty dQ^2 \, Q^4 \left[\Delta\Pi_{ud}(Q^2) \right]_{D>6} ,$$

$$H_8(\mu) = \mu^6 [\Delta\Pi_{ud}(\mu)]_{D>6} . \qquad (9)$$

The spectral integrals, $I_{1,8}(\mu)$, of course require spectral data to arbitrarily large s, whereas data is available only up to $s \simeq m_\tau^2$. It turns out, however, that classical chiral sum rules allow one to constrain the contributions to the spectral integrals, $I_{1,8}(\mu)$, from the region $s > m_\tau^2$, where data is absent [5]. These sum rules are the two Weinberg sum rules,

$$\int_0^\infty ds\,\Delta\rho(s) = F_\pi^{(0)2}$$
$$\int_0^\infty ds\,s\,\Delta\rho(s) = 0 , \qquad (10)$$

and the sum rule for the EM pion mass splitting in the chiral limit,

$$\int_0^\infty ds\,s\ln\frac{s}{\Lambda^2}\,\Delta\rho(s) = -F_\pi^{(0)2}\frac{4\pi}{3\alpha}\Delta m_\pi^{(0)2} . \qquad (11)$$

In Eq. (10), $F_\pi^{(0)}$ is the pion decay constant in the chiral limit, while in Eq. (11) $\Delta m_\pi^{(0)2}$ is the pion squared-mass splitting in the chiral limit. The LHS of Eq. (11) is independent of Λ. Details of the proceedure for implementing these constraints, called the "Residual Weight Method" (RWM), may be found in the first of Refs. [5]. The accuracy of the constraints turns out to decrease with increasing μ. For $\mu = 4$ GeV, where $D > 6$ OPE contributions may be safely neglected, one finds, after evolving to $\mu = 2$ GeV, in the $\overline{MS}$-NDR scheme,

$$M_7(2 \text{ GeV}) = 0.16 \pm 0.10 \text{ GeV}^3,$$
$$M_8(2 \text{ GeV}) = 2.2 \pm 0.7 \text{ GeV}^3 . \qquad (12)$$

A significant part of the quoted error is associated with uncertainties in the input values for $F_\pi^{(0)}$ and $\Delta m_\pi^{(0)2}$. These uncertainties play a significantly reduced role for lower scales μ. To take advantage of this observation, and work at lower μ, however, a knowledge of the $D > 6$ OPE contributions, $H_{1,8}$, is required.

FESR's for $\Delta\Pi_{ud}$ are relations of the general form

$$\int_{s_{th}}^{s_0} ds\,\Delta\rho(s)\,w(s) = \frac{-1}{2\pi i}\oint_{|s|=s_0} ds\,\Delta\Pi_{ud}(s)\,w(s) \qquad (13)$$

where $w(s)$ is any function analytic in a region containing $|s| \le s_0$. For sufficiently large s_0, the OPE for $\Delta\Pi_{ud}$ is to be used on the RHS. In general, at intermediate scales, OPE breakdown is expected to occur in the vicinity of the timelike real s axis [9]. We therefore work with "pinched" weights (those satisfying $w(s_0) = 0$) which suppress contributions from this region. Such weights have been shown to efficiently remove duality violating contributions in sum rules involving the flavor ud vector and axial vector correlators at scales $s_0 \sim 2 - 3$ GeV^2 [10]. We employ weights with a double zero at $s = s_0$. For polynomial weights, a term y^k in $w(s)$ (with $y \equiv s/s_0$) allows access to a_D, with $D = 2k + 2$. Logarithmic contributions are suppressed by both an additional power of α_s, and by additional numerical factors which occur generically when working with pinched weights.

To optimize the extraction of a_6, we have used pinched FESR's (pFESR's) based on two independent weights of degree 3 [5]. This allows us to reduce fractional errors on the spectral integrals and to optimally suppress possible logarithmic contributions of $D > 8$. The first weight was chosen in such a way as to produce the smallest fractional errors for the spectral integrals from among the space of weights of degree 3. The second was chosen to weight the spectral function in a very different manner. This provides a non-trivial check on the absence of residual duality violation since the optimally-fitted OPE parameters should reproduce the full set of spectral integrals, over the full range of s_0 in the fitting window, only if the OPE representation is, indeed, reliable at these scales, and for the doubly pinched weights employed in the analysis.

We have also employed a second, extended set of pFESR's to extract the a_D with $D = 8, \cdots, 16$. The results turn out to be sufficient to allow us to determine $H_{1,8}(\mu)$ for μ down to 2 GeV. We find an excellent match between the various spectral integrals and the optimized OPE representation in the full analysis windows of all the pFESR's employed, and excellent consistency between the values of the a_D determined from different pFESR's. Further details may be found in Ref. [6]. That reference also contains a full discussion of the rationale for the various weight choices, as well as a detailed discussion of the various tests which have been performed to verify that no signs of any residual duality violation are present in the analysis.

The pFESR extraction of a_6 gives us, essentially, a determination of $\langle O_8 \rangle$ (with the RWM estimate for the two VEV's, the $\langle O_1 \rangle$ contribution to a_6 is at the few percent level of the $\langle O_8 \rangle$ contribution). The results of the extended analysis allow us to determine $H_{1,8}(2 \text{ GeV})$ with good accuracy and hence to obtain a second, largely independent determination of $\langle O_8 \rangle$, as well as a determination of $\langle O_1 \rangle$ with errors significantly reduced over those obtained using the RWM approach. This "hybrid" determination, in the case of $\langle O_8 \rangle$, is largely independent of the pure pFESR determination since $\sim 60\%$ of the full RWM result for $I_8(2 \text{ GeV})$ is associated with the input chiral limit values of f_π and the π EM self-energy, which enter the RWM constraints, but do not appear at all in the pFESR analysis. The various spectral integrals have been evaluated using both the ALEPH [7] and OPAL [8] spectral data. Separate analyses were performed for each to the two databases.

We find, at $\mu = 2$ GeV in the $\overline{MS}$-NDR scheme, and in units of GeV^3, [5]

$$
\begin{aligned}
[M_8]_{ALEPH} &= 1.40 \pm 0.28, \\
[M_8]_{OPAL} &= 1.68 \pm 0.32
\end{aligned}
\tag{14}
$$

from the pure pFESR analysis, and

$$
\begin{aligned}
[M_8]_{ALEPH} &= 1.55 \pm 0.52, \\
[M_8]_{OPAL} &= 1.68 \pm 0.32, \\
[M_7]_{ALEPH} &= 0.23 \pm 0.05, \\
[M_7]_{OPAL} &= 0.21 \pm 0.05,
\end{aligned}
\tag{15}
$$

from the $\mu = 2$ GeV hybrid dispersive analysis. The M_7 hybrid results are in good agreement with those obtained by evolving the results of the high-scale ($\mu = 4$ GeV) dispersive analysis down to $\mu = 2$ GeV using the known anomalous dimension matrix, but have significantly reduced errors. The hybrid and pure pFESR M_8 results are in similarly good agreement. As noted above, since $\sim 60\%$ of the hybrid result is associated with the input to the classical chiral sum rule constraints and only $\sim 40\%$ with the hadronic τ decay data, the consistency of the results obtained in the two approaches is highly non-trivial.

A number of other attempts have also been made to determine the $SU(3)$ chiral limit values of the EWP operator ME's [11]. A detailed discussion of these approaches, and their relation to our work, may be found in Ref. [6]. Figures 2 through 8 of that reference show that our solutions for the a_D in all cases provide a better OPE representation of, not only the spectral integrals for the weights employed in our analysis, but also the spectral integrals corresponding to weights employed in the other analyses.

The results above correspond to a rather large, negative contribution to ε'/ε in the $SU(3)$ chiral limit, $(-15.0 \pm 2.7) \cdot 10^{-4}$, a factor of ~ 2 larger than given by most models employed previously in the literature. A sizeable enhancement of the gluonic penguin contribution is thus required if the Standard Model is to explain ε'/ε. Such an increase in the value of the gluonic penguin matrix element would also have an impact on our understanding of the origins of the $\Delta I = 1/2$ rule.

Finally, we note that the results obtained above for M_7 and M_8 should provide a useful constraint for the development of reliable lattice techniques for calculating both the gluonic and electroweak penguin matrix elements for physical m_s.

ACKNOWLEDGMENTS

K.M. acknowledges the ongoing support of the Natural Sciences and Engineering Research Council of Canada. The work of V.C. was supported in part by MCYT, Spain under Grant number FPA-2001-3031, and by ERDF funds from the European Commission. The work of J.D. and E.G. was supported in part by the National Science Foundation under Grant PHY-9801875.

REFERENCES

1.   See, e.g., A.J. Buras, hep-ph/0307203 and earlier references therein.
2.   A.J. Buras *et al.*, Nucl. Phys. B400, 37 (1993); M. Ciuchini *et al.*, Nucl. Phys. **B415**, 403 (1994).
3.   V. Cirigliano, A. Pich, G. Ecker and H. Neufeld, hep-ph/0307030; C.E. Wolfe and K. Maltman, Phys. Lett. **B482** (2000) 77 and Phys. Rev. **D63** (2001) 014008
4.   J.F. Donoghue and E. Golowich, Phys. Lett. **B478**, 172 (2000).
5.   V. Cirigliano *et al.*, Phys. Lett. **B522**, 245 (2001); Phys. Lett. **B555**, 71 (2003).
6.   V. Cirigliano, E. Golowich and K. Maltman, hep-ph/0305118.
7.   R. Barate *et al.* (The ALEPH Collaboration), Z. Phys. **C76**, 15 (1997); Eur. Phys. J. **C4**, 409 (1998).
8.   K. Ackerstaff *et al.* (The OPAL Collaboration), Eur. Phys. J. **C7**, 571 (1999).
9.   E.C. Poggio, H.R. Quinn and S. Weinberg, Phys. Rev. **D13**, 1958 (1976).
10. K. Maltman, Phys. Lett. **B440**, 367 (1998); hep-ph/0209091.
11. M. Knecht, S. Peris and E. de Rafael, Phys. Lett. **B508**, 117 (2001); J. Bijnens, E. Gamiz and J. Prades, JHEP **0110**, 009 (2001); S. Narison, Nucl. Phys. **B593**, 3 (2001).

HADRONIC SPECTRA AND KALUZA-KLEIN PICTURE OF THE WORLD

A.A. Arkhipov

Institute for High Energy Physics, 142280 Protvino, Moscow Region, Russia

Abstract. A manifestation of Kaluza-Klein picture in hadronic spectra is discussed. We argue that the experimentally observed structures in hadronic spectra confirm the Kaluza-Klein picture of the world.

> *"... the simpler the presentation of a particular law of Nature,*
> *the more general it is ..."*
> Max Planck, Nobel Lecture, June 2, 1920

INTRODUCTION

Dear Colleagues.

It seems that here is just the place where we could remember one of the greatest physicists of the last XX Century, I mean German physicist Max Planck. My experience in science allows me to definitely share Max Planck opinion in the above written fragment of his Nobel Lecture. Following this opinion I'd like to present here new very simple and at the same time very general physical law concerning the structure of hadronic spectra.

Although the modern strong interaction theory formulated in terms of known QCD Lagrangian is commonly accepted, this theory does not allow us to make an appreciable breakthrough in the problem of calculating the masses of compound systems so far mainly because that problem is a significantly non-perturbative one. In other words, this means that our theoretical understanding of low-energy QCD spectroscopy is far from desired. Even the best currently performed lattice computations in QCD cannot help us to understand the exact nature of the real hadron spectrum.

All of you know that the strong interactions are characterized by multi-particle production. The dynamics of the multi-particle systems with a necessity contains the so called many-body forces. Many-body forces are fundamental forces which take place in the multi-particle systems where the number of particles is greater than two, and they are responsible for the dynamics of the production processes. For example, the three-body forces are responsible for the dynamics of one-particle inclusive reactions; see Ref. [1] and references therein. A description of the many-body forces requires the use of multidimensional spaces, and as a consequence we cannot construct the strong interactions theory in a self-consistent way without the use of multidimensional spaces. Therefore,

CP717, *Hadron Spectroscopy: Tenth International Conference,*
edited by E. Klempt, H. Koch, and H. Orth
© 2004 American Institute of Physics 0-7354-0197-7/04/$22.00

seems it would be naturally to formulate the strong interactions theory in a multidimensional space from the very beginning.

The idea to use the multidimensional spaces in fundamental physics is not new: famous works of Kaluza and Klein were the first ones where this idea has been elaborated. In the year 1921, Kaluza proposed a unification of the theory of gravity and the Maxwell theory of electromagnetism in four dimensions starting from the theory of gravity in five dimensions. The basic idea of the Kaluza-Klein scenario may be applied to any model in Quantum Field Theory. Performing harmonic expansion for the multidimensional field we can reduce the multidimensional theory to the effective four-dimensional one with an infinite set of KK modes, i.e. an infinite set of four-dimensional particles with increasing masses. For the masses of the KK modes one obtains

$$m_n^2 = m^2 + \frac{\lambda_n}{R^2},$$ (1)

and the coupling constant g of the four-dimensional theory is related to the coupling constant $G_{(4+d)}$ of the initial multidimensional theory by the equation

$$g = \frac{G_{(4+d)}}{V_d},$$ (2)

where V_d is the volume of the compact internal space of extra dimensions $\mathcal{K}_d$. Eqs. (1,2) represent the basic relations of Kaluza-Klein scenario.

Unfortunately in the frame of allowed space in the Proceedings we will restrict ourselves here only to shortest fragments of the Report. We refer the interested reader to the extended version [16] (hereafter referred to as Report) for the details.

ON GLOBAL SOLUTION OF THE SPECTRAL PROBLEM

According to Kaluza and Klein we suggest that the input (fundamental) space-time $\mathcal{M}_{(4+d)}$ is represented as

$$\mathcal{M}_{(4+d)} = M_4 \times \mathcal{K}_d.$$

Let λ_n are characteristic numbers of the Laplace operator on $\mathcal{K}_d$ with a characteristic size $R_\mathcal{K}$

$$\Delta_{\mathcal{K}_d} Y_n(y) = -\frac{\lambda_n}{R_\mathcal{K}^2} Y_n(y).$$

Let $\lambda_\mathcal{K}$ be the set of all characteristic numbers of the Laplace operator

$$\lambda_\mathcal{K} \equiv \left\{ \lambda_n : n \in \mathbb{Z}^d \equiv \underbrace{\mathbb{Z} \times \mathbb{Z} \times \cdots \times \mathbb{Z}}_{d} \right\}.$$ (3)

There is one-to-one correspondence

$$\mathcal{K} \quad \Longleftrightarrow \quad (R_\mathcal{K}, \lambda_\mathcal{K}).$$

Let us consider a compound hadronic system h which may decay into some channel

$$h \rightarrow a + b + \cdots + c. \tag{4}$$

We introduce the spectral mass function of the given channel by the formula

$$M_h^{ab\ldots c}(R_\mathcal{K}, \lambda_{n_a}, \lambda_{n_b}, \cdots, \lambda_{n_c}) = \sqrt{m_a^2 + \frac{\lambda_{n_a}}{R_\mathcal{K}^2}} + \sqrt{m_b^2 + \frac{\lambda_{n_b}}{R_\mathcal{K}^2}} + \cdots + \sqrt{m_c^2 + \frac{\lambda_{n_c}}{R_\mathcal{K}^2}}. \tag{5}$$

Now we build the Kaluza-Klein tower:

$$t_h^{ab\ldots c}(\mathcal{K}) \equiv t_h^{ab\ldots c}(R_\mathcal{K}, \lambda_\mathcal{K}) \overset{def}{=} \left\{ M_h^{ab\ldots c}(R_\mathcal{K}, \lambda_{n_a}, \lambda_{n_b}, \cdots, \lambda_{n_c}) : \lambda_{n_i} \in \lambda_\mathcal{K} \right\}, \tag{6}$$

$$(i = a, b, \ldots, c).$$

After that we build the Kaluza-Klein town as a union of the Kaluza-Klein towers corresponding to all possible decay channels of the hadronic system h

$$\mathcal{T}_h(\mathcal{K}) \equiv \mathcal{T}_h(R_\mathcal{K}, \lambda_\mathcal{K}) \overset{def}{=} \bigcup_{\{ab\ldots c\}} t_h^{ab\ldots c}(R_\mathcal{K}, \lambda_\mathcal{K}). \tag{7}$$

We state:

$$\boxed{M_h \in \mathcal{T}_h(\mathcal{K})}. \tag{8}$$

Let $\mathcal{H}$ be the set of all possible physical hadronic states. We build the hadronic Kaluza-Klein country $\mathbb{C}_\mathcal{H}(\mathcal{K})$ by the formula

$$\mathbb{C}_\mathcal{H}(\mathcal{K}) \overset{def}{=} \bigcup_{h \in \mathcal{H}} \mathcal{T}_h(\mathcal{K}). \tag{9}$$

The whole spectrum of all possible physical hadronic states we denote $M_\mathcal{H}$

$$M_\mathcal{H} \overset{def}{=} \left\{ M_h : h \in \mathcal{H} \right\}. \tag{10}$$

We state:

$$\boxed{M_\mathcal{H} \in \mathbb{C}_\mathcal{H}(\mathcal{K})}. \tag{11}$$

The formulae (8) and (11) provide the global solution of the spectral problem in hadronic spectroscopy.

We would like to make some clarifying comments. First of all, in the construction of the global solution among all possible decay channels of the hadronic system h there have to be taken into account only those channels which contain the fundamental particles and their different multi-particle compound systems in the final states, as it should be. An appearance of non-zero KK modes of the fundamental particles in the final states of the decay channels is forbidden by the construction. Non-zero KK modes of the fundamental particles living in a compound system define the main properties of a compound system such as the mass and the life time of the system. An interaction of KK modes is weak, therefore we can calculate with a high accuracy the mass of a compound system as a simple sum of the masses of KK modes. Moreover, weakly interacting KK modes result very narrow widths of the compound states, and this phenomenon is observed at the recent experiments.

The dynamics of the compound systems decays is physically transparent: Non-zero KK modes of the constituents make a transition to zero KK modes, and we observe zero KK modes as decay products. In fact, we present here quite a new look on the Kaluza-Klein picture as a whole.

COMPARISON WITH THE EXPERIMENTAL DATA

In papers [2, 3, 4, 6, 8, 12] we have verified the global solution on the set of experimental data; see Tables 1-12 in the Report [16].

The most impressive fact from the view point of our developed theoretical conception is that all new recently observed states have the masses which excellently agree with theoretically calculated ones including the first observation of a very narrow charmonium state with a mass of $3871.8 \pm 0.7(stat) \pm 0.4(syst) MeV$ which decays into $\pi^+\pi^- J/\psi$ [11, 13]. It has been stressed that the $\pi^+\pi^-$ invariant mass for the $M(3872)$ signal region concentrate near the ρ mass [13]. Such state really exists, and it lives just on the second storey in the Kaluza-Klein tower of KK excitations for the $\rho J/\psi$-system; see Table 12 in the Report. As it is seen from Table 12 there is a wonderful agreement of experimentally measured mass with theoretically calculated one.

MEIN RUF TO SEARCH NEW STATES

We have found out quite an interesting correspondence where $K\pi$-system looks like a system built from two-pion system by replacement of some one pion with a kaon; see details in the Report.

Concerning a three-pion system we have found out that

$$a_2(1311) \in M_{10}^{3\pi}(1309-1313), \qquad a_2(1311) \in M_{10}^{\rho\pi}(1310-1312).$$

Moreover, we predict the strange partner of a_2-meson which we would like to call as a_2^s-meson

$$\boxed{a_2^s(1520) \in M_{10}^{K2\pi}(1517-1523), \qquad a_2^s(1520) \in M_{10}^{K\rho}(1519-1522)}.$$

Apart of isospin $a_2^s(1520)$-meson may have the same quantum numbers as $a_2(1311)$-meson. We call up to search $a_2^s(1520)$-meson and other strange partners of the three-pion states experimentally observed till now [14]. In this respect it seems the factory with an intensive kaon beams would be a very good device to realize such programm. However, we would like to especially emphasize that recently observed states in $K_S^0 K_S^0$ system reported at this Conference by ZEUS Collaboration [10] seem indicate on the possibility to observe at HERA the $a_2^s(1520)$-meson in $K_S^0\pi^+\pi^-$ system where the invariant mass of $\pi^+\pi^-$ system concentrated near the ρ-peak. I am asking the physicists from ZEUS Collaboration at HERA to consider such possibility.

CONCLUSION

We have shown that one simple formula with one fundamental constant described more than 120 experimentally observed hadronic states. This is the most impressive fact in our developed theoretical conception [15]. We did not ascribed the quantum numbers to the predicted states because we have made only model independent predictions concerning the masses of the states which have related only with the existence of a compact internal extra space.

The performed analysis allows us to conclude with a confidence: **The experimentally observed structures in hadronic spectra reveal the existence of the extra dimensions and confirm the Kaluza-Klein picture of the world**.

I began my talk with the saying of Max Planck, and I would like to finish the talk with the saying of Max Planck as well.

For it fell to this (atom) theory to discover, in the quantum action, the long-sought key to the entrance gate into the wonderland of spectroscopy, which since the discovery of spectral analysis had obstinately defied all efforts to breach it.

Max Planck, Nobel Lecture, June 2, 1920

We could paraphrase Max Planck and say that discovery of the fundamental scale of internal extra space with its geometry and shapes provides the long-awaited key to the entrance gate into the wonderland of hadronic spectroscopy, which since the discovery of strong forces had obstinately defied all efforts to open it.

REFERENCES

1. A.A. Arkhipov, hep-ph/0211449 (2002); preprint IHEP 2002-44, Protvino, 2002, available at http://dbserv.ihep.su/~pubs/prep2002/ps/2002-44.pdf
2. A.A. Arkhipov, hep-ph/0208215 (2002); preprint IHEP 2002-43, Protvino, 2002, available at http://dbserv.ihep.su/~pubs/prep2002/ps/2002-43.pdf
3. A.A. Arkhipov, hep-ph/0302164 (2003).
4. A.A. Arkhipov, hep-ph/0302213 (2003).
5. J. Yonnet, B. Tatischeff et al., Phys. Rev. C**63**, 014001-1 (2000).
6. A.A. Arkhipov, hep-ph/0304014 (2003).
7. J. Kuhn, this Conference.
8. A.A. Arkhipov, hep-ph/0305167 (2003).
9. E. Fadeeva, this Conference.
10. M. Barbi (ZEUS Collaboration), this Conference; hep-ex/0308006 (2003).
11. P. Krokovny (Belle Collaboration), this Conference; hep-ex/0307052, hep-ex/0308019 (2003).
12. A.A. Arkhipov, hep-ph/0306237 (2003).
13. K. Abe et al. (Belle Collaboration), hep-ex/0308029 (2003).
14. A.A. Arkhipov, hep-ph/0308321 (2003).
15. A.A. Arkhipov, hep-ph/0309002 (2003).
16. A.A. Arkhipov, hep-ph/0309327 (2003).

Infrared Features of Lattice Landau Gauge QCD and the Gribov Copy Problem

Sadataka Furui[*] and Hideo Nakajima[†]

[*]School of Science and Engineering, Teikyo University, Utsunomiya 320-8551, Japan
[†]Department of Information science, Utsunomiya University, Utsunomiya 320-8585, Japan

Abstract. Infrared features of gluon propagator, ghost propagator, QCD running coupling and the Kugo-Ojima parameter in lattice Landau gauge QCD are presented. The framework of PMS analysis suggests that there appear infrared, intermediate and ultraviolet regions specified by $\Lambda_{\overline{MS}}$, β_0 and β_1. The propagators and the running coupling of $q > 1 GeV$ are fitted by the $\widetilde{MOM}$ scheme using the factorization scale as the scale parameter. The running coupling data of $q < 14 GeV$ are fitted by the contour improved perturbation method. The Gribov copy problem is studied in $SU(2)$, $\beta = 2.2$ samples by comparing the data gauge fixed by the parallel tempering method and the data gauge fixed by the straightforward gauge fixing. The Gribov noise effect turned out to be about 2%.

INTRODUCTION

In Landau gauge QCD, it is necessary to restrict the gauge configuration expressed by the link variable $U_{x,\mu}$ to $\Omega_L = \{U | -\partial D(U) \geq 0, \partial A = 0\}$ which is called Gribov region (local minima)[2], where D is covariant derivative. Zwanziger argued that the uniqueness of the gauge field is guaranteed in its subset $\Lambda_L = \{U | A = A(U), F_U(1) = \mathrm{Min}_g F_U(g)\}$, and called the Λ_L fundamental modular(FM) region[3]. Here the optimizing function $F_U(g)$ is defined in the case of $U-$linear $(A_{x,\mu} = \frac{1}{2}(U_{x,\mu} - U^{\dagger}_{x,\mu})|_{trless\ p.})$ and $\log U$ $(U_{x,\mu} = e^{A_{x,\mu}})$ as, $F_U(g) = \sum_{x,\mu}\left(1 - \frac{1}{3}\mathrm{Re\ tr}U^g_{x,\mu}\right)$, and $F_U(g) = ||A^g||^2 = \sum_{x,\mu}\mathrm{tr}\left(A^g_{x,\mu}{}^{\dagger}A^g_{x,\mu}\right)$, respectively[4].

An interesting quantity which is relevant to infrared hadron physics is the QCD running coupling. In Landau gauge, it can be extracted from three gluon coupling and/or ghost-gluon coupling. In terms of gluon dressing functiuon $Z_A(q^2)$ and ghost dressing function $G(q^2)$, which are measurable in lattice Landau gauge, renormalization group invariant quatity[5] $\alpha_s(q^2) = g^2 G(q^2)^2 Z_A(q^2)/4\pi$ can be measured.

Colour confinement in infrared QCD is characterized by the Kugo and Ojima parameter $u(0)$ which is defined as

$$\frac{1}{V}\sum_{x,y}e^{-ip(x-y)}\langle\mathrm{tr}\left(\lambda^{a\dagger}D_{\mu}\frac{1}{-\partial D}[A_{\nu},\lambda^b]\right)_{xy}\rangle = (\delta_{\mu\nu} - \frac{p_{\mu}p_{\nu}}{p^2})u^{ab}(p^2),$$

where $u^{ab}(p^2) = \delta^{ab}u(p^2)$, $u(0) = -c$. The sufficient condition of the colour confinement is $u(0) = -1$. We measure the running coupling and the Kugo-Ojima parameter on lattice and study the ambiguity due to Gribov copy.

CP717, *Hadron Spectroscopy: Tenth International Conference*,
edited by E. Klempt, H. Koch, and H. Orth

NUMERICAL RESULTS OF LATTICE LANDAU GAUGE QCD

The gluon propagator and the ghost propagator

The gluon propagator of colour $SU(n)$ gauge is defined as

$$D_{\mu\nu}(q) = \frac{1}{n^2-1} \sum_{x=\mathbf{x},t} e^{-ikx} Tr\langle A_\mu(x)A_\nu(0)^\dagger\rangle$$

$$= (\delta_{\mu\nu} - \frac{q_\mu q_\nu}{q^2})D_A(q^2) \tag{1}$$

and the gluon dressing function as $Z_A(q^2) = q^2 D_A(q^2)$. Our result is shown in Fig.1(a) and is consistent with[7, 8].

The ghost propagator, which is the Fourier transform of an expectation value of the inverse Faddeev-Popov operator $\mathcal{M} = -\partial D = -\partial^2(1-M)$

$$D_G^{ab}(x,y) = \langle \text{tr}\langle \lambda^a x|(\mathcal{M}[U])^{-1}|\lambda^b y\rangle\rangle \tag{2}$$

where the outmost $\langle\rangle$ denotes average over samples U, is evaluated as follows. We take plane wave for the source $\mathbf{b}^{[1]} = \lambda^b e^{iqx}$ and get the solution of Poisson equation $\phi^{[1]} = (-\Delta)^{-1}\mathbf{b}^{[1]}$. We calculate iteratively $\phi^{[i+1]} = M\phi^{[i]}(\mathbf{x})(i = 1,\cdots,k-1)$. The iteration was continued until $Max_\mathbf{x}|\phi^{[k]}(\mathbf{x})|/Max_\mathbf{x}|\sum_{i=1}^{k-1}\phi^{[i]}(\mathbf{x})| < 0.001 \sim 0.01$. The number of iteration k is of the order of 60, in $SU(2)$, 16^4 lattice, and of the order of 100 in $SU(3)$.

We define $\mathbf{\Phi}^b(\mathbf{x}) = \sum_{i=1}^k \phi^{[i]}(\mathbf{x})$ and evaluate $\langle\lambda^a e^{iqx},\mathbf{\Phi}^b(\mathbf{x})\rangle$ as the ghost propagator from colour b to colour a.

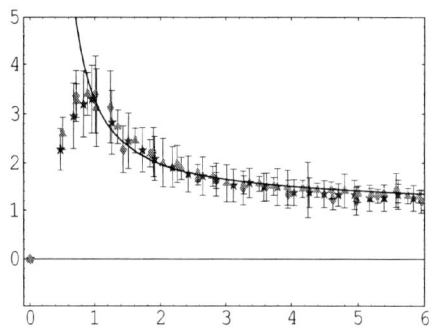 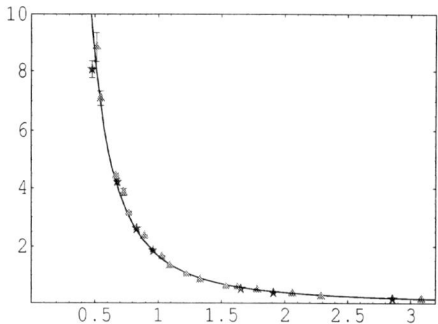

FIGURE 1. (a)The gluon dressing function as the function of the momentum $q(GeV)$. $\beta = 6.0$, 24^4(triangle), 32^4(diamond) and $\beta = 6.4$, 48^4(star) in $\log U$ version. The fitted line is that of the $\widetilde{MOM}$ scheme.
(b)The ghost propagator as the function of the momentum $q(GeV)$. $\beta = 6.0$, $24^4, 32^4$ and $\beta = 6.4$, 48^4 in $\log U$ version. The fitted line is that of the $\widetilde{MOM}$ scheme.

The gluon and the ghost dressing functions are fitted by using the framework of principle of minimal sensitivity(PMS)[10] or the effective charge method[11], in which the running coupling $h(q)$ is parametrized as

$$h(q) = y_{\overline{MS}}(q)(1+y_{\overline{MS}}(q)^2(\bar{\beta}_2/\beta_0 - (\beta_1/\beta_0)^2)$$
$$+ y_{\overline{MS}}(q)^3\frac{1}{2}(\bar{\beta}_3/\beta_0 - (\beta_1/\beta_0)^3) + \cdots. \tag{3}$$

The parameter $y_{\overline{MS}}(q)$ can be expressed by the parameter y defined as a solution of

$$\beta_0 \log\frac{\mu^2}{\Lambda^2} = \frac{1}{y} + \frac{\beta_1}{\beta_0}\log(\beta_0 y) \tag{4}$$

where Λ characterizes the scale of the system, and the function

$$k(q^2, y) = \frac{1}{y} + \frac{\beta_1}{\beta_0}\log(\beta_0 y) - \beta_0\log(q^2/\Lambda_{\overline{MS}}^2). \tag{5}$$

In PMS, y is treated as a function of q^2. However, in this work we fix the scale by the factorization scale $\mu = 1.97 GeV$ at which renormalizable quantity can be approximately factorizable into scheme independent and dependent parts[11, 8].
In order to be consistent with the $\overline{MS}$ scheme, we define the variable z as

$$z = -e^{(-1-bt/2c)} = -\frac{1}{e}(\frac{q}{\tilde{\Lambda}_{\overline{MS}}})^{-b/c}e^{iK\pi}$$
$$= -Z(q^2)e^{iK\pi} \tag{6}$$

where $t = \log(q^2/\tilde{\Lambda}_{\overline{MS}}^2)$, $c = \beta_1/\beta_0 = 51/22$, $b = \beta_0/2 = 11/2$, $\tilde{\Lambda}_{\overline{MS}} = (2c/b)^{-c/b}\Lambda_{\overline{MS}}$, $K = -b/2c$[10, 13]. When we fix y by the solution of eq.(4) with $\mu = 1.97 GeV$, ($\widetilde{MOM}$ scheme), we find that in the gluon dressing function the Landau pole at $z = 1/e$ remains, and another pole appears at $z \sim 0.17$. When y is chosen as q^2 dependent, the three regions $0 < z < 0.17$, $0.17 < z < 1/e$ and $1/e < z$ can be continuously connected[12], but there is a subtle problem of PMS in low energy[14] and leave the problem to bridge the three regions to the future.
The gluon dressing function $Z_A(q^2)$ and the ghost dressing function $G(q^2)$ in $\widetilde{MOM}$ scheme are plotted in Fig.1(a) and (b), respectively. They are singular at $q = \tilde{\Lambda}_{\overline{MS}} \simeq 0.25 GeV$ which should be washed away by the non-perturbative effects.

The QCD running coupling and the Kugo-Ojima parameter

The QCD running coupling $\alpha_s(q)$ turned out to have a peak of the order of 1 at $q \sim 0.5 GeV$ and decreases to a finite value at $q = 0$. We fitted the data by the contour improved perturbation series, which is a way to make a resummation of the series of coupling constant.
The effective running coupling in the $\overline{MS}$ scheme is expressed by the series of coupling constant $h^{(n)}$ as[12].

$$\mathscr{R}^n = h^{(n)}(1 + A_1 h^{(n)} + A_2 h^{(n)2} + \cdots + A_n h^{(n)n}) \tag{7}$$

The result of $\widetilde{MOM}$ scheme using $y = 0.01594$ is shown by the solid line in Fig. 2(a). The lattice data of $24^4, 32^4$ and 48^4 and the $\widetilde{MOM}$ scheme agree in $0.5GeV < q < 2GeV$ but slightly overestimates in $q > 2GeV$.

In contour improved perturbation method, physical quantities $\mathscr{R}$ are expressed in a series physical quantities $\mathscr{R}$ are expressed in a series

$$\mathscr{R}(q^2) = \mathscr{B}_1(q^2) + \sum_{n=1}^{\infty} A_n \mathscr{B}_{n+1}(q^2) \qquad (8)$$

$$\mathscr{B}_n = \frac{1}{2\pi} \int_{-\pi}^{\pi} \left(\frac{-1}{c(1 + W(Z(q^2)e^{iK\theta})} \right)^n d\theta \qquad (9)$$

Terms in the series (8) have alternating sign and A_3 is not known. Hence we tried a fit by choosing half of A_2 in the series(Fig.2(b))

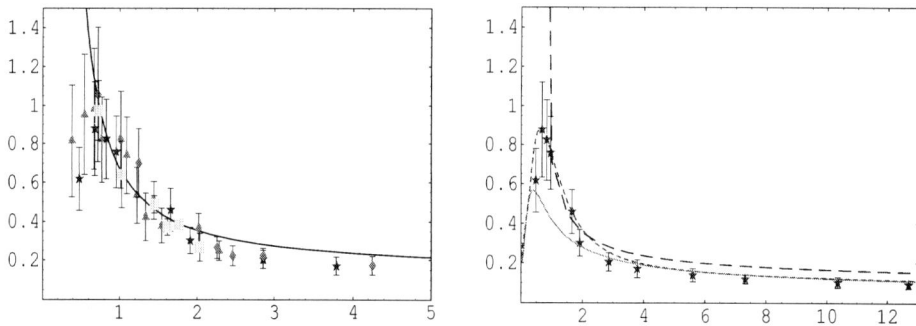

FIGURE 2. (a) The running coupling $\alpha_s(q)$ of $\beta = 6.0$, 24^4(box), 32^4(triangle), $\beta = 6.4$, 32^4(diamond) and 48^4(star) as a function of momentum $q(GeV)$ and the result of the $\widetilde{MOM}$ scheme.
(b) The running coupling $\alpha_s(q)$ as a function of momentum $q(GeV)$ of $\beta = 6.4, 48^4$ lattice. The solid line is the result of $\mathscr{R}^2$ using $\Lambda_{\overline{MS}} = 237MeV$. Dotted line is the result of $e^{70/6\beta_0}\Lambda_{\overline{MS}}$ and including half of A_2. Dashed line is the result of $\widetilde{MOM}$ scheme.

The Kugo-Ojima parameter depends slightly on $U-$linear or $\log U$ and c becomes larger as the lattice size becomes large. However the value saturates at about 0.8.

TABLE 1. The Kugo-Ojima parameter c, trace e/d and horizon function deviation h in $U-$linear and $\log U$ version. $\beta = 6.0$ and 6.4.

β	L	c_1	e_1/d	h_1	c_2	e_2/d	h_2
6.0	16	0.576(79)	0.860(1)	-0.28	0.628(94)	0.943(1)	-0.32
6.0	24	0.695(63)	0.861(1)	-0.17	0.774(76)	0.944(1)	-0.17
6.0	32	0.706(39)	0.862(1)	-0.15	0.777(46)	0.944(1)	-0.16
6.4	32	0.650(39)	0.883(1)	-0.23	0.700(42)	0.953(1)	-0.25
6.4	48				0.720(49)	0.982(1)	-0.26

THE GRIBOV COPY PROBLEM

We applied the parallel tempering(PT) method, which is used in finding global minimum in spin glass systems, to perform the FM gauge fixing of $SU(2)$ $\beta = 2.2, 16^4$ lattice configuration and compared with ensamble of 1st copy(Gribov copy obtained by the straightforward gauge fixing)[6]. The Kugo-Ojima parameter c of PT becomes smaller than 1st copy by about 2%. Gribov noise in the ghost propagator is similar to Cucchieri's noise[9] at $\beta = 1.6$, i.e. the infrared singularity of PT is weaker than 1st copy. The SU(2) gluon propagator, ghost propagator and running coupling in $q \geq 1 GeV$ region are consistent with results of[15]. The running coupling in our simulation has a peak at around $q = 1 GeV$ and is suppressed in the infrared.

We aimed to detect in the lattice dynamics, the signal of Kugo-Ojima confinement criterion derived in the continuum theory, formulated in use of Faddeev-Popov Lagrangian and BRST symmetry. We also noted Zwanziger's horizon condition, based on the lattice formulation, coincides with Kugo-Ojima criterion[3, 4]. However, our present data are not satisfactory to prove or disprove the confinement criterion. So far we do not find definite sign of global colour symmetry violation, but we are studying also colour off-diagonal ghost propagator, which is relevant to the generation of the mass of the gluon[16, 17].

ACKNOWLEDGMENTS

We are grateful to Daniel Zwanziger for enlightenning discussion. S.F. thanks Kei-Ichi Kondo, Kurt Langfeld, Karel van Acoleyen and David Dudal for valuable information. This work is supported by the KEK supercomputing project No.03-94.

REFERENCES

1. T. Kugo and I. Ojima, Prog. Theor. Phys. Supp. **66**, 1 (1979).
2. V.N. Gribov, Nucl. Phys. **B 139**1(1978).
3. D. Zwanziger, Nucl. Phys. **B 364** ,127 (1991), idem B **412**, 657 (1994).
4. H.Nakajima and S. Furui, Nucl. Phys. **B** (Proc Suppl.)**63A-C**,635, 865(1999), Nucl. Phys. **B** (Proc Suppl.)**83-84**,521 (2000), **119**,730(2003); Nucl. Phys. **A 680**,151c(2000), hep-lat/0303024,0305010.
5. L. von Smekal, A. Hauck, R. Alkofer, Ann. Phys.**267**,1 (1998), hep-ph/9707327;
6. H.Nakajima and S. Furui,Lattice2003 proceedings.
7. D.B. Leinweber, J.I. Skullerud, A.G. Williams and C. Parrinello, Phys. Rev. D**60**,094507(1999); ibid Phys. Rev. D**61**,079901(2000).
8. D. Becirevic et al., Phys. Rev. D **61**,114508(2000).
9. A. Cucchieri, Nucl. Phys. **B508**,353(1997), hep-lat/9705005.
10. P.M.Stevenson, Phys. Rev. D**23**, 2916(1981).
11. G. Grunberg, Phys. Rev. D**29**, 2315(1984).
12. K. Van Acoleyen and H. Verschelde, Phys. Rev. D**66**,125012(2002),hep-ph/0203211.
13. D.M. Howe and C.J. Maxwell, hep-ph/0204036 v2.
14. S.J. Brodsky and H.J. Lu, Phys. Rev. D**51**,3652(1995)
15. J.R.C.Bloch, A. Cucchieri, K.Langfeld and T.Mendes, hep-lat/0209040 v2.
16. D.Dudal,H. Vershelde, V.E.R. Lemes, M.S. Sarandy, S.P.Sorella, M.Picariello, A. Vicini and J.A. Gracey, JHEP06(2003)003.
17. K-I. Kondo and T. Shinohara, Phys. Lett. B**491**,263(2000).

Relativistic Faddeev approach to a non-local NJL model

Amir H. Rezaeian*, Niels R. Walet* and Michael C. Birse†

*Department of Physics,UMIST, PO Box 88, Manchester, M60 1QD, UK
†The Theoretical Physics Group, Department of Physics and Astronomy, University of Manchester, Manchester, M13 9PL, UK

Abstract. The diquark and nucleon are studied in a non-local NJL model. We solve the relativistic Faddeev equation and compare the results with the ordinary NJL model. Although the model is quark confining, it is not diquark confining in the rainbow-ladder approximation. We show that the off-shell contribution to the diquark T matrix is crucial for the structure of the nucleon: without its inclusion the attraction in the scalar channel is too weak to form a three-body bound state.

INTRODUCTION

The NJL model is a successful low-energy phenomenological model inspired by QCD [1]. It has also been applied to the nucleon, see Refs. [2]. However, the lack of confinement makes the model questionable in the baryonic sector. It has been shown that a non-local covariant extension of this model can lead to quark confinement for acceptable values of the parameters [3]. This occurs due to the fact that the quark propagator has no real poles, and consequently quarks do not appear as asymptotic states. There are several other advantages of the non-local version over the local NJL model: the non-locality regularises the model in a manner such that anomalies and gauge invariance are preserved and the momentum-dependent regulator makes the theory finite to all orders in the $1/N_c$ expansion. Finally the dynamical quark mass is momentum dependent in contrast to the ordinary NJL model and consistent with lattice simulations of QCD. As a result, the non-local version of the NJL model may have more predictive power.

Here we shall use the covariant quark-diquark formalism in a relativistic Faddeev equation for the nucleon, where we include only scalar diquark correlations. Due to the separability of the non-local interaction, the Faddeev equations can be reduced to a set of effective Bethe-Salpeter equations. This makes it possible to adopt the numerical method developed for such problems by Oettel et al. [4].

THE MODEL

We consider a non-local NJL model Lagrangian with $SU(2)_f \times SU(3)_c$ symmetry. There exists several versions of such non-local NJL models. Regardless of what version is chosen, a Fierz transformation allows one to rewrite the interaction in either the $q\bar{q}$ or qq channels. Here we truncate to the scalar $(0^+, T = 0)$ and pseudoscalar $(0^-, T = 1)$

CP717, Hadron Spectroscopy: Tenth International Conference,
edited by E. Klempt, H. Koch, and H. Orth

mesonic channels. To investigate the nucleon in the quark-diquark picture, one also needs to know the qq interaction. We truncate to the scalar $(0^+, T = 0)$ colour $\bar{3}$ qq channel,

$$
\begin{aligned}
\mathscr{L}_{I\pi} &= \frac{1}{2} g_\pi j_\alpha(x) j_\alpha(x), &\qquad \mathscr{L}_{Is} &= g_s \bar{J}_s(x) J_s(x), \\
j_\alpha(x) &= \int d^4 x_1 d^4 x_3 f(x - x_3) f(x_1 - x) \overline{\psi}(x_1) \Gamma_\alpha \psi(x_3), \\
\bar{J}_s(x) &= \int d^4 x_1 d^4 x_3 f(x_1 - x) f(x - x_3) \overline{\psi}(x_1) [\gamma_5 C \tau_2 \beta^A] \overline{\psi}^T(x_3), \\
J_s(x) &= \int d^4 x_2 d^4 x_4 f(x_2 - x) f(x - x_4) \psi^T(x_2) [C^{-1} \gamma_5 \tau_2 \beta^A] \psi(x_4).
\end{aligned}
\tag{1}
$$

Here $\Gamma_\alpha = (1, i\gamma_5 \tau)$ and m_c is the current quark mass of the u and d quarks. The matrices $\beta^A = \sqrt{3/2} \lambda^A (A = 2, 5, 7)$ project onto the colour $\bar{3}$ channel with normalisation $\text{tr}(\beta^A \beta^{A'}) = 3\delta^{AA'}$ and the τ_i's are flavour $SU(2)$ matrices with $\text{tr}(\tau_i \tau_j) = 2\delta_{ij}$. The object $C = i\gamma_2 \gamma_5$ is the charge conjugation matrix. Since we do not restrict ourselves to specific choice of interaction, we shall treat the couplings g_s and g_π as independent. For simplicity, we assume the form factor $f(x - x_i)$ to be local in momentum space, since it leads to a separable interaction in momentum space.

The dressed quark propagator $S(k)$ is now constructed by means of a Schwinger-Dyson equation (SDE) in rainbow-ladder approximation. Thus the dynamical constituent quark mass, arising from spontaneously broken chiral symmetry, is obtained as [the symbol Tr denotes a trace over flavour, colour and Dirac indices]

$$
m(p) = m_c + i g_\pi f^2(p) \int \frac{d^4 k}{(2\pi)^4} \text{Tr}[S(k)] f^2(k), \qquad S^{-1}(k) = \not{k} - m(k). \tag{2}
$$

Following Ref. [3], we choose the form factor to be Gaussian in Euclidean space, $f(p_E) = \exp(-p_E^2/\Lambda^2)$, where Λ is a cutoff of the theory. If one assumes that Λ is related to the average inverse size of instantons $1/\bar{\rho}$, its choice thus parametrises non-perturbative properties of the QCD vacuum. The choice (2) respects Poincaré invariance and leads to quark, but not colour, confinement, when the dressed quark propagator has no poles at real p^2 in Minkowski space. This occurs for

$$
\frac{m(0) - m_c}{\sqrt{m_c^2 + \Lambda^2} - m_c} > \frac{1}{2} \exp\left(-\frac{(\sqrt{m_c^2 + \Lambda^2} + m_c)^2}{2\Lambda^2} \right). \tag{3}
$$

For large enough values of the dynamical quark mass, the quark propagator has no real poles and quarks do not appear as asymptotic states. The propagator still has infinitely many pairs of complex poles, both for confining and non-confining parameter sets. This is a feature of such models and due care should be taken in handling such poles, which can not be associated with asymptotic states if the theory is to satisfy unitary.

Our model has four free parameters: the current quark mass m_c, the cutoff (Λ), the coupling constants g_π and g_s. We fix the first three to give a pion mass of 136.6 MeV with decay constant of 92.4 MeV, as the value of the zero-momentum quark mass in

TABLE 1. The model parameters for different sets, fitted as discussed in text. The resulting value of the dynamical quark mass $m(0)$ also is shown.

Parameter	set A	set B	set C
$m_0(0)$ (MeV)	250	350	400
$m(0)$ (MeV)	297.9	406.2	461.3
m_c (MeV)	7.9	14.4	17.6
Λ (MeV)	1046.8	723.4	638.1
g_π (GeV^{-2})	31.6	89.0	132.6

chiral limit $m_0(0)$. We analyse three sets of parameters, as indicated in table 1. Set A is non-confining and sets B and C are confining (i.e., they satisfy the condition Eq. (3)). The parameter g_s is not constrained by these conditions, which allows us to analyse the coupling constant dependence through the ratio $r_s = g_s/g_\pi$.

DIQUARK AND NUCLEON SOLUTION

In the rainbow-ladder approximation the diquark T-matrix can be written as

$$T(p_1, p_2, p_3, p_4) = f(p_1)f(p_3)[\gamma_5 C\tau_2\beta^A]t(q)[C^{-1}\gamma_5\tau_2\beta^A]f(p_2)f(p_4)$$
$$\times\delta(p_1 + p_2 - p_3 - p_4), \qquad (4)$$

where $q = p_1 - p_3 = p_4 - p_2$ and

$$t(q) = \frac{2g_s i}{1 + g_s J_s(q)}, \qquad (5)$$

$$J_s(q) = i\,\mathrm{Tr}\int \frac{d^4k}{(2\pi)^4}f^2(-k)[\gamma_5 C\tau_2\beta^A]S(-k)^T[C^{-1}\gamma_5\tau_2\beta^A]S(q+k)f^2(q+k). \quad (6)$$

In the above equation the quark propagators are the solution of the rainbow SDE Eq. (2). The diquark bound state is located at the pole of the T-matrix. The denominator of Eq. (5) is the same as that in pionic channel, thus one may conclude that at $r_s = 1$ the diquark and pion are degenerate. This puts an upper limit to the choice of r_s, since diquarks should not to condense in vacuum.

The relativistic Faddeev equation can be written as an effective two-body BSE between a quark and a diquark due to the separability of the two-body interaction in momentum-space. Using the the on-shell approximation this becomes an eigenvalue problem [4],

$$\int \frac{d^4k}{(2\pi)^4} K_{\alpha\gamma}(p,k;P)\psi_{\gamma\beta}(k,P) = \lambda(P^2)\phi_{\alpha\beta}(p,P), \tag{7}$$

$$\psi(p,P) = S(p_q)D(p_d)\phi(p,P), \tag{8}$$

$$K(p,k;P) = \frac{2}{g_{dqq}^2} X(P_1)S^T(q)\overline{X}(P_2). \tag{9}$$

Here $\psi(p,P)$ is the nucleon bound-state wave function related to the vertex function $\phi(p,P)$ by truncation of the legs. The parameters p and P are the relative and total momenta in the quark-diquark system, respectively. We define $p_q = \eta P + p$, $p_d = (1-\eta)P - p$ and $q = -p - k + (1-2\eta)P$. The relative momentum of quarks in the diquark vertices X and $\overline{X}$ are defined as $p_1 = p + k/2 - (1-3\eta)P/2$ and $p_2 = -k - p/2 + (1-3\eta)P/2$ respectively. (For η see below.)

These equations should be solved for the largest eigenvalue of $\lambda(P^2)$ with the constraint that $\lambda(P^2) = 1/g_{dqq}^2$ at $P^2 = M_N^2$, where M_N is nucleon mass. Here g_{dqq} denotes the diquark-quark interaction coupling, $g_{dqq}^{-2} = \left(\partial_{k^2}J_s\right)|_{k^2=M_d^2}$, where J_s is defined in Eq. (6). In the usual on-shell approximation the scalar diquark propagator is taken to be $D(p_d)^{-1} = -(p_d^2 + M_d^2)$ where M_d is the pole of the scalar diquark T-matrix. The functions X and its adjoint $\overline{X} = \gamma_0 X^\dagger \gamma_0$ are the vertex functions of two quarks within the diquark and can be obtained from Eq. (4),

$$X(p_1,k_d) = g_{dqq}\gamma_5 C\tau_2\beta^A f(p_1 + (1-\sigma)k_d)f(-p_1 + \sigma k_d). \tag{10}$$

The non-locality of our model naturally leads to an extended diquark, and provides a sufficient regularisation of the ultraviolet divergences for a diquark-quark loop. Above we have introduced the Mandelstam parameters η and σ, $\eta, \sigma \in [0,1]$. In principle, the eigenvalues should not depend on these parameters if the formulation is Lorentz covariant. However, in practise due to numerical errors and singularities in the propagators, one usually only finds that there are substantial plateaus where the results are η- and σ-independent. The nucleon vertex function $\phi(p,P)$ should correspond to a state of positive energy, positive parity and spin-$1/2$. This leads to the general structure

$$\phi(p,P) = \left(S_1 + \frac{i}{M_N} \not{p}S_2\right)\Lambda^\dagger, \tag{11}$$

where $\Lambda^\dagger = \frac{1}{2}(1 + \frac{\not{P}}{iM_N})$ is a positive energy projection for a nucleon with mass M_N, and S_1 and S_2 are scalar function. We now solve the equations in the rest frame of the nucleon. We expand the scalar functions in terms of Chebyshev polynomials of the first kind, which turns out to be very efficient for such problems [4].

RESULTS

The scalar diquark mass M_D is obtained as a pole of Eq. (5). We find that for a wide range of r_s, regardless of the parameter set and confinement, a bound scalar diquark

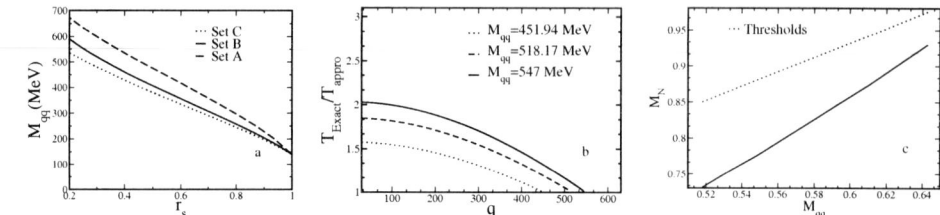

FIGURE 1. (a) Shows the diquark masses with respect to r_s for three sets of parameters. (b) Shows the discrepancy between the on- and off-shell approximation for three diquark masses (parameter set B). In (c) the nucleon mass is shown as a function of diquark mass for set B, the values are given in GeV.

exists. The diquark masses for various value of r_s for different parameter sets are plotted in Fig. 1a. For sets A, B, C no bound scalar diquark exists for $r_s < 0.110, 0.112, 0.146$, respectively. Please note that the confinement of the diquark for very small r_s is due to the screening effect of the ultraviolet cutoff and should not be associated to the underlying QCD confinement which stems from infrared divergence of gluon and ghost propagators. Having said that it is possible that diquark confinement may arise beyond the rainbow-ladder approximation [5].

In order to solve the Faddeev equation numerically, we have first used the on-shell approximation. We observe that one can not generate a three-body bound state in this model, without an artificial enhancement of the quark-diquark coupling of about 3. A similar feature has been observed in the ordinary NJL model as well [2], but the effect is even more severe here. As can be seen from Fig. 1 a decrease in r_s leads to a larger diquark mass, and an increase in the off-shell contribution to the qq T-matrix. It is this off-shell contribution that leads to a bound nucleon, and they are indeed substantial as can be seen in Fig. 1b. A preliminary result is shown in Fig. 1c. We also show a fictitious diquark-quark threshold defined as $M_{qq} + m_q^p$, where m_q^p is the real part of the first pole of quark propagator. Given this definition the diquarks in the nucleon are much more loosely bound in the non-local model than in the ordinary NJL model. Nonetheless, we obtain a bound nucleon solution near its experimental value.

The research of NRW and MCB was supported by the UK EPSRC; AHR acknowledges the award of an ORS studentship.

REFERENCES

1. J. Bijnens, Phys. Rept. **265**, 369 (1996).
2. A. Buck, R. Alkofer and H. Reinhardt, Phys. Lett. **B286**, 29 (1992); S. Huang and J. Tjon, Phys. Rev. **C49**, 1702 (1994); W. Bentz, H. Mineo, H. Asami, K. Yazaki, Nucl. Phys. **A670**, 48 (2000).
3. R. D. Bowler and M. C. Birse, Nucl. Phys. **A582**, 655 (1995); R. S. Plant and M. C. Birse, Nucl. Phys. **A628**, 607 (1998).
4. M. Oettel, L. Von Smekal, R. Alkofer, Comput. Phys. Commun. **144**, 63 (2002); M. Oettel, R. Alkofer, L. von Smekal, Eur. Phys. J. **A8**, 553 (2000); M. Oettel, G. Hellstern, R. Alkofer, H. Reinhardt, Phys. Rev. **C58**, 2459, (1998).
5. A. Bender, C. D. Roberts and L. V. Smekal, Phys. Lett **B380**, 7 (1997); G. Hellstern, R. Alkofer and H. Reinhardt, Nucl. phys. **A625**, 697 (1997).

Strong CP problem, Neutron EDM and Thermal QCD sum rules

Mohamed Chabab [1]

LPHEA, Physics Department, Faculty of Science Semlalia,
Cadi Ayyad University, P.O.Box 2390, Marrakech, Morocco

Abstract. The behaviour of the broken CP symmetry at finite temperature is examined. This is achieved through the investigation of the neutron electric dipole moment d_n induced by θ-term. By using thermal QCD sum rules, we find that below the critical temperature, the ratio $|\frac{d_n}{\theta}|$ slightly decreases but survives at temperature effects. This evolution implies that CP remains broken at finite temperature as required by Baryogenesis [1]

INTRODUCTION

Recently the finite temperature behaviour of symmetries has gained considerable interest. The question of symmetry restoration is a non trivial phenomenon, since it has been shown in [23, 24] that more heat does not necessarily imply more symmetry. Besides, the breaking of the symmetries has a profound implications in a particle physics and cosmology. The CP symmetry is certainly one of the most fundamental symmetries in nature. Besides its role in solving domain wall problem [25], it is a crucial ingredient to understand the matter-antimatter asymmetry [26].

In the three generation Standard Model, CP violation originates from the more obscure sector of the SM: the scalar part. It is parameterized, in the electroweak sector, by a single phase occurring in the Cabbibo-Kobayashi-Maskawa (CKM) quark mixing matrix [2]. CP violation could also originate from additional CP-odd four dimensional operator embedded in the following topological term "θ-term" in the QCD Lagrangian:

$$L_\theta = \theta \frac{\alpha_s}{8\pi} G_{\mu\nu} \tilde{G}^{\mu\nu},\qquad(1)$$

which breaks P, T and CP. $G_{\mu\nu}$ is the gluonic field strength, $\tilde{G}^{\mu\nu}$ denotes its dual and α_s is the strong coupling constant. The $G_{\mu\nu}\tilde{G}^{\mu\nu}$ quantity is a total derivative which contributes to the physical observables only through non perturbative effects, induced by instantons. A non zero value of θ may generate, in particular, a sizable neutron electric dipole moment (NEDM) which is related to the $\bar{\theta}$-angle by the following equation obtained within the framework of Chiral perturbation theory:

---

[1] E-mail address:mchabab@ucam.ac.ma

CP717, *Hadron Spectroscopy: Tenth International Conference,*
edited by E. Klempt, H. Koch, and H. Orth

$$d_n \sim \frac{e}{M_n}\left(\frac{m_q}{M_n}\right)\bar{\theta} \sim \begin{cases} 2.7 \times 10^{-16}\bar{\theta} & [3] \\ 5.2 \times 10^{-16}\bar{\theta} & [4] \end{cases} \qquad (2)$$

High precision Experiments have constrained the NEDM to $d_n < 1.1 \times 10^{-25} ecm$ [5], providing a stringent upper limit to $\bar{\theta} < 2 \times 10^{-10}$ [6]. The difficulty to explain the smallness of $\bar{\theta}$ in the standard model is usually known as the "strong CP problem". In this regard, several scenarios were suggested. The most elegant explanation is due to Peccei and Quinn [7], who identified $\bar{\theta}$ to the axion, a very light pseudo scalar boson arising from the spontaneous breaking of a global $U_A(1)$ symmetry. This particle may well be important to explain dark matter puzzle providing a peace of information on the missing mass of the universe [8].

Our aim in this work is to investigate the behaviour of the CP symmetry breaking at finite temperature and the thermal effects on the restoration of the strong CP problem. This is motivated by the possibility to restore some broken symmetries by increasing the temperature.

In section 2, we perform the the calculations of the the $\bar{\theta}$ induced NEDM using thermal QCD sum rules. Section 3 is devoted to the discussion and qualitative analysis of the thermal effects on the CP symmetry restoration.

NEDM FROM THERMAL QCD SUM RULES

In order to derive the NEDM through the QCD sum rules techniques [9, 10, 11], we consider a Lagrangian containing the following P and CP violating operators:

$$L_{P,CP} = -\theta_q m_* \sum_f \bar{q}_f i\gamma_5 q_f + \theta \frac{\alpha_s}{8\pi} G_{\mu\nu}\tilde{G}^{\mu\nu}. \qquad (3)$$

θ_q and θ are respectively two angles coming from the chiral and the topological terms while m_* is the quark reduced mass given by $m_* = \frac{m_u m_d}{m_u + m_d}$. The physical phase is $\bar{\theta} = \theta + \theta_q$. We usually start from the two points correlator in QCD background with a non-vanishing θ in the presence of a constant external electomagnetic field $F^{\mu\nu}$:

$$\Pi(q^2) = i \int d^4x e^{iqx} < 0|T\{\eta(x)\bar{\eta}(0)\}|0>_{\theta,F}. \qquad (4)$$

where $\eta(x)$ is the neutron interpolating current [12]:

$$\eta = 2\varepsilon_{abc}\{(d_a^T C\gamma_5 u_b)d_c + \beta(d_a^T C u_b)\gamma_5 d_c\}, \qquad (5)$$

and β is a mixing parameter. To select the appropriate Lorentz structure, $\Pi(q^2)$ is expanded in terms of the electromagnetic charge as:

$$\Pi(q^2) = \Pi^{(0)}(q^2) + e\Pi^{(1)}(q^2, F^{\mu\nu}) + O(e^2). \qquad (6)$$

The first term $\Pi^{(0)}(q^2)$ is the nucleon propagator which includes only the CP-even parameters, while the second term $\Pi^{(1)}(q^2, F^{\mu\nu})$ is the polarization tensor which may be expanded through Wilson OPE as: $\sum C_n < 0|\bar{q}\Gamma q|0 >_{\theta,F}$, where Γ is an arbitrary Lorentz structure and C_n are the Wilson coefficient functions calculable in perturbation theory [13, 14]. From this expansion, we keep only the CP-odd contribution part. The electromagnetic dependence of these matrix elements is determined in terms of the magnetic susceptibilities κ, χ and ξ [14], defined as:

$$< 0|\bar{q}\sigma^{\mu\nu}q|0 >_F = \chi e_q F^{\mu\nu} < 0|\bar{q}q|0 > \tag{7}$$

$$g < 0|\bar{q}G^{\mu\nu}q|0 >_F = \kappa e_q F^{\mu\nu} < 0|\bar{q}q|0 > \tag{8}$$

$$2g < 0|\bar{q}\tilde{G}^{\mu\nu}q|0 >_F = \xi e_q F^{\mu\nu} < 0|\bar{q}q|0 > \tag{9}$$

Moreover, the θ dependence of $< 0|\bar{q}\Gamma q|0 >_\theta$ matrix elements may be traced by considering the anomalous axial current [10]:

$$m_q < 0|\bar{q}\Gamma q|0 >_\theta = im_* \theta < 0|\bar{q}\Gamma q|0 > + O(m_q^2) \tag{10}$$

where the correction $O(m_q^2)$ is negligible since $m_\eta >> m_\pi$.

Putting altogether the above ingredients and after a straightforward calculation [11], the following expression of $\Pi^{(1)}(q^2, F^{\mu\nu})$ for the neutron is derived:

$$
\begin{aligned}
\Pi(-q^2) = &-\frac{\bar{\theta}m_*}{64\pi^2} < 0|\bar{q}q|0 > \{\tilde{F}\sigma, \hat{q}\}[\chi(\beta+1)^2(4e_d - e_u)\ln(\frac{\Lambda^2}{-q^2}) \\
&-4(\beta-1)^2 e_d(1+\frac{1}{4}(2\kappa+\xi))(\ln(\frac{-q^2}{\mu_{IR}^2})-1)\frac{1}{-q^2} \\
&-\frac{\xi}{2}((4\beta^2-4\beta+2)e_d+(3\beta^2+2\beta+1)e_u)\frac{1}{-q^2}...],
\end{aligned}
\tag{11}
$$

with $\hat{q} = q_\mu \gamma^\mu$.

The QCD expression (11) is confronted to the phenomenological parameterization $\Pi^{Phen}(-q^2)$ written in terms of the Neutron hadronic properties. The latter is given by:

$$\Pi^{Phen}(-q^2) = \{\tilde{F}\sigma, \hat{q}\}(\frac{\lambda^2 d_n m_n}{(q^2-m_n^2)^2} + \frac{A}{(q^2-m_n^2)} + ...), \tag{12}$$

where m_n is the neutron mass, e_q is the quark charge. the parameters A and λ^2, which originate from the phenomenological side of the sum rule, represent respectively a constant of dimension 2 and the neutron coupling constant to the interpolating current $\eta(x)$. This coupling is defined via a spinor v as $< 0|\eta(x)|n >= \lambda v e^{\alpha\gamma_5}$.

In the framework of QCD sum rules, the correlators at finite temperature are expressed in terms of the thermal Gibbs average of Wilson operator expansion [15, 18]. At relatively low temperature, where the system can be regarded as a non interacting gas of bosons, the thermal dependence of the vacuum condensates can be written as :

$$< O^i >_T = < O^i > + \int \frac{d^3p}{2\varepsilon(2\pi)^3} < \pi(p)|O^i|\pi(p) > n_B(\frac{\varepsilon}{T}) \tag{13}$$

where $\varepsilon = \sqrt{p^2 + m_\pi^2}$, $n_B = \frac{1}{e^x - 1}$ is the Bose-Einstein distribution and $< O^i >$ is the standard vacuum condensate (i.e. at T=0). In this approximation, we only kept the pion contributions, since in the low temperature region, the effects of heavier resonances ($\Gamma = K, \eta, ..etc$) are dumped by their distribution functions $\sim e^{\frac{-m_\Gamma}{T}}$ [17]. To compute the pion matrix elements, we apply the soft pion theorem given by:

$$< \pi(p)|O^i|\pi(p) >= -\frac{1}{f_\pi^2} < 0|[Q_5^a,[Q_5^a, O^i]]|0 > +O(\frac{m_\pi^2}{\Lambda^2}), \qquad (14)$$

where Λ is a hadron scale and Q_5^a is the isovector axial charge defined by:

$$Q_5^a = \int d^3 x \bar{q}(x) \gamma_0 \gamma_5 \frac{\tau^a}{2} q(x). \qquad (15)$$

Direct application of the above formula to the quark and gluon condensates shows the following features [16, 17]:
(i) Only $< \bar{q}q >$ is sensitive to temperature. Its behaviour at finite T is given by:

$$< \bar{q}q >_T \simeq (1 - \frac{\varphi(T)}{8}) < \bar{q}q >, \qquad (16)$$

where $\varphi(T) = \frac{T^2}{f_\pi^2} B(\frac{m_\pi}{T})$, $B(z) = \frac{6}{\pi^2} \int_z^\infty dy \frac{\sqrt{y^2 - z^2}}{e^y - 1}$ and f_π is the pion decay constant ($f_\pi \simeq 93 MeV$). The variation with temperature of the quark condensate $< \bar{q}q >_T$ results in two different asymptotic evolutions, namely:

$$< \bar{q}q >_T \simeq (1 - \frac{T^2}{8f_\pi^2}) < \bar{q}q >$$

for $\frac{m_\pi}{T} \ll 1$,

$$< \bar{q}q >_T \simeq (1 - \frac{\sqrt{\frac{\pi m_\pi}{2T}} T^2}{8f_\pi^2} e^{\frac{-m_\pi}{T}}) < \bar{q}q >$$

for $\frac{m_\pi}{T} \gg 1$.
(ii) The gluon condensate is nearly constant at low temperature and a T dependence occurs only at order T^8.

The determination of the ratio $\frac{d_n}{\bar{\theta}}$ sum rules at non zero temperature is now easily performed by applying Borel operator to both parameterizations of the Neutron correlation function shown in (11) and (12). Then finite temperature effects are introduced via the procedure discussed above. Finally, by invoking the quark-hadron duality, we deduce the final sum rules of the $\bar{\theta}$ induced NEDM at finite temperature:

$$\frac{d_n}{\bar{\theta}}(T) = -\frac{M^2 m_*}{16\pi^2} \frac{1}{\lambda_n^2(T) M_n(T)} (1 - \frac{\varphi(T)}{8}) < \bar{q}q > [4\chi(4e_u - e_d) - \frac{\xi}{2M^2}(4e_u + 8e_d)] e^{\frac{M_n^2}{M^2}}, \qquad (17)$$

where M represents the Borel parameter.

In order to get rid of the infrared divergence, the value of β has been set to 1 in (17). The Thermal evolution of the coupling constant and the mass of the neutron were determined from the thermal nucleon sum rules [17].

Within the dilute pion gas approximation, Eletsky has shown that the contribution induced the pion-nucleon scattering has to be considered [19]. It enters the nucleon sum rules through the coupling constant $g_{\pi NN}$, whose values lie within the range 13.5-14.3 [20].

Numerical analysis is performed with the following input parameters: the Borel mass has been chosen within the values $M^2 = 0.55 - 0.7 GeV^2$ which correspond to the optimal range (Borel window) in the $\frac{d_n}{\theta}$ sum rule at $T = 0$ [11]. For the χ and ξ susceptibilities we take $\chi = -5.7 \pm 0.6 GeV^{-2}$ [21] and $\xi = -0.74 \pm 0.2$ [22]. As to the vacuum quark condensate appearing in (17), we use its standard values [9].

ANALYSIS AND CONCLUSION

We have established the relation between the NEDM and $\bar{\theta}$ angle at non zero temperature from QCD sum rules. We find that the behaviour of the ratio $\frac{d_n}{\bar{\theta}}$ is connected to the thermal evolution of the pion parameters f_π, m_π and of $g_{\pi NN}$.

By analyzing the ratio as a function of T in the region of validity of thermal sum-rules $[0, T_c]$, we learn that $| \frac{d_n}{\bar{\theta}} |$ decreases smoothly with T (about 16% variation for temperature values up to 200 MeV) but survives at finite temperature. This means that either the NEDM value decreases or $\bar{\theta}$ increases. Consequently, for a fixed value of $\bar{\theta}$ the NEDM decreases but it does not exhibit any critical behaviour. Furthermore, if we start from a non vanishing $\bar{\theta}$ value at $T = 0$, it is not possible to remove it at finite temperature. We also note that $| \frac{d_n}{\bar{\theta}} |$ grows as M^2 or χ susceptibility increases. It also grows with quark condensate rising. However this ratio is insensitive to both the ξ susceptibility and the coupling constant $g_{\pi NN}$. We notice that for high temperatures, the analysis of $| \frac{d_n}{\bar{\theta}} |= f(\frac{T}{T_c})$ exhibits a brutal increase justified by the fact that for T beyond the critical value T_c, at which the chiral symmetry is restored, the constants f_π and $g_{\pi NN}$ become zero and consequently the ratio $\frac{d_n}{\bar{\theta}}$ behaves as a non vanishing constant. The large discrepancy between the values of the ratio for $T < T_c$ and $T > T_c$ may originate from the other contributions to the the spectral function which have been neglected, such as the scattering process $N + \pi \rightarrow \Delta$. These contributions, which are of the order T^4, are negligible in the low temperature region but become substantial for $T \geq T_c$. Moreover, this difference may also be due to the use of soft pion approximation which is valid essentially for low T ($T < T_c$). Therefore it is clear from this qualitative analysis, which is based on the soft pion approximation, that temperature does not play a fundamental role in the suppression of the undesired θ-term and hence the broken CP symmetry is not restored [1]. Indeed, some exact symmetries can be broken by increasing temperature [23, 24]. The symmetry non restoration phenomenon, which means that a broken symmetry at T=0 remains broken even at high temperature, is essential for discrete symmetries, CP symmetry in particular. Indeed, the symmetry non restoration is a crucial ingredient in solving the domain wall problem [25] and to create the baryon asymmetry in the early universe (BAU) [26].

ACKNOWLEDGMENTS

This work is partially supported by the convention de cooperation between CNRST-Morocco/GRICES-Portugal 681.02/CNR, and by the PROTARS III' grant D16/04.

REFERENCES

1. Chabab, M., El Biaze, N., and Markazi, R., J. Phys. **G27**, 2275 (2001).
2. Cabibbo, N., Phys. Rev. Lett. **10**, 531 (1963);
   Kobayashi, M and Maskawa, T., Prog. Theor. Phys. **49**, 652 (1973).
3. Baluni, V, Phys. Rev **D19**, 2227 (1979).
4. Crewther, R., Di Vecchia, P., Veneziano, G. and Witten, E., Phys. Letters **B88**, 123 (1979).
5. Barnett, R. M., and al, Phys. Rev. **D54**, 1 (1996).
6. Peccei, R. D., hep-ph/9807516.
7. Peccei, R. D., and Quinn, H. R., Phys. Rev. **D16**, 1791 (1977).
8. Lazarides, G., and Shafi, Q., "Monopoles, Axions and Intermediate Mass Dark Matter", hep-ph/0006202.
9. Shifman, M. A., Vainshtein, A. I., and Zakharov, V. I., Nucl. Phys. **B147**, 385 (1979).
10. Pospelov, M. and Ritz, A., Nucl. Phys. **B558**, 243 (1999).
11. Pospelov, M., and Ritz, A., Phys. Rev. Letters **83**, 2526 (1999).
12. Ioffe, B. L., Nucl. Phys. **B188**, 317 (1981);
    Chung, Y. and al, Phys. Lett.**B102**, 175 (1981); Nucl. Phys. **B197**, 55 (1982).
13. Shifman, M.A., Vainshtein, A. I., and Zakharov, V. I., Nucl. Phys. **B166**, 493 (1980).
14. Ioffe, B. L., and Smilga, A. V., Nucl. Phys. **B232**, 109 (1984).
15. Bochkarev, A. I., and Shaposhnikov, M. E., Nucl. Phys. **B268**, 220 (1986).
16. Gasser, J., and Leutwyler, H., Phys. Letters **B184**, 83 (1987);
    Leutwyler, H., in *QCD 20 years later*, edited by . P.M. Zerwas and H.A. Kastrup, World scientific Proceedings, Singapore, 1992. 1993).
17. Adami, C., and Zahed, I., Phys.Rev **D45**, 4312 (1992);
    Hatsuda, T., Koike, Y., and Lee, S. H., Nucl. Phys. **B394**, 221 (1993).
18. Mallik, S., and Mukherjee, K., Phys. Rev. **D58**, 096011 (1998).
19. Eletsky, V. L., Phys. Lett. **B245**, 229 (1990); Phys. Lett. **B352**, 440 (1995).
20. Blomgten, J., in *A critical issue in the determination of the pion nucleon decay constant*, Phys. Scripta **T87**, 53 (2000).
21. Belyaev, V. M., and Kogan, Y. I., Sov. J. Nucl. Phys. **40**, 659 (1984).
22. Kogan, Y. I., and Wyler, D., Phys. Lett. **B274**, 100 (1992).
23. Weinberg, S., Phys. Rev **D9**, 3357 (1974).
24. Mohapatra, R. N., and Senjanovic, G., Phys. Rev. **D20**, 3390 (1979) ;
    Dvali, G., Melfo, A., and Senjanovic, G., Phys.Rev. **D54**, 7857 (1996).
25. Zeldovich, Y. B., Kobzarev, I. Y., and Okun, L., JETP. **40**, 1 (1974);
    Kibble, T. W., J. Phys. **A9**, 1987 (1976); Phys. Rep. **67**, 183 (1980).
26. Sakharov, A., JETP Letters **5**, 24 (1967).

Quarks in Hadrons and Nuclei

G. Musulmanbekov

Join Institute for Nuclear Research, Dubna

Abstract. A dynamical quark model of hadron and nucleus structure is proposed. In the framework of the model, called the Strongly Correlated Quark Model, quarks and nucleons inside nuclei are arranged in crystal–like structure.

The majority of the physics community believes that the fundamental theory of strong interaction is QCD. However, the description of the dynamical structure of hadrons and nuclei in the framework of QCD is the unsolved problem so far. Phenomenological models of hadron structure can be divided into the current and constituent quark models. We believe that relationship between these models is based on chiral symmetry breaking.

Nuclear physicists apply three distinct models to describe the structure of nuclei: the shell, or independent particle, model, the liquid–drop, or collective, model and alpha–particle, or cluster, models. These three types of models are based on very different, in some cases, contradicting assumptions, yet all valid in their respective realms of application. In attempt to resolve this apparent contradictions, theoretical works suggesting that the nucleus may be in a solid phase have appeared from time to time in literature. Another important problem concerns with the role of quarks in forming of the nuclear structure: how are nucleons bound inside nuclei and do quarks manifest themselves explicitly in ground state nuclei?

In this paper we propose the model of nucleon structure, the so–called Strongly Correlated Quark Model (SCQM) [1, 2, 3], and utilize it to define the model of nuclear structure. It turns out that quarks and nucleons inside nuclei are arranged in a crystalline lattice. The ingredients of the model are the following. A single quark of definite color embedded in the vacuum polarizes its surrounding resulting in the formation of quark and gluon condensate. At the same time it experiences the pressure of the vacuum because of zero point radiation field or vacuum fluctuations which act on the quark tending to destroy the ordering of the condensate. Suppose that we place the corresponding antiquark in the vicinity of the first one. Owing to their opposite signs color polarization fields of the quark and antiquark interfere destructively in the overlapped space regions eliminating each other maximally in the space around the middle–point between the quarks. This effect leads to a decrease in condensates density in that space region and overbalancing of the vacuum pressure acting on the quark and antiquark from outer space regions. As a result an attractive force between the quark and antiquark emerges and the quark and antiquark start to move towards each other. The density of the remaining condensate around the quark (antiquark) is identified with the hadronic matter distribution. At maximum displacement in the $\bar{q}q-$ system corresponding to small overlapping of polarization fields, hadronic matter distributions

CP717, *Hadron Spectroscopy: Tenth International Conference,*
edited by E. Klempt, H. Koch, and H. Orth

have maximum extent and densities. The closer they come to each other, the larger is the destructive interference effect and the smaller hadronic matter distributions around quarks and the larger their kinetic energies. In that way quark and antiquark start to oscillate around their middle–point. For such interacting $\bar{q}q-$ pair located on the X axis at a distance $2x$ from each other, the total Hamiltonian is

$$H = \frac{m_{\bar{q}}}{(1-\beta^2)^{1/2}} + \frac{m_q}{(1-\beta^2)^{1/2}} + V_{\bar{q}q}(2x), \tag{1}$$

were $m_{\bar{q}}$, m_q are the current masses of the valence antiquark and quark, $\beta = \beta(x)$ is their velocity depending on displacement x, and $V_{\bar{q}q}$ is the quark–antiquark potential energy with separation $2x$. It can be rewritten as

$$H = \left[\frac{m_{\bar{q}}}{(1-\beta^2)^{1/2}} + U(x)\right] + \left[\frac{m_q}{(1-\beta^2)^{1/2}} + U(x)\right] = H_{\bar{q}} + H_q, \tag{2}$$

were $U(x) = \frac{1}{2}V_{\bar{q}q}(2x)$ is the potential energy of quark or antiquark. The quark (anti-quark) with the surrounding cloud (condensate) of quark – antiquark pairs and gluons, or hadronic matter distribution, forms the constituent quark. It is natural to assume that the potential energy of the quark (antiquark), $U(x)$, corresponds to the mass M_Q of the constituent quark:

$$2U(x) = C_1 \int_{-\infty}^{\infty} dz' \int_{-\infty}^{\infty} dy' \int_{-\infty}^{\infty} dx' \rho(x, \mathbf{r}') \approx 2M_Q(x) \tag{3}$$

where C_1 is a dimensional constant and the hadronic matter density distribution, $\rho(x, \mathbf{r}')$, is defined as

$$\rho(x, \mathbf{r}') = C_2 |\varphi(x, \mathbf{r}')| = C_2 \left| \varphi_Q(x' + x, y', z') - \varphi_{\bar{Q}}(x' - x, y', z') \right|. \tag{4}$$

Here C_2 is a normalization constant, φ_Q and $\varphi_{\bar{Q}}$ are density profiles of the condensates around the quark and antiquark located at distance $2x$ from each other. We consider by convention the condensates around quark and antiquark having opposite color charges. They have properties similar to compressive stress and tensile stress (around defects) in solids. Generalization to three–quark system in baryons is performed according to $SU(3)_{color}$ symmetry: in general, pair of quarks have coupled representations

$$3 \otimes 3 = 6 \oplus \bar{3} \tag{5}$$

in $SU(3)_{color}$ and for quarks within the same baryon only the $\bar{3}$ (antisymmetric) representation occurs. Hence, an antiquark can be replaced by two correspondingly colored quarks to get a color singlet baryon and destructive interference takes place between color fields of three valence quarks (VQs). Putting aside the mass and charge differences of valence quarks we may say that inside the baryon three quarks oscillate along the bi-sectors of equilateral triangle. Therefore, keeping in mind that the quark and antiquark in mesons and the three quarks in baryons are strongly correlated, we can consider each of them separately as undergoing oscillatory motion under the potential (3) in 1+1 dimension. The behavior of potential (3) evidently demonstrates the relationship between constituent and current quark states inside a hadron (Fig. 1). At maximum displacement quark is nonrelativistic, constituent one (VQ surrounded by condensate), since the influence of polarization fields of other quarks becomes minimal and the VQ possesses the maximal potential energy corresponding to the mass of constituent quark. At the origin of oscillation, $x = 0$, antiquark and quark in mesons and three quarks in baryons, being close to each other, have maximum kinetic energy and correspondingly minimum potential energy and mass: they are relativistic, current quarks (bare VQs). This configuration corresponds to so called "asymptotic freedom". In the intermediate region there is increasing (decreasing) of constituent quark mass by dressing (undressing) of VQs due to decreasing (increasing) of the destructive interference effect. This mechanism meets local gauge invariance principle. Indeed, destructive interference of color fields of quark and antiquark in mesons and three quarks in baryons depending on their displacements can be treated as phase rotation of wave function of single VQ in color space ψ_c on angle θ depending on displacement x of the VQ in coordinate space

$$\psi_c(x) \rightarrow e^{ig\theta(x)} \psi_c(x). \tag{6}$$

Phase rotation, in turn, leads to VQ dressing (undressing) by quark and gluon condensate that corresponds to the transformation of gauge field

$$A_\mu(x) \rightarrow A_\mu(x) + \partial_\mu \theta(x). \tag{7}$$

Here we drop color indices of $A_\mu(x)$ and consider each quark of specific color separately as changing its effective color charge, $g\theta(x)$, in color fields of other quarks (antiquark) due to destructive interference. Thus gauge transformations (7, 8) map internal (isotopic) space of colored quark onto coordinate space. On the other hand this dynamical picture of VQ dressing (undressing) corresponds to chiral symmetry breaking (restoration). Due to this mechanism of VQs oscillations the nucleon runs over the states corresponding to the certain terms of the infinite series of Fock space

$$|B\rangle = c_1 |q_1 q_2 q_3\rangle + c_2 |q_1 q_2 q_3 \bar{q}q\rangle + c_3 |q_1 q_2 q_3 g\rangle \ldots \tag{8}$$

Hereinafter we consider VQ oscillating along the $X-$ axis, with $Z-$ axis perpendicular to the plane of oscillation XY. Density profiles of condensates around VQs are taken in gaussian form. This choice is dictated by our semiclassical treatment of VQs motion. It has previously been shown [4] that the wave packet solutions of the time dependent Schrodinger equation for the harmonic oscillator move in exactly the same way as corresponding classical oscillators. These solutions are called "coherent states". This relationship justifies (partly) our semiclassical treatment of quantum objects.

We define the mass of constituent quark at maximum displacement as $M_{Q(\bar{Q})}(x_{max}) = \frac{1}{3}\left(\frac{m_\Delta + m_N}{2}\right) \approx 360 \; MeV$, where m_Δ and m_N are masses delta–isobar and nucleon correspondingly. The parameters of the model, namely, maximum displacement, x_{max}, and parameters of the gaussian function, $\sigma_{x,y,z}$, for hadronic matter distribution around VQ are chosen to be $x_{max} = 0.64 \; fm$, $\sigma_{x,y} = 0.24 \; fm$, $\sigma_z = 0.12 \; fm$. They are adjusted by comparison of calculated and experimental values of inelastic cross sections, $\sigma_{in}(s)$, and inelastic overlap function $G_{in}(s,b)$ for pp and $\bar{p}p-$ collisions [2]. The current mass of the valence quark is taken to be $5 \; MeV$. So far we have dealt with the scalar polarization

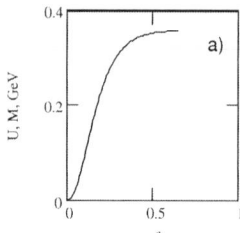 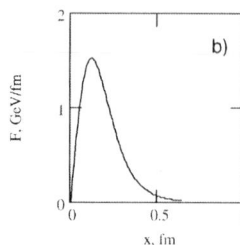

FIGURE 1. a) Potential energy of valence quark and mass of constituent quark; b) "Confinement" force.

field around VQ. In the framework of the model, as shown in a previous paper [3], the dominating contribution to proton spin comes from the orbital angular momentum of gluons and sea $\bar{q}q$–pairs circulating around the oscillating VQs. Now let us proceed to the many–nucleon problem. What would happen with oscillating quarks if we place a proton and a neutron nearby? Suppose that they aligned occasionally in such a way that a pair of quarks (one quark from the proton and another one from the neutron) with different flavor and color are nearest. Vacuum pressure acting from outside on these quarks decreases because of mutual influence (destructive interference, again) of the different color fields of these quarks. This effect results in attractive force between the nearest quarks from the proton and neutron. This force is near half in magnitude for that between the quark and antiquark in mesons and the three quarks in baryons at the same displacements. An additional attractive force comes from the flavor difference. To restrict these attractive forces quarks need to have parallel spins. As a result the potential (3) for adjacent quarks acquires an additional minimum at large quark displacements, small compared with the primary one. In that way the proton and nucleon form a bound state, namely, a deuteron. Now quarks of both nucleons oscillate around the deuteron center of mass interchanging their positions, with the quarks at free ends possessing maximal displacements. When all quarks pass the deuteron center of mass the adjacent quarks (at proton – neutron linkage) acquire angular orbital momentum ($l = 2$) that should be accompanied with spin flip of both quarks to conserve the total angular momentum of the deuteron.

Noting that three quarks inside nucleons are totally antisymmetric in the color space and two quarks from different nucleons at linkage are in antisymmetric color state ($\bar{3}$) having different flavors and parallel spins, we can construct more complex nuclei. The three nucleon system is formed by linkage of two quarks of each nucleon with quarks of two other nucleons according to the above rules. Three nucleon nuclei, namely ^3H and ^3He, represent a triangular configuration with three quarks at free ends. There is a drastic difference of quark oscillation pattern starting from the three nucleon system. The decrease of external vacuum pressure on quarks in each nucleon results in decreased attraction force between them that leads to the displacement of the origins of oscillations of each quark to the nucleon periphery and to the amplitude reduction of quarks oscillations, i.e. each quark having constituent mass oscillate near its own origin of oscillation. And so the suppression of current (bare) quark configurations inside the nuclear medium occurs. Completion of a four–nucleon system, ^4He, from a three–nucleon one, occurs by binding free quark ends in ^3H (^3He) with three quarks of an additional proton (neutron) again in accordance with the above rules. Since each quark in each of the four nucleons is coupled in a pair with a quark of an adjacent nucleon, current quark configurations are totally suppressed and only constituent quark configurations are realized in ^4He. Starting from ^4He all nuclei possess 3D-crystal – like structure. Indeed, planes of oscillations of two protons and two neutrons are located on opposite faces of an octahedron with common vertex. In this geometrical configuration four nucleons are in an $s-$ state that corresponds to the first s–shell of the shell model. Next, the $p-$ shell represents octahedron of bigger size with two $^3He-$ triangles instead of protons and two $^3H-$ triangles instead of neutrons. The triangles are located parallel to empty faces of the $^4He-$ octahedron, the free quark ends of these triangles coupled as in the $^4He-$ octahedron. This octahedron with the nested $^4He-$ octahedron represents the nucleus of ^{16}O. The next shell with principal number $n = 2$ is constructed in the same manner, extending triangles beforehand by adding a row of three protons to the row of two neutrons in ^3H and a row of three neutrons to the row of two protons in ^3He. Again, these triangles are located in couples on opposite faces of an octahedron parallel to unoccupied faces of the nested $p-$ octahedron. Construction of the next shells is performed in the same manner by extending triangles with new rows of neutrons and protons. When building the shells for $n > 2$ one needs to take into account the predominance of neutron number over proton number. It can be shown that at fixed distances between nucleon centers of mass (~ 2 *fermi*) the nucleons are arranged into a face–centered cubic lattice. It turns out that at nucleonic degrees of freedom our quark model of nuclear structure is identical to the lattice model formulated by N. Cook and V. Dallacasa more than twenty years ago [5] and called the FCC (face–centered–cubic)–lattice model. They demonstrated that it brought together shell, liquid-drop and cluster characteristics, as found in the conventional models, within a single theoretical framework. Unique among the lattice models, the FCC reproduces the entire sequence of allowed nucleon states as found in the shell model. Manifestation of a crystalline structure of nuclei has been observed in the diffraction pattern of scattering of α-particles on nuclei [6].

REFERENCES

1. G. Musulmanbekov, Nucl. Phys. Proc. Suppl. B 71 (1999) 117 and references therein.
2. G. Musulmanbekov, in Proc. Blois Workshop on Elastic and Diffractive Scattering, Eds. V. Kundrat and P. Zavada, World Scientific, 2000, p. 341.
3. G. Musulmanbekov, in Frontiers of Fundamental Physics 4, Ed. B.G. Sidharth, Kluwer/Acad. Press, 2001, p. 109.
4. E. Schrodinger, Naturwissenschaften, 14 (1926) 664; C.C. Yan, Am. J. Phys. 62 (1994) 147.
5. N.D. Cook and T. Hayashi, J. Phys. G: Nucl. Part. Phys. 23 (1997) 1109 and references therein.
6. D.A. Lednev, A.V. Yushkov, Izvestia RAN ser. fiz.(russian) 57 (1993) 107.

Gauge-invariant effective fields in QCD and the strong charge screening

D. S. Kuzmenko

Institute of Theoretical and Experimental Physics,
B.Cheremushkinskaya, 25, Moscow, Russia

Abstract. Gauge-invariant fields in hadrons are defined through the Wilson loop and written in terms of gauge-invariant field correlators. Dynamical equations for fields are considered. It is shown that at small distances in the leading order α_s aproximation they correspond to Maxwell equations with electric current and reproduce in particular the color-Coulombic OGE potential. At larger distances nonabelian triple field correlators lead to the emergence of circular effective (white) magnetic currents, and screening of the electric charge arises.

INTRODUCTION

As is known, it is rather difficult to treat the quantum chromodynamics in hadron region because it is formulated in terms of color degrees of freedom, while hadron states are colorless. We propose to handle the problem using white effective fields defined through the Wilson loop. Relying just on the QCD, we shall derive dynamical equations for these fields and expressions for effective electric and magnetic conserved currents. Then we shall analytically calculate distributions of effective fields and currents in the case of static quark and antiquark in the method of field correlators, and derive the relation between the scale QCD parameter and the string tension [1].

EFFECTIVE FIELDS AND DYNAMICAL EQUATIONS

The Wilson loop, written using the nonabelian Stokes theorem as the integral over the surface S bounded by the contour C, has a form

$$W[S,x_0](C) = {}_P \ \exp ig \int_S F_{\mu\nu}(z,x_0) d\sigma_{\mu\nu}(z), \tag{1}$$

where $F_{\mu\nu}(x,x_0) = \Phi(x_0,x) F_{\mu\nu}(x) \Phi(x,x_0)$ is covariantly shifted strength of the gluon field, $\Phi(x,y) = P \exp ig \int_x^y A_\mu^a(z) t^a dz_\mu$ is the parallel transporter, which transforms under gauge transformations as $\Phi(x,y) \to U^+(x)\Phi(x,y)U(y)$.

The gauge-invariant field is then defined as the derivative of the vacuum average of the trace of the Wilson loop over the element of the surface S,

$$_F J_{\mu\nu}(x) = \frac{\delta \ln \langle \mathrm{Tr}\, W[S,x_0] \rangle}{\delta \sigma_{\mu\nu}(x)} = \langle \mathrm{Tr}\, W(C = \partial S) \rangle^{-1} \langle \mathrm{Tr}\, ig F_{\mu\nu}(x,x_0) W[S,x_0] \rangle. \tag{2}$$

CP717, *Hadron Spectroscopy: Tenth International Conference,*
edited by E. Klempt, H. Koch, and H. Orth

One has to choose the surface S and trajectories connecting reference point $x_0 \in S$ to point x to complete the definition. Where confining fields are concerned, it is naturally to choose the minimal surface of Wilson loop and shortest trajectories. Then one arrives at the definition of "connected probe" [2]. Note that for the rest of this section the specification of surface and trajectories is inessential.

In the case of elecrodynamics the expression (2) defines the distribution of classical fields satisfying Maxwell equation with current $g^2 J_\mu(x) = g^2 \int_C dz_\mu \delta^{(4)}(z-x)$, and may therefore be interpreted as covariant generalization of the Biot-Savart law. In nonabelian case one can use the relation for differentiation of parallel transporters,

$$\partial_\mu \Phi(x,x_0) = ig A_\mu(x)\Phi(x,x_0) + ig \int_{x_0}^{x} dz_\beta \frac{\partial z_\alpha}{\partial x_\mu} \Phi(x,z) F_{\alpha\beta}(z)\Phi(z,x_0), \qquad (3)$$

where the second term arise due to the displacement of the integration contour, to obtain effective Maxwell equations with "magnetic current" k_μ^J,

$$\frac{1}{2} \varepsilon_{\mu\rho\alpha\beta} \frac{\partial}{\partial x_\rho} {}_{\scriptscriptstyle F} \bar{}_{\alpha\beta}(x) = k_\mu^J(x), \qquad \frac{\partial}{\partial x_\rho} {}_{\scriptscriptstyle F} \bar{}_\mu(x) = j_\mu^J(x), \qquad (4)$$

Gauge-invariant (colorless) currents k_μ^J and j_μ^J in (4) are expressed through vacuum averages of products of the Wilson loop, parallel transporters and covariantly shifted field strengths, see [1]. In the leading order of α_s the second term of (3) does not contribute, and using Bianchi identity and dynamical equations of QCD one finds that

$$j_\mu^J(x) = 4\pi C_F \alpha_s J_\mu(x), \qquad k_\mu^J = 0. \qquad (5)$$

This means in particular that the static color-Coulomb OGE potential is reproduced in terms of effective fields. The lowest (nonperturbative) contribution to the magnetic current is due to the triple correlator, $k^J \sim \left\langle g^3 E_i^a B_j^b E_k^c \right\rangle f^{abc} \varepsilon_{ijk}$. It is produced by the "displacement" term in (3).

DISTRIBUTIONS OF ELECTRIC FIELDS AND MAGNETIC CURRENTS FOR STATIC $Q\bar{Q}$ PAIR

Distributions of effective fields (1) for static quark and antiquark may be calculated in the method of field correlators [3], [4]. In this case the magnetic part of the effective field vanishes, and the electric field has the form

$$\varepsilon_i(\mathbf{r},\mathbf{R}) = n_k \int_0^R dl \int_{-\infty}^{\infty} dt \left(\delta_{ik}D(z) + \frac{1}{2}\frac{\partial z_i D_1(z)}{\partial z_k} \right) \equiv \varepsilon_i^D + \varepsilon_i^{D_1}, \qquad (6)$$

where $z = (\mathbf{r} - \mathbf{n}l, t)$; $\mathbf{R}$ is the vector directed from quark to antiquark, and $\mathbf{n} = \mathbf{R}/R$. The field (6) is defined through functions D and D_1, which parametrise the behavior of two-point gluon field strength correlators, see e.g. [5] and refs. therein. The function D

FIGURE 1. Vector distribution of magnetic currents **k** in bilocal approximation of field correlators method for $R = 2$ fm, $\sigma = 0.18$ GeV^2, $\lambda = 0.2$ fm. Positions of Q and $\bar{Q}$ are marked by circles.

decreases exponentially, $D(z) \sim \exp(-|z|/\lambda)$, where $\lambda = 0.2$ fm is the correlation length of gluon fields calculated, e.g., in [6]. The corresponding field ε_i^D has a form

$$\varepsilon_i^D(\mathbf{r}, \mathbf{R}) = n_i \frac{2\sigma}{\pi} \int\limits_0^{R/\lambda} dl \left| l\mathbf{n} - \frac{\mathbf{r}}{\lambda} \right| K_1 \left(\left| l\mathbf{n} - \frac{\mathbf{r}}{\lambda} \right| \right), \tag{7}$$

where K_1 is the McDonald function and $\sigma = 0.18$ GeV^2 is the string tension. At large $Q\bar{Q}$-separations $R \gg \lambda$ the distribution (7) forms the string with the universal profile $\varepsilon(\rho) = 2\sigma(1 + \rho/\lambda)\exp(-\rho/\lambda)$, where ρ is the distance to $Q\bar{Q}$-axes. The second part of the field (6), corresponding to function D_1, has both perturbative and nonperturbative contributions. The perturbative color-Coulomb field of one-gluon exchange is $\varepsilon_i^{D_1,p}(\mathbf{r}) = \varepsilon_i^{Coul}(\mathbf{r}) - \varepsilon_i^{Coul}(\mathbf{r} - \mathbf{R})$, where

$$\varepsilon_i^{Coul}(\mathbf{r}) = \frac{C_F \alpha_s r_i}{r^3} = -\nabla_i \frac{C_F \alpha_s r_i}{r}. \tag{8}$$

The nonperturbative part, ensuring the screening of the Coulomb charge and confinement of gluons, will be calculated in the following section. We apply now the Ampère law, $k_i = \varepsilon_{ijk}\nabla_j \varepsilon_k$ (see (4)), to expressions (6), (7) to calculate the distribution of magnetic current, and obtain circular currents shown in Fig. 1. Note that only ε_i^D contributes to magnetic current. The distribution of φ-component of the magnetic current in cylindrical coordinates, corresponding to the universal profile of electric field, has the form $k_\varphi(\rho) = -2\sigma\rho/\lambda^2 \exp(-\rho/\lambda)$. It satisfies the London's equation, $(\mathrm{rot}\,\mathbf{k})_z(\rho) = \gamma(\rho)\lambda^{-2}\varepsilon(\rho)$, where $\gamma(\rho) = (-2 + \rho/\lambda)/(1 + \rho/\lambda) \to 1$ at $\rho \gg \lambda$.

SCREENING OF THE STRONG COUPLING

Let us consider the remaining part of Maxwell equations (4) for static $Q\bar{Q}$, the Gauss law $\nabla_i \varepsilon_i = \rho$, for ε_i defined in (6)-(8), and recover the missing nonperturbative field $\varepsilon_i^{D_1,np}$, which ensures the confinement of gluons. In order to do the latter, we take the electric charge density ρ in the form (5), $\rho = 4\pi C_F \alpha_s (\delta(\mathbf{r}) - \delta(\mathbf{r} - \mathbf{R}))$, which leads

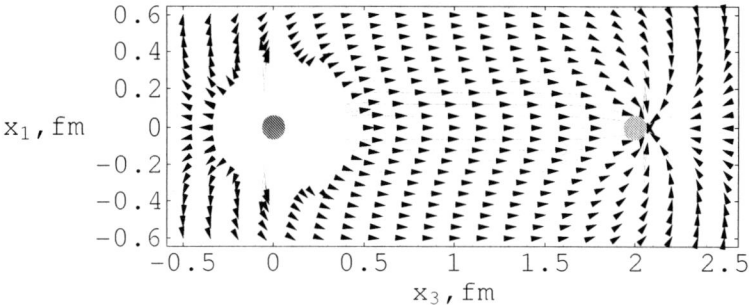

FIGURE 2. Vector distribution of total field $\varepsilon_i(x_1,0,x_3)$ for $\sigma = 0.18$ GeV^2, $\lambda = 0.2$ fm and constant $\alpha_s = 0.42$. Positions of quark and antiquark are marked by circles.

to the equation $\nabla_i\left(\varepsilon_i^{D_1,\mathrm{np}} + \varepsilon_i^{D}\right) = 0$. Solving this equation, we obtain two spherical distributions $\varepsilon^{D_1,\mathrm{np}}(r)$ around quark and antiquark, so that

$$\varepsilon^{D_1}(r) = \varepsilon^{\mathrm{Coul}}(r) + \varepsilon^{D_1,\mathrm{np}}(r) = \frac{C_f\alpha_s - \tilde{Q}(r)}{r^2} \equiv \frac{Q}{r^2}, \qquad (9)$$

where the "screening" charge $\tilde{Q}(r) = (2\sigma\lambda^2/\pi)\int_0^{r/\lambda} x^3 K_1(x)dx$ leads to the vanishing of the total charge at distances $r \gg \lambda$. The condition $Q(r)|_{r\to\infty} = 0$ determines the relation between parameters, $C_F\alpha_s|_{r\to\infty} = 3\sigma\lambda^2$. Let us stress that the central field (9) vanishes at $r \gg \lambda$ both outside and inside the string: the isotropic dielectric function ε, defined as $\varepsilon_i^{D_1}(r) = \varepsilon(r)\varepsilon_i^{\mathrm{Coul}}(r)$, falls exponentially at large distances,

$$\varepsilon(r)|_{r\to\infty} = \frac{\sqrt{\pi}}{2}\left(\frac{r}{\lambda}\right)^{5/2}\exp\left(-\frac{r}{\lambda}\right). \qquad (10)$$

Vector distribution of total field ε_i (6), (7), (9) for constant α_s is shown in Fig. 2. One can see the color-Coulomb central field near quark and antiquark, and confining string (tube) in between them.

We should note that distributions of gauge invariant fields and currents, similar to ones shown in Figs. 1 and 2, were calculated on the lattice in the method of abelian projection [7]. The resemblance is not surprising, since the effective field (1) presents a kind of projection related to the minimal surface of Wilson loop.

The behavior of charge $Q(r)$ is shown in Fig. 3 for classical (a) and quantum (b) cases. One can see the decrease of the charge, or the "screening". The quantum case is treated according to [8], where it was shown that confining fields in background perturbation theory lead to the modification of the logarithmic running coupling according to $\alpha_s(q^2) \to \alpha_B(q^2) = \alpha_s(q^2 + 1/\lambda^2)$. The background coupling α_B freezes at large distances, $\alpha_B(r)|_{r\to\infty} = \alpha_s(\lambda)$. This way one arrives at the relation between the scale parameter Λ_{QCD} and the string tension σ,

$$C_F\alpha_s(\lambda) = 3\sigma\lambda^2. \qquad (11)$$

As was discussed in [9], the value of $\Lambda_{\mathrm{QCD}} = 240$ MeV measured in lattice [10] in $\overline{\mathrm{MS}}$ regularizational scheme at $N_f = 0$ yields $\alpha_s(\lambda) = 0.42$, which satisfies (11) for physical

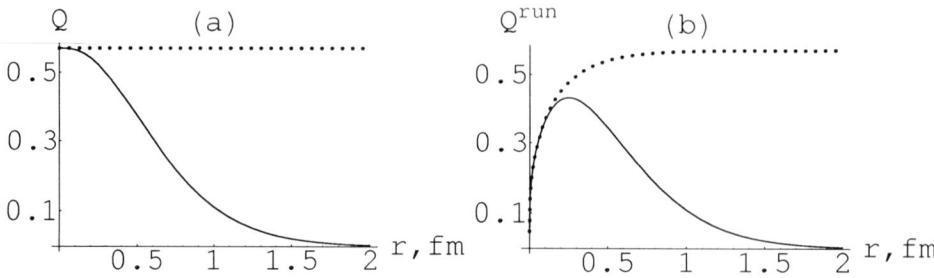

FIGURE 3. Effective charge $Q(r)$ (solid curves) and $C_f \alpha_s(r)$ (dotted curves) for constant coupling (a) and background coupling (b); $\sigma = 0.18 \text{ GeV}^2$, $\lambda = 0.2$ fm.

values $\lambda = 0.2$ fm and $\sigma = 0.18 \text{ GeV}^2$. Let us also note that phenomenon of the decrease of the QCD running coupling in infrared region was studied in lattice Landau gauge QCD and reported at this Conference [11].

SUMMARY

We have discussed in the talk that gauge invariant effective field, determined as surface derivative of the Wilson loop in gluodynamics with external currents, justifies Maxwell equations with classical current of Wilson loop at small distances. The mechanism of confinement in terms of effective fields is related to circular effective magnetic currents, which are expressed through three-point gluon field strength correlators, and the screening of the color-Coulomb part of the field. The phenomenon of the screening leads to the relation between the scale parameter of QCD and the string tension.

The author is grateful to Yu. A. Simonov and V. I. Schevchenko for many useful and inspiring discussions, as well as to A. M. Badalian and S. Furui for their interest to this work and valuable comments. This work has been supported by the Federal Program of the Russian Ministry of Industry, Science and Technology No. 40.052.1.1.1112, and grant NSh-1774.2003.2.

REFERENCES

1. D. S. Kuzmenko, hep-ph/0310017, submitted to Yad.Fiz.;
   D. S. Kuzmenko, V.I. Shevchenko, Yu. A. Simonov, Usp. Fiz. Nauk (2003) (in press).
2. A. Di Giacomo *et al* , Phys. Lett. B **236** (1990) 199; Nucl. Phys. B **347** (1990) 441.
3. L. Del Debbio, A. Di Giacomo and Yu. A. Simonov, Phys. Lett. B **332** (1994) 111.
4. D. S. Kuzmenko and Yu. A. Simonov, Phys. Lett. B **494**, 81 (2000); Yad. Fiz. **64**, 110 (2001).
5. D. S. Kuzmenko and Yu. A. Simonov, *Static potentials and confining field structure of baryons and 3g-glueballs*, this volume, hep-ph/0310034.
6. D. S. Kuzmenko, *Mass spectrum of heavy hybrid mesons in the QCD string model*, this volume, hep-ph/0310033.
7. V. Bornyakov *et al.*, Nucl. Phys. Proc. Suppl. **106**, 634 (2002); Y. Koma *et al.*, hep-lat/0302006.
8. Yu.A. Simonov, *Lecture Notes in Physics*, **479**, 139 (1996); Phys.At.Nucl. **58**, 107 (1995).
9. A.M. Badalian and D.S. Kuzmenko, Phys.Rev. D **65**, 016004 (2002).
10. S. Capitani *et al.*, Nucl. Phys. B **544**, 669 (1999).
11. S. Furui, H. Nakajima, *Infrared Features of Lattice Landau Gauge QCD and the Gribov Copy Problem*, this volume, hep-lat/0309166.

The QCD string in the chiral Lagrangian and the spectrum of pseudoscalar states

S.M. Fedorov* and Yu.A. Simonov*

*Institute of Theoretical and Experimental Physics, 117218, Moscow, B.Cheremushkinskaya 25, Russia

Abstract. Effective chiral Lagrangian is derived from QCD in the framework of Field Correlator Method. It contains the effects of both confinement and chiral symmetry breaking due to a special structure of the resulting quark mass operator. It is shown that this Lagrangian describes light pseudoscalar mesons, and Gell-Mann-Oakes-Renner relations for pions, eta and K mesons are reproduced. Spectrum of radial excitations of pions and K mesons is found and compared to experimentally known masses.

INTRODUCTION

QCD vacuum at low temperatures has two very nontrivial properties: confinement and chiral symmetry breaking (CSB). It is known from lattice calculations that phase transitions of deconfinement and chiral symmetry restoration take place at the same temperature [1, 2]. The fact that two critical temperatures coincide was not fully understood so far. It was shown in [3] that effective four-quark interaction leading to spontaneous chiral symmetry breaking, occurs in QCD due to confinement, and is associated with the QCD string. Thus, CSB is closely connected to confinement. In this approach the Effective Chiral Lagrangian (ECL) containing fields of light pseudo-scalar mesons is derived from QCD Lagrangian [3, 4, 5]. Confinement is taken into account through specific form of gluon-field correlators. As a result, expanding in powers of (derivatives) of bosonic fields, one obtains the ECL similar to the celebrated Gasser-Leutwyler Lagrangian [6], however in the nonlocal form [3].

ECL leads to the standard Gell-Mann-Oakes-Renner relations. We demonstrate, that the vanishing of meson masses in the chiral limit is due to cancellation of two terms in Green's functions of mesons. Poles of Green's function corresponding to radial excitations of pseudoscalar mesons are displaced from the masses, obtained in Hamiltonian approach without CSB effects (see e.g. [7] and references therein), and are shifted down by less than 15 %.

EFFECTIVE CHIRAL LAGRANGIAN

We consider Euclidean partition function for quarks and gluons in the presence of external classical currents v_μ, a_μ, s and p

CP717, *Hadron Spectroscopy: Tenth International Conference,*
edited by E. Klempt, H. Koch, and H. Orth
© 2004 American Institute of Physics 0-7354-0197-7/04/$22.00

$$Z = \int DAD\bar{\psi}D\psi \exp\left[-(S_0 + S_1 + S_{\text{int}} + S_{\text{g.f.}} + S_{\text{gh}})\right],$$

$$S_0 = \frac{1}{4}\int d^4x \left(F_{\mu\nu}^a\right)^2, \quad S_1 = -i\int d^4x \bar{\psi}^f(\hat{\partial} + \hat{v} + \gamma_5\hat{a} + s + i\gamma_5 p)^{fg}\psi^g, \quad (1)$$

$$S_{\text{int}} = -\int d^4x \bar{\psi}^f g\hat{A}^a t^a \psi^f.$$

Here $f, g = 1, 2, 3$ are flavor indices, t^a are generators of color SU(3) group. $S_{\text{g.f.}}$ and S_{gh} are gauge fixing and ghost terms.

Using generalized contour gauge [8, 9] one can integrate out gluon field A_μ. Expressing the result in terms of field correlators, and restricting ourselves to gaussian approximation (i.e. taking into account only bilocal correlators), after performing bosonization and integrating out quark fields we obtain effective chiral Lagrangian, with the degrees of freedom being scalar and pseudoscalar fields:

$$Z = \int DM_s D\phi_a \exp[-S_{\text{ECL}}], \quad S_{\text{ECL}} = 2N_f \int d^4x d^4y \left(J(x,y)\right)^{-1} M_s^2(x,y) + W(\phi),$$

$$W(\phi) = -N_c \text{tr} \ln\left[i(\hat{\partial} + \hat{v} + \gamma_5\hat{a} + s + i\gamma_5 p) + iM_s(x,y)e^{i\gamma_5 t_a \phi_a(x,y)}\right],$$

$$J(x,y) = \frac{1}{N_c^2 - 1}\int_0^1 dsdt \frac{\partial z_\rho(s,x)}{\partial s}\frac{\partial z_\lambda(s,x)}{\partial x_\mu}\frac{\partial z_{\rho'}(t,y)}{\partial t}\frac{\partial z_{\lambda'}(t,y)}{\partial y_\mu} \times$$

$$\times \text{tr}\left\langle F_{\rho\lambda}\left(z(s,x),x_0\right)F_{\rho'\lambda'}\left(z(t,y),x_0\right)\right\rangle_A.$$

$$(2)$$

Here tr refers to flavor and spinor indices and to space coordinates. M_s is the effective quark mass operator, and ϕ_a are fields of pseudoscalar mesons (up to the dimensional factor $2/f$, f is the decay constant, $\phi_a = 2\pi_a/f$). Bilocal approximation results in contour dependence of effective action. In what follows the exact position of contours is unimportant for our analytical results, while for numerical estimates we will assume that contours are chosen to minimize the spectrum (and area of the string world sheet).

Classical equations of motion have the solution $\phi_a^{(0)}(x,y) = 0$ for meson fields, and

$$M_s^{(0)}(x,y) = \frac{N_c}{4N_f}J(x,y)\text{Tr}\left(S(x,y)\right), \quad S(x,y) = \langle x|\left(i\hat{\partial} + iM_s\right)^{-1}|y\rangle, \quad (3)$$

for $M_s^{(0)}$. The existence of a nontrivial solution of equation (3) is a manifestation of the chiral symmetry breaking. Equation (3) was considered in [12] for the special case of heavy-light mesons, and it was shown that $M_s^{(0)}$ has a confining scalar solution for large interquark distances, thus confinement and CSB occur spontaneously and simultaneously from the nontrivial solution of the system (3).

MASSES AND GREEN'S FUNCTIONS OF MESONS

We consider ECL (2) and expand it in powers of the field ϕ_a up to the second order, introducing current quark masses $\mathcal{M}_f \equiv \text{diag}(m_u, m_d, m_s)$.

Quadratic in ϕ_a terms at zero momentum give masses of mesons. Taking into account that $\langle \bar{\psi}\psi \rangle_M = -i\langle \bar{\psi}\psi \rangle_E = -(1/Z)\delta Z[v,a,s,p]/\delta s(x) = -N_c \operatorname{Tr}(iS(x,x))$, where $\langle \bar{\psi}\psi \rangle_M$ and $\langle \bar{\psi}\psi \rangle_E$ denote quark condensate in Minkovski and Euclidean space respectively, one arrives at well-known Gell-Mann Oakes Renner relations. For pion it reads

$$f^2 M_{\pi^{\pm}}^2 = 2\hat{m}|\langle \bar{q}q \rangle| + O(m^2). \tag{4}$$

Similar relations are true for all other light pseudoscalar mesons. Here $\hat{m} = (m_u + m_d)/2$. We have neglected differences between quark condensates for different flavors, corrections are of order of m_q^2. Small mixing of ϕ_3 and ϕ_8 states due to isospin symmetry breaking (proportional to $m_u - m_d$) is also reproduced, and it yields a correction $\varepsilon = |\langle \bar{q}q \rangle|(m_u - m_d)^2/(4m_s - 2(m_u + m_d))$ to π^0 and η meson masses.

ECL (2) leads to correct Gell-Mann-Oakes-Renner relations for all light pseudoscalar mesons, and thus has correct symmetry structure. Also, there has to be some cancellation which makes pseudoscalar mesons massless in the chiral limit. The exact mechanism of this cancellation is presented below.

Let us consider Green's functions of mesons, generated by the pseudoscalar currents:

$$\mathfrak{G}_{ab}(x,y) = \langle J_a^5(x) J_b^5(y) \rangle = \frac{1}{Z} \frac{\delta^2 Z}{\delta p_a(x) \delta p_b(y)}, \quad J_a^5(x) = \bar{\psi}(x)\gamma_5 t_a \psi(x). \tag{5}$$

From the ECL one finds [4, 13] for Green's function in the momentum space:

$$\mathfrak{G}_{\pi^+\pi^+}(k) = \frac{N_c}{2} G_{\pi^+\pi^+}^{(0)}(k) - \frac{N_c^2}{4} G_{\pi^+\pi^+}^{(0,M)}(k) G_{\pi^+\pi^+}^{\phi}(k) G_{\pi^+\pi^+}^{(0,M)}(k),$$
$$G_{\pi^+\pi^+}^{(0)}(x,y) \equiv \operatorname{Tr}(S_d(x,y)\gamma_5 S_u(y,x)\gamma_5), \tag{6}$$
$$G_{\pi^+\pi^+}^{(0,M)}(x,y) \equiv \operatorname{Tr}(S_u(x,y)M_s(y)\gamma_5 S_d(y,x)\gamma_5).$$

Pion propagator, which has pole at $k^2 = -M_{\pi^{\pm}}^2$, takes the form

$$G_{\pi^+\pi^+}^{\phi}(k) = \frac{2}{N_c} \frac{1}{G_{\pi^+\pi^+}^{(0,MM)}(k) - G_{\pi^+\pi^+}^{(0,MM)}(k^2 = -M_{\pi^{\pm}}^2)},$$
$$G_{\pi^+\pi^+}^{(0,MM)}(x,y) \equiv \operatorname{Tr}(S_u(x,y)M_s(y)\gamma_5 S_d(y,x)M_s(x)\gamma_5). \tag{7}$$

Other possible contributions to Green function include additional pion exchanges, which are suppressed as $1/N_c$. Kaon and η meson Green functions differ only in flavor indices.

As argued in [4], all three Green functions $G_{\pi^+\pi^+}^{(0)}$, $G_{\pi^+\pi^+}^{(0,M)}$, and $G_{\pi^+\pi^+}^{(0,MM)}$ have the same set of poles, which are the poles of the quark model (i.e. confined $\bar{q}q$ system without chiral symmetry breaking) in pseudo-scalar channel, and can be represented as

$$G_{\pi^+\pi^+}^{(0)}(k) = -\sum_{n=0}^{\infty} \frac{c_n^2}{k^2 + m_n^2}; \ G_{\pi^+\pi^+}^{(0,M)}(k) = -\sum_{n=0}^{\infty} \frac{c_n c_n^{(M)}}{k^2 + m_n^2}; \ G_{\pi^+\pi^+}^{(0,MM)}(k) = -\sum_{n=0}^{\infty} \frac{\left(c_n^{(M)}\right)^2}{k^2 + m_n^2}. \tag{8}$$

Here $c_n = \sqrt{\frac{m_n}{2}}\varphi_n(0)$, $c_n^{(M)} = \sqrt{\frac{m_n}{2}}M(0)\varphi_n(0)$; $\varphi_n(\mathbf{r})$ is the 3D spin-singlet wave function of $\bar{q}q$ system, and $M(0)$ is a constant related to mass operator M_s, evaluated in [5] through $\sigma = 0.18$ GeV^2 and $T_g = 1$ GeV^{-1} to be $M(0) = 148$ MeV. Thus one has for the pion Green function:

$$\mathfrak{G}_{\pi^+\pi^+}(k) = -\frac{N_c}{2}\frac{\Psi(k)}{(k^2+M_{\pi^\pm}^2)\Phi(k)},$$

$$\Psi(k) = \sum_{n,m=0}^{\infty}\frac{c_n^2\left(c_m^{(M)}\right)^2}{(k^2+m_n^2)(m_m^2-M_\pi^2)}, \quad \Phi(k) = \sum_{n=0}^{\infty}\frac{\left(c_n^{(M)}\right)^2}{(k^2+m_n^2)(m_n^2-M_\pi^2)}. \tag{9}$$

Clearly, the Green function (9) has pole at $k^2 = -M_{\pi^\pm}^2$, and all poles of quark model are cancelled, since the same set of poles appears in functions $\Psi(k)$ and $\Phi(k)$. The radial excitations of $\pi^\pm$ meson are given by zeros of the function $\Phi(k)$. Masses of K^0, $\bar{K}^0$ radial excitations can be found the same way with the exchange of π meson mass and reference spectrum with those for K mesons. It should be mentioned, that η meson requires separate consideration, because of it's mixing with isoscalar state η', which is different for mesons and their radial excitations.

SPECTRUM OF MESONS' RADIAL EXCITATIONS

Masses and wave functions of reference spectrum can be obtained from the QCD string Hamiltonian (first derived in [14, 15, 16], and improved to take into account quark self-energy in [17]):

$$H = \frac{m_1^2}{2\mu_1} + \frac{m_2^2}{2\mu_2} + \frac{\mu_1+\mu_2}{2} + \frac{p_r^2}{2\tilde{\mu}} + \sigma r - \frac{4}{3}\frac{\alpha_s}{r}. \tag{10}$$

Here we have put $L = 0$; m_1 and m_2 are current masses of quarks, μ_1 and μ_2 are einbein parameters, to be found from the eigenvalues of Hamiltonian (10) via $\partial\bar{M}_n(\mu_1,\mu_2)/\partial\mu_1 = 0$, $\partial\bar{M}_n(\mu_1,\mu_2)/\partial\mu_2 = 0$; $\tilde{\mu} = \mu_1\mu_2/(\mu_1+\mu_2)$, and p_r is the radial component of momentum. This Hamiltonian allows to find spin averaged masses and wave functions. Spin-spin interaction can than be taken into account as a perturbation.

Plugging in numbers ($m_u = 0.005$ GeV, $m_d = 0.009$ GeV, $m_s = 0.17$ GeV, $\sigma = 0.18$ GeV^2 and $\alpha_s = 0.3$), we finally get the following chiral shift of reference (quark model) spectra:

pions:

$$\pi(1S) \quad 0.51 \text{ GeV} \rightarrow 0.14 \text{ GeV (exact)}$$
$$\pi(2S) \quad 1.51 \text{ GeV} \rightarrow 1.25 \text{ GeV (exp : 1.3 GeV)}$$
$$\pi(3S) \quad 2.18 \text{ GeV} \rightarrow 1.98 \text{ GeV (exp : 1.8 GeV)}$$

K mesons:

$$K(1S) \quad 0.63 \text{ GeV} \rightarrow 0.49 \text{ GeV (exact)}$$
$$K(2S) \quad 1.57 \text{ GeV} \rightarrow 1.43 \text{ GeV (exp : 1.46 GeV)}$$
$$K(3S) \quad 2.21 \text{ GeV} \rightarrow 2.1 \text{ GeV (exp : 1.83 GeV)}$$

It can be seen, that masses of radial excitations are shifted by less than 15%, and the shifts are small for high excitations. It should be mentioned, that masses of higher excitations and the slope of radial Regge trajectory differ from the experimental. The reason is that Hamiltonian (10) does not take into account effects of string breaking, which are important for highly excited states, since they have large spatial extent. As was shown in [18] the inclusion of string breaking effects does not violate the linearity of radial trajectories, which is in agreement with experimental data.

ACKNOWLEDGMENTS

The authors are grateful to A.M. Badalian for valuable comments and discussions. This work is supported by NSh-1774.2003.2 grant and INTAS-00110 and INTAS-00366 grants.

REFERENCES

1. F. Karsch, arXiv:hep-lat/9903031.
2. J. M. Carmona, M. D'Elia, L. Del Debbio, A. Di Giacomo, B. Lucini and G. Paffuti, Nucl. Phys. Proc. Suppl. **106**, 607 (2002) [arXiv:hep-lat/0110058].
3. Yu. A. Simonov, Phys. Rev. D **65**, 094018 (2002) [arXiv:hep-ph/0201170].
4. Yu. A. Simonov, arXiv:hep-ph/0302090.
5. Yu. A. Simonov, arXiv:hep-ph/0305281.
6. J. Gasser and H. Leutwyler, Annals Phys. **158**, 142 (1984);
    J. Gasser and H. Leutwyler, Nucl. Phys. B **250**, 465 (1985);
    for a review see H. Leutwyler, arXiv:hep-ph/9406283.
7. Yu. A. Simonov, arXiv:hep-ph/9911237.
8. S. V. Ivanov and G. P. Korchemsky, Phys. Lett. B **154**, 197 (1985);
    S. V. Ivanov, G. P. Korchemsky and A. V. Radyushkin, Yad. Fiz. **44**, 230 (1986).
9. V. I. Shevchenko and Yu. A. Simonov, Phys. Lett. B **437**, 146 (1998) [arXiv:hep-th/9807157];
10. V. I. Shevchenko and Yu. A. Simonov, Phys. Rev. Lett. **85**, 1811 (2000) [arXiv:hep-ph/0001299].
11. A. Di Giacomo, H. G. Dosch, V. I. Shevchenko and Yu. A. Simonov, Phys. Rept. **372**, 319 (2002) [arXiv:hep-ph/0007223].
12. Yu. A. Simonov, Few Body Syst. **25**, 45 (1998) [arXiv:hep-ph/9712248].
13. S. M. Fedorov and Y. A. Simonov, arXiv:hep-ph/0306216.
14. Yu. A. Simonov, Phys. Lett. B **226**, 151 (1989).
15. A. Yu. Dubin, A. B. Kaidalov and Yu. A. Simonov, Phys. Lett. B **323**, 41 (1994).
16. A. Yu. Dubin, A. B. Kaidalov and Yu. A. Simonov, Phys. Atom. Nucl. **56**, 1745 (1993) [Yad. Fiz. **56**, 213 (1993)] [arXiv:hep-ph/9311344].
17. Yu. A. Simonov, Phys. Lett. B **515**, 137 (2001) [arXiv:hep-ph/0105141].
18. A. M. Badalian, B. L. Bakker and Yu. A. Simonov, Phys. Rev. D **66**, 034026 (2002) [arXiv:hep-ph/0204088].

The $D_s(2317)$ and $D_s(2463)$ Mesons as Scalar and Axial-Vector Chiralons in the Covariant Level-Classification Scheme

S. Ishida[1]

Research Institute of Quantum Science, College of Science and Technology, Nihon University, Tokyo 101-0062, JAPAN

Abstract. The new narrow mesons observed recently in the final states $D_s^+ \pi^0$ and $D_s^{*+} \pi^0$ are pointed out to be naturally assigned as the ground-state scalar and axial-vector chiralons in the $(c\bar{s})$ system, which would newly appear in the covariant hadron-classification scheme proposed a few years ago.

INTRODUCTION

(Covariant Classification Scheme and Chiral states/Chiralons)   A few years ago we have proposed a covariant level-classification scheme for hadrons, unifying the seemingly contradictory two, non-relativistic and extremely relativistic, viewpoints. (Its essential points are reviewed by our previous talk[1].) Here the framework is manifestly Lorentz-covariant and the space for the static symmetry is extended from that of non-relativistic (NR) scheme to

$$SU(6)_{SF} \quad \otimes \quad \overline{|SU(2)_\rho}| \otimes O(3)_L, \qquad (1)$$

where a new additional $SU(2)$-space for the ρ-spin (ρ- and σ- spin being Pauli-matrices in the decomposition of Dirac γ matrices: $\gamma \equiv \sigma \otimes \rho$) is introduced for covariant description[2] of hadron spin-wave function (WF). The spin WF for the meson systems are generally given by the Bargmann-Wigner (BW) spinors, and are represented as the bi-Dirac spinors $W_\alpha^\beta = u_\alpha \bar{v}^\beta$: $\alpha = (\rho_3, \sigma_3)$, $\beta = (\bar{\rho}_3, \bar{\sigma}_3)$, where $\alpha(\beta)$ denotes the suffices of Dirac spinors of Heavy quarks(Light-anti-quarks) represented by the eigenvalues of ρ-spin and σ-spin. In the HL meson system the states with $(\rho_3, \bar{\rho}_3) = (+,+)$ and $(+,-)$ are expected to be realized in nature, reflecting the physical situation that the HL meson system has the non-relativistic $SU(6)_s$ spin symmetry (the relativistic chiral symmetry) concerning the constituent Heavy quarks (Light quarks). Our relevant $D_s(2317)/D_s(2463)$ mesons are naturally assigned as the scalar/axial-vector chiralons in the $(c\bar{s})$ ground states with $(\rho_3, \bar{\rho}_3) = (+,-)$, and play the role of chiral partners of the already established Paulons D_s/D_s^*, the states with $(\rho_3, \bar{\rho}_3) = (+,+)$.

---

[1] Representing the collaboration group with M. Ishida, T. Komada, T. Maeda, M. Oda, K. Yamada and I. Yamauchi .

[2] It is to be noted that in our scheme the squared-mass spectra are globally $\tilde{U}(12)$-symmetric[2] and the mass spectra themselves are able to be reconciled with the broken chiral symmetry.

CP717, *Hadron Spectroscopy: Tenth International Conference,*
edited by E. Klempt, H. Koch, and H. Orth
© 2004 American Institute of Physics 0-7354-0197-7/04/$22.00

(Description of HL Mesons) The WF of HL mesons are described as

$$\Phi_A^{\;B}(x,y) \sim \psi_{Q,A}(x)\bar\psi^{q,B}(y) \tag{2}$$

$$A = (\alpha,a),\ B = (\beta,b);\qquad \alpha,\beta = (1\sim 4);\ a = (c\ \text{or}\ b),\ b = (u,d,s),$$

and are assumed to satisfy the master Klein-Gordon equation of Yukawa-type[3]. The squared-mass operator is assumed to contain no light-quark Dirac matrices $\gamma^{(q)}$ in the ideal limit, leading to the chiral symmetric global structure of squared-mass spectra. The WF is separated into the two parts, the one of plane-wave center of mass motion and the other of internal WF: The internal WF with definite total spin J is expanded in terms of respective eigen functions $W(P)$ on spinor-space and $O(P_N,r)$ on internal space-time, where $W_\alpha^\beta(P_N)$ and $O(P_N,r)$ are covariant tensors respectively, in the $\tilde U(4)_{D.S.}$ (pseudo-unitary Dirac spinor) space and the $O(3,1)_L$ (Lorentz-space).

Spin WF/BW-spinors   As the complete set of spinor-space eigen-functions we choose the BW spinors, which are defined as solutions of the (local) Klein-Gordon equations. For the HL-mesons we have the two physical solutions:

$$U_\alpha^{\;\beta}(P) \equiv u_\alpha^{(Q)}(P)\bar v_{(\bar q)}^\beta(P);\ C_\alpha^{\;\beta}(P) \equiv u_\alpha^{(Q)}(P)\bar v_{(\bar q)}^\beta(-P), \tag{3}$$

As is evidently seen from Eq. (3), through the chiral transformation on light anti-quarks $\bar v(P)\gamma_5 = \bar v(-P)$, the former is changed into the latter as $U(P)\gamma_5 = C(P)$. They are decomposed into the pseudo-scalars/vectors, and scalars/axial-vectors, respectively, as

$$U_\alpha^{\;\beta}(v) = 1/2\sqrt2\,(1 - iv\cdot\gamma)\left[i\gamma_5 P_s(P) + i\tilde\gamma_\mu V_\mu(P)\right],$$

$$C_\alpha^{\;\beta}(v) = 1/2\sqrt2\,(1 - iv\cdot\gamma)\left[S(P) + i\gamma_5\tilde\gamma_\mu A_\mu(P)\right],\ (P_\mu\tilde\gamma_\mu = 0,\ v_\mu \equiv P_\mu/M). \tag{4}$$

Internal space-time WF/Yukawa oscillators   As the complete set of space-time eigen-functions we choose the covariant, 4-dimensional Yukawa oscillator functions. By imposing the freezing relative-time condition they become effectively the conventional, 3-dimensional oscillators:

$$\langle P_\mu r_\mu\rangle = \langle P_\mu p_\mu\rangle = 0 \ \Rightarrow\ O(3,1)_L \approx O(3)_L\,. \tag{5}$$

MASS SPECTRA FOR LOW-LYING D AND D_S-MESONS

Since of the static symmetry (1) the global mass spectra are given by

$$M_N^2 = M_0^2 + N\Omega,\ N \equiv 2n + L, \tag{6}$$

leading to phenomenologically well-known Regge trajectories. The masses of ground state mesons, P_s, V_μ, S and A_μ are degenerate in the ideal limit, and they are split with each others between chiral partners (spin partners) by the bilinear scalar-quark condensates (the perturbative QCD spin-spin interaction) as

$$M_0(0^-/1^-) \lesssim M_0(0^+/1^+) < M_1(L = 1)\,. \tag{7}$$

As for the splittings between chiral partners we obtain, applying the linear SU(3) σ-model, the universal relation[4] as

$$\Delta M^\chi = M_0(0^+) - M_0(0^-) = M_0(1^+) - M_0(1^-) \tag{8}$$

within the same quark-configuration mesons; and the relation of those between the different light-quark configurations as $\Delta M^\chi(c\bar n)/\Delta M^\chi(c\bar s) = a/b$, where the (a,b) is defined by $\langle\sigma^i\lambda^i/\sqrt2\rangle \equiv \text{diag}\{a,a,b\}$, and $a \equiv f_\pi/\sqrt2$, $(a+b)/2 \equiv f_K/\sqrt2$ ($f_\pi(f_K)$ being the decay constant of $\pi(K)$) . Through the experimental values of $\Delta M^\chi(c\bar s) = 350\text{MeV}/c^2$ and $a/b = 1/1.44$, we predict $\Delta M^\chi(c\bar n) = 240\text{MeV}/c^2$. In Fig. 1 (a) and (b) we show, respectively, the low-lying D meson and D_s meson mass spectra, presently known and/or predicted through the above relations.

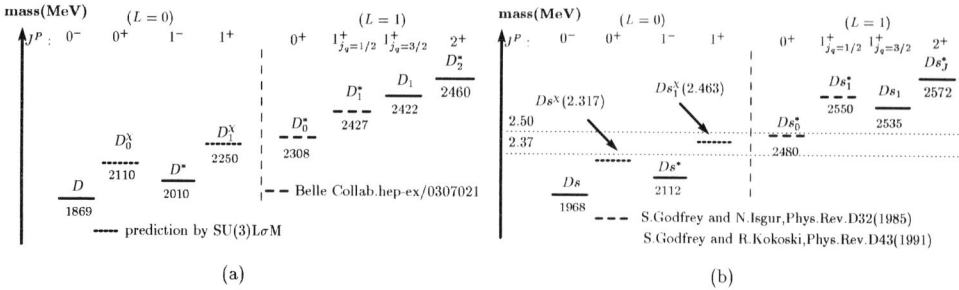

FIGURE 1. low-lying (a) D meson and (b) D_s meson mass spectra

DECAY PROPERTIES OF D_S-MESONS

The observed properties of D_s-mesons to be examined are as follows:

$$D_s(0^+;2.32) \rightarrow D_s(0^-;1.97)+\pi^0 \quad \text{observed}[5],$$

$$D_s(1^+;2.46) \rightarrow D_s(1^-;2.11)+\pi^0 \quad \text{observed}[6]. \tag{9}$$

$$R(0^+) \equiv \frac{Br(D^\chi_{s,0}(2.32) \rightarrow D_s^*\gamma)}{Br(D^\chi_{s,0}(2.32) \rightarrow D_s\pi^0)} < 0.078 \quad (\text{CLEO}[6]). \tag{10}$$

$$R(1^+) \equiv \frac{Br(D^\chi_{s,1}(2.46) \rightarrow D_s\gamma)}{Br(D^\chi_{s,1}(2.46) \rightarrow D_s^*\pi^0)} = 0.47 \pm 0.10 \quad (\text{Belle}[7]). \tag{11}$$

$$\Gamma_T[D_s(0^+;\ 2.32\ )],\ \Gamma_T[D_s(1^+;2.46)] < 7\text{MeV}. \tag{12}$$

(Pionic transitions) The observed processes (9) are iso-spin violating and considered to occur by the mixing of intermediate η meson with π-meson. From this picture and Eq. (8) we get the relation[4, 9],

$$\Gamma(D_s^+(0^+) \rightarrow D_s(0^-)\pi^0) = \Gamma(D_s^+(1^+) \rightarrow D_s(1^-)\pi^0), \tag{13}$$

which is consistent with the property (12). We can estimate phenomenologically the value of mixing parameter $\sin\theta$, by using the experimental branching ratio[8] of $D_n(c\bar{n})$ meson to the iso-spin violating decay channel as

$$(\sin\theta)^2_{\text{exp}} \approx \frac{Br(D_s^{*+} \rightarrow D_s^+\pi^0)(M_{D_s^*}^2/q^3)}{Br(D_s^{*+} \rightarrow D_s^+\gamma)} \bigg/ \frac{Br(D^{*+} \rightarrow D^+\pi^0)(M_{D^*}^2/q^3)}{Br(D^{*+} \rightarrow D^+\gamma)}$$

$$= (0.9 \pm 0.4) \cdot 10^{-3}. \tag{14}$$

This seems to be of reasonable order of magnitude as due to the virtual EM-interaction. In order to estimate the absolute magnitude of the width (13) in relation with those of the other HL-mesons, we shall set up the chiral symmetric interaction Lagrangian, applying the linear σ model in the framework of covariant oscillator quark model (COQM)[10], as

$$\mathcal{S}_Y^I = \int d^4x_1 d^4x_2 \mathcal{L}(x_1,x_2) \equiv \int d^4X \mathcal{L}_I(X),\ \mathcal{L} = \mathcal{L}^{ND} + \mathcal{L}^{AX} \tag{15}$$

$$\mathcal{L}^{ND} = g_{ND}\langle \Phi(x_1,x_2)M(x_2)\bar{\Phi}(x_1,x_2)\rangle.$$

$$\mathcal{L}^{AX} = g_{AX}\langle \Phi(x_1,x_2)(\partial_{2,\mu}^+ + i\sigma_{\mu\nu}\partial_{2,\nu}^-)[\partial_{2,\mu}M(x_2)]\bar{\Phi}(x_1,x_2)\rangle.$$

$$M \equiv s - i\gamma_5\phi \quad (s \equiv s^a\lambda^a/\sqrt{2},\ \phi \equiv \phi^a\lambda^a/\sqrt{2}),$$

$$\Phi \propto (1 - iv \cdot \gamma)(i\gamma_5 D + i\tilde{\gamma}_v D_v^* + D_0^\chi + i\gamma_5 \tilde{\gamma}_v D_v^{*\chi}),$$

$$\bar{\Phi} \equiv \gamma_4 \Phi^\dagger \gamma_4, \quad D = (D^0, D^+, D_s^+) \text{ etc. },\tag{16}$$

where only the Yukawa interaction of the scalar (s) and pseudo-scalar (ϕ) nonets with the light quarks in the HL-meson is taken into account. The interaction (15) consists of the two terms:

Firstly the g_{ND} term concerns with the mass-splitting between chiral partners, and gives dominant (compared to the g_{AX} term) contribution to the (quark-) spin non-flip processes. Accordingly, by fixing the coupling parameter g_{ND} from the experimental value of $\Delta M^\chi(c\bar{s}) = 350\text{MeV}$, we can predict the absolute values of the relevant pionic decay widths as

$$\Gamma(D_{n,0}^\chi \to D_n \pi) = \Gamma(D_{n,1}^\chi \to D_n^* \pi) = 133\text{MeV},\tag{17}$$

$$\Gamma(D_{s,0}^\chi \to D_s \pi^0) = \Gamma(D_{s,1}^\chi \to D_s^* \pi^0) = 122 \pm 54\text{keV},\tag{18}$$

where, in deriving Eq. (18), the estimated value Eq. (14) is used. The value of width (18) is consistent with the experiment (12).

Secondly the g_{AX} term in the interaction (15) corresponds to the extended, PCAC term, and concerns dominantly (compared to the g_{ND} term) to the spin-flip processes. Accordingly we can determine the coupling parameter g_{AX} by fitting the experimental decay width $\Gamma(D^{*+} \to D^0 \pi^+) = (96 \pm 23) \times 0.68\text{keV}$, and predict the π-(and/or σ-)mesonic decay widths of the other HL-mesons.

(Radiative decay) In order to treat systematically all the radiative transitions between the HL-mesons we shall set up the basic EM-interaction Lagrangian in the framework of COQM, as

$$\mathscr{S}_I^{EM} = \int d^4 x_1 d^4 x_2 \sum_{i=1,2} j_{i,\mu}(x_1, x_2) A_\mu(x_i) = \int d^4 X \sum_i J_{i,\mu}(X) A_\mu(X),$$

$$j_{i,\mu}(x_1, x_2) = -ie_i\left((m_1 + m_2)/m_i\right)\langle \bar{\Phi}_U(\partial_{i\mu}^- + ig_M \sigma_{\mu\nu}^{(i)} \partial_{i,\nu}^+)\Phi_U\rangle,\tag{19}$$

$$\Phi_U \equiv \Phi_U(-i\overleftarrow{v} \cdot \gamma), \quad \bar{\Phi}_U \equiv \bar{\Phi}_U(-i\overleftarrow{v} \cdot \gamma),$$

where Φ_U is the unitary correspondent of Φ, so defined as $\langle \bar{\Phi}_U \Phi_U\rangle \to \langle \Phi^\dagger \Phi\rangle$ at the rest frame. Here it is to be noted that our effective current $J_{i,\mu}(X)$ is obtained through the "minimal substitution" of $(\partial_{i,\mu} \to \partial_{i,\mu} - ie_i A_\mu(x_i))$, and accordingly it is conserved in the ideal limit.

Our effective current has also another remarkable feature due to the covariant nature of our scheme. The spin-current interaction (the second term in Eq. (19)) leads to the Hamiltonian

$$\mathscr{H}^{(i)\,spin} \equiv J_\mu^{(i)spin} A_\mu = \mu^{(i)}\sigma^{(i)} \cdot B + d^{(i)}\rho_1^{(i)}\sigma^{(i)} \cdot E, \quad \mu^{(i)} = d^{(i)} = e_i/2m_i.\tag{20}$$

This shows that our Hamiltonian contains the interaction through the "intrinsic electric dipole" $d\rho_1\sigma$ as well as the one through the magnetic dipole $\mu\sigma$. The "intrinsic dipole" gives contributions only for the transitions between chiralons and Paulons, while does none for the other transitions.

From the effective currents $J_{i\mu}$ in Eq. (19) we can calculate the relevant decay widths. In Table 1 we have given the predicted widths for all the radiative spin-flip transitions between ground state D_s mesons, and, for reference, the width for the transition of D_n meson, $D_n^{*,+}(1^-) \to D_n^+(0^-)\gamma$, in comparison with experiments. There, we have also shown the predicted values by the other chiral model.

TABLE 1. γ-decay widths for spin-flip transitions between the ground state D_s mesons. ($g_{c,n} = 1$, $m_c = M_\psi/2$, $m_{n(s)} = M_\rho(M_\phi)/2$).

	Processes	$P/\chi \to P/\chi$	Γ(keV) ours	Γ(keV) others [11]
(a)	$D_s(1^-) \to D_s(0^-)\gamma$	$P \to P$	0.33	0.43
(b)	$D_s(1^+) \to D_s(0^+)\gamma$	$\chi \to \chi$	0.26	0.43
(c)	$D_s(0^+) \to D_s(1^-)\gamma$	$\chi \to P$	22	1.74
(d)	$D_s(1^+) \to D_s(0^-)\gamma$	$\chi \to P$	82	5.08
	$D_n^{*,+}(1^-) \to D_n^+(0^-)\gamma$	$P \to P$	1.93(theor) $\leftrightarrow$ 1.54 $\pm$ 0.53(exp)	

From the results in Table 1 we see that our model gives the much larger widths for the transitions, (c) and (d), from chiralons to Paulons, compared to the other chiral model (, reflecting the above mentioned feature (20) of our currents,) while does the width of almost the same amount for transitions, (a) (and (b)), from Paulons(chiralons) to Paulons(chiralons). This difference is considered to come from the different identification of the relevant mesons in the two cases: The narrow D_s mesons are assigned as the conventional P-wave excited states in the other model, while they are the S-wave chiral states other than the P-wave Pauli-states in our scheme.

(Branching ratios between radiative to pionic decay widths)   From the predicted values of pionic (Eq. (10)) and radiative (Table 1) decay widths we obtain the ratios between them as follows:

$$R(0^+) \;=\; 0.18 \pm 0.10, \quad R(1^+) = 0.67 \pm 0.40, \tag{21}$$

which seems to be consistent with the experiments Eqs. (10) and (11).

CONCLUDING REMARKS

o The $D_s(2317)$ and $D_s(2463)$ mesons are shown consistently assigned as the chiralons with $J^P = 0^+$ and 1^+ in the $(c\bar{s})$ ground states.

o The decay width of $(D_n^\chi(1^+) \to D_n^*(1^-)\pi)$ is predicted as $\Gamma \simeq 130$MeV, and the radiative decay widths of chiral states into Pauli-states are predicted to be remarkably larger than those estimated in other works. These are to be checked experimentally.

o Further experimental search for chiralons are desirable.

REFERENCES

1. M. Ishida, in this proceedings.
2. S. Ishida and M. Ishida, Phys. Lett. **B539** (2002), 249.
   S. Ishida, M. Ishida and T. Maeda, Prog. Theor. Phys. **104** (2000), 785.
3. H. Yukawa, Phys. Rev. **91** (1953), 415, 416.
4. W. A. Bardeen and C. T. Hill, Phys. Rev. **D49** (1994), 409.
5. BABAR Collaboration, hep-ex/0304021 v1.
6. CLEO Collaboration, hep-ex/0305017 v1.
7. P. Krokovny, in this proceedings.
8. K. Hagiwara et al, Phys. Rev. **D66** (2002), 010001.
9. M. Ishida and S. Ishida, Prog. Theor. Phys. **106** (2001), 373.
10. S. Ishida, K. Yamada and M. Oda, Phys. Rev. **D40** (1989), 1497.
11. W. A. Bardeen, E. J. Eichten and C. T. Hill, hep-ph/0305049 v1.

Scalar Mesons and Chiral States

M. Ishida* and S. Ishida†

*Department of Physics, Meisei University, Hino 191-8506, JAPAN
†Research Institute of Quantum Science, College of Science and Technology, Nihon University,
Tokyo 101-0062, JAPAN

Abstract. The essential points and physical backgrounds of the covariant level-classification scheme, based on $\tilde{U}(12)_{SF} \otimes O(3,1)_L$, are reviewed: This scheme is extended from the non-relativistic $SU(6)_{SF} \otimes O(3)_L$ scheme by introducing the new $SU(2)$-spin (ρ-spin) degree of freedom, which is necessary for covariant description of composite hadrons. Our scheme predicts the existence of new type of chiral mesons and baryons (Chiralons) out of the conventional $SU(6)_{SF} \otimes O(3)_L$ scheme. The σ nonet is a typical example of chiralons to be assigned to the $(q\bar{q})$ relativistic S-wave state. The new narrow mesons $D_s(2317)/D_s(2463)$ are naturally assigned as the ground-state scalar and axial-vector chiralons in the $(c\bar{s})$ system.

INTRODUCTION

(Difficulty of Non-Relativistic Classification Scheme)   The non-relativistic (NR) hadron level-classification scheme with the approximate static-symmetry

$$SU(6)_{SF} \quad \otimes \quad O(3)_L \tag{1}$$

had been successful for these 4 decades, but recently the necessity for covariant classification has been strengthened: Theoretically the QCD, basic dynamics underlying the hadron physics, has the chiral symmetry, "maximally" relativistic symmetry. Phenomenologically the property of π-meson as a Nambu-Goldstone boson in the case of broken chiral symmetry has been wellknown. Moreover, the existence of light σ-meson as chiral partner of π-meson, $\sigma(600)$, which has been a controversial problem for many years, seems now to be confirmed. However, there is no suitable seat prepared for it in the NR scheme.

(Covariant Classification Scheme and Chiral states/Chiralons)   Correspondingly to this situation, a few years ago we have proposed[1, 2] a covariant level-classification scheme for hadrons, unifying the seemingly contradictory two, non-relativistic and extremely relativistic, viewpoints. Here the framework is manifestly Lorentz-covariant and the space for the static symmetry is extended from Eq. (1) to that of

$$SU(6)_{SF} \quad \otimes \quad \overline{SU(2)_\rho} \otimes O(3)_L, \tag{2}$$

where a new additional $SU(2)$-space for the ρ-spin (ρ- and σ- spin being 2 by 2 Pauli-matrices corresponding to the conventional decomposition of 4 by 4 Dirac γ matrices: $\gamma \equiv \sigma \otimes \rho$) is introduced for covariant description of hadron spin-wave function

CP717, *Hadron Spectroscopy: Tenth International Conference,*
edited by E. Klempt, H. Koch, and H. Orth
© 2004 American Institute of Physics 0-7354-0197-7/04/$22.00

(WF). As a result, in our scheme the squared-mass spectra become globally $\tilde{U}(12)$-symmetric[1, 3] and the mass spectra themselves are able to be reconciled with the broken chiral symmetry. The spin WF for meson and baryon systems are given by the Bargmann-Wigner (BW) spinors,[2] which are represented, respectively, as the bi-Dirac and tri-Dirac spinors

$$
\begin{aligned}
\text{mesons} \quad W_\alpha^\beta &= u_\alpha \bar{v}^\beta & &: \alpha = (\rho_3, \sigma_3),\ \beta = (\bar{\rho}_3, \bar{\sigma}_3), \\
\text{baryons} \quad W_{\alpha_1 \alpha_2 \alpha_3} &= u_{\alpha_1} u_{\alpha_2} u_{\alpha_3} & &: \alpha_i = (\rho_3^{(i)}, \sigma_3^{(i)}),
\end{aligned}
\tag{3}
$$

where $(\alpha, \beta, \alpha_i)$ denotes the suffices of Dirac spinors of (quark,anti-quark,i-th quark) represented by the eigenvalues of ρ-spin and σ-spin.

In light-quark(LL) meson systems the states with the combination of $(\rho_3, \bar{\rho}_3)$

$$
(+,+) \; ; \; (\text{Pauli} - \text{states}), \qquad (+,-), (-,+), (-,-) \; ; \; (\text{Chiral states}), \tag{4}
$$

are expected to be realized in nature, while in the heavy-light quark (HL) meson systems only the states with $(\rho_3, \bar{\rho}_3) = (+,+)$:(Paulons) and $(+,-)$:(Chiralons) are expected, reflecting the physical situation that the HL meson system has the non-relativistic $SU(6)_{SF}$ spin symmetry (the relativistic chiral symmetry) concerning the constituent Heavy quarks (Light quarks). The Pauli-states/Paulons Eq. (4) are also describable in NR scheme, while the chiral states/Chiralons Eq. (4) have appeared first in the covariant scheme and are out of the NR description. Similarly in the light-quark baryon systems, the states with $(\rho_3^{(1)}, \rho_3^{(2)}, \rho_3^{(3)}) = (+,+,+)$:(Paulons) and $(+,+,-), (+,-,-)$:(Chiralons) are expected to exist.

DESCRIPTION AND LEVEL STRUCTURE OF HADRONS

(Wave function and Wave Equation) The WF of mesons $\Phi_A{}^B$ and baryons $\Phi_{A_1 A_2 A_3}$ are described systematically as

$$
\begin{aligned}
\Phi_A{}^B(x,y) &\sim \psi_{q,A}(x)\bar{\psi}^{q,B}(y), \quad \Phi_{A_1 A_2 A_3} \sim \psi_{q_1,A_1}(x_1)\psi_{q_2,A_2}(x_2)\psi_{q_3,A_3}(x_3), \\
A &= (\alpha,a),\ B = (\beta,b); \quad \alpha,\beta = (1 \sim 4),\ a,b = (u,d,s,c,b),
\end{aligned}
\tag{5}
$$

and are assumed to satisfy the Klein-Gordon(KG) equation of Yukawa-type[4]

$$
\left[(\partial/\partial X_\mu)^2 - \mathscr{M}^2(r_\mu, \partial/\partial r_\mu \cdots ; \partial/i\partial X_\mu) \right] \Phi(X, r\cdots) = 0. \tag{6}
$$

The operator $\mathscr{M}^2$ is assumed to contain no light-quark Dirac matrices $\gamma^{(q)}$ in the ideal limit, leading to the chiral symmetric global structure of squared-mass spectra. The WF is separated into the two parts, the one of plane-wave center of mass motion and the other of internal WF, as

$$
\Phi(X, r\cdots) = \sum_N \sum_{\mathbf{P}_N, P_{N,0} > 0} \left[e^{iP_N \cdot X} \psi_N^{(+)}(P_N, r\cdots) + e^{-iP_N \cdot X} \psi_N^{(-)}(P_N, r\cdots) \right]. \tag{7}
$$

(Expansion of WF on [Spinor⊗Space-time] eigen-functions)   The internal WF with definite total spin J is expanded in terms of respective eigen functions on spinor-space and on internal space-time as

$$\psi_{J,\alpha\cdots}{}^{\beta\cdots}(P_N, r\cdots) = \sum_{i,j} c_{ij}^J W_{\alpha\cdots}^{(i)\beta\cdots}(P_N) O^{(j)}(P_N, r\cdots), \tag{8}$$

where $W_{\alpha\cdots}^{(i)\beta\cdots}(P_N)$ and $O^{(j)}(P_N, r\cdots)$ are covariant tensors respectively, in the $\tilde{U}(4)_{D.S.}$ (pseudo-unitary Dirac spinor) space and the $O(3,1)_L$ (Lorentz-space).

Spin WF/BW-spinors   As the complete set of spinor-space eigen-functions we choose the BW spinors, defined as solutions of the (local) KG equations

$$\left[(\partial/\partial X_\mu)^2 - M^2\right] W_{\alpha\cdots}^{\beta\cdots}(X) = 0 \xrightarrow{\text{Moment. Repr.}} \left(P_\mu^2 + M^2\right) W_{\alpha\cdots}^{(\pm)\beta\cdots}(P) = 0 \quad . \tag{9}$$

For the LL-mesons we have the four physical solutions:[2]

$$U_\alpha{}^\beta(P) \equiv u_\alpha^{(q)}(P)\bar{v}_{(\bar{q})}^\beta(P); \quad C_\alpha{}^\beta(P) \equiv u_\alpha^{(q)}(P)\bar{v}_{(\bar{q})}^\beta(-P), \tag{10}$$

$$D_\alpha{}^\beta(P) \equiv u_\alpha^{(q)}(-P)\bar{v}_{(\bar{q})}^\beta(P); \quad V_\alpha{}^\beta(P) \equiv u_\alpha^{(q)}(-P)\bar{v}_{(\bar{q})}^\beta(-P), \tag{11}$$

where U describes the Paulons with $(\rho_3, \bar{\rho}_3) = (+,+)$, while C, D and V describe the Chiralons with $(\rho_3, \bar{\rho}_3) = (+,-)$, $(-,+)$ and $(-,-)$, respectively. The U,V have $J^P = 0^-, 1^-$, while C,D have $J^P = 0^+, 1^+$. The eigen-functions of charge conjugation parity are obtained through the superpositions, $W = U \pm V$ and $C \pm D$. Their explicit forms are given by

$$\begin{array}{ccccccccc}
W_\alpha{}^\beta(v): & i\gamma_5 & i\tilde{\gamma}_\mu & -\gamma_5 v\cdot\gamma & -i\sigma_{\mu\nu}v_\nu & 1 & i\gamma_5\tilde{\gamma}_\mu & -v\cdot\gamma & -\gamma_5\sigma_{\mu\nu}v_\nu \\
J^{PC}: & 0^{-+} & 1^{--} & 0^{-+} & 1^{--} & 0^{++} & 1^{++} & 0^{+-} & 1^{+-} \\
\phi: & P_s^{(N)} & V_\mu^{(N)} & P_s^{(E)} & V_\mu^{(E)} & S^{(N)} & A_\mu^{(N)} & S^{(E)} & A_\mu^{(E)}
\end{array} \tag{12}$$

The above $W_\alpha{}^\beta(v)$ are the tensors of $\tilde{U}(4)(\supset SU(2)_\rho \otimes SU(2)_\sigma)$ symmetry, and correspond to all the 16 Dirac γ-matrices. By including the flavor SU(3), the LL mesons are classified as $\underline{144}(= 12 \times 12^*)$ representation of $\tilde{U}(12)_{SF}(\supset SU(3)_F \otimes SU(2)_\rho \otimes SU(2)_\sigma)$.[1, 3] In terms of static $SU(6)_{SF}(\supset SU(3)_F \otimes SU(2)_\sigma)$, the $\underline{144}$ includes four $36(= 6 \times 6^*)$.

As is evidently seen from Eq. (12) and from the chiral transformation on light quarks reducing $\gamma_5 u(P) = u(-P)$ and $\bar{v}(P)\gamma_5 = \bar{v}(-P)$, the $(P_s^{(R)}, S^R)$ and $(V_\mu^{(R)}, A_\mu^{(R)})$ ($R = N, E$) form linear representations of chiral symmetry.

For the HL-mesons we have the two physical solutions, U and C, given in Eq. (10). They are decomposed into the pseudo-scalars/vectors, and scalars/axial-vectors, respectively, as

$$U_\alpha{}^\beta \sim (1 - iv\cdot\gamma)\left[i\gamma_5 P_s + i\tilde{\gamma}_\mu V_\mu\right], \quad C_\alpha{}^\beta \sim (1 - iv\cdot\gamma)\left[S + i\gamma_5\tilde{\gamma}_\mu A_\mu\right]. \tag{13}$$

The U and C form $\underline{12}^*$ of $\tilde{U}(12)_{SF}$ symmetry. Through the chiral transformation, the former is changed into the latter as $U(P)\gamma_5 = C(P)$.

The light quark ground state baryons are assigned as $(12 \times 12 \times 12)_{Sym} = 364$ representation of $\tilde{U}(12)_{SF}$ symmetry. The $\underline{364}$ includes the baryons $(\underline{182_B})$ and anti-baryons $(\underline{182_{\bar{B}}})$. In terms of static $SU(6)_{SF}$, the $\underline{182_B}$ includes the $\underline{56_E}$(:Paulons) and $\underline{70_G}$, $\underline{56_F}$ (:Chiralons) with $(\rho_3, \rho_3, \rho_3) = (+,+,+)$ and $(+,+,-), (+,-,-)$, respectively. Both $\underline{56_E}$ and $\underline{56_F}$ include N-octets and Δ-decouplets with positive parity, while the negative parity $\underline{70_G}$ include N-octets with $J = 1/2$ and $3/2$, Δ-decouplet with $J = 1/2$ and Λ-singlet with $J = 1/2$.

Internal space-time WF/Yukawa oscillators   As the complete set of space-time eigen-functions we choose the covariant, 4-dimensional Yukawa oscillator functions. By imposing the freezing relative-time condition they become effectively the conventional, 3-dimensional oscillators:

$$\langle P_\mu r_\mu \rangle = \langle P_\mu p_\mu \rangle = 0 \quad \Rightarrow \quad O(3,1)_L \approx O(3)_L . \tag{14}$$

(Mass spectra for low-lying mesons and baryons)   Since of the static symmetry (2) the global mass spectra are given by

$$M_N^2 = M_0^2 + N\Omega, \ N \equiv 2n + L, \tag{15}$$

leading to phenomenologically well-known Regge trajectories.

The masses of ground state mesons and baryons are degenerate in the ideal limit, and they split with each others between chiral partners (spin partners) by the bilinear scalar-quark condensates (the perturbative QCD spin-spin interaction).

CANDIDATES FOR CHIRALONS/CONCLUDING REMARKS

(Experimental candidates of chiral particles)   In our level-classification scheme a series of new type of multiplets, chiralons, are predicted to exist in the ground and the first excited states. Presently we can give only a few experimental candidates or indications for them:

(LL-mesons)   One of the most important candidates is the scalar σ nonet to be assigned as $S^{(N)}(^1S_0)$: $[\sigma(600), \kappa(900), a_0(980), f_0(980)]$. The existence of $\sigma(600)$ seems to be established[5] through the analyses of, especially, $\pi\pi$-production processes. The firm experimental evidences for $\kappa(800\text{-}900)$ were reported in the production processes[7][8]. The properties of κ are consistent with those given formerly in $K\pi$ scattering phase shift[6].

In our scheme the two sets of P_s- and of V_μ-nonets are to exist: The vector mesons[9] $\rho(1250)$ and $\omega(1250)$, suspected to exist for long time, are naturally able to be assigned as the members of $V_\mu^{(E)}(^3S_1)$-nonet.

Out of the three established η, $[\eta(1295), \eta(1420), \eta(1460)]$ at least one extra, plausibly $\eta(1295)$ with the lowest mass, may belong to $P_s^{(E)}(^1S_0)$ nonet.

The two "exotic" particles $\pi_1(1400)$ and $\pi_1(1600)$ with $J^{PC} = 1^{-+}$ and $I = 1$, observed[10] in the $\pi\eta$, $\rho\pi$ and other channels, may be naturally assigned as the first excited states $S^{(E)}(^1P_1)$ and $A_\mu^{(E)}(^3P_1)$ of the chiralons.

(HL-mesons) The new narrow mesons $D_s(2317)/D_s(2463)$, observed recently in the final states $D_s^+\pi^0/D_s^{*+}\pi^0$, are pointed out[11] to be naturally assigned as the ground-state scalar and axial-vector chiralons in the $(c\bar{s})$ system. Their decay properties are well-explained in our scheme. The scalar/axial-vector chiralons are also expected to exist in $c\bar{n}$ system, denoted as D_{n0}^χ/D_{n1}^χ. Their masses are predicted around 2110/2250 MeV by using SU(3) linear σ model. Some results in search[12] for their existence are reported.

(qqq-baryons)　The two facts have been a longstanding problem that the Roper resonance $N(1440)_{1/2^+}$ is too light to be assigned as radial excitation of $N(939)$ and that $\Lambda(1405)_{1/2^-}$ is too light as the $L=1$ excited state of $\Lambda(1116)$. In our new scheme these two problems dissolve[13] in principle, because in the ground states there exist the two $\underline{56}$ with positive parity, $56_E^{\oplus}$ and $56_F^{\oplus}$, and one negative-parity 70_G. The other puzzle, that the predicted width $\Gamma(\Delta \to N\gamma)$ in the conventional treatment is much small compared with the experiment, may also be solved by considering the relativistic effect of the mixing between 56_E and 56_F. The other problem of extremely small width of $\Delta(1600) \to N\gamma$ is explained by the orthogonality between WFs of $\Delta(1600)$ and of $\Delta(1232)$ by assuming them to be ground states.[13]

(*Concluding remarks*)　We have summarized in this talk the essential points of the covariant level-classification scheme, which has, we believe, a possibility to solve the serious problem in hadron spectroscopy mentioned in Introduction. In this connection further investigations, both experimental and theoretical, for chiral states predicted in this scheme, are urgently required for new development of hadron physics.

REFERENCES

1. S. Ishida and M. Ishida, Phys. Lett. B**539** (2002) 249.
2. S. Ishida, M. Ishida and T. Maeda, Prog. Theor. Phys. **104** (2000), 785.
3. A. Salam, R. Delbourgo and J. Strathdee, Proc. R. Soc. London A**284** (1965) 146.
   B. Sakita and K. C. Wali, Phys. Rev. B**139** (1965) 1335.
4. H. Yukawa, Phys. Rev. **91** (1953), 415, 416.
5. N. A. Törnqvist, summary talk in proceedings of "σ-Meson 2000", KEK-proceedings 2000-4; NUP-B-2000-1; Soryusiron kenkyu **102** (2001) No.5.
6. S. Ishida et al., Prog. Theor. Phys. **98** (1997), 621.
   E. Beveren et al., Z. Phys. C**30** (1986), 651.
7. T.Komda in this proceedings. Wu Ning, hep-ex/0304001.
8. C. Gobel, in proc. of Nihon univ. and KEK symp., Ichigaya, Tokyo, Feb 24-26, 2003.
9. M. Oda, in this proceedings.
10. S. U. Chung, summary talk of Hadron99. V. Dorofeev; A. Popov, proc. of Hadron2001.
11. S. Ishida, in this proceedings. M. Ishida and S. Ishida, Prog. Theor. Phys. **106** (2001), 373.
12. I. Yamauchi, in this proceedings.
13. M. Ishida, in proc. of Nihon univ. and KEK symposium, 2003.

Chiral symmetry restoration in excited hadrons.

L. Ya. Glozman

Institute for Theoretical Physics, University of Graz, Universitätsplatz 5, A-8010 Graz, Austria

Abstract. The evidence, theoretical justification and implications of chiral symmetry restoration in excited hadrons are presented.

Different 3Q potential constituent quark models [1, 2, 3] predict a lot of baryon states at the mass region 2 GeV, which are not observed. A possible discovery of the strange pentaquark suggests that there should be in addition high-lying 5Q (and with higher amount of quarks) states that belong to different multiplets. Both constituent quark models as well as different pentaquark models rely crucially on spontaneous breaking of chiral symmetry (the very notion of constituent quark as a quasiparticle is related to chiral symmetry breaking in the QCD vacuum). A question then arises why all these states are not seen? Is it related to technical limitations to see them or it perhaps indicates that the physics of the high-lying hadrons is different as compared to the low-lying ones? Indeed, if one looks carefully at the nucleon excited states, see Fig. 1, one immediately notices regularities for high-lying states that definitely absent for the low-lying states. In particular, the nucleon (and delta) high-lying spectra show obvious patterns of parity doubling: starting from the 1.7 GeV region excited nucleons of the same spin but opposite parity are approximately degenerate. Is it accidental? If not, some symmetry should be behind this parity doubling. Then we have to understand why this symmetry is definitely absent for the low-lying states and persists only higher in the spectrum.

It has been suggested some time ago that this parity doubling reflects effective chiral symmetry restoration [4]. We know that the QCD Lagrangian posseses almost perfect $SU(2)_L \times SU(2)_R$ chiral symmetry, which is a symmetry of two independent (left and right) rotations of u and d quarks in the isospin space. If this symmetry of the QCD Lagrangian was intact in the QCD vacuum, then all hadrons would fall into representations of the parity-chiral group $SU(2)_L \times SU(2)_R \times C_i$, where the group C_i consists of elements identity and space inversion [6]. For baryons these multiplets are either parity doublets in N and Δ spectra, not related to each other, or quartets that contain degenerate nucleon and delta doublets. Either of these degeneracies should be fulfilled for any spin of baryons, because chiral symmetry of QCD does not constrain spin. From the low-lying nucleon and delta spectra we can conclude that there are no degeneracies of states of the same spin but opposite parity (even more, there is no one-to-one mapping of states of opposite parity with the same spin). It is this fact which was historically one of the most important arguments to conclude that chiral symmetry must be spontaneously broken. This spontaneous breaking leads to the appearance of quasiparticles (constituent

CP717, Hadron Spectroscopy: Tenth International Conference,
edited by E. Klempt, H. Koch, and H. Orth
© 2004 American Institute of Physics 0-7354-0197-7/04/$22.00

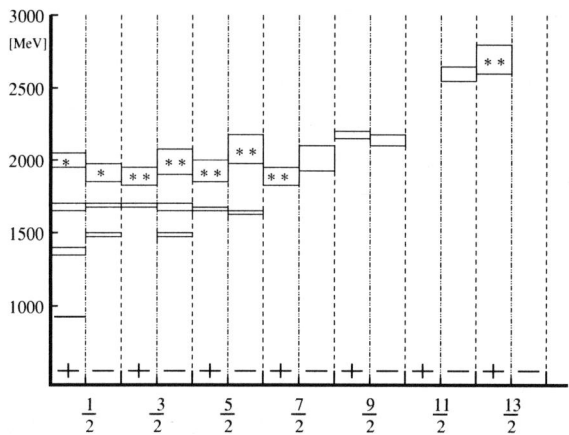

FIGURE 1. Excitation spectrum of the nucleon. The real part of the pole position is shown. Boxes represent experimental uncertainties. Those resonances which are not yet established are marked by two or one stars according to the PDG classification. The one-star resonances with $J = 1/2$ around 2 GeV are given according to the recent Bonn (SAPHIR) results.

quarks) and gives a basis for a constituent quark model. The latter one, supplemented by the effective interactions of constituent quarks which are also related to chiral symmetry spontaneous breaking [7, 2], is known to explain regularities of the low-lying baryons. However, higher in the spectrum the typical momenta of constituent quarks should increase, consequently the quasiparticle (constituent) mass of quarks should drop off and chiral symmetry will be effectively restored [4]. This is a plausible microscopical picture of symmetry restoration.

The systematic approach to the symmetry restoration based on QCD has been formulated in ref. [5, 6]. By definition an effective symmetry restoration means the following. In QCD the hadrons with the quantum numbers α are creared when one applies the local interpolating field (current) J_α with such quantum numbers on the vacuum $|0\rangle$. Then all the hadrons that are created by the given interpolator appear as intermediate states in the two-point correlator

$$\Pi = \imath \int d^4x \, e^{\imath qx} \langle 0|T\{J_\alpha(x)J_\alpha(0)\}|0\rangle, \tag{1}$$

where all possible Lorentz and Dirac indices (specific for a given interpolating field) have been omitted. Consider two local interpolating fields $J_1(x)$ and $J_2(x)$ which are connected by chiral transformation, $J_1(x) = UJ_2(x)U^\dagger$, where U is an element of the chiral group. Then, if the vacuum was invariant under chiral group, $U|0\rangle = |0\rangle$, it follows from (1) that the spectra created by the operators $J_1(x)$ and $J_2(x)$ would be identical. We know that in QCD one finds $U|0\rangle \neq |0\rangle$. As a consequence the spectra of two operators must be different. However, it may happen that the noninvariance of the vacuum becomes unimportant (irrelevant) high in the spectrum. Then the spectra of both operators become close al large masses (and asymptotically identical). This would mean that chiral symmetry is effectively restored. We stress that this effective chiral symmetry restoration does not mean that chiral symmetry breaking in the vacuum disappears, but only that the role of the quark condensates that break chiral symmetry in the vacuum becomes progressively less important high in the spectrum [5, 6]. One could say, that the valence quarks in high-lying hadrons decouple from the QCD vacuum.

Actually it is easy to prove that it must happen in QCD. At large space-like momenta $Q^2 = -q^2$ the correlator can be adequately represented by the operator product expansion, where all nonperturbative effects reside in different condensates [8]. The only effect that spontaneous breaking of chiral symmetry can have on the correlator is via the quark condensates of the vacuum. However, the contributions of all these condensates are suppressed by inverse powers of Q^2. This shows that even if the chiral symmetry is broken in the vacuum, at large space-like momenta the correlation function becomes chirally symmetric. In other words $\Pi_{J_1}(Q) \to \Pi_{J_2}(Q)$ at $Q^2 \to \infty$. The dispersion relation provides a connection between the space-like and time-like domains of the correlator. In particular, the large Q^2 correlator is completely dominated by the large s spectral density. Hence the spectral density at large s should be insensitive to the chiral symmetry breaking in the vacuum and must satisfy $\rho_1(s) \to \rho_2(s)$ at $s \to \infty$. If this chiral symmetry restoration happens in the regime where the spectrum is still quasidiscrete (i.e. it is dominated by resonances and the successive resonances with the given spin are separated), then these resonances must fill in representations of the parity-chiral group.

Clearly it is a matter of experiment to answer a question at which mass scale it happens. For example, the difference between the vector and axial-vector spectral densities, extracted from the weak decays of tau-lepton by ALEPH [9] and OPAL [10] collaborations is compatible with zero at masses of 1.7 GeV, though uncertainties are rather large. This difference is entirely from the spontaneous breaking of chiral symmetry and while it is large at ρ and a_1 masses, it gets strongly suppressed at the same mass scale where we see parity doublets in Fig. 1

A direct evidence [12] for chiral symmetry restoration can be infered from the recent results of partial wave analysis of proton-antiproton annihilation at LEAR [11]. In particular, the pions, $I, J^P = 1, 0^-$, and $\frac{1}{\sqrt{2}}(u\bar{u} + d\bar{d})$ f_0 mesons, $I, J^P = 0, 0^+$ would form $(1/2, 1/2)$ representations of the chiral group and would be degenerate level by level if the vacuum was chirally symmetric. Clearly it is not the case for the low-lying mesons, where effects of chiral symmetry breaking in the vacuum are strong. However, starting from 1.5 GeV mass we observe a pattern of chiral symmetry restoration, see Fig. 2.

The phenomenon of chiral symmetry restoration in hadron spectra rules out the non-relativistic potential description of high-lying hadrons in the spirit of the constituent quark model [12, 13, 14]. Consider, for instance, mesons. Within the nonrelativistic potential description of mesons the parity of the state is unambiguously prescribed by the relative orbital momentum of quarks. The states of opposite parity require different orbital angular momenta and hence different centrifugal repulsion. Hence they cannot be systematically degenerate within such a picture. Similar conclusions can be obtained for baryons [4]. Clearly the chiral symmetry restoration in high-lying hadrons is against relativistic scalar confinement potential description either, because scalar confinement

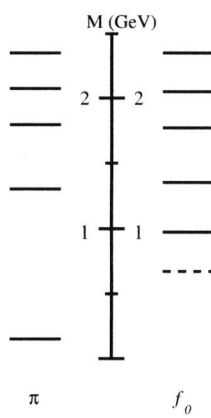

FIGURE 2. Pion and $n\bar{n}$ f_0 spectra.

manifestly breaks chiral symmetry. However, it does not contradict Lorentz-vector confining potential and this type of potential can be reconciled with parity doubling [15].

The most simple model of highly-excited hadrons compatible with the chiral symmetry restoration is a string (with the color-electric flux-tube in the string) and with bare quarks of definite chirality at the ends of the string [13]. Once the quarks posess a definite chirality, then all hadrons will form degenerate multiplets of opposite parity. For example, $n\bar{n}$ f_0 mesons and pions represent the following combinations of the right and left quarks

$$f_0 : \frac{1}{\sqrt{2}}(\bar{R}L + \bar{L}R), \tag{2}$$

$$\pi : \frac{1}{\sqrt{2}}(\bar{R}\tau L - \bar{L}\tau R). \tag{3}$$

Similar valence quark content relations can be written for other hadrons. Actually, parity partners represent different parity states of the same "basic" particle, energy of which is determined by the energy of the string. As a byproduct, this type of model automatically solves a famous spin-orbit problem of the constituent quark model: Since the chirality operator does not commute with the spin-orbit operator, there is no spin-orbit force once the chiral symmetry is restored.

Similar arguments about chiral symmetry restoration in meson spectra have been suggested in ref. [16, 17]. It has also been shown in ref. [18] that the string spectrum can be reproduced via the Salpeter-type or Dirac equations with vector confinement. Very interesting evidence for the chiral symmetry restoration has been obtained in lattice calculations [19], where it has been shown that the high-lying mesons decouple from the low-lying eigenmodes of the Dirac operator (which determine the quark condensate).

Clearly, the systematic experimental study of high-lying hadrons will be an interesting enterprise and will allow us to understand a lot about QCD in the confining regime. This program is just on the way at ELSA, JLAB and hopefully will be important for the future antiproton ring at GSI and for a new Japanese hadron facility.

REFERENCES

1. S. Capstick and N. Isgur, Phys. Rev. **D34** (1986) 2809.
2. L.Ya. Glozman, W. Plessas, K. Varga, R. F. Wagenbrunn, Phys. Rev. **D58** (1998) 094030.
3. U. Loring, B. S. Metsch, H. B. Petry, Eur. Phys. J. **A10** (2001) 447.
4. L. Ya. Glozman, Phys. Lett. **B475** (2000) 329.
5. T. D. Cohen and L. Ya. Glozman, Phys. Rev. **D65** (2002) 016006; hep-ph/0102206
6. T. D. Cohen and L. Ya. Glozman, Int. J. Mod. Phys. **A17** (2002) 1327.
7. L. Ya. Glozman and D. O. Riska, Phys. Rep. **268** (1996) 263.
8. M. A. Shifman, A. I. Vaistein and V. I. Zacharov, Nucl. Phys. **B147** (1979) 385.
9. R. Barate et al, Eur. Phys. J. **C4** (1998) 409.
10. K. Ackerstaff et al, Eur. Phys. J. **C7** (1999) 571.
11. A. V. Anisovich et al, Phys. Lett. **B491** (2000) 47; **B517** (2001) 261; preprint "Combined analysis of meson channels with I=1, C=-1 from 1940 to 2410 MeV"; preprint "I=0,C=-1 mesons from 1940 to 2410 MeV".
12. L. Ya. Glozman, Phys. Lett. **B539** (2002) 257.
13. L. Ya. Glozman, Phys. Lett. **B541** (2002) 115.
14. L. Ya. Glozman, Progr. Part. Nucl. Phys. **50** (2003) 247.
15. A. Le Yaouanc et al, Phys. Rev. **D31** (1985) 137.
16. S.R. Beane, Phys. Rev. **D64** (2001) 116010; hep-ph/0106022; see, however: M. Golterman and S. Peris, Phys. Rev **D67** (2003) 096001.
17. E. Swanson, hep-ph/0309296.
18. T. J. Allen et all, Phys. Rev. **D67** (2003) 054016.
19. T. DeGrand, "Excited states of hadrons and Dirac eigenmodes", to be published.

Hadrons in Matter

Hadrons in Nuclei

Ulrich Mosel

Institut fuer Theoretische Physik, Universitaet Giessen
D-35392 Giessen, Germany

Abstract. Changes of hadronic properties in dense nuclear matter as predicted by theory have usually been investigated by means of relativistic heavy-ion reactions. In this talk I show that observable consequences of such changes can also be seen in more elementary reactions on nuclei. Particular emphasis is put on a discussion of photonuclear reactions; examples are the dilepton production at ≈ 1 GeV and the hadron production in nuclei at 10 - 20 GeV photon energies. The observable effects are expected to be as large as in relativistic heavy-ion collisions and can be more directly related to the underlying hadronic changes.

INTRODUCTION

That hadrons can change their properties and couplings in the nuclear medium has been well known to nuclear physicists since the days of the Delta-hole model that dealt with the changes of the properties of the pion and Delta-resonance inside nuclei [1]. Due to the predominant p-wave interaction of pions with nucleons one observes here a lowering of the pion branch with increasing pion-momentum and nucleon-density. A direct observation of this effect is difficult because of the strong final state interactions (in particular absorption) of the pions. More recently, experiments at the FSR at GSI have shown that also the pion rest mass in medium differs from its value in vacuum [2]. This is interesting since there are also recent experiments [3] that look for the in-medium changes of the σ meson, the chiral partner of the pion. Any comparison of scalar and pseudoscalar strength could thus give information about the degree of chiral symmetry restoration in nuclear matter.

In addition, experiments for charged kaon production at GSI [4] have given some evidence for the theoretically predicted lowering of the K^- mass in medium and the (weaker) rising of the K^+ mass. State-of-the-art calculations of the in-medium properties of kaons have shown that the usual quasi-particle approximation for these particles is no longer justified inside nuclear matter where they acquire a broad spectral function [5, 6].

At higher energies, at the CERN SPS and most recently at the Brookhaven RHIC, in-medium changes of vector mesons have found increased interest, mainly because these mesons couple strongly to the photon so that electromagnetic signals could yield information about properties of hadrons deeply embedded into nuclear matter. Indeed, the CERES experiment [20] has found a considerable excess of dileptons in an invariant mass range from ≈ 300 MeV to ≈ 700 MeV as compared to expectations based on the assumption of freely radiating mesons. This result has found an explanation in terms of a shift of the ρ meson spectral function down to lower masses, as expected from theory (see, e.g., [7, 8, 9, 10]). However, the actual reason for the observed dilepton excess is far

CP717, *Hadron Spectroscopy: Tenth International Conference,*
edited by E. Klempt, H. Koch, and H. Orth
© 2004 American Institute of Physics 0-7354-0197-7/04/$22.00

from clear. Both models that just shift the pole mass of the vector meson as well as those that also modify the spectral shape have successfully explained the data [11, 12, 13]; in addition, even a calculation that just used the free radiation rates with their – often quite large – experimental uncertainties was compatible with the observations [14]. There are also calculations that attribute the observed effect to radiation from a quark-gluon plasma [15]. While all these quite different model calculations tend to explain the data, though often with some model assumptions, their theoretical input is sufficiently different as to make the inverse conclusion that the data prove one or another of these scenarios impossible.

I have therefore already some years ago proposed to look for the theoretically predicted changes of vector meson properties inside the nuclear medium in reactions on normal nuclei with more microscopic probes [16]. Of course, the nuclear density felt by the vector mesons in such experiments lies much below the equilibrium density of nuclear matter, ρ_0, so that naively any density-dependent effects are expected to be much smaller than in heavy-ion reactions.

On the other hand, there is a big advantage to these experiments: they proceed with the spectator matter being close to its equilibrium state. This is essential because all theoretical predictions of in-medium properties of hadrons are based on an equilibrium model in which the hadron (vector meson) under investigation is embedded in cold nuclear matter in equilibrium and with infinite extension. However, a relativistic heavy-ion reaction proceeds – at least initially – far from equilibrium. Even if equilibrium is reached in a heavy-ion collision this state changes by cooling through expansion and particle emission and any observed signal is built up by integrating over the emissions from all these different stages of the reaction.

In this talk I summarize results that we have obtained in studies of observable consequences of in-medium changes of hadronic spectral functions in reactions of elementary probes with nuclei. I demonstrate that the expected in-medium sensitivity in such reactions is as high as that in relativistic heavy-ion collisions and that in particular photonuclear reactions present an independent, cleaner testing ground for assumptions made in analyzing heavy-ion reactions.

THEORY

A large part of the current interest in in-medium properties of hadrons comes from the hope to learn something about quarks in nuclei. Indeed, a very simple estimate shows that the chiral condensate in the nuclear medium is in lowest order in density given by [10]

$$\langle \bar{q}q \rangle_{\mathrm{med}}(\rho, T) \approx \left(1 - \sum_h \frac{\Sigma_h \rho_h^s(\rho, T)}{f_\pi^2 m_\pi^2} \right) \langle \bar{q}q \rangle_{\mathrm{vac}} . \tag{1}$$

Here ρ_s is the *scalar* density of the hadron h in the nuclear system and Σ_h the so-called sigma-commutator that contains information on the chiral properties of h. The sum runs over all hadronic states. While (1) is nearly exact, its actual value is limited because neither the sigma-commutators of the higher lying hadrons nor their scalar densities are known. Only at very low temperatures these are accessible. Here $\rho_s \approx \rho_v \frac{m}{E}$ so that the condensate drops linearly with the nuclear (vector) density. This drop can be understood in physical terms: with increasing density the hadrons with their chirally symmetric phase in their interior fill in more and more space in the vacuum with its spontaneously broken chiral symmetry. Note that this is a pure volume effect; it is there already for a free, non-interacting hadron gas.

How this drop of the scalar condensate translates into observable hadron masses is not uniquely prescribed. The only rigorous connection is given by the QCD sum rules that relates an integral over the hadronic spectral function to a sum over combinations of quark- and gluon-condensates with powers of $1/Q^2$. It has been shown [17] that the QCDSR constrains the hadronic spectral function, but it does not fix it.

Thus models are needed for the hadronic interactions. The quantitatively reliable ones can at present be based only on 'classical' hadrons and their interactions. Indeed, in lowest order in the density the mass and width of an interacting hadron in nuclear matter at zero temperature and vector density ρ_v are given by (for a meson, for example)

$$m^{*2} = m^2 - 4\pi \Re f_{mN}(q_0, \theta = 0)\rho_v$$
$$m^*\Gamma^* = m\Gamma^0 - 4\pi \Im f_{mN}(q_0, \theta = 0)\rho_v \ . \tag{2}$$

Here $f_{mN}(q_0, \theta = 0)$ is the forward scattering amplitude for a meson with energy q_0 on a nucleon. The width Γ^0 denotes the free decay width of the particle. For the imaginary part this is nothing other than the classical relation $\Gamma^* - \Gamma^0 = v\sigma\rho_v$ for the collision width, where σ is the total cross section. This can easily be seen by using the optical theorem.

Note that such a picture also encompasses the change of the chiral condensate in (1), obtained there for non-interacting hadrons. If the spectral function of a non-interacting hadron changes as a function of density, then in a classical hadronic theory, which works with fixed free hadron masses, this change will show up as an energy-dependent interaction and is thus contained in any empirical phenomenological cross section.

DILEPTON PRODUCTION

Dileptons, i.e. electron-positron pairs, in the outgoing channel are an ideal probe for in-medium properties of hadrons since they – in contrast to hadronic probes – experience no strong final state interaction. A first experiment to look for these dileptons in heavy-ion reactions was the DLS experiment at the BEVALAC in Berkeley [19]. Later on, and in a higher energy regime, the CERES experiment has received a lot of attention for its observation of an excess of dileptons with invariant masses below those of the lightest vector mesons [20]. Explanations of this excess have focussed on a change of in-medium properties of these vector mesons in dense nuclear matter (see e.g. [11, 12]). The radiating sources can be nicely seen in Fig. 1 that shows the dilepton spectrum obtained in a low-energy run at 40 AGeV together with the elementary sources of dilepton radiation.

The figure exhibits clearly the rather strong contributions of the vector mesons – both direct and through their Dalitz decay – at invariant masses above about 500 MeV. If this strength is shifted downward, caused by an in-medium change of the vector-meson

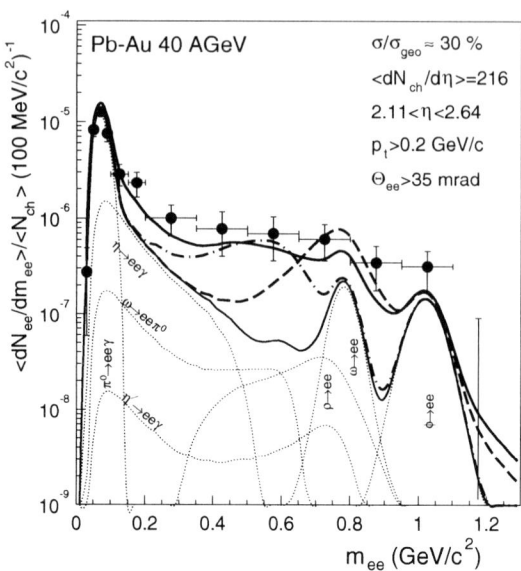

FIGURE 1. Invariant dilepton mass spectrum obtained with the CERES experiment in Pb + Au collisions at 40 AGeV (from [20]). The thin curves give the contributions of individual hadronic sources to the total dilepton yield, the fat solid (modified spectral function) and the dash-dotted (dropping mass only) curves give the results of calculations [13] employing an in-medium modified spectral function of the vector mesons.

spectral functions, then the observed excess can be explained as has been shown by various authors (see e.g. [21] for a review of such experiments).

As mentioned above such explanations always suffer from an inherent inconsistency: while the observed signal integrates over many different stages of the collision – nonequilibrium and equilibrium, the latter at various densities and temperatures – the theoretical input is always calculated under the assumption of a vector meson in nuclear matter in equilibrium. We have therefore looked for possible effects in reactions that proceed much closer to equilibrium and have thus studied the dilepton production in reactions on nuclear targets involving more elementary projectiles. It is not *a priori* hopeless to look for in-medium effects in ordinary nuclei: Even in relativistic heavy-ion reactions that reach baryonic densities of the order of 3 - 10 ρ_0 many observed dileptons actually stem from densities that are much lower than these high peak densities. Transport simulations have shown [21] that even at the CERES energies about 1/2 of all dileptons come from densities lower than $2\rho_0$. This is so because in such reactions the pion-density gets quite large in particular in the late stages of the collision, where the baryonic matter expands and its density becomes low again. Correpondingly many vector mesons are formed (through $\pi + \pi - > \rho$) late in the collision and their decay to dileptons thus happens at low baryon densities.

It is thus a quantitative question if any observable effects of in-medium changes of hadronic properties survive if the densities probed are always $\leq \rho_0$. With the aim of answering this question we have over the last few years undertaken a number of calculations for proton- [22], pion- [23, 24] and photon- [25] induced reactions. All of them have one feature in common: they treat the final state incoherently in a coupled channel transport calculation that allows for elastic and inelastic scattering of, particle production by and absorption of the produced vector mesons. We have also looked into the prospects of using reactions with hadronic [26] final states. In this case, the photoproduction of ϕ mesons on nuclei, our conclusion was that no in-medium signal from the ϕ could be observed due to the strong final state interactions of the outgoing kaons, the decay products of the ϕ meson. A semi-hadronic final state, such as $\pi^0\gamma$, as obtained in the photoproduction of ω mesons on nuclei looks more promising [27].

All the photonuclear calculations are done in a combination of coherent initial state interactions that lead to shadowing at photon energies above about 1 GeV and incoherent final state interactions. The shadowed incoming photon produces, for example, a vector meson which then cascades through the nucleus. The latter process we describe by means of a coupled-channel transport theory. The details are discussed in ref. [18]. A new feature of these calculations is that vector mesons with their correct spectral functions can actually be produced and transported consistently. This is quite an advantage over earlier treatments [21] in which the mesons were always produced and transported with their pole mass and their spectral function was later on folded in only for their decay.

A typical result of such a calculation for the dilepton yield – after removing the Bethe-Heitler component – is given in Fig. 2. Comparing this figure with Fig. 1 shows that in a photon-induced reaction at 1 - 2 GeV photon energy exactly the same sources, and none less, contribute to the dilepton yield as in relativistic heavy-ion collisions at 40 AGeV! The question now remains if we can expect any observable effect of possible in-medium changes of the vector meson spectral functions in medium in such an experiment on the nucleus where – due to surface effects – the average nucleon density is below ρ_0. This question is answered, for example, by the results of Fig. 3. This figure shows the dilepton spectra to be expected if a suitable cut on the dilepton momenta is imposed; with this cut slow vector mesons are enriched. In the realistic case shown on the right, which contains both a collision broadening and a mass shift, it is obvious that a major signal is to be expected: in the heavy nucleus Pb the ω-peak has completely disappeared from the spectrum. The sensitivity of such reactions is thus as large as that observed in ultrarelativistic heavy-ion reactions.

An experimental verification of this prediction would be a major step forward in our understanding of in-medium changes[1]. It would obviously present a purely hadronic base-line to all data on top of which all 'fancier' explanations of the CERES effect in terms of radiation from a QGP and the such would have to live.

---

[1] An experiment at JLAB is under way [28].

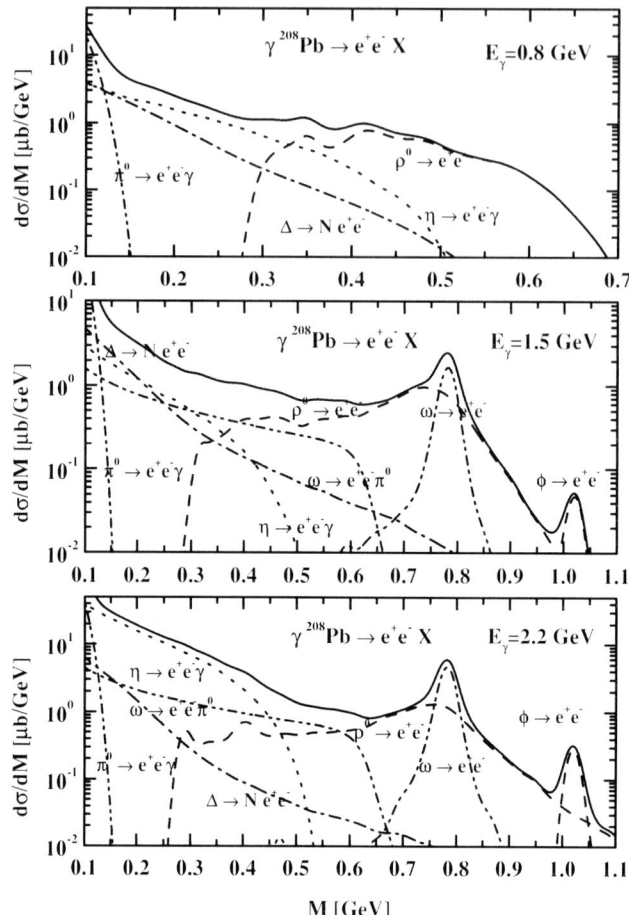

FIGURE 2. Hadronic contributions to dilepton invariant mass spectra for $\gamma + ^{208}Pb$ at the three photon energies given (from [25]). Compare with Fig. 1.

HADRON FORMATION

An in-medium effect different from the ones discussed so far happens when particles are produced by high-energy projectiles inside a nuclear medium. Then measurements of their yield can actually give information on the time it takes until the newly created

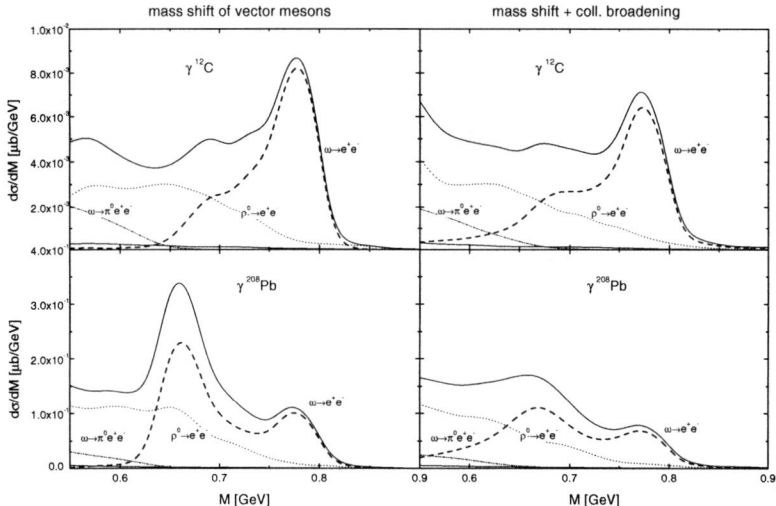

FIGURE 3. Dilepton mass yield with a dilepton-momentum cut of 300 MeV. Shown on the left are results of a calculation that uses only a shift of the pole mass of the vector mesons. On the right, results are given for a calculation using both mass shift and collisional broadening (from [18]).

particle has evolved into a 'normal', fully interacting particle. The longer this formation time is, the less absorption will take place. Thus nuclei serve as a kind of 'microdetector' for the determination of formation times. Obviously, this phenomenon is closely related to color transparancy. A particularly appealing probe are jets that emerge from the nuclei.

A major experimental effort at RHIC experiments has gone into the observation of jets in ultrarelativistic heavy-ion collisions and the determination of their interaction with the surrounding quark or hadronic matter [29]. Such experiments are obviously very sensitive to hadron formation times. In addition they can yield information on interactions while the final hadron is still being formed.

A complementary process is given by the latest HERMES results at HERA for photon-induced hadron production at high energies [30]. Here the photon-energies are of the order of 10 - 20 GeV, with rather moderate $Q^2 \approx 2$ GeV^2. Again, the advantage of such experiments is that the nuclear matter with which the interactions happen is at rest and in equilibrium.

In the high-energy regime the shadowing of the incoming photon, which is due to interference between interactions of the incoming bare photon and its hadronic components with the nucleons, becomes important. This coherence in the incoming state has to be combined with the incoherent treatment of the final state interactions in transport theory. For this purpose T. Falter has derived a novel expression for incoherent particle production on the nucleus [31] that allows for a clean-cut separation of the coherent initial state and the incoherent final state interactions which we again treat with our coupled-channel transport theory.

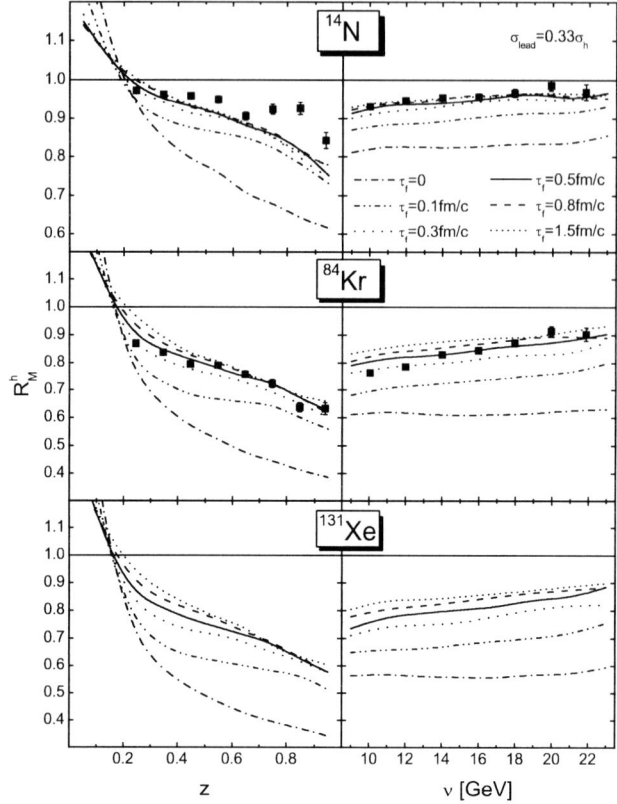

FIGURE 4. Multiplicity of produced hadrons normalized to the proton as a function of photon-energy ν (right) and of the energy of the produced hadrons relative to the photon-energy, $z = E_h/\nu$. The curves are calculated for different formation times given in the figure (from [32]).

An example of the results obtained is given in Fig. 4. The figure clearly shows that the observed hadron multiplicities can be described only with formation times $> \approx 0.3$ fm. The curves obtained with larger formation times all lie very close together. This is a consequence of the finite size of the target nucleus: if the formation time is larger than the time needed for the preformed hadron to transverse the nucleus, then the sensitivity to the formation time is lost. In [32] we have also shown that the z-dependence on the left side exhibits some sensitivity to the interactions of the leading hadrons during the formation; the curves show in Fig. 4 are obtained with a leading hadron cross section of $0.33\sigma_h$, where σ_h is the 'normal' hadronic interaction cross section.

CONCLUSIONS

In this talk I have illustrated with the help of two examples that photonuclear reactions can yield information that is important and relevant for an understanding of high density–high temperature phenomena in ultrarelativistic heavy-ion collisions. I have shown that the expected sensitivity of dilepton spectra in photonuclear reactions in the 1 - 2 GeV range is as large as that in ultrarelativistic heavy-ion collisions. I have also illustrated that the analysis of hadron production spectra in high-energy electroproduction experiments at HERMES gives information about the interaction of forming hadrons with the surrounding hadronic matter. This is important for any analysis that tries to obtain signals for a QGP by analysing high-energy jet formation in ultrarelativistic heavy-ion reactions.

ACKNOWLEDGEMENT

This talk is based on work done together with Wolfgang Cassing, Martin Effenberger, Thomas Falter and Kai Gallmeister. The work on which it is based has been supported by the Deutsche Forschungsgemeinschaft, the BMBF and GSI Darmstadt.

REFERENCES

1. T. Ericson and W. Weise, *Pions and Nuclei*, Clarendon Press, Oxford, 1988.
2. H. Geissel et al., *Phys. Rev. Lett.* **88**, 122301 (2002).
3. J.G Messchendorp et al., *Phys.Rev.Lett.* **89**, 222302 (2002).
4. A. Foster et al., nucl-ex/0307018
5. M.F.M. Lutz and E.E. Kolomeitsev, *Nucl.Phys.***A700**, 193 (2002).
6. L. Tolos, A. Ramos, A. Polls, *Phys.Rev.***C65**, 054907 (2002).
7. W. Peters et al., *Nucl.Phys.* **A632** 109 (1998).
8. M. Post et al., *Nucl.Phys.* **A689** 753 (2001).
9. M. Post et al., nucl-th/0309085, to be published.
10. J. Wambach, *Nucl. Phys. A* **715** 422c (2003).
11. W. Cassing et al., *Phys.Lett.* B363, 35 (1995).
12. R. Rapp, J. Wambach, *Adv. Nucl. Phys. 25* 1 (2000).
13. *Pramana* **60** 675 (2003), hep-ph/0201101
14. V. Koch et al., *Proc. Int. Workshop XXVIII on Gross Properties of Nuclei and Nuclear Excitations*, Hirschegg, Austria, Jan. 16-22, 2000, GSI Report ISSN 0720-8715.
15. T. Renk et al., *Phys. Rev.* **C66** 014902 (2002).
16. U. Mosel, *Progr. Part. Nucl. Phys.* **42** 161 (1999); *Proc. Int. Workshop XXVIII on Gross Properties of Nuclei and Nuclear Excitations*, Hirschegg, Austria, Jan. 16-22, 2000, GSI Report ISSN 0720-8715.
17. S. Leupold, U. Mosel, *Phys. Rev.* **C58** 2939 (1998); S. Leupold et al., *Nucl. Phys.* **A628** 311 (1998).
18. M. Effenberger, U. Mosel, *Phys. Rev.* **C62** 014605 (2000).
19. R.J. Porter et al., *Phys. Rev. Lett.* **79** 1229 (1997).
20. J.P. Wessels et al., *Quark Matter 2002 (QM 2002)*, Nantes, France, 18-24 July 2002, *Nucl. Phys.* **A715** 262 (2003).
21. W. Cassing, E.L. Bratkovskaya, *Phys. Rep.* **308** 65 (1999) .
22. E.L. Bratkovskaya, *Phys. Lett.* **B529** 26 (2002).
23. T. Weidmann et al., *Phys. Rev.* **C59** 919 (1999).
24. M. Effenberger et al., *Phys. Rev.* **C60** 027601 (1999).
25. M. Effenberger, E.L. Bratkovskaya, U. Mosel, *Phys. Rev.* **C60** 044614 (1999).

26. P. Muehlich et al., *Phys. Rev.* **C67** 024605 (2003).
27. P. Muehlich et al., to be published.
28. D. Weygand, private communication.
29. M. Gyulassy et al., Quark Gluon Plasma Vol. 3, ed. R.C. Hwa and X.N. Wang, World Scientific, Singapore, 2003.
30. V. Muccifora et al., *Nucl. Phys.* **A711** (2002) 254.
31. T. Falter, K. Gallmeister, U. Mosel, *Phys. Rev.* **C67** 054606 (2003), Erratum-ibid. **C68** 019903 (2003).
32. T. Falter, W. Cassing, K. Gallmeister, U. Mosel, nucl-th/0308073, to be published.

Medium Effects in Nuclear Interactions of Pions and Kaons

A. Gillitzer

Institut für Kernphysik, Forschungszentrum Jülich, D-52425 Jülich, Germany

Abstract. The objective of this contribution is to review the present status of experimental information on the properties of pions and kaons in nuclear matter. Medium effects on pions are discussed in the context of the experimental results on deeply bound pionic states. The repulsive in-medium potential for kaons has been determined in proton-nucleus collsions in line with results from studies of nucleus-nucleus collisions. The deduction of the depth of the potential for antikaons from heavy ion data is still affected by a large degree of model dependence. An experimental search for $\bar{K}$-nuclear bound states may clarify this question in the future.

INTRODUCTION

The question on the origin of the hadron masses as they are observed in nature is one of the basic topics in the physics of strong interaction. Hadrons are the only composed systems we know of to date that have a larger mass than their elementary building blocks. In case of the nucleon the bare quark masses account only for a tiny fraction of around 2%. Two basic properties of the strong interaction have been proposed to be responsible for the generation of hadron masses, namely confinement and spontaneous breaking of chiral symmetry in the ground state (i.e. the vacuum state) of QCD [1, 2, 3]. As a consequence of this spontaneous symmetry breaking mechanism a finite quark-antiquark pair condensate in the vacuum state and a gap in the vector meson and baryon mass spectrum is created. At finite nuclear density chiral symmetry has been predicted to be partially restored which should be manifest in a reduced value of the pairing gap and the quark condensate [4, 5, 6]. This should also have an influence on masses and other properties of hadrons inside the nuclear medium, which could therefore serve as probes to gain a better understanding of the generation of mass in strong interaction physics. A very general behaviour is expected from a simple scaling law according to which the masses of nucleons M_N, vector mesons m_V, scalar mesons m_s, and the pion decay constant f_π should scale in the same way as chiral symmetry is restored [7]. For pions and kaons the situation is different because of their special role as Goldstone bosons of spontaneous broken chiral symmetry [8], and a mass reduction according to this scaling is not expected. It is nevertheless vitally important to study the in-medium behaviour of pions and kaons. As will be discussed below, it is the s-wave pion-nucleon interaction that can be related to chiral symmetry breaking in nuclear matter. Kaons are particularly interesting also from the astrophysical point of view, because a sufficiently strong reduction of the antikaon mass in nuclear matter could result in kaon condensation in neutron stars.

CP717, *Hadron Spectroscopy: Tenth International Conference,*
edited by E. Klempt, H. Koch, and H. Orth
© 2004 American Institute of Physics 0-7354-0197-7/04/$22.00

Experimentally, different approaches are conceivable to study in-medium properties of mesons. The cleanest way to study a meson at rest inside nuclear matter is via the measurement of spectroscopic information from bound or quasi-bound states. This may be deeply bound states of quasi-atomic character, or nuclear states, exclusively or dominanantly bound by the strong interaction. The production of mesons in proton-nucleus collisions and in nucleus-nucleus collisions may be studied in the context of the influence of nuclear potentials on their yield, on their spectral shape, and (in AA collisions) on their flow pattern. The first method has been used for π^-, and very recently also for K^-, while K^+ and K^- production has been studied in proton-nucleus and nucleus-nucleus collisions.

DEEPLY BOUND π^- STATES AND THE IN-MEDIUM πN INTERACTION

The pion-nucleon and pion-nucleus interaction

The pion-nucleon interaction is characterized by a very weak s-wave interaction, and a p-wave interaction dominating already at rather low energies above threshold, governed by the coupling of the πN system to the Δ resonance. The s-wave interaction, as will be discussed in the next section, is particularly interesting due to its close relation to fundamental symmetry properties of the strong interaction. Recently, parameters of the s-wave πN interaction were determined to high precision in the measurement of strong level shifts in pionic hydrogen [9] and pionic deuterium [10]. From the measured $\pi^- p$ and $\pi^- d$ scattering lengths values for the πN isoscalar scattering length a^+ and the isovector scattering length a^- were deduced [11]:

$$a^+ = (-17 \pm 10) \times 10^{-4} \, m_\pi^{-1}, \quad a^- = (900 \pm 16) \times 10^{-4} \, m_\pi^{-1}, \tag{1}$$

the isoscalar scattering length being remarkably small, and almost compatible with zero.

Theoretical studies of the pion-nucleus interaction have a long history. It is traditionally modelled by an optical potential which is based on effective scattering length and volume parameters for s-wave and p-wave interaction, respectively, together with a low density approximation in which the potential is essentially linear in the nuclear density ρ [12, 13]. Its s-wave part is given by

$$2\mu U^{(s)}(r) = -4\pi[b(r) + \varepsilon_2 B_0 \rho^2(r)]. \tag{2}$$

where $b(r) = \varepsilon_1[b_0 \rho(r) + b_1 \Delta\rho(r)]$, $\rho(r) = \rho_n(r) + \rho_p(r)$, $\Delta\rho(r) = \rho_n(r) - \rho_p(r)$, $\varepsilon_1 = 1 + \mu/M = 1.147$, and $\varepsilon_2 = 1 + \mu/2M = 1.073$.

The effective s-wave isoscalar and isovector scattering lengths b_0 and b_1 can be related to the free πN scattering lengths $b_0^{free} = a^+$ and $b_1^{free} = -a^-$ by applying several in-medium corrections, the dominant of which is the correction for double πN scattering. For the isoscalar part this correction by far exceeds the free value [13, 14]

$$b_0 \simeq b_0^{free} - 2(b_1^{free})^2 \langle \frac{1}{r} \rangle \simeq b_0^{free} - 2(b_1^{free})^2 \frac{3p_F}{8\pi^2} \tag{3}$$

whereas the relative correction to the isovector part is small.

Pions as probes of chiral symmetry restoration in nuclei

To leading order in the pion energy ω the isospin even and odd πN T-matrix amplitudes at threshold are given by the low-energy theorem [15, 16]

$$T^{(+)}(\omega, \vec{q} = 0) = 0, \quad T^{(-)}(\omega, \vec{q} = 0) = \frac{\omega}{2f_\pi^2}, \tag{4}$$

well fulfilled by the measured values of the free πN scattering lenghts. The pion decay constant is related to the quark condensate through the Gell-Mann-Oakes-Renner relation [17]

$$m_\pi^2 f_\pi^2 = m_q \langle \bar{q}q \rangle_0. \tag{5}$$

It can be shown in a model-independent way that in the nuclear medium to leading order in nuclear density ρ the relation

$$\frac{\langle \bar{q}q \rangle_\rho}{\langle \bar{q}q \rangle_0} \approx 1 - \frac{\sigma_N}{m_\pi^2 f_\pi^2} \rho \tag{6}$$

holds [4, 5, 6], which with $f_\pi = 92.4$ MeV and $\sigma_N \approx 45$ MeV yields a reduction of the quark condensate by about 35%. It can be also shown that the relations (5) [18] and (6) [19] are still valid at finite density ρ to leading order, and therefore also the inverse proportionality of the s-wave isovector πN interaction strength and the square of the pion decay constant holds:

$$\frac{b_1^{\text{free}}}{b_1(\rho)} \approx \frac{f_\pi^*(\rho)^2}{f_\pi^2}. \tag{7}$$

Thus the isovector s-wave πN interaction in the nuclear medium reflects fundamental symmetry properties of the strong interaction, and it is therefore particularly important to determine the isovector strength parameter b_1 separately from the isoscalar parameter b_0.

Deeply bound pionic states

The closest possible approach to a situation where "real" pions at rest in nuclear matter may be studied experimentally is realized in pionic $1s$ states in heavy nuclei. These so-called deeply bound states exist as discrete states thanks to a subtle balance between the repulsive s-wave interaction and the attractive Coulomb interaction which allows the π^- to be accommodated in a potential pocket close to the nuclear surface [20, 21]. In Pb, the repulsive interaction reduces the binding energy of the $1s$ state from ~ 12 MeV, as due to the Coulomb interaction, by almost a factor of two to ~ 7 MeV. Thus the hadronic correction is of the same order of magnitude as the Coulomb binding. Due to the close vicinity to the nuclear surface, the pionic $1s$ and $2p$ states in heavy nuclei can

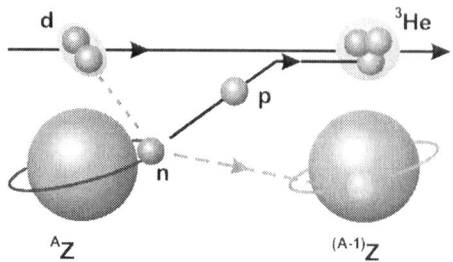

FIGURE 1. Schematic illustration of the formation of deeply bound pionic states in a $(d, {}^3\mathrm{He})$ transfer reaction. The population of quasi-substitutionl states with $l_\pi = l_n$ is strongly enhanced in recoilfree kinematics.

be therefore better characterized as nuclear pionic halo states, than as classical atomic states.

Using a standard pion-nucleus optical potential, the decomposition of the strong interaction into contributions from the s-wave part (repulsive), the p-wave part (attractive), and the imaginary part (repulsive) was theoretically studied for the pionic $1s$, $2p$, and $3d$ states, respectively, in ${}^{207}\mathrm{Pb}$ and ${}^{115}\mathrm{Sn}$ [22]. It was shown that the $1s$ shift is almost entirely determined by the s-wave part, and thus a measurement of the pionic $1s$ state allows to reliably deduce the πN s-wave interaction.

In the past pionic atoms have been formed with stopped negative pions, captured in high-lying states which subsequently deexcite in an electromagnetic cascade. This method, however, is not feasible in the case of deeply bound states, since the electromagnetic cascade due to strong absorption terminates before it can reach the lowest states. Toki *et al.* proposed [23] to populate deeply bound states in pionic atoms in pionic transfer reactions at recoilfree kinematics such as (n,d) or $(d, {}^3\mathrm{He})$, as sketched in Figure 1. In this reaction, where a neutron in the target nucleus is "replaced" by a π^-, the surface of the nucleus dominantly contributes. Therefore a strong relation of the angular momentum transfer and the transferred momentum is expected. At incident deuteron kinetic energies around 500 MeV the momentum transfer is close to zero, favouring the population of quasi-substitutional configurations with $l_\pi = l_n$.

Experimental method

Only a very brief description of the experiment is given here, more details can be found in Refs. [24, 26]. The GSI Fragment Separator (FRS) was used as high resolution spectrometer to study the $(d, {}^3\mathrm{He})$ reaction on Pb and Sn targets. In this two-body reaction, the momentum measurement of the emitted ${}^3\mathrm{He}$ nuclei allows to determine the pionic binding energy in the formation of bound pionic states. At the dispersive central focal plane of the FRS a pair of drift chambers is used to measure the ${}^3\mathrm{He}$ position and thus to determine its momentum, whereas the second part of the FRS from the central to the final focal plane is used to identify the ${}^3\mathrm{He}$ nuclei by ΔE and time of flight measured in scintillation counters. The two main objectives are to suppress high-rate

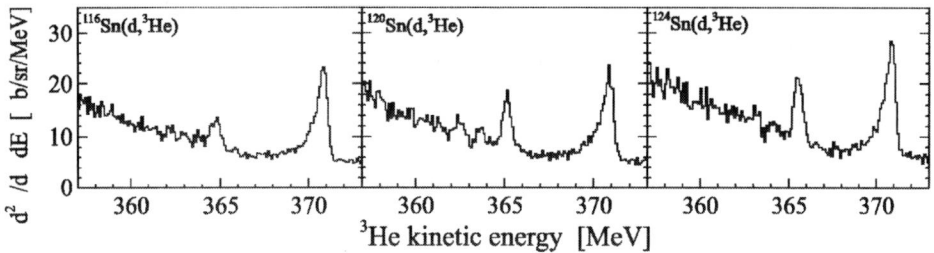

FIGURE 2. Double differential cross sections versus the ^{3}He kinetic energy measured in the ^{116,120,124}Sn$(d,^3$He$)$ reaction at incident deuteron kinetic energy $T_d = 250$ MeV/u.

background resulting from scattering and break-up processes of the d projectiles, and to guarantee a high-resolution measurement of the pionic binding energies at a level of $\sim 5 \cdot 10^{-4}$ relative to the beam energy. For energy calibration the $p(d,^3$He$)\pi^0$ reaction on polyethylene targets was used. In this reaction, within the FRS angular acceptance, quasi-monoenergetic ^{3}He nuclei are emitted at a momentum very close to that of ^{3}He nuclei that are produced in the population of deeply bound pionic states.

In the most recent experiment ^{112,116,120,124}Sn strip targets were irradiated with an intense high quality d beam at 503.388 MeV kinetic energy (corresponding to 250 MeV/u). In order to allow for good energy resolution, the targets had a thickness of only 20 mg/cm^2 and a strip width of 1.5 mm. The achieved resolution was $\delta E = 370$ keV (fwhm). On the downstream side the Sn targets had a thin polyethylene $((CH_2)_n)$ layer attached, in order to allow for a continuous online calibration and later corrections for longterm drifts.

Recent results

After the discovery of deeply bound pionic $2p$ and $1s$ states in ^{207}Pb in the ^{208}Pb$(d,^3$He$)$ reaction [24, 25, 26, 27], a better separation of $1s$ and $2p$ states was obtained in the study of the ^{206}Pb$(d,^3$He$)$ reaction [28]. In this presentation we concentrate on the most recent study of pionic Sn. Sn isotopes were chosen as targets for the study of pionic $1s$ states for two reasons: (*i*) Sn isotopes have $3s_{1/2}$ neutrons in orbitals with small binding energy, which allows to dominantly populate pionic $1s$ states in recoilfree kinematics at d kinetic energy of 500 MeV. (*ii*) Sn has a long chain of stable isotopes ranging from $A = 112$ to $A = 124$. This opens the possibility to measure the isotope shift of deeply bound pionic states, and to separate isovector and isoscalar parts in the s-wave pion-nucleus potential.

Figure 2 shows the double differential cross sections as a function of the ^{3}He kinetic energy at the dispersive focal plane of the FRS measured for the ^{116,120,124}Sn$(d,^3$He$)$ reaction. The skewed peak to the right side of the spectra is due to the $p(d,^3$He$)\pi^0$ reaction on the thin polyethylene backing. In the central part of the spectra a clear pionic $1s$ peak is seen for the three isotopes, corresponding to binding energies close to 4 MeV. To the left side more shallow pionic states and the free pion production continuum are

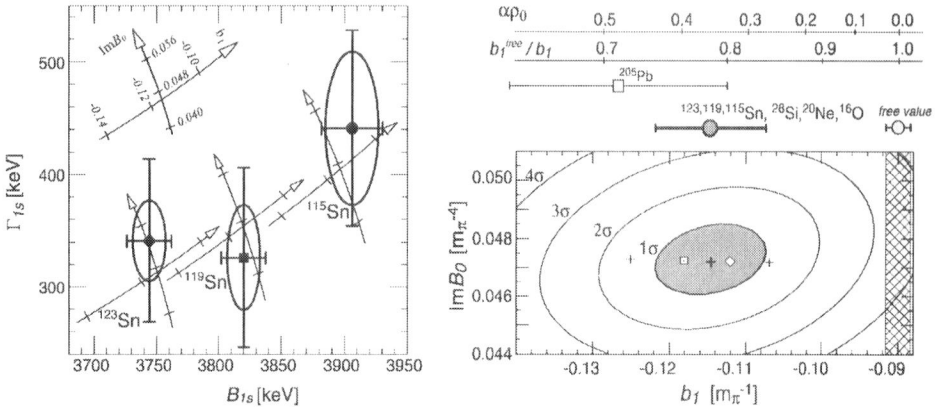

FIGURE 3. **Left:** Measured binding energies and widths of the pionic $1s$ states in [115,119123]Sn. The ellipses show the statistical errors, the horizontal and vertical bars are for the total errors. For each Sn isotope the theoretical relation of b_1 and ImB_0 with B_{1s} and Γ_{1s} as obtained with a common fit is shown. **Right:** Likelihood contours in the b_1-ImB_0 plane from a simultaneous fit of the binding energies and widths of the pionic $1s$ states in the three Sn isotopes and three light symmetric nuclei. The value for b_1 obtained from the earlier study of pionic [205]Pb is shown for comparison. See text for further explanation.

visible.

Figure 3 (left) shows the measured values [29] of the widths Γ_{1s} versus the binding energies B_{1s} for π^- bound to [115]Sn, [119]Sn, and [123]Sn. A clear monotonuous decrease of the binding energy with increasing neutron number is observed. The same trend is visible for the width. The isotope shift of the $1s$ pionic binding energy between [115]Sn and [123]Sn is 162 ± 30 keV.

The isotope dependence of the measured $1s$ binding energies and widths may be used to separate the isoscalar and isovector contribution to the pion-nucleus potential and thus to deduce constraints on the isovector effective scattering length b_1 in nuclear matter. Although an isotope effect has been clearly observed, the experimental errors would however result in a large uncertainty in the value of b_1 determined from the pionic $1s$ states in the Sn isotopes only. A more precise way to independently fix the isoscalar parameter b_0, as has been demonstrated in Ref. [30], is to use the known $1s$ binding energies in the symmetric light nuclei ^{16}O, [20]Ne, and [24]Mg. Based on the insensitivity of the $1s$ states to the p-wave part of the potential, p-wave parameters were adopted from global fits to the set of pionic atom data [31], and no attempt was undertaken to refit these parameters. In a simultaneous fit of B_{1s} and Γ_{1s} of the three Sn isotopes together with those of the symmetric light nuclei, best values for the potential parameters (see eq. (2)) b_0, b_1, ReB_0, and ImB_0 were determined. A likelihood plot in the $\{b_1, ImB_0\}$ plane is shown in Figure 3 (right). For the isovector s-wave parameter the fit value is $b_1 = -0.1149 \pm 0.0074\,\mathrm{m}_\pi^{-1}$, the absolute value of which is significantly enhanced compared to free πN scattering with $b_1^{\text{free}} = -0.0900 \pm 0.0016\,\mathrm{m}_\pi^{-1}$. Following the arguments outlined above, this can be interpreted as the consequence of a reduced in-medium chiral order parameter

$$\frac{f_\pi^*(\rho)^2}{f_\pi^2} \approx \frac{b_1^{\text{free}}}{b_1(\rho)} = 0.78 \pm 0.05. \tag{8}$$

This reduction by 22% is smaller than the reduction of $\sim 35\%$ predicted by eq. (6). However, one has to take into account that the $1s\,\pi^-$ effectively probes a nuclear density $\rho = 0.6\rho_0$ [32]. For normal nuclear density the observed ratio would then correspond to a ratio $b_1^{\text{free}}/b_1(\rho_0) \approx 0.64$ consistent with eq. (6). The same result is obtained by solving the Klein-Gordon equation with a density dependent parameter $b_1(\rho)$ according to the prescription $b_1(\rho) = b_1^{\text{free}}/(1 - \alpha\rho(r))$.

KAONS IN THE NUCLEAR MEDIUM

Theoretical predictions

There is a consensus in theoretical work on the general behaviour of kaons and antikaons in nuclear matter. In pioneering work, Kaplan and Nelson [33] demonstrated that kaons and antikaons couple attractively to the scalar nucleon density, whereas the vector coupling has opposite sign, and is repulsive for kaons and attractive for antikaons [34]. This corresponds to an increase of the K^+ mass and a decrease of the K^- mass in nuclear matter. Due to the attractive scalar interaction, the absolute mass shifts are expected to be larger for K^- than for K^+, reducing the K^+K^- theshold in nuclear matter as compared to the free threshold. The trend of the in-medium kaon and antikaon masses as a function of the density as predicted by some of the models [35, 36, 37, 38] is illustrated in Figure 4 (left).

The repulsive potential for K^+ is in line with K^+N scattering data [40] and K^+-nucleus scattering data [41]. For K^- the case is more complicated. The free isospin-weighted K^-p scattering length is repulsive due to the strong repulsive interaction in the $I = 0$ channel which overcompensates the attractive $I = 1$ interaction [42]. The values for the K^-p scattering length determined experimentally from K^-p scattering [42] and from the strong interaction shift in kaonic hydrogen [43] are in mutual agreement. The level shifts in heavier kaonic atoms are also repulsive [31]. This is understood as a consequence of the isospin singlet s-wave resonance $\Lambda(1405)$ which is commonly interpreted as a bound $\overline{K}N$ state around 30 MeV below the K^-p threshold. It is known that a subthreshold s-wave resonance results in a repulsive interaction at energies above threshold. On the other hand, inside nuclear matter the $\Lambda(1405)$ is expected to dissolve, and consequently the $\overline{K}N$ interaction to change from repulsive to strongly attractive [37].

Particularly strong attractive K^- potentials inside the nucleus $U_{\overline{K}} = -200 \pm 20$ MeV are obtained from a phenomenological study of kaonic atom data [39]. In contrast, microscopic models predict significantly shallower potentials with a depth in general in the range between -75 MeV and -140 MeV (see Ref. [44] and references therein). The chiral model of Ref. [44] finds a depth of only about -50 MeV for both the real and the imaginary part of the K^- potential. The dispersion approach of Ref. [45] predicts $U_K = +25$ MeV and $U_{\overline{K}} = -140$ MeV. Momentum dependent in-medium antikaon and hyperon spectral functions were theoretically studied in a selfconsistent chiral $SU(3)$ approach [47], which for vanishing K^- momentum predicts a broad double-humped K^- spectral distribution extending from ~ 0.35 GeV to ~ 0.55 GeV. In combining a simultaneous fit of kaonic atom and $\overline{K}N$ scattering data with a chirally motivated microscopic model $U_{\overline{K}} \approx -55$ MeV was obtained [48]. This brief overview on the theoretical work, though by no means complete, shows the disagreement in the predicted depth of the K^- potential, and reflects the uncertainties in the current understanding of the K^- properties in nuclear matter.

Kaon production in proton-nucleus collisions

Meson production in proton-nucleus collisions is affected by a nuclear potential in a twofold way. Attractive potentials (or mass shifts) will result in an enhancement of the yield, whereas a suppression is expected from repulsive potentials (or mass shifts). This effect will be most pronounced at subthreshold energies where the yield is extremely sensitive to the missing energy in the nucleon-nucleon system which has to be overcome by cooperative or multi-step processes. The spectral distribution of the produced mesons will also be affected by the in-medium potential, particularly at low momenta. For obvious reasons the experimental determination of a repulsive potential from the modification of the spectral shape is much more straight-forward than of an attractive potential. With a repulsive potential mesons produced at zero momentum will be observed at finite momenta after having propagated from the nuclear medium into free space, and thus a gap in the momentum spectrum at low momenta is created. In contrast, much more model dependence is involved in the interpretation of possible modifications of the spectral shape induced by an attractive potential.

This has been convincingly demonstrated by studies of K^+ production in the collision of protons with different target nuclei in recent measurements at the ANKE spectrometer [50] at COSY. The ratio of forward K^+ production on Cu, Ag, and Au targets to that on C targets was measured at proton beam energies between 1.5 and 2.3 GeV as a function of the kaon momentum p_K [51]. In this cross section ratio a strong suppression was observed at low kaon momenta $p_k < 200 - 250$ MeV/c which could not be explained by the effect of the Coulomb potential alone, which is larger for heavier nuclei. Best agreement with the experimental ratio is obtained with a repulsive K^+ potential $V_K^0 \approx 20$ MeV at the center of the nucleus [51, 52]. It is claimed that the experimental accuracy allows for a determination of V_K^0 to better than ± 3 MeV [51] in an almost model-independent way [52].

Kaon and antikaon production in nucleus-nucleus collisions

K^+ production [53] and K^- production [54] in relativistic nucleus-nucleus collisions was first observed in pioneering experiments at the BEVALAC. At GSI, subthreshold K^- production was measured at the Fragment Separator in Ne and Ni induced reactions [55]. At the KaoS spectrometer [56] a comprehensive program for the study of K^+ production was started and later extended to K^- production [57]. A systematic experimental study of K^+ and K^- production was also carried out at the FOPI spectrometer [58].

The observables most sensitive to the kaon and antikaon in-medium potentials were suggested to be the K^-/K^+ ratio, and the K^+ and K^- flow pattern. The K^-/K^+ ratio is expected to be enhanced by the negative in-medium shift of the K^+K^- threshold. The repulsive K^+ potential is predicted to result in an anti-flow for K^+ mesons, whereas a flow signal in line with the proton flow is expected for K^- mesons due to the attractive potential. However, for K^- mesons the enhancement of flow due to the potential may be compensated by the large absorption cross section due to reactions like $\overline{K}N \to \pi Y$ with Y denoting a Λ or a Σ hyperon. This problem is not present for K^+ mesons which have a large mean free path in nuclear matter, since only the elastic channel is open for the K^+N system at low momenta. In addition, due to the higher energy threshold for K^- production ($E_{lab} = 2.50\,A\,\text{GeV}$ given by the reaction $NN \to NNK\overline{K}$) than for K^+ production ($E_{lab} = 1.58\,A\,\text{GeV}$ given by the reaction $NN \to NK\Lambda$) the experimentally observable yield is much lower for K^- mesons than for K^+ mesons. Therefore both the experimental extraction of a K^- flow signal and its theoretical interpretation is very difficult, and the discussion is restricted to the flow pattern of K^+ mesons.

The KaoS collaboration has measured the azimuthal angular distribution of K^+ mesons in Au+Au collisions at $1\,A\,\text{GeV}$, and observed an azimuthal asymmetry in peripheral and semi-central collisions [59]. Kaons were found to be preferentially emitted perpendicular to the reaction plane. The ratio between out-of-plane to in-plane emission was found to be around 1.7 at mid-rapidity. In RBUU [60] and QMD [61] calculations this behaviour could be well reproduced with a repulsive K^+ nuclear potential, while flat distributions inconsistent with the experimental data are obtained without a K^+ potential.

The FOPI collaboration measured the K^+ and proton sideward flow in $1.69\,A\,\text{GeV}$ Ru+Ru and $1.93\,A\,\text{GeV}$ Ni+Ni collisions [62]. In both reactions it was found that the flow of kaons with low transverse momentum ($p_t < 0.35$ GeV/c) is anticorrelated with the flow of protons, whereas it is correlated for kaons with high transverse momentum ($p_t > 0.35$ GeV/c). The flow pattern is more expressed in the heavier system. The data were compared to RBUU transport calculations [63] with and without medium potential. The version without medium potential fails to describe the low p_t anti-flow phenomenon observed experimentally. In contrast, with a repulsive K^+ potential $U_K \approx 20$ MeV the data in both reactions are quantitatively reproduced. Good qualitative agreement was also found by an independent transport model including similar in-medium effects [64].

Comparative studies of K^+ and K^- production were carried out both in the KaoS and in the FOPI experiment. The KaoS collaboration measured K^+ and K^- yields in C+C [66] and in Ni+Ni [67] collisions. The projectile energies ranged from 0.8 up to $2.0\,A\,\text{GeV}$ for C+C, and up to $1.9\,A\,\text{GeV}$ for Ni+Ni. K^+ and K^- yields were compared at "equivalent" energies $\sqrt{s} - \sqrt{s_{thr}}$ where $\sqrt{s}$ and $\sqrt{s_{thr}}$ are the center of mass available energy and the threshold energy in the nucleon-nucleon system, respectively, as shown in Figure 4 (right). This presentation was chosen in order to minimize systematic uncertainties in the model description.

It came as a surprise that, as can be seen in Figure 4 (right), the subthreshold K^+ and K^- yields measured in both reactions per participant nucleon are practically identical.

kaon in-medium mass

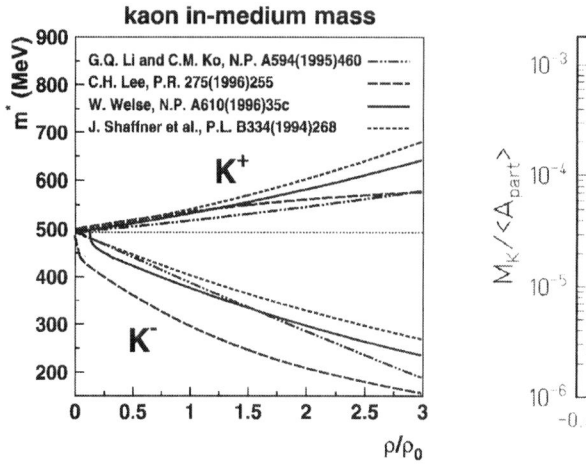

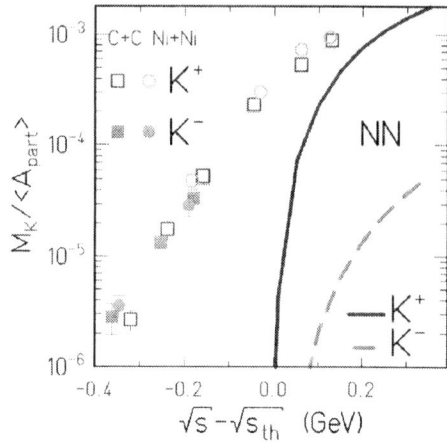

FIGURE 4. **Left:** Theoretical predictions [35, 36, 37, 38] for the kaon and antikaon mass in the nuclear medium as a function of the nuclear density (figure taken from Ref. [49]). **Right:** Subthreshold K^+ and K^- yields per participant nucleon as a function of the NN center of mass energy relative to the respective production threshold [65] measured in C+C [66] and Ni+Ni [67] collisions by the KaoS collaboration. The full and dashed curve are parametrizations to experimental data on elementary K^+ and K^- production in $p + p$ collisions, respectively (see text).

This is in striking contrast to the ratio of K^+ and K^- yields measured in proton-proton collisions at identical energies above the threshold, where the K^+ yield is between two and three orders of magitude larger than the K^- yield, depending on the the excess energy (see Figure 3 in Ref. [68]. This apparent contradiction was interpreted as an enhancement of subthreshold K^- production by a sufficiently strong attractive K^- potential that lowers the K^+K^- energy threshold inside nuclear matter. Transport calculations of the RBUU type [69] obtained K^- yields that are about a factor 5-7 too low if no reduction of the K^+K^- threshold due to an attractive K^- potential is taken into account, but are in agreement with both the K^- yield and the K^- momentum distribution, if an attractive K^- potential $U_K^- \simeq -100$ MeV at $\rho = \rho_0$ is included.

A similar conclusion was drawn from the K^-/K^+ ratio measured at the FOPI spectometer as a function of the rapidity and the center of mass kinetic energy in $1.93\,A\,$GeV Ni+Ni and $1.69\,A\,$GeV Ru+Ru collisions [70]. An increase of the K^-/K^+ ratio with decreasing center of mass kinetic energy ($0.1\,$GeV $< E_{cm}^{kin} < 0.4\,$GeV) and with increasing rapidity ($-1.25 < y^{(0)} < -0.5$ with $y^{(0)} = y^{lab}/y^{CM} - 1$) was observed. In the RBUU transport model of Ref. [69] the measured K^-/K^+ ratio and its dependence on the kinematical variables is well reproduced with a repulsive K^+ potential $U_K(\rho_0) = +30$ MeV and an attractive K^- potential $U_K^-(\rho_0) = -70$ MeV. In contrast, the computed K^-/K^+ ratio is clearly too small, if both potentials were set to zero, and it overshoots the data for the choice $U_K = +30$ MeV, $U_K^- = -120$ MeV. Qualitatively, the same result is obtained with the RBUU calculation of Ref. [71].

A completely different explanation for the enhancement of the K^- yield in nucleus-nucleus compared to proton-proton collisions is given by the authors of another transport model of the QMD type [72]. It is argued that the production channel $BB \rightarrow BBK^+K^-$ (B is a nucleon or a Δ), the only channel allowed for K^- production in pp collisons close to threshold, only marginally contributes in nucleus-nucleus collisions. In contrast, the mesonic $\Lambda(\Sigma)\pi \rightarrow K^-B$ channel is the dominant source of K^- mesons in this model, and thus via $NN \rightarrow NK\Lambda(\Sigma)$ reactions strongly links the K^- yield to the K^+ yield. As a consequence K^- mesons are created at a late stage of the reaction in the expanding system, and are therefore almost insensitive to the in-medium potential according to Ref. [72].

Recently, off-shell transport calculations were developed, in which the production and propagation of antikaons in the nuclear environment is described by dynamical spectral functions [73]. In these calculations $1.5\,A$ GeV Au+Au K^- data measured by the KaoS collaboration [74] are well described if medium effects on both antikaons and pions are taken into account, the K^- yields in $1.8\,A$ GeV C+C [66] and in $1.8\,A$ GeV Ni+Ni collisions [55, 67] are however underpredicted. The effect of the medium potential on the computed K^- yields is rather small, and difficult to be seen experimentally. The strongest enhancement is obtained at low K^- momenta.

Kaonic bound states

A way out of the ambiguities in extracting a nuclear K^- potential from K^- data in nucleus-nucleus collisions may exist in the experimental study of quasi-bound states of K^- mesons in nuclei. For sufficiently strong potential, nuclear bound states of antikaons will exist, as schematically illustrated in Figure 5. If these states could be found, the potential could be deduced in a model independent way from spectroscopic information. As Figure 5 shows, in this case three different classes of states will exist, namely (i) shallow atomic states with a small hadronic level shifts (the only ones known so far, since they can be populated in electromagnetic cascades), (ii) deeply bound atomic states, and (iii) nuclear states essentially bound by strong interaction.

For a moderate binding of the K^- meson by a few 10 MeV nuclear states of large widths of the order of 100 MeV are predicted. In this case nuclear states will proba-bly not be observable experimentally. However, if the K^- potential is deep enough, the K^- binding energy will be large enough ($B \simeq 100$ MeV) to close the $KN \rightarrow \pi\Sigma$ decay channel, and narrower widths are expected. As discussed above, calculations of kaonic bound states [75, 76, 77] based on chiral models [44] predict only moderate K^- bind-ing up to ~ 40 MeV and large width up to ~ 100 MeV. These predictions are in clear disagreement with the results obtained in a potential model for the $\Lambda(1405)$ resonance as bound $\overline{K}N$ state, using Brueckner-Hartree-Fock theory for the binding of $\overline{K}$ in nu-clei [78]. This model predicts a strong $\overline{K}$ binding in very light nuclei with energy levels below the $\pi\Sigma$ threshold, and widths between 20 MeV and 40 MeV. Additional binding is found in Ref. [79] as due to the mechanism of "superdense matter formation" around the K^- meson, which attracts the surrounding nucleons, and reduces their mean inter-nucleonic distance. With this model a particularly strong binding is predicted for the

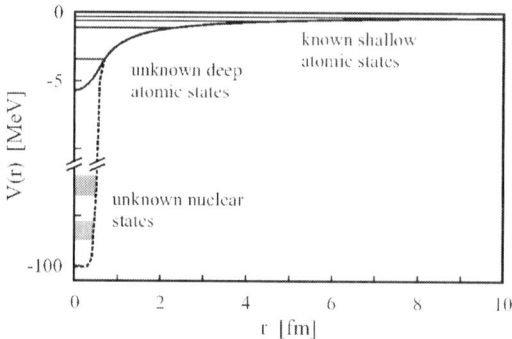

FIGURE 5. Schematic illustration of the K^--nucleus potential as a function of the distance. A short range nuclear potential attractive for antikaons in addition to the long range Coulomb potential may accomodate nuclear K^- states. So far only shallow atomic states are known.

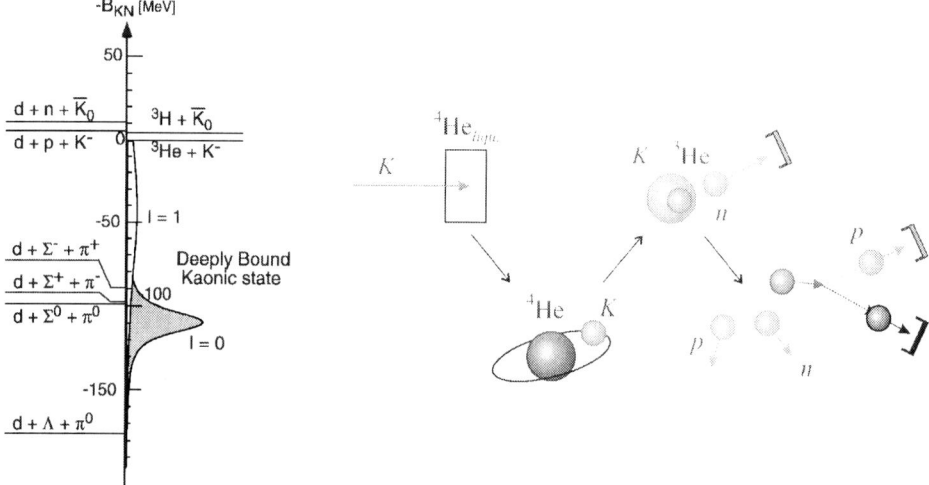

FIGURE 6. **Left:** Predicted [78, 79] energy levels of the bound $^3\text{He}K^-$ system in comparison to the energy thresholds for the possible three-baryon plus meson states with strangeness $S = -1$ and charge $Q = +1$. In the model of Ref. [78, 79] a strongly bound, relatively narrow $I = 0$ state is obtained, while for the $I = 1$ state the binding is weaker and the width is larger. **Right:** Schematic illustration of the $^4\text{He}(\text{stopped } K^-, n)$ reaction studied at KEK [81, 82].

$^3\text{He}K^-$ system with $B \approx 115$ MeV and $\Gamma \approx 22$ MeV, as shown in Figure 6 (left).

Different techniques were proposed for the experimental search for nuclear bound K^- states, as using the (K^-, N) knockout reaction on ^{12}C and ^{28}Si [80], or a $^4\text{He}(\text{stopped } K^-, n)$ reaction [81]. Based on the predictions of Refs. [78, 79], the latter reaction possibly forming a bound $^3\text{He}K^-$ state was recently studied at KEK. As schematically illustrated in Figure 6 (right), an identified K^- beam particle is stopped in a liquid ^4He target, captured into an atomic orbit, and at the end of the electromagnetic cascade undergoes a nuclear Auger effect with a neutron emitted. The measured time of flight reflects the kaonic binding energy. In addition cuts on the charged particle trajectories in the final state are applied in order to tag displaced decays of hyperons [81].

In this experiment very recently evidence was claimed for the observation of a strongly bound kaonic system of $K^- ppn$ with a binding energy $B = 173 \pm 4$ MeV and $\Gamma_{obs} = 14 \pm 4$ MeV [82]. Indications for such a bound system in agreement with the binding energy measured in Ref. [82] were also seen in a re-analysis of FOPI data with respect to possible $d\Lambda$ correlations [83].

Further studies are certainly needed in order to draw definite conclusions on the existence of discrete, strongly bound nuclear kaonic systems. In view of the limited intensity and quality of the existing K^- beams, and their uncertain availability in the future, the planned antiproton facility opens new opportunities to study nuclear kaonic states. At the High-Energy Storage Ring (HESR) an intense antiproton beam at momenta between 0.8 and 15 GeV/c with unprecedented quality will be available [84]. In the HESR ring a high-luminosity mode with $L = 2 \cdot 10^{32}$ cm^{-2}s^{-1} and $\delta p/p \simeq 10^{-4}$ based on stochastic cooling, as well as a high-resolution mode with $L = 10^{31}$ cm^{-2}s^{-1} and $\delta p/p \simeq 10^{-5}$ based on electron cooling, is planned. Antiproton induced reactions on proton and nuclear internal targets will be studied within a comprehensive diversified hadron physics program, using the universal, large acceptance detector system PANDA [84, 85].

Antiprotons are particularly suited for the study of slow hadrons in nuclear matter since the annihilation energy of almost 2 GeV can be converted into particle production without an associated momentum transfer. The most promising among the conceivable reactions producing slow K^- mesons inside a nucleus is considered to be the reaction $\bar{p}[p] \rightarrow \phi\phi$ with $[p]$ being a proton inside a nuclear target, as schematically shown in Figure 7 (left). A quite sizeable cross section $\sigma = 3.73 \pm 0.24\,\mu$b has been measured by the JETSET Collaboration at 1.4 GeV/c incident $\bar{p}$ momentum for the $\bar{p}p \rightarrow \phi\phi$ reaction in free space [86]. Due to the similar masses of the $\bar{p}$ projectiles and ϕ mesons, it is kinematically possible to create a pair of ϕ mesons, one of which takes the dominant fraction of the $\bar{p}$ momentum (called "fast" ϕ), and one of which has a very low momentum relative to the target nucleus (called "slow" ϕ). Since ϕ mesons are a source of low momentum K^+K^- pairs ($p_K = 127$ MeV/c in the ϕ rest frame), there will be a high chance for the K^- meson in the decay of the "slow" ϕ to be captured into a quasi-bound nuclear state if such a state exists. The kinematical situation is illustrated in Figure 7 (right) for an incident $\bar{p}$ momentum of 2 GeV/c with the center of mass polar angle of the "fast" ϕ confined within 30°. As can be seen in the figure, a dominant part of the populated phase space is associated with a momentum transfer less than 0.2 GeV/c. This is much smaller than the internal K^- momentum in a bound state even for a shallow K^- potential of a few 10 MeV. Experimentally, the detection of a fast K^+K^- pair with the Forward Spectrometer of PANDA, and a slow K^+ meson with the Target Spectrometer is required. From the measurement of the 4-momenta of these three particles the binding energy of the residual K^- meson can be unambiguously determined. As long as the annihilation process takes place on a single nucleon in the target nucleus, the missing mass region below the K^- mass is *only* populated in a reaction resulting in a quasi-bound state. Background reactions associated with Λ or Σ hyperons require at least two-step processes. Therefore we expect that the proposed reaction is particularly clean, and allows even to obtain conclusive information for the case that the K^- potential is shallow

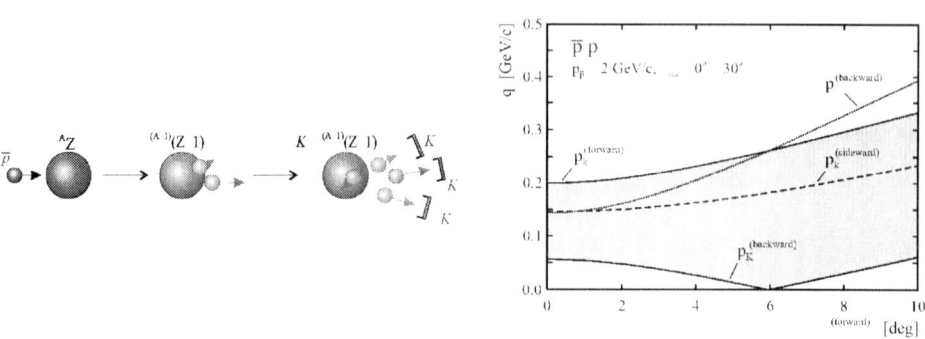

FIGURE 7. **Left:** Schematic illustration of the reaction ${}^{A}Z(\bar{p},(K^{+}K^{-})_{\phi}K^{+})[{}^{(A-1)}(Z-1)K^{-}]_{\text{bound}}$ proposed to be studied at HESR/PANDA. **Right:** Region of momentum transfer covered in the reaction ${}^{A}Z(\bar{p},(K^{+}K^{-})_{\phi}K^{+})[{}^{(A-1)}(Z-1)K^{-}]_{\text{bound}}$ as a function of the polar angle of the "fast" ϕ. The boundaries of the shaded region are given by the K^{+} from the decay of the "slow" ϕ emitted in forward and in backward direction, respectively, the dashed line is for sideward emission of the $K^{+}K^{-}$ pair. The momentum of the "slow" ϕ is also shown by the dotted line.

and and quasi-bound states have a large width.

Although the cross section for the proposed reaction is unknown, the achievable rates are expected to be sufficient for a good statistics measurement. Based on the measured cross section for the elementary reaction $\bar{p}p \rightarrow \phi\phi$ [86], on the polar angle cut $\theta_{\phi}^{cm} < 30^{o}$ (see Figure 7 (right)), and on an assumed effective proton number in the target nucleus $N_{\text{eff}} = 10^{-3}$ (which accounts for absorption effects and momentum mismatch, and seems to be conservative), a cross section $\sigma \simeq 0.5$ nb is obtained as a rough estimate. With a luminosity $L = 2 \cdot 10^{31}$ cm^{-2}s^{-1} this corresponds to ~ 900 events per day. Detailed simulations of an experimental study of the proposed reaction at PANDA are planned in the near future.

CONCLUSION

The study of deeply bound pionic states has added valuable information to the understanding of the properties of pions inside the nuclear medium, particularly on the s-wave isovector interaction which was previously not well defined. In the course of this experimental program, pionic $1s$ states have been identified in 5 nuclei. In the most recent study of pionic Sn, the isotope dependence of deeply bound pionic states was measured for the first time. From a common fitting of the pionic $1s$ states in light symmetric nuclei and in Sn isotopes, the isovector parameter b_{1} was deduced. It is found that the repulsive isovector interaction is significantly enhanced as compared to the πN interaction in free space, which can be understood in terms of a partial restoration of chiral symmetry in nuclear matter.

K^{+} production in proton-nucleus collisions, as well as K^{+} and K^{-} production in nucleus-nucleus collisions was experimentally studied in order to better understand the in-medium properties of these mesons. The repulsive nuclear potential for K^{+} mesons can be deduced in an almost model-independent way to be $U_K \approx +20$ MeV from the shape of the K^+ momentum spectrum measured in proton-nucleus collisions. Such a repulsive K^+ potential is consistent with the K^+ anti-flow observed in nucleus-nucleus collisions, as predicted by several transport models. In contrast, the interpretation of the experimentally found enhancement of the K^-/K^+ yield ratio in nucleus-nucleus collisions as compared to a much smaller ratio in elementary proton-proton collisions is much more model-dependent. Different transport models obtain a different sensitivity of the K^- yield to the nuclear K^- potential, and disagree in the depth of the potential, required to reproduce the measured K^- production cross sections.

This uncertainty may be resolved in an experimental study of nuclear bound states of antikaons, which are predicted to exist as discrete observable states for sufficiently deep antikaon potentials. First indications for the existence of nuclear systems with a bound K^- meson were recently seen. The future antiproton facility at GSI with the HESR ring and the PANDA detector opens new oppertunities for such studies. As particularly promising in populating bound nuclear K^- states, $\bar{p}$ annihilation on a proton in a nucleus into two ϕ mesons is proposed. From the fast K^+K^- pair and the slow K^+ meson in the final state, the 4-vectors of which are measured with the PANDA detector, the K^- binding energy can be uniquely reconstructed.

ACKNOWLEDGMENTS

The author would like to thank the members of the GSI-S236 Collaboration for helpful discussions, as well as M. Büscher, N. Herrmann, and P. Senger for their help in providing information on the ANKE, FOPI, and KaoS results.

REFERENCES

1. Y. Nambu and G. Jona-Lasinio, Phys. Rev. **122**, 345 (1961); **124**, 246 (1961).
2. T. Hatsuda and T. Kunihiro, Phys. Lett. B **185**, 304 (1987); Phys. Rep. **247**, 221 (1994), and references therein.
3. S. Klimt, M. Lutz, U. Vogl, and W. Weise, Nucl. Phys. A **516**, 429 (1990); U. Vogl and W. Weise, Prog. Part. Nucl. Phys. **27**, 195 (1991) and references therein.
4. E.G. Drukarev and E.M. Levin, Nucl. Phys. A **511**, 679 (1990); A **532**, 695 (1991).
5. T.D. Cohen, R.J. Furnstahl, and D.K. Griegel, Phys. Rev. C **45**, 1881 (1992).
6. M. Lutz, S. Klimt, and W. Weise, Nucl. Phys. A **542**, 52 (1992).
7. G.E. Brown and M. Rho, Phys. Rev. Lett. **66**, 2720 (1991).
8. W. Weise, Nucl. Phys. A **553**, 59c (1993).
9. H.C. Schröder et al., Phys. Lett. B **469**, 25 (1999).
10. P. Hauser et al., Phys. Rev. C **58**, R1869 (1998).
11. B. Loiseau, T.E.O. Ericson, and A.W. Thomas, Nucl. Phys. A **684**, 380 (2001).
12. M. Ericson and T.O.E. Ericson, Ann. Phys. **36**, 323 (1966).
13. T. Ericson and W. Weise, "Pions and Nuclei", Clarendon Press, Oxford, 1988.
14. J. Nieves, E. Oset, and C. García-Recio, Nucl. Phys. A **A554**, 509 (1993).
15. Y. Tomozawa, Nuovo Cim. A **46**, 707 (1966).
16. S. Weinberg, Phys. Rev. Lett. **17**, 616 (1966).
17. M. Gell-Mann, R.J. Oakes, and B. Renner, Phys. Rev. **175**, 2195 (1968).
18. V. Thorsson and A. Wirzba, Nucl. Phys. A **589**, 633 (1995).

19. E.E. Kolomeitsev, N. Kaiser, and W. Weise, Phys. Rev. Lett. **90**, 092501 (2003).
20. E. Friedman and G. Soff, J. Phys. G **11**, L37 (1985).
21. H. Toki and T. Yamazaki, Phys. Lett. B **213**, 129 (1988); H. Toki, S. Hirenzaki, T. Yamazaki, and R.S. Hayano, Nucl. Phys. **A501**, 653 (1989).
22. Y. Umemoto, S. Hirenzaki, K. Kume, and H. Toki, Phys. Rev. C **62**, 024606 (2000).
23. H. Toki, S. Hirenzaki, and T. Yamazaki, Nucl. Phys. **A530**, 679 (1991).
24. T. Yamazaki *et al.*, Z. Phys. A **355**, 219 (1996).
25. T. Yamazaki *et al.*, Phys. Lett. B **418**, 246 (1998).
26. H. Gilg *et al.*, Phys. Rev. C **62**, 025201 (2000).
27. K. Itahashi *et al.*, Phys. Rev. C **62**, 025202 (2000).
28. H. Geissel *et al.*, Phys. Rev. Lett. 88, 122301 (2002).
29. K. Suzuki *et al.*, to be published in Phys. Rev. Lett., preprint nucl-ex/0211023.
30. H. Geissel *et al.*, Phys. Lett. B **549**, 64 (2002).
31. C. Batty, E. Friedman, and A. Gal, Phys. Rep. **287**, 385 (1997).
32. T. Yamazaki and S. Hirenzaki, Phys. Lett. B **557**, 20 (2003).
33. D.E. Kaplan and A.E. Nelson, Phys. Lett. B **175**, 57 (1986).
34. A.E. Nelson and D.E. Kaplan, Phys. Lett. B **192**, 193 (1987).
35. G.Q. Li and C.M. Ko, Nucl. Phys. A **594**, 460 (1995)
36. C.H. Lee, Phys. Rep. **275**, 255 (1996).
37. T. Waas, N. Kaiser, and W. Weise, Phys. Lett. B **365**, 12 (1996); **379**, 34 (1996); W. Weise, Nucl. Phys. A **610**, 35c (1996); T. Waas and W. Weise, Nucl. Phys. A **625**, 287 (1997).
38. J. Schaffner, A. Gal, I.N. Mishustin, H. Stöcker, and W. Greiner, Phys. Lett. B **334**, 268 (1994).
39. E. Friedman, A. Gal, and C.J. Batty, Nucl. Phys. A **579**, 518 (1994).
40. B.R. Martin, Nucl. Phys. B **94**, 413 (1975).
41. D. Marlow *et al.*, Phys. Rev. C **25**, 2619 (1982).
42. A.D. Martin, Nucl. Phys. B **179**, 33 (1981).
43. M. Iwasaki *et el.*, Phys. Rev. Lett. **78**, 3067 (1997); T.M. Ito *et al.*, Phys. Rev. C **58**, 2366 (1998).
44. A. Ramos and E. Oset, Nucl. Phys. A **671**, 481 (2000).
45. A. Sibirtsev and W. Cassing, Nucl. Phys. A **641**, 476 (1998).
46. A. Gal, Nucl. Phys. A **691**, 268c (2001).
47. M.F.M. Lutz and C.L. Korpa, Nucl. Phys. A **700**, 309 (2002).
48. A. Cieplý, E. Friedman, A. Gal, and J. Mareš, Nucl. Phys. A **696**, 171 (2001).
49. N. Herrmann, private communication (2003); A. Devismes for the FOPI collaboration, talk presented at SQM01 (http://www-aix.gsi.de/ fopiwww/pub/fopi_www_conf.html).
50. S. Barsov *et al.*, Nucl. Instr. Meth. A **462**, 364 (2001).
51. M. Nekipelov *et al.*, Phys. Lett. B **540**, 207 (2002).
52. Z. Rudy, W. Cassing, L. Jarczyk, B. Kamys, and P. Kulessa, Eur. Phys. J. A **15**, 303 (2002).
53. S. Schnetzer *et al.*, Phys. Lett. B **49**, 989 (1982); Phys. Rev. C **40**, 640 (1989).
54. A. Shor *et al.*, Phys. Rev. Lett. **63**, 2192 (1989).
55. A. Schröter *et al.*, Z. Phys. A **350**, 101 (1994); A. Gillitzer *et al.*, Prog. Part. Nucl. Phys. **30**, 97 (1993); A. Gillitzer, Prog. Part. Nucl. Phys. **42**, 42 (1999).
56. P. Senger *et al.*, Nucl. Instr. Meth. A **327**, 393 (1993).
57. P. Senger and H. Ströbele, J. Phys. G **25**, R59 (1999).
58. A. Gobbi *et al.*, Nucl. Instr. Meth. A **324**, 156 (1993).
59. Y. Shin *et al.*, Phys. Rev. Lett. **81**, 1576 (1998).
60. G.Q. Li, C.M. Ko, and G.E.Brown, Phys. Lett. B **381**, 17 (1996).
61. Z.S. Wang, C. Fuchs, A. Faessler, T. Gross-Boelting, Eur. Phys. A **5**, 275 (1999).
62. P. Crochet *et al.*, Phys. Lett. B **486**, 6 (2000).
63. W. Cassing and E.L. Bratkovskaya, Phys. Rep. **308**, 65 (1999).
64. G.Q. Li and G.E. Brown, Nucl. Phys. A **636**, 487 (1998).
65. P. Senger, private communication (2003).
66. F. Laue *et al.*, Phys. Rev. Lett. **82**, 1640 (1999).
67. R. Barth *et al.*, Phys. Rev. Lett. **78**, 4007 (1997).
68. P. Moskal *et al.*, J. Phys. G **28**, 1777 (2002), and references therein.
69. W. Cassing, E.L. Bratkovskaya, U. Mosel, S. Teis, and A. Sibirtsev, Nucl. Phys. A **614**, 415 (1997).
70. K. Wiśniewski *et al.*, Eur. J. Phys. J. A **9**, 515 (2000).

71. G.Q. Li and G.E. Brown, Phys. Rev. C **58**, 1698 (1998).
72. Ch. Hartnack, H. Oeschler, and J. Aichelin, Phys. Rev. Lett. **90**, 102302 (2003).
73. W. Cassing, L. Tolós, E.L. Bratkovskaya, and A. Ramos, Nucl. Phys. A **727**, 59 (2003).
74. A. Förster for the KaoS collaboration, J. Phys. G **28**, 2011 (2002).
75. S. Hirenzaki, Y. Okamura, H. Toki, E. Oset, and A. Ramos, Phys. Rev. C **61**, 055205 (2000).
76. E. Oset, S. Hirenzaki, Y. Okamura, A. Ramos, H. Toki, and M.J. Vicente Vacas, Act. Phys. Pol. B **31**, 2285 (2000).
77. A. Baca, C. García-Recio, and J. Nieves, Nucl. Phys. A **673**, 335 (2000).
78. Y. Akaishi and T. Yamazaki, Phys. Rev. C **65**, 044005 (2002).
79. A. Doté, Y. Akaishi, H. Horiuchi, and T. Yamazaki, preprint nucl-th/0207085 (2002), submitted to Phys. Rev. Lett..
80. T. Kishimoto, Phys. Rev. Lett. **83**, 4701 (1999).
81. M. Iwasaki *et al.*, Nucl. Instr. Meth. A **473**, 286 (2001); Y. Akaishi and T. Yamazaki, Nucl. Phys. A **684**, 409c (2001).
82. M. Iwasaki *et al.*, preprint nucl-ex/0310018, submitted to Phys. Lett. B.
83. T. Yamazaki, talk given at the 2nd workshop "Challenges and Opportunities at the New International Accelerator Facility for Beams of Ions and Antiprotons at Darmstadt", GSI Darmstadt, October 14-17, 2003.
84. "An International Accelerator Facility for Beams of Ions and Antiprotons", Conceptual Design Report, GSI Darmstadt, November 2001; http://www.gsi.de/GSI-Future/cdr/.
85. P. Gianotti for the PANDA collaboration, these proceedings.
86. L. Bertolotto *et al.*, Phys. Lett. B **345**, 325 (1995).

759

Medium Modifications of Hadrons in Photon Induced Reactions

S. Schadmand *for the TAPS and A2 Collaborations*

II. Physikalisches Institut, Justus-Liebig-Universität Gießen

Abstract. Indications for in-medium modifications of hadron properties are reported from photoabsorption and meson production experiments. Strong medium modifications are observed in inclusive photoabsorption experiments and theoretical models investigate the in-medium dynamics of baryon resonances and their coupling to mesons. Recent experiments study the in-medium behavior of scalar and vector mesons where theoretical models expect in-medium modifications of the meson spectral functions that might be connected to partial restoration of chiral symmetry.

INTRODUCTION

Photoabsorption experiments demonstrate the complex structure of the nucleon and its excitation spectrum. On the free nucleon, several resonance regions contain the contributions of broad and overlapping nucleon resonances. The lowest resonance, the $\Delta(1232)$, is prominently excited by incident photons of 0.2–0.5 GeV. The energy regime of E_γ=0.5-0.9 GeV is called the second resonance region where $P_{11}(1440)$, $D_{13}(1520)$, and $S_{11}(1535)$ contribute. At even higher energies a third resonance region is visible.

The left panel of Fig. 1 shows the photoabsorption cross section [1] and provides a compilation of experimental cross sections for meson photoproduction from the proton, from MAMI, ELSA, GRAAL, and JLAB. The shapes of the meson cross sections reflect the resonance structure observed in photoabsorption. The curve indicates the sum of the meson production cross sections up to E_γ=800MeV and shows that the resonance structures can almost entirely be attributed to meson production. Single pion production dominates the region of the Δ resonance. Also, the three resonances in the second resonance region decay roughly to 50% via single pion emission. Above the two-pion production threshold ($E_\gamma \approx 0.4$ GeV), single π production looses in importance. The meson production channels involving charged pions are dominant as expected in electromagnetic excitation processes. The well-understood η photoproduction is characteristic for the $S_{11}(1535)$ resonance.

Nuclear photoabsorption and photofission experiments have provided rather different and intriguing results. The right panel of Fig. 1 shows a comparison of the photoabsorption cross sections on the free proton and on nuclei [2]. Overall, the cross sections scale with the atomic mass number due to the fact that the photon probe illuminates the entire nucleus. The figure displays the results as an average over the nuclear systematics. The Δ resonance is broadened and slightly shifted while the higher resonance regions seem to have disappeared. In fact, cross sections from nuclei heavier than ^{6}Li exhibit this behavior. The depletion of nuclear cross sections in the second resonance region has been

CP717, *Hadron Spectroscopy: Tenth International Conference,*
edited by E. Klempt, H. Koch, and H. Orth
© 2004 American Institute of Physics 0-7354-0197-7/04/$22.00

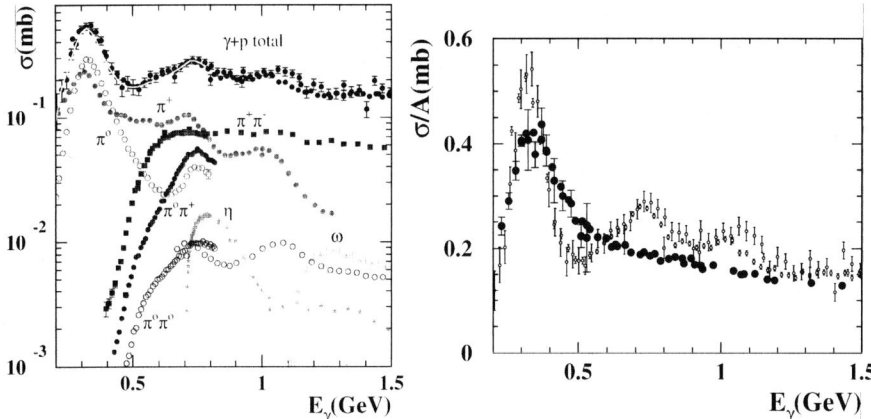

FIGURE 1. <u>Left</u>: Photoabsorption cross section on the free proton [1] and experimental decomposition into meson production channels MAMI, ELSA, GRAAL, and JLAB. The curve indicates the sum of the meson production cross sections and describes the total absorption. <u>Right</u>: Nuclear photoabsorption cross section (big symbols) representing an average over various nuclei compared to the elementary cross section. The plot is adapted from ref. [2].

taken as evidence for modifications of hadron properties in the nuclear medium. This phenomenon has not yet been explained in a model independent way.

It has been argued that an in-medium broadening of the $D_{13}(1520)$ resonance is a likely cause of the suppressed photoabsorption cross section [3]. The broadening could arise from a coupling of the resonance to the $N\rho$ final state since the ρ-meson itself is expected to broaden in the nuclear medium [4]. In another approach [5], the disappearance of the peak is modelled via a cooperative effect of the interference in double pion production processes, Fermi motion, collision broadening of the Δ and N^* resonances, and pion distortion in the nuclear medium. A deeper understanding of the situation is anticipated from the experimental study of meson photoproduction on nucleons embedded in nuclei in comparison to studies on the free proton.

MEDIUM EFFECTS

Experimental information on meson photoproduction off nuclei is scarce. Single charged pion cross sections have been measured below the double pion production threshold ($E_\gamma \approx 0.4$ GeV) [6]. The single pion cross sections exhibit a broad maximum between 0.2 and 0.4 GeV and explain the shape of the photoabsorption data indicating that the $\Delta(1232)$ resonance is a dominant producer of single pions. Single π° photoproduction has been studied into the second resonance region. The systematic study over a series of nuclei has not provided any hint for a depletion of the resonance yield [7]. The observed reduction and change of shape can be explained by trivial effects like absorption, Fermi smearing and Pauli blocking, and collisional broadening. A similar result is obtained in systematic studies of η photoproduction cross section from nuclei. In Fig. 2, the

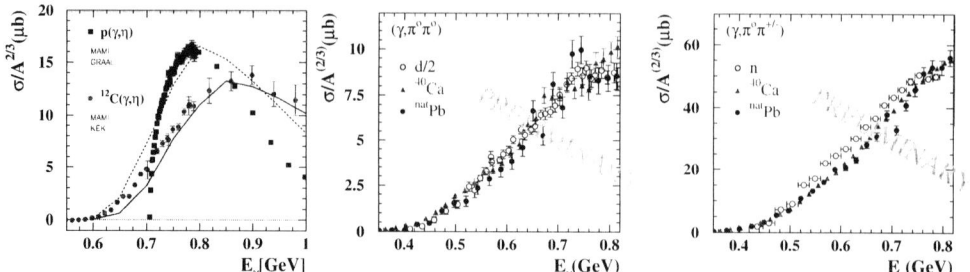

FIGURE 2. Left: η photoproduction cross section on the proton [9, 10] and on carbon [11, 12]. Middle and Right: Preliminary total cross sections for $\pi\pi$ photoproduction from lead [13] along with results from the deuteron [14, 15]. The nuclear cross sections are divided by $A^{2/3}$, the $\pi^\circ\pi^\circ$ deuteron cross section by 2.

carbon cross section, divided by $A^{2/3}$, agrees in overall magnitude with the proton cross section, which is a sign of strong η absorption in the final state. Around the η threshold, the influence of Fermi smearing is obvious. The comparison to the proton cross section indicates a mass shift of the resonance. This could be understood within the coupled channel BUU calculation of [8]. Here, a momentum dependent (solid curve) and a momentum independent (dashed curve) potential for nucelons and excited states are considered. In the case of a momentum dependent potential, the resonance with its relatively higher momentum would be less deeply bound, leading to a 'heavier' resonance. Still, the nuclear data clearly show the line shape of the $S_{11}(1535)$ resonance and not a depletion as observed in photoabsorption.

The middle and right panels of Fig. 2 show preliminary cross sections for $\pi^\circ\pi^\circ$ and $\pi^\circ\pi^\pm$ photoproduction on calcium and lead from a recent TAPS analysis. The nuclear cross sections are divided by $A^{2/3}$ and compared to results from proton and from nucleons bound in deuterons. Above the η threshold, background from from $\eta \to 3\pi$ has been subtracted. With the scaling with $A^{2/3}$, the nuclear data agree almost exactly with the cross sections on the nucleons. Thus, the total nuclear $\pi\pi$ cross sections do not seem to show any modification beyond absorption effects. It may be speculated that the strong 2π decay branch via Δ intermediate states ($N^\star \to \Delta\pi \to N\pi\pi$), together with the fact that the Δ resonance itself does not dramatically change in medium, dominate this behavior. Nevertheless, it should be noted that in the reaction $\pi^\circ\pi^\pm$, the two pions can stem from the decay of the ρ meson while the decay $\rho \to \pi^\circ\pi^\circ$ is forbidden. Accordingly, detailed studies of differential cross sections might reveal different modifications of the $\pi\pi$ correlations. In the following, first measurements of the invariant mass distributions of the pion pairs are presented.

DOUBLE PION PHOTOPRODUCTION AT THRESHOLD

Double pion production on nuclei is used as a tool to study the in-medium $\pi\pi$ interaction. The key issue is the behavior of scalar mesons, like the σ meson, in the nuclear medium. Some theoretical models expect a dropping of the σ meson mass as a function of nuclear density on account of partial restoration of chiral symmetry [16, 17, 18]. Other models interpret the σ meson as a scalar $\pi\pi$ scattering resonance [19]. Here, the meson-meson interaction in the scalar-isoscalar channel is studied in the framework of a chiral-unitary approach at finite baryonic density. The model dynamically generates the σ resonance, reproducing the meson-meson phase shifts in vacuum and accounts for the absorption of the pions in the nucleus. It qualitatively predicts a mass shift as observed in the data. The basic ingredient driving this shift is the p-wave interaction of the pion with the baryons in the medium, resulting in an in-medium modification of the $\pi\pi$ interaction.

In-medium modifications of the $\pi\pi$ interaction have been studied in pion-induced reactions on nuclei like $A(\pi^+, \pi^+\pi^-)$ [20] and $A(\pi^-, \pi^0\pi^0)$ [21]. However, pion-induced reactions occur at fractions of the normal nuclear density. This complicates the interpretation of the data in terms of medium effects. Photon-induced reactions can reach normal nuclear densities and should thus be more sensitive to in-medium modifications. A difficulty is the fact that distortion of differential spectra could occur from rescattering of the final state pions. For this reason, the experiments described in the following employ photon beam energies below 460 MeV where the pion momenta are moderate ($p_\pi \approx 100$ MeV) and thus their mean free path long (>10 fm) [22]. In this kinematic situation it may be assumed that the pions are created in the nuclear surface ($A^{2/3}$ scaling) and are emitted almost without final state interactions with the nucleons. A correlation between π° and $\pi^\pm$ on account of the ρ meson is not to be expected at these energies.

Fig. 3 shows the TAPS results for threshold $\pi\pi$ production on nuclei [23]. The

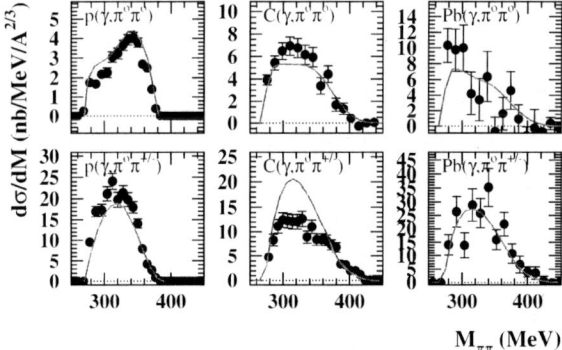

FIGURE 3. Differential cross sections of the reaction $A(\gamma, \pi^\circ\pi^\circ)$ (top row) and $A(\gamma, \pi^\circ\pi^\pm)$ (bottom row) with $A={}^1$H,^{12}C,^{nat}Pb for incident photons in the energy range of 400-460 MeV. Error bars denote only the statistical uncertainties. The figure is adapted from [23]. The curves are the prediction from [24].

curves are the prediction from [24]. In agreement with the prediction, the data indicate a dropping of the $\pi^\circ\pi^\circ$ mass (top row) in medium. This behavior is not observed for the $\pi^\circ\pi^\pm$ channel (bottom row) which can not stem from a σ meson. The comparison with the experimental data hints at the nature of the σ meson as a $\pi\pi$ resonance. It would be most desirable to confront this observation with QCD models which treat the σ as a $q\bar{q}$ state and explicitly take chiral-symmetry restoration into account.

CONCLUSIONS

Meson photoproduction on the free proton and on nucleons embedded in nuclei have been discussed. These studies are employed to investigate possible medium modifications of nucleon resonances by tagging on their characteristic meson decay. The experimental results from photoabsorption on nuclei at $E_\gamma <500$ MeV show clear signs of medium modifications by the fact that the resonances above the $\Delta(1232)$ seem to have disappeared. In meson photoproduction, the total cross sections do not seem to reflect this phenomenon. However, first detailed studies of double pion photoproduction from nuclei do indicate significant modifications of the $\pi^\circ \pi^\circ$ interaction in the nuclear medium.

ACKNOWLEDGMENTS

The $(\gamma, \pi\pi)$ data are part of the PhD work of S. Janssen, F. Bloch, and M. Kotulla. This work was supported by Deutsche Forschungsgemeinschaft, the U.K. Engineering and Physical Sciences Research Council, and Schweizerischer Nationalfond.

REFERENCES

1. http://pdg.lbl.gov/xsect/contents.html
2. V. Muccifora *et al.*, Phys. Rev. C **60**, 064616 (1999).
3. U. Mosel *et al.*, Prog. Part. Nucl. Phys. **42** (1999) 163
4. F. Klingl *et al.*, Nucl. Phys. **A624** (1997) 527
5. M. Hirata, *et al.*, Phys. Rev. C **66** (2002) 014612
6. J. Arends *et al.*, Z. Phys. A **305**, 205 (1982).
7. B. Krusche *et al.*, Phys. Rev. Lett. **86**, 4764 (2001).
8. J. Lehr, University of Giessen, private communication.
9. B. Krusche *et al.*, Phys. Rev. Lett. **74**, 3736 (1995).
10. F. Renard *et al.*, Phys. Lett. B **528**, 215 (2002)
11. M. Roebig-Landau *et al.*, Phys. Lett. B **373**, 45 (1996).
12. H. Yamazaki *et al.*, Nucl. Phys. A **670**, 202 (2000).
13. S. Janssen, PhD thesis, University of Giessen, to be published, 2002
14. V. Kleber *et al.*, Eur. Phys. J. A **9**, 1 (2000).
15. A. Zabrodin *et al.*, Phys. Rev. C **60**, 055201 (1999).
16. M. Lutz *et al.*, Nucl. Phys. **A542** (1992) 521
17. T. Hatsuda *et al.*, Phys. Rev. Lett. 82 (1999) 2840
18. R. Rapp et al., Phys. Rev. C 59 (1999) 1237
19. L. Roca, E. Oset and M. J. Vicente Vacas, Phys. Lett. B **541**, 77 (2002)
20. F. Bonutti *et al.* [CHAOS collaboration], Nucl. Phys. A **677**, 213 (2000)
21. A. Starostin *et al.* [Crystal Ball Collaboration], Phys. Rev. Lett. **85**, 5539 (2000).
22. W. Cassing *et al.*, Phys. Rep. **188** (1990) 363
23. J. G. Messchendorp *et al.*, Phys. Rev. Lett. **89**, 222302 (2002)
24. L. Roca, E. Oset and M. J. Vicente Vacas, Phys. Lett. B **541**, 77 (2002) and priv. commun.

Baryon Interactions in Nuclear and Hypernuclear Matter

H. Lenske, C. Keil, R.Shyam[†]

Institut für Theoretische Physik, Universität Giessen, D-35392 Giessen
[†]Saha Institute of Nuclear Physics, Calcutta, India

Abstract.
The production and structure of Λ hypernuclei are investigated in field theoretical models. The production in coherent p+A reactions is investigated by means of a resonance model. Results of exploratory calculations for associated strangeness production are presented for proton reactions on ^{40}Ca. The target single particle wave functions are obtained from DDRH theory with density dependent Dirac-Brueckner meson-baryon vertices. The dependence of the in-medium vertices on density is discussed.

INTRODUCTION

A major goal of hadron physics is to understand the interactions among the various members of the baryonic and mesonic flavor multiplets. For that aim an important class of processes is the implementation of a strange hadron into a nuclear environment by means of an appropriate reaction and investigations of the subsequent evolution of that system. Here we consider interaction within the lowest baryon flavor octet. We are especially interested in associated $K^+\Lambda$ strangeness production in hadronic reactions. As an interesting alternative to more common approaches using pion and kaon reactions [1] we consider strangeness production in exclusive $p+A \to_\Lambda (A+1)+K^+$ reactions at incident kinetic energies in the COSY energy regime. A field theoretical model is used describing associated strangeness production by the excitation of intermediate nucleon resonances decaying into the $K^+\Lambda$ channel. Since we are aiming at hypernuclei only those processes in which the hyperon in captured in a bound orbit are considered. Clearly, since these are highly selective reactions at large momentum transfer, a high sensitivity on nuclear wave functions has to be expected. We use DDRH theory [2] as a state-of-the-art relativistic field theory for the nuclear structure calculations, known to describe both normal nuclei [3] and hypernuclei [4] very accurately. This allows us to investigate the production cross sections for various nuclear states thus establishing the link from the production to the spectroscopy of the final hypernuclei and sampling their wave function at momenta which otherwise are not accessible.

CP717, *Hadron Spectroscopy: Tenth International Conference,*
edited by E. Klempt, H. Koch, and H. Orth
© 2004 American Institute of Physics 0-7354-0197-7/04/$22.00

ASSOCIATED STRANGENESS PRODUCTION IN PROTON-NUCLEUS REACTIONS

The production of Λ-hypernuclei with high intensity proton beams is an interesting alternative to more common pion and antikaon induced production scenarios [1] . In principle, the production process can proceed in a variety of reactions like $p + A(N, Z) \rightarrow {}_\Lambda B(N-1, Z) + n + K^+$, $p + A(N, Z) \rightarrow {}_\Lambda B(N, Z-1) + p' + K^+$, and $p + A(N, Z) \rightarrow {}_\Lambda B(N, Z) + K^+$ where N and Z are the neutron and proton numbers, respectively, in the target nucleus. Here, we study the last reaction [to be referred to as $A(p, K^+)_\Lambda B$] which is exclusive in the sense that the final channel is a two body system. In this reaction the momentum transfer to the nucleus is much larger than in (π^+, K^+) reaction, about 1.0 GeV/c as compared to about 0.330 GeV/c in forward direction.

The elementary production process is a two-nucleon mechanism (TNM) [6, 7] where the kaon production proceeds via a collision of the projectile nucleon with one of its target counterparts, thereby exciting intermediate baryonic resonances decaying in turn into a kaon and a Λ hyperon. The N^*1650), N^*1710), and N^*1720) states are espcially important [6]. The nucleon and the hyperon are captured into nuclear orbitals while the kaon is rescattered onto its mass shell. Three active bound state baryon wave functions are taking part in the reaction process allowing the large momentum transfer to be shared among the participants.

We use a field theoretical approach with effective Lagrangians for the nucleon-nucleon-pion $(NN\pi)$ and N^*-nucleon-pion $(N^*N\pi)$ vertices [6]. They are given by

$$\mathscr{L}_{NN\pi} = -\frac{g_{NN\pi}}{2m_N} \bar{\Psi} \gamma_5 \gamma_\mu \tau \cdot (\partial^\mu \Phi_\pi) \Psi. \tag{1}$$

$$\mathscr{L}_{N^*_{1/2}N\pi} = -g_{N^*_{1/2}N\pi} \bar{\Psi}_{N^*_{1/2}} i\Gamma \tau \Phi_\pi \Psi + \text{h.c.}, \tag{2}$$

$$\mathscr{L}_{N^*_{3/2}N\pi} = \frac{g_{N^*_{3/2}N\pi}}{m_\pi} \bar{\Psi}_\mu^{N^*} \Gamma_\pi \tau \cdot \partial^\mu \Phi_\pi \Psi + \text{h.c.}. \tag{3}$$

where m_N denotes the nucleon mass. The operator Γ (Γ_π) is either γ_5 (unity) or unity (γ_5) depending upon the parity of the resonance being even or odd, respectively. Following Ref. [7] we use a pseudovector (PV) coupling for the $NN\pi$ vertex and a pseudoscalar (PS) one for the $N^*_{1/2}N\pi$ vertex. The effective Lagrangians for the resonance-hyperon-kaon vertices are written as

$$\mathscr{L}_{N^*_{1/2}\Lambda K^+} = -g_{N^*_{1/2}\Lambda K^+} \bar{\Psi}_{N^*} i\Gamma \tau \Phi_{K^+} \Psi + \text{h.c.}. \tag{4}$$

$$\mathscr{L}_{N^*_{3/2}\Lambda K^+} = \frac{g_{N^*_{3/2}\Lambda K^+}}{m_{K^+}} \bar{\Psi}_\mu^{N^*} \Gamma_\pi \tau \cdot \partial^\mu \Phi_{K^+} \Psi + \text{h.c.}. \tag{5}$$

Here, $\Psi_\mu^{N^*}$ is the vector spinor for the spin-$\frac{3}{2}$ particle. Further discussions about the vertices and coupling constants involving such particles are found in Refs. [6, 7, 8]. The amplitude for graph 1b with spin-$\frac{1}{2}$ baryonic resonance, for example, is given by,

$$M_{1b}(N^*_{1/2}) = C_{iso}^{1b} \left(\frac{g_{NN\pi}}{2m_N} \right) (g_{N^*_{1/2}N\pi})(g_{N^*_{1/2}\Lambda K^+}) \bar{\psi}(p_2) \gamma_5 \gamma_\mu q^\mu$$

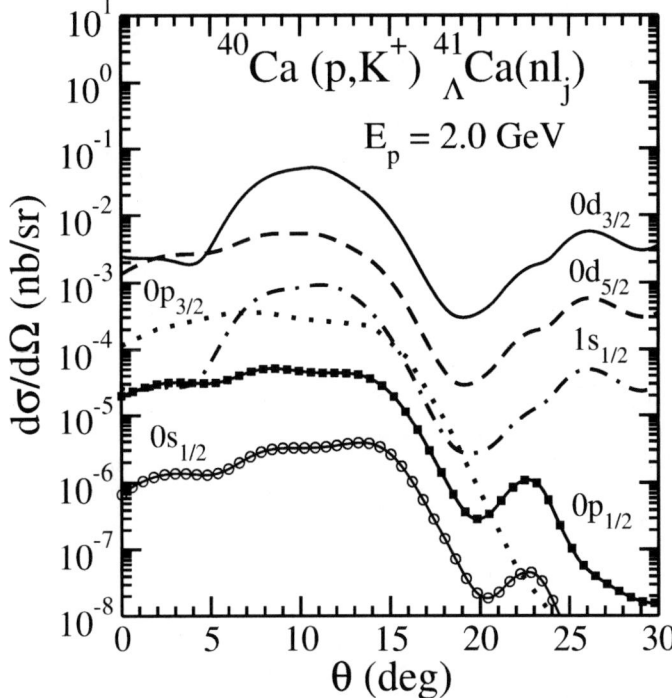

FIGURE 1. Angular distributions for production into various final Λ single particle orbitals.

$$\times \quad \psi(p_1)D_\pi(q)\bar{\psi}(p_\Lambda)\gamma_5 D_{N^*_{1/2}}(p_{N^*})\gamma_5$$

$$\times \quad \phi_K^{(-)*}(p'_K,p_K)\psi_i^{(+)}(p'_i,p_i), \tag{6}$$

where various momenta are as defined in Fig. 1b. For a more detailed discussion we refer to ref. [6].

Angular distributions for associated strangeness production in a (p,K^+) reaction at $E_p = 2$ GeV on a ^{40}Ca target are shown in fig. 1. Although initial and final state interactions are at present not included the magnitude of the cross sections in the nano- to picobarn range can be expected to be realistic, ranging at the lower end of the experimental feasibility. The shapes of the angular distributions are depending sensitively on the quantum numbers of the orbits into which the Λ is captured.

INTERACTIONS IN HYPERMATTER AND HYPERNUCLEI

The nuclear wave functions entering into the cross section calculation have been obtained by DDRH theory. This is a field theoretical approach accounting for the modifications of baryon-baryon interactions in matter with density dependent meson-baryon vertices. In [2] it was shown that a properly defined field theory, preserving covariance of the field equations and thermodynamical consistency, is obtained when vertex functionals depending on the field operators are used. The medium dependence of the vertices is derived from Dirac-Brueckner (DB) calculations and applied in relativistic mean-field calculations to nuclei [2, 3] and hypernuclei [4, 5]. Hence, we obtain an *ab initio* description once the free space baryon-baryon interaction is specified. In the strangeness sector, however, the information is sparse making it necessary to introduce the ratio of the scalar to the vector meson coupling constant as a free parameter which is determined from hypernuclear spectra [4]. Interestingly, we find significant deviations from the expectations of the naive quark model: While the latter predicts a reduction of the meson-Λ vertices by $1/3$ [1] the analysis of the existing data on Λ hypernuclei imply a reduction factors of about 50-60% [4]. The single Λ separation energies are well described by our calculations, especially for the heavier mass region, $A > 40$. Discrepancies for low masses, $A < 16$, seem to be related to additional dynamical self-energies coming from core polarization, thus supporting a conjecture of Polls et al. [9]

It is worthwhile to consider more closely the general properties of in-medium interactions for the $SU(3)_f$ flavor octet baryons. Denoting the in-medium vertices for meson $\alpha = \sigma, \omega$ by $\Gamma_{\alpha B}(\rho)$ the ratio of Λ to nucleon coupling constants is found to behave as [4]

$$R_\alpha = \frac{\Gamma_{\alpha\Lambda}}{\Gamma_{\alpha N}} = \frac{g_{\alpha\Lambda}}{g_{\alpha N}} + \mathcal{O}\left((\frac{k_F^\Lambda}{k_F^N})^2 \right) \tag{7}$$

showing that the in-medium vertices indeed reflect in leading order the free space properties of the couplings. The above relation indicates that the density dependence of baryons is most likely given by a common form factor depending on the Fermi momentum k_F^q of the flavor species q. In fact, inspection of the in-medium Bethe-Salpeter equation shows that medium dependencies are introduced primarily through Pauli-blocking which obviously is related to particles of the same flavor [10]. As an overall effect, the intrinsic density dependence of the meson-baryon vertices leads to a considerable variation of the coupling strength over the nuclear volume [4, 5], suppressing the coupling with increasing density. In fact, the in-medium interactions can be represented in terms of a susceptibility tensor and the free space interaction, at least in the ladder approximation [10]. In the DB vertices $\Gamma_{\alpha q}(k_F^q) = G_{\alpha q}F_{\alpha q}(k_F^q)$, obtained by solving the BS equation in ladder approximation, we may express the density dependence by form factors $F_{\alpha q}(k_F^q)$ and an overall strength $G_{\alpha q}$. From fig. 2 it is seen that the shape of the form factor is almost independent of the meson channel thus indicating universality.

SUMMARY AND OUTLOOK

The formation of hypernuclei in exclusive proton-nucleus reactions by associate $(K^+\Lambda)$ strangeness production has been discussed. The elementary vertices are described in a field theoretical model assuming a two-step type process where initially a nucleon is excited into a resonant state which subsequently decays into the $K^+\Lambda$ final state and the hyperon is captured by the target nucleus. Angular distributions for the outgoing K^+ were found to depend rather sensitively on the orbit occupied by the Λ. Hypernuclear

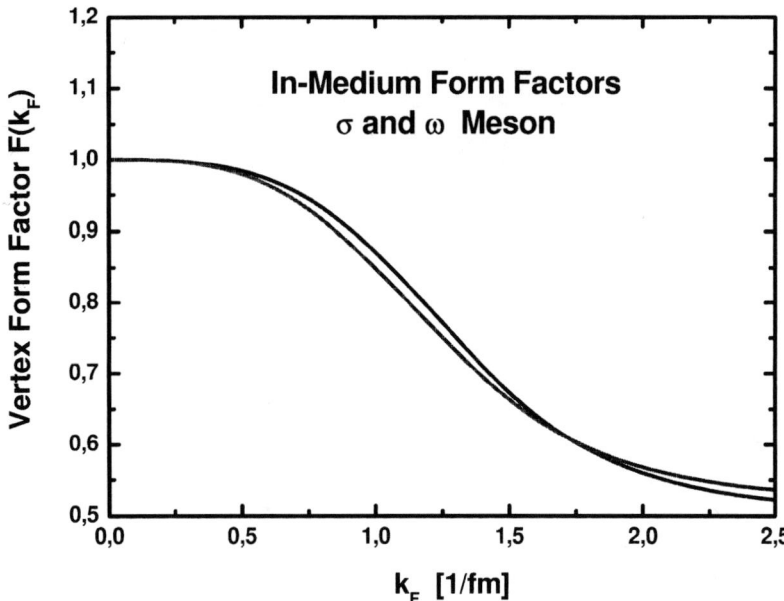

FIGURE 2. In-medium vertex form factors in the isoscalar meson channels describing the variation of the meson-baryon coupling strength with density.

structure was described by DDRH theory. A field theoretical approach to the density dependence of meson-baryon vertex functionals was discussed.

ACKNOWLEDGMENTS

This work was supported in part by FZ Jülich, grant COSY-066.

REFERENCES

1. R. E. Chrien and C. B. Dover, Annu. Rev. Nucl. Part. Sci. **39**, 113 (1989); M. May et al., Phys. Rev. Lett. **78**, 4343 (1997); H. Kohri et al., Phys. Rev. C. **65**, 034607 (2002).
2. H. Lenske and C. Fuchs: Phys. Lett. B **345**, 355 (1995); C. Fuchs, H. Lenske and H. H. Wolter: Phys. Rev. C **52**, 3043 (1995).
3. F. Hofmann, C. M. Keil, and H. Lenske: Phys. Rev. C **64**, 034319 (2001).
4. C. M. Keil, F. Hofmann, and H. Lenske, Phys. Rev. C **61**, 064309 (2000).
5. C. M. Keil and H. Lenske, Phys. Rev. C **66**, 054307 (2002).
6. R. Shyam, H. Lenske, U. Mosel, Phys. Rev. C 2003 (in print).
7. R. Shyam, Phys. Rev. C **60**, 055213 (1999); R. Shyam, G. Penner, and U. Mosel, Phys. Rev. C **63**, 022202(R) (2001).
8. G. Penner and U. Mosel, Phys. Rev. C **66**, 055211 (2002); **66**, 055212 (2002).
9. L. Vidaña, A. Polls, A. Ramos, M. Hjorth-Jensen, Nucl. Phys. A **644**, 201 (1998).
10. H. Lenske, to be published in *Spirnger Lecturer Notes*.

Nucleons in Nuclear Matter and the Transition to Quark Matter

W. Bentz[*], T. Horikawa[*], N. Ishii[†], H. Mineo[**], A.W. Thomas[‡] and K. Yazaki[§]

[*]Department of Physics, School of Science, Tokai University,
Hiratsuka-shi, Kanagawa 259-1292, Japan
[†]The Institute of Physical and Chemical Research (RIKEN),
Hirosawa, Wako-shi, Saitama 351-0198, Japan
[**]Department of Physics, National Taiwan University,
1 Roosevelt, Section 4, Taipei, Taiwan
[‡]Special Research Center for the Subatomic Structure of Matter, and
Department of Physics and Mathematical Physics, The University of Adelaide,
Adelaide, SA 5005, Australia
[§]Department of Physics, Tokyo Woman's Christian University,
Suginami-ku, Tokyo 167-8585, Japan

Abstract. We use an effective chiral quark theory to describe the single nucleon, the equations of state (EOS) of nuclear matter (NM) and quark matter (QM), and the phase transition from NM to QM at high density. We pay special attention to the effects of the nucleon quark structure on the EOS of NM, and of scalar diquark condensation (color superconductivity) on the EOS of QM.

INTRODUCTION

The possibility of a phase transition from nuclear matter (NM) to quark matter (QM) at high baryon densities is of great interest in connection with neutron stars and the possible existence of quark stars [1]. In order to describe such a phase transition, one needs a consistent theoretical framework which can account for the properties of single nucleons, NM and QM at the same time. It has been shown recently [2] that the Nambu-Jona-Lasinio (NJL) model [3] is a strong candidate for this purpose: Besides a covariant description of the nucleon as a quark-diquark bound state [4], it also allows a description of a saturating NM equation of state (EOS) [5] and of QM including the effects of color superconductivity [6]. It is therefore of interest to study the conditions under which a transition between NM and QM becomes possible. It is the purpose of this paper to show our results on this phase transition. For details of the model and the formulation we refer to Ref.[2]. The EOS for NM has also been used recently to describe the nuclear structure functions and the EMC effect[7, 8].

CP717, *Hadron Spectroscopy: Tenth International Conference,*
edited by E. Klempt, H. Koch, and H. Orth
© 2004 American Institute of Physics 0-7354-0197-7/04/$22.00

DESCRIPTION OF NUCLEAR MATTER AND QUARK MATTER

The NJL model is characterized by a chiral symmetric contact interaction between quarks. Because of its simplicity, the relativistic Faddeev equation for the nucleon can be solved in the ladder approximation, taking into account the interactions in the scalar and axial vector diquark channels[4]. For our present investigations at finite density, however, we will restrict ourselves to the scalar diquark channel, and to a simple approximation to the Faddeev equation, where the momentum dependence of the quark exchange kernel is neglected[5, 6]. Based on this quark-diquark description of the single nucleon, the EOS for NM can be constructed in the mean field approximation, and the phase transition to QM at high densities can be investigated.

The solution of the quark-diquark bound state equation leads to the nucleon mass $M_N(M)$ as a function of the constituent quark mass M. The effective (grand) potential for NM at zero temperature in the mean field approximation then has the form [5]

$$V^{(NM)} = V_{\text{vac}} + V_N + V_\omega, \qquad (1)$$

where V_{vac} describes the polarization of the Dirac sea of quarks due to the presence of the valence nucleons, V_N arises from the Fermi motion of the valence nucleons and depends on the function $M_N(M)$, and V_ω is the contribution of the mean vector field (ω_0) in NM . The conditions $\partial V^{(NM)}/\partial M = \partial V^{(NM)}/\partial \omega_0 = 0$ determine M and ω_0 for fixed chemical potential μ.

In effective field theories based on the linear realization of chiral symmetry, one often observes a collapse of the NM EOS because the σ mass decreases too rapidly as a function of the density. It has been shown in Ref.[5], however, that the saturation properties of the NM EOS can be described if the quark structure of the nucleon is taken into account, provided that one eliminates the threshold for the unphysical decay of the nucleon into quarks. This can be done, for example, in the proper time regularization scheme by introducing an infrared cut-off (Λ_{IR}) in addition to the ultraviolet one [9]. The elimination of the unphysical decay threshold then leads to a positive scalar polarizability of the single nucleon, and this in turn gives rise to an effective $NN\sigma\sigma$ interaction which raises the σ meson mass and prevents the collapse.

The EOS of color superconducting QM in the mean field approximation has the form [2]

$$V^{(QM)} = V_{\text{vac}} + V_Q + V_\Delta + V_\omega \qquad (2)$$

where V_{vac} describes the polarization of the Dirac sea of quarks due to the presence of the valence quarks, V_Q arises from the Fermi motion of the valence quarks without the effect of quark pairing, V_Δ arises from quark pairing in the scalar diquark channel and depends on the color superconducting gap (Δ), and V_ω is the contribution of the mean vector field (ω_0) in QM. The conditions $\partial V^{(QM)}/\partial M = \partial V^{(QM)}/\partial \Delta = \partial V^{(QM)}/\partial \omega_0 = 0$ determine M, Δ and ω_0 for fixed chemical potential μ.

After the calculation of the grand potential V, the EOS of NM and QM are obtained by applying the usual thermodynamic relationships, and the Gibbs criteria can be used to look for phase transitions. The chemical potential can be eliminated in favor of the baryon density ρ according to $\rho = -\partial V/\partial \mu$.

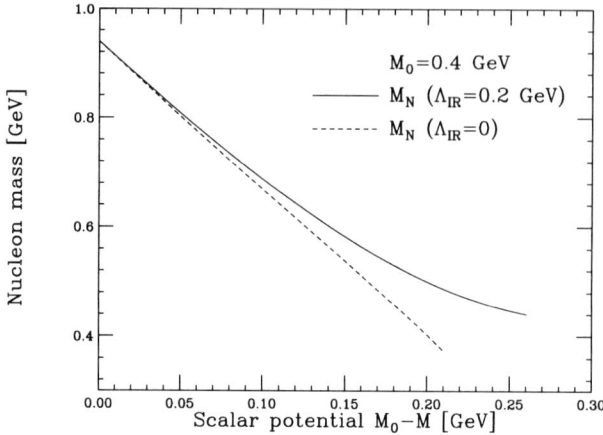

FIGURE 1. Nucleon mass as a function of the scalar potential for $\Lambda_{IR} = 0$ (dashed line) and $\Lambda_{IR} = 0.2$ GeV (solid line).

RESULTS

The function $M_N(M)$, which is the solution of the quark-diquark bound state equation for the nucleon, is shown in Fig. 1. The elimination of unphysical quark decay thresholds by an infrared cut-off Λ_{IR} leads to a scalar polarizability (curvature of the function $M_N(M)$), which is clearly seen in Fig. 1. This effect stabilizes the NM EOS, and leads to the saturation of the binding energy per nucleon as a function of density [5].

Since we can simultaneously describe the structure of the nucleon and stable NM, we can investigate the properties of a nucleon bound in the medium. Such an investigation has been carried out for the structure functions (the EMC effect) in Ref.[7, 8]. The main result is that the mean vector field has an important direct effect on the structure function, which leads to a successful description of the EMC effect.

We now ask the question whether there is a phase transition to QM at high baryon densities. Fig. 2 shows the EOS for NM (solid line), and for QM with several values for the strength of the pairing interaction in the scalar diquark channel (r_s). Curve 1 corresponds to normal, i.e., non-color superconducting, QM, and the other curves show the results for increasing strength of the pairing interaction.

It is clear from Fig. 2 that there is no phase transition from NM (solid line) to normal QM (dotted line) in our model. The scalar diquark condensation, however, gives rise to a substantial softening of the QM EOS, and to a phase transition from NM to QM at a transition density which decreases with increasing strength of the pairing interaction. The effective (constituent) quark mass for NM and QM is shown for the same cases as a function of the baryon density in Fig. 3. This figure demonstrates that the tendency toward a chiral phase transition is much stronger in QM than in NM.

We now assume a particular value of the coupling constant in the scalar diquark channel, which leads to reasonable transition densities, and investigate the nature of the phase transition more closely. Fig. 4 shows the resulting pressure of the ground state

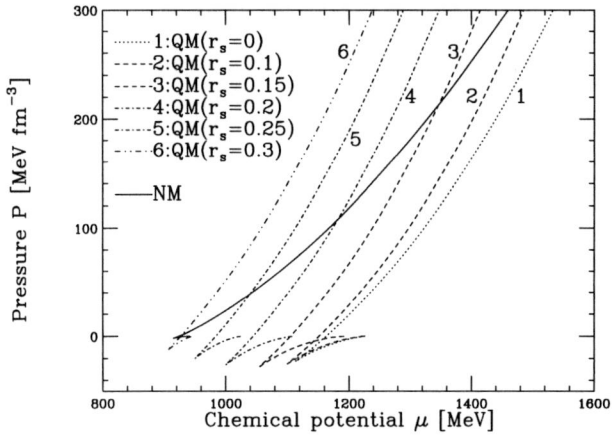

FIGURE 2. Pressure as function of chemical potential in NM (solid line), and in QM for several values of r_s, which is the ratio of the coupling constant in the scalar diquark channel to the one in the pionic channel.

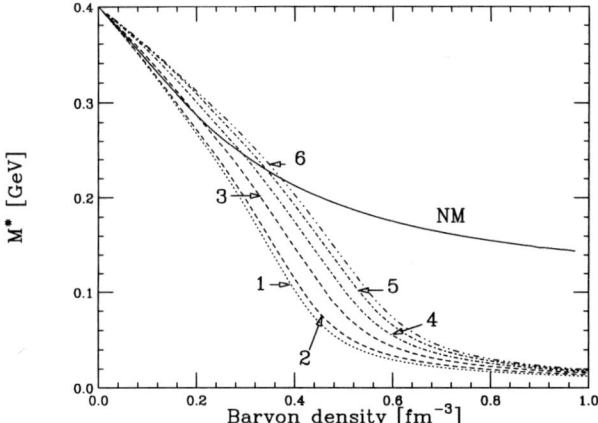

FIGURE 3. Effective quark mass as function of baryon density for NM (solid line) and QM. Labels 1 to 6 correspond to the same values of the pairing strengths in QM as in Fig. 2.

as a function of the baryon density for the case corresponding to $r_s = 0.2$ of Fig. 2.

We obtain first order transitions from the vacuum (VAC) to NM, where in both phases chiral symmetry is broken and color symmetry is intact, and from NM to color superconducting QM, where in the latter phase chiral symmetry is largely restored and color symmetry is broken. The present calculation gives large color superconducting gaps in the QM phase ($\Delta > 200$ MeV).

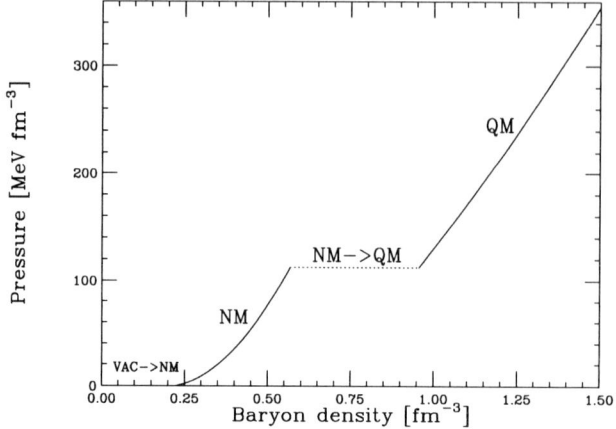

FIGURE 4. Pressure of the ground state of the system - vacuum (VAC), nuclear matter (NM) or superconducting quark matter (QM) - as function of the baryon density for $r_s = 0.2$. The mixed phases are also indicated.

SUMMARY

We used the NJL model as an effective chiral quark theory to describe the single nucleon, the saturation properties of normal NM, and high density QM. We have shown that there is a phase transition from NM to QM, provided that the effects of scalar diquark condensation (color superconductivity) are taken into account. This phase transition is characterized by the restoration of chiral symmetry and the spontaneous breaking of color symmetry in the high density region.

ACKNOWLEDGMENTS

This work was supported by the Grant in Aid for Scientific Research of the Japanese Ministry for Education, Culture, Sports, Science and Technology, Project No. C2-13640298, the Australian Research Council and The University of Adelaide.

REFERENCES

1. Proc. Int. Workshop on *Compact Stars in the QCD Phase Diagram*, **eConf C010815** (2002).
2. W. Bentz, T. Horikawa, N. Ishii and A.W. Thomas, Nucl. Phys. **A720** (2003), 95.
3. Y. Nambu and G. Jona-Lasinio, Phys. Rev. **122** (1960), 345; **124** (1961), 246.
4. N. Ishii, W. Bentz and K. Yazaki, Nucl. Phys. **A578** (1995), 617.
5. W. Bentz and A.W. Thomas, Nucl. Phys. **A696** (2001), 138.
6. M. Alford, K. Rajagopal and F. Wilczek, Phys. Lett. **B422** (1998), 247.
7. H. Mineo, W. Bentz, N. Ishii, A.W. Thomas and K. Yazaki, to be published.
8. H. Mineo, Contribution to this Conference.
9. D. Ebert, T. Feldmann and H. Reinhardt, Phys. Lett. **B388** (1996), 154.

Observation of K^+p and K^+d correlations from pA collisions

M. Nekipelov for the ANKE collaboration

Institut für Kernphysik, Forschungszentrum Jülich, 52425 Jülich, Germany

Abstract. In a series of measurements with the ANKE spectrometer at COSY-Jülich, the production of K^+-mesons accompanied by proton or deuteron at different beam energies, $T_p = 1.0, 2.0, 2.3$ GeV, has been investigated. These data supply direct evidence for the two-step reaction mechanism, with formation of intermediate pions, leading to kaon production. In addition, measurements with a deuterium target suggest that production cross section on neutron exceeds that on proton, and a corresponding ratio is close to 4.

K^+-meson production in proton-nucleus (pA) collisions at energies below the elementary $pN \rightarrow K^+ N\Lambda$ reaction threshold at $T_p = 1580$ MeV has attracted considerable interest for a long time. Irrespective of the large amount of data [1, 2, 3, 4, 5, 6, 7, 8, 9], which has been collected since the first measurements in 1988 [1], the nuclear kaon-production mechanism is far from being understood. Data on inclusive K^+ production can be equally well reproduced by different models that involve either multi-step processes [10, 11, 12, 13], cooperation between several target nucleons [14, 15], or high momentum components of the nuclear wave function [1, 7, 13, 16, 17]. All these mechanisms must be taken into account, and their relative contributions strongly depend on the reaction kinematics. It may turn out to be impossible to deduce the values of these weights from the conventional normalization to measured cross sections [8, 9]. Thus, new types of experiments, selectively exploring features of one or the other mechanism are mandatory, for example the measurement of (K^+p) and (K^+d) correlations.

In addition to the direct production on a single quasi-free target nucleon the K^+ mesons can also be produced in two-step processes with intermediate pion produced in the first chance collision, i.e a $pN_1 \rightarrow \pi X$ reaction, followed by $\pi N_2 \rightarrow K^+X$ on a second target nucleon. Depending on the beam energy, K^+-production may be due to both the direct and two-step reaction mechanisms. At low beam energies the two-step process is energetically favorable since the intrinsic nucleon motion can be utilized twice. Yet another possible scenario is when K^+ is produced on some sort of cluster. The K^+d final state can only be observed in case of the two-step reaction or cluster mechanism, while K^+p pairs come both from the two-step and direct production.

Another open question making the interpretation of kaon production from nucleus extremely hard is that kaon production from pn interactions close to threshold is unknown. Various theoretical models give diverse numbers for the factor between kaon production on proton and neutron ranging from 1 to 6. Measurements of correlated K^+ mesons and protons allow one to clarify the situation comparing resulting spectra with corresponding simulations.

CP717, *Hadron Spectroscopy: Tenth International Conference*,
edited by E. Klempt, H. Koch, and H. Orth
© 2004 American Institute of Physics 0-7354-0197-7/04/$22.00

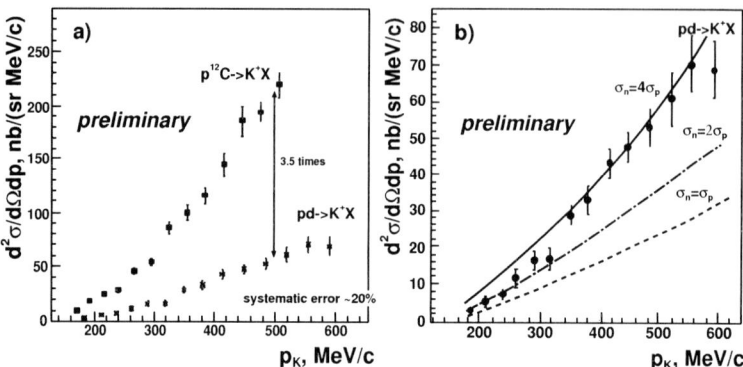

FIGURE 1. a) Double differential K^+-production cross sections, measured at $T_p = 2.021\,\text{GeV/c}$ with deuterium target and at $T_p = 2\,\text{GeV/c}$ with carbon target; b) K^+-production cross sections with deuterium target (as in a). The lines correspond to the simulations assuming a ratio of total production cross sections $\sigma(n)/\sigma(p)$: dashed - 1, dashed-dotted - 2 and solid - 4.

The ANKE facility ideally suites for such studies, since the spectrometer and its detectors have been optimized for K^+-meson detection [18]. The experiment was performed with carbon (foil) and deuterium (cluster-jet) targets at proton beam energies of T_p=1200 (C), 2021 (d) and 2300 MeV (C). Kaons with momenta between 175 and 610 MeV/c have been identified in the angular range $\vartheta < 12°$ ($\vartheta < 4°$ for deuterium target). Protons and deuterons, measured in coincidence with kaons, have been detected in a narrower solid angle cone of $\vartheta < 8°$.

Inclusive double differential K^+-production cross sections measured at $T_p = 2.021\,\text{GeV}$ with deuterium target are presented in Fig.1a together with the cross sections obtained with a carbon target at $T_p = 2.0\,\text{GeV}$ [19]. The factor of 3.5 between these two measurements can be understood in a framework of Glauber model [20], in which for the given beam energy incoming proton effectively sees in carbon 7.3 nucleons [21]. Additionally, theoretical calculations predict total direct-production dominance above free nucleon-nucleon threshold, therefore one might expect that at $T_p = 2.021\,\text{GeV}$ kaons are produced on independent nucleons, and no collective phenomena in the nucleus at this beam energy should be observed. Thus, within the accuracy of our experiment we assumed that $\sigma_d = \sigma_p + \sigma_n$. A comparison of measured cross sections and results of simulations using a ratio of production cross sections on neutron and proton as a free parameter is also shown in Fig.1b. Given the total cross sections for the individual channels considered (see table 1) as weights, simulated spectra have been corrected for the Fermi motion of nucleons in deuteron. The slight difference in the shape of the simulated and measured cross sections can be explained by the following reasons: first, this simplest approach does not include dynamics of the interaction in the different states, and second, there is no experimental data to check cross sections used as the weights for individual channels on the neutron. Within such a naive model the ratio of total K^+-production cross sections on the neutron and the proton is obtained as: $\sigma(n)/\sigma(p) = 4 \pm 1.5$.

As an independent check shape of the missing mass spectrum of (K^+p)-pairs has also

TABLE 1. Total kaon production cross section at $T_p = 2.021\,\text{Gev/c}$ calculated using equations from Ref. [22].

Reaction	$\sigma, \mu\text{b}$
$pp \rightarrow \Lambda K^+ p$	10.865
$pn \rightarrow \Lambda K^+ n$	19.420
$pp \rightarrow \Sigma^0 K^+ p$	1.361
$pn \rightarrow \Sigma^0 K^+ n$	1.680
$pp \rightarrow \Sigma^+ K^+ n$	0.522
$np \rightarrow \Sigma^- K^+ n$	2.776

been compared with the simulations, where the ratio σ_n/σ_p was varied from 1 to 10. As in the previous case a quasi-free moving nucleon, proton or neutron, has been chosen as a target. An example comparison for $\sigma_n/\sigma_p = 5$ is shown in Fig.2. It has been found that the shape of the missing mass spectrum is not very sensitive to this ratio, though the best agreement is reached at $\sigma_n/\sigma_p \approx 4.7$.

What concerns the K^+-production mechanisms from the nuclei, there are three possible sources of deuterons correlating with K^+ mesons: coalescent deuterons, the two-step production via $pN \rightarrow d\pi$ reaction, and the cluster mechanism. It is predicted that the $pN \rightarrow d\pi$ reaction provides a significant part of the total cross sections at low energies and reaches 30% at 1000 MeV [12]. It has also been shown that (K^+d) correlations can be observed in spite of distortions due to final-state interactions that the deuteron experiences on its way out of the nucleus. A rather flat background of coalescent deuterons should not be a problem below 1300 MeV, since two-step deuterons correspond to the kinematics of the $pN \rightarrow d\pi$ reaction, and a momentum distribution of those deuterons has a typical width of about 200–250 MeV/c.

The deuteron-momentum spectra measured in coincidence with the K^+-mesons at $T_p = 1.2\,\text{GeV}$ are shown in Fig.3a [23]. The one-peak fit to the experimental data yields a significantly wider distribution than theoretically predicted and is shifted to the low momentum region. Moreover, the peak-structure of shows a shoulder at the high-momentum tail. A better fit is obtained with three Gaussian peaks (Fig.3a). In this case, the maximum and width of the central peak are closer to the predictions according to Ref. [12], and the peak is assigned to correlated (K^+d) pairs, created via the two-step mechanism. The peak positions of two other contributions is in a qualitative agreement with phase-space calculations, in which K^+ mesons are produced on $(2N)$ clusters inside the nucleus [23].

In order to check whether experimental data are consistent with this conjecture, the kinematical features of both mechanisms were exploited. The beam energy of $T_p = 1200$ MeV is far above the threshold of the $pN \rightarrow d\pi$ reaction. Forward emitted pions have an energy of about 830 MeV (neglecting the intrinsic nucleon momenta), weakly varying with the emission angle. This pion energy is also well above the threshold of the $\pi N \rightarrow K^+ \Lambda$ reaction in the second step and produces kaons with momenta of 450–500 MeV/c for kaon emission angles less than 15°. Deuterons, produced in the first step have a rather wide angular distribution (more than 35°) and the shape of the deuteron peak weakly depends on the deuteron emission angle within 10–15°. Thus, the intensity

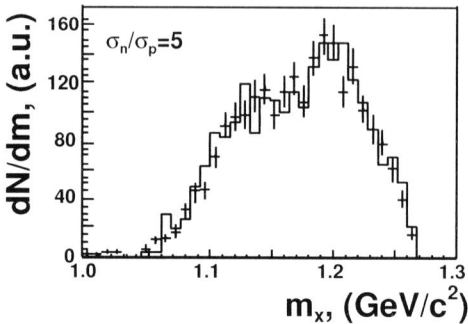

FIGURE 2. Missing mass spectrum of (K^+p)-pair, measured at $T_p = 2.021\,\text{GeV/c}$ with deuterium target. Solid line corresponds to the simulations assuming that production on neutron exceeds that on proton by a factor of 5.

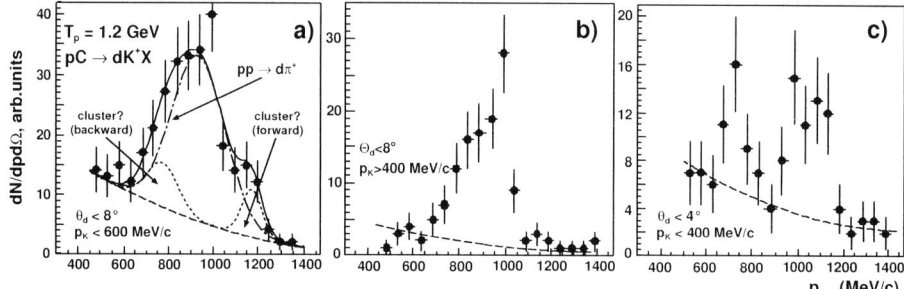

FIGURE 3. Deuteron-momentum spectra for pC collisions at 1200 MeV and different cuts on the deuteron emission angles and kaon momenta. The dashed lines indicate the fitted background distribution.

of the central peak should be proportional to the solid angle covered by the deuteron detector and strongly depend on the part of the kaon momentum spectrum accepted for correlations with deuterons. For possible $(2N)$ cluster mechanisms, the beam energy of 1200 MeV is less than 60 MeV above the threshold for the $pd \rightarrow dK^+\Lambda$ reaction. Since the binding energy of a $(2N)$ cluster is at least as large as for two individual nucleons, the excess energy should be even smaller. As a result, the deuterons from the cluster mechanism must have a much narrower, strongly forward peaked angular distribution $(< 10°)$. They correlate with kaons of momenta less than 425 MeV/c, the kinematical upper limit for kaons from the $2N$-cluster mechanism. A cut on the low-momentum part $(<400\,\text{MeV/c})$ of the kaon spectrum and on correlated deuterons within a reduced angular acceptance ($4°$ instead of $8°$) should thus reduce the deuteron peak from the two-step mechanism by a factor of 8. The experimental results after applying such cuts are presented in Fig.3b,c) and confirm our naive kinematical considerations.

In summary, double differential K^+-production cross sections with the deuterium target at $T_p = 2.021\,\text{GeV}$ have been measured indicating a strong difference in production on neutron and proton. Comparison of the experimental data (the values of the cross sections and the shape of missing mass spectrum of (K^+p) pair) with simulations give

$\sigma_n/\sigma_p \sim 4...5$. It has been shown, that the particular two-step mechanism $pN_1 \rightarrow d\pi$, $\pi N_2 \rightarrow K^+ \Lambda$ contributes to about 30% of the total kaon yield at $T_p = 1.2\,\text{GeV}$ [23]. The data also indicate that a significant fraction of the K^+-mesons maybe produced on two-nucleon clusters, i.e. in the reaction $p(2N) \rightarrow (dK^+)\Lambda$.

REFERENCES

1. V.P. Koptev et al., JETP **67**, (1988) 2177.
2. S. Schnetzer et al., Phys. Rev. C **40**, (1989) 640.
3. M. Büscher et al., Z. Phys. A **335**, (1996) 93.
4. M. Debowski et al, Z. Phys. A **356**, (1996) 313.
5. A. Badala et al., Phys. Rev. Lett. **80**, (1998) 4863.
6. Yu. Kiselev et al., J. Phys. G **25**, (1999) 381.
7. A.V. Akindinov et al., JETP Lett. **72**, (2000) 100.
8. V. Koptev et al., Phys. Rev. Lett. **87**, (2001) 022301.
9. M. Büscher et al., Phys. Rev. C **65**, (2002) 014605.
10. N. A. Tarasov et al., JETP Lett. **43**, (1986) 274.
11. W. Cassing et al., Phys. Lett. B **238**, (1990) 25.
12. A.A. Sibirtsev and M. Büscher, Z. Phys. A **347**, (1994) 191.
13. Yu. Paryev, Eur. Phys. J. A **5**, (1999) 307.
14. H. Müller and K. Sistemich, Z. Phys. A **344**, (1992) 197.
15. A. Bonasera and T. Maruyama, Prog. Theor. Phys. **90**, (1993) 1155.
16. A.A. Sibirtsev, Phys. Lett. B **359**, (1995) 29.
17. A.A. Sibirtsev, W. Cassing and U. Mosel, Z. Phys. A **358**, (1997) 357.
18. S. Barsov et al., Nucl. Instr. Meth. A **462** (2001) 364.
19. M. Nekipelov, V. Koptev, Proc. 5th Int. Conf. on Nuclear Physics at Storage Rings (STORI 02), 16 - 20 June 2002, Uppsala, Sweden; Physica Scripta **T104**, (2003) 40.
20. V. Franco, R. J. Glauber, Phys. Rev. **142**, (1966) 1195.
21. N. K. Abrosimov et al., Preprint PNPI **1146**, (1985) 1.
22. K. Tsushima et al., Phys. Rev. C, **59**, (1999) 369.
23. V. Koptev et al., Eur.Phys.J. A **17**, (2003) 235.

The mass of charmonium in nuclear matter

Su Houng Lee

Department of Physics and Institute of Physics and Applied Physics, Yonsei University, Seoul 120-749, Korea

Abstract. The masses of charmonium states immersed in nuclear matter are calculated in LO QCD and in QCD sum rules. While the mass shift for J/ψ are found to be less than -10 MeV, those for the $\chi_{0,1,2}$ and $\psi(3686)$ and $\psi(3770)$ are found to be more than -40 MeV. We investigate the feasibility of observing such mass shifts in the future accelerator project at GSI.

INTRODUCTION

Understanding hadron mass changes in nuclear medium and/or at finite temperature can provide valuable information about the QCD vacuum[1, 2, 3]. While the mass shifts for hadrons made of light quarks are sensitive to the restoration of the spontaneously broken chiral symmetry breaking[2, 4, 5, 6], those for the the heavy quark systems are sensitive to the changes of the non-perturbative gluon fields in nuclear matter. For the J/ψ, which consists of a charm and anticharm quark pair, both the QCD sum rules analysis [7, 8] and the LO perturbative QCD calculation [9, 10] show that its mass is reduced slightly in the nuclear matter mainly due to the reduction of the gluon condensate $(\langle \frac{\alpha_s}{\pi} G^2 \rangle)$ in nuclear matter, which is expected to decrease by 6% at normal nuclear matter density. However, the changes are much larger for excited charmonium states, due mainly to larger color dipole size of these excited states.

In this report, we summarize the expected mass shift for charmonium states in nuclear matter and study the feasibility of observing such mass shift in the future accelerator project at GSI[11].

The lowest dimensional QCD operators that characterizes the non perturbative nature of the QCD vacuum are the quark and gluon condensate. These condensate are estimated to have the following large non-perturbative expectation values in the vacuum [12],

$$\langle \frac{\alpha_s}{\pi} F_{\mu\nu}^2 \rangle \sim 1.5 \text{ GeV}/\text{fm}^3,$$
$$\langle \bar{q}q \rangle \sim 2 \text{ fm}^{-3}. \tag{1}$$

The gluon condensate can be written as the difference between the magnetic $B^2 = F_{ij}^2$ and electric $E^2 = \frac{1}{2}F_{0i}^2$ condensate, which respectively contribute to half of the zero temperature gluon condensate,

$$\langle \frac{\alpha_s}{\pi} B^2 \rangle = -\langle \frac{\alpha_s}{\pi} E^2 \rangle = \frac{1}{2}\langle \frac{\alpha_s}{\pi} F_{\mu\nu}^2 \rangle. \tag{2}$$

CP717, *Hadron Spectroscopy: Tenth International Conference*,
edited by E. Klempt, H. Koch, and H. Orth

The above relation follows naturally from the Euclidean formulation of QCD at zero temperature, such as in the lattice QCD, where the Euclidean space electric (magnetic) condensate is defined with a minus (plus) sign relative to its Minkowski space counterpart[12]. Due to the symmetry in the time and space directions in the 4 dimensional Euclidean space, the Euclidean space electric condensate is expected to have the same expectation value as the magnetic one[13, 14]. Hence the relation among the Minkowski space condensate in Eq.(2) follows.

At nuclear matter, the non perturbative quark and gluon field configuration are expected to change appreciably, such that the average gluon and quark condensate values decrease by 6% and 30%, respectively. These model independent results are obtained from the linear density approximation and the nucleon expectation values of the quark and gluon condensate, which are respectively known from the experimentally measured pi-N sigma term and from taking the nucleon expectation value of the trace anomaly relation[15]. The electric and magnetic part of the gluon condensate at nuclear matter can be estimated separately, by using the twist-2 gluon operator,

$$\langle N(p)|_{\text{S T}} F_\mu^\alpha F_{\alpha\nu}|N(p)\rangle = \left(p_\mu p_\nu - \frac{1}{4}m_N^2 g_{\mu\nu}\right) 2A_2(g),\tag{3}$$

where $A_2(g)$ is the second moment of the gluon distribution in the nucleon and is around 0.45, when the renormalization scale is in the of order 1 to 2 GeV. Using this and the linear density approximation, we have,

$$
\begin{aligned}
\langle\frac{\alpha_s}{\pi}E^2\rangle_{n.m.} &= \left(\frac{4}{9}m_N m_N^0 + \frac{3}{2}m_N^2\frac{\alpha_s}{\pi}A_2\right)\frac{\rho_{n.m.}}{2m_N},\\
\langle\frac{\alpha_s}{\pi}B^2\rangle_{n.m.} &= -\left(\frac{4}{9}m_N m_N^0 - \frac{3}{2}m_N^2\frac{\alpha_s}{\pi}A_2\right)\frac{\rho_{n.m.}}{2m_N}.
\end{aligned}\tag{4}
$$

Here, $m_N^0 \sim 0.75$ GeV is the mass of the nucleon in the chiral limit[16], which comes from taking the nucleon expectation value of the trace anomaly relation $T_\mu^\mu = -\frac{9}{8}\frac{\alpha_s}{\pi}F_{\mu\nu}^2$. As can be seen from Eq.(4), due to the additional factor of $\frac{\alpha_s}{\pi}$ in the second terms, the changes are dominated by the contribution from the first terms.

CHARMONIUM MASS SHIFT FROM QCD

The mass shift of charmonium states in nuclear medium can be evaluated in the perturbative QCD when the charm quark mass is large, i.e., $m_c \to \infty$. In this limit, one can perform a systematic operator product expansion (OPE) of the charm quark-antiquark current-current correlation function between the heavy bound states by taking the separation scale (μ) to be the binding energy of the charmonium [9, 17, 18]. The forward scattering matrix element of the charm quark bound state with a nucleon then has the following form:

$$T(q^2 = m_\psi^2) = \sum_n \frac{C_n}{(\mu)^n}\langle_o\ n\rangle_N.\tag{5}$$

781

Here, C_n is the Wilson coefficient evaluated with the charm quark bound state wave function and $\langle \circ_n \rangle_N$ is the nucleon expectation value of local operators of dimension n.

For heavy quark systems, there are only two independent lowest dimension operators; the gluon condensate ($\langle \frac{\alpha_s}{\pi} G^2 \rangle$) and the condensate of twist-2 gluon operator multiplied by α_s ($\langle \frac{\alpha_s}{\pi} G_{\alpha\mu} G_\nu^\alpha \rangle$). These operators can be rewritten in terms of the color electric and magnetic fields: $\langle \frac{\alpha_s}{\pi} E^2 \rangle$ and $\langle \frac{\alpha_s}{\pi} B^2 \rangle$. Since the Wilson coefficient for $\langle \frac{\alpha_s}{\pi} B^2 \rangle$ vanishes in the non-relativistic limit, the only contribution is thus proportional to $\langle \frac{\alpha_s}{\pi} E^2 \rangle$, similar to the usual second-order Stark effect. We shall thus calculate the mass shift of charmonium states due to change of the gluon condensate in nuclear medium by the QCD second-order Stark effect [10].

The mass shift of charmonium states to leading order in density is obtained by multiplying the leading term in Eq.(5), by the nuclear density ρ_N. This gives,

$$\Delta m_\psi(\varepsilon) = -\frac{1}{9} \int dk^2 \left| \frac{\partial \psi(k)}{\partial \mathbf{k}} \right|^2 \frac{k}{k^2/m_c + \varepsilon}$$
$$\times \left\langle \frac{\alpha_s}{\pi} E^2 \right\rangle_N \cdot \frac{\rho_N}{2m_N}. \tag{6}$$

In the above, m_N and ρ_N are the nucleon mass and the nuclear density, respectively; $\langle \frac{\alpha_s}{\pi} E^2 \rangle_N \sim 0.5$ GeV^2 is the nucleon expectation value of the color electric field obtained from Eq.(4) and $\varepsilon = 2m_c - m_\psi$. In Ref.[9], the LO mass shift formula was derived in the large charm quark mass limit. As a result, the wave function $\psi(k)$ is Coulombic and the mass shift is expressed in terms of the Bohr radius a_0 and the binding energy $\varepsilon_0 = 2m_c - m_{J/\psi}$. This might be a good approximation for J/ψ but is not realistic for the excited charmonium states as Eq.(6) involves the derivative of the wave function, which measures the dipole size of the system. We have thus rewritten in the above the LO formula for charmonium mass shift in terms of the QCD parameters $\alpha_s = 0.84$ and $m_c = 1.95$, which are fixed by the energy splitting between J/ψ and $\psi(3686)$ in free space[9]. Furthermore, we take wave functions of the charmonium state to be Gaussian with the oscillator constant β determined by their squared radii $\langle r^2 \rangle = 0.47^2, 0.74^2, 0.96^2$, and 1 fm^2 for J/ψ, $\chi_{0,1,2}$, $\psi(3686)$, and $\psi(3770)$, respectively, as obtained from the potential models [19]. This gives $\beta = 0.52, 0.43, 0.39$, and 0.37 GeV if we assume that these charmonium states are in the $1S$, $1P$, $2S$, and $1D$ states, respectively. Using these parameters, we find that the mass shifts at normal nuclear matter density obtained from the LO QCD formula Eq.(6) are -8, -40, -100, and -140 MeV for J/ψ, $\chi_{0,1,2}$, $\psi(3686)$, and $\psi(3770)$, respectively[20].

Although the higher twist effects on the charmonium masses are expected to be nontrivial, the result for J/ψ is consistent with those from other non-perturbative QCD studies, such as the QCD sum rules [7, 8] and the effective potential model [21, 22, 23], which are all based on the dipole interactions between quarks in the charmonium and those in the nuclear matter. The QCD sum rule results can also be applied for the $\chi_{0,1,2}$ states, and the results from the leading order gluon condensate is summarized in Table. 1.

Higher twist effects can be estimated in some calculations. For QCD sum rules for J/ψ, the corrections coming from dimension 6 operators are less than 30% of the leading order results[8]. The contributions from the $D\bar{D}$ meson loops in the $\psi(3686)$ and $\psi(3770)$ are also found to be less than 30% of the LO QCD result[20].

All the results are summarized in table 1.

TABLE 1. Charmonium Mass shift in nuclear matter in MeV

Charmonium	J^{PC}	QCD 2nd order Stark Effect	QCD sum rules	Effects of $D\bar{D}$ loop
η_c	0^{-+}	- 8 MeV	-5 MeV	No effect
J/ψ	1^{--}	-8 MeV	-7 MeV	< 2 MeV
$\chi_{0,1,2}$	$0,1,2^{++}$	- 40 MeV	-60 MeV	No effect on χ_1
$\psi(3686)$	1^{--}	-100 MeV		< 30 MeV
$\psi(3770)$	1^{--}	-140 MeV		< 40 MeV

OBSERVABILITY

Since the mass shift of the heavy quark system reflects the changes of the Gluon field configuration in the vacuum, it would be interesting to observe such effects in experiment.

Consider an anti-proton with incoming four momentum $(\omega, 0, k)$ annihilating a proton at rest $(m_N, 0, 0)$ and creating a charmonium moving with velocity v. The required incoming momentum k to create a charmonium state are summarized in Table I.

TABLE 2. Required momentum to create Charmonium with outgoing velocity v

	η_c	J/ψ	χ_0	χ_1	χ_2	$\psi(3686)$	$\psi(3770)$
k (GeV/c)	3.7	4.1	5.2	5.5	5.7	6.2	6.5
ω (GeV)	3.8	4.2	5.3	5.6	5.8	6.3	6.6
v (c)	0.78	0.8	0.83	0.84	0.85	0.86	0.87

Hence, the required incoming energy of the anti-proton to produce the charmonium state range from 4 to 6 GeV. In these energy region, the absorption cross section $\sigma_{\bar{p}-p} \sim 50$ mb. Hence, the anti-proton would be absorbed after travelling less than 1 fm in the nuclear matter. Moreover, once a charmonium state is created, their speed would be less than 0.9 c, which means that it will have to travel more than 10 fm/c to pass the diameter of a nucleus of A=125. Hence, considering the increased width of charmonium due to nuclear absorption[24], the charmonium is expected to decay inside the nucleus.

The cross section for the production of charmonium states and its subsequent decay into dileptons or $J/\psi + \gamma$ states are given by the following Breit-Wigner formula

$$\sigma_{BW}(E) = \frac{2J+1}{(2s_1+1)(2s_2+1)} \frac{\pi}{k^2} \frac{B_{in}B_{out}\Gamma_{Total}^2}{(E-E_R)^2 + \Gamma_{Total-medium}^2/4}, \quad (7)$$

where k is the c.m. momentum, E is the c.m. energy, B_{in} and B_{out} are the branching fractions of the resonance into the entrance and exit channels. The $2s + 1$ are the spin multiplicities of the incident spin states and J the spin of the charmonium. Also $\Gamma_{Total-medium} = \Gamma_{Total} + \Gamma_{medium}$.

783

If we substitute the medium mass shift and increase in width (due mainly to collision broadening), the cross sections are in the order of one to few hundred pbarn

The expected luminosity at the anti proton project at GSI is $2 \times 10^{32} cm^{-2} s^{-1}$. Therefore if the cross section is 1pb, it would corresponds to about 17 events per day.

TABLE 3. Measurable decay channel and expected event rate at GSI future accelerator.

Charmonium	$\Gamma_{Total} + \Gamma_{medium}$	Final state	cross-section to final state	events per day
$J/\psi(3097)$	87 KeV+ 20 MeV	$e^+ + e^-$	6pb	100
$\psi(3686)$	300 KeV+ 20 MeV	$e^+ + e^-$	0.6 pb	10
$\psi(3770)$	23.6 MeV +20 MeV	$e^+ + e^-$	1 pb	17
$\chi_{c0}(3417)$	16.2 MeV+ 20 MeV	$J/\psi + \gamma$	200 pb	3400
$\chi_{c1}(3510)$	0.92 MeV+20 MeV	$J/\psi + \gamma$	80 pb	1360
$\chi_{c2}(3556)$	2.08MeV+20 MeV	$J/\psi + \gamma$	350 pb	5950

Hence, the mass shift will be observable in the anti-proton project at the future accelerator facility at GSI.

ACKNOWLEDGMENTS

I would like to thank C.M. Ko and A. Gillitzer for useful discussions.

REFERENCES

1. G.E. Brown and M. Rho, Phys. Rev. Lett. **66**, 2720, (1991).
2. T. Hatsuda and S.H. Lee, Phys. Rev. C **46**, r34 ,(1992).
3. M. Post, S. Leupold, U. Mosel, nucl-th/0309085.
4. S.H. Lee, Nucl. Phys. A, **638**, 183c, (1998).
5. A. Hayashigaki, Phys. Lett. B, **487**, 96, (2000).
6. K. Tsushima, D.H. Lu, A.W.Thomas, K. Saito, and R.H. Landau, Phys. Rev. C **59**, 2824, (1999).
7. F. Klingl, S. Kim, S.H. Lee, P. Morath, and W. Weise, Phys. Rev. Lett. bf 82, 3396, (1999).
8. S. Kim and S.H. Lee, Nucl. Phys. A **679**, 517, (2001).
9. M.E. Peskin, Nucl. Phys. B **156**, 365, (1979).
10. M.E. Luke *et al.*, Phys. Lett. B **288**, 355, (1992).
11. See http://www.gsi.de/GSI-future.
12. M. A. Shifman, A. I. Vainshtein and V.I. Zakharov, Nucl. Phys. **147**, 385, (1979); *ibid.* **147** (1979), 448.
13. A. Di Giacommo and G.C. Rossi, Phys. Lett. B **100**, 481, (1981); ; A. Di Giacommo and G. Paffuti *ibid.* **108** 327, (1982).
14. S.H. Lee, Phys. Rev. D **40**, 2484, (1989).
15. M. A. Shifman, A. I. Vainshtein and V.I. Zakharov, Phys. Lett. B **78**, 443, (1978).
16. B. Borasoy and U.G. Meissner, Phys. Lett. B **365**, 285, (1996).
17. G. Bhanot and M.E. Peskin, Nucl. Phys. B **156**, 391, (1979).
18. Y. Oh, S. Kim, and S.H. Lee, Phys. Rev. C **65**, 067901, (2002).
19. E. Eichten *et al.*, Phys. Rev. D **17**, 3090, (1978). *ibid.* **21**, 203, (1980).
20. Su Houng Lee, Che-Ming Ko, Phys. Rev. C **67**, 038202, (2003).
21. S.J. Brodsky, et.al., Phys. Rev. Lett **64**, 1011, (1990).
22. D.A. Wasson, Phys. Rev. Lett **67**, 2237, (1991).
23. F.S. Navarra and C.A.A. Nunes, Phys. Lett. B **356**, 439, (1995).
24. K. Martins, Prog. Part. Nucl. Phys., **36**, 409, (1996); hep-ph/9601314.

Vector meson properties in nuclear matter

D. Cabrera*, E. Oset* and M.J. Vicente Vacas*

*Departamento de Física Teórica and IFIC, Centro Mixto Universidad de Valencia-CSIC,
Institutos de Investigación de Paterna, Apdo. correos 22085, 46071, Valencia, Spain

Abstract. In this talk, we report on the properties of vector meson resonances in the nuclear medium. First, the ϕ meson spectrum due to the kaon channel is obtained in terms of the ϕ selfenergy provided by a chiral $SU(3)$ dynamics model. The medium effects are considered in the properties of K and $\bar{K}$ mesons which are renormalized by $S-$ and $P-$wave interactions with the nucleons, based on the lowest order meson-baryon chiral effective lagrangian. Within this scheme the mass shift and decay width of the ϕ meson in nuclear matter are studied. Second, the ρ meson is studied by looking at the pion-pion scattering amplitude in the vector-isovector channel, following a chiral unitary framework in free space and in the medium. To account for the medium corrections, the pion propagators are modified with a $P-$wave selfenergy, driven by the excitation of $p-h$ and $\Delta-h$ components. In addition, the ρ is allowed to couple to baryonic resonances, particularly to the $S-$wave $N^*(1520)$.

ϕ MESON MASS AND DECAY WIDTH IN NUCLEAR MATTER

In this work we deal with the coupling of the ϕ to $K\bar{K}$ channels, which in vacuum accounts for 85 % of the total decay width. To describe the interactions of the ϕ meson with pseudoscalar mesons and baryons, we follow the model developed in refs. [1, 2]. The Lagrangian describing the coupling of the ϕ meson to kaons reads

$$\mathcal{L}_{\phi,kaons} = -ig_\phi \phi_\mu (K^-\partial^\mu K^+ - K^+\partial^\mu K^- + \bar{K}^0\partial^\mu K^0 - K^0\partial^\mu \bar{K}^0)$$
$$+ g_\phi^2 \phi_\mu \phi^\mu (K^-K^+ + \bar{K}^0 K^0) . \tag{1}$$

The calculation of the ϕ meson selfenergy in nuclear matter, Π_ϕ^{med}, proceeds by a dressing of the kaon propagators including both $S-$ and $P-$wave selfenergies, as well as vertex corrections demanded by gauge invariance. To aisle the medium effects, we subtract the vacuum selfenergy and our results will be presented in terms of this in-medium subtracted selfenergy, $\Delta\Pi_\phi^{med} = \Pi_\phi^{med} - \Pi_\phi^{free}$.

$S-$wave and $P-$wave kaon selfenergy; Vertex corrections

The interactions of K^+, K^0 and $\bar{K}^0$, K^- with the nucleons of the medium are rather different and it is necessary to treat them separately. The KN interaction is smooth at low energies. We take the approach of [3, 4] where a $t\rho$ approximation is used and a constant selfenergy is obtained.

CP717, *Hadron Spectroscopy: Tenth International Conference,*
edited by E. Klempt, H. Koch, and H. Orth

The $\bar{K}N$ interaction, however, is dominated at low energies by the excitation of the $\Lambda(1405)$ resonance, which appears just below the $\bar{K}N$ threshold. We follow the coupled channel chiral unitary approach to S−wave $\bar{K}N$ scattering developed in ref. [5]. The effective in medium interaction was obtained in [3] solving a coupled channel Bethe-Salpeter equation including Pauli blocking on the nucleons, mean-field binding potentials for the baryons involved and the medium modifications of π mesons and $\bar{K}$ mesons themselves, leading thus to a selfconsistent calculation. The inclusion of the S−wave selfenergy accounts for decay channels of the ϕ meson of the type $KYMh$, the major contribution coming from $Y = \Sigma, M = \pi$ [3, 5].

We include as well P−wave kaon selfenergies accounting for the excitation of Λh, Σh and $\Sigma^* h$ pairs ($\Sigma^* = \Sigma^*(1385)$). Because of strangeness conservation, only direct terms are permitted for the $\bar{K}$ excitations. Conversely, the K selfenergy arises from the crossed terms. Since the excited hyperon is far off-shell in the crossed kinematics, we expect the K to be barely modified by the P−wave selfenergy and we shall not include it in the calculation. The $\bar{K}NY$ vertices involved are derived from the lowest order chiral Lagrangian coupling the octet of pseudoscalar mesons to the $1/2^+$ baryon octet [3] and then treated in a non-relativistic approach. The P−wave $\bar{K}$ selfenergy can be written in terms of the Lindhard functions, including only direct kinematics, for the Yh excitations. We consider relativistic recoil corrections for the $\bar{K}NY$ vertices, to properly account for possible higher components of kaon momenta in the loops, and we have included static dipolar form factors in these vertices.

The introduction of the ϕ coupling to kaons and $\bar{K}NY$ interactions in a gauge vector description of the ϕ meson generates additional $\phi\bar{K}NY$ contact interaction vertices, leading to a set of vertex correction diagrams. In addition, we consider other contributions in which the ϕ meson directly couples to the hyperons, which turn to be very small.

Results and discussion

In Fig. 1 (left) we show the imaginary part of the subtracted ϕ selfenergy for different densities from $\rho_0/4$ to ρ_0 [6]. The resulting width of the ϕ meson grows as a function of the density, and after adding the free width it reaches the value of around 30 MeV at the vacuum ϕ mass for normal nuclear density [7]. Most of the ϕ decay channels contributing to the in-medium width have a smooth behaviour at the energies under consideration, except for the $\phi \to K\Sigma^* h$ channel which, neglecting the Σ^* width, has a threshold at 940 MeV. This threshold moves to higher energies as density grows because of the repulsive potential felt by the kaons.

In Fig. 1 (right) we show the real part of $\Delta\Pi_\phi^{med}$ which we find mildly attractive up to energies around 1.1 GeV. The change in the ϕ mass at $\sqrt{s} = M_\phi$ is approximately 8 MeV to lower energies at $\rho = \rho_0$. This small correction is in agreement with previous works [8, 9], although disagrees with naive scaling expectations.

In summary, we have studied the ϕ selfenergy in a cold symmetric nuclear medium, by considering the medium effects over the tadpole and $K\bar{K}$ decay diagrams which dominate the ϕ vacuum selfenergy. Our main input has been the dressing of the kaon propagators with S− and P−wave selfenergies. The S−wave antikaon selfenergy is obtained in a

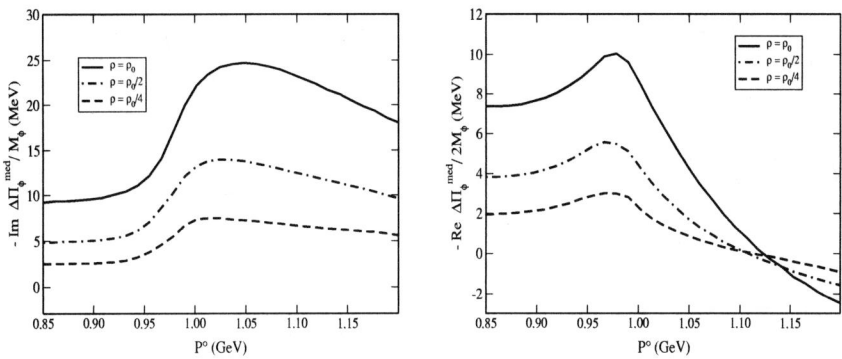

FIGURE 1. Imaginary (left) and real (right) part of the subtracted ϕ meson selfenergy in nuclear matter for several densities.

selfconsistent coupled channel unitary calculation based on effective chiral Lagrangians. The $P-$ wave selfenergy is driven by the coupling to hyperon-hole excitations. The main medium effect on the ϕ meson is the growth of its width by almost one order of magnitude at normal nuclear density. Nevertheless, the ϕ stays narrow enough to be visible as an isolated resonance and then give a clear signal in possible experiments measuring dilepton or K^+K^- spectra.

THE ρ MESON IN NUCLEAR MATTER

Meson-meson scattering in a chiral unitary approach

We study the ρ propagation properties by obtaining the $\pi\pi \to \pi\pi$ scattering amplitude in the $(I,J) = (1,1)$ channel. The model for meson-meson scattering in vacuum is fully explained in ref. [10]. The tree level amplitudes in the isospin basis ($|\pi\pi\rangle$, $|K\bar{K}\rangle$ $I = 1$ states) are obtained from the lowest order χPT lagrangians [11] including explicit resonance fields.

The final expression of the T matrix is obtained by unitarizing the tree level scattering amplitudes, in a coupled channel way. To this end we follow the N/D method, which was adapted to the context of chiral theory elsewhere. The free parameters of the theory are chosen to obey the low energy chiral constraints, by matching the expressions of the electromagnetic form factors calculated in this approach with those of one loop χPT.

The model successfully describes π, K electromagnetic vector form factors and the scattering amplitudes up to $\sqrt{s} \approx 1.2$ GeV with no free parameters. The results for the $\pi\pi \to \pi\pi$ scattering amplitude show that the inclusion of the $K\bar{K}$ channel introduces minimum changes compared to the calculation including only pion loops. Keeping this in mind, the calculations in nuclear matter are performed ignoring the contribution of kaon loops.

$(I,J) = (1,1)$ $\pi\pi$ scattering in the nuclear medium

The basic input for the calculation in nuclear matter is the pion selfenergy. It is written as usual in terms of the Lindhard functions accounting for both $p-h$ and $\Delta-h$ excitations [12]. Short range correlations are also accounted for with the Landau-Migdal parameter g', set to 0.7. We use a monopole form factor for the πNN and $\pi N\Delta$ vertices with the cut-off parameter set to $\Lambda = 1$ GeV. The Δ energy dependent decay width is explicitly kept, and recoil corrections have been considered in the $\pi N\Delta$ vertex.

As requested by the gauge invariance of the theory, a ρ-meson-baryon contact term must be considered [13], leading to a set of medium correction graphs. In addition to these terms, there are other medium corrections that arise if one sticks to the gauge vector field formalism and generates interactions via minimal coupling scheme [14]. Because of this a set of diagrams involving ρNN and $\rho\Delta\Delta$ vertices also contribute to the ρ meson selfenergy in nuclear matter. Also the $\rho\rho\pi\pi$ tadpole term gives a contribution because of the renormalization of the pion propagator.

A step forward is done by considering the coupling of the ρ meson to baryonic resonances. A much detailed work along these lines has been done in ref. [15], where many resonances are included. Here we focus on the $N^*(1520)$, which couples to the ρ in S-wave and therefore survives at zero ρ three-momentum. This correction manifests as an extra selfenergy term in the ρ propagator, driven by the excitation of a N^*-h pair. The lagrangian describing the interaction, $\iota_{N^*N\rho} = -g_{N^*N\rho}\bar{\Psi}_N S_i \vec{\phi}_i \vec{\tau} \Psi_{N^*} + h.c.$, is borrowed from ref. [16] and the contribution of the corresponding selfenergy diagram can be easily written in terms of a Lindhard function for the N^*-h excitation, which explicitly considers the N^* energy dependent decay width to $N\pi$, $\Delta\pi$, $N\rho$.

Results and discussion

In order to test the model dependence on the phenomenological parameters we performed variations of Λ in the range 0.9-1.1 GeV, and of g' in the range 0.6-0.8. The results are rather stable under both the variations, showing the kind of uncertainties in the present results which we can expect from uncertainties in the pion selfenergy.

We calculated in this approach the real and imaginary parts of T_{22}, the $\pi\pi \to \pi\pi$ scattering amplitude matrix element (Fig. 2). The imaginary part shows a clear broadening of the resonance and the peak position is slightly shifted upwards by about 30-40 MeV at $\rho = \rho_0$, which is also observed in the position of the zero of the real part. As a whole, the width of the ρ peak is increased to about 200 MeV at normal nuclear density. The coupling to the N^*-h components manifests as a visible bump at lower energies. As a result much strength is spread below the resonance mass [17].

In summary, we have analyzed the problem of the ρ properties in a nuclear medium following a chiral tensor formulation of $\pi\pi$ scattering in the ρ region together with unitarity in coupled channels. We have put together the different in medium mechanisms of previous works, leading to a substantial increase of the ρ width at normal nuclear density, accompanied by a small shift of the peak to higher energies, moderately dependent on the uncertainties of the pion selfenergy. An important source of strength appears at

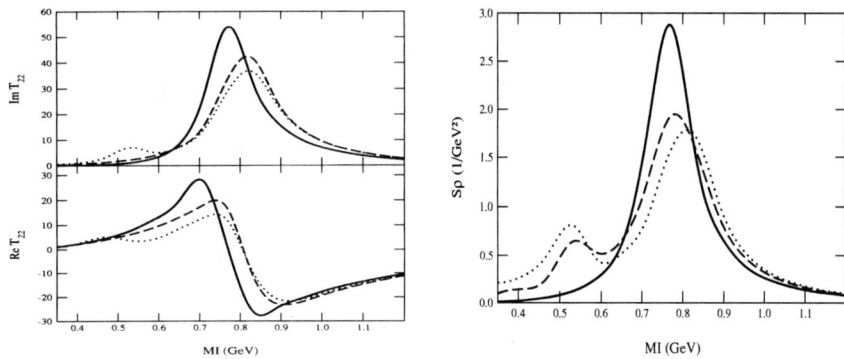

FIGURE 2. Left: Real and imaginary parts of the T_{22}. Long dashed (dotted) lines correspond to the results without (with) the coupling ρNN^* at $\rho = \rho_0$, and solid lines stand for the free case. Right: Spectral function for several values of nuclear density. Solid, dashed and dotted lines stand respectively for $\rho = 0$, $\rho_0/2$ and ρ_0.

low energies as a consequence of the coupling to $N^* - h$ excitations. We find, however, less overlap between the ρ peak and the N^* structure than in other works.

ACKNOWLEDGMENTS

I would like to thank the organizers for their kind invitation to the Hadron 2003 International Conference. This work is partly supported by DGICYT contract BFM2000-1326 and E.U. EURIDICE network contract HPRN-CT-2002-00311.

REFERENCES

1. F. Klingl, N. Kaiser and W. Weise, Z. Phys. A **356** (1996) 193 [arXiv:hep-ph/9607431].
2. F. Klingl, N. Kaiser and W. Weise, Nucl. Phys. **A624** (1997) 527 [hep-ph/9704398].
3. A. Ramos and E. Oset, Nucl. Phys. A **671** (2000) 481 [arXiv:nucl-th/9906016].
4. N. Kaiser, P. B. Siegel and W. Weise, Nucl. Phys. A **594** (1995) 325 [arXiv:nucl-th/9505043].
5. E. Oset and A. Ramos, Nucl. Phys. A **635** (1998) 99 [arXiv:nucl-th/9711022].
6. D. Cabrera and M. J. Vicente Vacas, Phys. Rev. C **67** (2003) 045203 [arXiv:nucl-th/0205075].
7. E. Oset and A. Ramos, Nucl. Phys. A **679** (2001) 616 [arXiv:nucl-th/0005046].
8. H. Kuwabara and T. Hatsuda, Prog. Theor. Phys. **94** (1995) 1163 [arXiv:nucl-th/9507017].
9. F. Klingl, T. Waas and W. Weise, Phys. Lett. B **431** (1998) 254 [arXiv:hep-ph/9709210].
10. J. A. Oller, E. Oset and J. E. Palomar, Phys. Rev. D **63** (2001) 114009 [hep-ph/0011096].
11. J. Gasser and H. Leutwyler, Nucl. Phys. **B250** (1985) 465, 517, 539.
12. E. Oset, P. Fernandez de Cordoba, L. L. Salcedo and R. Brockmann, Phys. Rep. **188**, 79 (1990).
13. G. Chanfray and P. Schuck, Nucl. Phys. **A555** (1993) 329.
14. M. Urban, M. Buballa, R. Rapp and J. Wambach, Nucl. Phys. **A641** (1998) 433 [nucl-th/9806030].
15. W. Peters, M. Post, H. Lenske, S. Leupold and U. Mosel, Nucl. Phys. **A632** (1998) 109 [nucl-th/9708004].
16. J. A. Gomez Tejedor, F. Cano and E. Oset, Phys. Lett. **B379** (1996) 39.
17. D. Cabrera, E. Oset and M. J. Vicente Vacas, Nucl. Phys. A **705** (2002) 90 [arXiv:nucl-th/0011037].

Structure function in nuclear matter in the NJL model

H. Mineo*, W. Bentz†, N. Ishii**, A.W. Thomas‡ and K. Yazaki§

*Department of Physics, National Taiwan University,
1 Roosevelt, Section 4, Taipei, Taiwan
†Department of Physics, School of Science, Tokai University,
Hiratsuka-shi, Kanagawa 259-1292, Japan
**The Institute of Physical and Chemical Research (RIKEN),
Hirosawa, Wako-shi, Saitama 351-0198, Japan
‡Special Research Center for the Subatomic Structure of Matter, and
Department of Physics and Mathematical Physics, The University of Adelaide,
Adelaide, SA 5005, Australia
§Department of Physics, Tokyo Woman's Christian University,
Suginami-ku, Tokyo 167-8585, Japan

Abstract. We present a consistent description of the structure function of a free nucleon, the equation of state of nuclear matter, and the structure function of a nucleon bound in nuclear matter, in the framework of the Nambu-Jona-Lasinio (NJL) model. The important role of the mean vector field in the nuclear medium to explain the EMC effect is emphasized.

INTRODUCTION

In this work we will be concerned with the medium modifications of the spin independent nuclear structure functions measured in deep inelastic scattering of leptons, that is, the EMC effect [1, 2]. It is known [3, 4] that the Fermi motion and binding effects on the level of nucleons cannot explain the observed reduction of the nuclear structure function in the range of Bjorken x between 0.4 and 0.8. It is therefore a challenging task to describe the nuclear systems in terms of nucleons with internal quark structure, and to investigate whether the binding effects on the quark level can account for the observations. For this purpose, we will use the Nambu-Jona-Lasinio (NJL) model [5] as an effective chiral quark theory. It has been shown [6] that the quark-diquark description of the single nucleon, which is based on the relativistic Faddeev approach [7], can be combined successfully with the mean field description of the nuclear matter equation of state. Since the description of the nucleon is completely covariant, off-shell effects can be investigated unambiguously, and the Ward identities for baryon number and momentum conservation can be incorporated rigorously from the outset, which is particularly important for the description of structure functions. This framework is therefore a powerful tool to investigate the origin of the EMC effect in terms of binding on the level of quarks. This is the purpose of our present work, where we will consider the case of infinite nuclear matter as a first step. For simplicity, we will limit ourselves to a valence quark description, that is, the effects of sea quarks will not be considered in this paper.

CP717, *Hadron Spectroscopy: Tenth International Conference,*
edited by E. Klempt, H. Koch, and H. Orth
© 2004 American Institute of Physics 0-7354-0197-7/04/$22.00

QUARK LIGHT CONE MOMENTUM DISTRIBUTIONS IN NUCLEAR MATTER

The light cone momentum distribution of quarks per nucleon in a nucleus with mass number A is defined as

$$f_{q/A}(x_A) = \frac{P_-}{A^2} \int \frac{dw^-}{2\pi} e^{iP_- x_A w^-/A} < A, P|\overline{\psi}(0)\gamma^+ \psi(w^-)|A, P >. \tag{1}$$

Here P^μ is the total 4-momentum of the nucleus and P_- its "minus component" [1], and ψ is the quark field. Since we will refer to the case $N = Z$, we need only the isospin symmetric combination $f_{q/A} \equiv f_{u/A} + f_{d/A}$. The quantity x_A is the Bjorken variable of the nucleus, which is equivalent to the fraction of the total P_- carried by a quark, multiplied by A. In the rest system of the nucleus (nuclear matter) we have $P_-/A = \varepsilon_F/\sqrt{2}$, where ε_F is the Fermi energy (that is, the mass per nucleon or the chemical potential) of the nucleons.

To calculate the distribution (1), we make use of the familiar one-dimensional convolution formula

$$f_{q/A}(x_A) = \int dy_A \int dz \, \delta(x_A - y_A z) f_{q/N}(z) f_{N/A}(y_A), \tag{2}$$

where $f_{q/N}(z)$ and $f_{N/A}(y_A)$ are the light cone momentum distributions of quarks in the nucleon, and of nucleons in the nucleus (per nucleon), respectively. (The definitions of these distributions can be found in Ref.[8].) The variable z is the fraction of the minus component of the nucleon momentum carried by a quark, and y_A is A times the fraction of the total P_- carried by a nucleon.

For the evaluation of the quark distribution in the nucleon ($f_{q/N}$), we describe the nucleon as a bound state of a quark and a scalar ($J^P = 0^+, T = 0$) diquark in the NJL model. The distribution $f_{q/N}$ can then be obtained by a straight forward Feynman diagram calculation [8, 9]. The presence of the nuclear medium is taken into account via scalar and vector mean fields which act on the quarks in the nucleon. In order to determine these mean fields for a given density, and also in order to evaluate the nucleon distribution function $f_{N/A}$, we make use of the successful mean field description of the nuclear matter ground state in the NJL model, which takes into account the quark substructure of the nucleons [6].

The most important relation of this approach, which shows the direct effect of the vector mean field on the quark distribution function, is as follows:

$$f_{q/A}(x_A) = \frac{\varepsilon_F}{E_F} f_{q/A0}(x_A' = \frac{\varepsilon_F}{E_F} x_A - \frac{V_0}{E_F}). \tag{3}$$

Here the distribution without the explicit effect of the mean vector field is denoted as $f_{q/A0}(x_A')$, and the nucleon Fermi energy has the form $\varepsilon_F = E_F + 3V_0$, where $E_F =$

---

[1] Our notations for light cone variables are $a^\pm = (a^0 \pm a^3)/\sqrt{2}$, $a_\pm = (a_0 \pm a_3)/\sqrt{2}$. The normalization of the nuclear state vector is $< A, P|\overline{\psi}\gamma^+ \psi|A, P >= 3A$.

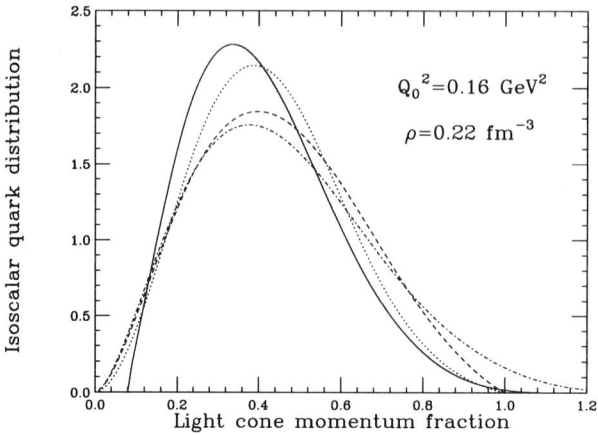

FIGURE 1. Sum of the valence up and down quark light cone momentum distributions at the low energy scale ($Q_0^2 = 0.16\,\text{GeV}^2$). For explanation of the lines, see text.

$\sqrt{M_N^2 + p_F^2}$ (M_N is the effective nucleon mass), and V_0 is the vector field acting on the quarks in the nucleon. The actual calculation of $f_{q/A}$ therefore proceeds as follows: First we calculate the distribution for a free nucleon in the quark-diquark approach, then we replace the quark, diquark and nucleon masses by the effective ones according to the nuclear matter equation of state, then we include the effect of the Fermi motion of nucleons with effective mass M_N, and finally we perform the scale transformation (3) to include the direct effect of the vector mean field. The actual calculations are carried out in the proper time regularization scheme which avoids unphysical quark decay thresholds for the nucleon, see Ref.[8] for details.

RESULTS

The medium modifications of the isoscalar valence quark distribution at the saturation density of our nuclear matter equation of state are shown in Fig.1 for the low energy scale, for which we use the value $Q_0 = 0.4\,\text{GeV}$. The dotted line shows the distribution in a free nucleon, which is consistent with the empirical parametrizations [10]. The dashed line shows the result when all masses are replaced by the effective ones, the dot-dashed line shows the result including the Fermi motion of nucleons without the direct effect of the mean vector field, and finally the solid line is obtained from the dot-dashed one by the scale transformation (3). It is clear from this figure, and can also be shown from Eq.(3), that the direct effect of the mean vector field is to squeeze the quark distribution from both the small and the large side of the Bjorken variable. The mean vector field is therefore essential to describe the depletion in the valence quark region, and also leads to an enhancement for smaller values of the Bjorken variable.

Fig. 2 shows our results for the structure function per nucleon in isospin symmetric nuclear matter (solid line), in comparison to the isoscalar structure function of a free

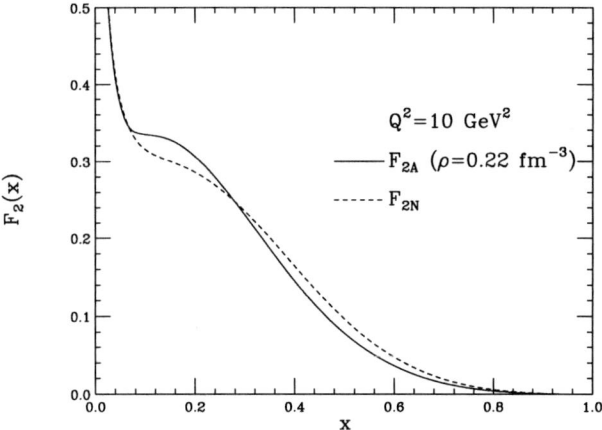

FIGURE 2. Structure function per nucleon in isospin symmetric nuclear matter (solid line), and isoscalar free nucleon structure function (dashed line) at $Q^2 = 10\,\text{GeV}^2$.

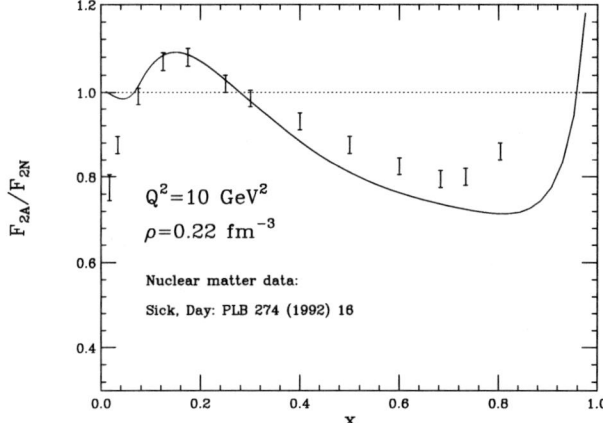

FIGURE 3. Ratio of the structure function per nucleon in isospin symmetric nuclear matter to the free isoscalar nucleon structure function at $Q^2 = 10\,\text{GeV}^2$.

nucleon (dashed line). These curves were obtained by evolving the quark distributions shown by the solid and dotted lines in Fig. 1 up to $Q^2 = 10\,\text{GeV}^2$ by using the Q^2 evolution code of Ref.[11]. Fig. 2 clearly shows the suppression of the structure function at large x and the enhancement at smaller x.

>From the two curves of Fig. 2, we obtain the EMC ratio between the nuclear and the nucleon structure functions shown in Fig. 3. The data points in this figure are the extrapolations of nuclear EMC data to the nuclear matter case. We see that the calculation can reproduce the main features of the EMC data, namely the suppression at large x and the enhancement at smaller x.

SUMMARY

In this paper we discussed the modifications of the structure function of a nucleon bound in the nuclear medium. For this purpose, we used an effective chiral quark theory which can account for the quark substructure of a single nucleon, and for the saturation properties of the nuclear matter binding energy. Our description of nuclear matter was done in the mean field approximation, which is characterized by mean scalar and vector fields which couple to the quarks in the nucleons. Our aim was to assess the effects which arise from the quark structure of the bound nucleon.

It is well known that the effects of the scalar and vector mean fields on the level of nucleons, including the Fermi motion, cannot explain the EMC data. In our approach, we investigated these effects on the level of quarks, and our most important result can be summarized as follows: The mean vector field in the medium influences the form of the quark light cone momentum distributions directly, besides its indirect influence through the equation of state of the system. This direct modification of the quark distribution is expressed by Eq.(3), and is the principal agent to explain the EMC effect in the framework of a mean field description of nuclear matter. While the effect of the vector field on the nucleon momentum distribution alone tends to cancel the effect of the scalar field [4], there remains an appreciable net effect if the description is done on the quark level, and the result is consistent with the EMC data.

ACKNOWLEDGMENTS

This work was supported by the Grant in Aid for Scientific Research of the Japanese Ministry for Education, Culture, Sports, Science and Technology, Project No. C2-13640298, the Australian Research Council and The University of Adelaide.

REFERENCES

1. J. Aubert *et al.*, Phys. Lett. **B 123** (1982) 275.
2. D.F. Geesaman, K. Saito and A.W. Thomas, Ann. Rev. Nucl. Part. Sci. **45** (1995) 337.
3. R.L. Jaffe, in *Proceedings of the 1985 Los Alamos School on Relativistic Dynamics and Quark Nuclear Physics*, edited by M.B. Johnson and A. Pickleseimer (Wiley, New York, 1985).
4. G.A. Miller and J. R. Smith, Phys. Rev. **C 65** (2002) 015211.
5. Y. Nambu and G. Jona-Lasinio, Phys. Rev. **122** (1960) 345; **124** (1961) 246.
6. W. Bentz and A.W. Thomas, Nucl. Phys. **A 696** (2001) 138;
   W. Bentz, T. Horikawa, N. Ishii and A.W. Thomas, Nucl. Phys. **A 720** (2003) 95.
7. N. Ishii, W. Bentz and K. Yazaki, Nucl. Phys. **A 578** (1995) 617.
8. H. Mineo, W. Bentz, N. Ishii, A.W. Thomas and K. Yazaki, Quark distributions in nuclear matter and the EMC effect, preprint (Oct. 2003).
9. H. Mineo, W. Bentz and K. Yazaki, Phys. Rev. **C 60** (1999) 065201.
10. A.D. Martin, R.G. Roberts, W.J. Stirling and R.S. Thone (MRST2002), Eur. Phys. J. **C 28** (2003) 455.
11. M. Miyama and S. Kumano, Comput. Phys. Commun. **94** (1996) 185.

Reactions

QCD and the Pomerons[1]

A. Donnachie

Department of Physics and Astronomy
University of Manchester
Manchester M13 9PL
England

Abstract. The phenomenology of the soft pomeron of hadronic physics is described and different models in nonperturbative QCD are outlined. A second pomeron, the hard pomeron, is introduced as a means of resolving a problem in the evolution of $F_2(x, Q^2)$ at small x. The two-pomeron approach gives an excellent description of $F_2(x, Q^2)$ at small x and the phenomenology can be understood in terms of a dipole model. Perturbative QCD successfully describes the evolution of the hard-pomeron component and allows the gluon density to be calculated. This is somewhat larger than usually supposed, but leads to a clean pQCD description of the charm structure function.

INTRODUCTION

The conventional treatment[1, 2] of pQCD evolution expands the DGLAP splitting matrix in powers of $\alpha_s(Q^2)$. This is almost certainly unsafe[3] at small x as it induces singularities which are unlikely to be present in the exact matrix. An analysis[4] of the theoretical uncertainties in conventional pQCD evolution leads to the conclusion that parton distributions are safe only over a restricted kinematic domain, $Q^2 \gtrsim 10$ GeV^2 and $x \gtrsim 0.005$. Regge Theory and pQCD should co-exist, not compete. When the two approaches are combined[5, 6, 7] the problem at small x is partially solved and provides a successful description of the proton structure function $F_2(x, Q^2)$ and its charm component $F_2^c(x, Q^2)$. This requires two pomerons, the soft pomeron well-known in hadronic reactions, and a new pomeron, by definition the hard pomeron. The concept of there being more than one pomeron pole is embedded in the BFKL equation. A running coupling leads[8] to isolated poles for the Mellin transform of the pomeron amplitude instead of the cut obtained in the case of a fixed coupling. In the two-pomeron approach the term which causes the rapid rise of $F_2(x, Q^2)$ with increasing $1/x$ is present already at small Q^2, so that in contrast to the conventional approach, pQCD evolution does not generate this term, but merely makes it become more prominent as Q^2 increases. This is apparent in the HERA data[9, 10, 11] for the photoproduction and electroproduction of charm, which exhibit the striking property[12] that at each fixed Q^2 they vary with the same power $x^{-\varepsilon_0}$ with $\varepsilon_0 \approx 0.4$. This behaviour, which is indicated by the thin line in

---

[1] Figures are reproduced with permission from *Pomeron Physics and QCD* by Sandy Donnachie, Guenter Dosch, Peter Landshoff and Otto Nachtmann (Cambridge University Press, 2002). See website http://www.damtp.cam.ac.uk/user/pvl/QCD

CP717, *Hadron Spectroscopy: Tenth International Conference,*
edited by E. Klempt, H. Koch, and H. Orth
© 2004 American Institute of Physics 0-7354-0197-7/04/$22.00

Fig.3, is not widely appreciated as the data are usually shown on a log-linear plot rather than a log-log plot.

THE SOFT POMERON

The soft pomeron is phenomenologically understood[13, 14] as a Regge pole with a linear trajectory $\alpha_1(t) = 1 + \varepsilon_1 + \alpha'_1 t$. The rise of total cross sections at high energies gives $\varepsilon_1 \sim 0.08 - 0.1$ and the energy dependence of the pp and $\bar{p}p$ differential cross section fixes $\alpha'_1 \sim 0.25\,\mathrm{GeV}^{-2}$. At small t, cuts arising from multiple pomeron exchange are weak, although they become important at larger t. Comparison of $\pi^\pm p$ and pp total cross sections supports the idea of quark additivity, that is the pomeron couples to single valence quarks with a γ^μ coupling. Rather direct evidence[15] supporting the hypothesis that the pomeron interacts with only one quark in the hadron is provided by the reaction $pp \to p + \Lambda\phi K$. In the $\Lambda\phi K$ centre-of-mass the Λ is produced preferentially in the direction of the dissociating proton, the K is produced backwards and the ϕ, which contains no valence quarks, is produced centrally. These data are consistent with a valence quark being back-scattered by the pomeron and a valence diquark going forward. If the soft pomeron is indeed a Regge pole then physical states, glueball states, should lie on the pomeron trajectory for $t > 0$. There is some evidence[16] for a 2^{++} glueball candidate at 1926 MeV, exactly the right mass to lie on the soft-pomeron trajectory. Confirming this, or otherwise, is a challenge for meson spectroscopy.

There are alternative models for the pomeron, involving uncorrelated gluon exchange. The exchange of two nonperturbative Abelian gluons interacting with the valence quarks, initially proposed by Low[17] and Nussinov[18] and developed in its final form by Landshoff and Nachtmann[19], reproduces many of the features of the phenomenology. It leads rather naturally to the γ^μ coupling and provides an explanation of quark additivity. Terms with the gluons coupling to different quarks are suppressed relative to those with the gluons coupling to the same quark. The suppression $\sim a^2/R^2$ where $a \sim 0.3\mathrm{fm}$ is the gluon correlation length in the vacuum and R is a typical hadronic radius, $\sim 1.5\mathrm{fm}$. In such a model the total cross section is constant, but reggeisation may be possible at the quark level[20], leading to a rising cross section. This approach is essentially based on quark-quark scattering and takes no explicit account of the bound state.

When the Abelian gluons are replaced by non-Abelian gluons in this approach, serious problems arise which are associated with neglecting quark confinement. It becomes necessary to consider directly hadron-hadron scattering. This leads to the Heidelberg model[21, 22, 14], in which the hadrons are represented by Wilson loops interacting via the gluonic vacuum. In this approach the whole hadron takes part in the interaction, not just the valence quarks. Quark additivity is lost and σ_T now depends on the size of the interacting hadrons. However as the ratio of the pion and proton radii is about 0.65 the results of quark additivity are reproduced. A bonus is that the model predicts $\sigma_T(Kp) \approx 0.55\sigma_T(pp)$, in agreement with experiment. As before, σ_T is constant. However the model allows the calculation[23, 14] of $d\sigma/dt$ which is qualitatively correct for pp scattering.

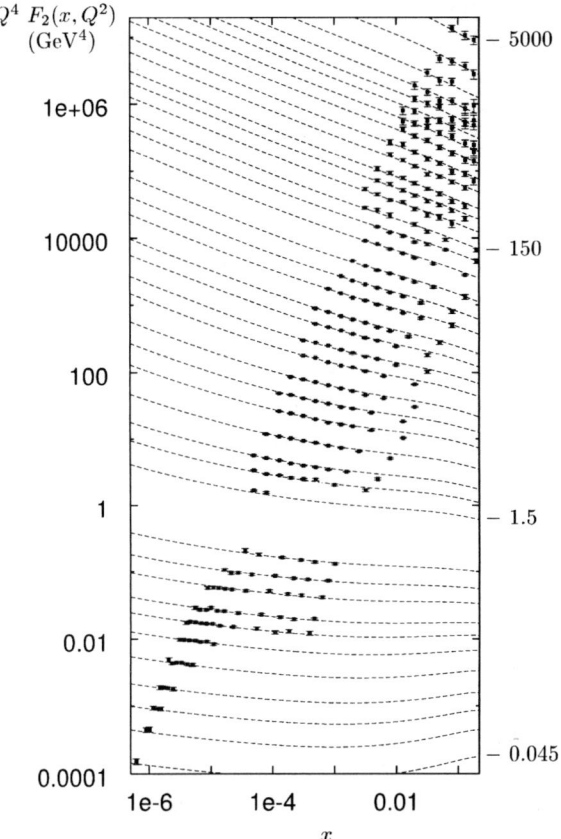

FIGURE 1. Data from H1[24] and ZEUS[25] with two-pomeron fit. Q^2 ranges from 0.045 to 5000 GeV^2.

THE HARD POMERON

If the soft pomeron is indeed a Regge pole then its parameters, namely its intercept at $t = 0$ and the slope of the trajectory, are universal and cannot be changed by pQCD evolution. This requires that the contribution of the soft pomeron to $F_2(x, Q^2)$ should be of the form $f_1(Q^2)x^{-\varepsilon_1}$. Within the Regge framework a second pomeron is required to describe the large rise of $F_2(x, Q^2)$ with increasing $1/x$. This suggests parametrising $F_2(x, Q^2)$ at small x as[5]

$$F_2(x, Q^2) = f_0(Q^2)x^{-\varepsilon_0} + f_1(Q^2)x^{-\varepsilon_1} \qquad (1)$$

at each Q^2 for which there are data, including $Q^2 = 0$. Fitting with the soft-pomeron intercept fixed at its classical value, $\varepsilon_1 = 0.08$, determines the hard-pomeron intercept to be in the range $0.35 \le \varepsilon_0 \le 0.5$. While varying ε_0 through this range has little effect on the shape of the hard-pomeron coefficient function $f_0(Q^2)$, the large-Q^2 component of the soft-pomeron coefficient function $f_1(Q^2)$ changes markedly from being approximately constant at large Q^2 to decreasing with increasing Q^2 beyond $Q^2 \approx 10\,\text{GeV}^2$. The constant behaviour at large Q^2 is obtained for $\varepsilon_0 \approx 0.4\, f_1(Q^2)$. Assuming this behaviour, all H1[24] and ZEUS[25] data for $x \leq 0.001$ are well fitted by

$$f_0(Q^2) = A_0 \frac{(Q^2)^{1+\varepsilon_0}}{(1+Q^2/Q_0^2)^{1+\varepsilon_0/2}}$$
$$f_1(Q^2) = A_1 \frac{(Q^2)^{1+\varepsilon_1}}{(1+Q^2/Q_1^2)^{1+\varepsilon_1}} \tag{2}$$

with $\varepsilon_0 = 0.4075$, $Q_0 = 2.88\,\text{GeV}$ and $Q_1 = 0.768\,\text{GeV}$ and with χ^2/N_{data} significantly less than 1. It is an extremely economical fit. The photoproduction data largely fix A_1, so there are effectively only four free parameters. Including a factor $(1-x)^7$ as a *crude* way of ensuring that $F_2(x,Q^2) \to 0$ as $x \to 0$, and adding terms for f_2 and a_2 exchange, the fit agrees well with data at larger x even up to $Q^2 = 5000\,\text{GeV}^2$. The fit is shown in Fig.1.

It is found that the charm data[11] are consistent with the hard-pomeron term only and further, that the hard pomeron is flavour blind:

$$F_2^c(x,Q^2) = 0.4 f_0(Q^2) x^{-\varepsilon_0} \tag{3}$$

where $f_0(Q^2)$ is defined in equation (2). The factor 0.4 is $\frac{4}{9}/(\frac{4}{9}+\frac{1}{9}+\frac{1}{9}+\frac{4}{9})$.

A DIPOLE MODEL

The model[26] is based on extending the Heidelberg model to incorporate both a soft and a hard pomeron. The parameters for the wave functions and for the dipole cross section are fixed by the Heidelberg model. It is assumed that large dipoles couple to the soft pomeron and small dipoles couple to the hard pomeron. This introduces one free parameter, the scale R_c for the hard/soft boundary. This can be fixed by fitting to $F_2(x,Q^2)$ and it is found that $R_c \approx 0.22\text{fm}$, in conformity with the usually accepted value. The fit to $F_2(x,Q^2)$ is good and the results are transportable to any reaction for which the wave functions of the participating particles are known. The predictions for the γp total cross section, for $\gamma p \to J/\psi p$, for F_2^c and F_2^L, for the $\gamma\gamma$ total cross section and the photon structure function F_2^γ, and for deep virtual Compton scattering, $\gamma^* p \to \gamma p$, are in excellent agreement with experiment.

For both $F_2^c(x,Q^2)$ and $F_L(x,Q^2)$ the photon wave function is concentrated at small distances. In the case of charm this is a consequence of the mass of the charm quark in the photon wave function. For the longitudinal structure function it is a consequence of the wave function of the longitudinal photon. Thus the hard pomeron is already dominant at moderate Q^2. The strong suppression in this model of the soft pomeron relative to the hard pomeron in $F_2^c(x,Q^2)$, which is purely a wave-function effect, is notable and provides an explanation for the almost-complete flavour-blindness of the hard pomeron commented on above. For the longitudinal structure function the increase of the short range (hard part) with increasing Q^2 is not as strong as for the transverse structure function, again a consequence of differences in the relevant photon wave functions. Nonetheless as the long range (soft part) of the longitudinal structure function is even more suppressed at large Q^2 relative to its contribution to the transverse structure function, the hard pomeron is dominant sooner in $F_L(x, Q^2)$ than in $F_2(x, Q^2)$.

THE DGLAP EQUATION

The singlet parton densities are defined by

$$\mathbf{u}(x, t) = \begin{pmatrix} x\Sigma_f(q_f + \bar{q}_f) \\ xg(x, t) \end{pmatrix}, \qquad t = \log(Q^2/\Lambda^2) \tag{4}$$

Take their Mellin transform with respect to x:

$$\mathbf{u}(N, Q^2) = \int_0^1 dx \, x^{N-1} \mathbf{u}(x, Q^2) \tag{5}$$

Then the DGLAP equation is

$$\frac{\partial}{\partial t} \mathbf{u}(N, Q^2) = \mathbf{P}(N, \alpha_s(Q^2))\mathbf{u}(N, Q^2) \tag{6}$$

where $\mathbf{P}(N, \alpha_s(Q^2))$ is the Mellin transform of the splitting matrix.

The normal procedure is to expand $\mathbf{P}(N, \alpha_s(Q^2))$ in powers of $\alpha_s(Q^2)$. However this is *illegal* when N is close to 0 as the terms in the expansion of $\alpha_s(Q^2)$ have singularities at $N = 0$ which are not present in $\mathbf{P}(N, \alpha_s(Q^2))$ itself. At any given Q^2, expanding $\mathbf{P}(N, \alpha_s(Q^2))$ in powers of $\alpha_s(Q^2)$ is invalid when one goes to sufficiently small x. However if we introduce the two-pomeron parametrisation the situation can be partially rescued as a fixed-power behaviour $\mathbf{u}(x, t) \sim x^{-\varepsilon}$ corresponds to

$$\mathbf{u}(N, Q^2) \sim \frac{\mathbf{f}(Q^2)}{N - \varepsilon} \qquad \mathbf{f}(Q^2) = \begin{pmatrix} f_q(Q^2) \\ f_g(Q^2) \end{pmatrix} \tag{7}$$

Now rewrite the DGLAP equation in terms of $\mathbf{f}(N, Q^2)$:

$$\frac{\partial}{\partial t} \mathbf{f}(N, Q^2) = \mathbf{P}(N = \varepsilon, \alpha_s(Q^2))\mathbf{f}(N, Q^2) \tag{8}$$

This cannot be applied to the soft-pomeron term as we need $\mathbf{P}(N, \alpha_s(Q^2))$ too close to $N = 0$. But for the hard-pomeron term we need $\mathbf{P}(N, \alpha_s(Q^2))$ for $N \approx 0.4$ where the expansion should be valid. One can then solve the DGLAP equation at small x for $f_q(Q^2)$ and $f_g(Q^2)$, with some starting value Q_0^2 of Q^2. A value $Q_0^2 = 20$ GeV^2 was taken. At sufficiently large x, conventional DGLAP evolution should not be influenced significantly by the singularities, and it was assumed to be correct down to $x = 0.01$ at

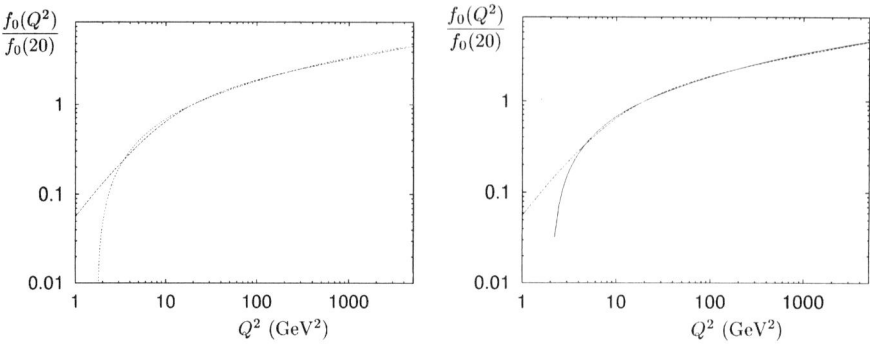

FIGURE 2. LO and NLO calculations of the hard-pomeron coefficient function, together with the phenomenological fit.

Q_0^2. The dominance of $F_2^c(x, Q^2)$ by the hard pomeron implies that the gluon structure function $g(x, Q^2)$ at small x is entirely hard-pomeron exchange, so we can use its value at $x = 0.01$, $Q^2 = 20$ GeV^2 to give $f_g(Q^2 = 20)$, and $f_q(Q^2 = 20)$ is known from the fit to $F_2(x, Q^2)$.

The result of the evolution, using four flavours, is almost the same for LO and NLO as there is very little difference[27] in the splitting matrix at LO and NLO for $N \approx 0.4$. The solution for $f_q(Q^2 = 20)$ is in astonishing agreement with the phenomenological fit as can be seen in Fig.2. At small x, the proton's gluon density is given by

$$xg(x, Q^2) = f_g(Q^2)x^{-\varepsilon_0} \tag{9}$$

At large Q^2 and small x this is larger than the conventional one: at $Q^2 = 200, x = 0.0001$ it is twice as large as MRST or CTEQ. This is significant for experiments at LHC.

CHARM AND LONGITUDINAL STRUCTURE FUNCTIONS

At small Q^2 $F_2^c(x, Q^2)$ should be calculated[28, 29, 30] from photon-gluon fusion, $\gamma^* g \to c\bar{c}$, at some fixed order in α_s and depends on the charmed-quark mass m_c. At large Q^2 a resummation to all orders in α_s is needed because of factors of powers of $\log(Q^2/m_c^2)$. This is achieved by changing at large Q^2 to the output from DGLAP evolution, where m_c can be neglected. The two calculations have to be matched at some value of Q^2, usually of order m_c^2, and is sensitive to exactly what value is chosen. In the two-pomeron approach the two calculations match rather well, as can be seen in Fig.3. The LO calculation of photon-gluon fusion requires $m_c \approx 1.3$ GeV and the NLO one requires $m_c \approx 1.6$ GeV.

The two-pomeron picture also gives a good description of J/ψ photoproduction[14, 12, 31].If the hard pomeron is just another glueball trajectory, then fitting σ^T and $d\sigma/dt$ gives its slope:

$$\alpha'_H \approx 0.1 \text{GeV}^{-2} \tag{10}$$

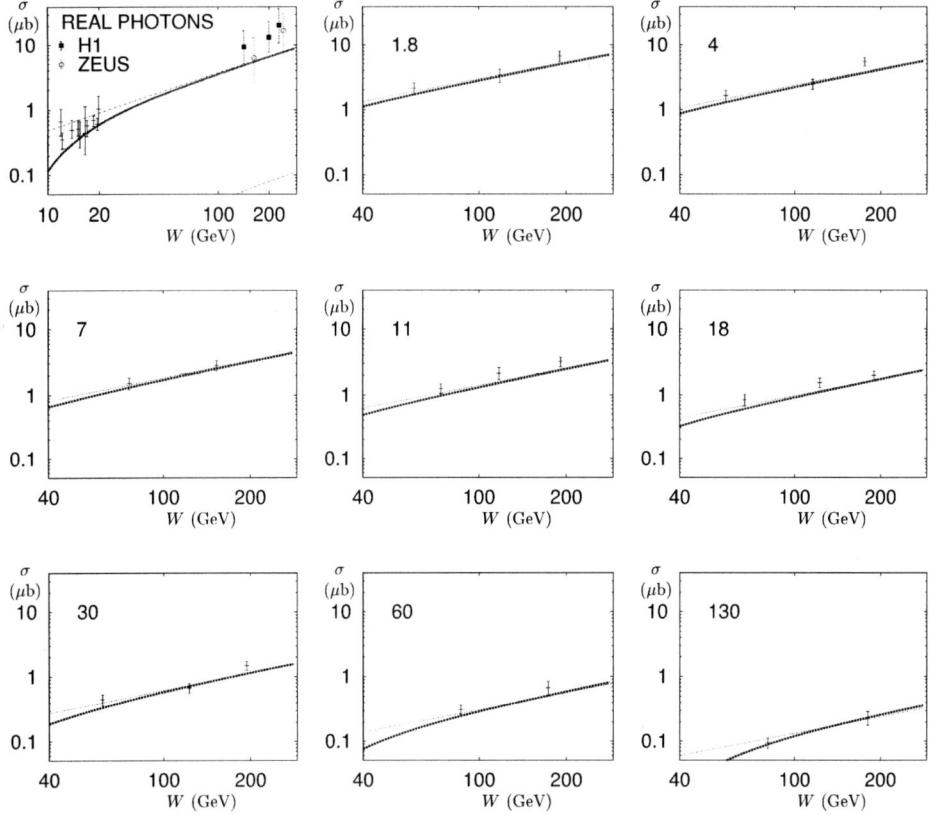

FIGURE 3. Data[9, 10, 11] for $\sigma^c(W)$. The thin lines correspond to (3), which coincides with the output from DGLAP evolution, and the thick lines are from NLO photon-gluon fusion.

Although the soft-pomeron term is appreciably smaller than the hard-pomeron term at the highest energies for which there are data, interference gives it an important role. There is a very distinctive and characteristic change in energy dependence in going from low energy, dominated by the soft pomeron, through the interference region and ultimately to hard-pomeron dominance.

The longitudinal structure function $F_2^L(x, Q^2)$ has the most sensitivity to the gluon distribution. There is some dependence on the charm-quark mass, principally at low Q^2, but it is much less than for $F_2^c(x, Q^2)$. The two-pomeron prediction is in good agreement with data. However the data depend on an assumed parametrisation to separate F_L from F_2 and errors are large[32]. So that although there is a clear difference between the hard-pomeron prediction and MRST, the data are not yet good enough to distinguish.

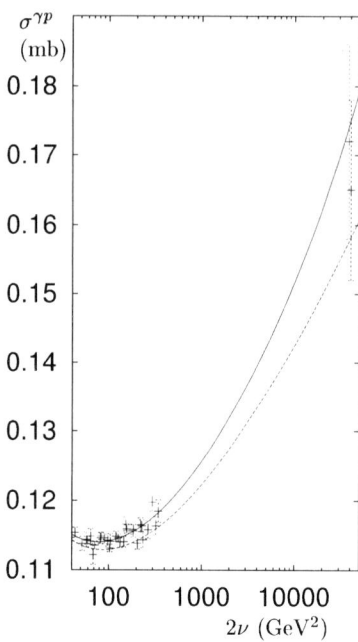

FIGURE 4. γp total cross section. The upper curve is the extrapolation to $Q^2 = 0$ of the fit to $F_2(x, Q^2)$. The lower curve omits the hard-pomeron term.

DISCUSSION

As scattering amplitudes are analytic functions of their variables, a singularity present at large Q^2 survives at $Q^2 = 0$. So if the hard pomeron is present at large Q^2 it should also be present at $Q^2 = 0$. This is reinforced by the charm-production data, as the W-dependence at $Q^2 = 0$ is the same as at higher Q^2. There is a small contribution from the hard pomeron to the γp total cross section, but it is masked by the much-larger soft-pomeron term and the systematic errors on the H1 and ZEUS data are too large to test whether the soft pomeron is adequate or whether an extra component is needed. Figure 4 shows the data for the photoproduction cross section and the extrapolation[14] to $Q^2 = 0$ of the fit to $F_2(x, Q^2)$. The soft-pomeron contribution does not differ much from fitting[14, 13] the γp total cross section directly with no hard-pomeron term. However there is possible evidence for an additional component in the $\gamma\gamma$ total cross section. The OPAL[33] and L3[34] data are shown in Fig.5. The final cross sections are rather sensitive to the Monte Carlo model used, different Monte Carlo simulations producing different results. The resulting uncertainty is included in the errors on the OPAL data. The L3 data are shown with the use of two Monte Carlo simulations. The curve in Fig.5 is the result of assuming factorisation of the soft-pomeron and reggeon terms in the γp and pp total cross sections. This is compatible with the OPAL data but the energy dependence is not sufficiently strong to match that of the L3 data. This may be an indication that a more rapidly-rising component must be added.

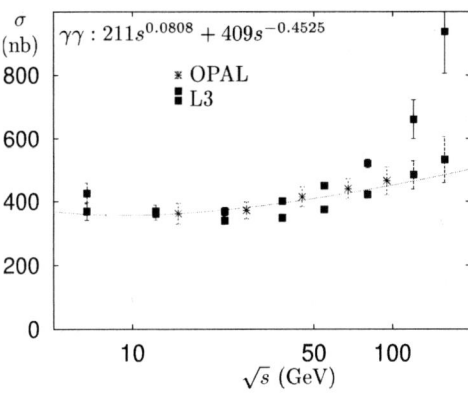

FIGURE 5. Data from OPAL[33] and L3[34] for the $\gamma\gamma$ total cross section. The curve is the prediction assuming factorisation of the soft-pomeron plus reggeon fits to the pp and γp total cross sections

The high-energy γp and $\gamma\gamma$ total cross sections are of crucial importance for the two-pomeron picture. If the hard pomeron is present in the total cross section for photon collisions, is the same true for pp collisions? As the photon wave function has a short-range component, one would expect that that the hard pomeron will couple to it with a larger relative strength than to a large object such as the proton, in the spirit of the dipole model outlined above. But even hadronic wave-functions have a high-momentum component, albeit small, so there is some prospect that the hard pomeron will show up at LHC. The dipole model predicts[26] that the hard-pomeron component could be as much as 15% of the pp total cross section at the LHC.

The sharp rise in $F_2(x, Q^2)$ at small x is certainly a consequence of gluon exchange. At one time there was a hope that the power ε_0 of $1/x$ i.e. the hard-pomeron term, might be calculated from the BFKL equation and therefore that it is a perturbative effect. At present it is far from clear that pQCD can be used to calculate the intercept $1 + \varepsilon_0$. The soft-pomeron trajectory surely cannot be calculated from pQCD, but lattice-gauge calculations of the glueball masses, or their direct measurement, may provide the trajectory.

SUMMARY

The soft pomeron of hadronic physics is well understood phenomenologically as a Regge pole, but it is still not fully explained by QCD models. It is essential to obtain glueball masses, and hence trajectories, to confirm it as a Regge pole.

The conventional approach to evolution needs modifying at small x. It can be corrected if it is combined with Regge theory, but only partly as we can only treat the hard-pomeron part. This is enough to extract the gluon distribution, which is larger at small x than conventionally supposed. The two-pomeron approach gives a good description of a wide variety of data, including charm photo- and electroproduction.

Key tests are the γp and $\gamma\gamma$ total cross sections, and the proton longitudinnal structure function, for all of which good data are required.

If the hard pomeron is a factorising Regge pole then it could be significant in the pp total cross section at the LHC.

REFERENCES

1. A. D. Martin, R. G. Roberts, W. J. Stirling and R. S. Thorne, *Eur.Phys.J.* **C18**, 117(2000)
2. J. Pumplin et al, *JHEP* **0207**, 012 (2000)
3. S. Catani and F. Hautmann, *Nucl.Phys.* **B427**, 475 (1994)
4. A. D. Martin, R. G. Roberts, W. J. Stirling and R. S. Thorne, hep-ph/0308087
5. A. Donnachie and P. V. Landshoff, *Phys.Lett.* **B518**, 63 (2001)
6. A. Donnachie and P. V. Landshoff, *Phys.Lett.* **B533**, 277 (2002)
7. A. Donnachie and P. V. Landshoff, *Phys.Lett.* **B550**, 160 (2002)
8. J. R. Forshaw and D. A. Ross, *Quantum Chromodynamics and the Pomeron*, Cambridge University Press, 1997, pp.113-138
9. M. Derrick et al, *Phys.Lett.* **B349**, 225 (1995)
10. S. Aid et al, *Nucl.Phys.* **B472**, 32 (1996)
11. J. Breitweg et al, *Eur.Phys.J.* **C12**, 35 (2000)
12. A. Donnachie and P. V. Landshoff, *Phys.Lett.* **B470**, 243 (1999)
13. A. Donnachie and P. V. Landshoff, *Phys.Lett.* **B296**, 227 (1992)
14. A. Donnachie, H. G. Dosch, P. V. Landshoff and O. Nachtmann, *Pomeron Physics and QCD*, Cambridge University Press, 2002. See www.damtp.cam.ac.uk/user/pvl/QCD/
15. A. M. Smith et al, *Phys.Lett.* **B163**, 267 (1985)
16. S. Abatzis et al, *Phys.Lett.* **B324**, 509 (1994)
17. F. E. Low, *Phys.Rev.* **D12**, 163 (1975)
18. S. Nussinov, *Phys.Rev.Lett.* **34**, 1286 (1975)
19. P. V. Landshoff and O. Nachtmann, *Zeits.Phys.* **C35**, 405 (1987)
20. E. Meggiolaro, *Nucl.Phys.* **B602**, 261 (2001)
21. O. Nachtmann, *Ann.Phys.* **209**, 436 (1991)
22. A. Kraemer and H. G. Dosch, *Phys.Lett.* **B272**, 114 (1991)
23. E. Berger and O. Nachtmann, *Eur.Phys.J.* **C7**, 459 (1999)
24. C. Adloff et al, *Nucl.Phys.* **B497**, 3 (1997)
25. J. Breitweg et al, *Phys.Lett.* **B407**, 432 (1997); S.Chekanov et al, *Eur.Phys.J.* **C28**, 175 (2003)
26. A. Donnachie and H. G. Dosch, *Phys.Rev.* **D65**, 014019 (2002)
27. R. K. Ellis, W. J. Stirling and B. R. Webber, *QCD and Collider Physics*, Cambridge University Press, 1997, p115
28. M. A. G. Aivazis, J. C. Collins, F. I. Olness and W-K. Tung, *Phys.Rev.* **D50**, 3102 (1994)
29. R. S. Thorne and R. G. Roberts, *Phys.Lett.* **B421**, 303 (1998)
30. M. Buza, Y. Matiounine, J. Smith and W. L. van Neerven *Eur.Phys.J.* **C1**, 301 (1998)
31. A. Donnachie and P. V. Landshoff, *Phys.Lett.* **B478**, 146 (2000)
32. C. Adloff et al, *Eur.Phys.J.* **C21**, 33 (2001)
33. G. Abbiendi et al, *Eur.Phys.J.* **C14**, 199 (2000)
34. S. Abatzis et al, *Phys.Lett.* **B519**, 33 (2001)

Electroproduction of Pseudoscalar Mesons on Nuclei at HERMES

Z. Akopov (on behalf of the HERMES Collaboration)

Yerevan Physics Institute, Yerevan, Armenia

Abstract. Results from unpolarized electroproduction of mesons at HERMES are presented, including production in a nuclear environment and its effect on the multiplicity of identified pions and kaons. The hadron attenuation, i.e. the reduction of the multiplicity on nuclei relative to that on deuterium has been measured as a function of two kinematic variables, using a 27.6 GeV positron beam. Also, results from electroproduction of neutral pions via the Primakoff mechanism are presented, using 12 GeV positron data.

INTRODUCTION

An incoming high energy positron or electron interacts with a nuclear target via the exchange of a virtual photon with the constituents of the nucleons - the quarks. Kinematic variables used to describe semi-inclusive deep inelastic scattering (SIDIS) are: the virtuality of the photon Q^2, the energy transfer of the virtual photon to the target nucleon v, and the fraction z of this energy transferred to the produced hadron:

$$-Q^2 = q^2 = (p - p')^2$$
$$v = E - E' \tag{1}$$
$$z = E_h / v$$

In SIDIS on a heavy nucleus (Fig. 1), the latter acts essentially as an ensemble of nucleon targets for either the struck quark or for the produced hadron that has been formed inside the nucleus. While the hadron still moves inside the nucleus, it is scattered on surrounding nucleons with a known hadron-nucleon cross-section, hence the formation time can be determined. The hadron formation time is the time between the initial hard interaction and the moment that the full hadron is formed. However, before the hadron is produced, the struck quark propagates over a certain distance loosing energy by radiating gluons. This is another source for the reduction of the hadron yield. By measuring the hadron yield and comparing this yield to the one obtained on a deuterium target, information on the hadron formation time can be obtained. This information is crucial for the interpretation of relativistic heavy ion collisions and high energy proton-nucleus and lepton-nucleus interactions.

HERMES data on semi-inclusive deep inelastic lepton-nucleus collisions provide a unique opportunity to study these effects. Experimental results are presented in terms of the hadron multiplicity ratio R_M^h, also called attenuation, defined as follows:

CP717, *Hadron Spectroscopy: Tenth International Conference,*
edited by E. Klempt, H. Koch, and H. Orth

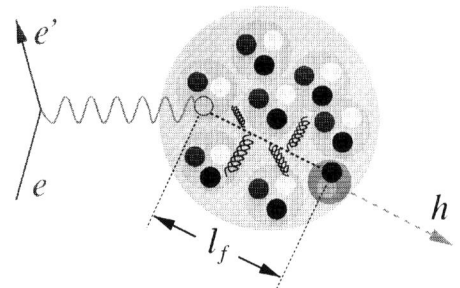

FIGURE 1. Nucleus acts as an ensemble of targets

$$R_M^h(v,z,Q^2) = \frac{\left(\frac{N_h(v,Q^2,z)}{N_e(v,Q^2)}\right)_A}{\left(\frac{N_h(v,Q^2,z)}{N_e(v,Q^2)}\right)_D} \qquad (2)$$

Here A denotes a heavy target, and D is deuterium, while N_h and N_e denote the number of produced hadrons and DIS positrons, respectively.

In the past, semi-inclusive leptoproduction of unidentified hadrons from nuclei was studied at SLAC with electrons [6] , and at CERN and FNAL with high-energy muons (EMC [7] and E665 [8] collaborations). HERMES has reported more precise data on the production of charged hadrons, as well as identified pions, kaons and protons/antiprotons ([4] [5]).

The measurements here reported were performed at the HERMES experiment using a 27.6 GeV positron beam stored in the HERA ring at DESY. The spectrometer consists of two identical halves located above and below the beam pipe. Both the scattered positrons and produced hadrons were detected with an angular acceptance of ±170 mrad horizontally, and ±(40-140) mrad vertically. The data was collected using various gas targets. Positron identification was performed using a TRD, scintillator preshower, and electromagnetic calorimeter. Identification of charged pions and kaons was done using a RICH in the momentum region between 2.5 and 15 GeV. The electromagnetic calorimeter provided neutral pion identification by detection of two clusters originating from two decay photons. Neutral pions were analyzed in the same momentum range.

RESULTS ON HADRON ATTENUATION

The multiplicity ratio has been determined as a function of z and v, while integrating over all other kinematic variables. The results for ^{84}Kr and ^{14}N targets are shown in Figure 2. The multiplicity ratios seem to be similar for π^+ and π^- mesons, indicating no charge dependence in the formation time of pions. However, there is quite a noticeable difference between distributions for positive and negative kaons which is possibly caused by the cross-section differences for the K^+N and K^-N processes. A similarly large

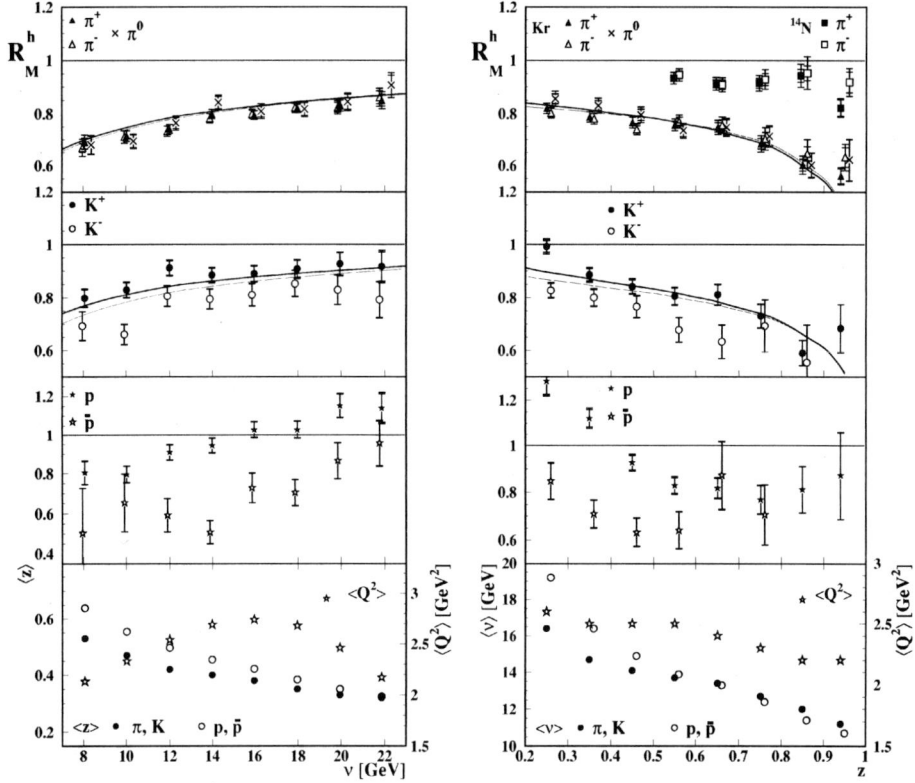

FIGURE 2. Multiplicity ratios vs. v and z for ^{84}Kr and ^{14}N. Lower panels show the average Q^2 values for each bin.

difference is observed for the proton ratios in comparison to those of the antiprotons. Overall, a significant decrease of the hadron multiplicity ratio is observed for Krypton.

The results are compared to calculations from Ref. [1], performed within the fragmentation function modification model, taking into account the absorption of hadrons formed inside the nucleus. A good description for the π^+, π^- and K^+ data is observed, while the prediction somewhat underestimates the K^- data.

Recently HERMES has measured the multiplicity ratios also for Neon and Helium targets, in an extended kinematic region using the TOF method in addition to the RICH to extend the particle identification to a larger kinematic range. These results, shown in Figure 3, show the nuclear dependence of the multiplicity ratio. It is observed that the heavier the nucleus the stronger is the hadron attenuation. Its dependence on kinematic variables shows that the attenuation becomes strongest at low values of v and at high values of z.

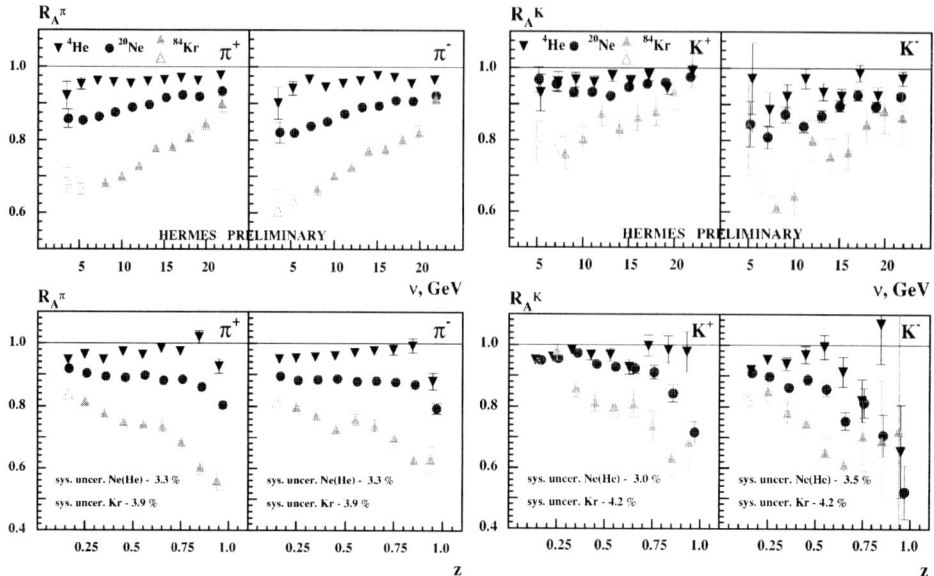

FIGURE 3. Nuclear dependence of multiplicity ratios. Left and right panels show the results for pions and kaons, respectively. Upper panels show the v dependence, while lower panels show the z dependence of the multiplicity ratios

PRIMAKOFF ELECTROPRODUCTION

The photoproduction of π^0 mesons in the Coulomb field of a nucleus, originally suggested by Primakoff [3], has been studied extensively during the sixties ([11] [12]). Recently, new interest in this subject was initiated by the attempts to measure the π^0 lifetime [13] because the cross section of the Primakoff process is inversely proportional to the π^0 lifetime. The importance of this process is also due to the unique opportunity to measure the Q^2 dependence of the neutral pion form factor provided in Primakoff electroproduction of pions. In this case both photons are spacelike and insight into the nature of the off-shell two-photon-pion coupling becomes accessible. Previously, measurements of the Q^2 dependence of the π^0 form factor were performed by the CELLO and CLEO collaborations in the high Q^2 region ([9] [10]).

HERMES has performed a measurement of Primakoff electroproduction, where the virtual photon interacts with the Coulomb field of the nucleus. The distribution of the angle between the virtual photon and the produced π^0 meson (the Primakoff angle) has been determined for Krypton and Deuterium targets. The result is shown in Figure 4. A clear enhancement can be seen in the low angle region for Krypton, as expected from the Primakoff mechanism. Once the analysis is finished, the HERMES data will allow to provide a data point for the π^0 form factors in the low Q^2 region, an area completely unexplored up to date.

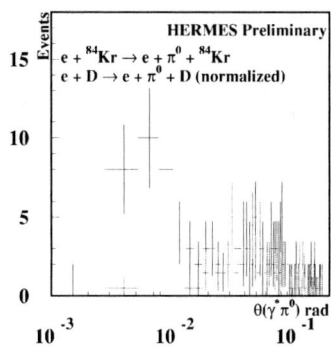

FIGURE 4. Primakoff angle distribution. In black and grey there is visible distinction between the Krypton and Deutron data

CONCLUSIONS

HERMES results on hadron multiplicity ratios on D, ^{3}He, ^{14}N, ^{20}Ne and ^{84}Kr targets have been presented. The attenuation of identified pions and kaons has been measured. For the first time results show that while multiplicities for pions are similar on a given target, they are different for positive and negative kaons. The results are in good agreement with the fragmentation function modification model. A significant nuclear dependence of the hadron attenuation has been observed, based on measurements on light and heavy nuclei.

For the first time, Primakoff electroproduction of neutral pions has been studied. The data will make it possible to extract the Q^2 dependence of the π^0 form-factor in the unexplored low Q^2 region.

REFERENCES

1. A. Accardi, V. Muccifora, H.J Pirner, Nucl.Phys A720, 131 (2003)
2. K.Ackerstaff et al.(HERMES Coll.), Nucl.Instrum. Methods A 417, 230 (1998)
3. H. Primakoff, Phys. Rev. 181, 899 (1951)
4. A. Airapetian et al., Eur. Phys. J. C20, 479 (2001)
5. A. Airapetian et al., Phys. Lett. B (in press), hep-ex/0307023
6. L.S. Osborne et al., Phys. Rev. Lett. 40, 1624 (1978)
7. EMC Coll., J. Ashman et al., Z. Phys. C52, 1 (1991)
8. E665 Coll., M.R. Adams et al., Phys. Rec. D50, 1836 (1994)
9. H.J. Behrend, el al., Z. Phys. C49, 401 (1991)
10. D.M. Asner et al., CLEO CONF95-24, EPS0188 (1995)
11. G.Morpurgo, Nuovo Cimento 31, 569 (1964)
12. G.Belletini et al., Nuovo Cimento 90A, 1139 (1965)
13. The PrimEx Collaboration, "Precision Measurements of the Electromagnetic Properties of Pseudoscalar Mesons at 11 GeV via the Primakoff Effect, (2000)

QCD studies at RHIC in polarized pp, dAu and AuAu collisions

B. Surrow

Massachusetts Institute of Technology, Department of Physics, Cambridge, MA 02139, USA

Abstract.
The Relativistic Heavy-Ion Collider (RHIC) at Brookhaven National Laboratory (BNL) is a unique facility which allows to study the collision of different nuclei and polarized protons at variable energy.

The spin physics program at RHIC at BNL focuses on the collision of polarized protons at a center-of-mass energy of $200 - 500\,\mathrm{GeV}$ to gain a deeper understanding of the spin structure of the proton in a new, previously unexplored territory. Several polarized fixed-target experiments have been conducted in the past to gain a deeper understanding of the spin structure of the proton. Those experimental efforts have been restricted to large values of Bjorken x. The role of the gluons to make up for the missing proton spin is currently only very poorly constrained from scaling violations in fixed target experiments. The need for a new generation of experiments to explore the spin structure of the proton is clearly apparent. The first polarized proton run in 2002 is the beginning of a multi-year experimental program which aims to address a variety of topics related to the nature of the protons spin such as the gluon contribution to the proton spin, the flavor decomposition of the quark and anti-quark polarization and the transverse spin dynamics of the proton.

Recent results at RHIC show that the collision of relativistic nuclei at a center-of-mass energy of $200\,\mathrm{GeV}$ per nucleon pair, can be interpreted as hard scatterings among quarks and gluons that make up the incoming nuclei. The collision of p+p and d+Au show a clear back-to-back topology of two jets as expected from the underlying hard scattering. In head-on Au-Au collisions on the contrary, the formation of the 'away-side' jet is strongly suppressed. One possible explanation of the missing jet is that a quark traveling through a medium of free quarks and gluons, commonly known as the Quark-Gluon Plasma, would interact strongly and thus loose energy. This phenomena is called jet quenching and was predicted as a potential signature for the formation of the Quark-Gluon Plasma.

Recent results on the spin and relativistic heavy-ion program at RHIC will be discussed.

INTRODUCTION TO RHIC SPIN PROGRAM

The first polarized proton run from December 2001 until January 2002 (RUN 2) at RHIC at BNL is the beginning of a multi-year experimental program which aims to address a variety of topics related to the nature of the proton spin such as: 1. spin structure of the proton (gluon contribution of the proton spin, flavor decomposition of the quark and anti-quark polarization and transversity distributions of the proton), 2. spin dependence of fundamental interactions, 3. spin dependence of fragmentation and 4. spin dependence of elastic polarized proton collisions. A recent review and status of the RHIC spin program can be found in [1].

The principle approach to study spin effects is to measure an asymmetry (A) which quantifies the normalized difference of measured yields for different initial-state spin configurations. Ultimately, any combination of beam polarization, i.e. either longitudi-

CP717, *Hadron Spectroscopy: Tenth International Conference,*
edited by E. Klempt, H. Koch, and H. Orth

nal (L) or transverse (T), will be possible at RHIC to access different aspects of the proton spin structure. A crucial fact to remember is that the statistical significance of double spin asymmetries varies as $P^4 \int L dt$ whereas for single spin asymmetries it varies as $P^2 \int L dt$. Thus, the demand on high polarization is particularly important for the measurement of double spin asymmetries. A focus of the first polarized proton run was the measurement of a transverse single-spin asymmetry, A_N. Non-zero values for A_N have been observed at the FNAL E704 [2] experiment for $\vec{p} + p \rightarrow \pi + X$ at $\sqrt{s} = 20\,\text{GeV}$ and $0.5 < p_T < 2.0\,\text{GeV}$. Theoretical models that explain the E704 data also predict non-zero values for A_N for pion production at RHIC. Qiu and Sterman [3] attribute the measured asymmetry to a higher-twist pQCD effect. The group of Anselmino and Leader perform a global analysis of semi-inclusive DIS data from HERMES [4] and E704 data. This approach involves transverse $k_\perp$ effects in the quark distribution functions ('Sivers effect') [5] as well as in the fragmentation functions ('Collins effect') [6] to account as possible explanations for the measured asymmetries. Besides the theoretical interest in measuring A_N, it will serve as a potential candidate to monitor the RHIC beam polarization at a particular experiment ('local polarimeter').

THE POLARIZED PROTON COLLIDER RHIC

The first collisions of polarized protons occurred in December 2001, ushering in a a new era to complement the ongoing relativistic heavy-ion program. RHIC is the first accelerator to accelerate and collide polarized protons, ultimately at high luminosity, at a center-of-mass energy of up to 500 GeV.

The key to maintain the proton polarization through acceleration despite its large anomalous magnetic momentum, is to perform a rotation of the proton spin by 180° in the horizontal plane around a particular axis. This manipulation is performed by helical dipole magnets, known as 'Siberian snakes', which have been used for the first time at a proton collider. With two Siberian snakes installed in each ring, cumulative tilt effects of the proton spin are canceled, thereby eliminating the influence of depolarizing spin resonances. Besides the installation of Siberian snakes, the PHENIX and STAR experiments are equipped with spin rotator magnets to allow for the precession from transverse to longitudinal polarization and thus to collide longitudinal polarized proton beams. These magnets have been successfully commissioned during the RHIC run in 2003 (RUN 3) and allowed the first collisions of longitudinally polarized protons at $\sqrt{s} = 200\,\text{GeV}$.

The first polarized proton run at RHIC (RUN 2) was carried out at a center-of-mass energy of 200 GeV. Each ring was loaded with 55 bunches of alternating polarization resulting in a a bunch crossing-time of 214 ns. A transverse polarization of about 20% was achieved at the injection energy of 24.6 GeV and was approximately maintained when the proton beams were accelerated to 100 GeV.

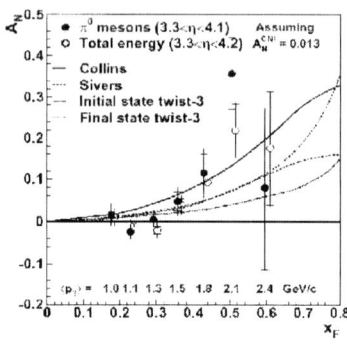

FIGURE 1. *Measurement of the transverse single-spin asymmetry of forward π^0 production, A_N, as a function of x_F by the STAR collaboration for $1.1 < p_T < 2.5\,GeV/c$ in comparison to pQCD model predictions evaluated at $p_T = 1.5\,GeV/c$ [7].*

FIRST POLARIZED PP RESULTS FROM STAR

The STAR collaboration has measured during the first polarized proton run (RUN 2) the transverse single-spin asymmetry, A_N, for forward π^0 production at $x_F \simeq 0.2 - 0.6$ and $p_T \simeq 1 - 3\,\text{GeV}$ [7]. A_N is extracted from :

$$A_N = \frac{1}{P}\frac{N^\uparrow - R \cdot N^\downarrow}{N^\uparrow + R \cdot N^\downarrow} \tag{1}$$

which requires three independent measurements: 1. the spin-dependent yields ($N^{\uparrow(\downarrow)}$) of forward π^0 production, 2. the relative luminosity $R = L^\uparrow/L^\downarrow$ and 3. the actual beam polarization P. The latter is the focus of a dedicated effort at RHIC to obtain a fast (relative) polarization measurement using pC elastic scattering at very small $|t|$ values. This Coulomb Nuclear Interference (CNI) polarimeter [8] will ultimately be calibrated to pp elastic scattering for a polarized hydrogen gas-jet target [9]. An upgrade program at the STAR experiment was performed with the installation of a beam-beam counter (BBC) [10] and a prototype forward-pion detector (FPD) [7]. In addition, a spin scaler system was commissioned to account for the beam polarization reversals every bunch crossing of 214 ns.

The STAR BBC consists of a hexagonal scintillator array structure at $\pm 3.5\,\text{m}$ from the nominal interaction point with full azimuthal coverage. The BBC is the main device to make the relative luminosity measurement and to provide a trigger to distinguish pp collision events from beam related background events by means of timing requirements. The FPD prototype system which is installed at $7.5\,\text{m}$ from the nominal interaction point facing the RHIC 'Yellow beam', consists of three lead-glass electromagnetic calorimeter modules together with a lead-scintillator calorimeter, which is a prototype module of the STAR endcap calorimeter. The latter device allows the reconstruction of π^0 mesons from their decay products ($\pi^0 \rightarrow \gamma\gamma$) by measuring the total energy and the transverse

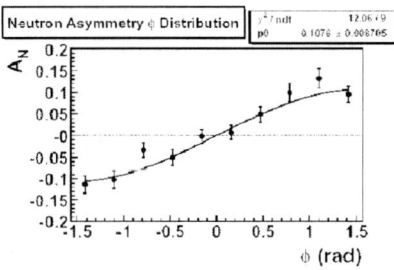

FIGURE 2. *Preliminary result of the measured neutron asymmetry as a function of the reconstructed azimuthal angle ϕ by the Local Polarimetry (PHENIX) collaboration using a Lead-Tungstate (PbWO$_4$) crystal electromagnetic calorimeter [13].*

shower profile. It consists of 12 independent towers, preshower detectors and a shower-maximum detector to perform a transverse shower profile measurement. This prototype module has been extensively studied using high energy electron test beams at SLAC. Its testbeam performance is well reproduced by a GEANT simulation.

Figure 1 shows the analyzing power, A_N, as a function of x_F measured by the STAR collaboration for $1.1 < p_T < 2.5\,\text{GeV/c}$. The systematic uncertainty has been estimated to be $\delta A_N \sim 0.05$ which does not include the normalization uncertainty from the RHIC beam polarization measurement. The analyzing power of the CNI polarimeter at 100 GeV has not been measured yet and is assumed to be the same as the measured analyzing power at 24.6 GeV to determine the RHIC beam polarization [11, 12]. A_N is found to increase with x_F and is similar in magnitude to the measurement of A_N performed by the E704 experiment at $\sqrt{s} = 20\,\text{GeV}$. These results are compared to pQCD model predications evaluated at $p_T = 1.5\,\text{GeV/c}$ involving different mechanism as discussed in the introduction, to account for the sizable observed transverse single-spin asymmetry A_N for forward neutral pion production.

FIRST POLARIZED PP RESULTS FROM PHENIX

The Local Polarimetry (PHENIX) collaboration installed as part of the PHENIX 'local polarimeter' development for neutral particle production a detector system located 1800 cm upstream and downstream of the RHIC IP12 collision point [13]. An electromagnetic calorimeter (EM-Cal) which consists of sixty (5×12 array) Lead-Tungstate (PbWO$_4$) crystals ($2.0 \times 2.0 \times 20.0\,\text{cm}^3$) was installed on one side of the RHIC IP12 interaction region facing the RHIC 'blue beam'. The length of this 5×12 array corresponds to about 1 interaction length. Sets of scintillator counters before and after the Lead-Tungstate array were used to define trigger conditions for photon and neutron samples. Simulation studies yield a purity of 98% and 89% for photons and neutrons, respectively. The systematic uncertainty has been estimated to be about 16%. The hadron calorimeter (H-Cal) which is a sandwich tungsten/optical-fiber calorimeter is installed on the other side of the RHIC IP12 interaction region which faces the RHIC 'yellow beam'. Its total length of 23 cm corresponds to about 2 interaction lengths. A postshower counter which consists of five $PbWO_4$ crystals provides a horizontal position measurement. Both calorimeter modules have an angular coverage of approximately 3 mrad around zero degrees. A scintillator hodoscope has been setup around the RHIC IP12 interaction region to suppress beam related background events. Figure 2 shows the measured transverse single-spin asymmetry for forward neutron production, A_N, as a function of the reconstructed azimuthal angle which shows the expected $\sin\phi$-type azimuthal dependence. A fit to this dependence allows to extract the underlying transverse single-spin neutron asymmetry. A_N is found to be -0.112 ± 0.007 with $\chi^2/ndf = 1.7$. The average measured A_N value for positive x_F amounts to -0.109 ± 0.007 and -0.110 ± 0.015 for the EM-Cal and H-Cal polarimeters, respectively. Both results agree within statistical uncertainties.

The PHENIX collaboration has measured the invariant differential cross section for inclusive neutral pion production, in $p + p$ collisions at $\sqrt{s} = 200\,\text{GeV}$ at mid-rapidity ($|\eta| < 0.35|$) as a function of p_T [14]. This analysis is mainly based on the PHENIX beam-beam counters and the PHENIX electromagnetic calorimeters. The PHENIX beam-beam counters cover a pseudo-rapidity range of $3.0 < |\eta| < 3.9$ with full azimuthal coverage. It was used to define a minimum-bias trigger which required the collision vertex to be within 75 cm of the nominal interaction point. A more restrictive vertex requirement of 30 cm was applied during the offline data analysis. The PHENIX electromagnetic calorimeters consist of two subsystems: two lead-glass sectors and six lead-scintillator sectors. This system is located at a radial distance of 5 m and spans a pseudorapidity range of $|\eta| < 0.35$ and an azimuthal interval of $\sim 22.5°$ per sector. It provides a fine $\eta - \phi$ segmentation of $\Delta\eta \times \Delta\phi \sim 0.01 \times 0.01$ which allows to reconstruct both decay photons up to at least 20 GeV/c in p_T of the neutral pion. A high-p_T trigger was used to enhance the sample of reconstructed high p_T neutral pions. This trigger is based on a threshold requirement of analog sums among 2×2 groupings of adjacent EMCal towers. The efficiency of this 2×2 high-p_T trigger was determined from the minimum-bias trigger sample as a function of p_T and was found to be flat in p_T for $p_T > 3\,\text{GeV/c}$ at the level of 0.78 ± 0.03. The integrated luminosity was determined from the number of minimum-bias triggered events using an absolute calibration of the respective trigger cross section based on a 'van der Meer scan'. The uncertainty of the luminosity normalization was estimated to be 9.6%. The minimum-bias trigger sample of 16 million triggered events corresponds to an integrated luminosity of $0.7\,\text{nb}^{-1}$ whereas the 2×2 high-p_T sample of 18 million triggered events refers to an an integrated luminosity of $39\,\text{nb}^{-1}$. The measured invariant differential cross section for inclusive neutral pion production for the minimum-bias triggered sample and the 2×2 high-p_T sample agree within statistical uncertainties in the range of overlap in p_T. The result of this analysis is shown in Figure 3. The measured differential cross section covers a p_T range of $\sim 1 - 14\,\text{GeV}$. Those results are compared to NLO pQCD calculations using the CTEQ6M [15] set of parton distribution functions and two sets of fragmentation functions, 'Kniehl-Kramer-Pötter' (KKP) [16] and 'Kretzer' [17], which mainly differ in the gluon-pion fragmentation function, D_g^π. Overall, good agreement is found between these NLO pQCD calculations and the measured results. At low p_T which is dominated by gluon-gluon and quark-gluon interactions, the calculations based on the KKP fragmentation functions which includes a larger gluon-pion fragmentation function, D_g^π, than in the case of 'Kretzer', is in better agreement with the data.

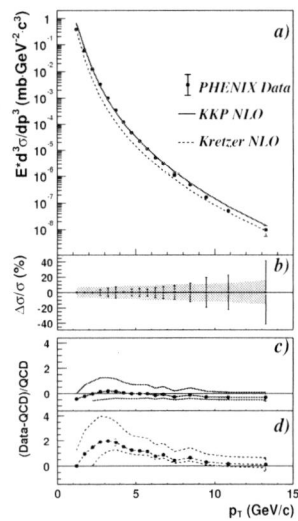

FIGURE 3. *Invariant differential cross section measurement by the PHENIX collaboration for inclusive neutral pion production in comparison to NLO pQCD calculations with equal renormalization and factorization scale of p_T using the 'Kniehl-Kramer-Pötter' (KKP) (solid line) and 'Kretzer' (dashed line) sets of fragmentation functions (a). The relative statistical errors (points) and point-to-point systematic errors (band) (b). The relative difference between the measured cross-section and the NLO pQCD calculations using the KKP (c) and the 'Kretzer' (d) set of fragmentation functions [14].*

SUMMARY AND OUTLOOK OF THE RHIC SPIN PROGRAM

The first polarized proton run at RHIC started a new era at BNL of exploring the spin structure of the proton. The main focus of the STAR experiment during the first polarized proton run (RUN 2) of transverse polarization was the measurement of a transverse single-spin asymmetry of forward π^0 production. In preparation of the first polarized proton run, an upgrade of the STAR experiment was performed with the installation of a beam-beam counter and a prototype forward pion detector (FPD) system, besides the commissioning of a spin scaler system. A new FPD system based on Pb-glass array calorimeters has been installed for the RHIC run in 2003 (RUN 3) [18]. First results on transverse single-spin asymmetries have been obtained and compared to model predictions. It will require more precise and differentiated measurements to discriminate among the competing models, e.g. a measurement of an azimuthal dependence of forward produced π^0 mesons around the reconstructed jet axis would point to a mechanism for the large measured transverse single-spin asymmetries as proposed by Collins as a final state effect. This would then open the potential to access transversity which is the least unmeasured parton distribution function. An upgrade of the existing electromagnetic FPD system by hadronic calorimetry would aid this effort to reconstruct the jet axis. The STAR detector will undergo major upgrade programs with the installation of the endcap calorimeter which is the principal device to explore the gluon polarization of the proton and the barrel calorimeter.

Large transverse single-spin asymmetries have been observed by the Local Polarime-try (PHENIX) collaboration for forward production of neutrons. The PHENIX collab-oration has measured the invariant differential cross section for inclusive neutral pion production in $p + p$ collisions at $\sqrt{s} = 200\,\text{GeV}$ at mid-rapidity ($|\eta| < 0.35|$) as a func-tion of p_T. Those measurements are found to be in good agreement with NLO pQCD calculations even at low p_T. This provides a solid basis for the planned polarized gluon density measurements with polarized protons at RHIC over a wide range in p_T [1].

HIGHLIGHTS FROM THE RHIC HEAVY-ION PROGRAM

One of the main goals of the Au-Au program at RHIC at BNL is to explore the formation of a new state of matter, often referred to as the Quark-Gluon Plasma, in the collision of Au-ion beams at a center-of-mass energy up to 200 GeV. Several signatures on the formation of the QGP have been suggested. A review of those signatures can be found in [19].

Signatures involving partonic collisions to probe a new state of matter play an im-portant role at RHIC in distinction to previous fixed-target heavy-ion programs. It is predicted that energetic partons propagating through matter lose energy through gluon radiation with a magnitude of this energy loss that is strongly dependent on the color charge density. Partonic energy loss has been widely discussed as a potential sensitive probe for the formation of a new state of matter formed in the most violent collisions of two Au nuclei at RHIC [20, 21]. A clean signature for partonic collisions in high-energy proton-proton collisions is the formation of back-to-back jets. The direct measurement of jets in Au-Au collisions is rather difficult. Probing the formation of a new state of matter through partonic energy loss can be studied through azimuthal correlations and inclusive spectra of high p_T hadrons in comparison to proton-proton collisions. Those experimental results will be presented in the following for the four RHIC experiments which are studying the collisions of Au nuclei.

Nuclear effects in Au-Au and d-Au collisions have been studied at RHIC in compar-ison to proton-proton data measured at RHIC using the following ratio usually referred to as the nuclear modification factor:

$$R_{AB}(p_T) = \frac{d^2N/dp_T d\eta}{T_{AB}d^2\sigma^{pp}/dp_T d\eta} \tag{2}$$

where $d^2N/dp_T d\eta$ is the differential yield per event in the nuclear collision $A + B$, $T_{AB} = <N_{AB}>/\sigma^{pp}_{inel}$ describes the nuclear geometry, and $d^2\sigma^{pp}/dp_T d\eta$ for $p + p$ inelastic collisions is determined from the measured $p + p$ differential cross section.

Figure 4 shows $R_{AB}(p_T)$ from the STAR Collaboration for minimum bias and central $d + Au$ collisions which reflects a clear enhancement in contrast to the central $Au + Au$ collisions, exhibiting large suppression in hadron production at high p_T [22].

Figure 5 shows experimental results in $Au + Au$ and $d + Au$ collisions as obtained by the PHENIX Collaboration [23]. The top panel of Figure 5 shows $R_{dAu}(p_T)$ for inclusive charged particles $(h^+ + h^-)/2$ compared to R_{AuAu} observed in $Au + Au$ collisions. The

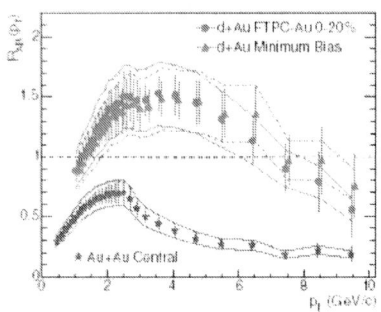

FIGURE 4. $R_{AB}(p_T)$ *for minimum bias and* $d+Au$ *collisions, and central* $Au+Au$ *collisions as obtained by the STAR collaboration. The bands show the normalization uncertainty. The error bars represent the quadratic sum of statistical and systematic errors [22].*

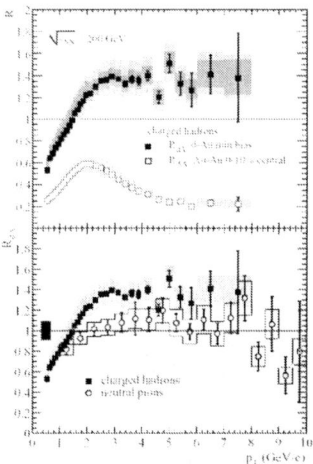

FIGURE 5. *Nuclear modification factor* R_{dAu} *for* $(h^+ + h^-)/2$ *in minimum bias* $d+Au$ *collisions compared to* R_{AuAu} *in the 10% most central* $Au+Au$ *collisions as obtained by the PHENIX collaboration. Inner bands show systematic errors. Outer errors bands include in addition normalization uncertainties (top). Comparison of* R_{dAu} *for* $(h^+ + h^-)/2$ *and the average of the* π^0 *measurements in* $d+Au$ *collisions. The bar at the left indicates the systematic uncertainty in common for the charged and* π^0 *measurements [23].*

lower panel compares the $(h^+ + h^-)/2$ result for $R_{dAu}(p_T)$ to the π^0 result in $d+Au$ collisions. The data clearly indicate that there is no suppression of high p_T particles in $d+Au$ collisions which is in contrast to the results obtained in $Au+Au$ collisions. An enhancement of inclusive charged particle production in $d+Au$ collisions is found at $p_T > 2\,\mathrm{GeV}/c$.

The BRAHMS measurement for R_{dAu} in comparison to $Au+Au$ results is shown in Figure 6 [24]. No suppression is observed in R_{dAu}, rather an enhancement at high p_T.

The nuclear modification factor R_{dAu} as obtained by the PHOBOS Collaboration is

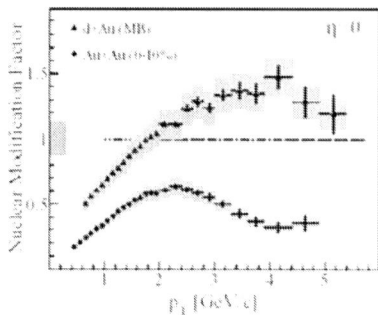

FIGURE 6. *Nuclear modification factor measured for minimum bias collisions of d + Au data at $\sqrt{s_{NN}} = 200\,GeV$ compared to central Au + Au collisions as obtained by the BRAHMS collaboration. Error bars represent statistical errors. Systematic errors are denoted by the shaded bands [24].*

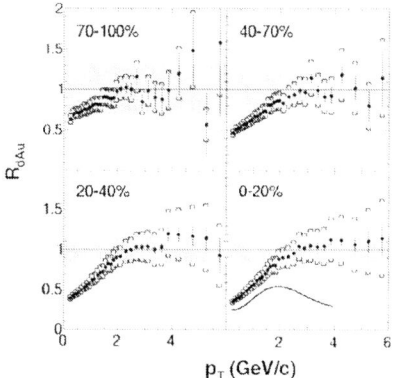

FIGURE 7. *Nuclear modification factor as a function of p_T for four bins of centrality as obtained by the PHOBOS collaboration. For the most central bin, the spectral shape for central Au − Au data relative to the UA1 $p\bar{p}$ reference data set is shown for comparison. The shaded area and brackets shows the uncertainty in R_{dAu} due to various systematic uncertainties [25].*

shown in Figure 7 as a function of p_T using the UA1 $p\bar{p}$ reference data set [25]. For comparison, the result from central $Au + Au$ collisions is also shown. For central $Au + Au$ collisions, the ratio of the spectra to the UA1 $p\bar{p}$ reference data set rises rapidly up to $p_T \approx 2GeV$ GeV/c, but falls off at higher p_T. This is in striking contrast to the behavior seen for central $d + Au$ collisions.

In summary, all four RHIC experiments are consistent in their findings of a strong suppression of the inclusive hadron yields [22, 23, 24, 25] in addition to the back-to-back correlations [26] in central $Au + Au$ collisions in comparison to $p + p$ collisions. If the result of this suppression would be an initial state effect, e.g due to gluon saturation phenomena [27], it should be also observed in $d + Au$ collisions. However, no suppression in $d + Au$ collisions is observed. This has been consistently found among all four

820

RHIC experiments. The inclusive hadron yields are found to be enhanced at high p_T in comparison to $p + p$ collisions which suggests that the Cronin effect [28] plays an important role in $d + Au$ collisions. The strong suppression seen in $Au + Au$ collisions are attributed to final-state interactions with the hot dense medium created in $Au + Au$ collisions [20, 21].

It will be important to explore various other potential signatures suggested for the formation of a Quark-Gluon Plasma to make a firm statement about the nature of the new state of matter created in the collision of $Au + Au$ nuclei at RHIC. In addition, it will be crucial to examine the region of higher gluon density, kinematically accessible in the forward region at RHIC, to explore the role of gluon saturation phenomena in the RHIC energy region.

REFERENCES

1. G. Bunce et al., *Ann. Rev. Nucl. Part. Sci.* 50 (2000) 525;
   L.C. Bland, in '15th International Spin Physics Symposium (SPIN 2002)', AIP Conf. Proc. **675** (2003) 98.
2. D. Adams et al. (E704 Collaboration), *Phys. Lett.* B261 (1991) 201; *Phys. Lett.* B264 (1991) 462.
3. J. Qiu and G. Sterman, *Phys. Rev.* D59 (1998) 014004.
4. A. Airapetian et al. (HERMES Collaboration), *Phys. Rev.* D64 (2001) 097101.
5. M. Anselmino et al., *Phys. Lett.* B362 (1995) 164; *Phys. Lett.* B442 (1998) 470.
6. M. Anselmino et al., *Phys. Rev.* D60 (1999) 054027; *Phys. Lett.* B442 (1998) 442.
7. J. Adamas et al. (STAR Collaboration), submitted to *Phys. Rev. Lett.*, hep-ex/0310058.
8. K. Kurita, in '15th International Spin Physics Symposium (SPIN 2002)', AIP Conf. Proc. **675** (2003) 812;
   O. Jinnouchi, in '15th International Spin Physics Symposium (SPIN 2002)', AIP Conf. Proc. **675** (2003) 817.
9. A. Bravar, in '15th International Spin Physics Symposium (SPIN 2002)', AIP Conf. Proc. **675** (2003) 830.
10. J. Kiryluk, in '15th International Spin Physics Symposium (SPIN 2002)', AIP Conf. Proc. **675** (2003) 424.
11. C.A. Allgower et al., Phys. Rev. D65 (2002) 092008.
12. H. Spinka, in '15th International Spin Physics Symposium (SPIN 2002)', AIP Conf. Proc. **675** (2003) 807.
13. Y. Fukao, in '15th International Spin Physics Symposium (SPIN 2002)', AIP Conf. Proc. **675** (2003) 584.
14. S.S. Adler et al. (PHENIX Collaboration), Phys. Rev. Lett.91 (2003) 241803.
15. J. Pumplin et al. (CTEQ Collaboration), *High Energy Phys.* 0207 (2002) 012.
16. B.A. Kniehl et al., *Nucl. Phys.* B597 (2001) 337.
17. S. Kretzer, *Phys. Rev.* D(2000) 054001.
18. A. Ogawa, in '15th International Spin Physics Symposium (SPIN 2002)', AIP Conf. Proc. **675** (2003) 407.
19. B. Müller and J. Harris, Ann. Rev. Nucl. Part. Sci. 71 (1996) 46.
20. M. Gyulassy and M. Plümer, *Phys. Lett.* B243 (1990) 432.
21. X.N. Wang and M. Gyulassy, *Phys. Rev. Lett.68 (1992) 1480.*
22. J. Adams et al. (STAR Collaboration), Phys. Rev. Lett.91 (2003) 072304.
23. S.S. Adler et al. (PHENIX Collaboration), *Phys. Rev. Lett.91 (2003) 072303.*
24. C. Adler et al. (BRAHMS Collaboration), Phys. Rev. Lett.91 (2003) 072305.
25. C. Adler et al. (PHOBOS Collaboration), *Phys. Rev. Lett.91 (2003) 072302.*
26. C. Adler et al. (STAR Collaboration), Phys. Rev. Lett.90 (2003) 082302.
27. D. Kharzeev et al., *Phys. Lett.* B561 (2003) 93.
28. J. W. Cronin et al. *Phys. Rev.* D11 (1975) 3105.

Hadron formation in electron induced reactions at HERMES energies

T. Falter*, W. Cassing*, K. Gallmeister* and U. Mosel*

*Institut fuer Theoretische Physik, Universitaet Giessen, D-35392 Giessen, Germany

Abstract. We investigate meson electroproduction off complex nuclei in the kinematic regime of the HERMES experiment using a semi-classical transport model which is based on the Boltzmann-Uehling-Uhlenbeck (BUU) equation. We discuss coherence length and color transparency effects in exclusive ρ^0 production as well as hadron formation and attenuation of charged pions, kaons, protons and anti-protons in deep inelastic lepton scattering off nuclei.

High energy meson electroproduction off complex nuclei offers a promising tool to study the physics of hadron formation. The relatively clean nuclear environment of electron induced reactions makes it possible to investigate the timescale of the hadronization process as well as the properties of hadrons immediately after their creation. In addition one can vary the energy and virtuality of the exchanged photon to examine the phenomenon of color transparency (CT).

In previous works [1–3] we have developed a method to combine the quantum mechanical coherence in the entrance channel of photonuclear reactions with a full coupled channel treatment of the final state interactions (FSI) in the framework of a semi-classical transport model. This allows us to include a much broader class of FSI than usual Glauber theory.

In our approach the lepton-nucleus interaction is split into two parts: 1) In the first step the electron emits a virtual photon which is absorbed on a nucleon of the target nucleus; this interaction produces a bunch of particles that in step 2) are propagated within the transport model. The virtual photon-nucleon interaction itself is simulated by the Monte Carlo generator PYTHIA v6.2 [4] which well reproduces the experimental data on a hydrogen target. Instead of directly interacting with a quark inside the target nucleus the virtual photon might fluctuate into a vector meson ($\rho^0, \omega, \Phi, J/\Psi$) or perturbatively branch into a $q\bar{q}$ pair before the interaction. While the latter is very unlikely in the kinematic regime of the HERMES experiment as we have shown in Ref. [3] the vector meson fluctuations become important at low Q^2 and clearly dominate the exclusive vector meson production measured at HERMES. The coherence length, i.e., the length that the photon travels as such a vector meson fluctuation V can be estimated from the uncertainty principle:

$$l_V = \frac{2\nu}{Q^2 + m_V^2}.$$

(1)

Here ν denotes the energy of the photon, Q^2 its virtuality and m_V the mass of the vector meson fluctuation. If l_V becomes larger than the internucleon distance in the nucleus

CP717, *Hadron Spectroscopy: Tenth International Conference,*
edited by E. Klempt, H. Koch, and H. Orth

the interactions triggered by the vector meson component V get shadowed in nuclear reactions [2, 3].

A direct photon interaction or a non-diffractive interaction of one of the hadronic fluctuations leads to the excitation of one or more hadronic strings which finally fragment into hadrons. The time, that is needed for the fragmentation of the strings and for the hadronization of the fragments, we denote as formation time τ_f in line with the convention in transport models. For simplicity we assume that the formation time is a constant τ_f in the rest frame of each hadron and that it does not depend on the particle species. We recall, that due to time dilatation the formation time t_f in the laboratory frame is then proportional to the particle's energy

$$t_f = \gamma \cdot \tau_f = \frac{z_h V}{m_h} \cdot \tau_f. \tag{2}$$

Here m_h denotes the hadron's mass and z_h is the energy fraction of the photon carried by the hadron. The size of τ_f can be estimated by the time that the constituents of the hadrons need to travel a distance of a typical hadronic radius (0.5–0.8 fm).

The formation time also plays an important role in the investigations of ultra-relativistic heavy ion reactions. For example, the observed quenching of high transverse momentum hadrons in $Au + Au$ reactions relative to $p + p$ collisions is often thought to be due to jet quenching in a quark gluon plasma. However, the attenuation of high p_T hadrons might also be due to hadronic rescattering processes [5] if the hadron formation time τ_f (in its rest frame) is sufficiently short.

We assume that hadrons, whose constituent quarks and antiquarks are created from the vacuum in the string fragmentation, do not interact with the surrounding nuclear medium within their formation time. For the leading hadrons, i.e. those involving quarks (antiquarks) from the struck nucleon or the hadronic components of the photon, we assume a reduced effective cross section σ_{lead} during the formation time τ_f and the full hadronic cross section σ_h ($h = \pi^\pm, K^\pm, p, \ldots$) later on. The hadrons with z_h close to one are predominantly leading hadrons and interact directly after the photon-nucleon interaction. Particles that emerge from the middle of the string might escape the nucleus due to time dilatation. However, about 2/3 of these intermediate z_h hadrons (mainly pions) are created from the decay of vector mesons that have been created in the string fragmentation. Because of their higher mass m_h (0.77 – 1.02 GeV) these vector mesons may form (or hadronize) inside the nucleus (see Eq. (2)) and thus be subject to FSI. The effect of the FSI, finally, will depend dominantly on the nuclear geometry, i.e. the size of the target nucleus.

The FSI are described by a coupled-channel transport model based on the Boltzmann-Uehling-Uhlenbeck (BUU) equation. For the details of the model we refer the reader to Ref. [6]. The important difference to a purely absorptive treatment of the FSI is that the particles resulting from the $\gamma^* A$ reaction do not have to be created in the primary $\gamma^* N$ interaction. In a FSI with a nucleon a hadron might not only be absorbed but also be decelerated in an elastic or inelastic collision. Furthermore, it may in addition produce several low energy particles. In the case of electroproduction of hadrons this finally leads to a redistribution of strength from the high z_h part of the hadron energy spectrum to lower values of the energy fraction z_h.

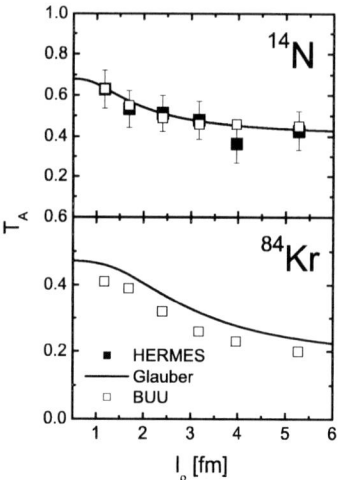

FIGURE 1. Nuclear transparency ratio T_A (3) for ρ^0 electroproduction plotted versus the coherence length (1) of the ρ^0 component of the photon. The data is taken from [7]. The solid line represents our Glauber result from Ref. [3]. For each transparency ratio calculated within our transport model (open squares) we used the average value of Q^2 and v of the corresponding data point.

In Fig. 1 we show the transparency ratio

$$T_A = \frac{\sigma_{\gamma^* A \to \rho^0 A^*}}{A \sigma_{\gamma^* N \to \rho^0 N}} \tag{3}$$

for exclusive ρ^0 production as a function of the coherence length (1) in comparison with the HERMES data [7]. The solid line displays the result that one gets if one uses our Glauber expression from Ref. [3].

The result of the transport model is represented by the open squares. For each data point we have made a separate calculation with the corresponding values of v and Q^2. For the N target the Glauber and the transport calculation are in perfect agreement with each other and the experimental data. This demonstrates that, as we have discussed in Ref. [2], Glauber theory can be used for the FSI if the right kinematic constraints are applied.

After applying all of the experimental cuts from Ref. [7], nearly all of the detected ρ^0 stem from diffractive ρ^0 production for which we assume zero formation in the calculations. The N data seems to support the assumption that the time needed to put the preformed ρ^0 fluctuation on its mass shell and let the wave function evolve to that of a physical ρ^0 is small for the considered values of Q^2. Furthermore, the photon energy is too low to yield a large enough γ factor to make the formation length exceed the internucleon distance and make CT visible. This conclusion is at variance with that reached in Ref. [8] where the authors also stress that one might see an onset of CT when investigating the transparency ratio as a function of Q^2 for fixed coherence length.

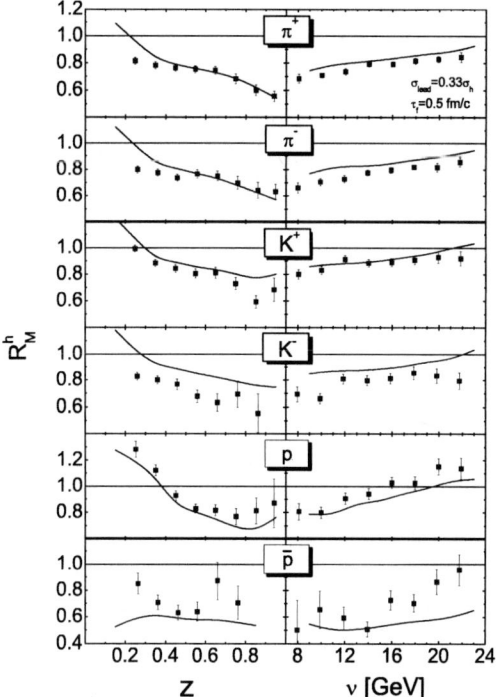

FIGURE 2. Calculated multiplicity ratios of π^+, π^-, K^+, K^-, p and $\bar{p}$ for Kr using a fixed leading hadron cross section $\sigma_{lead} = 0.33\sigma_h$ and formation time $\tau_f = 0.5$ fm/c. The experimental data has been taken from Ref. [16].

We now turn to Kr where we expect a stronger effect of the FSI. Unfortunately there is yet no data available to compare with. As can be seen from Fig. 1 the transport calculation for Kr gives a slightly smaller transparency ratio than the Glauber calculation, especially at low values of the coherence length, i.e. small momenta of the produced ρ^0. There are two reasons for this: About 10% of the difference arises from the fact that within the transport model the ρ^0 is allowed to decay into two pions. The probability that at least one of the pions interacts on its way out of the nucleus is about twice as large as that of the ρ^0. The other reason is that in the Glauber calculation only the inelastic part of the $\rho^0 N$ cross section enters whereas the transport calculation contains the elastic part as well. Thus all elastic scattering events out of the experimentally imposed t-window are neglected in the Glauber description. It is because of this t-window that also elastic $\rho^0 N$ scattering reduces the transport transparency ratio shown in Fig. 1. Both effects are more enhanced at lower energies and become negligible for the much smaller N nucleus.

In Ref. [9] we investigated the energy ν and fractional energy $z_h = E_h/\nu$ dependence of the charged hadron multiplicity ratio

$$R_M^h(z,\nu) = \left(\frac{N_h(z,\nu)}{N_e(\nu)}\right)_A \bigg/ \left(\frac{N_h(z,\nu)}{N_e(\nu)}\right)_D \tag{4}$$

in DIS off nuclei and compared with the N [10] and Kr [11] data from the HERMES collaboration. In Eq. (4) $N_h(z, v)$ represents the number of semi-inclusive hadrons in a given (z, v)-bin and $N_e(v)$ the number of inclusive DIS leptons in the same v-bin.

In Ref. [12] the observed R_M^h spectra were interpreted as being due to a combined effect of a rescaling of the quark fragmentation function in nuclei due to partial deconfinement as well as the absorption of the produced hadrons. Furthermore, calculations based on a pQCD parton model [13, 14] explain the attenuation observed in the multiplicity ratio solely by partonic multiple scattering and induced gluon radiation neglecting any hadronic FSI. It has already been pointed out by the authors of Ref. [12] that a shortcoming of the existing models is the purely absorptive treatment of the FSI. We avoid this problem by using the coupled-channel transport model.

In Ref. [9] we used the high z_h part of the charged hadron data from Ref. [11] to fix the leading hadron cross section to $\sigma_{\text{lead}} = 0.33\sigma_h$ during the formation time. The data for N and Kr could then be well described using a formation time $\tau_f > 0.3\text{fm/c}$ for all hadrons. This value is compatible with the analysis of antiproton attenuation in $p + A$ reactions at AGS energies [15].

In Fig. 2 we show the results for the calculated multiplicity ratio of π^-, π^+, K^-, K^+, p and $\bar{p}$ for Kr in comparison with the experimental data [16]. In our calculations we use the kinematic cuts of the HERMES experiment and take the detector geometry into account. We use a constant formation time of 0.5 fm/c and again scale all leading hadron cross sections with the same factor 0.33 during the formation time. Without further fine tuning we get a satisfying description of all the data meaning that the formation times of mesons, baryons and antibaryons are about equal.

The authors acknowledge valuable discussions with A. Accardi, N. Bianchi, A. Borissov, C. Greiner and V. Muccifora. This work was supported by DFG and BMBF.

REFERENCES

1. M. Effenberger and U. Mosel, Phys. Rev. C **62**, 014605 (2000).
2. T. Falter and U. Mosel, nucl-th/0202011; T. Falter and U. Mosel, Phys. Rev. C **66**, 024608 (2002).
3. T. Falter, K. Gallmeister, and U. Mosel, Phys. Rev. C **67**, 054606 (2003).
4. T. Sjostrand, P. Eden, C. Friberg, L. Lonnblad, G. Miu, S. Mrenna, and E. Norrbin, Comput. Phys. Commun.**135**, 238 (2001); T. Sjostrand, L. Lonnblad and S. Mrenna, hep-ph/0108264.
5. K. Gallmeister, C. Greiner, and Z. Xu, Phys. Rev. C **67**, 044905 (2003).
6. M. Effenberger, E. L. Bratkovskaya, and U. Mosel, Phys. Rev. C **60**, 044614 (1999).
7. A. Airapetian *et al.*, HERMES Collaboration, Phys. Rev. Lett. **90**, 052501 (2003).
8. B. Z. Kopeliovich, J. Nemchik, A. Schaefer, and A. V. Tarasov, Phys. Rev. C **65** 035201 (2001).
9. T. Falter, W. Cassing, K. Gallmeister, and U. Mosel, nucl-th/0303011; T. Falter and U. Mosel, nucl-th/0308073.
10. A. Airapetian *et al.*, HERMES Collaboration, Eur. Phys. J. C **20**, 479 (2001).
11. V. Muccifora, HERMES Collaboration, Nucl. Phys. **A711**, 254 (2002); E. Garutti, HERMES Collaboration, Act. Phys. Pol. B **33**, 3013 (2002).
12. A. Accardi, V. Muccifora, and H. J. Pirner, Nucl. Phys. **A720**, 131 (2003).
13. X. Guo and X. N. Wang, Phys. Rev. Lett. **85**, 3591 (2000); X. N. Wang and X. Guo, Nucl. Phys. **A696**, 788 (2001); E. Wang and X. N. Wang, Phys. Rev. Lett **89**, 162301 (2002).
14. F. Arleo, hep-ph/0306235.
15. W. Cassing, E. L. Bratkovskaya, and O. Hansen, Nucl. Phys. **A707**, 224 (2002).
16. A. Airapetian *et al.*, HERMES Collaboration, hep-ex/0307023.

Time-like Compton scattering and related exclusive processes in proton-antiproton annihilation at PANDA

Michael Düren

II. Phys. Inst. Univ. Giessen, 35392 Giessen
Michael.Dueren@uni-giessen.de

Abstract. Exclusive $p\bar{p}$ annihilation into two photons at large s and t can be described in terms of the handbag diagram. The process is separated into a 'soft' part which is parametrised by General Parton Distributions and a 'hard' part which describes the annihilation of a quasi-free $q\bar{q}$ pair into two photons. The process is of special interest due to its relation to the time-like formfactor and to Generalised Parton Distributions. It is proposed to measure this and related exclusive annihilation processes with a scalar meson, a vector meson or a lepton pair in the final state at the PANDA experiment at GSI. Estimates of the expected count rates based on a model and a previous experiment are compared.

GENERALISED PARTON DISTRIBUTIONS AND DVCS

The theoretical framework of Generalised Parton Distributions (GPDs) [1, 2], which has been developed only a few years ago [3, 4], caused a lot of excitement in the field of understanding the structure of the nucleon in the terms of QCD. The excitement is twofold. First of all, the GPDs unify the theoretical description of different processes like deep inelastic scattering (DIS), deeply virtual Compton scattering (DVCS), hard exclusive meson production (HEMP), wide angle Compton scattering (WACS), as well as its crossed channel, the time-like Compton scattering which is subject of this paper. In the GPD framework the well known formfactors and the polarised and unpolarised parton distributions turn out to be just limiting cases of the same GPDs. Secondly, the GPDs provide a broader insight into the nucleon than any of the previously studied functions. For the first time it is possible to provide detailed information about the localisation of partons inside the nucleon [5] and to access their orbital angular momentum [4].

The geometrical interpretation is illustrated in Figure 1 (left). The GPDs describe at the same time the longitudinal momentum xP of a parton wave packet (quark or gluon) and its transverse impact parameter b_T. Here, P is the longitudinal momentum of the nucleon and x is the Bjorken scaling variable which is interpreted in the infinite momentum frame as the fraction of the momentum carried by the parton. The transverse spatial distribution of charge in a non-relativistic picture is simply the Fourier transform of the nucleon formfactor. This illustrates the link between the formfactors and the GPDs. In a naive classical picture, it is intuitively clear that the GPDs can also access the orbital angular momentum of quarks and gluons which is nothing but the cross product of the longitudinal momentum and the transverse position of the parton.

CP717, *Hadron Spectroscopy: Tenth International Conference,*
edited by E. Klempt, H. Koch, and H. Orth
© 2004 American Institute of Physics 0-7354-0197-7/04/$22.00

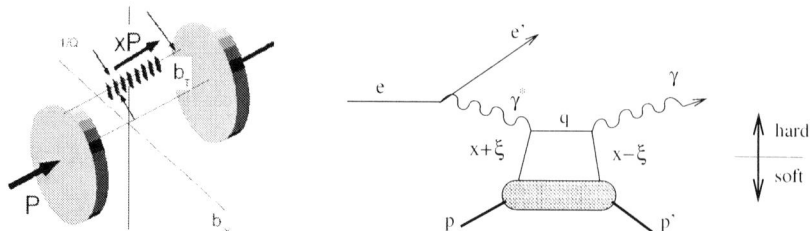

FIGURE 1. Left: The GPDs allow to measure at the same time the longitudinal momentum xP and the transverse impact parameter b_T of a quark wave packet. Right: In DVCS a quark with momentum fraction $x + \xi$ is taken out of the proton and put back with another momentum fraction $x - \xi$. The proton stays intact. The lower 'soft' part of the diagram is described by GPDs, the upper 'hard' part by pertubative QCD and QED.

DVCS is the prototype process for measuring GPDs. It is described by the handbag diagram as shown in Fig.1 (right). The process is divided into two parts. The upper 'hard' part of the diagram can be calculated pertubatively in QCD and QED. The hard scale in this process is defined by the virtuality of the photon γ^*. The lower 'soft' part of the diagram cannot be calculated pertubatively and is parametrised by the GPDs. A quark with momentum fraction $x + \xi$ is taken out of the proton and put back with another momentum fraction $x - \xi$. The proton stays intact. In the limiting case of the skewedness parameter $\xi = 0$ and mandelstam $t = 0$, the GPDs are identical to the ordinary parton distributions which describe the probability to find a quark with momentum fraction x in the nucleon. First measurements of DVCS processes have been performed by HERMES [6], CLAS [7], H1 and ZEUS [8]. In a similar way, the hard exclusive meson production can be described by GPDs. In this case the final state photon is replaced by a meson.

CROSSED CHANNEL WIDE ANGLE COMPTON SCATTERING

It has been shown that not only DVCS and hard exclusive meson production but also WACS can be described by the handbag diagram using GPDs [9, 10]. In WACS both photons are real. The hard scale, which is required for the pertubative description of the process, comes from the large transverse momentum transfer of the photon to the parton.

We proposed to study the crossed channel of WACS, the annihilation of $p\bar{p}$ into two photons at large transverse momentum at the PANDA experiment at GSI [11]. A priori it is not clear what is the dominant production mechanism for this process. It can be argued that the emission of the two photons from two independent quark lines (Fig. 2 a) is suppressed as quasi-free partons cannot emit a single hard photon. At very large energies ($\sqrt{s} \to \infty$) the annihilation is accompanied by the exchange of additional hard gluons (Fig. 2 b) [12]. The additional gluons transfer the large energy loss to the individual annihilating partons. It has been shown recently, that at intermediate energies ($s \approx 10 \text{ GeV}^2$) the process can indeed be described by the crossed-channel handbag diagram and the corresponding timelike GPDs (Fig.2 c) [13, 14]. In this case most of the energy is carried by the photon-emitting quark (i.e. x has to be large) and the

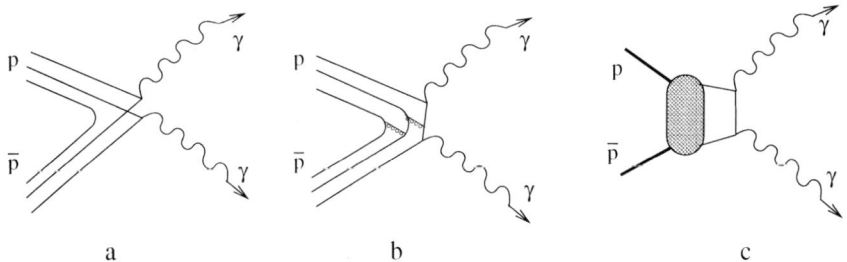

FIGURE 2. The crossed channel of wide angle Compton scattering is the annihilation of $p\bar{p}$ into two photons at large transverse momentum. The emission of two photons from two quasi-free partons is suppressed (a). At large energies hard gluons are required to account for the energy loss of the individual annihilating partons (b). At intermediate energies the handbag diagram is dominant (c).

momentum of the other partons is soft and of the order of the typical hadronic intrinsic momenta. It has been argued that in addition to QCD arguments, quark-hadron duality provides a further justification for the effective dominance of the handbag diagram even in kinematic regions where factorisation arguments based on large momentum scales alone can not assure its dominance [15].

CROSS SECTIONS AND EXPERIMENTAL REQUIREMENTS

First estimates of cross section σ and counting rates have been made using a simple model calculation [13]. The differential cross section for process 2 (c) is

$$\frac{d\sigma}{d\cos\theta} = \frac{2\pi\alpha_{em}^2}{s} \frac{R_V^2(s)\cos^2\theta + R_A^2(s)}{\sin^2\theta} \tag{1}$$

where θ is the emission angle of the photons in the CMS system, α_{em} is the fine structure constant and R_V and R_A are generalised vector and axial vector formfactors which can be obtained from double integrals of double distributions, which are a particular representation of GPDs. Using simple models for the double distributions as described in [13] the results as shown in fig. 3 are obtained. A cut of $45° < \theta < 135°$ assures that the process has large transverse momentum and can be described using the handbag diagram. Assuming for example an energy of $\sqrt{s} = 3.2$ GeV with a luminosity of $2 \cdot 10^{32}\text{cm}^{-2}\text{s}^{-1}$ for the proposed PANDA experiment, this calculation predicts a count rate of a few thousand events per month.

The experiment E670 at Fermilab has measured already the process $p\bar{p} \to \gamma\gamma$ at various angles, however with limited precision and energies up to $\sqrt{s} < 3.7$ GeV [16]. Comparing the model calculation of [13] with the measurement of [16] at $\sqrt{s} = 3.5$ GeV, one obtains the result that the measured cross section is a factor of about 35 above the model prediction. The reason for the discrepancy with the model calculation is likely the large contamination of the measured cross section by events with more than 2 photons (e.g. $p\bar{p} \to \pi^0\pi^0$) where the additional photons are not identified in the detector. The large possible background contamination requires a detector at PANDA which

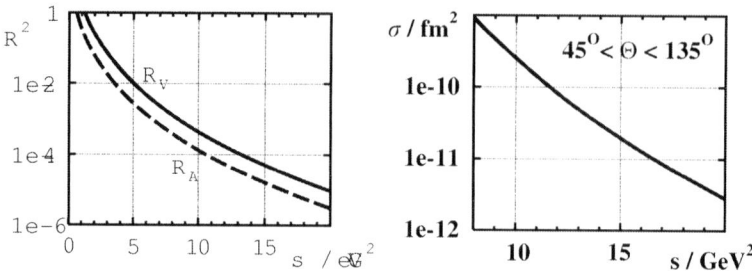

FIGURE 3. The generalised vector and axial vector formfactors (left) and the corresponding cross section (right) of $p\bar{p} \to \gamma\gamma$ as function of s as obtained from a model calculation.

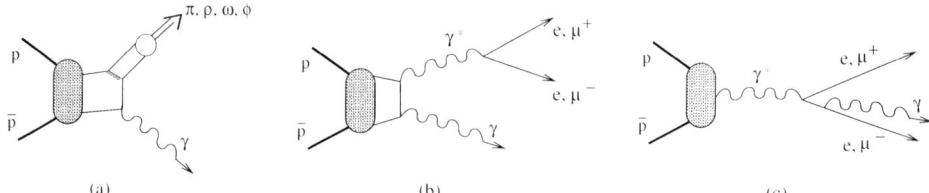

(a)   (b)   (c)

FIGURE 4. When the production of a single hard meson at large transverse momentum is described by the handbag diagram, an additional hard gluon is needed to form the final state meson (a). The production of a virtual photon (b) interferes with the corresponding BH-type diagram (c).

is superior to the one at the experiment E670 to allow for an additional suppression of multi-photon events. This requires a good energy resolution and hermeticity of the calorimeter.

Cross section measurements of the inverse process $\gamma\gamma \to p\bar{p}$ [17] are about a factor 3 above the estimate of [13] but in general confirm the handbag ansatz [14].

CROSSED CHANNEL HEMP AND DVCS

The production of real photons $p\bar{p} \to \gamma\gamma$ is not the only interesting process in connection with GPDs. In analogy to similar processes at electron scattering experiments, also the hard exclusive meson production (fig. 4 a) and the production of virtual photons which decay into lepton pairs (fig. 4 b) are of interest. For these processes the direct relation to GPDs has not yet been proven, but it can be hoped that these processes can also be understood in terms of the handbag diagram.

From the experimentalist point of view all these processes can be regarded as the annihilation of $p\bar{p}$ into a final state where one real photon is tagged at large transverse momentum on one side and a second particle $(\gamma, \gamma^*, \pi, \rho, \omega, \phi, ...)$ is detected at the other side, balancing energy and all three components of momentum. This gives a clear experimental signature. The comparison of the differential cross sections for the various final state particles will give new insights in the production mechanisms. If the handbag diagram is dominant for the production of hard mesons as in fig 4 (a), it is clear that for example the production of ϕ mesons is strongly suppressed, as the annihilating quark has to be at large x and we know from DIS experiments that there are no intrinsic strange quarks at large x. In addition to the sensitivity of the different processes to individual flavours of quarks, one obtains sensitivity to the helicity of quarks in the production of vector mesons and of lepton pairs. As in DVCS, the production of a virtual photon interferes with the corresponding BH diagram (fig. 4 c) which is related to the time-like proton form factor. In the production of a virtual photon one can distinguish the case where the virtuality q^2 is small and the hard scale comes from the transverse momentum as in the real photon production, and those where the virtuality $q^2 \geq 1$ GeV^2 is large and the hard scale comes from the virtuality as in DVCS.

CONCLUSIONS AND ACKNOWLEDGEMENTS

The exclusive annihilation of $p\bar{p}$ into two photons at large transverse momentum is described by GPDs and allows to extract information which is complementary to the results from electron scattering. The estimated count rates for the PANDA experiment are on the order of a few thousand $p\bar{p} \to \gamma\gamma$ events per month. The production of a virtual photon or different types of mesons in coincidence with a large p_T photon allows for a deeper understanding of the production mechanisms in exclusive $p\bar{p}$ annihilation.

I want to thank the BMBF for funding and C. Weiss, M. Hartig, R. Jakob, P.Kroll, M. Diehl for interesting discussions and new insights in the understanding of this matter.

REFERENCES

1. K. Goeke, M.V. Polyakov, M. Vanderhaeghen, Prog. Part. Nucl. Phys. 47 (2001) 401.
2. M. Diehl, habil. thesis, Univ. Hamburg (2003), hep-ph/0307382.
3. D. Müller *et al.*, Fortsch. Phys. 42 (1994) 101;
   A.V. Radyushkin, Phys. Lett. B **380** (1996) 417.
4. X.-D. Ji, Phys. Rev. Lett. **78** (1997) 610, hep-ph/9603249.
5. M. Burkardt, Phys. Rev. D **62**, 071503 (2000), hep-ph/0005108;
   J.P. Ralston, B. Pire, Phys.Rev. D **66** (2002) 111501.
   M. Diehl, Eur. Phys. J C **25** (2002) 223.
6. HERMES Collab., A. Airapetian *et al.*, Phys. Rev. Lett. **87** (2001) 182001;
   F. Ellinghaus for the HERMES Collab., Nucl. Phys. A **711** (2002) 170.
7. CLAS Collab., S. Stephanyan *et al.*, Phys. Rev. Lett. **87** (2001) 182002.
8. L. Favart, Nucl. Phys. A **711** (2002) 165.
9. A.V. Radyushkin, Phys. Rev. D **58** (1998) 114008, hep-ph/9803316.
10. M. Diehl, Eur. Phys. J C **8** (1999) 409, hep-ph/9811253.
11. H. Gutbrod *et al.* (eds.) "An International Accelerator Facility for Beam of Ions and Antiprotons", GSI report, Darmstadt (2001).
12. G.P. Lepage and S.J. Brodsky, Phys. Rev. D **22**, (1980) 2157.
13. A. Freund, A.V. Radyushkin, A. Schäfer, Ch. Weiss, Phys. Rev. Lett. **90** (2003) 092001.
14. M. Diehl, P. Kroll, C. Vogt, Eur. Phys. J. C **26** (2002) 567;
    M. Diehl, P. Kroll, C. Vogt, Phys. Lett. B **532** (2002) 99.
15. Close, Zhao, Phys. Lett. B **553**, 2003, 211.
16. T.A. Armstrong *et al.*, Phys. Rev. D **56** (1977) 2509.
17. e.g. G. Abbiendi *et al.*, Eur. Phys. J. C **28** (2003) 45.

Photo-Production of Proton Antiproton Resonances

Paul Eugenio and Burnham Stokes
for the CLAS Collaboration

Department of Physics, Florida State University, Tallahassee, FL, USA

Abstract. Preliminary results are reported on the reaction $\gamma p \to pp\bar{p}$. The data were obtained at the Thomas Jefferson National Accelerator Facility utilizing the CLAS detector and a tagged photon beam of 4.8 to 5.2 GeV incident on a liquid hydrogen target. The focus of this study is to search for possible intermediate resonances which decay to $\bar{p}p$. Both final state protons were detected in CLAS whereas the antiproton was identified via missing mass. General features of the accepted data are presented.

Introduction

The proton-antiproton system has had a rich history spanning more than thirty years. Initially, the $p\bar{p}$ system had much interest due to theoretical predictions of exotic matter. These predictions included: nucleon-antinucleon states that are loosely bound in a molecule-like structure called quasi-nuclear baryonium, and tighly-bound multi-quark baryonium ($qq - \bar{q}\bar{q}$) which have favored decays to nucleon-antinucleon final states.

Arround 1970, there were claims of a unusually-narrow meson resonance with a mass of 1.93 GeV/c^2 [1][2], and it was believed that this particle was not an ordinary meson and that it would couple to the proton-antiproton system. There were then claims that experiments found the narrow resonance in proton-antiproton scattering experiments [3][4][5][6]. Also, in the late 1970s there were claims of additional higher mass narrow resonances at 2.02 and 2.20 GeV/c^2 in the proton-antiproton system [7][6][8]. However follow up experiments did not make such claims[9][10]. And until recently, the debate had died out.

In 1997, CERN refuted their own earlier claims of the 1.93 and 2.02 GeV/c^2 resonances. Yet in 1999, a reanalysis of the CERN data confirmed the existence of the 2.02 and 2.2 GeV/c^2 resonances. Presently, the only well-known particle that decays to proton-antiproton is the J/ψ particle, with a mass of 3.097 GeV/c^2 [11]. Most of the past experiments involved proton-antiproton scattering or pion production. Recently Jefferson Laboratory has provided the first look at the proton-antiproton system through photoproduction.

In 1999 JLAB observed approximately 2000 accepted events of photoproduction of proton-antiproton($\gamma p \to pp\bar{p}$). This data was particularly exciting due to the fact that the yield was an order of magnitude larger than any previous photoproduction data set. Yet this data was still rather limited in statistics and preliminary studies of the proton-antiproton system were inconclusive. The experiment which followed, JLAB E01-017,

CP717, *Hadron Spectroscopy: Tenth International Conference,*
edited by E. Klempt, H. Koch, and H. Orth
© 2004 American Institute of Physics 0-7354-0197-7/04/$22.00

produced an order of magnitude greater number of $p\bar{p}$ events, and it is this data that is the main focus of the present work. With such a data set, it should be possible for the first time to search for the existence of both narrow and broad resonances via the use of partial wave analysis of the proton-antiproton system.

Experimental Results

The data were obtained using a photon beam incident on a liquid hydrogen target in the CEBAF Large Angle Spectrometer (CLAS). The particle detection system consists of drift chambers to determine the trajectories of charged particles, gas Čerenkov detectors for particle identification, scintillation counters for measuring time-of-flight (TOF) and particle identification, and electromagnetic calorimeters to detect neutral particles [12]. These detectors are designed to provide as much coverage of the 4π solid angle as possible.

For this experiment, the trigger required that a photon of an energy between 4.8 and 5.5 GeV be detected, it required that at least two of the timing counters surrounding the target measure hits, and it required that at least two of the six downstream TOF sectors measure hits.

After filtering events by particle identification, initial cuts were applied to the data set. These selections include beam energy, vertex position, and timing requirements. The photon energy was determined using the electron beam tagger. While the trigger required that a photon with an energy in the range of 4.8 to 5.5 GeV be identified, additional low energy photons could also be measured during the time window allowed to acquire the event. This would lead to an ambiguity in which photon beam particle was associated with the event measured in CLAS. Tight timing requirements as well as energy conservation cuts were used to take this ambiguity into account. Events were required to have a beam energy in the range of 4.8 to 5.5 GeV, and events below this energy are excluded from further analysis.

Nearly five thousand exclusive events were observed where all final state particles were identified in the CLAS spectrometer. However in CLAS, there are detector regions where particles can go unmeasured. For example, the CLAS toroidal magnetic field bends negatively charged particles back toward the beam. Quite often, these particles end up going back into the beam-line, and are lost. To increase the exclusive data yield, the anti-proton was allowed to be indetified via the missing mass.

Figure 1 shows the missing-mass-squared of events containing two identified protons. There is a prominent peak at a mass squared of $0.880(GeV/c^2)^2$, which is consistent with a missing antiproton. Selecting the events consistent with a missing antiproton ($0.85(GeV/c^2)^2 \leq MM^2 \leq 0.91(GeV/c^2)^2$) yields approximately 17,100 $\gamma p \rightarrow pp(\bar{p})$ events. Yet not all of these events are $\gamma p \rightarrow pp\bar{p}$ events as seen by the nearly flat background of non-antiprotons in the missing-mass squared distribution. Efforts are underway to further clean up and understand the background events under the antiproton signal.

Possible production mechanisms which describe the photoproduction of a proton-antiproton pair are diffraction/meson exchange, baryon exchange, and antibaryon ex-

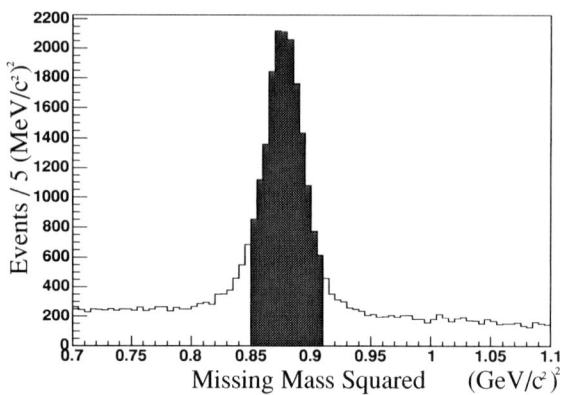

FIGURE 1. Missing Mass Squared off two protons.

change. In each process, an intermediate resonance may be produced. In meson exchange the photon transfers very little momentum to the target, but interacts, causing the photon to produce a resonance that decays to a fast forward-going proton-antiproton pair. In baryon exchange, the photon interacts with an exchange baryon converting it to a fast forward-going proton leaving behind a slow moving resonance at the target vertex which decays to a proton-antiproton pair. For antibaryon exchange, the photon interacts with an exchange antibaryon converting it to a fast forward-going antiproton, leaving behind a resonance at the target vertex which decays to two protons.

The distinction of meson exchange and baryon exchange production is clouded by the two identical protons. Without information identifying which is which, the two mechanisms are nearly indistinguishable. For antibaryon exchange it does not matter since both protons are at the same decay vertex. The two proton invariant mass is shown in Fig. 2. No obvious peaks or features are observed.

A simple way to distinguish the protons is by storting on the proton momentum. In the cases of meson and baryon exchange, one proton should be moving fast and in the forward direction. Whereas the other proton being produced at or near the target vertex receives very little momentum transfer from the beam and is expected to be slow going. Therefore, one can use the momentum of the two protons on an event by event basis and associate a $p_{fast}\bar{p}$ resonance with meson exchange and a $p_{slow}\bar{p}$ resonance with baryon exchange.

In Fig. 3, the proton momentum distribution is shown. The proton with the greatest magnitude of momentum is defined as the fast proton, and the other proton is defined as the slow proton. The light histogram shows the event by event distribution for the slow protons whereas the dark histogram is that for the fast protons. In addition, there are other kinematic variables that may help distinguish production mechanisms. These observables are currently under investigation.

The invariant mass of $p_{slow}\bar{p}$ is shown in Fig. 4. The distribution has some interesting structures, with a sharp rise at threshold and a possible narrow peak or dip near 2.0 GeV

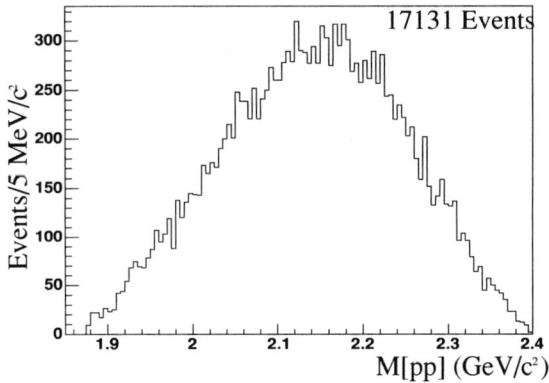

FIGURE 2. Accepted invariant mass distribution of the two protons.

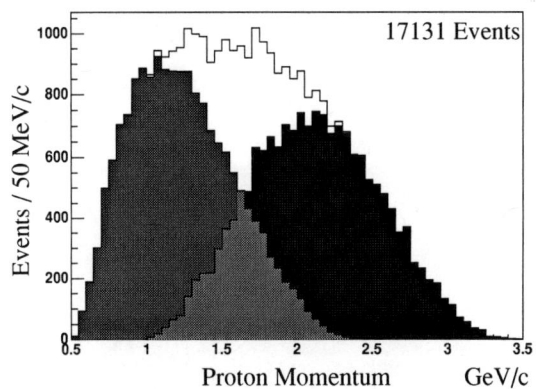

FIGURE 3. Proton momentum distributions. In each event, whichever proton has less momentum is placed in the "slow" distribution (left) and the other goes into the "fast" distribution (right)

and broader peak at 2.04 GeV. In the invariant mass distribution of $p_{fast}\bar{p}$ (Fig. 5) there are no obvious structures.

Future analysis plans include understanding background events, exploring kinematical features of the data, analyzing angular distributions, performing Monte Carlo simulations, and performing a partial wave analysis to search for resonant behavior.

REFERENCES

1. M.N. Focacci *et. al.*, Phys. Rev. Lett. 17, 890(1966).
2. D. Cline *et. al.*, Phys. Rev. Lett. 17, 1268(1968).

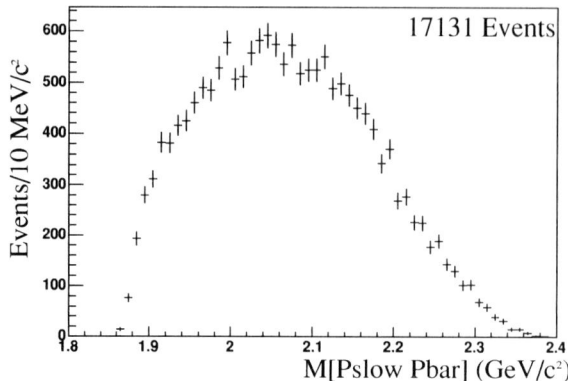

FIGURE 4. The invariant mass of the slow proton with the antiproton.

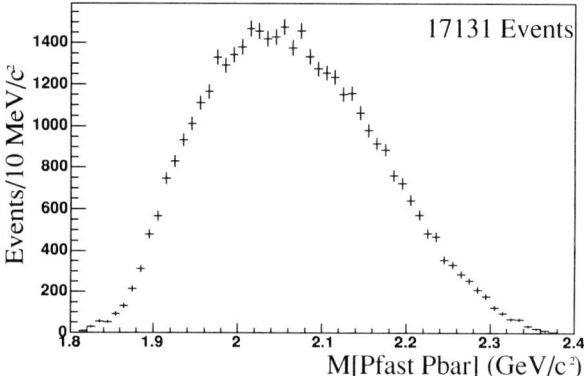

FIGURE 5. The invariant mass of the fast proton with the antiproton.

3. A.S. Carroll *et. al.,* Phys. Rev. Lett. 32, 247(1974).
4. T.E. Kalogeropoulos and G.S. Tzanakos Phys. Rev. Lett. 34, 1047(1975)
5. V. Chaloupka *et. al.,* Phys. Lett. 61 B, 487(1976).
6. P. Benkheiri *et. al.,* Phys. Lett. 68 B, 483(1977).
7. J. Bodenkamp *et. al.,* Phys. Lett. 133 B, 275(1983).
8. B.G. Gibbard *et. al.,* Phys. Rev. Lett. 42, 1593(1979).
9. R. Bizzarri *et. al.,* Phys. Rev. D 6, 160(1972).
10. J. Bensinger *et. al.,* Phys. Rev. D 23, 1417(1983).
11. M.W. Eaton *et. al.,* Phys. Rev. D 29, 805(1984).
12. B.A. Mecking *et. al.,* NIM A503, 513(2003).

Photoproduction of η'-mesons from the Proton

Ch. Elster*, A. Sibirtsev†, S. Krewald† and J. Speth†

*Institute of Nuclear and Particle Physics, Ohio University, Athens, OH 45701, USA
†Institut für Kernphysik, Forschungszentrum Jülich, D-52425 Jülich, Germany

Abstract. The presently available data for the reaction $\gamma p \rightarrow \eta' p$ are analyzed in terms of a model in which the dominant production mechanism is the exchange of the vector mesons ω and ρ. To describe the data at photon energies close to the production threshold we introduce a resonance contribution due to the well established $S_{11}(1535)$ resonance. Finally we study the contributions due to nucleon exchange to the η' photoproduction and find, that those contributions can be seen at large angles in the differential cross section.

INTRODUCTION

The reaction $\gamma p \rightarrow \eta' p$ was studied by several, collaborations, namely ABBHHM [1], AHHM [2] and SAPHIR [3, 4]. While ABBHHM and AHHM measured only the total cross section for the photoproduction of the η' meson at photon energies $1.67 \leq E_\gamma \leq 5$ GeV, the SAPHIR collaboration at ELSA studied the angular spectra of the η' mesons produced at photon energies $0.9 \leq E_\gamma \leq 2.6$ GeV. In addition, the photoproduction of η' mesons from the nucleon is presently experimentally investigated by the CLAS Collaboration at TJNAF [5] and by the Crystal Barrel Collaboration at ELSA.

Currently experimental and theoretical [6, 7, 8, 9] studies of the reaction $\gamma p \rightarrow \eta' p$ are motivated by the possibility to investigate excited baryons coupled to the η' meson. Assuming that the resonance production is the dominant contribution, an isobar analysis [3] of the SAPHIR data shows that the η'-photoproduction at photon energies $0.9 \leq E_\gamma \leq 2.6$ GeV can be described by the coherent excitation of two resonances, $S_{11}(1897)$ and $P_{11}(1986)$. However, the presently available data [1, 2, 3, 4] for the η'-photoproduction are quite limited. Thus, an evaluation of resonance properties from these data should be considered with caution. It is also clear that the contribution from the meson exchange current to the reaction $\gamma p \rightarrow \eta' p$ can not be completely neglected, as was shown in the analysis of the reaction $pp \rightarrow pp \eta'$.

Here we investigate up to what extent present data for the reaction $\gamma p \rightarrow \eta' p$ can be described by the exchange of vector mesons in the t channel and the nucleon exchange in the s and u channels. We do not consider contributions from baryonic resonances with masses above the production threshold. If possible resonances will significantly contribute to the production mechanism, then a discrepancy between our calculation and data should indicate this.

CP717, *Hadron Spectroscopy: Tenth International Conference,*
edited by E. Klempt, H. Koch, and H. Orth
© 2004 American Institute of Physics 0-7354-0197-7/04/$22.00

PRODUCTION VIA VECTOR MESON EXCHANGE AND RESONANCE CONTRIBUTIONS

One of the well established results in particle physics is that for very high energies, i.e. $E_\gamma > 5$ GeV, Regge trajectories provide the dominant processes for peripheral reactions. At energies well below 5 GeV, but still clearly above the resonance region the photoproduction of mesons can be described the t channel meson exchanges. In our case the photoproduction of the η' meson is dominated by the exchanges of the ρ and the ω meson. It is obvious, that already at the reaction threshold the η' meson photoproduction probes large $|t|$. For the energies available at ELSA and TJNAF, namely $E_\gamma < 2.5$ GeV, the photoproduction of η'- mesons probes $|t|$ up to $\simeq$-3 GeV^2. Thus, the ρ and ω meson exchanges should be very sensitive to the choice of the form factors at the nonlocal interaction vertices. Furthermore, it is important to notice that uncertainties in the t-channel contributions due to unknown coupling constants and form factors can be eliminated through data collected at energies slightly beyond the resonance region. Here the t-channel contributions dominate at the low four momentum transfer squared.

The effective Lagrangian densities used for the evaluation of the vector meson (V) exchange amplitudes are given as [10]

$$\mathscr{L}_{VNN} = g_V \bar{N} \gamma_\mu N V^\mu + \frac{g_T}{2m_N} \bar{N} \sigma_{\mu\nu} N V^{\mu\nu} \; ; \; \mathscr{L}_{V\eta'\gamma} = \frac{e g_{V\eta'\gamma}}{m_{\eta'}} \varepsilon_{\mu\nu\alpha\beta} F^{\mu\nu} V^{\alpha\beta} \eta'. \quad (1)$$

Here $m_{\eta'}$ stands for the mass of the η' meson, m_N for the mass of the nucleon. The vector meson field tensor is given by $V_{\mu\nu} = \partial_\nu V_\mu - \partial_\mu V_\nu$, and $F^{\mu\nu} = \partial_\nu A_\mu - \partial_\mu A_\nu$ with A^μ being the photon field. The vertices are dressed with dipole form factors with a cutoff mass of 2.1 GeV. The coupling constants are adopted from the Bonn potential model, $g_{\rho NN} = 3.9$ and $g_{\omega NN} = 10.6$, and the ratio of the tensor to vector coupling is 6.1 for the ρ meson and zero for the ω meson.

The couplings for the $\rho\eta'\gamma$ and $\omega\eta'\gamma$ vertices can be evaluated from the partial decay widths of the reactions $\eta' \to \gamma\rho$ and $\eta' \to \omega\rho$. The two quantities are related by

$$\Gamma_{\eta' \to \gamma V} = \frac{e^2 g_{V\eta'\gamma}^2}{32\pi} \frac{(m_{\eta'}^2 - m_V^2)^3}{m_{\eta'}^5}, \quad (2)$$

where $e^2 = 4\pi\alpha$, with α being the electromagnetic coupling. With $\Gamma_{\rho \to \gamma\eta'} = 60 \pm 5$ keV and $\Gamma_{\omega \to \gamma\eta'} = 6.1 \pm 0.8$ keV we obtain $g_{\rho\eta'\gamma} = 1.36$, and $g_{\omega\eta'\gamma} = 0.4$.

Because of the large uncertainties in the experimental results and the very limited number of experimental points [3, 4] presently available for the reaction $\gamma p \to \eta' p$, we prefer to introduce as few resonances into our calculation as possible. It is also clear, that large uncertainties in the selection of the parameters of potential resonances, i.e. mass and width, the coupling to the η' meson as well as our lack of knowledge about relevant form factors at the resonance vertices might provide quite a large freedom in describing the present data by isobar contributions. Having this in mind and considering that the vector meson exchange mostly underpredicts the present data close to threshold, we only have room for additional contributions at photon energies $E_\gamma < 1.7$ GeV, or at invariant collision energies below 2 GeV.

To study the influence of s-wave resonances on the behavior of the differential cross section close to the production threshold, we prefer to select only the well known $S_{11}(1535)$ resonance and investigate, whether its contribution is sufficient to account for the difference between our calculation based on vector meson exchange alone and the data for photon energies $E_\gamma < 1.7$ GeV.

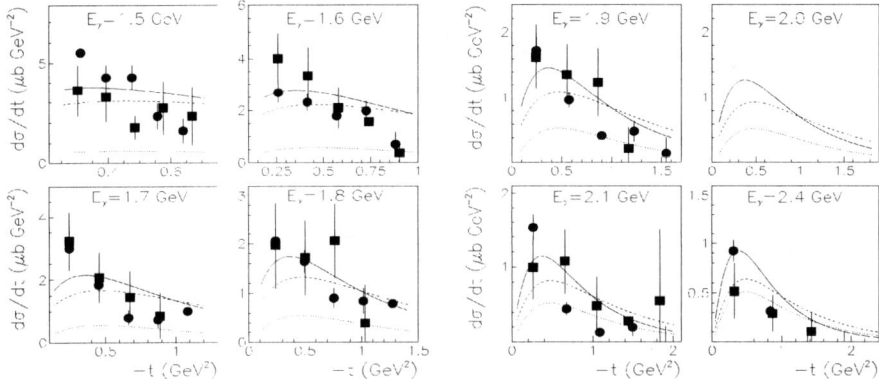

Fig. 1: The differential cross section for the reaction $\gamma p \to \eta' p$ as a function of the four momentum transfer squared t. The squares represent the first SAPHIR results given in Ref. [3] with absolute normalization. The circles stand for new SAPHIR data reported in Ref. [4] in relative normalization, which we multiply by a factor of 2.9 at all given photon energies E_γ. The dotted line show the contribution from vector meson exchange alone calculated with the dipole form factors at both VNN and $V\eta'\gamma$ vertices, while the dashed lines are our results with an additional inclusion of $S_{11}(1535)$ contribution. The solid lines show the calculations with $S_{11}(1535)$ and vector meson exchange contributions with exponential form factor at $V\eta'\gamma$ vertex.

In Fig. 1 the differential cross section for the reaction $\gamma p \to \eta' p$ is displayed. The dotted lines represent our results obtained with a dipole form factors at both, the VNN and $V\eta'\gamma$ vertices, and they reproduce the data for photon energies $E_\gamma \geq 2.1$ GeV quite reasonably. The calculations including the contribution of the $S_{11}(1535)$ together with the vector meson exchange are shown as dashed line. The solid lines represent the corresponding calculations with the monopole form factor at VNN vertex and exponential one at $V\eta'\gamma$ vertex. The resonant contribution dominates around the η' meson photoproduction threshold and vanishes with increasing the photon energy. We want to emphasize that our calculations do not prove that resonant contributions to the η' meson photoproduction are not necessarily only those from the $S_{11}(1535)$ resonance. Large uncertainties in selection of the coupling constants to resonances as well as in cutoff parameters allow sufficient freedom to consider more resonant contributions. However, in this work, we did not prefer to include any other resonance than $S_{11}(1535)$.

CONTRIBUTION OF THE NUCLEON EXCHANGE

In general, the contribution from the nucleon exchange current can be measured at small u or at backward angles in meson photoproduction reaction. Obviously, the size of the contribution will depend on the strength of the coupling of the η' meson to the nucleon. In fact, such an increase of the differential cross section at small u was detected in the photoproduction of π and ω mesons [11, 12]. Thus, it may also be observed in the photoproduction of η' mesons.

The effective Langrangians for the γNN and $\eta'NN$ interaction can be written as [13]

$$\mathscr{L}_{\gamma NN} = -e\bar{N}\left(\gamma_\mu \frac{1+\tau_3}{2}A^\mu - \frac{\kappa^S+\kappa^V\tau_3}{4m_N}\sigma_{\mu\nu}F^{\mu\nu}\right)N \; ; \; \mathscr{L}_{\eta'NN} = -ig_{\eta'NN}\bar{N}\gamma_5 N\eta'. \quad (3)$$

Here κ^S and κ^V are the isoscalar and isovector anomalous magnetic moments of the nucleon, $\kappa_p=\kappa^S+\kappa^V=1.79$ and while $\kappa_n=\kappa^S-\kappa^V=-1.91$ stand for the proton and neutron, respectively.

In principle, we should consider formfactors at the interaction vertices, since the nucleons are off shell in the intermediate state. We introduce the form factors similar to Ref.[14, 15] with a cutoff parameter $\Lambda_N=700$ MeV.

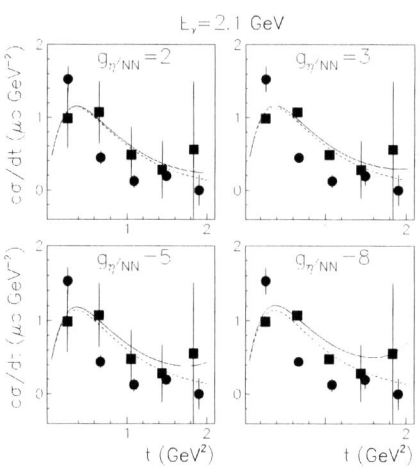

Fig. 2: The differential cross sections for the η' meson photoproduction as a function of four momentum transfer squared t at photon energies $E_\gamma=2.1$ GeV. The squares show the first SAPHIR results given [3] with absolute normalization. The circles are new SAPHIR data reported [4] in relative normalization, which we multiply by a factor of 2.9 at all given photon energies E_γ. The solid line represents our calculation including vector meson exchange, the contribution from the $S_{11}(1535)$ resonance as well as the contribution from the nucleon exchange with different $\eta'NN$ coupling constant. For the calculation shown as dashed line, the contribution from the nucleon exchange was omitted.

For our investigation it will be interesting to study if there are visible effects in the differential cross section depending on the strength of the $\eta'NN$ coupling. For this investigation we choose a fixed photon energy, $E_\gamma=2.1$ GeV, and vary the coupling constant $g_{\eta'NN}$ within the limits given above. The solid lines in Fig. 2 show those calculations, which are based on the contribution from the vector meson exchange, the $S_{11}(1535)$ resonance and the nucleon exchange current with different values for $g_{\eta'NN}$. From these figures we see that the differential cross section for η' meson photoproduction at small $|u|$ or equivalently at large scattering angles is sensitive to the $\eta'NN$ coupling constant.

SUMMARY AND CONCLUSIONS

Our goal was to investigate the importance of different reaction mechanisms in the reaction $\gamma p \rightarrow \eta' p$ from threshold to photon energies $E_\gamma = 2.4$ GeV. First we studied the contribution from vector meson exchange in the t channel to the η' meson photoproduction cross section. With this, the presently available data from the SAPHIR collaboration at ELSA [3, 4] can be well reproduced for photon energies $E_\gamma \geq 1.8$ GeV to 2.4 GeV.

The discrepancy between our calculation based on vector meson exchange alone and the data at small photon energies, $1.5 \leq E_\gamma < 1.7$ GeV, may be attributed to contributions of resonances. An analysis of the data in terms of the energy dependence of the extrapolated forward differential cross section $d\sigma/dt$ supports the assumption that this additional contribution results from a resonance with a mass below the η' meson production threshold. Thus we include the well established $S_{11}(1535)$ nucleon resonance into our calculation. Finally we can describe the available data for $1.5 \leq E_\gamma \leq 2.4$ GeV quite well. However we note, that with the presently available data we can not conclude that the $S_{11}(1535)$ resonance is the only one that contributes to the reaction $\gamma p \rightarrow \eta' p$.

As last step we investigated the contribution from the nucleon exchange to the reaction $\gamma p \rightarrow \eta' p$. The aim is to see whether this contribution can be detected experimentally. Our calculations show that the nucleon exchange contribution dominates at small $|u|$ for photon energies $E_\gamma \geq 1.8$ GeV. Our speculation is that an experimental measurement under the above kinematical conditions can be used for a better evaluation of the contribution from the nucleon exchange, which in turn allows to obtain the $\eta' NN$ coupling constant.

ACKNOWLEDGMENTS

This work was performed in part under the auspices of the U. S. Department of Energy under contract No. DE-FG02-93ER40756 with the Ohio University. The authors appreciate very useful discussions and communications with M. Dugger, J. Ernst, E. Pasyuk and B.G. Ritchie.

REFERENCES

1.   ABBHHM Collaboration, Phys. Rev. **175**, 1669 (1968).
2.   W. Struczinski et al., Nucl. Phys. B **108**, 45 (1976);
3.   R. Plotzke et al., Phys. Lett. B **444**, 555 (1998).
4.   J. Barth et al., Nucl. Phys. A **691**, 374 (2001).
5.   M. Dugger, Ph.D. Thesis, ASU, 2001.
6.   J.F. Zhang, N.C. Mukhopadhyay and M. Benmerrouche, Phys. Rev. C **52**, 1134 (1995).
7.   Z.P. Li, J. Phys. G **23**, 1127 (1997).
8.   B. Borasoy, E. Marco and S. Wetzel , Phys. Rev. C **66**, 055208 (2002).
9.   W.T. Chiang, S.N. Yang, L. Tiator, M. Vanderhaeghen and D. Drechsel, nucl-th/0212106.
10. see e.g. M. Benmerrouche and N.C. Mukhopadhyay, Phys. Rev. D **51**, 3237 (1995).
11. R.L. Anderson et al. Phys. Rev. Lett. **23**, 721 (1969)
12. R.W. Clifft et al. Phys. Lett. B **64**, 213 (1976).
13. F. Gross, J.W. Van Orden and K. Holinde, Phys. Rev. C **41**, 1909 (1990).
14. R.M. Davidson and R. Workman, Phys. Rev. C **63**, 025210 (2001).
15. H. Haberzettl, Phys. Rev. C **56**, 2041 (1997).

Recent Results from $2\pi^0$ Photoproduction off the Proton

Martin Kotulla, for the TAPS and A2 Collaborations

Department of Physics and Astronomy, University of Basel, CH-4056 Basel (Switzerland)

Abstract. The reaction $\gamma p \to \pi^0 \pi^0 p$ has been measured using the TAPS BaF$_2$ calorimeter at the tagged photon facility of the Mainz Microtron accelerator in the beam energy range from threshold up to 820 MeV. Close to threshold, chiral perturbation theory (ChPT) predicts that this channel is significantly enhanced compared to double pion final states with charged pions. The strength is attributed dominantly to pion loops in the $2\pi^0$ channel - a finding that opens new prospects for the test of ChPT. Our measurement is the first which is sensitive enough for a conclusive comparison with the ChPT calculation and is in agreement with its prediction. The data are also in agreement with a calculation in the unitary chiral approach.

In the second resonance region, a recent model interpretation of new GRAAL data claimed a dominance of the $P_{11}(1440) \to \sigma N$ reaction process. We present very accurate invariant mass distributions of $\pi^0\pi^0$ and $\pi^0 p$ systems, which are in contrast to the σN intermediate state and which show a dominance of the $\Delta \pi$ intermediate state.

INTRODUCTION

The description of the low energy properties of the nucleon as well as the study of nucleon resonances remain a long-standing task of hadronic physics. The unique features of the $2\pi^0$ channel - the strong suppression of the direct production (Δ Kroll-Rudermann, Born terms,...) - open new prospects to improve the knowledge in both fields.

Chiral Perturbation Theory

In the low energy regime where properties of the lowest lying baryons and mesons are studied, an approach exploiting the approximate Goldstone boson nature of the pion has been developed: chiral perturbation theory (ChPT) [1, 2]. This effective field theory has been extended to the nucleon sector (HBChPT[1]) [3, 4]. In general, it turns out, that ChPT is in good agreement with experiment in describing $\pi - N$ scattering [5]. From the study of $\pi\pi$ production processes, complementary information to the study of the single pion photoproduction channels can be gained. ChPT predicts that the $\pi^0\pi^0$ photoproduction channel is strongly enhanced due to chiral (pion) loops [6] which appear in leading (non vanishing) order q^3. This is a counter-intuitive result, since in the case of single pion production the cross sections for charged pions are considerably larger than the ones

---

[1] ChPT is used in this paper as a synonym for HBChPT

CP717, *Hadron Spectroscopy: Tenth International Conference*,
edited by E. Klempt, H. Koch, and H. Orth
© 2004 American Institute of Physics 0-7354-0197-7/04/$22.00

with neutral pions in the final state. In a calculation up to order M_π^2, the pion loops at order q^3 are responsible for two thirds of the total cross section [7]. This fact makes this channel unique, because unlike in other channels where the loops are adding some contribution to the dominant tree graphs, here they are absolutely dominating. In [7], the following prediction for the near threshold cross section was given:

$$\sigma_{tot}(E_\gamma) = 0.6 \text{ nb}((E_\gamma - E_\gamma^{thr})/10 \text{ MeV})^2 \tag{1}$$

where E_γ denotes the photon beam energy and E_γ^{thr} the production threshold of 308.8 MeV. The largest resonance contribution at order M_π^2 comes from the $P_{11}(1440)$ resonance via the $N^* N \pi \pi$ s-wave vertex. Actually, the uncertainty of the coupling of the $P_{11}(1440)$ to the s-wave $\pi\pi$ channel was a limiting factor for the accuracy of the ChPT calculation [7]. Therefore, for the most extreme case of this coupling, an upper limit for this cross was given in addition by increasing the constant in Eq. 1 from 0.6 nb to 0.9 nb.

Completing the overview of theoretical calculations of the reaction $\gamma p \to \pi^0 \pi^0 p$ close to threshold, this channel is also described in a recent version of the Gomez Tejedor-Oset model [8]. This model is based on a set of tree level diagrams including pions, nucleons and nucleonic resonances. In a recent work, particular emphasis was put on the re-scattering of pions in the iso-spin $I=0$ channel [9]. Double pion photoproduction via the Δ Kroll-Rudermann term is not possible for the $2\pi^0$ final state. In case of a $\pi^- \pi^+$ Kroll-Rudermann term, the charged pions can re-scatter into two neutral pions generating dynamically a $\pi\pi$ loop. This effect is doubling the cross section in the threshold region and is regarded by the authors as being a reminiscence of the explicit chiral loop effect described before.

In the past, two measurements of the reaction $\gamma p \to \pi^0 \pi^0 p$ below 450 MeV beam energy have been carried out [10, 11]. The second experiment showed an improvement in statistics by almost a factor 30. Nevertheless, in the threshold region the cross section still suffered from large statistical uncertainties (see Fig. 1).

Reaction Mechanisms in the second Resonance Region

Nucleon resonances are studied in a variety of experiments in an attempt to obtain information on the structure of the nucleon by comparison to quark model calculations. Most information has been gathered through πN scattering and π photoproduction. A complementary access is the double π production where the $2\pi^0$ channel turns out to be the most selective one. Because of the vanishing charge of the π^0, Born terms as well as direct production terms (Δ-Kroll-Rudermann, Δ-pion pole) are very much suppressed. Previously, two measurements of the reaction $\gamma p \to \pi^0 \pi^0 p$ were intensively studied in order to extract information on nucleon resonances. The MAMI results [11] were interpreted by the Valencia model [8] and gave a strong indication for a dominance for the $D_{13}(1520) \to \Delta\pi$. In a recent paper, the GRAAL collaboration reported on a measurement of the $2\pi^0$ channel from 650 MeV up to 1500 MeV [12]. These data were interpreted by an extention of the Laget-Murphy model [12]. Despite the bad coverage of the $P_{11}(1440)$ resonance with the incident beam energy, the authors emphasized that the

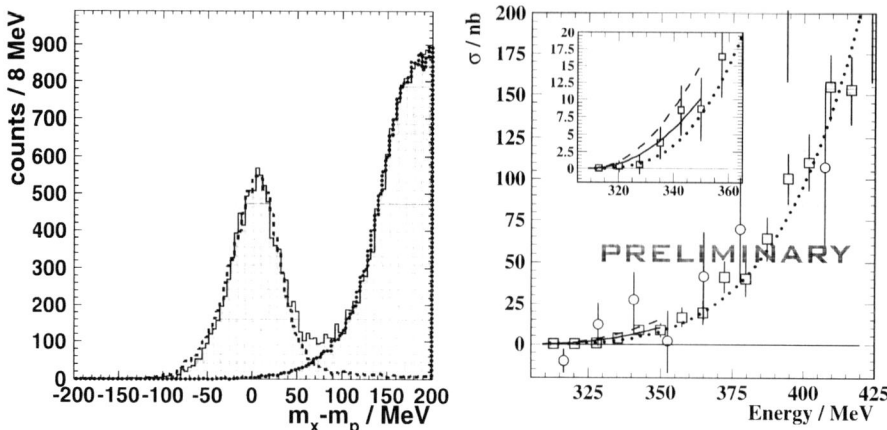

FIGURE 1. Left: Missing mass $M_X - m_p$ for two detected π^0 mesons for beam energies 780-820 MeV MeV (gray data, dashed $2\pi^0$ sim., dotted η sim.). Right: Total cross section for the reaction $\gamma p \to \pi^0 \pi^0 p$ (full squares) at threshold in comparison with [11] (open circles). The prediction of the ChPT calculation [7] is shown (solid curve) together with its upper limit (dashed curve) and the prediction of Ref. [9] (dotted curve).

data could be only explained by a dominance of the $P_{11}(1440) \to \sigma N$ reaction process. We present and discuss in this paper new and very precise invariant mass distributions of the $\pi^0 \pi^0$ and $\pi^0 p$ systems.

EXPERIMENTAL SETUP AND DATA ANALYSIS

The reaction $\gamma p \to \pi^0 \pi^0 p$ was measured at the electron accelerator Mainz Microtron (MAMI) [13, 14] using the Glasgow tagged photon facility [15, 16] and the photon spectrometer TAPS [17, 18]. The photon energy covered the range 285–820 MeV with an average energy resolution of 2 MeV. The TAPS detector consisted of six blocks each with 62 hexagonally shaped BaF_2 crystals arranged in an 8×8 matrix and a forward wall with 138 BaF_2 crystals arranged in a 11×14 rectangle. The six blocks were located in a horizontal plane around the target at angles of $\pm 54°$, $\pm 103°$ and $\pm 153°$ with respect to the beam axis. Their distance to the target was 55 cm and the distance of the forward wall was 60 cm. This setup covered $\approx 40\%$ of the full solid angle. The liquid hydrogen target was 10 cm long with a diameter of 3 cm. Further details of the experimental setup can be found in ref. [19].

The $\gamma p \to \pi^0 \pi^0 p$ reaction channel was identified by measuring the 4-momenta of the two π^0 mesons, whereas the proton was not detected. The π^0 mesons were detected via their two photon decay channel and identified in a standard invariant mass analysis from the measured photon momenta. Events were selected, were both of the two photon invariant masses fulfilled simultaneously the following cut: $110 MeV < m_{\gamma\gamma} < 150 MeV$. Furthermore, the mass M_X of a missing particle was calculated (see Fig. 1). In case of the reaction $\gamma p \rightarrow \pi^0 \pi^0 p$ the missing mass M_X must be equal to the mass of the (undetected) proton m_p. Above the η production threshold of 707 MeV, the $\eta \rightarrow 3\pi^0$ decay is a potential background source for the $2\pi^0$ channel via events where only two of the three π^0 mesons are detected by TAPS. A Monte Carlo simulation of the $2\pi^0$ and η reactions using GEANT3 [20] reproduces the line shape of the measured data. A cut corresponding to an interval of $-2.5\sigma \ldots min\{+2.5\sigma, 40MeV\}$ width of the simulated line shape has been applied to select the events of interest. The $\eta \rightarrow 3\pi^0$ background was estimated from these simulations to be below 2% for the highest beam energy of 820 MeV (compare Fig. 1) and was subtracted for the cross section determination. Background originating from random time coincidences between the TAPS detector and the tagging spectrometer was subtracted in the usual way, using events outside the prompt time coincidence window [16].

The cross section was deduced from the rate of the $2\pi^0$ events, the number of hydrogen atoms per cm^2, the photon beam flux, the branching ratio of the π^0 decay into two photons, and the detector and analysis efficiency. The geometrical detector acceptance and the analysis efficiency due to cuts and thresholds were obtained using the GEANT3 code and an event generator producing distributions of the final state particles according to phase space. The acceptance of the detector setup was studied by examining independently a grid of the four degrees of freedom for this three body reaction (azimuthal symmetry of the reaction was assumed). In a grid of total 1024 bins the acceptance is 100% for the beam below 410 MeV and above greater than 95% for the energy up to 820 MeV. The average value for the detection efficiency is 0.4%. The systematic errors of the efficiency determination are small, because the shape of the measured distributions are reproduced by the simulation. The systematic errors are estimated to be 8% and include uncertainties of the beam flux, the ätarget length and the efficiency determination.

RESULTS AND DISCUSSION

Chiral Perturbation Theory

The measured total cross section at threshold for the reaction $\gamma p \rightarrow \pi^0 \pi^0 p$ is shown in Fig. 1 as a function of the incident photon beam energy. The results are in agreement within the rather large error bars with a previous experiment [11]. The present data are compared with the prediction of the ChPT calculation [7] and is in agreement with it [21], although up to 20 MeV above threshold the data are somewhat lower than the ChPT prediction. The ChPT prediction using the upper limit of the coupling of the P$_{11}$(1440) to $(\pi\pi)_{s-wave}$ can be excluded. In the future, this might be exploited to establish a better constraint on this coupling by using our result as an input. The total cross section is also compared to the calculation with the chiral unitary model [9] and shows a good agreement with this latter calculation.

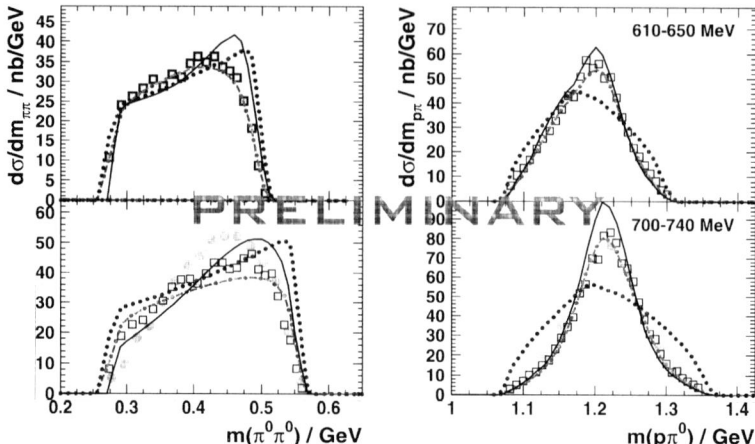

FIGURE 2. Invariant mass of $\pi^0\pi^0$ and $\pi^0 p$ for different bins of beam energy (full squares). The GRAAL data is shown by the full circles. The curves show σN phase space (dotted), $\Delta\pi$ phase space (dashed dotted) and the model calculation [8] (full curve).

Reaction Mechanisms in the second Resonance Region

The invariant mass distributions for two beam energies are shown in Fig. 2. They are compared to a $\Delta\pi$ phase space and a σN phase space simulations and to the Valencia model calculation [8]. For the σ a Breit-Wigner with a pole and a width of 800 MeV according to the Laget-Murphy model was assumed [22]. The GRAAL data around 720 MeV beam energy agrees very well with our $m_{\pi^0 p}$ data, whereas in the case of $m_{\pi^0 \pi^0}$, the agreement is worse. In the $m_{\pi^0 p}$ distributions, the $\Delta\pi$ intermediate state dominates starting already at 600 MeV beam energy. The other phase space distributions can not describe the data. In the $m_{\pi^0 \pi^0}$ distributions the differences between the different reaction processes is much less discriminative. The dominance of the $\Delta\pi$ intermediate state in the $2\pi^0$ production channel seems to be the more obvious explanation, although no interference effects are taken into account in this simplified comparisons. This observation is in contradiction to the claimed σN dominance in the Laget Model [12]. A future partial wave analysis has to clarify this discrepancy, and the presented data will provide strong constraints for solutions in the second resonance region.

Summary

In summary we have measured the total and differential cross sections for the reaction $\gamma p \to \pi^0\pi^0 p$. The prediction of the ChPT calculation [7] is in agreement with our measured data [21]. The upper limit quoted for this prediction can be excluded. This finding might be exploited for a better constraint on the $P_{11}(1440)$ to s-wave $\pi\pi$ coupling. Further on, the cross section is also well reproduced by another calculation [9], where pion loops are dynamically generated.

Secondly, invariant mass distributions for $\pi^0\pi^0$ and $p\pi^0$ are presented. In the second energy region, σN phase space and hence a dominant contribution of the process $P_{11}(1440) \to \sigma p$ seems to be unlikely. The differential cross sections are very well described with a $\Delta\pi$ intermediate state. This is supported by the Valencia model, where the dominant contribution stems from the $D_{13}(1520) \to \Delta\pi$ process and is in contradiction to the Laget model [12]. In a future partial wave analysis this data could provide strong limitations on the resonance parameters up to the second resonance region.

ACKNOWLEDGMENTS

We thank the accelerator group of MAMI as well as many other scientists and technicians of the Institut fuer Kernphysik at the University of Mainz for the outstanding support. This work was supported by Schweizerischer Nationalfond, DFG Schwerpunktprogramm: "Untersuchung der hadronischen Struktur von Nukleonen und Kernen mit elektromagnetischen Sonden", SFB221, SFB443 and the UK Engineering and Physical Sciences Research Council.

REFERENCES

1. Weinberg, S., *Physica (Amsterdam)*, **96A**, 327 (1979).
2. Gasser, J., and Leutwyler, H., *Annals Phys.*, **158**, 142 (1984).
3. Jenkins, E., and Manohar, A., *Phys. Lett. B*, **255**, 558 (1991).
4. Bernard, V., et al., *Nucl. Phys. B*, **383**, 442 (1992).
5. Fettes, N., and Meissner, U.-G., *Nucl. Phys. A*, **676**, 311 (2000).
6. Bernard, V., et al., *Nucl. Phys. A*, **580**, 475–499 (1994).
7. Bernard, V., Kaiser, N., and Meissner, U., *Phys. Lett. B*, **382**, 19–23 (1996).
8. Tejedor, J. G., and Oset, E., *Nucl. Phys. A*, **600**, 413 (1996).
9. Roca, L., Oset, E., and Vacas, M. V., *Phys. Lett. B*, **541**, 77–86 (2002).
10. Haerter, F., et al., *Phys. Lett. B*, **401**, 229 (1997).
11. Wolf, M., et al., *Eur. Phys. J. A 9 (2000) 5-8* (2000).
12. Assafiri, Y., et al., *Phys. Rev. Lett.*, **90**, 222001 (2003).
13. Walcher, T., *Prog. Part. Nucl. Phys.*, **24**, 189–203 (1990).
14. Ahrens, J., et al., *Nucl. Phys. News*, **4**, 5–15 (1994).
15. Anthony, I., et al., *Nucl. Instr. Meth.*, **A 301**, 230–240 (1991).
16. Hall, S., et al., *Nucl. Instr. Meth.*, **A 368**, 698 (1996).
17. Novotny, R., *IEEE Trans. Nucl. Sci.*, **38**, 379–385 (1991).
18. Gabler, A., et al., *Nucl. Instr. Meth.*, **A 346**, 168–176 (1994).
19. Kotulla, M., *Prog. Part. Nucl. Phys.*, **50/2**, 295–303 (2003).
20. Brun, R., et al., *GEANT3 Users Guide*, CERN, Data Handling Division DD/EE/84-1 (1986).
21. Kotulla, M., et al., *Phys. Lett. B*, **accepted** (2003).
22. Murphy, L., and Laget, J.-M., *DAPHIA-SPhN-96-10* (1996).

Measurement of the Spin Rotation Parameter A in the Elastic Pion-proton Scattering at 1.43 GeV/c

I.G. Alekseev*, N.A. Bazhanov†, P.E. Budkovsky*, E.I. Bunyatova†,
V.P. Kanavets*, A.I. Kovalev**, L.I. Koroleva*, S.P. Kruglov**,
B.V. Morozov*, V.M. Nesterov*, D.V. Novinsky**, V.V. Ryltsov*,
V.A. Shchedrov**, A.D. Sulimov*, V.V. Sumachev**, D.N. Svirida*,
V.Yu. Trautman** and V.V. Zhurkin*

*Institute for Theoretical and Experimental Physics, Moscow, 117218, Russia
†Joint Institute for Nuclear Research, Dubna, Moscow district, 141980, Russia
**Petersburg Nuclear Physics Institute, Gatchina, Leningrad district, 188300, Russia

Abstract. The ITEP-PNPI collaboration presents new results of the measurements of the spin rotation parameter A in the elastic scattering of negative pions on protons at $P_{beam} = 1.43$ GeV/c. The results are compared to the predictions of the different partial wave analyses. The experiment was performed at the ITEP proton synchrotron, Moscow.

INTRODUCTION

We present the new data on spin rotation parameter A in the $\pi^- p$ elastic scattering at 1.43 GeV/c. This experiment is the latest one from the series of the polarization parameters measurements in the resonance region performed by ITEP-PNPI collaboration in last decade. The momentum range $(0.8 - 2.1)$ GeV/c available with our beam-line contains nearly 65 % of the known light quark resonances. There are three clusters of resonances in this region corresponding to peaks in the total cross section. The main goal of our experiment is to obtain the necessary information for the *unambiguous reconstruction* of the pion-proton elastic scattering amplitudes by partial wave analyses (PWA).

The physical interest to the spectroscopy of the light baryons is based on several experimental observations that do not fit well the constituent quark models.

They are:

- the existence of the resonance clusters with masses 1.7 and 1.9 GeV/c^2;
- the presence of the negative parity resonances in the peak at $\sqrt{s} = 1.9$ GeV;
- possible existence of the parity doublets in the peaks at $\sqrt{s} = 1.7$ and 1.9 GeV;
- "missing resonances" near $\sqrt{s} = 2$ GeV.

The status of the modern experimental light baryon spectroscopy is not fully satisfactory. The resonances presented in the present-day issue of PDG are based mainly

CP717, *Hadron Spectroscopy: Tenth International Conference,*
edited by E. Klempt, H. Koch, and H. Orth

on the two PWA: **CMB80** [2] and **KH80** [1]. The both were performed more than two decades ago. However more recent analyses by **VPI** group [3] did not reveal the resonances $D_{13}(1700), S_{31}(1900), P_{33}(1920)$ and $D_{33}(1940)$. In the recent years **GWU-VPI** group continues the PWA development [4]. Meanwhile the precise and detailed data on asymmetry in $\pi^- p$ elastic scattering [5] and several measurements of the spin rotation parameters in $\pi^+ p$ elastic scattering [6] were performed by ITEP-PNPI collaboration.

FORMALISM

The meaning of the A and R spin rotation measuremets is clear from expression of the observables via transverse amplitudes f^+ and f^-.
They are:

$$
\begin{aligned}
\sigma &= |f^+|^2 + |f^-|^2 \\
P \cdot \sigma &= |f^+|^2 - |f^-|^2 \\
A \cdot \sigma &= \mathrm{Re}(f^+ f^{-*}) \cdot \sin(\theta_{cm} - \theta_{lab}) - \mathrm{Im}(f^+ f^{-*}) \cdot \cos(\theta_{cm} - \theta_{lab}) \\
R \cdot \sigma &= \mathrm{Re}(f^+ f^{-*}) \cdot \cos(\theta_{cm} - \theta_{lab}) + \mathrm{Im}(f^+ f^{-*}) \cdot \sin(\theta_{cm} - \theta_{lab}),
\end{aligned}
\tag{1}
$$

the polarisation parameters obeying the relation:

$$
P^2 + A^2 + R^2 = 1
\tag{2}
$$

In turn the transverse amplitudes correspond to the amplitudes of the scattering matrix $M = g + ih(\vec{\sigma} \cdot \vec{n})$ by simple relations: $f^+ = g + ih, \quad f^- = g - ih$. It follows from (1), (2) that differential cross section and normal polarization allow us to reconstruct the *absolute values* of the transverse amplitudes, while their *relative phase* may be obtained only by measurement of the spin rotation parameters.

EXPERIMENT LAYOUT

The spin rotation parameter A is determined by the measurement of the polarization of the recoiled proton from the proton target polarized along the pion beam. In this case the transverse component P_v of the recoiled proton polarization in the scattering plane simply relates to the A parameter: $P_v = A \cdot P_t$, where P_t is the target polarization. The P_v component is determined through the measurement of the azimuthal asymmetry of the recoiled proton scattering on carbon nuclei.

The details of the experimental setup could be found in [6]. The main elements are: longitudinally polarized proton target inside the super-conductive solenoid, thick filter carbon polarimeter, sets of the wire chambers for the tracking of the incident and scattered particles and TOF system for the identification of the beam particles.

TABLE 1. Spin Rotation Parameter A at 1.43 GeV/c in $\pi^- p$ Elastic Scattering.

θ_{cm}, **deg.**		A
range	mean	
155.0 – 162.2	160.4	-0.26 ± 0.22
162.2 – 165.6	163.9	-0.46 ± 0.23
165.6 – 172.0	167.4	-0.14 ± 0.23

DATA PROCESSING

The steps of the data processing were: elastic scattering events selection, the selection of the proton-carbon scattering in angular interval $(3 - 20)^o$ and determination of the A and P parameters by the maximum likelyhood method. For the elastic events selection a single χ^2 criterion accounting for complanarity and kinematic correlation was used. Resulting sample has (6-8) % of background contribution. The maximum likelihood function accounted for the target polarization, polarimeter analyzing power, the fraction and polarization of the quasielastic background and variable parameters A and P. The overall statistics had 6777 selected events.

RESULTS

The main source of systematic errors in A parameter is the uncertainty in the polarimeter analyzer power. To minimize these errors we performed a dedicated experiment on the measurement of the proton-carbon analyzing power [7]. It resulted in analyzing power precision better than 3 % and small level of the fake asymmetry. To suppress the contribution of the fake asymmetry to the A parameter errors the target polarization sign was periodically reversed.

The measurements were performed in the region of backward scattering angles where the predictions of the various PWA have maximum discrepancy. The results of the experiment are presented in table 1 and shown in figure 1 together with the various PWA predictions. The data on A parameter agree well with **FA02** (the latest by **GWU-VPI** group) predictions and deviate from **KH80** predictions on three standard errors level.

We conclude that **CMB80** and partly **KH80** analyses do not reconstruct properly the relative phases of the transverse amplitudes for the scattering to the backward hemisphere. This conclusion is supported by previous measurements of the spin rotation parameters in $\pi^\pm p$ scattering at 1.00 and 1.62 GeV/c. These results impose serious doubts about the present-day spectrum and properties of the light baryon resonances.

ACKNOWLEDGMENTS

Our thanks to professor G. Höehler for interesting and fruitful discussion. We are grateful to the staff of ITEP accelerator for excellent beam quality.

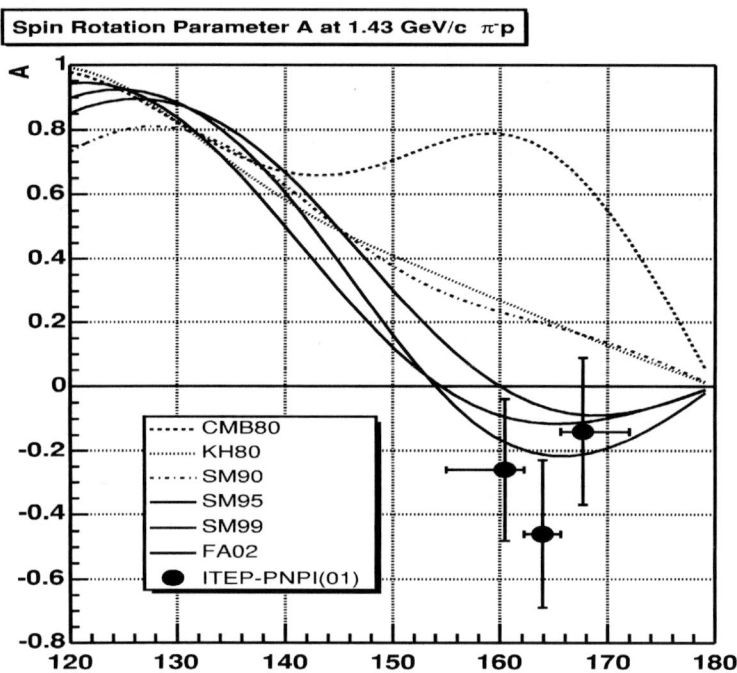

FIGURE 1. Spin Rotation Parameter $A(\theta_{cm})$ at 1.43 GeV/c in $\pi^- p$ Elastic Scattering.

This work was partly supported by Russian Fund for Basic Research grant 02-02-16121 and Russian Program "Fundamental Nuclear Physics".

REFERENCES

1. G. Hoehler, *Handbook of Pion-Nucleon Scattering, Physics Data*, No. **12-1**, Fachinformationzentrum, Karlsruhe, 1979.
2. R.E. Cutcosky et al., *Phys. Rev.* **D20** (1979), 2839.
3. R.A. Arndt et al., *Phys. Rev.* **C52** (1995), 2120.
4. http://gwdac.phys.gwu.edu/analysis/pin_analysis.html.
5. I.G. Alekseev et al., *Nucl. Phys.* **B348** (1991), 257.
6. I.G. Alekseev et al., *Phys. Lett.* **B351** (1995), 585;
   I.G. Alekseev et al., *Phys. Lett.* **B485** (2000), 32;
   I.G. Alekseev et al., *Eur. Phys. J.* **A12** (2001), 117.
7. I.G. Alekseev et al., *Nucl. Instr. Meth.* **A434** (1999), 254.

Energy dependence of the Λ/Σ^0 production cross section ratio in p–p interactions.

P. Kowina*, H.-H. Adam†, A. Budzanowski**, R. Czyżykiewicz‡,
D. Grzonka*, M. Janusz§, L. Jarczyk§, B. Kamys§, A. Khoukaz†,
K. Kilian*, P. Moskal*, W. Oelert*, C. Piskor-Ignatowicz§, J. Przerwa§,
T. Rożek*, R. Santo†, G. Schepers*, T. Sefzick*, M. Siemaszko¶,
J. Smyrski§, A. Täschner†, P. Winter*, M. Wolke*, P. Wüstner‖ and
W. Zipper¶

*IKP, Forschungszentrum Jülich, D-52425 Jülich, Germany
†IKP, Westfälische Wilhelms–Universität, D-48149 Münster, Germany
**H. Niewodniczański Institute of Nuclear Physics, PL-31-342 Cracow, Poland
‡M. Smoluchowski Institute of Physics, Jagellonian University, PL-30-059 Cracow, Poland,IKP,
Forschungszentrum Jülich, D-52425 Jülich, Germany
§M. Smoluchowski Institute of Physics, Jagellonian University, PL-30-059 Cracow, Poland
¶Institute of Physics, University of Silesia, PL-40-007 Katowice, Poland
‖ZEL, Forschungszentrum Jülich, D-52425 Jülich, Germany

Abstract. Measurements of the near threshold Λ and Σ^0 production via the $pp \rightarrow pK^+\Lambda/\Sigma^0$ reaction at COSY–11 have shown that the Λ/Σ^0 cross section ratio exceeds the value at high excess energies ($Q \geq 300$ MeV) by an order of magnitude. For a better understanding additional data have been taken between 13 MeV and 60 MeV excess energy.
Within the first 20 MeV excess energy a strong decrease of the cross section ratio is observed, with a less steep decrease in the higher excess energy range.
A description of the data with a parametrisation including $p - Y$ final state interactions suggests a much smaller $p - \Sigma^0$ FSI compared to the $p - \Lambda$ system.

INTRODUCTION

At the COSY–11 facility [1] measurements of the Λ and Σ^0 hyperon production were performed in the $pp \rightarrow pK^+\Lambda$ and $pp \rightarrow pK^+\Sigma^0$ reactions close to threshold [2] resulting in a cross section $\sigma(\Lambda)$ for the Λ production which is more then one order of magnitude larger than the cross section $\sigma(\Sigma^0)$ for the Σ^0 production.

Since the quark contents of these two hyperons are the same, based on the isospin relations only, the ratio of the cross sections $\mathscr{R} = \sigma(\Lambda)/\sigma(\Sigma^0)$ should be equal to three. In fact at high excess energies [3] a ratio of ~ 2.5 is observed in contrarry to the by more then one order of magnitude larger $\mathscr{R}$ at threshold.

In order to understand this behavior measurements [4] in the intermediate energy range (i.e. 13 MeV $\leq Q \leq$ 60 MeV), where the ratio $\mathscr{R}$ was expected to decrease from ~ 28 to ~ 2.5, have been performed.

CP717, Hadron Spectroscopy: Tenth International Conference,
edited by E. Klempt, H. Koch, and H. Orth
© 2004 American Institute of Physics 0-7354-0197-7/04/$22.00

EXPERIMENT

The measurements of the hyperon production were performed at the COSY–11 facility [1, 5] (see figure 1) at the Cooler Synchrotron COSY-Jülich [6].

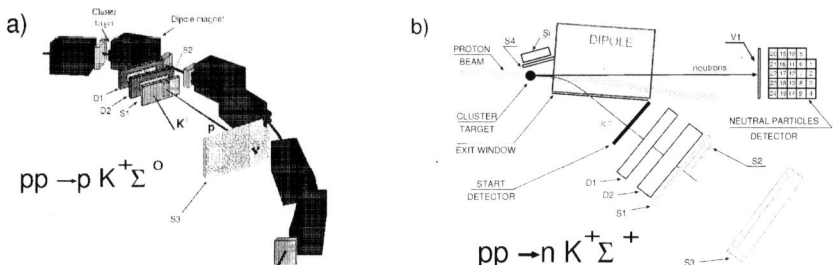

FIGURE 1. a) COSY–11 facility. b) Setup extended by the start and neutral particle detectors used in the measurements of the $pp \rightarrow nK^+\Sigma^+$ reaction.

One of the regular COSY dipole magnets serves as a magnetic spectrometer with a H_2 cluster beam target [7] installed in front of it. The interaction between a proton of the beam with a proton of the cluster target may lead to the production of neutral hyperons (Σ^0, Λ) via the reactions $pp \rightarrow pK^+\Lambda(\Sigma^0)$.

Events of the $pK^+\Lambda(\Sigma^0)$ production are selected by the detection of both positively charged particles in the exit channel (i.e. proton and K^+). The unobserved neutral particle is identified via the missing mass method.

Positively charged ejectiles are directed from the circulating beam by the magnetic field of the dipole towards the inner part of the COSY ring, where they are registered in a set of two drift chambers D1 and D2 for the track determination. Their momenta are reconstructed by tracking back the particles through the well known magnetic field to the assumed interaction point. The velocities of the ejectiles are given by a measurement of the time of flight between the S1(S2) start and the S3 stop scintillator hodoscopes from which in combination with the momentum the invariant mass of the particle is given. Therefore, the four-momentum vectors for all positively charged particles are known and the four-momentum of the unobserved neutral hyperon is uniquely determined.

To avoid systematical uncertainties as much as possible, COSY was operated in the "supercycle mode" i.e. the beam momenta were changed between the cycles, such that for example 10 cycles with a beam momentum corresponding to the excess energy $Q = 20\,\mathrm{MeV}$ above the Σ^0 threshold were followed by one cycle with the same Q above the Λ production threshold. The ratio of the number of the cycles was chosen inversely proportional to the ratio of the expected cross sections for the Λ and Σ^0 production. Thus, both cross sections were measured under the same conditions and possible changes in the detection system did not influence the data taking procedure, especially for the determination of the cross section ratio.

The extention of the detection system by an additional neutral particle detector (see figure 1b) allows for the measurements of neutrons in the exit channel, what is needed particularly in the $pp \rightarrow nK^+\Sigma^+$ reaction. To extend the acceptance an additional start detector for K^+ was installed in the system.

RESULTS

The hyperon production via $pp \rightarrow pK^+\Lambda(\Sigma^0)$ has been studied in the excess energy range between 13 and 60 MeV. In figure 2a) and figure 2b) the excitation functions and the energy dependence of the cross section ratio are shown, respectively. The most drastic decrease of the cross section ratio is observed between 10 MeV and 20 MeV following by a less steep decrease at higher Q-values.

The first published close–to–threshold data [2] have triggered many theoretical discussions. The results of available calculations are shown in figure 2b) and are briefly discussed in the following section.

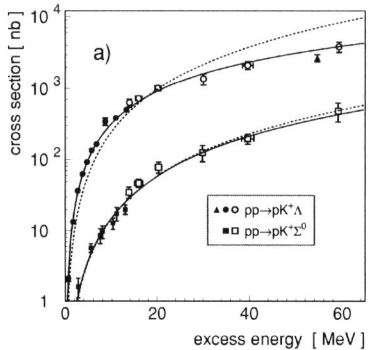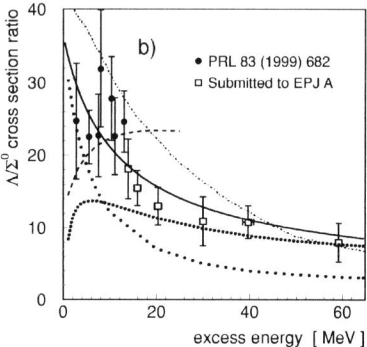

FIGURE 2. a) Total cross sections for the $pp \rightarrow pK^+\Lambda$ and $pp \rightarrow pK^+\Sigma^0$ production (full symbols [2, 8], open symbols [4] and triangle [9]). b) Cross section ratio for the $pp \rightarrow pK^+\Lambda$ and $pp \rightarrow pK^+\Sigma^0$ reactions.

Comparison with theoretical predictions

Presently different theoretical calculations with various dominant production mechanisms are available which reproduce at least the trend of the data, see figure 2b).

Calculations by Sibirtsev, Tsushima et al. [10, 11] were performed within two different models. In the first one – Boson Exchange model – (dense dotted line) pion and Kaon exchange is considered as the most important mechanism of the hyperon production [12]. The second model (dotted line) bases on the assumption that the hyperon is produced in the decay of N^* resonances excited via the exchange of π, η and ρ mesons [11, 13].

Hyperon production via N^* resonances was also investigated by Shyam et al. [14] (dashed-dotted) where $N^*(1650)$, $N^*(1710)$, $N^*(1720)$ were assumed to be excited by the exchange of the π, ρ, ω and σ mesons. The authors state that, at least close to the threshold, the dominant contribution to the hyperon production is the $N^*(1650)$ resonance.

Gasparian et al. [15] performed calculations within the Jülich Meson Exchange Model, where π and K-exchange was assumed including the interference between these two amplitudes (dashed line). It is found by the authors, that in the case of the Λ production K-exchange is dominant and constructive or destructive interference between π and K-exchange give similar results. For the Σ^0 production the strength of the contributions from π and K exchange are comparable resulting in a strong reduction of the Σ^0 production with a destructive interference by which the observed cross sections at threshold are reproduced. Within their calculations [15] the energy dependence of the cross section for other isospin channels are predicted like the Σ^+ production in the reaction $pp \rightarrow nK^+\Sigma^+$. Here, the predicted behavior of the cross sections for destructive and constructive π and K-exchange is opposite to that observed in Σ^0 production. For a destructive interference the cross section for $pp \rightarrow nK^+\Sigma^+$ is expected to be a factor of three higher and for constructive interference a factor of three lower than the cross section for $pp \rightarrow pK^+\Sigma^0$.

Data in the other isospin channels will help to extract the dominant mechanisms in the threshold hyperon production.

Measurements of the $pp \rightarrow nK^+\Sigma^+$ have been already performed at the COSY–11 facility. The data are presently under analysis [16].

Effective range parameters

The final state interactions of a two body subsystem in a 3-body final state like pK^+Y influence the excitation function in the threshold region and its analysis allows to extract information on the effective range parameters (for review see ref. [17]). A parametrisation of the cross section which relates the shape of the threshold behavior to the effective range parameters is e.g. given by Fäldt and Wilkin [18]:

$$\sigma = const \cdot \frac{V_{ps}}{F} \cdot \frac{1}{\left(1 + \sqrt{1 + \frac{Q}{\varepsilon'}}\right)^2} = C' \cdot \frac{Q^2}{\sqrt{\lambda(s,m_p^2,m_p^2)}} \cdot \frac{1}{\left(1 + \sqrt{1 + \frac{Q}{\varepsilon'}}\right)^2}. \quad (1)$$

The phase space volume V_{ps} and the flux factor F are given by [19]:

$$V_{ps} = \frac{\pi^3}{2} \frac{\sqrt{m_p\, m_{K^+}\, m_Y}}{(m_p + m_{K^+} + m_Y)^{\frac{3}{2}}} Q^2, \qquad F = 2\,(2\pi)^{3n-4}\,\sqrt{\lambda(s,m_p^2,m_p^2)}. \quad (2)$$

with the triangle function $\lambda(x,y,z) = x^2 + y^2 + z^2 - 2xy - 2yz - 2zx$.

The results of χ^2 fits using the Fäldt and Wilkin formula are presented in figure 2a) by the solid lines (the dotted lines correspond to pure S-wave phase space distributions). The parameter ε', which is related to the strength of the $p - Y$ final state interaction, and the normalization constant C' were extracted by the fits performed for each reaction separately resulting in:

$$C'(\Lambda) = (98.2 \pm 3.7)\ \text{nb/MeV}^2 \qquad \varepsilon'(\Lambda) = (5.51\,^{+0.58}_{-0.52})\ \text{MeV}$$

$$C'(\Sigma^0) = (2.97 \pm 0.27)\ \text{nb/MeV}^2 \qquad \varepsilon'(\Sigma^0) = (133\,^{+108}_{-44})\ \text{MeV}.$$

Assuming only S-wave production, the $p - \Lambda(\Sigma^0)$ systems can be described using the Bergman potentials [20], where scattering length $\hat{a}$ and effective range $\hat{r}$ are given by:

$$\hat{a} = \frac{\alpha + \beta}{\alpha\beta}, \qquad \hat{r} = \frac{2}{\alpha + \beta}, \tag{3}$$

with a shape parameter β, and $\varepsilon' = \alpha^2/2\mu$ where μ is the reduced mass of the $p - Y$ system [20]. The negative value of α is chosen since (at least for $p - \Lambda$) an attractive interaction is expected [21, 22].

The parameters $\hat{a}$ and $\hat{r}$ are interdependent (see Eq. (3)) and only a correlations between them can be deduced. In figure 3 the correlations obtained for the $p - \Sigma^0$ and $p - \Lambda$ systems are presented by solid and dashed lines, respectively. The errors in ε' are reflected in the error ranges and shown in the figure by the thinner lines. The cross symbol represents the averaged value of the $p - \Lambda$ effective range parameters extracted from a FSI approach in threshold Λ production [23].

It seems that the $p - \Sigma^0$ FSI are much smaller than the FSI for $p - \Lambda$ system.

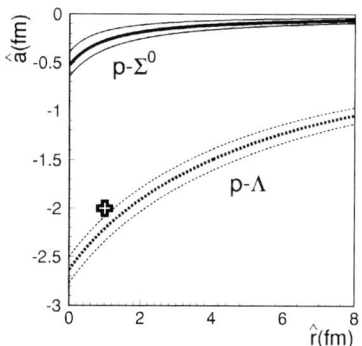

FIGURE 3. Correlation between the $p - \Sigma^0$ (solid lines) and $p - \Lambda$ (dashed lines) effective range parameters.

SUMMARY

The measurements of the energy dependence of the total cross sections for the $pp \rightarrow pK^+\Lambda$ and $pp \rightarrow pK^+\Sigma^0$ production performed at COSY–11 facility in the excess energies between 14 and 60 MeV showed that the cross section ratio strongly decreases in the excess energy range between 10 and 20 MeV.

Different theoretical models are able to describe the data within a factor of two with more or less the same quality, even though they differ in the dominant contribution to the production mechanism. Data in the hyperon sector are still too limited to put constrains on the existing models.

The new data suggest that the final state interactions in the $p - \Sigma^0$ channel are much weaker than in the case of the $p - \Lambda$ system.

Measurements of the hyperon production in the other isospin channels like $pp \rightarrow nK^+\Sigma^+$ reaction measured at COSY–11 will help to disentangle the production mechanisms in the threshold region.

REFERENCES

1. Brauksiepe, S., et al., *Nucl. Instr. & Meth.*, **A376**, 397–410 (1996).
2. Sewerin, S., et al., *Phys. Rev. Lett.*, **83**, 682–685 (1999).
3. Baldini, A., et al., *Total Cross–Sections for Reactions of High–Energy Particles*, Landolt–Börnstein, *New Series I/12*, Springer, Berlin, 1988.
4. Kowina, P., et al., *Submited to EPJ A* (2003).
5. Moskal, P., et al., *Nucl. Instr. & Meth.*, **A466**, 448–455 (2001).
6. Maier, R., *Nucl. Instr. & Meth.*, **A390**, 1–8 (1997).
7. Dombrowski, H., et al., *Nucl. Instr. & Meth.*, **A386**, 228–234 (1997).
8. Balewski, J., et al., *Phys. Lett.*, **B420**, 211–216 (1998).
9. Bilger, A., et al., *Phys. Lett.*, **B420**, 217–224 (1998).
10. Sibirtsev, A., et al., *e-Print Archive, nucl-th/0004022* (2000).
11. Tsushima, K., Sibirtsev, A., and Thomas, A. W., *Phys. Rev.*, **C59**, 369–387 (1999), erratum-ibid.C61:029903,2000.
12. Sibirtsev, A., *Phys. Lett.*, **B359**, 29–32 (1995).
13. Tsushima, K., Sibirtsev, A., and Thomas, A. W., *Phys. Lett.*, **B390**, 29–35 (1997).
14. Shyam, R., Penner, G., and Mosel, U., *Phys. Rev.*, **C63**, 022202 (2001).
15. Gasparian, A., et al., *Phys. Lett.*, **B480**, 273–279 (2000).
16. Rożek, T., and Grzonka, D., *COSY Proposal*, **117** (2002).
17. Moskal, P., Wolke, M., Khoukaz, A., and Oelert, W., *Prog. Part. Nucl. Phys.*, **49**, 1–90 (2002).
18. Fäldt, G., and Wilkin, C., *Z. Phys.*, **A357**, 241–243 (1997).
19. Byckling, E., and Kajantie, K., *Particle Kinematics*, John Wiley & Sons Ltd., 1973, iSBN 0 471 12885 6.
20. Newton, R. G., *Scattering Theory of Waves and Particles*, Springer-Verlag, New York, 1982.
21. Holzenkamp, B., Holinde, K., and Speth, J., *Nucl. Phys.*, **A500**, 485–528 (1989).
22. Rijken, T. A., Stoks, V. G. J., and Yamamoto, Y., *Phys. Rev.*, **C59**, 21–40 (1999).
23. Balewski, J., et al., *Eur. Phys. J.*, **A2**, 99–104 (1998).

Analysis of the η meson production mechanism via the $\vec{p}p \to pp\eta$ reaction

R. Czyżykiewicz[*], H.-H. Adam[†], A. Budzanowski[**], D. Grzonka[‡],
M. Janusz[§], L. Jarczyk[§], B. Kamys[§], A. Khoukaz[†], K. Kilian[‡], P. Kowina[‡],
P. Moskal[‡], W. Oelert[‡], C. Piskor-Ignatowicz[§], J. Przerwa[§], T. Rożek[‡],
R. Santo[†], G. Schepers[‡], T. Sefzick[‡], M. Siemaszko[¶], J. Smyrski[§],
A. Täschner[†], P. Winter[‡], P. Wüstner[∥] and W. Zipper[¶]

[*]M. Smoluchowski Institute of Physics, Jagellonian University, PL-30-059 Cracow, Poland,IKP,
Forschungszentrum Jülich, D-52425 Jülich, Germany
[†]IKP, Westfälische Wilhelms–Universität, D-48149 Münster, Germany
[**]H. Niewodniczanski Institute of Nuclear Physics, PL-31-342 Cracow, Poland
[‡]IKP, Forschungszentrum Jülich, D-52425 Jülich, Germany
[§]M. Smoluchowski Institute of Physics, Jagellonian University, PL-30-059 Cracow, Poland
[¶]Institute of Physics, University of Silesia, PL-40-007 Katowice, Poland
[∥]ZEL, Forschungszentrum Jülich, D-52425 Jülich, Germany

Abstract. Polarisation observables constitute a powerful tool for establishing the production mechanism of the η meson and for infering the presence of higher partial waves in the final system. Measurements of the proton analysing power for the $\vec{p}p \to pp\eta$ reaction have been performed by the COSY-11 group at three different excess energies: Q=10, 37 and 40 MeV. Data at Q=40 MeV indicate that the η meson is probably produced in partial waves higher than s wave.

INTRODUCTION

Despite the fact that the discovery of the η meson took place over fourty years ago [1], its production mechanism still remains an open question. Based on the close-to-threshold total cross section measurements for the $pp \to pp\eta$ reaction [2, 3, 4, 5, 6, 7], investigations on differential cross sections for this reaction [8, 9, 10, 11, 12, 13, 14, 15] and recently performed measurements of the proton analysing power for the $\vec{p}p \to pp\eta$ reaction [16, 17], there is a consensus that in the NN collisions the η meson is produced in a two-step process, where in the first stage exchange of one of the pseudoscalar or vector mesons excites the S_{11} resonance and subsequently this resonance decays into a Nη pair. The $S_{11}(1535)$ resonance seems to play an important role as an intermediate state since it has a large width covering the threshold energy for the $pp \to pp\eta$ reaction, and it couples strongly to the Nη system with a branching ratio corresponding to 30-55 % [18].

Although there exists a variety of models based on different assumptions concerning the mechanism of the η production, these models are in quite good agreement with the existing data on close-to-threshold total cross sections for the $pp \to pp\eta$ reaction [19]. As it was shown in reference [20], the excitation function for the $pp \to pp\eta$ process

CP717, *Hadron Spectroscopy: Tenth International Conference,*
edited by E. Klempt, H. Koch, and H. Orth
© 2004 American Institute of Physics 0-7354-0197-7/04/$22.00

can be equally well described by the intermediate excitation of the S_{11} via the exchange of either pseudoscalar or vector mesons. This implies that more limitations have to be added to the models in order to extract the way the η meson is really being created. One solution would be the verification of different models by means of the polarisation observables. At present there exist two models that predict the energy dependence of the proton analysing power for the $\vec{p}p \rightarrow pp\eta$ reaction [20, 21]. Theoretical forecasts for the two different excess energies Q=15 MeV and Q=40 MeV are shown in fig.1. There are significant differences between the models visible in both: the relative sign (at lower Q values) and in the magnitude of the proton analysing power. Measurements of this observable might therefore help to establish the adequacy of mechanisms of η meson production.

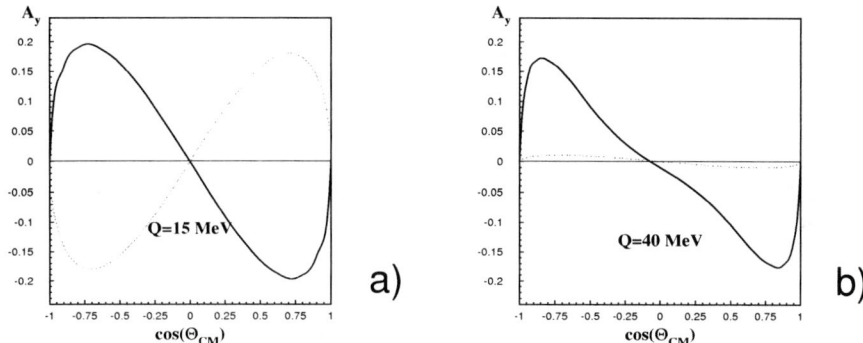

FIGURE 1. Predictions of the angular dependence for proton analysing power at (a) Q=15 MeV and (b) Q=40 MeV. The solid lines are the predictions of reference [20], whereas the dotted lines represent the calculations according to reference [21]. θ_{CM} is the centre-of-mass polar angle of the η meson.

EXPERIMENTS

There have been, so far, performed three measurements of proton analysing power for the $\vec{p}p \rightarrow pp\eta$ reaction at different excess energies. All these experiments made use of the COSY-11 installation [22], which is an internal detection setup mounted inside the COSY ring[1] [23] in the Research Centre Jülich. Experiments were performed in three separate runs in January '01 (at excess energy Q=40 MeV), September '02 (Q=37 MeV) and April '03 (Q=10 MeV). Data taken during the first run have been analysed and published [16] whereas the data from the two last runs await the full analysis. In the following the results from measurement at the excess energy Q=40 MeV will be presented as well as some preliminary estimations for the other runs.

The mentioned COSY-11 detection setup is displayed schematically in fig. 2a. Due to the short lifetime, the η meson cannot be registered in any of the detectors available at COSY-11. Its identification, however, can be performed using the missing mass

---

[1] Accelerator's name 'COSY' is an acronym of COoler SYnchrotron.

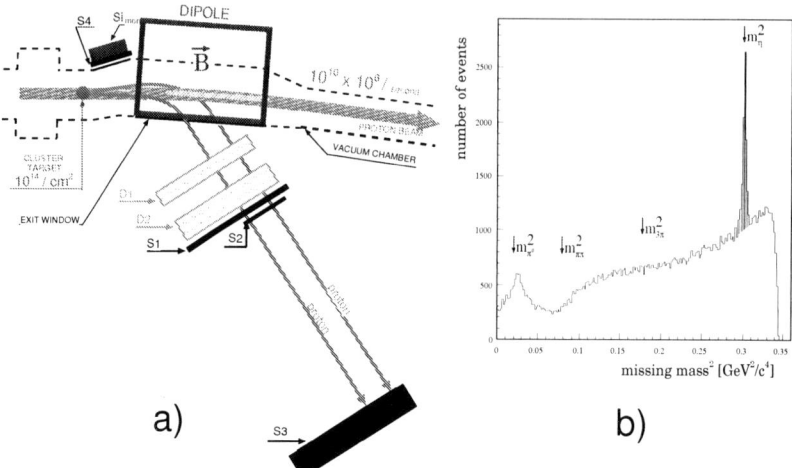

a)

b)

FIGURE 2. (a) Schematic view of the COSY-11 detection setup. D1 and D2 denote the two stacks of drift chambers with 6 and 8 detection planes respectively. S1, S2, S3 and S4 are the scintillator detectors. The time of flight is measured between the S1 scintillator hodoscope and the S3 scintillation wall. Having known the velocity of the particle from the time of flight measurement and the particle's momentum from the track reconstruction in the magnetic field inside the dipole magnet one is able to reconstruct the particle's mass. Si_{mon} are the silicon detectors used for the detection of elastically scattered protons. For more details the reader is refered to [22, 24]. (b) Square of the missing mass for events with two protons in the final state. Values indicated by arrows are the literature masses [25] of individual particles or systems of particles. Figure is adapted from [17].

method. Having known the beam and target's four-momenta, denoted as P_{beam} and P_{target} respectively, and having reconstructed the four-momenta of the two protons in the exit channel (P_1 and P_2), it is possible to determine the square mass of the unregistered particle or system of particles according to the formula:

$$m_X{}^2 = (P_{beam} + P_{target} - P_1 - P_2)^2. \tag{1}$$

The missing mass spectrum for events with two protons in the exit channel, obtained from the whole measurement at Q=40 MeV is shown in figure 2b. A high peak at the value of the η mass square is clearly seen. with about 6000 events inside the peak above the background. The background mainly originates from multi-pion production. The increase of the background in the higher-energy part of the spectrum is a reflection of the increasing acceptance of the detection system in this region. The mass resolution achieved during this measurement was equal to $\sigma_{m_\eta} = 1.6$ MeV/c^2.

In order to determine the analysing power[2] one has at the same time to monitor both the luminosity and the beam polarisation. During the first two measurements we made use of the EDDA experimental setup [26] in order to extract the beam polarisation,

---

[2] For detailed calculation of the analysing power, which takes into account the efficiency corrections, the reader is referred to [16].

whereas in the last two runs we monitored the polarisation by means of the COSY internal polarimeter [27] and our own monitoring system, allowing to measure elastically scattered protons in the horizontal and vertical planes.

RESULTS

The analysing power values for the $\vec{p}p \to pp\eta$ reaction as obtained from the measurement at an excess energy Q=40 are presented in figure 3. The range of the η centre-of-mass polar angle has been divided into four bins, with a width of 0.5 in $\cos(\theta_{CM})$. There are about 1000-1500 events in each bin. Figure 3a shows the results of analysis within the framework of the vector meson exchange model [28] for the $\vec{p}p \to pp\eta$ reaction. The dashed line in this figure represents the η $s+p$ partial wave contribution, the $s+p+d$ waves are represented by the dashed-dotted line, whereas the full model calculations, taking into account also the higer partial waves are shown as a solid line, which is close to the dash-dotted line. It is worth to note that the pure s wave should force A_y to be equal to 0. However, at present level of the measurement's accuracy we are not able to infer any quantitative conclusions.

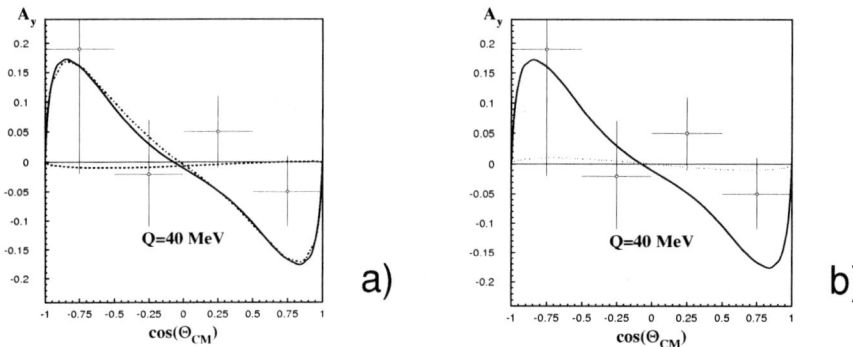

FIGURE 3. (a) Analysing power for the $\vec{p}p \to pp\eta$ reaction at Q=40 MeV. Curves in this figure are adapted from [28]. (b) Comparison of experimental data with predictions of different models. The labelling of curves has been explained in the caption of figure 1.

The interference term between Pp and Ps waves (denoted as G_1^{y0}, where we follow the notation from [29]) as well as the sum of $(Pp)^2$ and $(SsSd)$ interference terms $(H_1^{y0} + I_1^{y0})$ have been calculated and are equal to [16]:

$$G_1^{y0} = (0.003 \pm 0.004) \, \mu b,$$

$$H_1^{y0} + I_1^{y0} = (-0.005 \pm 0.005) \, \mu b.$$

The vanishing value of G_1^{y0} may suggest that there is no interference between Pp and Ps waves in the final state system. Unfortunatelly, during the first measurement the precision of data was not good enough to disentangle the sum of H_1^{y0} and I_1^{y0}.

OUTLOOK

As can be seen from figure 3b both model predictions lie within 2σ distance from experimental data, therefore at the level of accuracy obtained during the first measurement we are not able to distinguish between two different hypotheses of the η production. Thus further investigations were necessary. During the last two measurements the beam polarisation was increased from about 50% (which was the averaged polarisation during the first run) up to 75-80%. This nice feature, together with the fact that the luminosity integrated over the whole measurement's period was larger by a factor 1.5, should result in about two times better accuracy of the last two measurements. Moreover, the difference between predictions of two mentioned models are the largest at the excess energy Q=10 MeV, studied in present experiments, which should additionally help to discriminate one of the reactions models.

REFERENCES

1. Pevsner, A., et al., *Phys. Rev. Lett.*, **7**, 421–423 (1961).
2. Bergdolt, A. M., et al., *Phys. Rev.*, **D 48**, 2969–2973 (1993).
3. Chiavassa, E., et al., *Phys. Lett.*, **B 322**, 270–274 (1994).
4. Calén, H., et al., *Phys. Lett.*, **B 366**, 39–43 (1996).
5. Calén, H., et al., *Phys. Rev. Lett.*, **79**, 2642–2645 (1997).
6. Hibou, F., et al., *Phys. Lett.*, **B 438**, 41–46 (1998).
7. Smyrski, J., et al., *Phys. Lett.*, **B 474**, 182–187 (2000).
8. Calén, H., et al., *Phys. Lett.*, **B 458**, 190–196 (1999).
9. Tatischeff, B., et al., *Phys. Rev.*, **C 62**, 054001 (2000).
10. Moskal, P., et al., *πN Newsletter*, **16**, 367–369 (2001).
11. Moskal, P., et al. (2002), e-Print Archive: nucl-ex/0208004.
12. Moskal, P., et al. (2002), e-Print Archive: nucl-ex/0210019.
13. Abdel-Bary, M., et al., *Eur. Phys. J.*, **A16**, 127–137 (2003).
14. Moskal, P., et al., *Nucl. Phys.*, **A721**, 657 (2003).
15. Moskal, P., et al. (2003), e-Print Archive: nucl-ex/0307005.
16. Winter, P., et al., *Phys. Lett.*, **B 544**, 251–258 (2002), erratum-ibid. **B553**,339 (2003).
17. Winter, P., *Erste Messung der Analysierstärke A_y in der Reaktion $\vec{p}p \rightarrow pp\eta$ am Experiment COSY-11*, Diploma thesis, Rheinische Friedrich-Wilhelms-Universität Bonn (2002), ,IKP Jül-3943.
18. Caso, C., et al., *Eur. Phys. J.*, **C3**, 1–794 (1998).
19. Moskal, P., Wolke, M., Khoukaz, A., and Oelert, W., *Prog. Part. Nucl. Phys.*, **49**, 1 (2002).
20. Nakayama, K., Speth, J., and Lee, T. S. H., *Phys. Rev.*, **C 65**, 045210 (2002).
21. Faldt, G., and Wilkin, C., *Phys. Scripta*, **64**, 427–438 (2001).
22. Brauksiepe, S., et al., *Nucl. Instr. & Meth.*, **A 376**, 397–410 (1996).
23. Maier, R., *Nucl. Instr. & Meth.*, **A 390**, 1–8 (1997).
24. Moskal, P., et al., *Nucl. Instr. & Meth.*, **A 466**, 444–451 (2001).
25. Groom, D. E., et al., *Eur. Phys. J.*, **C 15**, 1–878 (2000).
26. Bisplinghoff, J., et al., *Nucl. Instrum. Meth.*, **A329**, 151–162 (1993).
27. Bauer, F., Annual Report 2001, Forschungszentrum Jülich (2001).
28. Nakayama, K., Haidenbauer, J., Hanhart, C., and Speth, J., *Phys. Rev.*, **C 68**, 045201 (2003).
29. Meyer, H. O., et al., *Phys. Rev.*, **C63**, 064002 (2001).

Isospin symmetry breaking as a tool for particle physics investigations

A. Magiera*, S. Abdel-Bary†, P. Hawranek†*, J. Ilieva**, K. Kilian†,
D. Kirilov‡, St. Kistryn*, S. Kliczewski§, W. Klimala†*, D. Kolev¶,
M. Kravčiková‖, T. Kutsarova**, J. Lieb††, H. Machner†, G. Martinská‡‡,
L. Pentchev**, N. Piskunov‡, P. von Rossen†, B.J. Roy†§§, I. Sitnik‡,
R. Siudak§, J. Smyrski*, R. Tsenov¶, M. Uličný†‡‡, J. Urbán‡‡ and
A. Wrońska†*

*Institute of Physics, Jagellonian University, Kraków, Poland
†Institut für Kernphysik, Forschungszentrum Jülich, Germany
**Institute of Nuclear Physics and Nuclear Energy, Sofia, Bulgaria
‡Laboratory for High Energies, JINR, Dubna, Russia
§Institute of Nuclear Physics, Kraków, Poland
¶Physics Faculty, University of Sofia, Sofia, Bulgaria
‖Department of Physics, Technical University Košice, Slovakia
††Physics Department, George Mason University, Fairfax, Virginia, USA
‡‡Department of Nuclear Physics, University Košice, Slovakia
§§Nuclear Physics Division, BARC, Bombay, India

Abstract. The importance of the isospin symmetry breaking in extracting various particle physics parameters is discussed. In particular the measurement of isospin symmetry breaking in the reaction $pd \rightarrow {}^3H\pi^+/{}^3He\pi^0$ is presented. The energy dependence of the measured cross sections ratio indicate isospin symmetry breaking effects. Within a simple model based on $\pi^0 - \eta$ meson mixing the mixing angle of these mesons is extracted.

INTRODUCTION

On the quark level the isospin symmetry is broken due to u and d quark mass difference and to their electromagnetic interaction. Since the origin of the quark masses is still unknown this symmetry was very often regarded to be accidental. Despite that the isospin symmetry is approximate its applications were very helpful in various nuclear and particle physics applications. However, isospin symmetry breaking plays also an important role in extracting quantities important in particle physics. First of all it allows to access current quark masses, which normally are not accessible directly due to the confinement. Presently the u and d current quark masses are extracted from the static properties of hadrons. The chiral perturbation theory allows to estimate the quark mass ratio m_u/m_d, while the lattice calculations deliver value of the average quark mass $(m_u + m_d)/2$. The mass limits obtained from this approach are shown in Fig. 1, where the unhatched area show the allowed region. The analysis, concerning the meson masses, their mixing and decays delivers the information on difference of the quark masses $m_d - m_u$ [1], the results are presented in Fig. 1 as the area between the two dotted lines.

CP717, Hadron Spectroscopy: Tenth International Conference,
edited by E. Klempt, H. Koch, and H. Orth

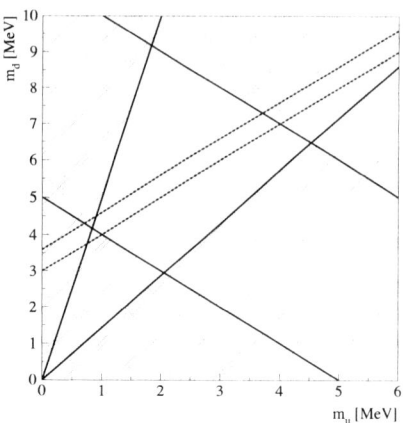

FIGURE 1. Unhatched area show allowed region for u and d quark masses deduced from the analysis complied in [2] (figure adopted from [2]). Downward going lines shows the limits from $(m_u + m_d)/2$ and the rising from the origin lines represent the limits from m_u/m_d. Two dotted lines show the limits for the $m_d - m_u$ obtained from the analysis of Ref. [1].

Usually it is not easy to relate observed isospin symmetry breaking effects for hadrons with those for quarks. One of the basic quantities that may be useful in quark mass determination is the magnitude of the meson mixing e.g. for $\pi^0 - \eta$ mesons. Starting with the isospin eigenstates of the SU(3) flavor symmetry, $| \tilde{\pi}^0 >$, and $| \tilde{\eta} >$ it is easy to show that the QCD Hamiltonian containing quark masses leads to nonzero probability for the transition between the states of different isospin. Therefore the physical states of π^0 and η mesons contains admixtures of both $| \tilde{\pi}^0 >$ and $| \tilde{\eta} >$ pure isospin states, what may be represented by:

$$
\begin{aligned}
| \pi^0 > &= \cos\theta_m \, | \tilde{\pi}^0 > - \sin\theta_m \, | \tilde{\eta} > \\
| \eta > &= \sin\theta_m \, | \tilde{\pi}^0 > + \cos\theta_m \, | \tilde{\eta} > .
\end{aligned}
\tag{1}
$$

The magnitude of isospin symmetry breaking is described by the meson mixing angle θ_m which may be easily related to u and d quark mass difference. Therefore a measurement of the $\pi^0 - \eta$ mixing angle may deliver further constraints on quark masses. Magnitude of this angle plays an important role in various processes where pions appear in the intermediate or final states. The $\pi^0 - \eta$ mixing may be important in the analysis of CP violation sources leading to the enhancement of ε'/ε [3]. The isospin symmetry breaking due to $\pi^0 - \eta$ mixing has strong implications on the analysis of $B \to \pi\pi$ decays [4] and influences the extracted value of $\sin 2\alpha$ which is related to certain CKM matrix elements.

The value of the $\pi^0 - \eta$ mixing angle is still not known precisely. Most of the theoretical calculations lead to similar values of 0.014-0.015 rad (see [5] for references). There are, however, some calculations delivering values out of this range e.g. 0.034 rad [6] or 0.010 rad [7]. Experimentally the value of the mixing angle is not easily accessible. The non-zero difference of the analyzing power found in the elastic $\bar{n}p$ and $n\bar{p}$ scattering [8] may be interpreted in terms of $\rho - \omega$ meson mixing, which is however, hidden in the charge symmetry breaking nucleon-nucleon potential. Recently reported observation of forward-backward asymmetry in the np→dπ^0 reaction [9] may be interpreted in terms of $\pi^0 - \eta$ mixing entering via transition potential [10]. The observed charge symmetry breaking in dd→$\alpha\pi^0$ [11] has no theoretical interpretation so far. In any case, it would need advanced four body calculations, that again contain some phenomenological potentials containing meson mixing.

A direct observation of the $\pi^0 - \eta$ mixing angle is possible only in some special cases. One of the possibilities is the comparison of the meson decay width for isospin symmetry forbidden and allowed decays. In Ref. [12] it was shown that from the comparison of the $\psi(2S) \rightarrow J/\psi + \pi^0$ and $\psi(2S) \rightarrow J/\psi + \eta$ decays the quark mass difference may be extracted directly. Other possibility are the studies of π^0 and η meson production in various reactions close to the η threshold. Such studies were performed for π^+d→ppη and π^-d→nnη reactions, which, interpreted within a model for η production, delivered the $\pi^0 - \eta$ mixing angle of 0.026±0.007 rad [13]. As shown in Ref. [5] the $\pi^0 - \eta$ meson mixing may play an important role also in the pd→^{3}Hπ^+/^{3}Heπ^0 reactions at beam momentum close to the η production threshold. It was proposed to study the ratio R of the cross sections for these reaction close to the η threshold for relative $p - \pi^0$ angle equal 180^0. Assuming that the isospin symmetry holds R should be equal 2. However, the pd→^{3}Heπ^0 reaction may proceed on two competing ways. In the first step the isospin pure states $| \tilde{\pi}^0 >$ and $| \tilde{\eta} >$ are produced. Than they couple to physical state $| \pi^0 >$ with the probabilities proportional to 1 and θ_m, respectively. That leads to strong interference effect, and the predicted deviations of R from 2 are about 20% [5].

According to these predictions we have performed the experiment measuring the cross sections and their ratio for pd→^{3}Hπ^+/^{3}Heπ^0 reactions. The beam momentum dependence has been measured over a narrow momentum region close to the η production threshold.

EXPERIMENTAL PROCEDURE

The experiment was performed at the COSY accelerator in Jülich using an extracted beam with an intensity of $5 \cdot 10^8$ protons/s. The ratio of differential cross sections was measured for five beam momenta. A liquid deuterium target with the thickness of 1 cm was used, yielding a luminosity of about $3 \cdot 10^{31}$ s^{-1}cm^{-2}. Since the expected isospin symmetry breaking effect is small the simultaneous detection of both reactions is necessary in order to reduce the systematic errors. This was achieved by the use of the upgraded Big Karl spectrometer, as shown schematically in Fig. 2. In the normal operation mode the spectrometer bend the particles to the focal plane. This location was used for the ^{3}He detection with the detector system containing a set of drift chambers for the track reconstruction and two layers of the hodoscopes for particles identification according to their energy loss and time of flight. Since the magnetic rigidity of ^{3}H is two times larger than ^{3}He, they are less bent in the first dipole reaching the position at dipole D1 window. There additional detection system consisting also of a set of drift chambers and scintillation hodoscopes was mounted and used for ^3H tracking and identification. Such experimental configuration allowed simultaneous detection of both pd→$^3H\pi^+$/$^3He\pi^0$ reactions. More details on the experimental procedure and on data evaluation can be found in Ref. [14]; below only the most important features are presented.

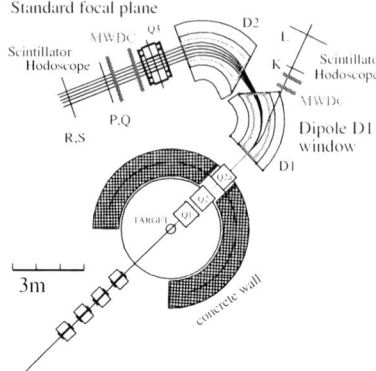

FIGURE 2. Upgraded Big Karl spectrometer. The ^3He ejectiles are detected in the focal plane while the ^3H particles are registered in the detection system located at dipole D1 window.

For both detection systems the particle track data are used to determine the particle momentum and angle at the target. The energy losses and the time of flight information are used for particle identification. All these data are sufficient to unambiguously identify the investigated reactions as presented in Fig. 3. Well known optic of the spectrometer at the focal plane allows to construct the missing mass spectrum shown in Fig. 3a. A clear peak located at 135 MeV/c^2, underlaid by only a small background, corresponds to the pd→$^3He\pi^0$ reaction. A bump at larger missing mass values is due to ^3He production associated with two pions. The missing mass resolution of 0.18 MeV originates mainly from the ^3He energy loss in the target. For identification of the pd→$^3H\pi^+$ reaction the information from the reconstructed tracks was used to determine ^3H momentum. Then the cuts on the proper momentum bands were applied and additionally it was requested that the energy loss exceeds some threshold value appropriate for the ^3H ejectiles. Finally, the time of flight spectrum was constructed. Its typical example is presented in Fig. 3b where the peak corresponding to the pd→$^3H\pi^+$ reaction is visible together with the deuteron peak originating from the tail of deuterons from the pd→dpπ^0/dnπ^+ reactions. The time of flight resolution after applying pulse height and position corrections is about 300 ps. Such spectra were used for integrating the events registered in the focal plane and in the dipole D1 window.

Large background in the ^3H spectrum comes from the secondary particles produced by the beam hitting the first dipole yoke close to the detection system at the first dipole exit. The background was reduced by introducing steerer magnets in the beam line and a special dipole magnet, located close to the target. This results in a non-zero beam incidence angle on the target. The beam incidence angle on the target was calibrated using the pp→dπ^+ reaction registered with the standard focal plane detection system. In all measurements the beam incidence angle on the target was kept constant at the

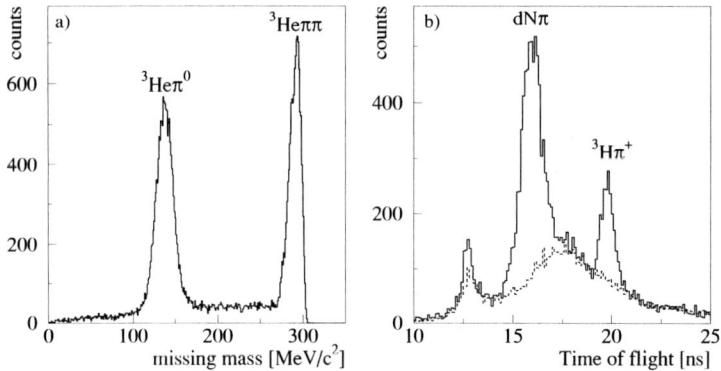

FIGURE 3. Panel a) shows the missing mass spectrum for ^{3}He particles detected at the focal plane for the beam momentum of 1.571 GeV/c. The peak at 135 MeV/c^2 clearly identifies the pd→^{3}Heπ^0 reaction. A bump at larger missing mass values corresponds to opening of the multi-pion production channels. Panel b) shows the time of flight spectrum for events selected by the applied momentum gate, for the beam momentum of 1.571 GeV/c. The peak corresponding to the pd→^{3}Hπ^+ reaction is well separated from other reactions. Also a prominent peak of deuterons from the pd→dpπ^0/dnπ^+ reactions is visible. The dashed histogram indicates background of particles not originating from the target.

value of 15.3 ± 0.5 mrad. The background was measured by selecting the events not originating from the target and it is shown in Fig. 2b as the dashed line.

The relative acceptance measurement was also performed using the pp→dπ^+ reaction for the beam momentum of 1.206 GeV/c. At this beam momentum this reaction precisely simulates the detection of ^3He and ^3H since the ratio of the rigidities for deuteron and pion is equal 2. Therefore kinematical coincidences of the outgoing deuterons and pions were used for the relative acceptance A determination, leading to the value A=18.8 ± 2.0.

EXPERIMENTAL RESULTS

The number of events for the pd→^{3}Heπ^0 reaction was obtained from the missing mass spectra by fitting the Gaussian distribution together with a linear background. The fitted Gaussian distribution and the background function were integrated within $\pm 3\sigma$ range. The final number of events for ^{3}He was obtained as the difference of these two integration results and the statistical error contains the statistics of the peak and the background.

The number of events for the pd→^{3}Hπ^+ reaction was extracted from the time of flight spectra. For the background subtraction the measured background of the events not originating from the target was used. The final number of ^{3}H events was obtained by integrating the fitted to the triton peak Gaussian distribution within $\pm 3\sigma$ region. The background distribution was also integrated within the same region and the total statistical uncertainty of the final number of ^{3}H events contains the error of the peak and the background.

The obtained numbers of events for ^{3}H and ^{3}He were corrected for the corresponding detection efficiencies. The total efficiency originates from the drift chambers efficiency and the events rejection probability due to double hits in a single scintillator paddle. The drift chambers efficiencies were measured for each run leading to values of 85%-98%. The two hit probability of 2%-3% was important and applied only for the triton detection system.

Using the numbers of events for ^{3}H and ^{3}He and the relative acceptance of the two detection systems the individual cross sections as well as their ratio were calculated. The results, together with their statistical errors, are presented in Fig. 4. They are compared with the ratios of the cross sections calculated using the data from Ref. [15] show large deviation from the value of 2 expected for isospin symmetry. Therefore it seems that for those data their quoted systematic error of 8% is underestimated. The systematic errors in our measurement comes mainly from the inaccuracy of the relative acceptance determination and is equal to 10%. This systematic error is the same multiplicative factor for all beam momenta. More details about the data evaluation and the results may be found in Ref. [16].

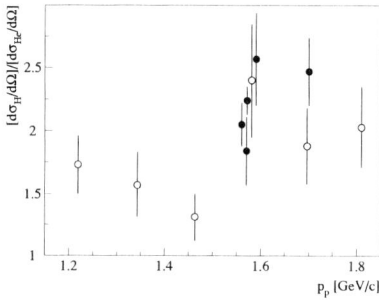

FIGURE 4.   Beam momentum dependence of the cross sections ratio for the pd→ ^{3}Heπ^0 and pd→^{3}Hπ^+ reactions for the relative proton-pion angle of 180^0. Full dots represent the results of this experiment. Open circles represent the cross sections ratio calculated using those results from Refs. [15], which were obtained at the same beam momenta.

DISCUSSION AND CONCLUSIONS

The measured values of the cross sections ratio changes rapidly close to the η production threshold. This indicates a presence of isospin symmetry breaking effects due to $\pi^0 - \eta$ meson mixing as predicted in Ref. [5]. The cross sections ratio is larger than 2 at beam momenta far above the η threshold. This may be due to systematical uncertainty of the data. It might be also caused by isospin symmetry breaking effects due to the different electromagnetic interaction in the two studied outgoing channels or due to the differences in the wave functions of ^{3}H and ^{3}He nuclei. Estimation of these effects was performed in Ref. [17]. Similar estimation for the present data shows that most important is the wave function effect, that may lead to increase of the ratio by about 10%. However, such effects in the narrow beam momentum range close to the η threshold will lead only to a constant shift of the cross sections ratio. The systematic uncertainty of the data and wave function effect may be accommodated into the model of Ref. [5] by introducing an overall normalization factor N. Than by fitting the model predictions to the data this normalization factor and the $\pi^0 - \eta$ mixing angle θ_m may be found. This results in $N = 1.15$ and $\theta_m = 0.006 \pm 0.005$ rad with the χ^2/n=1.13.

A more advanced model based on K-matrix formalism has been recently applyed to our data [18]. It uses additionally the data for the pd$\rightarrow^3$Heη and π^{-3}He$\rightarrow^3$Hη reactions. Fitting the most important K-matrix parameters and the mixing angle to all the data results in $\theta_m = 0.010 \pm 0.005$ rad. The mixing angle results which are extracted from our data are substantially lower then the value obtained from the analysis of π^+d$\rightarrow$ppη and π^-d$\rightarrow$nnη reactions [13]. Our data deliver mixing angle that is rather closer to the lower bound predictions of the QCD based models.

REFERENCES

1. B.M.K. Nefkens, G.A. Miller and L. Slaus, *Comm. Nucl. Part. Phys.*, **20**, 221-239 (1992).
2. K. Hagiwara *et al.*, Particle Data Group Collaboration, *Phys. Rev.*, **D66**, 010001-010974 (2002).
3. G. Ecker *et al.*, *Phys. Lett.*, **B477**, 88-92 (2000); S. Gardner and G. Valencia, *Phys. Lett.* **B466**, 355-362 (1999).
4. S. Gardner, *Phys. Rev.*, **D59**, 077502-077502 (1999).
5. A. Magiera and H. Machner, *Nucl. Phys.*, **A674**, 515-523 (2000).
6. B. Bagchi, A. Lahiri, S. Niyogi, *Phys. Rev.*, **D41**, 2871-2876 (1990).
7. T. Meissner and E.M.Henley, *Phys. Rev.*, **C55**, 3093-3099 (1997).
8. S.E. Vigdor *et al.*, *Phys. Rev.* **C46**, 410-448 (1992); R. Abegg *et al.*, *Phys. Rev. Lett.* **75**, 1711-1714 (1995).
9. A.K. Opper *et al.*, *nucl-ex/0306027*.
10. J.A. Niskanen, *Few-Body Systems*, **26**, 241-249 (1999).
11. E.J. Stephenson *et al.*, *nucl-ex/0305032*.
12. J. Gasser and H. Leutwyler, *Phys. Rep.*, **87**, 77-169 (1982).
13. W.B. Tippens *et al.*, *Phys. Rev.*, **D63**, 052001-052016 (2001).
14. J. Bojowald *et al.*, *Nucl. Instr. and Meth.*, **A487**, 314-322 (2002).
15. P. Berthet *et al.*, *Nucl. Phys.*, **A443**, 589-600 (1985); C. Kerboul *et al.*, *Phys. Lett.*, **B181**, 28-32 (1986).
16. M. Abdel-Bary *et al.*, *Phys. Rev.*, **C68**, 021603-021607 (2003).
17. H.S. Köhler, *Phys. Rev.*, **118**, 1345-1350 (1960).
18. A.M. Green and S. Wycech, *nucl-th/0308057*.

Hadron Production in COMPASS

Lars Schmitt for the COMPASS Collaboration

TU München, Physik-Department, D-85747 Garching, Germany

Abstract. COMPASS is a fixed target experiment running with a 160 GeV muon beam at CERN's SPS accelerator. It is aimed at the study of the nucleon spin structure using a polarised muon beam and a polarised ⁶LiD target. In a second phase it will investigate charmed hadrons and hadrons with gluonic degrees of freedom and study Primakoff and diffractive scattering. We present the status of analysis of the production of hadrons by muons.

PHYSICS OF COMPASS

The COMPASS collaboration was founded in 1996 to perform a number of measurements in hadron physics ranging from polarised structure functions examined with deep inelastic muon scattering to topics like light meson spectroscopy and the study of exotic hadrons [1]. After the first phase of the COMPASS experiment focusing on the contribution of gluons to the polarised structure function of the nucleon, a second phase is planned to address more topics with hadron beams.

Physics with the Muon Beam

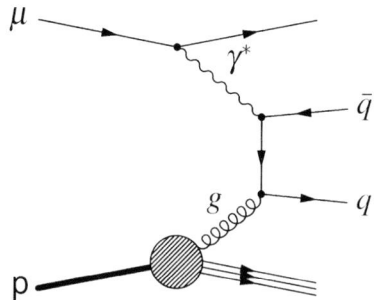

FIGURE 1. Photon-gluon fusion process.

In experiments at SLAC, DESY and CERN the spin structure of nucleons has been explored [2]. It was observed, that the spin carried by quarks inside the nucleon only contributes about one third to the total nucleon spin. This surprising result leads to the conclusion that the remaining part should come from gluons or from angular momentum. The first main objective of COMPASS is to measure this contribution from gluons.

In deep inelastic scattering gluons in the target nucleon can couple to the scattering lepton via the photon-gluon fusion process. For charm quarks this process is the dominant mode of production and therefore measuring asymmetries of open charm production in deep inelastic scattering is a clean way to measure the gluon spin in the nucleon.

In case of lighter quarks being produced, the photon-gluon fusion process can be enriched by requiring two hadrons or jets with high transverse momentum.

CP717, *Hadron Spectroscopy: Tenth International Conference,*
edited by E. Klempt, H. Koch, and H. Orth

The measurement is being performed in COMPASS by scattering a polarised muon beam of 160 GeV on a polarised target. As target material ^{6}LiD is used to measure the nucleon spin structure, because of its low dilution factor at nevertheless high values of polarisation. The material is polarised by dynamic nuclear polarisation in a magnetic field with microwaves at a temperature of a few mK.

As dominant signature COMPASS is looking for D^0 mesons in the decay $D^0 \to K^-\pi^+$. To reduce background the neutral D's are tagged by requiring them to come from the decay $D^{*+} \to D^0\pi^+$. COMPASS' goal is to measure the gluon polarisation to about 15% in open charm production [1]. Independently a similar accuracy can be reached from high p_T hadron pairs, however with a higher systematic error from the model dependence of the background description [3].

In addition COMPASS studies the flavour decomposition of the quark helicity distributions and the transverse spin distribution.

Physics with Hadron Beams

Primakoff Scattering. Primakoff scattering can be seen as Compton scattering in inverse kinematics: a particle is scattered in the Coulomb field of a heavy nucleus. For unstable particles for which Compron scattering would not be possible, the Primakoff process gives access to their electric and magnetic polarisabilities. For these quantities predictions are made by chiral perturbation theory. More details on the physics of Primakoff scattering are discussed elsewhere [4].

COMPASS will be able to measure the polarisability of the pion with an error of about 5% and for the first time could measure it for kaons as well.

Study of Gluonic Systems. Central collisions of a proton beam on a proton target provide a gluon-rich environment in the double pomeron exchange mechanism. This is assumed to efficiently produce glueballs and exotic non-$q\bar{q}$ mesons predicted by QCD. A candidate for the scalar glueball ground state 0^{++} is the f_0 at 1500 MeV/c observed by the experiments Crystal Barrel and WA102 [5].

COMPASS will be able detect many decay modes of this state and in addition access the mass region above 2 GeV where the first excited state, the tensor glueball 2^{++} is predicted around 2.3 GeV by lattice QCD.

COMPASS will have a very good acceptance owing to its two-stage spectrometer and the large acceptance target recoil detector. In addition a kinematical filter for glueballs is possible by requiring low p_T hadron jets [6]. COMPASS will also be able to contribute significantly to the search for exotic mesons.

Doubly Charmed Baryons. A further goal of the hadron program, to look for doubly charmed baryons (DCB), has been studied in recent reviews [7, 8]. Because of their particular structure decisive predictions are made by HQET [9]. Their dynamics should be similar to heavy mesons, where one light quark is coupled to one heavy partner. The decays of DCB offer a unique way to study spectator effects, since the contribution of each light quark flavour can be observed separately. The lifetimes of DCB are expected in the range of 100–200 fs [10] or even up to 1 ps for Ξ_{cc}^{++} [11].

Recently Fermilab experiment Selex (E781) has made observations of DCB in decay channels to Λ_c^+ plus mesons [12]. The production process seems to favour baryon beams and is very strong in the forward direction, where the largest part of SELEX's acceptance is. Extrapolating from these measurements COMPASS could observe up to 17,000 DCB in one year of beamtime. This would open up the possibility of detailed spectroscopy and precise tests of theoretical predictions. Conservative estimates assuming the double charm production probability to drop with the square of the single charm fraction would still yield up to 200 reconstructed decays.

OVERVIEW OF THE EXPERIMENT

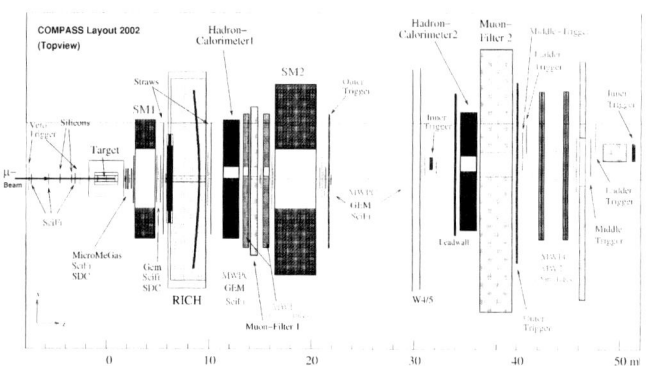

FIGURE 2.   General setup of the COMPASS spectrometer.

Spectrometer.   COMPASS uses a double magnetic spectrometer with tracking, electromagnetic and hadronic calorimetry and particle identification in each section: the first stage detects low momentum particles at large angles (± 180 mrad). High momentum particles are analyzed in the second part (± 25 mrad) using a large lever arm and a higher magnetic field than in the first. In this way an angular resolution down to 10 μrad (for small scattering angles) and a transverse resolution down to 7 μm can be achieved. A schematic view of the spectrometer is given in fig. 2. For the hadron beam programmes the gap of the first spectrometer magnet can be reduced to provide a higher field for the higher beam energy and also a smaller stray field in the tracking zones.

The tracking system is built up from a set of **L**arge **A**ngle **T**rackers (**LAT**) covering the outer region with lowest track density and **S**mall **A**ngle **T**rackers (**SAT**) for the inner regions. So-called *tracking stations* are distributed all over the spectrometer and consist of three different detector types staggered, each smaller one having finer granularity and rate capability and covering with some overlap a central hole in the next larger one. For the innermost part, directly in the beam, silicon detectors [13] and scintillating fibers are used. The LAT are small cell size drift chambers, straw chambers [14] and further downstream MWPC and large cell size drift chambers. The very important inner trackers between the beam region and the LAT consist of novel gaseous strip detectors, i.e. Micromegas [15] before the first magnet and GEM detectors [16] after that.

For identification of charged particles a RICH detector with a 3 m long C_4F_{10} radiator between the two spectrometer magnets is used [17]. Further downstream hadronic calorimeters provide the energy measurement needed to help forming a first level trigger. For the reconstruction of γ and π^0 an electromagnetic leadglass calorimeter placed in front of the second HCAL is available since 2003. Behind the calorimeters two muon filters allow to identify muons. Hodoscopes at the second muon filter serve as main trigger detectors and are correlated by means of fast lookup tables to provide kinematic selection and target pointing.

Data Taking in 2002. In 2002 COMPASS had its first major physics data taking period. After a long setup period, where many new detectors were commissioned and calibrated, 57 days were spent measuring with the target being longitudinally polarised (1.2 fb^{-1}), 19 days with transverse polarisation (0.3 fb^{-1}). In total about 5×10^9 events and 260 TB of data were recorded. The ^{6}LiD target achieved a record polarisation of up to 57%.

FIRST DATA ANALYSIS

Only a brief account of the status of the data analysis can be given here. The analysis is still in progress and being the first year with physics results the optimisation of precise calibrations and refined reconstruction algorithms is still going on. In particular detailed systematic studies still have to be performed for the major part of results. Therefore all presented results have to be considered as preliminary and subject to change.

Exclusive Production of ρ and ϕ

The exclusive production of vector mesons in COMPASS was studied on the base of about 1/6 of the data by reconstructing the decays $\rho \to \pi^+\pi^-$ and $\phi \to K^+K^-$ [18]. To ensure exclusive production a cut on the missing energy of $-2 < \Delta E < 2.5 \text{ GeV}$ was applied. The kinematic range was $10^{-3} < Q^2 < 10 \text{ (GeV/c)}^2$ and $7.5 < W < 16 \text{ GeV}$. In this sample 1.3 million ρ^0 and 42,000 ϕ were reconstructed.

The dependence on t and the angular distributions of ρ were studied. An approximate validity of s-channel helicity conservation was noted. The final statistics will allow improved measurements of both unpolarised and polarised observables for these channels.

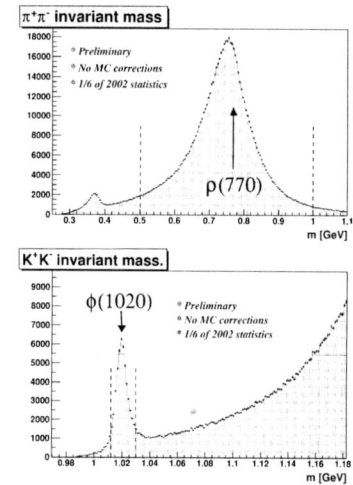

FIGURE 3. Mass distributions of ρ and ϕ.

Production of J/Ψ

In about 20% of the 2002 data production of $J/\Psi \to \mu^+\mu^-$ was studied. Being mostly elastically produced the signal is not useful for the measurement of $\Delta G/G$. A signal of 213 ± 17 J/Ψ at a mass of $m = 3103 \pm 6$ MeV was observed.

It has to be noted that the observation of J/Ψ was suppressed in 2002 by the DIS trigger. For the data taking in 2003 a dedicated J/Ψ trigger was set up.

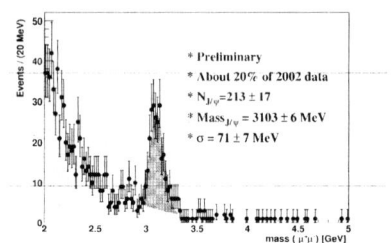

FIGURE 4. Mass distribution of $J/\Psi \to \mu^+\mu^-$.

ACKNOWLEDGMENTS

We thank the German Ministry of Education and Research, BMBF, for its support.

REFERENCES

1. The COMPASS Collaboration, COMPASS proposal, CERN /SPSLC 96-14, SPSC/P297 (Geneva, March 1996); CERN/SPSLC 96-30 (Geneva, May 1996).
2. K. Abe *et al.* [E143 coll.], *Phys. Rev.* D **58** (1998) 112003; A. Airapetian *et al.* [HERMES coll.], *Phys. Lett.* B **442** (1998) 484; B. Adeva *et al.* [SMC coll.], *Phys. Rev.* D **58** (1998) 112001.
3. A. Bravar, D. von Harrach and A. Kotzinian, *Phys. Lett.* B **421** (1998) 349, hep-ph/9710266.
4. M. A. Moinester, Intl. Workshop on Hadron Structure and Hadron Spectroscopy, Trieste, Italy, Feb. 2002, hep-ex/0206038. M. A. Moinester, V. Steiner, Proceedings of the COMPASS Summer School, Charles U., Prague (CZ), Aug. 1997, TAUP-2473-98, hep-ex/9801011
5. C. Amsler et al. [Crystal Barrel coll.], *Phys. Lett.* B **342**, 433 (1995). D. Barberis et al. [WA102 coll.], *Phys. Lett.* B 474 (2000) 423-426.
6. F.E. Close, A. Kirk, *Phys. Lett.* B **397**, 333-338 (1997)
7. M. A. Moinester, Z. Phys. A **355**, 349 (1996), and references therein.
8. L. Schmitt et al., contribution to the Workshop on Future Physics at COMPASS, CERN, Geneva, September 26-27 2002, to appear as CERN Yellow Report, hep-ex/0310049.
9. M. J. Savage, M. B. Wise, Phys. Lett B **248** (1990) 177. R. Roncaglia, et al., *Phys. Rev.* D **51**, 1248 (1995); *Phys. Rev.* D **52**, 1722 (1995); *Phys. Lett.* B **358**, 106 (1995); *Phys. Rev.* D **47**, 4166 (1993)
10. S. Fleck, J. M. Richard, Prog. Theor. Phys. **82** (1989) 760; J.M. Richard, *Nucl. Phys. Proc. Suppl.* **50** (1996) 147; J. M. Richard, *Nucl. Phys.* A **689** (2001) 235
11. B. Guberina, B. Melic, H. Stefancic, hep-ph/9911241, in Proc. of XVII Autumn School, QCD: Perturbative or Nonperturbative?, Lisbon (P), 29 Sep. - 4 Oct. 1999
12. M. Mattson et al., [SELEX Collaboration], *Phys. Rev. Lett.* **89** (2002) 112001. J.S. Russ et al., [SELEX Collaboration], Proceedings of ICHEP 2002, Amsterdam (NL), Jul. 2002, hep-ex/0209075.
13. H. Angerer et al., *Nucl. Inst. Meth.* A **512**/1-2, (2003) 229-238.
14. V.N. Bychkov et al., *Particles and Nuclei Letters*, **2** (2002) 111.
15. Y. Giomataris et al., *Nucl. Inst. Meth.* A **376**, 29 (1996);
16. F. Sauli et al., *Nucl. Inst. Meth.* A **386**, 531 (1997); B. Ketzer et al., *IEEE Trans. Nucl. Sci.* 49 (2002) 2403.
17. E. Albrecht et al., *Nucl. Inst. Meth.* A **502** (2003) 112.
18. A. Korzenev, Proceedings of DIS03, St. Petersburg (RU), May 2003.

Longitudinal polarization of Λ and $\bar{\Lambda}$ hyperons in DIS at COMPASS

M.G.Sapozhnikov [1]
on behalf of the COMPASS Collaboration

Joint Institute for Nuclear Research, Dubna 141980, Russia

Abstract. Λ and $\bar{\Lambda}$ production in deep-inelastic scattering of 160 GeV/c polarized muons on a polarized ^{6}LiD target is under study in the COMPASS experiment. Preliminary results from data collected during 2002 beam period are presented.

The study of polarization of $\Lambda(\bar{\Lambda})$ hyperons in the deep-inelastic scattering (DIS) can provide an information on the fundamental properties of the nucleon, such as polarization of the strange quarks in the nucleon [1] and to determine the mechanism of spin transfer from polarized quark to a polarized baryon [2]-[6].

The polarized nucleon intrinsic strangeness model [7, 8] predicts negative longitudinal polarization of Λ hyperons produced in the target fragmentation region ($x_F < 0$). Main assumption of the model is the negative polarization of the strange quarks and antiquarks in the nucleon. The meson cloud model [9] predicts that in the target fragmentation region the polarization of Λ should be anticorrelated with the target polarization. Therefore, it is expected that $P_\Lambda \sim 0$ for production on unpolarized target.

The measurement of the longitudinal Λ polarization in the current fragmentation region $x_F > 0$ is traditionally connected with investigation of spin transfer from quark to hadron [2]-[6]. According to the naive quark model the spin of Λ is carried by the s quark and the spin transfer from the u and d quarks to Λ is equal to zero. It means that the longitudinal polarization of Λ produced in DIS is $P_\Lambda \sim 0$. In [2], using $SU(3)_f$ symmetry and experimental data for the spin-dependent quark distributions in the proton, the contributions of u and d quarks in the Λ spin was found to be negative and substantial, on the level of 20% for each light quark. Contrary to this conclusion the $SU(6)$ based quark-diquark model [10] predicts a large positive polarization of the u and d quarks in the Λ at x>0.3. In [8] it was shown that the beam energies of the current experiments on Λ production is too small to study real spin transfer from the quark to baryon. It turns out that even at the COMPASS energy of 160 GeV most Λ, even in the $x_F > 0$ region, are produced from the diquark fragmentation. It is predicted that in the COMPASS kinematics the longitudinal Λ polarization is $P_\Lambda = -0.004, -0.07$, depending on the fragmentation model.

It is important to note that a significant part (up to 30-40%) of the observed Λ is from

---

[1] E-mail address:sapozh@sunse.jinr.ru

decays of heavy hyperons. such as Σ^0 and Σ^*. More clear situation is with analysis of the $\bar\Lambda$ production, however up to now the statistics of $\bar\Lambda$ produced in the DIS experiments was marginal.

The experimental situation with measurements of Λ and $\bar\Lambda$ longitudinal polarization $P_\Lambda(P_{\bar\Lambda})$ in DIS is summarized in the Table 1.

TABLE 1. Summary of experimental measurements of $\Lambda(\bar\Lambda)$ longitudinal polarization in DIS. Sign of polarization is given with respect to virtual photon momentum.

Reaction Exp.	$< E_b >$ (GeV)	$< x_F >$	N_Λ	P_Λ	$N_{\bar\Lambda}$	$< x_F >$	$P_{\bar\Lambda}$
$\bar\nu_\mu Ne$ WA59[11]	40	-0.47 >0	403 66	-0.59 ± 0.14 -0.11 ± 0.45			
μN E665 [12]	470	0.15 0.44	750	1.2 ± 0.5 0.32 ± 0.7	650	0.15 0.44	-0.26 ± 0.6 -1.1 ± 0.8
$\nu_\mu N$ NOMAD [13]	43.8	-0.36 0.21	5608 2479	-0.21 ± 0.04 -0.09 ± 0.06	248 401	-0.2 0.18	0.23 ± 0.20 -0.23 ± 0.15
eN HERMES [14]*	27.5	0.30	10568	$S_\Lambda =$ 0.06 ± 0.09	1687		

* The results were presented in terms of the spin transfer $S_\Lambda = \frac{P_\Lambda}{P_b D}$, where P_b is the beam polarization and D is the depolarization factor.

One can see that in the target fragmentation region the Λ polarization is indeed negative. The spin transfer for current fragmentation region seems small, however the experimental data are quite scarce, especially for $\bar\Lambda$.

We have studied Λ and $\bar\Lambda$ production by polarized μ^+ of 160 GeV/c on a polarized ^{6}LiD target of the COMPASS spectrometer constructed in the framework of CERN experiment NA58. A detailed description of the COMPASS experimental setup is in the talk of L.Schmitt at this Conference. This analysis uses data collected during the 2002 run. Total amount of the acquired data is about 260 TB. The analysis comprises about $1.7 \cdot 10^8$ DIS events, which correspond to 60% of the total 2002 statistics.

The V^0 events ($V^0 \equiv \Lambda$, $\bar\Lambda$ and K_s^0) were selected by requiring the outgoing muon tracks together with two hadron tracks of opposite charge. The primary vertex should be inside the target. The polarized ^{6}LiD target consists of two oppositely polarized 60 cm long cells. The data presented here are averaged on target polarization.

The secondary vertex must be downstream of the both target cells. The angle between vector of V^0 momentum and vector between primary and V^0 vertices should be $\theta_{col} < 0.01$ rad. Cut on transverse momentum of the decay products with respect to the direction of V^0 particle $p_t > 23$ MeV/c was applied to reject e^+e^- pairs from the γ conversion seen as the band at the bottom of the Armenteros plot shown in Fig.1.

The typical elliptical bands from the K_s^0, Λ and $\bar\Lambda$ decays are seen in Fig.1. Both Λ and $\bar\Lambda$ signals stand out clearly. The large number of produced $\bar\Lambda$ is a specific feature of the COMPASS experiment.

The standard DIS cut $Q^2 > 1$ (GeV/c)^2 and $0.2 < y < 0.8$ have been used. After background subtraction the experimental sample contains about 10800 Λ and 5900 $\bar\Lambda$.

The kinematical characteristics of produced Λ, $\bar\Lambda$ and K_s^0 are shown in Fig.2 and Fig.3. The x_F and Q^2 experimental distributions are compared to Monte-Carlo ones in

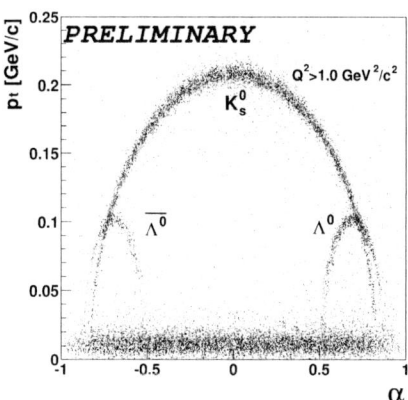

FIGURE 1. The Armenteros plot: p_t is the transverse momentum of the V^0 decay products with respect to the direction of V^0 momentum, $\alpha = \frac{p_L^+ - p_L^-}{p_L^+ + p_L^-}$, where p_L is the longitudinal momentum of the V^0 decay particle.

Fig. 2. One can see that we are able to access mainly current fragmentation region. The averaged value of x_F is $< x_F >= 0.21$, whereas for the Bjorken scaling variable x it is $< x >= 0.02$.

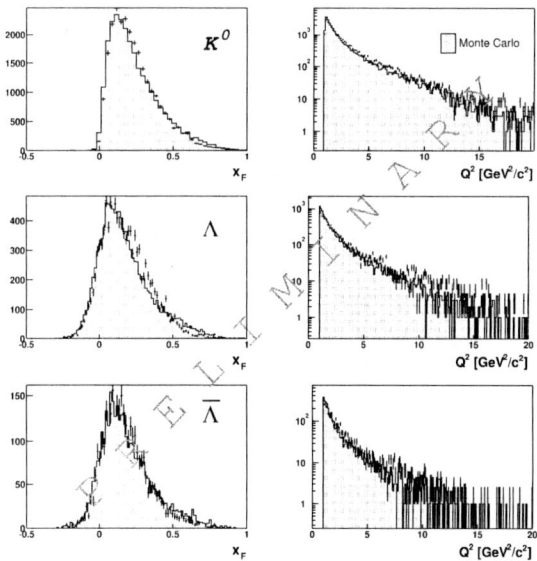

FIGURE 2. x_F (left column) and Q^2 (right column) distributions for K_s^0 (upper row), Λ (middle row) and $\bar{\Lambda}$ (lower row). The experimental data points are shown together with results of Monte-Carlo simulations (histograms).

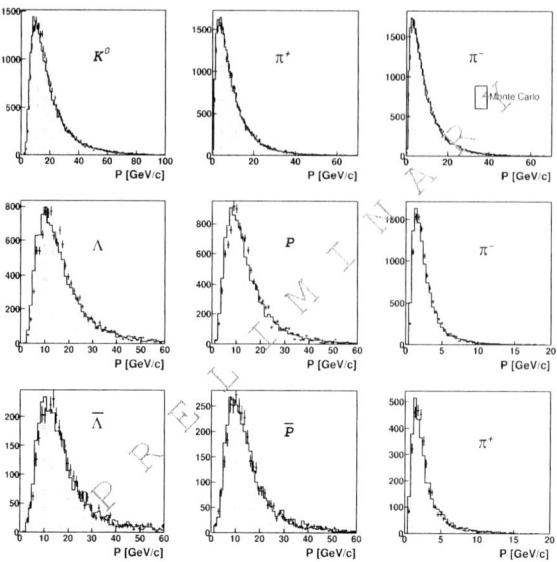

FIGURE 3. The momentum spectra for strange particles and their decay products are shown: K_s^0, π^+ and π^- in upper row; Λ, proton and π^- in middle row; and $\bar{\Lambda}$, $\bar{p}$ and π^+ in lower row. The experimental data points are shown together with results of Monte-Carlo simulations (histograms).

The momenta of V^0 particles and of their decay products are shown on Fig.3. The mean Λ momentum is 12 GeV/c, while decay pion momentum is 2 GeV/c. Both figures shows reasonable agreement between the experimental data and the Monte-Carlo simulations.

Λ ($\bar{\Lambda}$) hyperon polarization can be measured via the angular asymmetry of positive particle in $\Lambda \to p\pi^-$ ($\bar{\Lambda} \to \bar{p}\pi^+$) decays. Angular distribution of the positive particle in Λ ($\bar{\Lambda}$) rest frame is

$$\frac{d\sigma}{d\Omega} = N_0\left(1 + (-)\alpha P_i \cos\theta_i\right) \tag{1}$$

where N_0 is a normalization constant, $\alpha = 0.642 \pm 0.013$ is the decay asymmetry parameter, P_i, $i = x, y, z$ are the components of the polarization vector $\mathbf{P}$. The angles θ_i are between the direction of the positive decay particle and the axes of the coordinate system, which was defined using directions of virtual photon ($\mathbf{e}_\gamma$) and target nucleon ($\mathbf{e}_T$) in the Λ rest frame:

$\mathbf{n}_x = \mathbf{e}_\gamma$
$\mathbf{n}_y = \mathbf{e}_\gamma \times \mathbf{e}_T / |\mathbf{e}_\gamma \times \mathbf{e}_T|$
$\mathbf{n}_z = \mathbf{n}_x \times \mathbf{n}_y$

The analysis was performed slicing each angular distribution in 10 bins and fitting the invariant mass distribution of the V^0 decay products to determine the number of V^0 events in each bin. Fig. 4 shows measured angular distributions for K_s^0, Λ and $\bar{\Lambda}$ decays, corrected on the acceptance. The acceptance was determined by the Monte Carlo simulation of unpolarized $\Lambda(\bar{\Lambda})$ decays.

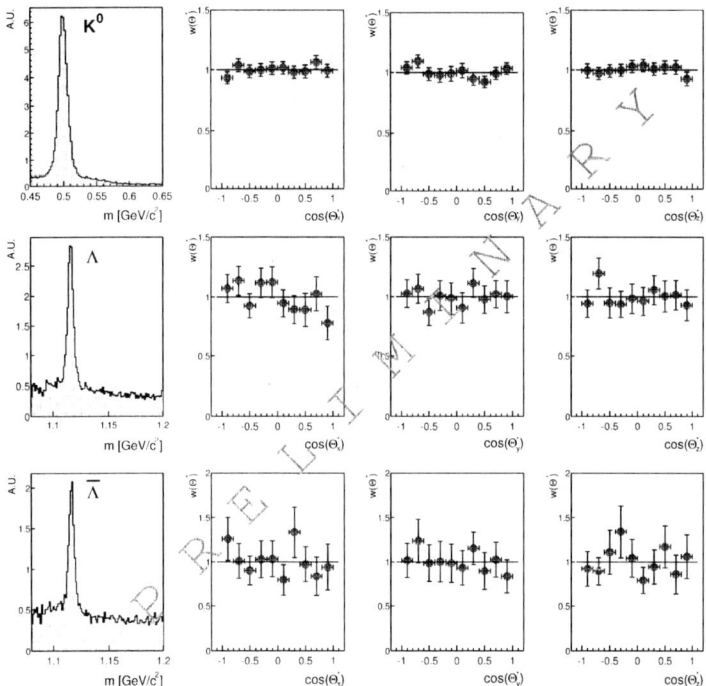

FIGURE 4. The $\cos\theta_x$ (second column) $\cos\theta_y$ (third column) and $\cos\theta_z$ (forth column) distributions for K_s^0 (upper row), Λ (middle row) and $\bar{\Lambda}$ (lower row). The first column shows invariant mass distributions for K_s^0, Λ and $\bar{\Lambda}$, respectively.

One could see that all angular distributions for K_s^0 decays are flat as expected. It means that no significant bias is introduced by the apparatus and by the analysis procedure.

The invariant mass distributions of K_s^0, Λ and $\bar{\Lambda}$ are shown in the first column of Fig.4 for a part of analyzed events. The level of the background events is not negligible and the procedure of the bin-by-bin fitting for determination of the angular distributions is essential to correct rejection of the background events.

The experimental $\cos\theta_i$ distributions are flat. It means that the Λ and $\bar{\Lambda}$ polarizations, averaged over the full kinematical range of COMPASS, are small.

Work on the determination of the polarizations and their systematic errors is continuing. The results from our 2002 data presented here demonstrate the good potential of COMPASS for measurements of Λ and $\bar{\Lambda}$ polarizations in DIS.

REFERENCES

1. J.Ellis et al., Phys.Lett., **B353**, 319 (1995); Nucl.Phys., **A673**, 256 (2000).
2. M. Burkardt and R. L. Jaffe, Phys. Rev. Lett. **70**, 2537 (1993).
3. M. Anselmino et al., Phys. Lett. **B509**, 246 (2001).

4. A. M. Kotzinian, A. Bravar and D. von Harrach, Eur. Phys. J. **C2**, 329 (1998).
5. C. Boros and L. Zuo-Tang, Phys. Rev. **D57**, 4491 (1998).
6. B. Q. Ma, I. Schmidt, J. Soffer and J. J. Yang, Phys. Lett. **B488**, 254 (2000).
7. J. Ellis, D. Kharzeev and A.M. Kotzinian, Z.Physik **C69**, 467 (1996).
8. J. Ellis, A.M. Kotzinian and D.V. Naumov, Eur. Phys. J. **C25**, 603 (2002).
9. W.Melnitchouk and A.W.Thomas, Z.Phys. **A353** 311 (1996).
10. J. J. Yang, B. Q. Ma and I. Schmidt, Phys. Lett. **B477**, 107 (2000).
11. S. Willocq et al., Z.Phys. **C53**, 207 (1992).
12. M. R. Adams et al., Eur. Phys. J. **C17**, 263 (2000).
13. P. Astier et al., Nucl. Phys. **B588**, 3 (2000);
    P. Astier et al., Nucl. Phys. **B605**, 3 (2001).
14. A. Airapetian et al., Phys. Rev. **B64**, 112005 (2001); S. Belostotski, IXth Workshop on High-Energy Spin Physics, Dubna, Russia, Aug 2 - 7, 2001.

Associated Strangeness Production at COSY-TOF*

Marc Wagner for the COSY-TOF collaboration

Physikalisches Institut, Universit‰Erlangen-N‚ rnberg, Erwin-Rommel-Str. 1, 91058 Erlangen
(supported by the German BMBF and Forschungszentrum J‚ lich)*

Abstract. The associated strangeness production in elementary hadron-hadron collisions is investigated exclusively in reactions of the type $pp \rightarrow KYN$ (Y=Λ,Σ) at the external COSY beam using the time-of-flight spectrometer COSY-TOF. The design of the apparatus allows a complete track reconstruction of all charged particles leading to the extraction of the total cross sections and differential observables of the mentioned reactions. Especially the Dalitz analysis of the reaction $pp \rightarrow K^+\Lambda p$ shows a strong contribution of N* resonances with varying strengths depending on the excess energy. Moreover, a strong evidence of a possible exotic penta quark resonance has been observed in the reaction $pp \rightarrow K^0\Sigma^+p$ at a significance level of about 5σ.

INTRODUCTION

The main goal in the investigation of the associated strangeness production close to threshold in reactions of the type pp $\rightarrow$ KYN (Y=Λ,Σ) is the insight into the reaction dynamics regarding strange-anti-strange-production. The simultaneous measurements of different reaction channels allows a direct comparison of the production ratio which is a powerful tool to test the related models due to vanishing parameters. The meson exchange model acts as an appropriate tool to describe the production process at energy regions near threshold including and combining various contributions as the exchange of strange and non-strange mesons as well as the excitation of resonances and the inclusion of final state interaction (FSI) between the produced hyperon and nucleon. Especially the measurement at various beam momenta helps to distinguish between the various resonance contributions and the influence of the FSI.

Another important topic to be supplied with results from strangeness production measurements in threshold region is the soliton-based prediction of an exotic pentaquark state which should be observed in the K^0p and K^+n subsystems induced in hadron-hadron collisions. Reasonable reactions for the search are $pp \rightarrow K^0\Sigma^+p$ as well as $pp \rightarrow K^+\Sigma^+n$. This resonance should be a member of an anti-decuplet assuming the N*(1710) resonance as a further prominent state. The penta quark mass is predicted around 1530 Mev/c^2 with an unusual narrow width of about 15 MeV/c^2 [1].

CP717, *Hadron Spectroscopy: Tenth International Conference,*
edited by E. Klempt, H. Koch, and H. Orth
© 2004 American Institute of Physics 0-7354-0197-7/04/$22.00

EXPERIMENT

The external experiment COSY-TOF is a large-angle, non-magnetic spectrometer with various start and stop detector components providing time-of-flight and geometrical track information of the charged particles. The detector combines high efficieny and acceptance at a moderate energy and momentum resolution. The whole arrangement together with a tiny liquid hydrogen target is installed inside a vacuum vessel (see Fig. 1). This ensures a rather precise definition of the interaction point and a strongly reduced contamination from background reactions in air. The outer detector consists of a huge cylindrical segmented scintillator barrel at approximately three meters length and some three meters in diameter and a circular endcap in the forward direction. The endcap consists of two separate hodoscopes, a central one (ëQuirlí) at a diameter of one meter and a ring-like one at an outer diameter of some three meters. Both consist of three segmented scintillator layers, one consisting of wedge like segments and two of left and right bended, helix formed elements. Apart from the track information the outer detector delivers also the stop timing.

The inner detector, shown in Fig. 2 together with a typical $K^+\Lambda p$-event, is optimized for track and vertex reconstruction. It consists of the ëStarttorteë, made of two layers of thin scintillators providing the start timing, a highly granulated doublesided silicon microstrip detector close to the target and two scintillating fibre hodoscopes at a distance of 10 cm and 20 cm, respectively. This system covers the full angular range of the reaction products for the investigated hyperon production from threshold up to the COSY limit at 3.4 GeV/c. It allows the complete reconstruction of the $pp \rightarrow K^+\Lambda p$ events including the precise measurement of the vertex of the delayed decay of the Λ-hyperon into two charged particles.

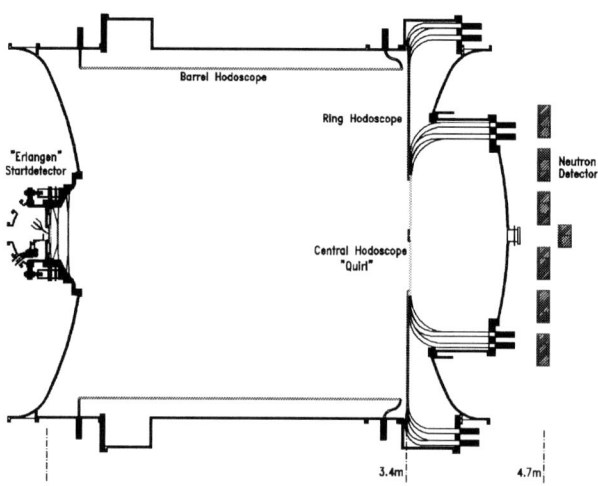

FIGURE 1. Setup of the TOF-Experiment.

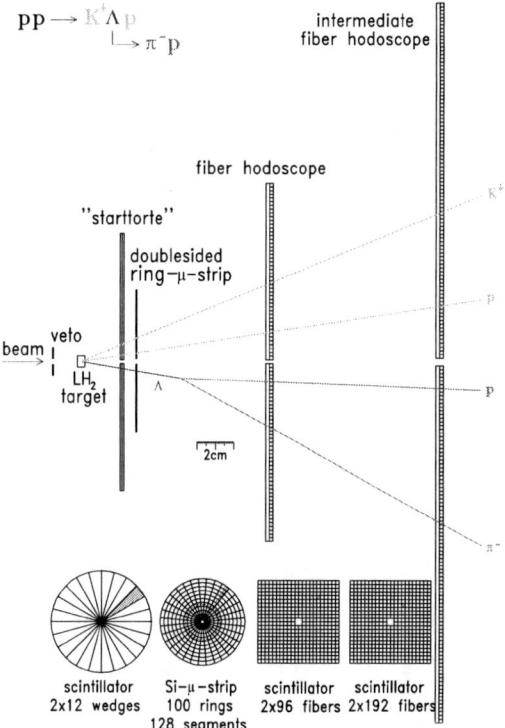

$pp \rightarrow K^+ \Lambda p$
$\qquad\quad \hookrightarrow \pi^- p$

intermediate
fiber hodoscope

fiber hodoscope

"starttorte"

doublesided
ring−μ−strip

veto
beam
LH₂
target

Λ

2cm

K^+

p

p

π^-

scintillator
2x12 wedges

Si−μ−strip
100 rings
128 segments

scintillator
2x96 fibers

scintillator
2x192 fibers

FIGURE 2. Scheme of the Start detector together with an event of the type $pp{\rightarrow}K^+\Lambda p$ with its characteristic delayed decay of the Λ-hyperon into pπ⁻.

RESULTS AND DISCUSSION

Λ Production

The reaction $pp \rightarrow K^+ \Lambda p$ has been investigated in detail between 2.5 GeV/c and 3.3 GeV/c beam momentum. For all measurements clean event samples on a very low background have been extracted. To demonstrate this, the missing mass spectra of the reconstructed Λ–hyperons are shown in Fig. 3 for the beam momenta of 2.95 and 3.2 GeV/c, respectively. Since the TOF apparatus covers the full phase space of the reaction a detailed Dalitz analysis is possible which is demonstrated in the left part of Fig. 4 at a beam mometum of 2.85 GeV/c≤ As one can see there are strong deviations from a homogeneous phase space distribution. From our previous investigations [2] and theoretical arguments it is most likely that the observed un-isotropy has its origin in the influence of the FSI and/or N* resonances. To get more information the data are compared with a model parametrization prepared by Sibirtev [3]. A satisfactory description of the data can only be achieved by a coherent addition of the resonances and the FSI together with a phase space contribution. The best agreement is obtained with a strongly dominating contribution of the N*(1650) resonance. The analysis of the Dalitz distributions at 2.95 GeV/c, 3.2 GeV/c and 3.3 GeV/c show an increasing contribution of the N*(1710) resonance which slightly dominates at 3.3 GeV/c which is in good agreement with calculations of Shyam et. al. [4]

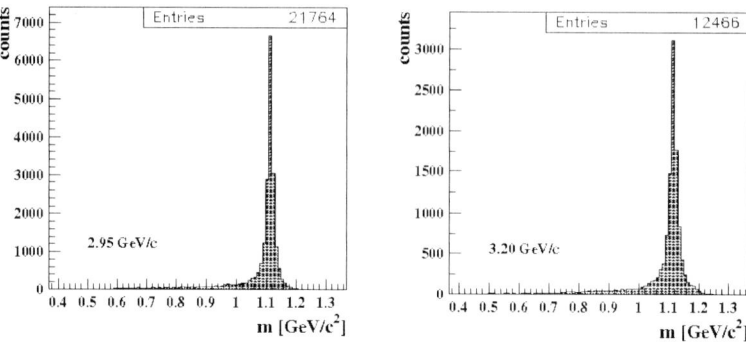

FIGURE 3. Λ missing mass distributions at 2.95 GeV/c (left) and 3.2 GeV/c (right).

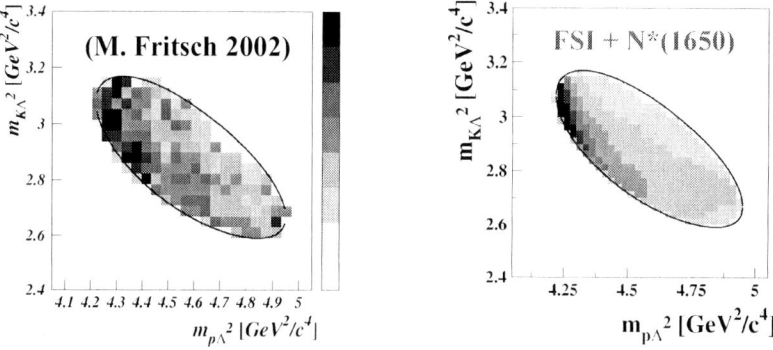

FIGURE 4. Dalitz plot of the reaction pp→KLp at 2.85 GeV/c. The parameterization of the model calculation (right) has been adjusted on the data (left).

Σ^+ Production

Since the assembly of the double-sided microstrip detector close to the target the measurement of the short range decaying Σ^+ hyperon has become possible. The access to Σ^+ production occurs via the two reaction channels $pp \to K^0 \Sigma^+ p$ and pp $\to$ K$^+\Sigma^+$n. As demonstrated in Fig. 5 (left) for the beam momentum at 2.95 GeV/c a very clean missing mass spectrum could be extracted from the reaction $pp \to K^0 \Sigma^+ p$, leading to the the total cross section at the beam momenta 2.85, 2.95 and 3.2 GeV/c, which represent the very first results of this channel at threshold regions. They are shown in Fig. 6 together with older results taken from bubble chamber measurements. The experimental results at threshold region are located at the upper limit of the expectations.

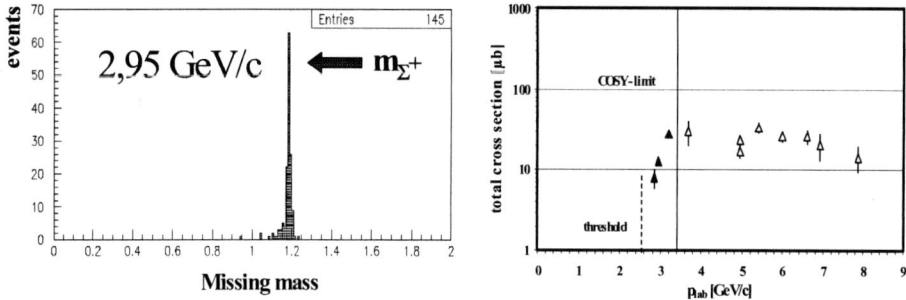

FIGURE 5. Missing mass spectrum (left) and total cross section (right) of the reaction pp→K⁰Σ⁺p in threshold region (filled symbols) together with older bubble chamber measurements (open symbols).

Within the measurement program of this reaction channel an investigation regarding a predicted exotic pentaquark state decaying into a Kp-system leads to a clear enhancement of about 5σ in the invariant mass spectrum of the K^0p-subsystem in the data at 2.95 GeV/c at a mass around 1530 MeV/c$\lesssim$ which is slightly lower than the value reported by the LEPS collaboration at Spring8 [5]. A first estimate of the total cross section of the related reaction $pp \rightarrow \Sigma^+ \theta^+$ roughly agrees with theoretical calculations [6].

ACKNOWLEDGEMENTS

We gratefully acknowledge the support of the German BMBF and the Forschungszentrum J¸lich, as well as the COSY crew for excellent cooperation.

REFERENCES

1. M.V. Polyakov et. al., Eur. Phys. J. A9 (2000)
2. Bilger et. al., Phys. Lett. B420 (1998) pp. 217-224
3. Private communications
4. Shyam et. al., Phys. Rev. C 63 (2001)
5. T. Nakano et. al., hep-ex/0301020
6. W. Liu and C. M. Ko, nucl-th/0308034

Diffraction at HERA

Armen Bunyatyan

Max-Planck-Institut für Kernphysik,
Saupfercheckweg 1, 69117 Heidelberg, Germany
and Yerevan Physics Institute, Armenia
E-mail: bunar@mail.desy.de

Representing the H1 and ZEUS Collaborations

Abstract. Recent measurements of inclusive processes, vector meson production and hadronic final states in diffractive interactions at HERA are presented. The data are used to investigate the factorization properties and study the partonic structure of colour singlet exchange.

INTRODUCTION

In its first running period, which ended in 2000, the ep collider HERA delivered more than $100pb^{-1}$ of integrated luminosity for each of the two experiments H1 and ZEUS. In this report we present some recent results on diffractive physics.

The diffractive dissociation of the photon $\gamma p \rightarrow Xp$ has characteristics similar to diffractive hadron–hadron interactions and can be described by Regge phenomenology. Within this framework, diffractive interactions at high energies are dominated by the exchange of the Pomeron, an object with vacuum quantum numbers. On the other hand, perturbative QCD (pQCD) calculations can be made for diffractive processes in which a hard scale is present: processes with high virtuality Q^2 or high four-momentum transfer squared $|t|$, or processes where the high p_T jets or heavy quarks are produced. In pQCD, the vacuum exchange can be modelled as the exchange of a colour singlet gluon ladder between the proton and the virtual photon, which fluctuates into a $q\bar{q}$ or $q\bar{q}g$ state before the interaction [1]. In this approach, the cross section is factorized into the square of the effective dipole wave function and the square of the cross section for diffractive scattering of the dipole off the proton.

At HERA, it is possible to vary the scale with which the Pomeron structure is probed in diffractive interactions, to study its partonic content and the transition from soft to hard interactions.

DIFFRACTIVE VECTOR MESON PRODUCTION

Vector meson production has been extensively studied at HERA both in the photoproduction ($Q^2 \sim 0$) and DIS regimes ($Q^2 > 1~GeV^2$). Figure 1 presents the exclusive photoproduction cross section for different vector mesons as a function of the γp center of mass energy, $W_{\gamma p}$ [2, 3]. The cross sections for light vector mesons (ρ, ω, ϕ) follow

CP717, *Hadron Spectroscopy: Tenth International Conference,*
edited by E. Klempt, H. Koch, and H. Orth
© 2004 American Institute of Physics 0-7354-0197-7/04/$22.00

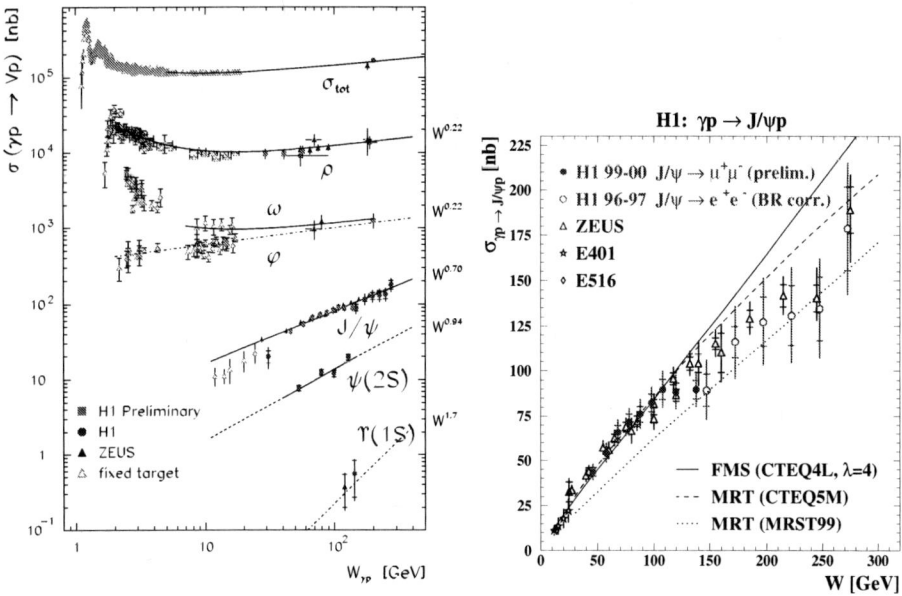

FIGURE 1. a) Vector meson cross sections as a function of the total CM energy of the photon-proton system $W_{\gamma p}$ in photoproduction ($Q^2 = 0$); **b)** The J/ψ meson cross section as a function of $W_{\gamma p}$ in photoproduction ($Q^2 = 0$) compared with pQCD calculations.

the soft behaviour of the total photon–proton cross section parameterized as $\sim W^{0.2}$ in agreement with the predictions of the Regge approach [4]. In contrast, the J/ψ cross section exhibits much harder energy dependence ($\sim W^{0.8}$) [5, 6]. Here, the mass of the charm quark sets the hard scale for pQCD calculations, and the steep energy dependence, which is related to the rise of the gluon density in the proton at small values of Bjorken-x, is quite well described by pQCD predictions [7].

Figure 2 shows the ρ cross section for different values of Q^2 as a function of W [8]. The energy dependence of the cross section becomes steeper as Q^2 increases. For light vector mesons, Q^2 plays the role of the hard scale, similar to the mass of the charm quark in the case of J/ψ photoproduction.

INCLUSIVE DIFFRACTION IN DIS

The kinematics of diffractive deep inelastic scattering, $ep \rightarrow eXp$, can be expressed in terms of the photon virtuality Q^2, the squared four momentum transfer t, Bjorken x, the fraction of the beam proton's momentum carried by the Pomeron, $x_{I\!P}$, and the fraction of the Pomeron's momentum carried by the struck quark, β. The cross section for the diffractive DIS process can be described by an expression analogous to that used for

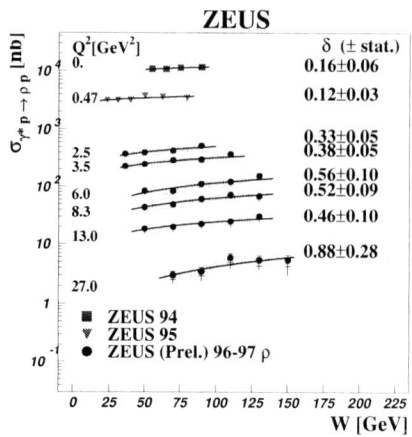

FIGURE 2. The ρ meson cross section as function of $W_{\gamma p}$ at different values of Q^2.

inclusive DIS:

$$\frac{d\sigma^D}{d\beta dQ^2 dx_{I\!P}} = \frac{2\pi\alpha}{\beta Q^4}(1 - y + \frac{y^2}{2})(F_2^D - \frac{y^2}{1+(1-y)^2}F_L^D),$$

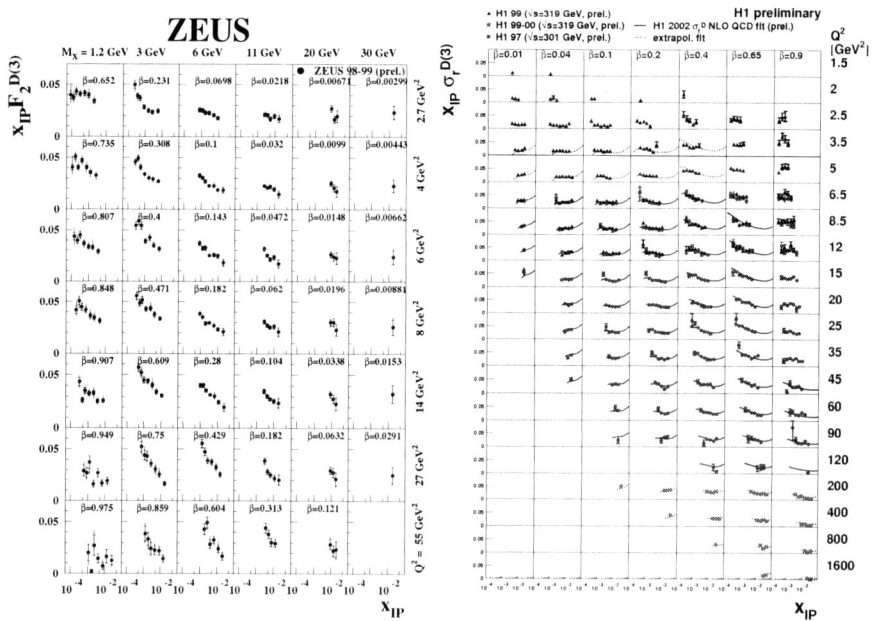

FIGURE 3. Diffractive structure function F_2^D and the reduced cross section $\sigma_R^D \equiv F_2^D - \frac{y^2}{1+(1-y)^2}F_L^D$ as a function of $x_{I\!P}$ for fixed β and Q^2.

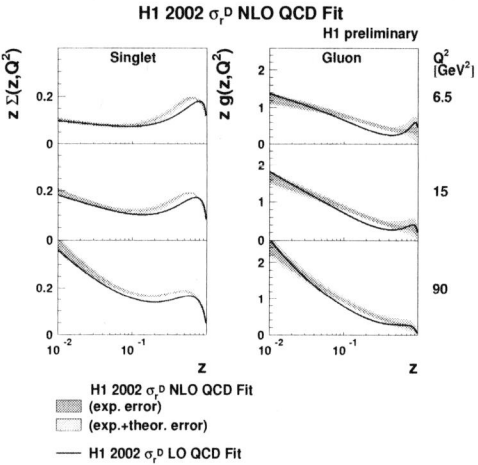

FIGURE 4. Pomeron parton densities. The left hand side shows the singlet quark distribution, the right hand side shows the gluon density.

In Fig.3 the new high precision H1 and ZEUS measurements of diffractive structure function are shown as a function of $x_{I\!P}$ for different Q^2 and β bins [9, 10]. The H1 measurements are used in a DGLAP QCD analysis in order to extract diffractive parton densities. These parton densities are universal for diffractive DIS processes [11]. In addition, Regge factorization between the $(x_{I\!P}, t)$ and (x, Q^2) dependences is assumed, according to which the Pomeron PDF is independent of $x_{I\!P}$ and t.

The extracted NLO singlet quark $\Sigma(z, Q^2)$ and gluon densities $g(z, Q^2)$, together with their uncertainties, are shown in Fig.4 [10]. Here the variable z corresponds to the momentum fraction carried by an individual parton. The comparison of quark and gluon contributions shows that up to 75% of the Pomeron's momentum is carried by gluons. It is also apparent that due to the indirect extraction of $g(z, Q^2)$ from the scaling violation, the gluon densities at high z have large uncertainties.

DIFFRACTIVE DIJET AND CHARM PRODUCTION

According to the QCD factorization theorem [11], the parton distribution functions extracted from the QCD fits to diffractive structure functions can be used to describe hadronic final states in diffractive DIS. Of particular interest are measurements of open charm, $\gamma^* q \to c\bar{c}$, and diffractive dijet production since the boson–gluon fusion process provides a direct probe of the gluon content of the Pomeron, in contrast to inclusive measurements. The presence of a hard scale provided by the high transverse momentum of the jets or the mass of the charm quark allows the testing of different QCD based models.

Figure 5 shows the cross sections for diffractive dijet and diffractive D^* meson production in DIS as a function of $z_{I\!P}$, which represents the longitudinal momentum fraction

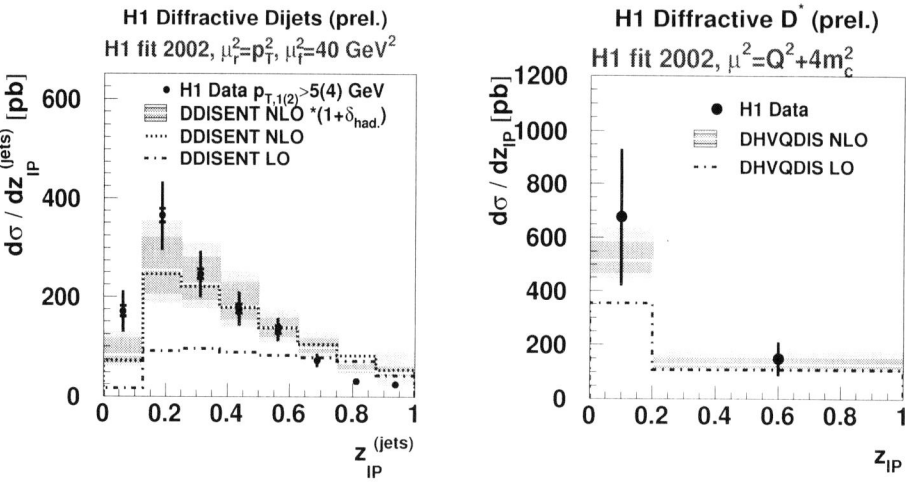

FIGURE 5. Diffractive dijet and diffractive D^* production cross sections compared with the DISENT and HVQDIS calculations.

of the diffractive exchange entering the hard scattering [12]. These measurements are compared with LO and NLO calculations using the DISENT [13] and HVQDIS [14] programs which use PDFs extracted from inclusive diffractive DIS data. The NLO calculations, corrected for hadronization effects, provide within the theoretical and experimental uncertainties a reasonable description of the shape and normalization of the measured cross section. The results are thus consistent with QCD factorization in diffractive DIS.

Dijet production in diffractive interactions was also studied in the photoproduction regime [15]. Here, the resolved and direct processes are distinguished by reconstructing the variable x_γ, defined as the fractional momentum of the quasi-real photon entering the dijet system. For direct processes $x_\gamma = 1$, while in the case of resolved processes $x_\gamma < 1$ will hold. Figure 6 displays the measured cross sections for diffractive dijet photoproduction as a function of x_γ^{jet} together with the prediction based on the diffractive parton densities from the QCD fit. The sum of resolved and direct processes in the model gives a good description of the data both in normalisation and in shape throughout the x_γ^{jet} range.

In contrast to this result, a large discrepancy was observed between the predictions and the measurements made by the CDF collaboration in the process $p\bar{p} \rightarrow pX$ at $\sqrt{s} = 1800\ GeV$ [16, 10]. The overestimation of the measured cross section by a factor of 5–10 indicates a breakdown of factorization. This has often been interpreted as being due to additional spectator interactions which lead to the suppression of the diffractive cross section. There is however no evidence for any suppression of the diffractive cross section in the H1 data in the region dominated by resolved photons, as might be expected on the basis of the CDF measurements.

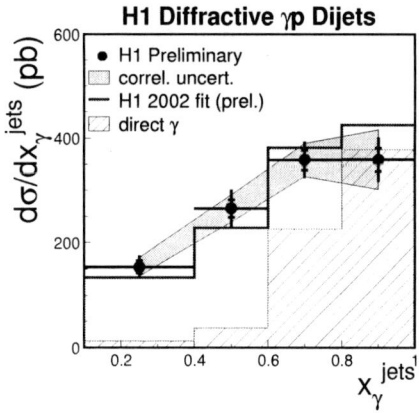

FIGURE 6. Diffractive dijet cross section in photoproduction compared with predictions using the QCD fits to diffractive DIS data.

CONCLUSIONS

HERA is an ideal facility for the study of the transition from soft to hard diffraction, explored using different hard scales ($Q^2, M_q, p_T^{jet}, ...$).

The measurements of inclusive diffraction and the hadronic final state in diffractive DIS can be described within a consistent picture. The results support QCD factorization with diffractive parton densities which are evolved according to the DGLAP equations and are dominated by the gluonic contribution.

REFERENCES

1. For a review see M. McDermott, DESY 00-126 (2000).
2. H1 Collaboration, *Nucl. Phys.* **B 463** (1996) 3; *Phys. Lett.* **B 541** (2002) 251.
3. ZEUS Collaboration, *Phys. Lett.* **B 377** (1996) 259; *Z. Phys.* **C 73** (1996) 73;
   Eur. Phys. J. **C 2** (1998) 247; *Phys. Lett.* **B 437** (1998) 432.
4. A. Donnachie and P. Landshoff, *Phys. Lett.* **B 296** (1992) 227.
5. H1 Collaboration, contributed paper 108 to EPS03, Aachen.
6. ZEUS Collaboration, *Eur. Phys. J.* **C 24** (2002) 345; *Phys. Lett.* **B 437** (1998) 432.
7. A.D. Martin, M.G. Ryskin and T. Teubner, *Phys. Rev.* **D 62** (2000) 14022;
   L. Frankfurt, M. McDermott and M. Strikman, *Journal High Energy Phys.* **103** (2001) 45.
8. ZEUS Collaboration, contributed paper 594 to EPS2001, Budapest.
9. ZEUS Collaboration, contributed paper 538 to EPS03, Aachen.
10. H1 Collaboration, contributed papers 089, 090 to EPS03, Aachen.
11. J.C. Collins, *Phys. Rev.* **D 57** (1998) 3051; *Phys. Rev.* **D 61** (2000) 019902.
12. H1 Collaboration, contributed paper 113 to EPS03, Aachen.
13. S. Catani, M.H. Seymour, *Nucl. Phys.* **B 485** (1997) 29;
    F. Hautmann, *Journal High Energy Phys.* **210** (2002) 25.
14. B.W. Harris, J. Smith, *Phys. Rev.* **D 57** (1998) 2806;
    L. Alvero, J.C. Collins, J.J. Whitmore, hep-ph/9806340.
15. H1 Collaboration, contributed paper 087 to EPS03, Aachen.
16. CDF Collaboration, *Phys. Rev. Lett.* **84** (2000) 5043.

Light antinuclei production in proton-proton, proton-nucleus, and antiproton-proton collisions

K. Protasov, R. Duperray, M. Buénerd

Laboratoire de Physique Subatomique et de Cosmologie, 53, Avenue des Martyrs,
IN2P3-CNRS, UJFG, F-38026 Grenoble Cedex, France

Abstract. The experimental data on the antideuteron, antitritium, and antihelium-3 production in proton-proton and proton-nucleus collisions are well reproduced within a simple model based on the diagrammatic approach to the coalescence model. First quantitative estimations of the antideuteron production cross section in antiproton-proton collisions are presented.

INTRODUCTION

The increasing interest in the study of production of light antinuclei in proton-proton and proton-nucleus collisions is mainly motivated by the presence of anti-nuclei in cosmic rays which has potentially important implications on the matter-antimatter asymmetry of the universe. >From this point of view, it is important to determine the amount of anti-matter which can be produced in the galaxy through the interaction of high-energy protons with the interstellar gas. A new generation of experiments (AMS [1], PAMELA [2], BESS [3]) should be able to measure the flux of anti-matter in a near future.

Furthermore, a possibility to perform experiments with antideuteron beams was discussed recently [4].

First estimations of the antideuteron flux in cosmic rays [5] have used the usual coalescence model. This model supposes that the nucleons, produced during the collision of a beam and a target, fuse into light nuclei whenever the momentum of their relative motion is smaller than a coalescence radius p_0 in the momentum space, which is a free parameter of the model, usually fit to the experimental data. A simple diagrammatic approach to the coalescence model developed in [6] provided a microscopic basis to the model. In this approach, the parameter p_0 is expressed in terms of the slope parameter of the inclusive nucleon production spectrum and of the wave function of the produced nucleus.

This diagrammatic approach has been generalized to describe the $\bar{\text{d}}$, $\bar{\text{t}}$, and $\overline{^3\text{He}}$ production cross sections [8, 9]. This approach can reproduce most existing data without any additional parameter in energy domains where the inclusive antiproton production cross sections are well known.

An additional contribution to the antideuteron flux in cosmic rays may come from the $\bar{\text{p}}\text{p} \to \bar{\text{d}}\text{X}$ reaction which implies the production of a nucleon-antinucleon pair at least. To our knowledge, this reaction has never been investigated either experimentally or theoretically. In this contribution, first quantitative estimations of this reaction cross section are presented.

CP717, *Hadron Spectroscopy: Tenth International Conference,*
edited by E. Klempt, H. Koch, and H. Orth
© 2004 American Institute of Physics 0-7354-0197-7/04/$22.00

DIAGRAMMATIC APPROACH OF THE COALESCENCE MODEL

The main ideas of the diagrammatic approach of the coalescence model for nuclear fragment production are reminded here for the reader's convenience [6]. The simplest Feynman diagram of Fig. 1 corresponding to fusion of two nucleons is considered as a basis for the coalescence model.

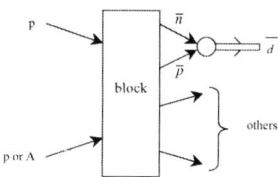

FIGURE 1. The simplest Feynman diagram corresponding to the coalescence of two antinucleons into an antideuteron.

The physical picture behind this diagram is quite simple: the nucleons produced in a collision (block) are "slightly" virtual and can fuse without any further interaction with the nuclear field. This diagram is not the only possible contribution to the full transition amplitude. However, as was shown in [7], where various diagrams were considered, there are mutual cancellations of a number of diagrams. As a result, at sufficiently large deuteron momenta the diagram of Fig. 1 is the dominant one, and at this stage the other diagrams can be neglected.

This diagram can be calculated directly using usual diagram technique. The cross section for $\bar{d}$ and $\bar{t}$ production can then be written (see [8, 9]) as

$$E_{\bar{d}}\frac{d^3\sigma_{\bar{d}}}{dp^3} = \frac{12\pi^3}{m\sigma_{inel}}R_{\bar{d}}\left|\int M(\mathbf{p}_1)M(\mathbf{p}_2)\varphi_d(\mathbf{q})\frac{d^3q}{(2\pi)^3}\right|^2;$$

$$E_{\bar{t}}\frac{d^3\sigma_{\bar{t}}}{dp^3} = \frac{96\pi^6}{m^2\sigma^2_{inel}}R_{\bar{t}}\left[\int M(\mathbf{p}_1)M(\mathbf{p}_2)M(\mathbf{p}_3)\,\phi_t(\mathbf{p},\mathbf{q})\frac{d^3p}{(2\pi)^3}\frac{d^3q}{(2\pi)^3}\right]^2.$$

In these expressions, $M(\mathbf{p})$ is the antiproton inclusive production amplitude related to the corresponding cross section by

$$E_{\bar{p}}\frac{d^3\sigma_{\bar{p}}}{dp^3} = |M(\mathbf{p})|^2;$$

φ_d and φ_t are correspondingly the deuteron and tritium wave functions, σ_{inel} the proton-nucleus inelastic total reaction cross section, m the nucleon mass. To take into account the threshold effects, a phenomenological correction factor $R_{\bar{d}}$ (or respectively $R_{\bar{t}}$) defined as $R_{\bar{d}}(x) = \Phi(x; 3m_p)/\Phi(x; 3 \times 0)$ is introduced here. Here Φ is the n particles phase space and $x = \sqrt{s + m_d^2 - 2\sqrt{s}E_{\bar{d}}}$. The denominator contains the high energy limit of the phase space to ensure R to be dimensionless and to do not change the value of the cross section out of the space phase boundary.

LIGHT ANTINUCLEI PRODUCTION IN PP- AND PA-COLLISIONS

The existing experimental data on the antideuteron production in pp- and pA-collisions (known from the authors) can be successfully described within this approach without any parameter (except the parameters used to describe the antiproton inclusive spectrum). As an example, Fig. 2 shows the results for a proton-proton collision data from CERN [11] and proton-nucleus data from FNAL [12].

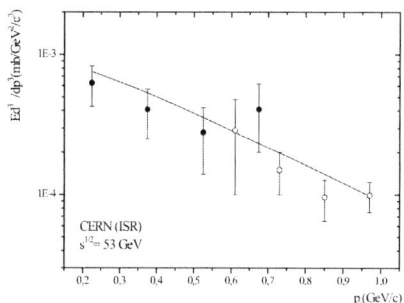 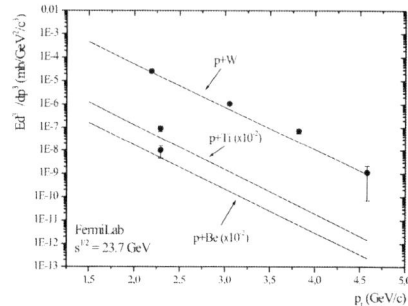

FIGURE 2. Left: Inclusive differential cross section of antideuteron production as a function of the transversel momentum p_t compared to the present calculations. The data are taken from [10] (black circles) and [11] (open circles). Right: Inclusive differential cross section for antideuteron production on Be, Ti, and W target as a function of the transversal momentum p_t compared to the present calculations. The data are taken from [12].

As an example of the $\bar{t}$ and $\overline{^3He}$ production the results for data from [13, 14] are presented in Fig. 3. On the average, data and calculations are within one order of magnitude. This result should be considered as a success in account of the numerous sources of uncertainities of the calculations and of the limited accuracy of the measurements. Note also that the $\bar{t}$ and $\overline{^3He}$ production cross section, measured at the same momentum, are expected to be close to each other (as it is clearly seen in the same experiment for t and ^3He production) whereas, in this experiment, they are quite different.

ANTIDEUTERON PRODUCTION IN $\bar{P}P$ COLLISIONS

The $\bar{p}p \rightarrow \bar{d}X$ which can give a non-negligible contribution to the antideuteron flux. To our knowledge, the cross section of this reaction was never investigated either theoretically or experimentally. This cross section can be estimated, at least for the order of magnitude, whithin our approach. The major difficulty, however, comes from the absence of any information about $\bar{p}p \rightarrow \bar{p} X$ reaction cross section which is an important ingredient of the present model. Nevertheless, an additional hypothesis can be made to overcome this difficulty by assuming that the $\bar{p}$ inclusive production cross section in the $\bar{p}p \rightarrow \bar{p}X$ reaction is the same as for p one in the reaction $pp \rightarrow pX$ (this cross section can be evaluated from the data [15]) and the $\bar{n}$ inclusive production cross section in the

894

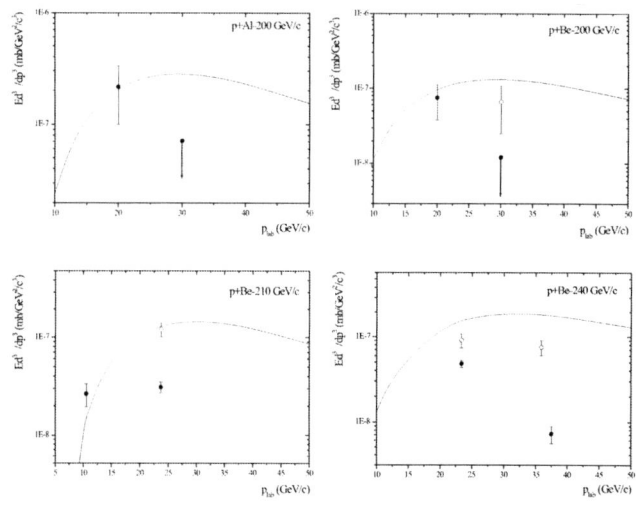

FIGURE 3. Inclusive differential cross section of t̄ (black circles) and $\overline{^3\text{He}}$ (open circles) production on Be and Al target as a function of the laboratory momentum p_{lab} compared to the present calculations. The data are taken from [13] and [14].

$\bar{p}p \to \bar{n}X$ reaction is the same as for $\bar{p}$ in the reaction $pp \to \bar{p}X$ (this cross section was recently parametrized in [16]). The results of these calculations are presented on the left part of the Fig. 4. On the right part of this figure, one can see a contribution to the antideuteron flux from this reaction. This contribution can be non-negligible especially at low energies where one expects to observe a contribution of antideuterons coming from exotic sources (SUSY particles decay [5] or primordial black holes evaporation [18]). The propagation of antideuterons was calculated whithin the usual leaky-box model [17]. To see more clearly the contribution of the reaction $\bar{p}p \to \bar{d}X$, a repopulation of low energy spectrum because of non-annihilation processus of antideuterons in a interstellar medium is not presented on this figure [19].

The future GSI antiproton facility [20] will be the ideal experimental facility to measure these cross sections.

CONCLUSIONS

The experimental data on the light antinuclei production in pp- and pA- collisions are well reproduced within a simple model based on the diagrammatic approach to the coalescence model. First quantitative estimations of the antideuteron production cross section in antiproton-proton collisions are presented. This reaction can be measured at the future GSI antiproton facility.

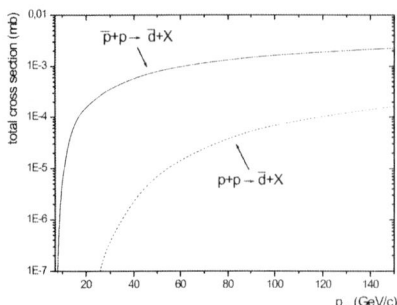 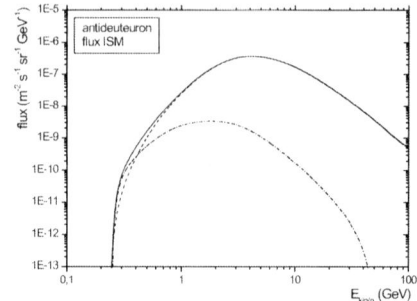

FIGURE 4. Left: Total cross section for antideuteron production in proton-proton (dash-dotted line) and antiproton-proton (solid line) collisions as a function of the laboratory momentum p_{lab}. Right: Antideuteron interstellar flux as a function of the kinetic energy per nucleon. The antideuteron non-annihilation inelastic collisions are not taken into account (see text). The dashed line showns the contribution of the pp, p He, He p→ $\bar{d}$ X reactions to the total flux, while the dash-dotted line corresponds to the $\bar{p}$p, $\bar{p}$He → $\bar{d}$ X reactions, the solid line corresponds to the total flux.

REFERENCES

1. M. Aguilar et al., Phys. Rep. **366**, 331 (2002);
   See section IV in Nucl. Phys. B (Proc. Suppl.) 113 (2002).
2. O. Adrani et al., Nucl. Instr. Meth. **A478**, 114 (2002).
3. Y. Asaoka et al., Phys. Rev. Lett. **58**, 051101 (2002).
4. F. Iazzi, Nucl. Phys. **A655**, 371c (1999).
5. P. Chardonnet, J. Orloff, P. Salati, Phys. Lett. **B409**, 313 (1997);
   F. Donato, N. Fornengo, P. Salati, Phys. Rev. **D62**, 043003 (2000).
6. V.M. Kolybasov, Yu.N. Sokol'skikh, Phys. Lett. **B225**, 31 (1989); Sov. J. Nucl. Phys. **55**, 1148 (1992).
7. M.A. Braun and V.V. Vechernin, Sov. J. Nucl. Phys. **44**, 506 (1986); **36**, 357 (1982).
8. R.P. Duperray, K.V. Protasov, A.Yu. Voronin, Eur. Phys. J., **A16**, 27 (2003).
9. R.P. Duperray, K.V. Protasov, L. Dérome, and M. Buénerd, to appear in Eur. Phys. J. (2003) ; nucl-th/0301103v1.
10. B. Alper et al., Phys. Lett. **46B**, 265 (1973).
11. W.M. Gibson et al., Lett. Nuov. Cim. **21**, 189 (1978).
12. J.W. Cronin et al., Phys. Rev. **D11**, 3105 (1975).
13. W. Bozzoli et al., Nucl. Phys. **B144**, 317 (1978).
14. A. Bussière et al., Nucl. Phys. **B174**, 1 (1980).
15. E.W. Andersson, et al., Phys. Rev. **19**, 198 (1967);
    M.A. Abolins, et al., Phys. Rev. Lett. **25**, 126 (1970);
    L.C. Tang and L.K. Ng, J. Phys. G: Nucl. Phys. **9**, 1289 (1983).
16. R.P. Duperray, C.Y. Huang, K.V. Protasov, and M.Buénerd, to appear in Phys. Rev. **D** (2003) ; astro-ph/0305274
17. M. Simon et al., Astro. Phys. J. **499** 250 (1998).
18. A. Barrau et al., A& A **288** 676 (2002).
19. B. Baret, R. Duperray, G. Boudoul, A. Barrau, D. Maurin, L. Derome, K. Protasov, M. Buénerd, Contribution to the ICRC'2003 Conference; astro-ph/0306221
20. www-new.gsi.de/zukunftsprojekt/

QCD analysis of polarized structure functions in next-to-leading-order, using improved valon model

Ali N. Khorramian [1][*][†], S. Atashbar Tehrani[**][†] and A. Mirjalili[‡][†]

[*]Physics Department, Semnan University, Semnan, Iran
[†]Institute for Studies in Theoretical Physics and Mathematics (IPM),
P.O.Box 19395-5531, Tehran, Iran
[**]Physics Department, Persian Gulf University 75168, Boushehr, Iran
[‡]Physics Department, Yazd University, Yazd, Iran

Abstract. Polarized parton distribution functions (PPDF) has been calculated in the improved valon model using next-to-leading order approximation. The Bernstein polynomial method has been applied to do direct fits. Our predictions for next-to-leading order of polarized parton distributions and proton structure function have been compared with leading order.

INTRODUCTION

Next-to-Leading Order(NLO) for unpolarized parton distributions in framework of valon model and a leading-order(LO) QCD analysis of polarized case has been performed as well in this frame [1, 2]. The valon model to *polarized* case in NLO approximation in QCD are employed. Here the distribution of unpolarized valons were calculated firstly by Hwa [3] which is a successful model in describing the internal hadronic structure. Afterward a new set of parameters which describe unpolarized valon distributions in proton were determined [4]. It is shown that polarized parton distribution functions are in connection with unpolarized ones [5]. According to this view we supposed the same relations between U and D polarized valon distribution functions and unpolarized ones. In this connection some unknown parameters are appeared which can be determined by using the Bernstein averages fitting method [6].

MOMENTS ANALYSIS OF POLARIZED VALON AND RELATED PARTON DISTRIBUTIONS IN NLO

In Ref.[4] the unpolarized valon distribution have been introduced and calculated in the following form:

$$G_{U/p}(y) = 72.49y^{1.75}(1-y)^{3.8}, \quad G_{D/p}(y) = 38.69y^{1.05}(1-y)^{4.51}, \quad (1)$$

---

[1] E-mail address:khorramiana@theory.ipm.ac.ir

CP717, *Hadron Spectroscopy: Tenth International Conference,*
edited by E. Klempt, H. Koch, and H. Orth

where $G_{U,D/p}(y)$ are the probability of finding unpolarized U, D-valon with momentum fraction y in unpolarized hadron. According to [2] which is corresponded to LO calculations, we can define as well in the NLO the polarized valon distributions from the unpolarized valon ones in the following form

$$\delta G_{j/p}^{NS}(y) = \delta F_j^{NS}(y) \times G_{j/p}(y) , \quad \delta G_{j/p}^{S}(y) = \delta F_j^{S}(y) \times G_{j/p}(y) , \tag{2}$$

where the subscript j refer to U and D valon type, and $G_{j/p}(y)$ in Eq. (2) is the probability of finding a j-valon with momentum fraction y in a polarized proton. In Eq. (2) we assumed the functional form of $\delta F_j^{NS}(y)$ and $\delta F_j^{S}(y)$ as follows

$$\delta F_j^{NS}(y) = N_j y^{\alpha_j} (1-y)^{\beta_j} (1 + \gamma_j y + \eta_j y^{0.5}) , \tag{3}$$

$$\delta F_j^{S}(y) = \delta F_j^{NS}(y) \times (\kappa y^{0.5} + \lambda y + \mu y^{1.5} + \nu y^2 + \rho y^{2.5} + \tau y^3) . \tag{4}$$

Let us define the Mellin moments of polarized valon distribution functions, $\delta G_{j/p}^{NS,S}$, as in below:

$$\delta M_{j/p}^{NS,S}(n) \equiv \int_0^1 y^{n-1} \delta G_{j/p}^{NS,S}(y) \, dy . \tag{5}$$

By inserting the Eq.(2) in above equation, the final results as a function of n which involves the unknown parameters of Eqs. (3,4) will be appeared. To calculate the NLO evolutions of the polarized parton distributions in the valon, we used its moments. The non-singlet(NS) part evolves according to

$$\Delta M_{NS\pm} = \left(1 - \frac{\alpha_s(Q^2) - \alpha_s(Q_0^2)}{2\pi} (\delta d_{NS\pm}^{(1)n} - \frac{2\pi b'}{b} \delta d_{qq}^{(0)n}) \right) L^{\delta d_{qq}^{(0)n}} , \tag{6}$$

where $L(Q^2) \equiv \frac{\alpha_s(Q^2)}{\alpha_s(Q_0^2)}$. The evolution in the flavor singlet and gluon sector are governed by 2×2 the anomalous dimension matrix with the explicit solution given by

$$\begin{pmatrix} \delta M_S \\ \delta M_{gq} \end{pmatrix} = \left(L^{\delta \hat{d}^{(0)n}} + \frac{\alpha_s(Q^2)}{2\pi} \hat{U} L^{\delta \hat{d}^{(0)n}} - \frac{\alpha_s(Q_0^2)}{2\pi} L^{\delta \hat{d}^{(0)n}} \hat{U} \right) \begin{pmatrix} 1 \\ 1 \end{pmatrix} , \tag{7}$$

where δM_{gq} is the spin dependent quark-to-gluon evolution function. All associated functions and quantities in above equations have been defined in Ref.[5].

NLO MOMENTS OF PROTON PPDF'S AND STRUCTURE FUNCTION

Having obtained the moments of polarized valon distributions, the determination of the moments of parton distributions in a proton are straightforward. Distributions that we shall calculate are δu_v, δd_v, $\delta \Sigma$ and δg. Their moments are denoted respectively by:

$\delta u_v^n(Q^2)$, $\delta d_v^n(Q^2)$, $\delta \Sigma^n(Q^2)$ and $\delta g^n(Q^2)$. Therefore the moments of polarized u and d-valence quark in a proton can be indicated by:

$$\delta u_v^n(Q^2) = 2\delta M_{U/p}^{NS}(n) \times \delta M_{NS+}(n, Q^2)\,, \tag{8}$$

$$\delta d_v^n(Q^2) = \delta M_{D/p}^{NS}(n) \times \delta M_{NS+}(n, Q^2)\,, \tag{9}$$

the factor 2 in Eq.(8) backs to existence of 2-U type valons. The moment of polarized singlet distribution (Σ) and gluon distribution are as follows:

$$\delta \Sigma^n(Q^2) = (2\delta M_{U/p}^{S} + \delta M_{D/p}^{S}) \times \delta M_S(n, Q^2)\,, \tag{10}$$

$$\delta g^n(Q^2) = (2\delta M_{U/p}^{NS} + \delta M_{D/p}^{NS}) \times \delta M_{gq}(n, Q^2)\,. \tag{11}$$

In Eq.(10) Σ symbol indicates $\sum_{q=u,d,s}(q+\bar{q})$, thus by having Σ contribution, and all valence quark in moment space, the contribution of $\delta \bar{q}$ can be specified directly. Also in Eq. (11), $\delta M_{gq}(n, Q^2)$ is the quark-to-gluon evolution function. As we know in the NLO contributions to $g_1(x, Q^2)$ we can use directly its moment in following form

$$g_1^n(Q^2) = \frac{1}{2}\sum_q e_q^2 \{(1 + \frac{\alpha_s}{2\pi}\delta C_q^n)[\delta q^n(Q^2) + \delta \bar{q}^n(Q^2)] + \frac{\alpha_s}{2\pi}2\delta C_g^n \delta g^n(Q^2)\}\,, \tag{12}$$

here $\delta q^n(Q^2)$, $\delta \bar{q}^n(Q^2)$ and $\delta g^n(Q^2)$ are moments of polarized parton distributions in a proton. Also δC_q^n, δC_g^n are the n-th moment of spin-dependent Wilson coefficients given by Ref.[5]. Since now the moments of polarized parton distributions has been determined, we can obtain the moment of polarized proton structure function in NLO by inserting the required distributions function in Eq. (8-11). According to Eqs. (3,4), which involves 16 unknown parameters and that they inter in sequence relations (8-11) and finally Eq. (12), it is obvious that the final version for $g_1^n(Q^2)$, includes as well these unknown parameters. If we can calculate the unknown parameters then the computation of all moments of polarized parton distributions and structure function, $g_1^n(Q^2)$, are possible.

QCD FITS TO EXTRACT POLARIZED VALON DISTRIBUTIONS IN NLO

Because for a given value of Q^2, only a limited number of experimental points, covering a partial range of values x, are available, one can not use the moments directly. A method devise to deal to this situation is that to take averages of structure functions with Bernstein polynomials [6]

$$g_{n,k} = \frac{(n-k)!\Gamma(n+2)}{\Gamma(k+1)\Gamma(n-k+1)} \sum_{l=0}^{n-k} \frac{(-1)^l}{l!(n-k-l)!} g_1^{(k+l)+1}(Q^2)\,, \tag{13}$$

where

$$g_1^{(k+l)+1}(Q^2) = \int_0^1 x^{(k+l+1)-1} g_1(x, Q^2)dx\,. \tag{14}$$

899

The Eq. (13) represent averages of function $g_1(x,Q^2)$ in the region $[\bar{x}_{n,k} - \frac{1}{2}\Delta x_{n,k}, \bar{x}_{n,k} + \frac{1}{2}\Delta x_{n,k}]$ [7]. To obtain experimental averages $g_{n,k}$ from the E143 and SMC data for xg_1 [8], we fit $xg_1(x,Q^2)$ for each bin in Q^2 separately, to the convenient phenomenological expression

$$xg_1^{(phen)} = \mathscr{A}x^{\mathscr{B}}(1-x)^{\mathscr{C}} . \tag{15}$$

Using Eqs.(13,14) the Bernstein averages $g_{n,k}(Q^2)$ can be written in terms of odd and even moments

$$g_{2,1}(Q^2) = 6\left(g_1^2(Q^2) - g_1^3(Q^2)\right) ,$$
$$g_{2,2}(Q^2) = 3\left(g_1^3(Q^2)\right) ,$$
$$\vdots$$

We shall use the result of Eq.(12) for the QCD prediction of $g_1^n(Q^2)$. The basic unknown fit parameters will be, $N_U, \alpha_U, \beta_U, \gamma_U, ..., \rho, \tau$. Thus there are 16 parameters to be simultaneously fitted to the experimental $g_{n,k}(Q^2)$ averages. After obtain 16 unknown parameters according to Eqs. (3,4), we are in situation to introduce polarized valon distributions which defined before in Eq. (2). In Fig.(1) we plotted $2\,y\delta G_{U/p}^{NS} + y\delta G_{D/p}^{NS}$ and $2\,y\delta G_{U/p}^{S} + y\delta G_{D/p}^{S}$ as a function of y in LO and NLO approximations.

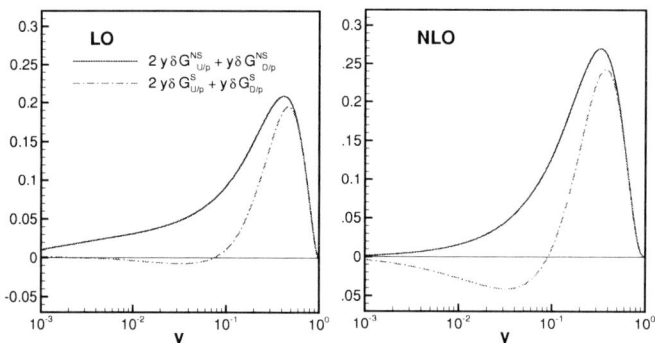

FIGURE 1. The plots of $2\,y\delta G_{U/p}^{NS} + y\delta G_{D/p}^{NS}$ and $2\,y\delta G_{U/p}^{S} + y\delta G_{D/p}^{S}$ as a function of y.

X-SPACE PPDF'S AND POLARIZED PROTON STRUCTURE FUNCTION

By using convolution integral as following

$$\delta q_{i/p}(x,Q^2) = \sum_j \int_x^1 \frac{dy}{y} \delta G_{j/p}(y)\delta f_{i/j}\left(\frac{x}{y},Q^2\right) , \tag{16}$$

where the summation is over different types of valons, we can obtain the PPDF's in the proton in x-space. In this equation we need in addition to $\delta G_{j/p}(y)$, to know the corresponding polarized i-parton distributions in a j-valon, $\delta f_{i/j}(z = \frac{x}{y}, Q^2)$. To obtain the z-dependence of parton distributions we can employ the inverse Mellin transform technique [5]. Now we can calculate all polarized parton distributions in a proton as a function of x, Q^2 by using convolution integral. We presented in Fig.(2) the results of $xg_1^p(x, Q^2)$, for $Q^2 = 2, 5$ (GeV^2) and compared it with experimental data Ref.[8].

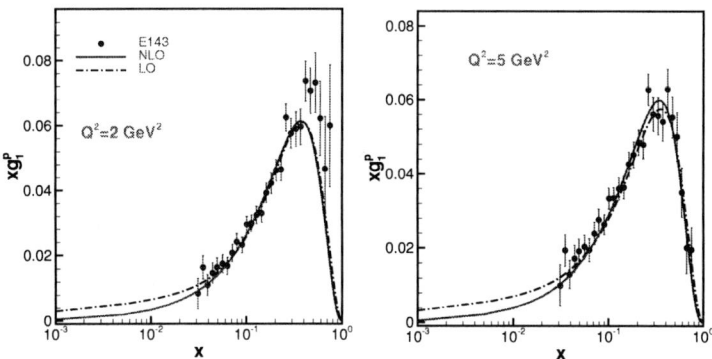

FIGURE 2. Polarized proton structure function xg_1^p as a function of x which is compared with LO and NLO approximations.

ACKNOWLEDGMENTS

We are grateful to R. C. Hwa for giving us his useful and constructive comments. A.N.K thanks from Semnan university for partial financial support to do this project. We acknowledge from Institute for Studies in Theoretical Physics and Mathematics (IPM) to support financially this project.

REFERENCES

1. F. Arash and Ali N. Khorramian, Phys. Rev. C **67** (2003) 045201; hep-ph/0303031.
2. Ali. N. Khorramian, A. Mirjalili and S. Atashbar Tehrani, Contribution to International workshop on QCD, Italy, 14-18 June 2003 hep-ph/0309258.
3. R. C. Hwa, Phys. Rev. D **22**,(1980) 759. ; R. C. Hwa, Phys. Rev. D **22**,(1980) 1593;R. C. Hwa and M. S. Zahir, Phys. Rev. D **23**, (1981) 2539.
4. R. C. Hwa and C. B. Yang, Phys. Rev. C **66** (2002)025204; R. C. Hwa and C. B. Yang, Phys. Rev. C **66** (2002) 025205.
5. G. Altarelli and Parisi, Nucl. Phys. B**126**, 298 (1977); B. Lampe and E. Reya, Phys. Rept. **332**,(2000)1.
6. C. J. Maxwell and A. Mirjalili, Nucl. Phys. B **645** (2002) 298.
7. J. Santiago and F. J. Yndurain, Nucl. Phys. B**563** (1999) 45 ; J. Santiago and F. J. Yndurain, Nucl. Phys. B**611** (2001) 447.
8. E143 Collaboration, K. Abe *et al.*, Phys. Rev. **D58**,(1998) 112003; Spin Muon Collaboration, D. Adams *et al.*, Phys. Rev. D **56**,(1997) 5330; Airapetian *et al.*, Phys. Lett. B **442**,(1998)484 .

Renormalization and factorization scales and scheme dependence for some QCD observables

A. Mirjalili[1*†], Ali N. Khorramian [**†] and S. Atashbar Tehrani[‡†]

*Physics Department, Yazd University, Yazd, Iran
†Institute for Studies in Theoretical Physics and Mathematics (IPM),
P.O.Box 19395-5531, Tehran, Iran
**Physics Department, Semnan University, Semnan, Iran
‡Physics Department, Persian Gulf University 75168, Boushehr, Iran

Abstract. The "Complete RG-improvement (CORGI)" serves to separate the perturbation series into infinite subsets of terms which when summed are renormalization scheme(RS)- Invariant. Crucially all ultraviolet logarithms involving the dimensionful parameter, Q, on which the observable depends are resumed, thereby building the correct Q-dependence. We used from this idea and indicated that in NNLO calculation as well as NLO calculation for when we have just one renormalization scale or two renormalization and factorization scales as it is appeared for moment of structure functions, all unphysical parameters can be completely eliminated from perturbative QCD predictions provided that all the ultraviolet logarithms involving the physical energy scale Q are completely resummed.

INTRODUCTION

The problem of the renormalization scale (scheme) dependence of fixed-order perturbative QCD predictions continues to frustrate attempts to make reliable determinations of the underlying dimensional transmutation parameter of the theory , Λ_{QCD}. Whilst a number of proposals for controlling or avoiding this difficulty have been advanced [1, 2, 3, 4] no consensus has been reached. To predominate this difficulty the idea of "Complete RG-improvement (CORGI)" is introduced [1]. In this approach it was proposed that one should perform a resummation to all-orders of *all* renormalization group (RG)-predictable terms at each order of perturbation theory. This procedure automatically organizes the series into infinite subsets of terms which are separately renormalization scheme (RS)-invariant, and crucially also, by completely resumming ultraviolet logarithms, generates the correct asymptotic dependence on the single dimensionful parameter 'Q' on which the observable depends.

In extension the above argument to involve factorization of operator matrix elements and coefficient functions, we show that in NNLO approximation on resumming all the ultraviolet logarithms, the μ renormalization and M factorization scale dependence will be disappeared.

---

[1] E-mail address: mirjalili@ ipm.ir

CP717, *Hadron Spectroscopy: Tenth International Conference,*
edited by E. Klempt, H. Koch, and H. Orth
© 2004 American Institute of Physics 0-7354-0197-7/04/$22.00

STRUCTURE FUNCTION MOMENTS

For the nth moment of a non-singlet structure function $F(x)$, which is defined by

$$M_n(Q) = \int_0^1 x^{n-2} F(x)\,dx\,,\tag{1}$$

we can have the following factorized form [5]

$$M = A\left(\frac{ca}{1+ca}\right)^{d/b} \exp(I(a))\,(1+r_1\tilde{a}+r_2\tilde{a}^2+r_3\tilde{a}^3+\ldots)\,,\tag{2}$$

where $I(a)$ is the finite integral

$$I(a) = \int_0^a dx\,\frac{d_1+(d_1c+d_2-dc_2)x+(d_3+cd_2-c_3d)x^2+\ldots}{b(1+cx)(1+cx+c_2x^2+c_3x^3+\ldots)}\,,\tag{3}$$

which can be readily evaluated numerically. We shall use $\tilde{a}$ to stand for $a(\mu)$ and a for $a(M)$. For simplicity we shall from now on suppress the n-dependence of terms in equations as we have done in Eq.(2). The coupling $a(\tau)$ itself, where $\tau\equiv b\ln(\mu/\tilde{\Lambda})$, is obtained as the solution of the transcendental equation [4]

$$\frac{1}{a} + c\ln\frac{ca}{1+ca} = \tau - \int_0^a dx\left[-\frac{1}{B(x)}+\frac{1}{x^2(1+cx)}\right]\,,\tag{4}$$

where $B(x)\equiv x^2(1+cx+c_2x^2+c_3x^3+\ldots)$.

PARAMETERIZING THE COEFFICIENTS

We Recall from [4] that for the single scale case of a dimensionless observable R (Q) with perturbation series

$$\text{R}\,(Q) = a+r_1a^2+r_2a^3+\ldots+r_na^{n+1}+\ldots\,,\tag{5}$$

the RS can be labelled by the non-universal coefficients of the beta-function $c_2, c_3,\ldots$, and by τ, which can be traded as a parameter for r_1 since [4, 6, 7, 8]

$$\tau - r_1 = \rho_0(Q)\equiv b\ln(Q/\Lambda_{\text{R}})\,,\tag{6}$$

is an RS-invariant. Using the self-consistency of perturbation theory- that is that for instance the derivative of the n-th order approximant R^n with respect to reorganization scale μ is of higher order than the approximant R^n itself, one can derive expressions for the partial derivatives of the perturbative coefficients with respect to the scheme parameters as follows

$$\frac{\partial r_2}{\partial r_1} = 2r_1+c,\quad \frac{\partial r_3}{\partial r_1}=c_2+2cr_1+3r_2,\quad \frac{\partial r_4}{\partial r_1}=c_3+2c_2r_1+3cr_2+4r_3,\,\ldots$$

$$\frac{\partial r_2}{\partial c_2} = -1,\quad \frac{\partial r_3}{\partial c_2}=-2r_1,\quad \frac{\partial r_4}{\partial c_2}=-\frac{a^3c_2+9a^3r_2}{3a^3},\,\ldots$$

$$\frac{\partial r_2}{\partial c_3} = 0,\quad \frac{\partial r_3}{\partial c_3}=-\frac{1}{2},\quad \frac{\partial r_4}{\partial c_3}=-\frac{-a^3c+6a^3r_1}{6a^3},\,\ldots$$

$$\vdots,\qquad\qquad \vdots,\qquad\qquad\qquad \vdots,\,\ldots\,.\tag{7}$$

on integration one finds

$$r_2(r_1,c_2) = r_1{}^2 + cr_1 + X_2 - c_2, \; r_3(r_1,c_2,c_3) = r_1{}^3 + \frac{5}{2}cr_1{}^2 + (3X_2 - 2c_2)r_1 + X_3$$

$$-\frac{1}{2}c_3, \; r_4(r_1,c_2,c_3,c_4) = \frac{4c_2^2}{3} - \frac{c_4}{3} - c_3r_1 + \frac{3c^2r_1^2}{2} + r_1^4 + 6r_1^2X_2 - 3c_2(cr_1 + r_1^2 + X_2),$$

.... $\hspace{11cm}$ (8)

where $X_i(n)$ are representing the RG-unpredictable part of $r_i(n)$ which are Q-independent and RS-invariants.

General Structure is as follows

$$r_n(r_1,c_2,\ldots,c_n) = \hat{r}_n(r_1,c_2,\ldots,c_{n-1}) + X_n - c_n/(n-1) \hspace{3cm} (9)$$

$\hat{r}_n$ is RG-predictable. X_n are unknown unless a complete N^{n}LO calculation has been performed. In generalization to moment problem, we need to parameters which label the scheme in which shows both dependence on renoramalization and factorization scales. In this case we have the r_n dependence as $r_n(\mu,M,c_2,\ldots,c_n,d_1,d_2,\ldots,d_n)$ [5]. As before M, μ can be traded, in this case for $r_1(M)$ and $\tilde{r}_1 \equiv r_1(M=\mu)$. Partially differentiating Eq.(2) with respect to μ,M,c_2,c_3,d_1, d_2,d_3, and demanding for consistency that it be for our case O(a^9) , so that the coefficients of a,a^2 and $a^3, \ldots a^8$ vanish.

In order to do calculation at NNLO approximation, we need to following partial derivatives with respect to μ and m as follows:

$$\mu\frac{\partial r_1}{\partial\mu} = 0, \; \mu\frac{\partial r_2}{\partial\mu} = br_1, \; \mu\frac{\partial r_3}{\partial\mu} = b(cr_1 + 2r_2), \; \mu\frac{\partial r_4}{\partial\mu} = b(c_2r_1 + 2cr_2 + 3r_3), \ldots$$

$$m\frac{dr_1}{dm} = -d, \; m\frac{dr_2}{dm} = -d_1 - dL - dr_1, \; m\frac{dr_3}{dm} = -d_2 - 2d_1L - dL^2 - d_1r_1 - dLr_1 - dr_2,$$

$$m\frac{dr_4}{dm} = -d_3 - 3d_2L - 3d_1L^2 - dL^3 - d_2r_1 - 2d_1Lr_1 - dL^2r_1 - d_1r_2 - dLr_2 - dr_3 , \ldots \hspace{1cm} (10)$$

Here we have defined for convenience $L \equiv b\ln(M/\mu)$. In above expressions X_n are analogous factorization and renormalization scheme (FRS) invariants. Consistently integrating the partial derivatives of r_1 yields

$$r_1 = \frac{d}{b}\tau_M - \frac{d_1}{b} - X_1(Q) , \hspace{4cm} (11)$$

where $\tau_M \equiv b\ln(M/\tilde{\Lambda})$ and $X_1(Q)$ is an FRS-invariant, analogous to $\rho_0(Q)$ for the single scale problem defined in Eq.(6). Consistently integrating the remaining partial derivatives and using Eq.(11) to recast the M and μ dependence in terms of r_1 and $\tilde{r}_1$, one obtains an explicit dependence of r_2 and $r_3 \ldots r_8$ on the FRS parameters r_1 and $\tilde{r}_1$. In below the results for r_2, $r_3 \ldots$ as a function of μ and m are written.

$$r_2(m,\mu) = \frac{1}{2d}(-br_1(r_1 - 2\tilde{r}_1) + d(r_1^2 + 2X_2))$$

$$r_3(m,\mu) = \frac{1}{6d^2}(2b^2r_1(r_1^2 - 3r_1\tilde{r}_1^2) - 3bd(r_1^3 - 2r_1^2\tilde{r}_1 - 4\tilde{r}_1X_2) + d^2(r_1^3 + 6r_1X_2 + 6X_3))$$

$$r_4(m,\mu) = \frac{1}{24d^3}(-6b^3r_1(r_1^3 - 4r_1^2\tilde{r}_1 + 6r_1\tilde{r}_1^2 - 4\tilde{r}_1^3) + b^2d(11r_1^4 - 36r_1^3\tilde{r}_1 + 36r_1^2\tilde{r}_1^2$$

$$+72\tilde{r}_1^2X_2) - 6bd^2(r_1^4 - 2r_1^3\tilde{r} + 2r_1^2X_2 - 12r_1\tilde{r}_1X_2 - 12\tilde{r}_1X_3) + d^3(r_1^4\tilde{r}$$

$$+2r_1^2X_2 - 12r_1\tilde{r}_1X_2 - 12\tilde{r}_1X_3) + d^3(r_1^4 + 12r_1^2X_2 + 24r_1X_3 + 24X_4))$$

$$\vdots \hspace{10cm} (12)$$

As in the single scale case there are constants of integration X_n representing the RG-unpredictable part of r_n. They are Q-independent and FRS-invariant.

RESUMMATION OF ULTRAVIOLET LOGARITHMS

To generalize the argument of [5] and to indicate that in NNLO approximation for single case in doing a resummation on all ultraviolet logarithms the μ dependent will be avoided, we need to keep the terms in Eq.(8) which involves the X_2 coefficients, since these coefficients show the contribution of r_n's in NNLO approximation. If we add all NNLO contributions and use the simplicity $c = 0$, $c_2 = 0$,..., then we will get the following result

$$NNLO \quad : \quad = X_2 a^3 + (3r_1 X_2)a^4 + (6r_1^2 X_2)a^5 + (10r_1^3 X_2)a^6$$
$$+(15r_1^4 X_2)a^7 + \cdots = X_2(\frac{a}{1-r_1 a})^3 \tag{13}$$

By substituting the expressions $r_1 = b\left(\ln\frac{\mu}{\Lambda} - \ln\frac{Q}{\Lambda_R}\right)$, and $a(\mu)=1/b\ln(\frac{\mu}{\Lambda})$ [4] we will arrive at

$$X_2(\frac{1}{b\ln(\frac{Q}{\Lambda_R})})^3 \tag{14}$$

in Eq.(13) which is as we expected, independent of renormalization scale-μ.

In order to show that in two scales case and in NNLO approximation, once again in doing the resummation all μ and m parameters will be disappeared, we back to expression of r_2... r_8 in Eq. (12) and keep just their NNLO contributions (the terms which involves the X_2 terms). Careful consideration of these contributions, make us clarification that we can sort these contribution in matrix form as below where adding the terms of each row indicate the NNLO contribution of r_2,... r_8,... and finally we can find the following relations between each columns as it is indicated in below

$$\begin{pmatrix}
\tilde{a}^2 & 0 & 0 & \cdots \\
\frac{1}{d^4}\tilde{a}^3 b(2d^4)\tilde{r}_1 & \tilde{a}^3 r_1 & 0 & \cdots \\
\frac{1}{d^5}\tilde{a}^4 b^2(3d^3)\tilde{r}_1^2 & \frac{1}{d}3\tilde{a}^4 br_1\tilde{r}_1 & -\frac{1}{2d}\tilde{a}^4(b-d)r_1^2 & \cdots \\
\frac{1}{d^5}\tilde{a}^5 b^3(4d^2)\tilde{r}_1^3 & \frac{1}{d^2}6\tilde{a}^5 b^2 r_1\tilde{r}_1^2 & -\frac{1}{d^2}2\tilde{a}^5 b(b-d)r_1^2\tilde{r}_1 & \cdots \\
-\frac{1}{d^5}\tilde{a}^6 b^4(5d)\tilde{r}_1^4 & \frac{1}{d^3}10\tilde{a}^6 b^3 r_1\tilde{r}_1^3 & -\frac{1}{d^3}5\tilde{a}^6 b^2(b-d)r_1^2\tilde{r}_1^2 & \cdots \\
\vdots & \vdots & \vdots & \ddots
\end{pmatrix} \tag{15}$$

The result for first column, second column and ... can be written in compact forms as follows:

$$s_1 = \frac{X_2\tilde{a}^2}{(1-\frac{\tilde{a}b\tilde{r}_1}{d})^2}, \quad s_2 = \frac{X_2\tilde{a}^3 r_1}{(1-\frac{\tilde{a}b\tilde{r}_1}{d})^3}, \quad s_3 = \frac{X_2\tilde{a}^4 r_1^2(1-\frac{b}{d})}{2(1-\frac{\tilde{a}b\tilde{r}_1}{d})^4}, \quad \cdots . \tag{16}$$

In adding all columns we will obtain:

$$C = \frac{X_2\tilde{a}^2}{(1-\frac{\tilde{a}b\tilde{r}_1}{d})^2}(1+\frac{\tilde{a}r_1}{(1-\frac{\tilde{a}b\tilde{r}_1}{d})}+\frac{\tilde{a}^2 b(-1+\frac{d}{b})r_1^2}{2d(1-\frac{\tilde{a}b\tilde{r}_1}{d})^2}+$$
$$\frac{\tilde{a}^3 b^2(-2+\frac{d}{b})(-1+\frac{d}{b})r_1^3}{6d^2(1-\frac{\tilde{a}b\tilde{r}_1}{d})^3}+\frac{\tilde{a}^4 b^3(-3+\frac{d}{b})(-2+\frac{d}{b})(-1+\frac{d}{b})r_1^4}{24d^3(1-\frac{\tilde{a}b\tilde{r}_1}{d})^4}+...) \tag{17}$$

which is in fact the Taylor expansion of following expression

$$C = a^{\frac{d}{b}} \frac{\tilde{a}^2 X_2}{(1 - \frac{\tilde{a}b\tilde{r}_1}{d})^2} (1 + \frac{b}{d}(\frac{\tilde{a}r_1}{(1 - \frac{\tilde{a}b\tilde{r}_1}{d})}))^{d/b} \tag{18}$$

By substituting the following relation for $\tilde{r}_1$ and $\tilde{a}$ [2]

$$\tilde{r}_1 = (\frac{d}{b\tilde{a}} - d\ln(\frac{Q}{\Lambda_m})), \ \tilde{a} = \frac{1}{b\ln(\mu/\Lambda_m)} \tag{19}$$

we will get the final result as follows

$$C_f = \frac{X_2(\frac{1}{\ln(\frac{Q}{\Lambda_m})})^{2 + \frac{d}{b}}}{b^2} \tag{20}$$

which is the NNLO contribution of moment of structure function and as we expected are independent of μ and M scales.

DISCUSSION AND CONCLUSIONS

In [5] it is shown that in NLO approximation if one can perform a resummation on all ultraviolet logarithm terms, all renormalization scale (one scale case) and all renormalization and factorization scales (two scales case-moment of structure function) will be avoided. We extended here this idea and indicated that in NNLO approximation we could rid of unphysical parameters in one and two scale cases as well if we do a proper resummation and this is confirmed the CORGI approach which claims that we can separate the perturbation series into finite subsets which when summed are RS-invariant.

ACKNOWLEDGEMENTS

A.M. thanks from Yazd University for partial financial support to do this project. A.N.K, A.M and S.A.T acknowledge from Institute for studies in Theoretical Physics and Mathematics (IPM) to support financially this project.

REFERENCES

1. C.J. Maxwell, hep-ph/9908463; C.J. Maxwell, Nucl. Phys. Proc. Suppl. **B86** (2000) 74.
2. H. David Politzer, Nucl Phys **B194** (1982) 493.
3. P.M. Stevenson and H. David Politzer, Nucl. Phys. **B277** (1986) 758.
4. P.M. Stevenson, Phys. Rev. **D23** (1984) 2916.
5. C.J. Maxwell and A.Mirjalili, Nucl. Phys **B577** (2000) 209.
6. G. Grunberg, Phys. Rev. **D29** (1984) 2315.
7. A.L. Kataev, N.V. Krasnikov, and A.A. Pivovarov, Nucl. Phys **B198** (1982) 508.
8. G. Grunberg, Phys. Lett. **B95** (1980) 70.

Production of η and η' mesons via the quasi-free proton-neutron interaction

P. Moskal [1*], H.-H. Adam[†], A. Budzanowski[**], R. Czyżykiewicz[*],
D. Grzonka[*], M. Janusz[‡], L. Jarczyk[‡], T. Johansson[§], B. Kamys[‡],
A. Khoukaz[†], K. Kilian[*], P. Kowina[*], W. Oelert[*], C. Piskor-Ignatowicz[‡],
J. Przerwa[‡], T. Rożek[*], R. Santo[†], G. Schepers[*], T. Sefzick[*],
M. Siemaszko[‡], J. Smyrski[‡], A. Strzałkowski[‡], A. Täschner[†], P. Winter[*],
M. Wolke[2*], P. Wüstner[¶] and W. Zipper[‖]

[*]IKP, Forschungszentrum Jülich, D-52425 Jülich, Germany
[†]IKP, Westfälische Wilhelms–Universität, D-48149 Münster, Germany
[**]Institute of Nuclear Physics, PL-31-342 Cracow, Poland
[‡]M. Smoluchowski Institute of Physics, Jagellonian University, PL-30-059 Cracow, Poland
[§]Uppsala University, S-75121 Uppsala, Sweden
[¶]ZEL, Forschungszentrum Jülich, D-52425 Jülich, Germany
[‖]Institute of Physics, University of Silesia, PL-40-007 Katowice, Poland

Abstract. A comparison of the close-to-threshold total cross sections for the η' meson production in both the $pp \to pp\eta'$ and $pn \to pn\eta'$ reactions should provide insight into the flavour-singlet (perhaps also into gluonium) content of the η' meson and the relevance of quark-gluon or hadronic degrees of freedom in the creation process. The excitation function for the reaction $pp \to pp\eta'$ has been already established. At present, experimental investigations of the quasi-free $pn \to pnX$ reactions are carried out at the COSY-11 facility using a beam of stochastically cooled protons and the deuteron cluster target. A method of measurement and preliminary results from the test experiments of the $pn \to pn\eta$ reaction are presented in this report.

INTRODUCTION

Close-to-threshold production of η and η' mesons in the nucleon-nucleon interaction requires a large momentum transfer between the nucleons and occur at distances in the order of ~ 0.3 fm. This implies that the quark-gluon degrees of freedom may play a significant role in the production dynamics of these mesons. Therefore, additionally to the mechanisms associated with meson exchanges it is possible that the η' meson is created from excited glue in the interaction region of the colliding nucleons [1, 2], which couple to the η' meson directly via its gluonic component or through its SU(3)-flavour-singlet admixture. The production through the colour-singlet object as suggested in reference [1] is isospin independent and should lead to the same production yield of the η' meson in the $pn \to pn\eta'$ and $pp \to pp\eta'$ reactions after correcting for the final

---

[1] email: p.moskal@fz-juelich.de
[2] present address: The Svedberg Laboratory, Thumbergsvågen 5A, Box 533, S-75121 Uppsala, Sweden.

CP717, Hadron Spectroscopy: Tenth International Conference,
edited by E. Klempt, H. Koch, and H. Orth
© 2004 American Institute of Physics 0-7354-0197-7/04/$22.00

and initial state interaction between the nucleons.

Investigations of the η-meson production in collisions of nucleons allowed to conclude that, close to the kinematical threshold, the creation of η meson from isospin $I = 0$ exceeds the production with $I = 1$ by about a factor of 12. This was derived from the measured ratio of the total cross sections for the reactions $pn \to pn\eta$ and $pp \to pp\eta$ ($R_\eta = \frac{\sigma(pn \to pn\eta)}{\sigma(pp \to pp\eta)}$), which was determined to be $R_\eta \approx 6.5$ in the excess energy range between 16 MeV and 109 MeV [3]. The large difference of the total cross section between the isospin channels suggests the dominance of isovector meson (π and ρ) exchange in the creation of η in nucleon-nucleon collisions [4, 3].

Since the quark structure of η and η' mesons is very similar we can – by analogy to the η meson production – expect that in the case of dominant isovector meson exchange the ratio $R_{\eta'}$ should also be about 6.5. If, however, the η' meson was produced via its flavour-blind gluonic component from the colour–singlet glue excited in the interaction region, the ratio $R_{\eta'}$ should approach unity after corrections for the interactions between the participating baryons.

Figure 1 demonstrates qualitatively the fact that the production of mesons in the proton-neutron collisions is more probable than in the proton-proton interaction if it is driven by the isovector meson exchanges only. This is because in the case of the proton-neutron collisions there are always more possibilities to realise the exchange or fusion of the isovector mesons than in the case of the reaction of protons.

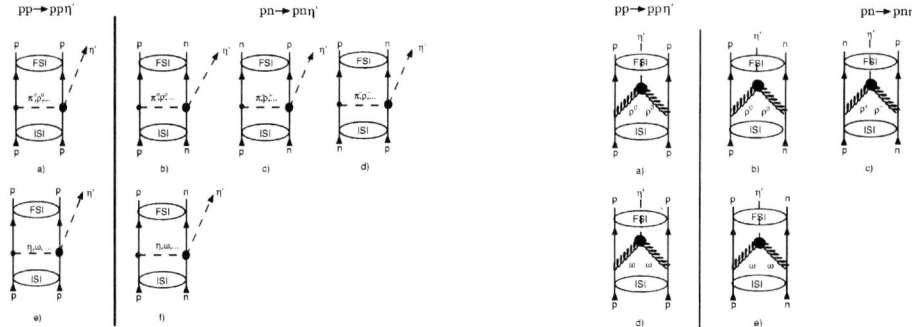

FIGURE 1. (**left**) Example of diagrams with the isovector and isoscalar meson exchange leading to the creation of the meson η' in the proton-proton and proton-neutron collisions. (**right**) Fusion of the virtual ω and ρ mesons emitted from the colliding nucleons.

The close-to-threshold excitation function for the $pp \to pp\eta'$ reaction has been already determined [5, 6, 7, 8] whereas the total cross section for the η' meson production in the proton-neutron interaction remains unknown. As a first step towards the determination of the value of $R_{\eta'}$ the feasibility of the measurement of the $pn \to pn\eta'$ reaction by means of the COSY-11 facility was studied by means of Monte-Carlo [9, 10]. As a second step, a test experiment of the $pn \to pn\eta$ reaction – suspected to have by at least a factor of 30 larger cross section than the one for the $pn \to pn\eta'$ reaction – was performed. In this test measurement, using a beam of protons and a deuteron cluster target, we have proven the ability of the COSY-11 facility to study the quasi-free creation of mesons via the $pn \to pnX$ reaction. Appraisals of simulations and preliminary results of the measurements of the quasi-free $pn \rightarrow pn\eta$ reaction performed using the newly extended COSY-11 facility [9, 10] will be presented in the next section.

TEST MEASUREMENT OF THE $PN \rightarrow PN\eta$ REACTION

As a general commissioning of the extended COSY-11 facility to investigate quasi-free $pn \rightarrow pnX$ reactions, we have performed a measurement of the $pn \rightarrow pn\eta$ process at a beam momentum of 2.075 GeV/c. The experiment, carried out in June 2002, had been preceded by the installation of a spectator [11] and neutron detectors, and by a series of thorough simulations performed in order to determine the best conditions for measuring quasi-free $pn \rightarrow pn\eta$ and $pn \rightarrow pn\eta'$ reactions [9, 10]. Figure 2 presents the COSY-11 detection facility with superimposed tracks of protons and neutron originating from the quasi-free $pn \rightarrow pnX$ reaction induced by a proton beam [12] impinging on a deuteron target [13]. The identification of the $pn \rightarrow pn\eta$ reaction is based on the measurement of the four-momentum vectors of the outgoing nucleons and the η meson is identified via the missing mass technique. The slow proton stopped in the first layer of the position sensitive silicon detector (Si_{spec}) is, in the analysis, considered as a spectator without interaction with the bombarding particle and it is moving with the Fermi momentum as possessed at the moment of the collision.

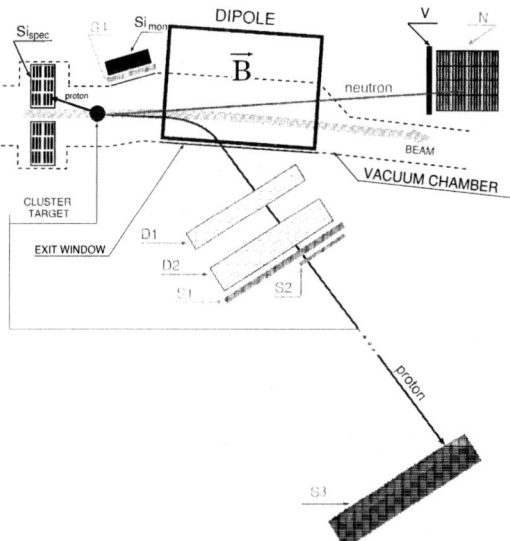

FIGURE 2. Schematic view of the COSY-11 detection setup [14]. Only detectors needed for the measurements of the reaction $pd \rightarrow p_{sp}pn\eta(\eta')$ are shown.
D1, D2 denote the drift chambers; S1, S2, S3, S4 and V the scintillation detectors; N the neutron detector and Si_{mon} and Si_{spec} silicon strip detectors to detect elastically scattered and spectator protons, respectively.

From the measurement of the momentum vector of the spectator proton $\vec{p}_{sp}$ one can infer the momentum vector of the struck neutron $\vec{p}_n = -\vec{p}_{sp}$ at the time of the reaction

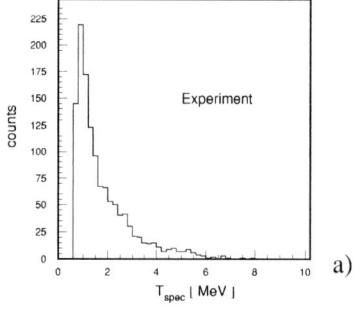 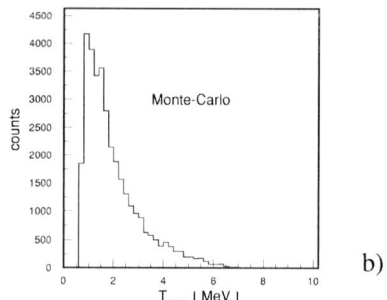

a)

b)

FIGURE 3. Distributions of the kinetic energy of the spectator protons.
(a) Experiment, (b) Monte-Carlo simulations taking into account the acceptance of the COSY-11 detection system and an analytical parametrization of the deuteron wave function [15] calculated from the PARIS potential [16].

and hence calculate the total energy of the colliding nucleons for each event. In the approximation that the struck neutron is treated as a free particle we can assume that the matrix element for quasi–free meson production off a bound neutron is identical to that for the free $pn \rightarrow pnMeson$ reaction [3]. In figures 3a and 3b the measured and expected distribution of the kinetic energy of the spectator proton is presented. Though still very rough energy calibration of the detector units one recognizes a substantial similarity in the shape of both distributions.

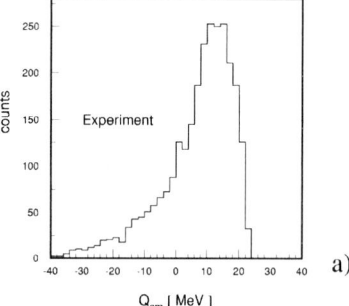 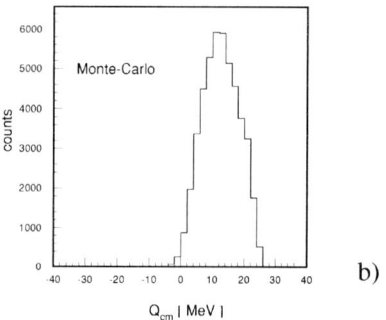

a)

b)

FIGURE 4. Distributions of the excess energy Q_{CM} for the quasi-free $pn \rightarrow pnX$ reaction, determined with respect to the $pn\eta$ threshold. (a) Experiment. (b) Simulation.

Figure 4 shows spectra of the excess energy in respect to the $pn\eta$ system as obtained in the experiment (4a) and the simulation (4b) for the $pn \rightarrow pn\eta$ reaction. The remarkable difference between the distributions comes from the fact that in reality additionally to the $pn \rightarrow pn\eta$ reaction also the multi-pion production is registered. The η and multi-

---

[3] For more comprehensive discussion of this issue the reader is referred to reference [17]

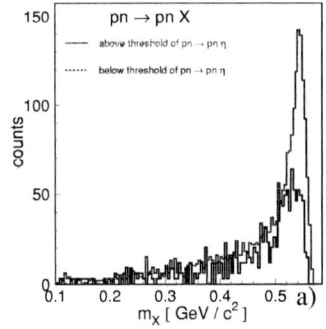 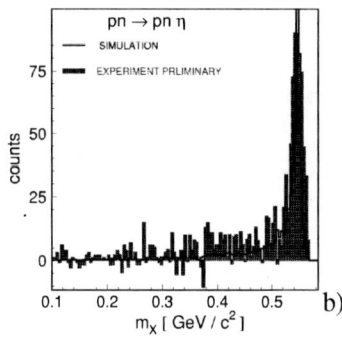

FIGURE 5. Missing mass spectra as obtained during the June'02 run:
a) Event distribution for $Q < 0$ (black line) and for $Q > 0$ (gray line).
b) Histogram represents the difference between number of events above and below threshold for the $pn \rightarrow pn\eta$ reaction, and the line corresponds to the Monte-Carlo simulation.

pion production cannot be distinguished from each other on the event-by-event basis by means of the missing mass technique. However, we can determine the number of the registered $pn \rightarrow pn\eta$ reactions from the multi-pion background comparing the missing mass distributions for Q values larger and smaller than zero. Knowing that negative values of Q can only be assigned to the multi-pion events we can derive the shape of the missing mass ditribution corresponding to these events. This is shown as the solid line in figure 5. A thorough evaluation of the background is in progress, however, rough comparison of events for positive and negative Q yields the promising results with a clear signal from the $pn \rightarrow pn\eta$ reactions, as can be deduced by inspection of figures 5a and 5b.

REFERENCES

1. S. D. Bass, Phys. Lett. **B 463** (1999) 286.
2. S. D. Bass, e-Print Archive: hep–ph/0006348.
3. H. Calén et al., Phys. Rev. **C 58** (1998) 2667.
4. G. Faldt, T. Johansson, C. Wilkin, Phys. Scripta **T 99** (2002) 146.
5. P. Moskal et al., Phys. Rev. Lett. **80** (1998) 3202.
6. P. Moskal et al., Phys. Lett. **B 474** (2000) 416.
7. F. Hibou et al., Phys. Lett. **B 438** (1998) 41.
8. F. Balestra et al., Phys. Lett. **B 491** (2000) 29.
9. P. Moskal, Schriften des FZ-Jülich: Matter & Material **11** (2002) 27; e-Print Archive: nucl-ex/0110001.
10. R. Czyżykiewicz, Diploma Thesis, Jagellonian University (2002), Berichte des FZ-Jülich, Jül-4017.
11. R. Bilger et al., Nucl. Instr. & Meth. **A 457** (2001) 64.
12. D. Prasuhn et al., Nucl. Instr. & Meth. **A 441** (2000) 167.
13. H. Dombrowski et al., Nucl. Instr. & Meth. **A 386** (1997) 228.
14. S. Brauksiepe et al., Nucl. Instr. & Meth. **A 376** (1996) 397.
15. M. Lacombe et al., Phys. Lett. **101B** (1981) 139.
16. M. Lacombe et al., Phys. Rev. **C21** (1980) 861.
17. P. Moskal, M. Wolke, A. Khoukaz, W. Oelert, Prog. Part. Nucl. Phys. **49** (2002) 1, hep-ph/0208002.

Associated strangeness production in pp collisions near threshold

P. Winter for the COSY-11 collaboration

IKP, Forschungszentrum Jülich, D-52425 Jülich, Germany

Abstract. Motivated by the ongoing discussion concerning the nature of the scalar resonances $f_0(980)$ and $a_0(980)$, the COSY-11 collaboration has taken exclusive data on the $pp \rightarrow ppK^+K^-$ reaction near the production threshold. A first total cross section $\sigma = (1.80 \pm 0.27^{+0.28}_{-0.35})$ nb for the excess energy $Q = 17$ MeV has been determined. In contrary to the η, ω, and η' single meson production studies which clearly show the strong pp final state interaction (FSI), the cross section values obtained at COSY-11 and DISTO can be both described by a fit with a four-body phase space including the proton-proton final state interaction as well as with one-meson exchange calculations neglecting FSI effects. Therefore, one might think about a compensation of the strong pp interaction through a pK^- FSI effect or an additional degree of freedom caused by the four-body final state. In the latter case, strong FSI effects can be expected at Q-values very close to the K^+K^- production threshold. Such a motivation triggered – in combination with the investigation of the $K\bar{K}$ interaction being relevant to the structure of the $f_0(980)$ – further measurements at the excess energies $Q = 10$ and $Q = 28$ MeV at COSY-11.

INTRODUCTION

Meson production close to threshold in nucleon-nucleon collisions offers an excellent tool in order to study meson-meson and baryon-meson interactions. Due to the low excess energies, the outgoing particles have a low relative momentum and hence final state interactions are more pronounced giving the possibility to derive e. g. scattering parameters. Additionally, because of the few relevant partial waves, the theoretical description of the reaction is simplified. Furthermore, studies of these elementary production processes enable to learn about the underlying production mechanisms. Last but not least, the high momentum transfer in such reactions probes the short range components of the hadronic interactions.

Over the last years, there have been several experimental investigations on the close to threshold meson production at different accelerators covering a mass range from the π- up to the ϕ meson. Recently, first results of the challenging studies of the close to threshold cross sections for the production of the broad $f_0(980)$[1] and a_0^+ resonances via the proton-proton collisions have been reported [1, 2]. The COSY-11 collaboration studies – besides other reactions – the K^+K^- production in proton-proton collisions. A first cross section $\sigma(Q = 17\,\text{MeV}) = (1.80 \pm 0.27^{+0.28}_{-0.35})$ nb has been published [3].

---

[1] In the following, f_0 and a_0 shall be regarded as an abbreviation for $f_0(980)$ and $a_0(980)$, respectively.

CP717, *Hadron Spectroscopy: Tenth International Conference,*
edited by E. Klempt, H. Koch, and H. Orth
© 2004 American Institute of Physics 0-7354-0197-7/04/$22.00

Subsequently, the measurements have been extended to excitation energy values of $Q = 10$ and $Q = 28$ MeV.

EXPERIMENT

The internal experiment COSY-11 [4] at the COoler SYnchrotron COSY [5] is shown in Figure 1. A hydrogen cluster target [6] is mounted in front of one of the regular COSY dipoles which acts as a magnetic spectrometer. The positively charged ejectiles are bent to the inner part of the ring and detected by a set of drift chambers followed by a time of flight (TOF) measurement via two scintillator detectors S1 and S3. The momentum determination is performed by tracing back the reconstructed trajectories through the known magnetic field to the interaction point. Together with the velocity calculable from the TOF measurement the four momentum of the positively charged particles are determined. Therefore, an identification of the ppK^+-system via the invariant mass of each track enables to identify the K^- by means of the missing mass method. Additionally, in

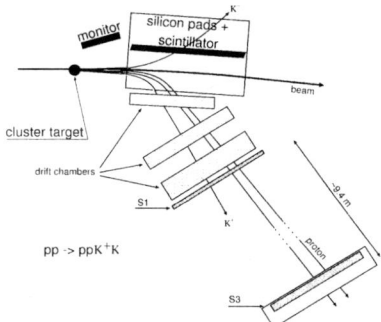

FIGURE 1. Experimental setup of the COSY-11 detection system.

the inner part of the dipole gap, a silicon pad detector in combination with a scintillator detector allows for the detection of negatively charged particles which is essential for a nearly 100% background reduction as will be shown later.

The luminosity is determined by the simultaneous measurement of the elastic pp scattering [7]. The detectction of the second proton is performed with another silicon pad detector, the so called monitor detector in Figure 1 close to the target.

PHYSICS MOTIVATION

The $K\bar{K}$ interaction has been under discussion for several years not only in the context of the structure of the isoscalar meson f_0. In the framework of a one-boson exchange model several theoretical groups have been working in this field. The strength of the $K\bar{K}$ interaction is crucial for the energy dependence of the cross section $\pi^+\pi^- \to K\bar{K}$ calculated by Krehl, Rapp and Speth [8]. Figure 2 (left part) shows the cross section for both $K^0\bar{K}^0$ and K^+K^- production including the $K\bar{K}$ interaction (solid lines) and without (dashed lines). It is obvious that the cross section rises much steeper for the inclusion of the kaon anti-kaon interaction. A similar behaviour can be expected for a proton-proton initial state although complete calculations for this case were not yet performed.

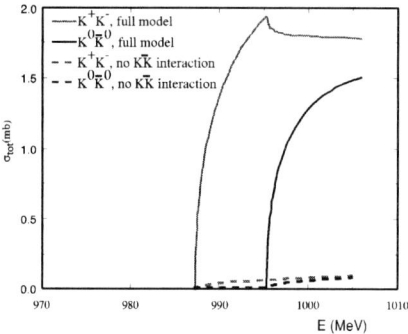 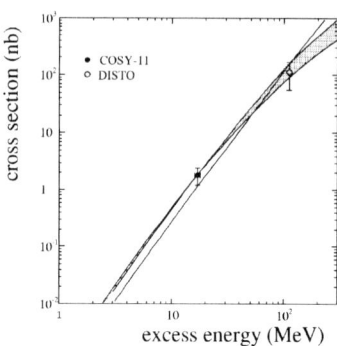

FIGURE 2. *Left picture*: Total cross section calculated within the Jülich meson exchange model [8] for the $\pi^+\pi^- \to K\bar{K}$ reaction. *Right picture*: Total cross section for $pp \to ppK^+K^-$. The data are taken from [9, 3] while further information on the lines are given in the text.

The knowledge of the $K\bar{K}$ interaction strength will certainly contribute to the understanding of the structure of the f_0 meson. Just to give an idea on the still open question concerning the f_0 we will mention some models without claiming any completeness.

The interpretation of this resonance as a normal $q\bar{q}$ state within the isoscalar nonet seems to be disfavoured [10, 11]. The predictions for the decay width of such states as well as the coupling to pseudoscalar final states ($f_0 \to \pi\pi$, $a_0 \to \pi\eta$) are more than one order of magnitude higher than the experimental observations. The authors of reference [11] find in their unitarity-enforcing analysis within a Jost function representation of the S-matrix a clear preference for the standard Breit-Wigner description, while in one of their former papers [12] the outcome of the analysis was a two resonance description of the f_0 meson. The changed result can be attributed to the inclusion of new data on J/ψ and D_s decays. Another prescription has been given by R. J. Jaffe [13] where the f_0 is assigned as a member of the lightest cryptoexotic $q^2\bar{q}^2$ nonet. The calculations are based on a semiclassical approximation to the MIT bag theory [14]. Such a $(qq\bar{q}\bar{q})$ configuration rather predicts a rich spectrum of experimentally not confirmed states. This problem does not occur in the potential model by Weinstein and Isgur [10] in which the f_0 is found to be a weakly bound $K\bar{K}$ system. Similar findings are given in the framework of the Jülich meson exchange model for $\pi\pi$ and $\pi\eta$ interactions [15, 8]. While the list of such different descriptions could be carried on, it should have become obvious that the structure of the isoscalar mesons is still barely known and that the $K\bar{K}$ interaction plays a crucial role in several models.

Finally, measurements on the reaction $pp \to ppK^+K^-$ close to threshold also open the possibility to study final state interactions in the proton-kaon and kaon-kaon systems. Figure 2 (right side) shows the two existing data points on the total cross section in $pp \to ppK^+K^-$ from the DISTO- and COSY-11 collaborations [9, 3]. The dashed line is a fit of a four body phase space with inclusion of the strong pp FSI. The solid lines are fits of phase space via an intermediate ppf_0 state also including the pp FSI. The shaded area stems from the uncertainty of the width of the f_0. The present available data do not allow to discriminate between the resonant and nonresonant production. It should be

914

mentioned that also a calculation by Sibirtsev et al. [16] without any FSI can reproduce the actual cross sections. This led to the speculation that a partial compensation of the pp and pK^- interaction takes place or an additional degree of freedom in the four body final state is responsible for the absence of those FSI effects.

STATUS OF THE ANALYSIS

In order to obtain further insight into the understanding of the $K\bar{K}$ interaction, the COSY-11 collaboration has extended their measurements on $pp \rightarrow ppK^+K^-$ to excess energies $Q = 10$ and $Q = 28$ MeV. The missing mass spectrum for the excess energy of $Q = 28$ MeV is shown in the left part of figure 3. The selected events include two identified protons and a positive kaon in the final state. Besides a clear peak at the

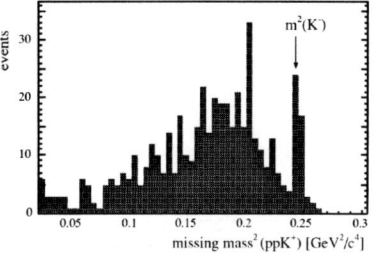 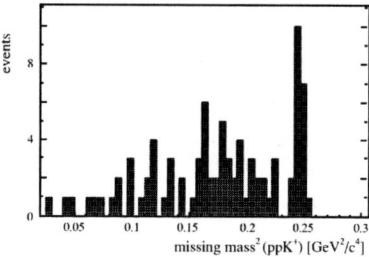

FIGURE 3. *Left picture*: Squared missing mass of the identified ppK^+ system. *Right picture*: The same events like in the left picture but with an additional hit in the silicon pad detector.

corresponding K^- mass a broad background is observed which is understood in terms of the intermediate excitation of the $\Sigma(1385)$ and $\Lambda(1405)$ resonances and misidentified pions stemming from $pp \rightarrow pp\pi^+X$ reactions [3]. It should be emphasized that in the area of the K^- peak, the contamination from the background is very small. The same events with the additional demand for a hit in the silicon pad detector mounted in the dipole gap are shown in the right part of figure 3. While the total amount of $pp \rightarrow ppK^+K^-$ events reduces as expected from Monte Carlo simulations, this cut drastically influences the background structure. Already at this stage the events within the K^- peak are nearly 100% background free.

A last step in this analysis will be not only to demand a hit in the silicon pad detectors but also to compare the hit position with the reconstructed one. Since the four momenta of the 2 protons and the K^+ are known, one can calculate the missing four momentum and reconstruct the expected trajectory of the K^- meson in the magnetic field [17]. Since this procedure was not yet applied to the data set at $Q = 28$ MeV, we will depict its efficiency with the already published data at $Q = 17$ MeV [3]. The comparison between the measured and expected hit position in the silicon pads for identified ppK^+ events is shown in figure 4 (left side). The black dots indicate those events which were assigned, according to the experimental resolution, to the K^- peak and the open circles all others. Whereas all events in the kaon signal scatter slighty around the expected correlation, the other events distribute quite randomly over the plot. The applied cuts (dashed lines) were extracted from Monte Carlo simulations in order to simultaneously minimize the loss of

K^- events and to maximize the reduction of the background. The result of applying this

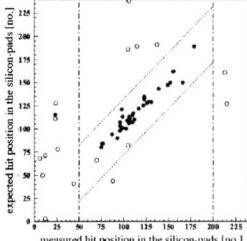 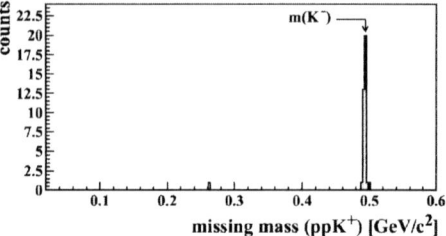

FIGURE 4. *Left picture*: Comparison of the measured hits in the silicon pad detectors for identified ppK^+ events. *Right picture*: Missing mass spectrum for $Q = 17\,\text{MeV}$ for events with two identified protons, one identified negative kaon and a hit in silicon pads within the cut explained in the text.

last cut to the selected events is a essential background free missing mass spectrum as is shown in the right part of figure 4 for $Q = 17\,\text{MeV}$. It is expected that the same supression of the background will be achieved for the two data sets at $Q = 10$ and $28\,\text{MeV}$.

SUMMARY

The COSY-11 collaboration extended their measurements on the elementary production of K^+K^- in pp collisions with new data at $Q = 10$ and $Q = 28\,\text{MeV}$. From a former measurement at $Q = 17\,\text{MeV}$ it is known, that this can be achieved background free. The analysis of both data sets is still in progress but for the higher excess energy a clear kaon signal is already seen. The new information on the excitation function in this reaction will help to understand the underlying process and the strength of the $K\bar{K}$ interaction. This might help to clarify the open question on the structure of the isoscalar meson f_0.

REFERENCES

1.   Moskal, P., et al., *J. Phys.*, **G 29**, 2235–2246 (2003).
2.   Kleber, V., et al., *e-Print Archive: nucl-ex/0304020* (2003).
3.   Quentmeier, C., et al., *Phys. Lett.*, **B 515**, 276–282 (2001).
4.   Brauksiepe, S., et al., *Nucl. Instr. & Meth.*, **A 376**, 397–410 (1996).
5.   Maier, R., *Nucl. Instr. & Meth.*, **A 390**, 1–8 (1997).
6.   Dombrowski, H., et al., *Nucl. Instr. & Meth.*, **A 386**, 228–234 (1997).
7.   Moskal, P., et al., *Nucl. Instr. & Meth.*, **A 466**, 448–455 (2001).
8.   Krehl, O., Rapp, R., and Speth, J., *Phys. Lett.*, **B 390**, 23–28 (1997).
9.   Balestra, F., et al., *Phys. Rev.*, **C 63**, 024004 (2001).
10. Weinstein, J. D., and Isgur, N., *Phys. Rev.*, **D 41**, 2236–2257 (1990).
11. Morgan, D., and Pennington, M. R., *Phys. Rev.*, **D 48**, 1185–1204 (1993).
12. Au, K. L., Morgan, D., and Pennington, M. R., *Phys. Rev.*, **D 35**, 1633–1664 (1987).
13. Jaffe, R. L., *Phys. Rev.*, **D 15**, 267 (1977).
14. Chodos, A., Jaffe, R. L., Johnson, K., and Thorn, C. B., *Phys. Rev.*, **D 10**, 2599–2604 (1974).
15. Janssen, G., Pearce, B. C., Holinde, K., and Speth, J., *Phys. Rev.*, **D 52**, 2690–2700 (1995).
16. Sibirtsev, A. A., Cassing, W., and Ko, C. M., *Z. Phys.*, **A 358**, 101–106 (1997).
17. Quentmeier, C., *Untersuchung der Reaktion $pp \to ppK^+K^-$ nahe der Produktionsschwelle*, Dissertation, Westfälische Wilhelms-Universität Münster (2001).

Summary Talks

The End of the Constituent Quark Model?

F.E. Close

Department of Theoretical Physics, University of Oxford,
1 Keble Rd., Oxford, OX1 3NP, United Kingdom;
f.close@physics.ox.ac.uk

Abstract. In this conference summary talk at Hadron03, questions and challenges for Hadron physics of light flavours are outlined. Precision data and recent discoveries are at last exposing the limitations of the naive constituent quark model and also giving hints as to its extension into a more mature description of hadrons. These notes also pay special attention to the positive strangeness baryon $\Theta^+(1540)$ and include a pedagogic discussion of wavefunctions in the pentaquark picture, their relation with the Skyrme model and related issues of phenomenology.

Introduction

My brief was to concentrate on light hadrons; but where do heavy hadrons end and light begin? I shall focus on what heavy flavours can teach us about light, and vice versa. The possible discovery of an exotic and metastable baryon with positive strangeness, the $\Theta^+(1540)$, has led to an explosion of interest in recent months and throughout this conference. There was a dedicated discussion session about it, which highlighted much confusion. In the hope of clarifying some of the issues, I have decided to devote a considerable part of this summary to a pedagogic description of wavefunctions and a review of some of the emerging literature that drew comment at the conference.

Light Hadron Spectroscopy and Dynamics: Present and Future

As regards the future of light hadrons experimentally: we have heard of several examples of innovative methods involving high energy machines. In particular, we have electron positron machines designed as B-factories, which turn out to access lower energies as a result of the initial state radiation[1]. They also provide copious data on $\gamma\gamma \rightarrow$ light hadrons. At HERA we have vector mesons produced diffractively but also now charge conjugation positive states produced apparently in the rapidity gap between the photon and the target proton[2]. Then there is the new opportunity for central production in proton-proton collisions at STAR and the proposal for this at HERA-g [3]. Exploiting the dk_T filter and ϕ dependences may separate $q\bar{q}$ states from those with gluonic admixtures or S-wave substructure (see later)[4].

Then we have heard about B and D decays into light hadrons from BaBar, BELLE and FOCUS and novel ideas from Ochs about $B \rightarrow K^* + 0^{++}$ as an entree into the light scalars[5]. There is also $D_s \rightarrow \pi(s\bar{s})_{0^{++}}$ which gives an entree to the scalar $s\bar{s}$ sector. The decays $\psi \rightarrow \gamma\gamma V$, where $V \equiv \rho, \phi$ have been reported from BES[6] and will be

CP717, Hadron Spectroscopy: Tenth International Conference,
edited by E. Klempt, H. Koch, and H. Orth
© 2004 American Institute of Physics 0-7354-0197-7/04/$22.00

pursued by CLEO-c [7]. So far these data have been applied to the $\iota(1440)$[6] region but eventually promise to provide information on the flavour content of any $C = +$ mesons, through $\psi \to \gamma R(C = +) \to \gamma\gamma V$. In particular for $R \equiv 0^{++}$ this can give essential information on the flavour contents, and hence mixings with the glueball of lattice QCD, of the various scalar mesons[8].

When high statistics data are available at CLEO-c and BES-III we can study $\chi \to \pi + R$ where $R \equiv$ light hadrons and complement the old $p\bar{p}$ data from LEAR. The bonus will be that $\sqrt{s} \sim 3.5\text{GeV}$ and that the overall J^{PC} is known in the subsequent partial wave analyses. Of particular interest here could be $\chi \equiv 1^{++}$ where in S-wave the recoil system $R \equiv 1^{-+}$, which is the exotic channel favoured for hybrid mesons. So there are reasons to be optimistic about sorting out light hadron spectroscopy and dynamics and solving whether and how the gluonic degrees of freedom are manifested in the strong QCD regime.

As G.W.Bush and T. Bliar might summarise the search for missing glueballs and hybrids: we know they exist; they are hidden but we will find them; give us time - we have only been searching for 20 years.

We have also heard[6] how ψ decays can give novel insights into baryon resonances in the timelike region through $\psi \to \bar{N}N^*$ or $\bar{\Delta}\Delta^*$. This selects isospin states apart from a background due to the intermediate $\psi \to \gamma^*$ channel, and gives complementary information to that from the maturing data from Jefferson Laboratory[9]. Finally we have the degeneracy of the hybrid candidate $\pi(1800)$ and $D(1865)$. The Cabibbo suppressed decays of the latter[10] share common channels with the strong decays of the former. Disentangling this is a significant overlap between heavy and light flavours of a pragmatic nature let alone the interesting potential implications for novel physics.

A problem in light hadron dynamics is that not all of the data can be correct. Swanson[11] has shown a nice figure listing all of the mesons as a function of J^{PC} from the PDG which is clearly overpopulated. This leads me to two requests: one to theorists and one to experimentalists.

Theorists: beware of taking your favourite random model; finding the J^{PC} states that agree with it and then ignoring, excusing or tweaking the model to apologise for those that do not not. It is important to keep the big picture in mind if we are to progress. Focus on the wood not the trees. There is information on more than spectroscopy. We have decays and also production dynamics that can provide essential constraints on models and interpretation.

Experimentalists: what am I supposed to regard as "official" data? Is a conference presentation, which does not get refereed for a peer reviewed journal "official"? At this Hadron series over the years we have seen reports of measurements on say phenomenon H1. Nothing is seen in a peer reviewed journal. Two years later the same group might report the phenomenon as H2; or they might say nothing and only when questioned by those in the know reveal that they no longer see any signal. However, this is not reported as an "official" withdrawal. It is hard enough trying to interpret the data without the added pollution of work in process. In particular, it is extremely important that states which are claimed, and then go away, be reported as withdrawn.

I would not want to stifle the presentation of preliminary data, as such creates discussion that can be mutually informative. However, when it is written up for the proceedings I would urge that a clear statement be made up front as to the status of the report: e.g.

on what timescale will a version be prepared for "official" publication? If the data are even more preliminary, I would suggest that no written summary be produced for the proceedings; or that a clear disclaimer be made that this is a report on an individual's analysis. I would hope that any such presentation has received the endorsement of the group, but suspect that this is not always the case, since at this conference I have heard at least one parallel session where group members appeared to be hearing of some analysis for the first time. This is fine for enabling the "critical filtering" that produces the best analyses, but dangerous nonetheless if associated "health warnings" are not prominent.

As regards the (scalar) glueball and hybrid states it is time to move on from simple ideas that such states exist in some pure sense. We have heard many times here statements on the line of "The $f_0(xxxx)$ is the scalar glueball", where you are invited to insert your number of choice out of $970, 1300, 1500$ or 1700. And for the hybrid: "The $\pi_1(yyyy)$ is exotic, ergo it is hybrid" (where $yyyy$ is 1400, 1600 or 1800). The only pattern seems to be that the former set involve odd and the latter even numbers in their first two places.

What I offer here is not a solution, but needs to be taken into account when seeking the solution. The real world contains thresholds for hadronic channels with the same J^{PC} as these objects and will involve mixings with those as well as between the primitive glueball and $q\bar{q}$ flavoured states. So the scalar mesons with I=0 in the PDG will be mixtures of glueball and flavoured $q\bar{q}$ at least. (Hence the interest in $\psi \to \gamma\gamma V$ alluded to earlier to help disentangle this mixing). Likewise with the π_1 states. The non-relativistic quark model was built, in part, on the absence of such exotic J^{PC} combinations. Now we have three being claimed. This is too much of a good thing and the presence of S-wave thresholds such as πb_1 and πf_1 around 1400MeV surely plays some essential role. Another school of thought has been presented here: could some of the π_1 states be evidence for $qq\bar{q}\bar{q}$ in $10\pm\overline{10}$ configurations[12]? Possibly, but beware the dog that didn't bark in the night: invoking multiquarks to accommodate one or two awkward states also implies the existence of whole multiplets of associated states. The failure to see them also needs to be explained in such models.

There is rather general agreement now that qualitatively the scalar mesons sector contains a scalar glueball degree of freedom[13] in the data, the question now is to quantify it. In this regard there seem to be two broad schools[14, 15] and data need to be able to distinguish between these as a minimum before we can claim the glueball as proven. First their common features: there is a scalar glueball present in the mass region up to around 1700MeV, which mixes with and disturbs the "simple" isoscalar $q\bar{q}$ sector. Now for their details. One[14] is that the mesons above 1GeV, $f_0(1700; 1500; 1370)$ are the I=0 states of $q\bar{q}$ mixed with the G, and that $K(1430)$ and $a_0(1450)$ are the other members of the extended nonet; in this scenario the mesons below 1GeV, in particular the $f_0(980)$ and $a_0(980)$ have a $qq\bar{q}\bar{q}$ or dimeson dynamical structure[16, 17]. The other[15] is that the $f_0(980)$ and $a_0(980)$ are in the nonet with the $f_0(1500)$ and $K(1430)$ (the $f_0(980)$ and $f_0(1500)$ having an interesting "inversion" of properties in $SU(3)$ flavour to the pseudoscalar η and η'); the $f_0(1370)$ is not recognised as a real resonance state, the $a_0(1450)$ if it exists is in some other nonet (perhaps with the $f_0(1700)$); the κ is not resonant and the $\sigma(600)$ is part of a very broad scalar glueball whose effects are felt throughout an extended energy range.

As I have shares in one of the above pictures, the following observations on the novel meson states reported at this conference might be distorted by my prejudices, but with that caveat in mind I offer them for consideration nonetheless.

When does the quark model work?

There is general agreement that the NRQM is a good phenomenology for $b\bar{b}$ and $c\bar{c}$ states below their respective flavour thresholds. Taking $c\bar{c}$ as example we have S states ($\eta_c, \psi, \psi(2S)$), P-states ($\chi_{0,1,2}$) and a D-state ($\psi(3772)$), the latter just above the $D\bar{D}$ threshold. Their masses and the strengths of the E1 radiative transitions between $\psi(2S)$ and χ_J are in reasonable accord with their potential model status. In particular there is nothing untoward about the scalar states.

Do the same for the light flavours and one finds clear multiplets for the 2^{++} and 1^{++} states (though the a_1 is rather messy); it is when one comes to the scalars that suddenly there is an excess of states. An optimist might suggest that this is the first evidence that there is an extra degree of (gluonic) freedom at work in the light scalar sector. But there is more: there is a clear evidence of states that match onto either $qq\bar{q}\bar{q}$ or correspondingly meson-meson in S-wave.

Such a situation is predicted by the attractive colour-flavour correlations in QCD[16, 17]. Establishing this has interest in its own right but it is also necessary to ensure that one can classify the scalar states and then identify the role of any glueball by any residual distortion in the spectrum. It is in this context that discoveries this year of narrow states in the heavy flavour sector provide tantalising hints of this underlying dynamics elsewhere in spectroscopy. If this is established it could lead to a more unified and mature picture of hadron spectroscopy.

The sharpest discoveries this year have involved narrow resonances: $c\bar{s}$ states, probably $0^+, 1^+$, lying just below DK, D^*K thresholds; and $c\bar{c}$ degenerate with the $D^o D^{*o}$ threshold. These are superficially heavy flavour states and out of my remit, but their attraction to these thresholds involves light quarks and links to a more general theme which I shall develop below.

First note how we have been reminded here by Barnes[18] that the $c\bar{c}$ potential picture gets significant distortions from the DD threshold region, such that even the $c\bar{c}$ χ states can have 10% or more admixtures of meson pairs, or four quark states, in their wavefunctions. Also Cahn[19] reminded us that the simple potential models of the D_s states are inadequate to explain the 2.32GeV and 2.46GeV masses of these novel states as simply $c\bar{s}$ in some potential. Furthermore, Davies[20] showed that the lattice seems to prefer the masses to be higher than actually observed, though the errors here are still large. In summary there is an emerging picture that these data on the D_s sector (potentially 0^+ and 1^+ and the S-wave DK and D^*K thresholds) and the $c\bar{c}$ sector (with the S-wave $D^o D^{*o}$ thresholds) confirm the suspicion that the simple potential models fail in the presence of S-wave continuum threshold(s).

Now let's examine this in the light flavoured sector. The multiplets where the quark model works best are those where the partial wave of the $q\bar{q}$ or qqq is lower than that of the hadronic channels into which they can decay. For example, the ρ is S-wave $q\bar{q}$ but P-wave in $\pi\pi$; as S-wave is lower than P-wave, the quark model wins; by contrast the σ is P-wave in $q\bar{q}$ but S-wave in $\pi\pi$ and in this case it is the meson sector that wins and the quark model is obscured.

A similar message comes from the baryons. The quark model does well for the Δ (S-wave in qqq but P-wave in πN); at the P-wave qqq level it does well for the D_{13}, which as its name implies is D-wave in hadrons, but poorly for the S_{11} which is S-wave in $N\eta$. The story repeats in the strange sector where the strange baryons with negative parity would be qqq in P-wave: the $D_{03}(1520)$ is fine but the $S_{01}(1405)$ is the one that seems to be contaminated with possible KN bound state effects.

As an exercise I invite you to check this out. It suggests a novel way of classifying the Fock states of hadrons. Instead of classifying by the number of constituent quarks, list by the partial waves with the lowest partial waves leading. Thus for example

$$0^{++} = |0^-0^-(qq\bar{q}\bar{q})\rangle_S + |q\bar{q}\rangle_P + \dots.$$

while

$$1^{--} = |q\bar{q}\rangle_S + |0^-0^-(qq\bar{q}\bar{q})\rangle_P + \dots.$$

or

$$\Delta(1230) = |qqq\rangle_S + |\pi N(qqqq\bar{q})\rangle_P + \dots$$

This holds true for

$$2^{++} = |q\bar{q}\rangle_P + |0^-0^-(qq\bar{q}\bar{q})\rangle_D + \dots.$$

the relevant S-wave vector-meson pairs being below threshold. For the remaining P-wave $q\bar{q}$ nonet with C=+ we have a delicate balance

$$1^{++} = |q\bar{q}\rangle_P + |0^-1^-\rangle_S + \dots.$$

where the $\pi\rho$ S-wave distorts the $q\bar{q}$ a_1, as is well known; the $f_1(1285)$ is protected because the two body modes are forbidden by G-parity; for the strange mesons the $K^*\pi$ and $K\rho$ channels play significant roles in mixing the 1^{++} and 1^{+-} states while the $s\bar{s}$ state is on the borderline of the KK^* threshold.

Chiral models which focus on the hadronic color singlet degrees of freedom are thus the leading effect for the 0^{++} sector but subleading for the vectors. An example was presented[21] where the N_c dependence of the coefficients of the chiral Lagrangian was studied. In the $N_c \to \infty$ limit it was found that $\Gamma(\rho) \to 0$, like $q\bar{q}$ whereas $\Gamma(\sigma) \to \infty$, like a meson S-wave continuum. Thus there appears to be a consistency with the large N_c limit selecting out the leading S-wave components.

Conversely, the "valence" quark model can give the leading description for the vectors or the Δ but there will be corrections that can be exposed by fine detail data. The latter are now becoming available for the baryons from Jefferson Laboratory; the elastic form factors of the proton and neutron show their charge and magnetic distributions to be rather subtle, and the transition to the Δ is more than simply the M1 dominance of the quark model. There are E2 and scalar multipole transitions which are absent in the leading qqq picture. The role of the πN cloud is being exposed; it is the non-leading effect in the above classification scheme. As we shall see later, the $\Theta^+(1540)$ as a pentaquark inspires novel insights into a potential pentaquark - or $N\pi$ cloud - component in the N and Δ.

The message is to start with the best approximation - quark model or chiral - as appropriate and then seek corrections.

Bearing these thoughts in mind it highlights the dangers of relying too literally on the quark model as a leading description for high mass states unless they have high J^{PC} values for which the S-wave hadron channels may be below threshold. It also has implications for identifying hadrons where the gluonic degrees of freedom play an explicit role and cannot simply be subsumed into the collective quasi-particle known as the constituent quark. Such states are known as glueballs and hybrids.

The lightest glueball is predicted to be scalar[13] for which the problems arising from the S-wave thresholds have already been highlighted. At least here, by exploiting the experimental strategies outlined at the start of this talk, we are possibly going to be able to disentangle the complete picture. For the 2^{++} and 0^{-+} glueballs above 2 GeV there are copious S-wave channels open, which will obscure the deeper "parton" structure. Little serious thinking seems to have been done here.

For the exotic hybrid nonet 1^{-+} we have a subtlety. In the flux-tube models abstracted from lattice QCD, the $q\bar{q}$ are in an effective P-wave[22, 23], which we may describe by $|q\bar{q}g\rangle_P$. There is a leading S-wave 0^-1^+ meson pair at relatively low energies, such that

$$1^{-+} = |0^-1^+\rangle_S + |q\bar{q}g\rangle_P + ...$$

The S-wave thresholds for πb_1 and πf_1 are around 1400MeV, which is significantly below the predicted 1.8GeV for lattice or model hybrids and tantalisingly in line with one of the claimed signals for activity in the 1^{-+} partial wave. All is not lost however; a $q\bar{q}$ or $q\bar{q}g$ nonet will have a mass pattern and decay channels into a variety of final states controlled by Clebsch-Gordan coefficients whereas thresholds involve specific meson channels. These can in principle be sorted out, given enough data in a variety of production and decay channels, but it may be hard.

SKYRMION MEETS THE QUARK

In the above we have discussed where components beyond the leading $q\bar{q}$ or qqq may obscure the simple quark model. We now come to a case where the **leading** component involves five constituents. If this discovery is confirmed it will make a sobering reminder that there can be phenomena latent in data that have been overlooked perhaps for decades.

In the textbooks, one of the major planks in establishing the constituent quark model is the absence of baryons with strangeness +1. The announcement of such a particle, and with a narrow width is therefore startling, if confirmed[24]. It is easy to accommodate positive strangeness; you just allow an extra $q\bar{q}$ to be present, e.g. $uudd\bar{s}$. The problem though is that such a state would be expected to fall apart so rapidly that its width would be broad. A narrow width, signaling metastability, therefore implies the existence of some inhibiting factor. Its parity is undetermined and that could itself discriminate among models. The state was predicted in the Skyrme model[26] where it is a member of a $\overline{10}$ with $J^P = \frac{1}{2}^+$. This is already an interesting conundrum for a quark model where the naive expectation is that the lightest state of a pentaquark $uudd\bar{s}$ has all constituents in a relative S-wave, hence $J^P = \frac{1}{2}^-$. However, this is true only when all the quarks are treated symmetrically. There is a considerable literature that recognises that ud in colour $\bar{3}$ with net spin 0 feel a strong attraction, which might even cause the S-wave combination to cluster as $[udu][d\bar{s}]$ which is the S-wave KN system, while the P-wave positive parity exhibits a metastability such as seen for the Θ. Two particular ways of realising this are due to Karliner and Lipkin[28] and Jaffe and Wilczek[29], which I will come to shortly.

A challenge for all quark models is the metastability of the state. The historical stability of the strange hadrons was due to their strong decays being forbidden; the first attempt to describe the Θ as a pentaquark[30] built on this idea by proposing that Θ be an isotensor resonance with states ranging from $uuuu\bar{s}$ with charge +3 to $dddd\bar{s}$ with charge -1. This can give a narrow width as there is no simple decay path that preserves isospin. The colour structure of such a pentaquark system is well defined. The I=2 flavour-space is totally symmetric and so is totally antisymmetric in colour-spin. This forces either $6_c \times (S = 0)$ or $3_c \times (S = 1)$. Only the latter can combine with the $\bar{s}$ in $\bar{3}_c$ to make the colour-singlet baryon. This leads to overall $J^P = \frac{1}{2}^-$ or $\frac{3}{2}^-$. The price, or excitement, is that there is a multiplet of states ($\Theta^{+++}....\Theta^-$) to be found. This may be already ruled out if the ELSA[24] data are confirmed as they find the Θ^+ but have no evidence for any partner, and thus suggest it is I=0.

Models with I=0 suggest that it is at the pinnacle of a flavour $\overline{10}$, which is where the original Skyrme prediction would place it. Thus there is also the interesting question of whether or under what circumstances there is any correspondence between the Skyrme and quark pictures.

Attempts to describe this as a pentaquark have been criticised in some quarters on the lines that it is meaningless to describe a hadron as made from a fixed number of quarks or antiquarks. Let's first make some obvious pedagogic remarks in order to accommodate some suggestions that I shall make later.

When the proton is viewed at high resolution, as in inelastic electron scattering, its wavefunction is seen to contain configurations where its three "valence" quarks are accompanied by further quarks and antiquarks in its "sea". The three quark configuration is thus merely the simplest required to produce its overall positive charge and zero strangeness. The question thus arises whether there are baryons for which the minimal configuration cannot be satisfied by three quarks.

A baryon with positive amount of strangeness would be an example; the positive strangeness requires an $\bar{s}$ and $qqqq$ are required for the net baryon number, making what is known as a "pentaquark" as the minimal "valence" configuration.

Hitherto unambiguous evidence for such states in the data has been lacking; their absence having been explained by the ease with which they would fall apart into a conventional baryon and a meson with widths of many hundreds of MeV. It is perhaps this feature that creates the most tantalising challenge from the perspective of QCD: why does Θ have width below 10MeV, perhaps no more than 1MeV[31].

If the data comprising the evidence as presented at this conference are being correctly interpreted, they suggest that the Θ is being produced with probability similar to that of the negative strangeness $\Lambda(1520)$. This suggests that the Θ is produced by the strong interaction between KN. However, such a strength seems to be at odds with the implied feeble decay strength implied by a 1MeV width into KN, unless perhaps Θ is produced by the strong decay from some state Θ^*, which is produced strongly by KN and has width of $\geq O(100)$MeV. The other possibility is that the production cross section of Θ is $O(10-100)$ smaller than that of $\Lambda(1520)$ (there emerged some hints after the conference that this might indeed be the case[25])

However, it may be premature to seek radical solutions given the nature of the current evidence[24]. The most immediate concern must be to establish not simply the spin and parity of the Θ, or other examples like it, but to verify that it indeed exists and is not some combination of statistical fluctuations, some complex novel dynamical background effect that has been overlooked, or psychological desire to be attracted by small positive signals while arguing away any compensating negative results. In the immediate term, a dedicated high statistics experiment involving photoproduction at Jefferson Laboratory, planned to take data in 2004 may help to settle some of these questions.

Whether or not it turns out to be real, the stimulus to theory has already reinvigorated interest in the Skyrme model (which even predicted that such a state should exist, at such a mass, though admittedly not with a width so small) and the pentaquark dynamics of the quark model. The Skyrme model and the quark model are both rooted in QCD though their relation has been obscure. Considerable theoretical attention into their relation has been stimulated by the Θ studies. (Following the conference there has appeared a paper which suggests[34] that the exotic Θ is an artifact of the rigid rotator approach to the Skyrme model, and that in the $SU(3)_F$ limit the $\overline{10}$ does not form.)

Skyrme's model, when extended to incorporate strangeness, implied that the lightest baryon families consisted of $(8\frac{1}{2}^+)$ which includes the nucleons, and a $(10\frac{3}{2}^+)$ which includes the Δ, Ω^-. This far its predicted pattern is like that of the quark model based on three quarks interacting with QCD forces and also as seen in the data. However, it was noticed that in this Skyrme model, there is a further family of ten (transforming like a "ten-bar", of SU(3)-flavour) with $J^P = \frac{1}{2}^+$. This is the family that can not be formed from three quarks and requires the pentaquark as a minimum configuration.

Initially it was thought that the pentaquark would lead to negative parity for the lightest states, in contradiction to the Skyrme model prediction of positive parity. However, the color magnetic forces of QCD, when combined with constraints on flavor and spin required by fundamental symmetries (such as Bose symmetry and the Pauli exclusion principle) cause the lightest observable states plausibly to contain one unit of internal angular momentum and thereby have positive parity [29, 28].

However, there does appear to be a significant potential difference between the models, which should be experimentally testable. Both predict that there are two further exotic members of the "ten-bar" family: they have strangeness minus two, like the familiar Ξ baryons, but whereas the familiar Ξ states have electric charges 0 or -1, these can have 0,-1 and also +1 or -2. Positively charged or doubly negatively charged baryons with strangeness minus two are hitherto unknown.

And this is where the potential difference arises. In the formulation of the Skyrme model for broken SU(3) in[27], the mass gap between the Θ and these Ξ has to be **larger** than that in the conventional ten, spanned by the $\Delta(1236)$ and $\Omega^-(1672)$. This appears to be unavoidable if the $\overline{10}$ masses are to be above those of the familiar decuplet. Indeed, they predicted this gap in the "ten-bar" to be some 540MeV leading to a mass for the Ξ

exceeding 2GeV. In the pentaquark picture, by contrast, one need only pay the price for one extra strange mass throughout the $\overline{10}$. This implies a relatively light mass for the Ξ ~ 1700MeV with the possibility that these states also could be relatively stable.

I will now describe the wavefunctions of the pentaquark in more detail to show that there is **no** simple mapping onto the Skyrme model as initially presented in [27].

$\overline{10}$ Wavefunctions

To get a feeling for a $\overline{10}$, first recall the most familiar decuplet of baryons. This forms a large inverted triangle with the Ω^- at its pointed base and $\Delta^{++};\Delta^-$ at the two extremes of its "shoulders"; the strangeness spans 0 to -3. Now consider the corresponding antibaryons, making a $\overline{10}$. Now we will have the (anti- Ω)$^+$ at the pointed head of the triangle and (anti-Δ)$^{--}$ and (anti-Δ)$^+$ at the extremes of its base; the strangeness spans +3 to 0. Note the electric charges of these states. The $\overline{10}$ of interest in the present story is like this but with the magnitudes of strangeness being two units less throughout than the antibaryon one just described. Thus instead of the (anti-Ω)$^+$ ($S = 3$) at the pointed head of the triangle we have $\Theta^+(S=1)$. In place of the (anti-Δ)$^{--}$ and (anti-Δ)$^+$ ($S = 0$) at the extremes of its base we have the exotic ($S = -2$) $\Xi^{--};\Xi^+$.

Thus we see the presence of three exotic correlations of strangeness and charge. The Θ^+ is what is claimed to have been discovered; the $\Xi^{--};\Xi^+$ are a remaining challenge.

We all know how to write the wavefunctions for a $\overline{10}$ made of three antiquarks. However, there appears to be some confusion about the analogous wavefunctions for a $\overline{10}$ made of pentaquarks. In particular the form quoted in the discussion session here is not a $\overline{10}$. Given this confusion I will describe here in a heuristic way, how to build them. This will immediately expose essential differences with the Skyrme model and suggest further novel implications in the baryon spectrum.

I am going to view the $qqqq\bar{q}$ as two diquarks qq-qq accompanying an antiquark. To form the wavefunctions and take care of their symmetries note first how the diquarks transform under SU(3)$_f$. Define the antisymmetric diquark states cyclically under $u \to d \to s$ so that (apart from normalisations)

$$(ud) \equiv (ud - du) \to \bar{s}; (ds) \equiv (ds - sd)\bar{u}; (su) \equiv (su - us) \to \bar{d}$$

Then take the traditional wavefunctions for antibaryons, retain one antiquark and replace the others by the corresponding diquark.

The Θ state $(ud)^2\bar{s}$ is thus seen immediately to be symmetric and analogous to the $\bar{\Omega}^+$. The analogues of the $\bar{\Delta}^{--}$ and $\bar{\Delta}^+$ are then respectively $(ds)^2\bar{u}$ and $(su)^2\bar{d}$. These form Ξ states with strangeness = -2 in our $\overline{10}$.

Before writing wavefunctions note immediately that there is only **one** extra strange mass in the Ξ states relative to the Θ. Thus in the pentaquark model one necessarily has low lying exotic Ξ states around 1700MeV if one identifies the $\Theta(1540)$ to set the scale. This is different from the Skyrme model as originally presented in [27].

This is an important fact worthy of some comment in view of the prediction[27] of the Θ in a $\overline{10}$ in a version of the Skyrme model. However, it was critical in that prediction that the mass gap from Θ to Ξ is **three** units of $\Delta(m_s - m_d) \sim 150$MeV, as for the conventional (anti)decuplet of (anti)Δ − (*anti*)Ω. In a pentaquark picture the mass gap is only a single unit.

The difference comes from the way that[27] implemented flavour symmetry breaking. A crucial assumption was that the SU(3) breaking for $m_s \neq m_d$ depends linearly on the hypercharge Y=B+S such that $M(Y) = M_0 - cY$ where $c > 0$. For the familiar baryon **10** this is equivalent to counting the number of strange quarks. However, this is not a general axiom. It does not work for mesons, for example, where $m(K^+) \equiv m(K^-)$ and $m(\omega) < m(\phi)$, nor for the octet baryons where $m(\Sigma) > m(\Lambda)$. The origin of these masses are immediately obvious in the quark model with hyperfine interactions.

The reason for the difference is that s and $\bar{s}$ contribute equally to the strange mass content, but cancel out in the hypercharge. In the $\overline{\mathbf{10}}$ of interest here, the simple correspondence familiar in the non-exotic **10** is lost. The mass gap from shoulder to toe of the 10, or from tip to base of the pyramid in the $\overline{\mathbf{10}}$ is given by the difference in **moduli** of the respective strangeness. Thus for the familiar 10 or $\overline{\mathbf{10}}$ which run from strangeness 0 to ±3 we have **three** units of strange mass, whereas for the case here which runs from strangeness +1 to -2 the modular difference is only one.

Ref.[27] forced the interval between the Theta and the N to be 1710-1540=170MeV and thereby inflate the mass splittings. As we already commented, the mass gap is $\frac{1}{3}m_s$ per stage in the $\overline{\mathbf{10}}$ for the pentaquark whereas ref[27] chose numbers with more like one m_s per unit gap. Now their model at first sight appears to hide beneath parameters $\alpha\beta\gamma$ (eq 16-18 in hep-ph/9703373). However, this is not really so. Critical is the mass gap per unit of strangeness in table 1 of ref.[27] which gives the mass gaps per unit strangeness to be $1/8\alpha + \beta - 5/16\gamma$ for normal 10 and $1/8\alpha + \beta - 1/16\gamma$ for the novel $\overline{\mathbf{10}}$. Hence in their convention where $1/8\alpha + \beta < 0$ then if $\gamma < 0$ (see later) the mass gap per unit of strangeness in their $\overline{\mathbf{10}}$ must be BIGGER than in the conventional 10. The only way to get it smaller, as in the pentaquark picture would be for $\gamma > 0$.

So, what can one say about γ in general?

First see eq 18 of ref[27] and the comment at end of section 2: "$I_1 > I_2$ so that the $\overline{\mathbf{10}}$ is heavier than the familiar decuplet". This tends to force γ negative and toward $2\beta/3$ (which is indeed in accord with their actual numbers of $\gamma \sim$-107MeV and $\beta \sim$-156MeV in their eq 27). So there appears to be an inherent distinction between the Skyrme picture of [27] and the pentaquark, so long as m($\overline{\mathbf{10}}$) > m(**10**).

There are other differences between the two pictures. To motivate these, we need first to look more carefully at the wavefunctions for the other states in the multiplet. The wavefunctions can be obtained by applying the U-spin lowering operator to the Θ. U_- changes $d \to s$ or $\bar{s} \to -\bar{d}$. U_- commutes with the Casimir operators of SU(3), and so under its operation one remains in the $\overline{\mathbf{10}}$. Thus for example

$$|p\rangle = \frac{1}{2\sqrt{3}}\left([(ud - du)(su - us) + (su - us)(ud - du)]\bar{s} + (ud - du)^2\bar{d}\right)$$

or more succinctly

$$p = -\sqrt{\frac{2}{3}}[(ud)(su)_+]\bar{s} - \sqrt{\frac{1}{3}}(ud)^2\bar{d}$$

928

We can expose the hidden $s\bar{s}$ or $d\bar{d}$ heuristically, though at the expense of suppressing the above symmetries, by writing this in the "shorthand" form

$$p(\bar{10}) = uud \left(\sqrt{1/3}|d\bar{d}> + \sqrt{2/3}|s\bar{s}> \right).$$

In similar fashion

$$|\Sigma^{+}(\bar{10}) >= U_{-}|p > \rightarrow uus \left(\sqrt{2/3}|d\bar{d}> + \sqrt{1/3}|s\bar{s}> \right)$$

These are like familiar baryons with extra hidden strangeness or hidden $d\bar{d}$ in a specific weighted combination for the $\bar{10}$. This immediately allows one to count the total number of $s+\bar{s}$ in each state. In the N, for example, you get: $(1/3) \times (0) + (2/3) \times 2 = 4/3$. For the Σ: $(2/3) \times 1 + (1/3) \times 3 = 5/3$. Hence one sees explicitly the equal mass rule but with $m_s/3$ per unit change of strangeness, consistent with our earlier observation that the total mass interval between the Θ and the $\Xi^{+} \equiv -(us)^2\bar{d}$ feels only **one** extra strange contribution.

Now we come to the interesting features, namely those states that are not at the corners of the $\bar{10}$. These can also form octet representations, whose wavefunctions are orthogonal to the above; they are

$$p(\mathbf{8}) \rightarrow uud \left(\sqrt{2/3}|d\bar{d}> - \sqrt{1/3}|s\bar{s}> \right) \tag{1}$$

$$\Sigma^{+}(\mathbf{8}) \rightarrow uus \left(\sqrt{1/3}|d\bar{d}> - \sqrt{2/3}|s\bar{s}> \right) \tag{2}$$

Counting the number of $s+\bar{s}$ one gets for the relative strange mass content to the mass pattern in the octet $N : \Sigma : \Xi = 2/3 : 7/3 : 2$.

Photoproduction of the $\bar{10}$ is interesting since the photon has $U = 0$ and so cannot cause transition from $p(\mathbf{8})(U = 1/2)$ to $p(\bar{10})(U = 3/2)$. By contrast, the neutron is in a $U = 1$ multiplet for both $\mathbf{8}$ and $\bar{10}$, and hence $\gamma n(\mathbf{8}) \rightarrow \bar{10}$ is allowed. To see this with the above wavefunctions, let the photon convert to a $q\bar{q}$ with amplitude proportional to the charge e_q; form the transition amplitude by isolating the terms in the $\bar{10}$ wavefunction where $(q_i q_j)(q_k q_l)\bar{q}_l$ occur with the q and $\bar{q}$ of the same flavour adjacent to one another. Thus

$$p \rightarrow (ud - du)u \left[s\bar{s} - d\bar{d} \right]$$

exposes the coupling to the mixed-antisymmetric conventional octet proton uud state, and a $U = 1$ state. This explicitly shows that $p(\mathbf{8}) \rightarrow p(\bar{10})$ transforms as $\Delta U = 1$ whereby photoproduction is forbidden.

The analogous exercise for a neutron gives

$$n \rightarrow (ud - du)d \left[s\bar{s} - u\bar{u} \right]$$

where the $q\bar{q}$ piece now transforms as $V = 1$, or equivalently as a linear superposition of $I = 1$ and $U = 0$. The latter therefore allows $\gamma n(\mathbf{8}) \rightarrow \bar{10}$.

Thus photoproduction could be imagined as a way to distinguish whether the pentaquark p^* is in **8** or **10̄**. However, it is at this point, if not already, that one realises that the language of **10̄** and **8** is not really suitable. The symmetry breaking allows mixing between the two multiplets and depending on the dynamics this may tend toward the extreme which respectively maximises and minimises the net $s + \bar{s}$ content. Thus the mass eigenstates may be expected to tend toward the following (subscripts L and H for light and heavy):

$$N_L = (ud)^2\bar{d}; N_H = (ud)(us)\bar{s}$$

and

$$\Sigma_L = (ud)(ds)\bar{u}; \Sigma_H = (ds)^2\bar{s}$$

In this case we see that for the set of "light" states, there is an increase of order m_s per strange gap, while the same is true for their heavy counterparts until the final stage where the Ξ is lighter than the Σ, thereby preserving the ubiquitous rule that there is only one unit of "extra" strange mass between Θ and Ξ.

Thus if one identifies the $m(\Theta) \sim 1540$MeV, one might identify $m(N_H) \sim 1710$ (contrast the Skyrme model which identified the **10̄** with this state) and then have the prediction of a lighter state, perhaps $m(N_L) \sim 1400$MeV, which could be related to the Roper resonance[29].

In the Skyrme model the **10̄** has $J^P = \frac{1}{2}^+$; there is no accompanying octet, and hence no possibility of mixing. In the pentaquark model one might naively expect that the lowest lying states are $J^P = \frac{1}{2}^-$; however, when the interquark QCD spin dependent forces are taken into account one finds[29, 28] that octet and **10̄** emerge lightest with $J^P = \frac{1}{2}^+$. However, one also finds that they are partnered by $J^P = \frac{3}{2}^+$ multiplets too. Let's now look into this and assess experimental tests.

Diquark Cluster Models

Early evidence that mesons and baryons are made of the same quarks was provided by the remarkable successes of the Sakharov-Zeldovich constituent quark model, in which static properties and low lying excitations of both mesons and baryons are described as simple composites of asymptotically free quasiparticles with a flavor dependent linear mass term and hyperfine interaction,

$$M = \sum_i m_i + \sum_{i>j} \frac{\vec{\sigma}_i \cdot \vec{\sigma}_j}{m_i \cdot m_j} \cdot v^{hyp} \tag{3}$$

where m_i is the effective mass of quark i, $\vec{\sigma}_i$ is a quark spin operator and v_{ij}^{hyp} is a hyperfine interaction.

As first pointed out by Karliner and Lipkin[28], a single-cluster description of the $(uudd\bar{s})$ system fails because of the repulsive interaction between the pairs of the same flavor, which prevents binding. This leads one to consider dynamical clustering into subsystems of diquarks or/and triquarks, which amplify the attractive color-magnetic forces. There are two routes that emerge naturally; one is that of[28], the other of Jaffe and Wilczek[29]. These naturally lead to $J^P = \frac{1}{2}^+$ as the lowest mass states.

The first step is common and is based on the strong chromomagnetic attraction between a u and d flavour when the ud diquark is in the $\bar{3}$ of the color $SU(3)$ and in the $\bar{3}$ of the flavor $SU(3)$ and has $I = 0, S = 0$, like the ud diquark in the Λ.

Such an idea has a long history, being the source of the $\Lambda - \Sigma$ mass difference, a possible linkage with the dominance of $u(x \to 1)$ in deep inelastic structure functions and of the maximisation of the polarisation asymmetry in this same limit. Such attraction between quarks in the color $\bar{3}$ channel halves their effective charge, reduces the associated field energy and is a basis of color superconductivity in dense quark matter[33]. There is the implied assumption that such a "diquark" may be compact, an effective boson "constituent", which is hard to break-up and hard for its constituents to rearrange with other quarks or antiquarks in the bound state. I shall refer to this by $[(ud)_0]$, the subscript denoting its spin, and the [ ] denoting the compact quasiparticle.

JW consider the following subcluster for the pentaquark: $[(ud)_0][(ud)_0]\bar{s}$. KL also start with the $[(ud)_0]$ seed, but regard the remainder as a strongly bound "triquark" $[(ud)_1\bar{s}]$. This internal structure is chosen to give the minimum energy to the triquark system; the $(ud)_1$ is coupled to spin 1, colour 6; the $\bar{s}$ couples to the u or the d to net spin 0. Thus we see there is this difference in details between the two approaches. First I will describe their dynamics and see what consequences there are.

For JW the two (ud) must combine to make 3_c in order to neutralise the $\bar{s} = \bar{3}$; since $\bar{3} \times \bar{3} \to 3$ is antisymmetric in colour, and since the $(ud)_0$ boson pair must be symmetric overall this implies that they are in P-wave (spatially antisymmetric).This gives a negative parity that combines with the negative parity of $\bar{s}$ to give an overall positive parity system. Thus one has $J^P = \frac{1}{2}^+; \frac{3}{2}^+$ pentaquark systems. It is possible to identify the mass with the Θ (see later); the metastability can be accommodated by insisting that the quasiparticles in $[(ud)_0][(ud)_0]\bar{s}$ prevent simple rearrangement to overlap with $[(ddu)][u\bar{s}]$, which are the NK colour singlet hadrons.

Whereas JW take the other ud diquark also to be in this configuration and then put the two diquarks in relative P-wave (by Bose symmetry after the colour is taken account of), KL by contrast took the remaining $ud\bar{s}$ and looked for the configuration in color and spin which would optimize the total (five-body) hyperfine interaction.

Karliner and Lipkin divide the system into two color non-singlet clusters which separate the pairs of identical flavor. The two clusters, a ud diquark and a $ud\bar{s}$ triquark, are separated by a distance larger than the range of the color-magnetic force and are kept together by the color electric force. Therefore the color hyperfine interaction operates only within each cluster, but is not felt between the clusters. They associate the $[(ud)_0]$ with the non-strange piece of the $\Lambda(1110)$ baryon.

Within the $[(ud)_1\bar{s}]$ the strange subsystems $u\bar{s}$ and $d\bar{s}$ are assumed to be in spin 0, by colour-spin forces analogous to the way that the K is lighter than K^*. If the diquark and triquark are in relative S-wave, then the colour attractions act among all the constituents leading to a freeze out that would be a KN S-wave system. In P-wave, the two separate quasi-particles avoid the contact hyperfine forces (their generalisations to the Fermi-

Breit effects are not discussed). These identifications help them to argue that the mass of the system agrees with that of the Θ. Analogous to JW, the metastability can be accommodated by insisting that the quasiparticles in $[(ud)_0][(ud)_1\bar{s}]$ prevent simple rearrangement to overlap with $[(ddu)][u\bar{s}]$, which are the NK colour singlet hadrons.

KL consider also heavy analogues $[(ud)_0][(ud)_1\bar{Q}]$ where $Q \equiv s; c; b$. They do not consider $Q \equiv u, d$, possibly because this enables annihilation with the like-flavoured quark in the triquark, thereby destroying the stability and overlapping with conventional udu or udd baryons. However, we shall see that such states can have non-trivial consequences.

JW do consider $[(ud)_0][(ud)_0]\bar{Q}$ with $Q \equiv u, d$. The dynamical difference is that the quasi-particle nature of the separate $([(ud)_0])$ may suppress the annihilation with the like flavour, enabling the states $[(ud)_0][(ud)_0]\bar{Q}$ with $Q \equiv u, d$ to have an existence. Such states would be expected to lie $O(100 - 150)MeV$ below the Θ, and they identify them with the Roper $n; p(1440)$ nucleon resonances.

The mass of the exotic Ξ^+ depends rather sensitively on the effects of clustering. First one needs the effective mass of the diquarks. The mass difference $m(\Delta) - m(N)$ implies that

$$\Delta m[(ud)_0] = -150MeV; \Delta m[(ud)_1] = +50MeV$$

relative to their mean masses. The strange counterparts follow from $m(\Sigma^*) - m(\Sigma)$ and implies

$$\Delta m[(us)_0] = -96MeV.$$

Thus

$$\Delta m(\Xi^+ - \Theta^+) = \Delta m(\bar{s} - \bar{d}) + 2\Delta m([(us)_0] - [(ud)_0]) \sim 230 - 250MeV.$$

So 1750-1800MeV is the mass range one obtains with this level of approximation, which is still significantly below the original prediction in ref[27]. There is further uncertainty in estimating the mass in that the orbital excitation of the $[us][us]$ and $[ud][ud]$ will have different energetics and the $\vec{L} \cdot \vec{S}$ shifts are also dependent on the flavours. These could easily add further uncertainties of $\pm 50MeV$. If and when the $\Xi^{+(--)}$ are discovered, along with their $J^P = \frac{3}{2}^+$ counterparts, their masses will enable the systematics of the clustering to be determined by fitting the above.

Other pentaquark states

At first sight the narrowness of the Θ would seem to argue against any identification of the broad Roper resonance as a nucleon analogue. However, this need not be so. Some of the following thoughts emerged from discussion with Maltman at the conference[35].

For a simple attractive square well potential of range 1fm. the width of a P-wave resonance 100MeV above KN threshold is of order 200MeV[29, 35]. However, this has not yet taken into account any price for recoupling colour and flavour-spin to overlap the $(ud)(ud)\bar{s}$ onto colour singlets uud and $d\bar{s}$ say for the KN. In amplitude, starting with

932

the Jaffe-Wilczek configuration, the colour recoupling costs $\frac{1}{\sqrt{3}}$ and the flavour-spin to any particular channel (e.g. K^+n) costs a further $\frac{1}{4}$. This appears to be akin to the factor $1/2\sqrt{6}$ found in ref[35] for the isospin-spin-colour overlap in their conventions. They go further and consider the mixing with the Karliner-Lipkin configuration gives for the lowest eigenstate a suppression of $\frac{1}{25}$[35]. Hence a width of $O(1-10MeV)$ for $\Theta \to KN$ may be reasonable.

The decay involves tunneling through the P-wave barrier, which by analogy with α-decay is exponentially sensitive to the difference between the barrier height and the kinetic energy of the state. This can affect the width of a spin $\frac{3}{2}$ "Θ^*" partner, whose mass $> m(\Theta)$, arising from the $\vec{L}\cdot\vec{S}$ splitting effects in the pentaquark system. This splitting is model dependent[36] but might be as small as $\sim 30-80$MeV. Given the exponential sensitivity in a tunneling width, the Θ^* could be high enough in the well to be broad. In any event such a state should be sought. If the above is a guide, only $\Theta^* \to \Theta\gamma$ is kinematically allowed as a transition. If its mass exceeds 1820MeV it becomes possible for a strong decay width $\Theta^* \to \Theta\pi\pi$ to feed at least some of the Θ signal.

The exponential sensitivity could frustrate attempts to differentiate between the pentaquarks and (original) Skyrme model in connection with the exotic Ξ states. In Skyrme these are above 2GeV and relatively broad; in pentaquarks they are ~ 1700MeV and hence the possibility of being relatively narrow, perhaps only 50% broader than the Θ[29]. However, exponential dependence could cause even 1700 MeV to be high enough in the well to give a broad width to $\Xi\pi$. It would be galling if the Θ were the only sharp state. (See also ref.[34] and comments earlier about the potential non-binding of the $\mathbf{10}$ in the Skyrme picture.)

For the Roper, the naive mass of the $uudd\bar{d}$, by comparison with the Θ, would be ~ 1400MeV. Its physical mass of 1430-1470MeV[37] may therefore have it elevated in the potential, where the exponential behaviour drives its large width. These possibilities need more careful study in models.

However, for the non-exotic states the picture is not so simple. As stressed earlier, there is no absolute meaning to a pentaquark configuration when a qqq constituent state can carry the same overall quantum numbers. Thus the $uudd\bar{d}$ is at best the $SU(3)$ flavour state within the pentaquark wavefunction of the Roper, or for that matter, of the nucleon. Mass eigenstates will be mixtures of qqq, these pentaquark states and higher configurations. The nucleon may be qqq in leading order with its pentaquark components, which naively exist at higher mass scale, revealed with increasing q^2. For the Roper, the mass scales of the pentaquark and the excited qqq components may compete. Certainly both states require qqq presence in order to understand the -2:3 relative amplitudes for $\gamma n : \gamma p$ magnetic moment/transitions. Furthermore, the existence of the $\Delta(1660)$ as a potential partner of the Roper (analogous to the $\Delta(1230)$ for the nucleon) plays an essential role in disentangling these states. Improved data on its photo-excitation from Jefferson Laboratory could help here. Note there is no simple place for such a state if the Roper were pure pentaquark.

The Nucleon Sea

As stressed above: the number of quarks in a nucleon is not a meaningful quantity. As q^2 varies the nucleon's structure is probed on ever finer resolution and the sea of $q\bar{q}$ is exposed. All that we can say is that three is the minimum number of quarks required to satisfy the overall quantum numbers. Thus there are pentaquark, heptaquark and ad-inf.-quark components to the wavefunction of any state which transforms like a nucleon.

In the case of a positively charged positive strangeness baryon, the $uudd\bar{s}$ is the minimal configuration compatible. Thus for such a state there is meaning to the pentaquark as the minimum "valence" wavefunction. Within the pentaquark sector, which contains this state, there are configurations which transform like a nucleon and these are in general mixtures of octet and $\overline{10}$, as described above

$$uud\left[cos\theta(d\bar{d}) + sin\theta(s\bar{s})\right]$$

The QCD forces that have led to the Θ being the lightest pentaquark state will lead to the above as the lightest analogues with nucleon quantum numbers. Note in particular that it is the attractive forces between ud pairs that have favoured these and contrast this with the $uudu\bar{u}$ pentaquark, which is in a **27** of flavour SU(3), and is pushed to higher energy by the repulsive forces between the symmetric u quarks.

If indeed the lightest proton excitation is the $(ud)^2\bar{d}$, then this would be a natural candidate for the leading piece of the five body proton Fock state. This is of course what the folklore for failure of Gottfried sum rule requires; the asymmetry that leads to the "missing" $(ud)uu\bar{u}$ in this picture is because the 27 is pushed up relative to the $\overline{10}$-**8** mixture. The antiquark sea is also naively polarised "against the flow" too.

There is an interesting duality between this and other interpretations of this flavour asymmetry in the sea. One is to invoke Pauli blocking of the $u\bar{u}$ due to the extra u flavour in the valence proton. In effect, this is an essential feature subsumed within the arguments of [29, 28], which addressed the pentaquark configuration for the Θ.

Another approach has been to consider the $N\pi$ cloud of the nucleon. The essence is that $p \to n\pi^+(u\bar{d})$ feeds the $\bar{d}$ whereas the $\bar{u}$ is energetically disfavoured, requiring $p \to \Delta^{++}\pi^-(d\bar{u})$. The QCD forces that push the $m(\Delta) > m(n)$ are the same that distorted the pentaquark configurations, favouring the **8**-$\overline{10}$ over the **27**. So the role of multiquark configurations, and their mapping onto the meson-baryon sectors, are all pervading. It is a question of approximation as to whether one or other dominates, or whether both play competing roles. Whether the $\Theta(1540)$ will turn out to be the first real evidence for a state with a minimal pentaquark "valence" configuration, or merely a strange story to tell future generations, it has certainly raised challenging questions and is leading to some unexpected insights.

Postscript on Θ for future historians

A vote at the conference on whether Θ can be regarded as (i) an established resonance, (ii) jury still out, (iii) is not a resonance, was split approximately $\frac{1}{4}; \frac{1}{2}$ and $\frac{1}{4}$ respectively. The total number of votes cast was $O(100)$. People who were involved in any of the relevant experiments or had written theory papers on Θ were excluded from the vote. There appeared to be a slight tendency for senior experimentalists to vote in category (iii). Whether this is because of experience with the machinations of statistics in the past, or frustration at having overlooked a major discovery, is for psychologists to debate.

Acknowledgments

I am deeply indebted to Mrs Hiltscher and Bernd Lewandowski for their help in preparing this talk and congratulate the organisers for a superb conference. In addition to talking with many people at the conference, I have profited from discussions about pentaquarks and exotic hadrons with J.Dudek, R.Jaffe, M.Karliner and H.Lipkin.

REFERENCES

1. Initial State radiation at BaBar: E.Solodov and R.Stroili; $\gamma\gamma$ at Belle: S.Hou (parallel session Mesons 1)
2. M.Barbi:"Observation of $K_s K_s$ resonances at HERA" (parallel session Mesons3)
3. P.Schlein: plenary talk
4. F.E. Close, A. Kirk and G.Schuler, Phys.Letters **B477** 13 (2000)
5. W.Ochs: plenary talk
6. W.Li: plenary talk
7. H.Stock: plenary talk
8. A.Donnachie "Meson Radiative Decays - a new flavour filter" (parallel session Mesons3);F.E.Close and A.Kirk, Z.Phys. C **21** (2001)
9. D.Weygand: plenary talk
10. A.Reis (parallel session Scalars 1); S.Malvezzi (FOCUS) (parallel session Mesons2).
11. E.Swanson: plenary talk.
12. S.Chung and E.Klempt hep-ph/0306018, Phys.Lett. **B563** (2003) 83
13. SESAM Collaboration, G. Bali *et al.*, Nucl.Phys.Proc.Suppl. **63**, 209 (1997); IBM Collaboration, J. Sexton *et al.*, Phys.Rev.Lett. **75**, 4563 (1995); F.E. Close and M.J. Teper, Report no. RAL-96-040/OUTP-96-35P; C.J. Morningstar and M. Peardon, Phys.Rev. D **56**, 4043 (1997); D. Weingarten, Nucl.Phys.Proc.Suppl. **73**, 249 (1999); C. McNeile and C. Michael, Phys. Rev. D **63**, 114503, (2001)
14. C. Amsler and F.E. Close, Phys. Lett. B **353**, 385 (1995); F.E.Close and A.Kirk, Z.Phys. C **21** (2001)
15. W.Ochs and P.Minkowski Eur.Phys.J. **C9** 283 (1999)
16. R.L.Jaffe, Phys.Rev. D **15** 281; R.L.Jaffe and F.E.Low, Phys.Rev D **19** 2105
17. F.E.Close and N.Tornqvist, J.Phys.G. **28** R249 (2002)
18. T.Barnes plenary talk.
19. R.Cahn discussion panel
20. C.Davies plenary talk
21. J.R.Palaez, hep-ph/0310237
22. N. Isgur and J. Paton, Phys. Rev. D **31** 2910 (1985)
23. F.E.Close and J.J.Dudek, hep-ph/0304243, Phys. Rev.Letters **91** 142001-1 (2003)
24. T Nakano et al (LEPS Collaboration) hep-ex/0301020 V V Barmin et al (DIANA Collaboration) hep-ex/0304040, S. Stepanyan et al (CLAS Collaboration) hep-ex/0307018, J.Barth et al. ELSA hep-ex/0307083, K.Hicks, these proceedings.
25. SAPHIR report, unpublished, private communication from K Hicks October 2003
26. M.Polyakov, panel discussion; A.V.Manohar, Nucl.Phys. B**248** 19 (1984); M.Chemtob, Nucl.Phys. B **256** 600 (1985); M.Karliner and M.Mattis, Phys. Rev D **34**, 1991 (1986); M.Praszalowicz in Proc Cracow Workshop on Skyrmions and Anomalies, Mogilany, Poland, Feb 20-24, 1987, World Scientific 1987, p.112
27. D.Diakanov, V.Petrov and M.Polyakov hep-ph/9703373; Z.Phys. A **359** (1997) 305
28. M.Karliner and H.J.Lipkin, hep-ph/0307243
29. R.L.Jaffe and F.Wilczek, hep-ph/0307341

30. S.Capstick, P.R.Page and W.Roberts hep-ph/0307019
31. R.Arndt et al nucl-th/0308012; R.Cahn, remarks in panel discussion. See also S.Nussinov hep-ph/0307357
32. M.Karliner and H.J.Lipkin, hep-ph/0307343
33. K.Rajagopal and F.Wilczek hep-ph/0011333
34. N.Itzhaki et al. hep-ph/0309305
35. B.Jennings and K.Maltman, hep-ph/0308286; K.Maltman: personal discussions during the conference
36. J.Dudek, unpublished
37. Particle Data Group, Eur.Phys.J. C **15**, 1 (2000)

Heavy Quark Summary

David G. Cassel

Laboratory for Elementary-Particle Physics, Cornell University, Ithaca, NY 14053-5001, USA

Abstract. This is a summary of some of the reports on progress in heavy quark physics presented at the Hadron 2003 conference.

INTRODUCTION

Many important new results in Heavy Quark (HQ) physics were reported at this conference. In broad categories, the experimental reports include:

Charm Mesons
- BES: J/ψ, $\psi(2S)$, $\psi(3770)$, and η_c decays [1, 2, 3, 4, 5, 6]
- KEDR/VEPP-4M: precision measurements of the J/ψ and $\psi(2S)$ masses [7]
- E835: precision measurements of χ_c masses [8]
- Belle, BaBar, and CLEO: observation of the η_c' [9]
- FOCUS: resonances in three-body final states in D decay [10, 11]
- FOCUS: search for CP violation and $D\bar{D}$ mixing [12]
- BaBar, CLEO, Belle: discovery of narrow D_{sJ}^{*+} and D_{sJ}^{+} mesons [13, 14, 15, 16]
- Belle: observation of the wide D_0^{*0} and $D_1'^0$ mesons [16]
- Belle: observation of a narrow $J/\psi\,\pi^+\pi^-$ state [16]

Charm Baryons
- Belle: study of the single charm baryon Ω_c^0 [16]
- SELEX: observations of new double charm baryons Ξ_{cc}^+ and Ξ_{cc}^{++} [17, 18]

Beauty Mesons
- BaBar and Belle: B^0 and B^+ decays and CP violation [14, 16]
- CDF and DØ: robust B_s decay signals which can lead to accurate measurements of $B_s\bar{B}_s$ mixing [17]

Beauty Baryons
- CDF and DØ: robust Λ_b signals which can lead to substantial expansion of our knowledge of beauty baryons [17]

Charm and Beauty
- H1 and ZEUS: photo- and electro-production of charm and beauty [19]

Future Experiments
- CLEO-c: e^+e^- annihilation in the charm threshold region at LEPP, Cornell [20]
- BES III: e^+e^- annihilation in the charm threshold region at IHEP, Beijing [6]
- Panda: heavy quark production in $p\bar{p}$ interactions at GSI, Darmstadt [21]

CP717, *Hadron Spectroscopy: Tenth International Conference,*
edited by E. Klempt, H. Koch, and H. Orth
© 2004 American Institute of Physics 0-7354-0197-7/04/$22.00

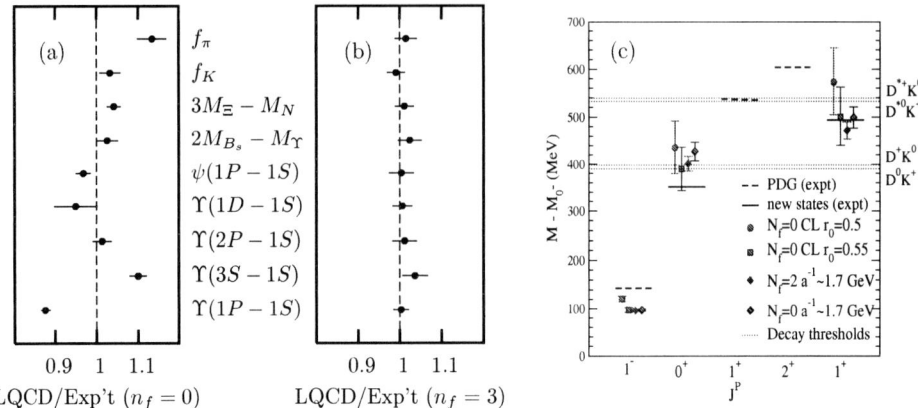

FIGURE 1. Results of LQCD calculations for Gold-Plated parameters divided by experiment measurements are shown for calculations (a) without quark loops, and (b) with three-flavor quark loops. The notation $\psi(1P - 1S)$ denotes the difference between the $\psi(1P)$ and $\psi(1S)$ masses. (c) illustrates the less satisfactory status of calculations of masses of unstable D_s states which are not Gold-Plated parameters.

Theoretical contributions included important new results on Lattice QCD calculations [22], and a variety of ideas for understanding the newly discovered narrow D^{*+}_{sJ} and D^{+}_{sJ} mesons [23, 24, 25, 26].

I had to make very difficult choices in order to keep this report to a reasonable length. I will discuss spectroscopy, strong decays to states including heavy quarks, and some theories. I am not able to include hadronic production [17], photo- and electro-production [19], weak decays to light quark states [10, 11], CP violation in B decay, [16, 14], search for CP violation in D decay and $D\bar{D}$ mixing [12], and details of future experiments [6, 20, 21]. Also, I will not be able to cover many theoretical reports, some not adequately and some not at all, [23, 25, 26, 27, 28, 29, 30]. I apologize in advance to all whose results and talks are not covered or are covered inadequately.

LATTICE QCD IN THE HQ SECTOR

Lattice QCD (LQCD) theorists are now poised to make precision (few %) calculations of many (but not all) important QCD parameters in the HQ sector, and a consortium of LQCD collaborations has recently made significant progress towards this goal [22, 31]. This consortium concentrates on "Gold-Plated" parameters, *i.e.*, certain parameters – particularly hadron masses and hadronic matrix elements with at most one hadron in the initial and final states – for hadrons that are at least 100 MeV below their strong decay threshold or with negligible decay widths.

With these restrictions the consortium will still be able to compute many non-perturbative QCD parameters that are crucial in HQ physics and measurements of CKM matrix elements. Gold-Plated parameters include: the weak decay constants

f_{D^+} and f_{D_s} that CLEO-c will measure [32], the weak decay constants f_{B^0} and f_{B_s} that are required to determine the CKM matrix elements V_{td} and V_{ts} from $B^0 \bar{B}^0$ and $B_s \bar{B}_s$ mixing, semileptonic decay form factors for D mesons that CLEO-c will measure, and semileptonic decay form factors for B mesons that are required to determine V_{cb} and V_{ub}. A major goal of this effort is to validate LQCD calculations in the D sector with CLEO-c measurements in order to utilize results for the B sector with confidence.

This success of this technique is based on technical advances in incorporating quark loops (dynamical quarks) in calculations. Most other LQCD calculations are quenched, $i.e.$, exclude quark loops. However, this approach is not yet suitable for all potential LQCD calculations of interest to the Hadron 2003 community, particularly calculations involving unstable particles! Davies reported first results from calculations which simulated 3 flavors of dynamical quarks. The consortium determined 5 free QCD parameters (bare $m_u = m_d$, m_s, m_c, m_b, and α_S), using M_π, M_K, M_{D_s}, M_Υ, and the $\Upsilon(2S)$–$\Upsilon(1S)$ mass difference. These parameters are then used in the calculations illustrated in Fig. 1. Figures 1(a) and 1(b) illustrate the importance of including quark loops as well as the success already achieved with these LQCD calculations. Clearly the consortium is already able to achieve agreement with experiment at the level of a few % for many well-measured HQ Gold-Plated parameters. On the other hand, the problems that arise with non-Gold-Plated parameters, $e.g.$, the masses of hadrons near strong decay thresholds are illustrated in Fig. 1(c). In order to have real theoretical predictions, the consortium is working hard to complete calculations of more Gold-Plated parameters before CLEO-c measures them.

CHARM BARYONS

Belle [16] is initiating a new era in charm baryon physics with a measurement of the Ω_c mass, x_p distribution, and semileptonic decay rates in their huge $\Upsilon(4S)$ data sample. The Ω_c^0, with quark content ssc (see Fig. 2) is the least well understood of the baryons with single c quarks. Belle detected Ω_c in the $\Omega_c \to \Omega^- \pi^+$ mode and observed 80.5 ± 10.8 events above a small background. This compares with about 50 events in multiple modes with large backgrounds from the E687 experiment [33] and 40 events from multiple modes from CLEO [34]. Belle obtained $M(\Omega_c) = 2693.9 \pm 1.1 \pm 1.4$ MeV, in excellent agreement with CLEO's value, $2694.6 \pm 2.6 \pm 1.9$ MeV but only in marginal agreement with the E687 result. Belle's results for the ϵ_P parameter in the Peterson parameterization of the x_p distribution and the cross section times branching fraction for events with $x_p > 0.5$ are also in good agreement with the CLEO results. This clearly demonstrates that there is a bright future ahead for Belle's charm baryon program.

SELEX previously reported a candidate for a doubly charmed (ccd) baryon $\Xi_{cc}^+(3520)$ [36]. Now [17, 18] SELEX has evidence for three more weakly decaying Ξ_{cc} states and for one strongly decaying excited Ξ_{cc} state. The signals appear after

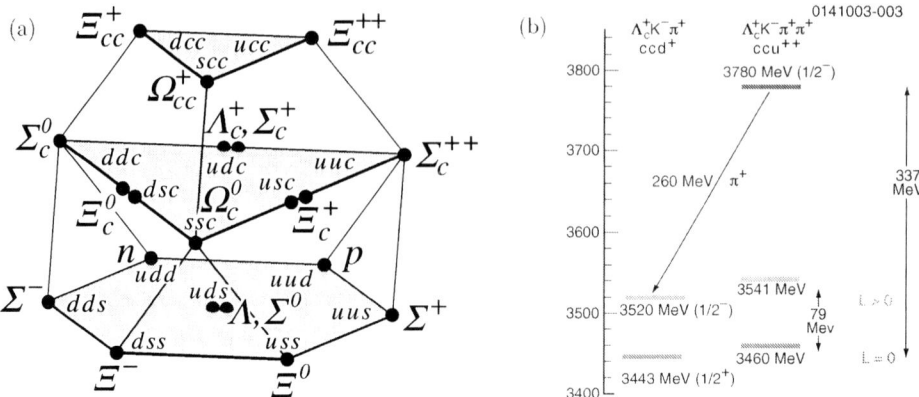

FIGURE 2. (a) The quark content of baryons that contain, 0, 1, or 2 charm quarks [35]. (b) Possible level scheme of the doubly charmed baryons from the SELEX experiment [17, 18].

isotropy and helicity cuts are used to reduce backgrounds or focus on $L > 0$ states. There are only a few events in each signal but backgrounds are also small. The possible spectroscopy arising from these signals is illustrated in Fig. 2. If the observed peaks are indeed doubly charmed baryons, the natural interpretation of the states in the $\Lambda_c K^- \pi^+$ column would be Ξ_{cc}^+ and of those in the $\Lambda_c K^- \pi^+ \pi^+$ column would be Ξ_{cc}^{++}. Two features of this spectroscopy are particularly interesting:

- Two of the new states are only about 80 MeV above the corresponding ground states, so all four of these states decay weakly.
- The strongly decaying excited state at 3780 MeV is about 340 MeV above the corresponding ground state. It is interesting to note that this mass difference is comparable to the mass difference between the lowest mass D_s and D_s^* states and the new states tentatively called D_{sJ}^{*+} and D_{sJ}^+ found by BaBar [13, 14, 37] and CLEO [15, 38].

Confirmation of these results should be forthcoming, since – at the SELEX values of production cross section times branching fraction – the COMPASS experiment should produce huge numbers of these doubly charmed baryons [18]. At this time there is no clear understanding of this spectroscopy, so serious theoretical effort will be required to understand this pattern of mass differences and its significance.

HEAVY/LIGHT QUARK MESON SPECTROSCOPY

The known states of mesons composed entirely of light quarks is given in Table 1. There are candidates for nearly all of the $q\bar{q}'$ states in the table, although evidence for some of them is not compelling enough for inclusion in the PDG [35] summary tables. The situation for mesons containing at least one heavy quark, given in Table 2, is very different. Less than one third of the possible states have been observed. Although much has been learned about the spectroscopy and decays of

Table 1. Mesons composed of the light quarks, u, d, and s, adapted from the PDG report [35]. Particles with names in boldface type are included in the PDG Meson Summary [35]; the others are not.

$N\,^{2S+1}L_J$	J^{PC}	$u\bar{d},\,u\bar{u},\,d\bar{d}$	$u\bar{u},\,d\bar{d},\,s\bar{s}$	$\bar{s}u,\,\bar{s}d$
		$I=1$	$I=0$	$I=1/2$
$1\,^1S_0$	0^{-+}	$\boldsymbol{\pi}$	$\boldsymbol{\eta},\boldsymbol{\eta'}$	$\boldsymbol{K}$
$1\,^3S_1$	1^{--}	$\boldsymbol{\rho}$	$\boldsymbol{\omega},\boldsymbol{\phi}$	$\boldsymbol{K^*(892)}$
$1\,^1P_1$	1^{+-}	$\boldsymbol{b_1(1235)}$	$\boldsymbol{h_1(1170)},\,\boldsymbol{h_1(1380)}$	$\boldsymbol{K_{1B}}$
$1\,^3P_0$	0^{++}	$\boldsymbol{a_0(1450)}^*$	$\boldsymbol{f_0(1370)},\,\boldsymbol{f_0(1710)}$	$\boldsymbol{K_0^*(1430)}$
$1\,^3P_1$	1^{++}	$\boldsymbol{a_1(1260)}$	$\boldsymbol{f_1(1285)},\,\boldsymbol{f_1(1420)}$	$\boldsymbol{K_{1A}}$
$1\,^3P_2$	2^{++}	$\boldsymbol{a_2(1320)}$	$\boldsymbol{f_2(1270)},\,\boldsymbol{f_2'(1525)}$	$\boldsymbol{K_2^*(1430)}$
$1\,^1D_2$	2^{-+}	$\boldsymbol{\pi_2(1670)}$	$\eta_2(1645),\,\eta_2(1870)$	$\boldsymbol{K_2(1770)}$
$1\,^3D_1$	1^{--}	$\boldsymbol{\rho(1700)}$	$\boldsymbol{\omega(1650)},\,\boldsymbol{\phi(1680)}$	$\boldsymbol{K^*(1680)}$
$1\,^3D_2$	2^{--}			$\boldsymbol{K_2(1820)}$
$1\,^3D_3$	3^{--}	$\boldsymbol{\rho_3(1690)}$	$\boldsymbol{\omega_3(1670)},\,\boldsymbol{\phi_3(1850)}$	$\boldsymbol{K_3^*(1780)}$
$1\,^3F_4$	4^{++}	$\boldsymbol{a_4(2040)}$	$\boldsymbol{f_4(2050)},\,f_4(2220)$	$\boldsymbol{K_4^*(2045)}$
$2\,^1S_0$	0^{-+}	$\boldsymbol{\pi(1300)}$	$\boldsymbol{\eta(1295)},\,\boldsymbol{\eta(1440)}$	$K(1460)$
$2\,^3S_1$	1^{--}	$\boldsymbol{\rho(1450)}$	$\boldsymbol{\omega(1420)},\,\boldsymbol{\phi(1680)}$	$\boldsymbol{K^*(1410)}$
$2\,^3P_2$	2^{++}	$a_2(1700)$	$f_2(1950),\,\boldsymbol{f_2(2010)}$	$K_2^*(1980)$
$3\,^1S_0$	0^{-+}	$\boldsymbol{\pi(1800)}$	$\eta(1760),\,\phi(1680)$	$K(1830)$

these states, there is clearly a very large amount physics waiting to be uncovered. Many parallels between light-quark and heavy-quark spectroscopies are evident in these tables, and the tables can be useful in orienting members of either the light-quark or heavy-quark community to the spectroscopy of the other community.

Heavy quark symmetry illuminates several aspects of the spectroscopy of charm-quark/light-quark ($c\bar{q}$) mesons [13, 39]. In the limit $m_c \to \infty$, the angular momentum ($j = l + s_q$) of the light quark q decouples from the spin of the charm quark s_c. For $l > 0$, spin-orbit forces split different j values, and for large finite m_c, spin-spin and tensor forces remove further degeneracies. This leads to the following spectroscopy when the light quark is $\bar{u}$ or $\bar{s}$:

- The $l = 0$, $j = \frac{1}{2}$ states combine with s_c to produce the 0^- (D) and the 1^- (D^*), the two lowest lying $c\bar{q}$ mesons, all four of which (D^0, D^{*0}, D_s^+, D_s^{*+}) are well known [35].
- The $l = 1$, $j = \frac{3}{2}$ states combine with s_c to produce 1^+ and 2^+ states. These states are above the $D\pi$ and $D^*\pi$ thresholds ($c\bar{u}$) or the DK and D^*K thresholds ($c\bar{s}$) but are relatively narrow because they decay via D waves. Candidates for these four states (called D_1^0, D_2^{*0}, D_{s1}^+, and D_{s2}^{*+}) have been observed [35] although the J^P of the particle I call D_{s2}^{*+} has not been established.
- The $l = 1$, $j = \frac{1}{2}$ states combine with s_c to produce 0^+ and 1^+ states, and this 1^+ state can mix with the 1^+ state from $j = \frac{3}{2}$. According to conventional wisdom, they could decay via S waves and would be above the $D\pi$ and $D^*\pi$

Table 2. Mesons composed of at least one heavy (c or b) quark, adapted from the PDG report [35]. Particles with names in boldface type are included in the PDG Meson Summary table [35]. The h_c, h_b, and η_b mesons have not been not found. The spin and parity assignments of the D_0^*, D_1', D_{sJ}^*, and D_{sJ} mesons are assumed.

		$c\bar{c}$	$b\bar{b}$	$c\bar{u},\ c\bar{d}$	$c\bar{s}$	$\bar{b}u,\ \bar{b}d$	$\bar{b}s$
$N\ ^{2S+1}L_J$	J^{PC}	$I=0$	$I=0$	$I=1/2$	$I=0$	$I=1/2$	$I=0$
$1\,^1S_0$	0^{-+}	$\eta_c(1S)$	$\eta_b(1S)$	$D(1865)$	$D_s(1969)$	B	B_s
$1\,^3S_1$	1^{--}	$J/\psi(1S)$	$\Upsilon(1S)$	$D^*(2010)$	$D_s^*(2112)$	B^*	B_s^*
$1\,^1P_1$	1^{+-}	$h_c(1P)$	$h_b(1P)$	$D_1(2420)$	$D_{s1}(2536)$		
$1\,^3P_0$	0^{++}	$\chi_{c0}(1P)$	$\chi_{b0}(1P)$	$D_0^*(2308)$	$D_{sJ}^*(2317)$		
$1\,^3P_1$	1^{++}	$\chi_{c1}(1P)$	$\chi_{b1}(1P)$	$D_1'(2427)$	$D_{sJ}(2463)$		
$1\,^3P_2$	2^{++}	$\chi_{c2}(1P)$	$\chi_{b2}(1P)$	$D_2^*(2460)$	$D_{sJ}^*(2573)$		
$1\,^1D_2$	2^{-+}						
$1\,^3D_1$	1^{--}	$\psi(3770)$					
$1\,^3D_2$	2^{--}						
$1\,^3D_3$	3^{--}						
$1\,^3F_4$	4^{++}						
$2\,^1S_0$	0^{-+}	$\eta_c(2S)$	$\eta_b(2S)$				
$2\,^3S_1$	1^{--}	$\psi(2S)$	$\Upsilon(2S)$				
$2\,^3P_2$	2^{++}		$\chi_{b2}(2P)$				
$3\,^1S_0$	0^{-+}		$\eta_b(3S)$				

thresholds ($c\bar{u}$) or the DK and D^*K thresholds ($c\bar{s}$). Hence, these states would be wide and hard to detect, so it was not thought to be surprising that these states had not been observed.

The $c\bar{u}$ spectroscopy and decays expected according to this picture are illustrated in Fig. 3(a). Belle observed [16, 40] two broad neutral states in B meson decay that are in accord with the $l=1$, $j=\frac{1}{2}$ $c\bar{u}$ states (D_0^{0*} and $D_1^{0'}$) in this picture.

However, BaBar shocked adherents to this conventional wisdom with the observation of a new narrow state tentatively named $D_{sJ}^*(2317)^+$, with mass below the threshold for decay to DK, that decays in the isospin violating channel $D_s^+\pi^0$ [13, 14, 37]. This was followed by CLEO's confirmation of this state and observation of another narrow state decaying to $D_s^{*+}\pi^0$, that CLEO tentatively names $D_{sJ}(2463)^+$ [15, 38]. Belle then confirmed both states [16, 41, 42]. This sequence of events was perhaps the most exciting recent development in heavy quark spectroscopy. Fig. 3(b) illustrates the spectroscopy of $c\bar{s}$ mesons after these remarkable discoveries.

OBSERVATION OF THE D_0^{*0} AND $D_1'^0$

Belle searched for excited $c\bar{u}$ states in $B^- \to D^{(*)+}\pi^-\pi^-$ decay [16, 40]. This environment is much cleaner than the usual hunting ground – continuum events

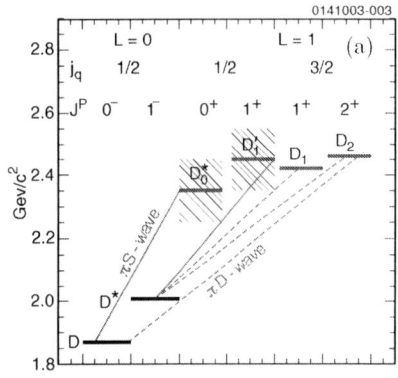
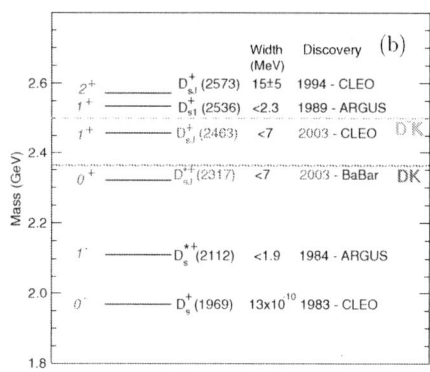

FIGURE 3. (a) the spectroscopy and decays of $c\bar{u}$ mesons after the Bell observation of D_0^{*0} and $D_1'^{0}$ states. The solid lines indicate S wave decays and the dashed lines indicate D wave decays. (b) the spectroscopy of $c\bar{s}$ after the discovery of the D_{sJ}^{*+} and D_{sJ}^{+} states. The thresholds for DK and D^*K decays are indicated by horizontal lines.

– for excited D states because, continuum events have large combinatorial backgrounds from non-resonant D and D^* candidates combined with multiple random pions. Belle observes robust signals for both $B^- \to D^+\pi^-\pi^-$ and $B^- \to D^{*+}\pi^-\pi^-$ decays. Belle's $B^- \to D^+\pi^-\pi^-$ signal was the first significant observation of this B^- decay mode [35].

Analysis of $D^+\pi^-\pi^-$ events is the simpler of the two analyses since only D_0^{*0} and D_2^{*0} should decay to this final state (see Fig. 3(a)). Belle then performs an unbinned maximum likelihood fit to the $D^+\pi^-\pi^-$ using sideband data to estimate combinatorial backgrounds. The fits are rather complex, requiring assumptions concerning the mass dependence of the decay widths, particularly for the wide D_0^{*0} state. Furthermore, the data prefer additional contributions from unknown higher mass virtual D and B states. Even with these contributions, the χ^2 of the fits with different versions of these assumptions are not excellent. However, the preferred fit does follow projections of the Dalitz plot onto the appropriate $M(D\pi)$ mass axis, and this fit generally follows the rather complex trends of projections of the data onto the $M(D\pi)$ axis in bins of helicity angle. The trends of the data and the fits in the neighborhood of $M(D\pi) = 2,540$ MeV in bins of helicity angle support the assumption that the contribution attributed to the D_2^{*0} is D wave. The D_0^{*0} and D_2^{*0} masses and widths from the preferred fit are given in Table 3, along with the PDG value for the D_2^{*0} mass and width.

The analysis of the $D^{*+}\pi^-\pi^-$ final state is even more complex since the $D_1'^{0}$ can decay to this final state via S wave decay so it is very broad, and the D_1^0 and D_2^{*0} can also reach this final state via D wave decays. The data are fit in a manner similar to the $D^+\pi^-\pi^-$ data, with the mass and width of the D_2^{*0} fixed to the value obtained in the $D^+\pi^-\pi^-$ fit. The results for the $D_1'^{0}$ and D_1^0 masses and widths are also given in Table 3. Agreement with PDG averages is excellent except for the

Table 3. Belle results for excited $l = 1$, $j_q = \frac{1}{2}$ and $\frac{3}{2}$ $c\bar{u}$ states. The errors for the Belle results are statistical, experimental systematic, and model dependent uncertainties in that order. The PDG masses and decay widths are from [35].

State	Belle		PDG	
	M (MeV)	Γ (MeV)	M (MeV)	Γ (MeV)
D_0^{*0}	$2,308 \pm 17 \pm 15 \pm 28$	$276 \pm 21 \pm 18 \pm 60$	---	---
$D_1'^{0}$	$2,427 \pm 26 \pm 20 \pm 15$	$384^{+107}_{-75} \pm 24 \pm 70$	—	—
D_1^0	$2,421.4 \pm 1.5 \pm 0.4 \pm 0.8$	$23.7 \pm 2.7 \pm 0.2 \pm 4.0$	$2,422.2 \pm 1.8$	$18.9^{+4.6}_{-3.5}$
D_2^{*0}	$2,461.6 \pm 2.1 \pm 0.5 \pm 3.3$	$45.6 \pm 4.4 \pm 6.5 \pm 1.6$	$2,458.9 \pm 2.0$	23 ± 5

D_2^{*0} width, for which there is another (unpublished) indication from FOCUS [43] that the width may be larger than the PDG average.

DISCOVERY OF THE $D_{sJ}^*(2317)^+$ AND $D_{sJ}(2463)^+$

The BaBar signal for the narrow state, tentatively called $D_{sJ}^*(2317)^+$, that decays to $D_s^+\pi^0$ [13, 14, 37] is illustrated in Fig. 4(a). The signal is very large and robust, and the new state appears to be the "missing" 0^+ ($l = 1$, $j = \frac{1}{2}$) $c\bar{s}$ state. However, contrary to conventional wisdom, the observed mass is below most (but not all [44, 45]) theoretical predictions for the 0^+ $c\bar{s}$ state and it is also below the DK threshold. Hence, this state should be narrow and the observed isospin violating strong decay $D_{sJ}^* \to D_s^+\pi^0$ could be favored. The low value of the D_{sJ}^* mass led to a flurry of theoretical activity [13, 23, 24, 25, 26, 46, 47, 48] including exotic, as well as relatively conventional, interpretations of the result.

The D_{sJ}^{*+} was also present in Belle and CLEO data, but nobody had looked for it before, because conventional wisdom suggested that it would be above the DK threshold, and therefore be very broad and very difficult to detect. CLEO confirmed the $D_{sJ}^*(2317)^+$ and provided evidence for another state tentatively called $D_{sJ}(2463)^+$ which decays to $D_s^{*+}\pi^0$ [38]. Belle confirmed the observations of both of these states by finding them in B decays [41] as well as continuum e^+e^- events [42]. CLEO data for the two signals are illustrated in Fig. 4(b) and Belle signals from their continuum e^+e^- data are illustrated in Fig. 5. These data exhibit clear signals in the $\Delta M(D_{sJ}^*) = M(D_s^+\pi^0) - M(D_s^+)$ and $\Delta M(D_{sJ}) = M(D_s^{*+}\pi^0) - M(D_s^{*+})$ mass difference distributions.

Analysis of these two new states is complicated because the two decay chains leading to $D^+\pi^0$ and $D^{*+}\pi^0$ final states differ only in the low-energy (~ 145 MeV) γ from $D_s^{*+} \to D_s^+\gamma$ decay, and the peaks in the two mass difference distributions $\Delta M(D_{sJ}^*)$ and $\Delta M(D_{sJ})$ are both near 350 MeV. Therefore it is possible for a D_{sJ} decay to appear in the D_{sJ}^* signal if the low energy γ from a D_s^* decay is not detected. Alternately, if a random low energy photon and the D_s from D_{sJ}^* decay reconstruct to the D_s^* mass, the combination will appear in the D_{sJ} signal. Separating the two signals requires careful analyses and cross checks. CLEO, BaBar, and

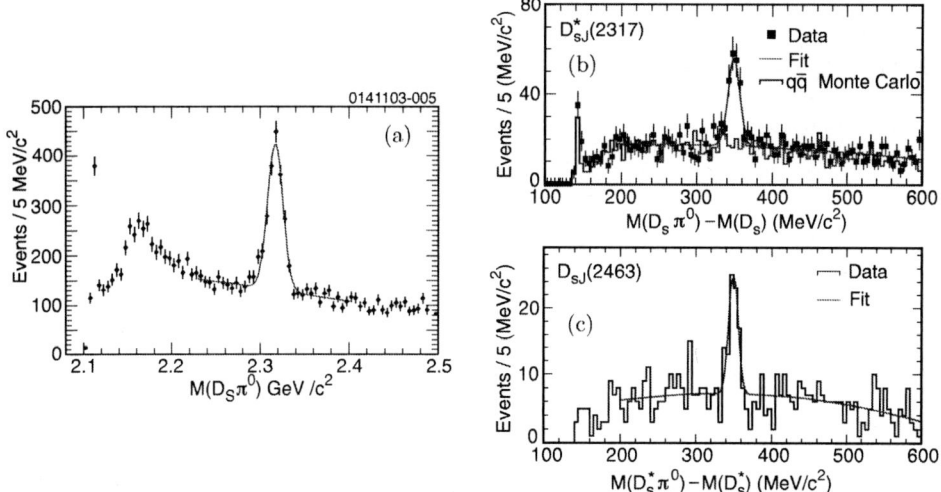

FIGURE 4. (a) The BaBar $D_{sJ}^*(2317)^+$ signal in the invariant mass distribution of $D_s^+\pi^0$ candidates. The narrow peak near 2.1 GeV is due to the isospin violating $D_s^{*+} \to D_s^+\pi^0$ decay. The $\Delta M(D_{sJ})$ (b) and $\Delta M(D_{sJ}^*)$ (c) mass differences from CLEO illustrating CLEO's $D_{sJ}^*(2317)^+$ and $D_{sJ}(2463)^+$ signals. The $q\bar{q}$ Monte Carlo data in (b) are normalized absolutely.

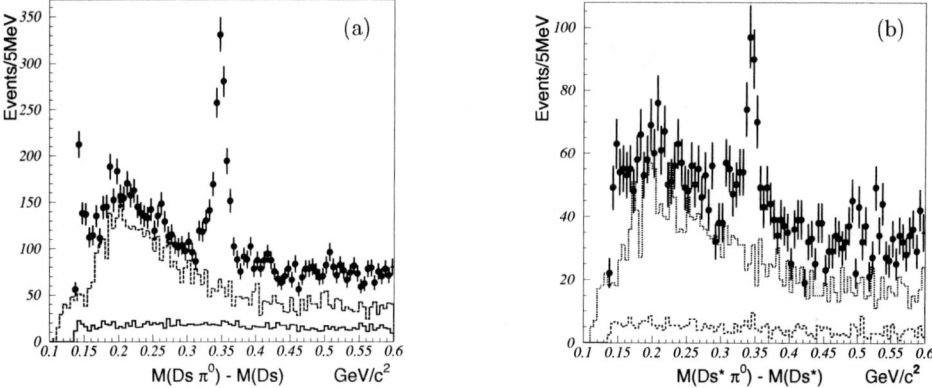

FIGURE 5. The Belle D_{sJ}^{*+} (a) and D_{sJ}^+ (b) signals from continuum data.

Belle used somewhat different techniques to disentangle the signals and demonstrate that neither signal is due entirely to misidentified events from the other. Each collaboration provides convincing evidence for the presence of both states.

The mass differences and decay widths obtained by BaBar, CLEO, and Belle are given in Table 4. The decay widths measured by all three experiments are consistent with zero. The $\Delta M(D_{sJ}^*)$ values are very consistent, but the measured $\Delta M(D_{sJ})$ mass differences are somewhat less consistent.

The evidence available so far is consistent with the spin-parity assignments of

Table 4. The mass differences, $\Delta M(D^*_{sJ})$ and $\Delta M(D_{sJ})$, and decay widths $\Gamma(D^*_{sJ})$ and $\Gamma(D_{sJ})$, measured by the BaBar, CLEO, and Belle collaborations. Belle (B) and Belle (C) refer to the Belle measurements in B meson and continuum data samples, respectively. The upper limits on decay widths are 90% CL ULs. The averages are naive weighted averages that do not take possible correlations into account.

Experiment	$\Delta M(D^*_{sJ})$ (MeV)	$\Gamma(D^*_{sJ})$ (MeV)	$\Delta M(D_{sJ})$ (MeV)	$\Gamma(D_{sJ})$ (MeV)
BaBar	$348.4 \pm 0.4 \pm 3.0$	< 10	$344.6 \pm 1.2 \pm 3.0$	
CLEO	$350.0 \pm 1.2 \pm 1.0$	< 7	$351.2 \pm 1.7 \pm 1.0$	< 7
Belle (B)	$351.3 \pm 2.1 \pm 2.0$		$346.8 \pm 1.6 \pm 2.0$	
Belle (C)	$348.7 \pm 0.5 \pm 0.9$	< 4.6	$344.1 \pm 1.3 \pm 1.1$	< 5.5
Average	349.2 ± 0.8		346.8 ± 1.1	

1^+ for the D^+_{sJ} and 0^+ for the D^{*+}_{sJ}:

Evidence that the D^+_{sJ} is 1^+ is reasonably compelling:

- Existence of the decay $D^+_{sJ} \rightarrow D^{*+}_s \pi^0$ and J^P conservation excludes 0^+.
- Belle observed $D^+_{sJ} \rightarrow D^+_s \gamma$ decays which excludes $J = 0$.
- Belle's angular distribution of $D^+_{sJ} \rightarrow D^+_s \gamma$ decays excludes 2^+ and favors 1^+.

Evidence that D^{*+}_{sJ} is 0^+ is less compelling:

- Existence of the $D^{*+}_{sJ} \rightarrow D^+_s \pi^0$ decay implies natural spin-parity.
- If the D^*_{sJ} is 0^+ the following decays, which are not seen, would be forbidden:
  - $D^{*+}_{sJ} \rightarrow D^+_s \gamma$ would be forbidden by J.
  - $D^+_{sJ} \rightarrow D^+_s \pi^+ \pi^-$ would be forbidden by J^P.

Theorists are also investigating consequences of molecular and tetraquark models [23, 25, 26]. Naively, one would expect that other charge states should exist if the D^*_{sJ} and D_{sJ} are $D\pi$ molecules. CLEO searched for signals in neutral and doubly charged states decaying to $D^{\pm}_s \pi^{\mp}$, $D^{*\pm}_s \pi^{\mp}$, $D^{\pm}_s \pi^{\pm}$, and $D^{*\pm}_s \pi^{\pm}$. No signals were observed and 90% CL ULs of $\sigma(\text{mode})\, \mathcal{B}(\text{mode}) \lesssim 10\%$ of the corresponding $D^+_s \pi^0$ or $D^{*+}_s \pi^0$ rate were obtained.

Efforts to quantitatively explain the pattern of excited $c\bar{u}$ and $c\bar{s}$ states with conventional spin-orbit and tensor forces have not been entirely successful [13, 39]. On the other hand, models in which the $D^{*+}_{sJ} - D^+_s$ and $D^+_{sJ} - D^{*+}_s$ mass differences arise from the spontaneous breakdown of chiral symmetry (SBCS) actually predicted the correct mass splittings at a time when conventional wisdom suggested much larger values [24, 44, 45, 46, 47]. Chiral multiplet models [24, 44, 45, 46, 47] predict, $M(1^+) - M(1^-) = M(0^+) - M(0^-)$, or $M(D^+_{sJ}) - M(D^{*+}_s) = M(D^{*+}_{sJ}) - M(D^+_s)$, if the D^{*+}_{sJ} and D^+_{sJ} are the 0^+ and 1^+ $c\bar{s}$ states, respectively. CLEO finds $\delta(\Delta M) = (351.2 \pm 1.7) - (350.0 \pm 1.2) = 1.2 \pm 2.1$ MeV.

Another possible consequence of chiral multiplet models generated substantial discussion at Hadron 2003. If the separations between the $l = 0$ and 1 $c\bar{s}$ states with $J = 0$ and 1 but opposite parities are due to SBCS, then SBCS may also separate the $l = 1$ and 2 states with $J = 1$ and 2 but opposite parities [24, 47]. This could then be repeated for other $Q\bar{q}$ systems and other l values.

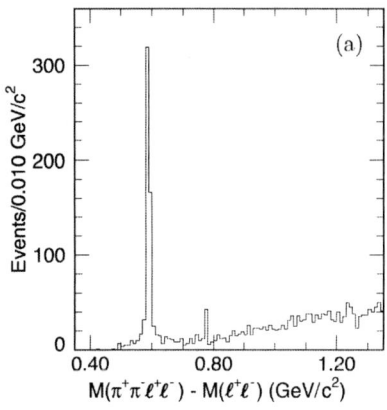 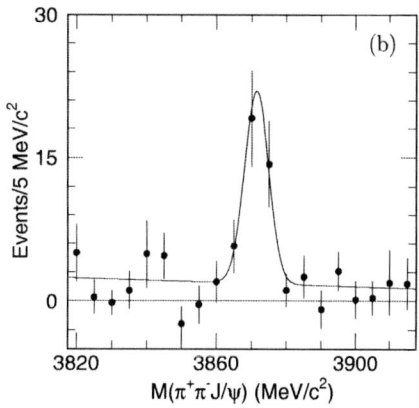

FIGURE 6. (a) Belle's distribution of $M(\ell^+\ell^-\pi^+\pi^-) - M(\ell^+\ell^-)$, with $M(\ell^+\ell^-)$ consistent with $M_{J/\psi}$. The large $\psi(2S)$ peak and the smaller – but significant – peak from the new state are clearly visible. (b) shows the $M(J/\psi\,\pi^+\pi^-)$ signal yields from unbinned maximum likelihood fits.

A NEW NARROW CHARMONIUM-LIKE STATE

The heavy quark community was surprised a second time this year with the Belle observation of a new narrow charmonium state that decays to $J/\psi\,\pi^+\pi^-$ [16, 49]. This is another example of Belle's highly successful program of utilizing its huge sample of B mesons as a source of particles containing charm quarks, either charmonium states or $c\bar{q}$ states. This is a particularly valuable source of charmonium states with quantum numbers other than $J^{PC} = 1^{--}$, the only $c\bar{c}$ states that can be produced directly in e^+e^- annihilation. The evidence for a new charmonium state decaying to $J/\psi\,\pi^+\pi^-$ comes from $B^\pm \to K^\pm J/\psi\,\pi^+\pi^-$ decays. The signals are illustrated in Fig. 6. The large peak in Fig. 6(a) is from $B^\pm \to K^\pm\psi(2S)$ decays, and is well-reproduced in Monte Carlo simulations. However, the mass spectrum in a Monte Carlo simulation is smooth above this peak and – in particular – does not reproduce the small peak near $\Delta M = 775$ MeV. Fig. 6(b) illustrates the $M(J/\psi\,\pi^+\pi^-)$ signal obtained from an unbinned maximum likelihood fit that includes the reconstructed B mass and the discrepancy between the measured center of mass energy of the B candidate and the beam energy in the center of mass. Using the $\psi(2S)$ peak in Fig. 6(a) to constrain the energy scale, Belle finds that the mass of the $J/\psi\,\pi^+\pi^-$ state is $M(X) = 3872.0 \pm 0.6 \pm 0.5 \pm 0.5$ MeV [49]. The measured decay width is $\Gamma(X) = 1.4 \pm 0.7$ MeV, or $\Gamma(X) < 2.3$ MeV (90% CL UL) [49]. This state is very near the $D^0\bar{D}^{*0}$ threshold at 3871 MeV, suggesting the possibility that it could be a loosely bound multiquark $D\bar{D}^*$ "molecular" state [16, 49]. It could also be a candidate for the missing $^3D_2$ charmonium state, although not everything observed so far is consistent with potential model expectations [49] for that state.

Table 5. KEDR/VEPP-4M measurements of the J/ψ and $\psi(2S)$ masses, compared to PDG averages of previous measurements.

Particle	KEDR/VEPP-4M (MeV)	PDG (MeV)
$M_{J/\psi}$	$3096.917 \pm 0.010 \pm 0.007$	3096.87 ± 0.04
$M_{\psi(2S)}$	$3686.111 \pm 0.025 \pm 0.009$	3685.96 ± 0.09

CHARMONIUM MASSES AND DECAY WIDTHS

New, high precision measurements of J/ψ and $\psi(2S)$ masses using the KEDR detector at VEPP-4M were reported at this conference [7, 50]. These measurements continue the long-term program of ever more precise measurements of the masses of $q\bar{q}$ and $Q\bar{Q}$ states at Novosibirsk. The absolute mass scale and precision of these measurements is achieved through the use of resonance depolarization to determine accurately the mean value of the storage ring beam energy. This measurement required a thorough understanding of and monitoring of many subtle accelerator effects, extremely careful experimental work, and close attention to a multitude of details. The results of this effort are given in Table 5 along with previous PDG averages [35]. The new measurements are compatible with, but about a factor of 3 more precise than the PDG averages.

The E835 Collaboration reported new preliminary precision measurements of χ_c masses [8]. These measurements are from the latest round of a series of precision measurements of charmonium masses at Fermilab. In these experiments, antiprotons are scattered on a gaseous H_2 target. The beam energy spread is very small $\sigma_E/E \sim 10^{-4}$ (comparable to the beam energy spread of VEPP-4M), but the mean beam energy is not known with high absolute precision, because resonant polarization is not an option with $\bar{p}$ beams. Therefore, the absolute energy scale is derived from other measurements of J/ψ and $\psi(2S)$ masses. Presumably the final results of these measurements will take into account the new KEDR/VEPP-4M measurements of the J/ψ and $\psi(2S)$ masses, so I am not including the preliminary values in this summary report. The preliminary new measurement of the $M(\chi_{c0})$ mass is substantially more precise than earlier measurements, while the precisions of the preliminary values of $M(\chi_{c1})$ and $M(\chi_{c2})$ are comparable to those of earlier measurements.

BES reported new measurements of branching fractions for η_c hadronic decays and accurate measurements of the η_c mass and width [3, 6, 51, 52]. BES reconstructed $J/\psi \to \gamma\eta_c$ events in five η_c decay modes: $K^+K^-\pi^+\pi^-$, $\pi^+\pi^-\pi^+\pi^-$, $K^{\pm}K^0_S\pi^{\mp}$, $p\bar{p}$, and $\phi\phi$. There are clean robust signals above reasonable backgrounds in all of these modes. Fig. 7 illustrates the signals in the first four of these modes. The precisions of the preliminary branching fractions measured for all of these modes are comparable to or better than the PDG averages [3, 6, 52]. The values of the η_c mass and decay width obtained by BES are given in Table 6. Except for the systematic error in the measured η_c decay width, the precisions of the new BES

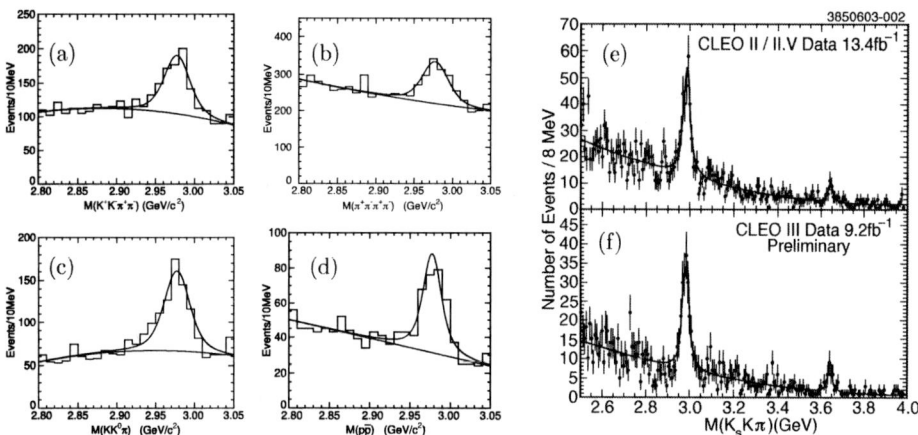

FIGURE 7. BES signals for (a) $\eta_c \to K^+K^-\pi^+\pi^-$, (b) $\eta_c \to \pi^+\pi^-\pi^+\pi^-$, (c) $\eta_c \to K^\pm K^0\pi^\mp$, and (d) $\eta_c \to p\bar{p}$. (e) and (f) CLEO signals for η_c' production in $\gamma\gamma$ fusions from two different configurations of the CLEO detector.

Table 6. Summary of new measurements of η_c and η_c' masses and decay widths. The BaBar and CLEO results are preliminary.

η_c	M (MeV)	Γ (MeV)	
BES	$2977.5 \pm 1.0 \pm 1.2$	$17.0 \pm 3.7 \pm 7.4$	
PDG	2979.7 ± 1.5	$16.0^{+3.6}_{-3.2}$	

η_c'	M (MeV)	Γ (MeV)	
Belle	$3654 \pm 6 \pm 8$	15^{+24}_{-15}	(< 55) 90% CL
BaBar	$3632.2 \pm 5.0 \pm 1.8$	$20 \pm 10 \pm 4$	
CLEO	$3642.7 \pm 4.1 \pm 2.0$	10^{+22}_{-10}	(< 46) 90% CL

measurements are comparable to those of the PDG averages. Uncertainties in the background shapes and the effect of a J/ψ veto dominate the large systematic error in this decay width.

In contrast to the η_c, the η_c' history is rather checkered, with one observation followed by a number of fruitless searches [9]. In fact, the η_c' does not appear in the 2002 PDG meson summary tables [35]. This situation changed dramatically with the Belle observation [53] of robust η_c' signals in $B^+ \to K^+\eta_c'$ with $\eta_c' \to K_S^0 K^\pm\pi^\mp$, followed by preliminary BaBar [54] and CLEO [55] observations of η_c' signals in $\gamma\gamma \to \eta_c'$, again with $\eta_c' \to K_S^0 K^\pm\pi^\mp$. The CLEO η_c' signals, observed in two data sets with substantially different detectors and software systems, are illustrated in Figs. 7(e) and 7(f). The masses and decay widths obtained by the three collaborations are given in Table 6. The results of the three experiments are in reasonable agreement.

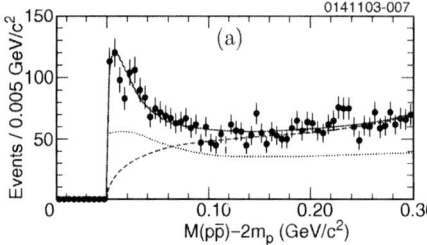 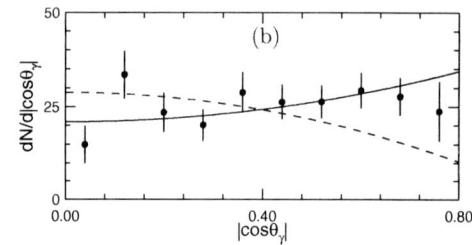

FIGURE 8. (a) The M($p\bar{p}$) mass distribution for $J/\psi \to \gamma p\bar{p}$ events. The points are the data, the solid line is the fit to the data, the dotted line is the acceptance, and the dashed line is the background estimated from $J/\psi \to \pi^0 p\bar{p}$ events. (b) The polar angle distribution for the photon in $J/\psi \to \gamma p\bar{p}$ decays. The solid line is a fit to $1 + \cos^2 \theta_\gamma$ and the dashed line is $\sin^2 \theta_\gamma$.

J/ψ DECAYS

The BES II data sample of 58 M J/ψ events is about a factor of eight larger than any other collection of J/ψ data. Many interesting results based on this data sample were reported at Hadron 2003 [3, 5, 6]. In the previous section I summarized the BES measurements of the η_c mass and width. In this section I will summarize an observation of a below-threshold $p\bar{p}$ resonance in $J/\psi \to \gamma p\bar{p}$ [5, 6, 56]. Some other J/ψ results are included in later sections.

The BES evidence for a $p\bar{p}$ resonance in $J/\psi \to \gamma p\bar{p}$ events is illustrated in Fig. 8(a). There is a clear signal above the background estimated from $J/\psi \to \pi^0 p\bar{p}$ events. The data are fit to the background function derived from $J/\psi \to \pi^0 p\bar{p}$ events and a Breit-Wigner resonance with an S-wave width. This fit is also illustrated in Fig. 8(a). The mass and width determined from this fit are $M = 1859^{+3}_{-10}{}^{+5}_{-25}$ MeV and $\Gamma < 30$ MeV at the 90% CL. This mass is about 18 MeV below the $p\bar{p}$ threshold, and the mass and width are not consistent with any established particle. If this is a $p\bar{p}$ resonance with $C = +1$, decaying via S-wave, it would have $J^{PC} = 0^{-+}$. The angular distribution for the photon in $J/\psi \to \gamma p\bar{p}$ decay would then be $1 + \cos^2 \theta_\gamma$. Fig. 8(b) shows that this is consistent with (and slightly favored by) the measured angular distribution, although $\sin^2 \theta_\gamma$ cannot be ruled out. The resonance could also be a $^3P_0$ state with $J^{PC} = 0^{++}$ that would decay via P-wave. A fit to the data with a P-wave width is almost as good as the S-wave fit and yields very similar M and Γ values. In either case, the angular distribution would be $1 + \cos^2 \theta_\gamma$, so the angular distribution cannot distinguish between these two possibilities. This state might be a bound $p\bar{p}$ system, an example of a system generally called baryonium. The idea that there might be a $p\bar{p}$ bound state just below threshold is not new [5, 6, 56].

EVIDENCE FOR $\psi(3770) \to J/\psi \, \pi^+ \pi^-$

BES-II accumulated about 20 pb^{-1} of data on and in the neighborhood of the $\psi(3770)$. This data sample is about a factor of 2 larger than the Mark III data

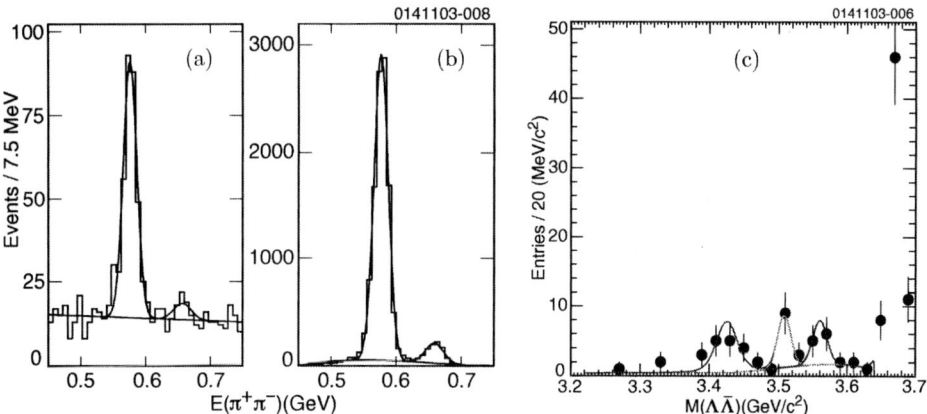

FIGURE 9. BES evidence for $\psi(377) \not\to D\bar{D}$. (a) the $\pi^+\pi^-$ energy spectrum for data and (b) a Monte Carlo simulation of the same spectrum. (c) distribution of $\Lambda\bar{\Lambda}$ invariant mass from $\psi(2S) \to \gamma\Lambda\bar{\Lambda}$ decay accompanied by a photon. The three fitted peaks are (from left to right) the χ_{c0}, χ_{c1}, and χ_{c3}.

sample. BES has used 8 of pb^{-1} of these $\psi(3770)$ data to search for events in which $\psi(3770) \not\to D\bar{D}$. Little is known about the $\psi(3770)$ except for its $D\bar{D}$ decays, and there is an apparent discrepancy between the cross section for $\psi(3770)$ production in e^+e^- collisions and the corresponding cross section for production of $D\bar{D}$ events [57]. Hence, searches for $\psi(3770) \not\to D\bar{D}$ decays are crucial for understanding the nature of the $\psi(3770)$. BES reported evidence for $\psi(3770) \to J/\psi\pi^+\pi^-$ decays [2, 6, 58], where J/ψ candidates are reconstructed in the e^+e^- and $\mu^+\mu^-$ modes. The results are illustrated in Figs. 9(a) and 9(b) which compare data to Monte Carlo simulations. In data and Monte Carlo simulations a large peak from $\psi(2S)$ radiative returns (initial state radiation) is clearly visible in the $E(\pi^+\pi^-)$ spectrum. The smaller peak near $E(\pi^+\pi^-) \sim 0.65$ GeV is attributed to $\psi(3770) \to J/\psi\pi^+\pi^-$ decay. The branching fraction obtained from this peak is $\mathcal{B}(\psi(3770) \to J/\psi\pi^+\pi^-) = (0.59 \pm 0.26 \pm 0.16)\%$. CLEO searched for the same decay mode in a smaller data sample (5 pb^{-1}) and found an upper limit of $\mathcal{B}(\psi(3770) \to J/\psi\pi^+\pi^-) < 0.26\%$ (90% CL) [59]. Clearly this is only the beginning of a program to understand the apparent discrepancy between the production cross sections for the $\psi(3770)$ and $D\bar{D}$ events.

$\psi(2S)$ DECAYS

The BES II sample of 14 M $\psi(2S)$ events is by far the largest in the world; it is more than a factor of three larger than the BES I sample, which was about a factor of two larger than the Crystal Ball sample. BES is investigating a large number of decay modes, is providing more precise measurements of branching fractions for known modes, and is finding evidence for modes not previously seen. At this conference [4, 6] BES presented results on the first evidence for $\chi_{cJ} \to \Lambda\bar{\Lambda}$ decay, the first observation of $\psi(2S) \to K_S^0 K_L^0$, and preliminary measurements of branching fractions for several $\psi(2S) \to VT$ (vector-tensor) decay modes.

Part of the motivation for studying $\chi_c \to \Lambda\bar{\Lambda}$ is to explore the validity of the color octet mechanism which appears to be required to describe χ_c decays [4, 6, 60]. The BES signals for $\chi_c \to \Lambda\bar{\Lambda}$ are illustrated in Fig. 9(c). The three χ_c peaks in the data are fit with Breit-Wigner functions with PDG [35] values of decay widths, convoluted with the detector mass resolution. The χ_c masses obtained from the fits are in good agreement with PDG averages. The branching fractions obtained from these fits are, $\mathcal{B}(\chi_{c0} \to \Lambda\bar{\Lambda}) = (4.7^{+1.3}_{-1.2} \pm 1.0) \times 10^{-4}$, $\mathcal{B}(\chi_{c1} \to \Lambda\bar{\Lambda}) = (2.6^{+1.0}_{-0.9} \pm 0.6) \times 10^{-4}$, and $\mathcal{B}(\chi_{c1} \to \Lambda\bar{\Lambda}) = (3.3^{+1.5}_{-1.3} \pm 0.7) \times 10^{-4}$. The agreement of these results with the color octet mechanism predictions is only marginal [4, 6, 60].

One motivation for measuring VT decay modes of the $\psi(2S)$ is to understand the "12% Puzzle" of Mark III and BES. QED and QCD suggest that the ratio of the decay widths for $\psi(2S) \to H$ to $J/\psi \to H$ (where H is any hadron state) should be roughly equal to the ratios of the semileptonic widths. This follows from the fact that lepton pair production and hadron production follow from the annihilation of the c and $\bar{c}$ quarks; to a virtual photon the first case and to three gluons in the second. The dominant difference between the $\psi(2S)$ and the J/ψ decay widths should the squares of the respective $c\bar{c}$ wave functions at the origin. This picture of charmonium hadronic decays, coupled with PDG [35] averages of leptonic branching fractions, leads to the prediction,

$$Q_H \equiv \frac{\mathcal{B}(\psi(2S) \to H)}{\mathcal{B}(J/\psi \to H)} \approx \frac{\mathcal{B}(\psi(2S) \to e^+e^-)}{\mathcal{B}(J/\psi \to e^+e^-)} \approx 12\%.$$

Many modes agree with this prediction within experimental errors, but some PV (pseudoscalar-vector) and VT modes do not. Measurement of as many modes as possible, in order to establish the pattern of successes and failures of the 12% Rule, is likely to be very useful for understanding the nature of the violations of the rule.

BES reported preliminary measurements of four $\psi(2S) \to VT$ decay modes for which only upper limits were available previously. These branching fractions along with the corresponding Q_H differences are given in Table 7. For these modes, the central values of Q_H are a factor of 2-4 below the 12% Rule prediction.

Fig. 10 illustrates the three Feynman diagrams that may hold the key to understanding violations of the 12% Rule [1]. In principle the three diagrams in this figure can contribute *interfering* amplitudes to the yield of any hadronic state in e^+e^- annihilation in the neighborhoods of the $\psi(2S)$ and J/ψ. The 12% Rule is derived by assuming that only the $a(ggg)$ amplitude contributes to hadronic decays and only the $a(\gamma^*c\bar{c})$ amplitude contributes to leptonic decays, so *ipso facto* interference is not an issue. On the other hand, it is well known that the interference between the $a(\gamma^*c\bar{c})$ and $a(\gamma^*)$ amplitudes is readily visible in the $e^+e^- \to \mu^+\mu^-$

Table 7. Preliminary BES-II measurements of $\psi(2S) \to VT$ branching fractions. The corresponding $J/\psi \to VT$ branching fractions are from the Particle Data Group.

VT Mode	$\mathcal{B}(\psi(2S) \to X)$ (10^{-4})	$\mathcal{B}(J/\psi \to X)$ (10^{-3})	Q_H (%)
ωf_2	$2.05 \pm 0.41 \pm 0.46$	4.5 ± 0.6	4.8 ± 1.5
ρa_2	$2.55 \pm 0.73 \pm 0.60$	10.9 ± 2.2	2.3 ± 1.1
$K^* \bar{K}_2^*$	$1.64 \pm 0.33 \pm 0.41$	6.7 ± 2.6	2.4 ± 1.2
$\phi f_2'$	$0.48 \pm 0.14 \pm 0.12$	$1.23 \pm 0.06 \pm 0.20$	3.9 ± 1.6

*0141003-001

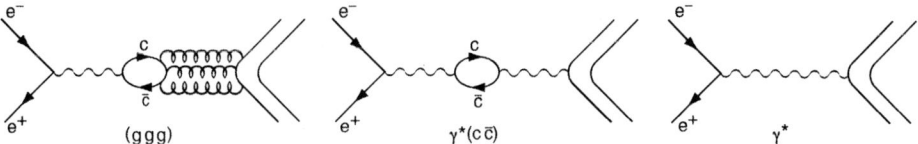

FIGURE 10. The three Feynman diagrams that can interfere and contribute to measurements of hadronic yields near the masses of charmonium states produced in e^+e^- annihilation.

cross section, so interference among two or three of these amplitudes should not be ruled out *a priori*. Analysis of these interference effects requires careful attention to cross sections, phases, and beam energy spreads. This approach has had some success [1, 61] in efforts to understand the violations of the 12% Rule in *PV* modes, *i.e.*, that destructive interference effects may explain some of the small Q_H values in these modes. Obviously establishing that these interference effects can explain the pattern of successes and failures of the 12% Rule will require a comprehensive understanding of the phases differences among the amplitudes contributing to each process. Measurements of Q_H in experiments in which e^+e^- annihilation is not the source of the charmonium (or Υ) states would be a significant step towards understanding whether or not the violations of 12% Rule are due these interference effects. Measurements in e^+e^- colliders with significantly different beam energy spreads can also illuminate the role of interference. In any event, we no longer have the luxury of assuming that the hadronic branching fractions of charmonium states produced in e^+e^- annihilation are entirely independent of the production mechanism(s).

Perhaps one of the more interesting violations of the 12% Rule is the observation of an *enhancement* of the Q_H value for the $K_S^0 K_L^0$ decay mode of the $\psi(2S)$. BES observes $K_S^0 K_L^0$ events by studying K_S^0 momentum spectra. These spectra, illustrated in Fig. 11 for both $\psi(2S)$ and J/ψ decays [4, 6, 62, 63] have large peaks at momenta corresponding to two body decays to $K_S^0 K_L^0$. The measured branching fractions are $\mathcal{B}(J/\psi \to K_S^0 K_L^0) = (1.82 \pm 0.04 \pm 0.13) \times 10^{-4}$ and $\mathcal{B}(\psi(2S) \to K_S^0 K_L^0) = (5.24 \pm 0.47 \pm 0.48) \times 10^{-5}$ [62, 63]. The corresponding Q_H value is $Q_H = (28.8 \pm 3.7)\%$, more than a factor of two above the 12% Rule.

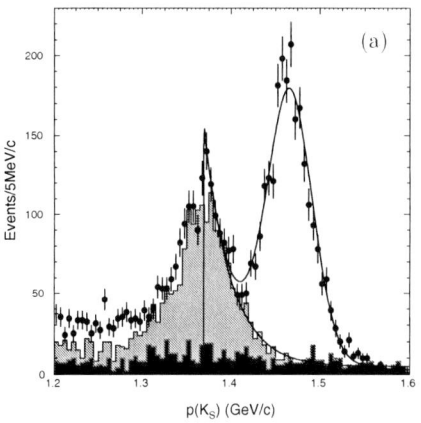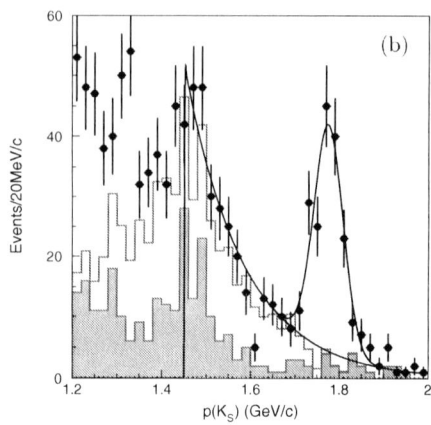

FIGURE 11. (a) the momentum spectrum of K_S^0 observed in J/ψ events, and (b) the momentum spectrum of K_S^0 observed in $\psi(2S)$ events. The points are data, the dark shaded histograms are from K_S^0 sidebands, the light shaded histograms are from Monte Carlo simulations of backgrounds. The unshaded peaks are the K_L^0 signals, and the curve is the best fit to the data.

Interference may be able to account for this enhancement above the 12% Rule [1, 64].

EVIDENCE FOR TWO-BODY
PSEUDOSCALAR-VECTOR DECAYS OF THE Υ

CLEO has accumulated by far the largest data samples at each of the bound Υ resonances: 1.2 fb^{-1} at the $\Upsilon(1S)$, 1.4 fb^{-1} at the $\Upsilon(2S)$, and 1.5 fb^{-1} at the $\Upsilon(3S)$. Additional data were accumulated in scans over the resonance peaks and in the near by continuum regions.

One of the first results from these data samples has been searches for $\Upsilon(nS) \rightarrow PV$ decays. One motivation for these searches is that no exclusive hadronic $\Upsilon(2S)$ and $\Upsilon(3S)$ decays, that do not involve transitions to other $b\bar{b}$ states, have been seen. Furthermore, $\Upsilon(nS) \rightarrow PV$ decays could illuminate the 12% Rule. The corresponding predictions for Q_H are 48% for the $\Upsilon(2S)$ and 72% for the $\Upsilon(3S)$ decays.

The preliminary results of these searches are given in Table 8 [65]. Several of the signals have at least a 3σ level of significance, and two of the signals – $\Upsilon(1S) \rightarrow \phi f_2'$ and $\Upsilon(1S) \rightarrow K_1(1400)\bar{K}$ – are significant by more than 5σ. These branching fractions are typically 2 or 3 orders of magnitude smaller than the corresponding charmonium branching fractions. This pattern of branching fractions adds another piece to the challenging puzzle of understanding the 12% Rule in charmonium decays.

954

Table 8. Preliminary branching fractions ($\mathcal{B}$), signal significance (N_σ), and 90% CL ULs (UL) in units of 10^{-6} for $\Upsilon(nS) \to PV$ decays.

Channel	$\Upsilon(1S)$			$\Upsilon(2S)$			$\Upsilon(3S)$		
	$\mathcal{B}$	N_σ	UL	$\mathcal{B}$	N_σ	UL	$\mathcal{B}$	N_σ	UL
$\rho\pi$	--	--	4	--	--	11	$9^{+7}_{-8} \pm 1$	1.5	22
$K^*\bar{K}$	$6^{+3}_{-2} \pm 1$	3.6	11	--	--	8	--	--	14
ρa_2	$9^{+4}_{-1} \pm 1$	3.0	19	--	--	24	$8^{+15}_{-6} \pm 1$	1.3	30
ωf_2	$3^{+2}_{-1} \pm 1$	2.6	7	--	--	11	--	--	8
$\phi f_2'$	$7^{+3}_{-2} \pm 1$	5.5	12	$6^{+6}_{-3} \pm 1$	3.0	17	--	--	14
$K^*\bar{K}_2^*$	$9^{+5}_{-4} \pm 1$	3.0	19	$11 \pm 8 \pm 2$	1.6	32	--	--	28
$b_1\pi$	$3 \pm 2 \pm 1$	2.9	8	--	--	12	$5^{+9}_{-4} \pm 1$	1.4	18
$K_1(1270)\bar{K}$	--	--	8	--	--	11	--	--	17
$K_1(1400)\bar{K}$	$14^{+4}_{-3} \pm 2$	5.6	23	$16^{+10}_{-7} \pm 2$	2.9	33	$7^{+10}_{-5} \pm 1$	1.5	22

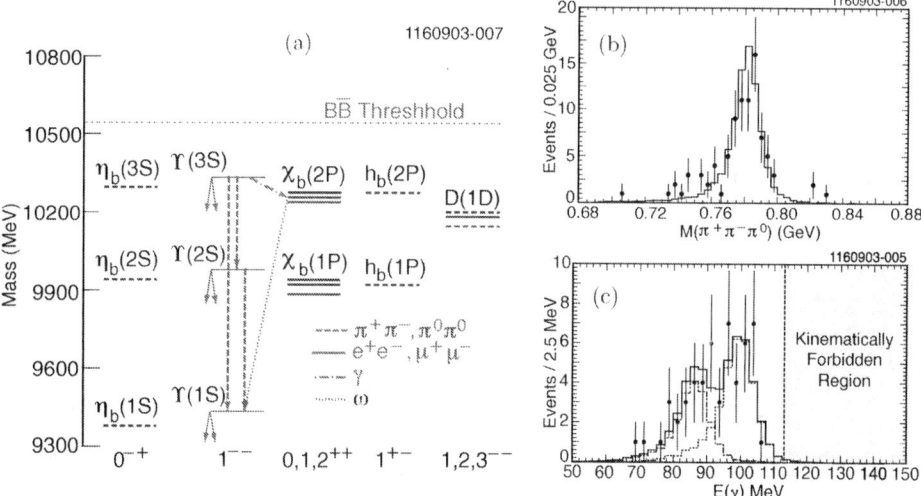

FIGURE 12. (a) Illustration of some of the transitions among the bound $b\bar{b}$ states that have been observed. (b) The distribution of $\pi^+\pi^-\pi^0$ invariant mass, $M(\pi^+\pi^-\pi^0)$, for $\Upsilon(3S) \to \gamma\pi^+\pi^-\pi^0 \Upsilon(1S)$ events. (c) The photon energy distribution $E(\gamma)$ for the same events.

OBSERVATION OF $\chi_B(2P) \to \omega\Upsilon(1S)$

Previously, $\pi\pi$ transitions were the only hadronic decays observed among the $b\bar{b}$ bound states. However – in principle – η and ω transitions via the strong interaction are also allowed. In a search for η and ω transitions, CLEO has observed $\chi_b(2P) \to \omega\Upsilon(1S)$ transitions, which are barely above threshold [66]. The observation of these transitions – which are highly suppressed by phase space – is quite curious.

Fig. 12(a) illustrates many transitions among the $b\bar{b}$ bound states. However, some observed or expected transitions, *e.g.*, those involving the $\chi_b(1P)$ and η_b states, have not been included for clarity. CLEO selected $\Upsilon(3S) \to \gamma\,\pi^+\pi^-\pi^0\,\Upsilon(1S)$ events where the $\Upsilon(1S)$ decayed to $\ell^+\ell^-$ for this analysis. The $M(\pi^+\pi^-\pi^0)$ distribution for these events, shown in Fig. 12(b), has an obvious ω peak with essentially no background. Fig. 12(c) illustrates the photon energy $E(\gamma)$ spectrum. The $\chi_{b1}(2P)$ and $\chi_{b2}(2P)$ states are above threshold for the decay $\chi_b \to \omega\Upsilon(1S)$ but the $\chi_{b0}(2P)$ state is not. The $E(\gamma)$ distribution is consistent with observation of both $\chi_{b1}(2P)$ and $\chi_{b2}(2P)$ as indicated by the histograms on the figure. CLEO obtains the following branching fractions for the two allowed decays:

$$\mathcal{B}(\chi_{b1}(2P) \to \omega\Upsilon(1S)) = (1.63^{+0.35\,+0.16}_{-0.31\,-0.15})\%$$

$$\mathcal{B}(\chi_{b2}(2P) \to \omega\Upsilon(1S)) = (1.10^{+0.32\,+0.11}_{-0.28\,-0.10})\%$$

THE FUTURE OF HEAVY QUARK PHYSICS

In lieu of trying to make a short summary of this summary of heavy quark physics, I will point out the likely sources of experimental heavy quark results in the future:

Immediate Future:

- BaBar and Belle have huge data samples and are already active in HQ physics.
- CDF and DØ are collecting large data samples and have a unique role to play in B_s and b baryons.
- We have seen many new results from BES II and expect more in the future.
- CLEO is in the midst of analyzing $\Upsilon(nS)$ data.
- H1 and ZEUS are entering heavy quark physics

Near future:

- The CLEO collaboration is starting to take CLEO-c (charm threshold) data. The first runs will be at the $\psi(3770)$.

Farther Future

- COMPASS will be able to collect huge samples of charmed baryons.
- BES III will take data in the charm threshold region with as much as an order of magnitude more luminosity than CLEO-c.
- PANDA will offer a whole range of new opportunities for heavy quark physics by producing heavy quark states in $p\bar{p}$ interactions.
- We should not forget that the ATLAS, CMS, and LHCb experiments at the LHC, and the BTeV experiment proposed for Fermilab, can be rich sources of information on heavy quark physics, as well as CP violation in B meson decay.

Although much has been learned about c and b quarks since the first signs of their existences in 1974 and 1977, respectively, there is still a large amount of excellent physics waiting to be uncovered. Heavy quark physics clearly holds the promise of a bright and exciting future!

ACKNOWLEDGEMENTS

I am delighted to acknowledge the many contributions of my CLEO and CESR colleagues to heavy quark physics and to my understanding of the subject. I am grateful to the many speakers at Hadron 2003 who provided me with preliminary versions of their talks and data, and who answered my incessant questions, in order to make it possible for me to produce a summary by the end of the conference. Discussions with T.-M. Yan and J. Rosner were very useful. B. Armstrong of the Particle Data Group kindly furnished TeX files from which I adapted Tables 1 and 2. Finally, I want to express my appreciation to the organizers of this delightful conference and to the members of their staff who have worked so hard and effectively to make our experience in the beautiful city of Aschaffenberg so productive and pleasant.

REFERENCES

[1] P. Wang in these proceedings,

[2] G. Rong in these proceedings.

[3] F. Lu in these proceedings.

[4] X. Mo in these proceedings.

[5] S. Jin in these proceedings.

[6] W. Li in these proceedings.

[7] A. Chamov in these proceedings.

[8] C. Patrignani in these proceedings.

[9] K. Seth in these proceedings.

[10] S. Malvezzi in these proceedings.

[11] A. Reis in these proceedings.

[12] G. Boca in these proceedings.

[13] R.N. Cahn in these proceedings.

[14] S. Spanier in these proceedings.

[15] D.G. Cassel in these proceedings.

[16] P. Krokovny in these proceedings.

[17] J. Russ in these proceedings.

[18] M. Moinester in these proceedings.

[19] J. Wagner in these proceedings.

[20] H. Stoeck in these proceedings.

[21] P. Gianotti in these proceedings.

[22] C.T.H. Davies in these proceedings.

[23] P. Bicudo in these proceedings.

[24] M. Nowak in these proceedings.

[25] K. Terasaki in these proceedings.

[26] I. Yamauchi in these proceedings.

[27] T. Feldmann in these proceedings.

[28] T. Komada in these proceedings.

[29] S. H. Lee in these proceedings.

[30] A.C. Torres in these proceedings.

[31] C.T.H. Davies *et al.*, hep-lat/0304004.

[32] CLEO-c Collaboration, R.A. Briere *et al.*, *CLEO-c and CESR-c: A New Frontier of the Weak and Strong Interactions*, Cornell Report CLNS 01/1742, Revised October 2001, available from a link at `http://www.lns.cornell.edu/`.

[33] P.L. Frabetti *et al.*, *Phys. Lett. B* **300**, 190 (1993) and *Phys. Lett. B* **336**, 106 (1994).

[34] CLEO Collaboration, R. Ammar *et al.*, *Phys. Rev. Lett.* **89**, 171803 (2002)

[35] Particle Data Group, K.Hagiwara *et al.*, *Phys. Rev. D* **66**, 010001 (2002).

[36] SELEX Collaboration, M. Mattson *et al.*, *Phys. Rev. Lett.* **89**, 112001 (2002).

[37] BaBar Collaboration, B. Aubert *et al.*, *Phys. Rev. Lett.* **90**, 242001 (2003).

[38] CLEO Collaboration, D. Besson *et al.*, *Phys. Rev. D* **68**, 032002 (2003).

[39] R.N. Cahn, and J.D. Jackson *Phys. Rev. D* **68**, 037502 (2003).

[40] Belle Collaboration, K. Abe *et al.*, hep-ex/0307021.

[41] Belle Collaboration, P. Krokovny *et al.*, hep-ex/0308019, (submitted to Phys.Rev.Lett.).

[42] Belle Collaboration, Y. Mikami *et al.*, hep-ex/0307052, (submitted to Phys.Rev.Lett.).

[43] F.L. Fabbri (FOCUS Collaboration) in *ICHEP 2000, Proceedings of the 30th International Conference on High Energy Physics*, Osaka, 2000, edited by C.S. Lim and T. Yamamaka (World Scientific, Singapore, 2001) v. I, p. 377.

[44] M.A. Nowak, M. Rho, and I. Zahed, *Phys. Rev. D* **48**, 4370 (1993).

[45] W.A. Bardeen and C.T. Hill, *Phys. Rev. D* **49**, 409 (1994).

[46] W.A. Bardeen, E. Eichten, and C.T. Hill, *Phys. Rev. D* **68**, 054024 (2003).

[47] M.A. Nowak, M. Rho, and I. Zahed, hep-ph/0307102.

[48] See Ref. [38] for additional references to theory papers.

[49] Belle Collaboration, S.-K. Choi *et al.*, hep-ex/0309032.

[50] KEDR Collaboration, V.M. Aulchenko *et al.*, *Phys. Lett. B* **573**, 63 (2003), hep-ex/0306050.

[51] BES Collaboration, J.Z. Bai *et al.*, *Phys. Lett. B* **555**, 174 (2003).

[52] BES Collaboration, J.Z. Bai *et al.*, hep-ex-0308073.

[53] Belle Collaboration, S.-K. Choi *et al.*, *Phys. Rev. Lett.* **89**, 102001 (2002).

[54] G. Wagner for the BaBar Collaboration, hep-ex-0305083.

[55] CLEO Collaboration, J. Ernst *et al.*, CLEO CONF 03-05, EPS 253.

[56] BES Collaboration, J.Z. Bai *et al.*, *Phys. Rev. Lett.* **91**, 022001 (2003), hep-ex-0303006.

[57] See Appendix A to Chapter III (p. 146) of Ref. [32] for a summary of previous data and references.

[58] BES Collaboration, J.Z. Bai *et al.*, hep-ex-0307028.

[59] T. Skwarnicki, Lepton Photon 2003, XXI International Symposium on Lepton and Photon Interactions at High Energy, Fermilab.

[60] BES Collaboration, J.Z. Bai *et al.*, *Phys. Rev. D* **67**, 112001 (2003), hep-ex-0304012.

[61] P. Wang, C.Z. Yuan, and X.H. Mo, hep-ph/0303144.

[62] BES Collaboration, J.Z. Bai *et al.*, hep-ex-0310023.

[63] BES Collaboration, J.Z. Bai *et al.*, hep-ex-0310024.

[64] P. Wang, C.Z. Yuan, and X.H. Mo, hep-ph/0305259.

[65] CLEO Collaboration, S.A. Dytman *et al.*, CLEO Conf. 07-03, LP-122 (unpublished).

[66] CLEO Collaboration, D. Cronin-Hennessy *et al.*, Cornell Report No. CLNS 03/1840, CLEO 03-12.

List of Participants

Achasov	Mikhail	Budker INP Novosibirsk, Russia
Akopov	Zaven	DESY Hamburg, Germany
Albertus	Conrado	Universidad de Granada, Spain
Amsler	Claude	Universität Zürich, Switzerland
Anisovich	Alexei	Universität Bonn, Germany
Anisovich	Vladimir	PNPI Gatchina, Russia
Arkhipov	Andrei	IHEP, Protvino, Russia
Baillon	Paul	CERN Geneva, Switzerland
Barbi	Mauricio	McGill University Montreal, Canada
Barnes	Ted	ORNL University of Tennessee, USA
Bartholomy	Olivia	Universität Bonn, Germany
Bashkanov	Mikhail	Universität Tübingen, Germany
Bayadilov	Dair	PNPI Gatchina, Russia
Bentz	Wolfgang	Tokai University, Japan
Bianchi	Nicola	INFN-Laboratori di Frascati, Italy
Bicudo	Pedro	CFIF Lisaboa, Portugal
Bloch	Frederic	Universität Basel, Switzerland
Bloise	Caterina	INFN-Laboratori di Frascati, Italy
Boca	Gianluigi	University of Pavia, Italy
Brinkmann	Kai-Thomas	TU Dresden, Germany
Bugg	David	Queen Mary, University of London,UK
Bunyatyan	Armen	MPI Heidelberg and Yerevan
Cabrera	Daniel	IFIC –Universidad de Valencia, Spain
Cahn	Robert	LBL Berkeley, USA
Cassel	David	LEPP, Cornell University
Chabab	Mohamed	Cadi Ayyad University, Marrakech, Marocco
Chamov	Andrei	Budker Institute Novosibirsk, Russia
Cherednikov	Igor	JINR Dubna, Russia
Chung	Suh-Urk	BNL Brookhaven, USA
Close	Frank	Oxford University, UK
Crede	Volker	Universität Bonn, Germany
Czyzykiewicz	Rafal	Jagellonian University Krakow, Poland
Davies	Christine	University of Glasgow
Denig	Achim	Universität Karlsruhe, Germany
Donnachie	Alexander	University of Manchester, UK
Duennweber	Wolfgang	Universität München, Germany

Düren	Michael	Universität Giessen, Germany
Elster	Charlotte	Ohio University, USA
Eugenio	Paul	Florida State University Tallahassee, USA
Fadeeva	Ekaterina	ITEP Moscow, Russia
Falter	Thomas	Universität Giessen, Germany
Fedorov	Sergei	ITEP Moscow, Russia
Feldmann	Thorsten	CERN Geneva, Switzerland
Fernandez	Francisco	Universidad de Salamanca, Spain
Fritsch	Miriam	Universität Bochum, Germany
Fuhrmann	Hermann	IMEP/ÖAW Wien, Austria
Furui	Sadataka	Teikyo University, Japan
Ganbold	Gurjav	JINR Dubna, Russia
Gianotti	Paola	INFN-Laboratori di Frascati, Italy
Gillitzer	Albrecht	IKP, FZ-Jülich, Germany
Glozman	Leonid	University of Graz, Austria
Goeke	Klaus	Universität Bochum, Germany
Gouz	Iouri	IHEP, Protvino, Russia
Hartmann	Olaf	GSI Darmstadt, Germany
Henner	Victor	Perm State University, Russia
Henning	Walter	GSI Darmstadt, Germany
Hicks	Kenneth	Ohio University Athens, USA
Hinterberger	Frank	Universität Bonn, Germany
Höistad	Bo	Uppsala University, Sweden
Hou	Suen	Academia Sinica Taipei, Taiwan
Illarionov	Alexei	JINR Dubna, Russia
Ishida	Muneyuki	Meisei University Hino, Japan
Ishida	Shin	Nihon University Funabashi, Japan
Jin	Shan	CAS Beijing, China
Johansson	Tord	Uppsala University, Sweden
Kalashnikova	Yulia	ITEP Moscow, Russia
Kaminski	Robert	Polish Academy of Science Krakow, Poland
Kanavets	Vadim	ITEP Moscow, Russia
Karl	Gabriel	University of Guelph, Canada
Khazin	Boris	Budker INP Novosibirsk, Russia
Khorramian	Alinaghi	IPM Teheran, Iran
Kienle	Paul	TU München, Germany
Kisiel	Jan	University of Silesia Katowice, Polen

Kleefeld	Frieder	CFIF Lisaboa, Portugal
Klein	Friedrich	Universität Bonn, Germany
Klempt	Eberhard	Universität Bonn, Germany
Koch	Helmut	Universität Bochum, Germany
Kolomeitsev	Evgeni	Niels Bohr Institute Copenhagen, Denmark
Komada	Toshihiko	Nihon University Funabashi, Japan
Kopf	Bertram	Universität Bochum, Germany
Korol	Alexander	Budker INP Novosibirsk, Russia
Kotulla	Martin	Universität Basel, Switzerland
Kowina	Piotr	IKP, FZ-Jülich, Germany
Krokovny	Pavel	Budker INP Novosibirsk, Russia
Kroll	Peter	Universität Wuppertal, Germany
Kuhn	Joachim	Carnegie Mellon University Pittsburg, USA
Kühn	Wolfgang	Universität Giessen, Germany
Kunihiko	Terasaki	Kyoto University, Japan
Kuzmenko	Dmitri	ITEP Moscow, Russia
Lee	Su-Houng	Yonsei University Seoul, Korea
Lenske	Horst	Universität Giessen, Germany
Lewandowski	Bernd	Universität Bochum, Germany
Li	Weiguo	CAS Beijing, China
Lu	Feng	CAS Beijing, China
Lundborg	Agnes	Uppsala University, Sweden
Lynen	Uli	GSI Darmstadt, Germany
Magiera	Andrzej	Jagellonian University Krakow, Poland
Maglich	Bogdan	HiEnergy Technologies, Inc.
Maltman	Kim	York University Toronto, Canada
Malvezzi	Sandra	INFN Milano, Italy
Matveev	Maxim	Universität Bonn, Germany
Meißner	Ulf	Universität Bonn, Germany
Metsch	Bernard	Universität Bonn, Germany
Meyer	Werner	Universität Bochum, Germany
Meyer-Wildhagen	Frank	Universität München, Germany
Mineo	Hirobumi	CAS & Taiwan University, Taiwan
Mirjalili	Abolfazl	Yazd University, Iran
Mo	Xiaohu	CAS Beijing, China
Moinester	Murray	Tel Aviv University, Israel
Mosel	Ulrich	Universität Giessen, Germany

Moskal	Pawel	IKP FZ-Jülich, Germany
Musulmanbekov	Genis	JINR Dubna, Russia
Nefediev	Alexei	ITEP Moscow, Russia
Nekipelov	Mikhail	IKP, FZ-Jülich, Germany
Nieves	Juan	Universidad de Granada, Spain
Nikolaenko	Vladimir	IHEP Protvino, Russia
Nowak	Maciej	GSI & Jagiellonian University Krakow, Poland
Noya	Hiroshi	Hosei University at Tama, Tokyo, Japan
Ochs	Wolfgang	MPI München, Germany
Oda	Masuho	Kokushikan University
Oelert	Walter	IKP FZ-Jülich, Germany
Orth	Herbert	GSI Darmstadt, Germany
Page	Philip	Los Alamos National Laboratory, USA
Patrignani	Claudia	Universita' and INFN Genova, Italy
Pelaez	Jose	Universidad Complutense Madrid, Spain
Peters	Klaus	Univeristät Bochum, Germany
Petrascu	Catalina	INFN-Laboratori di Frascati, Italy
Polyakov	Maxim	Univeristät Bochum, Germany
Poplawski	Nikodem	Indiana University Bloomington, USA
Protasov	Konstantin	IN2P3-CNRS Grenoble, France
Reis	Alberto	CBPF Rio de Janeiro, Brazil
Reyes	Marco	University of Guanajuato, Mexico
Rezaeian	Amir	University of Manchester, UK
Riska	Dan-Olof	University of Helsinki, Finland
Ritman	James	Universität Giessen, Germany
Roca	Luis	University of Valencia, Spain
Rohdjess	Heiko	Universität Bonn, Germany
Rong	Gang	CAS Beijing, China
Russ	James	Carnegie Mellon University
Santamarina	Cibran	Universität Basel, Switzerland
Sapozhnikov	Mikhail	JINR Dubna, Russia
Sarantsev	Andrey	PNPI Gatchina, Russia
Sasaki	Shoichi	University of Tokyo, Japan
Sassen	Felix	IKP FZ-Jülich, Germany
Sawada	Tetsuo	Nihon University Tokyo, Japan
Schadmand	Susan	Universität Giessen, Germany
Schlein	Peter	UCLA Los Angeles, USA

Schmitt	Lars	TU München, Germany
Schneider	Sonja	IKP, FZ-Jülich, Germany
Schwarz	Carsten	GSI Darmstadt, Germany
Seth	Kamal	Northwestern University Evanston, USA
Shklyar	Vitaliy	Universität Giessen, Germany
Sibirtsev	Alexander	IKP, FZ-Jülich, Germany
Sokolov	Andrei	GSI Darmstadt. Germany
Solodov	Evgeni	Budker INP Novosibirsk, Russia
Steinke	Matthias	Universität Bochum, Germany
Stoeck	Holger	University of Florida, USA
Stroili	Roberto	Universita' di Padova & INFN, Italy
Suh	Jun-Suhk	Kyungpook National University
Surrow	Bernd	MIT Cambridge, USA
Swanson	Eric	University of Pittsburgh, USA
Swat	Maciej	Indiana University Bloomington, USA
Teshima	Tadayuki	Chubu University Kasugai, Japan
Tornqvist	Nils	University of Helsinki, Finland
Tsuru	Tsuneaki	KEK Tsukuba, Japan
Uman	Ismail	Northwestern University Evanston, USA
Valcarce	Alfredo	Nuclear Physics Group, Spain
van der Heide	Jan	NIKHEF Amsterdam, Netherlands
Vereshagin	Alexander	St.Petersburg State University, Russia
Wagner	Jeannine	DESY Hamburg, Germany
Wagner	Marcus	Universität Erlangen-Nürnberg
Walz	Jochen	Max-Planck-Institut fuer Quantenoptik
Wang	Ping	CAS Beijing, China
Weinheimer	Christian	Universität Bonn, Germany
Weygand	Dennis	CEBAF Newport News, USA
Wiedner	Ulrich	DRS, Uppsala University, Sweden
Winter	Peter	IKP, FZ-Jülich, Germany
Wolke	Magnus	The Svedberg Laboratory, Uppsala, Sweden
Xiaobin	Ji	CAS Beijing, China
Yamauchi	Ichiro	TMCT Tokyo, Japan
Zheng	Hanqing	Peking University Beijing, China
Zou	Bingsong	CAS Beijing, China

D

Dalpiaz, P., 581
Davies, C., 615
Demiroers, L., 241
Denig, A. G., 83
Dimova, T. V., 60, 130
Donnachie, A., 110, 797
Donoghue, J. F., 675
Doroshkevich, E., 241
Druzhinin, V. P., 60, 130
Dshemuchadse, S., 241
Dünnweber, W., 388
Duperray, R., 892
Düren, M., 827
Dzheliadin, R., 155

E

Egger, J.-P., 175
Eidelman, S. I., 591
Ekström, C., 241
Elster, C., 837
Erhardt, A., 241
Eugenio, P., 832
Eyrich, W., 241

F

Fadeeva, E. A., 411
Falter, T., 822
Fedorov, S. M., 711
Feldmann, T., 533
Felix, J., 135
Fenyuk, A., 155
Fernández, F., 352
Filimonov, E. A., 270
Fransson, K., 241
Freiesleben, H., 241
Fritsch, M., 241
Fuhrmann, H., 175, 180
Furui, S., 685

G

Gallmeister, K., 822
Ganbold, G., 285

Gara, A., 135
Garzoglio, G., 581
Gavrilov, Y., 155
Gianotti, P., 525
Gillitzer, A., 241, 743
Glozman, L. Y., 726
Gollwitzer, K. E., 581
Golowich, E., 675
Golubev, V. B., 60, 130
Gottschalk, E. E., 135
Gouz, Y., 145, 155
Graham, M., 581
Groshev, V. R., 591
Grzonka, D., 852, 858, 907
Guaraldo, C., 175
Gustafsson, L., 241
Gutierrez, G., 135

H

Hahn, A., 581
Hartouni, E. P., 135
Hawranek, P., 863
Henner, V., 327
Hernández, E., 566
Hicks, K. H., 400
Höistad, B., 241
Horikawa, T., 770
Hou, S., 115

I

Iliescu, M., 175
Ilieva, J., 863
Ishida, M., 550, 721
Ishida, S., 550, 716, 721
Ishii, N., 770, 790
Ishiwatari, T., 175
Itahashi, K., 175
Iwasaki, M., 175

J

Jacewicz, M., 241
Janusz, M., 852, 858, 907
Jarczyk, L., 852, 858, 907
Ji, X., 231

Joffe, D., 581
Johansson, T., 241, 907

K

Kabachenko, V., 155
Kachaev, I., 155
Kalashnikova, Y. S., 317
Kamiński, R., 195
Kamys, B., 852, 858, 907
Kan, M. R., 270
Kanavets, V. P., 848
Karl, G., 290
Karnaev, S. E., 591
Karsch, L., 241
Karyukhin, A., 155
Kasper, J., 581
Keil, C., 765
Keleta, S., 241
Khokholov, Y., 155
Khorramian, A. N., 897, 902
Khoukaz, A., 852, 858, 907
Kilian, K., 241, 852, 858, 863, 907
Kirilov, D., 863
Kiselev, V. E., 591
Kistryn, S., 863
Kitamura, I., 150
Kleefeld, F., 332
Kliczewski, S., 863
Klimala, W., 863
Knapp, B. C., 135
Koch, I., 241
Kolev, D., 863
Kolomeitsev, E. E., 665
Komada, T., 337, 550
Konnopliannikov, A., 155
Kononov, S. I., 591
Konstantinov, V., 155
Korol, A. A., 60, 130
Koroleva, L. I., 848
Koshuba, S. V., 60, 130
Kostyuhin, V., 155
Kotov, K. A., 591
Kotulla, M., 842
Kovalev, A. I., 848
Kowina, P., 852, 858, 907
Kozlenko, N. G., 270
Kravchenko, E. A., 591
Kravčiková, M., 863

Kreisler, M. N., 135
Kremyanskaya, E. V., 591
Kress, J., 241
Krewald, S., 245, 280, 837
Krokovny, P., 475
Kroll, P., 451
Kruglov, S. P., 270, 848
Krupa, D., 357
Kuhlmann, E., 241
Kuhn, J., 377
Kullander, S., 241
Kupsc, A., 241
Kutsarova, T., 863
Kuzmenko, D. S., 421, 426, 706

L

Lasio, G., 581
Lauss, B., 175
Lee, S., 135
Lee, S. H., 780
Lenske, H., 765
Lepage, P., 615
Leśniak, L., 195
Levichev, E. B., 591
Li, L., 347
Li, W., 495
Lieb, J., 863
Loiseau, B., 195
Lopatin, I. V., 270
Lo Vetere, M., 581
Lu, F., 120
Lucherini, V., 175
Ludhova, L., 175
Lundborg, A., 431
Luppi, E., 581
Lutz, M. F. M., 665
Lysenko, A. P., 130

M

Machner, H., 863
Macrí, M., 581
Magiera, A., 863
Maglich, B. C., 37
Maltman, K., 675
Malvezzi, S., 77
Malyshev, V. M., 591

Przerwa, J., 852, 858, 907
Pyata, E. E., 60

Q

Qin, G. Y., 322

R

Rad'kov, A. K., 270
Reis, A., 312
Reyes, M. A., 135
Rezaeian, A. H., 690
Riska, D. O., 365
Robutti, E., 581
Roca, L., 160, 185
Roderburg, E., 241
Rong, G., 592
Rosen, J., 581
Roy, B. J., 863
Rożek, T., 852, 858, 907
Rumerio, P., 581
Rusack, R., 581
Russ, J. S., 507
Ryabchikov, D., 155
Ryltsov, V. V., 848

S

Santamarina, C., 170
Santo, R., 852, 858, 907
Santroni, A., 581
Sapozhnikov, M. G., 875
Sarantsev, A., 65
Sasaki, S., 416
Sassen, F. P., 190
Savinov, G. A., 591
Sawada, T., 200
Schadmand, S., 760
Schaller, L. A., 175
Schepers, G., 852, 858, 907
Schlein, P., 387
Schmitt, L., 870
Schneider, S., 245
Schönmeier, P., 241
Schröder, W., 241
Schulte-Wissermann, M., 241

Schultz, J., 581
Scobel, W., 241
Sefzick, T., 241, 852, 858, 907
Seki, R., 175
Semenov-Tian-Shanski, K., 260, 265
Seon-Hee, S., 581
Serednyakov, S. I., 60, 130
Seth, K. K., 543, 581
Shamov, A. G., 591
Shatilov, D. N., 591
Shatunov, Y. M., 60, 130
Shchedrov, V. A., 848
Shklyar, V., 275
Shusharo, A. I., 591
Shwartz, B. A., 591
Shyam, R., 765
Sibirtsev, A., 280, 837
Sidorov, V. A., 60, 130, 591
Siemaszko, M., 852, 858, 907
Silagadze, Z. K., 60, 130
Simonov, E. A., 591
Simonov, Y. A., 426, 711
Sirghi, D., 175
Sirghi, F., 175
Sitnik, I., 863
Siudak, R., 863
Skorodko, T., 241
Skovpen, Y. I., 591
Skrinsky, A. N., 130, 591
Smyrski, J., 852, 858, 863, 907
Solodkov, A. A., 155
Solodov, E., 49
Solovianov, O., 155
Soukharev, A. M., 591
Spanier, S. M., 485
Speth, J., 837
Stancari, G., 581
Stancari, M., 581
Starchenko, E., 155
Stepaniak, J., 241
Stöck, H., 515
Stokes, B., 832
Strasser, P., 175
Stroili, R., 55
Strzałkowski, A., 907
Sulimov, A. D., 848
Sumachev, V. V., 270, 848
Surovtsev, Y. S., 357
Surrow, B., 812
Svirida, D. N., 848